Encyclopedia of PVC

Encyclopedia of PVC

Second Edition, Revised and Expanded

Volume 2:
Compound Design and Additives

edited by

LEONARD I. NASS
L. I. Nass Company
Warren, New Jersey

CHARLES A. HEIBERGER
Plastics Industry Consultant
Princeton, New Jersey

MARCEL DEKKER, INC.   New York and Basel

Library of Congress Cataloging-in-Publication Data

(Revised for vol. 2)

Encyclopedia of PVC.

    2nd. ed., rev. and expanded.
    Includes bibliographies and indexes.
    Contents:　v. 1. Resin manufacture and properties --
v. 2. Compound design and additives.
    1. Polyvinyl chloride.　2. Nass, Leonard I.
I. Heiberger, Charles A.
TP1180.V48E5　1986　　　668.4'237　　　85-20753
ISBN 0-8247-7695-X (vol. 2)

MARCEL DEKKER, INC.
270 Madison Avenue, New York, New York 10016

Current printing (last digit):
10 9 8 7 6 5 4 3 2

PRINTED IN THE UNITED STATES OF AMERICA

Charles Heiberger left us on September 13, 1985, just as Volume 1 was being readied for final printing and publication. However, his presence as my coeditor has insured that he will always be with us, at least in memory. These volumes reflect his contributions to the series.

With Charles' death, I lost a collaborator who was very special. His many ideas were innovative and stimulating and his presence was inspiring. The reorganization of topics into our new format and the subsequent expansion of subject matter and of the series from three to four volumes result mainly from Charles' incisive thinking and powers of persuasion. It is also through Charles' efforts that we have so many new contributors working with us on this revised series. It was Charles who first expounded the thought that the new *Encyclopedia of PVC* would not necessarily make obsolete the original edition. What our new authors have done is to provide additional insight and complementary viewpoints to topics that were originally well prepared and documented some 10 years ago. Thus, both editions can rightfully stand together on one's shelves, side by side today, and this too is because of Charles' organizational abilities.

It will not be easy finishing this work without you, Charles. You were not just a staunch collaborator, but a good friend as well.

Leonard I. Nass

Foreword

Old soldiers never die—they just fade away. Old plastics never die—they hang on and stay.

There exist two outstanding examples that prove the validity of the metaphoric version of this sentence: the thermoset phenolics (Bakelite) and the thermoplastic polyvinyl chlorides (PVC). The former just celebrated the 75th anniversary of its appearance; the latter became commercial about 50 years ago. There is, however, a significant difference between these two classical synthetics: Bakelite has been—and still is—essentially restricted to those applications which need hard, thermostable materials; PVC, on the other hand, has been—and still is—expanding over all possible practical uses from soft rubbery shoe soles and garden hoses to hard and tough window frames and to giant sewer pipes. In 1983 the worldwide production and consumption of PVC was 12.5 million metric tons (27.5 billion pounds) and there is all reason to expect it will be about 17 million tons (37.4 billion pounds) in 1990.

At this time, it is extremely desirable that a new comprehensive presentation of all the many facets of this important material should be published in the form of four volumes of an encyclopedia where each chapter is contributed by a qualified expert to give an account of the development together with a detailed description of the present state of the art in synthesis, characterization, structure, properties, processing, and applications.

Such encyclopedic presentations with fact and figure collections can easily become monotonous and purely informative—like a telephone book—but they also may, beyond that, be educational and stimulating when the material is not presented as a dry enumeration of data and events but has suggestive concepts and ideas appropriately sprinkled throughout the text. The editors and authors have been remarkably successful in introducing into these volumes a refreshing flavor of progressive and imaginative contribution, which make them a source not only of reliable information but also of intellectual challenge. All chapters of the four volumes will convince the reader that PVC is an unusual polymer and will remain an important leader in the vast fields of its preferred uses for many more years.

PVC is technically not an "organic resin" because it contains normally 56.8 weight percent of chlorine and, in its after-chlorinated types, as much as 65 weight percent. Essentially, it is an inorganic-organic "hybrid" polymer and owes many of its interesting properties, such as flame retardancy, resistance to common solvents, polarity, compatibility with additives, and high softening characteristics, to the presence of its inorganic component. There exist several other inorganic-organic polymers: for example, polyvinylidene chloride (Saran), polytetrafluorethylene (Teflon), and the large families of silicones and phosphazenes. In all these cases, it is the inorganic component— silicon, phosphorus, fluorine, chlorine—which endows these materials with such valuable properties that all the enumerated polymers are commercially very successful even though in most cases the monomeric species are rather expensive and the methods of polymerization are not simple. In all these cases the inorganic component is a chemical constituent of each macromolecule. It is known, however, that the mechanical blend of truly organic polymers with inorgano-organic species also offers interesting new properties, and even the simple admixture of inorganic fillers to organic polymers is a field of widespread, successful activities, including the loading of PVC with finely powdered inorganics.

Any new large-scale demand for specific property combinations generates ideas for novel compositions and processing techniques, and the users of this encyclopedia will find their expectations for information, stimulations, and guidance fully and wholly satisfied.

Herman Mark
Polytechnic University
of New York
Brooklyn, New York

Preface to Second Edition

The original edition of this work was published in 1976—77 as a three-volume comprehensive reference work on PVC technology as it existed prior to about 1975. Since then it has become increasingly evident that an updated edition is necessary to cover a number of significant changes and advances that have occurred in PVC science and technology, among which may be mentioned:

Up-to-date handling methods for VCM; compliance with the EPA, OSHA, and FDA regulatory procedures in the United States

Computerized large reactor polymerization processes with improved resin quality and control, reduced environmental hazards, and better economics

Newly developed compounding additives plus improved compounding processes demonstrating better performance at lower cost, including fully automated continuous compounding methods

New processing techniques and equipment designs with computerized controls for faster throughput cycles with better quality control

New chlorinated-PVC materials and compounds with broad application in extrusion and molding markets

Renewed emphasis and attention to foamed, rigid PVC products in both Europe and the Western Hemisphere

The new *Encyclopedia of PVC* has now been expanded to four volumes with individual chapters grouped according to common interests: Volume 1, Resin Manufacture and Properties; Volume 2, Compound Design and Additives; Volume 3, Compounding Processes, Product Design, and Specifications; and Volume 4, Conversion and Fabrication Processes.

Safety and environmental concerns are now covered in two separate chapters: in Volume 1 for resins, and in Volume 3 for compounds and finished products. Likewise, the subjects of stabilization and plasticization have been divided: in Volume 1 when relating to resin properties, and in Volume 2 where individual stabilizer and plasticizer materials and formulations are more appropriately grouped with other additive classes. The earlier chapter on chemical modifications of PVC has been replaced by a more complete coverage of chlorinated PVC, which lately has become an important engineering plastic.

vii

New chapters covering the subjects of block and graft interpolymers, alloys, and polyblends; PVC foams; and testing of flexible PVC products have been added, emphasizing recent developments in these new or renewed-interest technologies.

Special efforts have been made to have a stronger international representation not only in the selection of individual contributors, but also in the consideration and coverage of significant foreign technology and methodology by all authors.

Each author is an expert in his field as well as knowledgeable in PVC technology as a whole. In all cases, individual contributions have been unified and enhanced by input from the editors and from other authors. Hopefully, the readers of this new edition will benefit greatly from this cross-fertilization of ideas.

Leonard I. Nass

Charles A. Heiberger

Preface to First Edition

Within a scant 35 years, polyvinyl chloride and its broad family of polymers and
compositions have grown from essentially nothing to what is today, by any standard of measurement, truly big business. Yet, in spite of this remarkable growth
record, there exists a paucity of hard-bound technical reference literature on the
subject. The present series of volumes attempts to correct that situation and to
provide the interested student and the serious research worker with a single
source reference of scientific background information and current "state-of-the-
art" technical data. Discussions are centered on all of the key raw materials and
on how to put them together, and on how then to process the resultant mixtures by
a host of conversion techniques. Environmental and disposal considerations are
covered, as well as the subjects of specifications, testing and quality control,
health and safety, operation of plants for production and profit maximization, and
numerous others. However, no attempt has been made to provide a "formulary"
of compound recipes. Further, except in isolated cases, only scant coverage is
devoted to discussion of individual products and applications, and markets or
economic factors. Rather, the emphasis has been placed on the fundamentals
surrounding formulation and processing philosophies in an attempt to instill in the
reader a sense of how to go about solving his problems himself, since it has been
repeatedly demonstrated that all such problems are unique and different, no
matter how similar to other problems they may first appear to be on the surface.

As editor, I have been most fortunate to have a group of eminently well-
qualified authors for the individual chapters. Each contributor is especially well
informed not only within the realm of his own particular discipline but on the
subject of PVC science and technology as a whole. As a result, these chapters
represent more than a collection of individual contributions since there was considerable direct collaboration among all authors. Hopefully, it is the readers
of this series who ultimately stand to benefit from this extensive cross fertiliza-
tion and pollination of ideas.

Leonard I. Nass

Contributors to Volume 2

Charles W. Fletcher, Synthetic Products Company, Cleveland, Ohio

*Melvin M. Gerson**, Podell Industries, Inc., Clifton, New Jersey

Saul Gobstein, Sa-Go Associates, Inc., Shaker Heights, Ohio

John R. Graff, Mobay Corporation, Dyes and Pigments Division, Haledon, New Jersey

L. G. Krauskopf, Exxon Chemical Company, Baton Rouge, Louisiana

Thomas C. Jennings, Synthetic Products Company, Cleveland, Ohio

Robert A. Lindner, Allied-Signal Corporation, Morristown, New Jersey

J. T. Lutz, Jr., Rohm & Haas Co., Bristol, Pennsylvania

V. E. Malpass, Ferro Corporation, Cleveland, Ohio

K. K. Mathur, Pfizer, Inc., Easton, Pennsylvania

Norman L. Perry, Colorite Plastics Co., Ridgefield, New Jersey

R. P. Petrich, Rohm & Haas Co., Bristol, Pennsylvania

D. B. Vanderheiden, Pfizer, Inc., Easton, Pennsylvania

Kurt Worschech, Neynaber Chemie GMBH, Division of Henkel, Loxstedt, West Germany

---

*Deceased

Contents of Volume 2

Contents of Other Volumes

Volume 4: Conversion and Fabrication Processes   (*Tentative*)

1

The Compounding of Polyvinyl Chloride

NORMAN L. PERRY

Colorite Plastics Co.
Ridgefield, New Jersey

I. POLYVINYL CHLORIDE, THE VERSATILE PLASTIC

Compounded polyvinyl chloride, more than any other plastic material, can be considered the "versatile plastic." Depending on how it is formulated, it may be a rubbery flexible material or a rigid high-impact product. It may be opaque or crystal clear. It can be formulated to be nontoxic, nonflammable, light-stable, and stain-resistant. Through proper formulation, polyvinyl chloride has been used in such diverse applications as electrical insulation, medical tubing, food wrap, outdoor furniture, swimming-pool liners, electrical conduit, pressure pipe, garden hose, house sidings, bottles, flooring, and clothing. As material for the ultimate product, compounded polyvinyl chloride can be formed by almost all the standard plastics processing techniques, including calendering, extrusion, injection and compression molding, blow molding, rotational molding, slush and dip molding, solution casting, electrostatic spray, and fluidized bed coating.

All references in the preceding paragraph refer to "compounded" or "formulated" polyvinyl chloride. Unlike most other plastics, polyvinyl chloride as sold in resinous form is unusable. Only by incorporating suitable additives can the chemist convert the resin to a form from which it can be processed into a useful end product. The total number of such additive materials that are being offered commercially for use with PVC are in the thousands. Many additional thousands of materials are not commercially available but have been proposed in the patent literature and in research reports. The formulator must select those components that will satisfy the requirements of end-use performance, processing characteristics, and cost.

II. COMPOUNDING FUNDAMENTALS

Compounding PVC essentially involves adding to the base PVC resin the components that will allow it to be processed into a finished product with desired properties, at minimum cost. The "families" of materials that will be chosen will probably fall into one of the following classifications, each of which is discussed in detail elsewhere in these volumes:

1. Plasticizers
2. Stabilizers
3. Lubricants
4. Impact modifiers
5. Processing aid resins
6. Fillers
7. Colorants
8. Miscellaneous (antistatic agents, ultraviolet absorbers, antiblocking agents, smoke control agents, flame-retardants, fungicides, odorants)

If one were to take PVC resin as it is received from the manufacturer and attempt to process it by normal conversion techniques (calendering, extrusion, injection molding, etc.), the resin would degrade during processing, giving off hydrogen chloride, discoloring, and sticking to the equipment. The results could be described only as disastrous. Adding to the PVC resin the proper type and amount of plasticizer, stabilizer, lubricant, pigment, and so on (as outlined above) results in a compounded mixture that can subsequently be processed on properly designed equipment into a satisfactory finished product. In the simplest sense, compounding can be defined as the mixing to homogeneity of PVC resin with the other components required for processing and performance. The homogeneous mixture of materials can subsequently be fed to processing equipment, providing a uniform, controlled product at maximum rates of production.

III. VINYL COMPOUND, PURCHASE OR MANUFACTURE?

The producer of vinyl products must decide whether to manufacture the vinyl compound internally (in any of the forms—dry powder, pellet, or fluid) or whether to purchase the compound, ready for the processing equipment, from a custom compounder or a resin supplier. This decision should be largely based on the use, volume of compound that is to be processed, the number of different compounds required, and the willingness of the manufac-

turer to invest capital for compounding equipment and capable employees.
(The various forms of compounds and the processes used in preparing compounded PVC are considered in Chapters 21—23.)

Internal compounding offers a potentially lower cost and provides the
manufacturer with the means for developing and marketing proprietary products in accordance with his technical ability. On the other hand, the manufacturer must be willing and able to make a substantial capital investment for
the necessary equipment. The internal compounder should be proficient in
the art of formulation and must set up internal quality control tests on incoming raw materials as well as on the compound.

At the time of preparation of this chapter, a pelletized vinyl processor
should be able to project a need for an annual use of about 8 million pounds
of compound in order to justify the capital investment required for internal
compound preparation. This would result in a 3- to 4-year payback of invested capital.

Purchase of compound puts the onus of performance on the supplier of
the compound. The processor can take advantage of the compounding expertise of his suppliers. Capital is not tied up in compounding equipment and
raw material inventory but can be used for operations or for finished product
plant expansion.

Assuming that the economics, availability of capital, and technical capabilities indicate the advisability of internal manufacture, the next factor that
must be considered is whether to make powder blend, pelletized compound, or
a fluid form of vinyl compound. The conversion process largely dictates the
most desirable form of compound feedstock. Fluid vinyl processes are generally separate and distinct from those that use powders or granulated pellets.
Formulating considerations for fluid vinyl processes are covered in Chapter
22. The discussion here is concerned primarily with processes that depend
on powders or pellets.

In general, the powder blending technique represents a lower capital investment than preparing a pelletized compound. Pelletized compound, however, is significantly more versatile than powder blend. Pellets are easier to
store, are not so moisture-sensitive, feed more readily into processing equipment, and can be processed by a much broader range of equipment types
than powder blend.

IV. PROCESSING EQUIPMENT LIMITATIONS

A fixed factor that the manufacturer must accept is the specific processing
equipment available and its suitability for producing a given item. Dolls'
heads will not be produced on a calender, nor can electrical conduit be readily produced by the plastisol technique. While these examples are obvious,
there may be other specific equipment limitations that are not so obvious and
may seriously impair the manufacturer's ability to use compounded PVC for a
desired end application.

Thus, a plastics processor may be expert at extruding polyethylene profiles; but an attempt to process formulated PVC on the same equipment might
run into difficulties. The processor might then go to the formulator or compound supplier and complain about the quality of the material supplied for the
job at hand. One supplier after another might be tried, in a search for the
magic PVC compound that would provide the desired product. The basic
problem, however, is the manufacturer's failure to recognize that the problem

is with the different plastic material and its peculiar processing requirements. Whereas formula modification may help in some instances (and a poorly formulated compound may not run on the best of equipment), difficulties can be expected, for example, if processing is attempted on an underpowered short-barreled extruder with unsuitable screw and die design. The most expert formulator cannot help in such a situation.

Although PVC has been described as a versatile plastic, it is not a very tolerant one. It will not tolerate poor die or screw design, sloppy temperature control, lack of internal streamlining in the equipment, or poor operator handling and processing techniques.

V. FORMULATING PLASTICIZED VINYL COMPOUNDS

In formulating for any given vinyl application, three factors must be considered: (1) suitability for application; (2) suitability for processing equipment; and (3) economics.

A. Choosing Raw Materials

To satisfy these requirements for a flexible PVC compound, the formulator chooses from the following groups of raw materials:

Resin	Plasticizer
Stabilizer	Filler
Lubricant	Colorant

1. Resin: The primary factors to be considered in choosing a resin for flexible PVC are molecular weight, bulk density, and plasticizer absorptivity. Other factors that can be important, depending on the specific process or application, are particle-size distribution, dry flow, gel and contamination levels, electrical characteristics, residual monomer content, heat stability, and clarity. For plastisol applications, which require a dispersion resin, rheological properties as measured by both low shear (Brookfield) and high shear (Severs) techniques must be considered.

All of the foregoing are discussed in detail in Chapters 3, 4, 6, and 7, and the reader is referred to them for a comprehensive discussion. Outlined below, however, are brief discussions of some of these properties and their importance in choosing the proper resin for a given application or process.

Resin molecular weight is probably the single most important property to be considered. Mechanical properties, such as tensile and tear strength, will improve with increasing resin molecular weight. On the other hand, resins of low molecular weight will provide lower melt viscosities and can thus be processed more readily. Therefore, the choice of resin molecular weight is often a compromise between desirable properties and ease of processing. In general, as plasticizer concentration decreases, or filler content increases, resin molecular weight should be decreased for optimum processing.

Certain forming processes require relatively low melt viscosities (e.g., injection molding) and in such instances, resins of lower molecular weight than what might be used for calendering or extrusion will be chosen.

Resin bulk density and plasticizer absorptivity are generally interrelated and affect processability. A resin with high bulk density offers the advantage of larger batch sizes during preblending as well as potentially high rates of extrusion from powder blend. Such resins may tend, however, to have relatively poor plasticizer absorptivity, so that difficulties in blending may occur with high plasticizer content or with polymeric plasticizers. Dry-blendability will also generally tend to worsen with decreasing resin molecular weight. The formulator will generally prefer the resin that has the highest bulk density that will still give the degree of dry-blendability required for a particular process.

If a finished product is to be extruded directly from powder blend, then particle-size and dry-flow characteristics become of interest. The percentage of "fines" in the resin should be controlled from lot to lot to ensure consistent fusion characteristics and thus uniform extrudability. Dry-flow rates should preferably be high to obtain maximum production rates and to prevent "bridging" in the hopper or conveying system.

Heat stability refers to a resin's response to a given stabilizer system or type. If the formulator is "locked in" to a given stabilizer system or type, he will look for a resin that will provide the best possible results with that system. It should be recognized that stabilizers can often be tailored to provide optimum response with various resins.

Considering electrical characteristics or clarity becomes important when the compounder requires such properties in the finished product. A resin's suitability for such applications is generally determined in a final formulation.

Summary of Resin Selection Techniques

1. Choose either a general-purpose or dispersion-grade resin, depending on processing procedure.
2. Choose optimum resin molecular weight. Use the highest molecular weight commensurate with satisfactory processability.
3. With general-purpose (suspension or bulk) resins, choose the resin of the highest bulk density that will still satisfy requirements for dry flow and plasticizer absorption.
4. Make certain the resin meets requirements for the finished product— clarity, gel content, electrical properties, etc.
5. With dispersion grade resins, choose proper viscosity characteristics—for either high shear(e.g., coating) or low shear (e.g., slush molding) applications.

2. Plasticizer: The type and amount of plasticizer used will have the most significant effect on the compound's properties and processing characteristics. See Chapters 10 and 11 for a complete discussion of plasticizers and the theory of their performance.

The most important groups of plasticizers with which the formulator will work are outlined below.

a. Phthalates: The most versatile and widely used group of plasticizers are the phthalates, produced by causing phthalic anhydride to react with alcohols. They provide an outstanding compromise of good compatibility, excellent processing behavior, low-temperature flexibility, low volatility, heat and light stability, and availability at relatively low cost. They may be considered the starting point in all vinyl formulation work.

The alcohols most commonly used to manufacture phthalate esters are butyl, hexyl, octyl, nonyl, decyl, undecyl, tridecyl, and benzyl. The aliphatic alcohols may be either straight-chain or branched. Mixtures of various isomers and alcohols are often used instead of a single alcohol.

Increasing either molecular weight or aliphaticity results in generally low-er volatility, poorer compatibility, slower resin solvation, improved resistance to water extraction, and poorer oil extraction properties. Increased aliphaticity improves low-temperature performance. Reducing molecular weight or increasing aromaticity within certain defined limits, usually increases volatility, improves compatibility and resin solvation, increases water sensitivity, and improves resistance to extraction by oils.

b. Aliphatic Diesters: Esters of aliphatic dicarboxylic acids (adipates, azelates, sebacates) give excellent low-temperature flexibility to PVC and are primarily used for that purpose. As with the phthalate series, volatility, compatibility, and solvation decrease as molecular weight of the acid increases from adipic to azeleic to sebacic. Resistance to water extraction improves as plasticizer molecular weight is incrased. Oil resistance and solvent resistance are generally poor for all aliphatic diesters. The least expensive low-temperature plasticizers tend to be the adipates; the sebacates are the most costly and have limited use. The most commonly used low-temperature plasticizer is dioctyl adipate.

c. Phosphates: Esters of phosphoric acid are used as plasticizers in applications requiring flame resistance. The commercially available and most commonly used phosphates are the triaryls (based upon aromatic structures such as isopropyl phenol and *tert*-butyl phenol), the trialkyls (based on aliphatic alcohols such as octyl or decyl), and mixed alkyl-aryl phosphates (based on combinations of the foregoing groups). As the plasticizer structure changes from the highly aromatic triaryl structure to the completely aliphatic trialkyl form, low-temperature flexibility improves, while chemical resistance and flameproofing efficiency decrease. The phosphates adversely affect the heat-stabilizing action of most barium-cadmium stabilizer systems and some increase or modification of the stabilizer system is often required. Aromatic phosphates are among the most fungus-resistant plasticizers.

d. Epoxides: Epoxy plasticizers are widely used for their beneficial effect on a vinyl composition's heat and light stability. They are stabilizing adjuncts, displaying a synergistic effect with most metallic stabilizers, especially those containing cadmium or zinc. The epoxidized oils (soybean and linseed) are considered nontoxic and are characterized by low volatility and permanence. The monoester types (alkyl epoxy stearate and tallate) provide good low-temperature flexibility.

e. Polymeric Plasticizers: Reaction of a dihydric alcohol (glycol) with a dicarboxylic acid results in a polyester of high molecular weight. Most such polyesters are chain-terminated with a monocarboxylic acid to provide the polymeric plasticizer.

Polymeric plasticizers are noted for their low volatility and resistance to extraction by various media. They are valuable in formulating high-temperature electrical insulation, in preparing vinyl compounds that will not mar or attack other polymers, and in preparing compounds that must resist oil or solvent extraction. Drawbacks of the polymerics include their high viscosity, low rate of solvation, poor low-temperature properties, and relatively high cost.

f. Secondary Plasticizers: In the vinyl field, secondary (or extender) plasticizers are defined as materials with relatively limited compatibility that are ordinarily used in combination with a primary plasticizer such as DOP. Besides limited compatibility, secondaries often provide other undesirable characteristics such as poor heat or light stability or high volatility; their effect on these properties should be carefully evaluated. Secondaries may be used for specific processing or end-use performance requirements, but they are mainly used for reducing cost.

g. Trimellitates: Commercially available trimellitate plasticizers are produced by the esterification of trimellitic anhydride with 2-ethylhexyl or isononyl alcohols. Trimellitates display exceptionally low volatility and are invaluable for preparation of high-temperature electrical compounds. They are also fairly resistant to extraction by both aqueous and nonaqueous media, and there is thus considerable interest in their use for critical medical applications.

h. Specialty Plasticizers: Many other plasticizers besides those outlined above have been developed over the years, and find application depending on economic factors, commercial availability of raw materials, and specific requirements of the vinyl industry. Examples of such materials are glycolate esters for food contact use and solid plasticizers (nitrile rubbers, polyurethanes, and ethylene-vinyl acetate copolymers) for permanence and chemical resistance.

3. *Stabilizer*: Under the effect of heat or light, unmodified polyvinyl chloride will degrade, resulting in loss of physical properties, discoloration, and evolution of hydrogen chloride. To minimize or eliminate these effects, a stabilizer must be added to the vinyl formulation. Hundreds of stabilizers and stabilizer combinations are available to the vinyl formulator. Although some are identified by their chemical composition, the large majority are sold as proprietary products or mixtures of materials.

If a stabilizer's only function were to provide stability during processing (that is, to prevent discoloration and hydrogen chloride evolution), the chore of the formulator in deciding which stabilizer to use would be relatively simple. The choice would depend simply on the relative cost and heat-stabilizing efficiency of the candidate materials, and the selection would become obvious. In practice, the compounder must consider the relative importance of short-term color control and long-term basic stability as well as secondary processing characteristics such as plateout tendencies, effect on melt rheology, lubrication characteristics, effect on rate of fusion, and possible chemical interaction with other formulation components. Also to be considered are the properties the stabilizer will impart to the finished product, such as opacity (or transparency), staining characteristics, light stability, odor, toxicity, and electrical properties.

Chapters 7, 8, and 13 give a complete discussion of the theory of PVC degradation, the chemistry of stabilization, and the effect of various stabilizer structures on processing and end-use properties.

The compounder will choose among the following families of stabilizers:

a. Alkyltin Mercaptides: On a pound-for-pound basis, organotin mercaptides represent the most powerful vinyl heat stabilizers available. They provide excellent long-term stability, good initial color, and crystal clarity. For these reasons they are the stabilizers of choice for most rigid applications. They are seldom used for stabilization of flexible vinyls because of their odor, poor light stability, and cross-staining characteristics.

b. Alkyltin Carboxylates: The nonmercaptide organotin stabilizers have limited use. Although organotin carboxylates provide outstanding clarity, they are expensive and relatively inefficient. They have largely been replaced by barium-cadmium combinations for flexible vinyls and by tin mercaptides for rigids.

    c. Barium-Cadmium Soaps: The barium-cadmium soaps generally consist of mixed or coprecipitated salts of barium and cadmium with a fatty acid such as lauric, myristic, or stearic acid. Many of the proprietary materials available also contain an antioxidant and a polyhydric alcohol to provide improved stability. The soaps are usually used in conjunction with a phosphite stabilizer. Soap systems generally provide excellent long-term stability, good color control, and moderate transparency. Their primary drawbacks are their tendency to show moderate staining in the presence of sulfides (formation of yellow cadmium sulfide) and in their tendency to plate out on equipment.

    d. Barium-Cadmium Liquids: Barium-cadmium liquids can be described as the workhorse of the vinyl industry for stabilization of flexible stocks. They provide excellent heat and light stability and can be used for the manufacture of crystal clears as well as opaques. The liquids are less prone than the soaps to plate out and sulfide-stain and they are generally less costly.

    e. Barium-Cadmium-Zincs: Barium-cadmium liquids or soaps are often combined with zinc organic salts. Such barium-cadmium-zinc combinations have proved especially effective for stabilizing highly filled stocks. Zinc, by its presence, can also provide almost complete protection against cadmium sulfide staining.

    f. Leads: Lead compounds are widely used for stabilizing electrical compounds and for processing techniques that require outstanding long-term heat stability. Use of lead salts is generally limited to opaque applications. Primary disadvantages are toxicity and severe tendency to sulfide-stain.

    g. Organic Stabilizers: Organic stabilizers are generally adjunct materials used in combination with metallic stabilizers. Organic phosphites are the most commonly used, and they display a synergistic effect with many metallic systems. They are an integral part of most barium-cadmium liquids and are generally recommended for separate addition with most barium-cadmium soap stabilizers.

Antioxidants are included in many of the proprietary stabilizer systems. They are sometimes added separately by the formulator to improve retention of physical properties at elevated temperatures.

Epoxides as plasticizers have been discussed previouly. Epoxides act synergistically with most metallo-organic systems to improve both heat and light stability. The epoxide group can be incorporated into the vinyl compound by adding the typical plasticizers (epoxidized soybean oil, octyl epoxy tallate, etc.) or by incorporating an epoxy resin (condensation product of bisphenol A and epichlorohydrin).

    h. Nontoxic Stabilizers: Markets exist for flexible PVC in the food packaging and medical fields. In the United States most nontoxic stabilizers used in flexible formulations consist of selected organic salts of calcium, magnesium, and zinc in combination with organic adjuncts such as antioxidants, polyols, and epoxides.

For rigid compounds, a few specific alkyltin mercaptides or carboxylates are generally used.

In all instances, for food packaging the stabilizer must be used in accordance with governmental regulations covering such applications. In the United States such authority is vested in the Food and Drug Administrations and the restrictions are codified in the Code of Federal Regulations.

4. Other Compounding Additives: In addition to the basic compounding materials—resin, plasticizer, and stabilizer—the formulator must be concerned with fillers, lubricants, and pigments and also such special-purpose additives as fungicides and flameproofing agents, smoke suppressants, antistats, UV absorbers, etc. (see Chapters 15 to 19).

Fillers are generally inorganic fine powders such as calcium carbonate, clay, talc, or barium sulfate added to reduce compound cost or to impart desirable characteristics. Important reasons for use of some commonly used fillers are as follows:

Calcium carbonate	Cost reduction
Clay	Improved electrical properties
Talc	Dry hand
Barium sulfate	High density, radiopacity
Aluminum trihydrate	Flame and smoke suppression

The lubricant system chosen is based on the other compound ingredients, processing conditions, and potential effect on end-use properties. Lubricants are generally fatty acids or their metallic or organic esters and are classified as internal or external in performance.

The pigment chosen must provide the desired color, be heat- and light-stable, and preferably be resistant to bleeding or migration.

Any additive incorporated to achieve a specific effect may also affect other properties. This may require further modification of the formulation to compensate for the additive or may limit the other properties that can be achieved.

B. Formulating Plasticized Vinyl for Desired Properties

We have indicated that compounded PVC is a versatile plastic; that is, properties can be varied over a wide range depending on the specific formulating components chosen. Among the properties that can be controlled and that are of interest to the formulator of flexible PVC are:

Hardness	Tensile strength
Modulus of elasticity	Elongation
Low-temperature performance	Volatility
Heat stability	Light stability
Electrical properties	Clarity
Toxicity	Chemical resistance
Permanance	Stain resistance
Flammability	Fungal and bacterial resistance
Barrier properties	Rheological properties
High-temperature performance	Outdoor weathering characteristics

1. Hardness: Hardness indicates the degree of softening that has been induced in a vinyl compound; it is related to the amount and type of plasticizer present. The property actually measured is the degree of indentation that a spring-loaded needle, of given dimensions, will produce when placed on the vinyl test specimen. The most commonly used piece of equipment for measuring hardness is the Shore Durometer, and the procedure is described in ASTM D2240.

The "A" type of Durometer provides a measure of the degree of softening of vinyl compounds containing the equivalent of about 35 phr (parts per hundred of resin) or more of di(2-ethylhexyl) phthalate (DOP). As shown in Figure 1, a compound with 35 phr DOP (in a medium-molecular-weight resin—I.V. of 1.0) has a Shore A hardness of 95, for a 10-sec reading. As DOP content increases, the A scale hardness decreases. Increasing plasticizer content results in greater penetration by the test needle and a correspondingly lower reading on the Durometer scale.

The maximum reading of the Shore A Durometer is 100, and the instrument cannot readily differentiate among compounds with a Shore A hardness greater than 95 (compounds with less than 35 phr of DOP or the equivalent). For such compounds, the Shore "C" or "D" Durometer is generally used. The principle and operation of these instruments are the same as in the A type, except that a more heavily loaded spring and a sharper needle are used to obtain penetration of the harder sample.

The relation between plasticizer concentration and Shore A hardness for a variety of commonly used plasticizers is shown in Figures 1 and 2.

Although plasticizer type and concentration are the predominant influences on compound hardness, other compound ingredients can have an effect on these properties. Liquid stabilizers tend to have some plasticizing effect and give a slightly lower reading than insoluble stabilizers do. Resin molecular weight (Fig. 3) can influence needle penetration to a small extent and can thus exert a minor effect on hardness readings.

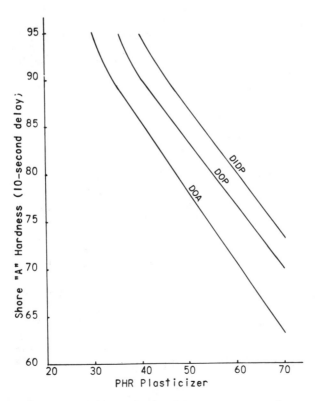

FIGURE 1 Effect of plasticizer concentration on Shore A hardness.

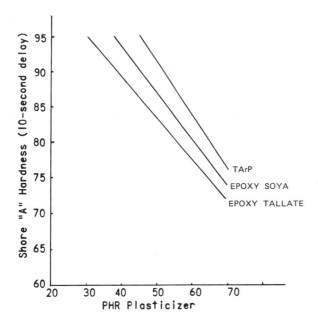

FIGURE 2 Effect of plasticizer concentration on Shore A hardness.

The effect of calcium carbonate on Shore A hardness is seen in Figure 4. Different fillers vary in their effect on hardness, depending on their plasticizer absorption characteristics.

To use the Shore hardness data effectively, one must recognize several factors. The first is that interlaboratory correlation under presumably iden-

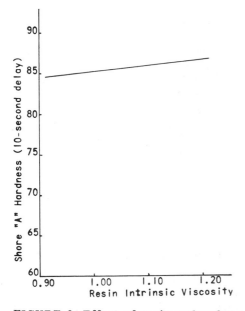

FIGURE 3 Effect of resin molecular weight on Shore A hardness.

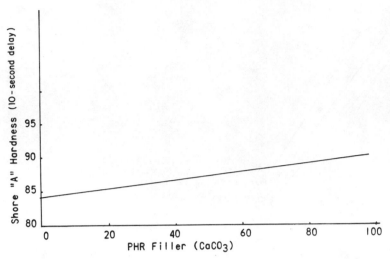

FIGURE 4 Effect of filler content on Shore A hardness.

tical conditions is no better than ±2 units. Second, Shore hardness is tem-
perature-dependent, and results obtained are valid only when a compound
has been properly conditioned under standard laboratory conditions. Third,
the hardness of a compound will increase with aging, especially in highly
plasticized formulations. Increases in hardness of up to 8 points within 1
week after processing are common. After that, hardness essentially levels
off to a constant value as long as temperature remains constant.

Attempts to use the rapid and simple Shore hardness tester as a plant
control tool must recognize the above limitations. Thus, an 80-Durometer
compound (as measured in the laboratory after proper conditioning) may
read 72 in the plant on a hot summer day, immediately after processing.

2. *Tensile Strength*: Tensile strength is the maximum force in tension
that a material is capable of sustaining before rupture occurs. It is expressed in terms of pounds per square inch or kilograms per square centimeter. Procedures used to measure tensile strength and also other physical
properties described below (modulus, elongation) are fully described in ASTM
D638. Test samples must be properly conditioned at the test temperature
prior to testing (at least 24 hr at 23°C has been widely accepted). Sample
geometry and rate of load applications will affect results, and they must be
specified if significant and comparative values are to be obtained.

A plasticized vinyl's tensile strength increases with

Decreasing plasticizer content
Increasing resin molecular weight
Increasing degree of fusion
Decreasing filler content

Figures 5, 6, and 7 show how plasticizer content and type, resin molecular
weight, and filler content affect tensile strength.

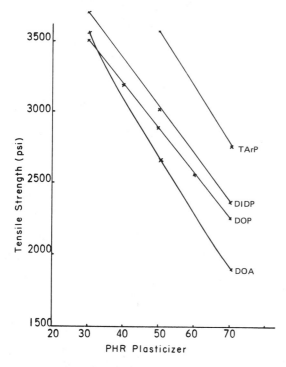

FIGURE 5 Effect of plasticizer concentration and type on tensile strength.

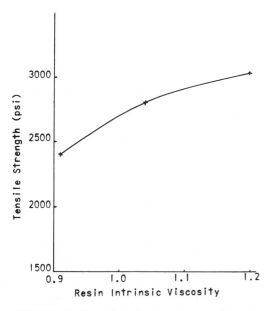

FIGURE 6 Tensile strength as a function of resin molecular weight.

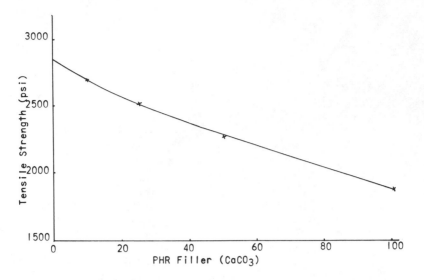

FIGURE 7 Effect of filler content on tensile strength.

    3. Modulus of Elasticity: A flexible vinyl product, as tensile force (stress) is applied to it, will tend to deform or stretch in the firection of force. The extent of deformation is defined as the strain. If stress is proportional to strain, measuring a series of such values and plotting a curve would yield a straight line, as shown in Figure 8. The slope of this curve (stress-strain) is defined as the modulus of elasticity and is expressed as pounds per square inch or kilograms per square centimeter.

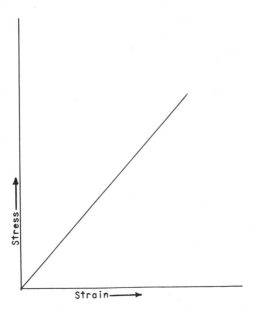

FIGURE 8 Ideal stress-strain curve.

The stress-strain curves of most plastics, however, yield not straight lines but curves. A plasticized vinyl, for example, might yield the type of stress-strain curve shown in Figure 9. Three distinct elastic modulus values can be obtained for a material showing the behavior depicted. Initial tangent modulus (line a) is the slope of the stress-strain curve at the origin; tangent modulus (line b) is the actual slope of the stress-strain curve at any specified stress or strain; secant modulus (line c) is the slope of the line drawn from the origin to any specified point on the curve.

The modulus of elasticity measured for most plasticized vinyl compounds, as reported in the literature, is the secant modulus measured at 100% elongation (compound at twice its original length), or strain = 1. The modulus is generally determined simultaneously with tensile strength measurements as described above and in ASTM D638.

One hundred percent elastic modulus varies with plasticizer content and type, and the relation is shown in Figure 10. For all plasticizers, the modulus decreases as plasticizer content increases.

4. Plasticizer Efficiency: Values of hardness and 100% elastic modulus as described above are often used in defining plasticizer efficiency, which refers to the amount of a given plasticizer that must be added to a formulation to achieve a desired degree of plasticization.

Figure 1 shows that a compound with 50 phr of dioctyl phthalate has a 10-sec Shore A hardness of 83. A compound with 50 phr of dioctyl adipate (DOA) has a Shore A hardness of 78. To achieve a hardness of 83, only 44 phr of dioctyl adipate need be used.

A triaryl phosphate TArP (see Figure 2) might be needed at a concentration of about 60 phr to achieve a hardness reading of 83. Using a phosphate at a level of 50 phr results in a compound with a Shore A hardness reading of 91.

Concentrations of 50 phr DOP, 44 phr DOA, and 60 phr phosphate all provide compounds with 10-sec hardness readings of 83 on the A scale. Dioctyl

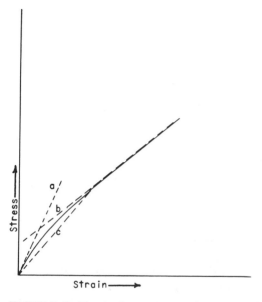

FIGURE 9 Typical stress-strain curve for plasticized vinyl.

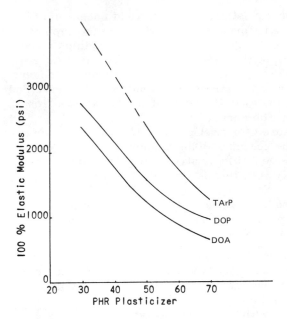

FIGURE 10 Elastic modulus vs. plasticizer content.

adipate, it can be said, is more efficient that DOP, since less DOA is required to achieve a required hardness. On the other hand, triaryl phosphate
is less efficient that DOP since more of it is needed to obtain a desired hardness value.

The same general relation can be seen in Figure 10, where it is noted
that a modulus of 1500 psi is obtained with 50 phr DOP, 44 phr DOA, and 64
phr phosphate. Again, DOA is more efficient than DOP, which is more efficient than a triaryl phosphate.

Plasticizer efficiency is a useful tool in formulating. In the above example, DOA would be given an efficiency rating of 44/50, or 0.88, compared
with DOP. If 10 parts of DOP were to be replaced by DOA to improve low-
temperature flexibility, only 8.8 parts of DOA should be added to preserve
equivalent modulus or hardness.

Efficiency values can vary, depending on whether modulus or hardness
is chosen for comparison. Thus, the efficiency value of a triaryl phosphate
(compared with DOP) is 1.28 at a modulus of 1500 psi but 1.20 at a Shore A
hardness of 83, even though the comparison in both cases is to 50 phr of
DOP. Efficiency values for some common plasticizers at several values of
modulus and hardness are summarized in Table 1.

5. *Ultimate Elongation*: Ultimate elongation is the maximum amount, expressed in percentage of original length, to which a vinyl specimen can be
stretched before rupture occurs. It is generally measured simultaneously
with tensile strength, as described previously. In general, ultimate elongation increases with increasing plasticizer content and resin molecular weight.
Elongation is reduced by filler content.

6. *Low-Temperature Flexibility*: All plasticized vinyl compounds become
stiffer as temperature is decreased, and they eventually become brittle at
some temperature. Depending on the types and amounts of plasticizer used,

TABLE 1   Efficiency Values for Some Common Plasticizers

	Modulus, psi			Hardness		
	1000	1500	2000	75	85	95
DOP	1.0	1.0	1.0	1.0	1.0	1.0
DOA	0.83	0.85	0.85	0.87	0.87	0.86
TArP	—	1.28	1.32	1.16	1.26	1.29

the rate at which flexibility decreases with decreasing temperature can be controlled. Thus, assume three compounds of the same flexibility at room temperature and plasticized, respectively, with DOA, DOP, and a triaryl phosphate. Examination of the stiffness of these compounds at a temperature significantly lower than room temperature (e.g., 0°C) will show that the DOA-plasticized compound is considerably more flexible than the one containing DOP, which in turn is more flexible than the one plasticized with phosphate.

Good low-temperature flexibility is usually obtained with low-viscosity aliphatic esters such as the adipates, azelates, sebacates, alkyl epoxy stearates, and trialkyl phosphates. Low-temperature performance becomes poorer with increasing plasticizer aromaticity or viscosity.

Various tests have been devised to differentiate among vinyl compounds in their ability to remain flexible and nonbrittle as temperature is reduced.

a. Clash and Berg Torsional Modulus: ASTM D1043 describes the Clash and Berg torsional modulus test. Torsional modulus is determined at varying subnormal temperatures, and a modulus-temperature curve is prepared. The compounds showing the smallest change in torsional modulus as temperature is reduced have have the best low-temperature flexibility. Single point measurements are often used for comparison. Thus, T_f is defined as the temperature at which a compound has a torsional modulus of elasticity of 135,000 psi. The lower the value of T_f obtained, the better the compound's low-temperature performance. Other values of torsional modulus can be used for such comparison.

b. Brittleness Temperature: Described in ASTM D746, the method of examining for brittleness establishes the temperature at which 50% of the test specimens show brittleness failure when subjected to a specified impact.

c. SPI or Masland Impact: The test for SPI is primarily used on calendered or extruded film and sheeting. The procedure measures the combined effect of basic compound low-temperature flexibility and the residual strains put into the product during processing. The film under test is folded into a loop of specified dimensions and is struck with a hammer of specified design at various temperatures. The brittleness temperature is defined as that at which 50% of the test specimens fail. This procedure has been accepted as part of the ASTM standards and is described in D1790. The SPI impact test is often used as a control test at a specified temperature on a go/no-go basis.

d. Mandrel Test: The procedure for the Mandrel test is generally performed on finished extruded items. The test specimen is conditioned for a given period at the test temperature and is subsequently wrapped about a mandrel of specified diameter. Any cracking that occurs in the test specimen generally signifies failure.

7. *Volatility*: With most vinyl compounds, it would be years before enough plasticizer was lost, due to plasticizer volatility at room temperature, to cause noticeable stiffening. Most of the plasticizers generally used in vinyls can be considered essentially low in volatility. As service temperature increases, however, or as surface area of the finished product as a function of its total weight increases, relative plasticizer volatility becomes important.

Volatility is, to a great extent, a function of plasticizer molecular weight. Generally, the higher the molecular weight, the lower the volatility of the plasticizer in the vinyl compound. Volatility is determined by exposing a test specimen to elevated temperature conditions over a specified period and determining the compound's weight loss. The rate of air circulation over the samples must be controlled if reproducible results are to be obtained.

ASTM D1203 is a widely used method for determining volatility characteristics. This procedure involves heat-aging the test specimen for 24 hr at an elevated temperature (usually 70° or 90°C) in proximity to, or in direct contact with, activated carbon (the function of which is to absorb the plasticizer vapors as they are produced). Although this procedure and other similar procedures can indicate basic compound volatility characteristics, results obtained do not necessarily indicate long-term life of the compound at elevated temperatures (see below).

8. *High-Temperature Service Characteristics*: It is obvious that a vinyl compound destined for high-temperature service applications should be formulated with low-volatility plasticizers. In addition, all formulation components should be chosen with the realization that they should not decompose into fragments of lower molecular weight and thus produce more volatile fractions during service. For example, experience has shown that highly branched plasticizers (those containing many tertiary carbon groups) are sensitive to such breakdown. To inhibit this tendency, high-temperature vinyl compounds containing such materials usually contain an antioxidant (e.g., phthalate).

In formulating for high-temperature service, one should remember the following:

1. Low-volatility plasticizers should be used.
2. The components should be stable at service temperatures or they should be stabilized to inhibit their breakdown.
3. The stabilizer system chosen should prevent or inhibit breakdown of the vinyl polymer at moderate temperature (as opposed to processing temperature).
4. Minor components (pigments, fillers, etc.) should not act as degradation catalysts for the plasticizers or resins.

The best criterion for determining high-temperature performance of a vinyl compound is exposure of a test specimen at conditions approaching as closely as possible the expected time and temperature of service. The greater the deviation from such service conditions in the interests of expediency, the less reliable the predicted behavior.

9. *Heat Stability*: The term heat stability refers to the ability of the compound to resist degradation and discoloration under conditions of processing. It can be measured by "static" or "dynamic" means (see Chapter 13). In the static test (ASTM D2115), samples are placed in a circulating-air oven at temperatures approximating those of normal processing and are

removed at regular intervals to determine degree of discoloration. In dynamic testing, the sample is processed on a two-roll mill and specimens are cut off at regular intervals. Discoloration and milling time before sticking occurs are observed. Torque rheometers such as the Brabender Plasticorder can also be used for dynamic testing.

All components of the formulation can affect heat stability. Differences in heat stability will often be observed between two presumably interchangeable resins from different manufacturers, or from different lots or recipes from the same manufacturer. These differences are often due to the concentration and type of polymerization residues present, the purity of the monomer used, and the degree of branching and unsaturation in the finished resin. As resin molecular weight increases, static heat stability will improve, all else being equal. Dynamic stability will often decrease, however, since higher internal stock temperatures will develop due to the greater frictional heat buildup resulting from increased melt viscosity.

Increasing plasticizer content generally improves both static and dynamic heat stability. Certain types of plasticizers, however, adversely affect heat stability; examples of such materials are the phosphate esters and the clorinated paraffins.

Most fillers tend to influence heat stability to some extent. Thus, even a relatively pure calcium carbonate (the most widely used type of filler) may require some modification in stabilizer system to compensate for its use. Impurities that may be present in the filler can have an important influence on stability and stabilizer requirements.

The formulator must also consider the potential effect on heat stability of such minor components of the formulation as the pigment and lubrication.

To summarize, every component of the vinyl formulation can affect heat stability and must be considered in choosing the stabilizer system and in the final decision about whether or not to use the desired additive. The effect of the additive can be determined by either a static or preferably a dynamic heat stability test. The stabilizer system can often be adjusted to compensate for the inclusion of the desired additives.

10. Light Stability: Like many organic chemicals, polyvinyl chloride compositions can undergo chemical reactions when exposed to ultraviolet light. When such reactions result in a loss of physical properties or a change in appearance, the material is said to undergo degradation. Such degradation is apparent from stiffening, embrittlement, spotting, and general discoloration. Changes that the vinyl compound will undergo can include chain scission of the base polymer, cross-linking of the polymer or of its breakdown fragments, and rearrangement or breakdown of the plasticizer molecules to a chromophoric, incompatible, or volatile form.

There has been relatively little agreement on significant tests for measuring light stability per se, since the term is somewhat ambiguous. Light exposure implies visible light. Yet visible light is not the primary cause of degradation of plastics; rather, it is the unseen ultraviolet portion that is an integral part of the sun's radiation and also many types of artificial illumination.

The intensity and wavelength distribution of ultraviolet light vary significantly both in sunlight (geographically and seasonally) and in various types of artificial illumination. The real question is not so much how good a compound's light stability is, as how well it resists degradation by the particular wavelength of light, visible or invisible, to which it is to be exposed in service. Thus, ability to withstand long exposures in an accelerated weathering device, where the light source is artificial, does not necessarily imply that the compound will display good resistance to sunlight, since the latter may have an entirely different ultraviolet spectrum.

The best measure of a compound's light stability involves exposure to the same type and intensity of radiation that it will experience in actual service. Since such exposures are usually impractical and time-consuming, a large number of accelerated tests based on artificial light have been developed to attempt to obtain light-stability information under controlled laboratory conditions in a relatively short period. Among such test devices have been the carbon arc Weather-Ometers and Fade-Ometers, various xenon testers, sunlamps, high-pressure mercury arc testers, fluorescent sunlamp/black light testers, and the germicidal lamp. Although these devices may give some indication of relative light stabilities, the results obtained should be treated cautiously until the instrument's performance has been calibrated and verified, using a wide variety of compositional variables. Care should be used in trying to extrapolate results of artificial-light aging studies to actual outdoor weathering performance without such prior verifications. Experience has been that the greater the degree of acceleration of a light aging test, the less reliable the results.

All components of the vinyl compound can affect light stability. Thus, care must be exercised in choice of resin, plasticizer, filler, and the stabilizer-lubricant system. Properly formulated flexible vinyls are among the most light-stable of polymeric materials.

For optimum light stability, a formulator will choose a base resin of relatively high molecular weight with minimal branching and little residual polymerization additive.

Satisfactory light stability can be attained with most plasticizers. Although light-stabilizing action has been claimed for octyl diphenyl phosphate, most phosphates have been found to worsen light stability. Chlorinated paraffins and highly aromatic secondary plasticizers should be avoided. If branched-chain alcohol esters are used, an antioxidant should be part of the stabilization system.

Epoxy plasticizers enhance light stability and should be incorporated at levels of 3 to 5 phr in almost all flexible vinyl compounds requiring good light aging.

Many fillers contain small quantities of various metallic oxide or salt contaminants, which can adversely affect light stability. Most of the commerically available calcium carbonate fillers are suitable, however.

For protecting clear or translucent compounds, certain additives function as light stabilizers by absorbing the bulk of the ultraviolet radiation, converting it to resonant intramolecular energy. Examples of such products are hydroxy-substituted benzophenones and benzotriazoles.

Excellent light stability can be attained with most of the commercially available barium-cadmium-zinc-phosphite combinations. Optimum results are obtained when these systems are used in combination with an epoxy plasticizer and an ultraviolet absorber.

A typical light-stable formulation would be as follows:

PVC resin	100
Ester plasticizer	30 to 80
Epoxy plasticizer	3 to 5
Calcium carbonate	0 to 50

Ba-Cd-Zn stabilizer	2 to 4
Phosphite stabilizer	0.5 to 1.0
UV absorber	0.1 to 1.0
Fatty acid or ester lubricant	0.3 to 1.0

Since ultraviolet radiation is one of the primary causes of vinyl breakdown, one way to protect the product is by screening out as completely as possible all ultraviolet radiation. This can be done through the use of high levels of highly opaque and light-stable pigments such as titanium dioxide and carbon black. This approach is used in formulating rigid compounds for outdoor exposure. Levels of 10 to 15% titanium dioxide are commonly used in formulating house siding, gutter systems, and vinyl windows.

11. Weatherability: The terms *weatherability* and *light stability* are often used interchangeably. Although the terms are interrelated, *weatherability* is the more inclusive since it describes the compound's ability to withstand protracted exposure to the elements and to the various processes related to weather conditions. A compound's weatherability, in addition to its light stability, can be affected by volatility (loss of plasticizer in hot climates), low-temperature flexibility (embrittlement in cold climates), resistance to fungi and bacteria (in hot, humid environments), resistance to water extraction (rainy areas), and the interrelation of the foregoing with each other and with ultraviolet light.

Weatherability essentially encompasses many potential causes of compound failure, and the formulator must consider the specific environment to which the product will be exposed if a life of maximum service is to be attained. Because of the complexities and interrelations among the various phenomena, no satisfactory accelerated test has been developed that will accurately predict total weatherability. All components and factors must be considered separately.

12. Electrical Properties: Electrical insulation is one of the most important outlets for plasticized polyvinyl chloride. The contributions of all compound components must be considered for their influence on insulation resistance, dielectric strength, and capacitance. The components used must generally be as free as possible from ionic impurities.

The compounder, besides being concerned over intrinsically electrical characteristics, must consider service conditions and maintenance of electrical and physical properties during use. The most restrictive nonelectrical requirement is maintaining physical properties at various elevated-temperature exposures. To effect such maintenance, the formulator will use low-volatility plasticizers and highly heat-stable formulations.

Choice of plasticizer will depend on the specific maximum service temperature to which the vinyl will be exposed on a continuous basis. As the service temperature increases, plasticizer volatility must be decreased and heat stability must be increased. Other exposure conditions that may have to be considered, depending on service conditions of the insulation, are oil resistance, resistance to fungi, low-temperature flexibility, light stability, flammability, and smoke generation.

Examples of plasticizers widely used for continuous service at varying temperatures are shown in the accompanying chart.

Service temperature	Plasticizer examples
60°C	Diisononyl phthalate
75°C, 80°C	Diisodecyl phthalate
90°C	Diundecyl phthalate
	Ditridecyl phthalate
	Polymerics
105°C	Polymerics
	Trioctyl and triisononyl trimellitates

Superior electrical properties are obtained when the compound is filled with a combination of clay and calcium carbonate, as is done with most primary insulation. Calcium carbonate alone may be used for jacketing applications.

Lead stabilizers such as tribasic lead sulfate, dibasic lead phosphite, and dibasic lead phthalate provide superior electrical properties and retention of physical properties for primary insulation. Barium-cadmiums may be used for jacketing.

13. Clarity: Plasticized polyvinyl chloride is intrinsically clear. Clarity is lost only through the incorporation of various compounding ingredients of limited solubility in the vinyl chloride resin. Such components may be contaminants (e.g., resin polymerization residues) or deliberate additives.

The resin used in manufacturing a clear vinyl must be specifically chosen for the application, and the plasticizer system should be of a high degree of compatibility. Choice of stabilizer is critical. Solubility within the basic compound is of prime importance here also, and best results have been achieved with organotin stabilizers and with specific barium-cadmium liquid compositions. Clear nontoxic compounds may be stabilized with calcium-zinc or some alkyltin stabilizers.

Lubricants that may be used in clear systems include stearic acid, stearamides, fatty acid esters, and mineral oil. Metallic fatty acid soaps and waxes are generally avoided where optimum clarity is desired.

Fillers and insoluble inorganic pigments are ordinarily not used in clear compounds. In exceptional situations when they are included, they are selected for refractive index match with the remainder of the vinyl formulation (see Chapter 17).

14. Toxicity: The environmental movement that began in the late 1960s has had its effect on the potential suitability of PVC for various applications. Environmental scientists raised many serious questions concerning the safety of PVC, residual monomer content, and worker and consumer exposure to monomer and compounding additives. The industry responded to these problems—modifying processes and instituting controls as required to ensure the safety of its workers and of the public.

The question of safety of the basic resin was raised during the early 1970s, when a number of cases of liver angiosarcoma were traced to frequent exposure to high levels of vinyl chloride monomer at a polymerization plant during the 1950s. A controlled animal study established the carcinogenic nature of VCM and the U.S. Occupational Safety and Health Administration (OSHA) reduced allowable worker exposure from 50 ppm to as low as 0.5 ppm of VCM per 8-hr weighted average (action level).

The PVC industry modified its operations to ensure the safety of its production employees and of its immediate plant environment. The resins manufactured from these modified processes were certified to contain less than 10 ppm residual VCM (RVCM) and most commercial resins sold by the late 1970s had less than 2 ppm. Extensive testing of flexible vinyl compounds manufactured from these low-RVCM resins showed no measurable levels of RVCM to a detection limit of 0.05 ppm. Monitoring of workers in PVC compounding and processing plants utilizing such resins has resulted in no findings of worker exposure above the action level of 0.5 ppm RVCM. There is little question today of the suitability of low-RVCM resins for food and medical applications. The potential safety of all the other components of a vinyl compound must be considered—from the standpoint of intended use as well as worker and consumer exposure.

In the United States specific regulations concerning additives intended for food packaging applications are promulgated by the U.S. Food and Drug Administration (FDA) and are codified under Title 21 of the Code of Federal Regulations. Similar regulations exist in most industrialized countries.

Vinyls for critical medical applications (e.g., intravenous or blood contact bags and tubing) are generally under the jurisdiction of the Bureau of Biologics of the FDA. However, for most medical applications, except for those mentioned above, specific government regulations do not exist in the United States. Most American producers of medical devices rely on FDA guidelines, on protocols for packaging of drugs as published in the U.S. Pharmacopoeia, and on their own testing for safety.

Most flexible compounds intended for food or medical use in the United States are plasticized with di(2-ethylhexyl) phthalate and epoxidized soybean oil and stabilized with calcium-zinc salts and organic esters. Rigid compounds are mostly stabilized with sanctioned alkyltin stabilizers.

In 1980 the National Institutes of Health reported a study showing carcinogenic effects of di(2-ethylhexyl) phthalate and adipate in white mice and rats at feeding levels of 3,000 to 12,000 ppm of their diets. At the time of this writing, significant additional test work on these plasticizers is in progress, and indications are that these esters present no apparent danger to humans when normally used in food and medical contact applications. It appears likely that no action will be taken to restrict use of these plasticizers.

Among the U.S. government agencies that are involved in regulating the use of PVC additives in the interest of consumer and worker safety are the FDA, OSHA, EPA, and the Consumer Safety Products Commission (CSPC). These agencies regularly promulgate regulations and continue to investigate potential problems in the use of plasticizers, stabilizers, and other additives, and the formulator's knowledge of regulations must remain current. Among materials presently regulated are lead stabilizers, various pigments, and some fillers.

15. Chemical Resistance: By the term *chemical resistance* is meant the ability of the vinyl compound to fulfill its function and maintain its properties in the presence of acids, alkalis, and organic chemicals. Because of the infinite number of materials that the vinyl may contact and the numberless time-temperature variables, there is no standard means of determining chemical resistance. A general test method appears under ASTM D543. In practice, the compounder must expose the compound to the specific chemical to be used at conditions approximating as closely as possible those of actual service. Physical or mechanical properties before and after exposure are determined to find the chemical's effect on the vinyl compound.

16. *Permanence*: The term *permanence* generally defines the ability of the plasticizer to remain within the vinyl and retain its plasticizing effectiveness under various exposure conditions. Thus, permanence encompasses volatility (discussed previously), the ability of the plasticizer system to resist extraction or migration into an adjoining material, and the stability of the plasticizer to its environment.

The resin-plasticizer relation is akin to the relation existing in a solid-liquid solution. The plasticizer is not bound chemically to the resin molecule, and it can be extracted from the vinyl by other materials that can act as solvents for the plasticizer. The extent of extraction that can occur is governed by the relative affinity of the plasticizer for the adjacent extractant (affecting extent of extraction), the viscosity of the plasticizer (affecting the ease with which the plasticizer can migrate from the vinyl), and the temperature of exposure (affecting rate of loss).

From a practical standpoint, the greatest problems of plasticizer extraction occur from water, soapy water, and organic oils and solvents. Best resistance to water and soapy-water extraction is shown by monomeric esters of high molecular weight, such as diisodecyl phthalate, diisodecyl adipate, and the trimellitates. Polymeric plasticizers display the greatest resistance to oil and solvent extraction. Many polymeric plasticizers do not stand up well to aqueous exposure because of their tendency to hydrolyze into fractions of lower molecular weight.

Vinyl plastics are often in intimate contact with other plastics. In such instances, it is generally undesirable for the plasticizer in the vinyl to migrate into the adjacent surface since this can cause stiffening of the vinyl and softening of the other polymer. Problems that come up in practice include migration into nitrocellulose, polystyrene, styrenic copolymers, and polycarbonates. Factors that dictate the plasticizer's tendency to migrate are its affinity for the other plastic and its mobility from the vinyl.

The highly viscous and relatively immobile polymerics are often used for applications requiring resistance to migration into other polymers. Thus, refrigerator gaskets, which may be in long-term intimate contact with polystyrene refrigerator bodies, are generally plasticized with polymeric materials. Similarly, vinyls intended for pressure-sensitive tapes, where they will be in long-term intimate contact with a rubber adhesive, will often contain a high percentage of polymeric plasticizer.

17. *Stain Resistance*: In consumer applications, the ability of the vinyl to maintain its original color and not accept stains from external agents is highly desirable. Determining the cause of a stain in a vinyl compound after service can require a fair degree of detective work and the cause is not always obvious. Among the causes to which various types of staining have been traced are moderate-temperature heat-aging, light-aging, migration into the vinyl of a component that is colored or can become discolored under various conditions, fungal attack, and industrial atmospheres.

Discoloration from moderate-temperature heat-aging can occur when different parts of a vinyl product are exposed to varying temperatures. An example of such discoloration is a vinyl floor that has become discolored under a radiator while the remainder of the floor retains its original color. This problem can often be minimized by careful selection of the stabilizer to protect against moderate-temperature heat discoloration. Satisfactory protection against discoloration under processing temperature conditions does not necessarily ensure protection against discoloration at moderate temperatures over a long period.

In a similar manner, different parts of a vinyl product may be exposed to varying amounts of light. As an example, a vinyl wall covering might be near a window, where it receives considerably more sunlight than other parts of the wall. To avoid such discoloration, components that will be light-stable should be chosen and the stabilizer system used should be one that will withstand such exposure. Secondary plasticizers should be avoided and lightfast colorants should be chosen.

Staining due to migration of an external component into the vinyl compound may be considered the opposite of plasticizer extraction. The question here is how poor a solvent is the plasticizer for external agents that may migrate into the vinyl and discolor it. Plasticizers that are relatively resistant to such staining have been developed.

Sometimes reformulation of the vinyl to resist external staining is not practical or desirable for economic or performance reasons. In those instances, it may be possible to change the external environment. An example of such an approach is in an application that may involve intimate contact between a rubber and a vinyl. Many rubber antioxidants can migrate readily into vinyl. Some of these antioxidants are light-sensitive and will subsequently discolor when the vinyl is exposed to ultraviolet light. The problem can be avoided by using nonstaining antioxidants in the rubber composition.

One of the most mysterious causes of vinyl discoloration has been attack by fungus. Fungi, attacking either the vinyl or a cellulosic substrate, can release a pink dyelike material that will cause random pink-spotting of the vinyl. The problem can be overcome by making both the vinyl and the substrate material resistant to fungal attack.

Industrial atmospheres may contain various reactive chemicals such as nitrous oxides and hydrogen sulfide. These may react with the pigments or stabilizers present, causing bleaching or discoloration of the product. This can be avoided by recognizing the possibility of exposure to such conditions and, if possible, choosing pigments and stabilizers that will not react with such gases. The hydrogen sulfide-staining problem can be anticipated by using nonstaining stabilizers, such as organotins or barium-cadmium-zincs (the zinc provides protection against sulfide staining), and by avoiding pigments based on lead or other metals that can form a colored sulfide.

Staining due to sulfide reaction with stabilizer or pigment can come from other sources besides the industrial atmosphere. Many foods, for example, contain sulfides, and so do cooking and heating gases. Cardboard and kraft paper, often used to package vinyls, may contain sulfides and can cause discoloration. Almost complete protection against sulfide staining can readily be obtained by careful selection of pigment and stabilizer.

18. Fungal and Bacterial Resistance: Although polyvinyl chloride is resistant to attack by microorganisms, most plasticizers are subject to attack and are a potential source of food to fungi and bacteria. Aliphatic plasticizers based on naturally occurring fats or oils are especially subject to such attack. Various tin, copper, mercury, arsenic, and organic compounds have been proposed and are being used as additives to inhibit attack of a vinyl compound by fungi and bacteria (see Chapter 16).

19. *Flammability*: Although polyvinyl chloride is completely self-extinguishing, most plasticizers are generally not. Resistance to flammability
can be achieved with phosphate plasticizers and they are widely used for this
purpose. Chlorinated paraffins are also often used for this reason, but it is
postulated that their primary effect is in reducing the concentration of other
normally flammable plasticizers in the vinyl composition.

Flameproofing can also be achieved by adding antimony trioxide to the
vinyl compound. This approach is satisfactory for opaque stocks in which
sulfide staining is not a potential problem.

Combinations of antimony trioxide with other additives such as aluminum
trihydrate and zinc borate are effective in minimizing flammability and sup-
pressing smoke development. Numerous proprietary materials have also been
developed to achieve this effect.

Many tests have been devised for determining relative flammability, and
some of these are described in ASTM D229, D568, D635, D757, D1433, D2843,
D2863, and D3801.

C. Techniques for Formulating Plasticized Vinyl

Various properties, then, can be achieved with flexible vinyls and also with
the formulation components that are available for attaining these requirements.

There are specific techniques the formulator can use, in a step by step,
systematic manner, to achieve the processing and product properties desired.
The number of different vinyl formulations designed to meet all possible requirements is infinite. No attempt can be made even to begin to list so-called
typical formulations. There is no such thing as a typical formulation; all
compounds are extremely specific in requirements and processing. Instead of
making a compendium of vinyl formulations, we show here by several specific
examples how a compounder can use the tools available to attain desired
properties.

Example 1

>   Product: General-purpose vinyl film
>   Production: Intensive hot fusion and calendering.
>   Requirements: Good all-around performance, translucent, 85 Durome-
>   ter, flame-retardant.

The first step is to determine (from Fig. 1) that an 85-Durometer compound can be achieved with 47 parts of DOP (the most commonly used plasticizer and the starting point in most vinyl formulations). Thus, Step 1 in
formulating gives

>   (1) PVC   100 parts
>   DOP   47

For the plasticizer content chosen, a medium-molecular-weight resin will
be suitable in most instances. With higher plasticizer content, resin molecular weight would be higher; for stiffer compounds, a resin of lower molecular
weight might prove the best.

For calendering, barium-cadmium liquids have proved effective and economical stabilizers; they also allow a clear or translucent end product. Lead stabilizers might contribute too much to opacity and they generally do not provide the degree of initial color control required for high-speed calendering. Tin stabilizers, due to their high cost and undesirable side effects, are not widely used in flexible PVC formulating. Thus, 2 parts of a barium-cadmium liquid are chosen, in a concentration typical of what is required for normal processing. Higher speeds and higher temperatures, however, might necessitate higher stabilizer concentrations.

Since the barium-cadmium liquids provide little lubricating action, a lubricant must be added. Stearic acid as a lubricant is effective and low in cost and also helps the heat-stabilizing effectiveness of many barium-cadmium liquids. About 0.5 part is typical for calendering.

The basic formulation has now been expanded to

(2) PVC (med. MW)   100 parts
    DOP   47
    Ba-Cd liquid   2
    Stearic acid   0.5

Because of the heat- and light stabilizing effectiveness of epoxy plasticizers, especially with barium-cadmium stabilizers, 3 to 5 parts of an epoxy plasticizer is added to the formulation, replacing an equivalent amount of DOP. An epoxidized soybean oil would be the most common choice.

(3) PVC (med. MW)   100 parts
    DOP   42
    Epoxidized soya oil   5
    Ba-Cd liquid   2
    Stearic acid   0.5

Since the compound should be flame-retardant, either a phosphate plasticizer or antimony trioxide should be incorporated. Since antimony salts are prone to sulfide staining, a generally undesirable trait in calendered film, a phosphate would be chosen. The most commonly used phosphates are the triaryl and alkyl diaryl esters. The choice depends largely on the relative importance of low-temperature flexibility and flameproofing ability. The triaryl phosphates provide a more effective flameproofing agent than the more aliphatic alkyl diaryl phosphate. The latter, however, provides significantly better low-temperature flexibility. Assuming that low-temperature flexibility is not an important requisite, a triaryl phosphate will be incorporated at a level of 10 to 15 parts. For this illustration, the 15-part level is chosen.

Since the product is to be translucent, a small amount of low-opacity filler can be added to reduce cost and to achieve translucency. Up to 10 phr of calcium carbonate will accomplish this without seriously affecting physical properties.

Through the use of an aromatic phosphate, low-temperature performance will be adversely affected. The formulator may wish to compensate partially for this (even though it is not part of the specification) by using an epoxy tallate (rather than an epoxidized soybean oil) in the compound or a straight-chain rather than a branched-chain phthalate. Such a choice might be based on economic considerations.

The formulation now reads

(4) PVC (med. MW)   100 parts
     DOP   27
     Phosphate   15
     Epoxy tallate   5
     Calcium carbonate   10
     Ba-Cd liquid   2
     Stearic acid   0.5

Examination of formulation 4 in the laboratory wil indicate that the various formulation steps taken to achieve some of the required properties may have modified other desirable characteristics. It will be noted that the compound is harder than it was originally (since the phosphate is less efficient that DOP). It will also be noted that incorporating a phosphate has worsened heat stability, perhaps to the extent that 2 phr of the barium-cadmium liquid will not provide sufficient stability for the processing.

To adjust the hardness, the DOP content should be increased, and an increase of about 4 parts of DOP is needed to bring the hardness reading back to 85.

The stability of phosphate-plasticized systems can be improved by adding a barium-cadmium soap to the formulation. The compound is now

(5) PVC (med. MW)   100 parts
     DOP   31
     Phosphate   15
     Epoxy tallate   5
     $CaCO_3$ 10
     Ba-Cd liquid   2
     Ba-Cd soap   1
     Stearic acid   0.5

While adding the barium-cadmium soap has improved heat stability, using a soap may now have raised two other problems. Since the soap is lubricating, the compound may now be overlubricated, and a reduction in stearic acid content may be necessary. In addition, barium-cadmium soaps have a tendency to plateout on calender rolls. To overcome plating out, a small amount of silica may be added. The final formulation is thus

(6) PVC (med. MW)   100 parts
     DOP   31
     Phosphate   15
     Epoxy tallate   5
     $CaCO_3$ 10
     Ba-Cd liquid   2
     Ba-Cd soap   1
     Stearic acid   0.3
     Silica   0.5

The preceding formulation should be prepared in the laboratory to check for exact Durometer required and degree of translucency and flameproofing obtained. Minor final adjustments in formulation can then be made.

Example 2

Product: Sheeting for outdoor upholstery use.
Production: High intensity powder mixing plus flat die extrusion.
Requirements: Low-temperature flexibility, resistance to water extraction, resistance to migration into clothing, opacity, light stability, resistance to fungi, 75 Durometer.

The required Durometer is obtainable with about 62 phr DOP (Fig. 1) so the starting point is

(1) PVC   100 parts
     DOP   62

One of the requirements of the formulation is resistance to migration into clothing, which could eventually cause stiffening of the vinyl. Experience has shown that the use of polymeric plasticizers will help to reduce such plasticizer loss. Polymeric plasticizers are expensive, however, and they generally exhibit poor low-temperature performance. Thus, the formulator will not want to use more than is necessary to do the job.

To determine the amount of polymeric needed to reduce migration to a satisfactory level, the formulator may devise accelerated migration tests into absorbent materials such as a silicate, or may prepare a controlled series of compounds and run use tests on various pieces of furniture in the field. An educated guess may be all that is necessary. The path chosen will depend on the test facilities available and the amount of time available to make a choice.

For this example, it can be assumed that the formulator has found that replacement of 20 parts of DOP by an equal efficiency concentration of a medium-molecular-weight polymeric plasticizer will give the migration resistance desired. Thus, the formulation is now

(2) PVC   100 parts
     DOP   42
     Polymeric plasticizer   24

Since the particular polymeric plasticizer chosen is less efficient than DOP, it is necessary to use 1.2 parts of polymeric to replace 1.0 part of DOP to maintain equal hardness.

Incorporating the polymeric plasticizer has now worsened low-temperature properties, perhaps to the point that will not allow the compound to pass whatever low-temperature specification exists. Thus, the plasticizer system must again be modified to improve low-temperature performance. Two routes are open: (1) use an adipate, an azelate, or a sebacate, or (2) replace the DOP with a more aliphatic phthalate.

Assuming the use of an aliphatic phthalate with an efficiency value of 0.95 compared with DOP, the formulation now reads

(3) PVC   100 parts
     $C_{7,9,11}$ mixed aliphatic
        phthalate   40
     Polymeric plasticizer   24

Since heat and light stability are necessary characteristics, an epoxy plasticizer should be used. An epoxidized soybean oil, which displays permanence properties similar to those of many polymerics, might be the best choice and could be used to replace an equal-efficiency amount of the polymeric. Assuming an efficiency value (compared with DOP) of 1.05 for the epoxy and 1.2 for the polymeric, 5 parts of epoxy would replace 6 parts of polymeric (to the nearest whole part).

Assuming that 20 parts of calcium carbonate filler can be used while still exceeding minimum tear and tensile strength requirements, the formulation now reads

(4)   PVC   100 parts
      $C_{7,9,11}$ mixed aliphatic
          phthalate   40
      Polymeric   18
      Epoxy soya   5
      $CaCO_3$ 20

Since the compound is fairly soft and the product requires good tear and tensile strength, a high-molecular-weight resin will be chosen. Since the sheeting is to be extruded from dry blend, and since it contains polymeric plasticizer and a relatively high plasticizer level, the resin should be one with high plasticizer-absorption properties.

A barium-cadmium or barium-cadmium-zinc stabilizer will probably be selected. In combination with the epoxy present, light stability will be satisfactory for outdoor use, especially in combination with highly opaque and light-stable pigments. Stearic acid will again be the lubricant of choice.

All the plasticizers used in this formulation are subject to attack by fungus. Thus, a fungicide must be added, since fungal resistance is part of the product requirement. Some fungicides adversely affect heat stability, so that it may also be necessary to modify the stabilizer system, perhaps by adding a phosphite stabilizer or increasing the barium-cadmium content, or both.

The final formulation, which must now be checked in the laboratory and adjusted if necessary, is

(5)   PVC (high MW, high absorp-
          tion)   100 parts
      $C_{7,9,11}$ mixed aliphatic
          phthalate   40
      Polymeric plasticizer   18
      Epoxidized soybean oil   5
      $CaCO_3$ 20
      Barium-cadmium liquid   2.5
      Phosphite stabilizer   0.5
      Fungicide   1.0
      Stearic acid   0.5

Choice of the proper type and amount of light-stable and nonmigratory pigments completes the formulation.

VI. FORMULATING RIGID VINYL COMPOUNDS

Rigid PVC occupies a unique position in the field of major thermoplastics. It can generally be considered as a low-cost engineering thermoplastic providing excellent physical properties, excellent aging characteristics, rigidity, chemical resistance, and high impact strength. Clear and nontoxic compositions can be achieved through proper choice of raw materials.

Rigid PVC is processed by the traditional means of extrusion, calendering, injection molding, blow molding, and compression molding. Like plasticized PVC, rigid PVC must be compounded before it can be used.

A. Materials Available

The materials that the formulator will work with are the following:

1. The basic PVC resin
2. A stabilizer system to prevent degradation during processing and use
3. Lubricants to facilitate processing
4. An impact modifier to enhance impact strength
5. A processing aid to facilitate fusion and provide ease of extrusion or molding
6. Pigments for proper color and to provide light stability (if required)
7. Fillers to reduce cost or to improve impact or certain other properties
8. Miscellaneous additives such as heat distortion improvers, light stabilizers, and antistatic agents
9. Plasticizers

Proper selection of compounding ingredients gives the formulator considerable latitude in the manufacture of rigids so that a vast range of property requirements and processing conditions can be met. Rigid PVC is used for potable-water pipe, electrical conduit, phonograph records, industrial parts, food containers, cosmetic containers, medical parts, house siding, storm windows, and rain gutter systems.

1. Types of Resin: Almost all PVC resin used for rigid applications is suspension polymerized homopolymer. The physical properties are enhanced as molecular weight increases, but lower-molecular-weight resins provide higher flow and are generally easier to process. Compounds intended for extrusion (pipe, siding, conduit, gutters, etc.) are usually based on medium-molecular-weight PVC resins (inherent viscosity 0.90 to 1.0). Injection molding or blow molding formulations will be based on resins with inherent viscosities of 0.75 to 0.90. Resins with inherent viscosities as low as 0.5 are chosen for injection blow molding applications or for injection molding of intricately shaped products. Copolymers of vinyl chloride with vinyl acetate, ethylene, or propylene are commercially available in some markets and are utilized for special applications. Vinyl chloride-vinyl acetate copolymers are generally the material of choice for high-resolution phonograph records. Vinyl chloride-olefin copolymers provide exceptionally high flow and excellent stability and are used in some injection molding applications. Copolymer resins are generally significantly more expensive than homopolymers.

Graft copolymers have been introduced in recent years, based on graft polymerization of vinyl chloride onto a rubberlike backbone such as ethylene-vinyl acetate, nitrile rubber (acrylonitrile-butadiene copolymer), or EPDM (ethylene-propylene dimer) rubber. Most such graft copolymers have not proved economically feasible and there has been relatively little commercial use. Much work has also been done in trying to develop internally plasticized copolymers by essentially the same approach, but with little economic success to date.

2. *Stabilizers*: To inhibit the degradation that unmodified PVC would undergo at processing temperatures, a stabilizer is added. Other purposes of the stabilizer are to react with any hydrogen chloride liberated and prevent the formation of color in the vinyl plastic as it is being processed. Stabilizers used in rigid compounding in the United States can be divided into six basic families:

Alkyltin mercaptides
Alkyltin carboxylates
Barium-cadmiums
Calcium-zincs
Leads
Antimony mercaptides

The alkyltin mercaptides are among the most efficient from the standpoint of pure heat-stabilizing action and processability. They can provide excellent clarity and do not cause flex-whitening. Drawbacks of the tin mercaptides are odor and poor light stability. Tin carboxylates are inefficient heat stabilizers, at least compared to the tin mercaptides. Clarity and flex-whitening resistance are excellent, and the tin carboxylates generally provide good light stability. Combinations of carboxylates and mercaptides are sometimes used in an effort to compromise between advantages and disadvantages of each group.

The use of certain alkyltin mercaptides and certain antimony mercaptides has been sanctioned by the National Sanitation Foundation for stabilization of potable-water pipe. The antimony mercaptides have provided suitable stabilization for this purpose at a somewhat lower cost than the tin salts.

The greatest advantage of the barium-cadmiums is the excellent weathering properties attainable with these materials when they are used in combination with epoxy plasticizer and ultraviolet absorber. Although clarity can be excellent, barium-cadmiums often cause flex-whitening of the final product. In addition, processability can be a difficult problem, and experience has indicated that lubricating the barium-cadmiums can be quite specific and sensitive. Dynamic heat stability obtained with barium-cadmiums is usually less than can be achieved with tin mercaptides. The odor level of barium-cadmium-stabilized rigids is usually extremely low.

Combinations of calcium, magnesium, and zinc organic salts with polyols, epoxides, and antioxidants are used for applications requiring nontoxicity. At low levels the calcium-magnesium-zincs provide fairly good initial color and clarity, but the processing stability provided to homopolymers is generally unsatisfactory. At higher levels processing stability may be satisfactory but initial color and calrity tend to be poor. In addition, compounds so stabilized tend to flex-whiten quite severely. Somewhat better success in working with calcium-magnesium-zinc nontoxic stabilizers has been achieved with some copolymers.

The use of certain alkyltin thyoglycolate esters and alkyltin maleates has been sanctioned by the Food and Drug Administration for food packaging applications. These stabilizers can provide excellent heat stability and clarity and are now widely used for stabilization of food and beverage containers as well as for medical devices. Use of these stabilizers is generally restricted to fairly low levels, which vary depending on the specific application. The reader is advised to consult the Code of Federal Regulations, Part 121, for more specific information.

The final group of stabilizers that could be used for rigid PVC stabilization are the lead salts. They provide efficient low-cost stabilization and are chosen where factors of opacity, toxicity, and staining characteristics are not important.

3. *Lubricants*: Lubricants used in rigid vinyl formulations generally are classified as being either internal or external in nature. The function of an internal lubricant is to facilitate the flow of the polymeric molecular network through the processing equipment. The function of an external lubricant is to provide a barrier between the polymer melt and the processing equipment, promoting flow and inhibiting adhesion.

The proper type and amount of lubricant depend on all the other compounding ingredients and the specific processing techniques used. To a great extent, proper lubricant selection has been a trial-and-error proposition. Among the materials used for lubrication of rigid PVC formulations are metallic stearates and laurates, stearic acid, glyceryl mono- and diesters, fatty alcohols, paraffin waxes, stearamides, montan wax esters, low-molecular-weight polyethylenes, mineral oils, and organic stearates. The selection and evaluation of lubricants for rigid PVC are considered in greater detail in Chapter 15.

4. *Impact Modifiers*: Choosing a suitable resin, stabilizer, and lubricant system will allow manufacturing of rigid vinyl product by most of the standard processing techniques. The product will be fairly brittle, however, To achieve better impact strength, certain rubberlike materials are often compounded into the formulation. Such materials generally have a limited degree of compatibility with PVC, forming a heterogeneous, two-phase system of vinyl and rubber (see Chapters 10 and 16). The most widely used impact modifiers are ABS or MBS terpolymers.

The ABS (acrylonitrile-butadiene-styrene) terpolymers are among the most efficient impact modifiers available. The MBS (methacrylate-butadiene-styrene) types can provide excellent clarity and relative freedom from flex whitening, and are thus widely used in clear applications. Because of their unsaturated backbone, both ABS and MBS modifiers are subject to oxidative and light degradation, so that impact properties will generally decrease under the effect of ultraviolet exposure and/or elevated-temperature aging.

Other modifiers that are used to provide impact resistance include ethylene-vinyl acetate copolymers, acrylics, and chlorinated polyethylenes.

Impact properties can be moderately improved through the addition of ultrafine calcium carbonates with particle sizes in the range 1 to 2 μm. This approach is generally not suitable where chemical resistance is required, although combined use of such fillers with ABS modifiers can produce low-cost high impact products.

Other approaches that have been taken to improve impact properties involve the use of graft or block copolymers of vinyl chloride with rubberlike polymers. The patent literature has many references, but commercial acceptability has been limited by economic considerations.

5. *Processing Aids*: Processing aids are designed to do what their name implies—provide ease of processing while having a minimum adverse effect on final product properties. In calendering they contribute to the development of a more uniform, free-flowing bank, and in extrusion they facilitate achievement of a smooth extrudate. Three basic types of processing aids are commercially available: the acrylics, the styrene-acrylonitrile copolymers, and the chlorinated polyethylenes.

The most widely used processing aid is basically a polymethyl methacrylate. It has little adverse effect on any desirable properties and it improves extensibility and hot tear strength. Styrene-acrylonitrile copolymers are somewhat less efficient than the acrylics as processing aids, but function in a similar manner.

Chlorinated polyethylenes can be beneficial to processing, depending on the molecular weight of the polyethylene used, its crystallinity, and its degree of chlorination. Some chlorinated polyethylenes, besides improving processability, will improve impact strength of the composition, thus fulfilling a dual function.

6. *Plasticizers*: In effect on processability of rigid PVC, plasticizers may be regarded as processing aids. They reduce melt viscosity and generally aid in producing a smooth end product. Unlike the processing aids mentioned above, however, plasticizers tend to have a seriously adverse effect on physical properties of the finished product, and hence their use in rigid applications should be avoided wherever possible. Their most pronounced deleterious effects will be seen in the reduction of heat distortion temperature, impact strength, and chemical resistance.

Using plasticizers in manufacturing flexible PVC is well established. If a normal plasticized PVC is exposed to progressively lower temperatures, a rapid blow will cause it to shatter in a brittle manner. The temperature at which a prescribed load will cause this shattering to occur is called the brittle temperature, or T_b. As the compounder reduces the total plasticizer content in a flexible PVC formulation, the value of T_b increases (i.e., resistance to embrittlement at subnormal temperatures worsens). The stiffness of the compound at room temperature, and also any other potential use temperature, increases.

As plasticizer level is continually reduced, compound stiffness at room temperature continues to increase, and eventually T_b reaches room temperature. At this plasticizer level the compound feels completely rigid. A sharp blow delivered at room temperature causes the product to shatter. At temperatures somewhat above room temperature, the product shows a slight degree of flexibility and the same blow does not cause shattering.

As plasticizer level continues to decrease, T_b becomes greater than room temperature, and even though some plasticizer is present the compound is rigid and highly brittle at room temperature.

Looking at the matter from the other direction, one finds that adding small amounts of plasticizer to a rigid compound will not make the compound flexible until enough plasticizer has been added to result in a reduction of the brittle point below room temperature (or below the temperature at which the compound is being evaluated). In an otherwise unmodified formulation (one that does not contain an impact modifier), the brittle temperature is closely related to the heat distortion temperature of the vinyl.

With the above in mind, it is not surprising to find that adding small amounts of plasticizer significantly decreases heat distortion temperature.

The extent of such reduction depends on the amount and type of plasticizer added. Each part of plasticizer reduces heat distortion temperature of a rigid compound from 1.5° to 3.0°C. The same reduction, incidentally, is noted with compatible liquid stabilizers, so that maximum heat distortion properties are achieved with nonsoluble solid stabilizers.

In addition, it has been noted that using small amounts of plasticizer (or compatible liquid stabilizers) causes a significant reduction in a compound's impact strength at room temperature. For example, as DOP is added to the compound, there is a continuing decrease in impact strength until a minimum is reached at a DOP concentration of about 10 to 15 phr. As DOP content increases beyond this level, a gradual improvement in impact strength is observed and the plasticizer begins to exert a flexibilizing effect.

The relative effect of plasticizer (DOP) addition on a rigid compound's impact strength is shown in Figure 11. Differences in effect between plasticizers have been observed and minimal effect on impact strength occurs with aliphatic plasticizers.

If the formulator recognizes and can live within the limitations imposed by plasticizer incorporation, he will obtain a significant improvement in processing behavior. If the plasticizer chosen is one of the epoxide type, it will significantly extend heat stability during processing and often improve outdoor weathering performance of the finished product.

Because of the above factors, plasticizers are used in rigid applications only where they are absolutely necessary and only in the minimum quantities needed to perform their function. If plasticizers are used, epoxides will generally be chosen as the preferred type.

7. Pigments: Pigments chosen will be governed basically by the color required. Heat stability, light stability, and resistance to migration should

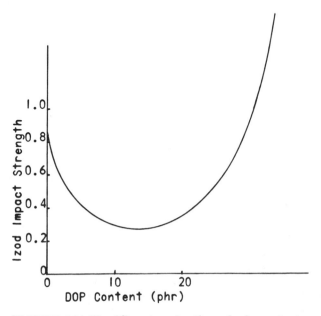

FIGURE 11 Significant reductions in impact strength on addition of small amounts of DOP plasticizer to a rigid compound.

be considered, as for plasticized stocks. It should be recognized that processing temperatures will generally be higher than with plasticized stocks so that a greater degree of heat stability will be needed. One additional point to recognize is that some pigments are based on metallic salts of lead, selenium, antimony, or copper. The metals in such pigments can react with the sulfide group in tin mercaptide stabilizers, leading to internal sulfide staining.

High concentrations of light-stable pigments (e.g., 10 phr TiO_2) markedly improve light stability and are often used in outdoor applications.

B. Performance and Evaluation of Rigid PVC Compounds

The formulator of rigid vinyl must be concerned with both processing behavior and product performance properties. Satisfactory processing behavior demands a melt viscosity suitable for the product being manufactured, adequate heat stability for the processing operation, and satisfactory rheological properties and lubrication in the melt stage so that a smooth product will be obtained.

Required product performance characteristics are dependent on the intended use of the rigid vinyl product. Properties that may have to be considered include heat distortion temperature, impact properties, stiffness, chemical resistance, permeability, clarity, light stability, toxicity, and any other specific requirements for the finished product. The rigid vinyl compound is compounded to take into consideration processing characteristics and product performance. How the formulator can choose among various available formulation components to meet the processing and end-use requirements is discussed below.

1. Melt Viscosity: A compound's melt viscosity must be controllable if the compound is to be processed satisfactorily (see also Chapter 29). Melt viscosity is principally determined by resin molecular weight but also by type and amount of lubricants, type of heat stabilizer, the presence of plasticizers or other viscosity-reducing additives, and process temperature.

Optimum melt viscosity depends on the specific processing technique used and the end product desired. Most extrusion operations utilize a medium molecular weight resin (inherent viscosity of 0.9 to 1.0) to optimize properties such as impact strength, which is generally required for applications such as rigid pipe, siding, and gutter systems. Most extrusion operations are designed to handle these medium molecular weight resins, and the molecular weight also helps in providing the melt integrity required for proper sizing of the extrudate.

In blow molding, however, too high a melt viscosity will prevent satisfactory fill of the mold during blowing, while too low a melt viscosity may cause excessive parison drip. For blow molding applications, resins with inherent viscosities ranging from 0.7 to 0.85 have proved satisfactory.

Still higher flow might be required for injection molding or injection blow molding, and resins with molecular weight as low as 0.5 inherent viscosity have been developed for these applications.

The addition of compatible liquids such as plasticizers and liquid stabilizers reduces melt viscosity. However, this approach can result in serious deterioration of desirable physical properties such as heat distortion, impact strength, and chemical resistance. Thus, the amount of liquid added to a rigid system should be minimized.

Obviously, lower melt viscosities are attained as melt temperature is achieved. The limiting factor is the heat stability of the vinyl compound. While heat stability can be improved by higher concentrations of stabilizers, this approach can adversely affect physical properties and generally increase cost.

2. *Heat Stability*: Unstabilized PVC cannot be processed without degradation. Thus, heat stabilizers as previously described are added to inhibit resin degradation during the processing operation. The type and amount of stabilizer added depend on required product performance characteristics as well as on the shear and temperature severity of the processing operation.

Extrusion operations on counterrotating twin-screw extruders generally permit significantly less stabilizer addition than if single-screw extruders were used. Temperature of an extrudate from a twin-screw extruder can be more closely controlled, and the polymer is subjected to significantly lower shear than with single-screw machines.

Proper design of gate and runners in an injection molding operation can also minimize the concentration of stabilizer required. However, if a significant amount of rework is to be used, higher stabilizer levels should be added than if minimal or no sprues and runners were to be recycled.

3. *Heat Distortion Temperature*: The heat distortion temperature of polyvinyl chloride is related to its second-order transition temperature, and the most common technique for this measurement is described in ASTM D648. In this test the heat distortion temperature is defined as that temperature at which a test specimen subjected to a fiber stress of 264 psi deflects 0.010 in. Completely unmodified and unplasticized PVC has a reported heat deflection temperature of approximately 78°C when tested by this procedure.

The heat deflection temperature is slightly reduced as the molecular weight of the base resin is lowered and as impact modifiers or nonliquid processing aids are added. The addition of plasticizer or liquid stabilizer has a major effect on heat distortion temperature and a reduction of 1.5° to 3.0°C per part of liquid can be expected.

Copolymer resins generally have lower heat distortion temperatures than homopolymers. Higher distortion temperatures have been obtained commercially with chlorinated polyvinyl chloride and have also been reported for stereospecific polymers manufactured by low-temperature polymerization. Other attempts have been made to improve heat distortion temperature by blending PVC with other polymers such as certain acrylic resins or SAN resins.

4. *Impact Properties*: Unmodified PVC has relatively poor impact resistance and tends to shatter when given a sharp blow at room temperature. The addition of plasticizer or liquid stabilizer makes this property even worse. Impact strength can be improved through the addition of certain rubber-type materials (as mentioned earlier in this chapter, and discussed in detail in Chapter 16). Typically 8 to 15% addition of certain types of acrylic, ABS, MBS, EVA, or CPE polymers will significantly improve impact characteristics and will change failure from a brittle to a ductile mode.

Many different tests have been developed to determine the impact properties of plastic materials and are fully described by the American Society for Testing and Materials. Among the tests that are used are ASTM D256, D1709, D3393, D2463, D3420, F725, and D4226.

One of the most widely reported tests is the Izod impact test as described in ASTM D256. A graph of impact modifer content versus Izod impact strength for a PVC formulation gives a typical curve such as that shown in Figure 12.

As incremental amounts of modifier are added, there is a relatively modest improvement in impact strength until a critical value is reached, beyond which there is a major improvement in Izod impact values. Thus it is possible for 10 parts of a modifier to give an Izod value of 2.5 ft-lb whereas 12 parts of the modifier may yield a value of 18. In actuality, quite a scattering of results will be obtained within this critical modifier concentration.

Examination of the type of break that occurs in the lower part of the curve will show essentially brittle-type failure. Breaks in the upper part of the curve are typically of the ductile type. Plotting a curve of this kind essentially indicates the level of impact modifier needed to reduce the value of brittleness temperature (T_b) to room temperature. In the example above, approximately 11 parts of modifier reduces brittle temperature to the temperature at which the test was conducted.

Alternatively, curves can be plotted of Izod impact strength versus temperature for two compounds that display, respectively, brittle (low Izod) or ductile (high Izod) break behavior at room temperature, as shown in Figure 13. It can be said that compound A, which shows high Izod values and ductile-type breaks at room temperature, has a T_b that is lower than room temperature. Conversely, the T_b of compound B is higher than room temperature.

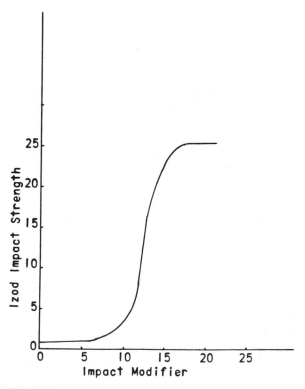

FIGURE 12 Impact strength vs. impact modifier content.

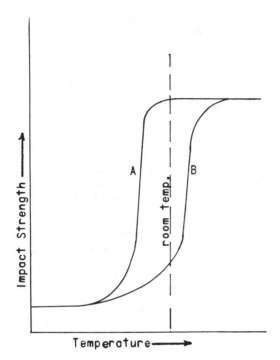

FIGURE 13 Typical effect of test temperature on impact strength of PVC.

Thus the brittleness temperature of a given compound may be more significant than any absolute Izod impact value determined at a single temperature. This value must certainly be considered for any PVC applications requiring reasonable impact properties at reduced temperatures.

Although it is obvious that the type and amount of impact modifier are very important in developing impact strength, other formulation components also have significant effects. Improved impact strengths are generally achieved with increasing resin molecular weight and minimization of liquid additives. Lubricants and fillers have significant effects, both beneficial and adverse, on impact strength. Finally, processing history will affect measured impact values, with impact strength reaching a maximum at what might be called optimum processing and then falling off as processing is continued.

5. *Stiffness*: Stiffness is usually measured by applying a force on a sample held as a cantilever beam and determining the angle of the bend (ASTM D747). Stiffness is decreased by incorporating impact modifiers and is increased by adding small quantities of plasticizing additives. Homopolymers are usually stiffer than copolymers.

6. *Chemical Resistance*: Rigid PVC has excellent chemical resistance. This basic resistance, however, is reduced by incorporating almost any processing or impact additive. To maintain maximum chemical resistance, PVC should contain only the minimum amount of stabilizers and lubricants required for processing.

To measure chemical resistance, the test specimen is exposed to a given chemical under predetermined conditions of temperature and time. After exposure, the weight gain (or loss) of the sample and changes in physical properties or required performance are measured.

7. *Permeability*: Like all plastics, rigid PVC displays certain permeability behavior. Materials encased within rigid PVC can migrate through the intramolecular interstices and escape from the package. On the other hand, environmental components (oxygen, carbon dioxide, water, etc.) can migrate through the container walls into the contents (see ASTM E96 and D1434). As with chemical resistance, lowest permeability is obtained with a material that contains a minimum of additives.

Permeability must be considered in determining whether a PVC container is a suitable package. One way to determine this is to fill a container from the test compound with permeant. The container is completely sealed and weight loss of the contents is determined as a function of time at room temperature and at a moderately elevated temperature.

8. *Clarity*: Unmodified, unstabilized, unlubricated rigid PVC displays excellent clarity, but it cannot be processed. Any lack of clarity in a finished formulation is due to the presence of an additive, and there are many more additives available that interfere with clarity than there are that preserve the resin's inherent clarity. Therefore, for optimum clarity, the type and quantity of all modifying materials must be carefully determined.

Certain groups of materials have generally proved satisfactory when clarity is required: organotin stabilizers, selected MBS, ABS, and acrylic modifiers, acrylic processing aids added in small amounts, and fatty acids and their organic esters as lubricants. The following are generally avoided when clarity is required: fillers, metallic soaps as lubricants or stabilizers, chlorinated polyethylenes, and hydrocarbon waxes. Even these materials, however, may be used in some instances in limited concentrations.

Maintenance of clarity under conditions of use must also be considered. There have been instances in which a perfectly clear PVC container became virtually opaque when filled with a product containing water or alcohol. Water hazing has often been traced to the stabilizer type, and opacification in the presence of alcohol has been traced to the type and concentration of processing aid and impact modifier.

9. *Light Stability*: Unless rigid PVC is properly formulated to provide light stability, it will deteriorate rapidly on outdoor exposure, especially in geographic areas where the ultraviolet content of sunlight is high.

Unfortunately, accelerated weathering tests in the laboratory do not always correlate with actual outdoor weathering results, and the vinyl compounder must depend on basic outdoor exposure data when formulating sunlight-resistant compounds. The following general comments related to sunlight resistance are based on outdoor weathering exposures the author conducted in Arizona.

1. Alkyltin mercaptides provide poor light stability in clear and translucent compounds and relatively little can be done to improve them.
2. Alkyltin carboxylates and selected barium-cadmium-phosphite stabilizers provide the bases of highly light-stable formulations.
3. Addition of an epoxy plasticizer and an ultraviolet absorber to alkyltin carboxylate or a barium-cadmium-phosphite can yield a clear or translucent material with excellent outdoor weathering performance.
4. Lead stabilizers, used in combination with a phosphite antioxidant, an epoxy plasticizer, and an ultraviolet absorber, provide moderate to good outdoor aging.

5. The impact modifier should be carefully selected. Many will interfere with weatherability, losing their impact-improving properties and actually accelerating degradation. MBS, ABS, and graft polymers will often present a problem. Best weathering is usually obtained with CPE, EVA, and all-acrylic modifiers.

6. High degrees of opacification with a light-stable pigment can provide excellent light stability, even with a base compound that has poor weatherability in the unpigmented state.

C. Formulating Techniques

In formulating for specific applications, the compounder must consider processing and economics. The finished compound must have suitable processing and stability characteristics consistent with the type of process being used; it must be suitable for its intended application; and, of course, this should be accomplished at the lowest possible cost.

Several typical formulations for rigid PVC applications and the rationale in their development are outlined below:

Example 1

Product: PVC bottle for hair shampoo,
Production: Blow molding.
Requirements: Good clarity, high impact strength, low permeability.

Experience has shown that the best blow molding characteristics are obtained with a resin whose inherent viscosity (IV) ranges from 0.7 to 0.85, and a resin in that molecular weight range is chosen.

Best clarity characteristics are obtained with a tin stabilizer, and specifically an alkyltin mercaptide will be chosen for its excellent heat stabilizing action. A methyl- or butyltin mercaptide will be chosen, depending on relative cost efficiencies, although odor levels with methyltins generally tend to be somewhat greater. Quantities required will range from 1.5 to 3.0 phr, depending on the type of processing equipment to be used and the amount of regrind that will be recycled.

The compound will be lubricated with 0.5 to 2.0 phr of a lubricant that will have a minimal effect on clarity, such as stearic acid, fatty acid alcohols, glycerol esters, montan wax esters, or stearamides. Generally a balance between external and internal lubrications will be developed.

A processing aid such as an acrylic polymer will be incorporated at levels of 1 to 3 phr to minimize melt fracture and thus provide a smooth parison. The acrylic modifier will also improve hot tear strength.

Impact strength will be obtained through the addition of an MBS modifier at a level of 8 to 15 phr. The highest degree of clarity and the lowest degree of flex whitening are generally obtained with the less efficient impact modifiers, which must be used at the higher levels to obtain adequate impact strength. If less than optimum clarity is accepted, lower quantities of a more efficient impact modifier would be considered, and such use would result in significantly lower cost. Lower levels of impact modifiers also generally provide lower permeability.

The addition of a tinting agent completes the formulation as follows:

PVC (0.75 to 0.9 IV)	100
Acrylic processing aid	1 to 5
MBS impact modifier	8 to 15
Organotin mercaptide stabilizer	1.5 to 3.0
Lubricants	0.5 to 2.0
Tinting agent	As required

Example 2

Product: PVC bottle for edible oil.
Production: Blow molding.
Requirements: Good clarity, high impact strength, low permeability, nontoxicity.

In formulating for a food packaging use, the same requirements that are outlined above for shampoo use must be met, with the additional requirements that all components must be sanctioned by the FDA for food packaging applications.

Major points of consideration, then, would be to use a resin with extremely low residual vinyl chloride levels so that no more than 5 parts per billion of residual vinyl chloride monomer remain in the finished product. In addition, the stabilizer used will probably be of an alkyltin mercaptide that has obtained with FDA sanction. Otherwise, the same general information techniques as outlined in Example 1 will apply.

Example 3

Product: Potable-water pipe.
Production: Extrusion from dry blend.
Requirements: Burst resistance, acceptable toxicity.

Best physical properties will be obtained with resins of medium molecular weight, and a resin whose inherent viscosity is 0.9 to 1.0 will be chosen.

Impact strength will generally be obtained with an ABS-type modifier, often used in combination with an ultrafine calcium carbonate filler. Levels of calcium carbonate are limited by its adverse effect on long-term burst resistance. A processing aid will be incorporated to provide a smooth extrudate and this will be of either the acrylic or the alpha methylstyrene type. Some chlorinated polyethylenes have also been suggested for this purpose, and chlorinated polyethylenes also function as impact modifiers.

The most widely used stabilizers in the United States for potable-water pipes are alkyltin or antimony mercaptides; levels of 0.3 to 2.0 phr are generally employed, depending on processing considerations.

A combination of external and internal lubricants will be used and combinations of calcium stearate and paraffin wax, sometimes with partially oxidized low-MW polyethylene, are often utilized for such purpose.

Pigments are used to provide the required color, and then the final formulation might be as follows:

PVC (0.9 to 1.0 IV)	100
Processing aids	1 to 5
Impact modifier	0 to 10

Calcium carbonate	0 to 10
Alkyltin or antimony mercaptide	0.3 to 2.0
Calcium stearate	0.5 to 1.5
Paraffin wax	0.5 to 1
Pigment	1 to 2

VII. CONCLUSIONS

Many procedures are involved in formulating both flexible and rigid PVC. Since new materials are constantly being developed and introduced, there will never be any single and best formulation for any given application. The vinyl compounder must stay constantly familiar with all products introduced that may help make formulating easier and less expensive, and that will open new avenues for the use of his materials.

Finally, the needs of the marketplace may dictate new or revised requirements which the formulator might not have been able to foresee or predict when originally developing the formulation, and these may necessitate formulation revisions or modifications. Likewise, the various governmental and other agency requirements with which the formulator must be cognizant frequently undergo changes, occasionally widespread in scope, and the formulator must be in a position to respond to these on short notice.

2

Actions and Characteristics of Stabilizers

THOMAS C. JENNINGS and CHARLES W. FLETCHER

Synthetic Products Company
Cleveland, Ohio

I. INTRODUCTION

In Chapter 8 (Volume 1) the theory of polyvinyl chloride degradation and stabilizer mechanisms was reviewed. In this chapter we deal primarily with the development, properties, and applications of PVC stabilizers per se.

The theory of PVC degradation and stabilizer mechanisms has been developed over the years along strict scientific lines and reflects many excellent scientific contributions [1–7]. The practical development and application of stabilizers, however, was in most instances a forerunner of stabilizer theory. The theory, in a general sense, explained phenomena and effects which were already being controlled by the PVC stabilizer compounder. To a large extent, the early stabilizers were developed as a result of the Edisonian type of trial-and-error pragmatism which was characteristic of the early vinyl compounders. Necessity was the mother of their inventions. Art more than science guided their hand. Dogged persistence and personal challenge were their driving forces.

Over 50 years have passed since Waldo L. Semon's plasticization patents gave birth to the U.S. vinyl industry in 1933 [8]. Since its inception, PVC has been nurtured and has grown by the development of specific stabilizers which have allowed resin to be used in a myriad of diverse industrial applications.

To all who contributed to the science and the art, the patents and the secrets, the theory and the magic...

To all who have been challenged by the need, the fascination, and the fun of formulating stabilizers...

To all who are yet to contribute and refuse to accept that the best stabilizers have already been invented...

To you, this chapter is dedicated.

A. The Necessity of Stabilization

One has only to observe the attempt to process PVC resin in the absence of stabilizer to become rapidly a believer in the necessity for stabilization. Heating unstabilized PVC above its fusion point initially gives rise to yellowing, followed quickly by gross discoloration, the evolution of hydrochloric acid, cross-linking, and ultimate charring to an infusible, unprocessable, corrosive black mass.

Polyvinyl chloride resin is inherently the most heat-sensitive of the major commercial thermoplastic resins. Nevertheless, virtually all vinyl compounding and fabrication techniques require that the resin be subjected to heat. Furthermore, many fabricated vinyl articles are exposed to varying degrees of heat and aging during their normal service life. Consequently, all PVC compounders worldwide, independent of location, application, or processing technique, have a need for PVC heat stabilizers.

The prolific growth of PVC during the past 50 years has been made possible to a large extent by the development of a wide variety of PVC heat stabilizers to protect the resin in a multitude of specific environments around the world.

B. Stabilizer Classification and World Markets

PVC heat stabilizers are classified as either primary or secondary stabilizers.

A primary stabilizer is a substance which, when employed as the sole stabilizer in PVC, imparts an acceptable degree of heat stability for commercial applications. Over the past 50 years, three dominant generic classes of primary heat stabilizers have emerged: lead stabilizers, mixed metal stabilizers, and organatin stabilizers [9].

A secondary stabilizer is a substance that cannot be employed as the sole stabilizer in a commercial PVC application, but can extend, complement, and synergistically improve the heat stability of PVC when used in conjunction with a primary stabilizer. There are two principal classes of secondary stabilizers: epoxidized oils and metallic soaps.

It is extremely difficult to assess accurately the size of the world market for the various stabilizer classes. The literature consists of numerous instances of conflicting data [10] and the problem is aggravated in certain cases where stabilizer packages consist of combinations of primary and secondary stabilizers in single packages.

Table I presents our best estimate of the 1983 world PVC stabilizer production by type and geographic region. This table indicates that the estimated volume of 415,000 metric tons can be further segmented into 56% primary stabilizer and 44% secondary stabilizer categories. Of the 56% or 229,000 metric ton primary PVC market; 54% is lead, 34% is mixed metal, and 12% is organotin—whereas the 186,000 metric ton secondary PVC stabilizer market is 83% epoxide and 17% metal soap. For further definition, the primary stabilizer category includes lead stabilizers such as inorganic lead salts, lead soaps, and all single packages (one packs) containing lead. It also includes primary mixed metal stabilizers such as all non-lead compounded stabilizer packages consisting of barium, calcium, cadmium or zinc stabilizers, all organotin stabilizers such as alkyltin, and finally, antimony compounds (including stabilizer/lubricant one packs). The secondary stabilizer category consists of epoxides such as epoxidized unsaturated natural oils,

TABLE 1 1983 Worldwide PVC Stabilizer Production Estimated by Region
(Thousands of Metric Tons)

Region	Primary lead	Primary mixed metal	Primary organotin	Secondary epoxide	Secondary soap	Total
United States	14	23	13	45	14	109
Western Europe	62	27	7	55	9	160
Japan	20	13	5	27	5	70
Other	29	14	2	27	4	76
Total	125	77	27	154	32	415

unsaturated esters, alkenes and resins, as well as secondary soaps such as
metallic stearates of alkali and alkaline earth metals.

The dominance of lead stabilizers in world primary stabilizer markets results from the extensive usage of lead stabilizers in rigid PVC pipe applications outside the United States, coupled with their nearly exclusive usage
in worldwide electrical wire and cable insulation applications.

Mixed metal stabilizers, the second largest volume primary stabilizer in
world markets, represent the stabilizers of choice in most flexible PVC applications. These stabilizers dominate in flexible calendered film and sheet,
nonelectrical flexible extrusions, and plastisols.

Organotins, the smallest of the primary stabilizer classes, find approximately 50% of their usage within the United States, where they are the stabilizers of choice in rigid PVC pipe, fittings, siding, and profile markets.
Worldwide, they dominate in clear rigid packaging and bottle applications.

Secondary stabilizers complement primary stabilizers. The epoxides are
mostly employed in combination with mixed metal stabilizers in flexible PVC
applications. The secondary soaps are principally employed in combination
with leads and organotins in rigid PVC applications.

Table 2 presents estimated stabilizer consumption according to PVC
fabrication technique. The extrusion process consumes the greatest portion
of stabilizer. In addition to being the largest fabrication technique for PVC,
the extrusion process subjects compound to extremely high temperature and
high shear stresses.

With regard to fabrication techniques, leads, organotins, and soaps are
for the most part employed in rigid extrusions. Mixed metals, leads, and
epoxides are used in flexible calendering, extrusion, molding, and coating
processes.

Table 3 presents stabilizer consumption by major application. Overall,
flexible film and sheet employ the greatest percentage of combined primary
and secondary stabilizers. This is the result of high epoxidized soybean
oil usage in these applications. Rigid pipe and fittings consume the second
largest percentage of the overall primary and secondary stabilizers. This
is the result of the stabilizer/lubricant one-pack concept finding wider
acceptance in world markets. It is further apparent from Table 3 that most
primary leads, organotins, and secondary soaps are employed in pipe and
fittings applications, whereas most mixed metals and secondary epoxides are
used in flexible film and sheet.

TABLE 2 1983 Worldwide PVC Stabilizer Consumption Estimated by Fabrication Technique (Thousands of Metric Tons)

Fabrication technique	Primary lead	Primary mixed metal	Primary organotin	Secondary epoxide	Secondary soap	Total
Extrusion						
Rigid	42	7	18	14	24	105
Flexible	45	17	<1	34	1	97
Total	87	24	18	48	25	202
Calendering						
Rigid	5	2	2	5	2	16
Flexible	14	28	2	55	<1	99
Total	19	30	4	60	2	115
Molding						
Rigid	12	<1	5	1	4	22
Flexible	7	23	<1	45	1	76
Total	19	23	5	46	5	98
Grand total	125	77	27	154	32	415

C. Properties of the Ideal Stabilizer

To understand the properties of an ideal stabilizer, it is necessary to consider two important aspects of stabilization. First, we must consider how the stabilizer is expected to function both theoretically and practically. Second, but equally important, we must consider the ancillary properties that the presence of the stabilizer per se contributes to the PVC compound.

To consider the functionality of a stabilizer from the theoretical standpoint, an understanding of the fundamental cause of thermal instability of PVC is needed. In Chapter 8 it was noted that the basic thermal instability of PVC is the direct result of a dehydrohalogenation or "unzippering effect" of adjacent hydrogen and chlorine atoms in the PVC chain. Most important, it was observed that the unzippering or degradation begins only at very specific sites on the PVC molecule. These sites are at labile chlorines, which represent an extremely small percentage of the total chlorine atoms present in the PVC molecule. These labile chlorines are allylic or tertiary in nature and occur only at branches, defects, and certain terminal points in the PVC chain [11].

The action of the ideal primary stabilizer is essentially a chemical reaction between the primary stabilizer and the PVC resin wherein the primary stabilizer molecule replaces a labile chlorine atom in the PVC resin molecule with a ligand that is less easily thermally displaced. This primary stabilization reaction is essentially subject to the basic rules of thermo-

TABLE 3 1983 Worldwide PVC Stabilizer Consumption Estimated by Major Application (Thousands of Metric Tons)

Application	Primary lead	Primary mixed metal	Primary organotin	Secondary epoxide	Secondary soap	Total
Rigid PVC						
Pipe and fittings	38	1	10	2	18	69
Profile	11	7	8	14	9	49
Packaging	6	1	5	2	1	15
Other	5	<1	2	1	1	9
Total	60	9	25	19	29	142
Flexible PVC						
Film and sheet	12	32	1	68	1	114
Wire and cable	33	2	<1	5	1	41
Plastisol	8	25	1	44	1	79
Other	12	9	<1	18	<1	39
Total	65	68	2	135	3	273
Grand total	125	77	27	154	32	415

dynamics, kinetics, and mass action which govern all chemical reactions. Furthermore, a chemical change of both reactants occurs.

$$\sim\!\!-\overset{|}{\underset{|}{C}}\!-Cl \quad + \quad M\!-\!S \quad \longrightarrow \quad \sim\!\!-\overset{|}{\underset{|}{C}}\!-S \quad + \quad MCl \qquad [1]$$

| (unstable resin) | (primary stabilizer) | (stable resin) | (spent stabilizer) |

   In the case of the ideal stabilizer, it is essential that the spent stabilizer by-product of the primary stabilization reaction be a neutral species incapable of causing direct or catalytic dehydrohalogenation of stable PVC molecules.

   In actuality, many metallic chloride by-products of primary stabilization reactions (e.g., $ZnCl_2$, $CdCl_2$, $RSnCl_3$) are Lewis acids capable of promoting dehydrohalogenation of PVC. In these instances, secondary stabilizers such as epoxides, metallic soaps, and chelators are required to mitigate the effect.

   From a practical standpoint, the ideal stabilizer must completely prevent any discoloration or generation of incipient quantities of HCl that result in the formation of conjugated unsaturation. The formation of any unsaturation in the molecule inherently triggers instability by activating new allylic chloride atoms in the polymer chain.

The ideal stabilizer is considered by compounders to be a fugitive additive. They love it for what it does, hate it for what it costs, and constantly strive to employ the lowest levels consistent with achieving their ultimate goal—a high-performance vinyl compound. As a direct corollary of this goal, high efficiency and low unit cost become the parameters of prime consideration when selecting a stabilizer. Theoretically, the ideal stabilizer approaches, as a limit, zero concentration and zero cost.

When a stabilizer is incorporated at a finite concentration in a PVC compound, it can, in fact, affect many auxiliary properties of the compound. These include physical properties, chemical resistance, radiation resistance, electrical properties, optical properties, rheological properties, organoleptic properties, toxicological properties, and processability. An ideal stabilizer has no negative effect on any of the above properties.

From a handling standpoint, an ideal stabilizer is indefinitely shelf stable, readily dispersible, permanently compatible, environmentally acceptable, and perfectly lubricating.

If an ideal stabilizer were to exist, it would be extremely efficient, readily available, low in cost, homogeneous, shelf stable, readily dispersible, easily handled, colorless, odorless, tasteless, nontoxic, nonextractable, nonmigrating, nonplating, soluble and compatible, heat stable, light stable, chemically resistant, stain resistant, moisture resistant, oxidation resistant, nonvolatile, nonplasticizing, nonconductive, readily processable, and evenly lubricating for all applications.

There is no perfect stabilizer, and there never will be. Around the world dozens of companies continue to offer thousands of modifications of the primary generic stabilizer types. Some are old, some new. Each was custom designed and optimized at some point in time for a specific customer, a specific application, a specific formulation, a specific piece of processing equipment, or a specific competitive situation. Each one in some way represents a compromise.

D. Stabilizer Optimization: A Compromise

Since no single substance or material has been found to satisfy all of the requisites of the perfect stabilizer, selection of the optimum stabilizer for a given application becomes a compromise or trade-off of one property for another, with the final selection being highly dependent on specific performance requirements, economic considerations, past precedents, and regulatory statutes.

Table 4 is an attempt to characterize broadly the basic parameters which govern the selection of the major subclasses within the principal generic stabilizer classes. This table is intended to be used as a general guide only. It is not intended to be an endorsement for, or caveat against, the use of any particular commercial product or group of products. It should be emphasized that the ultimate suitability of any particular product in any specific commercial formulation can and should be determined by the specific situation.

II. HISTORICAL DEVELOPMENT OF STABILIZERS

A. The Edisonian Approach

The Edisonian approach is essentially the trial-and-error method. It was this approach that led to the first PVC heat stabilizers.

TABLE 4 Stabilizer Selection Parameters[a]

Stabilizer type	Relative efficiency	Relative cost	Ancillary property impairment							
			Optical properties	Electrical properties	Toxicological properties	Light stability	Process-ability	Chemical stain resistance	Moisture resistance	Volatility and odor
Lead										
Basic salt	M	L	X	—	X	X	—	X	—	—
Stearate	L	L	X	—	X	X	—	X	—	—
Phosphite	M	M	X	—	X	—	—	X	—	—
One-pack	L	L	X	—	X	X	—	X	—	—
Mixed metal										
BaCd powder	H	M	X	X	X	—	X	X	—	—

BaCdZn liquid	M	L	—	X	X	—	—	—	X	X
CaZn nontoxic	L	M	—	X	—	—	X	—	X	—
Organotin										
Mercaptide	H	H	—	X	—	X	—	—	—	X
Carboxylate	M	H	—	X	—	—	X	—	—	X
Antimony										
Mercaptide	H	M	—	—	—	X	—	X	—	X
Auxiliary										
Epoxide	L	L	—	—	—	—	—	—	—	—
Metal soap	L	L	X	X	—	—	—	—	—	—
Chelator	L	M	—	X	—	—	—	—	X	X

aCode: H, high; M, medium; L, low; X, property impaired.

History has not documented the exact moment when the first PVC heat stabilizer was invented. Early PVC technology, at least in the United States, came as an outgrowth of the rubber industry. As a consequence, it is reasonable to assume that when the first attempts to fabricate PVC resin failed because of roll sticking, discoloration, and the evolution of HCl, compounders resorted to rubber technology for a solution.

State-of-the-art rubber technology in the early 1930s employed lubricants such as mineral oil, waxes, and stearic acid to prevent roll sticking. In addition, it was common practice to employ the metallic oxides of lead, magnesium, calcium, and zinc for proper curing, scorch control, and stabilization [12].

Armed with these basic tools, their ingenuity, and a propensity to seek synergistic combinations of additives, the early compounders attempted to process PVC successfully by eliminating its tendency to stick to processing equipment and degrade with the evolution of HCl.

B. Early Stabilizer Types

In their early attempts to neutralize HCl, compounders turned to the incorporation of bases. Table 5 lists the neutralization equivalents of common bases. The neutralization equivalent is defined as the parts by weight of base needed to neutralize 1.0 part of HCl. From Table 5 it appears that, based solely on neutralization equivalents, the relative effectiveness of the bases as PVC heat stabilizers should be as follows

$$MgO>CaO>NH_4OH>NaOH>ZnO>KOH>PbO$$

In fact, this is not the case. Litharge, PbO, is the most effective. The reason is that the alkali and alkaline earth oxides and hydroxides, being strong bases, tend to dehydrohalogenate PVC in their own right. Marvel et al. [13] reported that PVC degrades to a carotenoid-type highly colored polyene when heated in the presence of strong base. Ammonium hydroxide, NH_4OH, is too volatile to function as an effective stabilizer. In the case of ZnO, the by-product of its reaction with HCl, namely $ZnCl_2$, is a strong Lewis acid, a dehydrohalogenation catalyst.

Litharge proved to be an effective stabilizer because of the combination of its basic strength, its fine particle size, and its high neutralization equivalent. When combined with mineral oil for lubrication, oiled litharge became the cornerstone for the development of all primary lead stabilizers.

The combination of litharge and stearic acid was also found to be an excellent stabilizer system for flexible PVC. It was further discovered that the combination of stearic acid and an alkali or alkaline earth metal base greatly enhanced the effectiveness of the base as a stabilizer. It was a short step from mixtures of oxides and stearic acid to metallic stearate soaps and later to combinations of metallic stearate soaps that became the forerunners of the mixed metal stabilizer class.

The principal disadvantage of the mixed metal soap approach to stabilization was that mixed metal soaps made it difficult to vary stabilizer concentration and lubricant concentration independently. This was exacerbated by the fact that at high stabilizer use levels, dictated by certain applications, the high soap level gave rise to haze formation, overlubrication, plateout,

TABLE 5 Neutralization Equivalents of Common Bases

Common name	Chemical formula	Molecular weight	Neutralization equivalent
Magnesia	MgO	40.0	0.55
Quick lime	CaO	56.0	0.77
Ammonium hydroxide	NH_4OH	35.0	0.98
Lye	NaOH	40.0	1.10
Zinc oxide	ZnO	81.7	1.12
Caustic potash	KOH	56.1	1.53
Litharge	PbO	223.0	3.08

and exudation—four conditions resulting from the limited compatibility of the metallic soaps.

As a result of the overlubrication of metallic soaps, a search was made for metallo-organic complexes with greater inherent compatibility with PVC resin. The result was the development of organoleads, which rapidly gave way to organotins, the third principal class of primary PVC heat stabilizer.

Common phenolic antioxidants employed in rubber compounding were the first class of organic synergists or secondary stabilizers. Druesedow and Gibbs [14] later studied the rate of dehydrohalogenation of PVC under different atmospheres. They found, as illustrated in Figure 1, that at constant temperature, HCl evolution was greater in the presence of oxygen. This was confirmed by Imoto and Otsu [15], who showed that in the presence of oxygen the rate of HCl evolution increased with increasing time and temperature. In nitrogen, the rate is relatively constant (see Figure 2).

It is generally concluded that after incipient dehydrohalogenation occurs, oxidation of the initial species results in a structure which dehydrohalogenates at a greater rate. In addition, HCl can attack process equipment and give rise to transition metal chlorides, which are strong dehydrohalogenation catalysts in the presence of oxygen. Hence, the three principal functions of all early organic synergists were to scavenge HCl, prevent oxidation, and chelate metal chlorides.

III. DOMINANT GENERIC HEAT STABILIZER CLASSES

A. Leads

Lead stabilizers represent the largest primary stabilizer class in the world today. The dominance of lead stabilizers results from two fundamental facts.

1. Lead stabilizers, subject to certain caveats, represent the most cost-effective means of stabilizing PVC.
2. Lead stabilizers are unequaled in the electrical properties which their use imparts to flexible PVC wire and cable compounds [16].

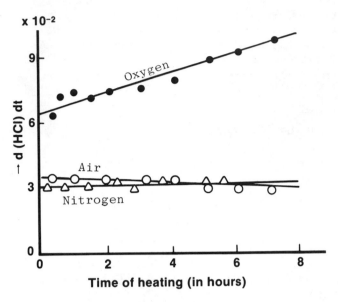

FIGURE 1 Thermal degradation of PVC in nitrogen, air, and oxygen.

Table 6 lists the approximate cost per pound of the key metals employed as primary PVC heat stabilizers.

All lead stabilizers can be considered to be derivatives of litharge, PbO. Litharge is effective as a stabilizer. It is particularly advantageous in electrical insulation applications, for four fundamental reasons:

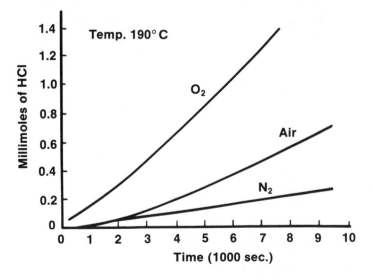

FIGURE 2 Rate of dehydrohalogenation of PVC as a function of heating time in oxygen, air, and nitrogen.

TABLE 6 1986 Stabilizer Metal Cost

Metal	Cost per pound	Source
Cadmium	$1.50	CdO
Zinc	0.41	ZnO
Lead	0.39	PbO
Antimony	1.35	Sb_2O_3
Tin	4.19	SnO

1. PbO is an excellent HCl scavenger because of its basicity and extremely fine particle size.
2. PbO is a weak base and will not dehydrohalogenate PVC per se.
3. $PbCl_2$, the reaction product of HCl and PbO, is not a strong Lewis acid and consequently does not catalyze spontaneous dehydrohalogenation of PVC.
4. $PbCl_2$ is one of the few metallic chlorides that is non-water-soluble and nonionizable. As a result, it will not reduce the electrical insulating properties of compounds on exposure to heat or moisture or on aging.

The improvements in lead stabilizers over the years can be viewed as improvements on the basic characteristics of litharge as a PVC heat stabilizer.

One of the most apparent disadvantages of litharge as a stabilizer is its yellow color. One of the first improvements over litharge was the use of white lead, that is, basic lead carbonate, $2PbCO_3 \cdot Pb(OH)_2$. White lead allowed the use of a lead stabilizer in white and pastel-colored compounds. However, its use introduced another potential problem, gassing. After approximately one-third of the stabilizer is consumed, white lead liberates CO_2, thereby causing gassing and blistering. This precludes its use in high-temperature applications.

The search for the perfect lead stabilizer continued. Early stabilizer formulators sought a white lead stabilizer that would not gas. The result was tribasic lead sulfate, $3PbO \cdot PbSO_4 \cdot H_2O$. This material became the cornerstone of lead stabilizers for rigid PVC applications. In plasticized PVC wire applications, however, the basic character of tribasic lead sulfate presented a problem at high temperatures because it saponified ester plasticizers during the long-term aging of insulation at 90° to 105°C.

The search for a compound that would be white, nongassing, and less reactive with polyester plasticizers in high-temperature wire applications turned toward the use of organic rather than inorganic acids. One material stands out in this area, namely dibasic lead phthalate, $2PbO \cdot Pb(OOC)_2C_6H_4 \cdot \frac{1}{2}H_2O$. Dibasic lead phthalate is unexcelled in its ability to impart long-term stability and retention of physical properties in Underwriters' Laboratories classes of wire at 90° and 105°C.

The weakness of dibasic lead phthalate, as well as all of the leads mentioned thus far, is lack of outstanding light stability and weather resistance. This shortcoming was finally overcome with development of dibasic lead phosphite, $2PbO \cdot PbHPO_3 \cdot \frac{1}{2}H_2O$. Dibasic lead phosphite approaches the ultimate in lead stabilizers. It is a white, nongassing, nonreactive, highly efficient stabilizer that imparts excellent weatherability. It is the only stabilizer known that will provide electrical properties and weather resistance in the absence of carbon black. It is finding increasing use in rigid extrusion applications such as vinyl window frames. Its ultraviolet resistance can be attributed to the antioxidant activity and high UV absorption of the phosphite anion.

All of the lead stabilizers previously mentioned are nonlubricating. In many applications it is advantageous to use a lead stabilizer that is lubricating. This need is fulfilled with two materials, dibasic lead stearate, $2PbO \cdot Pb(C_{17}H_{35}COO)_2$, and normal lead stearate, $Pb(C_{17}H_{35}COO)_2$. The dibasic salt is higher melting (250°C), less lubricating, and a better stabilizer than the normal salt. It also is an excellent light stabilizer and often is used in conjunction with dibasic lead phosphite for outstanding weatherability and processability. Normal lead stearate is low melting (110°C) and extremely lubricating. Its limited compatibility restricts its use to low levels in most applications.

Silicate-containing lead stabilizers represent a class of products that are formulated to retain plasticizer and long-term electrical properties in unfavorable environments. A family of products including normal lead orthosilicate, basic lead silicate, basic lead silicate sulfate, and basic lead chlorosilicate has been developed.

The principal lead stabilizer types and their major advantages are listed in Table 7. The principal advantages of lead stabilizers are their high efficiency, low cost, long-term heat stability, and electrical properties. Table 6 shows the tremendous cost advantage potential that lead-based stabilizers have over most other primary stabilizers by virtue of the low cost of lead metal.

Lead stabilizers, as a class, have several limitations. Because of their limited compatibility and pigmentlike characteristics, lead stabilizers cannot be used in clear applications. In addition, their inability to impart long term initial color-hold characteristics limits their use in pastel applications.

Lead stabilizers are prone to sulfide staining and cross-staining. Lead sulfide is black. As a result, lead-stabilized vinyl compounds are prone to discolor when exposed to H_2S, metallic sulfides, and several compounded products (e.g., rubber coatings that contain divalent sulfur). In addition, if a lead-stabilized vinyl stock is mixed, compounded, or placed in contact with PVC containing organotin or antimony mercaptides, discoloration due to sulfide cross-staining can occur.

Many lead stabilizers, being hydrated oxides, have a tendency to gas when processed at high temperatures. This results from the liberation of water or CO_2 and can give rise to bubbles or pinholes in vinyl stocks.

Perhaps the greatest deterrent to the use of lead stabilizers is their toxicity. There is no doubt that lead stabilizers are toxic. There are, however, three types of toxicology: scientific toxicology, political toxicology and emotional toxicology. The latter two are most difficult to address, and as a result the use of lead stabilizers has been restricted in many applications in many countries.

Lead stabilizers are both acutely and cumulatively toxic. As a result, extreme care must be exercised when they are handled. The principal potential entry routes of lead stabilizers into workers exposed to them are inhalation and ingestion.

TABLE 7 Principal Lead Stabilizers

Name	Formula	Specific gravity	Special property
Lead oxide	PbO	9.5	Low cost
Basic lead carbonate	$PbCO_3 \cdot Pb(OH)_2$	6.6	Low cost and white color
Tribasic lead sulfate	$3PbO \cdot PbSO_4 \cdot H_2O$	6.4	Overall cost/performance
Basic lead silicate sulfate	Complex	5.5	Low cost
Dibasic lead phthalate	$2PbO \cdot C_6H_4(COO)_2Pb$	4.6	Low reactivity
Dibasic lead phosphite	$2PbO \cdot PbHPO_3 \cdot \frac{1}{2}H_2O$	6.1	Outdoor weatherability
Dibasic lead stearate	$2PbO \cdot (C_{17}H_{35}COO)_2Pb$	2.0	Lubricity
Normal lead stearate	$(C_{17}H_{35}COO)_2Pb$	1.4	Lubricity
Lead 2—ethylhexoate	$(C_7H_{16}COO)_2Pb$	1.1[a]	Liquid product

[a]Sold as 60% solution in mineral spirits.

Lead stabilizer producers have gone to great lengths to improve the handling characteristics of their products so as to limit exposure in the workplace. Lead stabilizers are currently available in many forms to limit exposure [17].

Lead stabilizer pastes are available in plasticizers. Wetted lead powders containing mineral oil or plasticizer are often used.

Lead prills, granules, strands, and pellets are available where conventional lead stabilizers are formulated with waxes, resins, and binders. Collectively, these systems are often referred to as lead one-packs.

Another approach to limiting the exposure to lead stabilizers is through effective means of containerization. One of the most popular and most efficient methods of handling lead powders in bulk is the air pallet. A more pragmatic and simpler means of handling small quantities of lead powders is to supply the product in preweighed polyethylene or flexible PVC bags which can readily be charged to a Banbury mixer in total.

In the United States, the Occupational Safety and Health Administration (OSHA) requires that both the air level and blood lead level of the work force be monitored if lead exposure is above the action level [18].

There are two major applications for lead stabilizers worldwide: in flexible PVC wire and cable insulation, and in rigid PVC pipe. In PVC wire and cable insulation, no class of stabilizer can compete with leads for long-term retention of physical and electrical insulation properties. As metallic stabilizers are consumed, most generate ionizable or water-soluble chlorides (e.g., $BaCl_2$, $CdCl_2$, $CaCl_2$, $ZnCl_2$, R_2SnCl_2). Lead stabilizers, however, generate $PbCl_2$, which is neither ionizable nor water-soluble. As a result, lead stabilizers provide PVC wire and cable long-term aging and insulating properties that are unachievable with all other stabilizer classes.

The second major application of lead stabilizers worldwide (except in the United States) is in PVC pipe and profile extrusions. The pipe market, which includes potable water, drain, waste, and vent piping as well as electrical conduit and telephone duct, is the largest and most competitive market application for PVC resin. Pipe compounders throughout the world seek stabilizers with the best cost/performance ratio.

In most countries outside the United States, lead stabilizer one-packs are the stabilizers of choice for all rigid PVC pipe applications. Typical lead one-packs are admixtures of tribasic lead sulfate, lead stearate, paraffin waxes, and oxidized polyethylene wax. In the United States leads are not used in most pipe applications for many reasons, foremost among which is lack of sanction by the appropriate regulatory and certification authorities. This lack of approval stems from the adverse public reaction toward the use of lead in the workroom, consumer products, and articles of commerce.

Even with this, lead stabilizer producers in the United States are optimistic about the future use of lead stabilizers in PVC pipe, as exposure is reduced through lower use levels of highly efficient dustless one-packs and as the results of risk/benefit analyses of new lead stabilizer systems become available.

B. Organotins

Stabilizer efficiency increases as the compatibility, solubility, or degree of contact of stabilizer and resin matrix increases. The lead stabilizers, being pigmentlike, are precluded from achieving the degree of contact required to impart clarity to PVC resin and compounds.

The limited solubility of most metallic oxides, salts, and soaps in PVC caused early stabilizer chemists to investigate organometallic complexes as PVC heat stabilizers. It was theorized that the presence of the carbon-metal bond would enhance stabilizer solubility and therefore render organometallic compounds more efficient stabilizers. Tetraphenyllead and Tetraphenyltin were among the first materials to be effectively employed [19, 20].

Quattlebaum, considered by some to be the father of organotin stabilization, pioneered efforts which led to the early use of dialkyltin carboxylates as the first commercially successful organotin stabilizers for flexible PVC applications [21]. In the early 1950s, dibutyltin mercaptides and dibutyltin mercaptoacetate esters became the standards of the budding rigid PVC pipe and profile industry in the U.S. [22–24]. Today, methyltins, octyltins, estertins, mixed metal alkyltins, and alkyltin one-packs abound in the marketplace along with the butyltins.

To understand the basic chemistry, function, and application of organotin stabilizers, it is necessary to go back to their origins and trace the historical evolution of this remarkable class of stabilizers.

Tin compounds exist in two common valence states, stannous (+2) and stannic (+4). Simple inorganic tin compounds such as stannous carboxylates are of only limited commercial importance as PVC heat stabilizers, and stannic carboxylates are generally considered to accelerate the decomposition of PVC [25]. Both stannous and stannic chloride are extremely strong Lewis acids and readily dehydrohalogenate PVC.

The stannic oxidation state is the more stable state for tin. In this state tin is tetracoordinate. The Lewis acidity of stannic halides diminishes significantly as the tin is alkylated. Hence, with regard to Lewis acidity [26]:

$$SnCl_4 > RSnCl_3 > R_2SnCl_2 > R_3SnCl$$

The key to utilizing tin effectively as a PVC heat stabilizer is to diminish the strong Lewis acidity of the inorganic tin chlorides by alkylating tin. Accordingly, the first step in the development of effective stannic tin stabilizers is the alkylation of tin.

Tetraalkyltins can be prepared by one of three commercial processes: the Grignard, Wurtz, and aluminum alkyl routes.

Grignard:

$$SnCl_4 + 4RCl + 4Mg \rightarrow R_4Sn + 4MgCl_2$$

Wurtz:

$$SnCl_4 + 4RCl + 8Na \rightarrow R_4Sn + 8NaCl$$

Aluminum alkyl:

$$4R_3Al + 3SnCl_4 \rightarrow 3R_4Sn + 4AlCl_3$$

After its complete alkylation, tin is disproportionated by reaction with tin tetrachloride to produce near-quantitative amounts of the desired alkyltin chloride species, e.g.,

$$R_4Sn + SnCl_4 \rightarrow 2R_2SnCl_2$$

Di-*n*-butyltin dichloride emerged as the dominant alkyltin chloride intermediate for early organotin stabilizers. The exact reasons for this choice remain obscure. However, it is believed to be the result of at least four considerations; toxicology, volatility, economics, and the availability of *n*-butyl chloride.

From a toxicological standpoint, mono- and dialkyltins are considerably less toxic than trialkyltins. In addition, toxicity generally decreases as the chain length of the alkyl group increases [27].

From a volatility and odor standpoint, it is undesirable to volatize stabilizer by-products (alkyltin chlorides) during the processing, fabrication, or storage of vinyl products. Retention of stabilizer by-products is better assured with longer alkyl groups.

Lastly, the relative availability of high-purity, low-cost *n*-butyl chloride and its excellent reaction yields without rearrangement were pragmatic considerations in its choice as a preferred reactant for early alkyltin stabilizer intermediate synthesis.

As a result of the considerations mentioned above, the principal intermediate for the production of alkyltin stabilizers became dibutyltin dichloride:

$$
\begin{array}{l}
CH_3CH_2-CH_2-CH_2 \qquad\qquad Cl \\
\qquad\qquad\qquad\qquad\qquad Sn \\
CH_3CH_2-CH_2-CH_2 \qquad\qquad Cl
\end{array}
$$

Attention next focused on reacting dibutyltin dichloride with a wide variety of alcohols, phenols, and carboxylic acids in an effort to develop stabilizers, $(C_4H_9)_2Sn(Y)_2$, with the best overall properties. The carboxylates proved initially to be the most promising class of compounds to derivatize with the alkyltin dichloride. The linear saturated fatty acids proved to be best. When all optimization was completed, dibutyltin dilaurate emerged as the best compromise after giving due consideration to the ancillary properties of volatility, odor, physical form, lubricity, and compatibility.

Dibutyltin dilaurate is presumed to function by displacing labile chlorines on PVC with laurate ligands, e.g.,

Consequently, it is presumed to function by preventing the initiation of unzippering. In addition, the fatty character of the laurate anion contributes a lubrication function which prevents sticking of the PVC to hot processing surfaces.

The next major discovery of significant commercial utility was that maleic acid derivatives of dibutyltin oxide represented a class of stabilizers that were an order of magnitude better than the linear saturated simple carboxylic acid derivatives. Thereupon, dibutyltin maleate (polymer) and dibutyltin bis(alkyl maleate esters) and laurate/maleate combinations became widely accepted as the most efficient alkyltin carboxylate stabilizers.

It is currently held in some circles that the maleates and maleate esters confer extraordinary stability by preventing the initiation of unzippering by displacing labile chlorines and perform a secondary function of bleaching out color bodies which result from conjugated unsaturation by means of a Diels-Alder addition to polyenes in degraded PVC chains.

The maleate and maleate ester derivatives do not provide the excellent lubricity afforded by the laurates. In fact, maleates increase the tendency for vinyl to adhere to processing equipment. As a result, for optimum stability and lubricity a series of coreacted proprietary alkyltin carboxylates evolved which contained both laurate and maleate functionality.

The butyltin maleate-laurates were extremely successful for stabilizing flexible PVC and certain PVC copolymers. However, they lacked the necessary stability and processability to permit their use in rigid PVC extrusions.

In the early 1950s, it was discovered that the heat stability of alkyltin carboxylates was greatly enhanced by the addition of certain alkyl mercaptides. Shortly thereafter patents were issued on the use of dibutyltin dilaurylmercaptide as a heat stabilizer for rigid PVC applications. In 1953 patents were issued on the use of what proved to be the standard of the rigid PVC stabilizer industry for nearly two decades, dibutyltin bis(isooctyl mercaptoacetate) [24]:

$$CH_3-CH_2-CH_2-CH_2 \diagdown \underset{Sn}{} \diagup S-CH_2-\overset{\overset{O}{\parallel}}{C}-O-C_8H_{17}$$
$$CH_3-CH_2-CH_2-CH_2 \diagup \diagdown S-CH_2-\underset{\underset{O}{\parallel}}{C}-O-C_8H_{17}$$

This product was ideally suited for rigid PVC stabilization. It was highly efficient and as a result could be used at low levels, thereby having only a minimal effect on heat distortion temperatures. It was extremely compatible with PVC resins and essentially nonlubricating, which allowed it to be used in rigid clears. It was a water white, low-viscosity, nonvolatile liquid which could be handled with ease. In addition, it had good shelf stability and was not readily susceptible to oxidation or hydrolysis.

The emergence of dibutyltin bis(isooctyl mercaptoacetate) as a stabilizer in the 1950s was concurrent with the development of the U.S. rigid PVC market for pipe, profiles, and rigid PVC injection molding compounds. As a result, this material became the stabilizer of choice in most rigid formulations of the period.

Dibutyltin bis(isooctyl mercaptoacetate) has three undesirable qualities. It has poor odor, imparts poor light stability, and is relatively expensive.

The poor odor is an inherent property of mercaptide-based stabilizers and cannot be totally eliminated. Odor can be improved by reduction of free organotin chlorides and free mercaptan in the finished product or by inclusion of an odor mask.

The poor light stability imparted by dibutyltin bis(isooctyl mercaptoacetate) is also inherent in alkytin mercaptide-based stabilizers. It is believed that under the influence of sunlight the stabilizer undergoes a redox reaction which results in the formation of tin metal and tin sulfide. The poor light-stabilizing effects of organotin mercaptide stabilizers can be masked in rigid PVC weatherable compounds by the incorporation of large amounts of TiO_2 pigment. In clears, light stability can be improved marginally by the addition of selected UV absorbers.

The relatively high cost of dibutyltin bis(isooctyl mercaptoacetate) reflects the high cost of tin metal as well as the manufacturing costs inherent in the alkylation of tin.

In spite of its shortcomings, dibutyltin bis(isooctyl mercaptoacetate) completely dominated the U.S. rigid PVC stabilizer market from the early 1950s until the mid-1970s.

In the early 1970s, numerous improved versions of this stabilizer were introduced to the price-sensitive pipe market. These versions generally promised improved initial color at lower use level or at lower price.

High-efficiency alkyltins were marketed which contained between 24 and 27% tin metal. These generally included high-tin-containing additives such as dibutyltin sulfide, monobutyltin sulfide, or butylthiostannoic acid and anhydride [28, 29]. The latter compound received sanction in Germany for use in PVC food contact applications in 1966.

Despite their improved efficiency, the high-efficiency tins were not widely accepted in the U.S. pipe market because of their high unit cost. The U.S. pipe producers demanded lower-cost products. As a result, a series of improved butyltins were marketed with tin contents in the range 15 to 17%. These consisted of the basic dibutyltin stabilizer augmented by the addition of small amounts of monobutyltin moieties which enhanced initial color. Other products included the addition of small amounts of antioxidants, chelators, thioacids, zinc and stannous salts, and a myriad of other "secret additives" [30—32].

The improved butyltins achieved limited success. A significant break-through was the introduction of methyltins by Cincinnati Milacron Chemicals. Methyltins afforded the pipe, profile, and bottle compounder 10 to 15% higher tin content and higher efficiency at equivalent or lower price. These savings resulted from the direct alkylation of tin metal with methyl chloride to produce dimethyltin dichloride in a single high-yield step.

$$2CH_3-Cl + Sn \xrightarrow[\text{catalyst}]{\Delta} \quad CH_3 \diagdown \atop CH_3 \diagup Sn \diagup Cl \atop \diagdown Cl$$

Within a very short time, dimethyltin bis(isooctyl mercaptoacetate) displaced a significant portion of the dibutyltin bis(isooctyl mercaptoacetate) market.

$$CH_3 \diagdown \atop CH_3 \diagup Sn \diagup S-CH_2-\overset{O}{\overset{\|}{C}}-OC_8H_{17} \atop \diagdown S-CH_2-\overset{}{\underset{\|}{\underset{O}{C}}}-OC_8H_{17}$$

[dimethyltin bis(isooctyl mercaptoacetate)]

The methyltin technology initially took the form of a 21% tin stabilizer. However, in 1972 Synthetic Products Company introduced the concept of a low-cost, low tin content stabilizer to the U.S. pipe market. The product was a barium-tin-based stabilizer and the concept was a mixed metal organotin. The product consisted of a monobutyltin mercaptoacetate ester in combination with an overbased barium carbonate complex. It contained 7% tin and 12% barium and represented a low unit cost stabilizer for the pipe industry [33].

The theory behind the mixed metal stabilizer approach to pipe was that at a low stabilizer use level, as in pipe, organotin mercaptide is consumed by two competing reactions:

$$R \diagdown Sn \diagup SR' \atop R \diagup \diagdown SR' + Cl \atop Cl \rightarrow R \diagdown Sn \diagup Cl \atop R \diagup \diagdown Cl + R'S \atop R'S \qquad [2]$$

$$R \diagdown Sn \diagup SR' \atop R \diagup \diagdown SR' + 2HCl \rightarrow R \diagdown Sn \diagup Cl \atop R \diagup \diagdown Cl + 2R'SH \qquad [3]$$

In reaction (2), stabilizer is consumed in the primary stabilization mode. In reaction (3), stabilizer is consumed as an HCl scavenger. As a result, reaction (3) reduces the effective concentration of stabilizer available for the principal stabilizer function.

In a barium-tin stabilizer, the barium carbonate complex, due to its higher basicity, reacts with any free HCl, thereby preventing HCl from consuming primary tin stabilizer in a secondary reaction, e.g.,

$$BaCO_3 + 2HCl \rightarrow BaCl_2 + H_2O + CO_2$$

The barium carbonate is sacrificed to the organotin mercaptide with respect to the consumption of HCl in nonstabilizing side reactions.

It could be demonstrated that in many instances the mixed metal barium-tins and later mixed metal calcium-tins were one-for-one replacements for the classical 18 to 21% tin mercaptides.

The principal shortcoming of the mixed metal tins was their inability to yield quite as white a PVC pipe as straight standard tin stabilizers. The presence of the alkaline earth metal resulted in a product with a slightly yellow initial color compared to that produced with a conventional tin stabilizer.

The commercial success of the low-tin mixed metal barium and calcium buffered organotin mercaptides spurred organotin producers to develop a new class of internally buffered high monoalkyl straight tins cut back with mineral oil or other diluents to 5 to 12% tin. The internal buffering was achieved by sulfur bridging the organotin mercaptide, e.g.,

In this product, the Sn—S—Sn bond is sacrificial to the Sn—SR bond with respect to HCl in a manner not dissimilar to the function of the $BaCO_3$ in a mixed metal stabilizer. This allows the Sn—SR portion of the stabilizer to be consumed in the primary stabilization mode of displacing labile chlorines from the PVC chain while the Sn—S—Sn portion consumes HCl as a buffer, e.g.,

Sulfur-bridged, high-monobutyltin mercaptoacetates, when cut back to the same tin content as mixed metal barium tins, yield equivalent stability with superior initial color. Hence, they effectively checked the threat of mixed metal tins' displacement of straight tins from the pipe market.

After the success of the low-cost straight tins based on sulfur-bridged, high-monoalkyltin mercaptoacetates, the proponents of mixed metal tins turned to antimony technology to penetrate and displace tins from pipe while the basic tin producers concentrated on the development of more highly lubricating low-cost sulfur-containing liquids as well as solid stabilizer/lubricant one-packs.

The classical isooctyl mercaptoacetate ligand is highly compatible with PVC and, as a result, aids in rapid fusion. In multiscrew pipe extrusion, early fusion is not always desirable. Consequently, producers searched for low-cost, more externally lubricating ligands. The result was the development of "reverse ester" alkyltins based on mercaptoethanol esters rather than mercaptoacid esters. These esters are synthesized by the reaction of ethylene oxide, H_2S, and a carboxylic acid, e.g.,

$$CH_2\overset{\displaystyle O}{\overset{\diagup\ \diagdown}{-}}CH_2 + H_2S + RCOOH$$

$$\rightarrow HS-CH_2-CH_2-O-\overset{\displaystyle O}{\overset{\|}{C}}-R + H_2O$$

Of particular importance are the derivatives based on 2-ethylhexoic acid, neodecanoic acid, oleic acid, and tall oil fatty acid.

The structures of the nonlubricating isooctyl mercaptoacetate ligand and the lubricating mercaptoethyl oleate ligand are compared below:

$$-S-CH_2-\overset{\displaystyle O}{\overset{\|}{C}}-O-C_8H_{17}$$

(isooctyl mercaptoacetate)

$$-S-CH_2-CH_2-O-\overset{\displaystyle O}{\overset{\|}{C}}-C_{17}H_{33}$$

(mercaptoethyl oleate)

In addition to ligand modification to afford greater external lubricity in liquid organotin mercaptides for pipe, several suppliers have turned to the solid stabilizer/lubricant one-pack concept. In this concept, organotin mercaptide, calcium stearate, and paraffin wax are combined in the melt stage or manufactured in situ to prepare a one package stabilizer/stearate/wax combination product that can be sold in the form of a low-melting bead or flake. A typical composition consists of approximately 20 wt % organotin stabilizer, 20 wt % calcium stearate, and 60 wt % paraffin and oxidized polyethylene waxes.

Advocates of one-packs stress their convenience and claim synergistic effects of the combination of additives. Opponents of one-packs resist the lack of flexibility they give the compounder. In fact, the use of one-packs takes formulation decision making out of the hands of the pipe compounder and places it in the hands of the additive supplier, who may not have sufficient expertise with respect to the needs of the end PVC product or use.

Another class of tin stabilizer that was initially developed in Europe and reported to be of value in the area of improved toxicity, low odor, and improved weathering is the so-called ester tin [34]. These materials use acrylic esters to alkylate tin. A typical example of this class, derived from *n*-butyl acrylate, is:

$$
\begin{array}{c}
CH_3-CH_2-CH_2-CH_2-O-\overset{\overset{\displaystyle O}{\|}}{C}-CH_2-CH_2 \diagdown \\
\qquad\qquad\qquad\qquad\qquad\qquad\qquad\qquad Sn \\
CH_3-CH_2-CH_2-CH_2-O-\underset{\underset{\displaystyle O}{\|}}{C}-CH_2-CH_2 \diagup
\end{array}
\begin{array}{c}
\diagup SR \\
\\
\diagdown SR
\end{array}
$$

These materials are said to have the potential for competing with *n*-octyl-tins for food grade packaging applications. The food grade markets are dominated by two *n*-octyltin derivatives: di-*n*-octyltin maleate (polymer) and di-*n*-octyltin-*S*,*S'*-bis(isooctyl mercaptoacetate). New products are currently being tested to allow the inclusion of *n*-octyltin-*S'*,*S'*,*S''*-tris (isooctyl mercaptoacetate) in food contact applications to provide improved initial color at lower stabilizer use levels.

Table 8 lists the major alkyltin stabilizers of importance. Table 9 lists the estimated U.S. production of major stabilizer types [35]. Table 10 shows the relative toxicity of several key organotin stabilizers and stabilizer intermediates [27].

C. Mixed Metals

The historical connotation of the term "mixed metal stabilizers" refers to a class of non-lead-containing, non-organotin-containing stabilizers based on cadmium or zinc compounds, functioning as primary stabilizers, and augmented by the inclusion of alkali or alkaline earth metal compounds to prevent the premature formation of the strong Lewis acids, cadmium- and zinc chloride.

Even though the use of lead compounds and organotin compounds as PVC heat stabilizers predates the mixed metal approach, the mixed metals have achieved the largest volume of the three primary stabilizer classes in the U.S. market. This widespread use stems from their inherent flexibility in satisfying a wide variety of highly specialized end-use requirements for specific vinyl applications such as flooring, top coats, pool liners, wall coverings, roofing, upholstery, etc.

Generically, mixed metal products are classified as either "general-purpose" or U.S. Food and Drug Administration (FDA) grades. The general-purpose types most often consist of products based on barium-cadmium combinations and the FDA grades ordinarily consist of calcium-zinc types.

From a compositional standpoint, mixed metals take the form of powder, liquid, or paste. They are complex admixtures of as few as 2 or as many as 10 individual components. To the stabilizer producer, they are a formulator's dream and a plant manager's nightmare. Practically every product is custom-blended to address a particular customer application or competitive situation. They consist, in general, of salts, soaps, bases, antioxidants, chelators, plasticizers, solvents, and diluents.

In an effort to explain in an orderly manner a product line which has grown in a very disorganized fashion, it is essential, as with leads and organotins, to go back to the beginning and trace the evolution of the mixed metal stabilizer.

In the late 1930s, it was observed that the successful fabrication of PVC was contingent on the incorporation of additives that had the ability to absorb HCl and prevent sticking to processing equipment. Concurrent with the evolution of lead salts and organotin complexes, compounders began to

TABLE 8 Major Organotin Stabilizers

Dibutyltin dilaurate	$(n-C_4H_9)_2Sn(OOCC_{11}H_{23})_2$
Dibutyltin maleate	$[(n-C_4H_9)_2SnOOCH=CHCOO]_n$
Dibutyltin bis(isooctyl maleate)	$(n-C_4H_9)_2Sn(OOCCH=CHCOOC_8H_{17})_2$
Dioctyltin maleate	$[(n-C_8H_{17})_2SnOOCCH=CHCOO]_n$
Dibutyltin bis(lauryl mercaptide)	$(n-C_4H_9)_2Sn(SC_{12}H_{25})_2$
Dibutyltin bis(isooctyl mercaptoacetate)	$(n-C_4H_9)_2Sn(SCH_2COOC_8H_{17})_2$
Monobutyltin tris(isooctyl mercaptoacetate)	$n-C_4H_9Sn(SCH_2COOC_8H_{17})_3$
Dimethyltin bis(isooctyl mercaptoacetate)	$(CH_3)_2Sn(SCH_2COOC_8H_{17})_2$
Monomethyltin tris(isooctyl mercaptoacetate)	$CH_3Sn(SCH_2COOC_8H_{17})_3$
Dioctyltin bis(isooctyl mercaptoacetate)	$(n-C_8H_{17})_2Sn(SCH_2COOC_8H_{17})_2$
Dibutyltin bis(2-mercaptoethyl oleate)	$(n-C_4H_9)_2Sn(SCH_2CH_2OOCC_{17}H_{33})_2$
Monobutyltin tris(2-mercaptoethyl oleate)	$n-C_4H_9Sn(SCH_2CH_2OOCC_{17}H_{33})_3$
Dimethyltin bis(2-mercaptoethyl oleate)	$(CH_3)_2Sn(SCH_2CH_2OOCC_{17}H_{33})_2$
Monomethyltin tris(2-mercaptoethyl oleate)	$CH_3Sn(SCH_2CH_2OOCC_{17}H_{33})_3$
Monobutyltin sulfide	$(n-C_4H_9Sn)_2S_3$
Dibutyltin sulfide	$(n-C_4H_9)_2SnS$
Thiobis[monomethyltin bis(2-mercaptoethyl oleate)]	$CH_3\underset{S}{\overset{\mid}{Sn}}(SCH_2CH_2OOCC_{17}H_{33})_2$ $CH_3Sn(SCH_2CH_2OOCC_{17}H_{33})_2$

screen many different metallic soaps. This was done in the serendipitous and undocumented fashion that often characterizes fast-moving technology.

Cadmium compounds were initially patented for use as stabilizers in 1937 [36]. As compounders began to evaluate metallic soaps, they soon learned that cadmium and zinc soaps best prolonged the initial color of PVC but led to subsequent catastrophic blackening. By contrast, alkaline earth metal soaps, such as barium and calcium salts, prolonged long-term stability but did not impart good initial color to PVC compounds.

TABLE 9 Estimated U.S. Alkyltin Stabilizer Production (Millions of Pounds)

Year	Butyltin mercaptides	Methyltin mercaptides	Octyltin mercaptides	Alkyltin maleates
1966	5.6	0	0	0.2
1971	9.3	1.4	0.4	0.7
1976	10.1	4.5	0.6	0.3
1981	15.0	6.5	0.9	1.0
1986	18.0	8.0	1.2	1.5

The combination of barium and cadmium soaps gave the best overall performance properties for both initial color and long-term stability. This resulted from a synergistic effect which is the basis for the mixed metal approach to stabilization [37–39].

Simply stated, the cadmium soap is the primary stabilizer. It reacts with unstable PVC chains in a stabilizing mode by displacing labile chlorines with soap ligands. The by-product of the primary stabilization reaction is cadmium chloride, a strong Lewis acid, whose accumulation gives rise to the spontaneous dehydrohalogenation and concomitant catastrophic degradation of PVC.

The barium soap, while not an effective primary stabilizer, has the propensity to react with incipient cadmium chloride as it is generated, and simultaneously regenerating the cadmium soap primary stabilizer. The by-product of this primary stabilizer regeneration, barium chloride, is not a Lewis acid and is a relatively inactive species.

In a general sense, if one were to evaluate all of the common metals in the periodic chart as metallic soaps in PVC and measure their propensity to retard the discoloration of PVC as a function of time at an elevated temperature, each would generally follow either curve A or curve B in Fig. 3. The metal soaps giving type A behavior would generally be soaps of metals with ionic character. These are typically the alkali and alkaline earth metal soaps. The metal soaps giving type B behavior would generally be soaps of metals with covalent character. These are typically the heavy metals. When a type A soap is combined with a type B soap, the result follows the composite curve C.

If we assume that for every vinyl application there is some level of degradation, X, which is the maximum permissible level for that application, it can be seen that the level of degradation can best be prevented with a mixture of a type A and a type B soap.

This is a general graphic representation of the phenomenon which was discovered empirically over a number of years. It is the basis for the barium-cadmium synergism, the calcium-zinc synergism, and all other similar class A/class B metal synergisms in mixed metal stabilizers.

Over the years a wide variety of natural and synthetic fatty acids were evaluated for barium-cadmium soaps. In general, the linear saturated fatty acids from C_8 to C_{18} were found to be most desirable, with lauric acid (C_{12}) being considered by most authorities to be the optimum. Acids with lower chain length have a greater tendency to give rise to plateout, volatility,

TABLE 10 Acute Toxicity of Selected Organotin Compounds

Compound	LD_{50} (mg/kg)
Monomethyltin trichloride	575−1370
Monobutyltin trichloride	2200−2300
Monooctyltin trichloride	2400−3800
Dimethyltin dichloride	74−237
Dibutyltin dichloride	112−219
Dioctyltin dichloride	>4000−7000
Trimethyltin chloride	9−20
Tributyltin chloride	122−349
Trioctyltin chloride	>4000−29000
Monomethyltin tris(isooctyl mercaptoacetate)	920−1700
Monobutyltin tris(isooctyl mercaptoacetate)	1063
Monooctyltin tris(isooctyl mercaptoacetate)	3400−>4000
Dimethyltin bis(isooctyl mercaptoacetate)	620−1380
Dibutyltin bis(isooctyl mercaptoacetate)	500−1037
Dioctyltin bis(isooctyl mercaptoacetate)	1200−2100

and odor problems. Those with higher chain lengths tend to exude or migrate from plasticized PVC and are susceptible to attack by bacteria or fungi. Unsaturated fatty acids give rise to oxidation problems. Depending on the particular application and other formulation components, the weight ratio of the barium soap to the cadmium soap is typically between 2:1 and 1:2.

Recognizing that the large commercial applications for flexible PVC preceded those for rigid PVC in the United States by nearly two decades, it is not surprising that early mixed metals found their principal use in flexible PVC applications. It is, however, interesting to note that the mixed metals have continued to dominate flexible PVC applications in spite of the evolution of greatly improved lead and organotin products. The underlying reason for this continued dominance has been the discovery of organic synergists or secondary stabilizers that greatly extend the stability of the primary mixed metals.

The single most significant synergistic effect for mixed metals has been provided by use of epoxidized oils and epoxy esters. In flexible applications, they are generally employed in combination with the mixed metals at use levels of approximately twice those of the mixed metal stabilizer.

It was initially felt that epoxides functioned principally as HCl acceptors and scavengers. However, in the early 1970s, Anderson and McKenzie [40] studied in depth the synergism between epoxides and mixed metal stabilizers and concluded that the epoxy stabilization mechanism is quite complex and

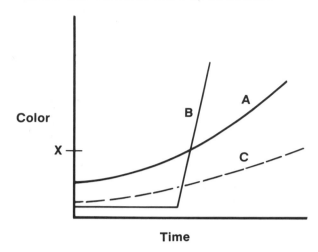

FIGURE 3 Stability patterns of metal soaps in PVC.

that there is strong evidence that, in fact, the epoxide functions as a primary stabilizer in the presence of catalytic amounts of cadmium or zinc chloride.

As PVC applications and processing equipment expanded, the limited compatibility and inherent lubricity of mixed metals, based on solid metallic soaps, made it difficult to vary stabilizer concentration and lubricity independently. In addition, high concentrations of stabilizer had a detrimental effect on clarity, plateout resistance, odor, printability, and heat sealability of the vinyl compositions. This gave rise to a need for mixed metal stabilizers with greater efficiency and compatibility with PVC.

The first discovery that was proposed to improve the compatibility of powdered mixed metal soaps was the incorporation with them of liquid organic phosphite esters as clarity enhancers [41]. It was initially believed that the principal function of the phosphite was to complex insoluble metal chlorides and thereby prevent haze. The current theory is that the function of the phosphite ester is much more complex. It is now recognized that the phosphite also prevents oxidation, aids in the dissolution of metallic soap micelles, and adds to double bonds, thereby breaking up the conjugated double bonds which give rise to yellowing in degraded PVC compounds. It has further been postulated that triaryl or aralkyl phosphites may engage in primary stabilization of PVC via an Arbuzov rearrangement reaction mechanism.

Since it was recognized that oxidation also contributed to the acceleration of PVC degradation, compounders began to evaluate and incorporate antioxidants in mixed metal PVC heat stabilizers. Phenolic antioxidants initially showed the most dramatic results, and preferred mixed metal systems became the combination of a cadmium soap, a barium soap, and a small amount of a phenolic antioxidant.

As the technology expanded, Lally et al. [42] discovered that the performance of powdered mixed metal stabilizers with a high cadmium or zinc content could be further improved by incorporation of polyhydric alcohols in combination with the soaps. It is generally agreed that the function of the polyhydric alcohol is to chelate cadmium or zinc chloride and thereby

TABLE 11 Typical Mixed Metal Stabilizer Systems

Component	Percent
General-purpose barium cadmium powder	
Barium stearate	60
Cadmium stearate	35
Bisphenol A	5
General-purpose barium cadmium zinc liquid	
Barium nonylphenate	25
Cadmium naphthenate	10
Zinc octoate	5
Decyl diphenyl phosphite	25
Glycol ether solvent	5
Mineral spirits	30
High-efficiency barium cadmium powder	
Cadmium laurate	50
Barium laurate	25
Pentaerythritol	20
BHT	5
Plastisol barium zinc stabilizer	
Barium neodecanoate	10
Zinc neodecanoate	5
Phenyl didecyl phosphite	20
Glycol ether solvent	5
Tall oil fatty acid	10
Octyl epoxy tallate	50
Nontoxic calcium zinc paste	
Calcium stearate	10
Zinc stearate	15
Tris(nonylphenyl) phosphite	25
Epoxidized soybean oil	50

prolong the activity of the stabilizer. Polyhydric alcohols such as glycerol, sorbitol, mannitol, trimethylolpropane, pentaerythritol, and dipentaerythritol are particularly effective. They are considered by some to function as primary stabilizers in the presence of catalytic amounts of cadmium or zinc.

The desire to control lubricity independent of stability, improve clarity and compatibility through the use of phosphites, and handle stabilizers more readily in bulk led to the inception and rapid growth of the liquid mixed metal stabilizers in the early 1950s.

Borrowing from their backgrounds in the paint drier industry, some stabilizer producers manufactured individual metallic soaps dissolved in mineral spirits, ketone, alcohol, or plasticizer. They subsequently combined the concentrated solutions with liquid phosphite esters to produce the first liquid mixed metal products. Initial products were based on combinations of cadmium, zinc, barium, strontium, and calcium salts of naphthenic, octoic, oleic, ricinoleic, and tall oil fatty acids.

Throughout the 1950s and 1960s, stabilizer producers concocted a myriad of mixtures, varying metals, metal ratios, carboxylic acid type, phosphite ester types, and solvent carriers.

Noteworthy developments during this period were the substitution of barium alkylphenates and overbased barium alkylphenates for barium carboxylates to improve solubility and plateout resistance; the use of benzoic acid and substituted benzoic acid derivatives to improve clarity of compositions; and the emergence of mixed alkyl-aryl phosphites in place of triphenyl phosphite as the optimum phosphite type for general-purpose PVC applications [43].

During the 1970s, very little new technology was introduced in the mixed metal stabilizer field. The products became mature and competition focused on services and price as the means of selling. During this period, a wide variety of solvent and plasticizer diluted liquids and filler-extended products were introduced. The compositions of some typical products are listed in Table 11.

The economic recession of 1980 had the positive effect of causing vinyl compounders to reassess the cost-effectiveness of their stabilizer systems for flexible PVC. This sense of austerity in conjunction with increased environmental awareness in the workplace gave rise to four criteria for the mixed metal stabilizer of the 1980s:

1. Reduction or elimination of cadmium salts.
2. Reduction of dust
3. Reduction of volatile solvents
4. Self-lubricating liquids

IV. EMERGING STABILIZER CLASSES

A. Antimony Mercaptides

The use of antimony mercaptides as PVC heat stabilizers was patented by Weinberg in 1954, concurrent with the development of dibutyltin mercaptide stabilizers [44]. For the next two decades, the commercialization of antimony tris(isooctyl mercaptoacetate) as a vinyl stabilizer was minimal, while dibutyltin bis(isooctyl mercaptoacetate) became the most widely used and most important PVC heat stabilizer in the U.S. rigid PVC market. Today,

over 20 years later, antimony stabilizers again challenge organotins in the
pipe market. This occurrence requires an explanation.

When antimony and dibutyltin mercaptides were first compared in the
early 1950s, the organotins were clearly the products of choice. At that
time, rigid PVC compounds required 1.5 to 3 phr of heat stabilizer to allow
processing from pelletized compound at high temperatures on single-screw
extruders. When organotin and antimony stabilizer analogs were compared,
the organotins were found to impart better heat stability, better light stability, and better processability. In addition, the antimony mercaptides
themselves exhibited poor storage stability.

When antimony and dibutyltin mercaptides were reevaluated in the mid-
1970s, many things had changed in the PVC pipe market. Pipe was processed from powder compound at lower temperatures on multiscrew extruders.
More important, stabilizer use level had decreased to less than 0.5 phr.
When organotin and antimony stabilizer analogs were compared under these
significantly different conditions, the performance of antimony derivatives
surpassed that of the conventional organotin derivatives. Concurrently,
Dieckmann [45] patented methods of preparing storage-stable antimony compositions through the incorporation of antioxidants.

To explain the improved performance of antimony stabilizers relative
to analogous organotin derivatives, it is essential to review the basic structure and properties of the two materials. The structures and molecular
weights of the analogous dibutyltin and antimony isooctyl mercaptoacetate
derivatives are depicted below:

$$CH_3-CH_2-CH_2-CH_2 \diagdown \qquad \diagup SCH_2COOC_8H_{17}$$
$$Sn$$
$$CH_3-CH_2-CH_2-CH_2 \diagup \qquad \diagdown SCH_2COOC_8H_{17}$$

dibutyltin-*S*,*S*'-bis(isooctyl mercaptoacetate)

molecular wt 639

(I)

$$\diagup SCH_2COOC_8H_{17}$$
$$Sb- SCH_2COOC_8H_{17}$$
$$\diagdown SCH_2COOC_8H_{17}$$

antimony-*S*,*S*',*S*"-tris(isooctyl mercaptoacetate)

molecular wt 731

(II)

It can readily be seen that from a theoretical standpoint the dibutyltin
derivative has two stabilizer ligands (isooctyl mercaptoacetate groups) per
molecule whereas the antimony derivative has three stabilizer ligands per
molecule. Therefore, the relative equivalent weights of the materials as
stabilizers are as follows.

Organotin:

Equivalent wt = 639/2 = 319.5

Antimony:

Equivalent wt = 731/3 = 243.8

The relative efficiency of antimony (II) versus organotin (I) as a stabilizer at infinite dilution is

$$\frac{\text{Equivalent wt Sb}}{\text{Equivalent wt Sn}} \times 100\% = \frac{639/2}{731/3} \times 100\% = 131\%$$

Therefore, one would expect that as the stabilizer use levels are reduced, the relative effectiveness of antimony approaches the point where it becomes 31% more effective than the organotin as a heat stabilizer. Dieckmann [46] reported that at use levels below 1 phr, antimony derivatives are superior to the corresponding dibutyltin mercaptides.

From mechanism considerations, if one assumes that the displacement of labile chlorines is a nucleophilic substitution with SN2 kinetics, one would consider the antimony stabilizer to be less sterically hindered than the corresponding dibutyltin derivative. This would result in more effective collisions at low concentrations and thereby explain the greater efficiency of antimony at low use levels.

The greater efficiencies of organotins at high use levels can be explained on the basis of two factors: the greater inherent compatibility of the organotins and the weaker Lewis acidity of their corresponding chlorides.

Organotins contain the carbon-tin bond. They are true organometallics. Antimony derivatives are not organometallic. As a result, the inherent compatability of the organotins ensures their effectiveness at high use levels, whereas the limited compatibility of the antimony derivatives decreases their effective concentration at high use levels.

Second, since antimony trichloride is a stronger Lewis acid than organotin chlorides, at high stabilizer use levels the concentration of the chlorides increases, and the antimony trichloride can impair heat stability of vinyl compounds. Antimony performs best against organotins at low use levels in pipe formulations.

It has been well established that organotin and antimony mercaptides do not give good light stability in rigid PVC applications. As a result, neither organotins nor antimony mercaptides are recommended for use in clear or translucent outdoor applications.

The poor light stability is thought to result from the autoxidation/reduction of these materials in the presence of sunlight. It is believed that the metallic cations are reduced to metal while the mercaptides are oxidized to sulfoxides or sulfones.

The inherently poorer light stability of antimony mercaptides compared to organotins is explained by the fact that in these derivatives antimony already exists in its lower oxidation state (+3, antimonous) whereas organotins exist in their higher oxidation state (+4, stannic). Muldrow [47] has demonstrated that pentavalent antimony mercaptides give better UV resistance in PVC.

In adequately pigmented and highly filled pipe compounds, Fletcher et al. [48] demonstrated that the weatherability characteristics of antimony- and organotin-stabilized pipe are not significantly different at low stabilizer use levels.

B. Stabilizer Lubricant One-Packs

A second emerging heat stabilizer class is the stabilizer lubricant one-pack. This stabilizer class is in the stage of rapid evolution, but more so in Europe and South America than in North America.

From the earliest days of PVC compound development and before the emergence of the lead, mixed metal, or organotin stabilizer classes, it was clearly recognized that successful processing and fabrication of PVC compounds required both a stabilization and a lubrication function.

During the period of rapid growth of PVC applications, the wide variety of processing techniques and the distinct differences in processing equipment dictated that stabilization and lubrication functions be considered separately. This permitted the optimization of each for a particular set of circumstances.

The ultimate end of optimization is standardization. As a result of decades of optimization, large-volume PVC applications, equipment, formulations, and operating conditions are becoming increasingly standardized. The best example of this can be observed in the PVC pipe business.

Once standardization occurs, the successful marketing of a one-pack additive becomes a distinct practical possibility. The one-pack combines several additives which would otherwise be added separately. The concept originated in Europe with lead stabilizer/lubricant packages for pipe. In addition to simplifying compounding, the lead one-packs provided an effective means of reducing the dust associated with handling lead powders. Lead one-packs are available in nondusting flakes, granules, and prills.

In the U.S. pipe market, where leads are not widely used, the one-packs developed to date combine organotin stabilizer, paraffin, calcium stearate and oxidized polyethylene wax. They offer the pipe compounder convenience and simplicity. They further reduce inventory demands. As a trade-off for the advantages, the one-pack concept tends to take quality control and formulation flexibility out of the pipe compounders' hands and places the responsibility upon the suppliers of one-packs. In return for this burden, it guarantees the one-pack supplier a greater volume of sales.

The ultimate one-pack will be the truly single-package additive which combines all additive functions. It will include stabilizer, lubricant, pigment, filler, reinforcement, flame retardant, UV absorber, and so forth. Products of this type will lose all of their generic identity and be geared to fabrication technique. These composite additives are expected to become standardized within the next decade for some applications and products. Prototypes are currently in the planning and evaluation stages. Examples of state-of-the-art one-packs are listed in Table 12.

V. ORGANIC SYNERGISTS IN STABILIZER COMPOUNDING

Many organic compounds synergistically enhance the heat stability of primary metallic stabilizers. Major classes of organic synergists and typical examples are listed in Table 13. The major classes are acids, phosphites, epoxides, polyhydric alcohols, antioxidants, and miscellaneous compounds.

A. Acids

Organic acids that enhance the heat stability of metallic stabilizers are generally weak acids that have the ability to react with free metallic bases

TABLE 12 Typical Composition: Stabilizer Lubricant
One-Pack

Component	Percent
Organotin system	
Organotin mercaptide	20
Calcium stearate	20
Paraffin wax	50
Polyethylene wax	10
	100
Lead system	
Tribasic lead sulfate	25
Dibasic lead stearate	50
Normal lead stearate	15
Wax	10
	100

or with the salts of still weaker acids. In so doing, they generate moieties
which have a greater propensity to displace labile chlorines via nucleophilic
displacements.

Stabilizer compositions which contain basic lead salts or barium alkyl-
phenylates are greatly enhanced by the addition of stearic acid. Lead and
barium stearates are believed to be formed in situ as a result of the following
reactions:

$$2RCOOH + PbX \cdot 2PbO \rightarrow (RCOO)_2Pb + PbX \cdot PbO \cdot H_2O \qquad [4]$$

$$2RCOOH + Ba(O-\bigcirc-R)_2 \rightarrow (RCOO)_2Ba + 2HO-\bigcirc-R \qquad [5]$$

In addition to the generation of primary carboxylate stabilizers as typified by reactions (4) and (5), the incorporation of a fatty acid builds inherent lubricity into the stabilizer. This generally enhances the dynamic
stability of PVC compounds by reducing frictional heat buildup during
processing.

Organic acids such as benzoic and substituted benzoic acids improve
the clarity of PVC containing mixed metal stabilizer systems. This is believed to be the result of the solubilization of equilibrium amounts of dispersed metallic oxides, hydroxides, and carbonates.

Thioacids such as mercaptopropionic, thiolactic, and thiomalic acids
enhance the initial color-hold properties of organotin mercaptides.

The activity of certain blue toners that are pH sensitive is also enhanced
by the addition of acids to a PVC compound.

TABLE 13 Typical Organic Synergists

Acids

   Stearic acid

   Oleic acid

   Benzoic acid

   Mercaptopropionic acid

   Thiodipropionic acid

Phosphites

   Triphenyl phosphite

   Tridecyl phosphite

   Diphenyl decyl phosphite

   Phenyl didecyl phosphite

   Tris(nonylphenyl) phosphite

Epoxides

   Epoxidized soybean oil

   Epoxidized linseed oil

   Isooctyl epoxy tallate

   1,2-epoxy dodecane

   Epichlorohydrin/bisphenol A condensate

Polyols

   Glycerol

   Sorbitol

   Mannitol

   Pentaerythritol

   Dipentaerythritol

Antioxidants

   2,6-Di-*tert*-butylphenol

   2,2'-Bis(*p*-hydroxyphenyl)propane

   2,2'-Methylenebis(4-methyl-6-*tert*-butylphenol)

   2,6-Di-*tert*-butyl-*p*-cresol

Miscellaneous

   2-Phenylindole

   Diphenylthiourea

   Dicyandiamide

   Melamine

   Aminocrotonate esters

   β-Diketones

Strong mineral acids have a deleterious effect on the performance of a stabilizer. These materials preferentially react with the metallic stabilizer to form metallic salts of the strong mineral acid; some of these salts may then function as strong Lewis acids.

B. Phosphite Esters

Phosphite esters were initially evaluated as plasticizers for PVC. Shortly thereafter they were found to improve markedly the clarity, heat stability, and light stability of PVC compounds stabilized with mixed metal stabilizers.

Neither lead, antimony, nor organotin stabilizers are particularly enhanced by the addition of phosphite esters. In state-of-the-art technology, phosphites are used only in combination with mixed metal stabilizers. The ultimate performance of a mixed metal stabilizer is generally achieved when it is used in combination with a phosphite ester, a phenolic antioxidant, an epoxide plasticizer and a small amount of fatty acid for lubrication.

In the synergistic stabilization trilogy of mixed metal/phosphite/epoxide, the phosphite ester is generally used at an overall level of 0.5 to 1.0 phr. In the case of liquid mixed metals, the phosphite is usually dissolved in the metal stabilizer solution. In powdered mixed metals, the phosphite is usually added as a separate component.

Phosphite esters can be trialkyl, triaryl, mixed trialkyl/aryl, or polymeric. The trialkyl phosphites are derivatives of alcohols. The low molecular weight species are too volatile for use in PVC. Tridecyl phosphite is widely used, but it is somewhat limited in effectiveness by its thermal instability and relatively low phosphorus content (6% P).

Triphenyl phosphite (10% P) was widely used in early mixed metal stabilizers. However, triaryl phosphites that are derivatives of water-soluble phenols are extremely sensitive to hydrolysis. When they are incorporated in liquid mixed metals, the shelf life of the products is significantly affected. When they are incorporated directly in PVC compounds, the shelf life of the vinyl dry blend prior to the melt compounding is affected.

The best overall performance in general-purpose PVC mixed metal stabilizers is achieved by the use of mixed, trisubstituted alkyl-aryl phosphites. Decyl diphenyl phosphite and didecyl phenyl phosphite have been found to yield optimum results.

Tris(nonylphenyl) phosphite may be used when USFDA acceptance or low toxicity is required. It is a common component of FDA-approved calcium-zinc paste stabilizers [49].

The effectiveness of a phosphite ester is generally a function of phosphorus content and degree of organic substitution. In monomeric phosphites, a high phosphorus content can be achieved only with low molecular weight materials, which give rise to volatility problems. In recent years, an increasing number of trisubstituted polymeric phosphite esters have been marketed. These materials are derivatives of glycols and higher polyols, diphenols, and polyphenols. They can be used to formulate stabilizer systems which give high efficiency, low volatility, and low extractables. They are usually significantly more expensive than the general-purpose monomeric phosphites and many of them also are prone to hydrolysis.

C. Epoxides

Epoxidized soybean oil is the most commonly used epoxide synergist. It is almost universally used in plasticized PVC in combination with mixed metal stabilizer systems at levels from 2 to 7 phr. It is believed to function as an HCl absorber and also as a catalyst or intermediate in the transfer of stabilizer ligands between mixed metals such as cadmium and barium. More recent evidence shows that under certain conditions, epoxides may also function as primary stabilizers [50, 51].

The synergistic stabilizer activity of an epoxidized oil is a direct function of its oxirane oxygen content. However, materials with an extremely high oxirane content tend to be either too volatile for practical use or less compatible with PVC and thus can give rise to exudation or accelerated aging, particularly when subjected to direct sunlight or high humidity.

Monomeric epoxy esters such as isooctyl-, butyl-, or 2-ethylhexyl-9, 10-epoxystearate have fair to good compatibility and impart low viscosity to plastisol formulations. Their inclusion also results in improved low-temperature flexibility (see chapter 14).

Other epoxides that find limited commercial use are 1-epoxydodecane, epoxidized chloroalkanes, and epichlorohydrin condensates with polyphenols such as bisphenol A.

D. Polyhydric Alcohols

Polyhydric alcohols or polyols are employed principally to enhance the performance of mixed metal stabilizers such as barium-cadmium, barium-zinc, and calcium-zinc types.

The mechanism by which polyols function is not well understood. It has been suggested that they complex metal chlorides and also serve as catalysts or intermediates in the transfer of stabilizer ligands in a manner not unlike the epoxides.

The relatively poor solubility of polyols in hydrocarbon solvents precludes their use in most liquid mixed metal stabilizers. They are employed principally in combination with mixed metal powdered stabilizers at levels of 10 to 50% by weight of stabilizer.

Pentaerythritol is the most common polyol used in barium-cadmium powders. Sorbitol and glycerol are the most common polyols used in calcium-zinc systems. Too high a use level of a polyol in a mixed metal stabilizer can give rise to severe moisture sensitivity and water blush of the final PVC compound. The compatibility of polyols can be significantly increased and their moisture sensitivity reduced by partially esterifying them. Glycerol monostearate is a widely used polyol derivative in calcium-zinc heat stabilizer systems.

E. Antioxidants

Substituted phenolic antioxidants are the most widely used antioxidants in PVC heat stabilizers. They are employed principally in mixed metal stabilizer systems at low use levels. Their synergistic effect appears to be catalytic rather than quantitative in areas where they are employed.

Low molecular weight substituted phenols are too volatile for use in PVC. 2, 2'-Bis(*p*-hydroxyphenyl)propane, commonly known as bisphenol A, and 2,6-di-*tert*-butyl-*p*-cresol, commonly known as BHT, are the most widely used materials. Where high-temperature applications demand a higher molecular weight, materials such as 2,2'-methylenebis(4-methyl-6-*tert*-butylphenol) are used.

Some phenolic antioxidants may give rise to staining with certain pigments and additives such as TiO_2, metal oxides, or hydroxides. A thorough understanding of the end-use compound must be considered before they are employed. Sulfur-containing antioxidants such as dilauryl thiodipropionate and thiolauric anhydride have found limited use in PVC heat stabilizers.

F. Miscellaneous Materials

A wide variety of organic materials have been used as stabilizers for particular resins or applications [52]. In Europe, until recently, PVC was often polymerized in alkaline emulsion systems. Resins produced in this manner contained substantial amounts of residual sodium carbonate. These resins were not particularly responsive to stabilization by metallic stabilizers. Two organic materials, diphenyl thiourea and α-phenylindole, were employed effectively to stabilize resins of this type. Currently, these stabilizers are no longer being used in significant quantities, as most Europeans have switched to suspension or mass polymerization processes.

In highly filled vinyl asbestos floor tile, dicyandiamide and melamine were used for many years as heat stabilizers. Their principal function was believed to be the complexing of iron or other contaminants of the asbestos fillers and subsequent prevention of a specific type of blue staining from occurring as a result of the interaction with plasticized PVC of asbestos containing iron contaminants. Currently, asbestos has been formulated out of most flooring applications. As a result, the use of dicyandiamide and melamine as additives has diminished.

Interest continues in both the United States and Europe in the applications of β-aminocrotonate esters and β-diketones. Both stabilizer types find use in combination with calcium-zinc stabilizers in situations where it is essential to prevent the formation of zinc chloride.

Butanediol-β-aminocrotonate has been shown to delay chloride formation to a greater extent than epoxides [52].

Stearoyl-benzoyl-methane has been shown to improve the initial color and color-hold characteristics of calcium-zinc stabilizer systems and has been suggested as a synergist for the use of calcium-zinc stabilizers in PVC bottle applications [53].

VI. LIGHT STABILIZERS

Most polymers, including PVC, are stable to visible light, which is the part of the electromagnetic spectrum between 400 and 700 nm [54].

All organic materials are degraded on prolonged exposure to ultraviolet light. Ultraviolet light at the earth's surface is generally considered to be the part of the electromagnetic spectrum between 290 and 400 nm. Fortunately, most of the ultraviolet light below 290 nm from solar radiation is screened out by the earth's atmosphere.

Depending on its structure, each organic substance exhibits a maximum sensitivity to a particular wavelength of UV light. For PVC, that wavelength is 320 nm.

The absorption of UV radiation by PVC excites PVC molecules. If the molecules remain in an excited state sufficiently long, the photolytic scission of certain C—Cl bonds occurs:

$$\text{UV}$$
$$\sim\!\!\text{CH}_2-\underset{\text{Cl}}{\text{CH}}\!\!\sim \;\rightarrow\; \sim\!\!\text{CH}_2-\underset{\bullet}{\text{CH}}\!\!\sim \;+\;\cdot\text{Cl}$$

This ultimately leads to conjugated unsaturation, discoloration, oxidation, cross-linking, embrittlement and loss of physical properties. The main difference between the photolytic scission and thermal fission of PVC is that the photolytic scission is principally a free-radical process.

The presence in PVC of hydroperoxide, carbonyl groups, olefinic groups, trace metals and polynuclear aromatics renders it even more susceptible to UV degradation. Hence, the heat history of the polymer prior to UV exposure has a significant effect on its ultimate UV resistance. Consequently, it is appropriate to consider that heat stabilizers have a significant effect on the ultimate light stability of PVC even though most of them are not light stabilizers per se. In the broadest sense, there are three types of light stabilizers: pigments, UV absorbers, and quenchers.

Pigments function principally by acting as light screens. Carbon black and TiO_2 are particularly effective in PVC. Zinc oxide is of limited use as a result of the zinc sensitivity of most PVC resins. Pigments are precluded in applications that demand clarity.

Ultraviolet absorbers are chemical additives that preferentially absorb harmful UV radiation more rapidly than PVC and dissipate it in the form of harmless lower-energy radiation. In this process, the UV absorber is not decomposed but rather regenerated.

The ideal UV absorber for use in PVC must satisfy the following criteria. It must absorb UV strongly at 290 to 400 nm. It must be colorless so that it does not discolor the compound. It must be compatible so that it does not affect clarity. It must be permanent and not exude on aging. It must be nonvolatile so that it is not lost during processing. It must be heat stable and not decompose during processing. It must be compatible with and unaffected by other compounding ingredients such as the heat stabilizer and antioxidants in the compound.

Substituted 2-hydroxybenzophenones and 2-hydroxyphenylbenzotriazoles are most widely employed. Substituted acrylonitriles and salicylate derivatives are of minor importance. General structures and specific examples are included in Table 14.

Another type of light stabilizer is known as a quencher. Quenchers do not function by absorbing UV radiation. Rather, they interact with excited PVC molecules and remove energy from them by an intermolecular energy transfer process.

Metal chelates, particularly nickel-organic compounds such as nickel dibutyldithiocarbamate (NBC), are extremely effective. NBC is not suitable in PVC, however, because of its green color and its propensity to interact with metallic heat stabilizers, particularly the organotin mercaptides.

A new class of hindered amine light stabilizers (HALS) are extremely effective in non-PVC applications. These materials are derived from piperidine. An example is bis-4-(2,2',6,6'-tetramethylpiperidinyl) sebacate. Unfortunately, this new class of materials, like all amine stabilizers, enhances dehydrohalogenation in PVC.

Ultraviolet light absorbers are of value principally in transparent PVC compositions. In highly pigmented systems, most of the incident UV energy is reflected by the surface. The portion absorbed is generally preferentially

TABLE 14 Principal UV Absorber Classes

2-Hydroxybenzophenones

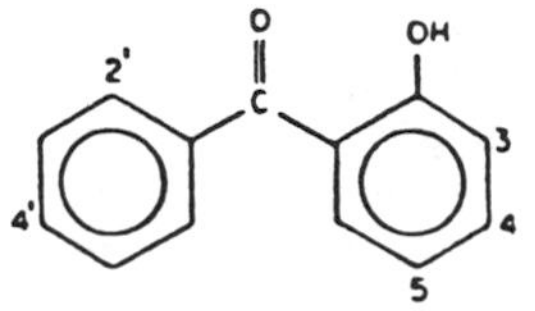

Examples: 2-Hydroxy-4-octyloxybenzophenone

2-Hydroxy-4-methoxybenzophenone

2-(2'-Hydroxyphenyl)benzotriazoles

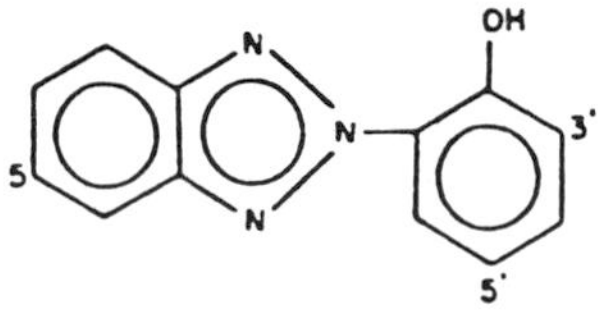

Examples: 2-(2'-Hydroxy-5'-methylphenyl)benzotriazole

2-(3',5'-Dialkyl-2'-hydroxyphenyl)benzotriazole

Substituted Acrylates

$$X - \overset{Y}{\underset{3}{C}} = \overset{Z}{\underset{2}{C}} - \underset{1}{COOR}$$

Examples: Ethyl 2-cyano-3,3-diphenylacrylate

2-Ethylhexyl 2-cyano-3,3-diphenylacrylate

Salicylates

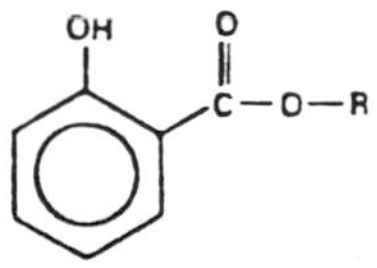

Examples: Phenyl salicylate

p-tert-Octylphenyl salicylate

absorbed by the relatively inexpensive pigment, which is present in greater quantities than the expensive UV absorber. As pigment concentration increases, diminishing returns are achieved by the use of a UV absorber. This is illustrated in Fig. 4 for two rigid PVC formulations, both containing 2-hydroxy-4-methoxybenzophenone. Both are heat stabilized, with a barium-cadmium system in one case and an organotin mercaptide in a second case [55].

In 1984 the UV light stabilizer market in the United States was estimated to be less than 3000 metric tons. Less than 10% of these materials are used in PVC.

Clear PVC finds few commercial applications that require weatherability. Most weatherable PVC compounds are highly pigmented with TiO_2 and are not augmented with UV absorber.

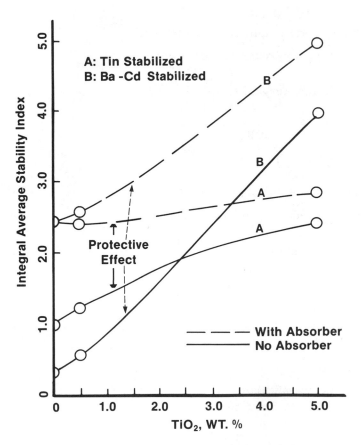

FIGURE 4 Ultraviolet absorbance in two rigid PVC formulations: A, tin stabilized, and B, Ba—Cd stabilized.

VII. STABILIZER SELECTION CRITERIA

A. Primary Stabilizer Contributions

Stabilizers make three primary contributions to PVC compounds:

1. Process heat stability
2. Processability
3. Long-term property retention

1. Process Heat Stability: The principal function of a stabilizer is to provide the PVC compound with sufficient heat stability to be fabricated without loss of its aesthetic properties. It is infrequent that a commercial stabilizer fails to perform this function in a field application. If this occurs, it is generally the result of a major mistake such as use of the wrong material or of a weighing error.

Most vinyl compounds tend to be overstabilized to compensate for slight variations in operating conditions and the processing of rework. Many PVC compounds are substantially overstabilized. Consequently, when a new stabilizer is evaluated in a plant situation, it is difficult to quantitate or measure the degree of insurance the stabilizer provides in a plant production environment.

In the laboratory, competitive stabilizers are best evaluated by comparing them with a control at several use levels. The stabilizer that gives the best performance at the lowest use level is the best stabilizer per se. In the final evaluation, stabilizers must be compared at equal cost levels since cost/performance is the ultimate selection criterion.

2. Processability: The processability contribution of a PVC heat stabilizer is a function of its inherent lubricity as well as its handling properties. Stabilizers range in consistency from plasticizer-like to filler-like substances.

If the inherent lubricity of a particular stabilizer is insufficient for a given application, the stabilizer may perform well in a static heat stability test but fail during processing applications. It is also important from a processability point of view that the product have excellent handling properties. Products with unusually strong odors, high viscosities, or dusty characteristics can irritate plant operating personnel.

3. Property Retention: Although it is frequently overlooked, one of the most important properties of a stabilizer is its own shelf stability. Most powdered stabilizers, including the leads and mixed metals, are indefinitely shelf stable. Liquid mixed metals and organotin stabilizers may occasionally precipitate with age and thereby lose their effectiveness. In addition, the mixed metals are prone to hydrolysis and oxidation and must be stored in closed containers to limit exposure to air and humidity. Liquid mixed metals over 1 year old begin to lose their effectiveness. Inventories should be rotated and unusually old material should be tested before use. Loss of efficiency occurs more rapidly in partially filled containers exposed to air.

In addition to providing process heat stability, a properly compounded stabilizer system should provide a reasonable degree of color retention to a finished vinyl compound during storage, during fabrication, and for the service life of the finished vinyl article. This requires that sufficient stabilizer of the proper type be employed to provide color retention during long-term exposure to moderate temperatures experienced during service life of the goods.

B. Stabilizer Side Effects

Most commercial stabilizers do provide the three primary stabilizer functions of process heat stability, processability, and property retention. As a result, the selection of a particular stabilizer is often made on the basis of the presence or absence of an ancillary property or stabilizer side effect.

1. Initial Color: The initial color provided to a PVC compound by a stabilizer is principally the result of the stabilizer's ability to initiate stabilization rapidly during the heat processing cycle; it is usually independent of the stabilizer's own color.

Stabilizers that are most effective in imparting good initial color are usually liquids or low-melting solids that are readily soluble in PVC resins. Organotins are ideal. Mixed metal liquids are fair to excellent. Powdered mixed metals are not as effective. Lead powders are least effective in maintenance of initial color.

Within the individual stabilizer classes, initial color can be increased in the following manner. For leads, it is improved by the inclusion of dibasic lead phosphite and lead carboxylates. Initial color is improved in organotin-stabilized compounds by increasing the ratio of monoalkyltin to dialkyltin in the derivatives. In mixed metal stabilizers it is achieved by increasing the ratio of cadmium to barium, or zinc to calcium and by the use of alkyl-aryl phosphite and/or polyhydric alcohol.

2. Ultimate Stability: Ultimate stability or long-term heat stability is achieved in stabilizer systems that have an extremely high neutralization capability for HCl and in systems that are incapable of generating strong Lewis acids as stabilizer by-products. Among the individual stabilizer classes, the leads are best, the dialkyltins are good, and the mixed metal systems are least effective.

Within the individual stabilizer classes, long-term stability can be achieved in the following manner. For lead stabilizers, it is achieved by increasing the PbO content of the stabilizer, usually through incorporation of a larger percentage of tribasic lead sulfate in the compounded product.

For organotins, long-term stability is improved by maximizing the alkyltin content and using mercaptide rather than carboxylate anions. Under these conditions, ultimate stability is directly proportional to tin and sulfur content.

For mixed metals, ultimate stability is improved by increasing the ratio of barium to cadmium, and calcium to zinc, and also by the addition of higher levels of epoxide, phosphite, and polyols as synergists.

3. Clarity: The clarity of a PVC compound with any given stabilizer is a direct function of the degree to which the stabilizer is soluble or dispersible in the compound. Clarity is inversely proportional to concentration for most powdered stabilizers.

Among the principal primary stabilizer classes, organotins impart the greatest degree of clarity. Mixed metal liquids give better clarity than powdered derivatives. Leads, being pigmentlike inorganic substances, impart the greatest degree of haze of the major stabilizer classes.

Within the organotin stabilizer class, clarity is generally improved by using mercaptoacetate rather than linear mercaptide anions in the case of sulfur-containing derivatives, and by using maleate esters rather than long-chain linear carboxylates in the case of nonsulfur tins. Small quantities of acidic additives have also been suggested to improve the clarity of organotins. These include hydroxy- and thiocarboxylic acids.

Liquid mixed metal stabilizers generally impart greater clarity than powdered soap-based systems. Clarity can be improved by maximizing the use of aromatic, branched-chain, and short-chain anions. Metallic soaps of long-chain linear saturated fatty acids, such as stearic acid, have limited compatibility in PVC. Clarity can also be improved by increasing the ratio of phosphite to metallic salt in these systems.

For lead systems, very little can be done to improve clarity. Neutral lead carboxylates based on aromatic acids can be employed in translucent applications. Lead salicylate has been suggested.

4. *Compatibility*: Compatibility or permanence is related to a stabilizer's propensity to migrate from a PVC compound over a period of time. Stabilizers which exude from PVC on aging give rise to adhesion problems, printability problems, dirt pickup problems, and toxicological concerns.

Generally, pigmentlike inorganic stabilizers, such as the leads, have no tendency to migrate from PVC. Organotins also tend to remain intact as a result of their inherent solubility and affinity for PVC resin. The mixed metal stabilizer systems, because of their intermediate solubility and potential lubricating nature, have the greatest propensity to lead to compatibility problems.

From a practical standpoint, most compatibility problems arise in plasticized PVC systems. Here, the use of high levels of epoxidized oils and excessive quantities of secondary plasticizers can result in the exudation of a combination of plasticizer and stabilizer. This effect is often delayed but can be accelerated by the presence of humidity and sunlight.

The compatibility of lead systems can be improved by avoiding the use of carboxylates and fatty acid lubricants. For organotins, compatibility is best achieved with mercaptoacid ester or short- or branched-chain carboxylate anions. In the case of mixed metals, the best way to ensure compatibility is to minimize the fatty soap content and overall use level. In addition, attention must be given to minimizing the content of other compounding ingredients such as secondary platicizers, aliphatic epoxide synergists, and fatty acid lubricants.

5. *Plateout*: Plateout is the term used to describe deposits of incompatible substances from the polymer melt onto processing equipment during the processing of PVC compounds. It commonly occurs on calender rolls and extrusion die lips.

When plateout deposits are analyzed, they are generally found to contain large amounts of particulate fillers and pigment employed in the compound. These by themselves are not the cause of plateout.

Plateout is caused by the formation of high-viscosity, sticky substances that are incompatible with the polymer melt. They may be stabilizer components, stabilizer by-products, or the reaction products of stabilizer components or stabilizer by-products with other compounding ingredients. When these substances are formed in the polymer melt, being both incompatible and sticky, they tend to deposit on the processing equipment. They subsequently cause filler and pigment particles from the polymer melt to adhere to the processing equipment. When plateout builds up to a high level, it generally causes streaking or poor surface on the calendered or extruded product. At that point one must either process an abrasive cleanup compound or shut down the equipment and physically remove the plateout deposits.

Inorganic lead compounds have little tendency to plate out. Organotin stabilizers per se do not plate out. However, reaction products of organotin mercaptide by-products with calcium stearate and wax have been reported to plate out in PVC pipe compounds.

Mixed metal products have the greatest tendency to plate out. When stabilizer solvents volatilize during flexible PVC processing, many of the salts employed in barium cadmium liquids become incompatible, sticky substances. This is particularly true of the barium or calcium components of these stabilizers.

For powdered mixed metals, it is the alkaline earth metal soap components that have the greatest tendency to plate out. This can be attributed to the physical form of the barium soaps at processing temperatures. At these temperatures, soaps are glassy liquid crystals with high melt viscosities.

Plateout of mixed metal stabilizers can generally be reduced by substituting barium nonylphenylates or overbased barium nonylphenylates for barium carboxylates in liquid stabilizers. In powdered stabilizers, plateout can be reduced by decreasing the barium component and incorporating small amounts of abrasives such as silica into the formulation as scrubbing agents. Acidic phosphite esters also reduce plateout by interacting with the basic barium soaps.

Plateout is directly proportional to the stabilizer concentration. It can often be reduced simply by cutting back the primary stabilizer and augmenting the reduced performance by increasing the concentration of a secondary stabilizer such as epoxidized soybean oil. It is the alkaline earth metal concentration of the stabilizer which contributes principally to plateout.

6. Volatility: Stabilizer volatility can result from loss of a stabilizer component such as solvent or antioxidant. In other instances, it can result from a stabilizer by-product such as alkyltin chloride, alkylphenols, H_2S, CO_2, or low molecular weight acids. It can also be the result of stabilizer hydrolysis or oxidation products such as alcohols and phenols. Stabilizer volatility leads to air pollution problems and odor problems in the finished products, as well as fogging in automative applications or condensation on processing equipment.

Lead stabilizers normally have no volatility or odor problems associated with their use. Organotin mercaptides and maleate esters generally have the worst odors during processing. These are due to volatile organotin chlorides, mercaptans, or maleic acid. In flexible PVC compounds, the residual odor of these moieties in the finished vinyl articles seriously restricts their use. This is not generally a problem in rigid PVC applications.

Liquid mixed metal stabilizer systems give rise to the greatest amount of volatiles. These arise from the evolution of solvents and from hydrolysis products of phosphite esters. State-of-the-art liquid concentrates have been developed that are essentially solvent-free and based on a combination of phenol-free and metal bases and polymeric phosphite esters. These products have a concentration of active components that is approximately twice that of early liquid stabilizers. This greatly reduces volatility and odor in processing and finished goods.

7. *Moisture Resistance*: Stabilizers that are sensitive to moisture are those which contain water-soluble components, hydrolysis-sensitive materials, or stabilization by-products that are water-soluble.

Moisture-sensitive stabilizers may yield finished vinyl products with poor properties. Bottles may water blush. Films may fog. Wire and cable may fail to meet insulation requirements. Floor tiles fail to maintain dimensional requirements. Food packaging fails extraction requirements. Flexible compounds lose aging stability under humid conditions.

Lead stabilizers are extremely moisture-resistant. The moisture resistance of organotins is fair to good. Mixed metal stabilizers are the most moisture-sensitive.

The moisture sensitivity of mixed metal stabilizers can be improved by using stabilizers with a low metal content and water-insoluble anions. The use of polyhydric alcohols should be avoided. Efforts should be made to use only hydrolysis-resistant phosphite esters such as high molecular weight trialkyls or tris(nonylphenyl) phosphites.

8. *Stain Resistance*: In the broadest sense, stabilizer-associated staining is a discoloration that results from the interaction of a vinyl stabilizer with air, moisture, fungus, sunlight, some chemical ingredients of the formulation, or an ingredient of a foreign surface in direct contact with the vinyl object.

The most common forms of staining associated with stabilizers are sulfur staining and sulfur cross-staining. These occur with metallic stabilizers that form colored sulfides. Lead sulfide is black, antimony sulfide orange, and cadmium sulfide yellow. Organotins and zinc stabilizers do not form colored sulfides. Zinc compounds often prevent cadmium from sulfur staining by preferential reaction with the sulfur. The source of sulfur which causes sulfur staining in vinyl stocks containing lead, cadmium, or antimony stabilizers can be H_2S in air, natural sulfides in mineral waters, sulfur in foods such as eggs and mustard, sulfide bleached kraft paper, and direct contact between vinyl and rubber articles compounded with sulfur. Sulfur cross-staining can occur when lead-, cadmium-, or antimony-stabilized PVC is compounded with rubber or vinyl rework or scrap that contains mercaptide stabilizers, metal sulfides, or sulfur-containing compounding ingredients.

Other common staining problems related to stabilizers are due to the reaction of stabilizer components with phenolic antioxidants, amines, UV absorbers, and metals.

Copper staining occurs when copper metal comes in contact with flexible PVC that is stabilized with strong alkaline stabilizer components such as barium nonylphenate. It produces a green discoloration in the vinyl.

Urethane staining occurs when PVC comes in contact with uncured or incompletely cured amine-cured urethanes. Highly colored chromophores result from the interaction of the amines with alkaline stabilizer components. Staining can be prevented only by use of a barrier coating between the vinyl and the urethane or by use of a precured urethane film.

Phenolic staining can occur when alkaline stabilizers are used in combination with phenolic antioxidants. Staining is due to the formation of quinoid groups.

Ultraviolet absorber staining can occur with benzophenone and benzotriazoles. These yellow stains are pH-dependent and can be reversed by the addition of acidic materials such as stearic acid to the formulation.

Titanium dioxide staining can occur when uncoated chloride-process TiO_2 with residual chlorides is used in combination with phenolic-containing stabilizers.

Fungal staining can occur under humid conditions when high quantities of fatty-based plasticizers or lubricants are employed. Zinc and some organotin stabilizers have been reported to reduce susceptibility of flexible PVC to fungal attack. Triaryl phosphate plasticizers are also helpful in this regard, but care must be taken with these materials since their inclusion will almost certainly necessitate a modification of the heat stabilizer system.

Ultraviolet staining occurs when stabilizers themselves undergo decomposition as a result of direct exposure to sunlight. This occurs with antimony mercaptides and certain organotin mercaptides. It is thought to be a result of a light-induced oxidation-reduction reaction of the stabilizer.

9. Mechanical Properties: Ideally, a stabilizer should not influence the mechanical properties of a PVC compound. Most stabilizers do have a measurable effect on both the mechanical properties and processability of PVC compounds. Lead stabilizers, being for the most part pigmentlike inorganic substances, have the least effect. Organotin mercaptoacetates, because of their extreme compatability, act as plasticizers and viscosity depressants for PVC. This behavior limits their use at high levels when high heat distortion temperatures are demanded in rigid PVC.

Mixed metal stabilizers fall between leads and organotins in their effect on mechanical properties. Their particular effect is dependent on the chemical structure of the specific ligand employed in their production.

10. Electrical Properties: PVC heat stabilizers have a marked effect on the electrical insulation properties of vinyl compounds. The presence in PVC of hydrophilic substances causes a loss of volume resistivity and other insulating properties. Consequently, the ideal stabilizers for these applications are hydrophobic species that are insoluble in water and non-ionizable. In addition, the by-products of stabilization (e.g., metal chlorides) preferably will be water-insoluble and nonionizable.

The only stabilizer class that meets these criteria fully is the lead stabilizers. As a result, leads are almost universally accepted as the stabilizers of choice in all wire and cable applications.

Calcium-leads, barium-leads, and nonmetallic synergists have been used with limited success where high volume resistivity is not key. Solid barium-cadmiums are used in applications requiring transparent vinyl insulation.

11. Rheological Properties: The rheological properties of stabilizers are of principal concern in the formulation of plastisols and organosols and in the processing of rigid PVC compositions.

In plastisols, stabilizer selection has a dramatic effect on viscosity and surface tension. As a result, initial viscosity, viscosity stability, air release characteristics, and cell structure in foamed vinyls are markedly affected by the choice of stabilizers. In general, liquid mixed metal stabilizers are preferred in plastisols.

Lead stabilizers are difficult to disperse and often require milling. They give rise to high viscosities and poor viscosity stability.

Organotin carboxylates are not especially effective heat stabilizers, and the sulfur-containing tins generally cause odor problems.

Mixed metal powders are difficult to disperse and often gel plastisols because of their grease-forming properties.

Liquid mixed metals provide low initial viscosities and some also provide good viscosity stability. In addition, they may facilitate air release. It is important that a plastisol stabilizer be neutral or acidic. Basic mixed metals give rise to viscosity drift due to interaction of the free basicity with residual emulsifier on plastisol resins.

Lead-, cadmium-, and zinc-containing stabilizers catalyze the decomposition of azodicarbonamide blowing agents and are often used to promote foaming at low temperatures.

Nonionic wetting agents, silicones and hydrocarbon diluents are often added to plastisol stabilizers to control air release or to maintain paste viscosity.

In processing rigid PVC compositions, basic lead stabilizers function in a pigmentlike manner rheologically, whereas lead stearates function as external lubricants.

As with the lead fatty acid soaps, mixed metal powdered stabilizers are highly lubricating. This characteristic makes it difficult to vary the stabilization and lubrication properties of a mixed metal stabilizer system independently in a rigid application. In addition, metal carboxylates give rise to high viscosities in PVC melts due to secondary cross-links formed between vacant d orbitals of the metal with p electrons of chloride atoms in PVC, as illustrated in the example below:

Organotin mercaptides are the most versatile type of stabilizer for rigid PVC processing. They exhibit the best rheological properties. The low melt viscosities experienced with the organotin mercaptides are attributed to the ability of sulfur p electrons to overlap the vacant tin d orbitals, thereby satisfying the orbital requirements of tin and preventing secondary cross-linking in the polymer melt [56].

VIII. STABILIZER SELECTION TECHNIQUES

The ultimate selection of a specific stabilizer for a particular end-use application involves four distinct decisions.

1. Selection of the generic class
2. Selection of the specific type
3. Optimization of the type
4. Determination of cost/performance property trade-offs

The selection of the generic class involves the choice of lead, organotin, or mixed metal. This choice can be dictated by governmental or other regulations, precedent, or the process of elimination.

Regulations in the United States prohibit the use of lead in potable water pipe but may require the use of lead in certain electrical applications.

Precedent influences the choice of organotins in weatherable rigid PVC compounds in the United States and the use of mixed metals, sometimes with leads, for similar applications in Europe.

The process of elimination of leads because of toxicity, opacity, or sulfur staining, and organotins because of cost, suggests that mixed metals will be the choice in flexible PVC calendered film and sheet applications worldwide. The elimination of organotins and mixed metals because of poor electrical properties is responsible for the use of leads in wire and cable worldwide. The opacity of many mixed metals and all leads suggests the use of organotins in crystal clear rigid packaging applications [57].

The preferred generic classes of stabilizer for major applications in the United States and the world are compared below:

Application	United States	World
Rigid PVC		
Pipe and fittings	Organotin or antimony	Lead
Extruded profile	Organotin	Mixed metals or lead
Bottles	Organotin	Organotin or calcium-zinc
Calendered goods	Organotin or mixed metals	Mixed metals or lead
Phonograph records	Lead or mixed metals	Lead
Flexible PVC		
Calendered film and sheet	Mixed metals	Mixed metals or lead
Extrusion and molding	Mixed metals	Mixed metals or lead
Flooring	Mixed metals	Mixed metals or lead
Plastisols	Mixed metals	Mixed metals or organotin
Wire and cable	Lead	Lead

It can be seen that organotins dominate rigid PVC applications in the United States and mixed metals dominate flexible PVC applications, with the exception of wire and cable, where lead is employed worldwide. In pipe, fittings, and extruded profile, the U.S. market currently uses organotin primarily. In most of the rest of the world, lead is used.

Once the generic class is determined, the selection of stabilizer type within the class is the next decision for the compounder.

In organotins the choice of type involves the nature of the alkyl group, the nature of the anion, and the physical form of the stabilizer. On an equal tin metal basis, there is little difference in performance between methyl-, butyl-, and octyltin derivatives that contain the same anionic ligands. Methyls and butyls are used in general-purpose applications. Some specific octyltins, butyltins, and methyltins are sanctioned for USFDA applications [58]. Sulfur-containing organotins are superior to nonsulfur organotins in heat stabilization. The sulfur-free carboxylate derivatives are superior to sulfur-containing organotins in light stabilization. Organotin liquids are available as either lubricating or nonlubricating types.

Within the mixed metal stabilizer class, the choice of type involves metal type and physical form. For general-purpose applications, barium-cadmium or barium-cadmium-zinc types are preferred. The inclusion of zinc improves sulfur stain resistance and initial color control in highly filled formulas. The optimum zinc level is determined by the zinc tolerance of the PVC resin, the plasticizer types and amounts, and the filler type and level.

Barium-zinc stabilizers are used in applications that have relatively moderate heat requirements, low rework requirements, or where the presence of cadmium is of toxicological concern.

For FDA and medical applications, nontoxic formulations based on calcium-zinc mixed metals are used in most plasticized applications. Nontoxic rigid PVC systems will use either FDA-sanctioned organotins or calcium-zinc types.

With regard to physical form, liquids are preferred where lubricity is not required or volatility is not of concern. For applications requiring low volatiles or a highly lubricating character, mixed metal powders are the products of choice.

In the lead stabilizer class, choice of type is among basic lead carbonates, sulfates, silicates, phthalates, phosphites, maleates, fumarates, or stearates. In addition, the compounder must decide whether to purchase a pure powder, a granule, or a complete compound one-pack stabilizer/lubricant system.

Sulfate or phosphite is used where heat stability is the major concern. Phthalate is used where low interaction with ester plasticizers is important. Phosphite with dibasic lead stearate is the choice for light stability. Stearate is used for lubricity. Carbonate is used for low cost.

The choice between pure powder, granule, or one-pack is determined by balancing cost versus environmental concerns, depending on individual plant use levels, equipment constraints, governmental regulations, and in-house technical skills.

The optimization of the stabilizer type within a class is determined by the cost/performance characteristics of competitive products. Stabilizer competitors are highly aggressive and constantly striving to produce superior products at lower costs for the customer.

As a rule, stabilizer performance is a function of metal content and, in mixed metal systems, the metal ratio. Stabilizer producers offer a wide variety of stabilizers with various metal contents and metal ratios. They are often referred to as "high efficiency," "standard," or "low cost." Price is generally related directly to metal concentration. Concentrates normally represent the best value, but not all formulators are equipped to use them judiciously.

Low-cost organotin liquids are generally diluted with mineral oils or plasticizers. Low-cost liquid mixed metals are diluted with mineral spirits. Low-cost powdered mixed metals are sometimes diluted with mineral fillers. Low-cost leads may be diluted with clays or silica. Typical percent metal ranges for liquid products are listed in Table 15.

After the selection of stabilizer class, type, and optimization of type has been completed, the compounder must address cost/performance property trade-offs. To do this effectively, the compounder must take into consideration the internal requirements of the conversion process and the expected life cycle of the object or product being produced.

TABLE 15 Typical Metal Content of Stabilizer Liquids

	Mixed metal				Organotin %Sn
	%Ba	%Cd	%Zn	%P	
Low cost	3−5	1−2	0−1	1−2	4−8
Standard	5−6	2−3	0−1	2−3	8−18
High efficiency	6−9	3−4	0−1	3−4	18−35

For products such as vinyl siding and flooring, it is not sufficient to stabilize only for processing. Stabilizer choice and stabilizer use levels must guarantee long-term service life of the finished vinyl article. Stabilizer selection is based on experience and results of long-term accelerated testing.

IX. INDIVIDUAL APPLICATIONS

A. Pipe

In the United States, the primary criteria for potable water pipe stabilizers are that they be low in cost, acceptable to the National Sanitation Foundation, and capable of producing white pipe consistently during continuous runs. Once these primary requisites are satisfied, stabilizers are compared according to the manner in which they perform in pipe extrusion with regard to wall thickness control, burn frequency, output rate, and frequency of downtime, corrosivity, and tooling wear. These can only be determined statistically over long periods of time.

Over the past two decades, the U.S. pipe market has switched from organotins with a high metal content to low-cost, low-metal types. Tin content of these stabilizers has fallen from 18 to 24% to 4 to 8%. In addition, a distinct portion of the market has switched to antimony or to low-cost tin one-packs that contain inexpensive waxes.

The active components of state-of-the-art products are either antimony tris(isooctyl mercaptoacetate) or mixed mono- and dialkyltin sulfur-bridged mercaptoethanol carboxylates.

In Europe, lead one-packs have been developed that have allowed the overall use levels in pipe to decrease from 2.5 to 1.2 phr. These materials are said to be based on combinations of lead sulfates, lead stearates, and proprietary waxes.

B. Fittings

The principal criteria for a good pipe fittings stabilizer are high-temperature stability, high-shear stability, and easy processibility.

Mixtures of mono- and dialkyltin isooctyl mercaptoacetates are ideal for this application because of their inherent compatability with PVC. They are used at levels of 1 to 2 phr. Leads are used in most of Europe for fittings applications.

C. Bottles

In bottles, initial color, clarity, and organoleptic properties are key.

For general-purpose bottles, special low-odor grades of methyl- and butyltin isooctyl mercaptoacetates have been developed for use at 1.5 to 2.5 phr.

For USFDA-sanctioned bottles, octyltins are used. The principal product is di-*n*-octyltin bis(isooctyl mercaptoacetate).

Mixed metal calcium-zinc types are used in some European bottle applications, principally because of their excellent low taste and odor properties and also because organotins are not approved for food contact in some European countries. In general, however, these materials do not process as well as the organotins.

D. Siding and Profile

In the United States, organotin stabilizers continue to be the choice for use in rigid exterior formulations as a result of their excellent performance in this application for the past two decades. The use of organotins in this application requires high levels of TiO_2 and limits the ability to produce certain dark colors.

In Europe, barium-cadmium-lead solid systems are used in outdoor applications. Lower TiO_2 levels in pastel colors are possible with these stabilizers. They are based on barium-cadmium laurates and lead phosphite-sulfite complexes with use levels between 3 and 4 phr.

E. Calendered Rigid Sheet

The calendering of rigid sheet requires stabilizers with lubricating character and low volatiles. Barium-cadmium laurates and stearates are ideal for this application. They are employed at use levels of 1 to 2 phr in combination with phosphite and epoxide synergists. Leads or organotins are also frequently used.

F. Phonograph Records

Records are generally stabilized with lead, barium-lead, or barium-cadmium powders. It is extremely important to use materials with a fine particle size and low volatiles to ensure excellent sound quality. The lubricating nature of these materials must be carefully adjusted and closely controlled for good mold release properties.

G. Calendered Flexible Film and Sheet

For flexible calendering, the principal stabilizer requirements are resistance to initial yellowing and long-term gradual color drift, excellent dynamic stability, low plateout, low odor, and low cost.

Most general-purpose grades are barium-cadmium-zinc liquids. For high-speed, thin-gauge film calendering, these are augmented by the addition of lubricating barium-cadmium powders. Use levels are 1.5 to 3 phr for liquids plus 0 to 1 phr of powder. Most applications include the addition of 3 to 5 phr of epoxy plasticizer as an auxiliary stabilizer.

Performance is generally a function of the cadmium, zinc, and phosphorus content of the stabilizer. Premium grades contain special antioxidants.

H. Flexible Extrusion and Injection Molding

Premium products for these applications are barium-cadmium-zinc powders and solvent-free barium-cadmium-zinc liquids. Use levels are 2 to 3 phr in combination with epoxy plasticizer.

I. Plastisols

The principal criteria for a good plastisol stabilizer are low volatiles, low odor, low viscosity, good air release, low plateout, and good mold release [59].

Neutral barium-zinc liquids with a high zinc content work well. Lead-, cadmium-, and zinc-containing stabilizers also decrease the blowing temperature of azodicarbonamide blowing agents. Potassium soaps are often added to control the cell structure of plastisol foams. Alkaline earth metal salts retard the decomposition of azodicarbonamide blowing agents. They are used only in high-temperature applications.

J. Flooring

Flooring stabilizers require good long-term moderate-temperature stability, good light stability, good stain resistance, and low toxicity.

Barium-zinc powders and liquids are preferred materials. They are used at levels of 2 to 3 phr in combination with 2 to 5 phr of epoxy plasticizer.

K. Wire and Cable

The stabilization of wire and cable insulation is completely dominated by lead stabilizers. Wire and cable must meet rigid requirements of the Underwriters' Laboratories in the United States, as well as British, Canadian, and other governmental standards for heat stability, electrical properties and mechanical properties [60].

Heat stability is a function of PbO content. Electrical properties arise from the nonconductivity of $PbCl_2$. Mechanical properties are retained by preventing plasticizer hydrolysis and volatility.

There are no universal lead stabilizers. The principal types used in various applications in the United States are listed below:

Application	Stabilizer
60° wire	Basic lead carbonate
80° wire	Basic lead silicate sulfate
90° wire	Dibasic lead phthalate
105° wire	Dibasic lead phthalate
Outdoor applications	Dibasic lead phosphite-stearate

X. TESTING AND EVALUATION OF STABILIZERS

The selection of the appropriate stabilizer for a vinyl compound is perhaps the most important decision a compounder makes in determining the final properties of the compound. Stabilizers influence the way a compound will process in manufacturing and the way it will ultimately perform in the application for which it is intended. Since stabilizers are such influential ingredients in vinyl compounds, they are subjected to a variety of tests unmatched by any other ingredient. Many of the tests have evolved empirically over the years and are meant to duplicate, in the laboratory, the environment to which a finished vinyl article will be subjected through its useful commercial life.

The purpose of this section is to familiarize the reader with the most common tests currently in use. Many of these tests have been developed over the years in the laboratories of stabilizer suppliers themselves. Other tests have been developed by regulatory agencies such as (in the United States) Underwriters' Laboratories, the National Sanitation Foundation, or the Plastics Pipe Institute, to provide performance information on vinyl compounds in specific applications. Finally, hosts of tests have been developed and standardized in the United States by the American Society for Testing and Materials (ASTM) and collected in a compendium of books on plastics testing. Most European testing is performed according to procedures laid down by the International Standards Organization (ISO). Table 16 lists the current ASTM procedures most applicable to PVC testing.

In many cases, the development of a desirable property is achieved at the expense of another property. By conducting a battery of tests, the vinyl compounder is able to balance the properties most critical to the end application of the vinyl article.

A. Color Development and Measurement

Many of the tests discussed in this section require visual examination of the test specimens to determine the suitability of a formulation for use. In many instances, formulations containing various additives are compared to one another. At other times, formulations may be compared to some previously established standard to determine acceptability.

The use of a visual technique to determine suitability of a specimen introduces operator-to-operator error. In addition, the human eye is sensitive only over a very narrow wavelength of light which we call the visible spectrum, between 380 and 700 nm. The human eye has difficulty distinguishing between various wavelengths that make up the visible spectrum and is not equally sensitive to all wavelengths along the spectrum. There are psychological as well as physiological problems with allowing operators to use a visual technique to determine suitability of products. Vinyl articles, the color of which may have drifted from a starting point because of exposure to heat, shear, or ultraviolet degradation will be judged with a different degree of sensitivity depending on the direction of color shift of the article.

In three-dimensional color space, as illustrated in Figure 5, color goes from light to dark by proceeding down the "L" scale, from red to green by proceeding along the "a" scale, and from yellow to blue by proceeding along the "b" scale. Shifts in colors of white articles which move into the blue or blue-green side of the color scale, even a very large distance, are normally determined by an operator to be much less indicative of poor stability than even very small moves into the red or yellow scale.

TABLE 16 Common Procedures for PVC Testing (*1983 ASTM Section 8 —
Volumes 8.01, 8.02, 8.03*)

Specifications for PVC resins and applications

D3915-80	PVC and related plastic pipe fitting compounds
D1755-81	PVC resins
D1784-81	PVC compounds and chlorinated rigid polyvinyl chloride (CPVC) compounds
D4216-83	Rigid PVC and related plastic building products compounds
D2474-81	Vinyl chloride copolymer resins
D2287-81	Nonrigid vinyl chloride polymer and copolymer molding and extrusion compounds
D1047-79	PVC jacket for wire and cable (see 1983 Vol. 10.02)
D2219-81	PVC insulation for wire and cable, 60°C operation (see 1983 Vol. 10.02)
D2220-80	PVC insulation for wire and cable, 75°C operation (see 1983 Vol. 10.02)

Test methods for thermal properties

D746-79	Brittleness temperature of plastics and elastomers by impact
D696-79	Coefficient of linear thermal expansion of plastics
D864-52 (1976)	Coefficient of cubical thermal expansion of plastics
D3801-80	Measuring comparative extinguishing characteristics of solid plastics in a vertical position
D621-64 (1976)	Deformation of plastics under load
D648-82	Deflection temperature of plastics under flexural load
D2843-77	Density of smoke from the burning or decomposition of plastics
D1870-68 (1978)	Elevated temperature aging using a tubular oven
D3364-74 (1979)	Flow rates for PVC and rheologically unstable thermoplastics
D1238-82	Flow rates of thermoplastics by extrusion plastometer
D1929-77	Ignition properties of plastics
D757-77	Incandescence resistance of rigid plastics in a horizontal position
D2863-77	Minimum oxygen concentration to support candlelike combustion of plastics (oxygen index)

TABLE 16 (Continued)

D568-77	Rate of burning and/or extent and time of burning of flexible plastics in a vertical position
D1433-77	Rate of burning and/or extent and time of burning of flexible thin plastic sheeting supported on a 45° incline
D635-81	Rate of burning and/or extent and time of burning of self-supporting plastics in a horizontal position
D3835-79	Measuring rheological properties of thermoplastics with a capillary rheometer
D4202-82	Thermal stability of PVC resin
D1525-82	Vicat softening temperature of plastics

Practices for thermal properties

D3045-74	Heat aging of plastics without load
D2115-67	Oven heat stability of PVC compositions

Test methods for mechanical properties

D1180-57	Bursting strength of round rigid plastic tubing
D1621-73 (1979)	Compressive properties of rigid cellular plastics
D695-80	Compressive properties of rigid plastics
D2236-81	Dynamic mechanical properties of plastics by means of a torsional pendulum
D790-81	Flexural properties of unreinforced and reinforced plastics and electrical insulating materials
D2583-81	Hardness, indentation, of rigid plastics by means of a Barcol impressor
D785-65 (1981)	Hardness, Rockwell, of plastics and electrical insulating materials
D3763-79	High-speed puncture properties of rigid plastics
D256-81	Impact resistance of plastics and electrical insulating materials
D3029-82a	Impact resistance of rigid plastic sheeting or parts by means of a tup (falling weight)
D3713-78 (1982)	Measuring response of solid plastics to ignition by a small flame
D1043-72 (1981)	Stiffness properties of plastics as a function of temperature by means of a torsion test
D747-70 (1981)	Stiffness of plastics by means of a cantilever beam
D2990-77 (1982)	Tensile, compressive, and flexural creep and creep rupture of plastics

TABLE 16 (Continued)

D1822-79	Tensile-impact energy to break plastics and electrical insulating materials
D638-82a	Tensile properties of plastics
D2289-81	Tensile properties of plastics at high speeds
D882-81	Tensile properties of thin plastic sheeting

Practices for mechanical properties

D2538-79	Fusion test of PVC resins using a torque rheometer

Test methods for optical properties

D523-80	Specular gloss
D2457-70 (1977)	Specular gloss of plastic films
D1746-70 (1978)	Transparency of plastic sheeting
D1925-70 (1977)	Yellowness index of plastics

Test methods for permanence properties

D543-67 (1978)	Resistance of plastic to chemical reagents
D1204-78	Linear dimensional changes of nonrigid thermoplastic sheeting or film at elevated temperature
D1042-51 (1978)	Linear dimensional changes of plastics
D1712-83	Resistance of plastics to sulfide staining
D2151-68 (1982)	Staining of PVC compositions by rubber compounding ingredients
D570-81	Water absorption of plastics
D1499-64 (1977)	Operating light-and-water-exposure apparatus (carbon-arc type) for exposure of plastics
D2565-79	Operating xenon arc-type (water-cooled) light-exposure with and without water for exposure of plastics
D794-82	Determining permanent effect of heat on plastics
D2299-68 (1982)	Determining relative stain resistance of plastics
D1435-75 (1982)	Outdoor weathering of plastics
G21-70 (1980)	Determining resistance of synthetic polymeric materials to fungi

TABLE 16 (Continued)

G22-76 (1980)	Determining resistance of plastics to bacteria
G23-81	Operating light-exposure apparatus (carbon-arc type) with and without water for exposure of nonmetallic materials
G23-77	Operating light exposure apparatus (xenon-arc type) with and without water for nonmetallic materials
G53-77	Operating light and water-exposure apparatus (fluorescent UV-condensation type) for exposure of nonmetallic materials
D2383-69 (1981)	Plasticizer compatibility in testing PVC compounds under humid conditions

Specifications for film and sheeting

D1927-81	Rigid PVC plastic sheet
D2123-81	Rigid poly(vinyl chloride-vinyl acetate) plastic sheet
D1593-81	Nonrigid vinyl chloride plastic sheeting
D1790-62 (1976)	Brittleness temperature of plastic film by impact
D3420-83	Dynamic ball burst (pendulum) impact resistance of plastic film
D1434-82	Gas transmission rate of plastic film and sheeting
D1922-67 (1978)	Propagation tear resistance of plastic film and thin sheeting by pendulum method
D2582-67 (1978)	Puncture-propagation tear resistance of plastic film and thin sheeting
D2838-81	Shrink tension and orientation release stress of plastic film and thin sheeting
D2457-70 (1977)	Specular gloss of plastic films
D1004-66 (1981)	Initial tear resistance of plastic film and sheeting
D1637-61	Tensile heat distortion temperature of plastic sheeting
D882-81	Tensile properties of thin plastic sheeting
D1746-70 (1978)	Transparency of plastic sheeting

Specifications for electrical insulating material

D922-80	Nonrigid vinyl chloride polymer tubing (see 1983 Vol. 10.01)
D150-81	AC loss characteristics and permittivity (dielectric constant) of solid electrical insulating materials

TABLE 16 (Continued)

D495-73 (1979)	Arc resistance of solid electrical insulation, high-voltage, low-current, dry
D257-78	DC resistance or conductance of insulating materials
D149-81	Dielectric breakdown voltage and dielectric strength of solid electric insulating materials at commercial power frequencies
D790-81	Flexural properties of unreinforced and reinforced plastics and electrical insulating materials
D229-82	Rigid sheet and plate materials used for electrical insulation
D374-79	Thickness of solid electrical insulation
D876-80	Nonrigid vinyl chloride polymer tubing used for electrical insulation (see 1983 Vol. 10.01)
D651-80	Tensile strength of molded electrical insulating materials (see 1983 Vol. 10.01)

Test methods for cellular plastics

D3576-77	Cell size of rigid cellular plastics
D1621-73 (1979)	Compressive properties of rigid cellular plastics
D1622-63 (1975)	Apparent density of rigid cellular plastics
D2237-70 (1980)	Rate-of-rise (volume increase) properties of urethane foaming systems
D2126-75	Response of rigid cellular plastics to thermal and humid aging
D1623-78	Tensile and tensile adhesion properties of rigid cellular plastics
D2842-69 (1975)	Water absorption of rigid cellular plastics

Practice for cellular plastics

D3748-78	Evaluating high-density rigid cellular thermoplastics

Test methods for analytical methods

D2124-70 (1979)	Analysis of components in PVC compounds using an infrared spectrophotometric technique
D1622-63 (1975)	Apparent density of rigid cellular plastics
D1045-80	Plasticizers used in plastics, sampling and testing
D793-49 (1976)	Short-time stability at elevated temperatures of plastics containing chlorine

TABLE 16 (Continued)

D1823-82	Apparent viscosity of plastisols and organosols at high shear rates by Castor-Severs
D1824-66 (1980)	Apparent viscosity of plastisols and organosols at low shear rates by Broodfield viscometer

Practices for color tests

D1729-82	Color differences of opaque materials, visual evaluation of (see 1983 Vol. 06.01)
E308-66 (1981)	Spectrophotometry and description of color in CIE 1931 system (see 1983 Vol. 06.01)

In order to overcome the inherent bias of human operators, color spectrophotometers or colorimeters may be used to give quantitative values to initial colors and color changes. The art of color comparison is quite complicated and mathematically involved. Over the years, a number of systems have been developed to identify quantitatively what the human eye sees. The major systems used today in the United States are the CIE L*a*b*, ANALAB 40, Hunterlab, and FMC2. Some examples of the methods of calculation are shown in Table 17 [61].

These equations are developed based on tristimulus colorimeter readings from XYZ values. The concept of XYZ values was developed in 1931 by asking a large number of human observers to rate color standards over a broad range of colors. Their responses to various different wavelengths of colors were plotted and the response of the CIE standard observer, that is, the average of all of the subjects, was plotted as a tristimulus value called $\bar{X}$, $\bar{Y}$, and $\bar{Z}$. The response curves are shown in Figure 6.

Upon these responses of the human eye through the various different wavelengths of color, are based all of the calculations used to express color numerically in three-dimensional space. A more complete discussion of color measurement is beyond the scope of this chapter, but many of the tests which are discussed in the following pages require a visual observation in comparing test results. Therefore, the reader is directed to the references cited at the end of this chapter [62, 63] and is cautioned to take into account the vagaries of various different methods of evaluating optical appearance. Whenever possible, it is suggested that numerical color data be generated and used as a method of evaluating test results in addition to visual evaluations.

When laboratory results for various different plastics formulations are being compared on the basis of a visual color inspection, it is important that the illumination source be closely controlled and that it can be varied in order to restrict metamerism [64]. Colors can play optical tricks on the human eye because of variations in the illumination source. Two samples compared to one another under a cool white fluorescent light source may appear quite similar, but in bright daylight illumination they will appear very different. Such mismatches are termed metamerism.

Laboratory evaluations of a sample under ordinary incandescent light or average fluorescent light often vary from the way the sample may appear

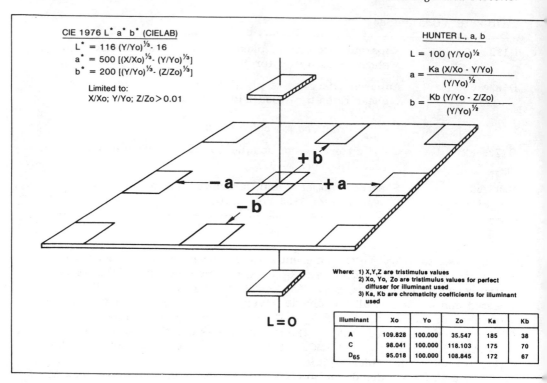

CIE 1976 L* a* b* (CIELAB)

$$L^* = 116 \ (Y/Yo)^{1/3} - 16$$
$$a^* = 500 \ [(X/Xo)^{1/3} - (Y/Yo)^{1/3}]$$
$$b^* = 200 \ [(Y/Yo)^{1/3} - (Z/Zo)^{1/3}]$$

Limited to:
X/Xo; Y/Yo; Z/Zo > 0.01

HUNTER L, a, b

$$L = 100 \ (Y/Yo)^{1/2}$$
$$a = \frac{Ka \ (X/Xo - Y/Yo)}{(Y/Yo)^{1/2}}$$
$$b = \frac{Kb \ (Y/Yo - Z/Zo)}{(Y/Yo)^{1/2}}$$

Where: 1) X,Y,Z are tristimulus values
2) Xo, Yo, Zo are tristimulus values for perfect diffuser for illuminant used
3) Ka, Kb are chromaticity coefficients for illuminant used

Illuminant	Xo	Yo	Zo	Ka	Kb
A	109.828	100.000	35.547	185	38
C	98.041	100.000	118.103	175	70
D_{65}	95.018	100.000	108.845	172	67

FIGURE 5 Three-dimensional color chart. (Courtesy of Hunterlab, Fairfax, Va.)

in natural daylight. Examples of different spectral distributions for various different light sources are shown in Figure 7. Since the spectral distributions of the various light sources vary greatly, it is easy to see why the sample appears so different under the light sources.

If optical appearance is a critical property of the finished part, the evaluations should be conducted under numerous light sources. This can be accomplished with the use of a light booth having three or four commonly used interchangeable light sources, as well as a light source meant to duplicate natural daylight. The best light source available is a filtered tungsten lamp. Fluorescent lamps never duplicate outdoor sunlight properly because they emit high-energy peaks of certain colors, even when attempts are made to balance the phosphors in the bulb. The human eye is misled by this high-energy output at certain wavelengths as opposed to the smooth curve generally characteristic of natural light. In addition to the illumination source, the background color can play optical tricks on the human eye, causing illusions which make different colors appear stronger or more subdued, and this too must be controlled.

Some formulations may fluoresce, and their ability to fluoresce may be affected by the type of stabilizer that is used. Some light boxes will allow one to run an ultraviolet source in combination with the other light sources and determine the effect of UV on optical appearance.

The standard lighting used to evaluate materials in a light box is the north sky daylight. This light source represents a natural, moderately overcast, north sky. Natural daylight is variable, but north sky daylight

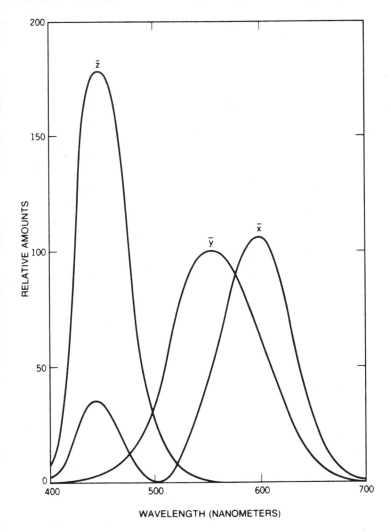

FIGURE 6 Analyzing appearance by measurements.

normally contains a little more blue than a standard daylight spectrum. This
can cause problems if color drift in a PVC formulation is in the direction of
the blue scale. Again, human observers generally equate good "white"
color with a blue-white color, and color may appear very good under north
sky daylight. However, if a stabilizer does not prevent the PVC formulation from developing conjugated unsaturation adequately as the polymer
degrades, other properties are lost. Stabilizers that cause the PVC compound or the pigment in the compound to remain relatively blue or to drift
toward the blue-green will be presumed to be better stabilizers. This may
or may not be the case, and it is one of the biases in human evaluations of
color change. The physical properties of the compound may be deteriorated
to a great degree although the color looks good; only further testing will
tell the whole story.

TABLE 17 Equations Used for Color Comparisons

Equation 1	Equation 2	Equation 4
CIEL*a*b*	ANLAB40	FMC2

Equation 1 — CIEL*a*b*

$$\Delta E_{CIE}(L^*a^*b^*) =$$

$$\left[(\Delta L^*)^2 + (\Delta a^*)^2 + (\Delta b^*)^2\right]^{\frac{1}{2}}$$

$$\Delta L^* = L^*_{sample} - L^*_{standard}$$

$$\Delta a^* = a^*_{sample} - a^*_{standard}$$

$$\Delta b^* = b^*_{sample} - b^*_{standard}$$

$$L^* = 25(100Y/Y_0)^{\frac{1}{3}} - 16 \quad 1<Y<100$$

$$a^* = 500\left[X/X_0\right)^{\frac{1}{3}} - (Y/Y_0)^{\frac{1}{3}}$$

$$b^* = 200\left[(Y/Y_0)^{\frac{1}{3}} - (Z/Z_0)^{\frac{1}{3}}\right]$$

Equation 2 — ANLAB40

$$\Delta E = \left[(\Delta C)^2 + (\Delta L)^2\right]^{\frac{1}{2}}$$

$$\Delta C = \left[(\Delta a)^2 + (\Delta b)^2\right]^{\frac{1}{2}}$$

$$\Delta L = L_{sample} - L_{standard}$$

$$\Delta a = a_{sample} - a_{standard}$$

$$\Delta b = b_{sample} - b_{standard}$$

$$L = 9.2Y_y$$

$$a = 40(V_x - V_y)$$

$$b = 16(V_y - V_z)$$

$$100 \frac{R}{R_{mgo}} =$$

$$1.2219V - 0.23111V^2 + 0.23951V^3 - 0.021009V^4 + 0.0008404V^5$$

Equation 4 — FMC2

$$\Delta E = \left[(\Delta C)^2 + (\Delta L)^2\right]^{\frac{1}{2}}$$

$$\Delta C = \left[(\Delta C_{RG})^2 + (\Delta C_{YB})^2\right]^{\frac{1}{2}}$$

$$\Delta L = 0.279K_2 (P\Delta P + Q\Delta Q)/aD$$

$$\Delta C_{RG} = K_1 (Q\Delta P - P\Delta Q)/aD$$

$$\Delta C_{YB} = K_1/b [S (P\Delta P + Q\Delta Q)/D^2] - \Delta S$$

$$P = 0.724X + 0.382Y - 0.098Z$$

$$Q = -0.48X + 1.37Y + 0.1276Z$$

$$S = 0.686Z$$

$$D = (P^2 + Q^2)^{\frac{1}{2}}$$

$$a^2 = (17.3)10^{-6}D^2/[1 + 2.73 P^2 Q^2/(P^4 + Q^4)]$$

$$b^2 = (3.098) 10^{-4} (S^2 + 0.2015Y^2)$$

$$\Delta X = X_{sample} - X_{standard}$$

$$\Delta Y = Y_{sample} - Y_{standard}$$

Equation 3

Hunter Lab

$$\Delta E = \left[(\Delta C)^2 + (\Delta L)^2 \right]^{\frac{1}{2}}$$

$$\Delta C = \left[(\Delta a)^2 + (\Delta b)^2 \right]^{\frac{1}{2}}$$

$$\Delta L = L_{sample} - L_{standard}$$

$$\Delta a = a_{sample} - a_{standard}$$

$$\Delta b = b_{sample} - b_{standard}$$

$$L = 10.0 Y^{\frac{1}{2}}$$

$$a = 17.5 \, (1.02X - Y)/Y^{\frac{1}{2}}$$

$$b = 7.0 \, (Y - 0.847Z)/Y^{\frac{1}{2}}$$

$$\Delta Z = Z_{sample} - Z_{standard}$$

$$\Delta P = P_{sample} - P_{standard}$$

$$\Delta Q = Q_{sample} - Q_{standard}$$

$$\Delta S = S_{sample} - S_{standard}$$

$$K_1 = 0.55669 + 0.049434Y$$
$$- (8.2575) \times 10^{-4} \, Y^2$$
$$+ (7.9172) \, 10^{-6} \, Y^3$$
$$- (3.0087) \, 10^{-8} \, Y^4$$

$$K_2 = 0.17548 + 0.027556Y$$
$$- (5.7262) \times 10^{-4} \, Y^2$$
$$+ (6.3893) \, 10^{-6} \, Y^3$$
$$- (2.6731) \, 10^{-8} \, Y^4$$

108

Jennings and Fletcher

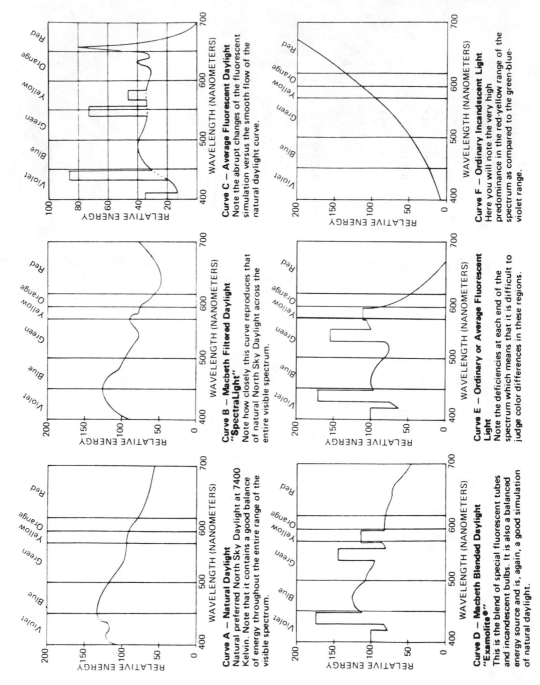

Curve A – Natural Daylight
Natural preferred North Sky Daylight at 7400 Kelvin. Note that it contains a good balance of energy throughout the entire range of the visible spectrum.

Curve B – Macbeth Filtered Daylight "SpectraLight"
Note how closely this curve reproduces that of natural North Sky Daylight across the entire visible spectrum.

Curve C – Average Fluorescent Daylight
Note the abrupt changes of the fluorescent simulation versus the smooth flow of the natural daylight curve.

Curve D – Macbeth Blended Daylight "Examolite®"
This is the blend of special fluorescent tubes and incandescent bulbs. It is also a balanced energy source and is, again, a good simulation of natural daylight.

Curve E – Ordinary or Average Fluorescent Light
Note the deficiencies at each end of the spectrum which means that it is difficult to judge color differences in these regions.

Curve F – Ordinary Incandescent Light
Here you will note the very high predominance in the red-yellow range of the spectrum as compared to the green-blue-violet range.

With the use of modern spectrophotometers and colorimeters, it is becoming a commonplace technique to evaluate color change based on the evaluation of ΔE's. The ΔE is the three-dimensional vector evaluation of a change in color space. It is equal to the square root of the sum of the squares of the color vectors, as in the following equation:

$$\Delta E = [(L*)^2 + (a*)^2 + (b*)^2]^{\frac{1}{2}}$$

The absolute magnitude of ΔE, however, is again not indicative of the way a human observer would evaluate the stability of a compound. The direction of the vector, that is, the quadrant into which the vector is moving, dramatically influences the evaluation of the stability of the compound by a human observer. But, in fact, the absolute color change, as measured by ΔE, is generally a good indication of how well a stabilizer is working to stop the dehydrohalogenation of a PVC compound and the formation of metal chlorides, both of which have a profound affect on the PVC's appearance. The smaller the ΔE, the greater the probability that degradation of the polymer has not occurred.

Modern laboratories rely more and more on quantitative methods of evaluation of color change than on the visual evaluations of numerous observers.

B. Heat Stability

1. Assessment of Color Change: The most widely used method of evaluating stabilizers is based on static heat stability. Vinyl compounds exposed at elevated temperatures over a period of time begin to degrade. The degradation can be measured visually by the color development or loss of physical properties of the compound. Under actual processing conditions, the compound is exposed to more than just heat; it is also exposed to the shear necessary to mold the part into the appropriate shape, as on a calendering roll, in an extruder, or in an injection molding machine. Laboratory tests have been developed to duplicate each of these processes. Laboratory testing is also used to measure stabilized compounds containing various proportions of other ingredients, such as fillers, impact modifiers, pigments, and lubricants, in order to screen the vast array of combinations needed to formulate the most useful compound for a particular application.

The type of testing done in the laboratory depends on the sophistication of the equipment available to do the testing. Often, because of economic constraints, a great deal of sophistication is not available. The following quick and simple tests have been accepted as standard in the industry for screening materials.

a. Static Heat Stability: In this type of testing, compounds are milled on a two-roll laboratory mill at a temperature ranging from 340° to 370°F for a period of about 5 minutes. The material is then sheeted off and cut into 11 small squares of about 1 in. by 1 in. The test squares from the formulations to be evaluated are then placed on 10 different panels, one sample from each formulation on each panel; the eleventh square is used as the control "off mill" sample. The panels are inserted into a circulating forced-air oven. They are removed at the desired intervals and the color changes among the various formulations are noted as a function of oven residence time.

The test appears simple, but a number of pitfalls may be encountered. The hot-air oven must be a high-velocity circulating air oven, with a small amount of outside air constantly bled into the oven to replenish the air inside it. This mitigates the amount of HCl buildup in the oven, which would otherwise accelerate the degradation of the vinyl compounds. The heat transfer between the plates and the samples must be uniform and the plates must not react with the samples. Stainless steel plates or steel plates covered with an aluminum foil that can be replaced after each test are recommended.

A uniform temperature gradient must be maintained throughout the oven. This is accomplished by the high air circulation rate and the use of a rotating carousel-type sample rack in the oven. Several designs of these carousels are available from oven manufacturers. The temperature response of the oven must be very fast. When the oven is shut down and the door opened to remove a sample panel, the oven must uniformly and quickly recover to the test temperature, or variations from one test to another may be seen. Samples placed on the trays must never be in contact with one another, particularly if the class of stabilizers being tested consists of types that will cross-stain with one another. Stabilizers or compounds that contain volatile materials that may cross-stain with the stabilizer or compounding ingredients of another formulation being tested at the same time must be avoided. If each of these factors is not controlled, the results obtained may be confounded and not indicative of the performance of the vinyl compounds being tested. This test procedure, as used for comparing PVC resins, is described in ASTM D2115. In addition, the ASTM has described two other methods of evaluating the static heat stability of vinyl compounds: ASTM D1870 and ASTM A4202. In these methods, samples are exposed to elevated temperature in either a tubular oven or a test tube. Both methods have significant drawbacks compared to the method described above. They are more time-consuming, heat transfer characteristics throughout the compound may not be as uniform as on a thin-milled sample placed on a plate, and the ability to remove HCl over the compound is extremely limited in both methods. The presence of HCl over the compound accelerates the compound's degradation. Although neither of these methods is as widely used as the circulating oven method, the tubular oven technique is considered the "standard" of the wire and cable industry. For sheet or film applications, however, the drawbacks mentioned above are real, and therefore the tubular oven is not used by these industries.

One of the problems with static heat testing is that potentially valuable stabilizers will be discarded on the basis of static heat stability comparisons alone. The compounder must remember that an individual property is developed at the expense of other properties, and if static heat stability is not the most critical aspect of product performance, a stabilizer should not be discarded solely because of performance in static heat stability testing. In addition, it is often difficult to evaluate which stabilizer really "wins" in a static test. Stabilizers evaluated in oven tests are classified in one of three ways: good initial color, good mid-term color, and good long-term stability. A stabilizer which offers good initial color, either "off the mill" or in the first several chips out of the oven, will often have very poor long-term stability. A good initial color is frequently developed at the expense of long-term stability, especially in the case of mixed metal-stabilized compounds, and long-term stability is most often developed at the expense of initial color. The selection criteria used to evaluate the most effective stabilizer often reside only in the eye and the mind of the person conducting the tests.

b. Press Stability: Milled samples subjected to static oven testing are also often subjected to a press stability test. In this test, the milled sheet taken from the two-roll laboratory mill is placed inside a 1/8- to 1/4-in.-deep cavity mold. Since the strips taken off the two-roll mill are normally 15 to 30 mils thick, a number of strips of the material are placed on top of one another until the desired thickness in the mold is obtained or slightly exceeded. Materials of different formulations can be piled adjacent to each other inside the mold until the cavity is full. The mold is then placed between two ferrotype plates to allow good heat conductance between the upper and lower surfaces of the heated platens on the press. The sandwich is put inside the press and subjected to a pressure of 10,000 to 60,000 psi. The platen temperature most often used is 360°F, but any temperature considered appropriate may be used as long as it is above the compound's glass transition temperature. The time interval of 1 to 15 minutes is arbitrary; it may be varied to provide the desired correlation with end-use application. At the end of the testing time, the platens in the press are water-cooled to avoid warpage when the mold is removed from the press. The mold is then removed and the samples are evaluated for color development, clarity, uniformity of flow, and fusion between the layers of plastic.

An aspect of this test is the fact that it is conducted largely in the absence of air. In addition, if a compound begins to degrade significantly, the test is really being conducted in the presence of entrapped HCl, which is considered to be a negative aspect of the test by some workers. This test is most often conducted because the thickness of the vinyl compound that can be built up in the press allows an evaluation of the development of opaqueness or haze in the thick sheets that cannot normally be seen in the thin strips evaluated in the oven test. Test samples prepared and exposed in this way are also often very good for compatibility testing because many ingredients will quickly migrate out of a thick sheet prepared in this manner. Quite often, the migration is due to incompatibility of lubricants, plasticizers, or stabilizers. Compounds specifically formulated for evaluation of incompatibility are also fabricated in a press. This is discussed further in a later section of this chapter.

c. Dynamic Test Conditions: The static heat stability tests outlined above generally provide a very good screening method for evaluating stabilizer additives in vinyl formulations, but they provide little information on how the stabilizer will work under the actual manufacturing conditions to which the compound will be subjected. To evaluate stabilizer performance more realistically, dynamic tests have been developed that reproduce the conditions the stabilizer would experience during processing. In PVC compounding and processing, the first exposure to heat and shear normally occurs in the blending operation. Whether a material is to be extruded, calendered, cast or molded, the vinyl formulation is preblended in some type of powder or paste mixer. Many of these operations introduce heat and shear, and several laboratory tests have been developed to evaluate these effects on a PVC compound.

The first is a high-intensity Henschel-type mixing test in which ingredients of the formulation are charged to a Henschel-type mixer or a Waring blender and mixed, while the rising temperature of the compound is noted and, frequently, recorded on a chart. One of two methods can be used to control the test, either a standard mixing time or a selected maximum temperature. Resin is charged to the mixer and stabilizer is immediately added so that the stabilizer can be present from the beginning of the time when the resin actually experiences heat and shear history. As the test goes on, fillers, pigments, and lubricants are all added at varying temperatures or times. Specific temperatures at which these ingredients are added should duplicate the temperatures at which they are added in the actual plant mixing operation. In many such procedures, a compound is mixed to 240°F maximum and samples of the resin are then taken out and examined by visual color evaluation. Samples from the various mixes are retained and compared to one another where various stabilizers are used. The problem with this method is that the order of addition of the individual ingredients is very important. Furthermore, the heat history is insufficient to induce gross changes in compound properties.

When the processing method in the plant is calendering, the most common method of dynamic stability evaluation is a two-roll dynamic milling test. Mill speed and temperature are selected to duplicate conditions that will be experienced in the production operation. The compound is allowed to continue milling and samples are taken at various selected time intervals and mounted on a heat stability chip card. The samples are normally run until the material sticks to the mill rolls or until catastrophic degradation is evidenced by blackening of the PVC. The mill is then cleaned, and the next stabilizer to be evaluated is compounded and run on the mill in the same manner. The chip cards then contain time-temperature comparisons among a group of similar compounds containing different stabilizers. In many cases, the lubrication and stabilization cannot be separated in a true dynamic situation, but the dynamic situation is that which best approximates the heat-shear history that the material will experience in actual manufacturing. The test is relatively easy to run and is often used on compounds that will be processed by means other than calendering. However, if a formulation is being developed for an injection molding or extrusion process, evaluation on a two-roll mill may give very misleading information. During the two-roll mill test, the formulation experiences a great deal of air exposure versus surface area, a condition that will never be experienced in an extruder or injection molding machine. Stabilizers that contain high levels of antioxidants might perform quite well on a two-roll mill, but the effectiveness of this type of stabilization mechanism may be unobserved when the material is processed in such a way that very little of the degradation is generated by oxidation. The test method, will, however, duplicate the type of oxygen-to-surface-area exposure that will be experienced in the calendering operation, and for this reason it is best used to evaluate compounds intended for calendering.

Formulations intended for extrusion or injection molding also must be evaluated for dynamic stability. The methods most commonly used for this employ either a torque rheometer instrument or an instrumented capillary rheometer. In torque rheometer testing, the sample is charged to a small mixing chamber, which is then closed to prevent the introduction of air. Rotor blades shear and mix the compound at an elevated temperature, and the processing stability of the compound is measured via rheologically induced force, by plotting mixer blade torque (in metergrams) versus time. A typical curve developed on a torque rheometer is shown in Figure 8. The evaluation of compounds with a capillary rheometer is discussed in detail in Chapter 29 (Volume 4 of this series).

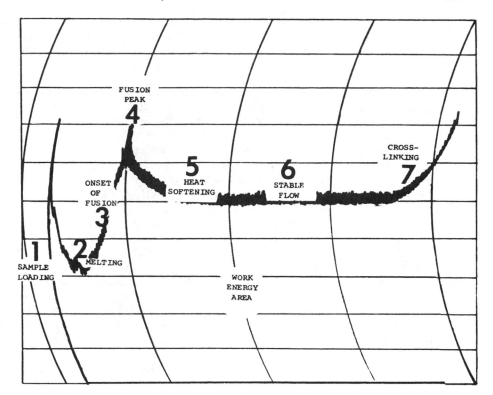

MINUTES · DYNAMIC SHEAR · MIXING

FIGURE 8 Torque rheometer mixing curve for PVC.

Several pieces of interesting data can be obtained by evaluating the torque curve. First, during sample loading (1*), a peak torque is developed which is an indication of torque that will be experienced in an extruder if the material is force-fed into the throat of the extruder. The material then experiences a heating and compaction region (2,3*), where it is gradually transformed from a powder to a viscous melt. As this happens, torque will build up to a point which is called the "fusion peak" (4*). At this point the material is totally viscoelastic but it is still a relatively cool melt. As it begins to heat up to the ultimate temperature of the mixing bowl, the torque drops as heat softening reduces the material's viscosity (5*). The material then establishes a steady-state condition which is called "stable flow" (6*). This level of torque represents an approximation of the viscosity within the metering section of an extruder at the test temperature and rate of shear. The test gives useful information on the power requirements that will be needed to process the material. The length of time between the peak fusion (4*) and the point at which viscosity again begins to increase or decrease (7*),

*These numbers refer to specific points indicated on Figure 8.

due to cross-linking or chain scission, is an indication of the inherent stability of the compound. This time is called the "dynamic stability time."
Compounds containing different stabilizers can be evaluated on the basis of
stability time as measured by a change in angle of more than 30° from stable
flow [65, 66].

The lubrication properties of the stabilizer can be evaluated by measuring the time between the sample loading peak and the fusion peak. This
is normally referred to as the "fusion time." Since stability is a function of
both the lubrication and the actual ability to stabilize the polymer, much can
be learned about the mechanism by which a stabilizer may be working. All
other things being equal, stabilizers that are very lubricating are expected
to give longer processing times. When the stabilizer is more lubricating,
the temperature developed through the shear heating, between the wall of
the mixing chamber and the resin is minimized. Since the heating is minimized, the viscosity is expected to be slightly higher because the resin is
at a slightly lower temperature. This sets up a complex system which makes
it difficult to evaluate the overall effectiveness of the stabilizer/lubricant
portion of the compound [67]. However, the total energy input is quanti-
fiable, and stabilizers can be compared with one another quantitatively on
the basis of energy absorption regardless of viscosity differences. Modern
torque rheometers can be equipped with a total torque integrator, which
records energy input by integrating the area under the stability curve; this
allows one to compare stabilizers on the basis of their ability to resist degradation to a specific amount of stress on the compound.

Another method used to develop dynamic stability information is an
injection molding technique. Parts are molded on a small laboratory molding machine over a number of cycles, and their color, appearance, and
physical properties can be measured. One stabilized compound versus another can be evaluated in this way. In addition, molded samples can be
ground up and remolded. The effectiveness of the stabilizer is then based
on the number of recycled shots that can be made with a compound before
significant color development occurs, or before sticking or shear-burning
occurs at the gates or within the runners of the mold.

Several test procedures have also been developed to measure the flowability of compounds. ASTM procedure D3364-74 defines a method for measuring flow rates with an extrusion plastometer. The plastometer is said to
be useful for predicting processability of compounds from changes in polymer
molecular weight, plasticizer content, stabilizer effectiveness, and lubricant
content. The major deficiencies of this method are the limited shear range
that can be achieved practically and the relative sensitivity of data on very
rigid compounds due to the low flow rates.

Another method commonly used to describe the flowability of PVC compounds has apparently been adapted from the thermosetting plastics industry.
For a number of years, thermoset molders have evaluated their compounds by
using ASTM spiral flow mold procedure D3123-72. Vinyl compounders have
used spiral flow molds to evaluate the effect of resin molecular weight,
processing aid, impact modifiers, and lubricants on the ability of a compound
to fill the cavities of complex molds. Most researchers have designed spiral
molds to match their own laboratory injection machines and to duplicate more
closely the shear rates to be encountered during processing on the production
molds. For a general discussion of mold design for forming test specimens of
plastic materials, a review of ASTM D647-68 is recommended, as well as ASTM
D3123 for spiral mold designs [68, 69].

The practice of grinding up extruded stock and reextruding a number of times is also quite commonly used to evaluate stabilizer performance in compounds designed for extrusion applications.

C. Light Stability and Weather Resistance

The ability of vinyls to resist change when exposed to natural sunlight becomes increasingly important as they find their way into a wide range of applications. Notably, vinyl-insulated wire, vinyl siding, and weatherable profiles for window and door frames bring increasing pressure for vinyl compounds to retain their appearance and physical properties over long periods of time outdoors. The subject of weatherability is one which, even today, is not fully understood. Much of the confusion arises from the fact that nature does not allow the luxury of controlled conditions that we can reproduce during laboratory evaluations. Therefore, a great deal of work has gone into the development of laboratory devices that can closely approximate the behavior of plastics in a natural environment but in an accelerated time frame.

Long-time exposure in a natural environment (i.e., weathering) is quite different from long-term steady-state light exposure. In nature, a plastic material experiences moisture in the form of rain, snow, and humidity, variability of temperature, exposure to oxidative gases or chemical pollutants, and biological attack. Mechanical abrasion and the detrimental effects of stresses induced in the plastic from frequent or severe changes in its physical dimensions, caused by expansion and contraction due to the thermal expansion coefficient of the material, also can cause degradation.

Developing weatherable compounds is as much an art as a science. Experiments to predict the long-term weatherability of plastic articles are always a series of compromises. To begin a development program, four questions must be answered:

1. How much time is available to conduct the testing?
2. Does the exposure site for testing adequately duplicate the commercial use locations and mode of exposure?
3. What properties are most likely to change over the commercial life of the article?
4. What are the cost constraints?

The answers to these questions will determine the amount and type of testing to be conducted. In general, one may elect to employ natural outdoor weathering, accelerated outdoor weathering, accelerated laboratory testing, or some combination of these approaches. Once a sample has been weathered, by whatever method is used, the effect of formulation variables on color and physical property retention can be determined [70, 71].

1. Natural Outdoor Weathering: Once the determination has been made to conduct long-term outdoor weathering, a number of options are available for the conditions of exposure of the samples. The rule here is, as with all other laboratory testing, "Attempt to duplicate the actual conditions that will be experienced during the commercial life of the article." This is an admirable rule, but one that is difficult to adhere to. Questions that are raised in outdoor testing are discussed below.

a. Test Locations: Is the site (or sites) selected representative of the conditions that will exist in the geographic area where the article is to be used? The degree of UV radiation in Puerto Rico, Arizona, or Florida, where many outdoor weathering tests are conducted, is significantly higher than that experienced in the midwestern United States or the New England states. The atmospheric conditions and the exposure to sulfur, sulfur dioxide, and nitrogen dioxide concentrations are also very different. Therefore, multiple test sites are usually necessary in order to have a good correlation with actual use conditions.

b. Sample Mounting: Will the mounting angle duplicate the commercial exposure? Most outdoor testing is conducted at 45° south. This is a somewhat accelerated condition compared, for instance, to a PVC siding hanging vertically on a house in a direction other than south. PVC roofing, on the other hand, is meant to lie horizontally or angled on a roof and experiences a great deal of UV exposure. Mounting angle is a very important consideration and is discussed further in Section X, C2 (Accelerated Outdoor Exposure).

c. Sample Preparation: Has the sample been processed on the same type of equipment and in the same way as the commercial article? Vinyl articles that are extruded or injection molded are processed in such a way that a certain orientation is given to the final article. Oriented PVC may react to stresses, such as UV or temperature/humidity differentials, in a far different manner than PVC processed so that the final product is isotropic. Many laboratories have gone to the use of small single- or twin-screw extruders rather than two-roll laboratory mills or molding presses for the preparation of test compounds designed to simulate extrusion or injection molding operations.

d. Sample Backing: Does the backing material adequately represent the type of insulation factor that will exist in use? Sample backings can affect heat buildup on the surface of the sample and also the amount of condensation that may form during dark or cool cycles.

e. Replication of Samples: Have enough sample replicates been placed on test to determine the sensitivity of the testing procedure at any given site? Have the samples been prepared in such a manner that if physical testing of a destructive nature is necessary, multiple samples can be taken at the appropriate time periods?

f. Frequency of Examination and Testing: For what length of time will the test be conducted? At what intermediate periods will testing of samples be done? In general, natural outdoor weathering presumes at least a 2-year testing cycle, with a greater frequency of sampling in the early months.

g. Criteria of Failure: What tests will be run on exposed samples? Will they be destructive in nature? Have the samples for destructive testing been prepared properly? If tensile testing or Izod impact tests are to be done, precut dumbell-shaped samples should be exposed so that no cutting of the exposed sample specimen need be done prior to the physical test. If sample specimens for color testing, physical testing, chemical immersion testing, etc. are needed in different shapes, multiple sample shapes will have to be prepared for each type of test.

h. Comparison with Accelerated Testing or Commercial Data: Is accelerated testing being conducted to predict what the long-term failure times

of particular formulas will be? It may be necessary to extend natural outdoor testing or speed up the frequency with which samples are inspected.
This can often be determined by failures in either accelerated testing or
commercial use.

i. Collection of Data: Data collected by natural outdoor weathering
should be determined on samples as soon as possible after removal from the
exposure site. In addition, testing programs should be set up to permit
continuity of the testing over the long testing period in the event that the
workers responsible for the program change. Data should be stored on
file cards, punch cards, or computer, but not by filing away the exposed
samples for later reference. Sample appearance will change with time in
the dark.

Many of the decisions made above will have to be made whether the
testing mode is natural, accelerated natural, or accelerated laboratory-type.
The answers to all of the questions should be obtained in a satisfactory manner before any testing has begun.

2. *Accelerated Outdoor Exposure*: Over the years, a number of methods have been used to accelerate the natural outdoor weathering exposure
of plastic samples in North America. Most of the testing is conducted in
Arizona, Florida, or Puerto Rico, each of which is an area of high UV saturation and, therefore, represents an accelerated mode of testing compared
to northern climates. As mentioned earlier, most of the exposures that are
conducted are done with a 45° south mounting, but other mountings are
sometimes used, most commonly 33° south, a variable (equatorial) mounting,
or a horizontal mounting.

Investigators have looked for additional ways to accelerate outdoor exposure to cut down on the time needed to do the testing and to lower the
cost. Failures in the field, caused by lack of understanding of long-term
weathering characteristics of their vinyl compounds, can have huge potential liability consequences for the compounder or manufacturer. Manufacturers therefore rely on some type of natural weathering to qualify their
products. In early methods of accelerated outdoor weathering, the test
specimen was mounted on a movable sample rack which was directed to follow
the sun through the course of the day. This increased the amount of exposure of the sample as the sun moved across the sky. Later, an apparatus
was developed that concentrated the sun's radiation on the sample specimen.
The apparatus is described as a Fresnel reflecting concentrator with a follow-
the-sun rack. It has 10 flat mirrors positioned so that the sun's rays strike
them at a near-normal angle while the apparatus is in operation. The mirror
design is such that each mirror is tangent to a theoretical parabolic trough
which directs the reflected sunlight uniformly onto the specimen target area.
An example of such an apparatus is shown in Figure 9.

The effect of concentrating sunlight onto a small sample area in this
manner is that the sample specimen receives energy having an intensity of
approximately eight suns. The difficulty with this method is the potential
for building up a great deal of heat through the absorption of ultraviolet
radiation and the reemission of thermal heat. To minimize this effect, some
sample devices are designed with an air deflector having the potential to
direct cooling air at a rate as high as 2000 CFM across the sample mount.
In natural outdoor weathering, the potential for thermal buildup on the
sample specimen is always present. The preparation of a backed or unbacked
sample is particularly critical when using accelerated weathering devices.

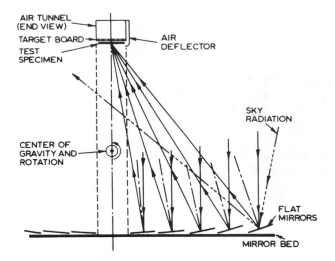

FIGURE 9 Optical system for EMMA(QUA) Fresnel reflecting concentrator. (Courtesy of DSET Laboratories, Inc.)

Table 18 shows a comparison of black- and white-panel temperatures for exposed test specimens, both backed and unbacked, for an accelerated Fresnel testing device versus a 45° south exposure. As can be seen from the data, large temperature variations occur between light and dark samples as well as between backed and unbacked sample mounts.

Test equipment can be operated in such a fashion that water spray cycles are optionally available to simulate natural environments and the dew formation that routinely occurs on plastic samples. The ASTM has established several procedures covering the operation of this type of apparatus. Standard E8338 is a test procedure that establishes the operating parameters for such devices. Standard D4141 establishes a procedure for nighttime water spray cycles. In addition, for better control of long-term outdoor accelerated weathering, the device is equipped with a precision solar scanning spectroradiometer that permits control of the exposure of the sample to ultraviolet radiation based on the amount of UV radiation deposited on the sample. Earlier radiation-controlled devices merely measured the amount of total radiation received by the sample or, in the case of normal outdoor testing, used a specified period of time. The amount of radiation that can be deposited in a specified period of time varies greatly, and therefore the reproducibility of the test varies greatly. By controlling the ultraviolet radiation exposure, one can set up test intervals that allow comparisons from year to year under reasonably controlled UV exposure modes. In addition, it is claimed that such devices can be set up to duplicate the ultraviolet exposure in other geographic locations. This is accomplished by calculating the average ultraviolet radiation to which a test specimen would be exposed in, for instance, Florida over 6 months, and then exposing it on the accelerated apparatus in Arizona until an equivalent amount of radiation has been deposited on the sample. Generally good results have been reported with this technique. Table 19 shows the type of acceleration obtainable with such a technique. Geographic locations that have vastly different atmospheric conditions may preferentially absorb certain wavelengths of solar radiation.

TABLE 18 Comparison of Black/White Panel Temperatures on EMMAQUA and 5° South[a]

	Backed		Unbacked	
	Black	White	Black	White
5° south	66°C	47°C	47°C	39°C
EMMAQUA	62°C	47°C	53°C	40°C

[a]Ambient temperature of 34°C on June 14, 1983.
Source: G. A. Ferlaut and M. L. Ellinger, Precision Spectral Ultraviolet Measurements and Accelerated Weathering, *J. Oil Chem. Assoc. 64*, 387–397 (1981).

The stability of polyvinyl chloride is particularly sensitive to radiation below 320 nm, and any differences in the frequencies of exposure in this range can cause differences in both the mode of failure and the rate of failure between the accelerated specimens and parts in commercial use. Zerlaut and Ellinger [72] have plotted typical ultraviolet spectra for the EMMA (QUA) test target versus more standard modes of exposure, as shown in Figure 10.

It is easy to see in Figure 10 that the EMMA significantly accelerates the amount of exposure in the wavelength region that we are particularly concerned with. Stabilizers containing antioxidant systems or UV quenchers that are particularly effective at reducing the effects of ultraviolet light on PVC will perform very well in this type of test. In long-term exposure, however, a number of other considerations may affect the performance of the vinyl compound, and stabilizers must be selected for their ability to inhibit not just mechanisms of ultraviolet degradation but a combination of degradation modes, the interplay of which is extremely complex. In order to understand better the stability of a compound subjected to various different degra-

TABLE 19 Pertinent Information for 5°S Florida and EMMA(QUA) Exposures

	Period		Exposure[a] (langleys)	Time[b]
Exposure	Begin	End		
5°S(F)	2/Feb/72	16/Aug/72	85,940	6 months
	29/Dec/72	29/Jun/73	79,870	6 months
	1/Aug/73	1/Feb/74	72,440	6 months
EMMA(QUA)	25/Jul/74	10/Aug/74	80,670	16 days
	14/Aug/74	1/Sep/74	82,490	18 days
	9/Sep/74	29/Sep/74	79,310	21 days

[a]Computed from monthly weather summaries for the region.
[b]Note the significantly shorter time needed to achieve the same radiation exposure with EMMA(QUA) exposures.

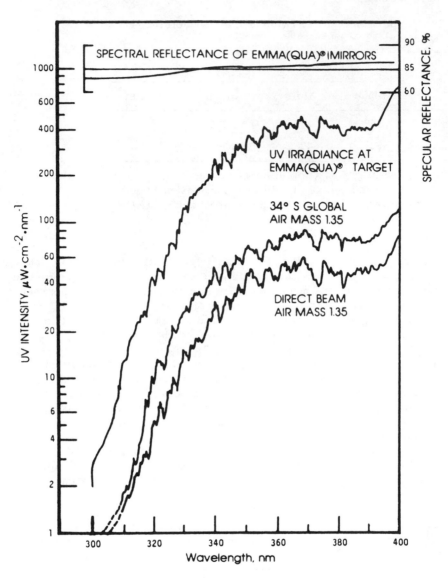

FIGURE 10 Typical ultraviolet spectra for midday, comparing UV at
EMMA(QUA) test target to 34° south global. (Courtesy of DSET Laboratories, Inc.)

dation modes, long-term weather testing is required. The EMMA test mentioned above, however, offers a quick and reliable screening method to
indicate the efficacy of a stabilizer with regard to ultraviolet stabilization
only; it can help to screen out stabilizers that are particularly poor for
applications requiring a high degree of UV stabilization.

    3. *Accelerated Laboratory Weathering*: In order to gain better control
and reproducibility in testing designed to determine the long-term weatherability of vinyl parts, much of the initial screening is conducted in laboratory
test equipment designed to duplicate the outdoor environment. Over the

years, a great deal of effort has gone into developing test equipment that will not only duplicate the effect of exposure to UV radiation, but also attempt to integrate the effects of abrasion, moisture, environmental pollutants, and light/dark cycles.

The most common types of machines currently in use employ xenon arcs, sunshine carbon arcs, violet carbon arcs, high-pressure mercury quartz lamps, fluorescent UV "black" lamps, and fluorescent sunlamps. Each type of lamp generates an entirely different spectral curve in the wavelength range 250 to 800 nm. No artificial light source accurately reproduces the spectral distribution of natural sunlight. In order to generate laboratory results in an accelerated fashion, most of them have a much higher energy distribution in the wavelength range between 250 and 350 nm which is most detrimental to PVC. This is helpful in developing stabilizers and additives that will be effective in commercial applications. Often, the final selection of an additive is not made until several years of weathering data in a natural environment have been collected, but a large number of stabilizer formulations can be screened very rapidly in the laboratory. Ultraviolet stability charts can be put together in the same way as heat stability charts and additives screened out in a timely manner. Formulations containing additives that are most effective are then compounded, processed, and sent out for the long-term outdoor weathering test. Additives used in these formulations can be presented to plastics producers for evaluation in their own compounds. Some examples of spectra of common light sources are presented in Figure 11. More complete comparisons of spectral curves can be found in the literature [73].

Figure 11 illustrates three different types of light sources—sunshine carbon arc, violet carbon arc, and xenon arc—versus a distribution for natural sunlight. These light sources, along with their recommended filtration, attempt to reproduce the spectral energy of natural sunlight between 280 and 320 nm (the range most detrimental to PVC) but they are not particularly high in energy output over these ranges. Therefore, degradation of PVC under these types of lamps can take a very long time. For instance, 2000, 4000, or even 6000 hr is not considered a long period of time for evaluating PVC in a xenon arc apparatus. Much work has been done by various industry groups in efforts to develop a standard testing method in the laboratory to duplicate outdoor weathering. The Vinyl Siding Institute of the Society of Plastics Industries in the United States has conducted numerous studies trying to identify a light source that will duplicate outdoor sunlight. They have also attempted to correlate results for different types of light sources and for the same type of light source in different machines throughout the industry. Their work is voluminous and ongoing and should be reviewed as the benchmark of today's technology [74, 75].

To greatly accelerate the UV exposure of a vinyl compound, the mercury quartz accelerated weathering device is recommended. This apparatus was originally developed by Charles Kuist at National Starch and Chemical Corporation. It is felt by some to represent such a highly accelerated test that the results cannot be correlated adequately with outdoor weathering.

An earlier development employed fluorescent sunlamps and ultraviolet black lights in an apparatus originally developed at American Cyanamid Company and later modified by others. ASTM procedure G53-77 covers testing under fluorescent radiation. An example of such a device is shown in Figure 12. The device has eight FS40 fluorescent lights which illuminate the

Atlas 6500 Watt Xenon Arc Lamp

Power Distribution Compared to Miami, Florida Daylight

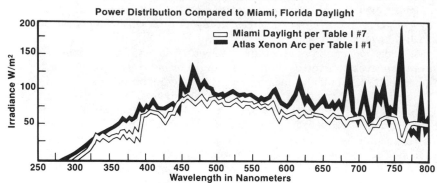

Atlas Sunshine Carbon Arc Lamp

Spectral Power Distribution

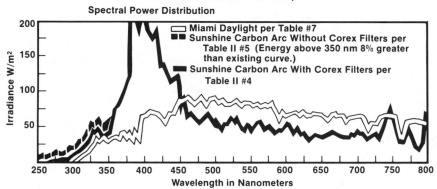

Atlas Enclosed Violet Carbon Arc Lamp

Spectral Power Distribution

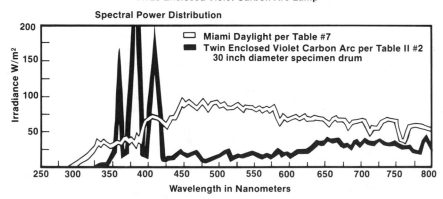

FIGURE 11 Energy distribution of synthetic light sources vs. natural sunlight. (Courtesy of Atlas Electric Devices Co.)

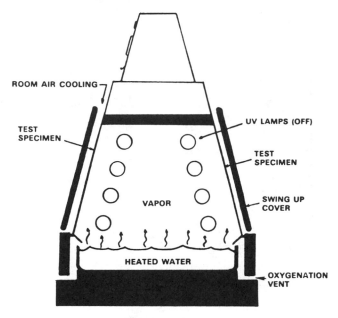

FIGURE 12 Accelerated weathering device conforming to ASTM G53-77. (Courtesy of Q-Panel Company.)

inside of the test chamber, with test specimen racks going up both sides. It can be programmed for varying periods of light and dark, with humidity available generally in the dark cycle. Examples of commercially used cycle times are:

> 4 hr UV/4 hr condensation
> 8 hr UV/4 hr condensation
> 2 hr UV/4 hr condensation

However, many other cycles are possible, and in some cases the machine may be run 24 hr a day, 7 days a week, without interrupting the light cycle. It has been found, however, that UV-induced degradation can actually be accelerated by interrupting the light cycle with brief dark humid periods. The ASTM procedure specifies what is known as a UV-B lamp, but three types of fluorescent UV lamps are generally available: types A, B, and C. Figure 13 shows the distribution of natural sunlight and the corresponding ultraviolet spectral regions of A-, B-, and C-type lamps.

The B-type lamp is specified in the ASTM procedure because UV-B lamps produce the lowest wavelengths occurring in natural sunlight. They have a peak intensity at about 313 nm and produce ultraviolet wavelengths which also occur in the A-type lamp. The lamp has a peak intensity at about 360 nm, which is generally too high to cause much degradation in PVC. A C-type lamp is typically not used in PVC testing because the wavelengths are far

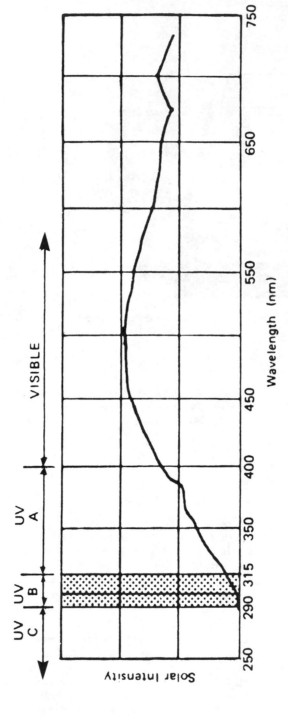

FIGURE 13 Division and classification of solar ultraviolet spectrum. (Courtesy of Q-Panel Company.)

below those normally occurring in natural sunlight reaching the surface of
the earth, since most of the shortwave ultraviolet radiation is absorbed in
the upper atmosphere. The spectral distribution of A- and B-type lamps
compared to natural sunlight is shown in Figure 14.

All the devices mentioned above can be equipped with irradiance detectors to display the spectral intensity at any given time and to integrate
the total spectral energy that has been deposited on a sample. This permits
one to control a certain amount of exposure given to a sample between test
intervals. This, again, helps to correlate the results with long-term outdoor
weathering. The most technically sophisticated of these detectors can monitor specific UV wavelengths.

All the devices can also circulate fresh air through the test chamber
to control the temperature buildup at the test specimen surface. This is
normally detected and controlled by the use of a black panel thermometer,
which displays the temperature on the surface of the black panel and is
used as a blank. As discussed earlier, the color of the sample and the particular backing can make a large difference in heat buildup on the surface
of the sample and can radically affect the rate of degradation of the polymer.
It is important to understand this for each individual series of tests conducted in an accelerated weathering device. Again, the rule "duplicate as
closely as possible the actual conditions to be encountered during the commercial service life of the compound" should be observed. When the temperature development at the surface of the test specimen varies by more than
5° to 7°C from that actually experienced in commercial use, the artificially
accelerated weathering data will most likely not be indicative of the failure
mode that will be encountered in use. Many vinyl compounds are particularly susceptible to long-term, low-temperature aging involving temperatures
slightly above 60°C. Much above this temperature, degradation mechanisms
other than ultraviolet degradation begin to come into play.

With the increasing use of vinyls in business machine housings, a new
type of long-term light stability has become important. Fluorescent light
systems in the interior of buildings in which business machine housings are
exposed (in many cases for 24 hr a day, 7 days a week) have become of
greater interest to the vinyl formulator. Test machines are available today
that duplicate conditions which might be encountered in an office atmosphere.
The typical machine uses superhigh output, cool white fluorescent lamps in
combination with FS40 fluorescent sunlamps. The fluorescent sunlamp is
generally filtered through a window glass-type filter in order to duplicate
light which might enter an office through windows in the building. This
type of testing is fairly new and a great deal of data has not been developed
or published on the stability of vinyl compounds under this type of lighting,
or whether the testing system itself adequately duplicates the degradation
mode encountered in office atmospheres.

Vinyl flooring and wall coverings that are typically displayed in showrooms and advertising booths that use fluorescent lighting undergo light
exposure that can vary from a very mild to quite extreme ultraviolet intensity. Radiant energy varies as the reciprocal square of the distance of the
light source from the exposed vinyl material. Display cabinets, even with
standard household lighting, will present an unrealistic exposure mode for
flooring or wall covering samples when compared to that which they will experience in commercial use. This is a problem with which additive suppliers
have had to cope. Only time and testing will ultimately determine whether
super-high output, cool white fluorescent testing is useful in developing vinyl compounds that will satisfy the marketing criteria of sample display cases.

Relative Spectral Energy Distribution

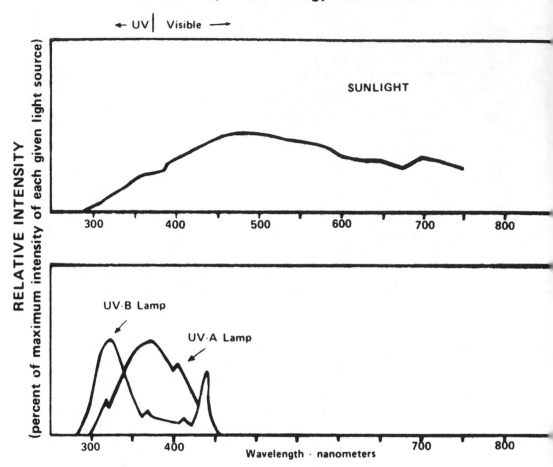

FIGURE 14   Relative spectral energy distribution.   (Courtesy of Q-Panel
Company.)

Finally, much work has been done using fluorescent sunlamps in conjunction with a black light.   Black lights are generally very short wavelength
ultraviolet lamps, similar to the UV-C lamp.   They are used in biological
laboratories as germicidal sterilization sources.   The combination of fluorescent sunlamps and black lights has been put together in a device called an
FS/BL testing unit.   Such a unit normally contains a rotating sample platform
mounted 4 to 6 in. away from the fluorescent sunlamp and black light combination (Figure 15).   The sample rack rotates so that ultraviolet energy from
the two lamp sources is evenly dispersed over the samples during the testing
periods.   In recent years, this type of testing has fallen out of favor with
vinyl compounders in the United States, but data derived by this test method
are available in the literature and a commercial unit with humidity control is
available in the United Kingdom [76].

FIGURE 15 Fluorescent sunlamp-blacklamp (FS/BL) exposure unit. (Courtesy of General Products Manufacturing Co., Cedar Knolls, N.J.)

The greatest utility of any accelerated laboratory device is as a screening tool for the rapid evaluation and discarding of materials, compounds, and additives that might be expected to fail in long-term outdoor testing. The accelerated method offers a quick and inexpensive route to do this, but the question still remains, how well does laboratory testing correlate with natural outdoor weathering? Kinmonth [77] has published a review of results of outdoor weathering versus laboratory testing covering a period from 1962 to 1977. Even with this wide-ranging intercorrelated review, along with the SPI data, it is difficult to draw a final conclusion as to which type of weathering device is most accurate and how well they work for various different polymers. Most formulators who work with specific compounds and who have many years of experience with indoor and outdoor weathering have established criteria that they feel is relevant to their own work. This seems to be the state of the art today and will probably continue for many years to come, with individual investigators feeling that they have some of the answers relative, at least, to their own problems concerning this very complex subject.

D. Compatibility Testing

During the processing and subsequent use of vinyl articles, it is desirable that all of the compounding ingredients be compatible with the vinyl resin and remain so throughout the commercial life of the article. The term used to describe this property with regard to additives is compatibility. Vinyl stabilizers are added at the earliest practical moment to PVC resin at or prior to the onset of processing of the formula. The ability of the stabilizer to disperse homogeneously throughout the matrix of the resin polymer and to remain compatible with the resin so as to better perform its function of stabilizing the compound is extremely important. The first stage at which stabilizer incompatibility manifests itself is during processing, in a form called plateout. Plateout is the buildup of certain compound ingredients on the metal surfaces of the processing equipment. The ability of a stabilizer to resist plateout is dependent on the original ingredients used to formulate the stabilizer, ingredient ratios, use level of the stabilizer in the compound, other compound ingredients present and their compatibility with the stabilizer, and the degree of degradation that takes place during processing. Each of these five factors contributes to incompatibility problems. As with other areas of testing that have been discussed, a complex system of interactions set up within the compound can contribute to compound failure. Although much is known about the performance of individual ingredients in a stabilizer that can contribute to plateout, very little is understood about the interreaction of stabilizers with other compound ingredients. The best way to evaluate the capability of a stabilizer to process well with little plateout is to run what is known as a plateout test. The test is conveniently conducted on a standard laboratory two-roll mill and is applicable to the processing of compounds in other types of equipment such as extruders, injection molding machines, thermoforming equipment, and plastisol molding or casting equipment.

Plateout test method[1]

Apparatus: Two-roll mill (standard speed 45 rpm), pyrometer, mixing beakers, scale, spatulas.

Procedure: (1) Prepare the following compounds:

(a) *Scrubbing compound*

Resin (IV = 0.8 to 0.9)	100.0
DOP	40.0
Epoxidized soya oil	6.0
Nonplating tin or lead stabilizer*	3.5
Stearic acid	0.5
Ground limestone	20.0
Cab-O-Sil (M-7)†	20.0
Carborundum No. 120	0.2

---

[1]Courtesy of Synthetic Products Co., "Stabilizer Test Procedures," Laboratory Standard 101-45.
*Synpron 357 or other suitable Ba/Cd stabilizer may also be used.
†Or 325 mesh silica.

Preblend on Hobart-type mixer or weigh individual components into beaker and mix. Additional carborundum may be added if mill is in very poor condition.

(b) *Cleanup compound*

Resin (IV = 0.8 to 0.9)	100.0
DOP	46.0
Hi-Sil 233*	4.0
TiO$_2$ (RA-50)	1.0
Stearic acid	0.3
Nonplating stabilizer (tin, lead, or Ba/Cd)†	2.0

Master batches of the above two compounds can be made and kept for extended work. All items should be added except for the stabilizer and mixed on a Hobart-type mixer or high-intensity mixer.

(c) *Standard plateout test compound*

Resin‡	100.0
DOP	67.0
Epoxidized soya oil	6.0
Stearic acid	0.5
Ground limestone	30.0
2-B red pigment	1.5
Stabilizer undergoing evaluation	At commercially used levels

Weigh ingredients in the order listed. Make this compound up fresh on the day the test is to be run; add stabilizer immediately before the run.

(2) Adjust roll temperature to 350°F (or other specified conditions).

(3) Run the scrubbing compound for approximately 5 minutes, strip off, and discard.

(4) Place the standard plateout test compound on the mill, start timer, adjust bank to pencil size, and allow to mill for 6 minutes after material is banded.

(5) Carefully strip plateout compound from mill without touching mill roll surfaces.

(6) Clean guides and area beneath guides with rag and stearic acid to remove compound that may have accumulated. Make sure none of this debris or stearic acid gets onto the roll surfaces.

---

*Or 325 mesh silica.
†Synpron 357 or other suitable Ba/Cd stabilizer may also be used.
‡The resin used here should be the same as that used in the production formulation.

(7) Without wiping rolls, place cleanup compound on mill and mill for 4 minutes with continuous mixing. Material should be of uniform color. Strip material, cool, and label.

(8) Clean the mill with a clean rag. A small amount of stearic acid may be used if necessary.

(9) Repeat steps 3 to 8 for each additional run. The swatches are compared at room temperature. The degree to which the white cleanup sheet becomes pigmented is indicative of the degree of plateout.

Note 1: At the mill setting used for this test, highly filled material will favor the back roll and should be allowed to band there. Unfilled material will usually favor the front roll. If meaningful results are to be obtained, all runs of a series should be done on the same roll.

Note 2: Be extremely careful to avoid contamination. Use paper-lined pan and change the paper each time the compound is changed.

Note 3: Uniformity of roll temperature is important. Temperatures lower than 350°F may cause changes in plateout tendencies.

The greater the degree of plateout, the greater the amount of 2-B red pigment that will adhere to the surface of the rolls along with the other compound ingredients causing the plateout. The cleanup compound, which is a mildly abrasive white compound, will then pick up the plated red pigment from the roll surfaces, and the degree of pink pigmentation of the white compound will be directly proportional to the amount of plateout that occurred in the test compound. This method is quite useful and appears to correlate very well with the commercial performance of compounds rated by this method.

E. Exudation and Migration

The fact that a stabilizer may process well and appear to be compatible with the formulation does not necessarily indicate that over a long period of time, in varying environments, the stabilizer will maintain its compatibility. If a stabilizer slowly migrates to the surface of a vinyl part, this migration is termed exudation or bloom, depending on the appearance of the deposit. If the deposit is wet or oily looking, it is normally termed exudation. If it is a dry, powdery, talclike deposit, it is called bloom. In either case, it occasionally consists of a mixture of the stabilizer with other formulation components, although many instances of exudation or bloom do not involve stabilizer at all. The stabilizer itself may not necessarily be at fault. In flexible formulations, the greatest potential for migration is from plasticizer, but stabilizers, except for leads, are generally highly compatible with the plasticizers and, therefore, will tend to migrate with them. A dry or powdery deposit generally indicates that the migrating stabilizer or plasticizer has picked up a high level of high-melting or solid-type product from the formulation, such as wax, filler (calcium carbonate), pigment (TiO_2), or any other solid component. Some lubricants migrate by themselves and result in a dry bloom.

A simple laboratory test has been devised that can indicate the long-term compatibility of a vinyl formulation. The test procedure is listed below.

Compatibility test method[1]

    Apparatus: Two-roll mill (standard speed 45 rpm), pyrometer, mixing beakers, scales, spatulas, forced air oven at 180°F, laboratory press at 350°F.

    Procedure:

Formulation no. 1 (normal level)		Formulation no. 2 (exaggerated level)
100.0	PVC*	100.0
40.0	DOP	40.0
0.1	Stearic acid	0.1
2.4	Stabilizer	10.0
1.0	Carbon black	1.0

    (1) Weigh individual components into beaker and mix.

    (2) Mill for 5 minutes at 350°F. Strip stock at approximately 40 gauge.

    (3) From each stock, cut out one 6 × 9 in. portion and press polish each material for 3 minutes at 3000 psi.

    (4) Cut each material in half, laying the two halves together.

    (5) Place the stock in 180°F oven, placing a 2-lb weight on the stock to ensure uniform contact.

    (6) Check the touching sides daily for evidence of bloom or exudation. This may be indicated by a wet, oily-looking surface or a hazy, bluish-type surface. The exaggerated level will normally exude first but is not an indication of a problem formulation; it merely shows that a tendency toward incompatibility is present.

The above method is useful for several reasons. First of all, the temperature at which the samples are aged accelerates the incompatibility of any ingredients from the vinyl compound, making this a quick screening test.

---

[1]Courtesy of Synthetic Products Co., "Stabilizer Test Procedures," Laboratory Standard 101-27.
*VC-100, Borden Chemical Co., Leominster, Mass., or equivalent grade.

Second, the carbon black used in the compound highlights the incompati-
bilities and helps them show up on the plastic surface when examined visually. Many times incompatibility, particularly in the form of the wet, oily
spew type of incompatibility, is difficult to see on a plastic surface. Often
it can be felt before it is seen and the carbon black helps to overcome this
problem. Third, both the stabilizer levels used and, in particular, the 10.0
phr stabilizer level in formulation #2 are generally an exaggeration of the
stabilizer levels used in commercial compounds. Therefore, formulation #2,
at the 180°F temperature using carbon black and the very high stabilizer
level, represents a greatly accelerated test for stabilizer-related exudation
or bloom.

F. Fogging and Volatility Testing

Vinyl compounds intended for film wrap applications require some unique
compatibility characteristics. When the vinyl film wrap is intended to come
into contact with foods, which often are at different temperatures at the
time of packaging or subsequently during storage, it must be able to resist
the formation of water droplets on the inside surface, which show up as
"fog." A test has been developed to evaluate both the hot and cold fog resistance characteristics of vinyl film.

Fog resistance test[1]

Objective:	To test stabilizer systems for their ability to prevent the formation of small water droplets (fog) on vinyl compounds intended for film wrap applications.
Apparatus:	600-ml beaker, deionized water, 8-oz. wide-mouth jars, hot plate, refrigerator at 40°F, two-roll lab mill at 350°F.
Formulation:	Commercial formulation to be tested. Milled 5 minutes after band at 350°F (or as specified). Sheeted off at (a) 30 mils or (b) 8 mils.
Procedure:	(a) *Hot fog*

A 30-mil press polished sheet containing the stabilizer
systems under evaluation in the test formulation is
placed over a 600-ml beaker full of water at 180°F.
After 30 minutes the respective stabilizer systems are
taken from the oven, cooled for 5 minutes, and then
rated against a control formula for their *prevention
of fog* on the exposed side of the vinyl.

(b) *Cold fog*

An 8 mil. film is placed over the mouth of a wide-
mouth 8 oz. jar filled with deionized water at room
temperature. The test film and the jar are then
placed in a refrigerator at approximately 40°F. The
test films are quickly evaluated every minute for
fog clearance for a total of 20 minutes. A very good
system may not develop any fog.

---

[1]Courtesy of Synthetic Products Co., "Stabilizer Test Procedures," Laboratory Standard 101-78.

In the case of antifogging formulations, a minor incompatibility of ingredients in the compound is sometimes necessary in order to pass the test. Certain vinyl-insoluble ester waxes that can act as wetting agents are used in vinyl compounds for the purpose of migrating to the surface and lowering the surface tension between the water and the vinyl compound so that droplets that begin to form do not have adequate surface tension to maintain the droplet shape. They then spread evenly over the surface of the vinyl and form a thin water film which does not appear as fog. The water is there, but it is not easily detected by the human eye and the problem is nonexistent in the consumer's mind.

In addition to fog stability, any incompatibility in the formulation that manifests itself with a high enough vapor pressure to cause an odor is generally unacceptable. This is usually tested for by volatility testing and can be done by placing a piece of vinyl film in a 8-oz. wide-mouth jar, loosely capped, in a 180°F oven overnight. A panel of judges is formed among testing personnel and the samples are evaluated each day by members of the panel for odor, which may slowly build up within the jar. This is generally indicative of volatilization of ingredients from the formulation. The stabilizer portion can have a great potential for causing odor, particularly if it is an organotin sulfur-containing material or a barium/cadmium stabilizer formulation that contains organic phosphites, solvents, free phenols, etc., capable of volatilizing out of the compound.

With electronic technology becoming more commonplace in the laboratory today, there is a more sophisticated method for volatility testing using a thermogravimetric balance. Curves of weight loss versus temperature and time can be developed for vinyl compounds. This, however, can be a costly and lengthy procedure, since the temperatures that are used generally should not exceed those to be encountered in the commercial application by much more than 20°F. At a greater temperature differential than this, the vapor pressures of liquid components are vastly exaggerated over what they are expected to be during the commercial life of the article. Volatility numbers can be generated that will never be encountered in the actual use of the vinyl article. If the testing temperature used is low and the incompatibility of volatile ingredients in the vinyl formulation is slight, testing can take quite a long time. A piece of testing equipment can be tied up for a very long period of time, and time in the laboratory represents money. Therefore, it is generally not practical to do long testing cycles on vinyl compounds themselves. Rather, individual ingredients that go into a vinyl compound, such as stabilizers, are tested on a thermogravimetric balance, where temperature is raised very rapidly and volatility of the stabilizer is plotted versus temperature. Stabilizer thermogravimetric curves can be compared to one another to give an indication of the potential volatility of one stabilizer compared to another.

G. Chemical Resistance Testing

1. Blush Resistance: With the increased activity in PVC for food packaging applications, the ability of a compound to resist absorption of liquids without a subsequent loss in its physical properties becomes very important. The principal material that most people are concerned with is water. Vinyl compounds intended for packaging of soft drinks, alcoholic beverages and other commercial solutions often experience a phenomenon known as water blush. This is the propensity of the compound to absorb water. This property can be radically affected by the stabilizer selected; solid, nontoxic stabilizers that contain any antioxidants or polyhydric alcohols that are soluble in water tend either to dissolve or to absorb water into the PVC compound. The result is a haziness in the compound. This phenomenon is generally looked for in thick sections of the PVC compound. The compound is milled, pressed into a $\frac{1}{4}$-in.-thick section, and placed in a low-temperature oven set between 110° and 180°F; the oven is kept at 100% relative humidity by leaving an open beaker of water inside the oven and continually checking that the beaker contains water. The normal exposure period is between 4 and 18 mr, but some test procedures specify checking the sample at 24-hr intervals for a period of up to 7 full days.

Another, more severe technique requires that the pressed sample actually be immersed in a water compound for 4 hr at 180°F and then checked visually for water blush. Most often, control samples are pressed into the same pressed plaque and visual comparisons are made. It is possible, however, to use an instrument such as a reflectometer or hazemeter for comparison. In either case, the sample to be tested must be compared to a control sample.

Solutions other than water can be used to evaluate a compound's blush resistance. A large market that has developed for bottles used in packaging mouthwash requires that the compound be resistant to water and alcohol solutions with alcohol contents between 13 and 22%. The blush testing can be conducted in the same fashion as above, except that, in the case of high alcohol content and 100% relative humidity testing, care must be taken that the alcohol cannot preferentially distill, leaving the vinyl compound exposed to water predominantly. Resistance of the compound to any other type of food product can easily be evaluated by the second immersion test method. Pieces of vinyl are immersed in the food intended for contact with the compound at an elevated temperature to accelerate the test and the sample is examined at intervals deemed to be appropriate to determine resistance.

In addition to the ability of the compound to resist blush, dimensional characteristics of a compound must be maintained in the presence of the food or solution to be contained. Testing can be done similarly to the immersion type of test or the high-humidity test with a specified sample size being exposed to the material of choice and the sample evaluated by comparing its dimensions and weight after testing to its original dimensions and weight. Dimensional ratios can be calculated and expressed as a percentage of the starting values. This property was particularly important in the days when vinyl asbestos flooring compositions were stabilized with amine-type stabilizers. When tiles were made of asbestos-filled formulations, it was particularly critical that they maintain dimensional stability so that they would fit together well and present the proper pattern on the floor over the whole commercial life of the floor. In addition, any "water growth," as it was termed, would cause the tiles to buckle against one another and cause the floor to release from the adhesive backings that were used to anchor the tiles. This is less of a problem today since most of the flooring formulations no longer contain asbestos and the stabilizer systems that are now used are not as sensitive to water growth as the amine types were.

2. *Staining*: Staining is a term used to describe a host of problems that can occur in the application of vinyl compounds. Three types of situations generally produce staining: exposure to atmospheric pollutants that

react with constituents of the compound, unwanted internal reactions between different compounding ingredients, and direct physical contact between the vinyl article and a material that can react with a constituent of the compound. The heavy metal salts contained in most stabilizers, as well as the inclusion of reactive chemical moieties such as carboxylic acids, phosphites and sulfur compounds, make the stabilizer a "hot" item ready to react not only with unstable ligands in the polymer but also with any material that comes in contact with the vinyl article.

Although staining occurs in all three of the situations listed above, experience has proved that the third type is generally the most common. In lead-, antimony-, barium/cadmium-, or barium/cadmium/zinc-stabilized articles, the most common type of staining is sulfur stain. Sulfur is present in the atmosphere and in most rubber compounds. Any vinyl compound that has the potential to come in contact with rubber compounds should be screened for sulfur stain resistance. Shoe soles, for example, are one of the major causes of sulfur staining on installed vinyl flooring.

Sulfur staining tests are conducted on milled vinyl stocks immersed in a saturated solution of hydrogen sulfide gas and water for 30 min up to 2 hr, then compared to a similarly treated control formulation which does not sulfide stain. Another technique is to expose a sample inside a closed container to an atmosphere containing hydrogen sulfide gas. Formulas stabilized with cadmium-containing stabilizers tend to form cadmium sulfide, which is yellow. Antimony-stabilized vinyls form orange stains. Compounds containing lead stabilizers turn dark brown or black on exposure to sulfide.

In addition to sulfur staining, several other types of staining occur frequently enough to be worthy of mention. They generally occur when a vinyl material is in contact with another plastic. The terms urethane staining, antioxidant staining and pinking are often used to describe these types of staining. In each case, a specific type of interaction is presumed to take place. In the case of urethane staining, the vinyl article is, of course, in contact with the urethane material. Staining between the free isocyanate radicals or amines used in the catalysis of the urethane compound, or sulfur-containing organotins used as urethane catalyst, and the lead, antimony, or cadmium stabilizer in the vinyl formulation is thought to occur. In general, it is difficult to identify the chemical moiety that causes the staining. In application, changes are made in the urethane resin formula to eliminate free isocyanate or polyol, or the catalyst is changed, and a staining test is conducted to determine the feasibility of putting the new urethane compound in contact with the vinyl compound.

Antioxidant staining may occur when vinyl is in contact with other materials that contain unusual antioxidants or with polyolefins that contain phenolic antioxidants not normally used in PVC. These antioxidants can react with ingredients in the vinyl formulation, particularly metals or free carboxylates from the stabilizer.

Phenolic antioxidants can undergo reactions with basic metal soaps or metal chlorides to form colored crystalline products [78]. In addition, the presence of metal soaps, bases, or salts can catalyze the transformation of phenolic additives into chromophoric compounds. An example is the effect of PbO_2 or $CdCl_2$ in oxidizing a cyclic phenol such as 2,6-di-*tert*-butyl-*p*-cresol (BHT) to 3,3',5,5'-tetra-*tert*-butyl-stilbene-4,4'-quinone as shown below. This oxidized BHT compound is red [79].

Quinone formed through biomolecular coupling of BHT

The migration of antioxidants out of highly plasticized vinyl compounds that have been formulated to be stable in the presence of the antioxidant, into filled or pigmented polymers never intended to contain such an antioxidant, can cause reactions between metals of the filler or pigment to form chromophoric compounds. In addition, phenolic materials are often quite volatile and tend to be absorbed by polymers containing carbon black.

Phenolic antioxidants are often used in conjunction with polyols in powdered stabilizers and can be condensed or inadvertently reacted to form undesirable esters with free carboxylates from the metal soaps. The phenolics are widely used, and therefore their properties are of interest in PVC formulation. Other antioxidant types cause many more problems in PVC and are therefore not used, but they do find their way into other polymers or elastomers. Widely used products such as hydroquinolines, naphthylamines, and p-phenylenediamines may be absorbed into a PVC article in contact with an elastomeric material and may undergo complex reactions to form color bodies, causing the vinyl to stain.

Lead-, antimony-, or cadmium-stabilized vinyl formulations often stain when they come in contact with rubber products containing various sulfur compounds. Besides the sulfur used to vulcanize the rubber, many of the accelerators and cross-linking agents used in rubber are thio compounds such as 2-mercaptobenzothiazole (MBT). The compound MBT is a primary accelerator for rubber but, due to problems such as odor, solubility and volatility, the MBT species is obtained by using MBT precursors in order to make more acceptable rubber compounds. Many of these precursors are very reactive with metals from the stabilizer. They may migrate between the polymer and elastomer in a flexible vinyl compound or stain at the point of contact between a rigid vinyl and an elastomer. Common reactive precursors or accelerators such as benzothiazyl disulfide (MBTS) or dipentamethylene thiuram hexasulfide can easily migrate and have available sulfide linkages capable of reacting with stabilizer components.

In addition, some antioxidants which are used in polyolefins to improve oxygen stability or UV stability have a particularly poor effect on the UV stability characteristics of the PVC compound. Highly flexible vinyl compounds which contain large amounts of plasticizers can induce migration of additives from one polymer to another and can accentuate staining characteristics of a vinyl compound in contact with another plastic.

Very light red to pink staining that occurs within a compound that contains naturally derived oils or acids is termed "pinking." In general, experience indicates that epoxy products may contain small amounts of natural organic contaminants, generally multi-ring compounds, which can react with the metals or acids present in a vinyl stabilizer to form chromophoric compounds that appear as pink stains. Changing the type or source of epoxide or the type or source of the stabilizer can often alleviate this problem.

Pinking also tends to occur in compounds that are susceptible to microbial attack. Formulations containing large proportions of plasticizer, particularly aliphatics, or highly plasticized compounds that contain epoxide appear to provide a good food source for microbes.

In each of the staining modes mentioned above, with the possible exception of microbial staining, colors are thought to develop by a heavy metal, such as cadmium, antimony, or lead, reacting with an amine, sulfur, or phenolic compound to form a chromophoric compound, whose appearance is identified as stain.

The term "cross-staining" is used to denote an interaction between compounds that may occur when certain scrap or regrind materials are blended with virgin compounds. The scrap material often does not contain the same stabilizer as the virgin. Where interactions among the stabilizers occur, the phenomenon is called cross-staining. Most of the time, cross-staining occurs when an antimony-, lead-, or cadmium-containing stabilizer interacts with a sulfur-containing material such as a sulfur-containing tin stabilizer. The most straightforward solution is to make the stabilizer types the same in both the virgin and scrap compounds so that staining does not occur.

The vinyl compounds that building wire in electrical wire and cable constructions comes in contact with represent yet another source of metal ion capable of reacting with compounding ingredients or stabilizers in the vinyl formulation. Copper wire, nickel-containing wire, and wire with a high content of iron can all react with carboxylates, phenols, or sulfur-containing materials in the vinyl compound to form a colored metal salt. This type of problem was more common in the early days of vinyl-clad copper wire. Stabilizer systems that are relatively free of carboxylates or phenolics are used in wire coating compounds today without any reaction with the copper wire. Enough is known about this type of formulation that very few, if any, problems are encountered today in vinyl-clad wire applications.

H. Quality Control Testing

Once a vinyl stabilizer has been selected for an application and is being purchased on a regular basis, some relatively simple tests are normally used to ensure a continuing level of quality in the incoming material. The main test that a laboratory should be prepared to perform on a vinyl stabilizer is that of heat stability. Generally, static oven heat stability will suffice, but in the case of injection molding or extrusion applications it is preferable to run a dynamic heat stability test such as a rheometer test.

Static heat stability on each incoming lot of stabilizer should be run against a control stabilizer sample not more than 3 months old, or against vinyl stock prepared from the control stabilizer sample.

In addition, specific gravity, color, refractive index, and viscosity are usually sufficient to show that a liquid stabilizer's composition is not varying. All of these tests can be run quickly with relatively inexpensive equipment in modern laboratories.

If a laboratory is sophisticated enough, the chemical tests to be performed on a stabilizer generally include metal analysis by atomic absorption or, in some cases, EDTA titration to determine that a stabilizer's metal ratio and overall metal concentrations do not vary outside a specified range. Spectral curves from an infrared or UV-visible spectrophotometer can often show that a stabilizer's composition is not varying. Although these are simple procedures to run, they can be very difficult tests to interpret. Modern PVC heat stabilizers are complex mixtures of many organic substances, hence their infrared or UV-visible spectra will be complex. It takes a highly trained spectroscopist to interpret variations in the spectra so that they can be related to performance attributes in the vinyl compound.

The keys to good stabilizer control are the performance tests, all of the tests mentioned previously: heat stability, dynamic stability, plateout, water blush, staining, low-temperature stability, and outdoor weathering characteristics. These are the tests that will be most applicable in determining the suitability of a stabilizer candidate throughout the intended service life of the vinyl article containing the stabilizer.

REFERENCES

1. Nass, L. I., ed., *Encyclopedia of PVC*, 1st ed., Dekker, New York, 1976.
2. Chevassus, F., and DeBroutelles, R., *The Stabilization of Polyvinyl Chloride*, St. Martin's, New York, 1963.
3. Penn, W. S., *PVC Technology*, Maclaren & Sons, London, 1966.
4. Gould, R. F., ed., *Stabilization of Polymers and Stabilizer Processes*, Advances in Chemistry Series, vol. 85, American Chemical Society, Washington, D.C., 1968.
5. Sarvetnick, H. A., *Polyvinyl Chloride*, Plastics Applications Series, Krieger, Publishing Company, Huntington, N.Y., 1977.
6. Kelen, T., *Polymer Degradation*, Van Nostrand Reinhold, New York, 1983.
7. Jellinek, H. H. G., ed., *Degradation and Stabilization of Polymers*, Elsevier Science Publishers, Amsterdam, 1983.
8. Semon, W. L. (to B.F. Goodrich), U.S. Patent 1,929,453 (1933).
9. Nass, L. I., in *Encyclopedia of Polymer Science and Technology*, vol. *12*, p. 725, H. Mark, N. V. Gaylord, and N. Bikales, eds., Wiley, New York, 1971.
10. U.S. International Trade Commission, Synthetic Organic Chemicals, U.S. Production and Sales 1978, 1979, 1980, 1981, 1982, Government Printing Office, Washington, D.C.
11. Guyot, A., Bert, M., Burille, M. F., and Michel, A., Trial for Correlation between Structural Defects and Thermal Stability in PVC, Third International Symposium on Polyvinyl Chloride, preprints Case Western Reserve University, 1980.

12. Morton, M., ed., *Rubber Technology*, 2nd ed., Van Nostrand Reinhold, New York, 1973.
13. Marvel, C. S., Sample, J. H., and Roy, M. F., *J. Am. Chem. Soc.*, *61*, 3241 (1939).
14. Druesedow, D., and Gibbs, C. F., Effect of Heat and Light on Polyvinyl Chloride, *Mod. Plast.*, *30*, 10, 123 (June 1953).
15. Imoto, M., and Otsu, T., *J. Inst. Polytech.*, *Osaka City Univ.*, *Ser. C*, 124 (April 1953).
16. White, E. L., in *The Encyclopedia of Basic Materials for Plastics*, H. R. Simonds and J. M. Church, eds., p. 433, Reinhold, New York, 1967.
17. Hurley, D. W., Lead Chemicals—Compliance with Environmental Regulations, *J. Vinyl Technol.*, *4*, 1 (1982).
18. Occupational Exposure to Lead, Final Standard, OSHA, Department of Labor, *Federal Register*, vol. 43, no. 220, Tuesday, November 14, 1978.
19. Smith, H. V., *Development of the Organotin Stabilizers*, Tin Research Institute of Great Britain, London, 1953.
20. Yngve, V. (to Carbide & Carbon Chemicals Corp.), U.S. Patent 2,219,463 (British Patent 497,879) (1940).
21. Quattlebaum, W., et al., U.S. Patents 2,344,002, 2,307,675, and 2,307,157.
22. Best, C. E., et al. (to Firestone), U.S. Patent 2,731,484 (British Patent 728,953) (1950).
23. Leistner, W. E., et al., U.S. Patents 2,726,277 and 2,726,254 (1950).
24. Weinberg, E. L., et al., U.S. Patent 2,648,650 (British Patent 719,733) (1951).
25. Caldwell, S. S., et al., U.S. Patent 2,629,700.
26. Brecker, L. R., Structure-Performance Relationships in Organotin Mercaptide Stabilizers, Third International Symposium on Polyvinyl Chloride, Case Western Reserve University, 1980.
27. Smith, P. J., Toxicological Data on Organotin Compounds, International Tin Research Institute Publ. No. 538.
28. Gottlieb, et al., U.S. Patent 3,424,717 (1969).
29. Larkin, W. A., et al., U.S. Patent 4,183,846 (1980).
30. Weisfeld, L. B., U.S. Patent 3,478,071 (1969).
31. Kauder, O., et al., U.S. Patent 3,565,930 (1971).
32. Brecker, L., U.S. Patent 3,565,931 (1971).
33. Jennings, T. C., et al., U.S. Patent 3,764,571 (1973).
34. Hampson, D., and Lanigan, D., paper presented at the 37th Annual Technical Conference, Soc. Plas. Eng., New Orleans, 1979.
35. Zuckerman, J. J., Organotins in Biology and the Environment, in ACS Symposium Series 82, *Organometals and Organometalloids, Occurrence and Fate in the Environment*, American Chemical Society, Washington, D.C., 1978.
36. Reed, M. C., et al., U.S. Patent 2,075,543 (1937).
37. Frye, A. H., and Horst, R. W., *J. Polym. Sci. 40*, 419 (1959).
38. Frye, A. H., and Horst, R. W., *J. Polym. Sci. 45*, 1 (1960).
39. Klemchuk, P. P., in *Stabilization of Polymers and Stabilizer Processes*, Advances in Chemistry Series, vol. 85, R. F. Gould, ed., American Chemical Society, Washington, D.C., 1968.
40. Anderson, D. F., and McKenzie, D. A., *J. Polym. Sci.*, *A-1* (8), 2905 (1970).
41. Leistner, W. E., et al., U.S. Patent 2,564,646 (1951).
42. Lally, R. E., et al., U.S. Patent 2,734,881 (1956).

43. Leistner, W. E., et al., U.S. Patent 2,716,092 (1955).
44. Weinberg, E. L., U.S. Patent 2,680,726 (1954).
45. Dieckmann, D. J., U.S. Patent 4,029,618 (1977).
46. Dieckmann, D. J., SPE Tech. Pap. XXII, 507 (1976).
47. Muldrow, C. N., U.S. Patent 4,287,118 (1981).
48. Fletcher, C. W., et al., The Effect of Thermal Stabilizers on the UV Stability of PVC Pipe Compounds, SPE Tech. Papers 27, 529 (1981).
49. Code of Federal Regulations, Title 21, Subpart E.
50. Michel, A., vanHoang, T. and Guyot, A., *Polymer Degradation and Stabilization*, 4, 427 (1982).
51. Anderson, D., and McKenzie, D., *J. Polym. Sci.*, *A-1* (8) (1970).
52. Michel, A., Recent Advances in PVC Stability with Metal and Organic Derivatives, Third International Symposium on Polyvinyl Chloride, Case Western Reserve University, 1980.
53. Gay, M., and Carette, M., Paper presented at SPE Polymer Modifiers and Additives Division 1st International Conference, Newark, N.J., November 6, 1985.
54. Nass, L. I., Actions and Characteristics of Stabilizers, in *Encyclopedia of PVC*, L. I. Nass, ed., vol. 1, chap. 9, Dekker, New York, 1976.
55. Ibid., p. 333.
56. Weisfeld, L. B., Thacker, G. A., and Hampson, D. G., *Vinyl Technology Newsletter* 5 (2), 47 (1968).
57. Karpel, S., PVC Sheet for Packing, in *Tin and Its Uses*, No. 139 (1984).
58. Code of Federal Regulations, Title 21, U.S. FDA.
59. Sarvetnick, H. A., *Plastisols and Organosols*, Van Nostrand Reinhold, New York, 1972.
60. Underwriters' Laboratories, Bulletin 83, *Thermoplastic Insulated Wire*, 4th ed., Melville, New York.
61. Osmer, D., Color Differences: Establishing and Interpreting Tolerance Limits, *Plast. Compounding*, 14 (Feb. 1983).
62. Judd and Wyszecki, *Color in Business, Science, Industry*, 3rd ed., Wiley, New York, 1975.
63. Robertson, A. H., How to Specify Color Differences, SPE RETEC, Cherry Hill, N.J. (1976).
64. Longley, W. V., The Visual Approach to Controlling Metamerism, *Color Research and Application 1*, 43, No. 1 (1976).
65. McCabe, C. C., Rheological Measurements with the Brabender Plastograph, *Chem. Can.*, 44 (Oct. 1960).
66. Blake, W. T., Predicting the Processability of Plastics, Tech. Bibliography No. 222, C. W. Brabender Instruments, Inc., 1964.
67. Rogers, M. G., Rheological Interpretation of Brabender Plasti-Corder (Extruder Head) Data, *Ind. Eng. Chem. Process Des. Dev.*, 9 (1), (1970).
68. Gomez, I. L., *Engineering with Rigid PVC*, p. 305, Dekker, New York, 1984.
69. Murrey, J. L., Injection Molding Rigid PVC Compounds, *SPE Tech. Papers 28*, 313 (1982).
70. Dunn, J. L., and Heffner, M. H., Outdoor Weatherability of Rigid PVC, *Tech. Papers*, 19, 483 (1973).
71. Kanal, M. R., Cause and Effect in the Weathering of Plastics, *Poly. Eng. Sci.*, 10 (2), 108 (1970).

72. Zerlaut, G. A., and Ellinger, M. L., Precision Spectral Ultraviolet Measurements and Accelerated Weathering, *J. Oil Colour Chem. Assoc.*, *64*, 387 (1981).

73. Kinmonth, R. A., and Scott, J. L., Playing the Numbers; an Inter-comparison of Radiant Exposure among Lamps and Daylight, *Atlas Sun Spots*, *14* (32) (1984).

74. Stoloff, A., Interim Report of the Vinyl Siding Division, Weatherability Committee, SPI Committee Meeting, January, 1980.

75. Stoloff, A., Weathering Committee Report, Vinyl Siding Institute of the PSI, September 1980.

76. Kinmonth, R. A., A Correlation Review—Published Results from 1962—1977, *Atlas Sun Spots*, 7, Issue 18, (1978).

77. Ibid.

78. Holtzen, D. A., Discoloration of Pigmented Polyolefins, *Plast. Eng.*, *33* (4), (1977).

79. Ibid.

3

Plasticizers: Types, Properties, and Performance

L. G. KRAUSKOPF

Exxon Chemical Company
Baton Rouge, Louisiana

I. INTRODUCTION

The objective of this chapter is to survey the materials used as plasticizers in vinyl compositions—their performance characteristics in vinyl compositions, manufacture, availability, properties, and identification.

A. Definition

The Council of the International Union of Pure and Applied Chemistry adopted the following definition of a plasticizer in 1951:

> A substance or material incorporated in a material (usually a plastic or an elastomer) to increase its flexibility, workability or distensibility. A plasticizer may reduce the melt viscosity, lower the temperature of a second order transition or lower the elastic modulus of the product.

The vinyl formulating chemist consequently evaluates the optimum plasticizer selection in light of the processing techniques used in the operation as well as the properties required for the end product. Furthermore, the chemist may select one or more materials from a wide variety of chemical substances to serve as the plasticizer. This is one of the primary reasons for the successful use of vinyl plastics in so many diverse applications. Plasticizer type and concentration must be recognized as the formulating variables that most dramatically alter the processability and end-use of flexible vinyl products.

B. History

Sears and Darby [1] published considerable information on the subject of plasticizer technology; they traced the history of plasticizer-like materials to very early civilization. The use of plasticizers in the modification of polymer properties began during the 19th century [2]. In 1868 the Hyatt brothers added camphor to nitrocellulose in order to make the polymeric material more easily moldable and less brittle in the finished form. The early references cite the use of tricresyl phosphate and dialkyl phthalates as partial replacements for camphor in nitrocellulose. In 1929 Ostromislensky [3] was granted a patent entitled "Polymer of Vinyl Chloride and Process for Making the Same," in which he related that "plastifiers" of low volatility might be mixed with his product. He speculated that the mechanism of "plastification" was probably a chemical reaction. During the 1920s the B. F. Goodrich Co. was seeking uses for the horny, rigid, and brittle polymer which they were able to make commercially from vinyl chloride; Waldo Semon [4] observed that boiling PVC particles in esters such as tricresyl phosphate and dibutyl phthalate gave a highly elastic and flexible adhesive which could also be molded into shapes. Thus, Semon is generally credited with being the first to plasticize PVC resins. Numerous patents were issued on the uses of plasticizers in the first half of the 20th century [5—7]. In 1933 Kyrides [8] was issued a patent on octyl alcohol esters; the claims specifically include di(2-ethylhexyl) phthalate, which became known as DOP [9].

Gresham [10], of B. F. Goodrich, patented the use of dioctyl phthalate and like materials as plasticizers for vinyl resins in 1943. DOP became commercially available in bulk quantities in 1940 [11] and held the leading position as a plasticizer for polyvinyl chloride for over 30 years. In the 1970s several events curtailed the commercial availability of DOP in the United States: (1) two leading producers, Union Carbide Corporation and Allied Corporation withdrew from the business, and (2) a major source of 2-ethylhexanol was shut down by Oxochem. Enjay, a predecessor of Exxon Chemical Americas, had introduced diisononyl phthalate in 1968, which was readily substituted for DOP in many applications. It is now recognized that alternatives such as diisononyl phthalate provide cost/performance benefits over DOP in general-purpose flexible PVC formulations whenever the two plasticizers are sold at the same price [12]. It should be noted that the versatility of PVC resin presents a multitude of formulating options to meet specified properties for flexible PVC applications. Therefore the ratio cost/performance is used in this chapter to emphasize that optimum commercial formulations provide minimized cost with the best balance of specified performance properties.

As commercialization of flexible PVC developed, many different phthalates and nonphthalate esters were found to be compatible with PVC and useful for imparting special properties. Thus the markets for flexible PVC were broadened to include end uses requiring exceptionally low volatility, extraction resistance, low-temperature properties, and numerous other characteristics not provided by "general-purpose" flexible PVC formulations.

C. Commercial Markets

The Modern Plastics Encyclopedia [13] lists over 400 materials classified as plasticizers that are available from over 50 suppliers. For all practical purposes, there are currently five major producers of vinyl plasticizers and about five minor producers in the United States; worldwide, there are about a dozen major suppliers.

Plasticizer production in the United States increased from 28.4 million pounds in 1940 to 170 million pounds by end of World War II in 1945. It peaked at slightly over 2 billion pounds in the late 1960s. Currently, the annual production of plasticizers in the United States is about 1.4 billion pounds, and free world production is approximately 4.5 billion pounds. The regional supplies of plasticizers for 1982−1983 were estimated as follows [14−16]:

Region	Million pounds	Metric kilotons
Western Europe	2024	920
United States	1445	657
Japan	825	375
Canada	123	56
	4417	2008

The market share for different types of plasticizers shows the worldwide predominance of phthalates. The U.S. market breakdown is representative [14]:

Plasticizer type	Approximate share (%)
Phthalates	71
Epoxides	7
Aliphatic dicarboxylic diesters	5
Polyesters	4
Phosphates	2
Extenders	2
Miscellaneous	9

Approximately 85% of the plasticizers produced worldwide are consumed as additives to impart flexibility to vinyl resins. The remainder are used as plasticizers for nonvinyl polymers, such as synthetic rubbers, cellulosics, and acrylics, and for nonplasticizer uses such as synthetic lubricants, hydraulic and dielectric fluids, and inert carriers. This chapter is limited to materials that have gained commercial acceptance as plasticizers for PVC resins.

The versatility of PVC is reflected in the variety of plastics fabricating processes utilized, as well as [in] the many end-use products made of plasticized PVC. This causes the traditional description of plasticizer market segments to be a confusing mixture of "processes" and end-use products. By combining market data from *Modern Plastics* [17] and SRI Chemical Economics Handbook [16], one may estimate plasticizer consumption according to both processes and end-use products.

Product	Plasticizer usage (%)
Calendered film, sheet, coated fabrics, and flooring	40
Extrusion-coated wire, substrates, and contours	35
Plastisol coatings, flooring, and molded shapes	20
Injection-molded parts and shapes	<5
Miscellaneous (coatings, latex, adhesives, etc.)	<5

The variety of end-use products made of plasticized polyvinyl chloride resins is limited only by human imagination and cost/performance comparisons with other materials. Considering costs, the SRI report [16] shows little change in plasticizer selling prices throughout the 1960s. Figure 1 shows that prices sharply increased in the early 1970s as a result of increased costs of hydrocarbon feedstocks caused by escalating crude oil prices. This has been historically true for worldwide plasticizer prices. Although European plasticizer prices are traditionally slightly lower than the U.S. market, the impact of crude costs is more dramatic in the plasticizer price swings [16a].

II. PLASTICIZER PERFORMANCE PROPERTIES IN PVC

A. What the Plasticizer Does

The major function of a plasticizer is to impart flexibility and workability. In so doing, both the plasticizer type and concentration significantly alter the glass transition and melting range of the polymer in the solid state [18–24]. Unfortunately, it is not possible to predict the performance characteristics of a plasticizer simply by defining the physical and/or chemical properties of the liquid plasticizers. Similarly, the mechanical properties of the plasticized polymer cannot be completely predicted from solvency characteristics. This is apparent from the mechanical properties of the polyvinyl chloride compositions as shown in the data tables contained in this chapter. It is possible, however, to predict performance characteristics with some degree of confidence by first classifying the plasticizer according to chemical structure and then relating it to other members of that chemical family whose performance characteristics are known. Although this has been done for most of the materials commonly used as plasticizers, a new or unidentified plasticizer must be assessed by evaluating its performance characteristics against known standards. A series of test formulations having a range of plasticizer levels are prepared and the properties of the test compounds are measured. This procedure provides absolute values in terms of hardness, modulus, extensibility, low-temperature properties, and so forth, which can be related to application requirements. These data also provide a baseline from which it is possible to predict the effects due to plasticizer interchanges and the use of plasticizer mixtures. One must be aware of the practical limits of mechanical properties that are attainable with vinyl compositions, the accuracy of the test methods employed in the measurement of these properties, and the effect of other formulating components besides plasticizer. Thus, the data base should be generated by using standard formulations in which all formulating variables are held constant except for the plasticizer. Methods used for sample preparation and measurement of properties must also be held constant.

Tables 1 and 2 list the mechanical properties of some typical commercial products made of polyvinyl chloride. The formulas have been oversimplified to demonstrate the effect of plasticizer. The properties in Table 1 typify those attainable from unplasticized to highly plasticized PVC compositions, and they show the result of varied levels of diisononyl phthalate (DINP) plasticizer. The values reported for the rigid polyvinyl chloride are approximations, because most of the test methods shown are designed to measure properties of plasticized vinyl compositions.

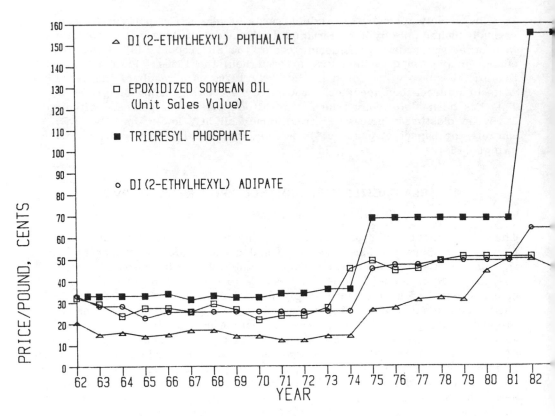

FIGURE 1   U.S. price of major plasticizers.   (From *Chemical Economics Handbook*, 579.5023 E, SRI International, Sept. 1984, reprinted with permission.)

The general effect of increased plasticizer concentration on mechanical properties is as expected, as shown in Table 1 and numerous publications [25−31]. However, the extremes of plasticizer concentration show some deviation from the pattern. At very high levels of plasticizer (greater than 90 phr*), the mechanical properties are altered only slightly as plasticizer concentration is increased [32]. At the low end (less than 30 phr), changes in mechanical properties are very dramatic as a function of plasticizer concentration. In fact, a phenomenon called antiplasticization occurs at plasticizer levels between 0 and about 20 phr. Extensibility decreases and brittleness increases to such a degree that commercial products containing such low levels of plasticizer are limited. As may be expected, this area has been the subject of numerous studies [1,33− 38]. Such studies have led to a better understanding of the morphological characteristics of plasticized polyvinyl chloride and the mechanism of plasticization.

Processing of PVC is enhanced by plasticizers, primarily because of the diffusibility and solvating action of the plasticizers on the high-molecular-weight PVC. Here again, small amounts of plasticizers acting as solvents have dramatic effects on processability. However, at very high plasticizer concentrations, the molten plasticized polymer has very little hot strength or elastic modulus. Thus, very highly plasticized PVC has processing problems that are quite different from those of unplasticized PVC. At in-

---

*phr = parts per hundred parts of resin, by weight.

TABLE 1 Typical Commercial Vinyl Products—Unfilled[a,b]

	ASTM	Rigid	Semirigid	Flexible	Very flexible	Highly flexible
DINP plasticizer level, by weight						
phr (parts/100 resin)		0	34	50	80	600
% of composition		0	25	33	44	86
Typical properties (approximate)						
Specific Gravity, @ 20/20°C	D-792	1.40	1.26	1.22	1.17	1.02
Durometer Hardness,	D-2240					
A scale		>100	95	83	66	<10
D scale		80	49	<35	<20	—
Tensile strength, psi	D-882	>6,000	3,500	3,000	2,100	—
Ultimate elongation, %		<15	285	330	385	—
Flexural stiffness, psi	D-747	>130,000	10,000	1,700	500	<100
T_4, °C	D-1043	>23	+20	+9	-12	—
T_f, °C	D-1043	+20	-1	-24	-45	—
Brittleness temperature, °C	D-746	>23	-16	-32	-46	—
(°F)		(>73)	(+3)	(-26)	(-51)	—
Typical products		Pipe Siding Records Blown bottles Credit cards	Floor covering Window shade Overlay films	Sheeting Waterbeds Pool liners Wall coverings Garden hose Tubing Upholstery	Coated cloth Playballs Shoe soles Highway markers Handle grips	Fishing lures

[a]Typical products.

[b]All properties are approximations.

termediate plasticizer levels, 20 to 60 phr, the plasticizer type (chemical structure) significantly influences the ease of PVC processing.

Table 2 is an overview of some typical commercial products made with filled, plasticized PVC. The types of plasticizers used reflect optimization of cost, end-use properties, and ease of processing. The effects of the various plasticizers will be reviewed as a function of chemical category (type) and specific structural effects. Bernardo and Burrell [39] have published a very readable introduction to the significant aspects of plasticization technology.

B. How the Plasticizer Works

Theoretical understanding of the mechanism of plasticization has gradually been improving. Sears and Darby [1,40] have prepared detailed treatments on the subject. Three favored theories are currently applied to the mechanism of plasticization (see also Chapter 10, Vol. 1 of this series):

1.   Lubricity theory—based on plasticizer reducing polymer-polymer interaction forces.
2.   Gel theory—loosely attached three-dimensional gel formed by polymer-plasticizer interaction.
3.   Free-volume theory—explains the fact that extrapolated curves of specific volume versus temperature do not intersect at the free volume of hypothetically pure and ordered system at absolute zero temperature.

Doolittle's [41] mechanistic theory does not separate the above but essentially includes them with thermodynamic concepts. Suffice it to point out here that one may consider plasticized polyvinyl chloride to be essentially a solid solution. Thus, there is a continuous dynamic state with respect to interactions of the following nature: polymer-polymer, polymer-plasticizer, and plasticizer-plasticizer.

In practical applications, the polymer-plasticizer solution contains many additional components; lubricants, stabilizers, diluents, dyes, pigments, fillers, degradation reaction products, absorbed entities from posttreatments and/or the application environment, etc. These are sometimes the cause of incompatibility or "spew" failure, which is a phase separation of plasticizer from the solid solution. The point is that most of the incompatibility problems found are not a result of incompatibility of the plasticizer per se. It is known, however, that most plasticizers have limits of compatibility with PVC. The plasticizer "works" when it is sufficiently compatible to maintain the desired polymer-plasticizer interaction to a greater extent than competing interactions, including the presence of other additives which influence the compatibility of the total plasticized PVC system.

III. PLASTICIZER EFFECTS ON PVC PROCESSING

The versatility of PVC becomes truly apparent when one examines the processes by which it is converted into end-use products. It is essentially adaptable to all processes used in molding and fabricating all other thermoplastic and elastomeric polymers, as well as those used for coatings and adhesives. The most commonly used processes for plasticized PVC include

TABLE 2 Typical Commercial Products of Filled Flexible Vinyl

	ASTM	Automotive upholstery	Waterstop	Building wire	Refrigerator gasket	Floor tile
		(a)	(b)	(c)	(d)	(e)
Composition, wt %*						
PVC		56	53	51	37	14
Plasticizer**		37	35	31	33	4
Filler		5	10	15	27	80
Stabilizers, lubricants, pigments, etc.			←———— $\sim 2-3$ ————→			
Properties						
Specific gravity	D-792	1.23	1.25	1.33	1.40	2.24
Hardness, Durometer A	D-2240	73	67	78	68	>95
Tensile, psi	D-882	2,200	2,000	2,500	1,800	900
Elongation, %	D-882	360	350	300	275	8
Flexural modulus, psi at 23°C	D-747	900	680	710	600	20,000
Brittleness, C	D-746	-37	-45	-29	-40	>23
(°F)		(-25)	(-49)	(-20)	(-40)	(>73)

*All compositions and properties are "typical;" plasticizer and filler types are varied to optimize desired properties.
**Typical plasticizer types: (a) Linear phthalates, DIDP, phthalate/adipate blends; (b) GP and linear phthalates; phthalate/adipate blends; (c) DINP, DIDP, DTDP, trimellitates; (d) Polymeric for PS contact, phthalate for ABS contact; (e) BBP, DHP, DIHP, DOP, DINP.

"hot compounding" as one of the steps in the process. In this step the premix of the PVC polymer with the various additives is first converted into a hot molten thermoplastic mass, followed by shaping and then cooling; the cooled shape is typically subjected to various posttreatment processes— printing, top coating, laminating, sealing, etc. Other processes involve preshaping of the formulated PVC premix followed by a fusion cycle which melts the particulate polymer into a continuous thermoplastic mass; this is then followed by cooling, and the composition develops increased strength as a function of time, as do the hot-compounded compositions. These processes utilize PVC premixes in a variety of forms: dispersions (plastisols and organosols), dry powders, solutions, and latices. The selection of specific grades of PVC resin is critical with respect to the process to be applied, as are many of the other additives used in the total formulation. Plasticizer type and concentration also have very significant effects on the process involved. All of the processes by which plasticizer and PVC resin are brought together and fused, however, follow the same stepwise pattern of plasticization:

1. Adsorption—the plasticizer is taken up by the PVC resin through adsorption on the particle surface and filling of intraparticle voids; adsorbed plasticizer may be removed from the PVC particles by centrifuging.
2. Absorption—plasticizer diffuses into the primary PVC particle, entering the amorphous regions intermolecularly; this is accompanied by a slight decrease in resin/plasticizer volume at the outset and is followed by the resin surfaces beginning to expand and entering a gel stage; at this point, the resin/plasticizer interaction is physically irreversible.
3. Fusion—following the gelation stage, a stepwise diffusion process takes place as polymer molecules swell and undergo disentanglement and separation; polymer-plasticizer homogenization takes place; ideally, resin grain boundaries begin to melt out and a continuous single phase plasticized PVC film is formed.
4. Toughening—after cooling there is a brief period (24 to 48 hr) of apparent increase in van der Waals' forces resulting in increased strength; beyond this point, the plasticized PVC may be considered a solid solution in which there is a continuous association and dissassociation between the polymer-plasticizer, polymer-polymer, and plasticizer-plasticizer molecules.

Obviously, the plasticizer type and concentration have significant effects on the processability of the formulated PVC composition, irrespective of which process is applied.

A. Hot Compounding

Hot-compounding techniques historically were derived from the processes normally used to compound elastomeric compositions. Equipment that was in place was modified for use at the high temperatures required for melting and shaping thermoplastic PVC (about 170° to 185°C or 340° to 370°F). Such equipment included Banbury mixers, two-roll mills, extruders, and calenders, which use mechanical energy and heat to convert the powdery premix into a molten plastic state, at which point it is shaped as desired.

Coaker and Bias [18] evaluated the effect of temperature on laboratory-prepared milled and molded specimens while comparing the effects of a strong solvating plasticizer, butylbenzyl phthalate, and a moderately solvating plasticizer, di(2-ethylhexyl) phthalate. Optimum temperatures ranged from 150° to 170°C but, as pointed out by Sears and Darby [1], in practice the optimum conditions are about 15°C above those measured in the work of Coaker and Bias. High levels of the strong solvating plasticizer result in hot melt rheology with poor hot strength, while a weak solvating plasticizer, ditridecyl phthalate (DTDP), had adequate hot strength and good elongation under processing conditions. Sears and Darby [1] studied Bargellini's work [19] and the work of Bergen and Darby [20] on the energy required to hot-compound plasticized PVC as a function of plasticizer type. Strong solvating plasticizers require more work for compounding as they enhance the swelling and undergo strong intermolecular association with the polymer, but the time required to accomplish optimum fusion is significantly shortened by the higher work energy associated with strong solvating plasticizers during the fluxing cycle. Schreiber [21] used a torque rheometer to determine the melt viscosity of plasticized PVC as a function of plasticizer concentration, shear rate, and temperature. He was able to ascertain a measure of the polymer-plasticizer interaction, thus a measure of plasticizer compatibility.

Although plasticizer type and concentration are generally selected to impart specified end-use properties, it is obvious that the solvating characteristics of the plasticizer system have significant effects on the processability during the mixing, compounding, and fabricating cycles.

B. Dry-Blending

In the 1950s Monsanto introduced a special porous grade suspension PVC resin and a technology referred to as dry-blending. This technology was designed to encourage the premixing of plasticizer with the PVC granular resin at slightly elevated temperatures (80° to 110°C) to bring about plasticizer take-up by the porous PVC resin particle while retaining the free-flowing particulate powder. The commercial effect was to circumvent the intensive energy requirement of batchtype mixing associated with Banbury mixers and two-roll mills. Thus, following the appropriate premixing cycle, it became possible to feed continuously a free-flowing dry blend of plasticized polymer into an extruder. In turn, the extruder could continuously convert the dry powder into a fluxed thermoplastic mass and extrude a continuous shape or coating. Commercial success first came in the wire and cable industry, where the practice was to extrude a plasticized PVC insulation over an electrical conductor. Thus a small cross-sectional area of extrudate allowed maximized back pressure and working in the extruder. Gradually PVC resin technology became refined to the point where today the majority of commercially available suspension grade PVC resins have the porosity required for good plasticizer take-up, yielding free-flowing dry blends prior to the fluxing operation. The dry-blend technology introduced some potential new ways to apply powdered PVC formulations as direct coatings [42,43]. Dry-blending is common practice in most of today's commercial hot-compounding processes. This enhances the ease of converting the dry premix to a fluxed thermoplastic mass and also enhances uniformity and product quality downstream of the compounding process.

Glass and Fields [44] reported the effects of porosity measured on commercial scale PVC resins and the means by which this could be accomplished; they observed that medium- to low-molecular-weight plasticizers were readily imbibed by the porous PVC resins, but high-molecular-weight plasticizers such as DTDP were more difficult to absorb into the resin pores and be completely dry-blended into the polymer matrix. McKinney [31] showed that in a copolymer resin the plasticizer absorption process is diffusion dependent; he demonstrated this by measuring changes in the glass transition temperature of the polymer as plasticizer moved into the amorphous regions of the resin grains. Thus, McKinney was the first to show that the resin grain actually becomes plasticized during the dry-blending operation. Park [28] reviewed the effects of plasticizer in the dry-blending process of PVC.

Uberreiter [46] showed that the dry-blending process was both kinetic and thermodynamic in nature. The filling of the particulate resin pores is essentially a kinetic process, controlled mainly by viscosity of the plasticizer, while the dry-blending process is not complete until the thermodynamics are satisfied with a compatible plasticizer. Wheeler [47] used a technique reported by Carleton and Mishuck [48] to characterize the state of plasticizer in PVC grains as a function of mixing time in commercial cycles with Henschel mixers and ribbon blenders compared to the laboratory method ASTM D2396. His work showed clearly that in the early stages of the mixing cycle, the plasticizer adsorption on resin particle surfaces occurs and can be reversed by centrifuging. Pore filling is also considered to be adsorption, but the centrifuge did not remove plasticizer from the resin pores or capillaries. Following adsorption, the second part of the cycle involves diffusibility of the plasticizer intermolecularly in the PVC amorphous regions. Wheeler's work showed that the absolute end of the dry-blending cycle occurs midway between the two significant breaks in the dry-blend curve as measured by ASTM D2396, which are characterized as "initial" and "final" dry-blend times. His work also gave a measure of the swelling that takes place in the resin grains as a function of plasticizer absorption. Defife [49] compared a number of techniques for measuring dry-blendability of PVC resins.

Overall, it should be recognized that the dry-blending process involves adsorption, including pore filling, followed by absorption which is the intermolecular imposition of plasticizer in PVC. Thus, the plasticizer choice is a significant variable in the dry-blending cycle due to the plasticizer's viscosity and solvating characteristics for PVC. Mixing temperature has a significant effect, but the dry-blending process is not complete until the plasticizer is imbibed intermolecularly with the PVC polymer. Commercial limitations remain today with systems that require very high levels of plasticizer and/or high-molecular-weight, weak solvating plasticizers.

C. Plastisols

A plastisol is a dispersion of PVC resin in plasticizer; if the plastisol contains a volatile diluent at more than 5% by weight, it is referred to as an organsol [50]. Plastisols and organosols are made by using special grades of nonporous, fine-particle-size PVC resins polymerized by emulsion or microsuspension processes. Although the surface area of these small particles is theoretically much larger than those of the suspension process PVC resins, the plasticizer is adsorbed and absorbed at a very

low rate, because of the low porosity of these particles. Commercial experience exposes one to complex rheological problems as the dispersion is
converted from liquid paste to a gel phase and ultimately to a molten thermoplastic mass. The rheological properties of these dispersions can be
critically influenced by every formulating ingredient, as well as by the
cycle used in mixing the paste. This chapter concentrates mainly on the
effects of plasticizer types and concentration in the area of plastisols. The
overall subject of plastisol/organsol technology is covered in Chapter 22
(Vol. 3).

Mixing and applications of plastisols entail concerns about several facets
of the dispersion. Every commercial application is concerned with one or
more of the following aspects of the plastisol: initial viscosity, viscosity
stability under environmental aging conditions, effects of shear (low- versus high-shear applications), gelation properties, and fusion of the complete formulation. In addition to basic properties, air release may be critical in order that inadvertent voids are not present in the finished product.
Conversely, deliberate entrainment of air is one way of making plasticized
PVC foams and is sometimes intentionally engineered into the formulation
(see Chapters 19 and 34). Plasticized PVC foams are generally prepared
by two different techniques: (1) chemically blown foam involves the use
of certain chemicals which decompose to form nitrogen, carbon dioxide, or
other gases, and (2) mechanically frothed foam involves the addition of
stabilizing agents which retain air as a gaseous foaming agent throughout
the gelation and fusion cycle of the plastisol. The technology of dispersions is also extended to include solid states, such as plastigels, which
are solid systems containing high levels of silica fillers having desirable
pseudoplasticity and high yield points. In addition, hydrogels are made
by dispersion technology, in which water plays a significant role as a
dispersing medium for silica fillers.

Obviously, the plasticizer is a significant variable in the study of PVC
dispersion technology. Plastisols have proved to be an excellent medium in
which the effects of plasticizer diffusibility and solvation characteristics
have been studied [51-56].

Plastisol viscosity is directly influenced by plasticizer viscosity. This
can be seen by comparing the viscosity of plastisols with direct substitu-
tions of low-viscosity plasticizers for high-viscosity plasticizers; in addition, if the plastisol viscosity/temperature curve is traced, one sees a reduction in viscosity at the initial stages of heating due to the reduction in
the viscosity of the continuous phase, namely the plasticizer per se. As
the heating continues, resin particles swell and the plastisol viscosity rises
sharply.

Plasticizer concentration directly affects the viscosity of the plastisol.
However, if one compares plastisol viscosity for different plasticizers at
an equal volume concentration of particulate solid (PVC resin), then it
appears that higher-viscosity, higher-molecular-weight plasticizers impart
lower plastisol viscosity. This is due to the peculiar nature of esters,
which have a lower specific gravity as molecular weight increases. Sears
and Darby [1] reviewed the aspects of rheology that apply to plastisols.
It is well for the plastisol designer to be familiar with the Mooney equation
relating the size, shape, and concentration of the particulate solids dispersed in liquid media:

$$\ln \eta_{rel} = \frac{K\phi}{1 - S\phi} = K\phi(1 - S\phi)^{-1}$$

where

η_{rel} = relative viscosity
K = size and shape factor of dispersed particulate
ϕ = volume concentration of dispersed phase
(particulate)
S = crowding factor, which is found to be approximately
equal to the reciprocal of the maximum volume concentration
attainable in the system

Lewis and Nielsen [57] published an excellent study in which they used glass beads dispersed in Arochlor to characterize the effects of particulate size and shape (K factor) as well as volume concentration on the rheological properties of dispersions. PVC resins readily agglomerate and deagglomerate under plastisol mixing and application conditions. This action causes significant variations in plastisol rheology, even when a given formulation is evaluated under apparently controlled conditions. These variations interfere with attempts to correlate laboratory and plant-scale performance. Plastisol testing permits generalizations only under rigorously controlled conditions of formulating reagents, mixing, and measurement practices.

Plastisol viscosity stability, gelation, and fusion properties are a function of the plasticizer diffusibility and solvency in much the same fashion as in dry-blending technology. It appears that the solvency predominates in controlling plastisol rheological and gelation characteristics [58]. Greenhoe [59] showed that there is no distinct correlation between gel temperature and fusion characteristics of plastisols, which suggests that diffusibility plays a role in the early part (gelation stages) of the plastisol curing cycle. Ideally, plastisol processors would like to have a plastisol based on a plasticizer system which is essentially inert to the PVC polymer during the storage and handling periods, but at the desired point of fabrication one would generally prefer a highly controllable and rapid gelation-fusion cycle. Plastisols designed for applications as chemically or mechanically blown foams require critically controlled rheology throughout the gelation and fusion cycles.

Graham and Darby [60] used microscopic measurements of plastisol cure cycles to determine solvency characteristics of plasticizers. They concluded that only plasticizers of high aromaticity produced plastisols capable of being fused at 150°C. At 170°C all plastisols developed sufficient physical properties to make performance comparisons, while at 200°C thermal decomposition of the polymer interfered with the measurement. As a generalization, they listed the following order of solvency for the various plasticizer families: phosphates > orthophthalates > isophthalates > terephthalates > adipates > azelates > sebacates. It is known, of course, that within these plasticizer families, the structure of the alcohol used in making the plasticizer is critical, as are other chemical and structural features. Frissell [61] arrived at a "semiquantitative" measure of solvency as contributed by specific chemical structures. Strong solvating action was contributed by oxirane oxygen, aromaticity, amide linkage combined with ester groups, and ester groups themselves. He pointed out that a diester plasticizer is significantly stronger in solvating power than a monoester and that adding a third ester group enhances the solvency characteristics of the plasticizer even more; he concluded that adding a fourth ester (tetraester) has a negative effect on solvency, however. Linear ether linkages and methylene groups were found to detract from solvency.

From the above, it is possible to separate the various chemical families according to general solvency characteristics and predict the effect of alcohol interchange within a family. The dialkyl phthalates (esters of an aromatic dibasic acid) show good solvency, which is reduced as molecular weight is increased above that of dihexyl phthalate. Butylbenzyl phthalate has exceptionally good solvency for PVC due to increased aromaticity and its solubility parameter δ which is about 8.5 to 9.9 compared to 9.6 to 9.7 for PVC [1].

Phosphates have reasonably good solvency, especially triaryl phosphates (CDP and TCP). Polyesters have fair to poor solvency, depending on molecular weight, chemical type, and impurities. Aliphatic hydrocarbon extenders are poor, and aromatic hydrocarbons are reasonably good in solvency, but naphthenic hydrocarbons are intermediate. Chlorinated hydrocarbons can vary significantly as a function of the starting hydrocarbon, purity, and degree of chlorination. Frissell [61] showed that chlorinated aliphatic hydrocarbons have maximum solvation for polyvinyl chloride at about 60% chlorination level, which incidentally is very close to the chlorine content of PVC—57%.

The relative solvency characteristics of the plasticizers discussed here were measured by the mimimum flux temperature test method [52] described in Section VII of this chapter (see Tables 13, 14, and 17—22). Although it is a static and empirical method for measuring plasticizer solvency, the minimum flux temperature test is a simple laboratory tool with good reproducibility, because the effects of essentially all influencing variables other than plasticizer structure are reduced to a minimum.

The solvency characteristics of plasticizer blends have been found to deviate from the calculated arithmetic average based on the separate components [62]. The deviation is shifted toward the stronger solvating components of the blend. Hence, plasticizer blends are commonly used for improving processability [63].

Critchley et al. [64] studied the effect of temperature on the fusion of plastisols as measured by the development of tensile strength. They examined several plasticizers ranging from phosphates and dibutyl phthalate as strong solvating types up to a poor solvator, ditridecyl phthalate. They found that it was necessary to define the end point at 90% of maximum tensile strength development in order to make valid and sharp comparisons; strong solvators such as tricresyl phosphate and dibutyl phthalate provided 90% fusion at 140°C, whereas ditridecyl phthalate required 165°C. Poppe [65] evaluated the factors of time and temperature while comparing the solvency properties of diisodecyl phthalate and diisononyl phthalate with those of DOP-plasticized plastisols in spread-coating applications. He found that for thin plastisol coatings, the duration of cure cycle (time) was not a critical factor, but a minimum critical temperature was required to attain adequate fusion properties. He concluded that DINP required a temperature about 2°C higher that DOP for fusion and DIDP required 7°C over the DOP control. Thus he concluded that commercial operations are primarily influenced by the temperature and heat convection characteristics of the oven as they affect heat transfer through the plastisol and substrate media. The plasticizer concentration did not appear to be a significant variable with respect to solvency or fusion temperature. Noting that most commercial ovens for DOP-plasticized systems are held at significantly higher temperatures than that required for DOP, he observed that DINP and/or DIDP can frequently be substituted at a cost benefit with little or no effect on the productivity cycle [66]. Werner [67,68] reported studies of various formulating additives which control plastisol rheology for coating of substrates, particularly with respect to controlling desired properties for porous substrates.

The subject of plastisol air release was studied in detail by Defife [69], who discusses the effects of plastisol viscosity as well as surface tension on the air release properties. He concludes that although higher plasticizer levels enhance air release, plastisol viscosity is not the primary controlling factor; the surface tension of the gas bubble formed as air is removed from the plastisol is the most critical factor. Daniels and Krauskopf [70] studied the effects of various air release and bubble stabilizing agents. They found that nonionic surfactants, such as polyethylene glycol monolaurate, increased deareation rates by about 20% when added to plastisols at 0.6 phr; silicone surfactants, however, significantly stabilized the plastisol air bubble when employed at 1 phr in the plastisol. It has also been found that diisononyl adipate appears to be a preferred low temperature/air release enhancing plasticizer compared to other commercially available adipates. Studies of air release effects of primary plasticizers have shown that diisononyl phthalate is traditionally equal to or better than di(2-ethylhexyl) phthalate with respect to enhancing air release [71]. Typical formulations for chemically blown and mechanically frothed foams are reported by Lasman and Scullin [72] and others [73,74].

D. Posttreatment

The versatility of plasticized PVC extends into the many posttreatment processes which may be applied during the fabrication of end-use products. One or more of the following processes are typically applied in posttreatment:

  Printing
  Top coating
  Plasma arc treating
  Cross-linking
  Adhesive coating
  Lamination
  Sealing—by means of solvent, heat, or ultrasonic energy

The criteria for selection of plasticizer with respect to posttreatments are that the plasticizer not interfere with the operation or affect end-use performance. This suggests maximized permanence of the plasticizer with the PVC polymer under the environment of the posttreatment operation and end use. Conversely, plasma arc treatment enhances the permanence of a plasticizer that would otherwise be transient [75—79]. Thus, the vinyl formulator considers the posttreatment process in view of the end-use requirements. The options, of course, are directed at selecting the optimized cost/performance plasticizing system.

Historically, the application of adhesive layers for pressure-sensitive vinyl tapes and adhesive-backed products, such as floor tile, demanded specific formulating to retard migration of the plasticizer into the adhesive. In particular, pressure-sensitive vinyl tapes, such as electrical tape, required the use of costly polymeric plasticizers [80,81]. Because of the significant technological advances in adhesive formulations and systems, it is now possible in some cases to use conventional dialkyl phthalate plasticizers in the flexible PVC formulation without encountering the traditional plasticizer migration which destroys the performance of the adhesive.

Solutions of vinyl/acrylic polymer blends have been applied as topcoats for many years. This technique provides a protective barrier to reduce plasticizer loss through rub-off mechanisms and has satisfied the needs for flexible PVC furniture and automotive upholstery.

The practice of laminating and sealing by means of heat or ultrasonic energy has been used commercially for a long time. It is preferred that the two materials be formulated to have similar melting temperatures so that there is an optimized condition for permanently sealing together the two different vinyl systems. The bringing together of dissimilar systems can lead to end-use difficulties even though fabrication by the sealing process is apparently successful. "Crinkly" bookbinder film is an example of end-use failure caused by the migration of dissimilar plasticizers from one film layer into the other. The relocation of the different plasticizers causes expansion of one film while the other contracts, as found by Ellis [82].

Other investigators have reported on techniques for controlling the problem of plasticizer migration when dissimilar systems are brought together [83—86]. The effects of plasticizer migration are discussed more completely in Section IV,B.

PVC can be cross-linked by electron beam or peroxide-catalyzed methods. A cross-linkable monomer, such as trimethylolpropane triacrylate, is incorporated into the PVC formulation to accomplish this. Telephone hook-up wire is an example of a commercial use. Cross-linking may also be applied to hot-compounded systems to improve dimensional stability, toughness, and thermal and solvent resistance.

IV. PLASTICIZER EFFECTS ON END-USE PERFORMANCE

Plasticizer type and concentration are the most critical formulating variables in optimizing cost and end-use performance properties of flexible PVC. This section reviews those options, recognizing that generalizations seldom apply to all situations. The general approach for the formulator is first to consider low-cost, general-purpose plasticizers in light of the end-use needs and the processing capabilities at hand. Plasticizer blends frequently meet the optimized needs, while in other cases the end-use limits the choice to a specific plasticizer. Some appreciation for relative demands may be gained by observing the market share satisfied by the various types of plasticizers. Phthalates are the most broadly used plasticizers, and nearly half of the phthalate volume is made up of those considered as general-purpose (GP) plasticizers.

A. Mechanical Properties

Mechanical properties are those which are measured under stress/strain conditions in order to establish the relationship between the applied strain and the resultant stress, or vice versa. In most cases, a resultant stress is measured as a function of an applied strain (deformation). Mechanical properties may be measured under three modes of applied strain: extension, compression, and shearing. But most measurements—and end uses—simultaneously impose all three of the modes to various degrees, which contributes to the complexity of mechanical properties. Rosen [310] has given an introductory discussion of polymer mechanical properties and their measurement.

Figure 2a represents the classic curve for the modulus of amorphous polymers as a function of time and temperature. Rosen [310] used a publication by Tobolsky and Catsiff to make the point that all amorphous polymers exhibit the four regions of state shown. The exact shape and position of the curve is a function of the polymer itself (chemical structure and morphology), and the time-temperature conditions of measurement. The usual practice is to maintain either temperature or time as a constant while measuring the dependent modulus. Figure 2a shows modulus as a function of time along with a temperature shift factor. Combining the two results gives a continuous curve of modulus as a function of time scale. It should be noted that impact tests represent very short time factors, whereas flexural and tensile tests generally involve lower load rates. In the case of plasticized PVC, the following three independent variables have effects that are characteristic of increased molecular free volume in the polymer matrix:

> Increased plasticizer concentration (phr)
> Increased temperature
> Increased time (i.e., a longer time span between the initial strain and
>   the point of stress measurement, or a lower rate of strain load)

Shift factors may be applied to the time effects, which results in a continuous modulus/temperature curve as shown in Figure 2a.

The classical work of Clash and Berg [311] investigated the effects of plasticizers in vinyl polymers by measuring apparent modulus of rigidity at constant load rates (time) as a function of temperature. Figure 2b was developed by Sears and Darby [1] from work by Clash and Berg [311]. It shows that increased plasticizer concentration shifts the modulus/temperature curve to the left on the temperature scale and changes the shape of the curve only slightly. Figure 2c shows that different plasticizer types impart significantly different shapes to the curve, i.e., low-temperature versus high-temperature effects, as well as the position of the modulus/temperature curve. Figure 2c shows that the plasticizing efficiency of tri(2-ethylhexyl) phosphate (TOP, or TOF) is significantly better than that of both DOP and tricresyl phosphate (TCP) at low temperatures. At room conditions, all three plasticizers appear to be similar in efficiency, as measured by stiffness moduli. Note, however, that the assessment of room temperature plasticizing efficiency by Durometer hardness shows TOP and DOP to be similar while TCP is significantly less efficient.

Figures 2b and 2c also show that brittleness temperatures, as determined by ASTM D746, do not correlate with T_f, the Clash-Berg low-temperature flexibility value. Clash and Berg made a somewhat arbitrary decision that the end of flexibility may be defined as the point at which the apparent modulus of rigidity is 135,000 psi, as measured by the procedure they used (ASTM D1043), but the curves do not indicate a distinct phase change at the T_f temperature. Brittleness point, as measured by ASTM D746, and T_f can only be correlated by developing appropriate shift factors reflecting the different stress rates applied in the two different test methods. Since this has not been done, the plastics industry has adopted the practice of characterizing low-temperature

properties by measuring both low-temperature flexibility, T_f, at a low
load rate and impact resistance, T_B, at a high load rate.

Both plasticizer type and concentration in polyvinyl chloride significantly alter the glass transition and melting temperature [87–91], which
are also relatable to free-volume concepts. Practical aspects of controlling
mechanical properties have been widely reported [1,92,93]. The chemical
structure of the plasticizer alters its effect on various mechanical properties, as well as the magnitude of the effect at any given concentration.
Table 3 and Figures 3 through 8 present comparisons of different plasticizers as a function of plasticizer concentration in commercial grade PVC.
Table 3 compares three different plasticizer types based on isononyl alcohol: phthalate, adipate, and trimellitate. Increased concentrations
(phr) cause softer, more flexible PVC with improved low-temperature
properties. The different types of plasticizers, with different organic
acid moieties, show different degrees of effectiveness in changing the
properties listed. The degree of plasticizer effectiveness is the characteristic defined as "efficiency."

Plasticizing efficiency must be defined quantitatively to facilitate the interchange of one plasticizer for another. Plasticizer efficiency may be defined as the amount of plasticizer required to accomplish a specified change
in a specified property of the PVC polymer. This definition reflects the
fact that the primary role of a plasticizer is to impart changes to the properties of the base, unmodified polymer. In practical usage, efficiency is
a means by which one can substitute one plasticizer for another on a quantitative basis while maintaining a specified desired property. Efficiency is
usually defined around certain mechanical properties, such as Durometer
hardness, at standard conditions. DOP is the traditional reference point,
hence the terminology "DOP equivalency." Tables 3 through 22 allow for
a comparison of plasticizing efficiency on the basis of Durometer A hardness. In most cases, performance properties are shown at an adjusted
level to provide Durometer A equal to 80, a typical value for many plasticized vinyl products. Some of the plasticizers are not shown at 80 Durometer A, but the performance characteristics at a concentration of 50 phr
provide some indication of relative plasticizing efficiencies. Durometer
hardness has been selected for evaluating plasticizing efficiency because
the test has reasonably good reproducubility and is relatively easy to conduct. Plasticizer efficiency may also be defined in terms of other properties such as modulus, elongation, damping action (effects on the dynamic
storage and loss moduli), or low-temperature properties.

The film and sheeting industry is noted for its empirical evaluation of
efficiency by "hand." The finite quantitative comparison of efficiency depends on the particular mechanical property being considered as well as
the overall plasticizer level used. Hence, the expression for efficiency
must include the definition of these parameters. Although it is possible
to make generalizations concerning plasticizing efficiencies, on occasion
it is necessary to restrict those generalizations to the specified property
being considered.

It is a common misconception that plasticizing "efficiency" correlates
with the solvating strength of the plasticizer. Two obvious examples that
refute this generalization are apparent by comparing di(2-ethylhexyl) adipate (DOA) with DOP, and TCP with TOP. Comparing Tables 13 and 18
shows that DOA has poorer solvating characteristics than DOP (difference
of 14°C as measured by minimum flux temperature), yet DOA is about 17%

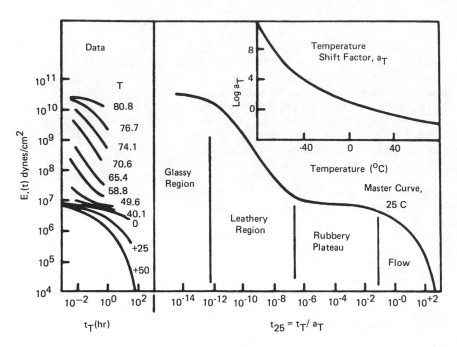

FIGURE 2a   Polymer modulus as a function of time-temperature superposition. (From S. L. Rosen, *Fundamental Principles of Polymeric Materials*, p. 273; reprinted with permission.)

more efficient than DOP in lowering the room temperature hardness of the vinyl composition. Table 19 shows that TOP is a poorer solvating plasticizer than TCP by about 39°C. Durometer hardness, however, shows that TOP is about 24% more efficient than TCP. On the other hand, the data tables contain numerous examples in which plasticizing efficiency and solvating characteristics appear to be correlated.

Performance characteristics of vinyl plasticizers are conveniently evaluated over a range of plasticizer concentrations, such as 25 to 90 parts by weight per 100 parts of resin (phr). Resultant properties can then be plotted as a function of plasticizer concentration. This provides further enlightenment, because it becomes apparent that the performance curves vary in slope as well as location on the grid as a function of plasticizer interchange. Hence, it becomes necessary to qualify the desired efficiency in such a manner as to indicate the general plasticizer level, that is "semi-rigid" to "very flexible." Table 3 and Figures 3 through 8 show the mechanical properties of vinyl compositions with various levels of commonly used plasticizers.

Of the classical plasticizer families, the diesters of aliphatic dibasic acids stand out as having unique effects on mechanical properties. This family, known as the "low-temperature" family, provides outstanding low-temperature flexibility and impact resistance to flexible vinyl, when formulated to specified room temperature hardness and modulus values. The other chemical plasticizer families have much less an effect on low-temperature properties, and the effect of alcohol structure is more apparent in the influence of their low temperature properties. For instance, linear alcohol

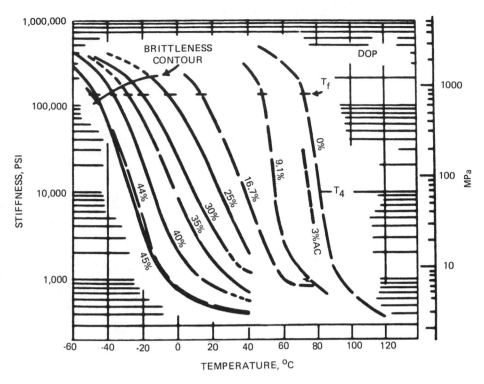

FIGURE 2b   Modulus temperature properties of PVC as a function of plasticizer concentration.   (From Ref. 1, p.309, with permission.)

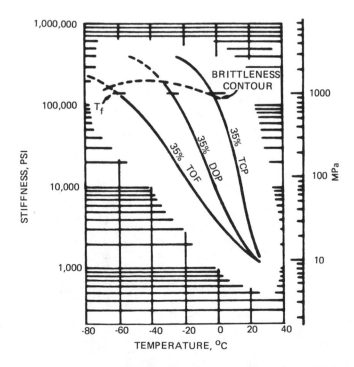

FIGURE 2c   Modulus temperature properties of PVC as a function of plasticizer type.   (From Ref. 1, p.310, with permission.)

TABLE 3   Effect of Plasticizer Concentration in PVC for Typical Plasticizers Based on Isononyl Alcohol

Property	Di(isononyl) phthalate			Di(isononyl)adipate			Tri(isononyl) trimellitate		
	phr-30	50	70	30	50	70	30	50	70
Compound specific gravity	1.271	1.221	1.185	1.250	1.194	1.154	1.274	1.225	1.189
Durometer A	96	84	72	95	81	68	>100	89	78
D	54	<35	<35	48	<35	<35	61	42	<35
100% Modulus, psi	3050	1820	1050	2720	1530	897	3330	2200	1400
Tensile strength, psi	3740	2970	2420	3710	2780	2220	4150	3050	2290
Elongation, %	279	331	377	286	345	383	275	327	370
Breaking energy, ft/lb	15	12	11	13	9	7	11	13	7
Clash-Berg, T_f, °C	+6	-24	-40	-14	-51	-70	+13	-17	-36
Brittleness, °C	-9	-33	-44	-39	-64	-73	-5	-27	-43
SPI volatility, wt %	0.6	1.4	1.8	9.9	10.0	14.7	0.4	0.5	0.6
Soapy water extraction, wt %	1.5	2.2	2.5	—	—	—	0.4	0.4	0.4

Source: G. D. Harvey, et al., unpublished works, Exxon Chemical Co., Baton Rouge, La. (1980–1985).

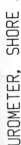

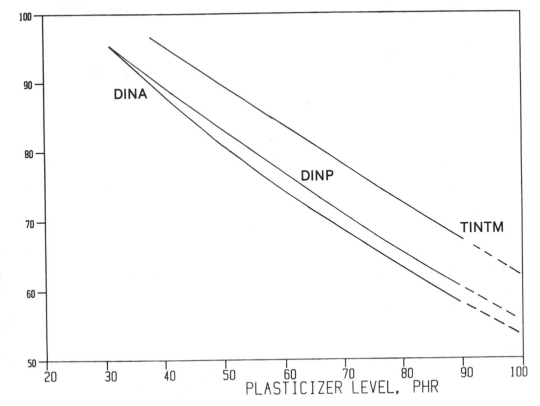

FIGURE 3 Hardness as a function of plasticizer level.

structures enhance both plasticizing efficiency and low-temperature properties. Higher-molecular-weight alcohols impart lower efficiency at room conditions, demanding more plasticizer to meet specified room temperature properties; this results in improved low-temperature properties by virtue of higher
plasticizer levels. This presents the classical choice to the formulator: which
is most cost effective, enhanced low temperature through the use of costly
aliphatic diesters, or higher concentrations of low-cost, less efficient plasticizers? The answer usually lies in a compromise between low-temperature
and specified room temperature properties.

Most commercial vinyl products contain a mixture or blend of plasticizers to provide the optimum compromise among desired performance
characteristics for given applications. Mechanical properties of the compositions using plasticizer blends fall within expectations, that is, intermediate between the properties provided by the separate components.
Thus, the mechanical properties provided by a plasticizer blend, or the
design of a plasticizer blend to provide desired properties, can be calculated provided that a given efficiency level is defined [94-96]. Tang
and Harris [97] successfully applied the technique of formulating with
plasticizer blends to a computerized system. Their work showed that the
calculated properties agreed well with laboratory-prepared samples based
on the computerized formulations.

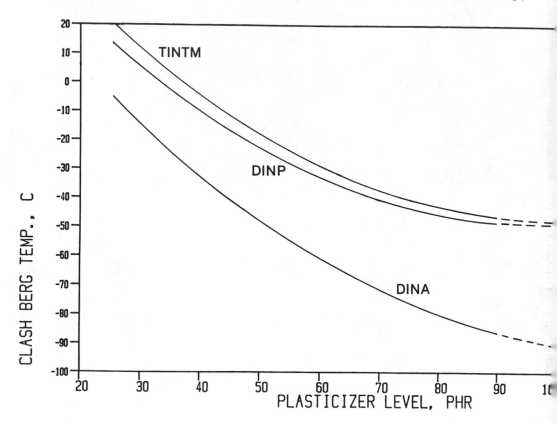

FIGURE 4 Clash Berg as a function of plasticizer level.

The effect of the molecular weight of the vinyl resin on the efficiency of a plasticizer is shown in Figure 9 [98]. Increasing molecular weight causes a general improvement in mechanical properties such as tensile strength, elongation, and low-temperature impact resistance. The efficiency correction, based on Durometer hardness, requires increased plasticizer concentration with increased molecular weight of the resin. This leads to flattening out of the tensile curve and further improvements of elongation and low-temperature impact resistance with the use of higher-molecular-weight polyvinyl chloride resins.

Copolymers, such as poly(vinyl chloride/vinyl acetate), lead to an apparent increase in plasticizing efficiency; that is, they require lower plasticizer levels to attain a specified hardness or modulus property at room conditions. This normally leads to a loss of low-temperature properties at the adjusted plasticizer level. Mechanical properties of plasticized polymers do not change in a straight line as a function of increased comonomer. In fact, depending on the comonomer, plasticizer incompatibility becomes apparent. Studies of dynamic modulus and stress relaxation of plasticized copolymers have led to the conclusion that the plasticizing mechanism through copolymerization is not the same as the mechanism with externally added plasticizers [99].

Frissell [100] has studied the effects of filler level on plasticizing efficiency. As expected, increased filler requires an increase in plasticizer

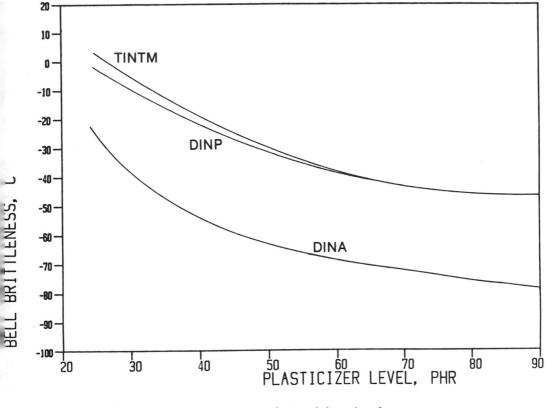

FIGURE 5 Brittleness as a function of plasticizer level.

level to retain mechanical properties of the unfilled vinyl composition. This becomes more difficult to accomplish as the filler level increases. The correction factor for the adjustment of plasticizer level due to filler concentration varies significantly, depending on the mechanical property being considered [101,102]. The correction factors in Table 4 show the additional plasticizer required, expressed as weight percent of the filler, to maintain the specified property of the unfilled formula. For instance, flexural and torsional modulus values are only slightly influenced by filler and can be maintained by additional plasticizer in the amount of 6 to 7% of the filler level (phr). Low-temperature impact resistance (brittleness temperature), however, requires 25 to 32% additional plasticizer, based on the filler weight used, and Durometer hardness requires 11 to 16 percent. Tensile strength decreases with increased filler loading. The required adjustment in plasticizer level is similar to that shown for brittleness failure, but in the opposite direction. This, then, is a formulating compromise that arises for filled flexible PVC products.

B. Permanence Properties

With very few exceptions, plasticizers do not chemically bond to the vinyl polymer. The plasticized vinyl composition is in a dynamic state in which

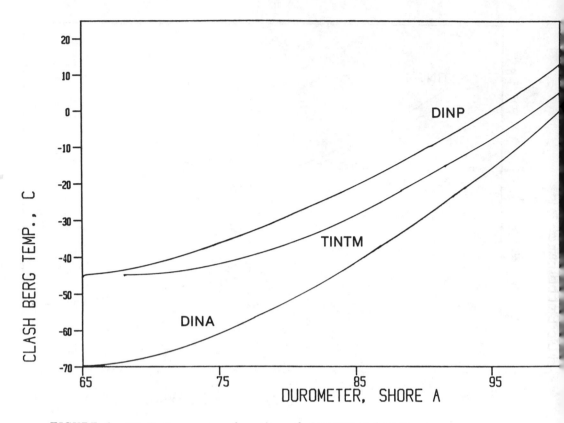

FIGURE 6   Clash Berg as a function of durometer hardness.

polymer and plasticizer molecules are continuously undergoing association
and separation.   Hence, the tendency of the plasticizer to be removed from
plasticized compositions is very real.   Although one actually measures the
fugitive nature of the plasticizer in the polymer, the subject is generally
referred to as permanence characteristics.   Permanence of a plasticizer is
critically affected by its chemical structure and the concentration used when
compounded into the vinyl polymer.   Other formulating variables, as well
as specimen dimensions and conditions to which the vinyl is exposed, significantly alter the plasticizer permanence.   Quackenboss [103,104] studied
the mechanism of plasticizer loss as a function of the conditions to which
the vinyl is exposed.   Others [105–108] have studied various plasticizers
as well as various concentrations and the types of losses experienced.

It is essential to understand the factors controlling the mechanism of
plasticizer loss in order to discuss the variables associated with it.   The
two primary factors controlling the loss of plasticizer are (1) the rate at
which the plasticizer can migrate to the surface from the inner part of the
vinyl, referred to as the diffusion rate, and (2) the rate at which the plasticizer leaves the surface.   To demonstrate, the sketch in Figure 10 represents an edgewise view of a piece of vinyl film.   The film faces are represented by the lines, and the space between the lines represents the interior of the vinyl.   The arrow labeled A represents the rate of plasticizer
diffusion through the vinyl composition, and the arrow labeled B represents

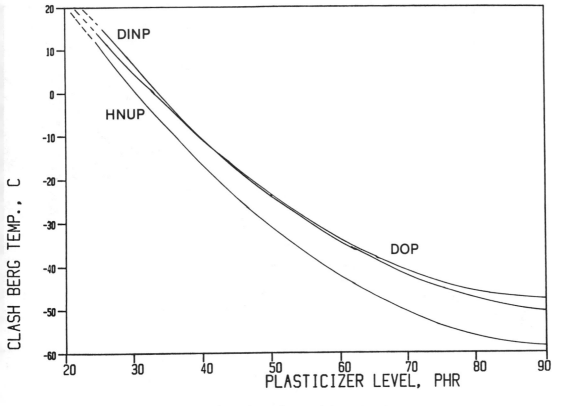

FIGURE 7   Clash Berg as a function of plasticizer level.

the rate at which the plasticizer leaves the surface of the film. The total loss of plasticizer from any vinyl composition is primarily influenced by either loss rate A or loss rate B. Obviously, just as in a chemical reaction, the determining rate will be the slower one. Thus, in exposure under which the rate of plasticizer loss from the surface is high (rate B), plasticizer removal is a diffusion-controlled type of loss. Plasticizer losses due to oil extraction and rub-off are typical examples of conditions under which the diffusion rate (A) is the controlling (lower rate) factor. This is the case because the plasticizer molecules are immediately removed either by the oil or by the rubbing medium as soon as they reach the surface of the film, and Fick's second law of diffusion (a second-order differential equation) holds:

$$\frac{dC}{dt} = D\frac{d^2C}{dx^2}$$

where

C = plasticizer concentration
t = time
D = diffusibility constant
x = length of path traveled by mobile plasticizer

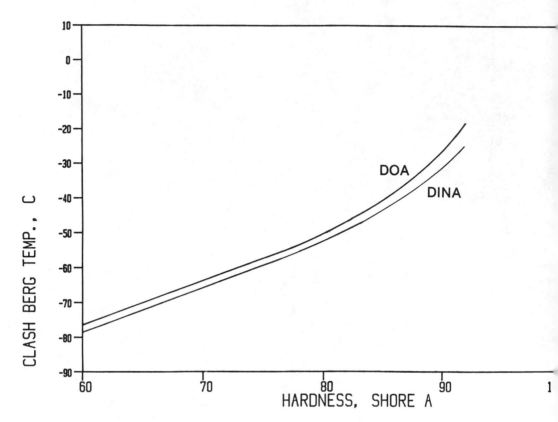

FIGURE 8   Clash Berg as a function of hardness.

It is possible to measure changes in the average values of C, t, and x in plasticized PVC specimens exposed to conditions where diffusibility is the controlling factor for plasticizer losses. From this, one can determine the diffusibility constant, D, as a function of plasticizer type and concentration. These values are shown as "K", in Tables 13, 14, and 17 through 22, as determined by the Reed et al. [107] oil extraction procedure at 50°C.

In exposures where rate A is very high or, conversely, the loss from the surface of the film is very low, the diffusion rate A is of minor significance and rate B is the determining factor. Examples are volatile losses and extraction by aqueous media. Thus, as Quackenboss has shown [103,104], plasticizer permanence characteristics can be classified as either diffusion-controlled (A) or surface-controlled (B) losses.

Plasticizer losses increase significantly as a direct function of plasticizer concentration, irrespective of the chemical type or the controlling loss mechanism. Wartman and Frissell [108] found that a deviation from expected loss rates was related to the compatibility of the plasticizer in the polymer.

Sears and Darby [1] review the numerous studies of the complexities associated with determining plasticizer permanence in polyvinyl chloride. This section will review the major parameters which significantly influence plasticizer permence and its measurement.

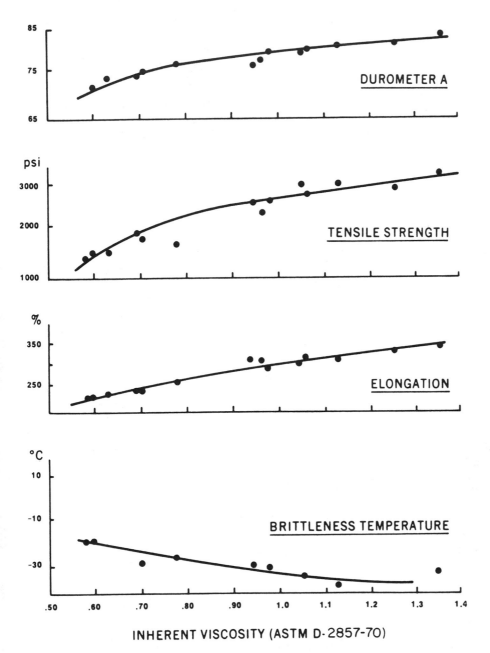

FIGURE 9 Mechanical properties vs. molecular weight of PVC at 50 phr DOP plasticizer. (From P.T. McCoy, Union Carbide Corp., unpublished work, 1964.)

TABLE 4   Plasticizer Adjustment Dependent on Filler Level

| | ASTM | Additional plasticizer required, wt % of filler level used | |
		DOP[a]	DINP[b]
Durometer hardness, A	D-2240	16	11
Tensile strength, psi	D-882	-36	-28[c]
Elongation, %	D-882	45	51
Flexibility, [E], T_4, T_f	D-747, D-1043	7	6
Low-temperature impact, °C	D-746	32	25

[a]From Ref. 308.
[b]From Ref. 309.
[c]For *tensile* there is a decrease with the addition of both filler and
plasticizer; therefore, to maintain a given tensile with added filler
it would be necessary to *remove plasticizer* from the compound. The
minus sign denotes removal of plasticizer

1. Compatibility: Compatibility has a major effect on plasticizer permanence because it is mainly controlled by the plasticizer-polymer interaction forces. This characteristic is a result of the fact that ideal plasticizers are only partial solvents for the polymer. For if the liquid plasticizer were a good solvent for the polymer, as is cyclohexanone for PVC, the
polymer-polymer and polymer-plasticizer forces would be sufficiently weak-
ened to preclude a gel structure, and the result would be a solution of polymer-solvent molecules.

The plasticizer-polymer interaction forces vary as a function of the
relative ratio of plasticizer to polymer and the conditions under which they
are measured. Thus these forces influence processability as well as permanence of the plasticizer-polymer combination. Predictors of plasticizer-
polymer compatibility generally rely on numerous attempts to apply indirect
measurement of properties that influence these forces. Unfortunately, in
the practical world of formulating plasticized PVC, the numerous additives
and reaction by-products incorporated in the PVC formulation significantly
interfere with the direct application of theoretical calculations and considerations.

Although the old adage of "like dissolves like" is somewhat useful with
respect to characterizing solvating groups of a plasticizer, the direct application of this concept is limited by the fact that plasticizers are large
molecules associated with macromolecular structures of polymers. It is
known that certain plasticizer-polymer solubilities show an adverse temperature effect, and likewise these interactions deviate significantly as a
function of the volume concentration of plasticizers in polymers. Wales [109]
pointed out that traditional phase diagrams are not completely useful because the plasticizer-polymer phase separation is a syneresis, that is, separation of a liquid from a gel by a contraction mechanism. Further, it is

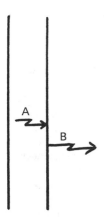

FIGURE 10 Edgewise view of vinyl film.

obvious that the solvating groups of a plasticizer are generally located internally rather than on the ends of the plasticizer molecule. Thus, the compatibility is controlled by the solvating groups along with the significant influence of the apolar groups providing a separation of polymer molecules.

Numerous investigators have concentrated on the measurement of plasticizer solvency and polarity as useful estimates of compatibility in PVC resins. Frissell [61] measured the effect of certain solvency groups in plasticizer structures. Anagnostopoulos et al. [110] reported the development of a microtest for measuring the compatibility of plasticizer in PVC. This method permitted quantitative measurement of a total plasticizer on a single particle of PVC resin as a function of temperature. Greenhoe [59] used a plastisol system to define the solvency as a function of different plasticizers; he recognized six transition points as the plastisol was heated to convert it from the liquid state to a fused flexible vinyl having useful mechanical properties. He showed that the final fusion states, elastomeric and fusion points, were significantly influenced as a function of plasticizer selection with respect to the temperature at which these states were reached; however, these points were not significantly different as a function of plasticizer concentration or time of exposure. His work suggests, however, that in the early part of the plastisol fusion curve, the effect of plasticizer is influenced significantly as a function of time and plasticizer concentration. Thus, the early part of the plasticizer take-up is a diffusion controlled mechanism. Su and coworkers [111] used gas-liquid chromatography (GLC) to determine thermodynamically the compatibility of di-n-octyl phthalate (DNOP) in PVC as a function of volume concentration over the temperature range 100° to 130°C. They concluded that at approximately 0.40 volume concentration (about 47 phr by weight); the heat of mixing was negative. Darby and co-workers [112] reported a practical technique for predicting plasticizer compatibility through the measurement of dielectric constant and solubility parameters. The Sears and Darby treatise [1] also shows how well a three-dimensional model of butylbenzyl phthalate fits in the network of a helical model of the PVC molecule; but in another part of that text it is reported that DOP shows better compatibility than butylbenzyl phthalate.

Many investigators have recognized gross incompatibility of plasticizers with PVC when they encountered difficulty in accomplishing fusion in a laboratory milling operation or difficulty in developing the ultimate mechanical strength properties even though the material appeared
to be appropriately prepared and fused. Gross incompatibility may also
be recognized by the appearance of haziness in the fused vinyl specimen.
From a practical point of view, one desires to predict compatibility of the
plasticized PVC matrix under end-use conditions. This is a difficult task.
Under most end-use conditions, a plasticized PVC system can remain compatible for a long time, but it may ultimately phase-separate. Laboratory
tests have been designed to accelerate the incompatibility. Unfortunately,
accelerating techniques require extension into an environment or set of
conditions which may not realistically predict those encountered in the
end-use application. Two such tests have been defined by ASTM as a
means of anticipating the effects of compression (ASTM D3291) and the
effects of high humidity (ASTM D2383).

The above procedures, as well as numerous variations thereof, are
useful for estimating the compatibility of a completely formulated vinyl
composition. Hence, a "go/no go" measure of compatibility is made on
the total plasticized PVC system, not knowing which of the formulating
variables is primarily responsible for observed incompatibility. Generally,
specimens that sustain the ASTM test conditions for 2 weeks without exhibiting spew are considered to have useful and acceptable compatibility.
Experience has shown that the elevated temperature and humidity test is
generally the most severe exposure for most plasticized vinyl compositions.
This is probably because the water upsets the balance of solubility parameters and hydrogen bonding and the polarity of the components employed
in the vinyl matrix under conditions realistically experienced in applications
of plasticized PVC.

Attempts to measure compatibility, or incompatibility, of fused vinyl
plaques quantitatively have not been too successful. Reed [113] used a
compressive force on a semispherical glass bead placed on top of a vinyl
specimen. His objective was to determine the total force (load × time)
to cause spew in test specimens. Poor reproducibility and absence of
correlation of test results with experience invalidated the approach. Hecker
and Perry [114] investigated the premise that compatibility reflected the
relative strength of the polymer-plasticizer interaction forces. They assumed that the more compatible plasticizers would cause greater swelling
of a polyvinyl chloride specimen immersed in the plasticizer, and they investigated a practical means for quantitatively measuring the compatibility
of various plasticizers. Apparently, competing mechanisms were interfering, for some results were obtained that were not entirely consistent with
experience. It is known that the resulting compatibility parameters (solubility, polarity, and hydrogen bonding) of mixtures of plasticizers may
or may not be significantly different from the parameters of the pure plasticizer components.

Basic studies of plasticizer compatibility with PVC have centered on the
Flory-Huggins approach of determining the solubility parameter [112,115,116]
Van Veersen and Meulenberg [117] showed that the ratio of apolar to polar
carbons provides a means of calculating the effect of chemical structure of
aliphatic esters on compatibility. This method is not effective, however,
for aromatic, epoxide, and phosphate plasticizers.

The dialkyl phthalate family of plasticizers is known for its good compatibility with vinyl resins over a wide range of concentrations. The alcohol selection does, of course, influence the degree of compatibility. Studies by Doty and Zable [118] and Walter [119] indicated that maximum compatibility of the phthalate plasticizer was found with dihexyl phthalate and dibutyl phthalate, respectively. Considering these works together, it appears that dipentyl phthalate represents the phthalate plasticizer of maximum compatibility. Compatibility decreases from this point as the alcohol chain length is decreased or increased. Ditridecyl phthalate represents the highest-molecular-weight phthalate that can be considered compatible at conventional use levels in typical commercial grade polyvinyl chloride. Dilauryl phthalate (normal C_{12} alcohol) is not compatible at conventional use levels. From this and other phthalate homologs, it is apparent that increased branching of the alcohol moiety provides increased compatibility when isomers are compared. Triesters, such as esters of trimellitic anhydride, show fairly good compatibility, but less than corresponding phthalates. Aliphatic diesters, monoesters, and plasticizer extenders have only limited compatibility with polyvinyl chloride. Chemical groups that enhance the solvency for polyvinyl chloride, that is, the plasticizer-polymer interaction, are generally believed to enhance compatibility, while groups that interfere with solvency reduce compatibility. Hence, aromatic, oxirane, ester, and halide groups enhance compatibility in polyvinyl chloride, but noncyclic ether linkages and hydroxyl and methylene groups reduce compatibility. Compatibility characteristics of plasticizers with copolymers of polyvinyl chloride are significantly influenced by the comonomer type. Polyvinyl butryal, for instance, is highly receptive to plasticizers with ether groups. [120]

The vinyl formulator may take advantage of plasticizers having limited compatibility with vinyl resins. While characteristics such as solubility parameter and dielectric constant may indicate poor compatibility for a given plasticizer, it is possible that the plasticizer can be used at some limited concentration to accomplish a desirable secondary effect. Aliphatic diesters, which can have poor compatibility as the sole plasticizer in polyvinyl chloride, are commonly used in admixture with phthalate plasticizers to provide improved low-temperature properties while maintaining a compatible composition. In some applications, the aliphatic diesters are used as the sole plasticizer and maintain sufficient compatibility because the total plasticizer level is low and the required life of the product is short.

2. Diffusibility, Permeability, and Extractability: Studies of fundamentals verify that plasticizer diffusibility, permeability, and extractability are a function of molecular mobility. Thus, the diffusibility is inversely influenced by strong plasticizer-polymer interaction, but it increases as a direct result of increased plasticizer concentration and temperature. Knappe [121] found that Fick's second law holds when the diffusibility is constant. This is the case over +20° to +50°C as shown by Kikkawa [122], but at >100°C his comparison of DOP, DOA, and tritolyl phosphate showed diffusibility to be nearly constant and independent of plasticizer type, and at 20° to 50°C the diffusibility was found to be significantly dependent on the type of plasticizer. Sears and Darby [1] studied the results reported by Chamberlain and Harrison [123], who measured the migration of various plasticizers from 50 phr plasticized PVC samples into an intimately contacted unplasticized PVC specimen. Plots show a straight line function of weight loss versus the square root of time, indicating diffusion-controlled migration. This study allows one to evaluate the effects of different plasticizers and the effect of alcohol moiety within plasticizer families. Of the 11 plasticizers studied, the linear dialkyl phthalates showed a diffusibility nearly twice that of DOP. Diisononyl phthalate (DINP) and diisodecyl phthalate (DIDP) have diffusibility constants that are 60 and 75% less that that of DOP, respectively; ditridecyl phthalate (DTDP) has a diffusibility rate 98% less than that of DOP. Tables 13, 14, and 17 through 22 compare the diffusibility of various plasticizers as measured by extraction with mineral oil according to the prodedure reported by Reed et al. [107] and Quackenboss [103,104]. When compared at a constant plasticizer level, 50 phr, the results in Tables 13 and 14 verify those reported by Chamberlain and Harrison. Table 14 also allows a comparison of diffusibility of various phthalates formulated to a constant Durometer hardness, that is, with the less efficient plasticizers used at higher concentrations to match a specified Durometer A of 80. Those comparisons show that the increased plasticizer concentration overwhelms the plasticizer structural effects within a given chemical family when the plasticizer loss is diffusion-controlled. Tables 13, 14, and 17 through 22 provide an opportunity to compare various chemical families. Low-temperature-type plasticizers, diesters of aliphatic dicarboxylic acids, are notoriously high in diffusibility, whereas more highly branched and bulky plasticizers such as polyesters have very low diffusibility as measured by the oil extraction technique. A comparison of these two families with dialkyl phthalates supports two of the base theories typically applied to understanding the mechanism of plasticized PVC. The lubricity theory suggests increased flexibility due to weak associations of the polymer-plasticizer molecules, whereas the gel theory suggests strong association between plasticizer and polymer in a three-dimensional network. Further comparisons can be made by observing the oil extractability within the families of epoxides and phosphates. In the epoxide family, the linear-like monoester of epoxidized ester of tall oil has a high diffusibility, whereas the bulky epoxidized soybean oil, a triester, has a low diffusibility constant; in the phosphate family, TOP imparts good low-temperature properties and also has a high diffusion constant, K, as compared to the more aromatic and stronger solvator TCP.

Experimental work involving practical applications indicates the complexities involved as a function of the environment and the time span during which the specimen is exposed. Under these conditions, it is more difficult to generalize about plasticizer effects due to the penetration of foreign substances, which, when given enough time, interrupts the plasticizer-polymer interactions and in general upsets the solubility relationships of the neat plasticized PVC system. Some reagents, such as aromatic hydrocarbons, which have a tendency to slightly solvate the PVC polymer, cause a swelling of the network, generating a new environment. Highly polar materials such as water or low-molecular-weight alcohols can emulsify the plasticizer and generate new diffusion patterns. Park and Van Hoang [124] report a method for measuring diffusion of additives and plasticizers with the use of a radioactive ingredient. Hasson et al. [125] studied the migration of plasticized PVC in contact with pitch and reported that plasticizer losses are not diffusion controlled, but rather follow the equation of William, Landel, and Ferry. Dutta and Graf [126] describe a method for measuring the increase in plasticizer concentration on the surface of flexible PVC, when exposed under static room temperature conditions, by the use of infrared spectroscopy; they report that their results are consistent with a diffusion-controlled mechanism.

Because plasticized PVC is applied in a diversity of end uses, plasticizer diffusibility and extractability have been studied widely in various end—use media. The studies include investigations of plasticized PVC designed for food processing equipment, food packaging, medical devices, and many reagents simulating these environs [127—138]. Most of these investigators acknowledge that plasticized PVC is quite hydrophobic, and in a static exposure to water it is difficult to measure quantitatively the absorption of water and/or extractability of plasticizers. Attempts to duplicate end-use conditions demonstrate the effects of time, turbulence, plasticizer emulsification, and combinations of different reagents simultaneously contacting the plasticized PVC surface. Some generalizations from these studies support the principles that plasticizer-polymer interactions reduce losses, as do steric effects and the potential for plasticizer solubility in the surrounding medium. Thus, if one accepts dialkyl phthalates as the general-purpose type, it is generallly found that the more highly branched phthalates are preferred over linear types when migration and extractibility resistance are of concern. Aliphatic diesters and other low-temperature plasticizers have very poor resistance to loss by diffusibility; polyester plasticizers are preferred for extraction resistance when entirely diffusion controlled, for instance, when the flexible PVC is intended for exposure to hydrocarbon-type oils. However, polymeric plasticizers have poor resistance to aqueous extractants.

PVC polymer alloys [136] have been reported to have exceptionally good resistance to extractability. This may be due to the interpenetrating networks of polymer alloys. Other workers have reported on extractability due to petroleum oils [137] and dry cleaning solvents [138].

Permeability of gases and water vapor is likewise controlled by diffusibility, as well as the solubility of the gaseous reagent in plasticized PVC matrix. Kummins and co-workers [139-141] reported extensive studies on moisture vapor permeability. Frissell [142] found that carbon dioxide is considerably more soluble than nitrogen in DOP- and phosphate-plasticized PVC films. Thus, carbon dioxide gas diffused through the films at a higher rate. He also found that DOP-plasticized PVC has a significantly higher degree of permeability than vinyls plasticized with solid polymers such as nitrile rubber.

3. *Chemical Resistance*: Vinyl compositions are susceptible to attack by certain organic solvents such as tetrahydrofuran, lower-molecular-weight ketones and esters, and chlorinated solvents. Increased plasticizer concentration significantly reduces resistance of flexible vinyls to chemical attack, although the interchange of various types of plasticizer shows only a minor effect.

Tables 5 through 7 summarize chemical resistance data measured according to ASTM D543 on unplasticized polyvinyl chloride and various plasticized compositions [143]. The plasticized compositions shown in Table 5 include two different materials with DOP plasticizer (one at a 90 Durometer A hardness, the other at a 60 Durometer A) and one material containing a medium-molecular-weight polyester plasticizer (60 Durometer A). Acetone dissolved the plasticized compositions and solvated the rigid vinyl specimens to the point of a gel. Ethyl acetate dissolved the 60 Durometer material and swelled the harder specimens. Hydrocarbon solvents showed no apparent attack on the rigid vinyl, but did show strong extractability characteristics for the monomeric DOP plasticizer, especially in the softer material

TABLE 5 Chemical Resistance of Vinyl Compounds: Percent Change in Weight[a]

Chemical substance	Plasticized vinyl			Unplasticized vinyl
	Polymeric plasticized (A)	Di-2-ethylhexyl phthlate		Polyvinyl (B) chloride
	60 Durometer A	60 Durometer A(C)	90 Durometer A(D)	
Acetic acid, 5%	—	+13	+6.0	+0.05
Acetone	—	Dissolved	Dissolved	Softened
Ammonium hydroxide, 10%	—	+0.36	+0.32	+0.10
28%	—	+0.65	+0.65	+0.18
Citric acid, 10%	—	+0.15	+0.12	+0.05
Clorox® (sodium hypochlorite, 5%)	+0.2[b]	+0.1[b]	+0.08[b]	c
Ethyl acetate	—	Dissolved	+24.0	Softened
Ethyl alcohol, 95%	—	-1.5	+0.16	c
Gasoline	-3.8	-17.0	-1.5	+0.08
Hydrochloric acid, 10% (3 N)	—	+0.5	+2.15	c

	A	B	C	D
Lighter fluid	—	—	—	c
Mineral oil at 23°C	+0.2	−4.5	−0.5	c
at 50°C	−1.0	−6.9	−1.5	+0.07
Motor oil (Veedol® 10-30)	—	−2.0	−0.32	+0.06
Wesson® oil	—	−4.0	−0.73	+0.06
Oleic acid, c.p.	—	−4.0	−0.91	c
Phenol, 5%	—	+7.0	+4.2	c
Sodium chloride, 10%	—	+0.10	+0.10	+0.08
25%	—	+0.15	+0.11	+0.07
Sulfuric acid, 25% (5.5 N)	—	+1.0	+0.6	c
95% (21 N)	Discolored	Discolored	+30.0	c

[a]ASTM D543-52T. Tested for 7 days at 23°C in various chemicals.

[b]Bleaching of color.

[c]Not a significant difference; measured results were less than 0.05%.

Materials: (A) Flexol® Plasticizer R2H (Union Carbide Corp.). (B) Composition of unplasticized PVC; 97% by weight Bakelite® Vinyl Resin QYSJ, inherent viscosity = 0.77 (Union Carbide Corp.); 3% by weight Mark® A stabilizer (Argus Chemical, Div. of Witco). (C) Bakelite VYNW—56%, DOP—42% (75 phr). (D) Bakelite VYNW—69%, DOP—29% (42 phr).

Source: E. E. Saunders, unpublished work, Union Carbide Corp., Bound Brook, N.J. (1956).

DOP—29%.

(60 Durometer A). The polymeric-plasticized composition, however, shows significantly improved resistance to plasticizer extraction by hydrocarbon solvents, compared with the DOP-plasticized counterpart. Although chlorinated hydrocarbon solvents are not shown, this type of solvent would be expected to show significant extraction of the plasticizer; the polymeric-plasticized composition would again show improvements over the DOP-plasticized compositions. However, even the rigid vinyl would be softened and swelled slightly on exposure to chlorinated hydrocarbon solvents.

All of the vinyl specimens were found to have good resistance to inorganic solvents. Sulfuric acid, however, seriously attacked the softer composition (60 Durometer A) irrespective of the interchange of DOP and polymeric plasticizers. Table 6 shows the resistance of plasticized vinyls formulated to a Durometer A hardness of 80 as a function of plasticizer interchange. In summary, only small differences occur as a function of the various plasticizer types when they are formulated to a given hardness. However, the alkyl epoxy tallate plasticizer appears to be exceptionally susceptible to concentrated sulfric and nitric acids, as would be expected.

Table 7 shows the effect of immersing various gasketing materials in DOP plasticizer [144]. Most elastomerics suffer some solvation or swelling when exposed to DOP, but some are only slightly affected. This typifies the attack of plasticizer-type esters on various gasketing materials commonly used in pipe and storage facilities.

Volatility: Plasticizer losses through volatilization are a concern under processing conditions as well as end-use applications. Poppe [66] reported a laboratory procedure for predicting plasticizer volatilization losses under commercial conditions of oven-cure cycles in plastisol spread coatings. His work shows the significant influence of molecular weight on plasticizer volatilization; diisononyl phthalate was found to be 45% less volatile than di(2-ethylhexyl) phthalate, while diisodecyl phthalate is 65% less volatile than di(2-ethylhexyl) phthalate. Beach [145] conducted extensive laboratory evaluations of volatile losses during hot compounding on a heated two-roll mill. His work showed diisononyl phthalate and diisodecyl phthalate to be 46 and 78% less volatile than DOP, respectively. He showed that the adipates of isononyl alcohol and 2-ethylhexanol were more volatile than their phthalate counterparts, but here again, diisononyl adipate was approximately half as volatile as DOA. Interestingly, diisononyl adipate was 40% less volatile than DOP. Both the Poppe and Beach studies were made with the appropriate correction factors for plasticizer efficiency and optimized processing temperatures with respect to the high-molecular-weight plasticizers versus the 2-ethylhexyl analogs. Beach also observed that significant volatilization occurs due to a thin-film coating of neat plasticizer on the bare roll. This effect essentially doubles the surface area from which the plasticizer is lost, although in a calendering operation the bare roll effect is even greater than twofold. However, the relative volatility of one plasticizer versus another was not influenced by this bare roll effect.

Frissell [146] concluded that there were three significant areas of concern with respect to plasticizer volatility when developing flexible vinyl formulations: (1) loss of plasticizer during processing operations, (2) optimum temperatures and time of exposure during the processing cycle, and (3) expected life of the product in the end-use application. He estimated that a typical commercial calendering operation in the manufacture of 4-mil-thick PVC film plasticized with di(2-ethylhexyl) adipate would lose

TABLE 6 Chemical Resistance[a] as a Function of Plasticizer Interchange in Flexible PVC[b]

Plasticizer	phr	10% NH_4OH	10% NaOH	30% H_2SO_4 (6.7 N)	10% HNO_3 (1.6 N)	10% HCL (3 N)	5% CH_3COOH	3% H_2O_2
Phthalates								
Di(2-ethylhexyl)	48	+0.7	0	0	+0.3	+1.1	+1.5	+0.1
Butyl 2-ethylhexyl	45	+0.6	0	+0.1	+0.3	+0.1	+1.3	+0.1
2-Ethylhexyl, isodecyl	49	+0.6	0	0	+0.3	+0.4	+1.6	+0.1
Diisodecyl	55	+0.6	0	+0.1	+0.4	+0.1	+1.8	+0.1
Epoxides								
Epoxidized soybean oil	60	+1.7	-1.3	-0.1	+0.5	0	+2.7	+0.4
2-Ethylhexyl epoxytallate	50	+1.0	-0.2	-10.9	-11.7	+0.1	+2.8	+0.1
Aliphatic diesters								
Di(2-ethylhexyl) adipate	40	+0.8	-0.1	+0.1	+0.9	+0.1	+2.5	+0.2
Diisodecyl adipate	52	+0.6	-0.1	0	+1.4	+0.1	+2.2	+0.2
Miscellaneous								
Tri(2-ethylhexyl) phosphate	47	+0.8	+0.1	+0.2	+3.4	+0.2	+3.6	+0.3
Di(2-ethylhexyl) isophthalate	51	+0.5	-0.1	0	+2.1	+0.1	+1.8	+0.1
Dipropylene glycol dibenzoate	49	+0.6	-0.1	-0.1	0	+1.1	+0.6	0
Polyethylene glycol di-2-ethylhexanoate	48	+1.4	-1.2	-0.8	-0.2	-1.1	+1.8	-0.9
Flexol® plasticizer R2H	64	+2.7	0	0	+0.9	+1.2	+2.2	+0.3

[a]Percentage weight change in 7 days at 23°C.
[b]All compounds 80 Durometer A hardness. All samples 0.025 in. thick. Test method ASTM D-543-52T. Base formula (by weight): PVC (inherent viscosity = 0.97), 100; plasticizer, as shown; dibasic lead phosphite, 2.0; dibasic lead stearate, 1.0.
Source: E. E. Saunders, unpublished work; Union Carbide Corp, Bound Brook, N.J., 1957.

TABLE 7 Chemical Resistance of Gasket Materials to DOP (Stamps Immersed in DOP for 105 days at 78°F)

Gasket material[a]	Hardness[b]			Volume change (%)		Wt. change, (%)[d]	Color solution[e]	Comments on specimen	Rating[f]
	Start	End[c]	48 hr[d]	End[c]	48 hr[d]				
Viton® A (1)	86	83	84	+14.8	+2.9	Nil	5	Slightly swollen	F
Neoprene® 7797	82	56	56	+40.7	+40.0	+46.3	Yellow	Swollen	X
Butyl K-53	66	66	66	+2.7	+1.9	+2.9	Yellow	No visible change	R
Thiokol® 3060 (2)	71	74	74	+1.9	+0.3	+0.4	Yellow	No visible change	R
Black rubber 3773	68	53	55	+41.0	+40.7	+58.3	Yellow	Swollen	X
Hycar®, D-24 (3)	63	50	52	+24.2	+21.4	+20.4	Yellow	Slightly swollen	F
Buna N (OR-25)	83	73	74	+22.8	+17.8	+19.5	Yellow	Slightly swollen	F
Hypalon® (1)	72	48	51	+60.2	+56.8	+108.0	Yellow	Swollen	X
Silicone No. 65	53	54	56	+8.1	+7.5	+3.5	5	No visible change	R

Buna S	46	39	39	+41.7	+39.3	+65.7	Yellow	Swollen	X
Natural rubber, gum	47	35	33	+50.4	+49.7	+103.0	Yellow	Swollen	X
Red rubber No. 107	81	65	65	+27.9	+27.2	+21.8	Yellow	Slightly swollen	F
Adiprene® L-100 (1)	45★	43★	43★	+14.9	+14.9	+15.3	Yellow	Slightly swollen	F
Saraloy® 300 (4)	64	64	67	+6.0	+4.6	+3.0	Yellow	No visible change	R
Garlock asbestos 7735	65★	51★	53★	+16.3	+12.0	+18.9	Yellow	Saturated with solution	F
U.S. asbestos No. 899	71★	49★	52★	+11.0	+10.2	+18.1	5	Saturated with solution	F

[a]Materials: (1) E. I. duPont de Nemours & Co., (2) Thiokol Chemical Corp., (3) B. F. Goodrich Chemical Co., (4) Dow Chemical Co.

[b]Hardness expressed as Durometer A except where ★ appears; asterisk indicates Durometer D hardness.

[c]Information obtained on the specimens immediately after removal from the solution.

[d]Information obtained on the specimens after drying 48 hr at 78°F.

[e]Pt-Co units; color of solution at start of test was 5 Pt-Co.

[f]Resistance rating: R, resistance good, recommended material; F, resistance fair, use of the material depends on specific service; X, unsuitable.

Source: L. S., van Delinder, and J. C., Canterbury, unpublished work; Union Carbide Corp., So. Charleston, W.Va.; 1962.

approximately 2% of the plasticizer in the typical manufacturing cycle; he projected that up to 2.5 lb of DOA fumes could be generated per hour of operation, which represents a significant engineering problem with respect to control of effluent vapors. He verified Quackenboss' [103] equation for predicting the end-use life, assuming that a 10% loss of plasticizer represents the end of the useful life of a plasticized PVC product:

$$\text{Life (hours to lose 10\% plasticizer)} = \frac{0.080}{\text{vapor pressure}}$$
$$\text{(mmHg at service temperature)}$$

Frissell also reported the weight change due to volatilization after selected 4-mil-thick plasticized PVC films were exposed in the laboratory for 2 1/2 years. Of the series of plasticizers he evaluated, only dibutyl phthalate had lost a significant amount of plasticizer, 18.4% by weight, after 2 1/2 years at ambient room conditions.

Tables 13, 14 and 17 through 22 show the percent weight loss of plasticized PVC films exposed to activated charcoal for 24 hr at 70°C (ASTM D-1203). Results are consistent with those in the literature, namely that vapor pressure, which is a function of molecular weight, is the primary factor in the loss of plasticizer due to volatilization. Thus, volatile plasticizer loss is a surface-controlled mechanism. Reed and coworkers [107, 147] showed that, under severe vacuum, volatile losses of plasticizer became diffusion-controlled because the plasticizer could leave the surface of the PVC film. High-molecular-weight plasticizers provide low volatile losses. Polymeric plasticizers which have been stripped of low-molecular-weight components are a preferred choice for the lowest volatility. Of the phthalates, the highest-molecular-weight homolog, diisotridecyl phthalate, is preferred. Trimellitate plasticizers are even higher in molecular weight than phthalates and therefore are preferred for low-volatility plasticized PVC.

PVC insulation for wire and cable applications is a classical end use that is limited by plasticizer volatility. Underwriters' Laboratories specifies the insulating material by the maximum allowable operating temperature, and this is defined through accelerated oven aging exposures. The classical 60° building wire insulation requires a specified retention of properties after 7 days exposure at 100°C.

General-purpose plasticizer DOP tends to be borderline in performance in this grade of building wire, while the higher-molecular-weight analog DINP imparts sufficiently low volatility to meet the performance specification with a substantial safety margin. New building wire specifications have been adopted with performance requirements to meet 90°C specifications; this requires that the PVC insulation withstand 7 days oven aging at 136°C. Under these conditions, plasticized PVC insulation usually will be formulated with trimellitates and blends with high-molecular-weight phthalates, such as ditridecyl phthalate, didodecyl phthalate, and dilinear undecyl phthalate.

5. *Degradation by Heat, Light, and Biological Effects*: Plasticizer permanence is, of course, influenced by changes that occur in the flexible vinyl composition. Other than changes due to penetration by extractants and their removal of plasticizer, the most commonly encountered changes are due to degradation by heat, light, radiation, and biological effects. Thermal degradation mechanisms involve dehydrohalogenation as well as oxidation. Photooxidation is the primary degradation mechanism under sunlight exposures. In outdoor exposures, the ultraviolet energy effects are

aggravated by many additional factors, such as rainwater extraction, corrosive reagents accumulating on the surface and biological attack. The problem of plasticizer permanence is compounded under these conditions, due to the presence of reactive intermediates such as free radicals. Many investigators have attempted to separate the contributing factors, while others have studied the effects of specific formulating variables in flexible PVC [148—151].

Plasticizers per se have good chemical stability, and their resistance to heat is such that there are few or no storage life problems commercially. However, variability in the heat and light stability of vinyl compositions as a function of plasticizer type and concentration is significant. Generally, heat and light stability of vinyl compositions are coincidently influenced as a function of plasticizer interchange. The degree of change, however, may be more apparent in one property than the other.

Stability tests can be classified into three types:

1. *Storage stability.* Tests are conducted on the plasticizer itself.
2. *Processing stability.* Dynamic tests are designed to simulate the thermal conditions under which vinyl compositions are processed.
3. *Aging stability.* Static exposures are designed to accelerate thermal or ultraviolet degradation of the vinyl product.

The storage stability tests usually entail exposing the plasticizer in a beaker to an elevated temperature. Stability is measured as a function of color, viscosity, or chemical changes. Iron or copper can be suspended in the heated plasticizer to simulate storage in heated tanks. Most plasticizers undergo a mild color change under such tests, although trace amounts of impurities can significantly influence the degree of color change. Epoxy plasticizers may undergo polymerization under certain conditions since the oxirane ring is a reactive site. When this occurs, measurable changes in oxirane level and viscosity are apparent. Chlorinated hydrocarbons undergo very significant color changes when heated in the presence of iron.

Processing heat stability is normally evaluated in dynamic tests [152, 153]. These tests abuse the vinyl composition mechanically as well as thermally under conditions that accelerate the degradation mechanism. They provide short-term empirical comparisons of performance, measured either by color changes or in the melt viscosity characteristics of the composition. In most cases, the plasticizer does not undergo chemical decomposition. Hence, the effect of plasticizer interchange on dynamic heat stability of vinyl compositions reflects the degree of lubricity the plasticizer imparts rather than the stability of the plasticizer itself. There is, however, an apparent improvement in dynamic heat stability with plasticizers that impart lubricating effects to the molten polymer system; the effect is actually due to a resultant reduction in heat history because of lubricity. Likewise, the volatility of a plasticizer may influence dynamic heat stability tests. A highly volatile plasticizer can cause an apparent decrease in heat stability; by virtue of its volatilization, the vinyl plastic matrix suffers increased heat history as compared to the more highly plasticized compositions.

Phosphate plasticizers are exceptionally poor in heat stability. Tri(2-ethylhexyl) phosphate is significantly inferior to the triaryl phosphates, apparently because of chemical breakdown in the presence of the decomposition products of polyvinyl chloride. Plasticizer extenders usually impart poor heat stability to vinyl compositions, the aromatic extenders being most severe. Impurities can cause a loss in heat stability. However, most commercial grade plasticizers are relatively free of impurities.

Epoxy plasticizers are unique in that the oxirane oxygen actively participates in the stabilization of polyvinyl chloride against thermal and ultraviolet degradation. Although epoxy plasticizers perform as adjuvant stabilizers in combination with certain metallic stabilizers, they are unsatisfactory as the sole stabilizer [154]. Tests with incrementally increased concentrations of epoxy plasticizer indicate that there is a level of optimum efficiency in the heat stabilizer process. This level depends on the other formulating variables in the composition. Levels of epoxy plasticizer-stabilizer greater than the optimum level show no further improvement in the heat stability of the vinyl composition. Theoretically, it should be possible to calculate the optimum level of epoxy stabilizer for any given composition based on the oxirane content. However, it would be necessary to know the exact number of labile chloride atoms contained in each given lot of PVC and the various chemical reactions that take place in the thermal stabilization process. Most commercial formulas for flexible PVC use epoxy stabilizer-plasticizer levels of 3 to 6 phr.

Static heat stability tests are commonly conducted to determine the ultimate stability of a vinyl product in an end-use application. These methods usually expose a fused vinyl specimen to elevated temperature in an oven or on a heated surface. Changes in color or mechanical properties (tensile or hardness characteristics) may serve as indicators of heat stability failure. Static heat stability tests are not reliable for predicting processing heat stability because the total effect of heat history is circumvented. However, it is generally found that results of static heat stability tests are reasonably similar to the results of the dynamic tests as a function of plasticizer interchange.

Light stability, or resistance to degradation by ultraviolet energy, is normally measured on test specimens of fused vinyl products. Extended outdoor exposures measure the effects of climatic elements in addition to sunlight. Laboratory tests accelerate the effects of sunlight by means of sunlamps or other types of simulators—fluorescent lamps, xenon arc sources, mercury vapor lamps, carbon arcs, etc. Numerous investigators have been unable to correlate accelerated radiant exposures with outdoor aging characteristics. Influential test variables included the position and location of the test specimen relative to the energy source as well as specimen thickness, color, clarity, composition, and heat history [87,155,156]. Comparison of the findings of various workers is complicated by the conditions on which conclusions are based. In most cases, the effect of sunlight degradation has been evaluated by noting changes in the appearance of the test specimen, that is, discoloration, spotting, crazing, or incompatibility. In some studies, the effect on mechanical properties of the vinyl composition was considered more significant than the change in appearance. Typical of the inconsistencies is the difference between the conclusion of Darby and Graham [156] that optimum weatherability is provided with a total plasticizer concentration of 35 to 40 phr, and the data in Table 8, which indicate an overall increase in light stability at higher plasticizer levels [157]. In general, the light stability characteristics of plasticized vinyl compositions follow a pattern similar to that of heat stability. It has been postulated that absorbed radiant energy is almost always translated into heat, which helps to explain similarities found in results of heat and light stability tests [158].

TABLE 8 Weathering Characteristics of Plasticized Vinyl as a Function of Plasticizer Concentration[a]

Plasticizer	phr	XIA Weather-Ometer failure		Florida failure	
		Hours	Type[b]	Hours	Type[b]
Di(2-ethylhexyl) phthalate	40	750	D,Sp,Stiff	750	D,Sp,Stiff
	50	1000	D,Sp,Br	750	D,Sp
	70	1000	D,Sp,Br	1000	D,T
Di(2-ethylhexyl) phthalate	40	750	D,Sp,Br	1000	D,Sp,Br
(duplicate specimen)	50	1000	D,Sp,Br	1500	D,Sp,Br
	70	1250	D,Sp,Br	2000	D,Sp,Br
Epoxidized soybean oil	40	750	D,Br	1500	Stiff, T
	50	750	D,Br	1500	Stiff,Sp,T
	70	750	D,Br,Sp	1500	Stiff,Sp,T
Di(2-ethylhexyl) adipate	30	750	D,Sp,Br	1800	D,Sp,Br
	40	750	D,Br	1500	D,Stiff,Sp
	50	750	D,Sp,Br	1500	D,Sp,Br
	70	1000	Sp,Br	2200	D,Sp,Br
Tri(2-ethylhexyl) phosphate	40	320	D,Stiff	320	D,T
	50	320	D,Stiff	500	D,T,Sp
	70	320	D,Sp	500	Sp,T
Octyl diphenyl phosphate	40	750	D,Sp,Br	500	D,Sp
	50	750	D,SP,Br	500	Y,Sp,T
	70	750	D,Sp,Br	250	T,Sp
Cresyl diphenyl phosphate	40	240	Y,Sp,Stiff	500	D,Sp,Stiff
	50	240	Y,Sp,Stiff	500	D,Sp,Stiff
	70	240	Y,Sp,Stiff	500	D,Sp,Stiff
Triethyleneglycol di(2-ethylhexanoate)[c]	40	500	D,Sp,Br	750	D,Sp,Stiff
	50	750	D,Sp,Br	750	D,Sp,Stiff
	70	750	D,Sp,Br	1000	Sp,Stiff
Flexol® plasticizer R2H	50	500	D,Sp,Br	500	D,Sp,Stiff
(Polyester)	70	500	D,Sp	500	D,Sp
	90	500	D,Sp	250	T,Sp

[a]Base formula (by weight); PVC (inherent viscosity = 0.97), 100; plasticizer, as shown; dibasic lead phosphite, 2.0; dibasic lead stearate, 1.0.
[b]D, darkening; T, tacky; Y, yellowing; Bch, bleached; Br, brittle; Sp, spotting; Stiff, stiffening.
[c]Triethylene glycol di-2-ethylhexanoate tested in 50/50 mixture with DOP.
Source: [159] E. E. Saunders; unpublished work; Union Carbide Corp., Bound Brook, N.J.; 1960.

Epoxy plasticizers show a stabilizing effect against the polymer degradation normally encountered under exposure to ultraviolet energy sources. The stabilizing effectiveness is altered by the plasticizer level in the vinyl composition and numerous other formulating variables, primarily the type of metallic stabilizer employed. The phosphates show some inconsistencies in sunlight resistance. Tables 8 and 9 show them to be noticeably inferior to the other types of plasticizers when used as the sole plasticizer. However, DeCoste et al. [87] found tricresyl phosphate to offer the best retention of physical properties when compared with other plasticizers in clear films. In addition, small concentrations (5 to 10 phr) of octyl diphenyl phosphate improved the outdoor weathering resistance of flexible films [156,160]. However, the improvements were not apparent in accelerated tests; higher concentrations of octyl diphenyl phosphate detracted from the light stability.

The effect of alkyl moiety within a given chemical family of plasticizers is difficult to measure in accelerated tests. However, field experience and prolonged outdoor exposures have indicated some general effects due to alcohol interchange. Plasticizers made with normal alcohols show a significant improvement in both heat and light stability over esters made with branched alcohols. Comparison of DOP with DIOP, DIDP, and NODP in Table 9 provides some indication of this effect.

Flexible PVC products are occasionally deliberately exposed to radiation to encourage cross-linking. Kojima and co-workers [161] studied the effect of plasticizer concentrations from 25 to 100 phr. Higher plasticizer levels encouraged increased cross-linking and decreased discoloration. Epoxy and polymeric plasticizers gave better cross-linking efficiency and less discoloration. But the aromatic plasticizers, phthalates and trimellitates, showed the opposite effects, apparently because the benzene ring acted as an electron trap.

Gamma radiation causes development of acidity in plasticizers, especially in the presence of PVC [162]. Dialkyl phthalates up to C_8 alkyl groups were the most stable. Szymanski and Smietanska [163] showed that the effectiveness of bis-aminocrotonate stabilizer was enhanced by calcium/zinc/epoxy stabilizer under exposure to gamma irradiation. In a later publication, it was concluded that only epoxy stabilizers containing benzene rings demonstrated antioxidative action [164]. Epoxy groups per se had no effect, but epoxidized oils enhanced the penetration of oxygen to the polymer due to their plasticizing action (free-volume effects).

Resistance to biological attack is of concern where plasticized vinyls are exposed to environmental conditions that are conducive to fungal growth. It is possible to formulate a fungus-inert vinyl composition, that is, a flexible vinyl composition that does not support fungal growth. Prolonged exposure, however, under atmospheric conditions that are conducive to growth of fungi usually results in fungal attack. This is generally due to accumulation on the film surface of foreign matter that will support the growth of fungi. Investigators have found that the diffusibility of fungicidal additives and plasticizers is critical in imparting resistance to fungal attack [165—168]. Thus, it is common practice to incorporate a small amount of fungicide in the formulation to inhibit fungal growth where application conditions suggest the need (see Chapter 19).

Burgess and Darby [169] have reported methods for determining fungal resistance that involve measuring gas absorption and weight changes.

TABLE 9 Weathering Resistance as a Function of Plasticizer Interchange[a]

	XIA Weather-Ometer failure		Florida failure	
	Hours	Type[b]	Sun Hours	Type[b]
Phthalates				
n-Butyl	500	D,Br,Sp,Stiff	—	—
2-Ethylhexyl (1st sample)	1000	D,Br,Sp	1000	D,Br,Sp
2-Ethylhexyl (duplicate)	1500	D,Br,Sp,Stiff	—	—
Isooctyl	1250	D,Sp,Stiff	1000	D,Sp,Stiff
Isodecyl	750	D,Br,Sp	750	D,Br,Sp
Tridecyl	—	—	1000	D,Sp
n-Hexyl, octyl,decyl	750	D,Sp,Stiff	—	—
n-Octyl, decyl	2250	D,Br,Sp,Stiff	—	—
Epoxides				
Soybean oil epoxide	750	D,Br	1500	D,Sp,Stiff
2-Ethylhexyl eopxy tallate	—	—	1500	D,Stiff
Diisodecyl-4,5-epoxy tetrahydrophthalate	—	—	1500	D,Bch,Stiff
Aliphatic diesters				
Di(2-ethylhexyl) adipate	750	D,Br,Sp	1500	D,Br,Sp
Diisodecyl adipate	500	D,Br	1250	D,Sp,Stiff
Di(2-ethylhexyl) azelate	1250	D,Br,Sp,Stiff	1500	D,Stiff,Bch
Phosphates				
Tri-2-ethylhexyl	240	D,Br,Sp	500	D,Sp,T
Octyl diphenyl	750	D,Br,Sp	500	Y,Sp,T
Cresyl diphenyl	240	Y,Sp,Stiff	500	D,Sp,Stiff
Tricresyl	90	Y	500	Y,T,Stiff
Miscellaneous				
Di(2-ethylhexyl) isophthalate	1000	D,Br,Sp	1000	D,Br,Sp
Flexol® R2H	500	D,Br,Sp	500	D,Sp,Stiff

[a]Base formula (by weight): PVC (inherent viscosity = 0.97), 100; plasticizer, 50; dibasic lead phosphite, 2.0; dibasic lead stearate, 1.0.
[b]D, darkening; T, tacky; Y, yellowing; Bch, bleached; Br, brittle; Sp, spotting; Stiff, stiffening.
Source: [159] E. E. Saunders, unpublished work, Union Carbide Corp., Bound Brook, N.J.; 1960.

Usually, fungal and bacterial resistance is evaluated either by measuring the amount of bacterial growth on inoculated plastic samples or by measuring a zone of inhibition when plastic samples are placed on inoculated agar and subjected to incubation conditions. ASTM D-1924 subjectively measures the degree of fungal growth on a plastic sample that has been inoculated and placed in a nutrient for incubation. In addition to subjective visual ratings, fungal attack may be measured by determining changes in mechanical properties. It has been found, however, that excessive fungal growth can be observed by the ASTM D-1924 method where no measurable change in tensile properties has taken place. Also, testing by outdoor exposure has shown that although fungal attack may be apparent, changes in tensile properties of flexible vinyl films primarily reflect changes due to weathering, that is, plasticizer loss and polymer degradation [170]. The zone of inhibition provides a measure of fungicidal activity and is useful in assessing the effect of fungicides rather than plasticizer selection. It may be determined according to American Association of Textile Chemists and Colorists (AATCC) Test Method 90-1962T.

Addition of fungicides reduces fungal susceptibility of plasticized vinyl effectively [171–173]. However, the effect of fungicides and additives other than plasticizers is ignored in the following discussion in the interest of assessing the response of fungi to plasticizer selection.

All matter can be separated into three classes with respect to fungal activity:

Fungicides: materials that are actively harmful to fungi.
Fungistats: materials that are not harmful to fungi but do not support fungal growth.
Food for fungi: materials that actively support colonization and fungal growth.

With the possible exception of tricresyl phosphate, none of the plasticizers demonstrates fungicidal characteristics.

Considerable work has been done on separating plasticizers into the last two classes [171,172,174–176]. Phthalates and phosphates generally fall in the class of fungistats. Aliphatic diesters, polyesters derived from aliphatic dicarboxylic acids, epoxidized oils and epoxy tallates are generally food for fungi. Derivatives of polyalkylene glycols and certain other specialty plasticizers are also foods. Changes in the alkyl moiety within a chemical family of ester-type plasticizers may significantly alter the resistance to fungal attack, however. Esters based on normal alcohols are more susceptible than those based on branched alcohols. Hence, it is found that di(n-octyl, n-decyl) phthalate is the most susceptible phthalate. Diisodecyl adipate, on the other hand, is sufficiently resistant to fungi to be classified as a fungistat rather than a food for fungi.

Klausmeier [177,178] studied the mechanism of fungal degradation of phthalate esters. He concluded that many microorganisms that fail to use plasticizers as the sole source of nutrient can cause biodegradation of plastics by action on the plasticizers when other nutrients are present. Hence, the ultimate growth of fungi on a fungus-inert vinyl composition can actually be due to the mechanism observed by Klausmeier rather than simply the accumulation of foreign matter on the film surface.

C. Special Properties

1. Electrical Insulation: The volume resistivity of plasticized polyvinyl chloride at 25°C is typically in the order of 10^{12} megohm-centimeters. Electric linemen's gloves and electrical insulating tape are made with relatively thin vinyl coatings. The most commonly known applications in which vinyl is used as electrical insulation are coated electrical conductors, namely building, appliance, and automotive wires. Numerous formulating variables besides the plasticizer have significant effects on the electrical insulating characteristics.

The dielectric constant of vinyl compositions with varied plasticizer concentrations has been determined in a series of samples containing both DOP and TCP plasticizers [179,180]. Both studies showed that the dielectric constant passes through a peak at a plasticizer level of 100 phr, that is, a 1:1 ratio of polyvinyl chloride and plasticizer. Fitzgerald and Miller [181] studied dimethylthianthrene in plasticized PVC containing plasticizer levels from 0 to 100%. They measured the dielectric constant over a wide range of temperatures. Comparing their findings with other published data, they concluded that the dielectric dispersion observed in a polar plasticizer and polar polymer combination is a result of a dipole of the plasticizer molecule together with dipole rotation of the polymer chain segments. Wartman [182] measured the volume resistivities of plasticized vinyls with varied plasticizers, to which he had intentionally added an ionic impurity. He showed that the conductance does not depend simply on either the compound stiffness or the plasticizer diffusion constant. He concluded that there were two factors controlling the electrical conducting characteristics of plasticized vinyl: (1) the dielectric constant of the composition involved, which would change with plasticizer level, and (2) the viscosity forces that impede the ion mobility. He observed that purification of a plasticizer was only partially effective in improving the electrical characteristics of the plasticized vinyl and envisioned the need for a plasticizer with a low conducting power to provide the ultimate in electrical insulation characteristics of flexible vinyl.

Deanin et al. [183] studied the effect of numerous variables on the volume resistivity of plasticized vinyls. They observed that purifying both the vinyl polymer and the plasticizer did not improve the insulation characteristics over the unpurified raw materials. However, ionic impurities in the base materials (either the resin or the plasticizer) could significantly impair the electrical insulation characteristics. Having experimented by adding various ionic stabilizers and impurities intentionally, the investigators concluded that electrical insulating characteristics are significantly impaired by the presence of an ionic contaminant that is soluble in the plasticized vinyl composition. They observed that lead stabilizers, and the degradation products that resulted from the stabilization of the vinyl polymer, actually improved the volume resistivity. Comparison of different types of plasticizers showed that a polymeric plasticizer and a DOP-plasticized composition provided similar insulating characteristics at equivalent flexibilities. Even nitrile rubber, a solid plasticizer, had little effect on the volume resistivity when compared with DOP-plasticized vinyls at equivalent moduli or flexibilities. Thus, they concluded that the mobility of the plasticizer itself was not a critical factor in promoting the mobility of ionic impurities. These investigators did show, however, that copolymers

of polyvinyl chloride and 2-ethylhexyl acrylate provided improved volume
resistivity over homopolymer resins when compared on an equal modulus
or flexibility basis.

Price [184] also found that ionic impurities that solubilize in the plasticizer significantly lowered the resistivity of highly purified DOP.

Although the works cited above show that the purity of the plasticizer
itself is not the chief factor in providing good electrical insulation properties, it is obvious that soluble ionic impurities, which may be introduced
through any of the raw materials, including the plasticizer, can impair the
electrical properties of plasticized vinyl. Hence, patents have been issued
on techniques for reducing the ionic impurities in plasticizers [185,186].

Electrical applications that use plasticized polyvinyl chloride usually
specify minimum insulating properties. However, the primary factor that
dictates plasticizer type is usually defined by requirements other than electrical ones. Such factors as processability or mechanical or permanence
properties define the plasticizer of choice. For instance, in the case of
linemen's gloves, plastisol rheology becomes a determining factor. In the
case of insulated wire and cable, the vinyl insulation is designed to meet
the limiting factor of oven aging resistance as defined by the performance
specification. Building wire rated at 60°F may be formulated with DINP
plasticizer, but 75°F-rated wire requires DIDP and plasticizers with lower
volatility to provide sufficient resistance to aging at elevated temperatures.
High-temperature appliance wire requires the more permanent type of plasticizer such as trimellitates or blends of trimellitates with high-molecular-
weight phthalates. In addition, antioxidants are generally added to the
vinyl formula to inhibit degradation of the polymer and plasticizer
[187−190]. Pressure-sensitive electrical tape must be formulated to inhibit migration of plasticizer into the adhesive [80,81].

It is obvious that although the electrical properties of the raw materials must be controlled, it is extremely unlikely that one could measure
the electrical properties of a liquid plasticizer and predict the performance characteristics for specific electrical properties on any given flexible vinyl composition.

2. Flammability Resistance: Polyvinyl chloride is approximately 57%
chlorine by weight and supports combustion only under extreme conditions.
Copolymers of polyvinyl chloride demonstrate some loss of flame resistance
as a function of increased level of comonomers (when the comonomers do
not contain halogen). Plasticized vinyls demonstrate a significant level
of combustibility. Film thickness, filler level, and other formulating additives which may act as flame retardants significantly alter the degree of
flammability, or burning rate. Increased plasticizer concentration promotes
higher flammability, but the plasticizer type is also a contributing factor.
In commercial practice, nonplasticizing additives are used to impart flame
resistance, which diminishes the adverse effects of the plasticizer.

Veprek and Frissell [191] found that all monomeric plasticizers except those containing chlorine and phosphorus caused high levels of
flammability. Polymeric plasticizers and epoxidized vegetable oils, such
as epoxidized soybean oil, cause slow to medium burning rates. This is
apparently due to the low vapor pressure of the higher-molecular-weight
plasticizers. All the phosphate plasticizers produced self-extinguishing
films when they were used as the sole plasticizer in flexible vinyl compositions. Within the phosphate family, the organic constituent signifi-

cantly influences the ability to provide flame retardance, as mentioned by
Veprek and Frissell. They used unfilled flexible vinyl films of 3-mil
thickness formulated to the equivalent of a Durometer A hardness of 80.
The compositions included powdered lead stabilizers and DOP was the
primary plasticizer. Flammability studies were conducted on experimental
films prepared by incrementally replacing portions of the DOP plasticizer
with the various phosphate plasticizers. Flammability was evaluated by
burning rate, as measured by the Society of Plastics Industires (SPI)
flammability test. The burning rate did not appear to change as a function
of incremental increases of the phosphate/DOP ratio. Each of the phosphate
plasticizers, however, imparts self-extinguishing characteristics when increased to a certain ratio of the plasticizing system. The ratio at which
the phosphate plasticizer provided self-extinguishment varied according to
the organic portion of the phosphate. Triaryl and alkyl-aryl phosphates
provided self-extinguishment when used at a level of 25% replacement of
the DOP. The tri(2-ethylhexyl) phosphate did not provide self-extinguishing properties until it was used at a level of 70% replacement of the DOP.

Keeney [192] studied the effect of phosphate structure on combustibility as measured by the limited oxygen index (LOI) test and smoke generation in the National Bureau of Standards (NBS) smoke test and the
Monsanto 2-ft tunnel test. The LOI test showed triaryl phosphate to be
significantly more burn-resistant than trialkyl phosphates. Smoke generation was poorest with triaryl phosphates, and trialkyl phosphates were
moderate in smoke generation. Alkyl-diaryl phosphates were the most resistant to smoke generation due to the formation of a phosphorous char
under combustion conditions. Dialkyl and aryl-alkyl phthalates appeared
to fall in a "moderate" category, similar to trialkyl phosphates with respect to smoke generation.

Chlorinated paraffins provided self-extinguishing characteristics when
used as partial replacements for DOP. [191] The level of chlorination on
the alphatic hydrocarbons significantly influenced the efficiency of flame
retardance. Experimentally prepared chlorinated paraffins with 60 to 70%
chlorine were more effective than those with a lower level of chlorine. The
60 to 70% chlorinated paraffins provided self-extinguishing properties when
used at approximately 50% replacement of the DOP.

Hoke [193] related the oxygen index test to the Motor Vehical Safety
Standard (MVSS) 302 burn rate test for flexible PVC-coated fabrics. He
also observed a synergism with the presence of phosphorous, chlorine,
and antimony in plasticized PVC coatings. Other investigators [194,195]
have studied the effects of various fillers and other formulating ingredients on the combustibility and smoke generation characteristics of plasticized PVC. Relatively little has been reported on the characteristics
as a function of non-phosphorous- and non-halogen-containing plasticizers [196].

The reader should recognize that the test results described in this
section were obtained in small-scale fire tests. These tests do not
necessarily reflect the performance or relative performance of the materials in a real fire.

Yarborough and Haskin [197] identified the combustion products of
pyrolyzed vinyl sheeting. Test specimens of flexible vinyl that were
pyrolyzed contained 55 phr of DOP plasticizer and 17 phr of titanium
dioxide with a small amount of tin stabilizer. The resin used was a
medium-molecular-weight nonsolvent polymerized vinyl chloride-vinyl

acetate copolymer of approximateky 3% vinyl acetate level. Test specimans
were pyrolyzed at approximately 1100°C under variously controlled levels
of oxygen. The following components were detected in the volatile products by mass spectrometric and chemical analyses: hydrogen chloride,
carbon dioxide, hydrocarbons, vinyl chloride, carbon monoxide, water,
and a trace of 1,1-dichloroethane or phosgene or both. It was established
that the dichlorinated products made up no more than 0.8 mole % of the
gaseous products. The primary constituents identified in the gaseous products varied as a function of oxygen level, as shown in Table 10. Clark
[198] evaluated technical publications up to 1972 and concluded that the primary gases resulting from combustion of PVC products are carbon monoxide,
carbon dioxide, hydrogen chloride, and water.

   3. *Stain Resistance*: Certain applications of plasticized vinyl require
resistance to staining. In all cases, the more highly plasticized vinyl compositions are more susceptible to any given staining agent than the less
plasticized compositions.

   Sulfide staining, which results from the formation of colored sulfide
salts when certain metallic stabilizers are used, is altered in severity by
the plasticizer level in the vinyl product, even though the plasticizer itself does not enter into the reaction in which the sulfide salts are formed.
As may be expected, plasticizers that contain the sulfur atom promote
severe sulfide staining in the presence of stabilizers, fillers, or pigments
containing heavy metals capable of forming colored sulfides. For this and
other reasons, sulfur-containing plasticizers have not gained wide commercial significance.

   The most common type of staining results from the absorption of a
staining reagent that may come in contact with the surface of the vinyl
product. Vinyl coverings for floors, walls, countertops, and upholstery
are typical applications that may require resistance to staining [199,200].

   Rigid vinyl products have good resistance to staining; this is reduced
significantly, however, with increased levels of plasticizer irrespective of
the plasticizer type. Water-soluble staining reagents are essentially non-
effective on plasticized vinyl. However, some colloidal substances cause
severe stains on vinyl surfaces when the water is allowed to evaporate.
The staining effect, then, is significantly altered according to the organic
constituents in the residual substance. Tomato paste staining is representative of this problem. Certain inks and other agents containing solvents
that are miscible with polyvinyl chloride are known to cause stains even
in rigid vinyl compositions.

   Pinner and Massey [201] studied stain resistance of flexible vinyl compositions as a function of plasticizer type by both thermodynamic and kinetic approaches. The change in stain resistance as a function of plasticizer concentration in all instances supported the kinetic explanation of
staining as a diffusion process. Various plasticizer types alter the degree
of staining (at constant levels of hardness or softness), however, indicating that other factors also influence stain resistance. The investigators found that plasticizers with a solubility parameter of 9.4 to 9.8
provided the best overall resistance to staining by oil base reagents.
Thus, it appears that stain-resistant vinyls are most readily attainable
with plasticizers that have solubility parameters of the same order as
that of polyvinyl chloride. Good stain resistance, however, was observed
for one exceptional material that had a solubility parameter of 8.7. This

TABLE 10 Combustion Products of Plasticized Vinyl[a]

	Mole percent gaseous products		
	Absence of O_2	Medium O_2 level	Very rich O_2
Benzene	3.0	6.0—11.0	—
Carbon dioxide	2.0	4.0—27.0	80.0
Hydrogen chloride	84.0	62.9—24.0	19.0
Ethylene	4.0	11.0—18.0	—
Acetylene	0.3	0.1—11.0	—

[a]From Ref. 197.

material was the isobutyrate/benzoate diester of 2,2,4,-trimethyl pen-tanediol-1,3. It was felt that further studies were necessary to elucidate the mechanism, believed to be surface volatility, to produce completely stain-resistant flexible vinyl compositions.

It has been suggested that a topcoat of rigid polyvinyl chloride, or some other stain-resistant film, is the most practical means of achieving stain resistance in flexible vinyl products [201,202]. It is also known that oven curing causes a plasticizer-deficient skin on the surface exposed to hot air; the result is that volatile plasticizers impart improved stain resistance [203]. However, if the plasticizer in the base material has a tendency to migrate into the surface layer, stain resistance will be degraded with time. Shah [204] found that plastisol resins with high residual soap levels impaired stain resistance whereas low-soap resins were very good.

Studies of cigarette smoke stainability confirmed the above generalizations [205] and showed that barrier coats of unplasticized acrylic/vinyl blends gave the best resistance to staining by cigarette smoke. Copolymers of vinyl chloride and ethylene were more resistant than plasticized homopolymer PVC, because less plasticizer was required to match the specified hardness. Plasticizer types compared as follows for short-term staining rate and long-term ultimate stain development:

	Short term	Long term
Best	Unplasticized top coat	
	Plasticized PVC/ethylene copolymer	
	Polymeric	Polymeric
	Phosphate	Phthalate
	Phthalate	Phosphate
Worst	Epoxy	Epoxy

In summary, practically all the commonly used plasticizers for vinyls are susceptible to oil-based stains. Specialty plasticizers containing ether linkages or a relatively high level of aromaticity (greater than that of low-molecular-weight dialkyl phthalates) show some degree of stain resistance when used below the 35 phr level. The stain resistance of vinyls containing plasticizer blends disproportionately reflects the effect of the more desirable component.

4. *Toxicological and Environmental Considerations*: The toxicological properties of plasticizers are considered highly critical for certain plasticized PVC application such as biomedical devices and the processing and packaging of foods. They are also important to workers involved in the manufacture, shipping, and storage of plasticizers and in plastic processing operations. The toxicological properties of plasticizers are evaluated by the procedures commonly used for other liquid organic chemicals. Plasticized PVC compounds are more difficult to characterize as a function of plasticizer effects because of interference by other formulating additives used besides the plasticizer. Standard practices have, nevertheless, been developed to evaluate the net effect of plasticized compounds against specific food and medical use requirements [128,206,207]. These methods and their relation to specific end uses are continually subjected to revisions, reflecting concern about subtle effects resulting from prolonged use or exposure.

Variability in human response or sensitivity to various reagents complicates the ability of the toxicologist to predict toxicological characteristics quantitatively. This is further complicated by variability in the test animals used in the evaluation procedures. Acute toxicological properties are however, quantified by statistically designed single-dose experiments [208]. The single oral dose required to cause death to 50% of the test species, expressed as LD_{50} in units of dose per kilogram of body weight, is the most commonly used criterion for characterizing toxicity. These values should be supplemented with data on the acute response to peritoneal injection, vapor inhalation, and contact with skin and eyes. Long-term (chronic) feeding tests are necessary to judge the potential of toxicological effects due to prolonged exposure. Even with all of the foregoing data available on toxicology, the interpretation and significance of test results in relation to specific conditions are uncertain. Hence, specific tests, such as administering plasticized PVC extracts to test animals [29,206] and extensively testing plasticizers for skin and oral toxicity [209], are commonly performed.

Plasticizers, as a class of materials, are known to have a very low order of acute toxicity. Relative toxicity, often related to chemical structure, can be defined by the single oral dose procedure as shown in Table 11. For the purpose of interpreting the degree of toxicity, the following standards are generally applied to data determined by single oral doses administered to albino rats:

LD$_{50}$	Relative toxicity
<1 mg/kg body wt	Dangerous
1—50 mg/kg	Serious
50—500 mg/kg	High
0.5—5 g/kg	Moderate
5—15 g/kg	Slight
>15 g/kg	Of extremely low order

Tricresyl phosphate was used in the early era of plasticizers without recognition of its potential toxicological effects. The "brownout" practices during World War II imposed ventilating constraints at sites where industrial processes had customarily been conducted with open windows. In England particularly, workers began to suffer from polyneuritis. After much investigation, it was learned that skin absorption of the ortho-cresyl phosphate isomer was the cause of the problem [210, 211]. Since that time, industrial cresols have been produced which are relatively free of *ortho*-cresol [212].

It is common practice for plasticizer suppliers to publish statements summarizing the toxicological and physiological properties of their products. Since the early 1970s, much concern has been expressed about the toxicity of plasticizers, particularly DOP and dibutyl phthalate (DBP). The issue became of concern when it was reported that measurable levels of phthalate were identified in whole blood that had been stored in plasticized PVC bags. Concern increased when it was found that the same plasticizer could be identified as being present in some humans and animals; in some cases there was no obvious source of the phthalate found in the subject's blood and tissues. To this date, there is no clinical evidence of a link between exposure to DOP or other phthalates and adverse effects on humans [213, 216].

In 1980 the National Cancer Institute (NCI) reported that high dose levels of DOP (called DEHP in their report) caused liver tumors in rats and mice. Earlier tests with lower dose levels had resulted in negative findings. The potential for harm caused great concern in spite of the historical use of DOP-plasticized PVC in food and medical applications without apparent harmful human health effects [217]. Since 1981 an industry-sponsored program has been conducted by the Chemical Manufacturers Association (CMA) on the health and environmental effects of phthalate plasticizers. The program is being carried out with the full review of responsible U.S. government agencies, principally the Environmental Protection Agency and the Food and Drug Administration. The program continues in progress at this writing, while many separate and distinct studies also continue. In July 1984 the CMA Phthalate Esters Panel issued a statement describing results from completed studies and expressing the opinion that the scientific data suggest that the tumor-inducing potential of DEHP is unique to rodents fed at high dose levels [218].

TABLE 11 Single-Dose Oral Toxicity of Plasticizers

Plasticizer	LD_{50}[a] (g or ml/kg)	Plasticizer	LD_{50}[b] (g or ml/kg)
Phthalates		Epoxides	
Dimethyl·CH_3	2.4—7.2	Soybean oil epoxide	23
Diethyl·C_2H_5	8.2	2-Ethylhexyl epoxy tallate	23
Dibutyl·C_4H_9 (normal and iso)	8—>20	Aliphatic diesters	
Di-n-hexyl· C_6H_{13}	30—39	di(2-Ethylhexyl) adipate	9.1
		Diisodecyl adipate	21
Dicyclohexyl·C_6H_{11}	30	di(2-Ethylhexyl) azelate	8.7

Diisooctyl·C_8H_{17}	23
Di-2-ethylhexyl·C_8H_{17}	26—34
Diisononyl·C_9H_{19}	≤10
Diisodecyl·$C_{10}H_{21}$	>64
Ditridecyl·$C_{13}H_{27}$	>64

Phosphates

Tri-2-ethylhexyl	40
Tricresyl	9.3

Miscellaneous

Triethylene glycol di-2-ethylbutyrate	8.4
Dipropylene glycol dibenzoate	9.8
Tetraethylene glycol di-2-ethylhexanoate	18
Triethylene glycol di-ethylhexanoate	31
Polyester Flexol® R2H	>63

[a]*Environmental Health Perspectives*, Experimental Issue no. 3, January 1973; U.S. Department of Health, Education, and Welfare.
[b]"Plasticizer Toxicology Studies," Union Carbide Corp.; unpublished works; 1945—1970.

The CMA testing program on environemntal effects investigated 14
commercially important (large volume) phthalates. The program was designed to determine (1) minimum levels at which effects on aquatic life
may occur, (2) the potential for bioconcentration in aquatic life, and (3)
the relative persistence of phthalates in the environment. Assessment of
test results by the CMA Phthalate Esters Panel indicates that all of the 14
phthalates studied have sufficiently high safety factors to demonstrate low
potential for adverse environmental effects [219].

Plasticized vinyl products intended for use in handling or packaging
foods and drugs or intended to serve as drug or medical devices in the
United States must meet the requirements of the federal Food and Drug
Administration as defined in the Code of Federal Regulations (CFR). Food
and Drug Administration regulations that cite plasticizers for these uses
are listed in CFR Title 21, Food and Drugs, parts 170 to 199, April 1,
1982:

175.105	Components of Adhesives
175.300	Resinous and Polymeric Coatings
175.320	Resinous and Polymeric Coatings for Polyolefin Films
176.170	Components of Paper and Paperboard in Contact with Aqueous and Fatty Foods
176.180	Components of Paper and Paperboard in Contact with Dry Food
177.1200	Cellophane
177.1210	Closures with Sealing Gaskets for Food Containers
178.3740	General-Purpose Usage of Plasticizers in Polymeric Materials (includes food wrap film)
181.27	Prior Sanction Food Ingredients—Plasticizers

The FDA has been evaluating results of CMA and other animal testing programs on plasticizers as they become available. As of this writing, there
has been no indication by FDA that they intend to further regulate these
products for food and medical uses.

D. Costs

General-purpose plasticizers are commodity products of the petrochemicals
industry. The selling price traditionally reflects the manufacturing costs
associated with such products. Figure 1 shows static pricing of the general-purpose plasticizers, except for dramatic increases in the early to middle 1970s which are clearly a result of increases in hydrocarbon feedstock
costs resulting from the oil embargo. Miscellaneous and specialty-type plasticizers do not benefit from the economics of highly automated and high-vol-
ume throughput associated with the commodity-type plasticizers. These
materials require selected and often relatively costly feeds in order to meet
the specific chemical structures needed to impart the desired specialty performance properties. Because of this, their selling prices are always
higher than those of the general-purpose plasticizers. Epoxy-type plasticizers, based on natural oils, are driven by a separate set of raw material feed cost factors, plus the fact that epoxidation is a costly and specialized operation. Hydrocarbon-type extenders, on the other hand, are
generally low-cost by-products of the petrochemicals industry. Since
their selling prices have remained relatively constant, the use of plas-

ticizer extenders reflects the pricing of the primary plasticizers; high costs for primary plasticizers justify increased use of low-cost extenders. Extenders are seldom used when primary plasticizer prices are low.

The astute formulator of plasticized PVC recognizes that pressures on feedstock and manufacturing costs, as well as the market demand, cause shifts in the pricing of raw materials. The primary need for general-purpose plasticizers is to impart a specified degree of flexibility, with less concern about secondary properties associated with that material. Hence, a calculation is all that is needed to evaluate the potential for substituting different plasticizers for optimized costing in the general-purpose market-place. Formulators of plasticized PVC purchase raw materials on a cost per weight basis, but must recognize that nearly all end-use products are sold on a volumetric basis. Even the costs associated with manufacturing items in the formulator's own plant are evaluated on a cost per volume basis. This volumetric cost exists in the automotive and appliance industries as well as the insulated wire and cable (cents per linear foot at specified wall thickness) and flooring markets (cents per square foot at specified thickness). Thus, it is insufficient to evaluate raw material costs on a cents per weight basis only; they must be evaluated on a cost per volume basis. This is accomplished by multiplying the cost of each raw material by its specific gravity (sp.gr), which leads to the awkward term known as "pound-volume cost." What this exercise actually does is to convert the raw material cost from a cost per mass to a cost per unit volume as shown below:

$$\text{Cost/mass (¢/lb)} \times \text{mass/volume (density or sp. gr.)} = \text{cost/volume (¢/vol)}$$

Three major ingredients make up the bulk of general-purpose plasticized PVC formulations. Their relative costs are shown below, expressed as both cents per pound (in 1984) and cents per volume:

	¢/lb	× sp. gr.	= ¢/vol
PVC, emulsion	50	1.40	70.0
PVC, suspension	30	1.40	42.0
Plasticizer	40	.972	38.9
Filler	6	2.71	16.2

Assuming that there is no significant deviation of the costs of other additives which are added in minor concentrations, stabilizer, lubricant, pigments, and so forth, the above tabulation shows that the filler is the lowest-cost ingredient when compared on either a cost per pound or cost per volume basis; thus it is always desirable to keep the filler concentration as high as possible consistent with meeting end-use performance and processing requirements. Plasticizer, on the other hand, is a relatively high-cost raw material on a weight basis, but it is significantly less costly than the typical suspension PVC resin on a volume basis. The significance is that, other than filler, the plasticizer is the next least costly ingredient in general-purpose flexible PVC formulations. Thus, the astute formulator recognizes the value of using a less efficient plasticizer, which can serve as a cost extender for the PVC resin. At this point, the formulator should evaluate the plasticizing efficiency of the alternative raw materials, recognizing that a less efficient plasticizer also permits an increase in the filler content in order to match the specific gravity (yield) of the product. This has been shown to convert into a significant savings to formulators and manufacturers of plasticized PVC compositions [12].

In addition to formulation raw material costs, one must consider costly penalties associated with processing losses. As reviewed in Section IV, B, 4, numerous investigators have quantified problems of plasticizer volatility during processing and fabrication. Current industrial practice involves containment of volatile effluents, an added cost. Poppe [66] made a comparison of energy versus raw material costs involving the substitution of lower-volatile DINP and DIDP for DOP in oven-cured, spread-coated plastisols on substrates. Acknowledging that slightly higher cure temperatures are required, he showed that the lower-volatile phthalates imparted a significant cost savings. His example used typical energy (heating) costs of European industry, which are much higher than those in the United States.

V.   CLASSIFICATION OF PLASTICIZERS

Classification of plasticizers facilitates an understanding of their performance characteristics and an intelligent selection of the optimum plasticizer or plasticizer mixture for any intended purpose. Hopff [272] arranges all plasticizers into three classes:

1. Internal (copolymers)
2. Vulcanizable (plasticizer is polymerized in situ or is chemically reacted onto the polymer)
3. Mixtures

The plasticizer industry is based primarily on materials falling in Hopff's third class. Doolittle [212] has separated solvents, plasticizers, diluents and extenders according to performance characteristics. He considers a plasticizer a nonvolatile solvent and defines an extender as a relatively nonvolatile diluent used as a partial replacement for the plasticizer without seriously affecting the mechanical properties of the plasticized polymer. As he points out, classification according to performance characteristics leads to many ovelapping and gray areas. In another publication, Doolittle [200] divides plasticizers extensively into chemical families and types, for uses with various polymers. Buttrey [221] has summarized the properties and performance characteristics of many plasticizers in vinyls, cellulosics, and other resins.

There is a tendency to refer to plasticizers as "primary" versus "secondary," which leads to many diverse and confusing definitions having to do with degrees of compatibility. Suffice it to say that a primary plasticizer for any given formula is the one that is added in the major amount to accomplish the most desired effect; secondaries are those used in lesser amounts to provide secondary desired effects, for instance, cost reduction, stabilization, low-temperature properties, etc.

Plasticizers for vinyl resins can be divided conveniently into the following seven chemical classes:

1. Phthalates—general purpose
2. Epoxides—stabilizers
3. Esters of aliphatic dibasic acids—low temperature
4. Phosphates—flame retardant
5. Polyesters—permanent
6. Miscellaneous
7. Extenders—low cost

Table 12 shows these materials and their associated chemistry. Plasticizer
synthesis usually involves the classical reaction of an alcohol with an organic acid to form an ester. Simple monoesters demonstrate limited compatibility with vinyls; the acid is usually a di- or trifunctional carboxylic
type, which is reacted with a stoichiometric amount of an alcohol. The R
and R' groups may be aliphatic or aromatic. Diesters that act as plasticizers can also be made by the reaction of 2 moles of monobasic acid with
1 mole of a diol.

Polyesters are formed by the reaction of polybasic acids with polyols.
In the case of epoxy plasticizers, naturally occurring glycerides or unsaturated monoesters are treated to add the oxirane group at unsaturated
sites on the fatty acid portion of the ester; this is also possible with synthetically derived unsaturated esters. Phosphates are based on the derivative of the inorganic phosphoric acid. Extenders are usually nonesters
and represent a variety of chemical types, such as napthenic, aliphatic,
aromatics or halogenated paraffins.

Analyses to identify an unknown plasticizer normally involve identification of the ester linkages or nonester functional sites; this may be
followed by saponification to identify the specific alcohol and acid involved. Esposito and Swann [222] have reported transesterification with
lithium methoxide in methanol to be a rapid and valid method for identifying both the alcohol and acid portions of plasticizers. For quick assessment, Seymour [223] measures physical properties: refractive index, density, and boiling point. Braun [224] has summarized four chromatographic
techniques for plasticizer separation and identification. Halsam et al. [225]
measured the saponification number for quantitative analysis. Techniques
for separating the plasticizer from the vinyl polymer have also been reported, along with chromatographic and infrared analytical methods [226—229].
Nuclear magnetic resonance allows identification of aromaticity and branched
versus normal alcohols in plasticizers.

Although all materials classified as plasticizers demonstrate performance
consistent with the definition adopted by the Council of the International
Union of Pure and Applied Chemistry, they differ markedly in degree of
plasticization and effect on secondary properties. Certain generalizations
can be made, however, about performance characteristics of each chemical
family and about certain structural variations within the families. Most of
the plasticizers are hydrophobic liquids and are generally miscible with
one another over a wide range of concentrations and temperatures. It
has become common practice in the vinyl industry to use plasticizer blends
to provide the optimum balance of properties desired in vinyl products
consistent with reasonable economics.

TABLE 12 Families of External Plasticizers

Type	Family	Chemical structure	Synthesis
General purpose	Phthalates	![benzene ring with two COOR groups] COOR, COOR	Esterification of phthalic anhydride and higher alcohols, ranging from C_4 to C13, and in some cases including benzyl chloride.
Stabilizer	Epoxides	R —$\overset{O}{\triangle}$— $(CH_2)_n COO$—CH_2 R —$\overset{O}{\triangle}$— $(CH_2)_n COO$—CH R —$\overset{O}{\triangle}$— $(CH_2)_n COO$—CH_2	Expoxidation of unsaturated natural oils, and esters of tall oil fatty acids.
Low temperature	Esters of aliphatic dibasic acids	$ROOC(CH_2)_n COOR$	Esterification of linear dicarboxylic acids such as adipic, azelaic, sebacic.

Type	Formula	Description
Flame retardant		
Phosphates	$PO(OR)_3$	Reaction of aliphatic alcohols and phenols with phosphorus oxychloride.
Permanent		
Polyesters	$R[(OOC(CH_2))_n COO(CH_2)_m CH_2]OOCR$	Esterification of a polycarboxylic acid with a polyhydric alcohol; usually, diacids and diols are used. Frequently chain-terminated with monobasic acids.
Miscellaneous	—	Typical examples are esters of; trimellitic anhydride, isophathalic acid, terephthalic acid, citric acid, glycols, alkane sulfonic acid, etc.
Extenders (low cost) Aliphatic hydrocarbons Naphthenic hydrocarbons Aromatic hydrocarbons Alkylated aromatics Chlorinated hydrocarbons	—	Petroleum fractions, refined distillates, alkylated aromatics, etc.

A. Phthalates—General Purpose

The phthalates provide the most desirable combination of performance
characteristics and low cost of any plasticizer family used in vinyl resins
and thus have become the most widely used. Approximately 70% of the
plasticizer market is made up of phthalates. Nearly half of this volume is
for general-purpose applications. These materials have been the "work-
horses" for almost 40 years, which testifies to their optimum combination
of cost, availability, and performance characteristics. DOP and DIOP have
traditionally served as the preferred general-purpose phthalates. Since
about 1970, the marketplace has been served with a variety of general-purpose phthalates, which offer potential for cost/performance optimization
for specific end uses [12]. Their outstanding features compare to those of
DOP as follows:

> DIHP, diisoheptyl—stronger solvating
> DINP, diisononyl—lower volatility
> Linears: HNUP-di(linear $C_7C_9C_{11}$) and NHDP-di(normal
> $C_6C_8C_{10}$)—low-temperature flex
> DOTP, di(2-ethylhexyl) terephthalate—F-2 nitrocellulose
>   mar resistance

All of the above have competed with DOP and DIOP at general-purpose
prices. However, in a commodity market, it must be recognized that
ultimately the product of choice will be that which is derived from the
strongest raw material position. These GP-type plasticizers are summarized
in Tables 13 and 14.

Lower-molecular-weight phthalates and aralkyl phthalates, such as DHP
(dihexyl), DBP (di-*n*-butyl), and BBP (butylbenzyl) are higher in cost
and volatility, which are disadvantages compared to DOP. The aralkyl
phthalate (BBP) is less efficient than DOP and is at a cost disadvantage
because of its specific gravity and traditional selling price, both of which
are significantly higher than those of DOP. Likewise, DBP, is not widely
used in the United States because of its high cost and volatility, but it
and DIBP are commonly used for PVC in the Far East and South America.

The designation BOP is commonly used for phthalate coesters made of
butanol and 2-ethylhexanol, where the butanol is usually 30% or less of the
alcohol blend. These coesters also enhance solvency, but volatility is quite
high because of the inherent presence of dibutyl phthalate.

Diisodecyl (DIDP) is considered by some as a general-purpose plasticizer because its cost is only a small premium over DOP, as is the case with
some linear phthalates. Its volatility is significantly lower than that of
DOP, but it may require slightly higher processing temperatures (about 7°
to 10° C) than the other general-purpose phthalates, because it has slightly less solvating strength.

Commercially available high-molecular-weight phthalates are not cost
competitive with the general-purpose phthalates. They provide exceptionally low volatility and are somewhat more difficult to process. This
group includes DTDP (ditridecyl), UDP (coester of branched $C_{11}C_{12}C_{13}$),
and DUP (di-linear undecyl), as shown in Table 15.

Tables 13 through 15 compare the physical properties of the phthalates
most commonly used in vinyls. It is of interest to note that the specific
gravity decreases as a function of increasing molecular weight in the
homologous series of dialkyl phthalates, which is a pattern opposite to

that found in most homologous series of organic compounds. The vapor pressure of these liquids behaves according to the Clausius-Clapeyron equation, as indicated by the straight-line relation of the natural logarithm of vapor pressure versus the reciprocal of absolute temperature. It has been found that blends of more than one plasticizer deviate only slightly from ideality [241].

Dibutyl phthalate is the lowest-molecular-weight phthalate typically used in polyvinyl chloride and similar resins. Even though it has a boiling point of 340°C at atmospheric pressure, it is generally too volatile to be used in so-called hot-compounding (calendering, molding, and extrusion) operations, where temperatures of 160° to 170°C are normally used. Also, vinyl films containing dibutyl phthalate plasticizer undergo significant changes in mechanical properties as a result of plasticizer loss at ambient room conditions. Hence, dibutyl phthalate is seldom used as the sole plasticizer for polyvinyl chloride.

Commercial utility of phthalates includes chemical species beyond those typically considered general-purpose types. The highest-molecular-weight phthalate which is used to plasticize PVC is ditridecyl phthalate; higher-molecular-weight phthalates generally lack compatibility with PVC.

The range of utility of the phthalate plasticizers provides an excellent series of chemical homologs from which the effect of alcohol structure can be studied. The performance characteristics of phthalate plasticizers commonly used in polyvinyl chloride are shown in Tables 13 through 15. The effects due to alcohol interchange that are found in the phthalate family are reasonably consistent with what is found in the other chemical families. Table 16 summarizes the effects of alcohol structure on performance characteristics of plasticizers. Some of the generalities reported in Table 16 are not discernible when one is considering the phthalate family alone, but they become apparent in other chemical families used as plasticizers.

Table 13 compares various isomeric C_8 phthalates. The comparison includes DOIP and DOTP, the 2-ethylhexyl esters of isophthalic and terephthalic acids respectively, because they are occasionally used on a commercial basis. Since they have the same molecular weight as the other dioctyl phthalates, their performance is somewhat similar except for properties associated with the solvency of PVC. As shown by the minimum flux temperature on Table 13, these two plasticizers are significantly less solvating than DOP, and they show a slight advantage in low-temperature properties of plasticized PVC; commercial experience has shown that these two plasticizers are also somewhat less compatible than the traditional dioctyl esters of *ortho*-phthalic acid. Pure DNOP (di-normal octyl phthalate) is not commercially available at present in the United States, but it has been found to be less solvating for PVC than DOP by +10°C, as shown by the minimum flux temperature, and has similar efficiency, as measured by Durometer A hardness. Low-temperature properties are 10° to 12°C better than those of DOP, which is a characteristic performance advantage for the straight-chain alcohol plasticizers versus those derived from branched-chain alcohols. Conversely, extraction by oil is approximately 80% higher than that of DOP and DIOP. This again is a typical characteristic of straight-chain versus branched-chain plasticizers.

Table 14 provides a comparison of the phthalates that are commonly employed as general-purpose types. Minor, but measurable, performance differences may be found with these materials versus the dioctyl plasticizers shown in Table 13. The higher-molecular-weight phthalates DINP

TABLE 13   Dioctyl Phthalate Plasticizers for Polyvinyl Chloride

Characteristics	ASTM	DOP (di-2-ethylhexyl)		DIOP (diisoctyl)
Typical physical properties[a]				
Refractive index, n_D^{23}		1.4839		1.4860
Specific gravity, 23°C/23°C		0.986		0.980
Viscosity, cps at 23°C		60		60
Pour point, °C		-46		-46
Freezing point, °C		-55		-45
Vapor pressure, mmHg at 200°C		1.3		1.0
Boiling point, °C at 5 mmHg		231		236
Flash point, °F		425		425
Typical color, Pt-Co		20		30
Typical performance characteristics[b]				
Minimum flux temperature, °C		106		104
Plasticizer concentration, phr		50	48	49
Durometer A (15 sec)	D-2240	79	80	80
Tensile strength, psi	D-882	3140	3200	3140
Elongation, %	D-882	325	320	325
Stiffness at 23°C, psi	D-747	1900	2150	1980
T_4, °C	D-1043	+7	+8	+10
T_f, °C	D-1043	-21	-20	-23
Brittleness temperature, °C	D-746	-24	-23	-24
SPI volatility (0.020 in. thick, 24 hr at 70°C), wt %	D-1203	1.0	1.0	0.9
Oil extraction, K at 50°C	SPI-VD-T13	1.9	1.8	2.1

[a]Typical physical properties from Ref. 123.
[b]Most typical performance properties taken from Ref. 67.

	DCP (dicapryl)		DNOP[a] (di-n-octyl)	DOIP (di-2-ethyl-hexyl isophthalate)		DOTP (di2-ethyl-hexyl terephthalate)	
1.4848			1.4800	1.4858		1.4873	
0.980			0.976	0.976		0.981	
60			40	63		64	
—			-41 (super-cooled)	-46		-46	
-60			-25	—		-48	
—			0.7	0.7		0.4	
—			—	241		258	
394			426	450		450	
30			—	40		20	
106			116	123		135	
50	52		50[b]	50	52	50	52
81	80		78	81	80	81	80
—	—		2790	3180	3120	—	—
—	—		280	289	295	—	—
2500	2100		—	1900	1820	2700	1350
+10	+9		0	+7	+6	+11	0
—	—		-35	-26	-28	-21	-31
-22	-23		-34	-28	-29	-28	-36
0.7	0.7		0.9	0.6	0.6	0.9	1.0
1.7	1.8		3.4	2.1	2.3	1.9	5.1

TABLE 14   General Purpose Phthalate Plasticizers for Polyvinyl Chloride

Characteristics	ASTM	BBP (butylbenzyl)		DBP (di-n-butyl)	
Typical physical properties					
Molecular weight		ca 312		278	
Saponification number		ca 359		402	
Refractive index, n_D^{23}		1.538		1.489	
Specific gravity, 23°C/23°C		1.126		1.046	
Viscosity, cps at 23°C		50		17	
Pour point, °C		-45		-40	
Freezing point, °C		<-35		<-35	
Vapor pressure, mmHg at 200°C		2.0		9	
Boiling point, °C/mmHg		370/760		340/760	
Flash point, °F		390		340	
Typical color, Pt-Co		30		20	
Typical performance characteristics					
Minimum flux temperature, °C		71		68	
Plasticizer concentration, phr		*50*	*44*	*50*	*36*
Durometer A (15 sec)	D-2240	74	80	70	80
Tensile strength, psi	D-882	3780	4000	3190	3900
Elongation, %	D-882	259	250	272	200
Stiffness at 23°C, psi	D-747	1050	1600	690	1400
T_4, °C	D-1043	+7	+14	0	+20
T_f, °C	D-1043	-12	-7	—	—
Brittleness temperature, °C	D-746	-10	-6	-22	-10
SPI volatility (0.020 in. thick, 24 hr at 70°C, wt %	D-1203	2.2	2.0	6.7	3.0
Water extraction (0.020 in. thick, 24 hr at 70°C), wt %	—	1.7	1.6	—	—
Oil extraction, K at 50°C	SPI-VD-T-13	2.0	1.1	2.0	0.5

DIHP (diisoheptyl)		NHDP (di-n-C$_6$, C$_8$,C$_{10}$)	HNUP (dilinear C$_7$,C$_9$,C$_{11}$)	DINP (diisononyl)		DIDP (diisodecyl)	
362		ca 418	ca 418	419		447	
315		ca 266	ca 266	266		251	
1.488		1.482	1.482	1.486		1.485	
0.993		0.971	0.971	0.970		0.966	
50		39	42	97		113	
-40		-30	<-50	-48		-37	
-50		-35	-38	-55		-53	
2.2		0.8	0.6	0.5		0.3	
220/5		235/4	252/10	252/5		261/5	
390		430	430	435		445	
20		25	20	20		25	
96		106	105	113		120	
50	48	53	50	50	53	50	55
77	80	79	80	82	80	84	80
3220	3380	3040	3050	3230	3210	3320	3260
312	290	358	310	325	327	325	333
1600	1960	2000	2300	1910	1820	1920	1480
+5	+11	0	+3	+10	+7	+12	+6
-18	-15	-32	-30	-24	-25	-26	-30
-23	-20	-37	-34	-24	-25	-24	-27
1.6	1.5	1.1	1.1	0.7	0.7	0.4	0.4
<1	<1	<1	<1	<1	<1	<1	<1
1.9	1.5	5.5	3.8	1.9	2.5	2.0	3.1

TABLE 15 High-Molecular-Weight Phthalate Plasticizers for Polyvinyl Chloride

Characteristics	ASTM	DTDP (ditridecyl)	UDP (diiso-$C_{11}C_{12}C_{13}$)	DUP (dilinear C_{11})
Typical physical properties				
Molecular wt.		531	503	475
Saponification number		211	224	238
Refractive index, n_D^{23}		1.483	1.486	1.481
Specific gravity, 23°C/23°C		0.951	0.956	0.957
Viscosity, cps at 23°C		304	190	56
Pour point, °C		-37	-40	-9
Freezing point, °C		—	—	+2
Vapor pressure, mmHg at 200°C		<.01	0.15	0.6
Boiling point, °C/at mmHg		242/2	275/5	262/10

	ASTM Method			
Flash point, °F		460	440	490
Typical color, Pt-Co		40	40	80
Typical performance characteristics				
Minimum flux temperature, °C		154	145 est.	132
Plasticizer concentration, phr		50	50	50
Durometer A (15 sec)	D-2240	93	90	88
D (15 sec)		41	40	35
100% modulus, psi	D-882	1950	2000	1740
Tensile Strength, psi	D-882	3020	3080	2800
Elongation, %	D-882	340	353	311
T_f, °C	D-1043	-27	-22	-37
Brittleness temperature, °C	D-746	-31	-33	-49
SPI volatility (0.010 in. thick, 24 hr at 70°C), wt %	D-1203	0.54	1.8	1.5

Source: G. D. Harvey, et al. (1985), unpublished work, Exxon Chemical Co., Baton Rouge, La.

TABLE 16   Generalized Effects of Alcohol Structure on Performance
of Plasticizers[a]

Characteristics	Increased molecular wt.	Increased branching	Increased Aromaticity
Compatibility	Decreases[b]	Increases	Increases
Solvency	Decreases	Increases	Increases
Efficiency (as measured by mechanical properties at room conditions)	Decreases	Decreases	Decreases
Low-temperature properties			
Equivalent concentration[c]	Independent	Impaired	Impaired
Adjusted for efficiency[d]	Improved	Impaired	Impaired
Permanence properties			
Diffusibility[e]			
Equivalent concentration[c]	Independent	Decreases	Decreases
Adjusted for efficiency[d]	Increases	Decreases	Decreases
Surface-controlled loss			
Equivalent concentration[c]	Decreases	Increases	Decreases
Adjusted for efficiency[d]	Decreases	Increases	Decreases

[a]Generalized effects hold in most cases within a homologous series (family)
of plasticizers; exceptions do exist.
[b]Increased molecular weight generally impairs compatibility. However, the
phthalates demonstrate maximum compatibility approximately at dipentyl
phthalate [118].
[c]Equivalent concentration 50 parts by weight per 100 parts PVC resin.
[d]Adjusted for efficiency—to equivalent hardness of Durometer A = 80.

and DIDP have lower plasticizing efficiency, which has a beneficial effect
on the low-temperature properties versus DOP when formulated to specified
room temperature properties such as hardness and modulus.   Lower volatile losses of the higher-molecular-weight phthalates are significant advantages over DOP even at the adjusted efficiency concentrations, as shown
by the SPI volatility test.   Table 14 includes the di(linear alcohol) phthalate
NHDP (normal C_6, C_8, C_{10} phthalate) and HNUP (linear C_7, C_9, C_{11} phthalate).
These two plasticizers are derived from commercial alcohols with very little
side-chain branching, hence the terminology "linear."   The alcohol for
NHDP is 100% normal, made by ethylene growth in the Ziegler process.   The
commercial phthalate is made from a mixed $C_{6,8,10}$ alcohol stream.   HNUP
designates the phthalate of alcohols derived from the oxonation of linear
alpha olefins, which is about 65:35 normal 1:2-methyl branched; the commercial HNUP phthalate is made from a mixed $C_{7,9,11}$ alcohol stream.   Both

of these commercial phthalates are coesters of mixed alcohols with an average carbon chain length of about $C_{8.9}$. The absence of the 2-methyl branch in NHDP shows slight, but not significant, differences in performance properties of NHDP versus HNUP. The linear phthalates NHDP and HNUP show improved low-temperature properties by about -10°C; oil extractability is, however, approximately three times higher than that of DOP, as measured by the oil extraction constant, K, at 50°C.

The lower-molecular-weight phthalates DIHP and BBP are noted for their stronger solvating characteristics than DOP, and this property enhances processability of the PVC compositions. This advantage is a compromise with volatility; note that DIHP is only slightly more volatile than DOP, whereas BBP is significantly more volatile and carries a significant penalty in low-temperature properties. Another material which is commercially available and falls in this category of strong solvating properties is dihexyl phthalate. DHP has solvating properties and volatility similar to those of BBP, but its low-temperature properties are similar to those of DIHP. Because of its lower specific gravity (1.01), DHP is commonly employed as an economical substitute for BBP (sp.gr. = 1.12) when strong solvating properties are required, while imparting improved low-temperature flexibility over BBP.

Table 15 compares the high-molecular-weight phthalates that are commercially available. This group of phthalates cannot be considered general-purpose types because of their higher costs, but they are a desirable type of plasticizer for applications requiring low volatility, such as high-temperature wire and cable insulation and fog-resistant automotive interiors. Because of their higher viscosity and lower solvating characteristics for PVC, this group of phthalates are somewhat more difficult to process than the lower-molecular-weight phthalates. The effects of alcohol molecular weight and structural characteristics are consistent with those in the low- and medium-molecular-weight phthalates. For instance, increased linearity in the alcohol structure gives a more efficient plasticizer when measured against specified room temperature properties and also enhances low-temperature flexibility by about 8° to 10°C. The high-molecular-weight linear phthalates, however, show poor resistance to migration and oil extraction of the plasticizer.

The presence of a phthalate plasticizer can be detected by infrared or nuclear magnetic resonance analyses. Gas chromatography (GC) gives a fingerprint of phthalate structures. It has been reported that thin-layer chromatography is capable of separating dibutyl and dioctyl phthalates [243]. High-performance liquid chromatography (HPLC) is now developed to the point of being able to separate members of the phthalate family. However, for stable esters such as phthalates, GC is the preferred technique. Specific analyses may include measuring saponification number or identifying the alcohol product of saponification [234].

B. Epoxides—Stabilization

Epoxy plasticizers are those that contain the three-membered epoxide ring. They were introduced in the vinyl industry in 1947 by the Rohm & Haas Corporation [244] and currently represent approximately 7% of the total plasticizer consumption. Epoxides are primarily used as adjuvant heat stabilizers [154]. They also enhance light stability in vinyl compositions. As compatible liquids, they also contribute plasticizing action to the vinyl composition, and this effect must be considered when they are used in vinyl formulations.

Epoxy plasticizers are most commonly made by epoxidation of natural oils. Epoxidized soybean oil is widely used in PVC formulating. Epoxidized linseed oil is used to a lesser extent, for economic reasons. The linseed oil permits a higher oxirane content, which improves its stabilizing efficiency slightly versus soybean oil epoxide. Epoxy tallates are monoesters with limited compatibility, but are sometimes used for stabilizing performance while coincidentally enhancing low-temperature properties. Epoxy plasticizers have also been produced from synthetically prepared unsaturated esters [245]. One such product, diisodecyl-4,5-epoxytetrahydrophthalate (EHDP), is shown in Table 17. In addition to serving as adjuvant stabilizers, the family of epoxy plasticizers permit some generalizations concerning structure/performance comparisons, as shown in Table 17. The epoxidized oils (soybean and linseed) are essentially tri(fatty acid) esters of glycerine having a molecular weight of about 1000. Their performance is similar to that of low-molecular-weight polyesters, with inferior low-temperature properties and sensitivity to aqueous reagents, but good resistance to volatility and oil extractibility. The epoxytallate, on the other hand, imparts improved low-temperature and poorer permanence properties. The performance characteristics of the unique EHDP plasticizer are consistent with the expectations from its chemical structure; it resembles its phthalate analog (DIDP). In addition, it imparts improved heat and light stability characteristics to the vinyl composition, in the manner of the epoxy plasticizers derived from natural oils. It is unique, in that it does not support fungal growth, whereas epoxidized oils are susceptible to fungal attack.

The compatibility characteristics of epoxy plasticizers have been a subject of great controversy in the vinyl industry for many years. Early use of epoxy plasticizers in high concentrations resulted in incompatible vinyl products. Although it had been known that residual unsaturation, due to incomplete epoxidation, causes reduced compatibility, Fath [246] demonstrated that the summation of epoxy linkages and residual unsaturation of epoxidized oils did not equal the unsaturation present in the original starting oil. He was able to account for the difference by measuring the hydroxyl level in the epoxidized oil. It was apparent that the unsaturated structures that had disappeared had been epoxidized and then reacted in situ to form vicinal dihydroxide by scission of the epoxide ring, which takes place under aqueous acidic conditions. The vicinal dihydroxide is then capable of forming a dimerized ether while splitting off a mole of H_2O. Thus:

Alkene   Oxirane   Vicinal   Dimerized
                                 dihydroxide   ether

Bauer [247] reported that the compatibility of epoxy plasticizers varied inversely with the hydroxyl and polyether content; this contributed to poor compatibility in addition to that associated with residual unsaturation. Further, epoxidized oils produced in the absence of aqueous media were found to have a very low hydroxyl-to-oxirane ratio and exhibited good compatibility when employed at primary plasticizer use levels in polyvinyl chloride.

The presence of oxirane oxygen can be detected by infrared spectroscopy if the epoxide group constitutes a major portion of the plasticizer molecule. Quantitative determinations are accomplished by adding hydrogen bromide through wet chemical methods similar to those described in ASTM D-1652-67 and American Oil Chemists' Society method Cd 9-57.

The stabilization reactions and characteristics of epoxy plasticizers are discussed in Chapters 8 and 13 of this series.

C. Esters of Aliphatic Dibasic Acids—Low Temperature

Esters of aliphatic dibasic acids represent about 5% of the plasticizer market. They are typically referred to as the low-temperature plasticizers because they impart improved low-temperature flexibility and impact resistance to vinyl compositions. Many aliphatic diesters are of limited compatibility and generally are used in admixture with a primary plasticizer from the phthalate family. Table 18 summarizes the properties and performance characteristics of the aliphatic diesters commonly used in plasticized vinyls.

Analytical identification can be accomplished by infrared, thin-layer chromatographic, HPLC, nuclear magnetic resonance, and GC techniques [225,226,229—244]. As with the phthalates, GC is the preferred technique. More specific information may be gained from saponification number, and the alcohol or acid salt products of saponification, or both, can be identified.

D. Phosphates—Flame Retardancy

Phosphates, widely used in nitrocellulose, were also among the first types of plasticizers used in vinyls. Today approximately 2% of the plasticizer market comprises phosphates; it is estimated that an equal volume of phosphates is applied in nonplasticizer applications. Phosphate plasticizers are commonly used in admixture with other monomeric plasticizers in vinyl, primarily because of their ability to impart flame retardance. Phosphorus-containing materials ranging from volatile liquids to waxy solids serve as plasticizers for vinyl [249—251]. However, availability and marginal performance characteristics have restricted the use of phosphate plasticizers in vinyl primarily to those shown in Table 19.

Physical properties differ somewhat as the chemical structure changes from trialkyl (2-ethylhexyl) phosphate to the alkyl aryl and triaryl phosphates. The performance characteristics in vinyl shift likewise. Plasticizing efficiency, as measured by Durometer hardness, is reduced as a function of increased aromaticity (TOP vs. CDP and TCP) even though the solvency characteristics increase as a function of aromaticity in the phosphate family. This comparison indicates that solvency is not the sole contributing factor to plasticizing efficiency, which might be erroneously concluded from examining the phthalate family, for example. Gamrath [252] patented the use of octyl diphenyl phosphate, pointing out the optimum

TABLE 17 Epoxy Plasticizers for Polyvinyl Chloride

Characteristics	ASTM	ESBO (epoxidized soybean oil)	EPT (2-ethylhexyl epoxytallate)	EHDP (diisodecyl tetrahydro-4,5-epoxyphthalate)
Typical physical properties				
Molecular weight		ca 900	ca 410	467
Oxirane oxygen, %		6.3–7.2	4.4–5.2	3.4
Refractive index, n_D^{23}		ca 1.471	ca 1.456	1.467
Specific gravity, 23°C/23°C		ca 0.993	ca 0.916	0.975
Viscosity, cps at 23°C		350–700	25–30	180
Pour point, °C		ca –5	—	–38
Freezing point, °C		20–25	–9	–40
Vapor pressure, mmHg at 200°C		—	0.3	0.3
Boiling point, °C at 1mmHg		Decomposes at boiling	213	220
Flash point, °F		315	232	232

		35–80	50–80	20–35
Typical color, Pt-Co		35–80	50–80	20–35
Typical performance characteristics				
Minimum flux temperature, °C		125	128	120
Plasticizer concentration, phr		50	50[a]	50
Durometer A (15 sec)	D-2240	80	80	81
Tensile strength, psi	D-882	3410	3010	3000
Elongation, %	D-882	324	378	300
Stiffness at 23°C, psi	D-747	2500	1500	3600
T_4, °C	D-1043	+14	-6	+14
T_f, °C	D-1043	-7	-37	-27
Brittleness temperature, °C	D-746	-16	-42	-21
SPI volatility (0.020 in. thick, 24 hr at 70°C), wt %	D-1203	0.2	0.6	0.4
Oil extraction, K at 50°C	SPI-VD-T13	0.7	6.9	1.6

[a]Incompatible at 50 phr.

TABLE 18 Esters of Dibasic Acids Plasticizers for Polyvinyl Chloride

Characteristics	ASTM	DOA (di-2-ethylhexyl adipate)		DNODA (di-*n*-octyl *n*-decyl adipate)	
Typical physical properties					
Molecular weight		371		ca 398	
Saponification number		302		ca 282	
Refractive index, n_D^{23}		1.445		1.447	
Specific gravity, 23°C/23°C		0.920		0.912	
Viscosity, cps at 23°C		11		16	
Pour point, °C		-60		0	
Freezing point, °C		<-75		-5	
Vapor pressure, mmHg at 200°C		2.4		0.8	
Boiling point, °C at 5 mmHg		214		235	
Flash point, °F		385		400	
Typical color, Pt-Co		25		50	
Typical performance characteristics[a]					
Minimum flux temperature, °C		120		125	
Plasticizer concentration, phr		*50*	*40*	*50*	*51*
Durometer A (15 sec)	D-2240	74	80	81	80
Tensile strength, psi	D-882	2880	3350	2780	2760
Elongation, %	D-882	395	340	370	372
Stiffness at 23°C, psi	D-747	1150	2600	2310	2240
T_4, °C	D-1043	-11	+2	-5	-5
T_f, °C	D-1043	-48	-37	-46	-46
Brittleness temperature, °C	D-746	-56	-48	-52	-52
SPI volatility (0.020 in. thick, 24 hr at 70°C), wt %	D-1203	2.8	2.4	2.2	2.2
Oil extraction, K at 50°C	SPI-VD-T13	9.3	4.8	31	32

[a]All of these materials demonstrate limited or poor compatibility at the reported concentrations.

DINA (diisononyl adipate)	DIDA (diisodecyl adipate)		DOZ (di-2-ethylhexyl azelate)		DBS (dibutyl sebacate)	
399	427		412		314	
281	263		272		357	
1.449	1.452		1.449		1.444	
0.917	0.915		0.916		0.934	
16	22		18		10	
-62	-64		-73		—	
-60	<-50		-105		-12	
1.5	0.6		—		—	
230	245		240		200	
405	425		415		356	
25	35		40		50	
134	148		143		—	
50	*50*	*53*	*50*	*47*	*50*	*38*
80	82	80	77	80	67	80
2910	2940	2850	2810	2980	—	—
380	365	370	350	330	—	—
2700	3000	2800	2070	2700	840	1600
-2	+3	-1	-12	-6	-25	-2
-47	-46	-49	-50	-47	-55	-34
-52	-48	-50	-54	-51	-60	-49
1.7	0.6	0.7	0.9	0.9	5.0	5.0
8.9	8.5	9.6	18.0	14.8	12.5	8.4

TABLE 19 Phosphate Plasticizers for Polyvinyl Chloride

Characteristics	ASTM	TOP (tri-2-ethylhexyl phosphate)	DDP (isodecyl di-phenyl phosphate)	CDP (cresyl diphenyl phosphate)	TCP (tricresyl phosphate)
Typical Physical Properties					
Molecular weight		435	390	340	368
Saponification number		386	460	493	457
Refractive index, n_D^{23}		1.442	1.506	1.563	1.551
Specific gravity, 23°C/23°C		0.924	1.075	1.195	1.131
Viscosity, cps at 23°C		11	24	48	215
Pour point, °C		-74	<-50	-38	-33
Freezing point, °C		<-100	—	<-38	<-35
Vapor pressure, mmHg at 200°C		1.9	0.5	1.0	0.5
Boiling point, °C/mmHg		220/5	Dec. at 245/10	255/5	265/5
Flash point, °F		405	465	450	470
Typical color, Pt-Co		100	50	30	70

Typical Performance Characteristics

		100		72		65		71	
Minimum flux temperature, °C		100		72		65		71	
Plasticizer concentration, phr		50	48	50	46	50	57	50	63
Durometer A (15 sec)	D-2240	78	80	76	80	85	80	90	80
Tensile strength, psi	D-882	2480	2620	2700	2950	3030	2940	3700	3300
Elongation, %	D-882	265	260	360	315	320	335	230	258
Stiffness at 23°C, psi	D-747	1750	2110	1650	2200	1900	1520	3520	2170
T_4, °C	D-1043	-8	-4	+3	+11	+15	+7	+20	+13
T_f, °C	D-1043	-53	-49	-15	-10	-3	-6	+8	-1
Brittleness temperature, °C	D-746	-58	-55	-20	-7	+2	-4	+10	+4
SPI volatility (0.020 in. thick, 24 hr at 70°C), wt %	D-1203	1.7	1.6	1.2	1.2	1.5	1.6	0.6	0.7
Water extraction (0.020 in. thick, 24 hr at 70°C), wt %		<1	<1	<1	<1	1.8	2.0	<1	<1
Oil extraction, K at 50°C	SPI-VD-T13	8.4	7.3	3.1	2.4	1.7	2.6	0.5	1.2

balance of properties it affords for plasticizing efficiency, low-temperature properties, volatility, and flammability resistance. The trialkyl phosphate TOP is exceptionally good in low-temperature properties but is high in volatility, whereas tricresyl phosphate is inferior in low-temperature properties but has lower volatile loss. Although all the phosphate esters impart flame resistance, effectiveness varies as a function of the organic moiety. The vapor pressure and flash point values indicate that tricresyl phosphate should provide better flame resistance than trialkyl phosphate. This has been verified in performance testing of vinyl films [191]. The phosphate plasticizers also have the common characteristic of impairing heat stability if used as the main constituents of the system. This fact, coupled with high cost, has limited the use of phosphates as primary plasticizers.

Phosphate plasticizers can be detected by infrared spectroscopy [253], although gas chromatography will give a fingerprint of the different types of phosphates. Wet chemical methods have been described for detecting *ortho*- and *meta*-cresols in the presence of other phenols [254].

E. Polyesters—Permanence

Polyesters constitute about 4% of the plasticizer market. They are referred to alternatively as polymeric, resinous or permanent-type plasticizers. A basic study of variations in starting materials and the resultant effect on the polyester plasticizer performance indicates the attendant comprises with the use of polyester plasticizers [255]. The most notable compromise of performance characteristics is in permanence versus low-temperature properties when polyester plasticizers are used. In this case, increased molecular weight generally improves permanence and impairs low-temperature properties. Frissell [256] has divided all polymeric plasticizers into three groups based on the viscosity range. The separation makes it possible to compare polyester plasticizers of low, medium, and high molecular weight.

Table 20 shows the properties and performance characteristics of typical commercial grade polyesters. The shift in plasticizing efficiency and permanence properties as a function of molecular weight is apparent. However, high average molecular weight of a polyester is no guarantee of good permanence, because residual low-molecular-weight ends may be present and they would be quite fugitive.

Polyester plasticizers have made possible the successful use of flexible vinyls in applications such as refrigerator gasketing, pressure-sensitive tapes, and oil-resistant electrical insulation for high-temperature appliance wiring, all of which require resistance to plasticizer loss due to migration, extraction, and/or volatilization.

Polyester plasticizers can be identified by infrared and chromatographic techniques [225,228], but normally a combination of techniques is required for specific identification.

F. Miscellaneous

Miscellaneous plasticizers are materials that do not fit in any of the foregoing chemical classes or that have been restricted to small-volume use primarily because of their cost and overall performance characteristics as plasticizers. They are used in order to take advantage of some unique characteristic or compromise of properties. Typical examples are shown in Table 21. Only the most significant types used in vinyl compositions are considered here.

1. *Iso- and Terephthalates*: Technically, these isomers of *ortho*-phthalate esters belong in the family of specialty plasticizers. However, as discussed earlier, plasticizing performance characteristics are similar in some respects to those of the phthalate family. The unique feature of iso- and terephthalates used as PVC plasticizers is their resistance to marring of F-2-type nitrocellulose lacquers [257,258]. DOTP is somewhat inferior to DOIP in terms of solvating strength (minimum flux temperature), as shown in Table 13.

2. *Other Phthalic Anhydride Derivatives*: Other plasticizers that use phthalic anhydride include triesters [259,261] and tetraesters [262]. These are prepared by reacting the phthalic half-ester with a monohydric ester to give triesters; or 2 moles of the half-esters can be reacted with a diol to give tetraesters. In addition, diaryl and di(chlorinated alkyl) phthalates have been assessed as vinyl plasticizers [263,264]. Other plasticizers with chemical structures and performance characteristics similar to those of phthalates but with starting materials other than phthalic anhyride have been reported [235,245]. Most of these materials are relegated to the class of specialty plasticizers, primarily because of their cost. Table 21 shows the properties and typical performance characteristics of a few specialty plasticizers derived from phthalic anhydride. Butyl phthalyl butyl glycolate is an example of a phthalate triester.

3. *Trimellitates*: Triesters of trimellitic anhydride have become increasingly important as vinyl plasticizers because of increased demand for high-temperature-resistant flexible PVC. For example, in 1985 a new grade of UL-rated PVC insulated building wire, NM-B, was specified to replace type T and TW grades. The NM-B grade is specified for 90°C maximum operating temperature. Low-volatility plasticizer systems composed of trimellitates blended with high-molecular-weight phthalates are finding use in this application.

Trimellitates are similar to the phthalates in processability and water resistance, while providing the low volatility typical of polyester-type plasticizers. Resistance to extraction by oil and hydrocarbon solvents, however, is inferior to that of their phthalate counterparts. Physical properties and typical performance characteristics of representative trimellitates are shown in Tables 3 and 21.

4. *Citrates*: The citrates have long been commercially available, but not widely used because of their cost. The trialkyl citrates contain a hydroxyl group, which impairs compatibility in vinyl resins. Thus, acylated citrates are the preferred type for use in vinyl resins.

$$
\begin{array}{c}
CH_2-COOH \\
|\\
HO-C-COOH \\
|\\
CH_2-COOH
\end{array}
\quad + \quad
\begin{array}{c}
3ROH \\
\text{Alcohol}
\end{array}
\longrightarrow
\begin{array}{c}
CH_2-COOR \\
|\\
HO-C-COOR \\
|\\
CH_2-COOR
\end{array}
\quad + \quad 3H_2O
$$

Citric Acid   Trialkyl Citrate

TABLE 20 Polyester Plasticizers for Polyvinyl Chloride

Characteristics	ASTM	Low molecular weight		Medium to high molecular weight			
		Plastolein® 9720 (A)	Admex® 515 (B)	Drapex® 409 (C)	Paraplex® G-54 (D)	Flexol® R2H (E)	Plastolein® 9765 (A)
Typical physical properties							
Molecular weight (approximate)		850	—	2,000	3,300	2,000	3,500
Refractive index, n_D^{23}		1.462	1.463	1.465	1.462	1.467	1.476
Specific gravity, 23°C/23°C		1.040	1.054	1.083	1.094	1.056	1.089
Viscosity, cps at 23°C		990	661	3,500	7,500	11,500	10,000
Pour point, °C		—	+2	+4	—	+5	—
Freezing point, °C		-8	—	—	0	<-10	+2
Vapor pressure, mmHg at 200°C		—	—	—	—	—	—
Boiling point, °C at 5 mmHg		—	—	—	—	—	—
Flash point, °F		500	475	530	550	—	530
Typical color, Gardner		5	1	<1	<1	5	4

Typical performance characteristics

Minimum flux temperature, °C		—	140	140	—	147	—
Plasticizer concentration, phr		50	50	50	50	50	50
Durometer A (15 seconds)	D-2240	80	81	84	90	93	77
Tensile strength, psi	D-882	—	2,600	—	—	3,290	—
Elongation, %	D-882	—	340	—	—	215	—
Stiffness at 23°C, psi	D-747	—	3,300	2,600	2,800	9,000	2,200
T_4, °C	D-1043	+10	+11	+14	+17	+22	+12
T_f, °C	D-1043	—	-31	-8	-5	+18	-6
Brittleness temperature, °C	D-746	-18	-12	-12	-4	+4	-12
SPI volatility (0.020 in. thick, 24 hr at 70°C), wt %	D-1203	0.5	0.8	0.2	0.8	0.1	0.3
Water extraction (0.020 in, thick, 24 hr at 70°C), wt %		0.8	<1	0.6	1.7	<1	<1
Oil extraction, K at 50°C	SPI-VD-T13	0.4	1.6	0.4	0.2	0.1	0.5

Sources: (A) Emery Chemicals, Product Data Sheets 613B and 618B, Aug. 1985; (B) Nuodex Inc., technical literature; (C) Witco Chemical Corp., Argus Chemical Div., technical literature; (D) C.P. Hall Co., "Esters for Industry," Oct. 1985; (E) not commercially available, previously supplied as Flexol® R2H by Union Carbide Corp. and as Morflex® PR2H by Morflex Chemical Co.

Table 21 Miscellaneous Plasticizers for Polyvinyl Chloride

	ASTM	Dicyclohexyl phthalate (A) (H)	Di(butoxyethyl) phthalate(B)	Butyl phthalyl butyl glycollate(C)	Tri(2-ethyl hexyl) trimellitate(D)	Isobutyrate, benzoate di-ester of 2,2,4-trimethyl pentanediol -1,3(E)	Dipropylene glycol dibenzoate(F)	Triethylene glycol di-2-ethylhexanoate (G)	Acetyl tributyl citrate (H)	Dipenta-erythritol hexaester (I)	33-35% Acryloni-trile buta-diene Hycar® (J) 1012 × 41
Typical physical properties											
Molecular weight		330	366	336	546	320	342	347	403	ca 840	—
Saponification number		340	306	499	307	351	328	323	556	390	—
Refractive index, n_D^{23}		—	1.485	1.492	1.482	1.483	1.530	1.443	1.444	1.455	1.516
Specific gravity, $23°C/23°C$		1.198	1.063	1.105	0.987	1.020	1.127	0.968	1.050	1.013	0.974
Viscosity, cps at 23°C		Solid	36	55	255	28	190	12	35	135	>50000
Pour point, °C		—	− 48	<−40	− 50	—	− 19	− 65	− 75	− 51	—
Freezing point, °C		63 m.p.	− 55	<−35	—	− 41	− 40	—	—	—	—
Vapor pressure, mmHg, at 200°C		1.1	1.2	1.8	—	40	1.2	6	4	—	—
Boiling point, °C/mmHg		224/4	227/4	345/760	260/1	75/.01	235/5	196/5	173/1	143/4	—
Flash point, °F		420	407	390	505	325	290	385	399	572	—
Typical color, Pt-Co		100 (molten)	80	45	—	—	85	30	<50	<150	—
Typical performance characteristics											
Minimum flux temperature, °C		—	106	—	132	—	78	132	102	—	—

Plasticizer concentration, phr		54 (50/50 DCHP/DOP)	50	43	50	50	50	50	50	50	50
Durometer A (15 sec)	D-2240	76	79	87	84	87	79	78	75	85	95
Tensile strength, psi	D-882	2660	—	3090	2530	3100	3740	—	3480	3640	3410
Elongation, %	D-882	320	—	310	335	300	240	—	276	267	185
Stiffness at 23°C, psi	D-747	—	1390	—	5500	—	1250	2700	860	1250	15000
T_4, °C	D-1043	—	+ 4	—	+ 17	—	+ 18	− 6	+ 4	+ 13	+ 24
T_f, °C	D-1043	− 29	− 23	− 8	− 16	− 25	+ 3	− 42	− 18	− 13	+ 2
Brittleness temperature, °C	D-746	—	− 24	—	− 20	—	+ 8	− 46	− 18	− 8	0
SPI volatility (.020 in. thick, 24 hr at 70°C), wt %	D-1203	—	1.3	—	0.3	7.2	1.3	4.6	2.3	0.1	0.2
Water extraction (0.020 in. thick, 24 hr at 70°C), wt %		—	4.6	—	< 1	—	3.8	2.8	2.0	< 1	< 1
Oil extraction, K at 50°C	SPI-VD-T13	—	4.2	—	2.2	—	1.0	14.2	1.8	0.8	nil

Sources: (A) Nuodex Inc. (B) Not commercially available; previously marketed as Kronisol® by FMC Corp., Industrial Chemical Group. (C) Not commercially available; previously marketed by Monsanto Co. (D) Amoco Chemicals Corp.; Bulletin 2572-1-62. (E) Eastman Chemical Products Inc., Sub. Eastman Kodak Co. (F) Velsicol Chemical Corp.; Product Information Bulletin 35060; Oct. 1980. (G) Flexol® 3GO, Union Carbide Corp. (H) Morflex Chemical Co., Inc.; Technical Bulletin—Citroflex® Plasticizers; Dec. 1983. (I) Hercules Inc., Polymers and Additives; Jan. 1986. (J) B. F. Goodrich Chemical Group.

$$
\begin{array}{c}
\text{CH}_2\text{—COOR} \\
| \\
\text{HO—C—COOR} \\
| \\
\text{CH}_2\text{—COOR}
\end{array}
\quad\xrightarrow[\text{Acetic anhydride}]{(\text{CH}_3\text{CO})_2\text{O}}\quad
\begin{array}{c}
\text{CH}_2\text{—COOR} \\
| \\
\text{CH}_3\text{COO—C—COOR} \\
| \\
\text{CH}_2\text{—COOR}
\end{array}
\quad + \text{ CH}_3\text{COOH}
$$

       Trialkyl citrate   Acetyl trialkyl citrate

Most of the citrate esters are acceptable as plasticizers in food packaging material [265]. Acetyl tributyl citrate is shown in Table 21. It has found limited use in PVC because of inferior low-temperature properties and resistance to aqueous extractants.

    5. Diesters of Alkylene Glycols and Polyalkylene Glycols: Alkylene and polyalkylene glycols (diols) make diesters when reacted with 2 moles of monocarboxylic acids. These diesters are plasticizers for PVC, but have experienced limited use primarily due to cost.

    Increased chain length of the polyalkylene glycols results in repeated ether linkages, which reduce solvency and compatibility with the vinyl polymer and promote water sensitivity. The nature of the organic acid significantly alters the performance characteristics of the diester. Dibenzoate esters, for instance, show very good solvency for polyvinyl chloride and only fair low-temperature properties. The dialkanoate esters show limited compatibility and improved low-temperature properties in vinyl compositions [266]. Certain dialkanoate esters of poly(alkylene) glycols act as plasticizers for polyvinyl butyral resin, which is applied as the interlayer in laminated safety glass [267]. Table 21 shows the characteristics of dipropylene glycol dibenzoate and triethylene glycol di-2-ethylhexanoate.

    6. Other Miscellaneous Liquid Plasticizers: This group includes ester amides [268], ketones [269], ethers, sulfonamides, and halogenated compounds. These are discussed in detail in other publications [270–272]. They are not widely used in the United States, but in other countries availability of raw materials and different end-use requirements for the plasticized vinyl compositions are responsible for more common use of the plasticizers classified as specialty materials in this presentation.

    3. Solid Plasticizers: Solid plasticizers are not widely used, but are a unique group of materials that demonstrate plasticizing action for vinyls. Usually these materials tend to enhance permanence while impairing processability and mechanical properties of the vinyl composition. For instance, nitrile rubber imparts "nerve," or excessive elastic character, to the compound during milling or processing. However, it provides flexibility to the end product and is essentially nonmigratory in the end-use application [273,274]. The polar polymeric materials, such as nitrile rubber and chlorinated polyethylene [275], yield "alloys" or "polyblends" when compounded with PVC. These materials have unique properties that are technically different from those of plasticized PVC.

    Crystalline solids, such as dicyclohexyl phthalate, act as processing aids and give markedly improved flow characteristics of the vinyl polymer in the melt region, but high stiffness and brittleness are problems at room temperature. Table 21 shows the properties of dicyclohexyl phthalate and performance characteristics when it is used in a 50:50 admixture with DOP.

8. *Vulcanizable Plasticizers:* Vulcanizable or convertible plasticizers
represent an attempt to combine thermoset and thermoplastic materials.
The principle is to use a compatible liquid containing a chemically reactive
site, which is activated by application of heat or a catalyst or both during
the vinyl fusion cycle. The concept of vulcanizable plasticizers has been
considered for a long time [276,277]. However, it was not applied commercially until the 1950s, and the market remains small. Vulcanizable plasticizers have been based on diallyl phthalate [278—280], acrylic esters [1],
and monochlorostyrene [281]. The most widely used mode of application is
vinyl dispersion technology, because the dispersion is applied as a coating
or molded into a desired shape while it is in the liquid state. Some applications make it desirable to use a dispersion to form a finished product
that is rigid. This can be accomplished by polymerizing the plasticizer
after the vinyl dispersion has been applied. Besides being used in coatings,
polymerizable plasticizers are used in the molding of rigid heels in slush-
molded ladies' high-heeled shoe boots, or in PVC designed for end uses
demanding exceptional stability.

G. Extenders—Low Cost

Extenders are generally low-cost liquids having limited compatibility with
vinyl resins. They are normally used in admixture with primary plasticizers in the interest of reducing costs. The presence of impurities and
unsaturation can cause impairment of heat and light stability of vinyl compositions. Plasticizer extenders do not actively solvate the vinyl polymer,
which results in an overall impairment of properties of the plasticized vinyl,
especially permanence properties. Extenders are usually derived from low-
cost petroleum sources. They may be used as found or treated to enhance
their performance in vinyl compositions. Treatments may include distillation,
extraction, hydrogenation, alkylation, chlorination, polymerization, or sometimes nothing more than filtration. They may be separated into the following
types:

    Aliphatic hydrocarbons
    Chlorinated hydrocarbons
    Cyclic and alkylated cyclic hydrocarbons
    Polymerized products
    Miscellaneous reclaimed oils

Aliphatic and naphthenic hydrocarbons are usually found as by-products
in the petroleum industry. Sometimes they are distilled to select a desired
molecular-weight range. Useful molecular-weight range is limited to a carbon skeleton of more than 10 and fewer than 18 carbon atoms. The lower
molecular weights are excessively volatile, and the higher-molecular-weight
hydrocarbons have limited compatibility with PVC. The degree of branching
and size of branching groups also influence both volatility and compatibility. Aliphatic hydrocarbon extenders may cause only a mild sacrifice of
heat and light stability, but typical use levels are usually limited to less
than 20% replacement of DOP (primary plasticizer) level, due to compatibility limits. Primary plasticizers having more limited compatibility than
DOP cause the maximum use level of extenders to be lowered even further.
Chlorinated hydrocarbons are more costly than the aliphatic hydrocarbons
themselves. Heat and light stability of chlorinated hydrocarbons are some-

TABLE 22 Typical Plasticizer Extenders

Characteristics	ASTM	Aliphatic hydrocarbon (Escoflex® 150)(A)	Aromatic hydrocarbons, (Panaflex® BN-1)(B)	(HB-40)(C)	Chlorinated hydrocarbon, (Cereclor® 42)(D)
Typical physical properties					
Molecular weight		—	—	—	ca 530
Refractive index, n_D^{23}		1.490	1.570	1.580	1.506
Specific gravity, 23°/23°C		0.890	0.926	0.984	1.16
Viscosity, cps at 23°C		16	165	74	2500
Pour point, °C		-51	—	-26	—
Freezing point, °C		—	—	—	—
Vapor pressure, mmHg at 200°C		—	—	115	—
Boiling point, °C/mmHg		295/760	309/760	345/760	Dec. 260/760
Flash point, °F		—	—	345	>392
Typical color, Pt-Co		75	>500	90	130
Typical performance characteristics (E)		(4/1 DOP/E-150)	(4/1 DOP/BN-1)	(4/1 DOP/ HB-40)	(4/1 DOP/C-42)
Minimum flux temperature. °C		117	115	111	—

Property	ASTM method				
Plasticizer concentration, phr		*50*	*50*	*50*	*50*
Durometer A (15 sec)	D-2240	84	81	83	84
Tensile strength, psi	D-882	3310	—	—	3030
Elongation, %	D-882	230	—	—	250
Stiffness at 23°C, psi	D-747	2700	1600	2900	2200
T_4, °C	D-1043	+16	+9	+6	+15
T_f, °C	D-1043	-22	-17	-19	-19
Brittleness temperature, °C	D-746	-28	-18	-20	-18
SPI volatility (0.020 in. thick, 24 hr at 70°C), wt %	D-1203	6.2	4.3	4.7	1.0
Water extraction (0.020 in. thick, 24 hr at 70°C), wt %		<1	1.4	1.3	<1
Oil extraction, K at 50°C	SPI-VD-T13	2.4	0.7	1.1	1.6

Sources: (A) Escoflex 150 (aliphatic hydrocarbon), East Coast Chemicals Co., Technical Information 185E, April 1962; (B) Panaflex BN-1 (alkylated aromatic), Amoco Chemicals Corp., Bulletin N-2, July 1959; (C) HB-40 (partially hydrogenated terphenyl), Monsanto Co. Plasticizers, Section III, p. 50; (D) Cereclor 42 (42% chlorinated *n*-paraffin); Imperial Chemical Industries, Technical Bulletin 500; (E) Union Carbide Corp., unpublished works; 1962-1967.

what inferior to those of the unmodified aliphatic hydrocarbons, but compatibility and volatility are significantly superior. However, compatibility under sunlight exposure is notably poor with chlorinated hydrocarbons. Performance characteristics vary not only as a function of the carbon skeleton length but also as a result of the degree of branching and chlorination. In the past, chlorinated waxes were used as plasticizer extenders. Chlorinated normal paraffins are commonly available today [282]. The principal difference between these two types of chlorinated hydrocarbons is the low degree of branching in the normal paraffin skeleton versus random branching in the waxes; this results in advantages in heat stability and viscosity for the unbranched materials. The available products allow the vinyl formulating chemist to select materials of carbon chain length ranging from about 10 to as high as 22; the chlorination level is normally between 45 and 52%. These materials can be used at levels as high as 30—35% replacement of DOP. However, higher-molecular weight phthalates are less tolerant of chlorinated paraffin extenders, due to compatibility limitations.

Cyclic hydrocarbon extenders are usually residue products. They may be derivatives of naphthalene, or aromatic residues that may have been subjected to hydrogenation or alkylation [283]. In general, the cyclic extenders impart poor odor, color and heat and light stability, and also poor volatility; aromatic extenders impair the low-temperature properties of vinyl compositions. Compatibility, however, is sufficiently good to allow 35—40% replacement of the primary plasticizer, such as DOP. The starting material and the posttreatment significantly influence the nature and performance of the aromatic hydrocarbon extenders.

A summary of some typical plasticizer extenders is shown in Table 22.

VI. PLASTICIZER SYNTHESIS

A. Plant and Laboratory Scale [284]

Commercial production of plasticizers is carried out in relatively simple equipment. Materials of construction are typically stainless steel; essentially, the plant must have the facilities to add liquid or solid reagents or both to the esterification reactor. Relatively minor posttreatment is normally involved as the reaction products are removed, separated, and purified. Figure 10 shows a typical flow diagram of a commercial plasticizer plant. Ackerson [285] points out that this particular unit is designed to produce 95% of the commercial-type plasticizers. The required versatility is provided by the materials of construction, the heating and cooling facilities for the reaction vessel, and alternative posttreating facilities. Although storage facilities for raw materials and the finished products are not shown in the diagram, they are more extensive than normal in order to provide added versatility in this particular unit.

Since the most widely used plasticizers are phthalate esters, their preparation is considered in some detail. Because of the similarity in the preparation techniques employed for many different esters, the preparation of phthalate esters is considered as a general example. Cases of other esters which exhibit significant variations from the preparative procedure used to produce phthalates are noted separately.

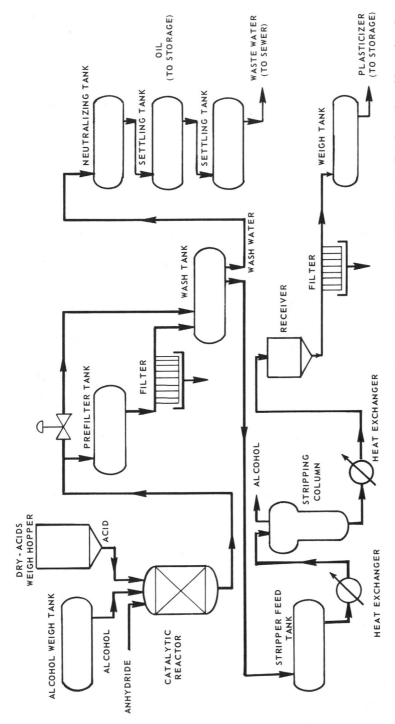

FIGURE 11 Commercial plasticizer syntheses—flow diagram. (Courtesy of *Chemical Engineering Magazine*.)

The basic chemistry involved in the preparation of phthalate esters takes place in two steps. In the first step of the reaction, a mole of alcohol and phthalic anhydride react to form a mole of monoester. This reaction is autocatalytic and relatively rapid, requiring only a few minutes at elevated temperature. The monoester then reacts with a second mole of alcohol to form the desired diester product. This second reaction is the rate-limiting step in the process and is accompanied by the evolution of a mole of water. A number of different catalysts, both acidic and metallic, have been used to enhance the rates of the esterification reaction.

The flow plan shown in Figure 11 is a typical design for carrying out this esterification on a commercial scale. Since the esterification reaction is an equilibrium one, removal of the water formed in the reaction is essential to drive the reaction to a desired level of conversion of the monoester to the diester. The reactor system and its associated equipment are specifically designed to accomplish this task. Occasionally, organic entrainers such as xylene or toluene are added to the reaction mass to aid in water removal.

Since phthalate esters typically must meet a specification on residual acid content, a neutralization and washing step is generally included in the process following completion of the reaction. Particularly in the case of acid-catalyzed esterifications, this step combines removal of the catalyst residues with removal of the unreacted phthalate monoester.

An excess of alcohol is normally employed during the reaction phase because of the equilibrium nature of the process. The presence of the excess alcohol helps by azeotroping the water and forces the reaction toward complete conversion of the phthalate monoester to the desired diester. Once the reaction is complete, however, this excess alcohol must be removed for the product to have the low light end content required to meet the specification normally imposed on phthalate plasticizers. This alcohol removal can be accomplished by any of a number of conventional stripping techniques. Steam stripping, nitrogen stripping, and vacuum distillation, with equipment such as wiped film evaporators, have all been employed for this step. Finally, the product may be subjected to finishing steps such as carbon treatment, filtration, and, in some cases, the addition of antioxidants prior to sale.

This process for making plasticizers is quite versatile, having been used in this general form for the production of adipates, trimellitates, isophthalates, citrates, and polyesters, among others. Naturally, reactor charges will be varied with these different products to reflect the different reaction stoichiometries that will be encountered. The versatility of the process is one of its strengths, since it allows the same plant to produce a number of different products. The number and type of products produced in the plant affect the nature and type of the auxiliary equipment, such as storage tanks, transfer lines, and loading facilities. Additional features, such as those to handle solid feeds, may be necessary for the production of some products.

The previous discussion was illustrated with a process flow scheme for a plant in which the esterification reaction is carried out in batch reactors. This type of operating mode is quite common in the industry and results in a plant with a considerable degree of operating flexibility. However,

when relatively few grades of products are to be produced, it has proved
practical to use a continuous reactor for the esterification section of the
plant [286,287].

Laboratory preparation of phthalate plasticizers is a straightforward
matter with the apparatus shown in Figure 12. In fact, once confidence
has been developed that laboratory procedures will produce esters with
reproducible colors, provided rigorous exclusion of air is accomplished,
laboratory esterification is the most reliable method of determining raw
material quality for the production of plasticizer [236–238]. Experimentally,
phthalates have been successfully prepared using olefins and alkyl groups
containing labile atoms in place of alcohols [239–241]. The traditional reaction of 2 moles of alcohol with one 1 of phthalic anhydride is depicted as
follows:

PHTHALIC ANHYDRIDE   MONOALKYL PHTHALIC ACID

DIALKYL PHTHALATE

Preparation of butylbenzyl phthalate [288] requires a procedure which
differs from the general process scheme described above. Application of
the general procedure to the preparation of mixed phthalate esters of two
different alcohols results in a statistical mixture of the possible reaction
products. In order to prepare a product containing only the desired
butylbenzyl phthalate without contamination from significant amounts of
either dibutyl phthalate or dibenzyl phthalate, a process is used in which
the reactants are phthalic anhydride, butyl alcohol, benzyl chloride, and
triethylamine. Commercial practice involves the introduction of all reagents
to the reactor at once. However, the process is conveniently viewed as
the sequential reaction shown in the following equation:

PHTHALIC ANHYDRIDE + C_4H_9OH (N-BUTANOL) + $(C_2H_5)_3N$ (TREITHYLAMINE) + BENZYL CHLORIDE → n–BUTYL HALF ESTER → TREITHYLAMINE SALT OF HALF ESTER → BUTYL BENZYL PHTHALATE + $N(C_2H_5)_3 HCL$ (TRIETHYLAMINE HYDROCHLORIDE) + H_2O (WATER)

Preparation of trimellitate plasticizer resembles that of phthalates, except that 3 moles of alcohol are added per mole of trimellitic anhydride:

TRIMELLITIC ANHYDRIDE + 3ROH (ALCOHOL) →(Δ, CATALYST) TRIALKYL TRIMELLITATE + $2H_2O$

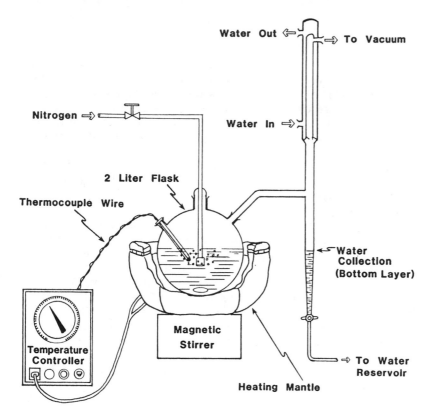

FIGURE 12 Laboratory esterification apparatus. (From R.W. Jenkins, Exxon Chemical Corp., unpublished work, 1975.)

Esters of aliphatic dibasic acids, low-temperature plasticizers, are synthesized similarly to the dialkyl phthalates [231,235,248]:

$$2ROH + HOOC-(CH_2)_n COOH \xrightarrow[\Delta]{H_2SO_4} ROOC(CH_2)_n COOR + 2H_2O$$

Adipic (C_6), succinic (C_4), glutaric (C_5), azelaic (C_9), and sebacic (C_{10}) acids have been used; however, adipic is most widely used because of its favorable economics. Alcohols commonly used in aliphatic diester plasticizers include butanol through isodecanol. In general, the shorter-chain alcohols are combined with the longer-chain dibasic acids and vice versa, so that the total number of carbon atoms is between 18 and 26. This affords a reasonable combination of compatibility and low volatility in vinyl compositions; any vinyl plasticizers not providing such characteristics have limited use.

Epoxy plasticizers are produced by adding oxygen across a double bond linkage, i.e.;

$$\text{C}=\text{C} \xrightarrow{[O]} \text{C}-\overset{O}{\text{C}}$$

Initially, commercial production of epoxy plasticizers commonly used hydrogen peroxide. The peroxide of an organic acid is generated in situ and epoxidation of the alkene proceeds in the oil phase while in contact with an aqueous layer. A commercial process that uses high-purity peracetic acid in an inert alkane [289] has been described. Absence of the aqueous medium results in a more efficient epoxidation process, yielding fewer undersirable side-reaction products.

Natural oils and other derivatives of natural products (tall-oil fatty acids) that contain a significant amount of unsaturation (iodine number 120 to 200) are commonly used as starting materials for producing epoxy plasticizers. Soybean and linseed oils are two of the most common, since they are readily available and have a high degree of unsaturation. These oils have a chemical structure typical of natural oils and can be represented as the tri(fatty acid) esters of glycerol. The 18-carbon fatty acid predominates, as indicated by the molecular weight of about 900.

Tall-oil acids are generated from a by-product of the paper industry, and they contain a significant level of unsaturated moieties, most of which are oleic and linoleic acids. These materials are first reacted with an alcohol, such as isooctanol or 2-ethylhexanol, to provide an unsaturated monoester structure, which is then subjected to epoxidation. The purity of the epoxy tallate esters can be expected to depend on the purity of the tall-oil acids. Commercial applications have shown that the performance characteristics of most commercially prepared epoxy tallates are equivalent to those of alkyl epoxy stearate, which uses commercial-grade oleic acid as the starting material.

Schematically, the treatment of the starting materials that provide epoxy plasticizers may be represented as follows:

Natural oils →[O]→ Epoxidized oils

Note: C_{18} acid predominates, but may range from C_{14} to C_{20} with varied C=C level

Unsaturated fatty acid, with C=C level dependent on starting material →ROH→ Ester of unsaturated fatty acid →[O]→ Epoxy tallate ester

Epoxy plasticizers have also been produced from synthetically prepared unsaturated esters [245]. One such product, diisodecyl-4,5-epoxytetra-hydrophthalate, is shown in Table 17. The performance characteristics of this plasticizer in vinyl are consistent with the expectations from its chemical structure; it resembles its phthalate analog (DIDP). In addition, it imparts improved heat and light stability characteristics to the vinyl composition, in the manner of the epoxy plasticizers derived from natural oils. It is unique, however, in that it does not support fungal growth, whereas epoxidized oils are susceptible to fungal attack.

Phosphates are produced by slow addition of phosphorus oxychloride to alcohol or phenol [290]:

$$3ROH + POCl_3 \longrightarrow P(OR)_3 + 3HCl$$
$$\overset{\parallel}{O}$$

Alcohol Phosphorus   Trialkyl Phosphate
   Oxychloride

The aliphatic alcohols are more readily reactive with phosphorus oxychloride than are the cresols. Thus, trialkyl phosphate is produced under cooling, whereas triphenyl and tricresyl phosphates require slightly elevated temperatures to accomplish synthesis. Mixed alkyl-aryl phosphates are readily synthesized by reacting phosphorus oxychloride with the desired ratio of aryl groups; then, at lowered temperatures, the alkyl alcohol is added [291].

Polyester plasticizers generally are reaction products of polyfunctional carboxylic acid and polyols [292-294]. Most commonly, dibasic acids, such as adipic or phthalic, are reacted with diols. If the reaction is conducted with a slight excess of organic acid, a monohydric alcohol is normally used as a chain terminator. Excess polyol results in residual free hydroxyl groups, which are normally terminated with acetic anhydride or some other monocarboxylic acid. The end product consists of a mixture of varied molecular weight polymers and is defined by average molecular weight values. In the case of polyesters, the content of the monofunctional material is the primary factor in controlling the average molecular weight, although reactants and reaction conditions are also factors. Progress of the reaction can be followed by measuring acid or hydroxyl numbers and viscosity. The reaction can be schematically represented as follows, where R represents alcohol terminated and R' represents acid terminated[*]:

$$x [HOOC(CH_2)_n COOH] + y[HOCH_2(CH_2)_m CH_2OH]$$

Dicarboxylic acid   Dihydric alcohol

$$\Delta \downarrow \quad \begin{array}{l} ROH \quad or \\ R'COOH \end{array}$$

$$R[OOC(CH_2)_n COOCH_2(CH_2)_m CH_2]_{x+y} OOCR' + [x + y + 1]H_2O$$

Polyester of varied molecular weight

---

[*]The example represents both alcohol and acid termination, but the two are not used simultaneously in practice.

B. Raw Materials

Plasticizer production consumes a wide range of raw materials derived from natural gas, oil, vegetable oils, and inorganic chemicals. These materials are subjected to numerous costly processes and treatments in the synthesis of plasticizer intermediates. Perhaps the most dramatic example is the treatment of raw materials such as natural gas and oil, which begins with the cracking operation to reduce the natural hydrocarbons to low-molecular-weight building blocks—alkenes. This is followed by separation and various reactions with the ultimate objective of building the higher-molecular-weight intermediates that are used to produce plasticizers. Vegetable oils or chemical by-products are usually subjected to relatively few treatments for the necessary purification or chemical alteration that will generate products that are useful as plasticizers.

The raw materials that the plasticizer industry uses can be separated into the following three classes: organic acids and acid anyhydrides, alcohols, and miscellaneous.

1. Organic Acids and Acid Anhydrides: Phthalic anhydride is the most commonly used cyclic dibasic acid anhydride because it is available at low cost in a pure state and its esters are highly compatible with polyvinyl chloride. Commercially, it is preferably used in the molten state, but may also be used in the flake form. It is produced by vapor-phase oxidation of *ortho*-xylene or naphthalene.

ORTHO — XYLENE

NAPHTHALENE

PHTHALIC ANHYDRIDE

The first synthesis of phthalic anhydride occurred prior to 1916, when, coincidentally, workers in Germany and in the United States separately carried out the reduction to practice [233]. Phthalic anhydride is used primarily in the synthesis of monomeric dialkyl phthalate plasticizers, although it is sometimes used for polymeric plasticizers.

The *meta* and *para* isomers of phthalic acid, known as isophthalic and terephthalic, respectively, are also useful for making plasticizers. Isophthalic acid became commercially available in the 1940s. It is made by liquid-phase oxidation of *meta*-xylene, or it can be found as a by-product in the commercial separation of isomer mixtures where dimethyl terephthalate is derived for production of polyester fiber [295].

Benzoic acid is the most common cyclic monobasic acid used in plasticizers. It is produced by oxidizing toluene:

TOLUENE   BENZOIC ACID

Benzoic acid is usually reacted in a 2:1 molar ratio with a dihydric alcohol to produce a diester plasticizer.

Trimellitic anhydride entered the plasticizer industry during the early 1960s. It is produced by the oxidation of pseudocumene, derived from aromatic petrochemical streams:

PSEUDOCUMENE   TRIMELLITIC ANHYDRIDE

Adipic acid is the aliphatic dibasic acid most commonly used in plasticizer synthesis. Low cost makes it the preferred dibasic acid for the synthesis of low-temperature plasticizers. It is made commercially by oxidizing cyclohexane, which is made by hydrogenation of benzene. Adipic acid is used to make low-temperature plasticizers and polyesters.

CYCLOHEXANE   ADIPIC ACID

Other aliphatic dicarboxylic acids used in the synthesis of plasticizers include succinic acid [$HOOC(CH_2)_2COOH$], azelaic acid [$HOOC(CH_2)_7COOH$], and sebacic acid [$HOOC(CH_2)_8COOH$]. Actually, the anhydride of succinic acid is used but azelaic and sebacic are used in the acid form. All of these acids are alcohol-soluble crystalline solids and may be used in making the same types of plasticizers as those for which adipic acid is used. Succinic is made by the hydrogenation of maleic anhydride. Azelaic and sebacic acids are derived from ozonolysis of natural fats and oils, which makes them more costly than adipic acid.

Monobasic acids are not widely used, since monoesters have limited compatibility in vinyls. Some monoesters are modified with chemical functionality, such as the oxirane group, that provides sufficient compatibility to allow their use at low concentrations in polyvinyl chloride. Monobasic acids may be reacted in a 2:1 molar ratio with diols to form diesters. The diesters of alkylene oxide-based diols find limited use in polyvinyl chloride.

Monobasic acids, such as isooctanoic, are also useful in polymeric plasticizer synthesis to esterify residual hydroxyl groups. Monobasic acids may be derived by ozonolysis of natural products or by oxidation of the corresponding alcohol.

2. *Alcohols*: Alcohols used in the synthesis of plasticizers range from methanol (CH_3OH) to tridecanol ($C_{13}H_{27}OH$), but the most commonly used plasticizer alcohols range from hexanol ($C_6H_{13}OH$) to decanol ($C_{10}H_{21}OH$). Both branched and unbranched (normal) alcohols are used in plasticizer synthesis; the structure of the alcohol depends on its source. Wickson [296] provides a historical review of the development of various types of plasticizer alcohols. The alcohol structure has significant effects on the performance characteristics of the plasticizers. Octanol, a common plasticizer alcohol, is a useful example for comparing the various methods of alcohol preparation and structural effects on plasticizer performance. The following grades of octanol have been used for plasticizers: isooctanol, 2-ethylhexanol, octanol-2, normal octanol.

1. *Isooctanol* is a member of a homologous series of randomly branched primary alcohols which range from C_6 to C_{13} and have found wide utility in plasticizers. Propylene and other low-molecular-weight olefins are oligomerized, providing a mixture of higher-molecular-weight olefins ranging from C_5 to C_{12} and containing methyl branches. The mixed oligomers are separated by distillation and subjected to oxonation [297,298], providing alcohols that have one or more carbon atom than the starting olefin. The oxonation, or hydroformylation, process yields 100% primary alcohols which have a specific molecular weight and are referred to as "iso" alcohols. Thus, isooctanol is made from heptene, isononal from octene, isodecanol from nonene, and so forth. Plasticizers made from the iso alcohols have performance properties similar to their specifically branched counterparts. For instance, DIOP and DOP are considered interchangeable, as are TIOTM and TOTM.

2. *2-Ethylhexanol* is a specific branched alcohol made by the aldol condensation of normal butyraldehyde, followed by dehydration and hydrogenation. Normal butyraldehyde is sourced from petrochemical streams or may be synthesized by hydroformylation of propylene. The commercial process leading to 2-ethylhexanol lacks the versatility associated with the oxo process, but does provide a product with a single molecular structure.

3. *Octanol-2* is a specific branched secondary alcohol. It is obtained as a by-product in the manufacture of sebacic acid from castor oil [299]. Limited availability and the poorer performance properties associated with all secondary alcohol structures have constrained the use of octanol-2 in plasticizers. DCP, dicapryl phthalate (technically a misnomer), has occasionally found use in plastisols, where it appears to impart desirable rheological properties.

4. *Normal octanol* is a primary unbranched alcohol that is commercially available in streams consisting of a mixture of normal alcohol homologs. The composition of these streams depends on the source and synthesis techniques. Historically, the only commercial source of normal alcohols used in the synthesis of plasticizers was coconut oil. Commercial production of synthetically produced normal alkanols was first accomplished by Continental Oil Company in 1962 by ethylene growth on a Ziegler-type catalyst [300], resulting in alcohols having even-numbered carbon skeletons. A mixture of C_6, C_8, and C_{10} normal alcohols is commonly used in plasticizer synthesis. The commercially available mixtures of normal alcohols have varied distributions with regard to carbon chain length. This becomes significant to the plasticizer consumer because plasticizers based on hexanol (C_6) are quite volatile, whereas plasticizers based on decanol (C_{10}) have low volatility but measurably less plasticizing efficiency. Since the phthalate of the alcohol mixture derived from coconut oil gave good overall performance, the synthetically produced mixtures of normal C_6, C_8, and C_{10} alcohols were designed to match the coconut oil-derived alcohols. However, improvements have been achieved with the synthesized normal alcohol mixtures by controlling the C_6:C_8:C_{10} ratio.

The normal alcohols are primarily used in manufacturing aromatic and aliphatic diester plasticizers. They contribute a very significant improvement in the low-temperature properties of vinyl, compared with similar esters based on branched alcohols. Plastisol rheology and heat and light stability are also improved somewhat but compatibility in polyvinyl chloride is somewhat more limited. Variations in the composition of the normal alcohol blends will change performance characteristics, most notably in processability and volatility.

Commercial grades of "linear" alcohols were introduced in the late 1960s. Fusco et al. [298] coined the term "linear" to describe primary alcohols which are produced from alpha olefins and contain more than 50% normal alcohols, the remainder being single methyl substituted on the 2 carbon. This is done by the use of carefully selected catalyst complexes, which encourage hydroformylation addition at the 1 carbon of alpha olefins. Commercially significant plasticizers that have been developed from these alcohols are HNUP (phthalate of linear C_7, C_9, C_{11} alcohol), shown in Table 13, and DUP (di-linear undecyl phthalate), shown in Table 14.

Projections have suggested that normal butylenes will become a low-cost feed from which plasticizer grade alcohols may be derived [301,302]. The *n*-butylenes may be dimerized to form octenes, with limited branching. These olefins may then be hydroformylated to a semilinear grade alcohol. The resulting dinonyl phthalate reportedly has performance properties that lie between those of DOP and DIDP [303], similar to those of commercial grade diisononyl phthalate.

Polyols, or polyhydric alcohols, are organic compounds having more than one hydroxyl group. Diols (containing two hydroxyl groups) are most commonly used in polymeric plasticizer synthesis, although materials

containing more than two hydroxyl groups have been used. Dipentaerythritol, for instance, contains six hydroxyl groups and is
used to make hexaesters, which are unique high-molecular-weight monomeric
plasticizers for vinyls [304,305].

Ethylene glycol, the simplest polyol, results from adding 1 mole of
water to 1 mole of ethylene oxide:

$$H_2C \overset{O}{\frown} CH_2 + H_2O \longrightarrow \overset{HO \quad OH}{\underset{H_2C-CH_2}{| \quad |}}$$

Ethylene oxide   Ethylene
                                                  glycol

If the ethylene oxide is encouraged to add further to ethylene glycol,
the reaction product continues to increase in molecular weight with
regularly spaced ether linkages, but the product remains a diol:

$$\overset{HO \quad OH}{\underset{H_2C-CH_2}{| \quad |}} \quad H_2C \overset{O}{\frown} CH_2 \longrightarrow \overset{HO \qquad\qquad\qquad OH}{\underset{H_2C-CH_2-O-CH_2-CH_2}{| \qquad\qquad\qquad\quad |}}$$

Ethylene   Diethylene glycol
glycol

$$n(CH_2 \overset{O}{\frown} CH_2)$$

$$\overset{HO \qquad\qquad\qquad\qquad\qquad\qquad OH}{\underset{H_2C-CH_2(O-CH_2-CH_2)_n-O-CH_2-CH_2}{| \qquad\qquad\qquad\qquad\qquad\qquad\quad |}}$$

Polyethylene glycol

Propylene oxide undergoes similar reactions to make a similar adduct, the
difference being a methyl group on alternate carbons along the chain.

Diols derived from condensation reactions do not contain the ether
linkage, which is generally not desired in vinyl plasticizers because it
has an undesirable effect on compatibility and water sensitivity. These
diols are commonly used in the manufacture of polyester plasticizers. In
addition, the diols may be used for producing monomeric diester plasticizers when reacted with 2 moles of monoacids.

$$2CH_3C\overset{H}{\underset{O}{\diagdown}} \xrightarrow[\text{(2) [H]}]{\text{(1) Aldol}} CH_3-\underset{\underset{OH}{|}}{CH}-CH_2-\underset{\underset{OH}{|}}{CH_2}$$

Acetaldehyde   Butanediol-1,3

Miscellaneous: *Organic halides* can be used in place of alcohols to produce esters. Although not commonly done on a commercial scale, it is possible to make dialkyl phthalates from alkyl halides, which are prepared from olefins [240, 241]. Benzyl chloride is an organic halide used in the commercial synthesis of pure esters such as butylbenzyl phthalate. It is prepared by passing chlorine over boiling toluene [306]:

$$\text{Toluene} + Cl_2 \longrightarrow \text{Benzyl chloride} + HCl$$

Toluene — Chlorine → Benzyl chloride

Phosphorus oxychloride is the main inorganic compound of significance to the plasticizer industry. It is used in manufacturing phosphate plasticizers. It is derived from phosphorous trichloride, phosphorus pentoxide, and chlorine:

$$PCl_3 + P_2O_5 + 3Cl_2 \longrightarrow 3POCl_3 + O_2$$

Vegetable oils such as soybean and linseed oils are essentially tri-(fatty acid) esters of glycerine with a reasonably high degree of unsaturation. In the naturally occurring state, they are not sufficiently compatible to be used as vinyl plasticizers. However, epoxidation of the unsaturated sites provides a unique heat stabilizer-plasticizer with good compatibility for polyvinyl chloride. As mentioned earlier, by-products of processes involving natural oils, such as tall oil, are used in producing epoxy monoester plasticizers.

VII. DESCRIPTIVE TEST METHODS

1. *Pour point is the temperature at which viscosity = 50,000 centistokes.*

2. *Flash point is determined by Cleveland Open Cup, ASTM D-92.*

3. *Minimum flux temperature is the minimum temperature for accomplishing fusion on a gradiently heated steel bar, using 100-phr plasticizer in emulsion-polymerized polyvinyl chloride by the following procedure.*

Procedure for Determining Flux Temperature [52]: The gradient hot bar is used to measure the solvating characteristics of plasticizers under static conditions. The apparatus consists of a steel bar resting on a series of hot plates. It is commonly referred to as the hot bar. The bar is cold rolled steel of size $1 \times 3 \times 26$ in. The top side has a smooth finish.

The heat source is a "vari-heat," which is supplied by Precision Scientific Company, Chicago catalog number 65500. The heaters are adjusted to provide a gradient temperature on the surface of the bar ranging from 40° to 200°C.

The minimum fluxing temperature test measures the temperature at which a plastisol develops the minimum tensile strength necessary for cohesion. The test composition is prepared according to the following formulation by spatula mixing in a beaker.

Component	By weight
Emulsion PVC (inherent viscosity 1.30)*	100
Plasticizer	100
Dibutyl tin dilaurate	1.0

To run the test, one lubricates the surface of the bar lightly with silicone stopcock grease, wiping the surface to leave an ultrathin film. The plastisol is then poured along the length of the bar on a uniform stream about 1/4 in. wide. After 20 minutes, the fused end of the strip is lifted and is gently and steadily pulled away from the bar, maintaining the specimen at right angles to the bar. The strip will break at the point where fusion is not quite sufficient for cohesion to have developed. The temperature at this point is measured with a contact pyrometer and called the minimum fluxing temperature. Reproducibility is of the order of ± 2°C.

4. *Mechanical and permanence properties were determined on compression-molded plaques with the formulation shown below.*

Component	By weight
Suspension PVC (inherent viscosity 0.97)*	100
Plasticizer	As shown
Dibasic lead phosphite	2.0
Dibasic lead stearate	1.0

5. *Oil extraction, according to M. C. Reed et al. [107].*

$$K = \frac{W_1 - W_2}{AT^{1/2}} = \frac{\text{weight Loss (g)} \times \text{time}^{-1/2} \text{ (h)}}{\text{surface area (m}^2)}$$

where

  K = diffusion constant
  W_1 = original specimen weight, g
  W_2 = exposed specimen weight, g
  A = total surface area of specimen (including edges), m^2
  T = exposure time, hr

*Inherent viscosity determined in accordance with ASTM D-2857-70.

6. *Procedure used to generate performance data at Exxon Chemical Co.*
[307]. The following are laboratory sample preparation and testing
procedures for plasticizers in PVC.

Formulations by weight:

	Suspension		Plastisol	
Component	By weight	Component	By weight
Susp. PVC (IV = 1.02)	100	Emul. PVC (IV = 1.20)	100
Plasticizer	25,35,50	Plasticizer	50,70,90
Mark 7101	2.0	Mark 7101	2.0
Stearic acid	.25	Stearic acid	—

Sample Preparation: Suspension PVC

1. Weigh and mix formula reagents in laboratory Hobart
   mixer (model N-50) at room conditions; mix all dry
   ingredients 1 minute at speed 1; add all liquids; mix
   5 minutes at speed 1 at room conditions.
2. Compound into fluxed homogeneous sheet on 8 × 16 in.
   equal speed, two-roll mill at 28 rpm for 5 minutes at
   specified temperature*; sheet off mill at 0.045 in. thickness;
   allow to cool.
3. Compression mold 6 × 6 in. test plaques to specified thickness,
   using Wabash press (model 75-184-4 STMAC):
   Preheat cycle: 15 minutes at 10 psi, at specified temperature*
   Mold cycle: 2 minutes at 500 psi, at specified temperature
   Cool cycle: 15 minutes with cold water on platens, 550 psi

---

Specified temperature for milling and molding is varied as a function
of plasticizer type:

Type	Milling temp.	Molding temp.
All aliphatic acid diesters and phthalates ≦ DIDP molecular weight	166°C (330°F)	171°C (340°F)
Phthalates > DIDP molecular weight	171°C (340°F)	177°C (350°F)
All trimellitates and polyesters	177°C (350°F)	188°C (370°F)

4. Die cut appropriate test specimens.
5. Condition test specimens under unstressed conditions for 7 days minimum at 23° ± 1°C (73° ± 1.8°F) 50 ± 3% relative humidity.

Sample Preparation: Plastisol

1. Weigh all dry ingredients—resins, fillers, etc.—directly into stainless steel mixing pot. Plasticizers and other liquid additions (stabilizers, etc.) are weighed into a suitable size glass beaker.
2. If more than one dry ingredient is used, these powders are mixed at speed 1 (slow) for 1 minute (Hobart Mixer model N-50).
3. For unfilled formulations, the liquid portion is added to approximately 43 phr by weight (to meet dough stage) at speed 1.
4. Mix dough at speed 2 (medium) for 5 minutes.
5. Reset mixer to speed 1. Add balance of liquids. The mixer is stopped and the beaker, pot, and stainless steel mixing blade are scraped with a semirigid spatula.
6. Reset mixer to speed 2 and mix for 8 additional minutes.
7. Stop mixing; record time of lab clock; scrape down blade and pot. Place vacuum desiccator and apply vacuum.
8. The sample will foam as the vacuum increases, then "break" as the air is released. Note: It may be necessary to break the vacuum or rap the desiccator to cause bubble breakage.
9. After the break, a timer is started and deaeration continues for 10 minutes. The vacuum is broken and the pot removed from the desiccator.
10. The deaerated plastisol is transferred to a suitable container (wide-mouth jar which can be covered to eliminate contamination) for running viscosities and placed in a constant-temperature bath for aging tests.
11. The time interval for plastisol viscosity (initial is 2 hr) began with the recorded time in step 7.
12. Samples (≤ 0.010 in. thickness) are cast on release paper, pregelled in a suitable oven (Werner-Mathis) for 30 sec at 150°C. Thick specimens (>0.010 in.) are gelled 45 sec at 150°C.
13. Gelled samples are compression-molded according to specified temperatures and cycle in Section 3 of Suspension PVC sample preparation.

Mechanical Properties:

Property	ASTM
Durometer hardness (stack 0.070 in. specimens to 1/4 in. minimum)	D-2240-81
Tensile properties (0.040 in.)	D882-81
Clash-Berg, T_f (0.075 in.)	D1043-72
Brittleness, T_B (0.075 in.)	D746-79

Permanence Properties:

1. Oven aging, forced air 7 days at 100°C (0.040-in.-thick die-cut specimens); suspend exactly 40 specimens in rack, rotating at 4.5 rpm; air flow 100 fpm; ELCONAP electric oven, type 1005
2. Volatility, activated carbon, 24 hr at 70°C; ASTM D1203-67, Method A (specimens 0.010 × 2 in. diameter)
3. Soapy water, 1% Ivory flakes, 24 hr at 70°C; ASTM D1239-55 (specimens 0.010 × 2 in. diameter)

REFERENCES

1. Sears, J. K. and Darby, J. R., *The Technology of Plasticizers*, Wiley, New York, 1982.
2. Mellan, I., *The Behavior of Plasticizers*, Pergamon, New York, 1961, p. 3.
3. Ostromislensky, I., U.S. Patent 1,721,034 (July 16, 1929).
4. Semon, W. L., to B. F. Goodrich, U.S. Patent 1,929,543 (Oct. 10, 1933).
5. I. G. Farbenindustrie A.G., British Patent 478,965 (Jan. 27, 1938).
6. Robertson, H. F., to Union Carbide Corp., U.S. Patent 2,222,490 (Nov. 19, 1940).
7. Douglas, S. D., to Union Carbide Corp., U.S. Patent 2,349,412 (May 23, 1944).
8. Kyrides, L. P., to Monsanto Co., U.S. Patent 1,923,938 (1933).
9. ASTM-D1249-81.
10. Gresham, T. L., to B. F. Goodrich, U.S. Patent 2,325,951 (Aug. 3, 1943).
11. *Mod. Plast.* 23 (3), 169 (1945).
12. Krauskopf, L. G., *Plast. Compounding*, Vol. 6, No. 1, pp. 28–32 (Jan./Feb. 1983).
13. *Modern Plastics Encyclopedia, 1983–84*, McGraw-Hill, New York, 1983, vol. 60, pt. 10A, p. 636.
14. *Mod. Plast.*, 61, (9), pp. 74–75 (Sept. 1984).
15. Lauzen, M., and Mancini, M., *Can. Plast,*, 42, 3 (Apr. 1984).
16. Frey, H. E., Bakker, J., Freeman, I., and Kamatari, O; Chemical Economics Handbook—SRI International; "Plasticizers", part 579; (Sept. 1984).
16a. *Eur. Chem. News* (Oct. 3, 1975), p. 37.
17. *Mod. Plas.*, 61, (1), p. 58 (Jan. 1984).
18. Coaker, A. W. M., and Bias, C. D., 26th Annu. Tech. Conf., Society of Plastics Engineers (SPE), New York, 1968.
19. Bargellini, F., *Mater. Plast.*, 28, 372–380 (1960).
20. Bergen, H. S., and Darby, J. R., *Ind. Eng. Chem.* 43, (10), 2404 (1951).
21. Schreiber, H. P., *Polym. Eng. Sci.*, 9 (1), 311–318 (1969).
22. Schreiber, H. P., *Polym. Eng. Sci.*, 10 (1), 13–18 (1970).
23. Goodrich, J. E., and Porter, R. S., *Polym. Eng. Sci.* 17 (1), 45 (1967).
24. McKinney, P. V., *J. Appl. Polym. Sci.* 9, 3359 (1965).

25. Ali, M. D., et al., *Paint Mfg.*, *20* (*3*), 103 (1950).
26. Clash, R. F., Jr., and Rynkiewicz, L. M., *Ind. Eng. Chem.* *36*, 79 (1944).
27. Doolittle, A. K., *The Technology of Solvents and Plasticizers*, Wiley, New York, 1954.
28. Park, R. A., (unpublished), Effects of Plasticizers upon Dry Blending PVC Resins, Dec. 3, 1958.
29. Taylor, R. B., and Tobolsky, A., U.S. Department of Commerce, Office Tech. Service AD413,495,26 (1963).
30. Terry, B. W., *Mod. Plast.* *44* (*5*), 160 (1967).
31. McKinney, P. V., *J. Appl. Polym. Sci.* *11*, 193 (1967).
32. Frissell, W. J., unpublished work, Union Carbide Corp., 1959.
33. Fuchs, O., and Frey, H. H., *Kunstoffe 49*, 213 (1959).
34. Ghersa, P., *Mod. Plast.* *35*, 135 (1958).
35. Gruenwald, I. G., *Kunstoffe 50*, 381 (1960).
36. Horsley, R. A., *Br. Plast. Progr.*, 77 (1957).
37. Kessenikh, R. M., et al., *Izv. Tomsk. Politekh. Inst. 126*, 36 (1964).
38. Voskresenskii, V. A., and Shakirzyanova, S. S., *Izv. Vyssh. Uchebn. Zaved. Khim. Khim. Tekhnol.* *6*, (*4*), 643 (1963).
39. Bernardo, J. J., and Burrell, H., *J. Polym. Sci.*, *1*, 537—578 (1972).
40. Sears, J. K., and Darby, J. T., in *Encyclopedia of PVC*, Vol. 1 (L.I. Nass ed.), Dekker, New York, 1976.
41. Doolittle, A. K., in *Encyclopedia of Chemical Technology*, Vol. 10, Interscience Encyclopedia, New York 1953, p. 777.
42. Geon® Vinyl Resin for Powder Coating, G-49, 7908 GL, B. F. Goodrich, 1979.
43 Joyce, A. C., Geon Vinyl Powders for Rotational Molding, Association of Rotational Molders' Spring Seminar, Scottsdale, Ariz., March 7, 1983.
44. Glass, J. E., and Fields, J. W., *J. Appl. Polm. Sci.*, *16*, 2269—2290 (1972).
45. McKinney, P. V., *J. Appl. Polym. Sci.*, 9, 3359 (1965).
46. Ueberreiter, K., *Plasticization and Plasticizer Processes*, Advances in Chemistry Series 48, ACS, Washington, D.C., 1965, pp. 35—48.
47. Wheeler, M. E., 26th Annu. Tech. Conf., SPE, May 1980, pp. 398—402.
48. Carleton, L. T., and Mishuck, E., *J. Appl. Polm. Sci.*, *8*, 1221—1255 (1964).
49. Defife, J. R., *J. Vinyl Technol. 2* (*2*) 95—99 (June 1980).
50. Whittington, L. R., *A Guide to the Literature and Patents Concerning Polyvinyl Chloride Technology*, 2nd ed., Society of Plastics Engineers, Stamford, Conn., 1963.
51. Hoy, K. L., *Am. Chem. Soc. Div. Org. Coat., Plast. Chem. Prepr.*, *25* (*1*), 375 (1965).
52. McKenna, L. A., *Mod. Plast.* *35*, 142 (1958).
53. Bauer, W. H., *SPE J. 17*, (2), 174 (1961).

54. Chosi, Y., et al., *Kogyo Kagaku Zasshi 68* (*4*), 718 (1965).
55. Van Veersen, G. J., and Meulenberg, A. J., *Plastics 17*, (*9*), 440 (1964).
56. Cohen, M. A., *J. Vinyl Technol.*, *2* (*4*), 234–237 (Dec. 1980).
57. Lewis, T. B., and Nielsen, L. E., *Trans. Soc. Rheol.*, *12* (*3*), 421–443 (1968).
58. Darby, J. R., and Graham, P. R., *Mod. Plast.*, *32* (10), 145–154, 250 (1955).
59. Greenhoe, J. A., *Plast. Technol.*, *6* (*10*), 43–47 (1960).
60. Graham, P. R., and Darby, J. R., *SPE J.*, 17 (1), 91–95 (Jan. 1961).
61. Frissell, W. J., *Mod. Plast. 38* (*9*), 232 (May 1961).
62. Courtier, J. C., *Plast. Mod. Elastomers 17* (*9*), 111 (1965).
63. Alexander, C. H., to B.F. Goodrich, U.S. Patent 2,500,891 (March 14, 1950).
64. Critchley, S. A., Hill, A., and Pater, C., in *Plasticization and Plasticizer Processes* (R. F. Gould, ed.), Advances in Chemistry Series 48, American Chemical Society (ACS), Washington, D.C., 1965.
65. Poppe, A. C., *J. Vinyl Technol.*, *3* (*3*), 175–178 (Sept. 1981).
66. Poppe, A. C., *Kunstoffe*, *70*, 38 (1980).
67. Werner, A. C., *SPE J.*, *22* (*10*), 67–70 (1966).
68. Werner, A. C., *Tappi*, *50* (*1*), 79A–84A (1967).
69. Defife, J. R., *J. Vinyl Technol.*, 2 (4), 222–226 (Dec. 1980).
70. Daniels, P. H., and Krauskopf, L. G., 84IT L379, unpublished work, Exxon Chemical Co., 1984.
71. Krauskopf, L. G., 82IT L619, unpublished work, Exxon Chemical Co., 1982.
72. Lasman, H. R., and Scallin, J. P., in *Encyclopedia of PVC*, Vol. 2 (L.I. Nass, ed.), Dekker, New York, 1977.
73. Rutkowski, A. J., unpublished works, Exxon Chemical Co., 1975.
74. Krauskopf, L. G., 83IT L110, L111, Exxon Chemical Co., 1983.
75. Toray Industries Inc., Japan Kokai Tokkyo Koho, 80, 121, 049, Application 79/27,660 (March 12, 1979).
76. Toray Industries Inc., Japan Kokai Tokkyo Koho, 80, 121, 050, Application 79/27,661 (March 12, 1979).
77. Kimura, I., *Sci. Ind.*, *39*, 753–6 (Dec. 1965).
78. Kuriyama, K., Murakami, S., Sugawarra, R., and Honda, K., France Demande 2,462,457 (Feb. 13, 1981).
79. Imada, K., Uenos, S., Nishina, Y., and Nomura, H., Shin-Etsu Chemical Ind. Co., Ger. Offen. 3,039,853 (May 14, 1981).
80. Bond, H. M., and Smith, W. P., to Minnesota Mining and Manufacturing Co., U.S. Patent 3,129,816 (April 21, 1964).
81. Oace, R. J., et al., to Minnesota Mining and Manufacturing Co., U.S. Patent Re. 23,843 (June 29, 1954).
82. Ellis, D., unpublished work, 82IT B42, Essochem Europe Inc., 1982.
83. Webber, C. S., and Pennisi, J. C., Great Britain Patent Specification 1,063,324 (March 30, 1967).
84. Balinth, I. J., to Johnson & Johnson Products Inc, U.S. Patent 4,335,026 (June 15, 1982).

85. Ernes, D. A., Gart, B. K., and Williams, R. C., *J. Appl. Polym. Sci.*, *29*, 383–397 (1984).
86. Downey, R. E., Goodyear Tire & Rubber Co., European Patent Application 82,630,079 (Aug. 20, 1982).
87. DeCoste, J. B., et al., *Chem. Eng. Data Ser. 3 (1)*, 131 (1958)
88. Hata, N., and Tobolsky, A. V., *J. Appl. Polym. Sci.*, *12*, 2597–2613 (1968).
89. Czekaj, T., and Kapko, J., *Eur. Polym. J.*, *17*, 1227–1229 (1981).
90. Fried, J. R., and Lai, S. Y., *J. Appl. Polym. Sci.*, *27*, 2869–2883 (1982).
91. Roy, S. K., Brown, G. R., and St. Pierre, L. E., *Int. J. Polym. Mater.*, *10*, 13–20 (1983).
92. Makaruk, L., and Retko, I., *J. Appl. Polym. Sci. 53*, 89–93 (1975).
93. Decoste, J. B., Howard, J. B., Wallden, V. T., and Zupko, H. M., Plasticized PVC for Retractable Cords, Annual Convention of Wire Association, Atlantic City, N.J., Oct. 15, 1958.
94. Alexander, C. H., to B.F. Goodrich, U.S. Patent 2,193,662 March 12, 1940).
95. McBroom, J., *Mod. Plast. 43 (5)*, 145 (1966).
96. Wartman, L. H., *Mod. Plast. 32 (6)* 139 (1955).
97. Tang, Y. P., and Harris, E. B., *SPE J. 23 (11)*, 91 (1967).
98. McCoy, P. T., unpublished work, Union Carbide Corp., 1964.
99. Okuyama, M., and Yanagida, T., *Kobunshi Kagaku, 20 (219)*, 385 (1963).
100. Frissell, W. J., *Plast. Technol. 8 (12)*, 32 (1962).
101. Harris, E. B., unpublished work, Union Carbide Corp., 1967.
102. Hooton, J. R., and Brofman, C. M., unpublished work, 83IT L782, Exxon Chemical Co., Baton Rouge, La., 1983.
103. Quackenboss, H. M., *Ind. Eng. Chem. 46 (6)*, 1335 (1954).
104. Quackenboss, H. M., unpublished work, Union Carbide Corp., Boundbrook, N.J., (1956).
105. Col, B., *Rubber Age 72 (2)*, 220 (1952).
106. Frissell, W. J., *Ind. Eng. Chem. 48*, 1096 (1956).
107. Reed, M. C., et al., *Ind. Eng. Chem. 46 (6)*, 1344 (1954).
108. Wartman, L. H., and Frissell, W. J., *Plast. Technol. 2 (9)*, 583 (1956).
109. Wales, M., *J. Appl. Polym. Sci. 15*, 293–310 (1971).
110. Anagnostopoulos, C. E., Coran, Y. E., and Gamrath, H. R., *Mod. Plast. 43 (2)*, 141–144 (1965).
111. Su, C. S., Patterson, D., and Schreider, H. P., *J. Appl. Polym. Sci. 20*, 1025–1034 (1976).
112. Darby, J. R., Touchette, N. W., and Sears, K., *Polym. Eng. Sci.*, *7 (4)*, 295–309 (1967).
113. Reed, M. C., unpublished work, Union Carbide Corp., 1953.
114. Hecker, A. C., and Perry, N. L., *SPE Tech. Pap.*, *6*, 58 (1960).

115. Anagnostopoulos, C. E., et al., *Mod. Plast. 43 (2)*, 141 (1965).
116. Darby, J. R., et al., *Polym. Eng. Sci. 7 (4)*, 295 (1967).
117. Van Veersen, G. J., and Meulenberg, A. J., *Kuntstoffe 57*, 561 (1967).
118. Doty, P. M. and Zable, H. S., *J. Polym. Sci. 1*, 90 (1946).
119. Walter, A. T., *J. Polym. Sci. 13*, 207 (1954).
120. Blair, C. M., and Carruthers, T. F., to Union Carbide Corp., U.S. Patent 2,120,927 (June 14, 1938).
121. Knappe, W., *Z. Angew. Phys.*, *6 (3)*, 96—101 (1954).
122. Kikkawa, M., *Kobunski Kagaku, 15*, 9—13 (1958).
123. Chamberlain, R., and Harrison, A. C., *Polym. Age, 36*, 549—334 (1972).
124. Park, G. S., and Van Hoang, T., *Eur. Polym. J., 15*, 817—822 (1979).
125. Hasson, A. M., Husson, F. C., Merle, G., and Gole, J., *J. Macromol. Sci.—Phys., B14 (4)*, 553—564 (1977).
126. Dutta, P. K., and Graf, K. R., *J. Appl. Polym. Sci., 29*, 2247—2250 (1984).
127. Vergnaud, J. M., *Polym.-Plast. Technol. Eng., 20 (1)*, 1—22 (1983).
128. Till, D. E., Reid, R. C., Schwartz, P. S., Sidman, K. R., Valentine, J. R., and Whelan, R. H., *Food Chem. Toxicol.*, *20*, 95—104 (1982).
129. Wildbret, G., *Environ. Health Perspect.*, 29—35 (Jan. 1973).
130. Jansen, L. E., and March, J., *Arch. Pharm. Chem. Sci. Ed., 5*, 43-49 (1977).
131. Messadi, D., and Vergnaud, J. M., *J. Appl. Polym. Sci., 27*, 3945—3955 (1982).
132. Messadi, D., and Vergnaud, J. M., *J. Appl. Polym. Sci., 26*, 2315—2324 (1981).
133. Messadi, D., Vergnaud, J. M., and Hivert, M., *J. Appl. Polym. Sci., 26*, 667—677 (1981).
134. Kampouris, E. M., Regas, F., Rokotas, S., Polychronais, S., and Pantazoylu, M., *Polymer, 16*, 840—844 (Nov. 1975).
135. Kampouris, E. M., *Polym. Eng. Sci., 16 (1)*, 59—64 (Jan. 1976).
136. Tordella, J. P., *Mod. Plast.*, 43 (1), 64—66 (Jan. 1976).
137. Kampouris, E. M., *Eur. Polym. J., 11*, 705—710 (1975).
138. Wentz, M., and Andrasik, I. J., *J. Coat. Fabrics*, 3—15 (July 1973).
139. Kumins, C. A., et. al., *J. Phys. Chem., 61*, 1290 (1957),
140. Kumins, C. A., and Roteman, J., *J. Polym. Sci., 55* 683 (1961).
141. Kumins, C. A., and Waythomas, D, J., *Official Digest, 34 (445)*, 177 (1962).
142. Frissel, W. J., unpublished work, Union Carbide Corp., 1953.
143. Frissel, W. J., and Saunders, E. E., unpublished work, Union Carbide Corp., 1956.
144. Van Delinder, L. S., and Canterbury, J. C., unpublsihed work, Union Carbide Corp., May 1962.
145. Beach, L. K., unpublished work, Ref. No. 7005 882, Exxon Chemical Co., Dec. 1969.

146. Frissel, W. J., *Ind. Eng. Chem.*, *48*, 1096 (1956).
147. Reed, M. C., *Mod. Plast.*, *27* (*4*), 117 (1949).
148. Williams, G. E., *J. Polym. Sci. Polym. Chem. Ed.*, *21*, 1491—1504 (1983).
149. Biggin, I. S., Gerrand, D. L., and Williams, G. E., *J. Vinyl Technol.* 4 (4), 150—156 (1982).
150. De Coste, J. B., and Wallder, V. T., *Ind. Eng. Chem.*, *47* (*2*), 314—322 (Feb. 1955).
151. De Coste, J. B., and Hansen, R. H., *SPE J.*, *18* (*4*), 431—439 (1962).
152. Himmler, G. G., and Nissel, F. R., *Plast. Technol.*, *3*, 4 (1957).
153. Stepek, J., et al., *Plast. Mod. Elastomers*, *17* (*8*), 150 (1965).
154. Anderson, D. F., and McKenzie, D. A., *J. Polym. Sci.*, *A-1* (*8*), 2905—2922 (1970)
155. Darby, J. R., et al., *Polym. Eng. Sci.*, *7* (*4*), 295 (1967).
156. Darby, J. R., and Graham, P. R., 17th Annu. Tech. Conf., SPE, Greenwich, Conn., 1961, Vol. VII, Session 24-1, 1—6.
157. Ward, M. P., and Krauskopf, L. G., unpublished work, Union Carbide Corp., March 1962.
158. Nass, L. I., unpublished work.
159. Saunders, E. E., unpublished work, Union Carbide Corp., 1960.
160. Orem, J. H., and Sears, J. K., *J. Vinyl Technol.*, *1* (*2*), 79—83 (1979).
161. Kojima, K., Kumafuhi, H., and Ueno, K., *Radiat. Phys. Chem.*, *18* (*5—6*), 859—863 (1981).
162. Krylovs, S. V., Kulifova, A. E., Ovchinnikov, Y. V., Lyutova, T. M., and Kozenko, S. M., *Plast. Massy*, *5*, 23—24 (1976).
163. Szymanski, W., and Smietanska, G., *J. Appl. Polym. Sci.*, *23*, 791—795 (1979).
164. Lerke, I., and Szymanski, W., *J. Appl. Polym. Sci.*, *28*, 513—518 (1983).
165. De Coste, J. B., *Ind. Eng. Chem. Product Res. Dev.* 7 (4), 238—247 (1968).
166. Ovchinnokov, Y. U., Bockkareva, G. C., and Bobrov, O. G., *Plast. Massy*, *12*, 13—15 (1980).
167. Cadmus, E. L., and Brophly, J. F., *Kunstoffe*, *72* (*4*), 25—26 (1982).
168. Stuhlen, F., and Pommer, E. H., *Kunstoffe*, *73* (*1*), 32—35 (1983).
169. Burgess, R., and Darby, A. E., *Br. Plast.*, *37* (*1*), 32 (1964).
170. Krauskopf, L. G., unpublished work, Union Carbide Corp., 1965.
171. Merz, A., and Dolezel, B., *Kunstoffe*, *57* (*9*), 726 (1967).
172. Merz, A., et al., *Kunstoffe*, *57*, 728 (1967).
173. Staff, C. E., to Union Carbide Corp., U.S. Patent 2,311,259 (Feb. 16, 1943).
174. Berk, S., et al., *Ind. Eng. Chem.* 49 (7), 1115 (1957).
175. Berk, S., *ASTM Bull.*, *168*, 53 (1950).
176. Scullin, J. P., et al., *Rubber Plast. Age*, *46*, 267 (1965).
177. Klausmeier, R. E., *Soc. Chem. Ind. Monogr.* *23*, 232 (1966).

178. Klausmeier, R. E., and Jones, W. A., *Dev. Ind. Microbiol.*, *2*, 47 (1960),
179. Fuoss, R. M., *J. Am. Chem. Soc.*, *61*, 2334 (1941).
180. Pendleton, J. W., and Engle, D. L., unpublsihed work, Union Carbide Corp., 1950.
181. Fitzgerald, E. R., and Miller, R. F., *J. Colloid Sci.*, *8 (1)*, 148 (1953).
182. Wartman, L. H., *SPE J.*, *20 (3)*, 254 (1964).
183. Deanin, R. D., et al., *Plasticization and Plasticizer Processes*, Advances in Chemistry Series 48, ACS, Washington, D.C., 1965, p. 140.
184. Price, R. E., *Polym. Eng. Sci.*, *6 (1)*, 75 (1966).
185. Farrar, M. W., and Johnston, K. G., to Monsanto Co., U.S. Patent 3,293,282 (Dec. 20, 1966).
186. I. C. I. Ltd., French Patent 1,361,041 (May 5, 1984).
187. Fischer, W. F., et al., to Standard Oil Development Co., U.S. Patent 2,593,420 (April 19, 1952).
188. Phillips, I., *Br. Plast.*, *37 (5)*, 261 (1964).
189. Phillips, I., *Br. Plast.*, *37 (6)*, 325 (1964).
190. Pugh, D. M. P., and Davis, B. I. D., *Plastics*, *30 (329)*, 103 (1965).
191. Veprek, K. J., and Frissell, W. J., unpublished work, Union Carbide Corp., 1954.
192. Keeney, C. N., *J. Vinyl Technol.*, *1*, 3 (Sept. 1979).
193. Hoke, C. E., 31st Annu. Tech. Conf., SPE, 1973, Preprints, pp. 548-552.
194. Mathis, T. C., and Hinchen, J. D., 31st Annu. Tech. Conf., SPE, 1973, Preprints, pp. 343-348.
195. Price, R. V., *J. Vinyl Technol. 1*, 2 (June 1979).
196. Brueggeman, B. G., oral consultation, Exxon Chemical Co., 1984.
197. Yarborough, V. A., and Haskin J. F., unpublished work, Union Carbide Corp., 1950.
198. Clark, C. A., 30th Annu. Tech. Conf., SPE, 1972, Preprints, pp. 623-627.
199. Monsanto Co., British Patent 970,596 (Sept. 23, 1964).
200. Van Etten, H. A., to E. I. du Pont de Nemours & Co., U.S. Patent 2,453,052 (Nov. 2, 1948).
201. Pinner, S. H., and Massey, B. H., *Br. Plast.*, *36 (10)*, 574 (1963).
202. Van Bramer, P. T., and Whitley, M. R., 21st Annu. Tech. Conf., SPE, Greenwich, Conn., 1965, 12, 1-5.
203. Godwin, A. D., oral consulation, November 1984.
204. Shah, A. C., 42nd Annu. Tech. Conf., SPEI, May 1984, p. 854.
205. Congdon, W. F., 32nd Annu. Tech. Conf., SPE, May 1974, p. 432.
206. USP XX, The United States Pharmacopoeial Convention (July 1, 1984), pp. 950-953.
207. U.S. Food and Drug Adminstration, *Toxicological Principles for the Safety Assessment of Direct Food Additives and Color Additives Used in Food*, Washington, D.C., 1982.
208. Weil, C. S., Carpenter, C. P., and Smyth, H. F., Jr., *Am. Ind. Hyg. Assoc. Q.*, *14*, 200 (1953).

209. Mallette, F. S., and van Haam, E., *Arch. Ind. Hyg. Occup. Med.*, *6*, 231 (1952).
210. Hunter, D., et al., *Br. J. Ind. Med.*, *1*, 277 (1944).
211. Piekacz, H., *Rocz. Panstwowego Zakladu Hig.*, *16* (*3*), 281 (1965).
212. Doolittle, A. K., *The Technology of Solvents and Plasticizers*, Wiley, New York, 1954, p. 862.
213. *Environmental Health Perspectives, Experimental Issue No. 3*, DHEW Publ. NIH 73-218, U.S. Dept. of HEW, Public Health Service, NIH, Jan. 1973.
214. Flexible Vinyls and Human Safety: An Objective Analysis, RETEC, Society of Plastics Engineers, March 20—22, 1973.
215. Conference on Phthalates, National Toxicology Program/Interagency Regulatory Liaison Group, Washington, D.C., June 9—11, 1981.
216. Group, E. F., *J. Vinyl Technol.*, *6* (*1*), 28—34 (March 1984).
217. Harris, G. W., Conference on Phthalates, National Toxicology Program/Interagency Regularity Liaison Group, Washington, D.C., June 9, 1981.
218. Moran, E. J., Position of the Phthalate Esters Panel on the Safety of Phthalate Esters, Ref. No. PE-80G, CMA, July 8, 1984.
219. Group, E. F., International Conference on Phthalic Acid Esters, University of Surrey, Guilford, England, Aug. 7, 1984.
220. Doolittle, A. K., in *Encyclopedia of Chemical Technology*, Vol. 10, Interscience Encyclopedia, New York, 1953.
221. Buttrey, D. N., *Plasticizers*, 2nd ed., Franklin Publ. Co., Hillsdale, N.J., 1960, p. 1.
222. Esposito, G. C., and Swann, M. H., *Anal. Chem.*, *34*, 1048 (1962).
223. Seymour, R. B., *Plast. World*, p. 58 (Sept. 1963).
224. Braun, D., *Plasticization and Plasticizer Processes*, Advances in Chemistry Series 48, ACS, Washington, D.C., 1965, p. 95.
225. Haslam, J., et al., *J. Appl. Chem.*, *1*, 122 (1951).
226. Criddle, W. J., *Br. Plast.*, *36* (*5*), 242 (1963).
227. Haslam, J., *Identification and Analysis of Plastics*, Van Nostrand, Princeton, N.J., 1965, p. 264.
228. Ligotti, I., et al., *Rass. Chim.*, *16* (*1*), 3 (1964).
229. Meise, W., and Ostromow, H., *Kunstoffe*, *54*, (*4*), 213 (1964).
230. Distillers Co., Ltd., and Tuerck, K. H. W., British Patent 675,581 (July 16, 1952).
231. Fraizer, R. B., to Union Carbide Corp., U.S. Patent 2,075,107 (March 30, 1937).
232. Staff, C. E., to Union Carbide Corp., U.S. Patent 2,311,261 (Feb. 16, 1943).
233. Stirton, R. I., in *Encyclopedia of Chemical Technology*, Vol. 10, Interscience Encyclopedia, New York, 1953, p. 586.
234. Wandel, M., and Tengler, H., *Kunstoffe*, *55* (*8*), 555 (1965).
235. Wickert, J. N., to Union Carbide Corp., U.S. Patent 1,972,579 (Sept. 4, 1934).
236. Coenen, A., in *Plasticization and Plasticizer Processes*, Advances in Chemistry Series 48, ACS, Washington, D.C., 1965, p. 76.
237. Dean, F., and Imperial Chemical Industries, British Patent 684,334 (Dec. 17, 1952).

238. Haslam, J., U.S. Patent 2,820,806 (Oct. 26, 1956).
239. Anastasiu, S., et al., *Mater. Plast.*, *2* (*6*), 351 (1965).
240. Monsanto Co., Netherlands Patent Application 6,503,262 (June 27, 1966).
241. Raether, L. O., and Gamrath, H. R., in *Plasticization and Plasticizer Processes*, Advances in Chemistry Series 48, ACS, Washington, D.C., 1965, p. 66.
242. Thinius, K., et al., *Plaste Kaut.*, *12* (*5*), 265 (1965).
243. Piekacz, H., *Bocz, Panstwowego Zakladu Hig.*, *16* (*3*), 281 (1965).
244. Ligotti, I., et al., *Rass. Chim.*, *16* (*1*), 3 (1964).
245. Mullins, D. H., et al., to Union Carbide Corp., U.S. Patent 2,924,582 (Feb. 9, 1960).
246. Fath, J., *Mod. Plast.*, *37* (*8*), 135 (1960).
247. Bauer, W. H., 19th Annu. Tech. Conf., SPE, Greenwich, Conn. 1963, 9, II-3.
248. Wickert, J. N., to Union Carbide Corp., U.S. Patent 2,032,679 (March 3, 1936).
249. Johnson, F., to Union Carbide Corp., U.S. Patent 2,668,800 (Feb. 9, 1954).
250. Lanham, W. M., to Union Carbide Corp., U.S. Patent 2,610,978 (Sept. 16, 1952).
251. Lanham, W. M., to Union Carbide Corp., U.S. Patent 2,909,559 (Oct. 20, 1959).
252. Gamrath, H. R., to Monsanto Co., U.S. Patent 2,504,120 (April 18, 1950).
253. Mellan, I., *Industrial Plasticizers*, Pergamon, New York, 1963.
254. Wehle, H., *Pharmazie*, *20* (*7*), 405 (1965).
255. Brice, R. M., et al., *SPE J.*, *19*, 984 (1963).
256. Frissell, W. J., *Resin Rev.*, *16* (*1*), 17 (1966).
257. Smith, P. V., Jr., et al., to Standard Oil Development Co., U.S. Patent 2,628,027 (Feb. 10, 1953).
258. Union Carbide Corp., British Patent 851,753 (Oct. 19, 1960).
259. Kyrides, L. P., to Monsanto Co., U.S. Patent 2,073,938 (March 16, 1937).
260. Levy, J., and Lighthipe, C. H., to Nopco Chemical Co., U.S. Patent 2,537,595 (Jan.9, 1951).
261. Monsanto. Co., British Patent 931,781 (July 17, 1963).
262. Mills, R. H., to Monsanto Co., U.S. Patent 3,155,714 (Nov. 3, 1964).
263. Carruthers, T. F., and Blair, C. M., to Union Carbide Corp., U.S. Patent 2,302,743 (Nov. 24, 1942).
264. Darby, J. R., to Monsanto Co., U.S. Patent 2,502,371 (March 28, 1950).
265. *Federal Register Amendment*, 31 F.R. 2897, Title 21, chap. 1, subchap. B, part 121, Food Additives, subpart E, p. 1, Feb. 18, 1966.
266. Doolittle, A. K., to Union Carbide Corp., U.S. Patent 2,285,420 (June 9, 1942).
267. Blair, C. M., and Carruthers, T. F., to Union Carbide Corp., U.S. Patent 2,120,927 (June 14, 1938).
268. Johnston, F., and Hensley, W. H., to Union Carbide Corp., U.S. Patent 2,472,900 (June 14, 1949).

269. Armour & Co., British Patent 518,027 (Feb. 15, 1940).

270. Buttrey, D. N., *Plasticizers*, 2nd ed., Franklin Publishing Co.,
      Hillsdale, N.J., 1960, p. 1.

271. Doolittle, A. D., *Encyclopedia of Chemical Technology*, Vol. 10,
      Interscience Encyclopedia, New York, 1953.

272. Hopff, H., in *Plasticization and Plasticizer Processes*, Advances in
      Chemistry Series 48, ACS , Washington, D.C., 1965, p. 87.

273. Imoto, M., et al., *Nippon Gomu Kyokaishi, 38* (*8*), 657 (1965).

274. Reed, M. C., *Mod. Plast.*, *27* (*4*), 117 (1949).

275. Merkel, R. R., and Searl, A. H., 21st Annu. Tech. Conf., Vol XI,
      Session XIII-4, 1–9, SPE, Stamford, Conn., 1965, 12,13,4,1.

276. Doolittle A. K., *Encyclopedia of Chemical Technology*, Vol. 10,
      Interscience Encyclopedia, New York, 1953, p. 862.

277. Garvey, B. S., Jr., to B. F. Goodrich, U.S. Patent 2,155,590
      (April 25, 1939).

278. Andrews, C. W., *Federation d'Association de Techniciens des Indus-
      tries des Peintures...de l'Europe Continentale*, 7, 133 (1964).

279. Hirzy, J. W., to Monsanto Co., U.S. Patent 3,336,362 (Aug. 15,
      1967).

280. Millard, J. W., et al., to Howards of Ilford, Ltd., British Patent
      1,074,467 (July 5, 1967).

281. Rubens, L. C., and Urchick, D., to Dow Chemical Co., U.S.
      Patent 3,275,713 (Sep. 27, 1966).

282. Rotenberg, D. H., in *Plasticization and Plasticizer Processes*,
      Advances in Chemistry Series 48, ACS, Washington, D.C., 1965
      p. 108.

283. Jenkins, R. L., to Monsanto Co., U.S. Patent 2,364,719 (Dec. 12,
      1944).

284. Lyford, J. IV., oral consultation, Exxon Chemical Co., Nov. 1984.

285. Ackerson, E. R., *Chem. Eng.*, *76* (*25*), 212 (1969).

286. Steinkkopff, D., *Verlag Darnstadat*, 1972, p. 78.

287. Eastman Chemical Division, Vol. 87 (16) *Chem. Eng.* 43 (Aug. 11, 1980).

288. Mills, R. H., to Monsanto Co., U.S. Patent 3,483,247; (December 9,
      1969).

289. Union Carbide Corp., British Patent 735,974 (Aug. 31, 1955).

290. Carruthers, T. F., to Union Carbide Corp., U.S. Patent 2,406,802
      (Sept. 3, 1946).

291. Van Wazer, J. R., in *Encyclopedia of Chemical Technology*, Vol, 10,
      Interscience Encyclopedia, New York, 1953, p. 506.

292. Carruthers, T. F., to Union Carbide Corp., U.S. Patent 2,929,827
      (March 22, 1960).

293. Hoaglin, R. I., et al., to Union Carbide Corp., U.S. Patent
      2,909,499 (Oct. 20, 1959).

294. Newton, L. W., to Union Carbide Corp., U.S. Patent 2,603,616
      (July 15, 1952).

295. Drexler, J., et al., *Plaste Kautschuk, 10* (*4*), 205 (1963).

296. Wickson, E. J., *ACS Symp. 159*, March 1980, pp. 181–195.

297. *International Encyclopedia of Chemical Science*, Van Nostrand, N.J.,
      1964, p. 862.

298. Fusco, E. J., et al., in *Plasticization and Plasticizer Processes*,
      Advances in Chemistry Series 48, ACS, Washingotn, D.C., 1965,
      p. 61.

299. Noller, C. R., *Textbook of Organic Chemistry*, Saunders, Philadel-
     phia, 1951, p. 541.
300. Lobo, P. A., et al., *Chem. Eng. Progr.*, *58* (*5*), 85 (1962).
301. *Chemical Marketing Reporter*, (Sept. 23, 1979), pp. 3, 23.
302. *Kagaker Kogyo Nippo* (July 16, 1980).
303. Lützel, G., and Holzmann, J., *Kunstoffe*, *74* (*4*), 222–224 (1984).
304. Fritz, F. A., to Hercules Powder Co., U.S. Patent 3,135,785
     (June 2, 1964).
305. Hayden Newport Chemical Corp. (J. P. Scullin and B. H. Silverman),
     German Patent 1,170,391 (May 21, 1964).
306. Stirton, R. I., in *Encyclopedia of Chemical Technology*, Vol. 10,
     Interscience Encyclopedia, New York, 1953, p. 596.
307. Harvey, G. D., unpublished work, 84IT L595, Exxon Chemical Co.,
     1984.
308. Krauskopf, L. G., in *Encyclopedia of PVC*, Vol. 1 (L. I. Nass,
     ed.), Dekker, New York, 1976, p. 521.
309. Hooton, J. R., unpublished work, 83IT L782, Exxon Chemical Co.,
     1983.
310. Rosen, S. L., *Fundamental Principles of Polymeric Materials*, Wiley,
     New York, 1982, p. 273.
311. Clash, R. F., Jr., and Berg, R. M., *Mod. Plast.*, *21* (*11*), 119
     (1944).

4

Lubricants for PVC

ROBERT A. LINDNER

Allied-Signal Corporation
Morristown, New Jersey

KURT WORSCHECH

Neynaber Chemie GMBH
Loxstedt, West Germany

I. INTRODUCTION

The two most important additives in rigid PVC compounds are stabilizers and lubricants. Without either, processing rigid PVC products successfully is impossible. It is desirable to know how important lubricants are, but it is critical to know how much to use. Use of excessive amounts is the problem most often encountered in the PVC area. Since these ingredients are so important for processing and physical properties, one would expect to find many references that explain lubricant usage. Little information is available and what is available is often very general. Statements about lubrication are general because of the nature of the problems. Lubricants interact with the other ingredients of a formulation, and a change in any one ingredient may affect the solubility of the lubricant in the entire system. Process machinery uses different ratios of heat and shear to achieve the same end product. A lubricant package designed for a low-heat, high-shear machine will fail dramatically in a high-heat, low-shear machine. A lubricant package is therefore usually designed for one specific set of conditions based on previous experience with the lubricants, the components of the compound, the machinery, and the process conditions. New lubricants should first be evaluated in the laboratory at different levels, temperatures, and shear rates.

Factorial interaction experimental designs are usually used in the laboratory to determine the practical range of usage. The results are then tested on full-size machinery. No one laboratory test has demonstrated complete correlation with production machinery, but a series of tests helps to eliminate materials that will not work. The problem of lubricant selection is made even more complex by the many different applications of PVC. A highly filled, plasticized floor covering compound requires a lubricant to meet different needs than those for a clear, rigid PVC food-grade bottle compound, or a high-impact, weatherable siding compound.

The task of organizing the existing knowledge in the field is an ongoing process. There is disagreement on how to classify lubricants as well as disagreement on the functionality of the lubricants within the classification systems. Many views have been proposed on the basis of specific tests. Time will sort out the various theories.

This chapter covers the function of lubricants, the mechanism of lubrication, a general discussion of molecular structure, attempts at the classification of lubricants, properties affected by lubricants, the types of materials available as lubricants and their advantages and disadvantages, lubricant selection for various applications, tests used for evaluation of lubricants, and typical lubricant formulation packages for various applications.

II. FUNCTION OF LUBRICANTS

The dictionary definition of the term lubricant does not fully encompass the role a lubricant plays in PVC, but if the interpretation of the key words "reducing friction on surfaces" is expanded to include intramolecular effects, a more complete picture of the function of lubricants in PVC is seen. The word lubricant is often used to describe a material whose purpose is to reduce adhesion between the polymer melt and the metal surfaces of processing equipment. We find that a material that reduces adhesion between metal and polymer usually reduces adhesion between polymer grains. In doing so, the lubricant also reduces interparticulate friction and changes the rate of fusion. Some lubricants reduce resistance to polymer chain slippage and allow one polymer molecule to slide over another, thus promoting flow. Lubricants can also be applied directly to the metal surfaces of the molds.

Usually, the term lubricant is used to describe internally compounded additives, whereas spray-on lubricants are called "release agents." In the PVC industry, assuming proper internal compounding of materials, directly applied release agents can be avoided. It is important to realize that in the context of the use of these materials in PVC, controlled application and controlled release are the key words. Lubricant level must be adjusted for optimum performance. Overuse of lubricants can lead to problems as severe as those experienced with insufficient lubrication. For example, in injection molding, if excessive amounts of external lubricants are used, cooled rings of melt are released prematurely from the mold surfaces, slide with the melt, and stack against each other. They do not knit together because they have already been cooled below their adhesion temperature. On heating, or on impact, the part opens up or shatters. This phenomenon is called delamination or splay by the manufacturer. It is due to uncontrolled release of the melt in the mold. In calendering operations, air blisters will appear between the film and the metal surfaces of the roll, causing imperfections in the surfacing. This is due to premature release from the roll due to overlubrication.

In extrusion processes, premature release of the material from the screw results in surging and erratic output. The extruder screw spins inside the melt and cannot drive material forward at a constant rate.

A lubricant must be only partially soluble in the PVC polymer. If the material has zero solubility, it will bloom to the surface. If it has too much compatibility, it will become a plasticizer and adversely affect the flexibility, hardness, or rigidity of the product. The materials most often used are long-chain hydrocarbons with a few polar groups to make them compatible. Some examples of these materials are natural waxes (e.g., carnauba, castor oil, hydrogenated castor oil), low-molecular-weight polyolefins, paraffin waxes, oils, fatty esters, fatty amides, and metal soaps, usually based on fatty acids and alkaline earth metals.

Lubricant additives are usually used in rigid PVC, where they are necessary to reduce melt viscosity and allow the machinery to process at high shear rates. Because they reduce frictional heat buildup and often contain costabilizing functional groups, lubricants can significantly improve the stability of a compound. In plasticized PVC, lubricant additives are used almost exclusively for their metal release properties, but in many cases they are not needed since the plasticizer system sometimes fulfills this role. The necessity of using lubricants increases as the plasticizer level decreases. Plasticizers reduce internal friction and promote flow. In applications where high shear occurs and the plasticizer level is low, lubricants are necessary. They are required for specialized areas such as food wraps, overlay films, wire and cable, and blown films. Lubricants function as processing aids. Because of the multitude of possible ingredients and the wide variety of processing equipment, there is no one universal lubricant that can provide all the needed properties for all formulations. Care must be taken to select lubricants and levels thereof which provide a proper combination of flow, metal release, and cost effectiveness. The best philosophy is to keep lubricant concentrations to a minimum. The old adage "If a little is good, twice as much is twice as good" does not apply here. Overlubrication is a problem that results in scorching, low output, surging, poor impact strength, and poor surface.

III. MECHANISM OF LUBRICATION

The mechanism of lubrication can be discussed in terms of two general classes of lubricants based on function. These are external lubricants and internal lubricants. They are very broad classifications and no one material functions entirely as an internal or external lubricant, but usually exhibits a combination of the properties of the two with one being predominant.

A. External Lubricants

These are responsible for two phenomena: metal release and particulate flow in the melt. The mechanism by which metal release is obtained is twofold. Some lubricants have a high affinity for metal surfaces and form a lubricant film, while others have low compatibility with the polymer and form a film at the interface between the polymer melt and the metal as they exude. Two types of lubricants, metal soaps and organic acids, appear to belong to the first group. Because of their polar nature, they have an affinity for metal and coat the metal surfaces. The polar ends of the molecules are attracted to the metal and the polymer, while the organic ends are attracted to themselves, forming a reduced-friction surface over which the polymer will slide. Esters of low polarity, polyethylenes, and aliphatic organic waxes represent the second group, which is incompatible with the polymer and will form a film on the polymer surface. In summary, the metal soaps and acids work by reducing the adhesion of the polymer to the metal, while the low-polarity organics work by providing a film that the polymer can slide on.

It is interesting to note that a synergistic effect occurs when members of the two groups are used together. This can be visualized as follows. The polar ends of the group 1 external lubricants are attracted to the metal and the polymer. The aliphatic tails are attracted to the group 2 lubricants. This allows the group 1 lubricant to act as a surfactant for the group 2 lubricant, spreading the droplets into an even film that the polymer can glide over. Controlled release is the key word, and this mechanism of metal release controls the properties of roll tracking in calendering, delamination in molding, and mixing in the extrusion process.

The second property imparted by external lubricants is "particle flow." Lubricants form a molecular film coating on the resin particles as well as the metal surfaces. They reduce friction and adhesion between resin particles in the polymer mass. If one thinks of resin particles as ball bearings lubricated by machine oil and tries to extrude them, the lubricated ball bearings would offer less resistance than unlubricated ones because frictional forces would be reduced to allow sliding as well as rolling (Figure 1). The total effect in extrusion is that lubrication of the resin particles slows fusion, which increases powder conveying over melt conveying. This phenomenon reduces the amperage demand on the machine and reduces temperature override problems in processing due to high viscosity. The effect can be observed in the Brabender fusion test. An analysis of the Brabender curve shows that the fusion peak is somewhat blunted where particulate flow dominates over melt flow. It should be noted that full melt homogenization is not necessary for development of full physical properties. What is necessary is rapid defusion of polymer chain segments across the polymer particle interfaces. To achieve this condition, it is necessary to obtain controlled fusion retardation such that particulate flow is maintained as long as possible for ease of processing but fusion is obtained before exit from the die.

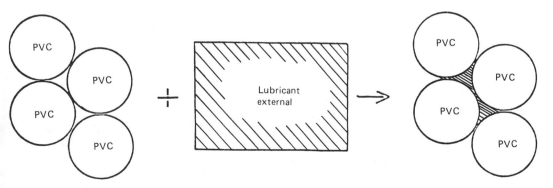

FIGURE 1 External lubricant.

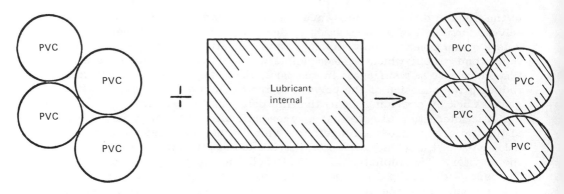

FIGURE 2 Internal lubricant.

     The controlling factor is proper lubricant selection. Lubricant must be
chosen to give the proper balance between particulate flow and molecular
flow. This will yield the optimum rheological properties for the material and
process. If good processing latitude or compound forgiveness is desired,
it is best to select lubricants with intermediate properties, which can control
particle flow versus melt flow over the widest range of conditions. When the
optimum balance is achieved, the PVC compound will have good physical
properties.

B. Internal Lubricants

These materials promote intermolecular flow via a chain slippage mechanism.
Such materials are characterized by being somewhat polar and semicompatible.
They are usually lower in molecular weight than the externals and demonstrate pseudoplasticizing effects. A plasticizer causes embrittlement at very
low concentrations, but internal lubricants are not polar enough to change
flexibility or hardness at room temperature. Internal lubricants function at
processing temperatures (Figure 2). The polar nature of these materials
allows them to be accepted into the polymer matrix at high temperature. The
van der Waals forces between chains are reduced by the polar lubricant,
which slips in between the polymer chains. This allows the chains greater
ease of movement because it decreases attraction between molecules. As rigid PVC cools, internal lubricant precipitates and becomes nonfunctional at
room temperature. The clarity is only slightly reduced because the lubricant
is very finely divided in the polymer matrix. On the other hand, external
lubricants adversely affect the clarity because of incompatibility and organize
themselves into large droplets.
     These pseudoplasticizers can adversely affect the heat distortion temper-
ature by lowering the glass transition temperature of the polymer. To test
for the properties of lubricants, a high-shear rheometric head attached to a
laboratory extruder can be used. With this device, calculations of viscosity
versus shear rate at various shear rates can be obtained. The die pressure
and output from the extruder give comparative information about the internal
properties of the lubricants. In injection molding, spiral-flow data allow
measurement of differences between internal lubricants based on length of

the spiral. Torque rheometers cannot usually be used to test internal lubricants. Since these materials are normally functional only at high shear, the
shear rates generated with older models of torque rheometers may be insufficient to discern differences at steady-state conditions. The data obtained
with microprocessor-controlled torque rheometers are more complete and
reproducible, and differences are more evident.

In real terms, internal lubricants reduce internal friction, which reduces
shear burn in injection molding applications, back pressure in extrusion, and
roll-parting forces in calendering. They give better gauge control in films,
sheets, and pipes, shorten cycles, and promote better mold filling. Internal
lubricants at typical concentrations have minimal adverse effects on fusion
time, nor do they contribute significantly to metal release. If the concentration of the internal lubricant exceeds its solubility constant in the polymer,
the internal will act as an external.

There are practical applications that take advantage of this phenomenon.
In bottles, an internal lubricant used at high concentrations gives high clarity and flow and allows use of a higher molecular weight resin in the formulation. Another application is in high-flow, rigid injection molding, where
maximum shear-burn resistance is required. This is best evaluated by a
shear-burn resistance test.

It has been known for some time that internal and external lubricants
can interact to produce enhanced effects. This is readily demonstrated by
the graphs in Hoechst's U.S. Patent 3,640,828 [2] (Figures 3–8), where
output is recorded at a given head pressure on an extruder as compounds
are processed that contain different ratios of polyethylene wax to ester waxes
and partially saponified waxes. One such case is illustrated in Figure 9. As
the ratio of wax OP to a polyethylene wax of molecular weight 2000 increases,

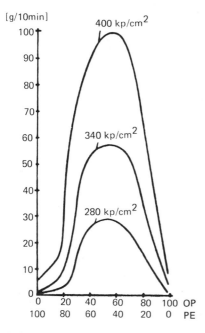

FIGURE 3 Hoechst-Wachs OP, PE wax MW 2000.

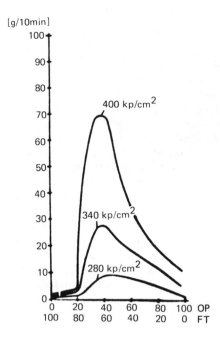

FIGURE 4 Hoechst-Wachs OP, FT paraffin.

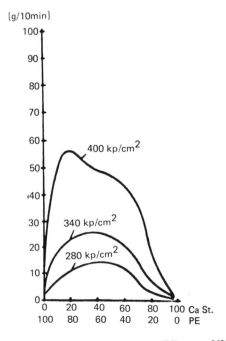

FIGURE 5 Ca Stearate, PE wax MW 9000.

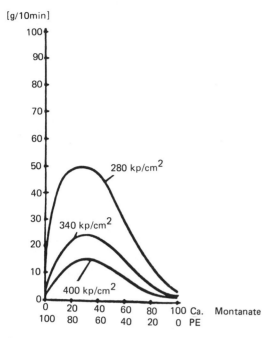

FIGURE 6 Ca Montanate, PE wax MW 9000.

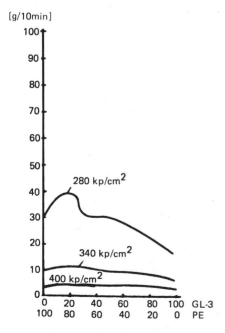

FIGURE 7 Hoechst-Wachs GL-3, PE wax MW 2000.

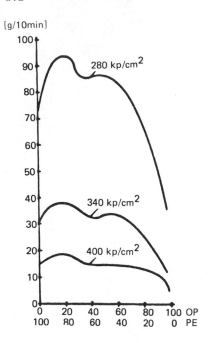

FIGURE 8 Hoechst-Wachs OP, PE wax MW 9000.

the output increases until it reaches a maximum at a ratio of 50:50, and then declines again as wax OP approaches 100%. Other graphs show similar results, but the results vary with the individual waxes evaluated and the internal-external balance within each wax and within the system. These effects appear to be due to the fact that the resin particle is partly invaded by and also surrounded by the internal lubricant, which comes in contact with the film of external lubricant. Therefore, the adhesion of the particle to the external lubricant film and of the external lubricant to the other resin particles is reduced. If each resin particle has a film of internal lubricant around it, then what is being registered is the adhesion between the internal lubricant film and the external lubricant film, which is less than the adhesion of the particles to the latter. The presence of internal lubricant also reduces the capacity of the resin to absorb external lubricant. Therefore, the external lubricant solubility is reduced and increased external lubrication will affect the results (Figure 10). The classical application of this is that equivalent or improved release effects can be achieved without increasing the adverse effects on fusion or without decreasing hot strength, resulting in a decreased demand for processing aid. This systematic approach contributes to good processing latitude and improved resistance to overlubrication.

IV. EFFECTS OF MOLECULAR STRUCTURE ON LUBRICANT FUNCTIONALITY

Many different materials can be used as lubricants, and an understanding of the molecular structure of lubricants and its effects on functionality is necessary for proper selection and utilization of lubricants. Variations in chain

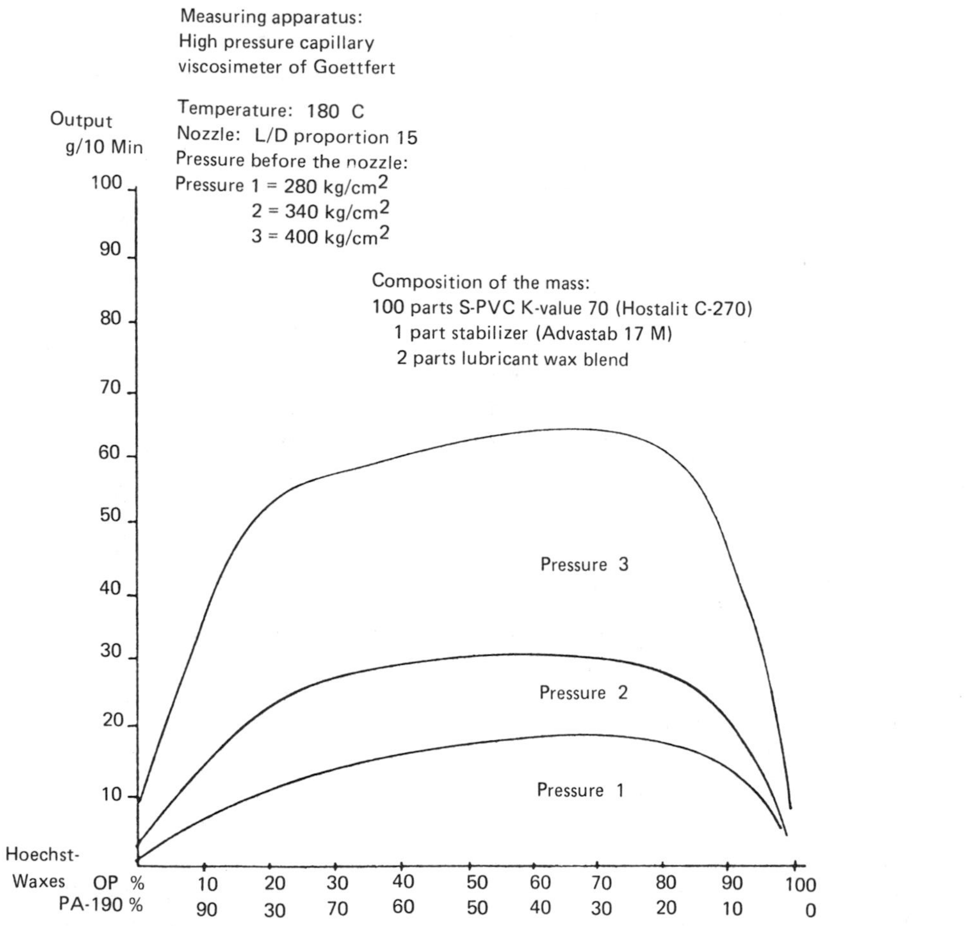

Measuring apparatus:
High pressure capillary
viscosimeter of Goettfert
Temperature: 180 C
Nozzle: L/D proportion 15
Pressure before the nozzle:
Pressure 1 = 280 kg/cm2
2 = 340 kg/cm2
3 = 400 kg/cm2
Composition of the mass:
100 parts S-PVC K-value 70 (Hostalit C-270)
1 part stabilizer (Advastab 17 M)
2 parts lubricant wax blend
Output
g/10 Min
100
90
80
70
60
50
40
30
20
10
Pressure 3
Pressure 2
Pressure 1
Hoechst-
Waxes OP %
PA-190 %
10 20 30 40 50 60 70 80 90 100
90 30 70 60 50 40 30 20 10 0

FIGURE 9 Flow behavior of hard PVC with a lubricant mixture of wax OP/PA-190.

PVC
PVC
PVC
PVC
Lubricant
internal/external
PVC
PVC
PVC
PVC

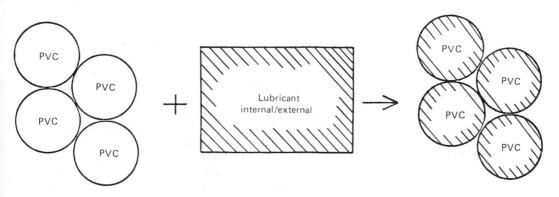

FIGURE 10 Internal/External lubricant.

TABLE 1 Comparison of Fusion Times of C_{16-18} and C_{22} Alcohols[a]

Lubricant	Fusion time (min)	Torque (mkp sec^{-1})	
		Maximum	15 min after maximum
C_{16-18} alcohol	4.6	3.59	2.56
C_{22} alcohol	7.3	3.25	2.51

[a]Measuring apparatus: Plasticorder PLV 151 (Brabender, Duisburg). Conditions: temperature, 165°C; mixer speed, 30 RPM; charge, 29 g. Composition of the mass: 100 parts S-PVC, K value 58; 2 parts tribasic lead sulfate; 0.2 part calcium stearate; 2.0 parts lubricant, as indicated.

length, polarity, and branching all change the solubility characteristics of the lubricant in the polymer.

The simplest examples are products that contain only one polar group. These materials can be fatty acids, fatty alcohols, simple esters, or branched or unsaturated materials. Long-chain alcohols are generally highly compatible flow promoters and function as internal lubricants. If formulations differing only in lubricant chain lengths are compared, an increase in the fusion time occurs from C_{16-18} to C_{22}, which corresponds to an increase in the external nature of the lubricant (Table 1). Although these materials would not ordinarily be classified as external lubricants, a trend toward external behavior is nevertheless established.

With fatty acids, the acid group is normally incompatible and orients itself toward the metal of the forming equipment. The acid coats the resin particle, causing a significant delay in fusion in most cases. The fusion time with fatty acids is longer than that with fatty alcohols, and as the molecular weight of the acid increases from C_{16-18} through C_{22}, the fusion time also increases.

Simple esters are made by the reaction of a fatty acid and a fatty alcohol. A typical member of this group is stearyl stearate. The polar ester group contributes to internal lubrication properties, but the molecule is about twice as big as the materials mentioned above and hence has a lower volatility. This structure contributes to a good balance of properties. Stearyl stearate offers the opportunity to study two effects: (1) the effect of position in the molecule of the ester group and (2) the effect of chain length. These effects were discussed by K. Worschech [1] in 1971:

Clearly the lubricating effect increases by spreading the C-chain lengths. It is obvious that this principle is also being maintained when the polar center is placed in the middle of the carbon chain. We would like to demonstrate this by the example of long chain linear composed wax esters in which the ester function represents the polar group. At first it is proved that the position of the ester group in the molecule within a certain range does not essentially influence the effectiveness of the lubricant (Table 2). This means that the lubricating properties of lauryl

TABLE 2 Comparison of Lauryl Stearate and Stearyl Laurate Lubricants[a]

Lubricant (esterwax)	Chemical structure	Fusion time (min)	Maximum torque (mkp sec^{-1})
Lauryl stearate M.P. 42.2°C	C_{12} —— O — C —— C_{17} (C=O)	5.5	2.80
Stearyl laurate M.P. 42.0°C	C_{18} —— O— C —— C_{11} (C=O)	6.0	2.84

[a]Measuring apparatus: Plasticorder PLV 151 (Brabender, Duisburg). Conditions: temperature, 165°C; mixer speed, 30 RPM; charge, 31 g. Composition of the mass: 100 parts S-PVC, K value 58; 2 parts tribasic lead sulfate; 0.3 part calcium stearate; 2.0 parts lubricant, as indicated.

stearate and stearyl laurate as well as their physical properties are almost identical; thus, we can simply add the chain length of the alcohol and the acid components in order to estimate the effectiveness of the wax ester.

Table 3 shows the data for molecules of palmityl stearate (G30), stearyl stearate (G32), stearyl behenate (G47) and behenyl behenate. As chain

TABLE 3 Effect of Chain Length on External Lubrication[a]

Lubricant (ester wax)	Carbon chain length	Composition 1[b]		Composition 2[c]	
		Fusion time (min)	Maximum torque (mkp sec^{-1})	Fusion time (min)	Maximum torque (mkp sec^{-1})
G-30	32	2.0	4.0	2.0	3.54
G-32	35	7.3	2.98	8.7	2.68
G-47	39	15.0	2.46	17.8	2.20
Behenyl behenate	43	42.5	2.34	—	—

[a]Conditions: temperature, 165°C; mixer speed, 40 RPM; charge, 32 g.
[b]Composition 1: 100 parts S-PVC, K value 58; 1.5 parts tin stabilizer (Irgastab 17 M); 0.2 part calcium stearate; 2.0 parts lubricant, as indicated.
[c]Composition 2: 100 parts S-PVC, K value 58; 2.0 parts tribasic lead sulfate; 0.3 part calcium stearate; 2.0 parts lubricant, as indicated.

TABLE 4 Plastification Time of Internal and External Loxiol Lubricants[a],[b]

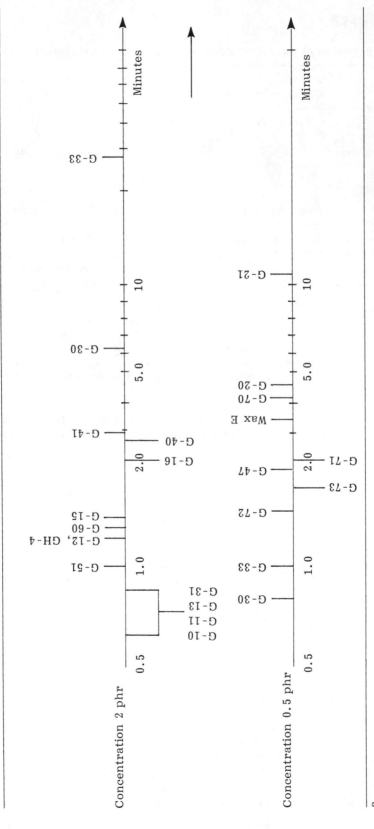

[a]Base formula: S-PVC, K value 55, 100 parts; lead sulfate, basic, 2 parts; calcium stearate, 0.3 part.
[b]G-31, Butyl stearate; G-51, stearyl alcohol; G-21, stearic acid; G-30, palmityl stearate; G-47, behenyl behenate.

length increases, so does fusion time. This illustrates that external lubrication increases with chain length and total molecular weight. Furthermore, by evaluating hexadecyl stearate versus palmityl stearate, one can estimate the effects of branching on materials of equivalent molecular weight. In Table 4 the fusion time with hexadecyl stearate is shorter than with palmityl stearate. Branching reduces the linearity, that is, chain length, and contributes to more internal lubricant nature. A comparison of an ester with total chain length equivalent to that of a fatty acid and fatty alcohol gives the results shown in Table 4, based on torque rheometer fusion time data. Butyl stearate, stearic acid, and stearyl alcohol are three commercial materials with approximate equal molecular weights. Table 4 shows that butyl stearate is slightly more internal than stearyl alcohol; stearic acid is so external that is does not appear on the 2-phr line. Stearyl stearate, which contains one polar group with roughly 34 carbons, is near the end of the 2-phr line. At the 0.5-phr line, stearyl stearate is near the beginning with stearic acid farther along. Stearic acid is much more external than either butyl stearate or stearyl alcohol.

The representative molecules discussed above have only one polar group. With materials that contain more than one polar group, changes in position of the polar groups and chain lengths have an effect on the lubricating properties. A general rule is that as chain length increases the external lubricating effect also increases. This is similar to the case of simple esters. Table 5 shows that as the chain length of fatty alcohols increases from dilauryl phthalate through distearyl phthalate, the fusion time and the external lubricating properties increase. A second principle observed is that if polar groups are added to a lubricant molecule without changing the chain length, the increased polarity increases compatibility and therefore increases the internal lubricating effect of the molecule. As shown in Table 6, addition of hydroxyl groups to the fatty acid chains of glycerol tristearate makes the molecule more compatible. The fusion time is reduced and the internal lubricating properties are increased. A third effect seen is due to the lin-

TABLE 5 Effect of Chain Length on Lubricating Properties of Fatty Alcohols[a]

Lubricant	Fusion time (min)	Torque (mkp sec^{-1})	
		Maximum	15 min after maximum
Dilauryl phthalate	3.2	3.56	2.6
Distearyl phthalate	14.5	2.86	2.5
No lubricant	2.0	4.07	2.9

[a]Measuring apparatus: Plasticorder PLV 151 (Brabender, Duisburg). Conditions: temperature, 165°C; mixer speed, 20 RPM; charge, 29 g. Composition of the mass: 100 parts 5-PVC, K value 58; 2.0 parts tribasic lead sulfate; 0.3 part calcium stearate; 2.0 parts lubricant, as indicated.

TABLE 6   Effect of Addition of Polar Groups on Internal Lubrication[a]

Lubricant	Fusion time (min)	Maximum torque (mkp sec^{-1})
Glyceryl tristearate[b]	19.0	2.3
Glyceryl tris (12-hydroxystearate)[b]	1.2	4.8
Soya bean oil[c]	11.7	3.3
Epoxidized soya bean oil[c]	2.2	3.8

[a]Measuring apparatus: Plasticorder PLV 151 (Brabender, Duisburg).
[b]Conditions: temperature, 165°C; mixer speed, 40 RPM; charge, 32 g. Composition of the mass: 100 parts S-PVC, K value 58; 2.0 parts tribasic lead sulfate; 0.3 part calcium stearate; 2.0 parts lubricant.
[c]Conditions: temperature, 165°C, mixer speed, 20 RPM; charge, 29 g. Composition of the mass: 100 parts S-PVC, K value 58; 2.0 parts tribasic lead sulfate; 0.3 part calcium stearate; 2.0 parts lubricant.

earity of the molecule. In Table 7, distearyl phthalate esters are evaluated. Fusion time increases from ortho through meta through para. The external lubricating nature therefore increases from ortho through meta through para, as the molecule appears to become more linear. Rotating the ester group around the benzene ring seems to result in an increase in the total length of the molecule, which changes the lubricant functionality. A fourth effect is produced by spreading the polar groups along the chain length of a molecule. In Table 8, two polar groups are spread farther apart in the diester without increasing the total molecular weight. The chain length of the internal diol increases, resulting in decreased interaction of the polar groups (decrease in melting point). The fusion time on the torque rheometer at 20 rpm decreases from 7.7 min with 1,2-ethylene distearate through 5.5 to 5.1 min with 1,10-decanediol dimyristate. This indicates that as the ester groups are spread along the chain, the compounds become more compatible and internal lubricating effects are increased.

The last subject in this section is the effect of the various chemical groups on the lubricating properties of these materials. Table 9 shows estimates of the properties produced by the various chemical groups listed. The double bond, epoxide, hydroxyl group, ester group, and ketonic group all contributing to internal lubricating properties, whereas the amide, amine, fatty acid, ether, and metallic ions all contribute to external lubricating properties. Tables 10 and 11 provide information on comparative lubricating properties of many of the materials available commercially.

V. CLASSIFICATION OF LUBRICANTS

Lubricant function varies widely among materials, depending on the great variety of structural variations possible. No lubricant is totally external or

TABLE 7 Effect of Linearity of the Molecule on Internal Lubrication[a]

Lubricant (chemical structure)	M.P. (°C)	Fusion time (min)	Maximum torque (mkp sec^{-1})
Distearyl-*o*-phthalate	54	4.2	3.88
Distearyl-*m*-phthalate	51	6.8	3.62
Distearyl terephthalate	85	7.3	3.53
No lubricant	–	0.6	5.54

Distearyl-*o*-phthalate

Distearyl-*m*-phthalate

Distearyl terephthalate

[a]Measuring apparatus: Plasticorder PLV 151 (Brabender, Duisburg). Conditions: temperature, 165°C; mixer speed, 30 RPM; charge, 30 g. Composition of the mass: 100 parts S-PVC, K value 58; 2 parts tribasic lead sulfate; 0.3 part calcium stearate; 2 parts lubricant.

internal; each lubricant exhibits a gradation of the above properties. This dual functionality gives rise to difficulty in classification because differences are not clear-cut. Many authors have attempted to classify lubricants according to criteria such as fluidity, molecular structure, haze, friction control, interaction solubility parameters, glass transition temperature, surface wetting, fusion time, roll release, and parting force. A good understanding of these methods will help the reader to understand the effects of lubricants.

A. Fluidity

The first empirical means of classifying lubricants was described by Marshall [5], using a system of fluidity measurements. In this method, compound was milled for a given amount of time at a specified temperature and formed into a 2-mil-thick sheet. Samples measuring 5 cm^2 were cut from the sheet

TABLE 8 Effect of Separation of Polar Groups Along the Molecule on Internal Lubrication[a]

Lubricant	Chemical structure	M.P. (°C)	Measurement I[b]		Measurement II[c]	
			Fusion time (min)	Maximum torque (mkp sec^{-1})	Fusion time (min)	Maximum torque (mkp sec^{-1})
Ethyleneglycol distearate	C_{17}—$\overset{O}{\overset{\|}{C}}$—O—$C_2$—O—$\overset{O}{\overset{\|}{C}}$—$C_{17}$	73.7	1.5	4.21	7.7	3.02
1,6-Hexandiol dipalmitate	C_{15}—$\overset{O}{\overset{\|}{C}}$—O—$C_6$—O—$\overset{O}{\overset{\|}{C}}$—$C_{15}$	56.0	1.2	4.47	5.5	3.14
1,10-Decandiol-dimyristate	C_{13}—$\overset{O}{\overset{\|}{C}}$—O—$C_{10}$—O—$\overset{O}{\overset{\|}{C}}$—$C_{13}$	54.5	1.0	4.64	5.1	3.24

[a]Measuring apparatus: Plasticorder PLV 151 (Brabender, Duisburg). Composition of the mass: 100 parts S-PVC, K value 58; 2 parts tribasic lead sulfate; 0.3 part calcium stearate; 1.0 part diol ester C_{38}.
[b]Measurement I: temperature, 165°C; mixer speed, 30 RPM; charge, 31 g.
[c]Measurement II: temperature, 165°C; mixer speed, 20 RPM; charge, 29 g.

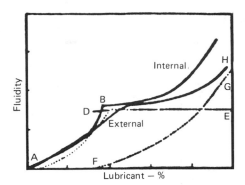

FIGURE 11 Make up of the observed plot for fluidity against lubricant concentration. Line AB and its continuation is the fluidity for an external type lubricant. and line AH for an internal lubricant.

and pressed under controlled conditions. The resulting percentage increase in surface area over a nonlubricated sample was measured and termed "fluidity." The equation $(Y - X)/X$ multiplied by 100, where X is the surface area of the unlubricated sample and Y is that of the lubricated sample, was used to calculate this term. It appears that the preheat time, temperature of the press, time allotted for pressing, and pressure are critical and difficult to maintain, and it is difficult to determine the surface area of a sample that is not circular. Therefore, the percentage of error in such an evaluation is high. Even so, the comparison of fluidity numbers for various lubricants may have some practical applications (Table 12). It is interesting to note that the most external lubricants exhibit the highest fluidity values. If fluidity is plotted against lubricant concentration, interesting results are obtained. Figure 11 shows that there is a rise in fluidity with increasing lubricant concentration until the compatibility limit is reached; at this point an inflection of the curve occurs, after which the fluidity continues to rise. This curve is a composite of the plots for internal and external lubrication with the external superimposed on the internal. In Figure 11, the composite fluidity curve A,B is the summation of the internal lubrication curve A,D,E and the external lubrication curve F,G. Internal lubricant has little effect after point D, and the external behavior is initiated prior to the limit at point F. It can be seen that the curve continues to rise until excessive lubrication prevents gelation. It can further be seen how these curves can be used to classify lubricants. In Figure 11, line A,B and its continuation is characteristic of an external lubricant. Its compatibility limit is reached at a lower concentration than that of sample A,H and its fluidity curve continues to increase at a greater rate. Sample A,H reaches its compatibility limit at a much higher concentration and its fluidity curve does not rise as rapidly as A,B. It is concluded that sample A,B is more external than sample A,H.

B. Clarity

One of the simplest ways to test for lubricant functionality is to determine the transmittance and haze values imparted to a clear PVC compound. Inter-

TABLE 9 Effects of Chemical Group on Internal and External Lubrication Properties[a]

Lubricant	Dosage of lubricant						Trade name
	0.2 phr		0.5 phr		1.0 phr		
	Fusion time	Max. torque	Fusion time	Max. torque	Fusion time	Max. torque	
Hexadecane	1.3	5.05	1.5	4.55			—
Stearic acid chloride	1.0	5.38	1.6	4.71			—
Stearic acid amide	1.6	5.06	2.9	4.40			—
Stearic acid	1.8	4.71	10.5	3.76			—
Stearic acid anhydride	1.8	4.97	7.5	3.81			—
Low-M.W. polyethylene	2.9	4.20	9.1	3.50			AC-polyethylene
Hydrogenated castor oil	1.1	5.46	1.6	4.98			—
Fischer-Tropsch paraffin	3.4	4.21	7.2	3.55			Loxiol G-22

Wax	1.2	5.15	2.4	4.47	—	—	Lupolen 1800 H
Ethylene bis-stearamide	2.0	4.88	6.6	4.07	—	—	Irgawax 280
Stearyl alcohol	—	—	1.7	4.92	2.4	3.84	Stenol PC
Octadecane	—	—	1.2	4.98	2.1	4.33	
Stearyl phthalate	—	—	1.5	5.08	2.8	3.70	Loxiol G-60
Monooleate	—	—	1.8	4.60	2.6	4.40	Loxiol G-10
Palmityl stearate	—	—	2.0	4.75	3.6	3.90	Loxiol G-30
Hexadecyl ether	—	—	6.3	3.67	20.9	3.62	—
Distearyl ketone	—	—	1.8	4.84	4.5	3.94	—
Ethylene glycol dimontan wax	—	—	5.5	3.85	—	—	Hoechst Wax E
Stearic acid nitrile	—	—	1.4	5.05	1.5	4.71	
No lubricant	0.8	5.52	0.8	5.52	0.8	5.52	

aMeasuring apparatus: Plasticorder PLV 151 (Brabender, Duisburg). Conditions: temperature, 150°C; mixer speed, 10 RPM; charge, 29 g. Composition: 100 parts S-PVC, K value 55; 1.5 parts tin stabilizer; 0.5 part calcium stearate; lubricant as shown.

TABLE 10 Fusion Times of External Lubricants[a,b]

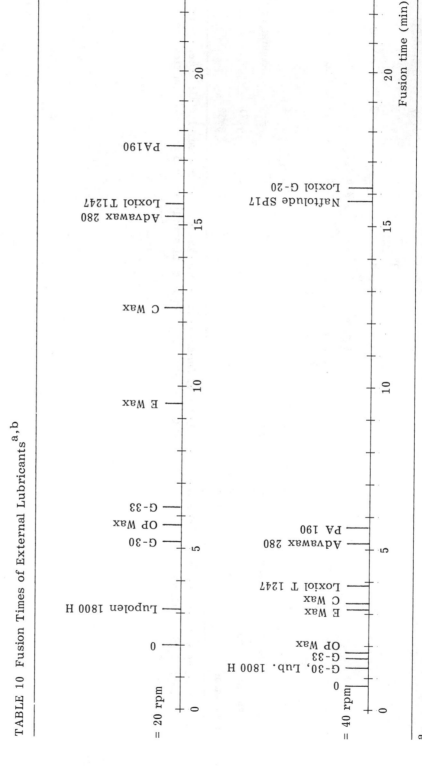

[a]Base formula: S-PVC, K valu 55, 100 parts; tribase, 2 parts; calcium stearate, 0.3 part; lubricant, 1 part.
[b]Plastograph conditions: temperature, 165°C; n = 20 rpm, 40 rpm; E = 29 g.

TABLE 11 Glyceride Lubricants for PVC: Structure versus Function[a]

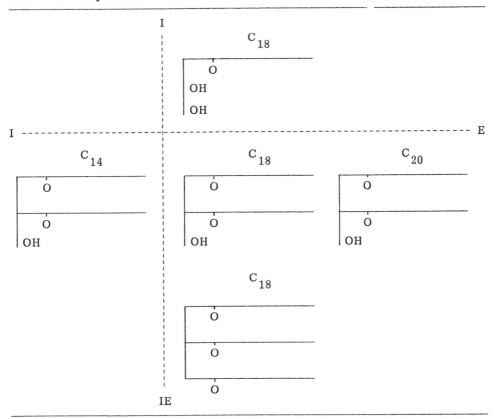

[a]Key: I, internal; IE, intermediate; E, external. X axis: effect of chain length; as chain length increases, external lubricating function increases. Y axis: as degree of esterification increases, external lubrication function increases.

nal lubricants have considerable compatibility over a wide range of concentrations and external lubricants become incompatible at very low concentrations. Once the solubility product constant of the solvent, PVC, is exceeded by the solute, lubricant, the solute begins to precipitate out, forming microdroplets approximately 1 to 4 μm in diameter embedded in the matrix of the PVC. The result is the formation of haze, since the lubricant droplets and the PVC have different refractive indices and the light entering the PVC is scattered. As the concentration of the lubricant increases above the solubility limit, more and larger droplets are formed, increasing the degree of haze. It must be noted that sample preparation is very important. If one sample cools slower than another, it may exhibit more haze, because the lubricant will have the opportunity to flow together, resulting in larger droplets and leading to erroneous results. Gale [4] prepared PVC strips by extrusion and measured the sheet clarity versus lubricant concentration.

TABLE 12   Comparative Physical Properties for Lubricants[a]

Lubricant	Type	Formula	M.P.	C.L.	0.25 phr	0.5 phr	1.0 phr	2.0 phr	L.B.	Supplier
DBP	Plasticizer	Dibutyl					12		I	—
Octadecane	Hydrocarbon	$C_{18}H_{38}$	28	3.5		8	13	23	I	—
Paraffin wax	Hydrocarbon	Straight-chain paraffin M.W. 500	64	1.0	6	14	20		I.E.	—
Sasolwax	Hydrocarbon	Synthetic hydrocarbon M.W. 800	100	0.5	12	18	23	32	E	7
A. Wax 3500	Hydrocarbon	Polythene wax M.W. 5000	104	0.5	8	16	26	40	E	2
Lauric acid	Fatty acid	$C_{11}H_{23}COOH$	44	3.8		8	15	20	I	—
Stearic acid	Fatty acid	$C_{17}H_{35}COOH$	70	1.2	10	20	40	48	I.E.	—
Wax S	Fatty acid	C_{24}–C_{30} acids (Montanic acid)	82	0.6	12	25	40	NG	E	2
Calcium stearate	Metal soap	$(C_{17}H_{35}COO)_2Ca$	148	1.25		8	25	45	I.E.	—
Lead stearate	Metal soap	$(C_{17}H_{35}COO)_2Pb$	103	0.4	10	17	30	48	E	—

Trade name	Type	Composition	M.P.	C.L.					L.B.	Source
Stennol PC	Fatty alcohol	C_{16}–C_{18} alcohols	50	3.0	5	12	25	50	I	6
Stearyl alcohol	Fatty alcohol	$C_{17}H_{35}CH_2OH$	58	2.6	5	12	24	45	I	—
Abril Wax 10DS	Fatty amide	Alkylated stearamide	140	>1.0			35		I.E.	1
Acrawax C	Fatty amide	Alkylated stearamide	140	1.25		5	13	25	I.E.	4
Butyl stearate	Fatty ester	$C_{17}H_{35}COOC_4H_9$		>3.0						
Loxiol G-30	Fatty ester	Alkylated fatty acid	45	>2.0	4	8	12	16	I	3
Loxiol G-31	Fatty ester	Alkylated fatty acid	15	1.5		9	16	22	I	3
Wax E	Fatty ester	Totally esterified montanic acid	80	0.5		22	38	NG	E	2
Loxiol G10	Partial ester	Alkylated unsaturated fatty acid	-4	>4.0		5	14	30	I	3
Wax OP	Partial ester	Part saponified butyl montanate	100	0.5		20	42	NG	E	2
Chlorooctadecane	Fatty halide	$C_{17}H_{35}CH_2Cl$	19	6.0		6	10	15	I	—
Octadecylamine	Fatty amine	$C_{17}H_{35}CH_2NH_2$	60	>0.5		16	30		I	—

[a]M.P., Melting point (°C); C.L., compatibility limit (%); E, external; L.B., lubricant behavior; NG, no gelation; I, internal. [b]1, Bush Beach & Segner Bayley Ltd. (Abril Industrial Waxes Ltd.); 2, B.A.S.F.; 3, O. Neynaber; 4, Rex Campbell & Co. Ltd.; 5, Joseph Crosfields & Sons Ltd.; 6, J. H. Little & Co. Ltd. (Henkel Int. GmbH); 7, Petrowaxes Ltd. (Sasol Marketing Co. S.A.). Where no supplier is given, the lubricant can be obtained from more than one source.

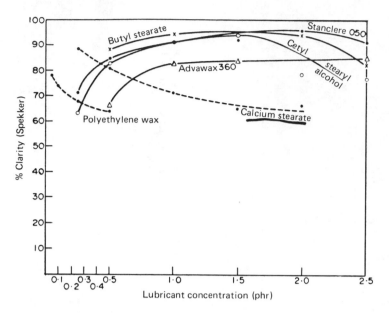

FIGURE 12 Effect of lubricant on clarity of extruded strip.

The data (Figure 12) for butyl stearate at concentrations from 0.5 through 2.5 phr show that clarity is maintained or slightly increased from the lower to the higher concentration; at 2.0 phr, it reaches its maximum and begins to lose clarity. The lubricant starts to become incompatible at a level of about 2 phr. Because it maintains clarity over a wide range, it is an inter-

TABLE 13 Clarity of Extruded Strips

Classification	Lubricant	% Clarity	
		1.0 phr	2.0 phr
Internal	Stanclere 1050	91	96
	Glycerol monoricinoleate	—	95
	Butyl stearate	94	94
	Loxiol G-12	90	95
External	Calcium stearate	71	67
	Stearamide/stearic acid	81	62
	Hoechst wax E	55	—
	Polyethylene wax	74 at 0.1 phr	64 at 0.5 phr

TABLE 14 Clarity and Stress Whitening

Lubricant(s)	Modified			Unmodified		
	% WL[a]	% Haze	CWR[b]	% WL[a]	% Haze	CWR[b]
None	86	5	7	91	4	0
Octadecane	82	8	>9			
Stearyl alcohol	84	8	7—8	90	5	0
Stearic acid	75	10	5	89	4	0
12-Hydroxy stearic acid	81	9	8			
Sodium stearate	73	14	>9			
Potassium stearate	71	83	>9			
Calcium stearate	64	42	8			
Butyl stearate	82	8	7	91	4	0
Glycerol monostearate	82	6	7—8	90	4	0
N,N'-Ethylene bis-stearamide	79	9	8			
Wax S	77	9	7—8	54	74	>9
Wax E	82	6	7			
Wax OP	80	7	3—4	79	10	9
Paraffin wax	79	10	7—8	20	100	9
AC-629A	72	15	7—8			
Paraloid K-175	86	5	7—8	90	4	0
K-175/Calcium stearate— 1.0/0.3	84	6	4	73	27	9
Glycerol monostearate/wax OP—0.7/0.3	83	6	8	90	4	3
Wax OP/wax S—0.8/0.2	80	9	4			
Wax OP/wax E—0.8/0.2	80	7	4			
Stearyl alcohol/AC-629A— 0.8/0.2	83	8	7			
Butyl stearate/calcium stearate—0.7/0.3	81	8	6—7			

[a]% white light transmittance.
[b]Crease whitening resistance.

nal lubricant, and only at a higher dosage does it become incompatible and exhibit an external lubricant nature. For cetyl-stearyl alcohol, one sees the same phenomenon at a concentration of 1.5 phr. The compatibility limit is reached and the material begins to exceed the solubility product constant and becomes external or incompatible. With calcium stearate, as concentration increases, clarity decreases. This indicates that calcium stearate is an external lubricant even at low concentrations. With polyethylene wax, the same phenomenon occurs and one can conclude that polyethylene is an even more external lubricant than calcium stearate. The deviations observed at low concentrations of butyl stearate and stearyl alcohol are due to insufficient lubrication, and the decrease in clarity is due to surface imperfection caused by sticking to the roller mill.

Table 13 shows that 2 phr of materials such as glyceryl monostearate, glyceryl monoricinoleate, and butyl stearate maintain good clarity, whereas materials such as calcium stearate, stearamide, montanic acid ester, and polyethylene have an adverse effect on clarity even at 1 phr. Further technical data on clarity imparted by various lubricants are shown in Table 14.

Table 15 compares montan ester derivatives to complex esters and to an internal lubricant. It can be seen that even at 5 phr an internal lubricant does not adversely affect clarity and that both montanic acid esters, with and without partial saponification, have more effect on clarity than do complex esters. It is particularly noteworthy that complex esters that contain unsaturation contribute higher clarity than the saturated ester, indicating increased compatibility of products that contain double bonds. It is equally interesting that montan derivatives containing metal soaps are sub-

TABLE 15 Effects of Montan Ester Derivatives, Complex Esters, and an Internal Lubricant on Clarity[a]

	Dosage (phr)							
	0	0.2	0.5	1.0	1.5	2.0	3.0	5.0
Loxiol G-70/G-74	98.4	—	98.5	92.3	63.4	—	—	—
Loxiol G-71	98.4	—	98.0	92.1	80.6	—	—	—
Loxiol G-72	98.4	—	98.4	98.6	97.2	—	—	—
Loxiol G-73	98.4	—	98.2	96.3	—	74.8	63.4	—
Montanic acid ester, partly saponified	98.4	99.0	94.3	60.4	42.0	—	—	—
Montanic acid ester	98.4	—	97.0	—	57.3	—	—	—
Loxiol G-10	98.4	—	—	—	—	—	98.7	96.4

[a]The degree of transparency is stated below the various testing levels, i.e., 0 to 5.0 phr. These figures enable us to indicate the amounts of high-molecular esters to be used for the production of clear PVC products: Loxiol G-70, 0.5 to 1.0 phr; Loxiol G-71, 0.5 to 1.0 phr; Loxiol G-72, 1.5 to 2.0 phr; Loxiol G-73, 1.0 to 1.5 phr; and Loxiol G-74, 0.5 to 1.0 phr. Clarity is achieved with a dosage of up to the lower amount. With increased dosage distinct haziness develops.

TABLE 16 Clarity Imparted by Simple Esters[a,b]

		Dosage (phr)								
	0	0.5	0.75	1.0	1.5	2.0	2.5	3.0	3.5	4.0
Loxiol G-40	98.4	—	—	—	—	—	—	98.5	90.4	45.6
Loxiol G-41	98.4	—	—	—	—	98.8	98.4	73.1	—	—
Loxiol G-30	98.4	—	—	—	98.6	96.7	47.5	38.8	—	—
Loxiol G-32	98.4	99.3	99.5	99.4	88.5	—	—	—	—	—
Loxiol G-47	98.4	99.2	—	97.3	45.7	—	—	—	—	—
Behenyl behenate	98.4	98.9	88.9	—	—	—	—	—	—	—

[a]Transparency measurements made on 5-mm-thick rigid PVC sheets. Measuring device, Gardner. (The transparency measurements were made against a black background; the results were represented in percent light transmission values.)
[b]Test formula: PVC, suspension polymer, K value 60, 100 parts; tin stabilizer, 1.5 parts.

stantially less compatible than those without such soaps, indicating that calcium montanate is a very efficient external lubricant. The montan ester containing partial saponification is more external than the straight montanic acid ester.

Table 16 shows the clarity imparted by various simple esters, which are listed in order of molecular weight. As molecular weight increases, the degree of clarity imparted decreases, indicating that external lubricating nature increases with molecular weight. Also, in a comparison of tridecyl stearate and stearyl stearate, the branching increases compatibility and provides more internal nature than the linear structure. For the variation of percent transmittance and percent haze over a wide range of typical lubricants, see Table 14.

C. Brabender Fusion Time

The most widely used method of classifying lubricants is by their fusion time in the typical C. W. Brabender PlastiCorder test. The assumption made is that internal lubricants are much more compatible than external lubricants and as temperature rises they are solubilized in the resin and do not form a lubricating film at particle interfaces. External lubricants form a strong film between the resin particles, inhibiting fusion and increasing the time to coalescence.

A plot of fusion time and torque for ingredients at the same concentration shows great differences in the effects of various lubricants. To compare internal lubricants, excessively high concentrations must be used. The more external effect the lubricant exhibits, the less internal effect it can have. In the Brabender fusion time classification the external nature of each

TABLE 17 Data for Lubricants in the Metal Release Test[a]

		Time to tack (in minutes) at 350°F with 300-g sample size at 0.040-in. gap												
Lubricant package (phr)	No lubri- cant	VPG576	G15 70% WG1563 30%	HOB 7119	E.B.S.	HOB 7108	HOB 7107	HOB 7107 80% CaST 20%	CaST	St. Acid	Ethylene glycol Montan ester	Partially sapon- ified Montan ester	Poly- glyc- erol	G-15
0.25	9.0	9.0	11.0	9.5	12.0	14.5	20.0	23	10.5	13.0	10.5	14	13	11
0.5	9.0	9.5	21.0	17.0	12.5	44.5	43.5[b]	51[b]	14.0	14.5	15.5	18	14	14
0.75	9.0	9.5	34.5	17.5	13.5	50[b]	48.5[b]	53[b]	21.0	15.0	31.0	26.5	21	14.5
1.0	9.0	12.5	46.0[b]	16.5	13.0	50[b]	48.0	58[b]	49.0[b]	15.5	44.5[b]	53.0[b]	22	17.0
Stability time at 1.0 phr			46.0	31.5	25.5	50.0	48.0	58	49.0	28.0	44.5	53.0	47	

[a]Base formulation: Formosa 614 (PVC), 100.0 phr; BTA-3F (modified), 8.0 phr; Thermolite 831, 2.0 phr; Thermolite 813, 0.2 phr.
[b]Compound degraded before it stuck.

lubricant is used to obtain a comparative estimate of its internal lubricating properties. All lubricants, no matter how internal, have some effect on fusion. Externals have a significant effect on fusion at typical use levels. Comparing the position of a fatty alcohol stearate ester (G-33) on the top chart to the same material on the bottom chart in Table 4, one sees that the entire top chart can be compressed into 1 min on the bottom chart. This indicates the highly internal lubricating effect of the materials at the top as compared to the typical external lubricants at the bottom. It should be noted that certain materials can be misclassified in this test. For instance, alkaline earth salts of fatty acids, such as calcium stearate, promote fusion but not flow and also contribute to good metal release.

D. Metal Release

A traditional and practical means of classifying lubricants is by their ability to promote release from metal surfaces. Internal lubricants do not provide a significant amount of release. Externals exude to form a lubricant film and thus give efficient release. Materials that are intermediate give some release as well as some internal lubricating properties. This allows the classification of materials as predominantly internal or external according to their metal release property on a heated roll mill. The particulars of this test are discussed in Section VI.D. From the data in Tables 17, 18, and 19 it is apparent that lubricants such as VPG 576 (i.e., glyceryl mono-

TABLE 18 Data for Lubricants in the Metal Release Test[a,b]

Lubricant concentration (phr)	Time to sticking (min)					
	Loxiol G-70	Loxiol G-71	Loxiol G-72	Loxiol G-73	Wax E	Ca stearate
0	5	5	5	5	5	5
0.1	40	40	15	25	10	15
0.3	70	95	75	70	25	35
0.5	95[c]	105[c]	100[c]	95[c]	40	55
0.75	105[c]	105[c]	100[c]	95[c]	95	85
1.00	—	—	—	—	110[c]	

[a]Base formulation: SPVC B.F.G. type 106, No. 00497-3, 100.00 phr; tin stabilizer Advastab TM-181, 1.50 phr; external lubricant (as indicated), 0.1 to 1.00 phr.
[b]Test conditions: two-roll dynamic mill; Brabender torque rheometer; transparency; Vicat softening point. Two-roll mill conditions: rolls 8 x 16 in.; roll surface temperatures, back roll 310°F, front roll 310°F; roll speeds, back roll 23 rpm, front roll 18.5 rpm; roll friction ratio, 1:1.25; stock temperature, 358°F.
[c]Compound burns fast to mill rolls due to lack of thermal stability rather than any loss of lubricating efficiency.

TABLE 19 Data for Lubricants in the Metal Release Test[a],[b]

Lubricant concentration (phr)	Fusion time (min)	Maximum torque at fusion (mg)	Torque (mg) 10 min after max. torque	Stock temperature (°C) 10 min after max. torque
No lubricants	1.625	1750	1500	161
Loxiol G-70 0.25	3.5	1700	1450	161
Loxiol G-70 0.50	8.75	1625	1425	162
Loxiol G-70 0.75	44.0	1550	1400	161
Loxiol G-70 1.0		No fusion		
Loxiol G-72 0.25	2.0	1750	1450	161
Loxiol G-72 0.50	2.5	1725	1450	162
Loxiol G-72 0.75	2.75	1625	1425	163
Loxiol G-72 1.0	4.0	1625	1425	162
Wax E 0.25	3.0	1625	1475	161
Wax E 0.50	4.25	1575	1437	161
Wax E 0.75	8.0	1500	1412	160
Wax E 1.0	17.75	1425	1375	160

[a]Base formulation: SPVC B.F.G. type 106, lot No. 00497-3, 100.00 phr; stabilizer Advastab TM-181, 1.50 phr; lubricant calcium stearate, 0.20 phr; internal lubricant Loxiol G-16, 1.00 phr; external lubricant (as indicated), 0.25 to 1.0 phr.
[b]Brabender Plastograph conditions: apparatus, Plasticorder PL-V300; roller, type 6 (electric), capacity 30 ml; loading chute, B.F.G. type (cold); ram weight, 15 kg; rotor speed, 30 rpm; charge weight, 28 g; bowl temperature set, 160°C.

myristate) are quite internal and give very little metal release, whereas complex fatty ester types, or polyethylene, give considerable ease of release. Montanate esters, complex fatty esters, oxidized polyethylene, and metal fatty acid soaps are normally classified as "release agents," a subgroup in the huge family of external lubricants.

E. Structure

With the development of the test procedures outlined above, investigators were able to correlate molecular structure with functionality. Illmann's work (Table 20) sums up the basic principles of estimating lubricant function from an assessment of structure. Figure 13 depicts some esters of glycerol. Increased degree of esterification of glycerol, from the top toward the bottom of the figure, leads toward a more external lubricant. The molecular weight has increased without increasing the polarity; therefore, the material takes on a more external function. Looking at the horizontal axis, when the chain

TABLE 20 Lubricant Classification

No lubricants	Plasticizers and processing aids such as acrylates, e.g., acryloid K 120		
Internal lubrication Soluble in PVC	Monoglycerin esters		O=, RO— Chain C_{14-18}
Clear	Stearyl alcohol	Polar groups	HO— Chain C_{14-18}
Polar	Stearic acid (technical)		O=, HO— Chain C_{14-18}
	Soaps: stearates etc.		Chain C_{16-18} — [polar / O-Me-o / Center] — Chain C_{16-18}
			Chain C_{16-18} — [polar / Amide Groups / Center] — Chain C_{16-18}
Balance between internal and external lubrication	Wax GL-3		Long chain — [polar / Center] — Long chain
	Wax OP (derived from montan wax)		Long chain C_{28-32} — [polar / Soap-Ester-Groups / Center] — Long chain C_{28-32}
	Wax E		Long chain C_{28-32} — [polar / Ester Groups / Center] — Long chain C_{28-32}
	Wax S	Polar group	O=, HO— Long chain C_{28-32}
External lubricants: (these must be adjusted carefully to avoid overlubrication)	Paraffin oil, hydrocarbons		Short branched chain
	n-Paraffin, FT-paraffin		Long straight chain
	Wax PA-520, polyethylene wax		Long somewhat branched chain

Source: Ref. 3.

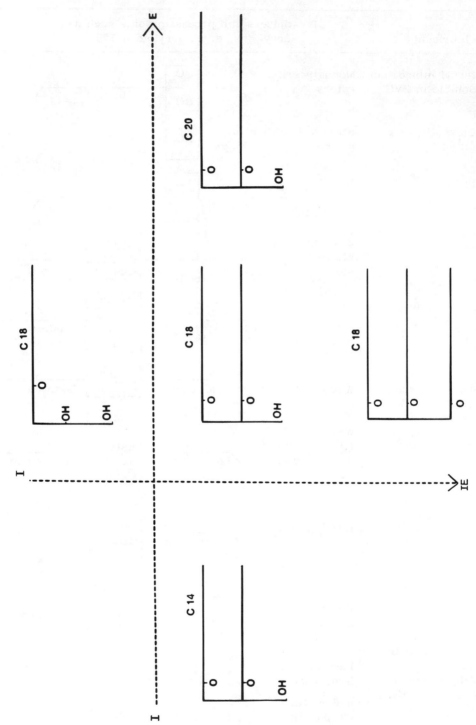

Figure 13 Relationshnp of ester structure to function of lubricants for PVC. Along the X axis: as the chain length increases, the external lubricating function increases. Along the Y axis: as the degree of esterification increases, the external lubrication function increases. I: Internal; IE: Intermediate; and E: External.

length is increased without increasing or affecting the polarity by maintaining the same degree of esterification, the polar nature of the molecule decreases; therefore, its trend is toward a more external functioning lubricant. For example, glyceryl monobehenate with a C_{22} chain length is more external than glyceryl monostearate but more internal than glyceryl dibehenate. A diagram like this can be drawn for esters of almost any alcohol-acid combination. The changes in molecular structure that can be effected can be estimated. If it is known how one of the members functions, one can then estimate how the rest of the members of the group will act.

F. Solubility

Another means of classifying lubricants is by their solubility. Internal lubricants should be soluble in the PVC at elevated temperatures but, unlike plasticizers, not soluble at room temperature. External lubricants are insoluble in PVC. An evaluation of their ability to penetrate, wet, and swell PVC resin particles can give some idea of their functionality, particularly as internal lubricants. In Figure 14 tricresyl phosphate, calcium stearate, n-butyl stearate, and dimethyl phthalate are evaluated by plotting the average change in diameter of a PVC particle versus increasing temperature. It is apparent that at low temperatures the plasticizers tricresyl phosphate and dimethyl phthalate readily wet and penetrate the PVC particle, resulting in large increases in volume of the PVC molecule. Note that there is a rapid decrease in volume of the dimethyl phthalate as temperature increases, probably because of volatilization. Next note that n-butyl stearate is an internal lubricant. It invades the resin particles only as temperature increases near

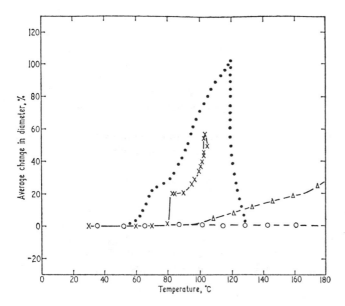

FIGURE 14   Change in PVC particle diameter with temperature:
        xxx PVC(D60/11) in tricesyl phosphate
        ooo PVC(D60/11) in calcium stearate
        △△△ PVC(D60/11) in n-butyl stearate
        ●●● PVC(D60/11) in dimethyl stearate

the melting temperature of the rigid PVC, becoming soluble at high temperatures and not at low temperatures, a characteristic of internal lubricants. Calcium stearate does not wet or swell the particle of PVC, even above its melting temperature. It is not soluble and thus is classified as an external lubricant. It has been shown by Khanna [7] (Table 21) that contact angle measurements of liquids above their melting points will yield their adhesion tension and the advancing contact angle. The power of adhesion tension indicates the degree of lubricant wetting of the polymer; an advancing contact angle indicates the extent of wetting of the polymer by the lubricant. High adhesion and low contact angle indicate internal lubrication. Conversely, in low adhesion tension, a high advancing contact angle indicates incompatibility, thus an external lubricating nature. In Table 21, butyl stearate, glyceryl monostearate, dioctyl sebacate, and tricresyl phosphate have a very high adhesion tension and low contact angle. Glyceryl monostearate and butyl stearate do not have as low a contact angle as dioctyl sebacate and tricresyl phosphate; this shows that they are intermediate between dioctyl sebacate and tricresyl phosphate, which are true plasticizers, and lead stearate and oxidized polyethylene, which are external lubricants. Calcium stearate and monoglycerides of behenic acid are totally incompatible, having high contact angles and low adhesion tension.

G. Solubility Parameters

Solubility parameters furnish another means of comparing the relative activity of potential lubricant molecules. Lubricants may be categorized as external or internal by calculating the solubility parameter constants of these materials. Thus one can estimate the solubility of various substances in various resins, such as lubricants in PVC. The old axiom "like dissolves like" is applicable here. Materials that have similar solubility parameters are soluble in one another and therefore act as solvents or plasticizers. Those that have solubility parameters not too far from that of the polymer act as internal lubricants in the polymer. Materials that have divergent solubility parameters are incompatible and therefore act as external lubricants. Immergut and Mark [8] expressed this very well in their paper on principles of plasticization that cohesive energy density is synonymous with solubility parameter:

> COHESIVE ENERGY DENSITY. The cohesive energy density (CED) was defined in 1931 by Scatchard [10] as the energy of vaporization per unit volume. Hildebrand and Scott [11] then proposed the square root of the cohesive energy density as a numerical value identifying the solvent power of specific solvents.
> Solubility will occur if the free energy of mixing, ΔF, is negative—i.e., $\Delta F = \Delta H - T \Delta S < 0$. Since ΔS, the entropy of mixing, is usually large and always positive for polymer-solvent systems, the sign of ΔF will be determined by the sign and magnitude of ΔH, the heat of mixing. For cases where $\Delta H > 0$, according to Hildebrand and Scatchard,

$$\Delta H = \phi \phi_s (\delta_s - \delta_p)^2$$

> where ϕ and ϕ_s are the volume fraction of solvent and polymer, respectively, and δ_s and δ_p are the solubility parameters of the solvent and the polymer. The solubility parameter, δ, is defined by

TABLE 21 Wetting Parameters of Additives to a PVC EP Suspension Polymer (ISO Viscosity Number 105)

Additive	Adhesion tension, $\gamma \cos \theta$ (dyn/cm)	Advancing contact angle, θ (deg)	$\cos \theta$	Melting point (°C)	Viscosity η (P)
Additives that wet the polymer					
n-Butylstearate	30.6	13	0.974	24	0.13
GMS	33.7	15	0.965	62	0.104
DOS	33.0	0	1.0	liquid	0.18
TCP	33.6	3	0.998	liquid	0.57
Additives that partly wet the polymer					
Low-molecular-weight polyethylene wax (PA-190)	28.1	23	0.92	129	
Lead stearate	25.8	31	0.857	98	
Additives that do not wet the polymer					
Calcium stearate	13	65	0.42	120	
Monoglyceride of a fatty acid (Coroxin TK1)	14.3	63	0.45	75	0.14
Water	24	69	0.35	liquid	0.01

$$\delta = (\text{DEIV})^{\frac{1}{2}} = (\text{CED})^{\frac{1}{2}}$$

For cases where there is strong interaction between solvent and polymer, such that ΔH is negative, the polymer will dissolve since ΔF will be zero. However, if there is no strong polymer-solvent interaction, ΔH being positive, the magnitude of ΔH determines whether or not the liquid will dissolve the polymer. This means that only liquids for which ΔH will be very small will be solvents for the polymer, and the smaller ΔH, the better the solvent. In fact, if $\Delta H = 0$, ΔF must be negative, and solution is certain. Thus, in selecting a suitable solvent, one tries to have $\Delta H = 0$ by trying to make the term $(\delta_s - \delta_p) = 0$. In other words, one attempts to choose a liquid with a solubility parameter (or cohesive energy density) as close to that of the polymer as possible.

Tables 22 and 23 show some representative values for polymers and organic liquids [8]. The values for liquids are calculated according to the relation

TABLE 22   Solubility Parameter δ of Polymers

Polymer	δ
Teflon	6.2
Polyethylene	7.9
Polyisobutylene	8.1
Polyisoprene	8.3
Polybutadiene	8.6
Polystyrene	9.1
Poly(vinyl acetate)	9.4
Poly(methyl methacrylate)	9.5
Poly(vinyl chloride)	9.7
Polyethylene glycol terephthalate	10.7
Cellulose nitrate	11.5
Poly(vinylidene chloride)	12.2
Nylon (6,6)	13.6
Polyacrylonitrile	15.4

$$\delta = \left(\frac{\Delta E}{v}\right)^{\frac{1}{2}} = \frac{\Delta H - RT}{M/d}$$

where   H = latent heat of vaporization
         T = absolute temperature
         R = gas constant
         M = molecular weight
         d = density

If ΔH is not known, it can be calculated fairly accurately from Hildebrand's equation: $\Delta H = 23.7 T_{bp} + 0.020 T_{bp}^2$ where T_{bp} is the boiling point in °K.

For polymers, δ cannot be calculated according to the above relations because the polymer is not volatile—i.e., it has no boiling point or ΔE of vaporization. There are two methods of overcoming this problem. The δ can be determined experimentally by determining solubility (or swelling of slightly cross-linked samples) in a series of liquids of known δ. The δ of the polymer is then the midpoint of the range of δ values of the liquids in which the polymer dissolves. For the cross-linked samples, one takes the δ value of the liquid in which maximum swelling occurs.

The alternative method is to calculate δ for the polymer from the structural formula and the density of the polymer by using the molar attraction constants tabulated by Small [9] (see Table 24). Assuming the additivity of these constants, δ can be calculated according to the relation

TABLE 23 Solubility Parameter δ of Common Solvents

Solvent	δ
Ethyl ether	7.4
Carbon tetrachloride	8.6
Xylene	8.8
Toluene	8.9
Ethyl acetate	9.1
Benzene	9.2
Methyl ethyl ketone	9.3
Cyclohexanone	9.9
Acetone	10.0
sec-Butyl alcohol	10.8
n-Butyl alcohol	11.4
Cyclohexanol	11.4
n-Propyl alcohol	11.9
Ethyl alcohol	12.7
Nitromethane	12.7
Ethylene glycol	14.2
Methanol	14.5
Glycerol	16.5
Water	23.4

$$\delta = \frac{d\Sigma G}{M}$$

where d is the density, ΣG is the sum for all the atoms and groupings in the polymer repeat unit, and M is the molecular weight.

For example, the calculation of δ for PVC is as follows [shown in Table 25] as compared to an experimental value of 9.7 ± 1.0. This approach is very useful for polymer-plasticizer systems, and especially for cases where polymeric plasticizers, which cannot be treated as volatile liquids, are being considered.

Since the system described above is useful for polymer solvents and for plasticizers, it could be extended to internal lubricants.

It appears that sufficient accuracy can be obtained from molecular structure calculations to estimate compatibility. Table 26, from Small [9], compares the solubility parameters obtained by calculation with those obtained experimentally. The parameters are sufficiently close that meaningful estimates can be obtained by calculation from molecular structure. It appears that this could be a useful technique for estimating the compatibility of materials in PVC.

TABLE 24 Molar Attraction Constants

Group	G	Group	G
—CH$_3$	214	—O—	70
—CH$_2$—	133	CO	275
—CH<	28	COO	310
>C<	-93	CN	410
CH$_2$≡	190	Cl	270
—CH=	111	Br	340
>C=	19	I	425
CH≡C—	285	CF$_2$	150
—C≡C—	222	S	225
—C$_6$H$_5$	735	SH	315
—C$_6$H$_4$—	658	ONO$_2$	440
		Si	-38

Source: Ref. 9.

TABLE 25 Calculation of δ for PVC

$d = 1.4$
$M = 62.5$

$$
\begin{array}{ll}
\begin{array}{c} \left[\!\!\left[\mathrm{CH_2{-}C} \right]\!\!\right] \\ \quad\quad\mid \\ \quad\quad\mathrm{Cl} \end{array}
& \begin{array}{l}
1\ CH_2 = 133 \\
1\ CH\ =\ 28 \\
\underline{1\ Cl\ \ =\ 270} \\
\Sigma G\ =\ 431
\end{array}
\end{array}
$$

$$\delta = \frac{1.4 \times 431}{62.5} = 9.66$$

Source: Ref. 8.

H. Heat Distortion Temperature

Another way to classify lubricants in rigid PVC formulations is to determine their effect on the compound's heat distortion temperature (HDT). The HDT is mainly a function of the polymer's glass transition temperature (T_g) plus any solvating or plasticizing effects that formulation additives may exert on the polymer. The magnitude of difference between the polymer's T_g and the compound's HDT is directly related to the degree of solubility of the additive in the polymer. Lubricants that depress the HDT most are the most soluble and the most compatible, and are thus classified as internal lubricants or plasticizers. Materials that have very little effect on the HDT (i.e., little solubility in the polymer) are external lubricants. This is seen in the data developed by Ceccorulli et al. [12] and King and Noel [13] on dry-blended PVC compound. Plotting T_g or HDT against R, where R is the total weight percentage of additives in the composition, allows the estimation of relative lubricating effects. In Figure 15 for example, pure PVC has a T_g of 82°C, but with 0.1% stearic acid the HDT is 2°C less, and with 0.1% butyl stearate the HDT changes by 15°C. Stearic acid is an external lubricant and butyl stearate is either an internal lubricant or a plasticizer, or both. King and Noel [13] evaluated a series of esters and other compounds to assess their effect on the HDT (Figure 16). Materials such as di-2-ethylhexyl phthalate (DOP), dibutyl phthalate (DBP), and butylbenzyl phthalate (BBP) continue to affect the HDT as their concentrations increase beyond the limit of the testing, indicating that these materials must be plasticizers. Materials such as methyl stearate and butyl stearate have an increasing effect on the HDT to a maximum depression of 20°C, indicating that these materials are internal lubricants or secondary plasticizers.

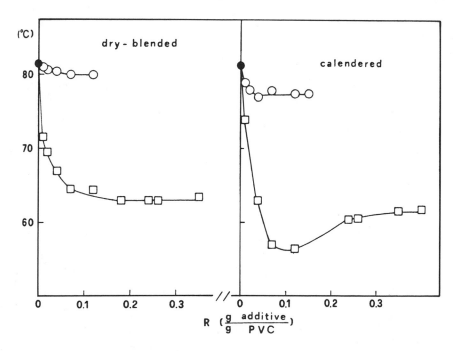

FIGURE 15 T_g vs. additive content: ○ HSt, □ BuSt, ● pure PVC.

TABLE 26 Calculated and Observed Solubility Parameters of Polymers[a]

Polymer	δ (cal/cm^3)$^{\frac{1}{2}}$ (calc.)	(obs.)
Polytetrafluoroethylene	6.2(1)	
Polyisobutylene	7.70	8.05(4)
Polyethylene	8.1(2)	7.9(5)
Natural rubber	8.15	7.9(6)
		7.98(7)
		8.35(4)
Polybutadiene	8.38	8.4–8.6(4)
Butadiene/styrene		
85:15	8.48	8.5(4)
75:15	8.54	8.09(7)
		8.6(4)
60:40	8.65	8.67(4)
Polystyrene	9.12	9.1(4)
		8.6–9.7(6)
Polystyrene/divinylbenzene		9.1(8)
Buna N (butadiene 75)	9.25	9.38(7)
(acrylonitrile 25)		9.5(4)

Polymer	δ (cal/cm^3)$^{\frac{1}{2}}$ (calc.)	(obs.)
Polymethyl methacrylate	9.25	9–9.5(9)
Neoprene GN	9.38	8.18(7)
Polyvinyl acetate	9.4	9.25(4)
Polyvinyl chloride	9.55	9.48(10)
		9.7(6)
Polyvinyl bromide	9.6	9.5(11)
Polymethyl chloroacrylate	10.1	
Cellulose dinitrate	10.48	10.56(6)
Polyglycol terephthalate	10.7(3)	
Polymethacrylonitrile	10.7	
Cellulose diacetate	11.35	10.9(6)
Polyacrylonitrile	12.75	

[a](1) Assumed liquid $(CF_2)_n$; (2) Assumed liquid $(CH_2)_n$; (3) Amorphous form; (4) Scott and Magat; (5) Richards; (6) Magat; (7) Gee; (8) Boyer and Spencer; (9) Alfrey, Goldberg, and Price; (10) Doty and Zable; (11) Edelson and Fuoss.

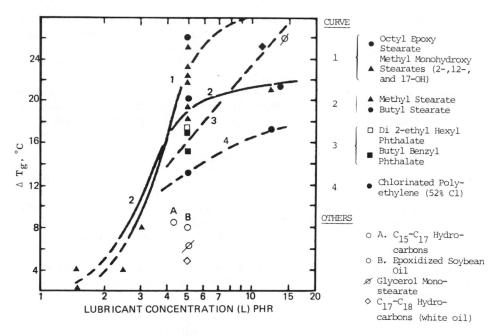

FIGURE 16 Effect of esters and other additives on T_g of PVC at various concentrations.

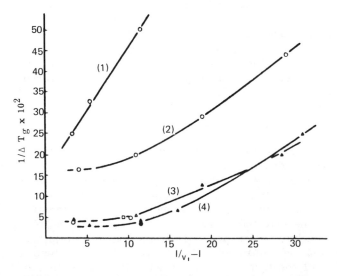

FIGURE 17 Variation of ΔT_g with volume fraction v_1 lubricant. (1) 12-hydroxy stearic acid (2) stearic acid (3) △, methyl stearate; ○, butyl stearate; □, 1-dodecanol; (4) ▲, methyl 12-hydroxy stearate; ○, octyl epoxy stearate.

The change in HDT indicates the degree of solubility or compatibility of the materials in the polymer. The change in HDT or glass transition temperature over the volume fraction is used to calculate an interaction parameter (Figure 17). Paraffin wax, lead stearate, and polyethylene have little or no effect on the HDT, and it appears that materials that have an interaction parameter range of 0 to 65 are external lubricants. Those in the range 60 to 160 are internal lubricants. Materials in the range 160 to 225 are secondary plasticizers, and materials between 225 and 325 are plasticizers (Table 27).

I. Metal Release

Lubricants can be classified according to nontack time. In this procedure, a two-roll mill is used to mill a given amount of compound. The PVC gradually degrades until a point is reached at which it overwhelms the lubricant film that is preventing adhesion to the metal roll. The relative efficiency of external lubricants can be estimated by this test. Internal lubricants give short metal release times.

Table 18 shows a comparison between Henkel complex esters, montanic esters, and calcium stearate. These materials, with the exception of wax E are all external lubricants.

J. Effect on Machine Conditions

A number of investigators have tried to use real-life tests to classify lubricants. Davis and Fraser [14], in 1973, attempted to classify lubricants into three categories: type 1, lubricants that reduce power torque in the feed zone; type 2, lubricants that control retardation of fusion induced by type 1 lubricants; and type 3, lubricants that reduce melt torque in the metering zone of the die. Their technique was to use a Brabender torque rheometer and program the heating cycle to increase at a given rate. The fusion time is spread out and the induction time, which is the onset of fusion, can be determined. Because internal lubricants are compatible, they depress the glass transition temperature; coalescence and fusion occur at a temperature lower than that of the resin alone. External lubricants inhibit fusion and fusion takes place at a temperature higher than that of the resin alone (see Tables 28 to 32). This difference in fusion temperature allows the determination of lubricant functionality.

K. Compaction Test

Another attempt to use real-life situations to classify lubricants was made by Fahey [15], who placed compounded dry blend in a heated, confined cavity and applied pressure and temperature to determine apparent density versus time. Materials segregate because pressure and temperature affect melting and deformation. In Figure 18, a powder with external lubricants densifies to about 1.1 1/cm^3 because the external lubricants allow the particles to slip by each other. The external lubrication prevents transition into homogeneous melt and a density plateau is maintained. In the example the plateau is maintained beyond the time of the test. Ultimately densification would continue until a density of approximately 1.4 g/cm^3 was obtained in PVC formulations. With an internal lubricant (Figure 19) the powder densifies to about

TABLE 27 Interaction Parameters of PVC Additives[a]

Additive	$\Delta Tg(°C)/v_1$
Paraffin wax (Advawax 165)	0
Lead stearate	0
Loxiol GE-1	0
Polyethylene, PA-190	0
9,10-Dihydroxy stearic acid	15
12-Hydroxy stearic acid	25
N,N'-Ethylene bis-stearamide (Advawax 280)	25
Advawaxes 135 and 136	25
Advawax 146 (glycol stearates-palmitates + free acid)	35
Tristearyl citrate	50
Octadecane	55
$C_{17}-C_{18}$ paraffins (white oil) (Marcol 52)	60
Stearic acid	60
Glycol esters of montanic acids (Wax E)	60–65
Glycerol monostearate	85
Advawax 121	90
Loxiol G-40	95
Loxiol GH-1	100
Epoxidized soybean oil (Paraplex G-62 or Epoflex 945)	100
$C_{15}-C_{17}$ paraffin-naphthene fraction (Mentor 29)	110
1-Eicosanol	120
1-Octadecanol	130
1-Hexadecanol	170
Chlorinated polyethylene (Cereclor S-52)	180
1-Tetradecanol	185
1-Dodecanol	210
2-Dodecanol	220
Methyl stearate	220
Butyl stearate	225
Di-2-ethyl hexyl phthalate	225
Butyl benzyl phthalate	230
Dibutyl phthalate	240

TABLE 27 (Continued)

Additive	$\Delta Tg(°C)/v_1$
Methyl-9,10-dihydroxy stearate	240
Methyl-2-hydroxy stearate	275
Methyl-17-hydroxy stearate	290
Methyl-12-hydroxy stearate	300
Octyl epoxy stearate (Drapex 3.2)	325

[a]Additive concentration: 0.08 ± 0.01 volume fraction (v_1).
Source: Ref. 13.

TABLE 28 Evaluation of Materials That Accelerate Fusion at a Level of 1 phr
in a Formulation Containing Tribasic Lead Sulfate and Dibasic Lead Stearate

Material	Fusion temperature (°C)	Induction temperature (°C)	Induction torque (m·kgs)	Torque at 180°C (m·kgs)
Blank	141	121	1.50	2.30
Tritolyl phosphate	132	120	2.50	2.30
Aromatic oil	140	122	1.60	2.20
Stearyl phthalate	132	119	2.00	2.15
Glyceryl monoricinoleate	128	115	1.90	2.10
"Rapivyl"	137	118	1.50	2.05
N-Butyl stearate	132	116	1.80	2.05
Tristearyl citrate	136	130	2.50	2.05
Lauryl alcohol	133	114	1.60	2.00
Glycerol monostearate	131	117	1.80	2.00
Isodecyl stearate	136	129	2.00	1.90
Stearyl alcohol	141	136	1.60	1.90
Ethyl palmitate	134	116	1.70	1.90
Dibutyltin maleate	128	116	2.00	2.30
Acrylic processing aid	128	118	2.50	2.30

Source: Ref. 14.

TABLE 29 Evaluation of Materials at 0.5 and 1.5 phr in a Formulation Containing Tribasic Lead Sulfate and Dibasic Lead Stearate

Materials	Fusion temperature (°C)	Induction temperature (°C)	Induction torque (m·kgs)	Torque at 180°C (m·kgs)
0.5 Tritolyl phosphate	136	119	1.65	2.30
1.5 Tritolyl phosphate	127	113	2.85	2.55
0.5 Glyceryl monoricinoleate	133	114	1.80	2.20
1.5 Glyceryl monoricinoleate	132	113	1.55	1.90
0.5 "Rapivyl"	130	120	2.10	2.25
1.5 "Rapivyl"	131	109	2.10	2.00
0.5 N-Butyl stearate	132	114	1.75	2.05
1.5 N-Butyl stearate	132	113	1.50	1.90
0.5 Isodecyl stearate	135	113	1.45	2.05
1.5 Isodecyl stearate	145	137	1.70	1.05
0.5 Glyceryl monostearate	131	113	1.60	2.05
1.5 Glyceryl monostearate	133	111	1.60	1.95
0.5 Tristearyl citrate	135	115	1.70	2.10
1.5 Tristearyl citrate	185	164	0.60	1.75
0.5 Stearyl alcohol	138	114	1.55	2.20
1.5 Stearyl alcohol	132	112	1.55	1.90

Source: Ref. 14.

TABLE 30 Evaluation of Fusion-Retarding Materials in a Formulation Containing Tribasic Lead Sulfate and Dibasic Lead Stearate

Material	Fusion temperature (°C)	Induction temperature (°C)	Induction torque (m·kgs)	Torque at 180°C (m·kgs)
12-Hydroxystearic acid	200	190	0.20	1.50[a]
Lauric acid	190	185	0.25	1.05[a]
Stearic acid	192	184	0.25	1.35[a]
Polyethylene wax	185	170	0.90	1.40[a]
Ethylene bis-stearamide	185	174	0.40	2.10[a]
Paraffin wax	180	172	0.70	1.30
Stearyl stearate	166	150	1.40	1.95
Paraffinic oil	161	150	1.25	1.90

[a]Extrapolated values.
Source: Ref. 14.

TABLE 31 Evaluation of Materials at 0.3 and 0.6 phr in a Formulation Containing Tribasic Lead Sulfate and Dibasic Lead Stearate

Material	Fusion temperature (°C)	Induction temperature (°C)	Induction torque (m·kgs)	Torque at 180°C (m·kgs)
0.3 Paraffin wax	185	147	0.95	1.80
0.6 Paraffin wax	186	173	1.00	1.60
0.3 Stearic acid	141	119	1.45	2.05
0.6 Stearic acid	149	114	1.35	1.60
0.3 Polyethylene wax	166	135	1.00	1.70
0.6 Polyethylene wax	179	140	0.80	1.60
0.3 Paraffinic oil	139	115	1.20	2.15
0.6 Paraffinic oil	140	119	1.65	2.10

Source: Ref. 14.

TABLE 32 Evaluation of Metal Soaps at 1 phr in a Formulation Containing Tribasic Lead Sulfate and Dibasic Lead Stearate

Metal soap	Fusion temperature (°C)	Induction temperature (°C)	Induction torque (m·kgs)	Torque at 180°C (m·kgs)
Magnesium laurate	137	115	1.25	2.05
Lithium 12-hydroxy stearate	195	185	0.35	1.30[a]
Aluminum stearate	193	182	0.35	1.40[a]
Cadmium stearate	189	177	0.40	1.40[a]
Lithium stearate	187	173	0.60	1.75[a]
Lead stearate	185	169	0.75	1.45[a]
Potassium stearate	176	153	0.40	1.80
Dibasic lead stearate	154	128	1.15	2.05
Barium stearate	144	127	1.40	1.95
Magnesium stearate	144	121	1.35	2.00
Calcium stearate	136	120	1.05	1.95

[a]Extrapolated values.
Source: Ref. 14.

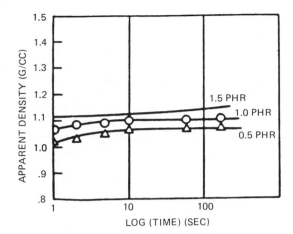

FIGURE 18 XL 355 wax (180 C).

0.8 to 0.9 g/cm^3. A dense pack is not formed because the particles do not have sufficient external lubrication to slip over themselves under pressure. Densification takes place when the particles deform and become a homogeneous melt. This lack of dense pack is typical of internal lubricants and the effect can be summarized as follows: the greater the tendency to produce a dense pack in this system, the longer the time to melting and complete densification as a homogeneous melt.

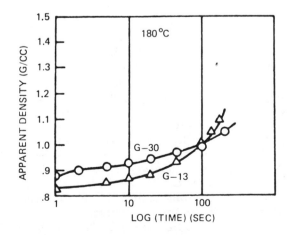

FIGURE 19 G-13/G-30 (1.5 phr).

L. Pressure Profile

Another system for real-life classification of lubricants has been designed
by Henkel. This technique utilizes an extruder with transducers implanted
at the feed, compression, and metering zones. Simultaneous pressure readings are recorded after a steady state is achieved while a sample is being
processed in the extruder. Internal lubricants exhibit high pressure in the
feed and compression zones. Frictional heat builds up and a drop in melt
viscosity is seen due to overheating in the metering zone. External lubricants inhibit fusion and exhibit low or no pressure in the feed and compression zones. With the delay in fusion, the melt would have some particulate
nature and exhibit higher melt pressures in the metering zone. The graph
of pressure against the position in the extruder zones with various lubricants (Figure 20) indicates that pressures with internal lubricants are high
in the first zone and decrease through the third zone. Graphs for external
lubricants are low in the feed zone and increase through the metering zone.
A well-balanced system shows low pressures in the feed zone and increases
in the compression zone and then a drop at the metering zone.

VI. EVALUATION OF LUBRICANTS

Evaluating lubricants is probably as much art and experience as it is science.
A number of tests are very important in making lubricant selections. The
following discussion covers tests used in evaluating the processability imparted by lubricants rather than effects on the performance of the finished
product. Finished product performance testing is discussed in the Formulations section.

A. Mixing Techniques

The first step is sample preparation. The preferred device for preparing
dry-blend samples for testing is a laboratory-size high-intensity mixer (Figure 21). Essentially, this mixer consists of a jacketed cavity with a high-
speed rotor at the bottom. The rotor speed can be varied between 1800 and
3600 rpm. A possible procedural technique for mixing is shown in Table 33.
    The important point is that resin should be heated to 150 to 160°F and
the stabilizers then added as soon as possible so that degradation is minimized during the mixing and blending procedure. The lubricants should not
be added at this point, because it is important for the stabilizer to penetrate
the resin rapidly. Another important point is that the temperature should
be raised above the boiling point of water so that any hygroscopic materials
can release adsorbed water, to prevent blistering during testing or sample
preparation. The lubricants can be added near the end of the heating cycle,
where they can melt and completely disperse in the dry blend. In this manner, they trap the stabilizer in the resin by encapsulating the resin particles, encouraging particulate flow in the semimelt. The dry blend should
be cooled to prevent long heat retention and ensuing compound degradation.
All samples should be aged at standard conditions for 24 hr before testing.
Stabilizers and internal lubricants require time to be absorbed into the resin
particles. If the material is tested without aging, the fusion test may indicate that the material is more externally lubricated than it is.

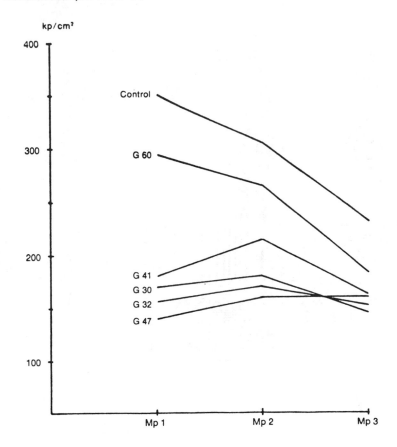

Measured values for torque, back pressure and output

Dosage in phr	Torque mkp	Back pressure kp	Output grams/min.
Control, no lubricant	21	1950	122
1.0 LOXIOL G 60	17	1500	124
1.0 LOXIOL G 41	14	1250	115
1.0 LOXIOL G 30	12.5	1100	123
1.0 LOXIOL G 32	10.5	1130	121
1.0 LOXIOL G 47	14	1130	125

FIGURE 20 Influences of various lubricants on the pressure diagram in the extruder.

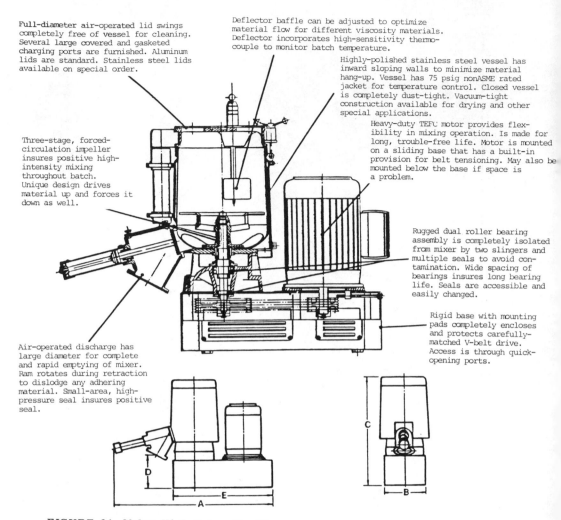

Full-diameter air-operated lid swings completely free of vessel for cleaning. Several large covered and gasketed charging ports are furnished. Aluminum lids are standard. Stainless steel lids available on special order.

Deflector baffle can be adjusted to optimize material flow for different viscosity materials. Deflector incorporates high-sensitivity thermocouple to monitor batch temperature.

Highly-polished stainless steel vessel has inward sloping walls to minimize material hang-up. Vessel has 75 psig nonASME rated jacket for temperature control. Closed vessel is completely dust-tight. Vacuum-tight construction available for drying and other special applications.

Heavy-duty TEFC motor provides flexibility in mixing operation. Is made for long, trouble-free life. Motor is mounted on a sliding base that has a built-in provision for belt tensioning. May also be mounted below the base if space is a problem.

Three-stage, forced-circulation impeller insures positive high-intensity mixing throughout batch. Unique design drives material up and forces it down as well.

Rugged dual roller bearing assembly is completely isolated from mixer by two slingers and multiple seals to avoid contamination. Wide spacing of bearings insures long bearing life. Seals are accessible and easily changed.

Rigid base with mounting pads completely encloses and protects carefully-matched V-belt drive. Access is through quick-opening ports.

Air-operated discharge has large diameter for complete and rapid emptying of mixer. Ram rotates during retraction to dislodge any adhering material. Small-area, high-pressure seal insures positive seal.

FIGURE 21 Welex High Intensity Mixer.

One of the pitfalls of this technique involves small quantities of liquids that are difficult to weigh out and to deliver accurately. A syringe or a beaker that is wet-tared should be used to increase precision. Materials that are added into the vortex may coat the rotor and may never be dispersed. Thermocouples in the scraper blade are inherently inaccurate because the heavy mass of the blade requires considerable time to heat or cool. The temperature of the powder may be 20 to 30°F hotter at any given moment during heating than the thermocouple in the heavy scraper blade registers. Occasionally, the mixer should be stopped and the walls scraped to ensure uniformity. In a 10-liter mixer, which is a good size for testing, 1000 to 2500 g of dry blend can be prepared at one time. Charges that are either too large or too small interfere with the mixing vortex.

As mentioned previously, alternative mixing techniques can be used. Many researchers prepare blends by using a different order of addition. Some fill the mixer with dry ingredients and add liquids when the blades start turning. As long as a consistent mixing method is used each time, internally consistent data can be obtained. A modification of the above proce-

TABLE 33 Henkel Mixing Technique

Turn on steam full, 20 lb, 240°F.

Charge resin.

Mix at 3600 rpm.

Add stabilizer at 150°F.

Add wax at 160°F.

Stop and scrape sides at 170°F.

Add pigment at 180°F.

Add processing aid or impact modifiers at 200°F.

Shut off steam at 220°F and turn on cold water and scrape sides of mixer.

Change to 1800 rpm and mix while cooling.

Drop batch at 150 to 160°F.

Note: may need to turn up to 3600 rpm to blow out mixer.

dure can be used for plasticized PVC. Steam can be turned on until the resin reaches 150°F and then turned off. The liquid stabilizer/plasticizer premix can then be added slowly to the side of the vortex. Frictional heat buildup will normally raise the temperature of the dry blend to 212°F. A small amount of emulsion-grade resin (approximately 3%) can be added, if desired, to absorb excess plasticizer and obtain a free-flowing powder blend. Without this step one occasionally sees a wet-blend that is not completely free-flowing. Shortly after the addition of the emulsion grade resin, lubricants can be added. Cold water can be fed into the jacket and the system cooled at a low mixer speed.

A dangerous practice sometimes used is to allow the plasticized material to fuse in the Henschel blender and then be chopped by the blades. The blades can snap off and tear through the sides of the machine, injuring anyone standing nearby. It is also not advisable to granulate for testing purposes, as granules do not give as much information as powder in the Brabender fusion test. Although plasticized PVC can be blended in a Henschel-type blender, a ribbon blender that is jacketed and can be heated with steam and cooled with cold water is preferred. The mixing procedure would not vary from that used in a high-intensity mixer. The materials would have to be heated continually because there is no frictional heat buildup with a ribbon blender. When a ribbon blender is used, it is not necessary to add an emulsion grade drying resin in the blending technique. Compounds should be stored at least 24 hr under standard conditions before testing, to allow all the internal lubricants and plasticizers to reach physical equilibrium with the resin.

B. Brabender Plasticorder Fusion Test

The most widely used test for evaluating lubricants in PVC is the Brabender Plasticorder fusion test. It is a very simple test and indicates how the material will perform in the extrusion operation. One should be aware that even though the test is simple to perform, it is very sensitive. Consistency of

TABLE 34 Brabender Plasticorder PVC Powder Fusion Test

Apparatus

   Brabender Plasticorder PL-V300 with 60-ml roller No. 6 head

Conditions

   Use 60-ml capacity roller mixing head No. 6.

   Set temperature as required (175°C)

   Roller speed 30 rpm

   Connector set as required (1:5)

   Indicator adjustment as required (5×)

   Air cooling pressure 60 lb/in.^{2}

   Slide set as required (0)

   Damper set 8 sec

   Charge 40 × SPG

   Loading chute cold

   Ram weight 15 kg

   Preheat time 1 min

Procedure

   Compound was prepared in Henschel mixer and aged overnight.

   Cold chute is inserted in throat of mixing head 1.0 min before test is
      started.

   Sample is pressurized with 15 kg free-falling weight on chute ram.

   Machine is stopped after 15 min, and fusion time and torque at fusion and
      10 min thereafter are recorded.

---

operation is necessary in order to obtain valid data. There are many variables that affect this test. Powder fusion is affected by pressure, temperature, and shear rate, any of which can be varied. It is important to modify
the test in a way that best simulates the application that is being evaluated.
See Table 34 for one company's typical procedure.

   The test procedure in Table 35 is best suited for evaluating internal lubricants or internal/external lubricant packages, for single-screw and injection-molding processes. To improve repeatability, a few rules should be
followed. First, the sample should be properly aged so that plasticizers and/
or internal lubricants have sufficient time to reach equilibrium with the resin.
Fresh samples tend to increase variance because the lubricants have not had
time to be absorbed; thus they appear more external or, if the resin is still
warm, appear more internal. Second, the sample size should be approximately 40 times the specific gravity of the compound. Ideally, one obtains den-

TABLE 35 Brabender Stability Test: Dynamic Early Color Development

---

Apparatus

    Brabender Plasticorder PL-V300 with 60-ml electrically heated No. 6 roller head. Mold pliers, 1 cm^2 or 1 cm diameter cavity, 0.5 mm deep.

Conditions

    Set temperature 190°C

    Roller speed 60 rpm

    Connector set 1:5

    Indicator adjustment 5×

    Air cooling pressure 60 lb/in.^{2}

    Slide set 0

    Damper set 8 sec

    Charge 40 × SPG

    Loading chute Cold

    Ram weight 15 kg

    Preheat time 1 min

Procedure

    Compound is prepared in a Henschel mixer and aged overnight.

    Cold chute is inserted in throat of mixing head 1 min before test is started.

    Sample size is 40 × SPG.

    Compound is poured into chute and allowed to preheat 5 min before the test is started.

    Sample is pressurized with a 15 kg free-falling weight on chute ram.

    Test is begun on pressurization.

    Machine is stopped every 5 min and a sample is extracted with mold pliers.

    Sample is mounted on card.

    Fusion time and torque, both at fusion and 10 min after, are recorded.

    Color and clarity, if applicable, are demonstrated by mounted sample.

Note: Any deviation from the above is recorded on the experimental sheet.

sification of the powder to melt and a mass that exactly fills the bowl. In many cases, PVC blends have a density of approximately 1.4 g/cm^3 and 56 g provides a sample size at which comparisons can be made. If one uses high levels of filler, high levels of plasticizers, or high levels of TiO$_2$, a calculated specific gravity is necessary to determine the optimum sample size. A larger sample size may be desired in order to model pipe compounds in conical twin-screw processing equipment.

A pressure factor may be introduced to duplicate the effects of processing equipment. With pipe compounds, high levels of external lubricants are usually used and pressurization is necessary to achieve any fusion at all. One may wish to calculate the compression ratio of the screws in order to determine the densification factor needed to optimize the test. High levels of external lubricants may also necessitate the use of higher temperatures and roller speeds to increase the solubility and shear rate and shorten fusion time. It is important that the rate of injection of the sample "pressurization in the bowl" be constant; a chute equipped with sufficient dead weight is suggested; when the weights are placed on the plunger and the load released, gravity acts as a constant force. Temperature is critical; the loading chute should be preconditioned in an oven so that it will not act as a heat sink, drawing heat away from the bowl. The chute should be introduced at the same point before the start of the test each time. It is suggested that the same operator perform the tests in a full set of experiments to minimize variance.

After the preliminary determination, fusion should be expected at 2 to 4 min with the control sample. Fusion time should be set so that negative and positive variations can be easily differentiated in the unknown materials. If the conditions are such that the fusion takes place for all samples in the range 0.5 to 1 min, the result would be within the limits of error of the machine. The differences in data would not be meaningfully interpretable. If the material does not fuse, temperature and shear rate may have to be changed to obtain meaningful results. It is easiest to change shear rate or temperature to increase the sensitivity of this test. Different lubricants act differently at different temperatures and shear rates, and crossovers in functionality can occur so that one lubricant may appear more external than another at a low temperature but less external at a high temperature. This phenomenon also occurs as a function of shear rate. The temperature should reflect the process conditions. The amount of lubricant used should be varied to obtain an optimum level for testing, even if that level would be impractical for actual production, to determine differences.

The shear rate of the Brabender machine is low compared to that of production machinery. This limits the amount of information that can be generated on internal lubricants. The roller speed can be varied and relative trends observed. The more effective internal lubricants will have a smaller effect on equilibrium torque with increasing shear rate when melt temperature is constant. At low shear rates, external lubricants will reduce equilibrium torque because of improved metal release, which allows the polymer mass to slip along the wall and between the kneading rotors. A torque rheometer should be used to estimate fusion time but not functionality of lubricants, unless other tests corroborate the results. Problems can arise with materials that are evaluated at shear rates different from those used in the production machinery. For internal lubricants, the Brabender extrusiograph equipped with a rheometric head for evaluating the melt viscosities and flow natures of materials is suggested to obtain a more complete picture of lubricant functionality.

Cleanout and reconditioning prior to the start of the next test is important in the fusion test. The old material must be ejected and material scraped from the blades. A scrubber is introduced. Ideally, this should be wood from which all rosins and residues have been chemically leached. Since this is impractical, we have used cedar chips which give a constant factor in the test. After approximately 5 min, the wood chips have been compacted in the bowl, have scraped off any residue, and have absorbed any residual lubricant film that was deposited during the test. Air is used to blast out all the wood chips or dust. The bowl is closed and allowed to reach thermodynamic equilibrium, which takes approximately 15 min. Cleaning and thermal equilibration of the equipment is critical and should be adhered to strictly. A bowl that has not come to thermal equilibrium registers erroneous fusion times. If a proper cleanout is not effected, the residual lubricant film from the previous test may interact with the new system and give an erroneous result as well. It is often wise to repeat the first sample at the conclusion of a test series to eliminate errors due to warmup and to check reproducibility. Ideally, all tests should be performed in a room held at constant temperature and humidity. It is not always true that external lubrication is related to speed of fusion; some lubricants can give false results in this test. Therefore this test should be used only in conjunction with other tests that verify and confirm the results. The main use of this test is as a simple method for comparing the fusion of known compounds to that of unknown compounds. A match can be made to the properties of new compounds so that they can be run safely in an extruder without unanticipated results. Experience is needed to properly interpret these results. Figure 22 illustrates some typical curves obtained in the fusion test.

C. Brabender Stability Test

It is important that the processing stability of a compound be evaluated in a dynamic test because lubricants affect the stability of the compound. High-viscosity formulations are subjected to more mechanical stress and will be less stable than low-viscosity materials. Lubricant stability and stabilizer and lubricant functionality are tested simultaneously in the Brabender stability test. Test conditions of 190°C and 60 rpm are arbitrarily chosen as constant conditions. Materials should be injected and treated as in the fusion test (discussed in the preceding section) and samples should be extracted every 5 min and mounted. In formulations with short stability time a sample may be extracted every 2 min to observe color development. An experienced operator will remove similar amounts of sample, and the variation will be insufficient to have a major effect on the test.

Many investigators run this test up to cross-linking time or degradation. It is also important that color development be evaluated (Tables 35 to 38). Most products are not acceptable due to color change long before any cross-linking takes place. In highly externally lubricated compounds, cross-linking time is not even perceptible because particulate flow predominates over melt flow. Immediately after the onset of cross-linking, the test should be terminated and cleanout begun. It is suggested that the cleanout technique used in the fusion test be used. Any small particles of degraded material left in the bowl will adversely affect the next sample.

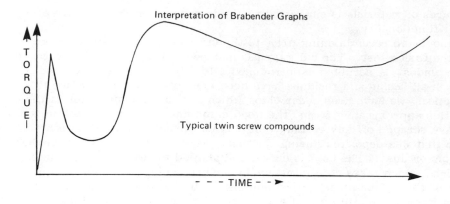

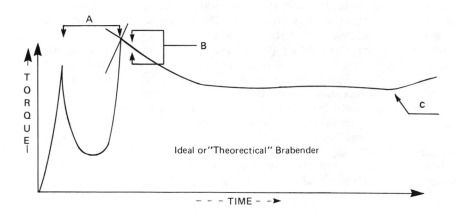

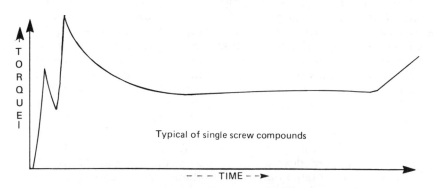

FIGURE 22 Interpretation of Brabender Graphs: A fusion time; B degree or curve amplitude indicates possible problem with die swell; C crosslinking.

TABLE 36 Brabender Stability Test Data: Formulations

Component	Formulation		
	A	B	C
GEON 110 × 334	100.0	100.0	100.0
Acryloid K 120 N	2.0	2.0	2.0
Dow cpe 0.56	2.0	2.0	2.0
TiO_2	1.5	1.5	1.5
Calcium stearate	1.0	1.0	1.0
E.B.S. wax	1.0	0	0
TM 181	1.6	1.6	1.2
Loxiol G-33	0	0.8	0.8
XL 165	0	0.2	0.2

1. Regrind stability: To get a complete picture of the processing stability and longevity-of-use stability of the compound, a static oven test is often used to evaluate regrind or end-use stability. This test consists of evaluating the stability of pieces of parts or material. It is important that the pieces all be similar in shape, size, and thickness. A hot-air circulating oven is used, in which samples are placed on a rotating table to ensure equal exposure conditions. Samples can be removed at various time intervals to observe color development.

TABLE 37 Brabender Stability Test Data: Results for Formulations A, B, and C

Measurement	Formulation		
	A	B	C
Brabender stability time at 190°C, 60 rpm	33 minutes	38 minutes	32.5 minutes
Fusion time at 190°C, 60 rpm	0.4 minutes	0.5 minutes	0.7 minutes
Torque (meter-grams)	2950	2700	2750
Torque after 10 min (meter-grams)	1450	1300	1350

TABLE 38 Color Development Evaluation

Formulation	Min 5	Min 10	Min 15	Min 20
A E.B.S. wax 1.0 TM-181 1.6	▨	▨		
B Loxiol G 30 .8 XL-165 .2 TM-181 1.6	▨	▨	▨	▨
C Loxiol G 30 .8 XL-165 .2 TM-181 1.2	▨	▨	▨	

D. Roll Release Test

This test evaluates metal release properties that are critical in both calenders and dies for the extrusion operation. No lubricant evaluation procedure is complete without this test. Materials that can be incorrectly evaluated in the Brabender fusion test reveal external functionality via metal release time.

The metal release test gives a numerical value to the ability of a lubricant to promote separation from metal surfaces. It should be noted that different metal surfaces have different coefficients of friction and adhesion with PVC melts. Since many calendar rolls used to process rigid PVC are chrome-plated, it is best to use a chrome-plated roll so that the metal release correlates with production machinery. Since shear rate is a factor, one should have a variable-speed mill so that the surface velocity of the rolls can be adjusted to approximately that of production equipment. It should also be noted that larger rolls are easier to work on and, since safety is a factor, the larger the roll, the safer the equipment. In order to maintain shear rates equivalent to those in production and to obtain reliable results, the gap between the two rolls should be carefully maintained. A smaller gap will introduce higher shear rates and greater frictional heat buildup. Lubricants are more soluble at higher temperatures, and this will affect the results of the test by shortening the time to sticking. A typical procedure is outlined in Table 39. The mill must be able to handle typical processing temperatures of rigid PVC—that is, up to 425°F. This capability requires a hot-oil-heated mill. The roll seals must be of a quality to retain a tight seal at that temper-

TABLE 39 Roll Release and Stability Test Procedure

Apparatus

  Two-roll rubber mill with oil heat capacity

  Roll dimensions 8 × 16 in.

  Friction ratio 1:1.25

  Rolls with chromed steel surface

Conditions

  Roll surface

    Back roll, −370°F

    Front roll, −370°F

  Roll speeds

    Back roll−23 rpm

    Front roll−18.5 rpm

    Mill gap−40 mils

Procedure

  1. Compounds are prepared in a Henschel mixer and aged overnight.

  2. Sample charge is weighed out at 180 g. Sample charge is then added to the mill and allowed to band.

  3. Once banding has occurred, timer is started.

  4. All excess cold compound not circulating in the nip is taken off and discarded.

  5. Every 5 min a 2-in. square sample is extracted and retained for mounting and the sample is mixed in for 1 min.

  6. When compound bonds to the roll surface, it is cut off and the test is terminated.

  7. Rolls are then cleaned with stearic acid and presurfaced by using a special cleaning compound containing silica, which cleans the surface and absorbs all excess stearic acid.

Report

  Time from banding to sticking is reported as a function of efficiency of external lubrication. Five-minute samples are mounted and observed for color change as a function of stability. Power consumption is reported as a function of internal lubrication and stock temperature is reported to ensure consistency of results.

ature. One drop of contamination by heating oil will affect the test. Contamination is a problem, and care must be taken not to introduce foreign matter.

Sample size should be selected such that no cold material hangs up in the nip and all the material is rolling in the bank or on the rolls. It is very important to maintain a temperature similar to or as close as possible to the production temperature, since materials can function differently at low than at high temperatures. To minimize erroneous results, one must test near the process temperature. In many cases, compounds are designed not to stick during their entire processing life; consequently, these compounds will run without sticking until they degrade. A test on these compounds would be meaningless. In order to compare lubricants effectively, changes in the compound must be made. One should reduce the concentrations of the lubricants without changing ratios to promote sticking of the material before it degrades. If a test is still not attainable with lower concentrations of the lubricants, stabilizer levels may be increased to lengthen stability until tack. This work should be done prior to the evaluation of a test set. Once meaningful conditions have been set up and a test achieved, one is next faced with the problem of cleanup. As soon as tack is achieved, immediately scrape the material from the roll and add a small amount of stearic acid to effect the release of the remaining stock material. This should be followed by the addition of a cleaning batch that contains silica abrasive and has a high filler content to loosen the burnt material and remove it from the roll. The roll should then be rubbed with stearic acid and sufficient time allowed for the stearic acid to volatilize, leaving the rolls absolutely clean and ready for the next test. An example of some data produced by this test is included in Table 18.

E. Capillary Rheometers

Capillary rheometers can range in complexity from simple melt indexers to microprocessor-controlled multivariant systems. Simple melt indexers give limited information because of the low shear stress usually employed. The more complex capillary rheometers give multipoint versus single-point data and allow evaluation of the polymer melt at shear rates generally prevailing in the processes of injection molding, extrusion, and calendering. A typical melt rheometer (Figures 23 and 24) is the Sieglaff-Mckelvey. Theoretical considerations are discussed in Chapter 29 of this series. Lubricants modify relationships between shear stress and shear rate. Internal lubricants shift melt fracture to higher shear rates and external lubricants reduce the shear stress per unit shear rate. Illmann [3] presents data indicating that combinations of lubricants are more effective than either internal or external lubricants alone. Output is related to the total quantity of lubricants used, their internal nature, and their external nature. The temperature at which testing is performed is a critical factor. Shah [16] demonstrated that viscosity is shifted by changing temperature (see Figures 25 and 26). Differences in lubrication should be tested under the process conditions of the material to optimize output.

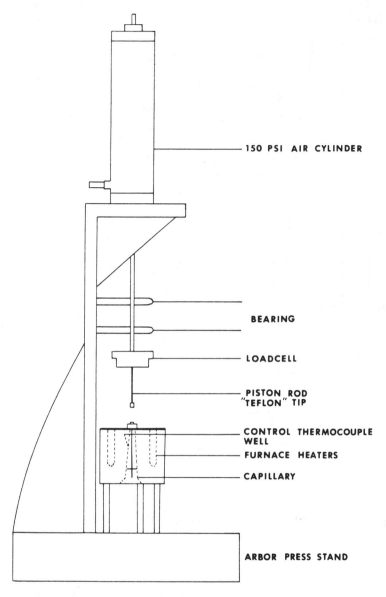

FIGURE 23 Schematic diagram of a piston rheometer.

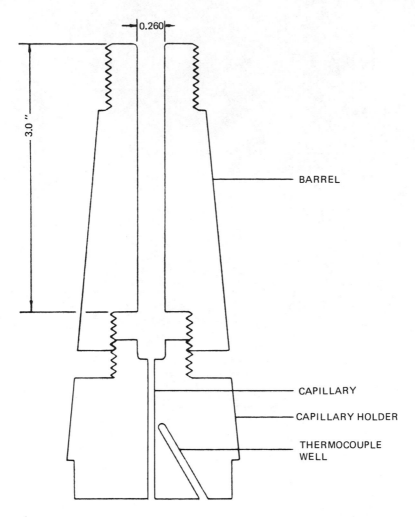

FIGURE 24 Diagram of a barrel and capillary assembly.

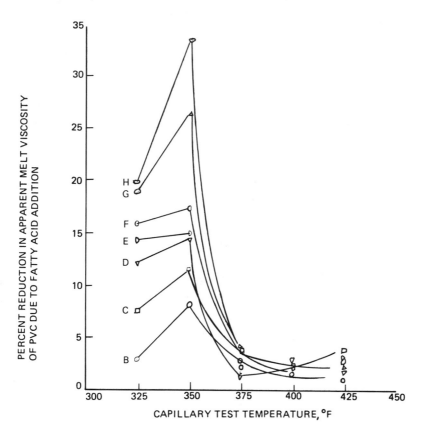

FIGURE 25 Effect of test temperature on percent viscosity reduction of PVC due to fatty acid addition. Shear rate: 100 sec^{-1}.

SAMPLE	ACID	NO. CARBON ATOM
B	BUTYRIC	4
C	CAPRYLIC	8
D	LAURIC	12
E	MYRISTIC	14
F	PALMITIC	16
G	STEARIC	18
H	DECOSANOIC	22

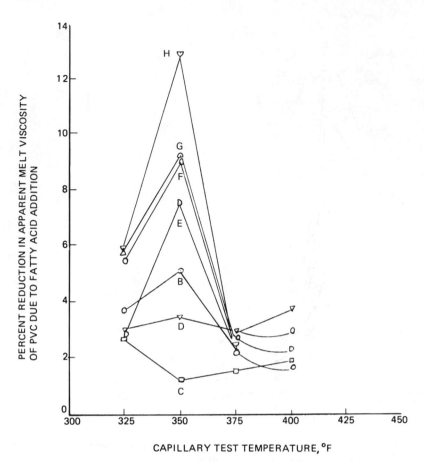

FIGURE 26 Effect of test temperature on percent viscosity reduction of PVC due to fatty acid addition. Shear rate: 1000 sec^{-1}.

SAMPLE	ACID	NO. CARBON ATOM
B	BUTYRIC	4
C	CAPRYLIC	8
D	LAURIC	12
E	MYRISTIC	14
F	PALMITIC	16
G	STEARIC	18
H	DECOSANOIC	22

F. Extrusiometer

This device is an extruder that feeds a rheometric head (Figure 27). Pressure drop at the orifice is obtained and the Poiseuille equations provide shear rate versus shear stress and shear rate versus viscosity data. Testing procedures are included (Tables 40 and 41). A screw is used to feed the rheometric die. This provides a mixture of both particulate and melt flow similar to that observed on production equipment. Transducers can be placed in the body of the extruder to aid in determination of the fusion point (see Figures 27 and 28). This is important because if fusion occurs too late in the extruder (i.e., the material is overlubricated), the particulate flow dominates in the orifice, leading to erroneous conclusions. Early fusion, which indicates underlubrication, explains discoloration due to sticking in the extruder. Only a properly lubricated compound can be evaluated for critical stabilizer level. Measurements of the swell ratio can be used to evaluate internal lubricants. The higher the swell ratio, the less the relaxation of van der Waals forces which occurs in the PVC melt. An alternative method for evaluating internal lubricants is to observe the shear rate at which melt fracture occurs. This can be done by variation of the speed of the extruder and the orifice diameter. The more data points obtained, the easier it is to determine the fracture point; for example, see Tables 42 and 43 and Figure 29. PVC formulations designed for twin-screw extruders contain higher amounts of external lubricants. To obtain meaningful results, evaluations should be performed on a twin-screw extruder with a rheometric head attached.

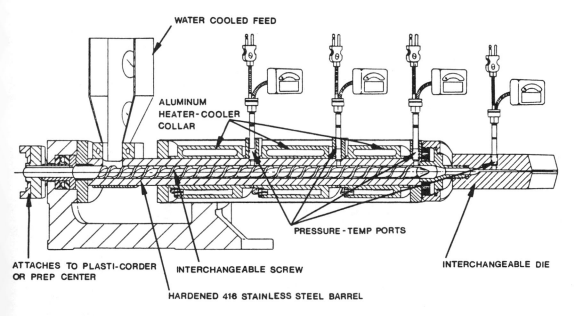

FIGURE 27 Extrusiograph schematic.

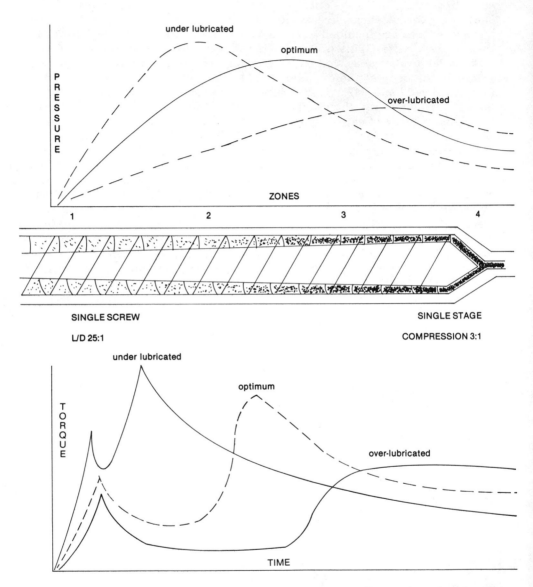

FIGURE 28 Comparison of Extrusiograph pressure profile and Brabender data.

TABLE 40 Test Procedure for Extrusiograph® Evaluation

Apparatus

    C. W. Brabender Extruder (attachment for Plasticorder)

        3/4-in. diameter

        25:1 length-to-diameter ratio

    Screw: Rigid type (three zone)

        Single stage

        Compression ratio 3:1

        Unvented

    Hopper with screw feeded

    Die: 2-mm rheometric rod die with pressure transducer and thermocouple

Test conditions

    Hopper-screw speed 30 rpm

    Extruder speed 120 rpm

    Torque lever arm settings (5X) 5:1 with 15-kg weight on arm lever

    Temperature settings

        Zone 1 (feed) 160°C

        Zone 2 (compression) 190°C

        Zone 3 (metering) 200°C

        Zone 4 (die) 190°C

Report

    Torque

    Die temperature and pressure

    Output

    Appearance of the extrudate

TABLE 41 Extrusiograph Rheometric Calculations Applying Classic Mathematical Formulas

---

Plots of output versus pressure will reveal the influence that length-to-diameter die ratios have on throughput. In order to generate such graphs,
a range of pressure values is produced at the entrance of the three capillary
lengths by varying screw speed. For each specific pressure, output is
measured and plotted on the ordinate. Pressure drop is plotted on the abscissa. The latter values are found by solving Poiseuille's equation for pressure drop, which is given here:

$$\Delta p = \frac{8Qnl}{\pi r^4}$$

where Q is volumetric flow rate (cm^3/sec), n is the viscosity coefficient
(poise), r is the radius of the capillary, and l is the length of the capillary
(cm).

However, it is first necessary to solve the equation given for viscosity (n):

$$n = \frac{(\Delta pr)/(2l)}{(4Q)/(\pi r^3)}$$

Since the viscosity coefficient (n) is equal to shear stress (t) divided by
shear rate (y), the same calculation provides the numerical values for shear
stress

$$t = \frac{\Delta pr}{2l}$$

and for shear rate

$$y = \frac{4Q}{\pi r^3}$$

For our purposes, it is important to evaluate and plot shear rate versus viscosity, as well as the quality of the surface of the extrudate, as these will
indicate the processability of any potential compound. These properties are
particularly good indicators of the efficiency of an incorporated internal lubricant, as the lower melt viscosity of equivalent shear rate will allow one
to evaluate comparative samples.

TABLE 42 Formulations Evaluated

Component	Formulation				
	A	B	C	D	E
PVC, K value 65	100	100	100	100	100
Phthalate plasticizer	30.0	30.0	30.0	30.0	30.0
Flame retardant	2.0	2.0	2.0	2.0	2.0
Tribasic lead sulfate	5.0	5.0	5.0	5.0	5.0
Lead stearate	0.15	0.15	0.15	0.15	0.15
Ethylene bis-stear-amide wax	1.5				
Partial ester, Loxiol[a] HOB 7121		1.5			
Simple ester, Loxiol[a] G-30			1.5		
Partial ester, Loxiol[a] HOB 7135				1.5	
Simple ester, Loxiol[a] G-40					1.5
Complex ester, Loxiol[a] G-71	0.7	0.7	0.7	0.7	0.7

[a]Loxiol is a registered trademark of Henkel Corp.
Source: Henkel

TABLE 43 Dynamic Heat Stability Results

Measurement	Formulation					
	A	B	C	D	E	F
Brabender stabilization time at 205°C, 150 rpm, min	28	61	38	54	39	39.5
Fusion time at 205°C, 150 rpm, min	————————————Instantaneous————————————					
Torque after 10 min, m·g	550	550	550	600	550	1100

TABLE 44 Compound Formulations Evaluated in the Extrusiometer Test

Component	Formulation							
	A	B	C	D	E	F	G	H
Medium-molecular-weight copolymer	100	100	—	—	—	—	—	—
VC-95PM[a] (medium-molecular-weight homopolymer)	—	—	90	90	90	90	—	—
VC-80[a] (low-molecular-weight homopolymer)	—	—	—	—	—	—	90	90
Paraloid KM-228[b]	—	—	10	10	10	10	10	10
Paraloid K-120N[b]	—	3	—	1	2	3	—	2
Thermolite 31[c]	3	3	3	3	3	3	3	3
Stearic acid	0.5	0.5	0.5	0.5	0.5	0.5	0.5	0.5

[a]Borden Chemical Company.
[b]Rohm & Haas Company.
[c]M & T Chemicals, Inc.
Source: Ref. 18.

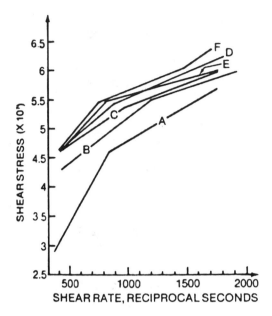

FIGURE 29 Shear stress vs. shear rate.

TABLE 45 Melt Fracture Effects with Different PVC Resins

Measurement	Compound—Test no.						
	G-3	G-5	G-4	G-1	G-2	G-6	G-7
Die diameter, in.	1/8	1/8	1/8	1/16	1/16	1/16	1/16
Shear rate, sec^{-1}	161	167	308	1030	2050	1050	1160
Stock temperature, °C	173	215	238	165	175	201	224
Melt fracture	Med.	None	Low	Med.	Med.	None	None
Swell ratio	1.12	1.07	1.08	1.23	1.18	1.23	1.19
Temperatures, °C							
Zone 1	125	142	147	140	157	138	144
Zone 2	180	177	185	175	185	180	182
Zone 3	180	193	180	180	187	183	194
Zone 4	180	193	175	173	175	176	196
Screw, rpm	31.5	31.5	63	31.5	63	31.5	31.5
Rate, g/min	30.5	31.6	58.3	24.0	47.8	24.4	27.1
Torque, m·g	4900	6100	6200	7140	6870	7200	6450

Source: Ref. 18

TABLE 46 Effects of Processing Aid Resins and Temperature

Measurement	Compound—Test no.											
	A-6	B-5	A-2	B-2	G-2	H-2	C-5	C-4	D-5	F-5	F-6	F-7
Die diameter, in.	1/8	1/8	1/16	1/16	1/16	1/16	1/8	1/8	1/8	1/8	1/16	1/16
Shear rate, sec^{-1}	106	115	1600	1630	2040	1870	1320	248	123	136	1010	1960
Stock temperature, °C	203	200	180	174	175	175	215	233	213	213	226	235
Melt fracture	High	None	High	Low	Med.	None	High	Med. High	Med. High	None	None	None
Swell ratio	1.18	1.56	1.30	1.62	1.18	1.50	1.17	1.19	1.21	1.27	1.49	1.51
Temperatures, °C												
Zone 1	130	137	125	133	157	163	148	145	148	145	145	150
Zone 2	182	186	173	173	185	185	187	190	182	188	186	195
Zone 3	184	186	192	187	187	190	193	183	194	193	193	204
Zone 4	176	180	175	175	175	175	195	175	194	194	194	195
Screw rpm	31.5	31.5	63	63	63	63	31.5	63	31.5	31.5	31.5	63
Rate, g/min	20.0	21.7	37.4	38.2	47.8	43.6	25.0	46.7	23.2	25.6	23.7	45.6
Torque, m·g	1750	2275	3700	4550	6870	6880	5000	5050	3950	5850	7370	7120

Source: Ref. 18.

An evaluation of a process aid additive by Mendham et al. [18] is shown in Tables 44 through 46 to illustrate the type of data obtained with the equipment described above. Another means of evaluating rheology is with a variable-speed mill equipped with pressure transducers to register the parting force of the rolls. This enables one to evaluate the pressure exerted by the material as it passes through the nip of a calender. The lower the force, the better the flow properties of the compound and the less roll bending that is needed to prevent crowning of the film.

G. Spiral Flow Test

In this test a small injection-molding machine is fit with a spiral mold. A typical procedure is shown in Table 47. The machine is set up so that the 24-in. mold (Figure 30) is filled approximately one-half to two-thirds with a stock control material. Injection is at a set pressure and screw profile. Lower-viscosity materials flow more readily than high-viscosity ones in a given amount of time and fill the mold to a greater extent in that time interval. Some typical data have been provided by Henkel (Tables 48A and 48B).

FIGURE 30 Spiral flow molding test.

TABLE 47 Spiral Molding Test Injection Molding Conditions

I.   Machine

      Arburg No. 200—All Rounder

      39 ton clamp

      2 oz. shot

II.   Heats

      A—Nozzle, 45%

      B—Front zone, 170°C

      C—Center zone, 170°C

      D—Rear zone, 165°C (cooling water on feed)

III. Times

      A—Injection, 10 sec

      B—Clamp, 30 sec

      C—Overall, 33 sec

IV.   Speed

      A—Screw, 50 to 60 rpm (50 except 60 on larger shots)

      B—Inject, 2.5 (on control knob)

V.   Pressures

      A—Injection, 1000 psi

      B—Back, 0

VI.   Mold

      Mold, 24-in. spiral

      Mold temperature, 100°F

TABLE 48A Formulations Used in the Spiral Flow Test

Component	Formulation					
	A	B	C	D	E	F
Formalon 614	100.0	100.0	100.0	100.0	100.0	100.0
Paraloid KM-611	3.0	3.0	3.0	3.0	3.0	3.0
Paraloid KM-120N	1.5	1.5	1.5	1.5	1.5	1.5
Supra Flex 100	3.0	3.0	3.0	3.0	3.0	3.0
TiO_2	1.0	1.0	1.0	1.0	1.0	1.0
Thermalite T-133	1.6	1.6	1.6	1.6	1.6	1.6
Ca stearate	0.7	1.4	1.6	1.6	0.7	1.3
Loxiol G-70	0.7					
Loxiol G-30	0.3					
Wax E		0.3				
AC 629A			0.1			0.1
XL 165				0.2		
Advawax 280					0.7	
WE-2					0.3	
Loxiol HOB						0.3

TABLE 48B   Results of Spiral Flow Test[a]

Measurement	Formulation					
	A	B	C	D	E	F
Brabender stability time at 190°C, 60 rpm	32	31	32	32.5	28.5	31
Fusion time at 190°C, 60 rpm	0.5	0.4	0.5	0.5	0.5	0.5
Maximum torque	3300	3400	3350	3300	3200	3300
Torque after 10 min	1900	1950	2000	1950	1950	2000
Izod impact (ASTM D256)						
Trial I	0.5	0.51	0.51	0.62	0.6	0.5
Trial II	0.65	0.55	0.51	0.61	0.5	0.55
Heat distortion temperature (ASTM D256), °C						
Trial I	72	72	75	74	73	74
Trial II	74	74	76	76	75	75
Spiral flow molding test, inches	15.2	12.7	12.6	13.8	15.3	13.6

[a]Summary of conclusions: A vs. B shows that calcium stearate at high levels reduced flow while not contributing significantly to improved stability, whereas G-30 improved flow significantly. A vs. B shows that Loxiol G-70 contributed to improved flow over montan esters; also, the Loxiol ester system has improved thermal stability over the system containing amide waxes. C vs. F shows that additive HOB-7121 improves the flow of high calcium stearate when used at low levels.

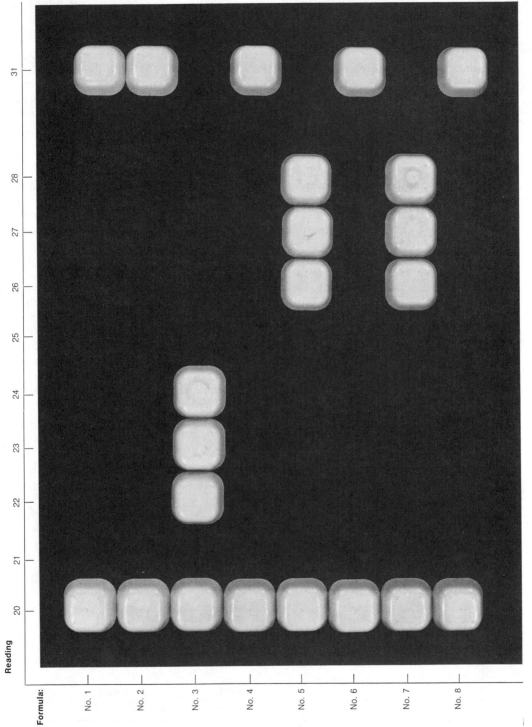

FIGURE 31 The scale ranges from 15 to 31, whereby 15 signifies the longest injection time and 31 the shortest. Readings between 23 and 31 have good flow behavior and processing length. Readings below 23 are unfavorable.

341

H. Shear Burn Test

This test consists of increasing the firing cylinder pressure in an injection-molding machine to cause the shot to be pushed into the mold more rapidly until shear burn appears at the sprue. The pressure at which shear burn occurs is recorded. With a known nozzle dimension, the shear rate at which the onset of shear burn occurs can be observed and the efficiency of internal lubricants determined (e.g., Table 49 and Figure 31).

I. Production Scale-Up

In the final analysis, all compounds must be run on production-scale equipment. To avoid the cost of large volumes of scrap of unusable material, preliminary testing should be performed on laboratory equipment. For example, to evaluate extrusion compounds a small single-screw extruder (length-to-diameter ratio 24:1) or a laboratory twin-screw extruder may be used. Either device should be equipped with pressure profile capability (Figure 27). (For examples of the data obtained, see Tables 50A to 50C.)

The compound should be tested for output rate and extrudate quality — for example, melt fracture, orange peel, color, gloss, die swell, and in the case of transparent stocks, clarity. Motor amperage and screw torque should be measured to estimate power consumption in full-size equipment. Laboratory equipment allows the elimination of candidate formulations that are unsuitable in production-size equipment. The final optimization study should be made on the actual type of equipment that will be used for production. Figures 32A, 32B, and 33 illustrate the types of information obtained from a production-size twin-screw extruder with a pressure transducer in the die.

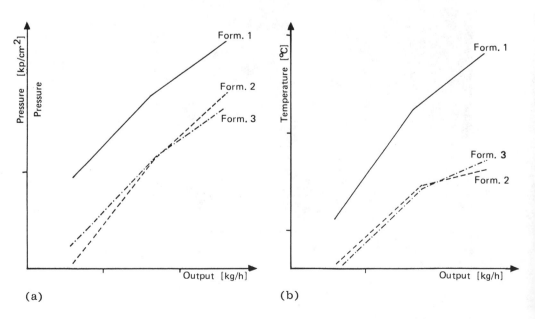

(a)   (b)

FIGURE 32 Typical diagram of pressure (a) and temperature (b) curves of trials on the twin screw extruder.

TABLE 49 Results of Injection Molding Tests[a,b]

Measurement	Formulation			
	I[c]	II[d]	III[e]	IV[f]
Cylinder temperature, °C	170/195	170/195	180/205	180/205
Head temperature, °C	195	195	210	210
Screw speed, rpm	59	59	59	59
Max. molding pressure, kp/cm^2	1200	1200	1200	1200
Max. molding time	2	2	2	2
Back pressure, kp/cm^2	4	4	3	3
Total cycle, sec	33.0	31.0	35.2	35.2
Temperature of compound, °C	238	235	228	243
Dosage, mm	61	61	61	61
sec	21.6	19.4	22.8	22.4
Assessment				
Shear combustion after adjustment	27	31	26	31
Blooming	Slight	None	Substantial	Very slight
Scaling	None	None	Slight	Very slight
Burning at screw tip	None	None	None	Slight
Falling distance (1 kp)				
Without breaking, cm	105	80	105	80
100% break, cm	150	150	180	120
Oven test, ½ hr, 160°C	Slight flaking	Slight flaking	Slight flaking	Very slight flaking
Vicat softening point in air, °C	84.5	85	86	86.5
Residual stability				
Initial color	Very light yellow	Very light yellow	Yellowish	Yellowish
Early color stability	120	105	90	90
Stability at finish	210	210	195	180

[a]Conditions: cylinder diameter, 45 mm; screw tip, spiral; die, open; die diameter, 4 mm; mold, disks; sprue, bar; sprue diameter, 4/5 mm; wall thickness, 1.8 mm (min.), 2.2 mm (max.); injected weight, 112 g.
[b]Starting-point formula: suspension PVC, K value 58, 100 parts; Winnofil S (ICI), 0.5 part.
[c]Tribasic lead sulfate, 3.0 parts; dibasic lead stearate, 0.3 part; Loxiol GS-2, 1.9 parts.
[d]Tribasic lead sulfate, 3.0 parts; dibasic lead stearate, 0.3 part; Loxiol GS-2, 1.9 parts; processing aid, 1.5 parts.
[e]Tribasic lead sulfate, 3.0 parts; Loxiol VGS 891, 1.6 parts.
[f]Tribasic lead sulfate, 3.0 parts; Loxiol VGS 891, 1.6 parts; processing aid, 1.5 parts.

TABLE 50A Formulations

Component	Formulation							
	A	B	C	D	E	F	G	H
PVC (BFG)	100.0	100.0	100.0	100.0	100.0	100.0	100.0	100.0
Durastrength 200	8.0	8.0	8.0	8.0	8.0	8.0	8.0	8.0
Superflex	5.0	5.0	5.0	5.0	5.0	5.0	5.0	5.0
TiO_2	15.0	15.0	15.0	15.0	15.0	15.0	15.0	15.0
Paraloid K 120 N	1.0	1.0	1.0	1.0	1.0	1.0	1.0	1.0
M&T T 137	2.0	2.0	2.0	2.0	2.0	2.0	2.0	2.0
Calcium stearate	1.6	2.4	1.6	3.0	1.6	1.6	1.6	1.6
XL 165	0.8	0.8	0.8		0.8			
Advawax 280			0.8	1.2				
Loxiol G-30	0.8							
Loxiol G-33						0.8		
Loxiol 7121							0.8	
Loxiol 9889								0.8
Exp. No. 304					0.8			

TABLE 50B Extrusion Tests on Formulations A to H[a]

Formulation	Cylinder temperature (°C)				rpm	Mass pressure				Stock temperature	Torque	Output
	1	2	3	4		1	2	3	4			
A	195	205	205	195	60/120	600	4100	3000	1800	200	5250	103.06
B	195	205	205	195	60/120	600	4400	3100	1900	200	5500	103.97
C	195	205	205	195	60/120	800	5200	3100	1900	200	5350	109.86
D	195	205	205	195	60/120	1200	6000	3000	1800	200	5750	111.99
E	195	205	205	195	60/120	500	3800	3100	1900	200	5050	101.53
F	195	205	205	195	60/120	500	4000	3100	1900	200	5050	102.50
G	195	205	205	195	60/120	700	4600	3100	1900	200	5200	104.16
H	195	205	205	195	60/120	600	4000	3100	1800	200	5000	100.65

[a]For conclusions based on tests, see footnote, Table 50C.

TABLE 50C Additional Tests on Formulations A to H[a]

Measurement	Formulation							
	A	B	C	D	E	F	G	H
Brabender stability time, 190°C, 60 rpm	39.5	44.0	35.5	38.5	35.5	39.0	38.0	38.0
Fusion time, 190°C, 60 rpm	1.2	0.7	1.2	0.6	1.3	1.2	0.8	1.2
Max. torque	2750	3000	2500	2900	2500	2600	2800	2600
Torque after 10 min	1850	2000	1800	1900	1900	1850	1900	1800
Brabender fusion time, 185°C, 30 rpm	4.1	2.6	3.5	2.0	3.2	3.8	2.9	3.4
Max. torque at fusion	2050	2500	2000	2450	1900	1950	2150	1900
Torque after 15 min	1700	2050	1750	1950	1650	1700	1700	1650

[a]Conclusions based on tests: Loxiol G-33 in formulation E exhibits best overall performance, lowest torque on extrusion, high flow, and good heat stability. Loxiol lubricants promote flow over high calcium stearate formulations (A, E, F, G, H vs. B and D). Higher calcium stearate acts as a costabilizer in formulations B and D but does not delay color in amide wax formulations.

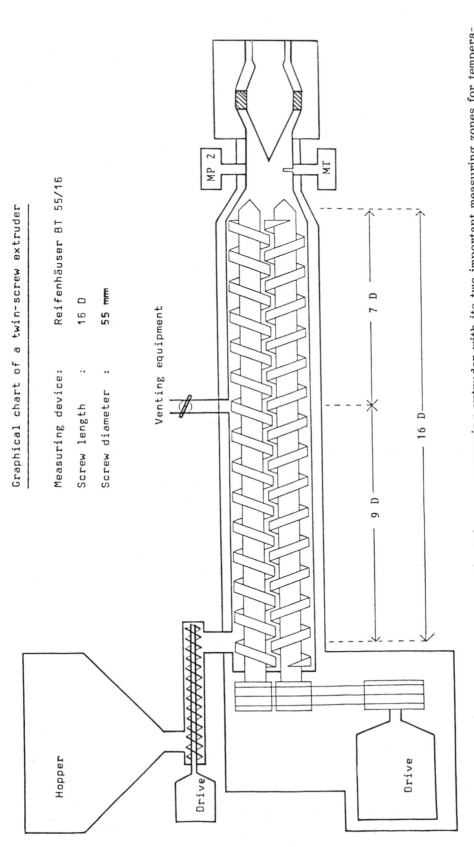

Graphical chart of a twin-screw extruder

Measuring device:		Reifenhäuser BT 55/16
Screw length	:	16 D
Screw diameter	:	55 mm

Venting equipment

Hopper

Drive

Drive

MP 2

MT

7 D

9 D

16 D

FIGURE 33 Shows a graphical structure of a twin screw measuring extruder with its two important measuring zones for temperature and pressure in the front of the screw tips. The typical demonstration of an analysis of a test series run on a twin is shown on the chart.

VII. PROPERTIES AFFECTED BY LUBRICANTS

Lubricants affect both processing properties and finished product quality. They begin to exert their influence on the processing before the compounds reach the forming equipment. Dry blends are influenced considerably. In the dry-blending process, the addition of lubricants controls the particle size density and the dispersion of other additives. Density and powder flow properties are influenced by the frictional heat buildup. The friction between particles caused by particle collision and ultimate agglomerate breakup influences the bulk density by influencing the particle size distribution and thus the packing. The higher the density, the greater the compression the screws will exert on the powder, thus promoting faster fusion, more frictional heat buildup, etc.

With powder conveying, two properties are required: resistance to caking and powder transport. Problems can occur with either high-melting or low-melting lubricants. Caking can occur in the throat of an extruder, in a hopper, or in storage bins. In most cases this is due to a lubricant that has been melted and dispersed over the resin during the blending process where the blended compound is then conveyed or delivered for storage at a temperature above the melting point of the wax lubricant. When the lubricant cools, it causes resin particles to adhere to each other, resulting in a caking problem. The caking may bridge in the hopper of the extruder or cause a solid mass to form in the storage bin. The solution is to cool the dry blend sufficiently before it is delivered to the bin or hopper. Caking can also occur when powder blends containing low-melting lubricants are placed in a hopper that has been allowed to heat by conduction from the extruder below. The heating melts the lubricant and causes the powder blend near the throat to become wet, resulting in a hangup or bridging. With dry blends using liquid lubricants or with plasticized PVC, the problem may be avoided by using higher blending temperatures, which allow the material to be more fully absorbed by the resin. In some cases, introduction of a drying resin or emulsion grade PVC resin is necessary to absorb the excess liquid materials in order to effect a free-flowing dry blend. In conveying, high-melting lubricants of a powdery nature can be sucked out of the dry blend and gather in the filters of the vacuum conveying system, eventually shutting down the system. Such lubricants may be responsible for the buildup of fine particle size material within the tubes which will choke off the flow. These problems occur because dry blends, during mixing, never exceed the melting point of the lubricants; therefore they do not melt over the resin and are not tackified to the larger resin particles. Calcium stearate and certain amide waxes are notorious for this type of problem. The easiest solution is to add a lower-melting wax lubricant, which acts as a tackifier to bond the amide or calcium stearate fine particles to the resins.

The extrusion process is controlled almost entirely by properties affected by the lubricants. First, the compound must have sufficient dynamic heat stability to be processed through the equipment. Lubricants such as metal soaps and materials that contain double bonds or hydroxyl groups will act as hydrogen chloride scavengers or secondary stabilizers as well as lubricants. They contribute significantly to the heat stability of a PVC compound. The second property is the time of fusion of the powder into a homogeneous melt that can be formed into a useful product. The speed and degree of fusion are critical aspects. External lubricants coat the resin particles and make them roll like ball bearings over each other. They inhibit fusion and alter partial fusion. External lubricants control the degree of particulate flow versus melt flow. The back pressure is influenced by the internal/external balance because of its effect on the melt viscosity and the degree of particulate flow versus melt flow. Where internal lubricant is insufficient, the melt is too viscous; frictional heat buildup occurs and causes temperature override. The amperage required in an extrusion process is also controlled by the internal/external balance since the total work done by the screw is influenced by the amount of melt being processed by the screw as well as by the melt viscosity.

Output, onset of surging, development of screw marks, and die swell are all influenced by lubricants. For optimum output, the external lubrication level must not be so high that there is slipping on the screw, preventing the screw from pushing the molten mass forward and out through the die. Surging is influenced by too high a level of external lubrication; the material catches on the screw and then slips off the screw at every revolution. Screw marks occur when two ribbons of molten material that come off the screw at the tip do not weld well together in the die because of overlubrication.

Die swell in profiles and gauge control in flat-sheet die extrusions are also influenced by the lubrication, particularly if the material in the flat-sheet die is not completely homogenized. Some materials exhibit more die swell than others because of differences in the amount of particulate flow versus melt flow and in the efficiency of the internal lubricants. Particulate flow has less die swell than melt flow, but melt flow with internal lubrication has less die swell than melt flow without internal lubrication. Resistance to plateout and die drip are also related to lubricants. Die drip is caused by externally lubricating materials that are very incompatible; high-melting paraffin waxes and some polyethylenes may actually be physically squeezed out of the polymer melts and form a lake in the bottom of the die. As the polymer continues to move through the die, some material in the lake is pushed out of the die, drips down the face of the die, and collects on the floor. This problem can be solved by reducing the external lubricant level or using more compatible lubricant.

Plateout is a complex phenomenon that usually involves an interaction between lubricant, stabilizer, pigment, and PVC. Incompatible materials are deposited out of the melt onto the metal surfaces. If these materials are not continually purged, they accumulate and stick to the metal surface, causing a coating to build up which eventually adversely affects the gauge control of the sheet or profile. The only way to remove it is to shut down and clean out the extruder. The cause of plateout is believed to be interaction of impurities and internal lubricants. These tend to bring the plateout to the surface, where it enters the incompatible layer of external lubricant and is deposited against the metal surface. Use of a more compatible external lubricant is one way to slow this process down; other ways involve using a less compatible or higher-molecular-weight internal lubricant to prevent the migration, or a scrubber lubricant that dissolves the plaque and carries it downstream.

The primary function of lubricants is to impart metal release. The most important property contributed by a lubricant is reduction of adhesion to metal surfaces. Several factors influence this other than the efficiency and molecular weight of the lubricant. Interactions can occur with other additives. Addition of alkyltin maleate, for example, increases the tack of the compounds and requires more external lubrication. Localized frictional heat buildup also influences release, because as the temperature rises, lubricant solubility increases. The lubricants are more soluble and less effective as external lubricants at higher temperatures. By reducing the temperature of the die, in effect cooling the material at the surface, one can cause the external lubricant to become more effective and give better metal release. Internal lubricants affect resistance to shear burn. They act as plasticizers at high temperatures, allowing the molecules to flow over each other with less frictional heat buildup and thus less polymer shear stress and burning.

A. Product Properties Affected by Lubricants

The product properties affected by lubricants are clarity, impact resistance, and appearance. Clarity can be adversely affected by external lubricants. Materials that are incompatible precipitate in the polymer matrix, where differences in the index of refraction cause a haze to develop in the film. In these cases, paraffins and metal soaps should be avoided in favor of esters. As with any solution, haze can be affected by rate of cooling in the process. The effect of lubricants on impact resistance is complex. Internal lubricants can plasticize slightly and improve the product's ability to absorb impacts by decreasing its brittleness. The compatibility of the lubricants with PVC and/or modifiers is an important factor because lubricants can disperse fillers and modifiers more evenly. The fillers can then act as microfracture sites and spread impacts over a wider area. Incompatible lubricants can weaken adhesion between the PVC and modifiers and adversely affect the impact resistance. The most prevalent problem is overlubrication that causes insufficient fusion and results poor spider line weld. Overlubrication inhibits adhesion of particle to particle. In the case of spider line welds, the wave fronts of polymer passing through the spider are coated with such a strong lubricant film that they fail to knit together and leave a weak section. In injection molding, the impact resistance of parts exhibiting splay problems is also poor because the material entering the mold releases, causing ring sections to float in the melt, and these sections stack up against one another as in a log jam. Since these sections are already cooled on the outside, they do not knit back together sufficiently to give impact strength to the part. Poor burst strength in pipes and fittings is a closely related problem; it is the result of polymer failing to knit together as the two melt flows meet in the back of the mold, because of overlubrication. Weak sections exist in both the pipe and the fittings and when water pressure is applied to the pipe, a rupture along the spider line knit or the weld line will occur.

The aesthetic properties of products can be influenced considerably by lubricants or effects attributed to lubricants. Burn streaks may be caused by materials stuck on the fins of the spider, on the screw tip, or along the edges of the sheet. This occurs because of insufficient external lubrication. Burning can also occur with excessive external lubrication when the material slips behind the screw or cylinder of the injection molder and has a long residence time in the equipment. Burning and sticking caused by too little external lubrication are often observed in the form of black carbonized flecks distributed here and there on the sheet or profile. This is caused by periodic detachment of burned particles from where they are sticking. Scorching caused by excessive lubrication usually results in a high-gloss surface and gradual yellowing of the product or yellow streaks. Shear burn is caused by insufficient internal lubrication and violent overheating as the material passes through the nozzle at high shear. The frictional heat buildup can make the material stick to the surface of the orifice as the high temperature increases the solubility of the lubricants and reduces the external lubricating effect. In molding, both can occur simultaneously. Part discoloration may be due to excessive external lubrication causing too long a dwell time in the machine as the material slips behind the screw during injection molding.

Many processors try to control gloss by using machine conditions and lubricants to increase particulate flow so that a low- or medium-gloss product results. This has proved unwise because particulate flow is closely related to insufficient fusion. The product becomes too variable in its impact resistance properties when particulate flow dominates sufficiently to produce a low-gloss product. The impact resistance properties are adversely affected by a few degrees variation in temperature. It is never a good idea to try to affect product surface gloss by juggling lubricants and temperatures because fusion can be very poor. Even though a hot die can be used to melt the surface and the surface is fused and has a high gloss due to the external lubrication, impact resistance will be very poor because inside the extruded profile, the particulate nature dominates and the material lacks integrity. Excessive lubrication may cause die drip, with a bright strip or streaks being formed on the profile.

Other aspects of appearance are pigment streaking and flocculation, both of which are associated with overlubrication. In the case of streaking, this is due to insufficient work in the compound; that is, overlubrication has caused insufficient plasticity with consequent inadequate dispersion of pigments in the compound. In rare cases, pigments are more soluble in the lubricants than in the polymer matrix. The external lubricants will bring pigments to the surface and streaking will occur due to pigment flocculation. Various colors of plateout may also result as the pigments become part of the plateout. This is most often seen with organic pigments and high-molecular-weight hydrocarbons such as paraffins and polyethylenes. Many other properties that are more specific to a particular product or product line are also influenced by lubricants. For instance, it is particularly important that heat distortion temperature be maintained in siding and pipe fittings. As temperature rises, internal lubricants become more soluble in the matrix and begin to act like plasticizers. At the temperature where they become functional they reduce the melting point of the resin and consequently its ability to maintain its rigidity. Some low-molecular-weight lubricants begin to border on plasticizers; therefore, they may adversely effect the tensile properties and increase the elongation properties of the product.

Sinks that form in complex injection moldings are influenced by relaxation of strains in the melt. Higher internal lubrication allows the material to be molded with less strain, and therefore fewer sinks are formed.

Paintability, printability, and laminability are all affected by external lubricants. Lubricants that are incompatible or volatile come to the surface and form a film, which will give poor bonding and allow the print or paint to be scraped away or the laminates to separate. The solution is to avoid extremely volatile or external lubricants and use higher concentrations of intermediate lubricants that have better compatibility. Complex esters and montans are other possibilities.

Various lubricants have an effect on the antistat, antifog, and antiblock properties of PVC. These characteristics are imparted by overloading the film with lubricant, which exceeds the solubility product constant, becomes external, and exudes to the surface. In the case of anitblock, a lubricant film on the surface prevents adhesion of one plasticized film to another. In a roll under pressure, amides prevent adhesion of the film at the pressure point, allowing it to be unwound without damage. In the antistat situation, lubricants with free hydroxyl groups come to the surface and the hydroxyl groups distribute static charge. This effect is often used in packaging films where automated machines are used for wrapping; it is essential that cut films do not stick because of static charge. It is also important in denesting, where static charges may cause the stacked films to adhere to one

another. Antifog additives appear to work in a similar way. They exude to
the surface and change its properties in such a way that water beads on the
surface. This is important for packaging foods having a large component
of water.

Crease-whitening is another property that is affected by lubricants,
although it is probably more influenced by impact modifiers. The edges may
turn white, showing incompatibility between the modifiers and PVC. This
incompatibility may be accentuated when excessive amounts of external lubrication are used in the formalization. It can be avoided by using high
concentrations of internal and or intermediate lubricant and getting sufficient release from these compatible materials by exceeding their solubility
constant so that they act as externals. Usually the proper choice of impact
modifier is the overriding factor.

Water blush is also effected by the lubrication systems. Lubricants that
contain hydroxyl groups and are sufficiently low in molecular weight to be
mobile allow water to penetrate the material by migration. This water probably reacts with metallic stabilizers to form metal hydroxides. The effect is
not due to hygroscopy because heating the product will not reverse the process. Water blush can be avoided by using lubricants that do not have active
hydroxyl groups or that are too high in molecular weight to be mobile.

Finally, the end-use stability or service life of the product can be affected
adversely by some lubricants. As polymer degradation occurs, the product
becomes more brittle and subject to damage by impact. Lubricants that are
thermally unstable in rigid compounds, particularly amide waxes, tend to
accentuate this problem.

B. Advantages and Disadvantages of Lubricants

1. Advantages of Internal Lubricants: In summary, the benefits of
using internal lubricants are

1. Increased molding speed without shear burn
2. Reduced die swell
3. Prevention of sinks
4. Improve die filling
5. Promotion of good weld line and spider line knit
6. Lack of adverse effect on paintability or printability
7. Good clarity
8. Lack of adverse effect on crease-whitening
9. Promotion of flow

2. Disadvantages of Internal Lubricants: Detrimental effects are:

1. Decreased heat distortion temperature
2. Sometimes decreased Izod impact resistance
3. Contribution to plateout if not of the highest quality
4. Relatively high dosages required for effective use

3. Advantages of External Lubricants: The advantage of using external lubricants is that they promote efficient release from metal surfaces.

4. Disadvantages of External Lubricants: Disadvantages (mainly
through usage at too high a concentration) are:

1. Promotion of splay and delamination
2. Promotion of poor weld line strength

3. Exudation causing poor printability, paintability, and laminatability
4. Adverse effect on crease-whitening
5. Adverse effect on fusion

VIII. TYPES OF MATERIALS AVAILABLE AS LUBRICANTS

A. Fatty Alcohols

Fatty alcohols are efficient internal lubricants in PVC and exhibit good com-patability. However, they are quite volatile and evaporate readily out of the polymer melt. Their use is confined primarily to low-molecular-weight PVC applications designed for either injection/blow molding or injection molding because they are too volatile for applications such as calendering and extrusion.

B. Fatty Acids

Stearic acid is probably one of the most widely used lubricants. It exhibits excellent external lubricating properties but volatility is a problem. This material is widely used in plasticized PVC because processing temperatures are much lower than in rigids and the volatility problem is not as great at lower temperatures.

C. Fatty Acid Esters

Because of the great variety of molecular structures that are possible, lu-bricants which are highly internal or highly external can be made. The possibility for structural rearrangement gives great versatility to the com-pounder. These materials are characterized primarily by their high efficien-cy, high degree of forgiveness, and wide processing latitude.

The first group of esters comprises the *partial esters of glycerol*, which are widely used in rigid compounds. Advantages of these materials are that they are moderately costabilizing, contribute to good flow, and are low in cost. Disadvantages are that they can contribute to plateout if not of the highest purity, are somewhat water-sensitive, and may contribute to water blush in bottle and sheet applications. They may lower heat distortion tem-peratures. Their flow is good but not necessarily superior.

The second group is *simple esters*, e.g., cetyl palmitate. Such mate-rials contribute to good flow, resist plateout, and are thermally stable. They are not water-sensitive, since they contain no hydroxyl groups. Since these materials are somewhat intermediate in lubricant functionality, they do not have as adverse an effect on heat distortion temperature as some other mate-rials. But they are not perfect, and since they lack hydroxyl groups they also lack a costabilizing effect. They find application primarily in rigid PVC injection molding and complicated profile.

The next group are the *fatty acid diesters*. The high-molecular-weight members of this group, e.g., ethylene dimontanate, are compatible external lubricants that contribute to good clarity. They resist plateout, are ther-mally stable, and since they contain no glycerol, they are not water-sensi-tive and do not affect heat distortion temperature adversely. Disadvantages are that they are not costabilizing, they are very expensive, and they are not really the most efficient externals. The low-molecular-weight members of the group, e.g., distearyl phthalate, are excellent internal lubricants with good compatibility and promote excellent flow. They are thermally stable, not water-sensitive, resist plateout, and do not affect heat distortion temperature too adversely. Disadvantages are that they are not costabilizing and they are moderately high in cost. Use of these materials is typically limited to rigid PVC. The high-molecular-weight ones are used as externals primarily for clear applications and the low molecular weight ones are used as internals for clear applications primarily.

D. Pentaerythrityl and Polyglyceryl Esters

Pentaerythritol, polyglycerols, and sorbitol form esters which function similarly to lubricants. The lower-molecular-weight materials of this class are normally internals and the higher-molecular-weight materials tend to be intermediates. The lower-molecular-weight materials contribute to superior flow and the intermediates give a good balance of properties. They resist plateout and have good thermal stability. Pentaerythrityl esters are not water-sensitive. The group as a class do not adversely affect heat distortion temperature or impact resistance significantly. Some members are mildly costabilizing. The lower-molecular-weight materials have a low threshold of effect and therefore are high-efficiency internals. This is a very versatile molecular structure and allows the chemist a wide range of selection. Most of the materials have good cost/performance ratios. Their application is primarily in rigid PVC, although sorbitan and polyglyceryl esters find use as internal lubricants and antifog and antistat agents in semirigid food packaging films.

E. Polyesters

Certain polyester polymers, partially based on long-chain fatty acids such as pentaerythritol adipate stearate polymers, function as external lubricants in PVC. They exhibit good metal release and good processing latitude or compounding forgiveness. In injection-molding applications they tend to resist delamination better than other lubricants. Since they are somewhat compatible, they contribute to superior clarity in film and bottle compounds. This unique blend of compatibility with metal release contributes to excellent printability and laminability properties compared to other externals. The only disadvantage is that they are not the most efficient external lubricants.

F. Fatty Acid Amides

The fatty acids amides function intermediately as lubricants and have a good balance of lubricating properties. Since they are not highly internal they have a minimal effect on heat distortion temperature and impact resistance. Their big disadvantage is they impart very poor thermal stability. This poor stability is well known, and they are seldom used outside the United States and Canada because of this problem. Investigators in the United States continue to use them because of their desirable properties. These materials are well suited for use as antiblock agents, particularly in plasticized films. Here, the problem of stability is not so important because plasticized materials can be processed at much lower temperatures than rigids. Use of amides may be necessary if denesting is a required function.

G. Metal Soaps or Metal Stearates

A typical example of a metal soap is calcium stearate. Such materials range in function from intermediate to external lubricants, depending on the chain length of the fatty acids. All materials that have chain lengths in the range of stearates are external lubricants. The advantages here are that they contribute to good release and are costabilizing. In fact, some are true stabilizers. Some of the members, calcium stearate particularly, cause a highly viscous melt and can be used as a process aid to decrease fusion time. Disadvantages are that the metal soaps are so incompatible that they produce cloud and cannot be used in clear formulas, and they contribute very little to internal lubricating properties. Calcium stearate contributes to such high melt viscosity that it can actually resist flow. Some of these materials have such high melting points that they cannot reach a melt in the compounding. These dust particles can separate readily during vacuum conveying and cause clogging in the system. Some members, particularly calcium stearate, interact with paraffins and stabilizer and thus can contribute significantly to plateout. This group of materials, particularly the lower-molecular-weight ones, are used as stabilizers in rigid and plasticized PVC. The higher-molecular-weight materials are used mostly in rigids as lubricants, release agents, and costabilizers.

H. Hydrocarbons

A typical hydrocarbon has a melting point of 165°F and a 30-carbon chain length. These products range functionally from intermediate (mineral oils) to highly external (165°F melting point paraffins). The high-molecular-weight materials exhibit good release and are low in cost. Disadvantages are poor processing latitude for the high-molecular-weight hydrocarbons and a tendency to cause die drip. Microcrystalline paraffins can also dissociate via oxidation and lose functionality. These materials can be used in plasticized through rigid formulations, but they are used primarily in rigid PVC compounds.

I. Polyethylenes

Both oxidized and unoxidized polyethylenes are used. Their advantage is that they give excellent release. A disadvantage is that the unoxidized materials are very concentration-sensitive, which leads to poor processing latitude. They sometimes promote pigment separation and they can contribute to plateout problems. They are used in a wide variety of applications, but most often in pipe compounds.

J. Lubricating Process Aids

Lubricating process aids give excellent high-temperature release and have little adverse effect on fusion. However, they may contribute to plateout in extrusion and calendering and to delamination in molding.

IX. INTERACTIONS OF OTHER COMPOUNDING INGREDIENTS WITH LUBRICANTS

Lubricants are one part of a multicomponent PVC compound, and each ingredient affects the solubility of the other ingredients in the compound. Lubricants can have synergistic reactions with other lubricants. They also act differently with various process aids, stabilizers, modifiers, and fillers. Because modifiers are a large component of a compound, they affect the solubility of lubricants in the entire compound. They can have a marked effect on the acceptance, solubility, and lubricant demand of the compound. Stabilizers, on the other hand, often consist of soluble materials which may act as internal lubricants in the compound. Other stabilizers may contain metal soaps and may act externally to upset the internal/external balance in the compound. Since process aids compensate for overlubrication, they are used to counteract the side effects of some lubricants.

A. Interactions with Processing Aids

Essentially, processing aids are low-melting resins, usually acrylics, that bind the PVC resin particles together and allow the extruder, through shear heating, to promote fusion and bring the compound rapidly to melt temperature through frictional heat buildup. One could consider them as fusion promoters. This may be better appreciated by an examination of the work of Wilson and Raimondi [19]. Table 51 compares poly-α-methylstyrene (PAMS) with and without an acrylic processing aid in a typical commercial pipe formulation. It is evident that the processing aids are fusion promoters and that PAMS is a more efficient promoter than the acrylic. It also appears that PAMS has some internal lubricating effects because the melt viscosity appears to be significantly lower with PAMS. Table 52 shows that as the concentration of PAMS increases from 0 to 3 phr, the fusion time of compounds containing ethylene bisstearamide (EBS) wax also increases. This shows that EBS wax is more incompatible in compounds containing PAMS

TABLE 51 Properties of a Commercial PVC Pipe Formulation with and without an Acrylic Processing Aid and with PAMS

Property	With 2 phr acrylic	Removal of process aid	Substitution of PAMS for acrylic, phr		
			2	4	6
Brabender data[a]					
Fusion time, min	3.0	3.35	2.6	1.5	0.5
Fusion torque, M-g	4000	3990	3600	3850	3725
Melt viscosity, M-g (torque 15 min after fusion)	2800	2840	2725	2690	2625
Max stock temp., °C	194	192	192	190	189
Physical Data					
Tensile strength at yield, psi	7750	7490	7610	7780	7680
Tensile strength at break, psi	7510	7810	7660	7890	7350
HDT-264 psi, °C	72	73	72	73	69
Izod impact ft-lb/in.	1.0	1.4	1.3	0.7	0.9

[a]Data were obtained with a No. 6 roller head with 8 sec damping at 60 rpm and 170°C jacket temperature.
Source: Ref. 19.

TABLE 52 Melt Viscosity and Fusion Parameters of Bis-Stearamide Formulations[a,b]

	phr Calcium stearate								
	0			0.75			1.5		
	No PAMS								
Bis-stearamide, phr	0	0.75	1.5	0	0.75	1.5	0	0.75	1.5
Fusion time, min	3.5	4.0	4.1	1.7	4.8	10.5	1.5	6.9	4.0
Equilibrium torque	2430	2415	2340	2485	2385	2070	2470	2160	2415
	3 phr PAMS								
Bis-stearamide, phr	0	0.75	1.5	0	0.75	1.5	0	0.75	1.5
Fusion time, min	0.6	7.6	13.6	1.7	6.5	11.0	1.4	4.8	9.4
Equilibrium torque	2305	2186	2180	2425	2145	2090	2370	2130	2015

[a]Torque rheometer conditions: 170°C at 60 rpm, 60-g charge.
[b]Formulation: PVC, 100; tin stabilizer (T31), 1.25; titanium dioxide, 2.0; calcium carbonate, 2.5; calcium stearate, varied as shown; bis-stearamide, varied as shown; PAMS, varied as shown.
Source: Ref. 19.

TABLE 53 Melt Viscosity and Fusion Parameters of Paraffin Wax Formulations[a],[b]

	No PAMS		
Paraffin wax, phr	0.6	1.0	1.4
Fusion time, min	6.5	14.0	18.9
Equilibrium torque	2370	2240	2020
	1.5 phr PAMS		
Paraffin wax, phr	0.6	1.0	1.4
Fusion time, min	4.2	8.0	13.5
Equilibrium torque	2335	2230	2025
	3.0 phr PAMS		
Paraffin wax, phr	0.6	1.0	1.4
Fusion time, min	3.0	5.5	9.5
Equilibrium torque	2310	2210	2015

[a]Torque rheometer conditions: 170°C, 60 rpm, 60-g charge.
[b]Formulation: PVC, 100; tin stabilizer, 1.25; titanium dioxide, 2.0; calcium stearate, 1.0; paraffin wax, varied as shown; PAMS, varied as shown.
Source: Ref. 19.

TABLE 54 Rupture of "Carina" 67-01: Draw Ratios of Various Melts at Break

Parts APA-1	Tested by	Temperature (°C)	At Break	
			Draw ratio	Stress (N/m^2)
0	ICI	180	1.4	1.1×10^5
	Shell	185	1.7	3.4
	Shell	190	4.9	8.1
	Shell	195	6.0	7.4
1	ICI	180	2.4	4.3
	Shell	185	2.0	4.1
	Shell	190	3.7	5.9
	Shell	195	6.0	6.1
3	ICI	180	6.4	12.1
	Shell	185	6.8	13.2
	Shell	190	7.9	12.0
	Shell	195	12.3	11.8

Source: Ref. 20.

processing aid. Table 53 shows the influence of increasing concentration of PAMS versus increasing concentration of paraffin wax. The fusion time decreases with addition of PAMS. Therefore, the paraffin must be more compatible with PAMS and less externally lubricating in this system than in a system without PAMS.

Processing aid has a marked effect on the overall lubricant demand of a compound. Considerable interaction takes place, and it is important to evaluate this. In many cases, if proper lubricating techniques are used, processing aids are virtually unnecessary. This is particularly true when large quantities of acrylic or acrylic copolymer impact modifiers are used. These materials are also low-melting resins, contribute to frictional buildup, and have draw properties. Additional processing aid is rarely necessary for compounds intended for twin-screw processing, where some fusion delay is desirable.

Processing aids are essential only where very little or no impact modification additives are being used, or complete homogenization is necessary for the development of clarity. This may be illustrated by work of Cogswell [20] yielding the results shown in Table 54. The draw ratios of various melts at break are shown. When the stock temperature is 195°C, a draw ratio of 6:1 is obtained with unmodified PVC; with PVC containing three parts of modifier at 180°C, a draw ratio of 6.4:1 is obtained. This implies that PVC on its own has sufficient melt strength to allow for a drawdown and machine setups, provided the PVC compound is fused and a homogenous melt is obtained. To do this, one must go to a higher temperature, which is readily possible. Since acrylic impact modifiers have similar properties of fusion promotion, it is not necessary to add processing aids when large quantities of acrylic modifiers are being used.

B. Interactions with Fillers

The interactions between lubricants and fillers are illustrated in Tables 55A to 55C where a series of compounds with two levels of filler are evaluated for fusion time and heat stability. It can be seen that as the filler level increases, the fusion delay and melt viscosity increase. This is not as easily seen in the Brabender fusion time as in the extrusiograph data, where die pressure and pressure profile can be seen. From the fusion delay and scorch data, it appears that the filler adds some lubricating qualities to the compound, and indeed this is possible because stearates are normally used to coat fillers. Uncoated materials in rigid PVC cause frictional heat build-up to a point that degradation becomes a problem. The filler has changed both the fusion time and the melt rheology so much that a drastic change in lubricants is required. The metal release properties are changed not only by the calcium stearate and stearic acid left on the coated filler particle, but also by the filler particle itself, which prevents intimate contact with metal surfaces and reduces the adhesion of the compound to the metal surfaces. Also, the filler particles act as a scrubber on the metal surfaces, and as the compound moves down the extruder or over the forming equipment, it has a continual scrubbing action, depending on the abrasiveness of the filler.

Cook [21] evaluated the extrusion torque of compounds lubricated with calcium stearate, stearic acid, or paraffin wax (Figure 34). In each case, he evaluated compounds containing no filler and 50 parts of filler. Several different interactions take place. The torque is higher as filler concentration increases, but when stearic acid concentration increases, the extrusion

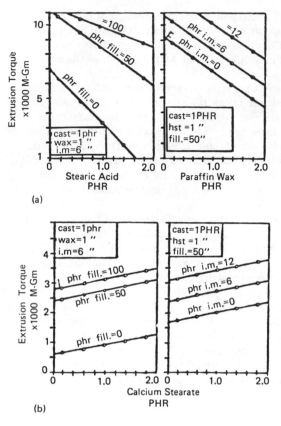

(a)

(b)

FIGURE 34 (a) Extrusion torque vs. stearic acid and paraffin wax; (b) extrusion torque vs. calcium stearate.

torque drops off. As calcium stearate concentration increases, the extrusion torque increases with filler content up to 50 phr and then 100 phr. This emphasizes that different interactions of lubricants and fillers can occur. Melt viscosity is increased by the filler particles jamming against each other during the movement of the melt.

Formulation changes can be made to compensate for high filler levels, as shown in Tables 56A and 56B. In formulation B versus D, when one changes from acrylic to a lubricating chlorinated polyethylene impact modifier, flow is significantly improved without loss of impact resistance. In A versus B, when a switch is made from a small amount of large-particle-size filler to a large amount of small-particle-size filler, the impact resistance is improved but stability is reduced significantly because of frictional heat buildup, from 28 min in A to 18 min in B. With lubricating impact modifier, B versus D, melt viscosity is reduced. The spiral flow mold test shows the improvement in the length of B versus D, an increase from 13.4 to 15.2 in., with a change in modifier. The next step is to increase the internal lubricant from 0.7 to 1.4 phr. In D versus F, we see an increase from 15.2 to 17.7 in. in the spiral flow.

The increase in melt viscosity that occurs when excessive amounts of filler are added can be corrected by the use of significant amounts of internal lubricants. In plasticized PVC there are also significant reactions with calcium carbonate fillers. Coated fillers are not as frequently used in these

TABLE 55A   Test Results from Filler Level Study

Measurement	Formulation								
	A	B	C	D	E	F	G	H	J
PVC (BFG)	100.0	100.0	100.0	100.0	100.0	100.0	100.0	100.0	100.0
Durastrength 200	8.0	8.0	8.0	8.0	8.0	8.0	8.0	8.0	8.0
Filler ($CaCO_3$)	0	—	—	—	—	—	—	—	—
Superflex[a]	—	5.0	—	—	—	20.0	—	—	—
OMYA 95T[b]	—	—	5.0	—	—	—	20.0	—	—
OMYA 90T[b]	—	—	—	5.0	—	—	—	20.0	—
Ultraflex[a]	—	—	—	—	5.0	—	—	—	20.0
Dupont TiO_2	15.0	15.0	15.0	15.0	15.0	15.0	15.0	15.0	15.0
K120N	1.0	1.0	1.0	1.0	1.0	1.0	1.0	1.0	1.0
T 137	2.0	2.0	2.0	2.0	2.0	2.0	2.0	2.0	2.0
CaSt	1.6	1.6	1.6	1.6	1.6	1.6	1.6	1.6	1.6
Loxiol G-30	0.8	0.8	0.8	0.8	0.8	0.8	0.8	0.8	0.8
XL 165	0.8	0.8	0.8	0.8	0.8	0.8	0.8	0.8	0.8

[a]Product of Pfizer, Inc.
[b]Product of Pluess-Staufer.

TABLE 55B Test Results from Filler Level Study[a]

Measurement	Formulation									
	A	B	C	D	E	F	G	H	J
Brabender stability time, 190°C, 60 rpm	42	37	39	42	35	37.5	39	42.5	35
Fusion time, 190°C, 60 rpm	1.2	1.4	1.5	1.2	1.2	1.7	1.6	1.3	1.7
Max. torque	2650	2650	2600	2650	2650	2700	2850	2700	2950
Torque after 10 min	1850	1850	1850	1800	1850	1950	1975	1850	2100
Brabender fusion time, 185°C, 30 rpm	3.0	3.3	3.5	3.1	3.2	4.4	4.1	3.2	4.6
Max. torque at fusion	2100	2000	2050	1950	2000	2050	2150	2050	2250
Torque after 15 min	1750	1725	1700	1650	1750	1750	1850	1750	2000

[a]For conclusions, see footnote, Table 55C.

TABLE 55C Additional Test Results from Filler Level Study[a]

Formulation	Cylinder temperature				rpm	Mass pressure				Stock temperature	Torque	Output
	1	2	3	4		1	2	3	4			
A	195	205	205	195	60/120	800	5000	3000	1800	200	5300	105.63
B	195	205	205	195	60/120	600	4400	3100	1800	200	5200	103.17
C	195	205	205	195	60/120	600	4600	3100	1900	200	5200	104.79
D	195	205	205	195	60/120	700	4400	3100	1800	200	5200	106.32
E	195	205	205	195	60/120	600	4400	3000	1800	200	5100	104.46
F	195	205	205	195	60/120	300	2800	3100	1900	200	4400	95.40
G	195	205	205	195	60/120	500	3800	2300	2000	200	5100	103.26
H	195	205	205	195	60/120	500	4200	3300	2000	200	5200	102.70
J	195	205	205	195	60/120	300	2800	3400	2000	200	4300	93.35

[a]Conclusions: Formulations F and J with 20 phr Superflex and Ultraflex were overlubricated; fusion was delayed into the metering zone of the screw. Low output was also evident. Heat stability was maintained with Pluess-Staufer 90-T but was reduced with other fillers; the most dramatic reduction was seen with Ultraflex. Fusion delay was observed at the 20 phr loading level for Pfizer Superflex and Ultraflex over Pluess-Staufer 90T and 95T; no delay was observed at the 5 phr level.

TABLE 56A Formulation Changes to Compensate for High Filler Levels:
Formulations Used

Component	Formulation (phr)					
	A	B	C	D	E	F
Formalon 614	100.0	100.0	100.0	100.0	100.0	100.0
Paraloid KM 611	3.0	3.0	0	0	0	0
Dow CPE .56	0	0	0	1.5	0	1.5
Paraloid K120N	1.5	1.5	1.5	1.5	1.5	1.5
HiFLEX 100	3.0	0	0	0	0	0
Superflex	0	20.0	20.0	20.0	20.0	20.0
TiO_2	1.0	1.0	1.0	1.0	1.0	1.0
M&T T-133	1.6	1.6	1.6	1.6	1.6	1.6
Calcium stearate	0.7	0.7	0.7	0.7	0.7	0.7
Loxiol G-30	0.7	0.7	0.7	0.7	1.4	1.4
Loxiol G-70	0.3	0.3	0.3	0.3	0.3	0.3

compounds, and the oil-absorption properties of the fillers become important.
The high surface area of small particles results in high absorption of liquid
additives, including plasticizers, lubricants, and stabilizers. Lubricants
can be used to seal the surface of the fillers and reduce oil absorption. They
can be used to increase flow, particularly in high-shear applications such as
wire and cable. The introduction of internal lubricants is useful for reducing
the internal friction and reducing the melt viscosity.

C. Interactions with Impact Modifiers

Lubricants also interact with impact modifiers. When an impact modifier is
added to a PVC compound, it forms an alloy with the PVC which may drasti-
cally alter the solubility parameters. Detweiler and Purvis [22] showed the
influences of these interactions. In comparing the percent white light transmission of several compositions, it is seen that the compatibility of the lubricants is influenced by the addition of modifier (Table 14). The sample
containing stearic acid allows 89% transmission in the unmodified formulation
and only 75% in the modified formulation. Wax OP, on the other hand, is
slightly more internal in the modified compound than in the unmodified compound. With Wax S, the pattern seen with OP is repeated—77% transmission
for modified and 54% for unmodified compound. The differences in solubility
of the lubricants in the alloys compared to the unmodified PVC compounds
also appear as differences in fusion rate on a torque rheometer and on a mill.
Table 57A shows that the fusion times of OP Wax and Wax S are different in
modified and unmodified compounds. Both promote more fusion delay in unmodified compounds and tend to appear more external in unmodified compounds. Lubricants can affect the impact strength of the compound significantly.

TABLE 56B Formulation Changes to Compensate for High Filler Levels: Results Obtained

Measurement	Formulation					
	A	B	C	D	E	F
Brabender stability time at 190°C, 60 rpm	28	18	23.5	27	25	21.5
Fusion time at 190°C, 60 rpm	0.3	0.2	0.6	0.5	0.4	0.5
Maximum torque	3300	3900	3250	3200	3400	3000
Torque after 10 min	2000	2250	2150	2050	2050	1950
Izod impact (ASTM D256)						
Trial I	0.5	0.85	0.95	0.85	0.6	0.5
Trial II	0.6	0.78	1.0	1.1	0.65	0.7
H.D.T. (ASTM D648), °C						
Trial I	71	72	73	72	70	71
Trial II	73	73	73	74	72	72
Spiral flow molding test, in.	15.2	13.4	14.7	15.2	16.9	17.7

TABLE 57A Fusion Behavior in Modified and Unmodified Compounds

| Lubricant(s) | Modified | | | Unmodified flux time (min), mill |
| | Flux time (min) | | Brabender flux torque | |
	Mill	Brabender		
None	1/4–1/2	3/4	4300	1/4–1/2
Octadecane	1/3	1	4050	
Stearyl alcohol	1/4–1/2	3/4	4250	1/2
Stearic acid	1/4	3/4	3940	1
12-Hydroxystearic acid	1/4	1/2	4200	
Sodium stearate	1/4	$4\frac{1}{2}$	3300	
Potassium stearate	$1\frac{1}{2}$			
Calcium stearate	4	3/4	4200	
Butyl stearate	1/4	1/2	4150	3/4
Glycerol monostearate	1/4	3/4	4350	1/2
N,N'-Ethylene bis-stearamide	1/4	1	4100	
Wax S	1/4	3/4	4000	
Wax E	1/4	1	4050	
Wax OP	1/4	3/4	4100	3/4
Paraffin wax	1/4	1	4150	$4\frac{1}{2}$
AC-629A	1/4–1/2	2	3350	
Paraloid K-175	1/2	1/2	4550	1/4
K-175/calcium stearate, 1.0/0.3	1/4–1/2	1	4400	1/4
Glycerol monostearate/ Wax OP, 0.7/0.3	1/4	3/4	3850	1/4
Wax OP/Wax S, 0.8/0.2	1/4	1/2	4150	
Wax OP/Wax E, 0.8/0.2	1/4	$1\frac{1}{4}$	3850	
Stearyl alcohol/AC-629A, 0.8/0.2	1/4	$1\frac{1}{4}$	3850	
Butyl stearate/calcium stearate, 0.7/0.3	1/4	$1\frac{1}{4}$	3850	

TABLE 57B   Impact Strength in Modified and Unmodified Compounds

Lubricant(s)	Izod impact strength	
	25°C	10°C
None	14.6	3.2
Octadecane	23.3	4.6
Stearyl alcohol	21.7	3.5
Stearic acid	20.2	4.2
12-Hydroxy stearic acid	22.8	4.1
Sodium stearate	27.2	22.4
Potassium stearate	26.6	11.6[a]
Calcium stearate	24.0	12.9[a]
Butyl stearate	24.5	13.4[a]
Glycerol monostearate	25.0	17.1[a]
N,N'-Ethylene bis-stearamide	26.0	18.3[a]
Wax S	21.3	11.4[a]
Wax E	22.6	4.2
Wax OP	19.3	4.1
Paraffin wax	24.9	5.7[a]
AC-629A	23.4	12.2[a]
Paraloid K-175	23.2	4.2
K-175/calcium stearate, 1.0/0.3	24.0	19.5
Glycerol monostearate/Wax OP, 0.7/0.3	23.4	4.6
Wax OP/Wax S, 0.8/0.2	22.5	16.2
Wax OP/Wax E, 0.8/0.2	21.2	17.6
Stearyl alcohol/AC-629A, 0.8/0.2	23.8	9.9[a]
Butyl stearate/calcium stearate, 0.7/0.3	13.6	9.3[a]

[a]Denotes scattered values—both hinge breaks and clean breaks.

In Table 57B, differences in impact can be seen at 10°C. Since the Izod is a crude method for measuring impact, no conclusions can be drawn about the mechanism of the differences. The overall result demonstrates conclusively that modifiers must be taken into account when evaluating lubricants.

D. Interactions with Stabilizers

Lubricants have a significant effect on the dynamic heat stability of PVC compounds. PVC is an unstable polymer. Stabilizers scavenge HCl liberated from the polymer and react with the polymer chain to form more stable compounds. Reduction of HCl evolution slows the rate of degradation. Lubricants can dramatically alter the heat stability of a PVC compound by various mechanisms. External lubricants can prevent sticking and the degradation associated with sticking and stagnation, including black specks and yellow scorched areas present in even highly stabilized materials with insufficient external lubricant. Internal lubricants can reduce shear stress on the polymer and subsequent shear burning. Lubricants with costabilizing groups can delay the degradation time and onset of color. Metallic lubricants can act as secondary stabilizers. External lubricants have the greatest effect on stability under dynamic conditions. Unless formulations have sufficient metal release, stabilizer will not prevent the eventual degradation of the PVC melt in a continuous process. The compound will stick and eventually burn, and the burned material will catalyze additional degradation. Additional stabilizer delays the onset of degradation but does not prevent it. External lubricants also lower the apparent shear stress at constant shear rate of the compound, thus reducing frictional heat buildup and extending stability. In static oven testing, the value of release and reduced shear stress are not apparent; dynamic testing is used to evaluate external lubrication. Internal lubricants break intermolecular attractions (van der Waals forces) and promote flow. This is seen in the change in slope of the shear stress versus shear rate curve. Internal lubricants lower the tendency for plug flow and subsequent viscosity heating, which allows the PVC to be processed at a lower temperature and reduces the need for stabilizer.

Some lubricants are costabilizers or stabilizers in their own right. They not only reduce melt viscosity but also have reactive chemical sites that scavenge HCl. The most effective lubricants are the soaps of group I metals, the alkalis, and group II, the alkline earth metals. Lubricants that contain hydroxyl, oxirane, or keto groups or double bonds also have costabilizing properties. To illustrate the costabilizing effect of hydroxyl groups, an oven stability test (Table 58) was performed on four compounds. The first contained glyceryl monostearate (GMS, Loxiol G-12); the second and third contained esters with no free reactive groups, Loxiol G-60 and Loxiol G-41; and the fourth contained no lubricant. The formulation with GMS was the only one to show any improvement in stability over a system with no lubricant. To test whether the extra stability might have been due to the rheological properties of the GMS, the two lubricants with no free reactive groups were chosen to bracket the functionality of the GMS in the fusion text (Table 59). The results indicated that the two free hydroxyl groups interact with the PVC and alkyltin stabilizer in such a way as to promote additional stability.

In lead-stabilized systems (Table 60), a pentaerythrityl ester gives considerably better stability than a simple ester. Not only are the functional groups on the lubricant important, but how they react with primary stabilizers is also important. Table 61, developed by Hecker and Cohen [23], shows a great difference in the fusion time of lubricants when they are used

TABLE 58 Thermostability Test[a,b]

Lubricant (phr)	Early color stability (min)
Loxiol G-12, 1.5	120
Loxiol G-60, 1.5	105
Loxiol G-41, 1.5	105
Without lubricant	105

[a]Conditions: roller mill, 170°C; oven, 180°C.
[b]Basic formulation: S-PVC, 100 parts; tin stabilizer, 1.5 parts.

TABLE 59 Plastograph Test[a,b]

Lubricant (phr)	Fusion time (min)
Loxiol G-12, 2.0	1.2
Loxiol G-60, 2.0	1.1
Loxiol G-41, 2.0	3.1
Without lubricant	0.2

[a]Conditions: temperature, 165°C; mixer speed, 40 rpm; charge, 29 g.
[b]Basic formulation: S-PVC, K value 58, 100 parts; tin stabilizer, 1.5 parts; calcium stearate, 0.3 parts.

TABLE 60 Thermostability Test[a,b]

Lubricant (phr)	Long-term stability (min)
Loxiol G-41, 2.0	105
P-ester, 2.0	120
P-ester, 1.0	105
Without lubricant	90

[a]Oven temperature, 180°C.
[b]Basic formulation: S-PVC, 100 parts; lead stabilizer, 1.0 parts; lead stearate, 0.5 part.

TABLE 61 Effect of Lubricants on Fluxing Rate of Various Stabilizer
Systems[a]

Lubricant	I[b]	II[c]	III[d]	IV[e]
None	10	6	2.5	6.0
Stearic acid	19	11	7.0	43
Wax E	13	7.5	45	80
12-Hydroxystearic acid	17	8	3	13
Advawax 280	15	7.5	5.5	56

[a]Base formula: medium-molecular-weight homopolymer resin, 100 parts;
lubricant, 1.0 parts.
[b]Mark WS, 3.0; Mark C, 1.0; Drapex 4.4, 2.0.
[c]Mark 99, 3.0; Mark C, 1.0; Drapex 4.4, 2.0.
[d]Mark 292, 2.0.
[e]Mark A, 2.0.
Source: Ref. 23.

with different stabilizers. Table 62, developed by Henkel, demonstrates the
same effect. With lead stabilizers, the lubricants seem to be much more external than with alkyltins or barium-cadmium. The lubricants also change
their relative functionality depending on the type of stabilizer employed.
G-47, which is the most internal lubricant in formulations stabilized by lead
and barium-cadmium, is nearly the most external lubricant in an alkyltin-
stabilized formula.

    These results are not unanticipated, because the addition of any ingre-
dient to a compound changes the solubility product constant of the compound.
Stapfer and co-workers [24] reported that there are considerable differences in how a stabilizer affects the stability and lubricating properties of
different lubricants, as seen by the fusion time (Table 63). It is important
to be aware of these interactions and to know that the lubricant packages
should be adjusted to compensate for changes in the stabilizer. Metal soaps
are in most cases external lubricants. Andrews et al. [25] indicated that as
various metal soaps are added, fusion time increases with some metal soaps
and decreases with others, but in all cases the dynamic stability time increases (Table 64). This indicates that the materials function as secondary
stabilizers. Those that shorten fusion time have the additional property of
acting as fusion aids.

E. Interactions between Lubricants

Lubricants can interact synergistically to produce enhanced effects. This
is often seen where internals react with externals to give synergistic release
effects. Some lubricants are compatible with each other and others are not.
This can result in improving or hindering the solubility of the lubricant
package components in the PVC polymer, with resulting improvements in
metal release or flow. Examples of interactions between internals and externals have been documented by Illmann [3] (Figure 9). Illmann evaluated
the output of a capillary viscometer at three pressures with compounds in
which the ratio of internal to external is varied along the entire range of
lubricants tested. Figure 9 shows that the output increases significantly

TABLE 62 Fusion of External Lubricants[a,b]

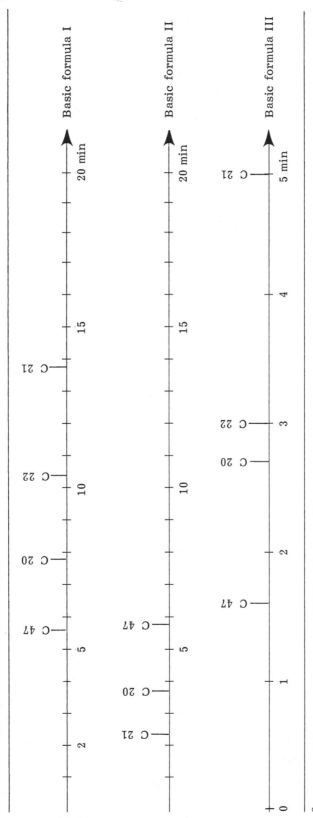

[a]Test conditions: temperature, 165°C; 40 rpm; 32 g.
[b]Basic formulas: (I) Solvic 223, 100 parts; Tribase, 2 parts; calcium stearate, 0.3 part; lubricant, 1 part. (II) Solvic 223, 100 parts; Advastab 17M, 1.5 parts; calcium stearate, 0.3 part; lubricant, 1 part. (III) Solvic 223, 100 parts, Mark WSX, 2.5 parts; MarkC, 0.5 part; lubricant, 1 part.

TABLE 63   Influence of Lubricants on Stability[a]

| Stabilizer (2.0 phr) | Lubricant (10 phr) | | | | | | | | | | | |
| | Glycol Distearate | | | Distearyl carboxylate | | | Distearamid | | | Paraffin | | |
	S	F	T	S	F	T	S	F	T	S	F	T
Liquid tin mercaptide	10.8	0.5	2050	9.1	0.3	2075	8.9	0.3	2000	9.4	1.4	2050
Solid tin mercaptide	9.9	6.3	2200	10.9	3.0	2190	7.9	6.3	2250	10.5	8.7	2010
Liquid tin carboxylate	6.1	0.3	2250	5.6	0.4	2225	7.6	0.5	2075	7.5	1.2	2350
Solid tin carboxylate	8.4	2.5	2190	8.1	6.1	2175	12.3	3.5	2350	9.4	7.1	2250
Liquid barium-cadmium	2.7	2.7	2500	3.1	0.5	2580	1.7	2.0	2575	2.0	8.5	2700
Solid barium-cadmium	6.1	8.5	2400	5.3	4.7	2275	3.6	6.0	2400		>30	
Liquid calcium-zinc	>20	9.3	500	>20	9.6	525		>30			>30	
Solid calcium-zinc	>20	1.9	620	>20	1.3	590		>30		>20	2.4	590

[a]S, Stability time (min); F, fusion time (min); T, minimum torque (m·g).

TABLE 64 Effect of Lubricant Type and Concentration on Brabender Fusion, Dynamic Heat Stability, Torque at 170 and 210°C, and Static Heat Stability of PVC Powder Blends Stabilized with a Butyl Thiotin Compound[a],[b]

Lubricant	phr in formula (x)	Brabender evaluation				Static heat stability (minutes to noticeable decomposition)
		at 170°C		at 210°C		
		Fusion time (min)	Torque (m·g)	Decomp. time (min)	Torque (m·g)	
Cadmium stearate	1	10.5	1500	15	1025	Not tested
	2	12.5	1400	20	850	90
	3	15.0	1200	29	700	Not tested
	4	14.0	1200	35	600	Not tested
Barium stearate	1	1.5	1650	15	1075	180
	2	1.25	1650	20	800	165
	3	1.00	1650	—	675	150
	4	0.75	1650	33	600	135
Magnesium stearate	1 (2)[c]	1.50	1700	13	1050	150
	2 (3)	1.25	1650	15	950	150
	3 (4)	1.25	1600	18	1000	180
	4 (5)	1.00	1600	17	950	180
Lithium stearate	1	2.75	1600	17	1000	150
	2	2.50	1600	24	875	165
	3	3.75	1650	35	700	150
	4	2.50	1600	45	500	150

TABLE 64 (Continued)

| Lubricant | phr in formula (x) | at 170°C | | at 210°C | | Static heat stability (minutes to noticeable decomposition) |
		Fusion time (min)	Torque (m·g)	Decomp. time (min)	Torque (m·g)	
Dibasic lead stearate	1	8.50	1600	16	950	
	2	10.00	1650	20.5	850	Not tested
	3	7.25	1700	31	700	
	4	11.00	1600	42	825	
Aluminum stearate	1	4.25	1650	15	1000	
	2	2.75	1600	15	950	Not tested
	3	2.25	1575	15	850	
	4	4.75	1550	16	1000	

Brabender evaluation

Hoechst wax OP	1	3.00	1700	13	1000	135
	2	3.00	1600	18.5	850	135
	3	3.00	1500	37	600	135
	4	4.75	1300	27	500	135
Hoechst wax E	1	4.25	1650	12	1300	150
	2	13.75	1350	15	925	150
	3	12.50	1175	20	600	150
	4	20.50	1000	37	425	150
Proprietary brand of Mixed lubricants E1	1	1.75	1700	11	1100	Not tested
	2	3.50	1650	15.5	975	
	3	7.50	1450	17.5	825	
	4	12.00	1150	—	—	
No added lubricant	Zero	2.75	1750	12.5	1200	Not tested

aAll trial blends based on a medium K value suspension PVC resin and the following stabilizer lubricant concentrations (phr): butyl thiotin stabilizer, 1; magnesium stearate, 1; stearic acid, 0.2; lubricant under investigation, x.
bBrabender conditions: at 170°C, 60 rpm rotor speed, 32-g charge via quick-loading shute; at 210°C, 70 rpm rotor speed, 32-g charge via quick-loading shute. Static heat stability at 180°c by method ISO R305.
cFigures in parentheses are actual quantities of magnesium stearate in the trial powder blends since there is already 1 phr in the base formula.

TABLE 65 Metal Release Properties: Loxiol G-15 with Varying Proportions of Mixtures of Esters

Total addition of lubricant (parts by wt. per 100 parts by wt. of PVC)	Addition of mixed esters (parts by wt. per 100 parts by wt. of PVC) (I)	(II)	Nonstickiness on the rolling mill at 185°C (min)
0	0	0	5
0.2	0	0.2	5
0.2	0.05	0.15	35
0.2	0.1	0.1	65
0.2	0.15	0.05	85
0.2	0.2	0	75

Source: Ref. 26.

as the more internal lubricant increases from 0 to 80%. The highest flow occurs with the more internal lubricant at a minimum level of 30%. The output decreases as the more internal lubricant level is increased beyond 70%. The changes in blend ratio are not directly proportional to the output and some interaction phenomenon takes place. It appears that below 30% of the more external lubricant, there is insufficient interaction reducing the output.

Henkel's technical group at Loxstedt, West Germany developed data [27] showing that a similar phenomenon occurs with metal release properties (Table 65). The concentration of intermediate lubricant, Loxiol G-15, is varied in proportion to a complex ester. With no lubricant the nontack time is 5 min; no improvement is seen with internal lubricant at 2 phr. The internal gives little metal release effect on its own. The nontack time increases as the proportion of external lubricant is increased up to about 75% and then decreases. Some synergistic effect occurs, because release is not directly proportional to the amount of external lubricant added to the system. A similar trend occurs with an internal material; for example, distearyl phthalate varied in a similar manner with the complex ester, as seen in Table 66. A second type of interaction was described by Hartitz in 1973 [27]. He observed that increasing the concentration of some lubricants caused increasing fusion delay, whereas with other lubricants it had little effect on fusion or promoted fusion (Tables 67 to 69). He then observed that if lubricants were evaluated in combination with calcium stearate or a microcrystalline wax, the fusion time could be completely altered (Tables 70 and 71). It is interesting that the materials that show a fusion time decrease with calcium stearate all contain superpolar groups, whereas the materials that exhibit fusion time decreases with microcrystalline wax are much more aliphatic. Combinations of a lubricant from one group with a lubricant from the other gave enhanced fusion time delay, indicating a reduction of surface tension of the group B lubricant by the group A lubricant in the PVC environment. Knowledge of interactions would allow more precise lubrication for each specific need. At present, lubrication is still as much art as science. Further studies will continue to define and direct the trend toward science.

TABLE 66 Metal Release Properties: Distearyl Phthalate with Varying Proportions of Complex Ester

Total addition of lubricant (parts by wt. per 100 parts by wt. of PVC)	Addition of lubricant (parts by wt. per 100 parts by wt. of PVC)	Addition of distearyl phthalate (parts by wt. per 100 parts by wt. of PVC)	Nonstickiness on rolling mill at 185°C (min)
0	0	0	5
0.2	0	0.2	5
0.2	0.05	0.15	15
0.2	0.1	0.1	75
0.2	0.14	0.06	85
0.2	0.2	0	75

X. LUBRICANT SELECTION AND TYPICAL LUBRICANT FORMULATIONS

As repeatedly emphasized in this chapter, lubricant selection is very important. Lubricants affect not only the processing of PVC but also the physical properties and the end-use performance of the product. As often as not, it is the end-use requirements of the product rather than processing properties that determine lubricant selection. We begin with PVC pipe. The United States and Canada differ from the rest of the world in formulating PVC pipe. At present, standards do not permit lead stabilizer usage in potable water pipe in the United States and Canada. North American technology differs, based on that limitation.

A. Pipe

Pipe is a commodity product and is processed mainly with twin-screw extruders, although some processing with single-screw extruders still occurs. Low cost and high volume are the requirements in the industry. Requirements are good gauge control, good rigidity resistance to long-term high-pressure exposure, sufficient impact resistance, and crush resistance. A typical U.S. formulation is as follows:

PVC (K value 65 to 70)	100 parts
TiO_2	1
Coated precipitated calcium carbonate filler	3 to 5
Low-cost methyl- or butyltin mercaptide stabilizer	0.4 to 0.3
Calcium stearate	0.8 to 0.6
Paraffin (melting point 160 to 165°F)	1.0 to 1.2
Oxidized polyethylene	0.1 to 0.15

TABLE 67 Fusion Times of Compounds Containing Single or Two-Phase Lubrication Systems

Product	Melting point (°C)	Composition	Fusion time of compound (min)							
			Lubricant A added alone at parts level:				2 parts lubricant A in combination with:			
							Ca stearate		FT 300	
			0.25	0.5	1	2	0.5	1	0.25	0.5
1. Calcium stearate	140–145	Ca salt of stearic acid	—	0.5	0.6	0.6	—	—	1.7	3.3
2. Barium stearate	162–167	Ba salt of stearic acid	—	—	1.2	1.3	1.6	1.1	4.2	15
3. Lithium stearate	210–215	Li salt of stearic acid	1.5	2.2	15	15	5.5	1.8	15	15
4. Stearic acid	65	Stearic acid	—	—	1.3	1.0	15	15	2.0	2.5
5. Cenwax A	78	Hydroxystearic acid	—	—	0.4	0.3	15	15	0.8	0.8
6. PE 18-18	50	Stearyl stearate	—	—	0.7	1.8	15	15	1.0	1.5
7. Wax OP	100	Montanic ester (partly saponified)	0.7	1.8	15	15	5.0	2.5	15	15
8. Wax E	82	Montanic ester	0.3	0.5	3.6	15	15	15	15	15
9. Durowax FT 300	110	Fischer-Tropsch wax	0.5	0.9	1.3	1.5	15	15	—	—
10. Microthene 510	113	Polyolefin	0.4	0.4	0.5	0.3	1.0	1.0	0.4	0.4
11. PA 190	126	Polyolefin	0.5	0.4	0.8	0.5	3.0	2.8	0.6	0.7
12. Advawax 280	149	N,N'-Ethylene-bis-stearamide	—	—	0.5	1.0	15	15	1.7	1.6
13. Kaydol	Liquid	Mineral oil	—	—	0.5	1.8	15	15	0.7	1.0
14. Silicone oil	Liquid	Silicone oil	1.7	2.3	(1.3)	2.7	15	15	3.2	3.3

Source: Ref. 27.

TABLE 68 Effect of Lubricant Concentration on Fusion Time

Lubricant	Fusion time (min)
2 pph Ba stearate	1.3
0.5 pph Ca stearate	0.5
2 pph Ba stearate 0.5 pph Ca stearate	1.6
0.5 pph FT 300	0.9
2 pph Ba stearate 0.5 pph FT 300	>15
2 pph Cenwax A	0.3
0.5 pph Ca stearate	0.5
2 pph Cenwax A 0.5 pph Ca stearate	15
0.5 pph FT 300	0.9
2 pph Cenwax A 0.5 pph FT 300	0.8

Source: Ref. 27.

TABLE 69 Classification of Lubricants with Respect to Behavior in a Two-Phase System

Group A Fusion time decreases or remains constant when adding Ca stearate and increases when adding FT 300	Group B Fusion time increases when adding Ca stearate and remains constant or decreases when adding FT 300
Calcium stearate	Stearic acid
Barium stearate	Cenwax A
Lithium stearate	Stearyl stearate
Wax OP	Advawax 280
	Mineral oil
(Stearic acid)	Silicone oil
	Durowax FT 300
	PA 190
	Microthene 500

Source: Ref. 27.

TABLE 70 Two-Phase Lubricant Systems with Synergistic Effect on Fusion Time: Comparison of Group A and Group B

Group A	Fusion time (min)		Group B	Fusion time (min)		Group A & B	Fusion time (min)		Torque (mkg)
Lubricant	1 part	2 parts	Lubricant	1 part	2 parts	Lubricant system	A B 0.5 + 0.5	A B 1.0 + 1.0	
Ca stearate	0.6	0.6	Cenwax A	0.4	0.3	Ca stearate + Cenwax A		>15	—
			Stearyl stearate	0.7	1.8	Ca stearate + stearyl stearate		2.4	4.98
			Advawax 280	0.5	1.0	Ca stearate + Advawax 280		>15	—
			Durowax FT 300	1.3	1.5	Ca stearate + Durowax FT 300		>15	—
Ba stearate	1.2	1.3				Ba stearate + Cenwax A		>15	—
						Ba stearate + stearyl stearate		4.3	4.98
						Ba stearate + Advawax 280		3.3	5.12
						Ba stearate + Durowax FT 300		>15	—
Li stearate	>15	>15				Li stearate + Cenwax A	4.4		5.30
	(2.2)[a]					Li stearate + stearyl stearate	(2.5)		5.16
						Li stearate + Advawax 280	3.2		5.36
						Li stearate + Durowax FT 300	>15		—
Wax OP	>15	>15				Wax OP + Cenwax A	2.9	>15	5.00
	(1.8)[a]					Wax OP + stearyl stearate	4.5		5.02
						Wax OP + Advawax 280	3.0		5.06
						Wax OP + Durowax FT 300	>15		—

[a]At 0.5 part.
Source: Ref. 27.

TABLE 71 Effect of Various Metal Stearates on Fusion Time

Lubricant	Softening range (°C)	$C_{17}H_{35}COO^-$ (%)	Residual stearic acid (%)	Fusion time when added at 0.5 pph (min)	Fusion time when added at 0.5 pph in combination with	
					0.5 phr Ca stearate	0.5 phr SP-9
Lithium stearate	210—215	97.61	0.4 (10.4)	2.2 (2.5)	3.7 (3.8)	>30 >30
Sodium stearate	210—225	92.50	0.3 (10.3)	3.2 (3.7)	2.5 (3.2)	>30 >30
Magnesium stearate	115—120	95.89	2.3 (10.3)	1.3 (1.9)	2.2 (2.3)	>30 >30
Barium stearate	162—167	80.50	8.7 (11.2)	1.5 (1.3)	2.5 (3.2)	3.6 4.0
Stannous stearate	80—85	82.69	0.3 (10.3)	1.4 (1.7)	2.9 (2.2)	>30 >30
Ferrous stearate	100—105	91.03	0.6 (10.6)	2.2 (3.0)	3.2 (3.8)	5—18[a] 6
Ferric stearate	90—95	93.84	10.3 (17.3)	2.1 (3.4)	4.5 (4.7)	5.3 5.0
Chromic stearate	90—105	94.24	17.6	5.0	>30	>30

[a]Reproducibility problems.
Source: Ref. 27.

A typical European formulation is

PVC (K value 65 to 70)	100 parts
Basic lead sulfate	0.8 to 1.2
Dibasic lead stearate	0.8 to 1.0
Calcium stearate	0.2 to 0.4
Simple fatty acid ester	0.4 to 0.6
Stearic acid	0.15 to 0.3
Microcrystalline wax (melting point 170 to 200°F)	0.15 to 0.25
Pigment TiO_2	1.0
Filler-coated precipitated calcium carbonate	3 to 5

The testing program includes measurements of sizing, methylene chloride resistance for fusion, ball drop impact, quick burst, static long-term pressure, and extraction studies for etiology.

Formulations that are extrudable from powder by single-screw technology by the two-stage single-screw approach are as follows.
U.S. formulation:

PVC (K value 65)	100 parts
TiO_2	1 to 1.5
Coated precipitated calcium carbonate filler	3 to 5
Acrylic processing aid	1.0 to 1.5
Low-cost methyl- or butyltin mercaptide stabilizer	1.0 to 1.5
Calcium stearate	1.5 to 1.8
Paraffin (melting point 160 to 165°F)	0.4 to 0.6

European formulation:

PVC (K value 65)	100 parts
Basic lead sulfate	1.5
Normal lead stearate	0.8 to 1.2
Calcium stearate	0.4
Dicarboxylic acid diester	0.6 to 0.8
Simple fatty acid ester	0.5 to 0.8
Fatty acid	0.3 to 0.5
Coated precipitated calcium carbonate	3 to 5
TiO_2	1 to 1.5

In the United States, pipe that is used for drain, waste, and vent has no pressure specifications; therefore, it is made with a less expensive formulation and most often on twin-screw machines.
A typical formulation is

PVC (K value 65)	100 parts
TiO_2	1
Acrylic process aid resin	0 to 1.5
Precipitated coated calcium carbonate filler	30
Low cost methyl- or butyltin mercaptide stabilizer	0.4
Calcium stearate	0.8 to 0.6
Paraffin (160 to 165°F melting point)	1.0 to 1.2
Oxidized polyethylene	0.15 to 0.3

B. Weatherable Profiles

The next largest group is weatherable profiles. These include siding, which is presently unknown in Europe and in the rest of the world but has a large market in the United States and Canada, and window profile, which is a small but growing market in the United States but a developed market throughout the world. These applications require good resistance to ultraviolet light exposure, good impact resistance, both at low temperature and at ambient temperature, and good gauge and sizing.

A typical pastel-colored formulation for siding and window profile is

PVC (K value 62 to 65)	100 parts
Impact modifier	6 to 8
TiO_2	10 to 12
Coated precipitated calcium carbonate	3 to 5
Butyltin mercaptide	1.5 to 1.8
Calcium stearate	1.2 to 1.5
Process aid resin	0 to 1.0
Simple ester	1.0 to 1.2
Oxidized polyethylene	0.4 to 1.0

The lubricant level is varied, depending on the process characteristics of the machinery. A typical formulation for window profiles based on the European technology with barium-cadmium systems is

PVC (K value 65)	100 parts
Barium-cadmium stabilizer (solid)	2.5 to 3
Alkyl aryl phosphite stabilizer	0.5
Epoxidized soya oil	1.0 to 1.5
Phthalate fatty alcohol diester	0.5 to 1.0
Simple ester	0.5
12-Hydroxystearic acid	0.2 to 0.4
Coated precipitated calcium carbonate	3 to 5
Pigment, TiO_2	3 to 5
Pigment, colorant	To suit

The trend is that the barium-cadmium stabilized systems are being modified for dark colors for the U.S. market. Some attempts are being made to use alkyltin maleate stabilizer systems in weatherable profiles in the United States. It should be noted that such systems are quite tacky and require considerably more external lubrication than typical alkyltin mercaptide systems.

The newest type of formulation for window profiles now in use in Europe has been developed to reduce the cadmium soap levels and is as follows:

PVC	100 parts
Acrylic impact modifier	6
Dibasic lead phosphite	2.5 to 3
Barium-cadmium stabilizer	1.0
Alkyl aryl phosphite stabilizer	0.3
Dibasic lead stearate	0.2 to 0.4
Calcium stearate	0.3
Simple ester of fatty acid and fatty alcohol	0.3 to 0.5
12-Hydroxystearic acid	0.2 to 0.3

Another market segment peculiar to Europe, particularly parts of Spain, Italy, Yugoslavia, and Greece, consists of window shutters. These products are now becoming popular in Florida.
A typical formulation is

Suspension PVC (K value 65)	100 parts
Dibasic lead phosphite	2.5 to 3
Lead stearate	0.5 to 1.0
Calcium stearate	0.4
Simple esters	0.5 to 0.3
Hydrocarbon or polyethylene wax	0.3 to 0.2

C. Pipe Fittings

Pipe fittings represent a universal application. In the United States, the systems are usually alkyltin-based and in Europe they are mostly lead-based. The specifications in the United States are that they pass both the Izod impact test (ASTM D256) and the ball drop impact test, a heat distortion test (ASTM D648), a quick burst test, and an oven test for splay.
A typical U.S. formulation is

PVC (K value 56 to 58)	100 parts
TiO_2	1.5
Impact modifier CPE or acrylic	0 to 5
Calcium carbonate	3 to 5
Processing aid resin	0.5 to 1.5
Methyltin or butyltin mercaptide stabilizer	1.0 to 1.5
Calcium stearate	0.7 to 1.0
Simple ester or partial ester of polyhydric alcohol	0.9
Oxidized polyethylene wax	0.1

A typical European formulation for high flow is

PVC (K value 58 to 60)	100 parts
Tribasic lead sulfate	3
Calcium stearate	0.3 to 0.5
Dicarboxylic acid diester	1 to 2
Simple ester	1
Calcium carbonate	1

For high-vicat compounds the typical formulation is

PVC (K value 58 to 60)	100 parts
Calcium carbonate	1.0
Acrylic fusion promoter	1.5
Tribasic lead sulfate	2
Calcium stearate	0.5 to 0.8
Lead stearate	0.2
Simple ester	1.0

D. Transparent Bottles

Transparent bottles represent a universal application. They are mainly of two types: those for general-purpose use and those for food packaging. Specifications are clarity (light transmittance), a drop burst test, low odor, low extraction, and manufacture from materials approved by local governing agencies in the case of food packaging. A typical U.S. formulation for food packaging is

PVC (K value 56 to 58)	100 parts
Acrylic impact modifier	10 to 12
Acrylic processing aid	1.5 to 2
Octyltin stabilizer	2.0
Mixed glycerol esters	1.2
Oxidized polyethylene wax	0.15 to 0.3

A typical U.S. bottle formulation for general-purpose application is

PVC (K value 56 to 58)	100 parts
Acrylic impact modifier	10 to 12
Acrylic processing aid resin	1.5
Butyltin stabilizer	2
Mixed glycerol esters	1 to 1.5
Oxidized polyethylene wax	0.15 to 0.3

The bottle-blowing technology in the United States is usually by an extrusion blow-molding process and is performed mainly from pelletized feedstocks. In Europe a much greater variety of processes and materials are used, which mainly operate from dry-blend feedstocks. More varied ingredients are allowed by the European food and drug regulators, and the distinction between food and general-purpose applications is usually not made. A typical European formulation for an alkyltin-stabilized system for an extrusion blow-molding process is

PVC, suspension grade (K value 58 to 60)	100 parts
Octyltin mercaptide	1.2 to 1.5
Methacrylate-butadiene-styrene impact modifier	5 to 15
Processing aid	1 to 1.5
Mixed glycerol esters	0.8 to 1.2
Oxidized polyethylene wax	0.2

In Europe a formulation for injection stretch blow molding is

PVC, suspension grade (K value 58 to 60)	100 parts
Octyltin mercaptide	1.2 to 1.5
Processing aid	0 to 1
Hydrogenated castor wax	0.6
Simple ester	0.6
Oxidized polyethylene wax	0.2

Some European processors also employ nontoxic systems based on calcium-zinc stabilizers, which are not as popular in the United States because of haze. French processors in particular have considerable experience in water bottles made with these stabilizers since alkyltins are not sanctioned in France. A typical formulation processable by extrusion blow molding is

PVC, suspension (K value 58 to 60)	100 parts
Methacrylate-butadiene-styrene impact modifier	5 to 15
Processing aid	1 to 1.5
Epoxidized soya oil	5
Calcium stearate	0.2
Zinc octoate	0.1
β-Diketone	0.2 to 0.3
Oxidized polyethylene wax	0.3
Mixed glycerol esters	0.8 to 1.2

For the injection-molding stretch technique without impact modifiers, a typical formulation is

Suspension PVC (K value 58 to 60)	100 parts
Processing aid	1
Epoxidized soya oil	5
Calcium stearate	0.2
Zinc octoate	0.1
β-Diketone	0.2
Hydrogenated castor oil	0.6
Simple ester	0.6
Oxidized polyethylene wax	0.15

E. Rigid Film and Sheet

Films are usually defined as products of calenders under 10 mils thick and sheets are usually produced by calendering or by extrusion with coat hanger dies. The sheet is usually 10 mils or more in thickness. Typical specifications call for film having sufficient impact resistance, good draw-down properties during vacuum forming, resistance to crease whitening, good denesting properties, and low odor.

A typical tin-stabilized film is as follows:

PVC (K value 56 to 58)	100 parts
Impact modifier	6 to 10
Processing aid	1.5
Tin stabilizer (octyltin for applications sanctioned by the U.S. Food and Drug Administration or butyltin for general-purpose use)	2.0
Mixed glycerol esters	1.0
Complex ester	0.3

For low crease-whitening effect on a clear extruded sheet, a typical formulation is

PVC (K value 56 to 58)	100 parts
Impact modifier (low crease-whitening type)	12 to 15
Acrylic processing aid	2 to 1.5
Tin stabilizer (octyltin for U.S. FDA-sanctioned applications and butyltin for general-purpose use)	2.0

| Mixed glycerol esters | 1.25 |
| Oxidized polyethylene wax | 0.25 |

For food-grade packaging, European systems are similar to U.S. systems:

Suspension-grade PVC (K value 60)	100 parts
Octyltin stabilizer	1.5
Methacrylate-butadiene-styrene (MBS) modifier	0 to 12
Mixed glycerol esters	0.8 to 1.0
Oxidized polyethylene wax	0.2 to 0.5

For rigid, stretch films for cellophane-type tapes, a special technique is used for manufacture. This is called the Luvitherm process, and the following formulation is used:

Emulsion-grade PVC (K value 70 to 75)	100 parts
Diphenyl thiourea stabilizer	0.3 to 0.4
Complex ester with partial saponification or montanic acid esters (same use level)	3.5 to 4.0

Phonograph record compounds: the most important specification is that there should be no surface staining or imperfections that would cause distortion in the sound of the product.

Vinyl chloride/vinyl acetate copolymer (88/12)	90 to 100 parts
PVC resin, K value 55 (optional)	10 to 0
Barium lead stabilizer	1.5
Processing aid	0 to 2
Lubricant (simple fatty esters)	0 to 0.5
Carbon black	0.25

European formulations differ from U.S. practice. A typical European phonograph record formulation is

Vinyl chloride/vinyl acetate copolymer (88/12)	100 parts
Carbon black	0.2
Alkyltin stabilizer	0.2 to 0.3
Calcium stearate	0.3
Butyl stearate	1.0 to 1.5
Complex esters	0.3 to 0.4

Copolymer-based sheet for credit card applications: a typical U.S. formulation is

Vinyl chloride/vinyl acetate copolymer (92/8)	100.0 parts
Tribasic lead sulfate	3.0
Lead stearate	1.2
Oxidized polyethylene	0.2
Acrylonitrile-butadiene-styrene (ABS) impact modifier	5 to 10
TiO_2	2.0
Mixed simple esters	1.0

F. Films for Vacuum Forming

In Europe special formulations are used for vacuum forming films. In order to get better flow and less crease whitening, a vinyl chloride/acetate copolymer is used. A typical formulation is

Suspension-grade PVC (K value 60)	50 to 80 parts
Vinyl chloride/acetate copolymer (90/10)	50 to 20
Flow modifier or process aid resin	1 to 2
Alkyltin stabilizer	1.5
Partly saponified complex ester	0.6 to 0.8
Mixed glyceryl esters	0.8 to 1.0
Ethylene bis-stearamide	0.4

G. Plasticized Film and Sheet

One large application is food-grade film wraps for meats, breads, and vegetables. European and U.S. formulations appear to be very similar. A typical formulation is

Suspension-grade PVC (K value 65)	100 parts
Dioctyl adipate	25
Epoxidized soya oil	5
Calcium-zinc stabilizer	2.5
Tris(isononylphenyl) phosphite	0.5
Sorbitan monooleate 20EO (antifog agent)	1.5
Glyceryl monooleate	1.5 to 2.5
Oxidized polyethylene wax	0.2 to 0.3
Polyethylene wax	0.05 to 0.1

In calendered films for furniture overlay, a typical European formulation is

Suspension-grade PVC (K value 65)	100 parts
Dioctyl phthalate	15
Epoxidized soya oil	3
Acrylonitrile-butadiene-styrene (ABS) modifier	5
Calcium carbonate	5
Barium-cadmium solid	0.7
Barium-cadmium liquid	1.5
Tridecyl stearate	1.0
12-Hydroxystearic acid	0.2 to 0.3

H. Wire and Cable

For 105°C wire, the European formulation is

PVC (K value 70)	100 parts
Trimellitate plasticizer	50
Epoxidized soya oil	3
Dibasic lead phthalate	10
Antioxidant	0.5
Tridecyl stearate	0.5
Kaolin filler	10

The U.S. formulation for wire and cable high-speed processing is

PVC resin (K value 62 to 65)	100.0 parts
Calcium carbonate filler (fine coated)	2.0 to 10.0
Calcined clay	10.0 to 30.0
Dioctyl phthalate	20.0 to 40.0
Tribasic lead sulfate	3.0 to 5.0
Dibasic lead stearate	0.2 to 0
Partial ester of polyhydric alcohol	1.0 to 1.5
Oxidized polyethylene wax	0.3 to 0.2

REFERENCES

1. Worschech, K. F., and Wolf, K., *Kunststoffe*, *61*, 645 (1971).
2. U.S. Patent 3,640,828, Walter Bruts et al. (to Hoechst), (February 8, 1972).
3. Illmann, G., *SPE J.*, *23*(6), 71 (1967).
4. Gale, G. M., *PVC Technol. (Proc. Symp.)*, *41*, 61 (1973).
5. Marshall, B. R., *Br. Plast.*, 70 (August 1969).
6. Detweiler, D. M., and Purvis, M. T., *SPE Tech. Pap.*, *19*, 647 (1973).
7. Khanna, R. K., *J. Plast. Inst.*, *42*, 158 (1974).
8. Immergut, E. H., and Mark, H. F., Principles of Plasticization, *ACS Adv. Chem. Ser.*, *48*, 1 (1965).
9. Small, P. A., Solubility of Polymers, *J. Appl. Chem. 3*, 71 (February 1953).
10. Scatchard, G., *Chem. Rev.*, *8*, 321 (1938); *ibid.*, *44*, 7 (1949).
11. Hildebrand, J. H., and Scott, R. L., *The Solubility of Non-Electrolytes*, 3rd ed., Reinhold, New York, 1950.
12. Ceccorulli, G., et al., *J. Macromol. Sci. Phys.*, *B20*, 4, 519 (1981).
13. King, L. F., and Noel, F., *Polym. Eng. Sci.*, *12*, 2 (1972).
14. Davis, P. J., and Fraser, S. J., *SPE Tech. Pap.*, *19*, 479 (1973).
15. Fahey, T. E., *J. Vinyl Technol.*, *4*, 3 (1982).
16. Shah, P. L., *Polym. Eng. Sci.*, *14*(11), 773 (1974).
17. Lindner, R. A., and Bohaczuk, R. A., *Plast. Compound.* (Jan./Feb. 1983).
18. Mendham, W. E., Jalbert, R. L., and Bosselman, J. E., *SPE Tech. Pap.*, *13*, 1078 (1967).
19. Wilson, A. P. et al., *Polym. Eng. Sci.*, *18*(11), 888 (1978).
20. Cogswell, F. N., *Pure Appl. Chem.*, *55*(1), 177 (1983).
21. Cook, P. J., *SPE Tech. Pap.*, *19*, 526 (1974).
22. Worschech, K. F., *SPE Tech. Pap.*, *23*, 219 (1977).
23. Hecker, A. C., and Cohen, S., *SPE Tech. Pap.* (Quebec Sect. SPE RETEC, Montreal, October 16, 1964).
24. Stapfer, C. H., Hampson, D. G., and Dworkin, R. D., *SPE Tech. Pap.*, *14*, 276 (1968).
25. Andrews, K. E. et al., *Br. Plast.*, *43*(11), 88 (1970).
26. Worschech, K. F. et al., U.S. Patent 3,875,069 (to Neynaber Chemie) (April 1, 1975).
27. Hartitz, E. J., *SPE Tech. Pap.*, *19*, 362 (1973).
28. Lindner, R., *Modern Plastics Encyclopedia*, McGraw-Hill, New York, Volume 59, 1982–83.

5

Polymeric Modifiers: Types, Properties, and Performance

V. E. MALPASS

Ferro Corporation
Cleveland, Ohio

R. P. PETRICH and J. T. LUTZ, JR.

Rohm & Haas Co.
Bristol, Pennsylvania

Polyvinyl chloride has become one of the largest-volume plastics, in large part due to the wide variety of applications for which it is used. The basic polymer has an excellent balance of properties, in particular excellent clarity, good physical strength, and low flammability. However, it is also burdened by inherently poor heat stability—PVC chains are susceptible to dehydrochlorination reactions which cause decomposition at relatively low temperatures. In addition, the base polymer is rather difficult to fuse (which we now know to be related to its ordered or partially crystalline structure), requiring the use of rather high processing temperatures. This combination of difficult fusion and facile decomposition results in a narrow range of acceptable processing temperatures. The commercial success of PVC is a tribute to the use of complicated formulations with many additives to overcome these processing deficiencies.

A wide variety of additives is required in PVC formulations: heat stabilizers, lubricants, colorants, etc. These additives, which have been discussed in other chapters, are primarily small molecules, covering a wide range of chemistries. In addition, polymeric additives are widely used in PVC formulations. The physical and processing properties of these polymeric additives are blended with those of the PVC matrix to better suit the fabrication or end-use requirements. However, these are not simply alloys of PVC with commercial polymers prepared for other purposes. Rather, these polymeric additives are very specialized polymers, customized to meet the demanding requirements of PVC formulators.

Three basic types of polymer additives for PVC are considered in this chapter: (1) processing aids, (2) impact modifiers, and (3) specialty modifiers.

I. PROCESSING AIDS

A. Mechanism of Operation

Although processing aids have been used commercially in PVC compounds for over 25 years, the mechanism of their action has still not been elucidated clearly. This is not very surprising considering the lack of understanding of PVC rheology which prevailed until recently. Some of the early proposals for processing aid mechanisms range from simple additivity of the melt viscosity of the high-molecular-weight processing aid with that of the PVC resin, to a complex proposal that the processing aid influences the melting and recrystallization kinetics of the PVC crystallites. Ryan [1] discussed some of these early hypotheses but could not find evidence to prove that any of them were correct. Today, much more is known about the morphology of PVC. We now know that PVC has a well-defined particulate structure which is very difficult to destroy at temperatures below the degradation point [2–6]. The rheological characteristics of PVC melts are determined by this particulate structure. With this understanding of PVC morphology, we can say today that the mechanism of action of processing aids is related to their effects on these particulate flow processes.

Processing aids are used in PVC formulations for many reasons, but they can all be grouped into two general categories of effects:

1. They accelerate and control the fusion process in PVC compounds
2. They strongly affect the rheological characteristics of the fully fused PVC melt

Any proposed mechanism for the action of processing aids must consider these two different categories.

If one considers first the mechanism by which processing aids improve the fusion characteristics of PVC powder blends, it is generally accepted that they accelerate the process by increasing the interactions between the PVC grains. Gould and Player [7] state that processing aids improve the fusion process by increasing particle-to-particle friction, hastening the breakdown of the PVC grains and the ultimate exchange of chains between particles. They also suggest that the processing aid molecules improve the heat transfer during this process by virtue of increasing contacts between grains. Menges [4, 5] argues that the PVC grains are bound together during the fluxing process by a "putty" consisting of PVC polymerization auxiliaries and minor additives in the PVC formulation. He considers the processing aid as one important component in this putty. Gould and Player attribute the particular effectiveness of acrylic processing aids in increasing interparticle friction to their high surface hardness. Bohme [8], on the other hand, bases his explanation of acrylic processing aid performance on their ability to increase the coefficient of friction between the PVC melt and the metal surfaces of processing machinery. He argues that the acrylic processing aid causes the compound to adhere better to the metal surface, thus improving the heat transfer and the shear between the compound and the machinery surfaces. Krzewki and Collins [9] also expressed the opinion that processing aids accelerate the fusion process due to increased interparticle and particle-metal friction.

1. Improvement of Fusion: One would first observe the effect of a processing aid on the processing of a rigid PVC powder blend during the densification/plastification/fusion process. Processing aids greatly shorten the time and the heat/shear history necessary to melt and homogenize the compound. In the early days of development of rigid PVC, this effect was first noted in laboratory experiments on two-roll mills. Addition of a processing aid shortened considerably the time required to flux the powder blend, form a homogeneous band with smooth edges, and develop a uniform rolling bank. With the more modern laboratory test equipment available today, especially instrumented kneader mixers such as the Brabender and Haake machines, one can more accurately determine the effects of processing aids on the fusion characteristics of PVC compounds. When running a predetermined quantity of PVC powder blend under a controlled temperature/speed condition, the recorded evolution of the torque needed to turn the blades as a function of time gives a direct measure of the fusion process. For example, the two curves shown in Figure 1 compare a clear, high-K-value PVC compound run under the same conditions without processing aid and with 1.5 phr of acrylic processing aid added. The point of maximum torque is generally accepted as representing the beginning of plastification. Comparing the times to maximum torque peak, we can see that the addition of processing aid reduces the plastification time from 50 to 27 sec. We can also see that the maximum torque is considerably increased, reflecting the increased interaction of the PVC particles with each other and with the chamber surfaces.

These instrumented kneader mixers can also be run in a programmed temperature mode, where the bowl temperature rises at a preset rate. Although the overall shapes of the curves obtained from these experiments are somewhat different, one can note in this case again a substantial shortening of the time required for fusion when a processing aid is included in the formulation (Figure 2).

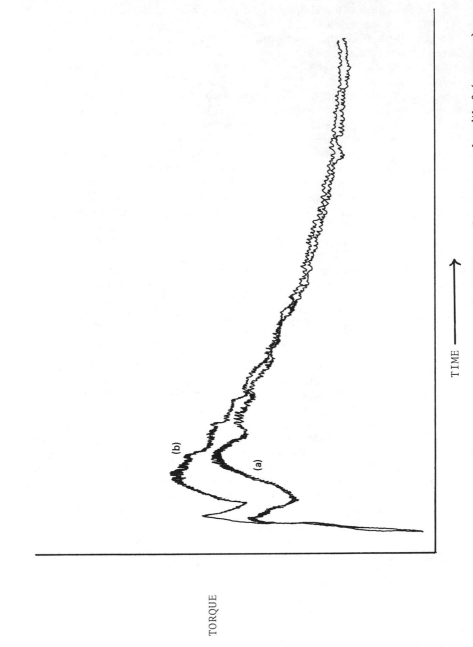

FIGURE 1 Gottfert kneader chamber fluxing curves for PVC powder compounds with 0 (curve a) and 1.5 (curve b) phr acrylic processing aid.

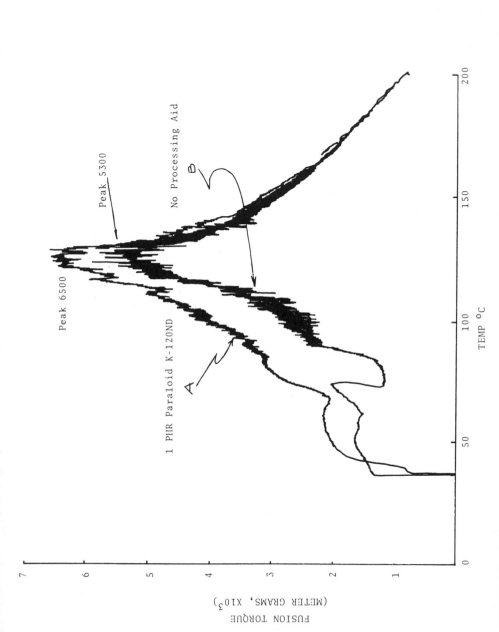

FIGURE 2 Haake programmed temperature rise kneader chamber fusion curves for PVC (K = 55) powder compounds with 0 (curve A) and 1 phr (curve B) acrylic processing aid. Temperature increase, 4°C/min.

TABLE 1 Effect of Processing Aid on Melt Pressure Distribution—Single-Screw Extrusion

Processing aid and use level (phr)	Melt pressure (bars)		Back pressure on screw (kN)
	10.5D	Die entry	
None	290	275	25.5
Standard acrylic			
2 phr	290	275	26.5
4 phr	325	305	28.5
"High-efficiency" acrylic			
3 phr	320	315	29.5

Processing studies on extruders fitted with instruments for continuous measurement of pressures and temperatures also demonstrate the effect of processing aids in improving the fusion of PVC powder blends. Table 1 contains data on the melt pressure distribution along the barrel of a single-screw laboratory extruder for the same compound run with and without acrylic processing aid. The compound with processing aid gives higher melt pressure readings in the first zones, showing that the fusion point for this compound is moved closer to the feed port than for the control without processing aid. Pazur and Uitenham [10] confirmed this effect with direct visual observations of the melting process along a screw that was quickly removed from the extruder. Their data, shown in Figure 3, indicate that the melt pool in the screw channel is at least twice as wide for a compound containing processing aid and impact modifier than for the unmodified control compound.

2. *Effects of Processing Aids on Rheology of PVC Melts*: The most important function of processing aids in rigid PVC is the improvement of the rheological characteristics—homogeneity, strength, and elasticity—of the melt after fusion and during fabrication. Unmodified rigid PVC compounds are notably poor in this respect, in particular when processed at low to moderate temperatures. Under such conditions, the crystalline regions in the primary PVC particles do not melt and the degree of molecular interdiffusion between particles is very low, so the interparticle strength is very poor. In this circumstance, the molecules of the processing aid function to bind together the particles and improve melt strength. In the case of the most widely used acrylic processing aids, their ability to perform this function is based on three inherent characteristics:

1. Excellent compatibility with PVC, even on the molecular scale
2. Flexible elastic characteristics of the acrylic polymer chains
3. Long, high-molecular-weight polymer chains

Even when processing temperatures are raised sufficiently to allow greater particle breakdown and interpenetration of PVC molecules from one particle

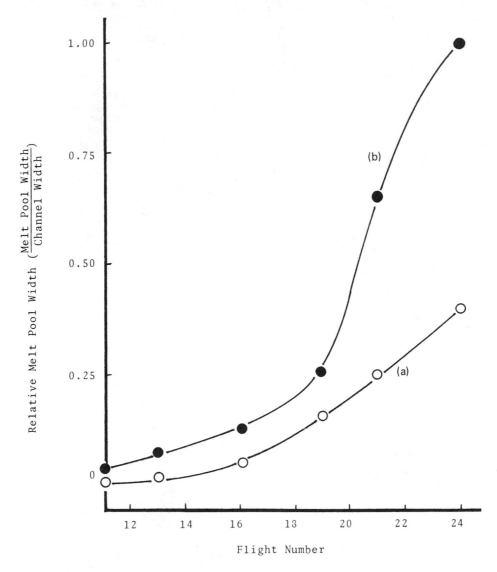

FIGURE 3 Melting rate vs. screw flight number for extrusion of unmodified PVC powder blend (curve a) and the same compound with processing aid and impact modifier added (curve b). (From Ref. 10.)

to another, these processing aids continue to provide a strong enhancement of melt strength, because they mix intimately with the shorter, stiffer PVC chains and provide a longer-range viscoelastic network. This network greatly assists in binding together the PVC domains, equalizing stress concentrations within the melt during fabrication and thus greatly reducing processing defects [7]. Evidence for these effects is presented below.

Homogenization: The homogenizing action of processing aids, tying together the semicrystalline PVC domains into a more uniform viscoelastic

(a)

(b)

FIGURE 4 Milling characteristics of rigid PVC compound with no processing aid (a) compared to the same formulations with 2 phr acrylic processing aid (b).

structure, can be observed quite easily even on laboratory equipment. Figure 4 shows the results of a laboratory roll milling experiment. The two photographs compare an unmodified, high-K-value PVC compound with an identical one containing 2 phr of acrylic processing aid. Each compound was milled for 5 minutes at 175°C, thus both are well beyond the point of fusion. It is evident that the compound containing the processing aid is much more homogeneous—the rolling bank is smooth, the edges are well knit, and the surface of the mill band is uniform and glossy. This effect is the key to successful production of a high-quality surface in high-speed rigid PVC calendering. In addition, the homogenization effect is important in extrusion processes for uniform delivery of constant quality and quantity of melt to the dies. Likewise, in injection molding, this homogenization effect is important in obtaining good quality surfaces and avoiding weakness at weld lines. In all types of processes, the processing aids assist in binding together the myriad formulation ingredients to avoid plateout, splay, and so forth.

Elasticity and Extensibility: Notwithstanding the fusion and homogenization effects discussed above, the most important reason for adding processing aids to rigid PVC compounds is to improve the elasticity and extensibility of the melt. This effect can be observed even in simple laboratory experiments such as milling. An experienced operator can easily sense by hand the improvement in tear strength and extensibility which occurs when a processing aid is added to the formulation. The earliest processing aids were developed using just this simple kind of test. However, today more sophisticated instrumentation is available to allow more accurate measurement of the effects. Various laboratory tests have been developed to measure the improved elasticity in extrusion processes. One approach is simply to increase extrusion haul-off speed until the point is reached where the extrudate begins to tear or crack [7, 10]. The data of Gould and Player, given in Table 2, demonstrate the dramatic improvement processing aids can make in the maximum drawdown ratio of extrudate emerging from the die.

If the take-off equipment is instrumented to allow direct measurement of force and speed ratio, additional information can be gained. The Gottfert Rheotens, which allows such measurements, has been used to characterize the effects of various processing aids on PVC melt elasticity. Petrich [11] used the Rheotens to compare the melt strength and elongation effects of (1) a standard acrylic processing aid, (2) a "high-efficiency" (specialized composition and higher molecular weight) acrylic processing aid, and (3) a

TABLE 2 Limiting Extruder Takeoff Rate as a
Function of Processing Aid Content

Processing aid content (%)	Maximum extrudate draw ratio without tearing
0	1.32
3.0	2.55

Source: Ref. 7.

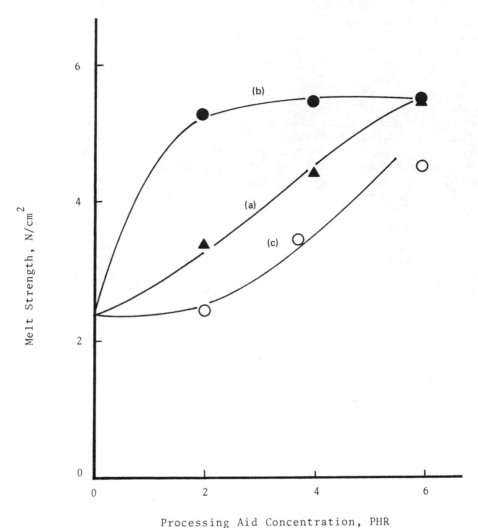

FIGURE 5 Melt strength of rigid PVC extrudates as a function of processing aid content. (a) Standard acrylic processing aid; (b) high-efficiency acrylic processing aid; (c) lubricating processing aid. (From Ref. 11.)

lubricating processing aid. Figure 5 illustrates a substantial, essentially linear improvement in the breaking stress of the molten extrudate as the concentration of standard acrylic processing aid increases (curve a). The high-efficiency processing aid (curve b) exhibits a much more dramatic effect, reaching a plateau value at only 2 phr which is equivalent to that reached by the other product at 6 phr. The lubricating processing aid (curve c), since it is a dual-purpose product, exhibits somewhat lower efficiency than the standard. The corresponding data on the elongation at break for these melt extrudates show some interesting effects (Figure 6). The standard processing aid causes a sharp improvement in melt elongation even at 2 phr, which remains relatively constant at higher concentrations.

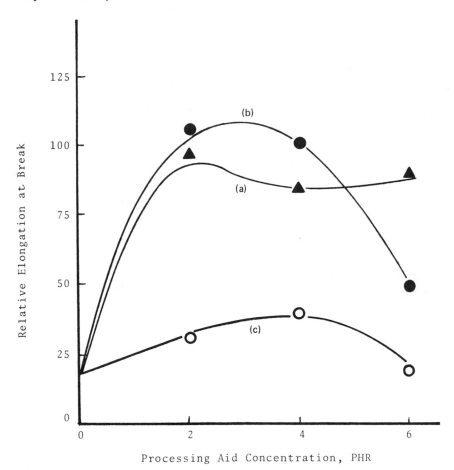

FIGURE 6 Breaking elongation of rigid PVC extrudates as a function of processing aid content. (a) Standard acrylic processing aid; (b) high-efficiency acrylic processing aid; (c) lubricating processing aid. (From Ref. 11.)

On the other hand, the high-efficiency product is essentially equal to the standard at 2 and 4 phr but shows a reduction in melt elongation at the high concentration of 6 phr. It was theorized that this very high molecular weight product requires higher melt temperatures for optimum elongation at such high concentrations. (In effect, this is not a great practical concern, since very few PVC compounds utilize such high processing aid levels.) Pazur and Uitenham used the Gottfert Rheotens to study the effect of processing aid on the melt extension properties of a twin-screw PVC pipe compound [10]. They showed (Figure 7) that the addition of processing aid progressively increases the tensile viscosity of the compound at any given elongation. In addition, their data indicate that a high-efficiency processing aid can outperform the standard type in this respect, even at much lower concentrations. They also observed that a certain minimum melt temperature and processing aid concentration are necessary for effective action. The

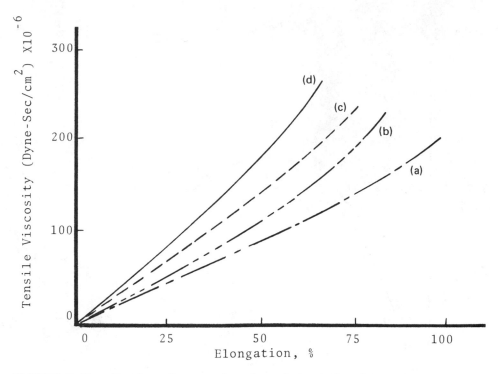

FIGURE 7 Tensile viscosity as a function of elongation for PVC melts containing varying concentrations of processing aid. (a) No processing aid; (b) 1.0 phr standard processing aid; (c) 1.5 phr standard processing aid; (d) 0.5 phr high-efficiency processing aid. (From Ref. 10.)

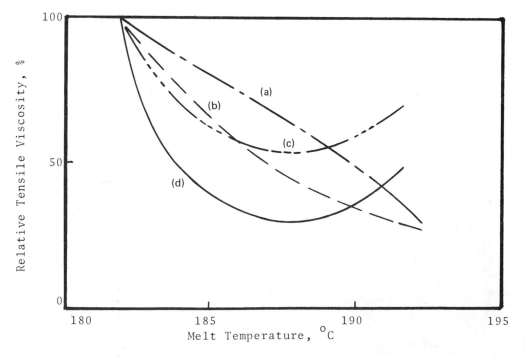

FIGURE 8 Relative tensile viscosity as a function of temperature for PVC melts containing varying concentrations of processing aid. (a) No processing aid; (b) 1.0 phr standard processing aid; (c) 1.5 phr standard processing aid; (d) 0.5 phr high-efficiency processing aid. (From Ref. 10.)

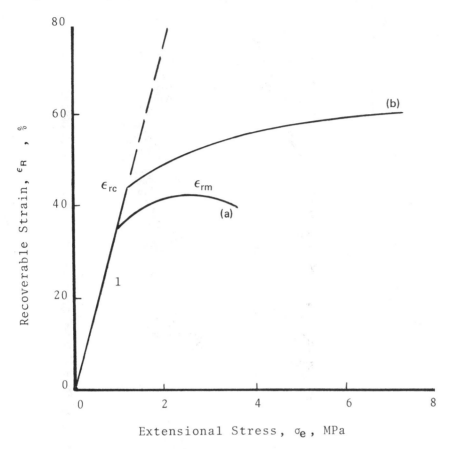

FIGURE 9 Relationship between recoverable strain and extensional stress
for PVC melts extruded through converging knife edge die. (a) Without
processing aid; (b) with 3 phr processing aid; (1) linear Hookean region;
ε_{rc}, critical strain; ε_{rm}, maximum strain. (From Ref. 7.)

data shown in Figure 8 indicate that the standard processing aid only begins
to improve the tensile viscosity of the base stock at temperatures above about
190°C. Like those of Petrich, these results show some tendency for higher
optimum processing temperatures for the high-efficiency processing aids.

Gould and Player [7] developed a more sophisticated test procedure,
measuring the postextrusion swell on the extrudate from a converging zero-
length die on a ram extruder. Measurements of swelling ratio and flow pressure allow the calculation of the recoverable strain ε_r and the average extensional stress σ_e. Plots of the reversible strain as a function of average
extension stress, as shown in Figure 9, indicate a substantial improvement
when processing aid is added to the formulation. The processing aid has no
effect on the initial, Hookean region of the curve. However, it increases
the value of ε_{rc}, the critical, recoverable strain point at which the curve
departs from linearity, and gives considerably higher recoverable strain
values over a wide range of stress in the nonlinear region. This increase
in the critical strain ε_{rc} is a key feature of processing aid action, since

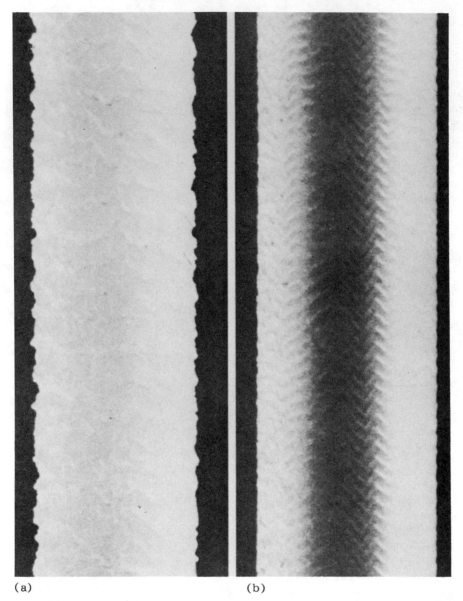

(a)   (b)

FIGURE 10 Effect of processing aids on surface roughness of rigid PVC
extrudates. (a) No processing aid, 40 rpm screw speed; (b) 1.5 phr standard acrylic processing aid, 40 rpm; (c) 1.2 phr high-efficiency acrylic processing aid, 40 rpm; (d) 1.2 phr high-efficiency acrylic processing aid,
60 rpm.

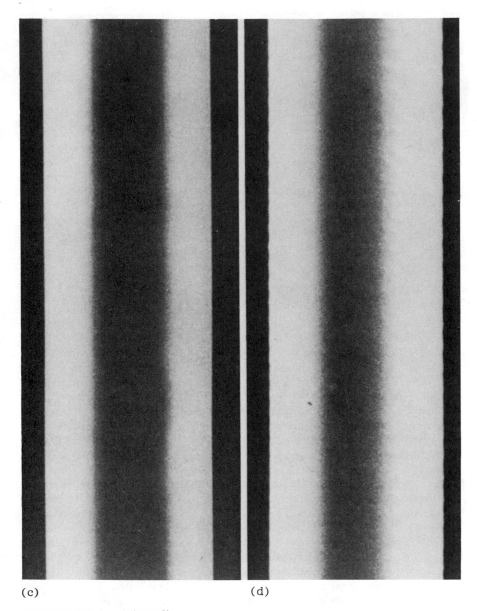

(c)   (d)

FIGURE 10 (Continued)

melt fracture is thought to begin to occur in the dies at ε_{rc}. The general
term melt fracture covers a wide variety of defects which occur during polymer processing, especially extrusion. The semicrystalline, particulate nature of PVC melts causes them to have poor extensibility and thus to be
especially susceptible to melt fracture problems. Addition of minor amounts
of processing aid is an effective solution to these problems. This is illustrated in Figure 10. The photomicrographs show the surface of extrudates at
varying speeds and with different concentrations of processing aids. The

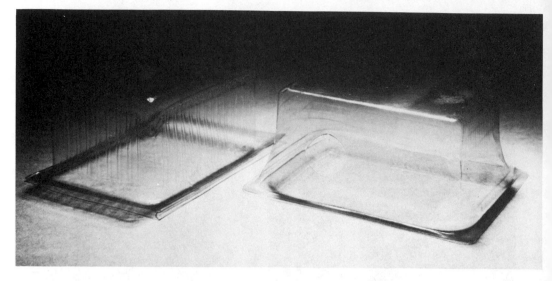

FIGURE 11 Improvement of thermoformability of rigid PVC film through the use of processing aids. The package on the left contains 2 phr acrylic processing aid; that on the right contains none.

control material (a), with no processing aid, shows very rough surface finish when extruded at a screw speed of 40 rpm. Compound (b), containing 1.5 phr standard acrylic processing aid, shows much less melt fracture at 40 rpm. Compound (c), containing 1.2 phr of the high-efficiency processing aid, gives a completely smooth surface finish under the same conditions. The bottom photo shows that this compound, with 1.2 phr high-efficiency acrylic processing aid, gives a satisfactory surface even at a screw speed of 60 rpm.

The improvement of melt strength and extensibility given by processing aids can also be illustrated by thermoforming experiments. Figure 11 shows a good example of the thermoforming benefits of a processing aid in commercial PVC film. The container on the left was made with a commercially calendered film containing 2 phr of acrylic processing aid. This film easily thermoformed into a good package with full corners, relatively uniform wall thickness, and good definition of decorative/functional details. The package on the right was made from the same film formulation but without processing aid. Under the same thermoforming conditions, the corners are not fully filled, and the film tore before drawing down to meet the sides of the mold. Because of these dramatic improvements in thermoformability, PVC calenderers frequently use more processing aid in their formulation than is necessary strictly for good processing on the calender, in order to facilitate the forming of the films by the end-user. Likewise, the improved melt strength and extensibility contributed by processing aids can also be useful in other postfabrication operations such as belling of pipe and welding of profiles.

Melt Viscosity: Although processing aids normally raise the melt viscosity of PVC compounds slightly, a key factor in their success is that this effect is far smaller than their beneficial effect on melt elasticity/extensibility. The increase in viscosity actually depends on the viscosity of the base formulation and on the test being used. Ryan [1] found that the melt viscosity of a high-molecular-weight PVC compound, measured in a capillary rheometer, increased only slightly over a broad range of temperatures and shear rates when an acrylic processing aid was added. Capillary rheometer measurements on lower-K-value PVC resins have shown a significant increase in melt viscosity. However, in practical torque rheometer tests at shear rates similar to those used in commercial processing equipment, the increase in melt viscosity associated with processing aid addition is usually no greater than 5 to 10%. These melt viscosity increases usually have no significant detrimental effect on commercial PVC fabrication processes.

B. Effect on Other Properties of PVC Products

Since modern processing aids exert their beneficial effects on PVC processing characteristics at levels as low as 1 to 3 phr, they have very little effect on the physical properties of rigid PVC materials. The polymers used as processing aids generally have fairly good physical properties themselves, related to their high molecular weight and high-T_g compositions. In addition, the compatibility of most processing aids with PVC avoids problems of inhomogeneities in the product, which could serve as initiation sites for failure.

Physical properties of compounds containing normal use levels of processing aids normally are within experimental error of the properties of control samples without processing aid. In fact, in many cases the physical properties of fabricated PVC products containing processing aids are *better* than those of unmodified controls due to the better fusion in compounds containing processing aid. Likewise, the optical properties of PVC compounds are not greatly changed by the addition of processing aids, and the improvements in surface quality, gloss, and so forth usually outweigh any changes in the actual clarity of the product. The heat stability of rigid PVC compounds containing processing aids, especially the most popular commercial acrylic processing aids, is at least as good as that of unmodified control compounds.

Overall, the effect of commercial processing aids on properties other than processibility can generally be assumed to be neutral.

C. Types

The use of plasticizers was the earliest attempted solution to the problems of inherently poor processing of PVC. A variety of low-molecular-weight chemical compounds were found to be sufficiently soluble and mobile to penetrate the PVC structure, allowing it to soften at considerably lower temperatures. Thus, the minimum acceptable processing temperature was lowered considerably. However, these plasticizers also caused a sharp reduction of the modulus and strength of PVC, even at low concentrations. Nevertheless, the resulting material had a useful balance of properties and found extensive commercial use.

The original commercial processing aids for rigid PVC were discovered as a result of a search for a "high-temperature plasticizer"—a material which would provide improved fluidity at processing temperatures but remain inert at use temperatures. During the 1950s it was found that some high-molecular, PVC-compatible polymers could function in this way. By the late 1950s the first commercial processing aid for rigid PVC, Paraloid®* K-120, was avail-

---

*Trademark, Rohm & Haas Company.

able to PVC compounders. Processing aids quickly became accepted as an essential component of rigid PVC formulations and combined with developments in improved stabilizers and processing machinery to allow a very rapid growth of rigid-PVC applications. In particular, processing aids provided great assistance in allowing the use of simpler, more economical fabrication techniques directly from powder blends. Thus, it could be said that the development of processing aids was one of the important factors in the establishment of PVC as a leading material for rigid pipe, siding, window profiles, and other extrusions, as well as clear films and bottles.

In those early days of development of processing aids, the first compositions reported in the literature as having processing aid functionality were copolymers of styrene with acrylonitrile [12] or methyl methacrylate [13]. Although such compositions have been made and used captively by PVC producers, they were not produced and sold on the open market purposely as processing aids for PVC. The first commercially promoted processing aids, introduced by Rohm & Haas Company in the late 1950s, were of the acrylic type, that is, methacrylate/acrylate copolymers with methyl methacrylate predominating. Today, over 25 years later, this type of composition remains dominant. This success is based on the inherent compatibility of the methacrylate polymers with PVC, their specific rheological properties, and their intrinsic stability to heat and light. Many acrylic monomers are available, and they can be combined in various ratios and structures, as well as in polymers of different molecular weights, so that a wide variety of acrylic processing aids can be tailored to any application desired. These various combinations have been the subject of many patents [14—19]. Outside the acrylic field, poly(2-methyl styrene) has been used somewhat in commercial PVC applications [20]. Newer compositions suggested, but not yet used in significant volume, include poly(neopentylene terephthalate) and poly(alkylene carbonate) [21, 22].

In addition to conventional processing aids, many multipurpose materials have been conceived. The most successful of these has been a class of lubricating processing aids, which combine very strong external lubrication effects with the conventional rheological improvements of acrylic processing aids [23—26]. These materials are typified by Rohm & Haas' Paraloid® K-175. Besides such commercial additives, similar principles have been applied to internal modification of PVC resins during their production to give "easy processing" lubricated resins [27, 28]. Other experimental multipurpose processing aids have been tested but without extensive commercial use to date. These include processing aids which reduce the viscosity of the PVC compound. Both acrylic [29, 30] and polyester [31] compositions have been described. Other compositions have been designed to improve the thermal stability of the PVC compound through incorporation of glycidyl methacrylate [32] or oxirane [33] functionality. Lastly, the possibility of raising the heat distortion temperature of PVC compositions has been suggested through the use of processing aids incorporating methacrylate copolymers with bicyclic side chains [34] or styrene/acrylonitrile/acrylamide terpolymers [35].

D. Applications

The use of processing aids in formulating PVC for various applications is covered in greater detail in a later section, but some general comments can be made here. The benefits of accelerated fusion, better homogenization, and improved melt strength/extensibility described above apply in almost all rigid PVC applications and fabrication processes. Therefore, processing aids are found very commonly in almost all rigid PVC formulations. Use levels can vary widely, depending on a particular formulation, processing equipment, and so forth, but the most common range is 1 to 3 phr.

II. IMPACT MODIFIERS

Impact modifiers are generally compounded into rigid PVC to improve the impact behavior of the PVC while retaining other critical performance properties necessary to achieve the demanding application requirements for clear and opaque rigid PVC.

The impact resistance of rigid PVC compounds is the key factor allowing their use in applications such as shatterproof clear bottles, crease-resistant clear packaging film, opaque pipe or house siding, and opaque flame-resistant injection-molded housings. Even though rigid PVC has inherent toughness due to its polarity and low-temperature molecular relaxation and is ductile over a range of use conditions, it would be brittle and have inadequate toughness under conditions of high deformation rate and concentrated stress without the incorporation of impact modifiers.

A. General Classes

Impact modifiers for rigid PVC may be considered as falling into three general classes:

1. Grafted particulate rubbery polymers such as methacrylate-butadiene-styrene (MBS), acrylate-methacrylate (all acrylic), acrylate-butadiene-methacrylate (modified acrylic), and acrylonitrile-butadiene-styrene (ABS), which all introduce an incompatible rubbery phase
2. Semicompatible plasticizing polymers such as chlorinated polyethylene (CPE) and ethylene-vinyl acetate (EVA)
3. Inorganics such as stearic acid-coated calcium carbonate

Nitrile rubbers (NBR) are also used for PVC impact modification and can fall into class 1 or 2 depending on the composition and level of cross-linking. Copolymers of vinyl chloride and insoluble rubber are also used in Europe as impact modifiers and generally fall into class 1.

B. Performance Requirements for Impact-Modified PVC

Incorporating rubbery particulate impact modifiers into rigid PVC by the process of powder blending followed by melt mixing enhances the natural toughness of PVC. The range of use conditions over which PVC can behave in a tough or ductile manner is extended up to higher rates of deformation or down to lower use temperatures. Impact modifiers raise the ductile-brittle strain rate and lower the PVC brittleness temperature. Impact modification reduces the notch sensitivity of PVC in that it can remain ductile

under the severe geometric constraints and stress conditions of the notched
Izod impact test.

Whereas impact improvement is critical to successful use performance
and market penetration, this improvement must be achieved with the retention of other key properties. For clear rigid PVC applications such as blow-
molded bottles and calendered or extruded film, the optical clarity, yellowness, resistance to crease whitening, and water blush must be controlled;
rigidity must be adequate to withstand the application loads applied; and
processing must be optimized to give a product of homogeneous quality at
acceptable production throughput rates. The effect of the impact modifier
on rate of fusion of the PVC is an important aspect of processing. Complete
fusion of the PVC particles is essential to achieve the high level of transparency and low haze for clear applications. However, for opaque applications
such as extruded PVC pipe, profile, or siding, complete fusion is not always
achieved and may not be necessary for optimum balance of part toughness
and processing throughput rate. Retention of rigidity, processing fluxing
and color stability, postextrusion forming including thermoforming, heat
stability, and weatherability are all important performance properties.
Weatherability is usually assessed as impact strength retention and degree
of color retention after outdoor exposure. Other opaque applications with
generally complete fluxing, but where toughness, processing flow, thermo-
oxidative stability, and part color retention are critical, are injection-molded
housings for appliances, business machines, and electrical appliances. Flame
retardance and low smoke emission are other properties which may have to
be considered to meet customer specifications for many of these opaque
applications. The inherent flame retardance of the PVC resin can be augmented by addition of other flame-retarding components such as halogen and
antimony oxides to achieve the most stringent Underwriters Laboratories
(UL) V-0 rating.

For clear, rigid PVC applications, graft polymers such as methyl methacrylate-butadiene-styrene have become the major impact modifiers in commercial use. Acrylonitrile-butadiene-styrene graft copolymers are also used.
The all-acrylic acrylate-methacrylate impact modifiers are widely used in
opaque, weatherable PVC applications due to their excellent weathering
characteristics. Modified acrylics offering good weatherability with improved
processing and flow are also used extensively. Other impact modifiers for
opaque PVC applications include ABS graft copolymers, chlorinated polyethylene, ethylene-vinyl acetate copolymers, grafted elastomers based on
vinyl chloride, nitrile rubbers, and stearic acid-coated calcium carbonate.

Recent developments in commercial impact modifiers in the United States
with guidelines for selection based on end-use requirements, processing
considerations, and cost/performance are described in a recent review
article [36].

C. Effects of Modifier Structure on Modified PVC Properties

Impact modifiers are usually particulate rubbery graft polymers or blends
of both hard and rubbery polymers. The rubber component of the modifier
is usually insoluble and incompatible with PVC and forms a distinct disperse
phase with a low glass transition temperature, T_g. The rubbery particulate
usually has a grafted outer glassy shell to achieve the desired controlled

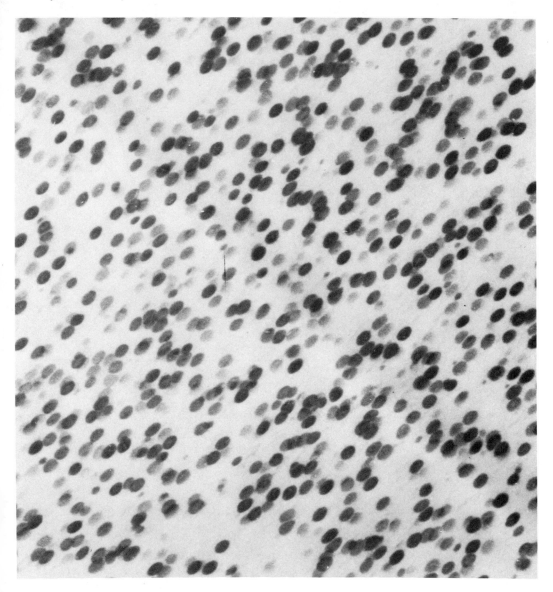

FIGURE 12 Transmission electron photomicrograph ($\times 50,000$) of well-dispersed MBS impact modifier in PVC, using OsO_5 for staining the rubber particles.

compatibility or wetting with the PVC matrix and good dispersion. An example of a well-dispersed uniform particle size MBS impact modifier is shown in the transmission electron photomicrograph in Figure 12. Some of the outer shell may be ungrafted and solubilize into the PVC glassy matrix.

Typically, the rubber component will have a T_g lower than the $-40°C$ β-relaxation of PVC to confer toughness at lower use temperatures. The

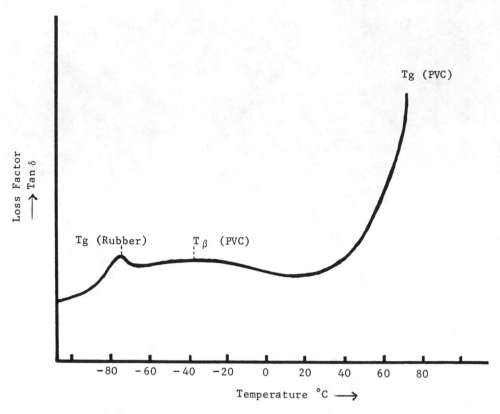

FIGURE 13 Dynamic mechanical loss spectrum of MBS impact-modified PVC measured at low frequency (~1 Hz) showing the T_g of the dispersed rubber phase as a shoulder on the broad β-relaxation region of PVC. Area under shoulder is proportional to rubber content.

dynamic mechanical loss spectrum of MBS-modified PVC identifying the SBR loss peak near -60°C is shown in Figure 13. Polybutadiene gives the greatest lowering of PVC brittleness temperature due to its very low T_g of -85°C (cis, vinyl 1,2 commonly used in ABS). The brittleness temperature of PVC is lowered progressively by addition of impact modifier. The rubber content and particle size distribution of the impact modifier as well as the composition and molecular structure of the graft polymer influence the effectiveness of the impact modifier in lowering the brittleness temperature of the PVC. The number and size distribution of rubber particles per unit volume of glassy PVC and the adhesion of the grafted rubber particles with the PVC rigid continuum are the major factors affecting the efficiency of the impact modifier in reducing the brittleness temperature or increasing the strain rate for embrittlement of the PVC resin. Figure 14 shows the major composition and structure effects on PVC brittleness temperature. Figure 15 summarizes the major effects of composition and structure on impact efficiency and effectiveness.

The distribution of rubber particle size governs the impact efficiency and balance of optical (transparency, haze) and impact properties. A

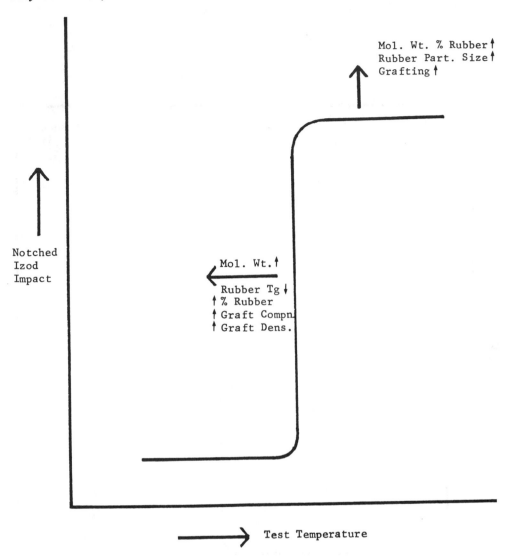

FIGURE 14 General effect of PVC molecular weight and impact modifier composition and structure on Izod impact brittleness temperature.

smaller particle size with narrow distribution and with cross-linking in the
rubber and graft stage is used to give very low stress whitening, high
transparency, and low haze with improved impact resistance. This type of
modifier is used in PVC packaging film, where resistance to crease whitening,
excellent transparency, and tear and puncture resistance are required.

Very high impact efficiency can be achieved at the expense of transparency and haze by using larger rubber particles for toughening. Stress
or crease whitening generally increases with rubber particle size. Opaque
PVC for injection-molding applications contains this type of modifier.

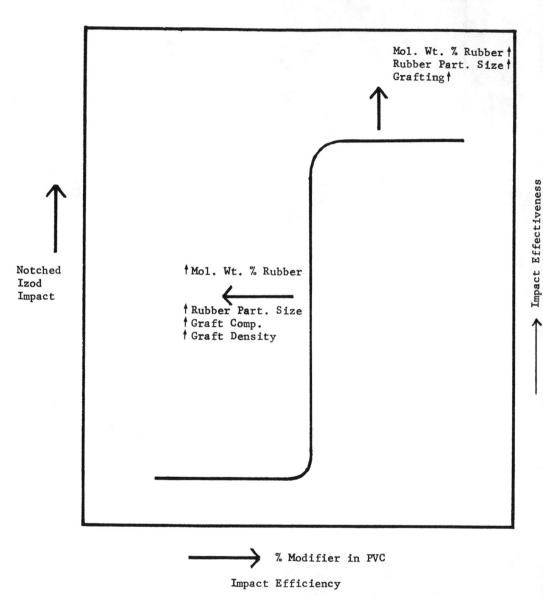

FIGURE 15 General effect of PVC molecular weight and impact modifier composition and structure on Izod impact efficiency and effectiveness.

An excellent balance of high impact efficiency and good optical properties is achieved with a controlled distribution of small and larger rubbery
particles. The rubber particle size used commercially varies from an
average of less than 800 to more than 2000 Å. For optimum clarity combined
with impact resistance, the dispersed rubber must be below the half-wave-
length of visible light (nonscattering) or have a close match of refractive
index with the PVC matrix containing solubilized processing and property
additives. The rubber particle size and distribution must be tightly controlled to ensure consistent impact modifying and optical performance. Also,
the individual grafted rubber particles must be very well dispersed to ensure
full utilization of the rubber toughening mechanisms and avoid processing
and optical problems due to particle agglomeration.

Grafting of each rubber particle is essential for attaching the rubber
particles to the surrounding glassy matrix. This allows the transfer of
strain and energy necessary for the rubber particles to dissipate impact
energy. Grafting is required for effective dispersion of the 1000-μm resin
powder particles on melt mixing in PVC to give uniformly dispersed grafted
rubber particles of the same 0.1- to 0.2-μm size as the original rubber latex.
Maximum impact efficiency with good optical properties and processing is then
realized.

The degree of grafting controls the amount of free or ungrafted rigid
copolymer which can be mixed or solubilized into the PVC. The melt flow of
the PVC can be improved in this manner, but careful control of molecular
weight of the ungrafted polymer must be maintained to control PVC processing and impact resistance.

Incorporating a rigid core in the rubber particle can offer the impact
efficiency of larger rubber particle size at lower rubber content combined
with the benefits of low stress whitening and good optical properties.

Figure 16 shows the effects of modifier particle size, particle size distribution, rubber cross-linking, and incorporation of rigid central core on the
impact-stress whitening balance.

D. Mechanics of Impact Modification

The toughness of a plastic material during impact is governed by the load
response of the material to the rapid rate of deformation under the temperature and geometry constraints imposed.

Because of the viscoelastic nature of polymers, their load response to
deformation is highly dependent on the rate of deformation. The higher the
rate of deformation, the more elastic (Hookean) is the load response, because
the viscous flow component of response (plasticity or cold flow) requires
more time to function and lags behind the elastic response. Reducing temperature has the same effect as increasing rate of deformation.

Impact resistance or toughness can best be described by measuring the
load response to deformation up to impact abuse rates of deformation over a
wide range of use temperatures and for different test geometries.

1. Uniaxial Tensile Deformation: The simplest form of deformation is
tensile (uniaxial) elongation. The fundamental mechanical parameters of
tensile stress, strain, and strain rate can readily be obtained from the tensile load-displacement trace. Tensile stress is load divided by cross-sectional area in the neck-down region of the specimen. Strain is elongation divided
by the deforming or gauge length of the specimen. Stress and strain at

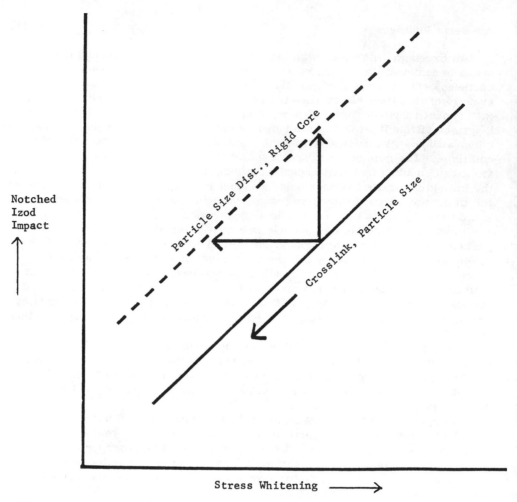

FIGURE 16 General effect of impact modifier composition and structure on balance of Izod impact energy and stress whitening of impact-modified PVC.

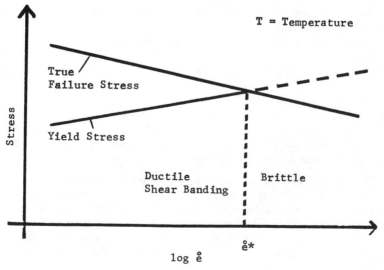

FIGURE 17 Response of yield and failure stress to strain rate $\dot{e}$; the ductile-to-brittle transition occurs at $\dot{e}*$. (From Ref. 38.)

engineering yield (maximum load) and fracture are readily determined from
the load-elongation curves. Ductility is demonstrated by the occurrence of
macroscopic plastic yielding or load decrease prior to break. Yielding is
associated with the onset of necking of the test specimen, which reduces
the cross-sectional area and load-bearing capability. The extent of the
load decrease at yield is attributable to the severity of local necking. Severe local necking is referred to as unstable neck formation [37]. Toughness is measured as the total work required to fracture the plastic material,
which is obtained from the area under the tensile stress-strain curve up to
the point of fracture (or point of unstable neck formation).

The response of plastic yielding or cold shear flow to rate of deformation
and temperature controls the envelope or limit of ductile failure. The ductile failure envelope is described by combinations of rate and temperature
which cause a transition from ductile (yielding) to brittle (elastic fracture
without yield) behavior. If the yield strength (stress) is greater than the
brittle strength, then as Vincent [38] pointed out, the plastic will fail in
a brittle manner under those conditions. Yield stress increases with strain
rate more than the brittle strength, so a material which is ductile at a low
strain rate (low yield stress) can become brittle at the rate where brittle
strength equals yield strength. This condition as shown in Figure 17 is the
ductile-to-brittle transition rate for that particular test temperature. As
temperature is lowered, the ductile-to-brittle strain rate decreases. Testing
at one rate and varying the temperature can similarly define a ductile-to-
brittle transition temperature at that particular rate. Deformation rate,
stress state (uniaxial, biaxial), and temperature must be taken into consideration to define the ductile-to-brittle behavior of a plastic material. Figure
18 represents the failure envelope dependence on strain rate and temperature.

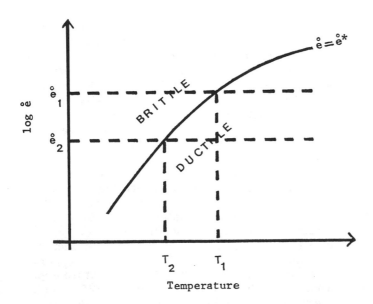

FIGURE 18 Dependence of ductile-brittle transition on temperature, T, and
strain rate, $\overset{\circ}{e}$.

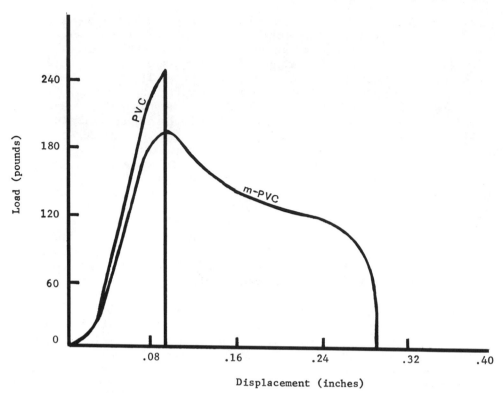

FIGURE 19 Typical room temperature tensile load-displacement dependence for PVC (K57 packaging) and 10 phr MBS-modified PVC measured at a test speed of 20,000 in./min.

As can be seen from the high-speed load-displacement traces in Figure 19, incorporation of rubbery particulate impact modifier into a continuous rigid PVC matrix induces bulk yielding and ductility at rates of deformation where the unmodified PVC would be brittle. This is achieved by lowering the aggregate engineering yield strength of rigid PVC to favor yielding with more stable neck formation over brittle fracture and allow extensive elongation (cold flow) beyond yield up to fracture. Whitening in the neck-down region is intensified and the ductile-to-brittle transition rate is increased by the addition of impact modifier. Similarly, the ductile-to-brittle transition temperature at impact rates is considerably lowered. The rubbery impact modifier particles have very low modulus and act to lower the bulk engineering yield stress of PVC by decreasing the effective load-bearing cross section while maintaining or increasing the elongation of PVC beyond yield. Obviously, if there are other matrix defects such as incomplete fusion of the PVC particulates, the impact performance will be affected.

The toughness or fracture energy of rigid PVC is considerably increased above the ductile-to-brittle transition rate of unmodified PVC by incorporating grafted rubbery particulate impact modifier. The tensile fracture energies of a rigid PVC (K57 packaging resin) and impact-modified PVC (10 phr MBS) are compared up to a test speed of 40,000 in/min in Figure 20. At

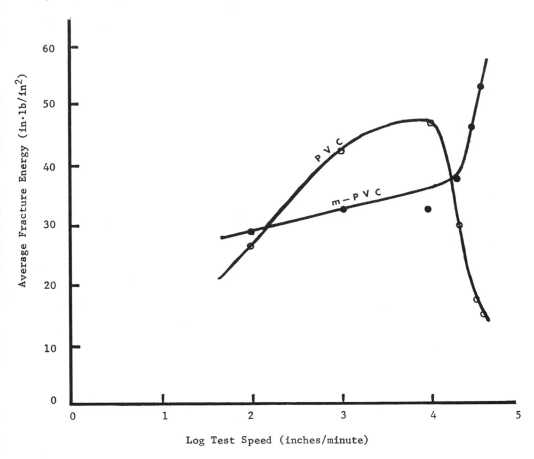

FIGURE 20 Effect of 10 phr MBS impact modifier on room temperature strain rate response of tensile fracture energy of a K57 packaging PVC formulation.

high rates of deformation where rigid PVC is brittle and has low fracture energy, impact-modified PVC continues to increase in toughness by virtue of its rapidly increasing tensile yield stress and associated higher stress levels up to break. Elongation beyond the engineering yield point causes extensive adiabatic heating, which can lead to healing of stress whitening and appreciable strain recovery [39]. This increase in plastic temperature may facilitate flow at high rates of deformation once the yield process has been initiated. Vincent [40] found that adiabatic heating at low rates of deformation appreciably reduced the cold drawing capability of PVC.

Increasing the molecular weight of the PVC resin gives the expected overall increase in toughness at high strain rates. High-speed tensile and Izod impact fracture energies of ABS-modified PVC increase with PVC resin molecular weight [41].

High-speed tensile testing has been found useful in simulating the hoop stress in PVC bottle walls during drop impact. High-speed tensile fracture energy has been shown by Matonis and Aubrey [42] and Malpass et al. [43] to be correlated with mean bottle failure height, both of which increase on addition of impact modifier. Fenelon [44] found that onset of unstable neck-

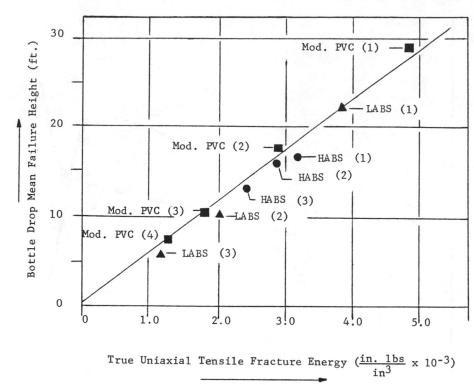

FIGURE 21 Correlation of bottle drop mean failure height with true uniaxial tensile fracture energy at 10,000 in./min for the materials indicated. (From Ref. 44, p. 542.)

ing with severe localized thinning occurs beyond the engineering yield point. Brown and Ward [45] had shown previously that different stress levels are required for yield initiation and propagation. The formation of an unstable neck gives rise to a complex stress state and defines the ultimate uniaxial stress and strain the PVC can withstand. By using this necking onset condition to obtain tensile fracture energy in conjunction with true stress and strain data (from high-speed photography), Fenelon [44] established a universal relationship between fracture energy and mean bottle drop height for different families of materials, including ABS and impact-modified PVC. This correlation, shown in Figure 21, between uniaxial tensile fracture energy and multiaxial bottle wall fracture strongly supports a strain energy criterion for failure analogous to the classical Von Mises-Hencky plasticity yield condition [46]. Figure 22 describes the actual configuration of the PVC specimens at different points along the tensile load-elongation trace.

2. *Multiaxial Impact Testing*: To evaluate toughness under conditions more closely simulating actual end use, falling-weight impact measurements on flat plates or film are performed. ASTM D1709 describes the standard test procedure for obtaining the energy required for 50% of the test specimens to fail by a visible crack or break. A free-falling tup can strike the specimen from various heights (variable impact energy and strain rate), or

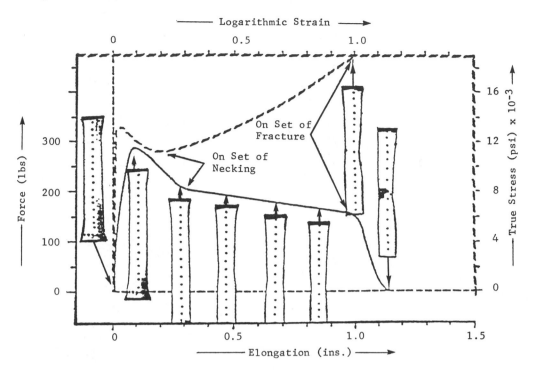

FIGURE 22 Force vs. elongation and true stress vs. true strain for Mod. PVC (2) at 10,000 in./min. Photographs of specimens at various stages during deformation are included. (From Ref. 44.)

the tup weight can be varied from a fixed drop height (variable energy at constant strain rate). The commonly used Gardner variable height impact test is described in ASTM D3027 and is a simplified falling-weight impact tester which does not clamp the specimen in place. The geometry of the impacting tup, the tup diameter relative to test specimen diameter, and the specimen thickness are all important geometric considerations which affect the impact values determined. This type of test procedure, although simple and inexpensive, requires the use of many specimens, and the failure criteria can be subjective. Also, depending on the exact sequence of tests performed, different 50% failure energies can be obtained. The familiar Bruceton staircase method locates an initial failure height and then proceeds with testing incrementally above and below that height to establish the height giving 50% failures. A preferred and more comprehensive procedure is to determine the percentage of failures at a given drop height and progressively increase the drop height, thereby obtaining a profile of impact toughness over a range of drop heights. This is called the Probit procedure [47]. Figure 23 shows the linear relationship between fraction broken and impact energy plotted on a normal probability scale for ABS sheet over a range of temperatures.

More recently, instrumented falling weight impact tests have been introduced [48]. Weight and drop height can be varied to change the impact

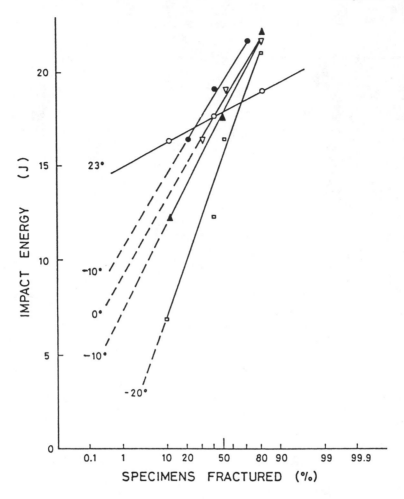

FIGURE 23 Probit curves from falling-weight tests on ABS over a range of
temperatures. Plate specimens 2 mm thick, injection-molded at 240°C. (From
Ref. 52.)

energy. Load and impact energy (area under the load-time trace) are
plotted against time and are recorded during the entire impact test from
initial impact up to failure. In addition to flat plate or film test specimens,
Izod and Charpy tests can be performed with the variable deformation rate
and load-time recording capabilities of the tester.

A major advantage of the instrumented falling-dart penetration impact
tester is that very few test specimens are required to define an average
impact value at a particular temperature. Temperature can be varied down
to low use temperatures to determine a practical brittleness temperature [49].

Figure 24 shows a typical load-time trace (2 msec) obtained for a milled
and compression-molded packaging PVC (K57) formulation containing 10 phr
of MBS impact modifier impacted with a hemispherical tup traveling 7 ft/sec.
The biaxial load maximum with subsequent deformation beyond this yield
point is clearly demonstrated. Unmodified PVC under these conditions has a
lower fracture energy with less deformation.

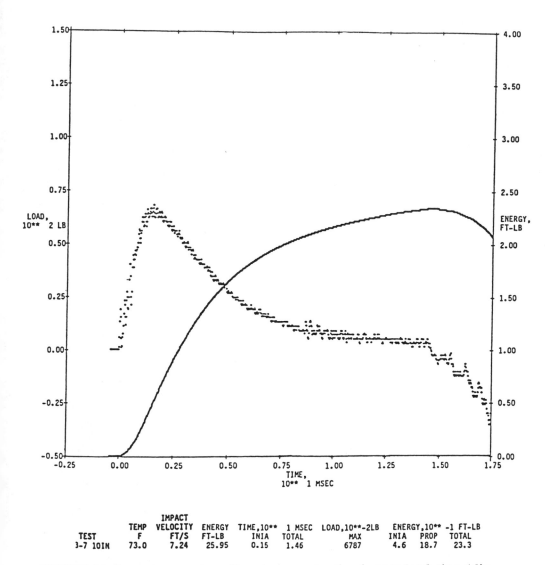

TEST	TEMP F	IMPACT VELOCITY FT/S	ENERGY FT-LB	TIME,10** 1 MSEC		LOAD,10**-2LB	ENERGY,10** -1 FT-LB		
				INIA	TOTAL	MAX	INIA	PROP	TOTAL
3-7 10IN	73.0	7.24	25.95	0.15	1.46	6787	4.6	18.7	23.3

FIGURE 24 Room temperature Dynatup penetration impact load-time (displacement) trace for K57 PVC modified with 10 phr MBS, showing integrated energy (area under curve) up to point of fracture.

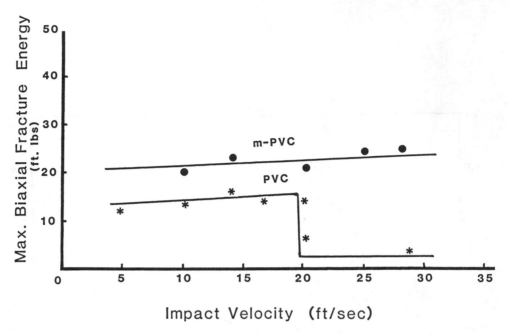

FIGURE 25 Room temperature Dynatup penetration impact energy to fracture
for a K57 packaging PVC and 10 phr MBS-modified PVC.

Figure 25 compares the fracture energies obtained over a wide range of
impact velocities for the K57 packaging PVC formulation and the same formulation containing 10 phr MBS impact modifier at room temperature. The
modified PVC exhibits tough impact behavior over the complete speed range
of the instrument (impact velocity from 2 to 28 ft/sec) with considerably
higher fracture energies than obtained for the unmodified PVC formulation.
A small number of brittle fractures are obtained over this range of speeds
with the modified PVC. However, for the PVC formulation without impact
modifier, only brittle failures were obtained above a test speed of 20 ft/sec;
that is, this was the ductile-to-brittle transition condition under the imposed
test conditions for unmodified PVC. The biaxial stress state significantly
lowers the fracture toughness capability of unmodified PVC compared to
the uniaxial tensile fracture energy shown earlier in Figure 20.

   3. Notched Impact Testing: Bucknall [50] provides a comprehensive
description of impact testing and the mechanics of the fracture process. By
far the most widely used in the U.S. plastics industry is the ASTM D256
notched Izod impact test. Notched Charpy impact testing according to ASTM
D256-81 is also common in the United States but is perhaps more prevalent
in Europe.
   PVC is a notch-sensitive material; that is, its impact fracture energy
as measured by notched Izod or Charpy impact testing falls rapidly with
decreasing notch radius. This is because a sharp notch reduces plastic
yielding by virtue of its higher local strain rate, stress concentration, and
triaxiality. Shoulberg and Gouza [51] performed a comprehensive evaluation
of the notch and temperature sensitivity of unmodified PVC, impact-modified
PVC, ABS, and PC. Figure 26 describes the notched Izod impact surface

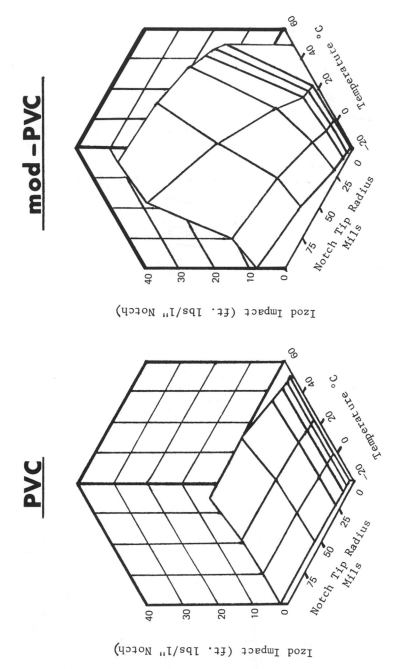

FIGURE 26 Notched Izod impact of PVC and impact-modified PVC, showing the effects of notch tip radius and test temperature. (From Ref. 51)

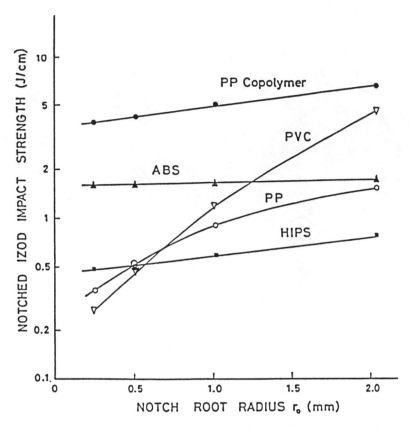

FIGURE 27 Relationship between Izod impact strength and notch root radius for toughened and untoughened polymers at 23°C. (From Ref. 52.)

responses that they generated for PVC and impact-modified PVC. Morris [52] also compared the severe notch sensitivity of PVC with other polymers such as PP homo- and copolymer, HIPS, and ABS, as shown in Figure 27.

Fracture mechanics measurements with sharply notched specimens show that the fracture surface energy, G_c, of unmodified PVC is low in comparison with that of modified PVC. In the case of PVC, PC, PMMA, HDPE, and MDPE, the zone of plastic deformation at the root of the notch is small enough to permit valid measurement of G_c according to the principles of linear elastic fracture mechanics (LEFM). However, for HIPS, ABS, and toughened PVC, yielding by multiple crazing or shear flow is too extensive to allow analysis by LEFM. Therefore, yielding fracture mechanics is employed. The critical crack opening displacement, δ_c, and the plastic work parameter, J_c, or J integral have been used to define the fracture toughness of ductile polymers. These two parameters are related to yield stress σ_y as follows:

$$J_c = \sigma_y \delta_c \qquad\qquad [1]$$

In the elastic case J_c equals G_c. The plastic work parameter has been used by Plati and Williams [53] to obtain the data in Table 3.

TABLE 3 Critical Strain Energy Release Rate Determined by Two Methods

Material	G_c (J/m^2)	
	Charpy	Izod
Polystyrene (GPPS)	0.83×10^3	0.83×10^3
PMMA	1.28×10^3	1.38×10^3
PVC (Darvic 110)	1.42×10^3	1.38×10^3
Nylon 66	5.30×10^3	5.00×10^3
Polycarbonate[a]	4.85×10^3	4.83×10^3
PE (medium density)[b]	8.10×10^3	8.40×10^3
PE (high density)[c]	3.40×10^3	3.10×10^3
PE (low density)	34.7×10^3	34.40×10^3
PVC (modified)	10.05×10^3	10.00×10^3
HIPS	15.8×10^3 (J_c)	14.0×10^3 (J_c)
ABS (Lustran 244)	49.0×10^3 (J_c)	47.0×10^3 (J_c)

[a]Specimens cut in the extrusion direction.
[b]Density = 0.940; MI = 0.2.
[c]Density = 0.960; MI = 7.5.

The J_c values of fracture energy can be estimated from the work required to break notched specimens at impact rates of deformation. The work or energy to break, w, is obtained from the area under the load-displacement trace up to break. The work required to fracture a notched specimen of a ductile plastic, W, can be related to the fracture mechanics parameter J_c by

$$W = \frac{w}{(d-a)t} = \frac{J_c}{2} \qquad [2]$$

where d is the specimen depth, a the initial crack or notch depth, and t the thickness. Crack initiation energy, W_i and crack propagation energy, W_p, can be similarly obtained from the load-displacement trace by assuming that the crack initiates at maximum load. Malpass and Gaggar [54] measured the W fracture energies for notched injection-molded Izod specimens of a low-molecular-weight injection molding grade of PVC and ABS-modified formulations. They used a high-rate tester to perform flexural three-point bend tests on Izod test specimens at test speeds from 10 to 11,000 in./min at room temperature and down to -40°C. The ABS modifiers were shown to raise progressively the strain rate at which the PVC ductile-brittle transition occurred. Varying the rubber grafting gave different increases in the ductile-brittle transition rate. The ductile-brittle transition rate also decreased with decreasing temperature. Figure 28 shows the Arrhenius relationship between ductile-brittle transition rate and temperature for PVC and PVC containing different levels of ABS modifier. An apparent activation

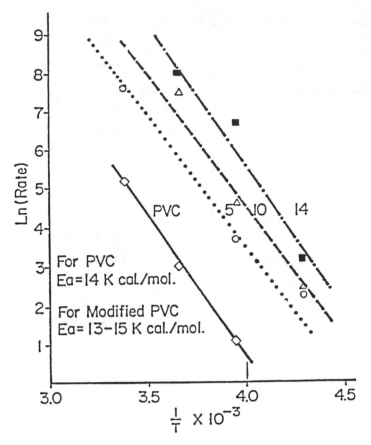

FIGURE 28 Log rate vs. reciprocal temperature plot for PVC and modified
PVC (5, 10, and 14 phr ABS modifier). (From Ref. 54.)

energy of ~14 kcal/mole was obtained for the ductile-brittle process for the
PVC and modified PVC formulation, which was consistent with the activation
energy for the B-relaxation process in PVC [55, 56]. Apparently, impact
modification did not change the distribution of relaxation processes, but
provided a stress bias to facilitate ductile behavior up to higher rates of
deformation.

Havriliak and Malpass [57] have also demonstrated that instrumented
variable-rate notched Izod testing of a K57 packaging PVC and MBS-modified
PVC can provide fracture energy values for initiation, propagation, and
total break and describe the ductile-brittle transition rates. As can be seen
from Figure 29, the unmodified PVC was brittle and had a very low fracture
energy down to very low impact velocities of 2 ft/sec. The 10 phr MBS-modified PVC remained ductile with a high fracture energy up to an impact velocity of 23 ft/sec, which is double the normal Izod impact test velocity of 11.3
ft/sec. The local strain rate at the root of the notch can be calculated from
simple beam theory [51]. For a notch of 10 mil radius at an impact velocity
of 11.3 ft/sec the strain rate would be 48,200 in./in./min. Different MBS
impact modifiers will provide different levels of fracture energy enhancement

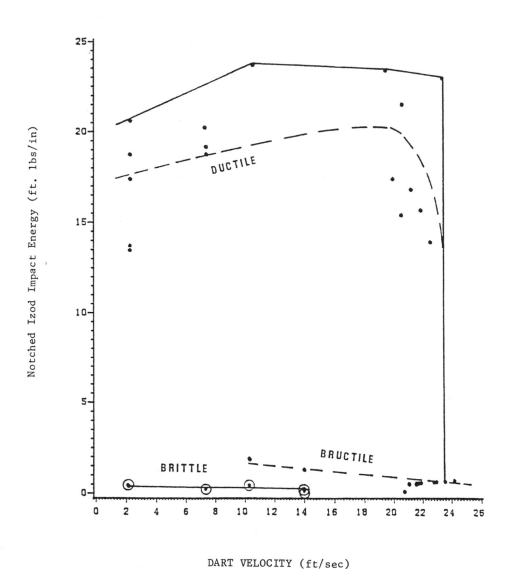

FIGURE 29 Dependence of notched Izod impact energy on dart velocity for modified and unmodified PVC. (From Ref. 57.)

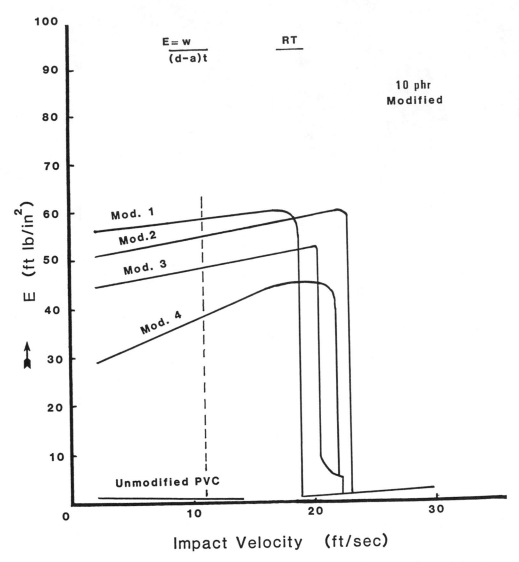

FIGURE 30 Dependence of fracture energy, E, on strain rate for notched specimens (Izod).

and will give different increases in the ductile-brittle transition rate of PVC. Figure 30 compares the fracture energies of some different MBS impact modifiers in the K57 PVC formulation at the 10 phr level.

E. Mechanisms of Rubber Toughening in PVC

1. Fully Fused PVC State: Various mechanisms have been proposed to explain toughening of glassy polymers by incorporation of rubber particles. Merz et al. [58] were the first to propose an energy absorbing mechanism due to stretching of the adhered rubber particle across a crack front, thereby holding the crack faces together and blunting the crack growth. Scattering of light from microcracks explained stress whitening, and opening of the microcracks provided a mechanism for large strain deformation.

Bucknall and Smith [59] calculated that the energy of stretching rubber particles across crack fronts would not explain the increase in energy required to fracture the rubber-toughened glassy matrix. They applied the earlier work of Kambour on glassy polymers [60] and proposed a multiple crazing mechanism. Crazes would initiate at the high dilatational stress region at the equator of a discrete rubber particle and develop in a direction perpendicular to the applied stress. Crazes would propagate from one rubber particle to another along the region of maximum dilatational stress. These crazes contained fibrillar matrix material stretched in the direction of deformation and hence were load-bearing along with the stretched rubber particles. Crazes were terminated when the stress concentration at the tip fell below the critical level for propagation or when a large rubber particle was encountered [59, 61]. In this way a large number of small crazes are formed. Dense crazing throughout a large deformed region accounts for the high energy absorption in tensile and impact tests. Stress whitening is caused by light scattering by the crazes. Optical microscope examination of thin sections of HIPS]59] and subsequent electron microscopic studies of crazed HIPS, ABS, and toughened PVC have confirmed multiple craze formation [62, 63].

Shear yielding in the dilatational high stress concentration region at the equator of the rubber particles was proposed by Newman and Strella [64] to describe cold flow and macroscopic yielding or necking. This matrix yielding mechanism was based on optical microscope studies of rubber particle distortion in ABS tensile specimens. The deformation was attributed to a local reduction in the T_g of the SAN due to triaxial tension. The failure envelope studies of Sternstein and Ongchin [65], however, require that triaxial tension promotes crazing with brittle fracture rather than shear yielding. Petrich [66] interpreted the toughening of PVC by MBS impact modification as cold flow in the PVC matrix resulting from local lowering of the yield stress around the rubbery particles. Electron microscopic examination of the stress-whitened region of fractured MBS-modified PVC showed no evidence of craze or void formation. Stress whitening was attributed to birefringence caused by mismatch of the refractive indices of oriented rubber particles and glassy matrix.

Sternstein et al. [67] analyzed the stress distribution around a soft spherical inclusion to determine the initial elastic conditions for craze formation and shear yielding. They applied Goodier's elasticity solution, which assumes perfect adhesion at the surface of the inclusion and infinite dilution. This solution applies only to small elastic deformation and does not account for viscoelastic matrix behavior at higher strains. Contours of the major principal (tensile) stress distribution controlling craze formation are shown in Figure 31. Stress levels are highest at the equatorial position relative to the direction of tensile deformation and lowest at the polar position in the direction of tensile stress. Crazes form along the equatorial plane. The maximum shear stresses are shown in Figure 32. The locus of maximum shear stress extends from the equatorial position out into the glassy matrix.

The effect of varying the inclusion from a soft rubber much lower in modulus than the glassy matrix to a hard inclusion of the same order of modulus has also been analyzed [67]. The maximum local tensile stress concen-

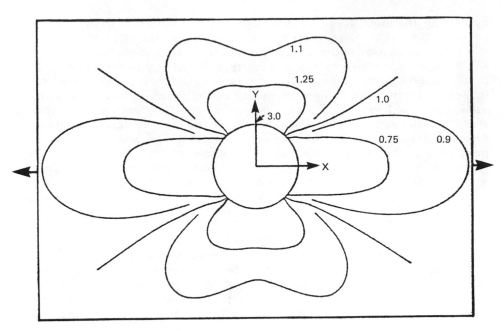

FIGURE 31 Major principal stress contours (σ_1) for an elastic solid containing a hole. Contour numbers are per unit applied tension with sample loading as shown. (From Ref. 67.)

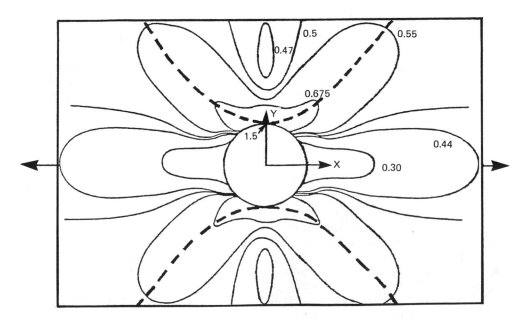

FIGURE 32 Maximum shear stress contours. Contour numbers are per unit applied tension. (From Ref. 67.)

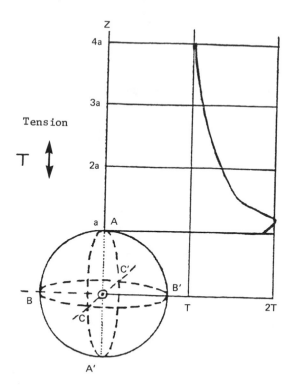

(a)

(b)

FIGURE 33 Tensile stress decay around a spherical rigid inclusion and
spherical cavity (rubber particle). (a) Tension at points along OZ for
rigid inclusions; (b) tension on equatorial plane for spherical cavity.
(From Ref. 67.)

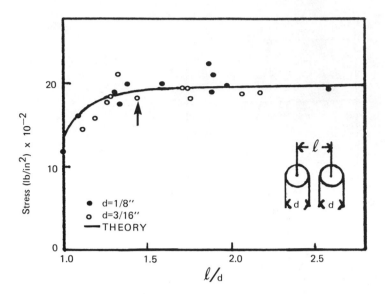

FIGURE 34 Stress distribution between adjacent rubber spheres with 1/8 and 3/16 in. diameters. (From Ref. 68.)

tration for a hard inclusion occurs at the poles of the particle and also acts parallel to the applied stress. Figure 33 shows the stress decay and direction of stress into the glassy matrix for a soft and a hard inclusion.

Overlapping of principal and shear stresses around adjacent rubber particles further increases the stress concentrations and thereby lowers the bulk stress required to initiate crazing or yielding. Matsuo et al. [68]

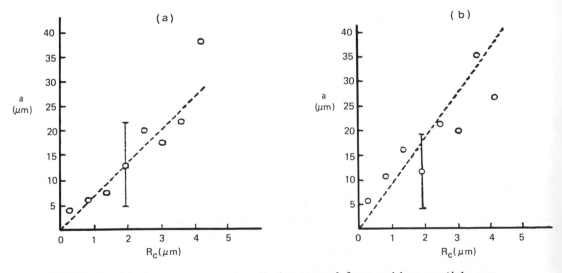

FIGURE 35 (a) Average craze length (measured from rubber particle centers) vs. equivalent particle radius (low applied strain). (b) Same as (a) at higher applied strain. (From Ref. 69.)

demonstrated this experimentally using polystyrene grafted rubber balls in a polystyrene matrix. Figure 34 shows the stress interaction, which becomes negligible when the center-to-center distance of the balls exceeds 1.45 times their diameter. The results showed no effect of particle diameter. Donald and Kramer [69], however, reported that in strained thin films of HIPS the length of the equilibrium craze as measured by electron microscopy depends on the size of the nucleating rubber particle. No effect of occluded matrix within the rubber particle was found. Figure 35 shows these craze length results at two strain levels.

Craze nucleation requires that the initial stress intensification at the rubber equator exceeds the stress concentration at a static craze tip. For craze propagation to occur, the equatorial stress must extend at least three craze fibril diameters from the surface of the rubber particle into the glass [69]. The distance that stress intensification extends into the glassy matrix increases with particle diameter. Therefore, larger particles allow principal stress overlap at greater separation.

The overlap of shear stress fields similarly facilitates shearing at approximately 45 to 58% to the direction of loading. Hobbs et al. [70] used a shear stress overlap approach to predict the shift in ductile-to-brittle transition rate for rubber-modified nylon and EPDM-modified polypropylene. Shear banding has been shown to be the predominating mechanism in these systems. Figure 36 illustrates the shear stress field overlap concept.

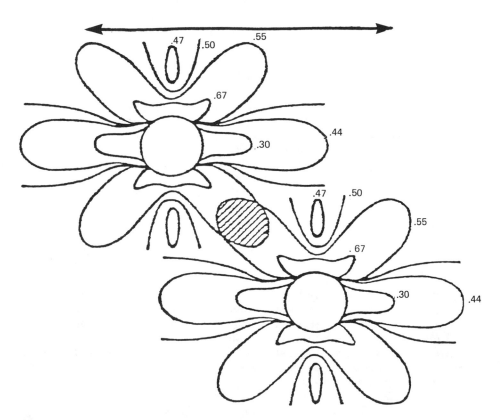

FIGURE 36 Overlapping shear stress field contours for neighboring particles in uniaxial tension. Principal stress is in the vertical direction. (From Ref. 70.)

Bucknall et al. [71] demonstrated that crazing and shear yielding can occur simultaneously in an HIPS/PPO blend. Differences in the contributions of the two mechanisms have explained differences in tensile behavior between different plastics such as HIPS and ABS [72—74]. Interaction between crazes and shear bands is important in ductile polymers such as PVC. Crazes tend to be short and terminate at shear bands which are formed well before general yielding [75,76] and orient parallel to the applied stress between adjacent rubber particles and are normal to the craze direction. Bucknall [50] has shown that PVC modified with 5% ABS undergoes very little volume increase during tensile creep and that shear yielding is the predominant deformation mechanism.

Yee [77] supported the interpretation of impact toughening in PVC and PC as due to a stress state being induced in the glassy matrix, favoring voiding and material flow around the modifier particles. His findings did not support any change in the matrix molecular relaxation mechanism. Malpass and Gaggar [54] suggested that impact modification of PVC could provide a stress bias to the PVC low-temperature β-relaxation process akin to increasing temperature. The ductile-brittle transition strain rate at room temperature was shown to conform to an Arrhenius rate activated flow process. Kramer [75] had previously shown that shear band propagation is a stress- and temperature-activated process which follows the Eyring flow equation.

Yee [77] also pointed out that at a sufficient concentration of rubber particles, the intervening local glassy matrix essentially undergoes tensile stretching or plane stress deformation and shear flow, as opposed to plane strain deformation, which only allows crazing.

Breuer, Haaf, and Stabenow [78, 79] determined by transmission electron microscopy of microtomed layers that cavitation of rubber impact modifier particles and shear banding were the major contributors to stress whitening of deformed impact-modified PVC. Crazing was not observed. The highest toughness was achieved when the outer stage grafting level was low to give a network of rubber particles throughout the PVC matrix. Stress whitening was assigned to cavities within the rubber particles aligned at 55° to 64° to the stress direction along shear bands. Siegmann et al. [80] showed a similar shear banding with rubber cavitation in CPE impact-modified PVC film by using an optical arrangement. Toughness can, therefore, be increased by small grafted rubber particles facilitating predominantly shear band formation. Cross-linking of the rubber particles can be used to reduce void formation and stress whitening in PVC and to maintain spherical particle shape after processing.

Donald and Kramer [81] also determined by electron microscopy of thin films of ABS polymers that rubber cavitation relieves the hydrostatic tension produced by local shear. The concentration of closely packed rubber particles gives a glassy matrix distance between nearest-neighbor particles of about 0.1 μm. This thickness is such that when rubber particles cavitate in a shear stress field, the glassy regions are everywhere under unrestricted plane stress conditions, which allows for shear flow. This shearing mechanism is not available in polystyrene. The impact toughness of high-impact polystyrene is governed by crazing only. ABS polymers allow both crazing and shear band formation.

Tough, ductile, high-impact performance of fully fused rigid PVC is achieved when a sufficient concentration of small, discrete rubbery par-

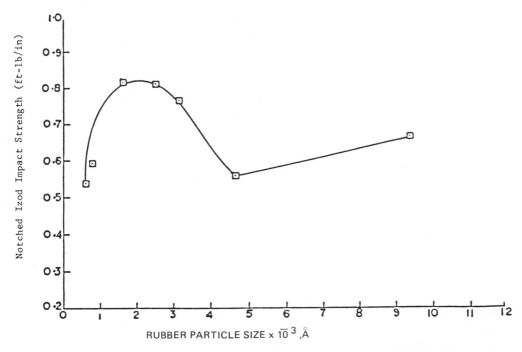

FIGURE 37 Effect of PBD particle size on impact strength of PVC (10 parts). (From Ref. 82.)

ticles is uniformly dispersed throughout the glassy PVC matrix with adequate wetting or adhesion to the PVC matrix to induce a stress field throughout it. Predominantly shear flow and bulk yielding are induced in the glassy matrix up to high rates of deformation and down to low temperatures. Some rubber cavitation may also occur. The stress field is a result of interacting tensile and shear stresses, which increase the stress concentration between adjacent rubbery inclusions and thereby lower the bulk stress required to initiate shear band formation and yielding. The resultant shear stress concentration between rubber particles depends primarily on the separation and hence concentration of the rubber particles and not on their size. However, larger particles are more efficient craze nucleators because the stress enhancement zone is above the three-fibril spacing distance required for craze formation. Crazing requires a longer time than shear to propagate and contributes less at higher impact rates. The rubber particles must have sufficient size (>800 Å) and low enough T_g to be effective as a discrete low-modulus phase up to high strain rates and down to low use temperatures. Morton et al. [82] found 2000 Å to be optimum for PVC shearing. Figure 37 shows their impact results with polybutadiene particle size varying from 700 to 4500 Å. Highly oriented shear bands formed in the thin glassy matrix between rubber particles terminate propagating crazes and may prevent craze formation. Shear bands can also initiate from the highly stressed tip of a craze, bringing it to a halt. These termination mechanisms deemphasize the need for large rubber particles to effectively terminate crazing in PVC.

TABLE 4 Particle States of Suspension PVC

Term	Approximate size in typical PVC	Origin or description
Grain	100 μm	Free flowing at room temperature
Agglomerate	10 μm	Formed during polymerization by merging of primary particles
Primary particle	1 μm	Formed from single polymerization site at conversions of 10 to 50%
Domain	100 nm	Presence not clearly proved, possibly formed by mechanical working within or from primary particles
Microdomains	10 nm	Crystallite or nodule?

2. Particulate PVC Structure: In high-molecular-weight opaque PVC formulations used for vinyl siding and window profiles, the state of fusion of the PVC particulate structure and associated disposition of the rubber impact modifiers has a profound effect on processing and impact performance.

In 1967 Berens and Folt [83] published their landmark finding that PVC could retain vestigial particulate structure through melt processing into the finished part. A temperature above 200°C was required to destroy those PVC primary particles. Rabinovitch and Summers [85] subsequently found that at about 215°C all the crystallites which maintain the integrity of the primary particles disappear and a continuous melt is formed.

Suspension-polymerized PVC is the most commonly used PVC for opaque extrusion applications. Each powder grain of suspension PVC is generally covered by a skin. The different particle states of suspension PVC are shown in Table 4. Table 5 compares the primary particle and grain size of mass, suspension, and emulsion polymerized PVC. Figure 38 shows PVC primary particles with impact modifiers and pigment distributed between them.

The extent of particulate breakdown not only affects the physical properties of the processed PVC part but also controls the processing melt rhe-

TABLE 5 Comparison of Primary Particle and Grain Sizes of PVCs

	Primary particle size (μm)	Grain size (μm)
M-PVC	0.5 to 1.5	50 to 100
S-PVC	0.5 to 1.5	75 to 250
E-PVC	0.15 to 0.40	50 to 150

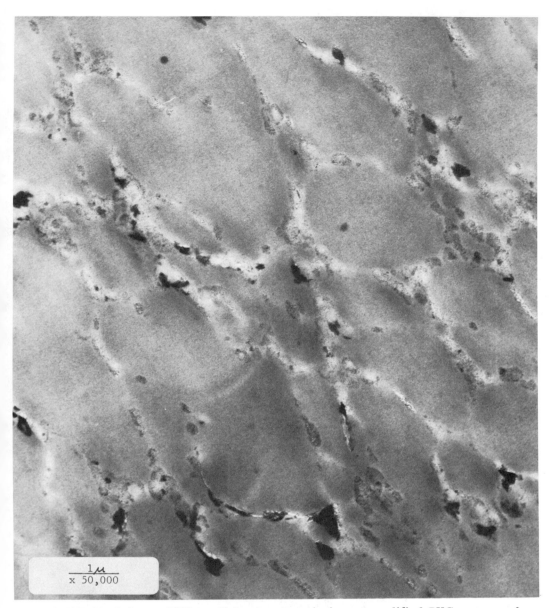

$$\frac{1\mu}{x\ 50,000}$$

FIGURE 38 Primary PVC particle structure in impact-modified PVC processed at 311°F (155°C) melt temperature at ×50,000 magnification (OsO_5 pre- and poststained transmission electron photomicrograph showing modified acrylic rubber dispersed between primary particles).

GEON Resin. 101EP F 24

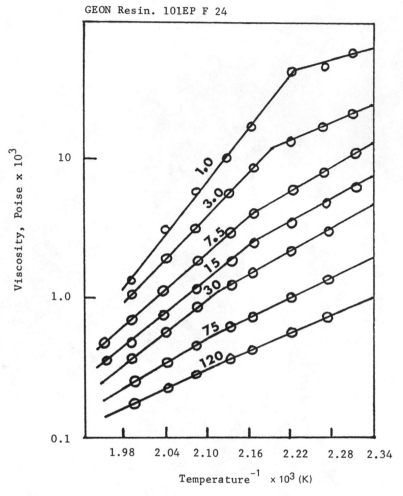

FIGURE 39 Change of viscosity-temperature dependence (activation energy) from particulate to fully fused state for Geon PVC resin. Shear rates (sec.$^{-1}$) are shown for each plot. (From Ref. 86.)

ology. Collins and Krier [86] have quantified these rheological effects as shown in Figure 39 in terms of the activation energy for melt viscosity. Faulkner [87] used a Haake Rheocord torque rheometer programmed for a constant rate of temperature increase to show differences in torque and hence viscosity associated with progressive PVC particle breakdown. Figure 40 shows a torque-temperature curve with the particle breakdown identified. Another melt elasticity technique for measuring relative particle breakdown has been described by Krzewski and Collins [88]. This method uses a zero-length capillary to give a simple direct measurement of melt elasticity. Melt elasticity and die swell increase as PVC particulate structure is destroyed, as shown in Figure 41.

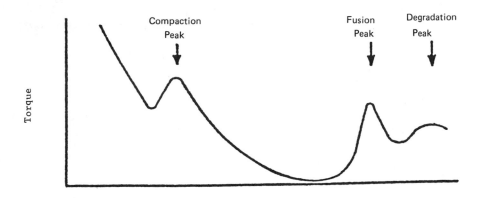

FIGURE 40 Temperature-torque relationship for PVC resins. (From Ref. 87.)

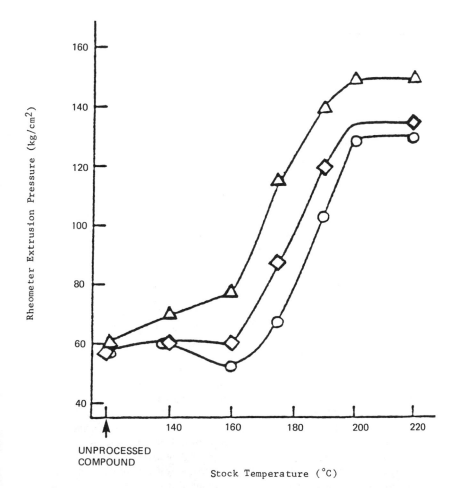

FIGURE 41 Standard fusion curves for compounds containing 0 (○), 2 (◇), and 5 (△) phr of processing aid. Percent fusion = $(P - P^*)/(P_{220} - P^*) \times 100$. (From Ref. 88.)

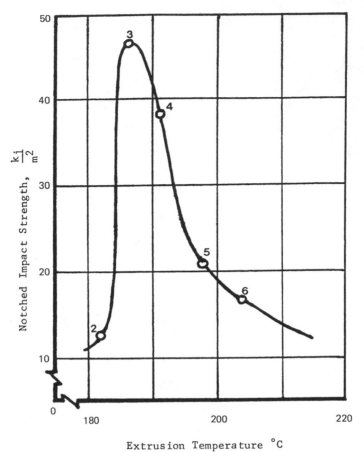

FIGURE 42 Notched Izod impact strength and morphology of high-impact
PVC. The numbers indicate samples examined by electron microscopy.
(From Ref. 89.)

     Menges et al. [89] demonstrated that the particulate structure of EVA-
modified PVC affects notched Izod impact. As shown in Figure 42, a maximum
was found at some optimum state of particulate structure. The EVA was
shown by electron microscopy to form a network structure around the primary PVC particles. The EVA then acts as a rubber cement between primary
particles, presumably operating to dissipate crack propagation energy. CPE
acts in a similar manner [90]. Both impact modifiers preserve a rubber mem-
brane as the continuous phase between 160° and 180°C. At higher temperatures, a phase inversion occurs and CPE or EVA becomes the discrete
particulate phase and PVC the continuum. Notched Izod impact is improved
when CPE or EVA is continuous. Other acrylic and modified acrylic modifiers affect PVC particulate structure in different ways depending on the
modifier particle size and integrity. Some intermediate microstructure with
partial breakdown of the PVC primary particle appears to be needed to obtain optimum notched Izod impact energy. For falling dart or penetration
impact, however, a state close to full PVC fusion is required with no micro-

defects [92]. Processing aids have similar important influences on rate of PVC particulate structure breakdown [91]. Shear flow with rubber particle stretching between PVC particles appears to be the preferred mechanism for good notched Izod impact under partial breakdown conditions. Effective distribution of the impact modifier and good adhesion to the PVC primary particles are essential for toughness under impact conditions. Stearic acid-coated ultrafine $CaCO_3$ particles would be expected to function as adhesion points at the surfaces of contacting PVC particles and could increase density, modulus, and dart impact strength under suitable processing conditions.

3. *Optical Properties*: The clarity and color of blow-molded PVC bottles and calendered or extruded film are critical to the aesthetic appearance and functional performance in displaying the contents of PVC bottles and film packaging.

Impact modifier distribution, dispersion, and refractive index of the dispersed modifier phase affect optical properties. The dispersion and solubility of other compounding ingredients in the PVC are also important to optical properties. Any discontinuities greater in size than about half the wavelength of visible light can act as scattering centers and cause haze and color formation. The compounded PVC has a refractive index controlled by the refractive indices of the PVC resin and all the compounded soluble ingredients such as processing aids, lubricants, and heat stabilizers. The closer the grafted impact modifier particle is in its refractive index to that of the compounded PVC continuum, the higher is the clarity of the impact-modified PVC. Perhaps more important is the latitude in grafted modifier particle size to achieve higher impact as the modifier refractive index approaches that of the PVC continuum.

A difference of as little as 0.0010 between the refractive index of the PVC continuum, which is typically 1.5400, and that of the modifier will cause visible light scattering, increase haze, and reduce white light transmission as measured by ASTM D1003. Therefore, the composition of the grafted modifier—that is, internal rubber composition, grafted outer shell(s), any occluded polymer in the rubber substrate, and any free or unattached polymer component of the modifier—must be tightly controlled to achieve the desired optical properties. Use of soluble polymeric components in PVC to adjust its refractive index to compensate for light scattering has been proposed [92]. The yellowness index as measured by ASTM D1925 also must meet critical customer requirements and usually must be less than 3.0. A negative value indicates a bluish tint.

A phenomenon which is directly attributed to the light scattering caused by the differential in refractive index between dispersed modifier and PVC continuum is that of color reversal [93]. This describes the observation of yellow color with transmitted white light on a white background and blue color when scattered light is viewed against a black background. This optical reversal effect can cause ingredients in a bottle to have undesirable oily-appearing hues. Figure 43 shows how color reversal changes as the composition of the impact modifier and its refractive index change.

The blueness of the scattered light increases as the refractive index of the modifier is increased above that of the PVC continuous phase. Conversely, the yellowness of scattered light increases as the refractive index of the dispersed modifier phase is lowered below the PVC. A similar effect is observed as temperature is changed for a given impact-modified PVC formulation where the refractive indices are closely matched at room temper-

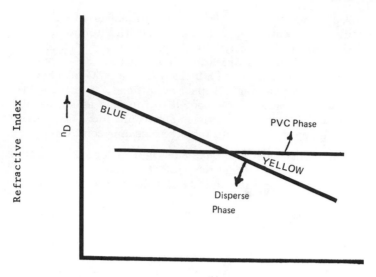

Modifier Composition

FIGURE 43 Effect of changing impact modifier composition and refractive index on color reversal (color of scattered light).

ature. The haze and blueness of scattered light increase as temperature is decreased below room temperature. Haze also increases above room temperature, but the color is observed to become yellow.

4. *Gel Particles/Dispersion*: Another important aspect of the optical quality of impact-modified PVC is the concentration of residual "gel" particles and other contaminants. The effect of the impact modifier on PVC fluxing is therefore critical to the achievement of a homogeneous fluxed state. Acceleration of the PVC fusion rate is generally obtained by incorporating rubbery particulate modifiers [94]. This is not the case with CPE or EVA, which tend to delay PVC fusion [90]. Processing conditions must be adjusted to recognize these changes and obtain a low gel count. The dispersion of the modifier is critical for optimizing optical and physical properties. Different degrees of modifier particle aggregation have been obtained by varying graft structures, which have given different impact-optical properties [95]. Other sources of optical defects in clear PVC film or bottles are typically fiber particles and assorted particulates. Any tendency toward thermal degradation during melt processing, particularly in extrusion, can cause defects due to small degraded PVC particles as well as discoloration.

5. *Melt Properties*: Adding impact modifier to PVC significantly affects the flux and melt elasticity of the PVC. Melt viscosity of fully fused PVC is not changed appreciably by the low weight and volume percents of modifier used.

Table 6 contains Brabender torque and flux time data for PVC modified with 15 wt % (20 vol %) ABS high-impact modifier [94]. The major effect of incorporating modifier into the dry blend which did not contain processing aid was to significantly reduce the time to flux of the PVC. Torque values

TABLE 6 Torque Rheometer Characteristics of Powder Blends

Material[a]	Time to peak torque (sec)	Torque (m-g)		
		Peak	Peak + 5 min	Peak + 15 min
PVC (K-69)	480	2300	2300	2300
85% K69 PVC/15% ABS-1	17	3650	2800	2620
85% K69 PVC/15% ABS-2	10	3600	2100	2000
PVC (K-57)	160	2100	1800	1700
85% K57 PVC/15% ABS-1	20	3400	1900	1750
15% K57 PVC/15% ABS-2	13	3500	2100	1875

[a]ABS-1 and ABS-2 are high-rubber graft copolymers with the same diene content but with differences in grafting density and molecular weight.

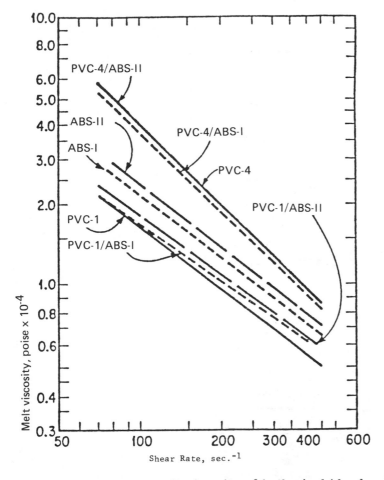

FIGURE 44 Apparent melt viscosity of both vinyl blends and ABS modifiers: logarithmic dependence on apparent shear rate. The blends are intermediate. (From Ref. 94.)

were generally slightly higher for the modified PVC. Melt viscosity of the
modified PVC resins was found to be approximately the average of the melt
viscosities of the blend components on a volume fraction basis:

$$\eta = \eta_1 V_1 + \eta_2 V_2$$

Figure 44 shows the effect of 15 wt % of two different ABS impact modifiers
on PVC melt viscosity.

   Figure 45 shows that the extrusion die swell, S, of PVC was reduced by
ABS modifier in proportion to the die swell, S_n, and volume fraction, V_n, of
the blend components. Table 7 contains experimental and calculated die
swell values for modified PVC according to the simple relationship

$$S = S_1 V_1 + S_2 V_2$$

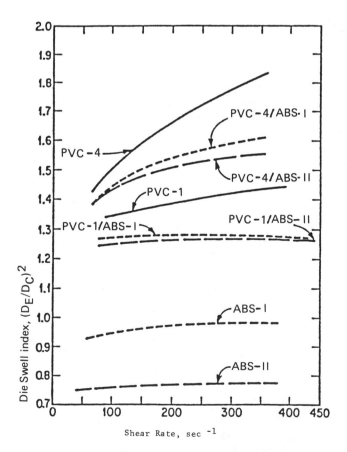

FIGURE 45 Die swell index dependence on shear rate for the modified vinyl
compounds as well as the ABS resins themselves. (From Ref. 94.)

TABLE 7 Extrudate Swell and Surface Roughness

Material	Shear rate (sec^{-1})	Shear stress (psi)	RMS roughness (10^{-6} in.)	Die swell index		
				Expt.	Calc.	Deviation (%)
K 69 PVC	100	57.42	600	1.50		
	200	57.81	600	1.56		
	300	58.98	900	1.72		
	400	56.25	900	1.77		
85% K69 PVC/	100	54.69	600	1.46	1.40	-3.8
15% ABS-1	200	55.47	600	1.54	1.52	-0.8
	300	54.30	600	1.59	1.62	+1.9
	400	54.20	600	1.62	1.68	+4.3
85% K69 PVC/	100	56.64	550	1.45	1.36	-5.9
15% ABS-2	200	57.81	500	1.51	1.48	-1.5
	300	55.47	550	1.56	1.58	+2.3
	400	55.31	600	1.57	1.64	+5.0
K-57 PVC	100	25.78	35	1.35		
	200	29.70	60	1.38		
	300	30.47	70	1.42		
	400	32.44	80	1.46		
K85% K57 PVC/	100	27.34	35	1.27	1.26	-0.8
15% ABS-1	200	29.70	45	1.28	1.30	+1.5
	300	31.65	60	1.28	1.33	+3.8
	400	33.59	55	1.28	1.35	+5.6
85% K57 PVC/	100	25.39	35	1.25	1.23	-2.1
15% ABS-2	200	34.38	55	1.27	1.26	-0.6
	300	35.16	85	1.27	1.29	+1.3
	400	34.77	115	1.29	1.31	+3.1

Similar results have been obtained for MBS and acrylic impact-modified PVC melts. Casale et al. [96] have also described the melt viscosity and die swell of ABS polymers in terms of the effective volume concentration of grafted butadiene rubber particles.

F. Types of Impact Modifier

1. Methacrylate-Butadiene-Styrene Polymers: MBS modifiers are graft polymers prepared by polymerizing methyl methacrylate or mixtures of methyl methacrylate with other monomers in the presence of polybutadiene or polybutadiene-styrene rubber [97]. These polymers are blended and melt-mixed with PVC resins to improve impact toughness, which is usually measured by notched Izod impact or falling dart impact testing. However, MBS modifiers also accelerate PVC resin fluxing and generally facilitate PVC processing.

MBS impact modifiers are available from a number of producers in the United States and Japan (see Appendix). They are also produced in Europe by the same primary manufacturers. MBS modifiers must contain approved antioxidant systems and be very low in residual monomer to be suitable for food grade applications. Sequenced grafting gives very high monomer conversion, and the isolation and drying processes are designed to give very low residual monomer.

SBR latex particle size is controlled by different processes. One approach is to develop larger particles by chemical microagglomeration of a precursor SBR emulsion with a smaller, narrow particle size distribution [98]. In this way, highly controlled particle size distributions can be achieved for balancing optical properties with high impact efficiency. Another approach is direct growth of rubber emulsion particles with a larger, broader particle size distribution.

The smaller particle size MBS modifier is used for clear PVC bottles and film with high clarity and sparkle. The larger particle size MBS is used for high-impact applications with good clarity control, such as blow-molded bottles with high drop height impact performance. The use of cross-linker in both the rubber latex and grafting stage [99] and multistaging the rubber latex using cross-linker in both the rubber latex and grafting stage [100] have been reported to achieve low crease whitening with good impact.

Cross-linked MBS modifiers with rigid core centers within the soft rubber particles have also been described [101, 102]. These modifiers offer very low stress whitening with high impact efficiency and good retention of mechanical properties. Lower rubber content for high impact efficiency is also an advantage in recovering and drying the modifier resin.

Dry Blending/Compounding: MBS modifiers are usually incorporated in PVC with a high-intensity powder blender. The order of addition of compound ingredients and the temperature at which each ingredient is added are important. They vary to some degree depending on the specific modifier being used. A typical blending procedure is shown in Table 8.

Melt Mixing/Compounding: Preblending is frequently followed by a melt mixing step to produce compounded pellets for subsequent processing into bottles, film or injection molding.

Melt mixing can be performed on a mill to give fluxed homogenized sheet for subsequent dicing. A continuous melt mixer can also be used to deliver melt to an extruder for strand pelletizing.

TABLE 8 Blending Procedure for PVC Impact
Modifier Mixtures

Component	Temperature when added (°F)
PVC	Room temperature
Stabilizer	180
Modifier	190
Lubricant	200
Mix to	200 to 240
Cool to	100

Typically the mill temperature is about 350°F and a minimum of 3 minutes milling after flux is used to achieve optimum dispersion of modifier and other ingredients.

Processing: PVC compounds containing MBS impact modifier can be extruded successfully from either powder blend or pellets. Extrusion blow molding of powder blend to produce clear PVC bottles is predominantly practiced in Europe. Extrusion stock temperatures should range from 370° to 390°F, depending on the stabilization system and melt rheology. Pelletized compound is generally used for extrusion blow molding, film extrusion, and injection molding in the United States. Calendering of PVC compounds modified with MBS is usually accomplished by fluxing the preblend, mill mixing, and then feeding the milled stock to the calender. Calendering temperatures usually vary from 330° to 360°F, depending on the formulation used.

MBS impact modifiers generally have little effect on the power-law capillary extrusion melt viscosity of PVC compounds. However, they have a major effect on the flux time, acting similarly to process aids in significantly accelerating the PVC fluxing process. They also contribute to the hot strength of PVC melts by increasing the degree of molecular chain entanglement. Processing conditions must be adjusted to accommodate the shorter flux times and obtain a gel-free, well-homogenized product.

Significant reduction in gel colonies in MBS-modified PVC has been achieved by blending the MBS with poly(ethyl acrylate). PVC containing 3% of a blend of MBS and poly(ethyl acrylate) had 93% fewer gels than PVC with the unblended MBS modifier [103].

Optical Properties: For clear bottles and film, excellent clarity, resistance to crease whitening, and resistance to water blushing are required. MBS impact modifiers are generally preferred for superior balance of optical properties with less sensitivity to processing conditions. Typically, with 12 phr MBS modifier in a general-purpose PVC master batch suitable for film or bottle, the optical properties shown in Table 9 are obtained.

Impact: The effect of incorporating MBS impact modifier into PVC resin of different molecular weights is shown in Figure 46. Increasing PVC molecular weight increases notched Izod impact over the entire modifier concentration range [104].

TABLE 9 Optical Properties of 12 phr MBS Compound[a]

Light transmission (%) ASTM D1003	87 to 89
Haze (%) ASTM D1003	3 to 4
Yellowness index ASTM D1925	3 to 5
Stress whitening	Very slight to definite
Color reversal	None to distinct

[a]The optical properties are dependent on the total formulation used and the processing history.

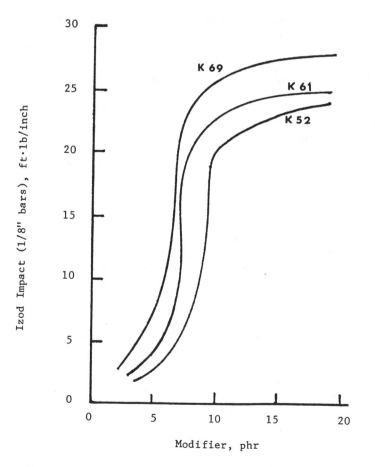

FIGURE 46 Effect of PVC molecular weight and impact modifier concentration on impact strength at 23°C. (From Ref. 104.)

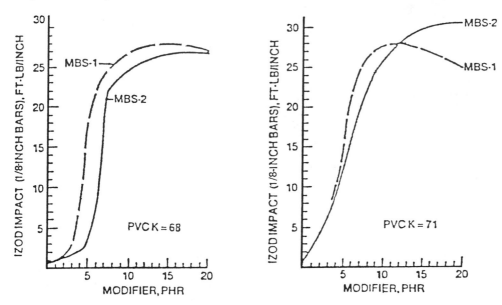

FIGURE 47 Impact efficiency vs. effectiveness at 23°C. (From Ref. 104.)

The impact improvement in PVC with two different MBS modifiers is illustrated in Figure 47. The curves differentiate between impact efficiency (concentration effect) and impact effectiveness (level of impact at a given concentration of modifier). MBS-1 is the more efficient and effective in this particular K68 formulation, but MBS-2 becomes more effective than MBS-1 at higher concentrations in the particular K71 PVC formulation. Figure 48 compares the impact-modifying performance of the range of available MBS modifiers in a general-purpose K57 PVC master batch. The 1/4 in. Izod values show the transition from ductile to brittle behavior in the typical modifier usage range of 9 to 15 phr. A recent study using the instrumented Dynatup Tester to vary strain has shown different impact toughness behavior for MBS impact modifiers in a K57 compound at fixed modifier concentration. Figure 30 compares modifier toughness and defines the ductile-brittle transition rates for the different modifiers.

As can be seen from the results in Figure 49, prolonged milling has little effect on the impact strength of MBS-modified PVC. Compared to ABS, they generally exhibit improved heat stability and clarity as well as room temperature impact efficiency [104].

Chemical Resistance: MBS-modified PVC stands up well to chemical exposure, as shown by the results in Table 10. Exposing MBS-modified PVC to a strong acid or base causes only minor changes in impact.

Free-Flow, Low-Dust MBS: Recently, there has been an emphasis on improving the resin powder flow and bulk handling behavior of MBS impact modifiers. MBS grades with improved powder flow have been introduced in Japan and Europe. The resin flow is usually measured by a funnel flow test. The improved powder flow resins have higher bulk density with less dust, fewer fines, and less tendency to cake when stacked in a bag. The lower

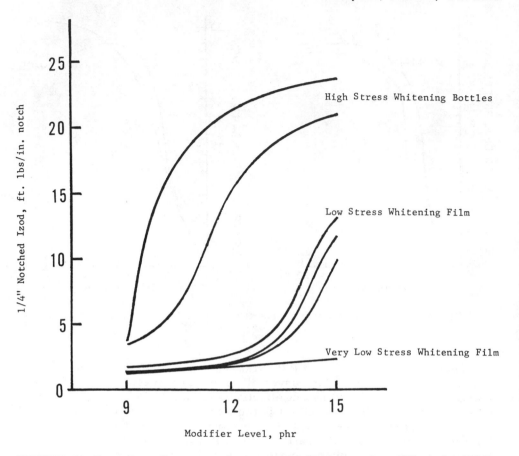

FIGURE 48 Impact performance of commercial MBS impact modifiers for PVC film and sheet in a representative packaging PVC formulation (room temperature and 10 mil radius notch).

TABLE 10 Effect of Chemical Exposure on Impact Strength

	Treatment			
	None	10% NaOH[a]	80% H_2SO_4 [a]	Hexane[b]
Izod impact strength				
1/8 in. bars	25.8	21.6	25.0	26.0
(ft-lb/in. notch)				

[a]Two weeks at 90°C.
[b]Two weeks at 80°C.

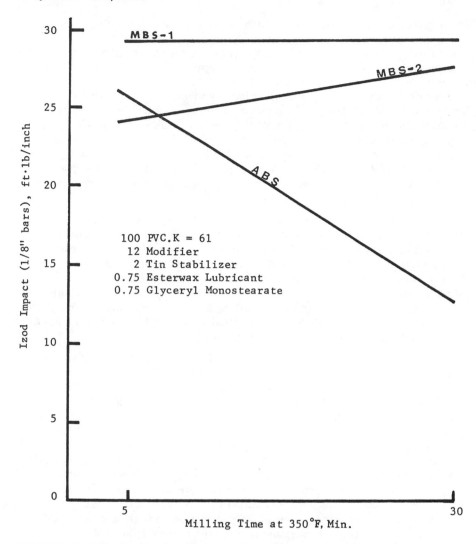

FIGURE 49 Effect of processing on impact strength. (From Ref. 104.)

dust level will reduce the explosion hazard from fine dust exposed to an ignition source. Some improvements in PVC compounding rate may also be possible. The improved flow versions are expected to offer advantages in bulk or large bag handling, pneumatic resin transfer, and automatic weighing.

The improvement in MBS powder flow can be achieved by using a spray-drying resin recovery process from emulsion [105]. A process for free-flow MBS has also been described which utilizes an additional emulsion polymer component coprecipitated with conventional MBS [106].

2. Acrylonitrile-Butadiene-Styrene (ABS) Polymers: A range of ABS resins are available in the form of free-flowing white powder for blending and compounding with PVC (see Appendix). Some grades of ABS are de-

signed to give clear PVC compounds with improved toughness, rigidity
retention, and good chemical resistance. Other grades give opaque PVC
compounds with improved to excellent toughness and improved processing.
ABS resins facilitate fluxing of PVC resins, improve hot melt strength, and
reduce melt elastic strain recovery or die swell [94].

ABS graft copolymers are prepared by chemically grafting styrene and
acrylonitrile monomers onto polybutadiene or styrene-butadiene rubber substrates in emulsion. The diene emulsion is prepared as a precursor to the
grafting polymerization. The composition, particle size, and particle size
distribution of the rubber emulsion are carefully controlled to ensure consistent impact-modifying performance and optical properties of the grafted
rubber modifiers in PVC. Other factors controlled in the final dried resin
are residual monomer and moisture content.

Powder Blend Procedure: Powder blending of the modifier, PVC resin,
and additives is similar to MBS modifier blending. Blending is usually performed in a high-intensity mixer/bowl cooler or ribbon blender system. Stabilizer is added to the PVC resin and blended at maximum speed for approximately 1 minute up to 125°F. ABS modifier is added and blended for an
additional 1 minute up to 180°F. Other ingredients (pigments and fillers)
are then introduced; lubricants are added at 190° to 200°F and blending continues up to 225°F; the blend is then dropped from the blender and cooled
to 112°F for storage or further processing.

Melt Mixing Procedure: The powder blend may be further heated and
fluxed in an intensive mixer until a temperature of 290° to 320°F is reached.
The melt is held for 2 to 4 minutes at a minimum temperature of 330°F. The
blend is transferred to a mill before a temperature of 380°F is reached. The
stock is milled for a further 2 to 4 minutes and then diced to give the final
pellet form. For some processes, such as calendering, the melt is transferred directly to the calender rolls to give the intermediate sheet or film
for subsequent thermoforming. The viscosity and elasticity of the modified
PVC during calendering are important for output rate and energy requirements. ABS modifiers accelerate PVC resin fluxing and reduce die swell
or strain recovery during melt processing while generally increasing the
hot melt strength. The appearance of ABS-modified PVC is smoother and
more cohesive during melt processing than that of unmodified PVC.

Extrusion of powder blend, usually to produce pipe or profile, generally
requires temperatures that are slightly higher than that for unmodified PVC.
This is attributed primarily to faster fluxing of the blend. Extruders
should be operated under conditions to achieve 350° to 380°F, depending
on the molecular weight of the PVC resin. Both single- and twin-screw
extrusion processes are used.

Types of ABS Modifier: ABS modifier resins are generally classified
according to the tensile modulus of the modifiers themselves [107]. This
roughly corresponds to rubber level in the modifier:

1. Polymers with a high tensile modulus (3×10^5 psi) and the lowest
rubber level are used primarily as processing aids for opaque applications
with some impact modification and hot strength improvement; these modifiers
have the least effect on the physical properties of rigid PVC.

2. ABS polymers with intermediate tensile modulus (2×10^5 to 3×10^5
psi) and rubber content are used for opaque PVC applications where hot

TABLE 11 Physical Properties of ABS/PVC Blends[a]

Property	High-modulus ABS	Intermediate-modulus ABS	Low-modulus ABS
Hardness (Rockwell R)	112	110	108
Tensile strength (psi)	7,300	7,100	6,600
Tensile modulus (psi)	390,000	370,000	370,000
Flexural yield strength (psi)	13,600	12,000	11,200
Flexural modulus (psi)	460,000	430,000	375,000

[a]Approximate K value of PVC, 69; ABS modifier concentration, 20%.

melt strength and embossing retention are essential and plastic flow and impact performance are required.

3. Polymers with a low tensile modulus (2×10^5 psi) and the highest rubber content are the most efficient as impact modifiers but are also very important in melt processing. It is noteworthy that modifiers for opaque and clear PVC fall in this grouping and applications include extrusion, calendering, and injection molding.

General Properties of ABS Modified PVC: Typical physical properties of 20% ABS-modified PVC shown in Table 11 demonstrate that the mechanical strength and modulus of PVC are improved as the tensile modulus of the modifier is increased.

Impact Behavior: Optimum impact is achieved by selecting the highest-molecular-weight PVC which can be consistently fluxed with an ABS modifier. For a PVC of a given molecular weight or K value, the impact strength will increase with ABS modifier concentration up through a maximum and then decrease. The modifier impact efficiency and level of toughness depend on the class of ABS impact modifier used. Figure 50 describes these toughening effects for the three classes of ABS and for PVC resins of three different molecular weights. Impact modifiers for opaque and transparent PVC applications are included in the low-modulus, high-impact modifier efficiency class.

The effect of styrene-acrylonitrile ratio on ABS impact modifier efficiency in PVC has been given an exhaustive fracture mechanics treatment [108, 109]. More than one maximum in impact was found for some ABS compositions. High-speed tensile and puncture measurements have confirmed the synergistic effect of modifying PVC with low-modulus ABS impact modifier [41, 110]. Figure 51 shows the improvement of PVC (K57) tensile fracture energy at high impact rates of deformation by incorporating high-rubber ABS impact modifier. Figure 52 shows the synergistic effect of ABS modifier on notched Izod and tensile fracture energy for PVC resins of three different molecular weights.

The effectiveness of a high-rubber ABS impact modifier in raising the brittleness strain rate of a low-molecular-weight injection molding PVC at

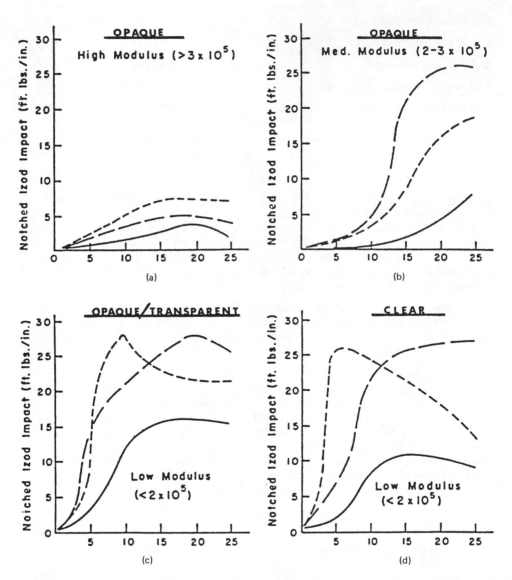

FIGURE 50 Impact efficiency of various ABS polymers in PVC. (From
Ref. 111.)

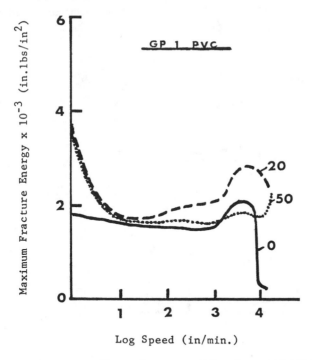

FIGURE 51 Effect of ABS modifier % on K57 PVC tensile fracture energy at test speeds up to 12,000 in./min. (From Ref. 41.)

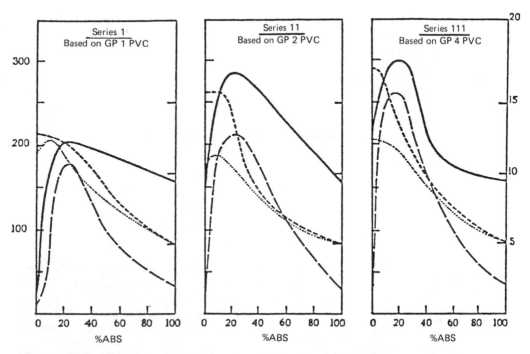

FIGURE 52 Comparison between (− −) 10 × Izod impact energy, (---) 12,000 in./min tensile yield stress, (—) 12,000 in./min tensile fracture energy, and (····) 12,000 in./min apparent shear strength, as a function of PVC/ABS blend composition. (From Ref. 110.)

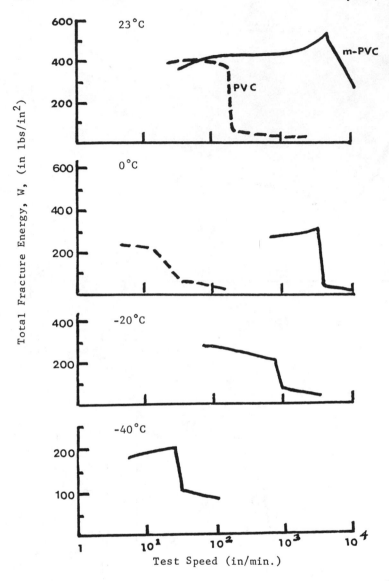

FIGURE 53 Total fracture energy of 14 phr ABS impact-modified PVC (——)
and unmodified PVC (- - -) from flexural notched bend testing. (From
Ref. 54.)

low use temperatures down to -40°C is demonstrated in Figure 53. Here,
three-point bend testing on notched Izod specimens was used to measure
the ductile-to-brittle transition strain rate for unmodified PVC and PVC
modified with increasing levels of ABS impact modifier [54]. This ductility
improvement down to very low use temperatures can be realized while maintaining the good melt flow and thermal stability during injection molding
essential for acceptable product performance. The data in Table 12 show

TABLE 12 Test Speeds at Which Ductile-Brittle Transitions Occur at Various Test Temperatures (in./min).

Material	23°C	0°C	-20°C	-40°C
PVC	180	20	2 to 5	0.5 to 1.0
5 phr modifier	2,000	95	40	9.5
10 phr modifier	6,000	2,000	100	8 to 16
14 phr modifier	10,000	3,000	800	25

the impact improvement realized by incorporating high-efficiency ABS modifier in a K55 PVC formulation for injection molding.

Chemical Resistance: ABS-modified PVC retains the chemical resistance characteristics typical of ABS resins. It is resistant to attack by aqueous solutions of inorganic salts, alkalis, and most mineral acids. Certain solvents, alcohols, and oils have little effect on ABS-modified PVC. Strong oxidizing agents such as concentrated sulfuric acid or chlorine solution will attack ABS-modified PVC. Glacial acetic acid, chlorinated hydrocarbons, and aromatic hydrocarbons have a considerable swelling action on these polymers. Table 13 shows properties of 15% ABS-modified PVC after 7 days immersion in some critical liquid reagents [111].

3. Acrylic Impact Modifiers: All-acrylic and modified acrylic impact modifiers are used primarily for opaque outdoor PVC applications requiring good retention of toughness, color, and appearance (see Appendix).

The all-acrylics have no unsaturation and therefore are least vulnerable to oxidation. They have been the principal impact modifiers used in high-molecular-weight PVC for weatherable applications such as window profile, house siding, gutters, downspouts, and other related PVC items. Extensive outdoor weathering tests and in-use behavior have confirmed their performance in stabilized and pigmented PVC [112−114].

TABLE 13 Chemical Resistance of ABS-Modified PVC[a]

	Unmodified PVC compound	High-modulus ABS	Intermediate-modulus ABS	Low-modulus ABS
30% H_2SO_4	0.03	0.21	0.04	0.06 to 0.28
10% NaOH	0.14	0.29	0.10	0.18 to 0.29
Ethyl alcohol	0.05	0.10	0.07	0.10 to 0.13
Benzene	51.9	61.1	63.8	64.5 to 67.1
Pentane	0.01	0.01	0.01	0.01

[a]Approximate K value, 69; wt % change after 7 days immersion.

Most acrylic modifiers are grafts of methyl methacrylate onto poly(alkyl acrylate) substrates such as poly(butyl- or 2-ethylhexyl acrylate). Core-shell acrylic modifiers have been described [115]. Also, acrylic impact modifiers combining impact improvement with processing and functionality are now available.

Modified acrylics are designed to offer improved low-temperature impact efficiency with good processability [116], which reduces the need for additional processing aids. These usually are based on an alkyl acrylate-butadiene copolymer with methyl methacrylate grafts [117—121].

Blending with PVC: Powder blending of acrylic and modified acrylic modifier with PVC resin is generally carried out in a high-speed mixer. The recommended order of addition of ingredients and temperatures at which the ingredients are added are similar to those generally used for MBS modifiers.

Melt Mixing: The same general conditions are used for compounding and pelletizing as for MBS modifiers. However, acrylic-modified PVC is widely processed direct from the powder blend into siding or profile.

Impact Strength: The impact improvement of high-molecular-weight opaque PVC is highly dependent on the particulate structure of the PVC [91]. The PVC particulate structure is sensitive to the extrusion processing temperature and to the fluxing efficiency of the processing aid in combination with the impact modifier [90].

Acrylic impact modifiers are used at levels down to 6 phr PVC resin to provide ductility at use temperatures. Up to 4 phr, processing aid is used in the PVC formulation to achieve satisfactory processing and finished part quality. At extrusion melt temperatures below 200°C, some particulate structure will remain. Very low die swell is obtained under these conditions. For notched Izod impact strength a preferred microstructure is obtained when partial breakdown of the PVC primary particle has been achieved. Falling dart impact energy increases to a maximum when destruction of the primary particles is nearly complete. Both single- and twin-screw extruders are used to extrude PVC powder blends for weatherable applications. Formulations and extrusion conditions must be optimized to achieve the desired impact and part quality.

Conventional notched Izod impact efficiency and low-temperature ductility of fully fused acrylic modifier in PVC (K62) are shown in Table 14. Pro-

TABLE 14 Impact Strength Efficiency of Acrylic Modifier in PVC (K=62)

Testing temperature (°C)	Izod impact strength of PVC containing acrylic impact modifier at		
	10%	7.5%	5%
23	23.0	20.7	14.4
16	10.1		
10	9.4		
0	2.1		

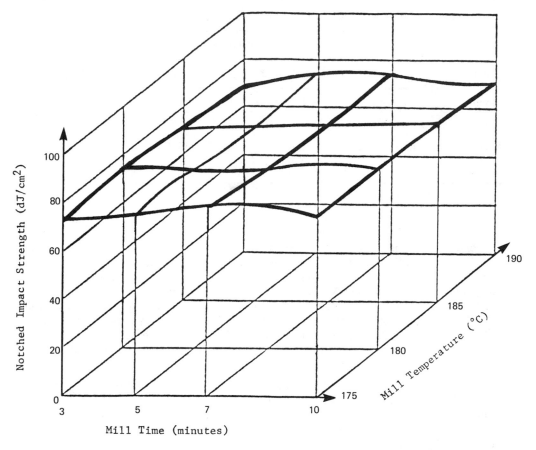

FIGURE 54 PVC modified with 8 phr of an acrylic impact modifier. (From Ref. 114.)

longed milling has little effect on impact of acrylic-modified PVC. EVA-modified PVC, EVA-grafted PVC, and CPE-modified PVC show more sensitivity to both the milling temperature and time [114]. Figure 54 shows the insensitivity of acrylic impact-modified PVC to milling temperature and time.

Weathering performance: The results in Table 15 show the retention of notched Izod impact strength and absence of color shift for a fully fluxed acrylic-modified PVC after outdoor exposure in Florida and Arizona. No measurable change in impact or color occurred after 5 years of exposure in Florida. Only moderate loss of impact occurred after 3 to 5 years of exposure in Arizona.

4. *Chlorinated Polyethylene Polymers*: Chlorinated polyethylene polymers are prepared by chlorinating high-density polyethylene. The properties of CPE polymers and CPE-modified PVC depend on molecular weight, degree of crystallinity, chlorine content, and distribution of chlorine atoms in the polymer chain [122, 123]. CPE with 30 to 40 wt % chlorine exhibits the low glass transition temperature of HDPE and low crystallinity. The level of compatibility with PVC is controlled largely by the chlorine content

TABLE 15 Effect of Outdoor Weathering on Acrylic-Modified PVC (85/15)

Location	Property	Years of exposure							
		0	1/4	1/2	1	2	3	4	5
Florida	Izod impact	12.6	16.2	14.1	11.9	11.1	12.8	14.1	12.5
	Color[a]	W	W	W	W	W	W	W	W
Arizona	Izod impact	15.5	19.8	16.8	15.4	11.7	6.6	8.3	9.7
	Color	W	W	W	W	SY	SY	W	W

[a]W, white; SY, slightly yellow.

and distribution of chlorine atoms on the polyethylene backbone [123]. Polymers containing 25 to 40% chlorine with a random distribution of chlorine atoms are the best impact modifiers for PVC; they have sufficient compatibility for adhesion to the PVC matrix without losing particulates and dispersibility. Polymers containing less than about 25% chlorine are incompatible with PVC and do not confer any property enhancement. Polymers with more than about 48% are highly compatible with PVC and become solubilized, acting as plasticizers. See the Appendix for available commercial grades, recommended uses, and special features.

Blending Procedure: A recommended room temperature blending procedure for achieving good dispersion of CPE modifiers in PVC is shown in Table 16.

Blending can also be performed in a high-speed mixer. Care must be taken to minimize heat buildup and avoid "caking" of the modifier. Low-speed operation of the mixer and a minimum cycle time should be used. Table 17 shows a typical high-speed mixing procedure.

Melt Mixing: Conditions similar to those used with MBS- and acrylic-modified PVC are used. Generally, CPE-modified PVC is extruded from a powder blend.

Impact Modification of Rigid PVC: Increasing the chlorine content of CPE from 25 to 36% improves the impact strength of CPE-modified PVC [124],

TABLE 16 Procedure for Blending at Room Temperature

Materials and procedure	Time (min) when added
PVC polymer and dry materials	At start
Lubricants	5
Liquid plasticizers (if required)	7
Fillers and other additives	12 to 15
Mix for total time of	15 to 25

TABLE 17 Procedure for High-Speed Mixing

Materials and procedure	Time (min) when added	Blend temperature (°F)
PVC resin, dry stabilizers, and lubricants	0.5	—
Continue blending; add liquids	1 to 2	100
Add all other solids	1 to 2	125

but with some sacrifice in processability. Lower impact strength but better clarity is obtained with a 42% chlorine level in CPE. Most rigid PVC compounds are modified with CPE containing about 36% chlorine because of the preferred balance of impact, processing, and dispersibility obtained.

The impact efficiency of CPE polymers has been compared with that of acrylic, MBS, ABS, and EVA modifiers [125]. Figure 55 shows the notched

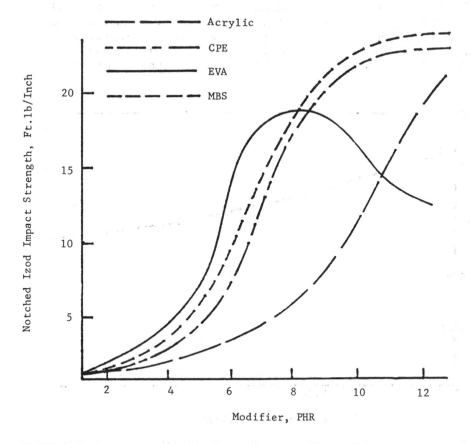

FIGURE 55 Izod impact strength of impact-modified PVC. (From Ref. 125.)

TABLE 18 Typical Physical Properties of CPE-Modified Rigid PVC

Property	Unmodified PVC	12% CPE-modified PVC
Tensile strength (psi)	8,200	6,500
Tensile modulus (psi)	430,000	330,000
Elongation (%)	30	60
Flexural strength (psi)	13,900	11,000
Flexural modulus (psi)	470,000	360,000
DTUL, 264 psi (°C)	72	72

Izod impact efficiency of a CPE modifier in a medium-molecular-weight PVC. Low-temperature ductility is also improved.

General Properties: Addition of an amorphous CPE with 40% chlorine to a high-molecular-weight PVC gives the typical physical properties shown in Table 18.

Outdoor Weathering: Color and impact are retained after prolonged outdoor exposure when a properly formulated blend is used. Such a formulation requires careful choice of stabilizer and pigment system. Accelerated weathering performance of a pigmented PVC compound containing 20 phr of CPE with 36% chlorine and an ultraviolet light stabilizer is shown in Figure 56.

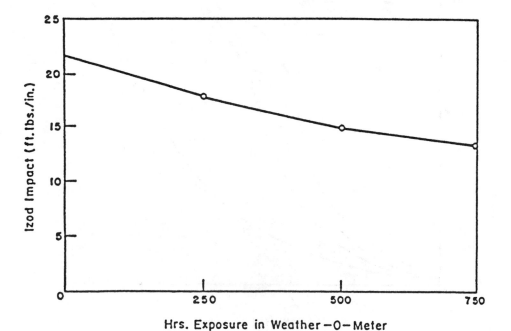

FIGURE 56 Accelerated weathering of CPE-modified PVC, 20 phr CPE. (From Ref. 112.)

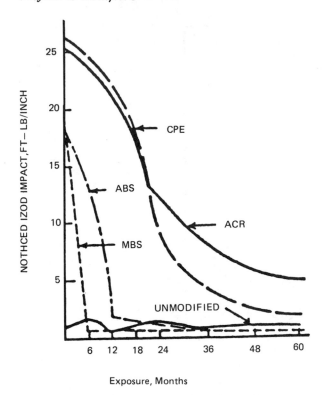

FIGURE 57 Weatherability of white PVC. (From Ref. 104.)

Outdoor weathering of CPE has been compared with that of acrylic, MBS, and ABS and shown to be comparable with all-acrylic modifiers [104]. Both MBS- and ABS-modified PVC contain unsaturated butadiene rubber and undergo major loss of impact and suffer severe color development during outdoor aging. Figure 57 compares the weatherability of white PVC containing different modifiers on exposure in Florida at 45°.

Transparent blends of CPE and suspension PVC have been described [126] which have good weathering, toughness, and blocking resistance.

5. *Miscellaneous Modifiers*: Other rubbery polymers have been used to impact-modify PVC and improve PVC processing. These modifiers are not widely used commercially in the United States, but some find broader usage in Europe.

Ethylene Copolymers:

Ethylene-vinyl acetate: PVC containing 5 to 15% EVA copolymer dispersed as microdomains exhibits good impact strength synergism [127]. These modifiers are ethylene-vinyl acetate copolymers with 30 to 50 wt % vinyl acetate. Impact strength and efficiency increase with vinyl acetate content, as do elongation and thermal stability. Modulus, heat distortion temperature, low-temperature flexibility, and melt flow are reduced. The most impact-efficient modifiers fall in the 40 to 50% range. Melt flow and thermal stability improve on EVA addition, but modulus, strength, and heat

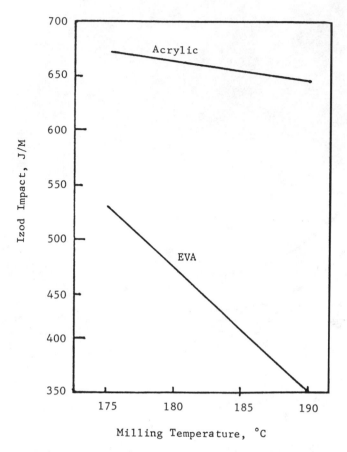

FIGURE 58 The effect of milling temperatures on impact strength of PVC modified with acrylic or EVA impact modifiers. (Ref. 176.)

deflection temperature decrease. When EVA reaches about 25 wt % it forms a second continuous phase interpenetrating the PVC matrix and acts as a rubbery plasticizer. Menges et al. [89] have shown that the maximum notched Izod impact occurs when the PVC grain structure is broken down into primary particles, with EVA forming a rubbery network structure around the primary particles which acts to blunt crack initiation and growth. A sharp drop-off in Izod impact occurs after this optimum as the EVA undergoes a phase inversion and becomes dispersed particulates as the PVC primary structure is destroyed. EVA and CPE are therefore very similar in their network formation around PVC primary particles to give optimum notched Izod impact. Figure 58 shows the sensitivity of EVA-modified PVC to milling temperature compared to acrylic modifier.

EVA-modified PVC offers good heat and light stability, improved processing, weatherability approaching that of acrylics, chemical resistance, and good low-temperature impact. Compared to CPE, EVA resins have lower specific gravity but slightly less chemical resistance. Alloys of EVA and PVC are opaque and have a lower flexural modulus than CPE-modified PVC because of EVA's strong plasticizing effect. EVA modifiers have been used in Europe for a number of years as impact modifiers for opaque PVC for outdoor applications such as window profiles.

EVA modifiers were originally supplied only in pellet form, which created difficulties in dry-blending and processing in PVC powder formulations. Powder forms have become available, but they tend to agglomerate. Suppliers are working to resolve these problems. See the Appendix for available grades and uses.

EVA terpolymers: Terpolymers of ethylene, vinyl acetate, and carbon monoxide are available for use as weatherable impact modifiers in PVC. Products are offered in both pelletized and ground powder form. These modifiers confer good weatherability in terms of impact and color retention [128].

Ethylene-alkyl acrylate copolymers: Ethylene copolymers containing 10 to 35 wt % alkyl acrylate function as impact modifiers in PVC. Ductile impact values of 20 ft-lb/in. have been reported [129].

Ethylene-carbon monoxide copolymers: These copolymers have been reported to give a threefold increase in PVC impact strength [130].

Ethylene-propylene grafted with acrylate: Ethylene-propylene copolymer has been grafted with methyl methacrylate in a twin-screw extruder, using peroxide initiator. At 7 phr this grafted copolymer gave a notched Izod impact strength of 24 ft-lb/in. in PVC [131].

Methyl methacrylate suspension grafting of EPDM with various initiators gave an efficient impact modifier for PVC. Ductile PVC Izod impact values have been reported. Similar MMA grafting on EPR gave a poor impact modifier [132, 133]. Grafting EPDM with vinyl chloride gave short graft chains which did not allow good adhesion with the PVC and did not function as an impact improver [133]. Higher notched Izod impact values of 10 to 18 ft-lb/in. are claimed for vinyl chloride-EPR copolymer mixtures using lauroyl peroxide for grafting [134].

Chlorinated ethylene-1-butene: PVC compositions modified with chlorinated ethylene-1-butene rubber containing 5-ethylidene-2-norbornene as comonomer have improved toughness at room temperature and low temperature and retain toughness on accelerated weathering [135].

Vinyl Chloride Grafted Polymers and Grafted PVC:

Vinyl chloride grafted impact modifiers: Vinyl chloride has been grafted onto EVA [136], EPR [134], BD [133], acrylic [137], and NBR [138] to provide impact modifiers with improved compactibility with PVC.

Vinyl chloride graft copolymers: In Europe, vinyl chloride copolymerized with elastomers such as acrylic or EVA has been used quite extensively, mostly for extrusion into window profiles and other outdoor applications. Grafting of vinyl chloride onto polyacrylate is claimed to give a polyacrylate-vinyl chloride copolymer which is very stable to processing

temperature and shear [139]. The size and distribution of the elastomeric
acrylate particles are fixed by the copolymerization process and are not
further changed for processing or subsequent fabrication. Polyacrylate-
vinyl chloride graft copolymers are claimed to be processable over a far
wider range of conditions than polyethylene vinyl acetate-vinyl chloride
graft polymers, which achieve a high degree of strength and toughness over
a relatively narrow processing range. Acrylic- and CPE-modified PVC
blends are also reported to have narrower processing windows than polyacrylate-vinyl chloride polymer to give consistent notched Izod impact strength,
and precise dimensional control.

Impregnating porous PVC slurry particles with elastomeric acrylate and
MBS polymers to give a high-impact, weatherable compound has been described [140–142].

Butadiene Copolymers:

Butadiene-acrylonitrile (NBR): Butadiene-acrylonitrile copolymer
nitrile rubbers have been used as modifiers for PVC toughness and processing improvement for many years. Matsuo [143] has described the morphology of different compositions of NBR modifiers after melt mixing with
PVC and how the dynamic mechanical and impact properties are affected.

An important continuing use has been to improve the vacuum forming
characteristics of impact-modified flexible vinyl sheet. Deep draws are
achievable with little thinning.

Addition of NBR to rigid PVC gives a rapid and continuous decrease in
room temperature modulus because of the partial solubility of the rubber in
the PVC matrix. Dispersion of the NBR in the PVC is governed by the
cross-link level of the NBR and the mixing conditions. NBR elastomers are
available as crumb or powdered resins. Specific dry-blending procedures
have been developed for NBR powdered modifiers [144].

The tensile strength, elongation, and impact strength of the PVC-NBR
compounds depend on the molecular weight and composition of the NBR used
[145]. Heat deflection temperature is lowered by NBR addition.

Other butadiene copolymers: Ductile impact performance of PVC has
also been achieved by adding 10 to 15 wt % butadiene-diethyl fumarate.
Notched Izod impact values of 23 to 28 ft-lb/in. have been reported [146].
Similarly, butadiene-dimethyl itaconate copolymer at 10 wt % raises PVC impact by a factor of 5 [147]. Adding only 5 wt % of butadiene copolymers
containing 2-vinylpyridine or 5-vinylpicoline produces PVC impact strengths
in the range 10 to 15 ft-lb/in. [148].

Methyl methacrylate grafting of butadiene rubber is reported to give an
efficient impact modifier for PVC. Ductile PVC impact was achieved with
an optimum monomer-to-rubber ratio of 0.75 [133].

Butyl and Chlorinated Butyl Rubbers: PVC blends containing 10%
isobutylene-isoprene rubber have been reported to have a notched Izod impact value of 14 ft-lb/in. [149]. This system has little sensitivity to the
mechanical work used to achieve satisfactory blending.

Isobutylene-isoprene rubber has been chlorinated to obtain one atom of
chlorine per double bond. Blending this chlorinated copolymer with PVC
gives a transparent compound which has five times the impact strength of
unmodified PVC [150].

Norbornene-Carboxylate: The impact resistance and processability of PVC are claimed to be improved without sacrificing transparency or mechanical properties by blending with 10 phr of a ring-opening polymer of norbornene in which less than 5% of the monomer units contain carboxylic acid groups and the remainder contain carboxylate alkyl ester groups. Impact strength was improved by a factor of 8 and melt flow was increased from 0.06 to 0.1 g/10 min [151].

Polyethylene Glycol or Ethylene Oxide Copolymer: Milling 8 phr of polyethylene glycol or ethylene oxide copolymer containing no pendant epoxy groups with PVC reportedly improves Izod impact strength from 1 to 21 ft-lb/in. Melt flow is also increased [152].

Polyurethane-Urea: Addition of thermoplastic polyurethane-urea elastomer has been reported to increase sharply the notched Izod impact strength of PVC. A maximum impact strength is reported at 7.5% elastomer. A significant decrease in impact strength is observed at higher elastomer loadings [153].

Cross-linked Polyester Modifiers: Solution blending of PVC, acrylic processing aid, stabilizer, lubricant, and 20 phr of cross-linked adipic acid-1,4-dihydroxy-2-butene-ethylene glycol-propylene glycol copolymer rubber gave a PVC formulation with 20.5 ft-lb/in. notched Izod impact strength at 23°C [154].

Polysiloxane Impact Enhancer: Addition of 0.5 part poly(dimethylsiloxane) fluid with a viscosity of 20 centistokes to CPE-modified PVC (10/90) gave compositions which had double the Izod impact and reduced shrinkage during thermal aging [155].

Inorganic Particulate Modifiers:

Precipitated calcium carbonate: Stearic acid-coated, precipitated calcium carbonates of ultrafine particle size (0.1 μm) have been found to reduce notch sensitivity and improve the low-temperature impact resistance [156] and UV stability of rigid PVC [157]. The high surface area of the $CaCO_3$ is believed to increase the capacity to neutralize the HCl responsible for yellowing. At 5 to 10 phr in rigid PVC, long-term impact is retained and minimum yellowing on weathering is observed. Reaction of 0.07 μm $CaCO_3$ with stearic acid has been shown to give improved notched Izod and drop weight impact strength with a range of PVC resin molecular weights. A concentration of 20 to 30 phr of coated 0.07 μm $CaCO_3$ in rigid PVC was comparable to 8 phr of acrylic impact modifier up to 0°C and had a higher modulus.

Notched Izod impact improvement was more pronounced for medium- and high-molecular-weight PVC compounds. The optimum loading was approximately 15% and depended on good melt mixing to achieve good dispersion.

In cases where higher levels of impact performance are required, the filler is used in conjunction with a polymeric impact modifier. Preblending 0.07 μm $CaCO_3$ with ethylene-vinyl acetate copolymer has given 14.1 ft-lb/in. notched Izod impact strength for 5 phr of a 50:50 blend of EVA with $CaCO_3$ [158]. Similar results have been reported for other impact modifier combinations [159].

Calcium carbonate has also been treated with other stearates. Ammonium stearate with dodecylbenzenesulfonic acid has been used as a surface coating.

A 20 phr loading in PVC gave a high Charpy impact strength of 19 kJ/m^2 at 23°C and no break for the unnotched Charpy impact test at -20°C [160]. Palmitic acid has also been used with stearic acid as a surface coating agent [161] with EVA or polyethylene.

An update on the use of $CaCO_3$ in PVC and other plastics can be found in McMurrer [162]. Available commercial products are described with details on type, particle size, surface treatment, price, applications, suppliers, and properties. This subject is discussed further in Chapter 17 of this series.

Fine particle size hydrated alumina: PVC compounds containing 6 to 12 phr Al(OH)$_3$ (particle size 1 μm) and an acrylic impact modifier have improved impact strength and improved weatherability compared to the PVC containing either ingredient alone. Torque rheometer and extrusion tests showed lower motor amperage and torque with increasing Al(OH)$_3$ concentration [163]. A notched Izod impact strength of 22.6 ft-lb/in. was reported for 6 phr Al(OH)$_3$ and 5 phr acrylic impact modifier compared to 14.3 ft-lb/in. when Al(OH)$_3$ was omitted.

III. SPECIALTIES

A. Polymeric Gloss-Reducing Agents

Conventional gloss-reducing agents are inorganic fillers. Most commonly used because of their effectiveness are the various expanded silicas and silicates. Their relatively fine particles of irregular shape protrude through the PVC surface to scatter reflected light, giving the PVC article a low gloss or matte finish. Adding inorganic fillers, especially those used for gloss reduction, has an adverse effect on physical properties of PVC and other notch-sensitive plastics.

There is a unique polymeric gloss-reducing agent currently marketed by only one company. This product Paraloid® KF-710, is an acrylic heteropolymer supplied as a fine granular powder. Like inorganic gloss reducers, it is effective in reducing the gloss of both plasticized and rigid PVC (Figure 59). The similarity ceases here, however, for the polymeric gloss reducer does not markedly impair physical properties (Figure 60).

When the polymer disperses in the PVC melt, it forms a dispersed acrylic phase that is optically and chemically compatible with the melt. As the PVC melt cools, the acrylic-rich surface of the melt shrinks preferentially in a fine, uniform "wrinkle" reminiscent of tung-oil varnish-based wrinkle finishes of yesteryear. Under a microscope (or to the unaided eyes) the surface of the polymeric "flatted" PVC appears very much like that of matte-roll embossed PVC. The surface remains intact, and there are no imperfections to concentrate stresses that result in physical property impairment.

When conventionally flatted PVC is reheated and postformed as in vacuum forming, gloss increase or burnishing frequently presents an unattractive lack of uniformity of appearance. For the reasons given in explaining the mechanism governing gloss reduction by the polymeric flatting agent, reheating and postforming have little or no effect on the gloss of the PVC containing that additive (Figure 61). In using any flatting or gloss-reducing agent, one must remember that a thermoplastic material such as PVC will tend to assume the physical structure of the surface in contact while hot (this is the basis for conventional embossing). Therefore PVC compounds

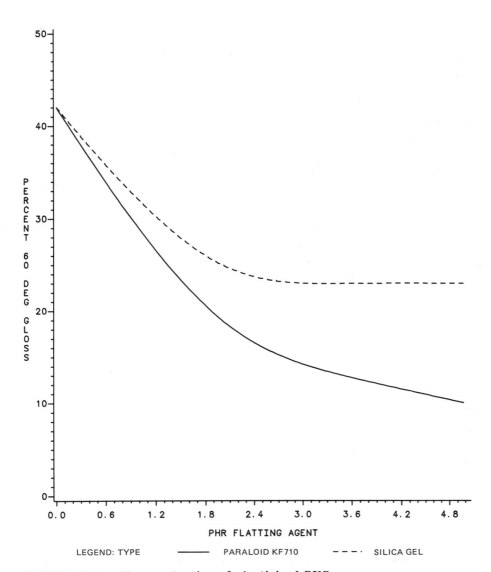

FIGURE 59(a) Gloss reduction of plasticized PVC.

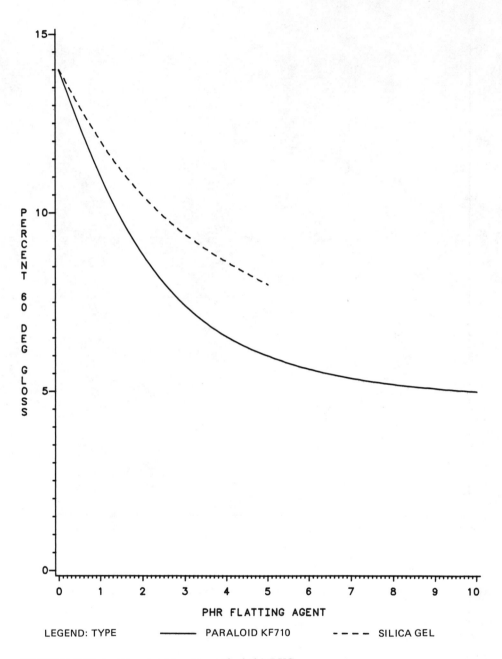

FIGURE 59(b) Gloss reduction of rigid PVC.

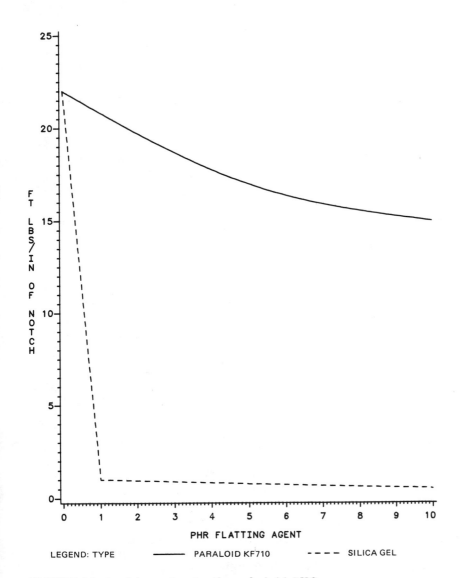

FIGURE 60 Izod impact retention of rigid PVC.

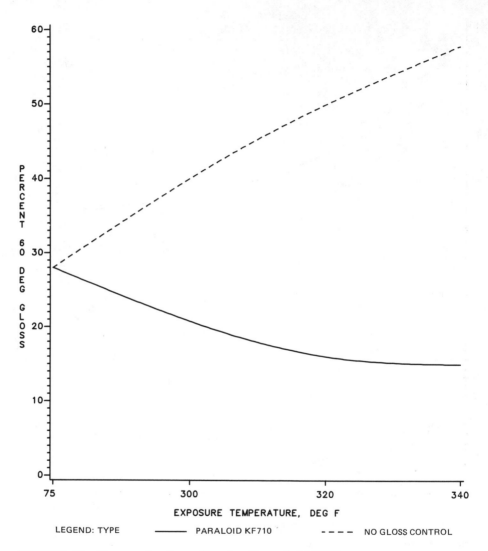

FIGURE 61 Gloss retention after heating rigid PVC.

TABLE 19 Contact Clarity of PVC Films

Plasticizer level	Flatting agent	Haze (%)
0	None	10
	Silica	100
	Acrylic	6
25 phr DOP	None	12
	Silica	100
	Acrylic	7
50 phr DOP	None	6
	Silica	98
	Acrylic	16

containing either inorganic or the polymeric flatting agent will have a glossy surface if they are processed at an elevated temperature in contact with polished metal.

PVC sheet and profiles are successfully made with both inorganic and polymeric flatting agents. Calendered goods frequently are mechanically embossed in addition to having gloss reduced by the flatting agents. The use of a polymeric flatting agent has the advantage of resulting in a nonabrasive composition, therefore extending the life of the embossing roll and minimizing the need for expensive recutting of the rolls. Due to their refractive indices and particle sizes, all inorganic gloss reducers impart opacity to the PVC formulation. Obviously this is no problem in pigmented systems, but there are shortcomings when low-gloss clears are desired. The polymeric gloss reducer does not impair the clarity of clear PVC compounds due to the refractive index and dispersibility of the all-acrylic polymer (Table 19). This unique feature allows the fabricator to second-surface print or decorate a low-gloss film that can be laminated to another substrate without losing definition or color of the pattern.

The formulations in Table 20 are intended to illustrate typical use levels of the polymeric flatting agent. They may require adjustments in flatting agent levels or other ingredients to meet specific production or product demands.

B. Polymeric Additives for Rigid PVC Foam

The acrylic processing aids are effective in promoting uniform fine-cell development in chemically blown PVC foams, both plasticized and rigid.

For many rigid PVC foams, fine cells and low density are only part of the requirements for producing serviceable foam profiles and pipe. Impact strength is impaired as thin cell walls become the structural component in foams. Many foamed products have been made successfully with combinations of acrylic processing aids, impact modifier, and lubricating-processing aids.

TABLE 20 PVC Formulations Containing Acrylic Flatting Agent

	Rigid PVC		Semirigid PVC		Flexible PVC	
		Control		Control		Control
PVC	100 (K=62)	100 (K=62)	100 (K=69)	100 (K=69)	100 (K=69)	100 (K=69)
Plasticizer (DOP)	—	—	25	25	50	50
Acrylic flatting agent	1 to 5	0	2	0	2	0
Acrylic processing aid	2	2	—	—	—	—
MBS impact modifier	12	12	—	—	—	—
Tin stabilizer	2	2	2	2	2	2
Glyceryl monostearate	0.75	0.75	—	—	—	—
Paraffin wax	0.75	—	—	—	—	—
Stearic acid	—	—	0.25	—	0.25	—
Contact clarity (% haze)	6	10	7	11	16	6
Gloss (% at 60°)	11 to 5	14	7	56	12	66
Izod impact (ft-lb/in.)	20 to 17	23	—	—	—	—
Hardness (Shore A, 10 sec)	—	—	88	88	75	70
Tensile strength (psi)	—	—	3570	3570	2800	300
Elongation (%)	—	—	270	290	360	350
Coefficient of friction (ASTM D1894)	—	—	—	—	3.92/2.53	7.63/7.53
Blocking force (lb/in. at 170°F)	—	—	—	—	0.11	0.32

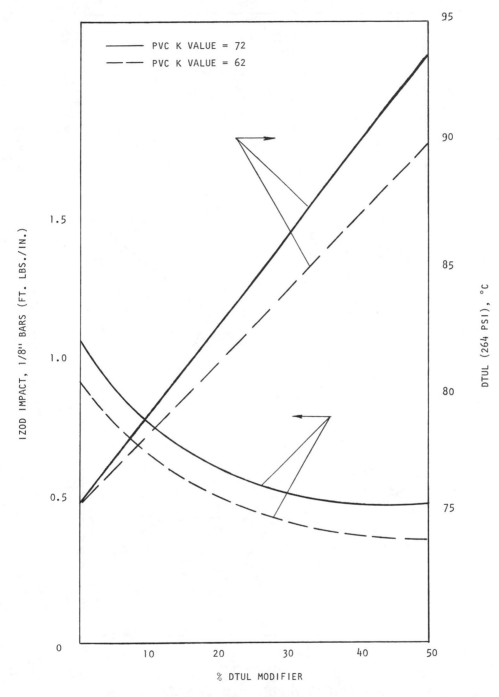

FIGURE 62 Effect of DTUL modifier concentration on impact strength and DTUL.

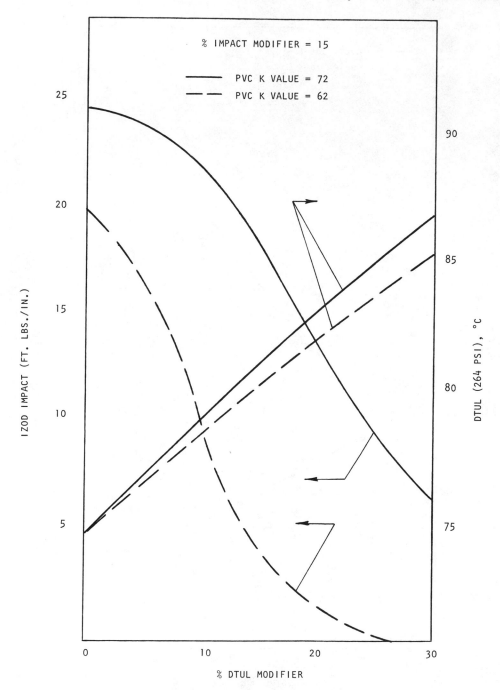

FIGURE 63 PVC-acrylic DTUL modifier impact modifier system.

TABLE 21 The Effect of DTUL Modifier on the Properties of PVC (K = 72)

Property	DTUL modifier in PVC (%)		
	0	30	50
Flexural stress (psi)	11,500	13,500	15,500
Flexural modulus (psi)	434,500	437,300	449,500
Tensile strength (psi)	9,320	9,000	9,400
Elongation at break (%)	156	120	11
Tensile modulus (psi)	420,100	435,800	438,200
Tensile impact strength (ft-lb/in.^{2})	199	46	25

To minimize handling inconvenience and to optimize the three functions, an all-acrylic polymer system is currently offered under the designation of Paraloid® KM-318F (see appendix). This unique product offers (1) the good hot melt strength (processing aid function) necessary to contain the foam-producing gas and prevent rupture on leaving the die, (2) the impact strength necessary to provide ductility and cracking resistance, and (3) the polymeric lubricating functionality required to give good slip through the die without disturbing the balance of fusion and rheological properties so critical for high-speed foam profile production.

C. Heat Distortion Temperature Improvers

Polymers having glass temperatures above that of PVC (about 87°C) have been used to raise the heat distortion temperature (DTUL, ASTM D648) of PVC compounds. Various polymers based on acrylonitrile or styrenics are used as combinations of processing aid/DTUL improver or impact modifier/ DTUL improver. Increase of DTUL per part of additive is relatively low, as one would expect when the mechanism is an additive rather than a reactive one.

In most cases, with the processing aid/DTUL improvers, tensile strength is not affected, stiffness is increased, and impact strength is impaired (Table 21; Figure 62). Additional impact modifier can restore impact to a practical level (Figure 63), or the impact modifier/DTUL combination may be used. Commercially significant (important) increases in DTUL in the range 10° to 20°C require concentrations of 30 to 50% modifiers, making the resulting compound more representative of an alloy than of modified PVC. DTUL improvers are available for clear or opaque formulations.

IV. PVC APPLICATIONS

A. Building Products

1. Pipe and Conduit: High-molecular-weight acrylic processing aids, acrylic-styrene, and α-methylstyrene processing aids continue to be the "aspirins" of the pipe industry that is predominantly supported by twin-

480 Malpass, Petrich, and Lutz

TABLE 22 PVC Pipe Conduit Formulations

	Twin-screw	Single-screw
Constant ingredients (for all pipe types)		
PVC resin (K=67, 68; IV=0.92)	100	100
Tin stabilizer	0.4	1.2
Calcium stearate	0.8	1.5
Paraffin wax (165°C)	1.2	0.8
TiO_2	1.0	1.0
Variable ingredients dependent on applications		
Pressure		
Acrylic processing aid (or high-efficiency acrylic processing aid)	0.75 (0.5)	1.0 (0.75)
Calcium carbonate	0 to 5	0 to 5
DWV		
High-impact efficiency MBS or ABS (or CPE)	1 to 2 (3)	1 to 2 (3)
Calcium carbonate	0 to 5	0 to 5
Sewer		
Calcium carbonate	0 to 5	0 to 5
Drain		
Acrylic lubricant	0.75	1.0
Calcium carbonate	20 to 40	20 to 40

screw extrusion of powder compounds. Unless the extrusion line is extremely sophisticated, the processing aids, especially the more cost-efficient acrylics, provide a means to minimize the batch-to-batch variations that are commonly experienced due to vagaries in formulation ingredients and mixing techniques. The more uniform fusion and improved hot melt strength that result allow for improved economics and properties due to closer tolerances (no need to overweigh to compensate for variations), higher extrusion rates, and more ductile compounds.

Rigid and modified PVC have become significant factors in the construction industry. Rigid PVC drain/waste/vent (DWV) pipes and rigid PVC sewer pipes are widely used in domestic and commercial construction. Rigid PVC potable-water pipe applications are more limited because of local code restrictions based on lack of confidence in long-term performance and misinformation about the safety of rigid PVC compounds. Rigid PVC conduit is increasingly used because it is a natural electrical insulator, is easier to install than metal, and is inherently corrosion-resistant and flame-retardant. The formulations in Tables 22 and 23 group the constant and variable ingredients for various pipe and conduit uses.

TABLE 23 PVC Pipe Formulations

Variable ingredients dependent on applications[a]

Application	Ingredient	Twin-screw	Single-screw
Telephone duct	Acrylic lubricant processing aid	0.75	1.0
	High-efficiency acrylic, mod. acrylic impact modifier (or CPE)	2.0 (3 to 5)	2.0 (3 to 5)
	Calcium carbonate	20 to 30	20 to 30
Electrical conduit	Acrylic processing aid (or high-efficiency acrylic processing aid)	0.75 (0.5)	1.0 (0.75)
(Small diameter)	High-efficiency acrylic mod. acrylic impact modifier	4.0	4.0
	(or high-efficiency MBS impact modifier)	3.0	3.0
	(or CPE)	(5 to 6)	(5 to 6)
	Calcium carbonate	10	10
Irrigation pipe	Acrylic processing aid (or high-efficiency acrylic processing aid)	0.75 (0.5)	
	High-efficiency acrylic or mod. acrylic impact modifier	3.0	
	TiO_2	7.0	
	Calcium carbonate	0 to 5	

[a]Constant ingredients are the same as in Table 22.

For *pressure pipe*, only the processing aid is used. A major physical property of pressure pipe is its ability to sustain internal pressure. Sustaining this pressure is a function of the tensile strength of the PVC compounds. Adding any kind of rubbery impact modifier will, of necessity, decrease the tensile properties. Since the impact requirements per se are marginal for pressure pipe, impact modifiers are not used. Processing aids are used to facilitate the production of high-quality pipe with a homogenous character and good ductility.

In some high-pressure pipe applications, usually in pumping high-pressure water in mines and other similar applications, the mode of failure is secondary to the amount of pressure or the length of time to fail. In these cases, catastrophic failure, that is, brittle failure, can pose significant hazards to human beings. Therefore, these specialty grade high-pressure pipes frequently contain 8 to 12 phr of high-efficiency MBS impact modifiers and are also made with thicker walls to compensate for compromises in burst strength.

In the *drain, waste, and vent pipe*, processing aids are used for the properties mentioned above, and a high-efficiency impact modifier is also used in low concentrations. Although the concentrations of 1 to 2 phr appear to be insignificantly small, experience has proved that the presence of these impact modifiers gives an added degree of toughness that protects the DWV pipe in shipping and rough handling and gives good serviceability.

For *sewer pipe*, processing aids and some calcium carbonate are used. Again, these are large-diameter, thick-walled pipes that have sufficient impact strength because of their thickness. It is important to use processing aids to make the melt homogeneous and the resulting product ductile.

In various *drain pipe* applications, large amounts of calcium carbonate are used to give the pipe stiffness and to lower the cost. Therefore, it is critical that a special acrylate processing aid/lubricant system be used to bind the calcium carbonate particles. The addition of very low levels of a product such as Paraloid® K-175 enables the pipe to retain adequate impact strength even though it is made from highly filled PVC compounds.

Telephone duct formulations frequently include weatherable impact modifiers because telephone duct is frequently exposed to ultraviolet light. In addition, telephone duct formulations often include significant quantities of calcium carbonate, again for stiffness and lower cost. In this regard, acrylic modifier systems perform in a superior manner in binding the calcium carbonate and maintaining good integrity and good impact strength in these highly filled PVC compounds.

In *electrical conduit*, the need for impact modifier depends on the diameter of the conduit. For small-diameter pipes, the geometry is such that impact modifiers are required at relatively low levels. For larger diameters, processing aids alone suffice. Calcium carbonate is present in significant quantities, and the use of all-acrylic systems is recommended.

Irrigation pipe formulations require the use of weatherable impact modifiers due to outdoor exposure of the pipe. Here again, levels of impact modifier are relatively low but have been shown to be sufficient to give good weatherable service. For any exterior application, higher levels of TiO_2 are usually incorporated to act as sunscreens to protect the PVC.

2. *PVC Injection Molded Pipe Fittings*: These may be made from precompounded pellets or, when used in large volume, injection molded directly from powder formulations. In Table 24, the type I and type I-improved formulations require very little impact modifier. The use of processing aid is critical to achieve the proper fusion in the relatively short residence time of the PVC powder compound in the injection molder. The use of a polymeric acrylic lubricant is desirable to minimize weld-line failure, shear burning, and splay. For type II compounds, which require somewhat more impact strength, the impact modifier level is somewhat higher. In either case, the compounder is served well by using high-efficiency impact modifiers.

B. Exterior Sheet and Profiles

1. *Vinyl Siding*: Vinyl has overtaken aluminum as the preferred siding material. Rigid, solid PVC and PVC-clad window systems represent the greatest growth potential for PVC in the next decade. Rain collection systems continue to be popular due to ease of installation and minimal mainte-

TABLE 24 PVC Injection-Molded Pipe Fitting Formulations

	Type I/type I improved (cell class 12454/13454)	Type II (cell class 14333)
PVC resin (K=57 to 61; IV=0.68 to 0.79)	100	100
High-efficiency MBS, ABS impact modifier; (CPE)	1.0 (3.0)	3.0 (5.0)
Acrylic processing aid (or high-efficiency acrylic processing aid)	1.0 (0.75)	1.0 (0.75)
Acrylic lubricant processing aid	0.5	0.5
Tin stabilizer	1.5	1.5 to 2.0
Calcium stearate	0.5	0.5
Paraffin wax (165°F)	0.5	0.5
TiO$_2$	1.5	1.5

nance requirements. More than 15 years experience in the United States and 25 years experience in Europe have demonstrated the viability of rigid PVC building products [164–171].

PVC siding is normally extruded from powder compounds. The powders are extruded on both twin-screw and single-screw extruders with equal success. In the formulations in Table 25 several critical factors should be noted. One is the use of a weatherable impact modifier. The use of impact

TABLE 25 PVC Siding Formulations

	Twin-screw	Single-screw
PVC resin (K=67, IV=0.92)	100	100
Acrylic, mod. acrylic, CPE impact modifier	5 to 6	5 to 6
Acrylic processing aid (or high-efficiency acrylic processing aid)	1 (0.5)	2 (1.5)
Acrylic lubricant processing aid	0.5	0.5
Tin stabilizer	1.6	2.0
Calcium stearate	1.3	1.0
Paraffin wax (165°F)	1.0	1.0
TiO$_2$	10 to 14	10 to 14

484

Malpass, Petrich, and Lutz

modifiers with olefinic unsaturation results in rapid degradation of the impact modifier and degradation of the PVC. The result is rapid discoloration and rapid loss of impact strength. Another factor to note is the proper use of stabilizer. Stabilizer levels in PVC siding and other exterior applications are higher than those normally used in interior applications. This high level is used to protect the PVC effectively from excessive degradation during processing, which would otherwise continue and accelerate on exposure to ultraviolet light. A third and important factor is the level of TiO_2, which is in the range 10 to 14 phr to give a high degree of opacity and shield the PVC resin from UV light degradation. The formulations presented are, obviously, for a white PVC compound. A problem frequently arising in the use of colored compounds is the inability to use enough TiO_2 to shield the PVC adequately from UV radiation. Special formulations for colored PVC must be developed to give successful outdoor durability. As in all compounds that are processed from powder, the inclusion of processing aid is necessary in order to convert the powder rapidly through the melt stage to a tough homogeneous melt for rapid extrusion with good appearance and toughness in the final article. Properly adjusted processing aid and lubricating processing aid levels yield high-quality surfaces that enable the processor to extrude at very high rates with good quality. As in all formulations that can be extruded on either twin-screw or single-screw extruders, the balance of processing aid and lubricant must be carefully considered. All single-screw formulations require relatively early fusion of the PVC. Therefore, the amount of processing aid is higher and the lubricant level is lower (especially external lubricants) than would be found in twin-screw formulations. A critical factor in twin-screw formulations is regulating the fusion point so that it occurs very close to the adaptor or in the adaptor and die block as the compacted/sintered powder is converted to a homogeneous melt.

2. *Vinyl Profiles: Window Lineals; Rain Delivery Systems*: *PVC window profile* formulations are somewhat different from those for PVC siding applications, although they are both considered to be exterior applications. Lubricant systems are somewhat different and TiO_2 levels are generally lower for window profiles. It has been argued that the use of a weatherable impact modifier in window profile formulations is not necessary. Experience has proved, however, that there is a need for reasonably high impact strength and ductility in PVC window profile formulations just as there is in PVC siding. In the postextrusion fabrication, there are several punching and cutting operations. If the PVC formulation is not ductile/tough, the part will be brittle and break during fabrication. Furthermore, the parts, whether siding or window profiles, are shipped during cold weather and receive a certain amount of abuse in transit. Other problems arise, especially in cold weather, in installation. Cutting and errant blows of hammers can cause severe problems in brittle PVC products. Table 26 presents a typical set of tin- and barium/cadmium-stabilized PVC window profile formulations. Although in 1987 there were very few barium/cadmium-stabilized rigid PVC systems in the United States, the use of barium/cadmiums and mixed lead/barium/cadmium systems has been widespread in Europe for many years. Many studies show that barium/cadmium is a superior stabilizer system, especially when used in combination with epoxy-containing materials, for outdoor durability. Therefore, the use of TiO_2 in barium/cadmium systems is generally lower.

TABLE 26 PVC Window Profile Formulations

	Twin-screw	Single-screw
Tin stabilized		
PVC resin (K=67; IV=0.90)	100	100
Acrylic, mod. acrylic, or CPE impact modifier	5 to 6	5 to 6
Acrylic processing aid	1	2.5
Acrylic lubricant processing aid	0.5 to 1.0	0.5 to 1.0
Tin stabilizer	1.5	2.0
Calcium stearate	1.2	1.0
Paraffin wax (165°F)	1.0	0.75
TiO_2	8 to 12	8 to 12
Barium/cadmium stabilized		
PVC resin (K=67; IV=0.90)	100	100
Acrylic, mod. acrylic, or CPE impact modifier	5 to 6	5 to 6
Acrylic processing aid	1	1.5
Acrylic lubricant processing aid	0.5	0.5
Ba/Cd stabilizer	3.0	4.0
Alkyl-aryl phosphite	0.5	0.5
Calcium stearate	1.0	—
Ester wax	0.5	—
Glyceryl monostearate	0.4	2.0
TiO_2	7.0	7.0
$CaCO_3$ (fine particle size, coated)	5.0	5.0

The window profile formulations can generally be considered as starting points for *rain delivery systems* as well.

The choice of type and concentration of TiO_2 and fillers merits some consideration at this point. Although color retention is proportional to the TiO_2 level in white exterior compounds, impact retention appears to respond to an optimum level of TiO_2 (Figure 64). Apparently, more than 10 phr TiO_2 begins to act more like filler than an ultraviolet shielding agent, with resulting loss in impact strength. Ferguson suggests that 10 phr TiO_2 gives a good balance of optical and impact properties on outdoor exposure [164]. The amount and type of TiO_2 required will depend on the stabilizer system and the type of impact modifier. Summers and Rabinovitch [172, 173] demonstrated the advantages of nonchalking TiO_2 where the photoreactivity rate is lowest due to a combination of rutile crystalline structure and alumina or silica coatings that apparently prevent the TiO_2 from coming in direct contact with the PVC polymer.

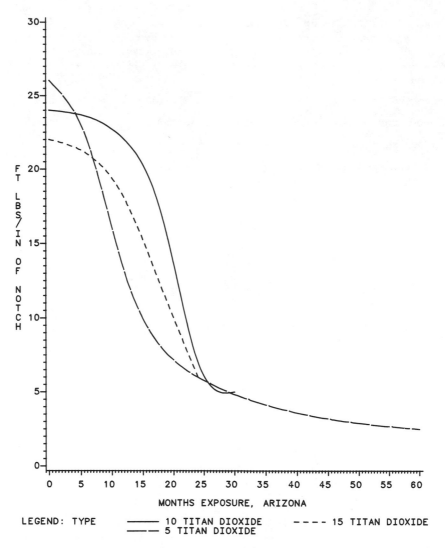

FIGURE 64 Weatherability of white PVC: effect of titanium dioxide on Arizona exposure test results.

Although chalking grades of TiO_2 have some advantage for whites where self-cleaning of surface erosion and bleaching due to greater photoreactivity are advantages, these "advantages" are contraindicated in many colored systems. In addition, surface erosion leads to rapid deterioration of impact strength due to notch-sensitizing defects.

The use of significant amounts of fillers in exterior PVC formulations remains controversial. Coated calcium carbonate and alumina trihydrate (ATH) with average particle sizes of less than 1 μm have been shown to enhance impact strength [173]. Effects of alumina trihydrate on weathering are primarily in retention of color. A patent has been applied for the use of fine particle size ATH in PVC [163]. Mathur and Tapper [174] reported that coated, precipitated calcium carbonate (CPCC) at a particle size of 0.5 μm can be used to enhance the impact retention of exterior PVC formulations. Further, the use of 0.07 μm CPCC at 5 to 10 phr is claimed to allow the reduction of TiO_2 levels by 30 to 40% in the formulation, because of the UV-shielding potential of the CPCC [175]. Claims are also made for both ATH and CPCC with respect to ease of processing (dispersion) and minimization of extruder barrel and screw [174, 175].

Interior PVC extruded profile formulations are made from prepelletized compound or directly from powder blends. With powder blends, both twin-screw and single-screw extruders have been used successfully. Most interior solid profiles receive postextrusion treatments similar to those imparted to PVC window profile lineals. In addition, PVC interior profiles are normally exposed to environments that receive impact and other abuse in service. For this reason, significant amounts of impact modifiers are used in these formulations. The choice of impact modifier is made on the basis of minimal contribution to die swell and high efficiency in imparting impact strength to the compound. Low contribution to melt viscosity or elasticity results in low die swell. When sophisticated and complicated profiles are being made, freedom from die swell is critical in order to maintain close tolerances. In the formulations presented in Table 27 for extrusion from powder blends, the processing aid and high-molecular-weight acrylic lubricant are suggested to facilitate fusion, impart good hot melt strength, and give excellent part appearance as well as rapid extrusion rates. An advantage of using an extremely efficient impact modifier is that lower levels can be used to obtain a desired level of effect, and with the lower level, less contribution to adverse melt viscosity and die swell will be achieved. Although weather resistance is not required, the low die swell characteristics of acrylic impact modifiers qualify them, as well as properly selected MBS and ABS modifiers, for this application.

TABLE 27 Interior PVC Profile Formulations

	Twin-screw	Single-screw
PVC resin (K=67; IV=0.90)	100	100
High-efficiency, low die swell MBS or acrylic impact modifier	6 to 10	6 to 10
Acrylic lubricant processing aid	1.0	1.0
Tin stabilizer	1.5	2.0
Calcium stearate	0.8	0.8
Paraffin wax (165°F)	1.2	0.7
Calcium carbonate	3 to 5	3 to 5
Pigments	As needed	

C. High-Flow PVC Injection-Molded Articles

Large, thin-walled, or moderately thick-walled parts can be made from prepelletized or powdered PVC compounds. Television cabinets and power tool, appliance, and business machine housings made from PVC have excellent aesthetics and toughness and are inherently fire-resistant. A critical factor in formulating for these applications is good melt flow, which is required to fill the rather large molds rapidly. Processing aids are needed to promote rapid fusion, reduce the time and temperature needed to fuse the compound, and impart high gloss and ductility to the part. Significant levels of impact modifier are needed to give the part adequate toughness. The impact modifier type and level are important because injection molding formulations are based on relatively low-molecular-weight PVC resins, which have inherently lower strength than the higher-molecular-weight PVCs normally used in extrusion. The high-molecular-weight acrylic lubricant is an important ingredient since it can prevent shear burning caused by the high injection pressures needed to fill the mold. Also important is the ability of the acrylic lubricant to assist in mold filling by reducing sticking to the mold and yet enhancing, rather than interfering with, weld-line strength when the melt fronts meet the mold. Higher stabilizer levels are required for protection at the higher temperatures and shear rates required to fill the mold. An efficient internal lubricant at high levels is also required to reduce the melt viscosity. Higher external lubricants are required, but their use must be carefully minimized in order not to interfere with aesthetics or weld-line strength. All of these concepts are embraced in the formulations of Table 28.

TABLE 28 High-Flow PVC Injection Molding Formulations

	General purpose	Exterior use
PVC resin (K=52; IV=0.57-0.60)	100	100
High-efficiency MBS or ABS impact modifier	10 (or 12)	—
(or low die swell MBS or ABS impact modifier)		
High-efficiency acrylic impact modifier	—	10 to 12
Acrylic processing aid	1.5 to 2.0	1.5 to 2.0
(or high-efficiency acrylic impact modifier)	(1.2 to 1.7)	(1.1 to 1.7)
Acrylic lubricant processing aid	1.0 to 1.5	1.0 to 1.5
Tin stabilizer	2.0	2.2
Glyceryl monostearate	2.7	2.7
Polyethylene wax	0.3	0.3
TiO_2	As needed	10 to 12

TABLE 29 Extruded PVC Sheet and Film (Single-Screw)

	Clear	Opaque	Non-stress-whitening clear
PVC (K=60 to 62; IV=0.76 to 0.80)	100	100	100
High impact efficiency MBS or ABS impact modifier	10 to 12	—	—
High impact efficiency MBS—crease whitening resistant	—	—	10 to 15
General-purpose MBS (or ABS) impact modifier	—	8 to 12 (6 to 10)	—
Acrylic processing aid (or high-efficiency acrylic processing aid)	2.0 (1.6)	2.0 (1.6)	2.0 (1.6)
Acrylic lubricating processing aid	1.0	1.0	1.0
Tin stabilizer	2.0	2.0	2.0
Montan ester wax	0.4	—	0.4
Glyceryl monostearate	0.75	—	0.75
Calcium stearate	—	2.0	—
Paraffin wax (165°F)	—	0.7	—
Blue toner	0.0004	—	0.0004
TiO_2	—	As needed	—

D. Extruded and Calendered Sheet and Film

A significant percentage of the clear and opaque sheet and film that is produced contains impact modifiers in varying concentrations, depending on specific applications. The formulations presented in Table 29 illustrate typical compositions for general-purpose clears, for opaques, and for non-stress-whitening clear formulations. Clear formulations are most frequently used in packaging applications. Non-stress-whitening clears require, in addition to impact modifiers and lubricant systems that give good clarity, freedom from stress or crease whitening. Most of the large-volume application compounds for extruded or calendered sheet are prepared from powder blends. For that reason, processing aids are a vital part of the formulation, not only for rapid fusion and ductility but also for the excellent hot strength that is required in downstream take-off operations. Postforming requires ductility for cold-breaking and hot strength for thermal vacuum forming. Acrylic processing aids contribute to these properties. The choice of impact modifier for opaque applications is much broader because of the lack of a requirement for clarity. Whether working in clear or opaque systems, the

TABLE 30 Calendered PVC Sheet and Film Formulations

	Clear	Opaque	Non-stress-whitening clear
PVC resin (K=60; IV=0.76)	100	100	100
MBS or ABS impact modifier	10 to 15	—	—
High-efficiency impact modifier	—	—	10 to 15
General-purpose MBS or ABS (or acrylic or high-efficiency MBS or ABS impact modifier)	—	8 to 12 (6 to 10)	
Acrylic processing aid (or high-efficiency acrylic processing aid)	2.0 (1.6)	2 (1.6)	2.0 (1.6)
Acrylic lubricant processing aid	0.5 to 1.0	0.5 to 1.0	0.5 to 1.0
Tin stabilizer	2.0	1.6 to 2.0	2.0
Glyceryl monostearate	0.5 to 0.8	0.5 to 0.8	0.5 to 0.8
OP wax (Hoechst)	0.2 to 0.4	0.2 to 0.4	0.2 to 0.4
E wax (Hoechst)	0.05 to 0.2	0.05 to 0.2	0.05 to 0.2
Blue toner (1% in PVC)	0.06	—	0.06
TiO_2		As needed	

choice of a high-impact-efficiency modifier is most advantageous, from both economic and final product property considerations. In calendered sheet, the amount of processing aid is adjusted carefully to give an optimum rolling bank and also to give good hot strength that allows the very high production rates required to make calendering of high-quality sheet and film a profitable operation.

E. PVC Containers

PVC bottles and containers are becoming increasingly popular in the United States for food and nonfood applications. The formulations of Tables 31 and 32 illustrate general-purpose bottle compounds such as would be used for non-food-packaging applications. Also presented are calcium/zinc formulations typical of those used in Europe, where bottle configurations are somewhat different from those in the United States, and typical calcium/zinc U.S.-type formulations. Calcium/zinc is preferred as the lowest-toxicity system for use in food grade containers, but several tin stabilizers now have FDA sanction. In formulating for food containers it is important that each of the ingredients be certified as FDA-approved.

TABLE 31 Blow-Molded Bottle Formulations

General-purpose A		Calcium/zinc (Europe) ribbed bottle	
PVC (K value 58)	100.0	PVC (K value 58)	100.0
Butyltin stabilizer	1.5	Calcium stearate or behenate	0.35
Acrylic processing aid	1.5	Zinc octoate (23% zinc)	0.10
Glyceryl monooleate	0.5	Epoxidized soya oil	5.5
OP wax (Hoechst)	0.2	Loxiol G15 (Henkel)	0.5
Toner (1% in PVC)	0.06	Stearoyl benzoyl methane	0.3
Clear MBS or ABS impact modifier	12.0	OP wax or PED 191 (Hoechst)	0.25
		Loxiol G78 (Henkel)	0.15
General-purpose B		Toners (1% in PVC)	0.065
PVC (K value 58)	100.0	Acrylic processing aid	0.3
Butyltin stabilizer	2.0	Clear MBS impact modifier	8.0
Glyceryl monostearate	0.5	Ca/Zn (U.S.) 16 oz. Boston round	
OP wax (Hoechst)	0.2	PVC (K value 58)	100.00
Acrylic processing aid	1.0	Ca/Zn package stabilizer	2.0
Acrylic lubricating processing aid	1.0	ESO	4.0
Toner (1% in PVC)	0.06	Acrylic processing aid	1.5
Clear MBS or ABS impact modifier	12.5	Stearoyl benzoyl methane	0.3
		Glyceryl monooleate	0.5
		OP wax (Hoechst)	0.2
		Tris(nonylphenyl) phosphate	0.5
		Clear MBS impact modifier	12.0
		Toner (1% in PVC)	0.07

F. Cellular Rigid PVC Profiles

As discussed earlier, rigid PVC foams can be made to have a broad range of properties. Table 33 illustrates starting formulations that give several choices of polymeric additive systems to suit end-use and processing requirements. Some formulators find the balance of properties of an additive such as Paraloid® KM-318F to be ideally suited to their equipment and end-product requirements. Others have formulated successfully around high-efficiency acrylic processing aids. Because cellular PVC articles may be exposed directly or indirectly to sunlight, it is advisable to use all-acrylic systems or mixtures of acrylic processing aids/lubricants with weatherable modifiers.

TABLE 32 Blow-Molded PVC Bottle Formulations

	General-purpose tin	Food grade octyltin	Food grade Ca/Zn
PVC resin (K=58; IV=0.72)	100	100	100
Clear MBS or ABS impact modifier	12 to 14	12 to 14	12 to 14
Acrylic processing aid	2.0	2.0	2.0
Acrylic lubricant processing aid	0.5 to 1.0	0.5 to 1.0	—
Tin stabilizer	2.0	2.0 to 2.2	—
Ca/Zn stabilizer	—	—	2.0
Glyceryl monostearate	0.5	0.5	0.5
Montan ester wax	0.2	0.2	0.2
Benzoyl stearoyl methane	—	—	0.3
Alkyl-aryl phosphite (food-approved)	—	—	0.5
Epoxidized soya oil	—	—	4.0
Blue toner (1% in PVC)	0.06	0.06	0.07

TABLE 33 Cellular PVC Profile Formulations

	Twin-screw	Single-screw
PVC resin (K=62; IV=0.80)	100	100
Paraloid® KM-318F[a] (Rohm & Haas)	8 to 9	8 to 9
Azodicarbonamide	0.5	0.5
Tin stabilizer	1.5	1.5
Calcium stearate	0.8	1.0
Paraffin wax (165°F)	1.0	0.8
Polyethylene wax	0.1	—
Calcium carbonate	5.0	5.0
Pigments	As needed	

[a]Alternative systems; depending on end-use requirements:

A— High-MW acrylic processing aid 6 phr
   Acrylic impact modifier 3
   High-MW acrylic lubricant 1

B—High-MW acrylic processing aid 6 phr
   CPE 5
   High-MW acrylic lubricant 1

C— High-MW acrylic processing aid 3 to 8 phr

V. COMMERCIAL IMPACT MODIFIERS FOR PVC

Company and Trade name	Recommended uses	Use level (phr)	Special features
	MBS		
Rohm & Haas			
Paraloid KM-611	Bottles, opaque film and sheet, pipe, fittings, molded parts	5 to 15	Low crease whitening, low temperature impact
Paraloid KM-641	Bottles, opaque film and sheet, pipe, fittings, molded parts	5 to 15	Low crease whitening, low temperature impact
Paraloid KM-653	Opaque film and sheet, pipe, fittings, molded parts	5 to 15	High impact efficiency, low temperature impact
Paraloid BTA-753	Opaque film and sheet, pipe, fittings, molded parts	5 to 15	Very high impact efficiency, low temperature impact. Very low odor
Paraloid KM-680	Opaque film and sheet, pipe, fittings, molded parts	5 to 15	Very high impact efficiency; low temperature impact, color
Paraloid BTA III F	Bottles, clear film, sheet	8 to 15	Excellent impact and clarity, low crease whitening, low water blush, chemical resistance
Paraloid BTA III N2	Bottles, clear film, sheet	8 to 15	Excellent impact and clarity, low crease whitening, low water blush, chemical resistance
Paraloid BTA-731	Bottles, clear film and sheet	5 to 15	Superior room temperature and low temperature impact— otherwise similar to Paraloid BTA III N2
Paraloid BTA-702	Clear film and sheet	8 to 15	Superior combination of clarity, crease whitening resistance, toughness, and chemical resistance

Company and Trade name	Recommended uses	Use level (phr)	Special features
Kaneka Texas (Kanefaguchi)			
Kane Ace B-11A	Film and sheet	5 to 20	Glasslike clarity without crease whitening
Kane Ace B-31	Film and sheet	5 to 20	Glasslike clarity without crease whitening, higher impact than B-11A
Kane Ace B-18A-1	Film, sheet bottles	5 to 20	Ideal balance of impact strength and clarity, natural color
Kane Ace B-22	General purpose	5 to 20	Ideal balance of impact strength and clarity, slightly fluorescent color
Kane Ace B-28	General purpose	5 to 20	High impact strength, natural color
Kane Ace B-56	Profile, film, and sheet injection molding	5 to 20	High impact strength down to low temperature, opaque and transluscent applications
M & T Chemicals			
Metablen C-223	Opaque interior	2 to 15	Very high impact efficiency
Metablen C-201/203	Clear bottles and sheet	8 to 15	High clarity and impact
Metablen C-110	Film and sheet	8 to 15	High clarity, very low crease whitening

ABS

Borg-Warner			
Blendex 101	Opaque sheet and profile	10 to 40	Improves processing, hot strength, and emboss retention for vacuum forming
Blendex 131	Opaque sheet and profile	10 to 40	Improves processing, toughness, hot strength, and emboss retention
Blendex 201	Opaque sheet and profile	10 to 40	Good balance of high modulus and impact. excellent vacuum forming

Commercial Impact Modifiers (continued)

Company and Trade name	Recommended uses	Use level (phr)	Special features
Blendex 305	Solid and foam profiles	10 to 20	Good impact resistance
Blendex 301	Solid and foam profiles	10 to 20	Controlled particle size (40 mesh) 305 for powder blending
Blendex 310	Opaque, pipe, fittings, sheet	8 to 15	High efficiency processing
Blendex 311	Opaque pipe, fittings, sheet	8 to 15	Controlled particle size 310 for powder blending
Blendex 336	Opaque and transparent sheet, film, conduit, injection molding	7 to 15	High impact efficiency with crease whitening for embossed lettering
Blendex 405	Transparent calendered film and sheet, bottles	7 to 18	High impact, clarity, low crease whitening
Blendex 401	Transparent calendered film and sheet, bottles	7 to 18	Controlled particle size 401 for powder blending
Blendex 702	Opaque profile, sheet, injection molding	10 to 50	Impact and high HDT

Kaneka Texas (Kanefaguchi)

Telalloy A-10	Pipe, profile, injection molding	20 to 80	Opaque applications, improves impact and HDT

Mobay

Novadure A95	Rigid PVC	25 to 35	Thermal stability

Acrylics

Rohm & Haas

Paraloid KM-323B	Weatherable PVC siding, window profiles	5 to 15	Excellent weathering, impact and processing
Paraloid KM-330	Weatherable PVC siding, window profiles	5 to 15	Excellent property retention after weathering

Appendix (continued)

Company and Trade name	Recommended uses	Use level (phr)	Special features
Paraloid KM-334	Weatherable PVC siding, window profiles, pipe fittings	4 to 15	Superior impact efficiency and retention after weathering; excellent processing
Paraloid KM-318F	Rigid foam, nailable profiles and sheet, foam pipe	5 to 10	Uniform foam density, excellent weathering
M & T Chemicals			
Durastrength 200 (modified acrylic)	Siding and window profiles	5 to 8	Excellent weathering, impact, and processing

CPE

Dow Chemical			
CPE 3614A	Pipe, sheet, siding, profile injection-molded products	2 to 12	Good impact, excellent processing, excellent retention of properties and color after weathering
CPE 3615	Pipe, siding, rigid and foam profiles, injection-molded products	2 to 12	Small uniform particle size, improved impact with excellent processing, excellent property retention after weathering
CPE 4213	Calendered and injection-molded products	2 to 12	High chlorine content, lower viscosity, improved flow, but lower impact
CPE 3623A	Profiles, molded goods	2 to 12	Intermediate between 3615 and 4213 in impact efficiency, low viscosity
XO-2243.5	Extruded and injection-molded products	2 to 12	Good low-temperature properties; excellent weatherability, high impact

Commercial Impact Modifiers (continued)

Company and Trade name	Recommended uses	Use level (phr)	Special features
	EVA		
E. I. DuPont			
Elvaloy (terpolymer)	Sheet and profile	4 to 5	Superior weatherability and color retention
Mobay			
Levapren 450N	Rigid PVC	4 to 8	Weather resistance
Levapren KL3-2450	Rigid PVC	4 to 8	Weather resistance
Levapren KL3-2451	Rigid PVC	4 to 8	Weather resistance

REFERENCES

1. Ryan, C., *Soc. Plast. Eng. J.*, *24*, 89 (1968).
2. Berens, A., and Folt, V., *Polym. Eng. Sci. 8*, 5 (1968).
3. Faulkner, P., *J. Macromol. Sci. Phys.*, *B11*, 251 (1975).
4. Menges, G., and Berndsten, N., *Kunststoffe*, *66*, 735 (1976).
5. Menges, G., et al., *Kunststoffe*, *69*, 562 (1979).
6. Rabinovitch, E., and Summers, J., *J. Vinyl Technol.*, *2(3)*, 165 (1980).
7. Gould, R., and Player, J., *Kunststoffe*, *69*, 7 (1979).
8. Bohme, K., *Angew. Makromol. Chem.*, *47*, 243 (1975).
9. Krzewki, R., and Collins, E., *Soc. Plast. Eng. ANTEC Prepr.*, 570 (1981).
10. Pazur, A., and Uitenham, L., *Soc. Plast. Eng. ANTEC Prepr.*, 573 (1981)
11. Petrich, R., Reprints, Plastics & Rubber Institute Conference on Processing of PVC, London, April 1978.
12. U.S. Patent 2,646,417 (1953), B. F. Goodrich.
13. U.S. Patent 2,791,600 (1957), B. F. Goodrich.
14. U.S. Patent 3,373,229 (1965), I.C.I.
15. U.S. Patent 3,764,638 (1973), Stauffer Chemical.
16. U.S. Patent 3,833,686 (1974), Rohm & Haas.
17. British Patent 1,378,434 (1974), Kanegafuchi.
18. U.S. Patent 3,673,283 (1972), Japanese Geon.
19. U.S. 3,874,740 Patent (1974), Tenneco.
20. Wilson, A., and Raimondi, V., Reprints, PRI Conference on Processing of PVC, London, April 1978.
21. U.S. Patent 4,105,624 (1978).
22. U.S. Patent 4,137,280 (1979), Air Products.
23. U.S. Patent 3,859,384 and 3,859,389 (1975), Rohm & Haas.
24. Petrich, R., *Mod. Plast.*, *49(8)*, 74 (1972).
25. Graham, R., *Am. Chem. Soc. Div. Org. Coat. Plast. Prepr.*, *34*, 172 (1974).
26. French Patent 2,324,660 (1974), Protex.

27. British Patent 848,153 (1960), BASF.
28. U.S. Patent 4.051,200 (1976), BASF.
29. German Patent 2,163,986 (1972), Mitsubishi Rayon.
30. U.S. Patent 3,867,481 (1975), Rohm & Haas.
31. German Patent 2,017,398 (1970), Rohm & Haas.
32. U.S. Patent 3,096,313 (1967), Du Pont.
33. U.S. Patent 3,284,545 (1966), Rohm & Haas.
34. U.S. Patent 3,485,775 (1969), Rohm & Haas.
35. U.S. Patent 3,584,079 (1971), Monsanto.
36. McMurrer, M. C., *Plast. Compounding*, 77–85 (Nov./Dec. 1983).
37. Haward, R. N., *The Physics of Glassy Polymers*, Applied Science, 1973, Chapter VI.
38. Vincent, P. I., *Polymer*, *1*, 7 (1960).
39. Watling, R. E., and Montgomery, R. A., *Soc. Plast. Eng. Annu. Tech. Conf.*, 1972, p. 143.
40. Vincent, P. I., *Plastics*, *27*, 115 (1962).
41. Malpass, V. E., *Soc. Plast. Eng. Annu. Tech. Conf.*, May 1967.
42. Matonis, V. A., and Aubrey, N. E., *Soc. Plast. Eng. J.*, *26*, 49 (1970).
43. Malpass, V. E., Paul, L. A., and Fenelon, P. J., *Soc. Plast. Eng. Annu. Tech. Conf.*, 1973.
44. Fenelon, P. J., *Polym. Eng. Sci.*, *15*(7), 538 (1975).
45. Brown, H. R., and Ward, I. M., *Polymer*, *14*, 469 (1973).
46. Bucknall, C. B., *Toughened Plastics*, Applied Science, 1977, p. 153.
47. Bucknall, C. B., *Toughened Plastics*, Applied Science, 1977, p. 287.
48. Dynatup Impact Tester, Effects Technology, Santa Barbara, Calif.
49. Szamborski, E. C., and Hutt, R. J., *Soc. Plast. Eng. Annu. Tech. Conf.*, Prepr. 884, 1984.
50. Bucknall, C. B., *Toughened Plastics*, Applied Science, 1977, p. 272.
51. Shoulberg, R. H., and Gouza, J. J., *Soc. Plast. Eng. J.*, *9*, 23–32 (1967).
52. Morris, A. C., *Plast. Polym.* *36*, 433 (1968).
53. Plati, E., and Williams, J. G., *Polym. Eng. Sci.*, *15*, 470 (1975).
54. Malpass, V. E., and Gaggar, S. K., *J. Vinyl Technol.*, *1*(2), 112 (1979).
55. Radon, J. C., *Polym. Eng. Sci.*, *12*, 431 (1972).
56. McCrum, N. G., et al., *Anelastic and Dielectric Effects in Polymeric Solids*, Wiley, New York, 1967.
57. Havriliak, S., Jr., and Malpass, V. E., *Soc. Plast. Eng. Annu. Tech. Conf.*, 1985, pp. 645–648.
58. Merz, E. H., Claver, G. C., and Baer, M., *J. Polym. Sci.*, *22*, 325 (1956).
59. Bucknall, C. B., and Smith, R. R., *Polymer*, *6*, 437 (1965).
60. Kambour, R. P., *Nature (London)*, *195*, 1299 (1962).
61. Sultan, J. N., and McGarry, F. J., *Polym. Eng. Sci.*, *13*, 29 (1973).
62, Michler, G., Gruber, K., Pohl, G., and Kaestner, G., *Plaste Kautsch.*, *20*, 756 (1973).
63. Matsuo, M., *Polym. Eng. Sci.*, *9*, 206 (1969).
64. Newman, S., and Strella, S., *J. Appl. Polym. Sci.*, *9*, 2297 (1965).
65. Sternstein, S. S., and Ongchin, L., *ACS Polym. Prepr.*, *10*(2), 1117 (1969).
66. Petrich, R. P., *Polym. Eng. Sci.*, *12*, 757 (1977).
67. Sternstein, S. S., Ongchin, L., and Silverman, A., *Appl. Polym. Symp.*, *7*, 175 (1968).

68. Matsuo, M., Wang, T., and Kwei, T. W., *J. Polym. Sci.*, *A2(10)*, 1085 (1972).
69. Donald, A. J., and Kramer, E. J., *J. Appl. Polym. Sci.*, *27*, 3729–3741 (1982).
70. Hobbs, S. Y., Bopp, R. C., and Watkins, V. H., *Polym. Eng. Sci.*, *23(7)*, 3 (1983).
71. Bucknall, C. B., Clayton, D., and Keast, W. E., *J. Mater. Sci.*, *7*, 1443 (1972).
72. Haward, R. N., Mann, J., and Pogany, G., *Br. Polym. J.*, *2*, 209 (1970).
73. Bucknall, C. B., and Clayton, D., *J. Mater. Sci.*, *7*, 202 (1972).
74. Fenelon, P. J., and Wilson, J. R., *Am. Chem. Soc. Div. Org. Coat. Plast. Prepr.*, *34(2)*, 326 (1974).
75. Kramer, E. J., *J. Polym. Sci. Phys.*, *13*, 509 (1975).
76. Kramer, E. J., *J. Macromol. Sci. Phys. B10*, 191 (1974).
77. Yee, A. F., *J. Mater. Sci.*, *12*, 757 (1977).
78. Breuer, H., Haaf, F., and Stabenow, J., *J. Macromol. Sci. Phys.*, *B14*, 387 (1977).
79. Haaf, H., Breuer, H., and Stabenow, J., *Angew. Makromol. Chem.*, *58/59*, 95 (1977).
80. Siegmann, A., English, L. K., Baer, E., and Hiltner, A., *Polym. Sci.*, *24(11)*, 877–885 (1984).
81. Donald, A. M., and Kramer, E. J., *J. Mater. Sci.*, *17*, 1765–1772 (1982).
82. Morton, M., Cizmecioglu, M., and Lhila, R., *Polym. Blends Comp. Multiphase Syst.*, *Am. Chem. Soc.*, *260*, 221–230 (1984).
83. Berens, A. R., and Folt, V. L., *Trans. Soc. Rheol.*, *11*, 95 (1967).
84. Rabinovitch, E. B., and Summers, J. W., *J. Vinyl Technol.*, *2(3)*, 165 (1980).
85. Summers, J. W., *J. Vinyl Technol.*, *3(2)*, 107 (1981).
86. Collins, E. A., and Krier, C. A., *Trans. Soc. Rheol.* *11*, 225 (1967).
87. Faulkner, P. G., *J. Macromol. Sci. Phys.*, *B11*, 25 (1975).
88. Krzewski, R. J., and Collins, E. A., *J. Macromol Sci. Phys.*, *B20*, 443 (1981).
89. Menges, G., Berndtsen, N., and Opfermann, J., *Plast. Rubber Process.*, *4*, 156 (1979).
90. Rosenthal, J., *Soc. Plast. Eng. Annu. Tech. Conf. Prepr.*, 1983, 606.
91. Rabinovitch, E. B., and Summers, J. W., *Soc. Plast. Eng. Tech. Pap.*, *27*, 516 (1981).
92. Ryan, C. F., and Jalbert, R. L., *Encyclopedia of PVC*, Vol. 2, ed. L. Nass, Dekker, New York, 1977, p. 628.
93. Ferguson, L. E., *Soc. Plast. Eng. J.*, *28(9)*, 33–37 (1972).
94. Malpass, V. E., *Soc. Plast. Eng. Tech. Pap.*, *15*, 55–59 (1969).
95. Purcell, T. O., *Polym. Prepr. Am. Chem. Soc.*, *13(1)*, 699 (1972).
96. Casale, A., Moroni, A., and Spreafico, C., *Am. Chem. Soc. Polym. Prepr.*, *15(1)*, 334 (1974).
97. Ohtsuka, S., Watanabe, H., and Amagi, Y., *Soc. Plast. Eng. Annu. Tech. Conf. Prepr.*, *13*, 707 (1967).
98. Japan Patent 54/135889 (1979), Kureha Chem. Ind.
99. Japan Patent 58/152039 A2 (1983), Kanegafuchi Chem. Inc.
100. European Patent Appl. EP 68357 A1 (1983), Kureha Chem. Ind.
101. Netherlands Patent Appl. NL 82/1149A (1983), Kureha Chem. Ind.
102. European Patent Appl. EP 50848A2 (1982), Kureha Chem. Ind.

103. U.S. Patent 4,440,905 A (1984), Rohm & Haas.
104. Lutz, J. T., *Plast. Compounding*, *34–44* (Jan./Feb. 1981).
105. British Patent GB 156476-A (1977), Kanegafuchi; Japanese Patent JP 52003637-A (1977).
106. European Patent Appl. EP 66382 A1 (1982), Rohm & Haas.
107. Carlson, A. W., Jones, T. A., and Martin, J. R., *Mod. Plast.* (May 1967).
108. Pavan, A., Ricco, T., and Rink, M., *Mater. Sci. Eng.*, *45(3)*, 201–209 (1980).
109. Rink, M., Ricco, T., and Pavan, A., *Angew. Makromol. Chem.*, *93(1)*, 221–224 (1981).
110. Malpass, V. E., *Appl. Polym. Symp.*, 5, 87–102 (1967).
111. Ryan, C. F., and Jalbert, R. L., *Encyclopedia of PVC*, Vol. 2, ed. L. Nass, Dekker, New York, 1977, p. 621.
112. Ryan, C. F., and Jalbert, R. L., *Encyclopedia of PVC*, Vol. 2, ed. L. Nass, Dekker, New York, 1977, p. 627.
113. Lutz, J. T., *Polym. Plast. Technol.*, *Eng.*, *16*, 58–80 (1978).
114. Rabinovic, I. S., *J. Vinyl Technol.*, *5(4)*, 179–182 (1983).
115. U.S. Patent 3,678,133 (1972), Rohm & Haas.
116. Stoloff, A., *Plast. Eng.*, *35(7)*, 29–31 (1979).
117. U.S. Patent 4,443,585 (1984), Rohm & Haas.
118. Japan Patent JP 57/165415 A2 (1982), Kanegafuchi Chem. Inc.
119. Japan Patent JP 55/7841 (1980), Kanegafuchi Chem. Ind.
120. Japan Patent JP 55/155009 (1980), Kanegafuchi Chem. Ind.
121. Japan Patent JP 55/166217 A2 (1981), Mitsubishi Rayon.
122. Young, W. L., and Serdynsky, E. D., *Soc. Plast. Eng.*, *Regional Tech. Conf.*, Ohio, 1965.
123. Blanchard, R. R., and Burnell, C. N., *Soc. Plast. Eng. J.*, *24(1)*, 74 (1968).
124. Deanin, R. D., and Shah, M. R., *Org. Coat. Plast. Chem.*, 44, 102–107 (1981).
125. Burnell, C. N., and Gunson, D. J., *Soc. Plast. Eng. Regional Tech. Conf.*, Oct. 27, 1981.
126. German Patent DE 3236514 A1 (1984), Hoechst A.G.
127. Deanin, R. D., and Shah, N. A., *J. Vinyl Technol.*, *5(4)*, 167–172 (1983).
128. Japan Patent JP 58/168646 A2 (1983), Mitsui Polychem.
129. British Patent 1,031,424 (1966), DuPont; French Patent 1,349,099 (1964).
130. British Patent 958,399 (1964), Union Carbide; U.S. Patent 3,156,744 (1964).
131. European Patent Appl. EP 33220 (1981), Rohm & Haas.
132. Pegoraro, M., and Beati, E., *Proc. IUPAC Macromol. Symp. 28th*, p. 649.
133. Beati, E., Pegoraro, M., and Briano, M., *J. Appl. Polym. Sci.*, *26(7)*, 2185–2195 (1981).
134. U.S. Patent 4,195,137 (1980), Hooker Chem.
135. Japan Patent JP 58/23844 A2 (1983), Mitsui Petrochem.
136. McGill, W. J., and Wittstock, T., *Plast. Rubber Process. Appl. 3(1)*, 77–83 (1983).
137. U.S. Patent 3,330,996 (1967), Pechiney.
138. U.S. Patent 3,327,022 (1967), Pechiney.
139. Menzel, G., *Plastverarbeiter*, *34(9)*, 816–821 (1983).

140. Japan Patent JP 57/98537 A2 (1982), Kanegafuchi Chem. Ind.
141. Japan Patent JP 57/98542 A2 (1982), Kanegafuchi Chem. Ind.
142. Japan Patent JP 57/98543 A2 (1982), Kanegafuchi Chem. Ind.
143. Matsuo, M., *Jpn. Plast.*, 7–16 (July 1968).
144. Woods, M. E., and Frazer, D. G., *Soc. Plast. Eng. Tech. Pap.*, *20*, 426–428 (1974).
145. Deanin, R. D., and Sheth, K. B., *Am. Chem. Soc. Div. Org. Coat. Prepr.* *43*, 23–26 (1980).
146. U.S. Patent 2,779,748 (1957), U.S. Rubber.
147. Netherlands Patent Appl. 6,502,232 (1966), Shell Internationale.
148. U.S. Patent 2,780,615 (1957), U.S. Rubber.
149. U.S. Patent 3,158,644 (1964), Monsanto.
150. U.S. Patent 3,090,768 (1963), Esso Research and Engineering.
151. Japan Patent JP 57/209949 A2 (1982), Mitsubishi Petrochem.
152. European Patent Appl. EP 12559 (1980), Rohm & Haas; U.S. Appl. 966,935 (1968).
153. Mokry, J., Stresinka, J., Pelzbauer, Z., and Marcincin K., *Chem. Zvesti*, *37(6)*, 783–790 (1983).
154. U.S. Patent 4,225,684 (1980), Borg-Warner.
155. European Patent Appl. EP 51770 A2 (1982), Bayer A.G.
156. Mathur, K. K., and Driscoll, S. B., *J. Vinyl Technol.*, *4(2)*, 81–86 (1982).
157. Mathur, K. K., and Tapper, M., *J. Vinyl Technol.*, *5(4)*, 173–178 (1983).
158. U.S. Patent 4,373,051 A (1983), National Dist. and Chem.
159. Simek, J., Stepak, J., and Kutova, J., *Sb. Vys. Sk. Chem. Technol. Praze*, *S9*, 205–215 (1983).
160. French Patent FR 2480771 A1 (1981), Rhone-Poulenc.
161. Czech. Patent CS 205243B (1983), to Dvorak, J., Fajtova, J., Laita, Z., Nejedly, E., Prochazkova, M., Rehorova, H.
162. McMurrer, M. C., *Plast. Compounding*, 88–96 (Jan./Feb. 1982).
163. Stofoff, A., and Skrada, K. A., *J. Vinyl Technol.*, *4(2)*, 76–80 (1982).
164. *Kunstoffe*, *55(9)*, 724 (1965).
165. *Rubber Plast. Age*, *46(1)*, 39 (1965).
166. *Plastics*, *31(347)*, 1158 (1966).
167. *Br. Plast.*, *40(10)*, 104 (1967).
168. *Mod. Plast.*, *45(3)*, 96 (1967).
169. *Mod. Plast.*, *44(3)*, 94 (1966).
170. Lutz, J. T., Jr., *Polym. Plast. Technol. Eng.*, *11(1)* (1978).
171. Lutz, J. T., Jr., *Soc. Plast. Eng. Annu. Tech. Conf.*, Montreal, 1977.
172. Summers, J., *J. Vinyl Technol.*, *5(2)* (June 1983).
173. Summers, J., and Rabinovitch, E., *J. Vinyl Technol.*, *5(3)* (Sept. 1983).
174. Mathur, K., and Tapper, M., *J. Vinyl Technol.*, *5(4)* (Dec. 1983).
175. *Plast. World* (Nov. 1982), pp. 55–57.
176. Lutz, J. T., *Polym. Plast. Technol. Eng.*, *21(2)* (1983).

6

Fillers: Types, Properties, and Performance

K. K. MATHUR and D. B. VANDERHEIDEN

Pfizer Inc.
Easton, Pennsylvania

I. INTRODUCTION

The concept of cost reduction by use of filling materials has been known throughout the ages. Since the earliest days, very significant advances have been made in this area in terms of fine particle size technology, tailoring particle morphology, beneficiation of natural materials to attain high purity, surface treatments for improved matrix compatibility, and the development of coupling agents to achieve polymer-to-filler bonding for improved mechanical properties.

In general, fillers are defined as materials that are added to the formulation to lower the compound cost. Such materials can be in the form of solid, liquid, or gas. By the appropriate selection and optimization of such materials, not only the economics but other properties such as processing and mechanical behavior can be improved. This chapter primarily covers the solid fillers used in PVC applications. Liquid fillers such as extender plasticizers and gaseous fillers such as blowing agents are covered in separate chapters.

For effective utilization of fillers in PVC, a complete understanding of the individual filler characteristics is essential. Each class of fillers appears to exhibit specific characteristics which make them especially suited for the given application, for instance, ultrafine precipitated calcium carbonates for improved notched Izod impact strength and calcined clays for electrical properties. Although these fillers retain their inherent characteristics, very significant differences are often seen, depending on the molecular weight of PVC, compounding technique, and the presence of other additives in the formulation. Therefore, once the basic property requirements are established, the optimum filler type and loading for cost/performance balance must be determined. Frequently, blends of two different fillers are used to attain a balance of properties. Special consideration must also be given to determining the acceptability of the processing properties. Such information will lead one to a complete understanding of the benefits to be derived from the use of filler in a given compound.

For effective utilization of fillers, it is critical that they be well dispersed in the polymer system. Some coarser fillers can be dry-blended and injection-molded at 1 to 10 phr without a major sacrifice in surface appearance and physical properties. Melt compounding is essential to disperse fine fillers for maximum impact strength improvement as well as retention of impact strength at high loadings. In a highly agglomerated state, small particle size fillers may behave like coarse fillers, causing undue weakness in the matrix which will be related to loss in notched Izod impact and low-temperature impact properties in PVC.

The addition of filler also requires a balance of formulation for optimum processing properties. For example, the use of stearate-coated calcium carbonates will require a reduction in lubricant level for optimum fusion torque and maximum extruder output in rigid PVC compounds. The stearate-treated fillers sometimes also act as costabilizers, which should be taken into consideration for formulation economics.

Therefore, before making a final decision on a filled PVC compound, it is critical to establish the following:

1. Optimum loading level for property/benefit
2. Optimum formulation for processing/production output
3. Economics of filled formulations

II. FILLER CLASSIFICATION

Fillers have been classified in many different ways, ranging from their shapes (e.g., spheres, rods, ribbons, flakes) to specific characteristics (e.g., conductivity, fire retardancy). For simplicity, PVC fillers can be classified in two categories: (1) extenders and (2) functional materials.

Although practically all of the fillers exhibit some functional property, the above classification is tied to the primary reason for using a filler. By definition, an extender filler primarily occupies space and is mainly used to lower the formulation cost. A functional filler, however, has a definite and required function in the formulation apart from cost; examples are antimony oxide for fire retardancy and pyrogenic silicas for rheology modification.

As with all attempts to classify, there are gray areas where overlap and poor definition exist. For example, some of the extender fillers, when reduced to a finer particle size and/or surface-treated, would be reclassified as functional fillers. The development of new surface treatment technologies has further broadened the ability to graft functional characteristics to extender fillers. As an even further complication, some functional fillers in rigid PVC may turn out to be merely extenders in plasticized PVC. Such factors have seriously complicated the task of establishing sharp boundary lines between extenders and functional fillers in terms of their generic composition; however, on a performance basis, the two can be separated as shown in Figure 1.

Therefore, the extender fillers basically lower the formulation cost and increase flexural modulus, while the functional fillers provide at least one specifically required function in the formulation, such as thixotropy, fire retardancy, opacity, color, or impact modification.

III. CHEMISTRY AND PROPERTIES OF PVC FILLERS

A. Extender Fillers

In general, an ideal extender filler should (1) be spherical to permit retention of anisotropic properties, (2) have an appropriate particle size distribution for particle packing, (3) cause no chemical reactivity with PVC or additives, (4) have a low specific gravity comparable to or lighter than that of PVC resin, (5) have a desirable refractive index and color, and (6) be low in cost. Thus far, no single product stands out in all of these specifications. Ground limestones come closest to the specifications and are most commonly used as extenders in both rigid and flexible PVC. Ground talcs and clays, even though they improve some electrical properties, are also classified as extenders for flexible PVC because they allow filling to high loadings without adversely affecting the physical properties, and they meet other extender qualifications.

1. Limestone: Limestone is a naturally occurring mineral. Chemically, it is $CaCO_3$ and may contain a small amount of $MgCO_3$ and possibly traces of other impurities such as SiO_2, Al_2O_3, and Fe_2O_3. In mineralogical terms, it is classified as a trimorphous mineral because it exists in three distinct crystal structures, namely calcite, aragonite, and vaterite.

Calcite, in its various geological forms, is one of the most widely occurring minerals. It is the chief constituent in all limestones and marbles and is found in sedimentary and metamorphic rocks. The fossiliferous form of

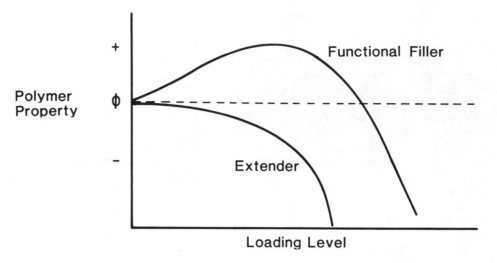

FIGURE 1 Characteristics of functional and extender fillers.

limestone exists in oyster shells, which find special applications. Its crystal
habits are shown in Figure 2.

   Limestone is a moderately soft mineral. On the Mohs hardness scale, it
is rated at 3. Because of this property, limestone can be ground to a very
fine particle size. Depending on the ore quality, dry or wet beneficiation

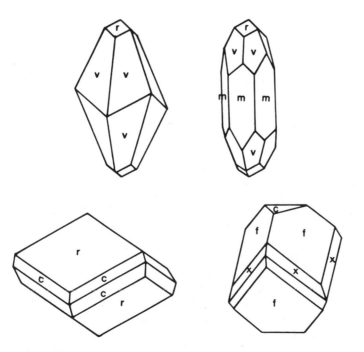

FIGURE 2 Crystal habits of calcite.

is often necessary to improve the color. The milling can be conducted in dry or wet form, often being dictated by the particular wet or dry benefi- ciation requirements. The milled product is then classified through a par- ticle classifier to achieve the desired top size.

In rigid PVC applications, smaller top size ground limestone is preferred for several reasons: (1) it permits greater impact strength retention up to higher loadings, (2) it minimizes abrasive wear on processing equipment, particularly extruders, and (3) it provides excellent surface appearance of parts.

Stearate surface-treated ground limestones offer many advantages over uncoated products, such as hydrophobicity, better powder flow properties (causing less hang-up in automated systems), the possibility of formulation cost reduction by lowering lubricant and stabilizer level, significant improve- ments in low-temperature impact strength, and reduced abrasivity. The stearate treatment chemically reacts with the limestone surface and is usually applied in sufficient quantity to form a monomolecular layer. The limestone and stearic acid reaction takes place as follows:

$$2C_{17}H_{35}COOH + CaCO_3 \rightarrow (C_{17}H_{35}COO)_2Ca + CO_2\uparrow + H_2O$$

In general, limestones of relatively fine particle size are used in rigid PVC as compared to plasticized PVC. In PVC plastisols and organosols, where thixotropy and plastic viscosity are critical, finer particle size ground lime- stones may be preferred. With the availability of a wide variety of particle sizes and suitable morphology, these materials are frequently blended in plastisols for improved particle packing.

In rigid PVC formulations, relatively fine particle size stearate-coated ground limestone fillers are most commonly used. Because of the fine par- ticle size and stearate coating, they offer improved processing as well as physical property advantages in the form of improved notched Izod and drop weight impact strength (Table 1).

In plasticized PVC, the finer ground limestones impart higher tensile strength and tensile modulus, with some reduction in ultimate elongation, as compared to the coarser limestone extenders (Table 2).

The dielectric properties of ground limestone, along with other commer- cially important minerals and PVC, are listed in Table 3 [1].

The stearate-coated ground limestones in PVC wire and cable formula- tions demonstrate increased volume resistivity, insulation resistance, and significantly reduced moisture sensitivity (Table 3).

In addition, a series of wet ground limestone fillers of very small par- ticle size (1.0 μm) have been introduced by a few manufacturers under the name of ultrafine ground limestone (UFGL). These products are sold com- mercially with and without the stearate treatment and provide improved properties over the coarser ground limestones. Key improvements are seen in impact strength retention in rigid PVC and reduced abrasivity.

In addition to the ground limestone, there are numerous manufactured (precipitated) grades of $CaCO_3$ which are also widely used; these are dis- cussed further in Section III.

2. *Kaolin Clay*: The mineral commercially known as kaolin is basically a hydrous aluminosilicate mineral with the chemical formula $Al_2O_3 \cdot 2SiO_2 \cdot 2H_2O$. Thus, a theoretical kaolin contains 46.5% silica, 39.5% alumina, and 14.0% water. The kaolin family covers three different species, namely nacrite, dickite, and kaolinite. The U.S. deposits are primarily of the kaolinite type.

TABLE 1 Effect of 20% Loading of Stearate-Coated 3-μm Ground Limestone
on Physical Properties of PVC[a]

Property	Unfilled rigid PVC	20% stearate-coated 3-μm ground limestone
Flexural modulus (psi)	510,000	650,000
Tensile yield stress (psi)	7,800	6,700
Elongation at yield (%)	5.0	4.8
Notched Izod impact strength (ft-lb/in.)	0.86	1.70
Drop weight impact (ft-lb)	2.3	4.1

[a]Formulation (in phr): PVC K-62, 100.0; organotin mercaptide, 1.6; acrylic
process aid, 1.5; lubricant A, 1.2; lubricant B, 0.5; and TiO_2, 1.0.

Kaolinite is a platy material. The crystal elongates parallel to the c-axis.
The alumina octahedral sheet shares the oxygen with silica tetrahedral
sheets, while the OH and O form the external layers (see Figure 3). The
basal spacing is perfect for cleavage. Therefore, by selection of the appropriate milling equipment, kaolin can be delaminated to fine sheets.

The molecular weight of kaolin is 258.09. The specific gravity of the
uncalcined product is 2.58, which increases to 2.63 on calcination.

The bound water begins to be liberated at 330°C. Calcination at 650°C
produces partial dehydration of the kaolinite lattice, resulting in its conversion to pseudocrystalline "meta-kaolinite." Complete dehydration is accomplished at 900°C. At this stage, a significant change in density and light
refractivity takes place.

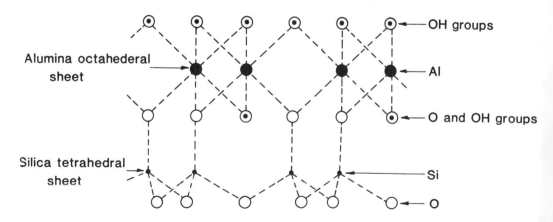

FIGURE 3 Idealized structure for the alumina/silica layers of hydrated
aluminum silicate (kaolin).

TABLE 2 Effect of Ground Limestone Particle Size on Properties of Plasticized PVC[a] at 12 phr Filler Level

Limestone particle size (μm)		BET surface area $(N_2 - m^2/g)$	Ultimate tensile strength (psi)	Tensile modulus at 100% elongation (psi)	Percent ultimate elongation
Top	Average				
15	3	4.3	3230	3210	110
25	5	3.8	2070	1880	140
41	7	2.7	1960	1760	150
31	6	1.9	1960	1610	150

[a]Formulation (in phr): PVC MHMW, 100.0; Sb_2O_3, 1.4; N-octyl-N-decyl phthalate, 67.0; octyl epoxytallate plasticizer, 5.5; Ba/Cd/Zn stabilizer, 2.2; lubricant, 0.2; and limestone filler, 12.0

TABLE 3 Effect of Stearic Acid-Treated and Untreated Ground Limestones in a PVC Wire and Cable Formulation

	Formulation	
	I	II
Component		
PVC	100	100
DOP	45	45
Lead stabilizer	4	4
Lead stearate	0.5	0.5
Calcined clay	10	10
Ground limestone (3 μm)	30	—
Stearate-treated ground limestone (3 μm)	—	30
Properties		
Modulus at 100% E (psi)	2060	1950
Ultimate tensile strength (psi)	2730	2740
Ultimate elongation (%)	260	280
Dielectric strength (volt/mil)	330	320
Volume resistivity (ohm-cm^{-1})	3.6×10^{14}	4.2×10^{14}
Insulation resistance (ohms)	2.3×10^{14}	4.1×10^{14}
Dielectric constant at 60 cps	5.78	5.59

Kaolin clays are commercially sold in air-floated, water-washed, calcined, and silane-treated forms. The color and purity of the products depend on the beneficiation technique and posttreatments. The calcined clays are primarily used in wire and cable jacketing and insulation formulations at 5 to 15 phr levels. Much higher levels cause increased stiffening and lowering of elongation. Because of the alignment of platelets in the extruder direction, the kaolin-filled PVC wire and cables possess even further improved electrical properties.

The platelet alignment increases the electron flow path considerably (see Figure 4). The critical electrical properties of kaolin and other fillers are given in Table 4. The low-cost air-floated and water-washed clays are commonly used in flooring, film, toys, and upholstery applications.

3. *Talc*:   Chemically, talc is a hydrated magnesium silicate. The chemical formula for pure talc is $3MgO \cdot 4SiO_2 \cdot H_2O$. In actuality, the ratio of MgO to SiO_2 may vary from 1:2 to 1:1. The structure of talc is shown in Figure 5.

Talc as found in nature can be associated with a wide variety of other materials. Montana talc is an exceptionally pure talc ore that very nearly approximates the theoretical formula of talc. New York talcs generally contain only 40 to 60% talc. Texas talc contains high amounts of dolomite ($CaCO_3 \cdot MgCO_3$), while Vermont talc has varying levels of magnesite. California and Nevada talcs vary in impurities with the particular ore source.

Talc is the softest of known minerals. On the Mohs hardness scale, it is rated 1. The specific gravity of talc is 2.7. Mining of talc is carried out by classical mining techniques. Although most of the mines are open pit, there are some active underground mines.

TABLE 4 Dielectric Properties of Selected Fillers and PVC

Typical fillers	Dry resistivity (ohm-cm)	Humid resistivity (ohm-cm)	Dielectric strength (volt/mil)	Dielectric constant[a] Ke (at 1 mc)	Dielectric loss (%)
$CaCO_3$	10^{11}	10^7	60 − 80	6.1	0.05
Kaolin	10^{13}	10^6	70 − 120	2.6	0.1
Kaolin (calcined)	10^{13}	10^8	60 − 100	1.3	0.06
Kaolin (calcined and surface-treated)	10^{13}	10^{12}	80 − 150	1.3	0.003
Kaolin, partially calcined	10^{13}	10^5	70 − 100	1.3	0.01
Talc	10^{14}	10^9	—	5.5 − 7.5	0.001
PVC	10^{16}	10^{16}	700 − 1300	2.9	0.01

[a]Measured at 1 MHz.

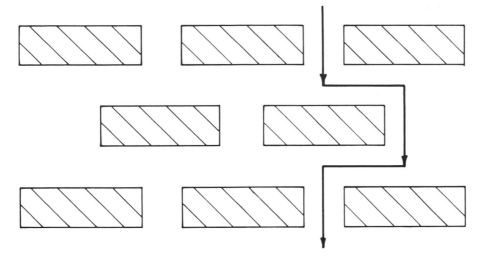

FIGURE 4 Flow of electrons in kaolin-filled PVC.

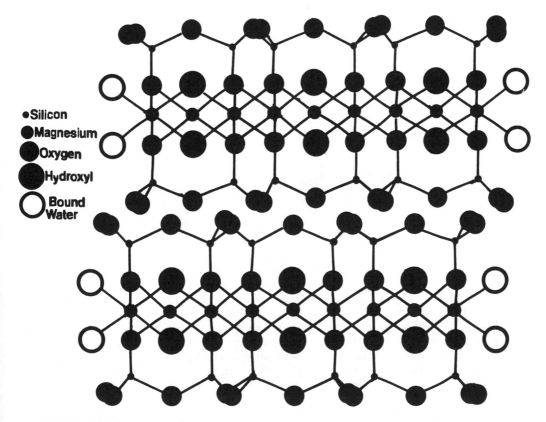

FIGURE 5 Structure of talc.

In a typical talc operation, the ore is crushed into ½ to 3 in. pieces with a jaw crusher, then water washed and sorted to remove impurities. It is then cone crushed and milled to varying particle sizes in a closed loop with pneumatic classifiers to obtain a product with desired particle sizes and distributions.

Talcs with high impurities require beneficiation by flotation, magnetic separation, bleaching, and so forth to attain a brighter and chemically purer product.

Because of the perfect basal spacing, talc can be exfoliated into perfect platelets when appropriate milling techniques are used.

In plastic processing, the platy particles align in the machine direction. In general, platy talcs improve hot strength, increased modulus, and tensile strength and are generally used in plasticized PVC applications, especially vinyl flooring. The fibrous New York talcs have also been used in this application in the past. A scanning electron micrograph of talc-filled PVC is shown in Figure 6.

In addition to their use as fillers, fine-ground talcs are also used as antiblocking agents in thin PVC films. The delaminated calcined talcs in general have properties similar to those of kaolin clay, as shown in Table 4.

Because of its high oil absorption, talc is also frequently used to increase PVC plastisol thixotropy and plastic viscosity.

The talc surface is highly active. Talc can be surface-treated to achieve a wide variety of useful properties. The most common surface-treating agents are polyethers and polyols, which give the talc improved properties for a variety of applications [2]. Because of the silicate surface, talc also reacts with silanes. In PVC such products show improved dispersibility.

B. Functional Fillers

1. Precipitated Calcium Carbonate: Precipitated calcium carbonates are among the most versatile functional fillers used in the PVC industry. They can be manufactured from lime by three basic technologies as follows [3]:

Lime/CO_2: $Ca(OH)_2 + CO_2 \rightarrow \underline{CaCO_3} + H_2O$

Lime/soda: $Ca(OH)_2 + Na_2CO_3 \rightarrow \underline{CaCO_3} + 2NaOH$

Solvay process: $NH_3 + H_2O + CO_2 + NaCl \rightarrow \underline{NaHCO_3} + NH_4Cl$

$Ca(OH)_2 + 2NH_4Cl \rightarrow CaCl_2 + 2NH_4OH$

$CaCl_2 + Na_2CO_3 \rightarrow \underline{CaCO_3} + 2NaCl$

As a result of the chemical processing, precipitated calcium carbonates have a chemical purity far beyond that normally found in the ground limestones. The chemistry and varied crystallography of calcium carbonate also permit the manufacture of products displaying a wide range of particle size, particle shape, and particle size distribution, as shown in Table 5.

Probably the most important value of precipitated calcium carbonate in rigid PVC today stems from the ability of surface-treated products in the particle size range 0.07 to 0.5 μm to increase substantially the impact strength of PVC (Table 6 and Figures 7 and 8) [4, 5]. This is accomplished with relatively low loading levels in the 5 to 30 phr range and is not accompanied by a loss in the heat stability of the compound (Table 6).

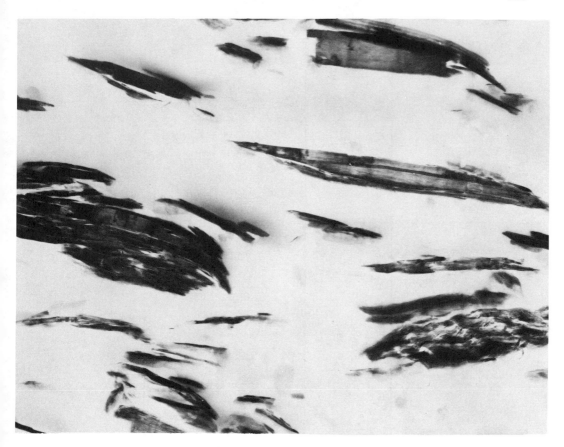

FIGURE 6 Microtomed section of talc-filled PVC under scanning electron microscope at ×35,000.

Improvements in ultraviolet (UV) light stability and exterior weatherability of PVC have also been reported, hypothesized to result from the HCl-scavenging ability of the precipitated calcium carbonate [4, 6]. Calcium stearate-coated precipitated calcium carbonate is also being used in some plasticized applications as a thixotrope and a secondary heat stabilizer.

Precipitated calcium carbonates also provide functionality in the fire resistance of PVC. Precipitated calcium carbonates of very small size (0.05 to 0.10 μm) and high surface area (~20 m^2/g) effectively scavenge HCl gas liberated from burning PVC, thereby reducing the corrosiveness of the generated smoke.

$$CaCO_3 + 2HCl \rightarrow CaCl_2 + CO_2 + H_2O$$

At the same time, the carbon dioxide and water generated in the neutralization reaction serve to keep the oxygen index low enough for self-extinguishing properties. Applications of calcium carbonate in PVC jacketing compounds for electrical cables have been promoted based on the HCl-scavenging functionality of the filler [7].

TABLE 5 Identification and Physical Properties of Surface-Treated Calcium Carbonate Products

Material	Product identification	Particle size via TEM analysis (μm)		BET surface area (m²/g)	Oil absorption ASTM D281-31 (lb oil/100 lb filler)	Dry brightness (%)	Percent wet out in water (%)
		Average	Top Size				
Surface-treated fine natural ground calcium carbonate	3μ GL	3	15	3.6	14	94	0
Surface-treated very fine precipitated calcium carbonate	0.5μ PCC	0.5	<1.5	6.0	26	94	0
Surface-treated ultrafine precipitated calcium carbonate	0.07μ PCC	0.07	<0.2	20	35	94	0

TABLE 6 Modifications of a Rigid PVC Compound with Surface-Treated Calcium Carbonates to Provide High Impact Resistance at Ambient and Low Temperatures

	Traditional approach using elastomeric impact modifier	Impact modification using surface-treated calcium carbonates	
	Unfilled rigid PVC control impact modified with 8 phr MBS	Rigid PVC with 0.5-μm PCC and 2.5 phr MBS	Rigid PVC with 0.07-μm PCC (no additional impact modifier)
Falling weight impact (ft-lb)			
at 23°C	25.2	25.8	25.4
at 0°C	25.0	24.9	25.0
at -30°C	21.8	18.9	22.0
Notched Izod impact (ft-lb/in.)	5.4	5.4	6.0
Tensile yield strength (psi)	7,300	6,900	7,200
Flexural modulus (psi)	400,000	490,000	490,000
Heat deflection temperature (°C)	64	64	64

2. Precipitated Silicas and Silica Gels: Precipitated silicas and silica gels are classified as wet process silicas to differentiate them from the thermally processed pyrogenic silicas. They are produced by the reaction of an aqueous alkali silicate solution with a mineral acid solution, e.g. [8],

$$Na_2SiO_3 + H_2SO_4 \rightarrow Na_2SO_4 + SiO_2\downarrow + H_2O$$

Precipitation can be conducted in three general ways, based on the order of addition of the reactants:

1. Alkali silicate solution added to acid solution
2. Acid solution added to alkali silicate solution
3. Simultaneous addition of alkali silicate and acid solutions into water or neutral salt solution

The first of these processes is the preferred method for producing silica gels. These products are characterized by very small primary particle sizes and high surface areas in the range 200 to 800 m^2/g. These physical properties and their relatively high cost promote their use in many applications also served by the pyrogenic silicas. These applications in PVC include: (1) prevention of plateout, (2) antiblocking of film, (3) flow control aid for resin powders, (4) viscosity control agent, (5) processing aid, and (6) selective adsorbent and moisture removal agent [9].

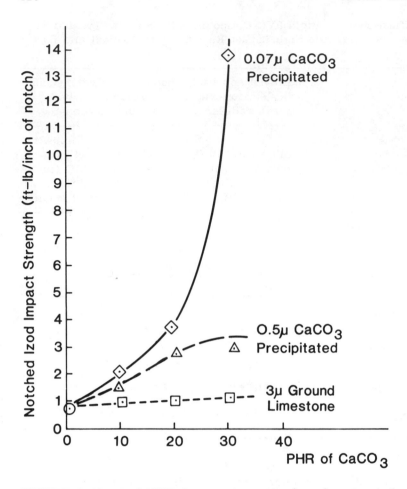

FIGURE 7 Notched IZOD impact strength of surface-treated calcium
carbonates in a rigid PVC compound.

     Precipitated silicas are generally produced by precipitation procedure
2 or 3 above [8]. Simultaneous addition is generally the preferred technique,
allowing variation and control of pH, precipitation time, temperature, and
electrolyte concentration to produce a range of precipitated silica products
(Table 7).
     Precipitated silicas offer a range of property improvements in PVC. In
a formula simulating a PVC floor covering, advantages reported included
(1) increased hardness, (2) improved elasticity, (3) increased resistance to
heat distortion, and (4) improved scratch resistance (Table 7) [10]. Where
transparency or translucency is desirable, large additions of synthetic silicas can be made with minimal effect. In plastisols used to coat textiles, precipitated silicas can provide improved mar resistance and flatting if desired.
They are also used in vinyl plastics as release agents to smooth out calendered sheets.
     3. *Precipitated Metallic Silicates*: Precipitated metallic silicates are pro-
duced by a process very similar to that just described for precipitated sil-

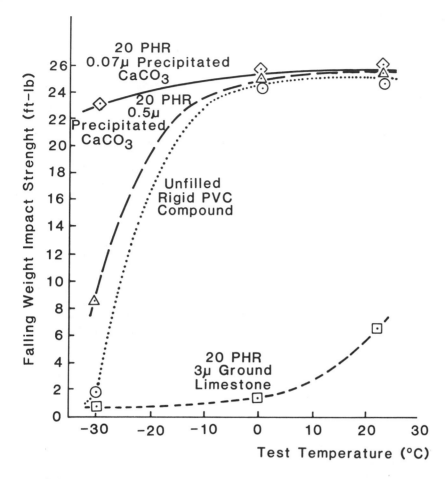

FIGURE 8 Low-temperature falling weight impact behavior of unfilled and 20 phr CaCO₃-filled rigid PVC.

icas. In the case of metallic silicates, the aqueous alkali silicate is reacted with a solution of a metal salt instead of a mineral acid. For example:

$$Na_2SiO_3 + CaCl_2 \rightarrow CaSiO_3 + 2NaCl$$

The most important metallic silicates still in use are calcium silicate, basic lead silicate, aluminum silicate, and barium silicate. Typical physical properties are reported in Table 8. Although the importance of the metallic silicates has waned significantly in recent years in favor of the synthetic silicas, they do complement the pigment group in that they have a significantly alkaline pH of 9 to 12 compared to pH 6 to 9 for the precipitated silicas and a highly acid pH (3.6 to 4.3) for the pyrogenic silicas.

Metallic silicates have been added to vinyl formulations for a wide variety of reasons. Advantages reported for their use include (1) improved electrical resistivity, (2) modified rheological properties, (3) improved hand and drape characteristics in vinyl fabrics, (4) prevention of plateout, (5) optical extension of TiO₂ in opaque PVC formulations, and (6) processing improvements in dry-blend extrusion operations.

TABLE 7 Effect of Precipitated Silica on PVC
Mechanical Properties [10]

Property	Unfilled	50 phr FK-320
Shore A hardness immediate	55	85
Modulus 100 kg/cm^2	25	144
Permanent elongation at breaking (%)	73	45
Permanent set 50°C (%)	62	57
Needle tear-out resistance (kg/cm)	27	28

4. Pyrogenic Silica: Pyrogenic silicas, also referred to as fumed silicas, comprise a very special class of functional silica fillers. The special properties of pyrogenic silicas result primarily from the unique manufacturing process, first practiced by Degussa in 1942, in which silicon tetrachloride is hydrolyzed in an oxygen-hydrogen flame [11].

$$2H_2 + O_2 \rightarrow 2H_2O$$

$$SiCl_4 + 2H_2O \rightarrow SiO_2 + 4HCl$$

$$2H_2 + O_2 + SiCl_4 \xrightarrow{1000°C} SiO_2 + 4HCl$$

In terms of physical properties, pyrogenic silicas are characterized by (1) exceptionally small discrete particle sizes ranging from 7 to 16 nm (0.007 to 0.016 μm) with attendant surface areas of 120 to 380 m^2/g, (2) low moisture levels, generally less than 1.5%, (3) acidic pH values in the range 3.6 to 4.3, (4) extremely high chemical purity, generally >99.8% SiO_2 on an ignited basis, and (5) very low bulk densities, ranging from 50 to 100 g/l (specially densified). These properties are compared with those of the precipitated silicas and silicates in Table 8 [8].

The most important application of pyrogenic silicas in PVC is related to their use as rheology modifiers in PVC plastisols and organosols. Very small amounts of fumed silicas provide increased viscosity coupled with pronounced thixotropy [12]. It is this thixotropic flow behavior that is essential where processing and/or application of the plastisol requires a higher degree of fluidity than can be tolerated once the product is in place in its ultimate use.

Although less important, many other PVC applications are served by pyrogenic silicas [10]. As a result of their purity and low moisture content, fumed silicas are very good insulators and highly suitable for improving the electrical properties of PVC cable compounds. Pyrogenic silicas are also used to reduce plateout in PVC processing. Surface-modified (hydrophobic) pyrogenic silicas are recommended to improve the dry flow (anticaking) properties of pure PVC powders while the standard grades are used to correct the dry flow deficiencies of dry blends (plasticized PVC powder).

TABLE 8 Physical Property Comparison of Silica-Based Pigments [9, 10]

Physical property	Pyrogenic silica	Precipitated silica	Precipitated metal silicate
Discrete particle size (nm)	7 − 16	15 − 100	20 − 50
Surface area (m^2/g)	100 − 500	40 − 250	35 − 180
Drying loss (%)	<1.5	4 − 7	5 − 9
SiO_2 (dry) (%)	>98.5	83 − 90	63 − 80
DBP absorption (g/100g)	−	175 − 285	165 − 220
pH (5% in water)	3.6 − 4.3[a]	6 − 9	9 − 12
Compacted apparent density (g/l)	50 − 100	150 − 250	100 − 250

[a]4% in water.

5. *Antimony Oxide*: Antimony oxide (Sb_4O_6) is a very effective fire retardant additive for plasticized PVC [13−15]. The pigment is produced by roasting the naturally occurring sulfide ore (Sb_4S_6) and then is purified by distillation [12]. The chemistry of antimony is much like that of its more familiar group V family member, phosphorus, and similar fire retardant materials have been developed around each element as a base. A review prepared in 1979 [17] lists over 200 references on the uses of antimony compounds as fire retardants.

In plasticized PVC, the principal means of achieving flame resistance starts with the substitution of a flame-retarding phosphate ester plasticizer for the more flammable varieties. The flame resistance can be significantly increased beyond this point by the addition of as little as 1 to 5 phr antimony oxide [18]. The combination of phosphate plasticizer and antimony oxide pigment generally produces acceptable retardancy while retaining good low-temperature properties but with a sacrifice in clarity of the resulting compound [19].

6. *Zinc Oxide*: Interest in zinc oxide for PVC revolves around its pigmentary properties as well as heat and light stabilization properties. Zinc oxide pigment exists in many grades, the basic properties of which are indicated in Table 9 [20]. Two primary pyrometallurgical processes are used for its manufacture, the French or indirect process and the American or direct method [21]. Zinc sulfide ore, the common starting raw material, is roasted, purified, and the zinc metal vapor burned directly to zinc oxide in the American process while the French process contains an intermediate condensation step to zinc metal.

Based on its primary pigment properties (white color, fine particle size, high opacity/UV absorption), zinc oxide has found uses in flexible and rigid PVC for exterior applications, although its use is small compared to that of TiO_2. In vinyl flooring, zinc oxide has been used often as a heat and light stabilizer, especially in formulations involving large amounts of limestone filler. Stabilization mechanisms involving synergism with alkaline earth salts have been proposed (see Chapter 13). Many proprietary stabilizer systems for vinyl floor tile are known to contain zinc oxide as an important ingredient.

TABLE 9 Typical Properties of Titanium Dioxide and Zinc Oxide

Property	Titanium dioxide		Zinc oxide
	Anatase	Rutile	
Average particle size (μm)	0.3	0.2 $-$ 0.3	0.2
Density (g/cm^3)	3.9	4.1	5.6
Refractive index	2.55	2.76	2.01
Tinting strength	1200	1600	210
Oil absorption (lb/100 lb)	18 $-$ 30	16 $-$ 48	10 $-$ 25
Hardness (Mohs)	5 $-$ 6	6 $-$ 7	4+

7. *Titanium Dioxide* (see also Chapter 18): The functional value of TiO_2 in PVC, as in most other application areas, lies in its unsurpassed opacifying (hiding) power among white pigments. It exists in two primary crystal forms, anatase and rutile, the basic properties of which are given in Table 9 [20]. Two basic manufacturing processes are presently in use for TiO_2. The sulfate process operates from an ilmenite or iron titanate raw material, a hydrous titanium dioxide being extracted with sulfuric acid, purified, and calcined to either anatase or rutile pigment grades. The newer chloride process starts with a natural rutile ore that is converted to titanium tetrachloride, purified, and then reacted with oxygen at about 1500°C to reform the rutile pigment. Both processes include numerous finishing steps including very important inorganic surface treatments plus milling, drying, and classification to produce the final pigment products.

The pigmentary properties of whiteness, brightness, and opacity are the prime reasons for the incorporation of TiO_2 in PVC. Although anatase was used in some early formulations where a blue tone whiteness was required, finer particle sized rutile specialty grades have now effectively replaced and eliminated anatase in most polymer systems. The blue tone rutile pigments develop much higher brightness and opacity and impart superior weathering characteristics compared to anatase. One area where the coarser anatase grades continue to be the product of choice is in plastisols [22].

Rutile TiO_2 is also an effective ultraviolet light absorber and is thus capable of protecting PVC in exterior applications. Numerous studies paralleling those in the paint area have shown that the inorganic-coated, nonchalking grades of rutile TiO_2 (rutile-exterior grades) provide the greatest degree of protection [23]. TiO_2 levels of 12 phr in flexible PVC appear to provide the best practical protection. It is certainly no coincidence that the 8 to 12 phr level is also the area of optimum cost/performance ratio on the basis of optical properties.

8. *Iron Oxide Pigments* (see also Chapter 18): Natural and synthetic iron oxides are frequently used as pigments in plasticized PVC [24], where the processing temperatures are relatively lower. Iron oxide pigments, in addition to excellent ultraviolet stability, offer good chemical and mildew resistance. The chroma and hue of these pigments frequently require toning

with brilliant organic pigments for brighter shades [25]. The blending of inorganic and organic pigments usually results in the most economical pigmentation package.

In rigid PVC, where processing temperatures may be as high as 195° to 200°C, the iron oxide yellow and iron oxide black pigments are not recommended. At these processing temperatures, the yellow iron oxide ($Fe_2O_3 \cdot H_2O$) will partially dehydrate to α-Fe_2O_3, giving a reddish yellow color. Similarly, the iron oxide black begins to oxidize to γ-Fe_2O_3, giving a reddish black shade. Several patents have shown the use of phosphate doping to increase the processibility of yellow iron oxide pigment [26]. In Europe, the iron oxide pigments are very frequently used in rigid PVC. This is accomplished by selecting appropriate stabilizers and lubricants and by maintaining lower processing temperatures.

Since the iron oxide pigments absorb low amounts of infrared, they are frequently used in PVC products designed for exterior applications. Because of the low infrared absorption, they exhibit low heat buildup on exposure to sunlight and could be especially suited for applications such as PVC siding, profiles, outdoor posts, tubular patio chairs, and so forth [27].

The iron oxide pigments used for audio, video, instrumentation, and disk recording PVC tapes require very special properties. The slow recording speed and high fidelity require ultrafine gamma ferrites. The magnetic inks for PVC credit cards and railway tickets use synthetic magnetite, which permits high concentration in the polymer system.

9. *Organic Pigments*: A wide variety of ultraviolet-stable organic pigments are being used for exterior PVC applications. The key requirements for such pigments in PVC are (1) thermal stability, (2) ultraviolet stability, (3) environmental fading stability, and (4) acid rain stability. Additional requirements are discussed in detail in Chapter 18.

10. *Carbon Black* (see also Chapter 18): Carbon black pigments can provide a variety of special property advantages in PVC, even though they do not provide the reinforcing characteristics commonly associated with their primary use in rubber. Carbon blacks are produced by three quite different processes which lead to substantial differences in final product properties [28]. By far the most important process is the furnace black process, in which a highly aromatic oil feedstock is decomposed in a furnace chamber containing the combustion products of an oil or gas flame. Over 90% of carbon blacks produced and sold are made by the furnace black process. The remaining products result from use of the far less important channel black and thermal black processes. General property ranges of carbon blacks made by the three processes are compared in Table 10 [28–30].

In PVC, functionality of the carbon blacks can be exceptionally broad [29, 20]. For pigmentation purposes, the relatively coarse furnace or thermal blacks are commonly used because of their relatively lower cost and ease of dispersion. Carbon black pigmentation also provides significant protection from UV and thermal degradation. Carbon black is capable of absorbing harmful ultraviolet radiation and at the same time scavenging the free-radical degradation products capable of catalyzing further degradation. Thus, carbon blacks are well-known stabilizers and screening agents for flexible and rigid PVC compounds for exterior applications.

Carbon blacks are also used to control the electrical properties of PVC. Depending on the carbon black chosen, high electrical conductivity or high electrical resistivity can be imparted. In applications where high conduc-

TABLE 10   Carbon Black Properties Related to Production Methods

Property	Furnace black	Channel black	Thermal black
Average particle size (nm)	13 — 75	10 — 30	150 — 500
Surface area (N_2 BET) (m^2/g)	25 — 560	100 — 1125	6 — 15
Oil absorption (cm^3/g)	0.70 — 1.85	1.0 — 5.7	0.3 — 0.5
pH	3.3 — 9.0	3.0 — 6.0	7.0 — 8.0
Percent volatiles	1.0 — 9.5	3.0 — 17.0	0.1 — 0.5

tivity is required, for example to reduce static charge buildup on molded parts, a carbon black with a fine particle size, high structure, and low volatile content is required. Where high electrical resistance is desired, a carbon black with a coarser particle size, low structure, and high volatile content will be most effective. The coarse size and low structure decrease the number of particle contacts through which electrons can flow and the surface volatiles act as insulators, further inhibiting conductivity in this case.

11. Asbestos: In terms of volume usage, asbestos was a very large volume filler for PVC tiles in years past. More recent health and safety studies have shown that inhalation of fibrous asbestos over a period of time can cause asbestosis, a nonmalignant fibrotic lung condition, bronchogenic carcinoma, and mesothelioma, a rare cancer of the chest lining and abdominal cavity. Because of these health risks, the PVC tile industry has in recent years reformulated their products [31].

In nature, asbestos occurs in two basic forms: chrysotile, also known as serpentine asbestos, and the amphiboles, which include tremolite, amosite, crocidolite, actinolite, and anthophyllite. These varieties differ chemically. The chemical composition from deposits varies depending on the associated impurities. The color varies from white to gray. The specific gravity varies from 2.48 to 2.56.

High-aspect-ratio asbestos exhibits excellent reinforcement and contributes to dimensional stability and impact strength, thereby allowing a greater total volume loading of filler for improved economics. Other high-aspect-ratio organic fibers lack the basic characteristics of asbestos. Organic fibers suffer from poor wetting in the PVC resin and a relatively lower hot strength, which is critical for continuous processing operations.

Both of these problems have been overcome to some degree by the industry. Overall, the industry has been quite secretive about the new formulations, but the recent literature and patents [32] from major manufacturers shed some light in this area.

The most direct patents have discussed the use of synthetic mineral wool, polyester, and cellulosic fibers to replace asbestos. A host of wetting agents, mixing equipment, and incorporation techniques have been tried and tailored to achieve maximum wetting of fibers.

IV. FILLER PROPERTIES AFFECTING THE PERFORMANCE OF FILLED PVC

A. Color

To the vinyl formulator, color of the filler (or, more precisely lack of color) is generally second only to cost in the ranking of filler selection criteria. Cost and color are inexorably related, however, because whiteness is a key factor on which filler value is established by the manufacturer, whether it is a product of natural origin or chemical synthesis.

Color is established by the light absorption of the pigment over the visible region. The ideal white pigment will show no absorption. The ideal is rarely if ever achieved, however, because very low levels of chemical impurities and other structural defects in the filler crystals can lead to extraordinary absorption and color effects in white fillers [33]. If the impurity exists as a separate phase or mixture in the filler, several tenths of a percent may be required to produce a significant color effect. If the impurity is well dispersed in the crystal lattice, however, only a few parts per million are needed to produce a major effect. For example, although the absolute iron content often provides a poor correlation with the color of kaolin, as little as 0.003% Fe_2O_3 in some clays can produce an intense color ranging from yellowish to reddish [34]. In another study of barium sulfate, calcium carbonate, and dolomite fillers, the brightness and color varied inversely with impurity level, particularly that of iron [35, 36]. There is certainly a general trend for the brightness to fall and the color intensity to increase as the iron content and other impurities increase, regardless of the type of filler.

In examining the filler's color in PVC, it is important that the color be measured in a fully wetted state, whether that is a paste with DOP or a fully compounded vinyl sheet. In the dry state, the actual color of the white filler is difficult to distinguish because of the enhanced scattering of the pigment in air. Visual screening in the dry state is wholly inadequate and even instrumental measurements have questionable value. In a DOP paste or compounded vinyl formulation, light scattering is minimized or eliminated and color differentiation is significantly enhanced. In this case, visual or instrumental assessments can easily be made and comparisons drawn among fillers of the same or closely similar refractive index.

Where the color of fillers of substantially different refractive indices must be compared, it is best to do the comparison in a formulation designed to eliminate the inherent scattering power differences of the fillers. This can be accomplished by using relatively high concentrations of the fillers (approximately 100 phr) to enhance their color contributions while adding a small amount (approximately 0.2 phr) of titanium dioxide to completely opacify the formulation and mask any filler scattering power differences. Such a technique can prove useful, for example, in comparing a silica (refractive index = 1.5) or talc (1.57) filler with a ground limestone (1.62), where the goal of the measurement is to distinguish the true color differences of the fillers apart from scattering. In nonopaque films, the color intensity is diluted by the whiteness effect of scattered light compared to the transparent film.

B. Refractive Index

From the historical perspective of unfilled systems, the present-day vinyl formulator is often looking for filled systems that perform and look as if they were

unfilled. To achieve this goal, the filler must be as transparent as possible
in PVC and display no inherent color. Both of these properties can be well
characterized in PVC according to the Kubelka-Munk theory [37, 38], which
relates all optical properties of pigments to the fundamental processes of
absorption (K) and scattering (S) of light.

The transparency of an insoluble filler in a homogeneous polyvinyl chlo-
ride/plasticizer binder depends on the complete absence of light scattering
in the heterogeneous filled system. The ability of a white pigment to scatter
light (M) depends on several factors, but by far the most important is the
index of refraction, as indicated in the following Lorentz-Lorenz equation
[39, 40]:

$$M = \frac{(n_p/n_0)^2 - 1}{(n_p/n_0)^2 + 1}$$

where n_p is the index of refraction of the pigment and n_0 the index of refraction of the surrounding medium. When $n_p = n_0$ and $M = 0$, there is no
scattering and a perfectly transparent filled PVC results. With the refractive index of PVC at 1.55 and most phthalate plasticizers in the range 1.48
to 1.50, a typical formulation with 100 phr DOP would have a refractive index of about 1.53. Most silica fillers with refractive indices ranging from
1.48 to 1.55 would be quite transparent in such a PVC system.

As the refractive index of the filler varies progressively from 1.53, the
optical transparency of the filled system progressively decreases. Fillers
with refractive indices in the 1.57−1.65 range including talc (1.57), clay
(1.57), mica (1.59), calcium silicate (1.59), calcined clay (1.62) and calcium
carbonate (1.65) generally produce translucent PVC compositions (M = 0.0174
to 0.0255). Fillers with refractive indices above 1.7 are generally considered
as hiding pigments and are added specifically for their opacifying power.
Totally opaque vinyl systems will generally include from 0.5 to 10 phr of ZnO
(R.I. = 2.01) or rutile TiO_2 (R.I. = 2.76). As with any such system, the
degree of transparency or opacity will also be affected by the concentration
of the filler and the film thickness.

In addition to index of refraction, particle size is also an important variable affecting the scattering power of a filler. Mitton [37, 38] has offered
the following empirical relationship of scattering power (S) to particle size
(d) for white pigments where λ is the wavelength of light and M and n_0 are
defined above.

$$S = \frac{\alpha M^3 \sqrt{\lambda}}{(\lambda^2/2d) + n_0^2 \pi^2 M^2 d}$$

Plots of S versus d for TiO_2 and $CaCO_3$ are given in Figure 9 for λ =
560 nm. Note the differing optimum particle sizes depending on the different
refractive indices for the two pigments. The optimum size is also affected
by changes in the refractive index of the surrounding medium (n_0). Finally,
the scattering power of all pigments progressively decreases as the particle
size falls below approximately 0.1 μm. This size is so much smaller than the
wavelength range of visible light that the scattering interaction is essentially

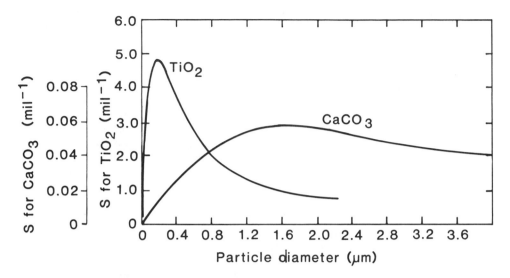

FIGURE 9 Empirical plot of scattering coefficient (S) versus particle size (PS) using a green filter (560 nm) for TiO_2 and $CaCO_3$ pigments.

lost. Thus, whereas a PVC filled with a 1-μm calcium carbonate might be translucent, the same formulation filled with a 0.07-μm calcium carbonate could be quite transparent.

When a filled part is bent or otherwise stressed, an obvious whitening generally occurs along the fold. This stress whitening also results from a discontinuity in refractive index. The mechanical stressing force produces a partial phase separation, producing small voids that are now filled with air having a refractive index of 1.0. These air/resin and air/filler interfaces refract and scatter light very efficiently because of the large differences in refractive index at the interface. The ease and extent to which stress whitening occurs generally increase in filled systems and are related to poor filler wetting and/or weak filler bonding to the polymer-plasticizer network.

C. Particle Size, Size Distribution, and Shape

Particle size, size distribution, and shape of fillers are key variables affecting their performance in PVC [41−44]. They are also key specification properties in the manufacture of most fillers because of their important contribution to overall production costs. In PVC, these filler properties strongly influence the mechanical properties (tensile strength, impact strength, modulus, and hardness), dimensional properties (shrinkage and creep), surface gloss, fire resistance, permeability, and air entrapment. The rheology and resulting processibility of the vinyl compounds are also primarily controlled by these three variables, whether the system is a liquid vinyl plastisol or a vinyl melt being processed via calendering, extrusion, or injection molding. These three properties play such a central role in the ultimate properties of the filled vinyl formulation that most filler producers will offer the same filler chemistry in several particle sizes, size distributions, and even particle shapes to give the formulator the greatest latitude in developing the desired properties in the ultimate product.

Although the selection of fillers is quite large in terms of these three properties, some generalizations can be drawn to help the PVC formulator. For mineral fillers of natural origin, particle shape is largely predetermined. Limestone, silicas, and other minerals ground to a specific size display an irregular particle shape characterized by the fracture surfaces resulting from the grinding process. Particle size distributions are largely determined by the milling and classification equipment used and are thus relatively similar from product to product. Clay and talc products are generally platy in shape with greater variations in shape and distribution based on ore source and process variations. In general, particle size distributions are broad, a characteristic that typifies the natural products.

For the synthetic minerals and pigments that have been made available, much wider varieties of particle shapes and sizes are possible. In general, the synthetic products display smoother, more regular crystal surfaces and much narrower particle size distributions. These characteristics generally offer PVC property advantages over the less costly natural products. Synthesis technology must generally allow optimization of particle shape, size, and distribution for a particular filler function (opacity, impact strength, gloss, etc.) in order to provide a property advantage or benefit commensurate with the added cost of the filler.

The importance of size, shape, and size distribution measurements to the filler manufacturer and user has resulted in the development of numerous measurement techniques and equipment. A general list of these techniques with appropriate references is provided in Table 11. Excellent books and references are also available covering this important area in greater detail [45−47]. Familiarization with the general techniques employed is necessary for the filler user to develop accurate comparisons of competitive materials.

In general, sieve analyses are relied on for particle analyses down to 44 μm (325 mesh). The methods are simple, fast, and convenient. From 44 to about 1 μm, sedimentation methods are most commonly employed with some use of microscopy, often as a reference technique. From 1 to about 0.1 μm, centrifugal sedimentation and electron microscopy techniques are commonly used. Below 0.1 μm, electron microscopy was the only available method until the recent release of equipment utilizing dynamic laser light scattering to provide particle size measurements from approximately 3 to 0.003 μm [48, 49]. It is important to note that sieve and sedimentation analyses provide particle statistics based on weight distributions while microscopy and the optical or laser light scattering techniques provide number distributions. Care must be taken that proper comparisons are made based on the technique and type of data generated.

D. Surface Area and Porosity

In addition to particle size, size distribution, and shape, surface area and porosity of the filler are key factors affecting the rheology and ultimately the mechanical properties and performance of the filled PVC system. Specific surface area (SSA) is the term most commonly used for powders and is defined as the area on a molecular scale that is exposed to a liquid or a gas by 1 g of the powder (common units are meters squared per gram. Porosity in this case is defined as the collection of surface flaws that are deeper than they are wide and can be further characterized in terms of pore size, pore volume, pore area, and pore shape.

TABLE 11 Types of Particle Size Analysis Techniques

	Applicable size range (μm)	Method or reference
Screen analysis	44	ASTM D-1921-63
		ASTM C-92-46
Elutriation		
Air	100−5	[50, 51], ASTM B293-60
Liquid	100−5	[52]
Gravitational sedimentation		
Pipette	50−2	[53]
Balance	50−2	[54]
Micromerograph	250−1	[55]
Divers	30−2	[56], ASTM D422-39
Turbidimetric (Extinction)	150−2.5	[57]
Centrifugal sedimentation		
Beaker centrifuge	3−0.05	[58]
Centrifuge pipette	2−0.1	[59]
Microscopy		
Visible	100−0.5	[60], ASTM E20-62T, D1366-65
UV	100−0.2	[60]
Electron	5−0.005	[61]

Specific surface area is clearly related to particle size and can be calculated for a host of particle geometries, assuming nonporous surfaces and well-defined size distributions. In the simple case of monodisperse spheres,

$$SSA = 6/dP$$

where SSA is the specific surface are, d the particle diameter, and P the true density of the powder. Herden [62] used this relationship to tabulate specific surface area as a function of particle diameter, as shown in Table 12.

This exercise is instructive in that it provides a general but useful frame of reference for interpreting surface area results. It also illustrates the sensitivity of surface area measurements to the fraction of fine particles in a powder sample. It should be noted that surface areas calculated from particle size data will at best establish the lower limits for the sample. True SSA values are often higher by a factor of 1000 or more based on variations in particle shape, surface irregularities, and porosity that always exist in the real filler sample.

TABLE 12 Relation between Specific Surface Area and Particle Diameter

Diameter	0.1 μm	1 μm	10 μm	100 μm	1 mm
SSA	60 m^2	6 m^2	6000 cm^2	600 cm^2	60 cm^2

Surface area can be determined by several methods based on adsorption [63], photoextinction analysis [64, 65], and permeability [66, 67]. Because of its simplicity and broad applicability, however, the Brunauer, Emmet, Teller theory employing gas adsorption is almost universally used for surface area measurements today [68, 69]. The equipment allowing full characterization of the gas desorption processes also provides the necessary data for characterization of the sample porosity in terms of pore size, pore volume, pore area, and pore shape [70].

For the vinyl formulator, surface area data on the filler(s) represent an important complement to particle size data in estimating reasonable loading levels for rheology and processibility of the filled system. While particle size data on a mass basis tend to overemphasize the coarse fraction of the filler and provide no measure of porosity, surface area data compensate for these shortcomings. The entire available filler surface must be properly wet for optimum properties and performance of the filled system. Since some fillers are capable of adsorbing and inactivating many formulation additives, such as plasticizers and liquid stabilizers and lubricants, increased additive levels are generally called for, based on the total filler surface area incorporated in the formulation. In a similar way, porosity (pore volume, size, area, shape) information is an essential complement to surface area in that it describes that part of the total surface (as determined by the adsorption of nitrogen) which may not be available to the larger polymer molecules. It is surface that should not be overlooked, however, because it can provide adsorption sites for smaller organic molecules (plasticizers, stabilizers, inhibitors, antioxidants), resulting in severe performance problems in the ultimate PVC formulation [71].

E. Surface Treatment

It is rare today to find a line of filler products that does not include several grades that are surface-modified for particular end-use properties or applications. In altering the filler surface chemistry, the energy and degree of interaction at the polymer-filler interface can be changed significantly, leading to surprising enhancement of mechanical and/or dimensional properties in filled vinyl formulations.

Surface treatment of a mineral filler can provide a host of useful benefits in the ultimate PVC formulation. The uncoated mineral surface, which is generally hydrophilic, can be made hydrophobic, thereby increasing its compatibility, dispersibility, and processibility in PVC. The coated surface is also less likely to adsorb and deactivate other formulation additives such as plasticizers, heat stabilizers, and/or antioxidants that are required for optimum PVC performance. Finally, the degree of filler-polymer loading or adhesion can be increased through the use of an appropriate coupling agent, thereby providing functional reinforcement from an otherwise nonfunctional filler. In this regard, silicone and silane coupling agents have been widely used on the various silica and silicate fillers with reasonable success [10, 11]. Their effectiveness with calcium carbonate has been somewhat limited. Fatty acids have generally proved to be effective dispersing/coupling agents for calcium carbonate fillers [74-76]. As always, the vinyl formulator should carefully evaluate the cost/benefit performance of the coated versus uncoated filler. The high cost of many potential coupling agents has limited the commercial viability of many coated fillers in vinyl formulations.

F. Dispersibility

Proper dispersion of the filler(s) in PVC is essential to the performance and ultimate success of the filled system (see Figures 10, 11). Responsibility in this important area is shared by the filler manufacturer and the formulation compounder. Lack of performance on the part of either contributor will almost surely result in inferior performance of the filled vinyl formulation.

Dispersibility of a filler is established primarily by (1) the chemical nature of the filler, (2) its surface characteristics, (3) its particle size, and (4) the process by which it is made. The chemical nature of the filler defines the bond strengths and many of the special surface characteristics that will be encountered in any given material. Filler manufacturers have generally gone to great lengths to define the chemical and surface characteristics of their materials. The physical and chemical properties of the filler surface will establish its tendency to aggregate and the strength of the clusters formed.

Particle size of the filler is also important to its ease of dispersion. In general, the surface energies of particles greater than 1 μm are such that the driving force for aggregation is minimal and the aggregates that do form are generally weak. As the particle size drops below 1 μm, and particularly as it enters the colloidal region below 0.1 μm, the tendency of the particles to aggregate in large clusters increases progressively and substantially. In this region, protective colloids or surfactants must also be employed to stabilize the particles from reagglomerating once the initial aggregates are broken during the dispersion process.

Finally, the closely guarded processes by which the fillers are produced have an all-important effect on the ultimate dispersibility of the product. It is in this area that substantial effort is expended to maintain the optimum filler performance at a minimum manufacturing cost. An incredibly extensive patent literature has grown out of an almost universal need by filler manufacturers to improve the ease of dispersion of their products [77]. Favorite techniques include (1) the use of various surface treatments to change the surface chemistry and reduce the forces of particle-particle attraction and bonding and (2) methods to circumvent the drying process or reduce the forces of agglomeration that occur during drying. Clearly, techniques to improve pigment dispersibility will continue to receive a great deal of attention from filler manufacturers and users.

In PVC compounding, three basic dispersion processes are commonly used: dry blending, melt shear, and liquid dispersion [78, 80]. In practice, a combination of these techniques is often employed, although there are few hard and fast rules. The experience of the compounder is still the primary factor determining the dispersion processes used. The dry-blending technique achieves dispersion primarily by impact and attrition grinding. Unless followed by another dispersion process, there is little opportunity

FIGURE 10 Agglomerated filler in matrix.

for effective wetting of the pigments by the resin. Melt shear dispersion is
most commonly employed in PVC compounding and works well in follow-up
combination with dry blending. The effectiveness of melt shear depends on
the ease of wetting of the filler in the PVC and a high enough viscosity to
allow high shear forces to be effectively transmitted to the filler aggregates.
Liquid dispersion processes for PVC plastisols involve predispersion of the
filler in the vinyl plasticizer. Equipment type can vary but the type of
dispersion forces is generally determined by viscosity, which is most often
controlled by the filler concentration. The choice of dispersion technique
usually is dictated by cost, the ultimate property requirements of the filled
composite, and, finally, the dispersibility of the fillers required to provide
those properties.

G. Abrasion and Hardness

The abrasivity of mineral fillers is generally recognized to depend on three
key factors: (1) filler hardness, (2) particle size, and (3) particle shape.
    When the filler particle contacts a softer surface (the surface being ab-
raded), the extent of the damage is controlled by the difference in hardness,
the energy of the interaction (which involves the mass and acceleration of
the particle), and, finally, the efficiency of the interaction (which involves
the shape of the particle including the existence of sharp points and edges).
    The hardness of mineral fillers is established by their comparative
ratings on the Mohs scale. The higher the Mohs hardness value, the harder
the mineral. The Mohs hardness scale is defined as follows:

1.	Talc	6.	Feldspar
2.	Gypsum	7.	Quartz
3.	Calcite	8.	Topaz
4.	Fluorite	9.	Corundum
5.	Apatite	10.	Diamond

The Mohs hardness is useful in establishing a general expectation of abrasivity for a synthetic or natural mineral product of high purity. When
dealing with ground natural fillers, however, it is important to assess the

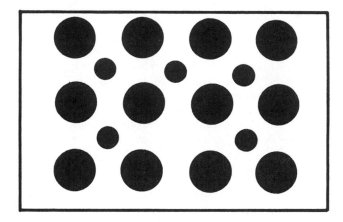

FIGURE 11 Filler dispersed to discrete size in matrix.

mineral purity. For soft minerals like talc, the abrasivity of the product may be established primarily by the level of impurities of higher hardness. Abrasivity of mixtures increases disproportionately with the level of the more abrasive component.

Abrasion studies conducted on synthetic mineral filler samples that were essentially monodispersed but of different average size showed a generally linear increase of abrasion with size [80]. Similar studies with ground natural products have tended to confirm this abrasion-size relationship, but exceptional care must be taken to control the particle size distribution and purity of each sample [80]. The coarse particle fraction of the total size distribution disproportionately controls the abrasivity of the sample. In addition, any harder mineral impurities tend to be more difficult to grind and thus are concentrated in the coarse fraction.

Particle shape also tends to affect abrasivity, although the effect appears to be somewhat smaller than that of hardness or particle size. Clearly, spherical shapes and relatively smooth surfaces decrease the overall abrasivity of the product. Particles with sharp crystal points or edges appear capable of more abrasion damage per impact than other less aggressive shapes. Once again, the irregular particle shapes created in grinding a natural product tend to be more abrasive than particles of the same mineral composition and size produced synthetically. It is admittedly difficult, however, to eliminate the effects of purity and particle size distribution in these comparative assessments.

To the PVC compounder, abrasion has two important effects. First, excessive wear on equipment, such as processing rolls, screws, cylinder walls, dies, molds, and mixers, has a decidedly negative economic impact on operations. Costs of equipment repair and replacement as well as lost production time should be considered when filler evaluation and choices are made. Unfortunately, wear is not related to fillers alone, and abrasion and wear are difficult to assess in tests of short duration. Second, excessive wear leads to fine metal contamination in the vinyl compound. This can be manifested as a discoloration of the vinyl compound and/or a more insidious premature heat stability failure of the compound resulting from catalytic degradation influenced by the metallic contamination [81]. Clearly, the abrasivity of potential fillers should be assessed early in the development of any filled vinyl formulations.

V. PROCESSING WITH FILLERS

Filler materials, as they are purchased, are generally found in an agglomerated state. The degree of agglomeration is dependent on processing steps, filler size, surface energy, surface coating, and so forth. The energy required to deagglomerate these particles is directly proportional to the energies holding the particle agglomerates together.

In general, the small particle size fillers (below 1 μm average size), because of their high surface area, contribute to multiplied surface and electrostatic charges, causing high amounts of agglomeration. The presence of a monomolecular layer of a surface coating such as a fatty acid satisfies some of the surface energetics, resulting in soft agglomerates that are easier to break down by low-level mechanical energy.

The filler aggregates form weak contact points in the matrix. Therefore, a dispersion to discrete particle size resulting in complete wetting of filler with the polymer is essential. Usually the presence of 2 to 3% agglomerates will cause measurable changes in impact strength of a filled PVC. This property change is further magnified on low-temperature impact strength testing.

The PVC industry has been using single-screw, twin-screw, and Banbury techniques to compound fillers in PVC. Melt-mix compounding prior to final shaping becomes highly desirable when the particle size of filler is very small and the loading level is above 5 phr. Some recent work has shown that a medium particle size precipitated $CaCO_3$ (0.5 μm) can be dry-blended up to 10 phr on a high-intensity mixer and injection-molded without sacrificing the impact strength property. In fact, injection molding equipment manufacturers are currently working on screw designs for better compounding on the molder to eliminate added melt compounding cost.

Both the single-screw and twin-screw extruders are excellent means of compounding and simultaneously forming materials. In general, when varying the filler size, surface treatment, and loading levels, a study to balance the lubricant package and stabilizer level is essential. Overall, the fine particle size fillers raise torque, promoting early fusion, with some reduction in compound stability. Such studies can be made on a torque rheometer and can be magnified by an increase in filler loading as well as the rotor speed for each formulation before use (Figure 12).

Similarily, the coated fillers generally reduce fusion time and fusion torque and greatly increase compound stability. The calcium stearate coating formed on the $CaCO_3$, because of its large size and high melting point, cannot penetrate the PVC molecule to function as an intramolecular lubricant. It basically acts as an external lubricant and a costabilizer, requiring an overall reduction in external lubricant for optimum processing (see also Chapter 15). For large runs, the production output of formulations containing coated (lubricating) and uncoated fillers can be optimized by designing a 2^3 factorial experiment to study the effect of key parameters (see Figure 13).

The extrusion index for single- or twin-screw extruders can be calculated as follows:

$$\text{Single screw} = \frac{(\text{output})^2}{(\text{amps})(\text{melt pressure})} \times 100$$

$$\text{Twin screw} = \frac{(\text{output})^2}{(\text{torque})(\text{thrust})} \times 100$$

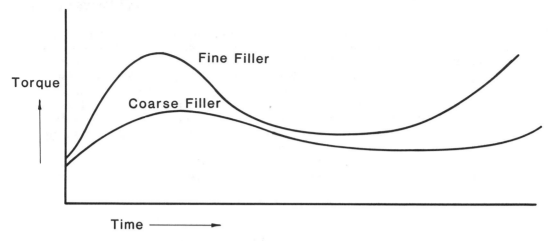

FIGURE 12 Effect of filler size on processing properties of PVC.

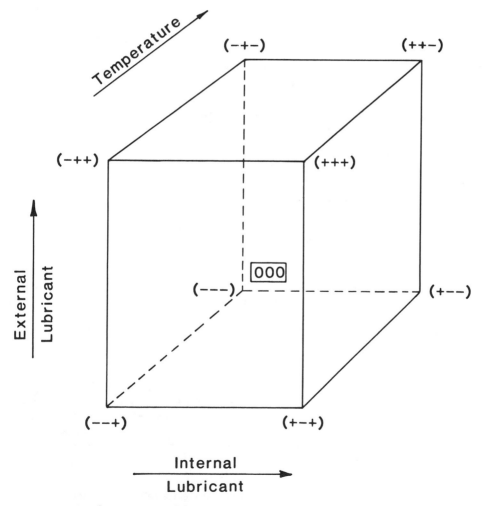

FIGURE 13 2^3 factorial design for extrusion performance study.

Several other studies [83, 84] have examined the processing efficiency
of filled systems which become necessary from an economic point of view.

Like rigid PVC, the plastisol formulations also require intense mixing
of fillers for optimum dispersion. This is particularly true for PVC plastisols
used in automotive undercoating, where stearate-coated ultrafine precipitated $CaCO_3$ (0.07 μm) is used as the thixotrope. Dispersion of such fine
$CaCO_3$ fillers requires medium- to high-intensity mixing to deagglomerate
the particles for maximum thixotropy, surface smoothness, and gloss properties.

In the coatings industry, plastisol dispersion is generally measured on
the Hegman gauge, which basically shows the maximum size of the particles
or agglomerates present in the dispersion.

VI. MECHANICS OF FILLED POLYMERS

The composite properties of filled PVC are primarily governed by the following filler factors: (1) particle size, (2) morphology, and (3) interfacial
adhesion with polymer.

Reduction in filler size leads to a decrease in the thickness of the polymer interlayer and consequently an increase in the proportion of polymer
at the filler boundary. This has a definite effect on mechanical properties
because of the increased surface contact with the polymer. The particle
morphology—platy, fibrous, spherical, and so forth—has a major effect on
the processing and reinforcement of composites. Interfacial adhesion affects
stress-strain behavior, thus affecting tensile properties.

When properly selected and compounded, there are significant benefits
to be obtained by using particulate fillers in PVC over the unfilled polymer.
The key advantages are:

1. Rheology and particle packing (plastisols)
2. Improved dimensional stability
3. Increased stiffness (modulus)
4. Increased toughness (impact strength) in rigids
5. Improved electrical properties
6. Reduced cost

Most of these factors have been studied in detail in special polymer composites. A discussion of some mathematical treatments of these factors now
follows.

A. Plastisol Rheology

The addition of filler materials has a most definite effect on the melt rheology
of rigid PVC as well as the "filled rheology" of plastisols and organosols.
The effect of fillers on the melt rheology of rigid PVC was discussed in the
previous section. This section deals with plastisol and organosol rheology.

Besides improving key polymer properties such as tear resistance, hardness, and creep resistance, the addition of appropriate filler material to
plastisols and organosols improves thixotropy, with the added possibility of
cost reduction.

1. Thixotropy: Many studies have been done to relate the theoretical
description of the rheology of dispersed systems, beginning with Einstein's
equation

$$\eta_s = \eta_0 (1 + 2.5\phi)$$

In this equation, η_s is the viscosity of the suspension, η_0 the viscosity of the liquid medium, and ϕ the volume fraction of suspended particles. This equation is based on the assumptions that (1) all the filler particles are smooth-surfaced spheres with low concentration and (2) there is no particle-particle interaction. Unfortunately, most plastisol and organosol filler materials have nonuniform shapes, sizes, and surface characteristics and the tendency is to load these fillers much above the range of Einstein's equation. The complexities of real-world systems notwithstanding, significant work has been reported explaining the rheological behavior of particle suspensions since Einstein.

A generalized Mooney equation [86] appears to give an excellent fit:

$$\ln(\eta_r) = \frac{K\phi}{1 - s\phi}$$

In this equation $\eta_r = \eta_s/\eta_0$ is the relative viscosity, ϕ is the volume fraction of filler, K is an adjustable parameter related to the size and shape of particles, and s is an adjustable parameter related to the space-filling properties of the particles. Thus, a practical user of filler would like to have ϕ as high as possible and η_r as low as possible.

2. Particle Packing: The definitive analysis of the packing of monodisperse spheres was done by Graton and Fraser [88]. They found that there are six possible ordered packing arrangements and, associated with each, a characteristic void content. These results are summarized in Table 13 [88].

A classical empirical study on the packing of spheres was done by McGeary [89]. He found that in random packing, a roughly orthorhombic arrangement having a void content of 37.5% predominates. By studying binary, ternary, and quaternary mixtures, he was able to derive relationships related to relative sizes and amounts of components and the observed packing density. His results are shown in Table 14.

For binary systems (see Figure 14) McGeary found that the packing density improved as the ratio of d_1 to d_s tended to infinity. The maximum density achievable was 86% of the theoretical maximum. Packing density improved rapidly as d_1/d_s increased from 1 to about 7 and then changed abruptly and increased gradually as d_1/d_s ranged from 7 to 100.

For three spheres in contact, as shown below, the largest diameter sphere which will just fit through the triangular pore is 0.154d or $d_1/d_s = 6.5$. Thus, if d_1/d_s is less than 6.5, the smaller spheres will be able to penetrate the interstitial area of the packing, but if d_1/d_s is greater than 6.5, infiltration of the smaller spheres may occur. McGeary's results indicate that very efficient packing can be obtained by a mixture of only four or five components.

For nonspherical fillers, the following equation has been used:

$$\eta = \eta_0 \exp \frac{\alpha F}{1/f - KF}$$

where α and K are shape factors of particles ($\alpha = 10.5$ to 24.8 and K = 1.35 to 1.90) and f is a magnitude determined by the ratio of thickness of the surface layer on the particle and its size.

TABLE 13 Packing of Spheres

	Points of contact	Volume of unit cell	Volume of unit void	Voids %
Cubic	6	$8.00\ r^3$	$3.81\ r^3$	47.6
Orthorhombic (two orientations)	8	$6.93\ r^3$	$2.74\ r^3$	39.5
Tetragonal	10	$6.00\ r^3$	$1.81\ r^3$	30.2
Rhombohedral (two orientations)	12	$5.66\ r^3$	$1.47\ r^3$	26.0

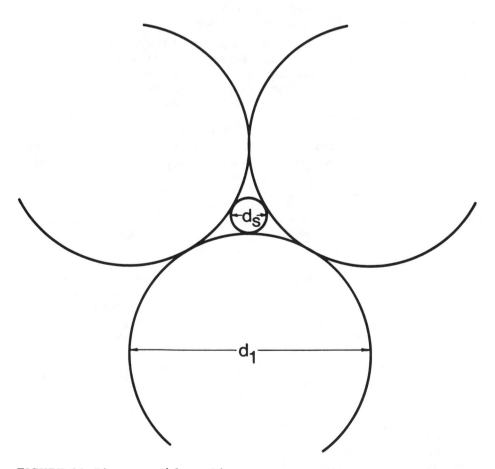

FIGURE 14 Binary particle packing.

TABLE 14 Packing Density for Mixtures of Spheres

Component	d Ratio	X_1	X_2	X_3	X_4	% Theoretical
1	1	1.00	—	—	—	60.5
2	7	.726	.274	—	—	86
3	38	.647	.244	.109	—	94
4	316	.607	.230	.102	.061	97.5

Milewski [90] has studied the particle packing of more complex systems such as spheres-flakes, spheres-fibers, and so forth. Because of the excessive costs involved in selecting appropriate lots for blending of materials for optimum packing, such systems have been mostly justified only in specialized applications thus far.

B. Moduli of Filled Systems

One of the most dramatic effects of adding a filler to the polymer matrix is seen by an increase in stiffness (modulus) of the composite.

Smallwood [91], on the basis of a hydrodynamic concept, proposed the following equation for calculation of the elastic modulus of filled composites, which is analogous to Einstein's equation for viscosity:

$$E_{fill} = E_{unfill}(1 + 2.5F)$$

where E_{fill} and E_{unfill} are the elastic moduli of filled and unfilled composites and F is the volume proportion of the filler. This equation has been found suitable only for low concentrations of filler in the composite.

Kerner's [92] equation, given as follows, is well recognized for calculation of this property in filled polymers:

$$\frac{E_{fill}}{E_{unfill}} = 1 + \left(\frac{ABF}{1 - BF}\right)$$

where $A = (7 - 5\nu)/(8 - 10\nu)$, $B = [(E_{fill}/E_{unfill}) - (E_{fill}/E_{unfill} + A)$, and E_{fill} is the elastic modulus of the filler.

A further modification of this equation is

$$\frac{E_{fill}}{E_{unfill}} = \frac{1 + ABF}{1 - B\psi F}$$

where ψ is a function depending on the maximum degree of filling F_m

$$\psi F = \left(1 + \frac{1 - F_m}{F_m^2} F\right) F$$

or

$$\psi F = 1 - \exp\left(\frac{-F}{1 - F/F_m}\right)$$

The versatile Halpin-Tsai equation for predicting the composite modulus is based on micromechanics theory and is fully discussed by others [93–95]. For a composite containing a flake or platelet oriented parallel to the stress direction, the modulus enhancement ratio is given by

$$\frac{\bar{E}}{E_m} = \frac{1 + \zeta \eta v_f}{1 - \eta v_f} \qquad\qquad [1]$$

where

$$\eta = \frac{E_f/E_m - 1}{E_f/E_m + \zeta} \qquad\qquad [2]$$

and $\zeta = 2(d/t)$ or $2 \times$ (aspect ratio).

The quantities are identified as:

$\bar{E}$ = tensile modulus of composite
E_m = corresponding tensile modulus of unfilled polymer (matrix)
E_f = corresponding tensile modulus of filler
v_f = volume fraction of filler in composite
ζ = geometric factor which takes into account the filler aspect ratio
d = average equivalent diameter of platelets
t = average thickness of platelets

For a composite containing spherical particles (a particulate-filled system), equations (1) and (2) are still applicable, where the aspect ratio now becomes equal to 1 and $\zeta = 2$.

1. Prediction of Composite Moduli for Spherical Fillers: Work reported by Radosta [96] basically shows that all that is needed to predict the composite modulus from equations (1) and (2) for a calcium-carbonate-filled polymer are the corresponding moduli for the unfilled polymer and the filler and the volume fraction of filler used. The actual polymer used for the matrix phase and the test temperature at which the values were obtained do not enter into the calculations but are taken into account by the value of the modulus used for the unfilled polymer. (Note: it is assumed that the modulus of a mineral filler does not change appreciably over the useful temperature range of most thermoplastics.) Therefore, if one were to vary the polymer matrix modulus either by using different polymers as the matrix phase or by varying the test temperature, and if the data were plotted as the modulus enhancement ratio versus the matrix modulus, then the data points should lie on one continuous curve. This is indeed the case, as seen in Figure 15. By using the data from the previous section, the matrix mod-

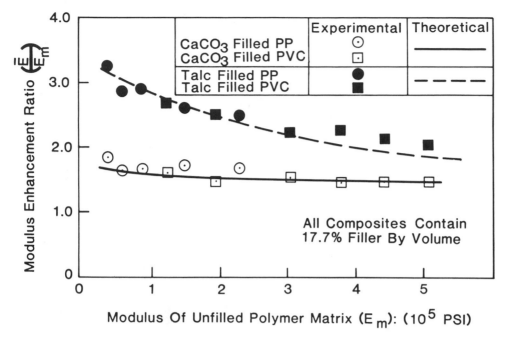

FIGURE 15 Comparison of experimental modulus data with the theoretical curves from the Halpin-Tsai equations.

ulus has been varied by changing both the test temperature and the matrix polymer (polypropylene and PVC). The solid line in Figure 15 represents the curve calculated by using equations (1) and (2) and a value of 5,000,000 psi for the elastic modulus of calcium carbonate. The agreement between the experimental data and the calculated curve is excellent. (Note that although the Halpin-Tsai equations were derived for tensile modulus, it has been shown that the modulus enhancement ratios for flexural modulus are equivalent to the ratios for tensile modulus and either can be used as long as one is consistent.)

   2. Prediction of Composite Moduli for Platy Fillers: Due to the platy nature of some fillers, the additional parameter of aspect ratio is needed to predict the composite modulus of these materials. If all the platelets are aligned parallel to the stress direction, we can use equations (1) and (2). However, examinations of sections obtained from injection-molded specimens show that the platelets are not perfectly aligned parallel to the mold surfaces, but are more randomly arranged throughout the specimen. At present, a good predictive equation for random three-dimensional orientation of platelets has not been established.

   For the purposes of this discussion, a constant K is introduced into equation (1) as a scale factor to adjust for the decrease in modulus due to deviations of the platelets from perfect parallel alignment. The constant K in equation (3) below is expected to vary between 1 (for perfect parallel alignment) and 3 (for a completely random arrangement).

$$\frac{\bar{E}}{E_m} = \frac{1 + \zeta \eta v_f}{K(1 - \eta v_f)} + \frac{K - 1}{K} \qquad\qquad [3]$$

As seen in Figure 15, the data points for the platy composites all lie on one continuous curve. As with the spherical filler data, the matrix modulus has been varied by changing both the test temperature and the matrix polymer. The dashed curve in Figure 15 represents the curve calculated by using equations (3) and (2) with values of 25,000,000 psi for the elastic modulus of the platy filler, 20 for the average aspect ratio of the platy filler, and 3.0 for the scale factor K. Although there is good agreement between the experimental data and the calculated curve, a theoretical equation derived to accommodate three-dimensional orientation of platelets would give a better fit and would not require the empirical scale factor K used here.

C. Stress-Strain Properties

The addition of fillers to PVC raises the flexural modulus, as discussed previously. Generally, such an increase in modulus is associated with some reduction in tensile strength and elongation. The addition of finer spherical fillers to rigid PVC contributes good flexural modulus and improved tensile strength, with some reduction in elongation. This effect is much more dominant when platy materials such as talc and clays are used.

The stearate-treated small particle size spherical fillers (e.g., precipitated calcium carbonate), which exhibit low adhesion (coupling) with the PVC matrix, enable the compounded PVC to retain greater tensile strength and elongation up to much higher loading levels. The filled PVC composites show a clear yield point. This is believed to be due to destruction in adhesion of filler with the PVC as well as filler-filler agglomerates. This is described as "dewetting."

The elongation of a filled composite can be described by

$$\varepsilon_B = \varepsilon_B^0 \, (1 - \phi 1/3)$$

where ε_B is elongation at break for filled PVC, ε_B^0 is elongation at break for unfilled PVC, and ϕ is the volume fraction of the filler. The change in elongation and cross-sectional area of filled PVC has been expressed in terms of Poisson's ratio:

$$V = \frac{C/C_0}{L/L_0}$$

where V is Poisson's ratio, C_0 the initial cross-sectional area, and L_0 the original length. Since the filled PVC increases in cross-sectional area, Poisson's ratio is less than 0.5.

The measurement of impact strength at various strain rates also provides very useful performance information. For a filler exhibiting perfect adhesion with the PVC matrix and following Hookean behavior, the impact strength is described as

$$I_s = \frac{E \varepsilon B^2}{2}$$

where I_s is the impact strength, E is Young's modulus, and εB is the elongation at break.

Work reported by Mathur and Driscoll [4] and Radosta [5] has clearly demonstrated that fillers with smaller particle sizes improve rigid PVC impact resistance dramatically. The surface treatment with fatty acid further improves the ductile-brittle transition.

The optimum particle size range appears to be about 0.05 μm. Fillers much smaller than 0.05 μm form tightly held agglomerates and are very difficult to disperse, thus offering no further advantage in terms of impact strength. The impact strength improvement is very significant in medium- to high-molecular-weight PVC. Such fillers, when added above 15 phr, act as impact modifiers. It is believed that the higher melt viscosity of medium- to higher-molecular-weight PVC permits greater torque buildup, which is required for good dispersion of fine fillers.

It is hypothesized that these high-surface-area coated fillers, when subjected to impact, go through "dewetting," exhibiting ductility up to much higher loadings and lower temperatures.

D. Diffusion in Filled Polymers

The addition of filler materials to PVC affects the permeability and diffusion in PVC compounds.

Nielsen's [97] basic model essentially assumes that diffusing materials have to go around the filler particle. Therefore, platy fillers aligned with the surface render a more lengthy path. The addition of platy talc and clays to PVC wire and cable compounds essentially improves the insulating efficiency due to these principles.

In general form, the diffusion of materials through a filled PVC compound can be described by:

$$\frac{P_{fill}}{P_{unfil}} = V_{pol}/t$$

Where P_{fill} and P_{unfil} are permeability of filled and unfilled compounds respectively, V_{pol} is volume fraction of polymer; t is tortuosity defined as the extended distance material has to travel to go through the film [98]. This equation assumes that all the filler platelets are oriented parallel to the surface.

Nielsen has further modified the equation to include the particle shape and orientation in the matrix.

$$\frac{P_{fill}}{P_{unfil}} = \frac{P_M}{P_{unfil} F^\eta + P_M(1 - F^\eta)} \cdot \frac{V1}{t} + \frac{V_{pol} + V2}{t}$$

where η is a constant varying from 0 to 1, characterizing average specific diffusion path. This constant depends upon the particle shape and orientation.

TABLE 15 Calculation of Pound-Volume Cost of a $CaCO_3$-Filled PVC Compound[a]

Ingredients	Formula weight	Specific gravity	Volume	Cost/lb	Pound cost
PVC resin	100.0	1.40	71.43	$0.35	$35.00
Lubricant A	1.2	0.85	1.41	$0.98	1.18
Lubricant B	0.4	0.82	0.48	$0.80	0.32
Stabilizer	2.0	1.02	1.96	$1.85	3.70
Stearic acid	0.5	0.85	0.58	$0.44	0.22
$CaCO_3$	12.0	2.71	4.43	$0.07	0.84
	116.1		80.29		$41.26

[a]Calculation:

$$\text{Cost per pound} = \frac{41.26}{116.1} = \$0.355$$

$$\text{Compound specific gravity} = \frac{116.1}{80.29} = 1.446$$

$$\text{Pound-volume cost} = \frac{\$0.355}{1.446} = \$0.245$$

1. Cost Reduction: In addition to modifying various properties as previously discussed, the selection of appropriate fillers is expected to lower the formulation cost.

Since fillers are sold on per-pound basis, their economic value comes from the volume of resin they replace. Therefore, it is essential to compare them on the basis of unit volume of resin they will replace (see Table 15). The normal cost units are cents per cubic inch. In order to calculate this, the cost of each material in cents per pound and the specific gravity of fillers under consideration is necessary. The cost per cubic inch can then be calculated from the following relationship, where 0.0361 is the specific gravity conversion factor:

$$\text{Cents/in.}^3 = \text{cents/lb} \times \text{sp. gr.} \times 0.0361$$

VII. ANALYTICAL METHODS AND FILLER QUALITY CONTROL

A. Chemical Properties

The key chemical properties of interest to a PVC compounder include:

Material purity
Percent acid insolubles
Iron or other metallic impurities
Water solubility
pH (water slurry)

TABLE 16 Physical Property Tests

Property	Test method
Particle size	ASTM D1366-55T
Surface area	ASTM E20-62F
Specific gravity	ASTM D153-54
Bulk density	ASTM D1895-65T
Oil absorption	ASTM D1483-60
Brightness—color	ASTM E313-67
Moisture content	ASTM C25-58

The mineral purity will vary from mine to mine. The presence of high amounts of foreign materials in the ore can cause multiple problems. High silica levels have been related to abrasivity of filler. High iron and other metallic impurities affect heat stability of the compound. The typical chemical analysis of a commercial grade ground limestone is reported by its producer as follows:

$CaCO_3$	96.00%
$MgCO_3$	1.50%
SiO_2	1.20%
Al_2O_3	0.30%
Fe_2O_3	0.08%
H_2O	0.25%

B. Physical Properties

The physical properties of prime interest to PVC compounders are as follows:

Particle size and distribution
Surface area
Dispersibility
Specific gravity
Bulk and tap densities
Dry brightness
Wet color
Oil absorption
Refractive index
Moisture content
Mohs hardness

These properties are tested by the ASTM tests indicated in Table 16.

REFERENCES

1. Conley, R. F., *PM&AD—RETEC Proceedings*, Nov. 6 and 7, 1985, p. 144.
2. Dientenfass, L., *Colloid Z.*, *155*, 121–130 (1957).
3. Standen, A. (ed.), *Kirk-Othmer Encyclopedia of Chemical Technology*, 2nd ed., Interscience, New York, 1964, pp. 7–11.
4. Mathur, K. K., and Driscoll, S. B., *SPE Annu. Tech. Conf.*, 912–917 (1982).
5. Radosta, J. A., *SPE Annu. Tech. Conf.*, *25*, *27*, 593–595, 1979.
6. Mathur, K. K., and Vanderheiden, D. B., in *Proceedings of the ACS International Conference on Polymer Additives*, March 1982, Plenum Publishing, pp. 371–389.
7. Leuchs, O., 19th International Wire and Cable Symposium, Atlantic City, 1970, p. 239.
8. *What Are "White Reinforcing Fillers"?*, Degussa Corp., Teterboro, N.J., 1980.
9. Marvel, D., *Silica Gel*, W. R. Grace & Co., Baltimore, Md., Tech. Bulletin, 1980.
10. *Synthetic Silicas and Silicates for PVC*, Tech. Bull. Pigment No. 51, Degussa Corp., Teterboro, N.J., 1978.
11. *Basic Characteristics and Applications of Aerosil®*, Tech. Bull. Pigment No. 11, Degussa Corp., Teterboro, N.J., Tech. Bulletin, 1980.
12. *Aerosil® for PVC Plastisols*, Tech. Bull. Pigment No. 41, Degussa Corp., Teterboro, N.J., Tech. Bulletin.
13. Furnivall, J. C., Kupfer, A. D., and Irvine, J. L., *SPE Annu. Tech. Conf. Proc.*, *27*, 541–544 (1981).
14. Augustyn, E. J., Schwarcz, J. M., *SPE Annu. Tech. Conf. Proc.*, *23*, 202–207 (1977).
15. Nelson, E. D., and Kaufman, S., *J. Fire Flammabil.*, *13*, 79–103 (April 1982).
16. Wang, C. Y., *Antimony—Its Geology, Metallurgy, Industrial Uses and Economics*, Charles Griffin & Co., London, 1952.
17. *Use of Antimony Compounds as Fire and Flame Retardants*, Bibliography FR-3, M&T Corp., Rahway, N.J., 1979.
18. Bhatnagar, V. M., in *Proceedings of 1975 International Symposium on Flammability and Fire Retardants*, Technomic Press, Westport, Conn., 1975, pp. 238–248.
19. Hindersinn, R., and Wagner, G. M., in *Encyclopedia of Polymer Science and Technology*, 1967, p. 7.
20. Patton, T. C., in *Pigment Handbook*, Vol. I, Wiley Interscience, New York, 1973.
21. Mathewson, C. H., *Zinc: The Metal and Its Alloys and Compounds*, *ACS Monogr. 142*, Reinhold, New York, 1959.
22. Komar, L. C., Titanium Dioxide in Polymers, *26th SPE Annu. Tech. Conf.*, 303 (1968).
23. DeCoste, J. B., and Wallder, V. T., Weathering of Polyvinyl Chloride, *Bell Telephone Syst. Tech. Publ. Monogr. 2376*.
24. Herman, E., Eisenoxidpigments in PVC, *Kunst. Plast.*, *10* (1963).
25. Vanderheiden, D. B., *Paint Varnish Prod.*, 19–24 (Sept. 1974).
26. U.S. Patents 3,652,334 (1972) and 4,053,325 (1984).
27. Mathur, K. K., and Kramer, K., *J. Vinyl Technol.*, *5*(*1*), 32–38 (1983).
28. Standen, A. (ed.), *Kirk-Othmer Encyclopedia of Chemical Technology*, 2nd ed., Interscience, New York, 1964, pp. 243–282.

29. Katz, H. S., and Milewsk, J. V. (eds.), *Handbook of Fillers and Reinforcements for Plastics*, Van Nostrand Reinhold, New York, 1978.
30. Grayson, M. (exec. ed.), *Kirk-Othmer Encyclopedia of Chemical Technology*, 3rd ed., Wiley-Interscience, New York, 1978, p. 643.
31. *Chem. Week*, 16 (May 3, 1978).
32. U.S. Patents 4,260,534 (1979), 4,250,064 (1974), 4,242,397 (1979), 4,193,841 (1977), 4,138,521 (1976), 4,097,644 (1972), 3,962,507 (1976), etc.
33. Mitton, P. B., in *Pigment Handbook*, Vol. III D-C, Wiley, New York, 1973.
34. Leoin, V. V., Danilova, D. A., and Shoets, L. V., *Bum. Prom.*, *4*, 15—16 (1972).
35. Cremers, M., *Polym. Paint Colour J.*, 852—862 (Nov. 3, 1976).
36. *Ibid.*, pp. 936—941 (Dec. 1, 1976).
37. Atkins, J. T., and Billmeyer, F. W., *Color Eng.*, *6(6)*, 40 (May—June), 1968.
38. Judd, D. B., Wyszecki, G., *Color in Business Science and Industry*, 3rd ed., Wiley, New York, 1975.
39. Mitton, P. G., Vejnosk, L. W., and Frederick, M., *Official Digest of the Federation of Societies for Paint Technology* (1961), p. 33.
40. *Ibid.*, *34(444)*, 73—89 (1962).
41. Nikaido, T., Japan Kokai 78, 82851 (1978).
42. Shanks, H. G., *J. Appl. Polym. Sci.*, *26(9)*, 3099—3102 (1981).
43. Mathur, K. K., Greenzweig, J., and Driscoll, S. B., *SPE Annu. Tech. Conf.*, *24*, 732—736 (1978).
44. Mathur, K. K., and Driscoll, S. B., *J. Vinyl Technol.*, *4(2)* (1982), pp. 81—86.
45. Allen, T., *Particle Size Measurement*, Chapman & Hall, London, 1974.
46. Jarrett, B. A., and Heywood, H., *Br. J. Appl. Phys.*, *Suppl. 3*, S21—S26 (1954).
47. Lloyd, P. J., Scarlett, B., and Sinclair, J., paper presented at the Particle Size Analysis Conference, Univ. of Bradford, Bradford, England, 1970.
48. Chu, B., *Laser Light Scattering*, Academic Press, New York, 1974.
49. Instrument companies: Leeds & Northrup Instruments (Microtrac), St. Petersburg, Fla.; Coulter Electronics Inc. (model N4), Hialeah, Fla.; Nicomp Instruments, Santa Barbara, Calif.
50. Roller, P. S., *Proc. ASTM 32*, 608 (1932).
51. Dalla Valle, J. M., *Micromeritics*, Pitman, New York, 1948, p. 72.
52. Andrews, L., *Industrial Engineering Symposium* (1947), p. 114.
53. Andreasen, A. H., Kolloid. Beith. 27, 349—358 (1928).
54. Odean, S., *Soil Sci.*, *19*, 1 (1925).
55. Eadie, F., and Payne, R., *Iron Age*, *174*, 99 (1954).
56. Berg, S., *Ingenioervidensk. Skr.*, Vol. 2 (1940).
57. Sharatt, E., Van Somersen, E., and Rollenson, E., *J. Soc. Chem. Ind.*, *64*, 63 (1945).
58. Martin, S., *Ind. Eng. Chem. Anal. Ed.*, *11*, 47 (1939).
59. Komack, H. J., *Anal. Chem.*, *23*, 844 (1951).
60. Green, J., *Industrial Rheology and Rheological Structures*, Wiley, New York, 1949.
61. Riedel, G., and Ruska, H., *Kolloid Z.*, *96*, 86 (1941).
62. Herden, G., *Small Particle Statistics*, 2nd ed., Butterworths, London, 1960.

63. Brunauer, S., Emmet, P., and Teller, E., *J. Am. Chem. Soc.*, *60*, 309 (1938).
64. Rose, H. E., *J. Appl. Chem.*, *2*, 80 (1952).
65. *ASTM Book of Standards*, Part 2, ASTM, New York, 1944, p. 47, C115—42.
66. Lea, F. M., and Nurse, R. W., *J. Soc. Chem. Ind.*, *58*, 278 (1939).
67. Carman, P. C., *J. Soc. Chem. Ind.*, *57*, 225 (1938).
68. Lowell, S., *Introduction to Powder Surface Area*, Wiley, New York, 1979.
69. Gregg, S. J., and Sing, K. S. W., *Adsorption Surface Area and Porosity*, 2nd ed., Academic Press, London, 1982.
70. *Gas Adsorption Equipment*, Micromeritics Instrument Corp., Norcross, Ga.
71. Mathur, K. K., Witherell, F. E., *SPE Annu. Tech. Conf.*, May 6, 1980.
72. Marsden, J. G., *J. Appl. Polym. Sci.*, Symp. 14.
73. Friedman, L. J., *SPE Tech. Pap.*, *15*, 287 (1969).
74. McCord, A., U.S. 3,333,980 (1967).
75. Solvay Cie, Fr. Demande, 2,231,695 (1974).
76. Shikata, T., Japan Kokai, 75,02,754 (1975).
77. Gutcho, M. H. (ed.), *Inorganic Pigment Manufacturing Processes*, Noyes Data Corp., Park Ridge, N.J., 1980.
78. Miller, D. B., *Color Eng.*, *6(4)*, 46—51 (July—August 1968).
79. Reeve, T. B., and Dills, W. L., *SPE Annu. Tech. Conf.*, New York (May 1970), Preprint 16, 574—576.
80. K. K. Mathur (Pfizer Inc.), unpublished work.
81. Levy, G. L., *Wire J.*, *8*, 39 (1971).
82. K. K. Mathur (Pfizer Inc.), unpublished work.
83. Baronin, G. S., Minkin, E. V., and Artemova, T. G., Deposited Doc. SPSTL 663, Khp - *D81*, 1981, Available from SPSTL.
84. Laukhin, V. Ya, *Khim. Neft. Mashinostr.*, (*11*), 14—17, 1982.
85. Einstein, A., *Ann. Phys.*, *17*, 549 (1905); 289 (1906); *34*, 591 (1911).
86. Mooney, M., *J. Colloid Sci.*, *6*, 162 (1951).
87. Simha, R., and Frisch, H. L., in *Rheology*, Vol 1, F. R. Eirich (ed.), Academic Press, New York, (1956).
88. Graton, L. C., and Fraser, H. J., *J. Geol.*, *43*, 785 (1935).
89. McGeary, R. K., *J. Am. Ceram. Soc.*, *44(10)*, 513 (1961).
90. Milewski, J. V., *RP/C 1974*, 29th Annu. Conf., SPI.
91. Smallwood, H., *J. Appl. Phys.*, *15*, 758 (1944).
92. Barrer, R. M., et al., *J. Polym. Sci. Part A*, *1*, 2565 (1963).
93. Aston, J. E., Halpin, J. C., and Petit, P. H., *Primer on Composite Materials*: *Analysis*, Technomic Publications, Stamford, Conn. (1969).
94. Kardos, J. L., *Crit. Rev. Solid State Sci.*, *3*, 419—450 (Aug. 1973).
95. Halpin, J. C., and Kardos, J. L., *J. Appl. Phys.*, *43*, 2235 (1972).
96. Radosta, J. A., *SPE Tech. Pap.*, *21*, 526 (1975).
97. Nielsen, L. E., *J. Macromol. Sci. Part A*, *1*, 929 (1967).
98. Barrer, R. M., et al., *J. Polym. Sci. Part A*, *1*, 2565 (1963).

7

Colorants: Types, Properties, and Performance

MELVIN M. GERSON*

Podell Industries, Inc.
Clifton, New Jersey

JOHN R. GRAFF

Mobay Corporation
Haledon, New Jersey

---

*Deceased

547

Color in vinyl plastics is a desirable feature for marketability and identification. While colors specified to fulfill these ends are based on esthetic principles, achieving the desired colors is based on physical, chemical, and engineering principles.

Colorants are defined as materials that will modify the light incident on a surface so that light of other wavelengths, at the same or lesser intensities, is reflected or transmitted [14]. Colorants are generally colored materials, although optical brighteners that reinforce color appearance may themselves be colorless. Colorants can be incorporated into the plastic mass or applied to the surface of the plastic as a coating or laminate.

There are basically two types of colorants: those that are soluble in the vinyl formulation (dyes), and those that are insoluble (pigments). Some pigments show a slight solubility in vinyl formulations, although such manifestation is generally undesirable and considered a defect. Solubility leads to bleeding or migration, plateout, and crocking, and can disqualify an otherwise acceptable pigment from use in vinyls.

I. DYES

Since solubility, by definition [1], refers to the dissociation of a chemical into individual molecules or ions, we can readily infer the properties of many dyes.

Advantages	Disadvantages
1. Excellent transparency	1. Migration or bleeding
2. Ease of incorporation into the formula	2. High degree of reactivity with other ingredients in the formula
3. Excellent tinting strength or coloring ability	3. High degree of reactivity with the degradation products of the formula
	4. Poor lightfastness
	5. Lack of heat stability

The disadvantages of dyes so outweigh their advantages, particularly in lightfastness and heat stability, that few dyes are used in flexible vinyl formulations. The development of solvent soluble dyes suitable for use in plastics [2] is progressing slowly, and selected members of this class of colorants are finding special uses where they can be incorporated into formulas in very low concentrations and where lightfastness is not a requirement in end use.

II. PIGMENTS

The predominant method for coloring vinyls is with *pigments*. For our purpose, we must define *pigments* as insoluble materials that have a particulate structure and that can selectively absorb and reflect the rays of incident light or confer opacity to the plastic, or both.   This definition includes all the colored pigments that can be used in concentrations as low as 0.001 phr and the white pigments, such as titanium dioxide, which are sometimes used at concentrations of 20 phr or higher.

Vinyl processing requires pigments with a special set of characteristics, compared with other colorant-using industries.   The special requirements result from the temperatures used to process vinyl formulations and the degradation products that form during the processing life of a vinyl compound. Pigments that in themselves will not withstand temperatures of 300° to 500°F for varying lengths of time obviously cannot be used.   Pigments that are sensitive to acids are generally unstable in vinyls, since HCl is a degradation product of vinyls.   Pigments with borderline sensitivity to acids can sometimes withstand the rigors of processing if an acid acceptor, such as an epoxy stabilizer, is present; such an addition would not help, however, if long-term lightfastness is required.

Pigments containing any metallic cations normally found in stabilizers must be used carefully.   In some cases, these pigments may enhance stability; in other cases, unexpected reactions occur and too much of a given cation may cause a loss of control over stability.

It is hard to formulate rigid rules about suitability of use in vinyls for any pigment.   As pigment technology improves and as vinyl formulations change, many pigments previously considered unsatisfactory become usable; for example, pigments that migrate in plasticized vinyl because of slight solubility in the plasticizer may be suitable for use in unplasticized rigid vinyl compounds.

Some pigments are unsuitable for use where electrical resistance is required since absorbed soluble salts that are difficult to wash out remain in the pigment matrix after drying; these are ionic in nature and electrically conductive.   The electrical resistance of such pigments can be enhanced by preparing a pigment dispersion by the technique known as flushing.   This process effects the transfer of the pigment precipitate from the water phase directly to an oil phase (plasticizer) without previous drying. The oil wets the surface of the pigment preferentially and becomes bound to it.   The water separates into a distinct phase (carrying the water-soluble salts with it) and is carried off during the decantation phase of this operation [14].

The technology of pigments and the science of color is growing at a very rapid rate; therefore descriptions of pigment use such as this chapter are only a review of what is known at present.   The latest information on the subject is best obtained from the technical literature of pigment manufacturers.

III.   PIGMENT DISPERSIONS

Pigments are furnished as dry powders.   Before they can be used in a vinyl formulation, they must be properly dispersed.   Dispersion of pigments is a three-step process: (1) disintegration of the agglomerates that have formed during drying, packaging, and storing; (2) wetting of the surface and replacing the pigment-air interface with a pigment-vehicle interface; and (3) stabilization of the dispersion to prevent reflocculation [3,4,14].

The key to using a pigment most efficiently for the best color value and end-use properties lies in its proper dispersion.   Money value derives from tinting strength in relation to cost. This property depends on the degree of dispersion—the greater the better in the effort to approach ultimate particle separation Pigment dispersion requires shear for its accomplishment when "flushing" is not possible. This shear is defined as the sliding of two adjacent or connecting parts or layers on each other so that they move apart in a direction parallel to the plane of their contact, that is, in opposite directions. While such shear occurs with a high-intensity mixer, such as a Banbury or similar type of continuous mixer, a two-roll mill, a calender, or an extruder, the time that vinyl stocks are exposed to shearing stress in such equipment is not enough to result in proper pigment dispersion. Specialized equipment such as three-roll ink mills, pebble mills, high intensity-high speed mixers, and impingement mills are required for this latter purpose. Wetting agents and surfactants, acting by physicochemical methods, can shorten and simplify this operation, but basically dispersion of pigments remains a mechanical operation [3,4,5,14]. The science of pigment dispersion has become a specialized technology, and as a result, the vinyl formulator has available a series of predispersed pigments in a variety of vehicles and forms. They may be prepared as pigment pastes (high-viscosity liquids) in any of the plasticizers that are required for specific formulations, or as chips, powders, or pellets, in compounded vinyl, which are handled like any other solid ingredient in the formula. The advantages of pigment dispersions, compared with dry pigments, are lack of dust, ease of handling, ability to be directly incorporated into the formula, and provision for standardization of shade and strength by the pigment dispersion manufacturer to a close tolerance. Testing and evaluating pigment dispersions are similar to testing and evaluating pigments themselves.

IV. PARTICLE SIZE, TINTING STRENGTH, AND TRANSPARENCY

The ultimate aim in using any pigment is to achieve the maximum in color value and in transparency or opacity. Pigments are manufactured to an optimum particle size for color value, tinting strength, and transparency-opacity characteristics. Agglomeration of these particles during manufacture tends to interfere with achieving this goal. Particle size is therefore a definite factor in the use of pigments and is controlled in the pigment manufacturing and dispersion operations. The desirable particle size varies from a low of about 20 nm with certain carbon blacks to 2 μm for iron oxides, chrome oxides, and others. A better understanding of the significance of particle size is available form studying titanium dioxide as an opacifying pigment. The opacity property of a white pigment directly depends on its refractive index [3,5] and inversely depends on its particle size, up to a point. The refractive index of a pigment is an inherent property that depends on its crystal structure and chemical composition. The refractive index of titanium dioxide as the highest known among colorless (white) substances, contributes to its excellent opacifying characteristics [7]. The other controlling factor is implied by the Mie theory [8,17]—that when the particle size of a pigment is reduced below one-half the wavelength of incident light, its ability to scatter light is reduced in proportion. This optimum particle size can vary with crystal structure, particle shape, and refractive index; it is of the order of 0.2 nm for titanium dioxide.

V. PROPERTIES OF PIGMENTS FOR VINYL PLASTICS

The capacity for color, transparency or opacity, and tinting strength of a pigment is generally inherent in the pigment and independent of the vinyl formulation (excluding solubility considerations). Every other property affecting pigment suitability for use in vinyl plastics—heat stability, crocking, and so on—is in some way affected by the choice of the other ingredients of a vinyl formulation or may affect this choice. Consequently, no one should evaluate a pigment (or pigment dispersion) without using the pigment in a valid vinyl formulation and comparing results from a similar formulation uncolored or pigmented with a similar pigment as a reference standard.

The most important properties of a vinyl formulation that can be affected or changed by the presence of a given pigment are listed below. Also described are the test techniques generally used for evaluating colorants on each of these properties.

A. Heat Stability

The colorant is incorporated into a vinyl compound (with or without white pigment) and subjected to heat for varying periods. It is compared with a similar vinyl compound, without the colorant, that has been treated in the same way. Pigments that fade or change color indicate a lack of inherent heat stability. Pigments that accelerate the browning or darkening of the compound indicate an undesirable reactivity.

B. Light Stability

The light stability of a pigment may vary with its use in masstone or in tint tone. This property is evaluated for compounded vinyl stocks in both forms and compared with similar unpigmented stocks. Exposure in accelerated light-aging or weathering devices is generally the method. (See "Outdoor Durability," Section K.) Interpreting the results is similar to judging heat stability. Loss of color indicates lack of inherent lightfastness. Premature browning or brittleness of the stock indicates reactivity with the vinyl formulation.

C. Sulfide Staining

Pigmented vinyl compound is exposed to an atmosphere rich in hydrogen sulfide fumes and high in humidity. Change in color (usually significant darkening) indicates sensitivity to this chemical, which is usually present in industrial fumes. Since certain stabilizers, fire retardants, and other additives will cause this darkening, unpigmented vinyl compound of the same formulation should be tested simultaneously for comparison.

D. Migration, Bleeding, and Blooming

Prepare a sheet or film of the vinyl formulation as required and place it in contact with a similar white or uncolored sheet of vinyl under a weight of 1 lb/in.^{2} This test is usually conducted for 5 hr at 180°F. Other times and temperatures may be used if they better represent end-use requirements. Transfer of color to the uncolored sheet indicates migration or bleeding. Pigments that migrate will usually bloom (appear on the surface as a fine powder that crocks easily) after a period of storage at room temperature. Pigments that are slightly soluble in the liquid ingredients of the vinyl formulation show this defect. Pigments that will migrate in flexible vinyl formulations may not migrate in rigid formulations. Pigments that sublime during heating may also bloom regardless of the presence of plasticizers.

E. Crocking (Dry or Wet)

Crocking refers to ruboff, and the procedures used are those specified for textiles in accordance with AATCC [9]. A piece of white, unsized fabric is rubbed across the vinyl surface under constant pressure. After the specified number of rubs, the fabric is examined visually for color pickup.

F. Ease of Incorporation; Texture

Evaluating ease of incorporation must be consistent with the methods of processing to be used. An excellent test is to band a white vinyl stock compound on a two-roll mill. Add the pigment or dispersion to be tested and mill for 3 minutes with constant stripping and rebanding. Strip the sheet, allow to cool to room temperature, and reband for 3 minutes. The sheet is examined for color streaks and other evidence of unincorporated pigment.

G. Electrical Characteristics

Because many pigments may be slightly ionic or polar or contain soluble salts as a result of the manufacturing operation, their electrical characteristics should be evaluated in accordance with NEMA procedures [11].

H. Plateout

Mill a colored sheet of vinyl on a two-roll mill for at least 5 minutes. Follow this immediately with a pure white vinyl compound on the same mill, and determine the degree of color imparted to the white sheet by pigment left on the rolls. When blooming, bleeding, or migrating pigments exhibit this phenomenon, they should be eliminated from the formula. Plateout of pigments can be eliminated by varying the amount of lubricant, changing lubricant, or incorporating a small amount of a finely divided absorptive extender pigment such as silica or aluminum hydrate.

I. Resistance to Acid, Alkali, and Soap

Two or three drops of the test solution are placed on the surface of the pigmented vinyl compound for periods up to 5 hr. The compound is washed clean and the exposed spot is examined for change in color.

J. Oil Absorption and Rheology

Flow characteristics of pigmented vinyls are particularly important in plastisol or extrusion technology. They may be evaluated by practical processing tests or by tests of their consistency over a period with a Brookfield Viscometer or Brabender Plasti-corder. The rheology of pigmented vinyl compounds is generally influenced by the oil absorption of the pig- ment. This property of a pigment is defined as the amount of vehicle re- quired to surround each particle of the pigment completely and fill its interstices [13]. It is measured by determining the amount of vehicle re- quired to change a pigment-vehicle mixture from a noncontinuous appear- ance to a homogeneous, wet-appearing mass. This will vary with the ve- hicle and its wetting ability, the pigment particle size and shape, the sur- face treatment applied during pigment manufacture, and the surface charges on the pigment [3].

K. Outdoor Durability

Outdoor durability includes lightfastness and also refers to chalking (for- mation of a fine powder on the surface of the plastic) and effect of weath- ering on tensile strength and embrittlement, and so on. These phenomena must be considered as the interrelation between the pigment particles and vinyl formulation, and the degradation products of the formulation on ex- posure [18]. While some evaluation can be made in accelerated weathering devices, actual continuous exposure for 12 or more months in Arizona or Florida is believed to represent the severest conditions of humidity, sun- light, ultraviolet light, and weather that vinyl plastics are likely to be subjected to in actual service [19]. The names and addresses of commer- cial weathering stations in Florida and Arizona are listed in reference 12.

VI. PIGMENT CHARACTERISTICS (GENERAL)

Color, tinting strength, tint tone, chemical resistance, lightfastness, heat stability, and so on are characteristics of the individual pigment, its particle size, its pigmentary shape, and any impurities that may be present as a result of the individual manufacturing process. Any general listing of pig- ments and their properties, as in the Color Index [2], are at best an av- erage of the properties of those pigments within any given family or class that are commercially available. Individual products, even from the same manufacturer, may vary in particle size and shape, or in impurities present because of raw materials used or of surface treatments to the pigments, such as additives to control oil absorption, decrease flocculation tendencies, in- hibit chalking (in the case of titanium dioxide), or improve transparency. These variations can result in marked differences in performance among the same pigments when they are evaluated in any of the performance tests described above. This is nowhere more apparent than in the electrical characteristics, rheology, weathering, and heat and light stability tests discussed. Most surfactants are innocuous in vinyl formulations but ac- count for the differences in ease of dispersion and rheology that can be found between pigments of the same type from different manufacturers.

   A common modification of pigments is accomplished by resination. Solubilized rosin is added to the precipitation tank before neutralization and precipitation of the pigment. When precipitation of the pigment occurs, the rosin is also precipitated with and on the pigment.

   Resination of pigments at less than 5% may be considered a surface treatment. Resination from 5 to 30% by weight of some pigments is used to modify the optical or working characteristics to a significant degree. Re- sination improves transparency and makes the color more brilliant. Un-

fortunately, resinated pigments are not as heat-stable as the base un-
resinated pigment and cannot be processed at equivalently high temperatures.

The designation "toner" refers to pigments that are almost 100% colorant.
"Lakes" refer to pigments precipitated in the presence of an inert, extender
pigment of very low refractive index, such as aluminum hydrate or barium
sulfate. The most common forms of these extended pigments are either
60% color pigment and 40% extender, or 40% color pigment and 60% extender.
They have the triple advantage of ease of dispersion, improved rheology,
and less sensitivity to weighing errors. They are commonly used with the
more expensive pigments such as cadmiums, vat oranges, vat yellows, and
vat blue. The FD&C pigments are all prepared by precipitating the approved FD&C dyestuff on such an extender (usually aluminum hydrate).
The characteristics of the extenders used are such that they have little
or no effect on the color or transparency of the pigment.

VII. PIGMENTS, COLOR, AND STRENGTH

The color of a pigment should be evaluated only when the pigment is used
in a representative vinyl formulation that is characterized by concentration
and thickness of film or sheet consistent with the end-use requirements
of the vinyl itself [14]. One of the significant tests must apply to cost
of the formulation. Pigments used in vinyls vary in price from a low of
about 15¢ per pound for some black pigments to $40 per pound for some
esoteric organic pigments derived from vat dyestuffs.

When a colorant is used in a formula and is observed in reflected light
(light is incident on the surface and reflected to the eye), the *masstone*
of the colorant is evaluated. When the same formulation is observed in
transmitted light (i.e., the light is observed through the pigmented sheet
or film) the *transparency* and *undertone* of the colorant is evaluated. When
the colorant is evaluated in a combination with a white pigment (generally
at a ratio of 5 parts of colorant to 95 parts of white pigment) the *tint tone*
of the colorant is evaluated. It must be understood that the undertone of
a pigment (when observed in transparency) is not necessarily the same as
the tint tone of the pigment when observed in combination with white [14].

Definitions of *color* and methods for numerical evaluation of the color
properties of a pigment are discussed in Section XVI,A.

VIII. CLASSIFICATION OF PIGMENTS

Two general methods useful in classifying pigments involve chemical composition and color. Any general study of pigments should begin with their
chemical composition, since this generally defines their performance properties. Individual studies can then be made, based on the color requirements
of the particular problem.

IX. INORGANIC PIGMENTS: OXIDES (HYDRATED OXIDES)

Many of the most durable pigments for use in vinyls are metallic oxides.
Since this class of compounds is generally unreactive, the properties that
most often are manifested in vinyls are excellent heat stability, light
stability, chemical inertness, lack of bleeding and migration, desirable

electrical characteristics, and very low absorption. These properties are improved by calcining, and those oxides that have been so treated are especially suitable for vinyl formulations.

A. Titanium Dioxide

The oxides of titanium are virtually the only white pigments found in vinyl formulations. They exist commercially in two crystal forms, anatase and rutile. Both crystals are tetragonal and in an octahedral pattern. In anatase, the octahedra are packed so that 4 of the 12 edges of each octahedron are shared with adjacent octahedra; in rutile, two such edges are shared [7]. See Figures 1 and 2.

Rutile pigments have the highest refractive indices of white pigments and are much more resistant to chalking than anatase. Anatase pigments have generally a slightly bluer shade (and therefore appear whiter); they are slightly lower in refractive index and are slightly easier to disperse.

Surface treatments on titanium dioxide pigments may consist if zinc oxide, aluminum oxide, silicon dioxide, or other oxides deposited in monomolecular layers by vapor-phrase calcination. In many instances, these surface treatments can affect the heat and light stability of vinyl formulations.

Titanium dioxides, because of their excellent tinting strength and very high opacity, are the whitest pigments known. The older white pigments such as zinc sulfide, lead carbonate, and lithopone can be used for other purposes in vinyls but are rarely used as white colorants.

B. Antimony Oxide

Antimony oxide is used because of the fire retardancy it imparts. Since it will cause a whitening effect and loss of transparency, it must also be considered a white pigment. As a fire retardent, antimony oxide is discussed further in Chapters 17 and 19.

C. Oxides of Lead, Zinc, and Silicon

Lead oxides, zinc oxides, and silicon oxides are all of little value because of their refractive indices (i.e., poor opacity), low tinting strength, and reactivity in vinyl formulations. Silicon oxides are so low in refractive index that they are classed as inert, extender, or filler pigments. (See Chapter 17.)

Lead oxide and zinc oxide, when used at the concentrations required to color vinyls, are so reactive that it is difficult to provide suitable heat and light stabilization for them. Special grades of zinc oxide have recently been developed that reportedly are inert. They have been recommended for use as light and weathering stabilizers in PVC rather than as pigments [10].

D. Iron Oxides and Hydrates

The compounds in the class containing oxides and hydrates of iron vary in color from light yellow, through red, to a dark brown and black. The iron oxides that are mined often contain many impurities (such as magnesium, manganese, zinc, and aluminum in ionic form) that may be too reactive for vinyl use. When these metallic ions are calcined, they are con-

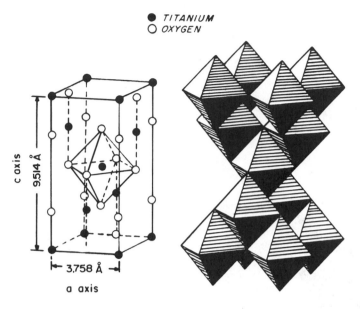

● *TITANIUM*
○ *OXYGEN*

FIGURE 1   Crystal structure of anatase titanium dioxide.   (Courtesy of Titanium Pigments Division, NL Industries, Inc.)

verted to their oxide form and are quite unreactive at temperatures below 400°F.   Hydrated iron oxides, which are yellow, are also stable to temperatures below 400°F.   As vinyl-processing temperatures increase, improved stabilization techniques are required.   Red iron oxide pigments mostly are manufactured synthetically and consist of at least 98% Fe_2O_3. Synthetic manufacture results in a high degree of purity, low reactivity, and a wide variety of shade and undertones.

Brown iron oxide pigments, burnt siennas and burnt umbers, are calcined natural pigments that are combinations of Fe_2O_3 with significant amounts of manganese dioxides, silicon oxides, or aluminum oxides.   Synthetic brown oxides are also prepared by calcining iron oxide, so that a blend of FeO and Fe_2O_3 forms in the proper proportion and assumes a desired crystal form.   Black iron oxide is prepared in a similar fashion. Overheating, or too strenuous oxidation in any of these calcination processes, will cause a reversion of the desired pigment to the red Fe_2O_3.

Iron oxides are desirable because of their excellent durability, inertness, and low cost.   Their chief disadvantages lie in their high specific gravity, relative coarseness or large particle size (when compared with organic pigments), low tinting strength, and lack of brilliance (see paragraph on colorimetry, Section XVI,A).   The yellow iron oxides are more properly characterized as tan pigments.   The red iron oxides form brick red, brown, or brown maroon colors, in contrast with the brighter cadmium salts and organic red pigments.

E.   Chromium Oxides and Hydrates

The only other oxide pigments that are commercially important are the green oxides and hydrates of chromium.   Chromium oxide, Cr_2O_3, is

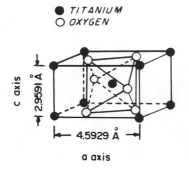

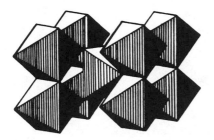

FIGURE 2   Crystal structure of rutile titanium dioxide.   (Courtesy of
Titanium Pigments Division, NL Industries, Inc.)

probably the least reactive and most lightfast green pigment known.   It is
low in brightness, opacity, and tinting strength, and very difficult to disperse.   Hydrated chromium oxide (chrome hydrate) is $Cr_2O_3 \cdot 2H_2O$.   This
pigment has all the stability characteristics of chromium oxide (except that
it will not withstand ceramic processing temperatures), is much easier to
disperse, is more transparent, has higher tinting strength, and is much
more brilliant in color.   While hydrated chromium oxide is more expensive,
it is a much more practical pigment in vinyl formulations than chromium
oxide is.

F.   Miscellaneous Colored Oxides

Almost all other colored oxides have been evaluated for use as pigments.
Many of them, such as the cobalt oxides, have found use in ceramic processes in which their ability to resist temperatures to 1500°F are of value.
They are, however, deficient in brilliance and tinting strength when compared with other pigments available for use in vinyls.

X.   ELEMENTAL PIGMENTS

A.   Carbon Blacks

The generic term carbon black refers to a large series of colorants based on
the formation of elemental carbon during the controlled, incomplete combustion of natural gas or oil.   Bone black (animal black or drop black)
results from the controlled combustion of animal bones, with the organic

matter forming elemental carbon on a calcium carbonate-calcium phosphate substrate [20,22,23].

Three main forms of carbon black pigments are available [21], depending on the feedstocks and method of manufacture, characterized as follows:

Channel blacks[a]	Furnace blacks	Thermal blacks
Small particle size	Larger particle size	Largest particle size
High jetness	Low jetness	Lowest jetness
Acidic surface	Alkaline surface	—
High oil absorption	Low oil absorption	Low oil absorption
Brown undertone	Blue undertone (recommended for tinting)	Bluest undertone

[a]No longer manufactured in the United States, but available through imports.

The older forms of carbon black, often referred to as pigment blacks, lamp blacks, acetylene blacks, and so on, have largely been supplanted by these newer, more economical forms of black pigments.

Because of the variety of conditions subject to control during the manufacture of carbon blacks, these pigments can vary in particle size, adsorbed gases and moisture, porosity, structure (linking of carbon atoms into chains), and surface area. This results in a wide variation in masstone and tint-tone color, electrical conductivity, oil absorption, ease of dispersion, and rheological properties. Channel carbon blacks are recommended for UV stabilization for outdoor exposure [14,18,21] because of their ability to form and react with free radicals and also their ability to absorb ultraviolet radiation and render it harmless.

Carbon black pigments are so varied in their properties and are the subject of so much investigation [24] that they are best described by their individual characteristics (Table 1) and by an explanation of their choice for some specific uses. Jetness of black pigments can be estimated by a nigrometer reading (the lower the reading, the higher the jetness). As jetness increases in masstone, its tinting strength decreases, and oil absorption, ease of dispersion, rheological problems, and cost increase. Consequently, more than one black pigment is required for industrial vinyl operations.

An example of the compromises that must be made can be found in considering pigments for the two separate applications, phonograph records and electrical insulation. Phonograph records use carbon black as a filler for its low pound-volume cost; and the very small particle size of channel black helps to minimize scratches and other surface noise. The Regular Color Channel Black (Table 1) traditionally has been used because of its low cost (Compared with the smaller sizes), relatively high jetness, and high electrical conductivity, shown by its ability to resist the formation of electrostatic charges. However, owing to problems in connection with air pollution, channel black pigments have been phased out of production and have been replaced with furnace blacks of approximately equal surface characteristics and particle size.

The carbon black best designed for use in electrical insulation (highest electrical resistance) is the Long Flow Furnace Black (Table 1). Two related characteristics contribute primarily to the excellent electrical proper-

TABLE 1 Typical Characteristics of Some Black Pigments

Pigments	Nigro-Meter	Surface area (m²/g)	Particle diameter (μm)	Oil absorption (lb oil/100 lb pigment)	Tinting strength	Volatile content (%)	Fixed carbon (%)	pH	Characteristics
High color channel	64	850	12	375	165	13.0	87.0	3.0	Jettest black in common use
Medium color channel (A)	71	380	16	160	175	5.0	95.0	5.0	Economical jet black
Medium color channel (B)	74	320	17	155	185	5.0	95.0	5.0	Economical jet black for UV stability. Low electrical conductance
Regular channel	81	130	22	125	182	5.0	95.0	5.0	General-purpose black. Phonograph records
Long flow channel	84	295	28	88	170	12.0	88.0	3.5	Easy to disperse; very low oil absorption; highest electrical resistance
Conductive furnace	94	190	29	250	140	2.0	98.0	7.5	Most electrically conductive black; for floors, for antistatic effects
Oil furnace (A)	84	86	25	85	227	1.0	99.0	7.5	Excellent low-cost channel black substitute; high tint strength and low oil absorption
Oil furnace (B)	90	42	42	72	187	1.0	99.0	8.0	High tint strength, blue tone, low oil absorption; used for tinting
Gasfurnace	99	23	80	70	100	1.0	99.0	9.5	Very blue tone; very low oil absorption. Most popular black for tinting.
Thermal	110	6	470	33	35	0.5	99.5	9.5	Lowest oil absorption; bluest tone; lowest cost

ties of the latter pigment, its ease of dispersion and its low structure. Ease of dispersion relates to the pigment's ability to be wetted by a vehicle (resin or plasticizer) and to have its envelope of air completely removed. Since organic vehicles are much less conductive than air and have a much lower tendency to ionize under high voltage, this prohibits formation of paths for electrical charges. Long Flow Furnace Black is also described as a "low-structure" black. The individual carbon particles are discrete and not connected (Figures 3 and 4). This results in easier separation of the pigment particles during dispersion and prevents formation of long pigment chains, which can transmit electrical current. On the contrary, the most conductive carbon black pigments has very high oil absorption and the lowest amount of adsorbed materials on its surface.

These grades of channel black are also being replaced, principally by furnace blacks having similar characteristics and properties.

B. Bone Blacks

Bone blacks are still used in masstone in vinyls to achieve high jetness with a blue undertone. The presence of calcium carbonate and calcium phosphate gives them a color effect that other black pigments cannot duplicate. The presence of other salts and impurities makes it difficult to stabilize formulations that contain bone black and to control for color during processing. Bone black should therefore never be used for tinting.

C. Metallic Pigments

The other types of pigments based on elemental chemicals are the metallic flake pigments. Although many metallic elements are hammered into flakes and used in the paint, ceramic, and printing ink industries [25,20], the vinyl plastics industry limits itself to aluminum, colored aluminum, copper, and bronze flakes (Table 2). All the metallic pigments are prepared by hammering the metal into thin sheets in the presence of a lubricant such as stearic acid. The most desirable effect is achieved when the flakes orient themselves in a plane parallel to the surface of the film, and the reflectance is similar to the specular, or mirrorlike, reflectance from the metal itself. Using coarser flakes (approximately 200 mesh), leafing grade powders (the leafing characteristic is enhanced by adding a maximum of lubricant), and slow processing at slightly higher than normal roll temperatures (on a calender) or die temperature (in an extruder) will enhance this metallic effect.

The copper alloys were first used for gold effects in vinyls. Since the metals of these alloys are reactive, the stability of these compounds is limited. The metals tend to react with the vinyl compound during processing and with the degradation products of vinyls when they are exposed to light and humidity. Aluminum flakes are much less reactive, and normally show only a slight dulling on long-term exposure to light and humidity. One can render both types of flakes less reactive by coating them with a very thin coating of a thermosetting resin. Although this enhances the stability of the copper alloys, it does not permit them to equal the aluminum pigments in durability. Durability of the aluminum pigments is also enhanced by this treatment.

A whole gamut of metallic colors and tones can be obtained by using transparent pigments in combination with aluminum flakes. Gold, copper,

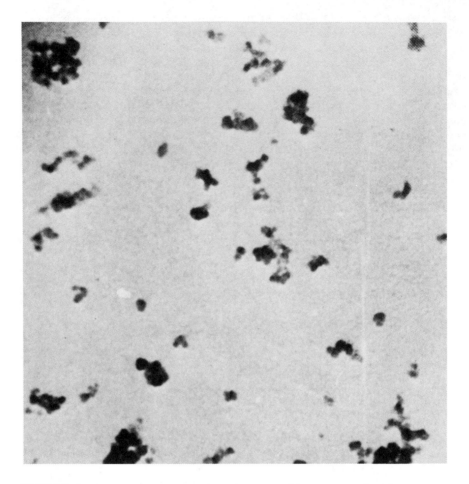

FIGURE 3   Low-structure carbon black.   (Courtesy of United Carbon Co.)

brass, and bronze colors of any degree of lightfastness and heat stability
can be prepared, depending on whether one uses low-cost transparent
yellows with or without an azo red.   More lightfast formulations will result
with the higher-cost, more durable organic pigments.   The effect created is
a yellow, gold, or bronze color distributed throughout a film.   Although
the aluminum flakes can be incorporated into the vinyl film to give a metallic effect colored to the degree desired, the optical effect is not quite
the same as that obtained from bronze powders.   The increase in durability
and processability may be worth the compromise, however.   Another type
of metallic effect is obtained by coating very thin (less than 0.001 in.)
aluminum foil on both sides with a thermoplastic or thermosetting resin that
has been pigmented with transparent, durable, nonbleeding pigments.   The
coated foil is then cut into very small flakes.   They are available in two
grades, "Glitter," from 0.005 to 0.010 in. in diameter, and "Flitter," from
0.015 to 0.100 in. in diameter.   Flitters and glitters can be incorporated
into clear, colored, transparent, pastel or white film to give unusual color
effects.   They are seldom used as the sole colorant of a film, but to
give a speckled effect, especially in floor coverings and other decorative
vinyls.

TABLE 2 Elemental Metal Pigments

Commercial	Metal composition
Aluminum bronze	Aluminum
Copper bronze	Copper
Pale gold bronze	Cu, 92%; Zn, 6%; Al, 2%
Rich pale gold bronze	Cu, 90%; Zn, 9.25%, Al, 0.75%
Rich gold bronze	Cu, 77%; Zn, 22%; Al, 1%
Green gold bronze	Cu, 68.75%; Zn, 31%; Al, 0.25%

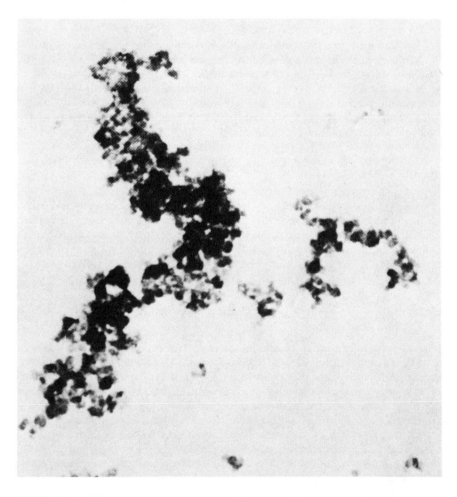

FIGURE 4 High-structure carbon black. (Courtesy of United Carbon Co.)

XI. METALLIC SALTS

These brightly colored inorganic pigments, metallic salts, are used because of their excellent heat stability, lack of tendency to migration and crocking, and good to excellent lightfastness. Their chief deficiencies are their very high specific gravity (high pound-volume cost), relatively large particle size, and low tinting strength, when compared with organic pigments. A technique for improving the chemical resistance and the lightfastness of these pigments has been devised that consists of coating each pigment particle with a coating of fumed silica. This permits the use of some pigments which have had borderline lightfastness and resistance to soap, alkali, and acid.

A. Iron Blues and Chrome Greens

The iron-containing salts such as Prussian blues and chrome greens (coprecipitates of ferric ferrocyanide and lead chromate) contain the reactive Fe cation and significantly affect heat and light stability.

B. Chrome Yellow and Orange Pigments

The chrome yellow and orange pigments compose a series based on lead chromate. Medium chrome yellow is lead chromate, with the lighter shades being mixtures of lead chromate and lead sulfate. Chrome oranges are basic lead chromate and are too sensitive to acids for use in vinyls. The chrome yellows vary in shade from light greenish yellow to a reddish yellow. They are attacked by acids and alkali and will turn black in the presence of hydrogen sulfide.

C. Molybdate Oranges

Molybdate oranges are complex mixtures of lead chromate, lead sulfate, and lead molybdate. They have excellent tinting strength and heat and light stability. One of their chief uses is in combination with high-cost organic red and maroon pigments for formulations that are bright in color, have good stability except to acids and alkalis, have high opacity, and are relatively low in cost. They will turn black in the presence of hydrogen sulfide.

D. Cadmium Pigments

Cadmium pigments vary in shade from the bright yellow of cadmium sulfide, through the reds of the cadmium sulfoselenides, to the deep maroon of cadmium selenides. Since they do not darken in the presence of H_2S, they are used for outdoor exposure (except in yellow tints) and where the maximum in heat stability is required. They are available as CP (chemically pure) and lithopone extended grades. In this case, the lithopone refers to barium sulfate, which is inert in vinyl formulations. While simple chemical formulas can be written for their composition, x-ray diffraction studies indicate that the term *coprecipitation,* frequently used to describe their preparation, implies a complex crystal or solid solution that provides for unexpected chemical stability [27].

E. Mercury-Cadmium Pigments

A newer series of pigments based on coprecipitation of red mercury sulfide with cadmium sulfide and selenides is analogous in properties and similar in shade to the cadmium orange, red, and maroon series. Their chief advantage is a lower cost than the cadmium series [27]. In vinyl formulations, their stability equals the cadmium pigments except where exposure to moisture or to acids is a problem. These pigments are also available in the CP grades and lithopone grades.

F. Ultramarine Blue

Ultramarine blue, an inorganic pigment made from calcining sulfur, clay and a reducing agent at high temperatures, has never been identified chemically. It has a bright, reddish blue color. Ultramarine blues are chemically stable except to acids. Since HCl has been identified as a heat and light degradation product of PVC, special stabilization is required for processing ultramarine blue formulations. This pigment is not recommended where lightfastness is a main requirement.

G. Mineral Violet

Manganese violet, permanent violet, and mineral violet are synonyms for manganese ammonium phosphate. The chemical stability of this substance is good, except to alkali and moisture. Its lightfastness is excellent in masstone and tint tone for interior use, but not good enough for exterior use. It has largely been supplanted by the newer organic violet pigments, which have better money value (color strength per unit cost).

H. Miscellaneous Chemical Salts

Strontium chromate, zinc chromate, cobalt salts, colored silicates, and other colored chemical salts are used as pigments in other industries besides the production of polyvinyl chloride. They are generally low in money value and brightness and are chiefly recommended for the ceramic industries. The exceptions are the various zinc chromate salts that are used in vinyl copolymer solution coatings and vinyl plastisol-organosol coatings to provide corrosion resistance for steel, aluminum, and magnesium metal substrates.

The characteristics of some inorganic pigments suitable for vinyl compounding are given in Table 3.

XII. ORGANIC PIGMENTS

Organic pigments compose a group of colorants that embraces a wide variety of chemicals having an equally wide variety of costs and properties. They are brilliant in color and have good tinting strength. Vesce [34] has categorized these pigments by chemical class. New classes must be continually added to this list because of new developments [2] (Table 4).

Organic pigments are classified according to chemical structure. Similarity of structure generally implies a similarity of properties. Table 4 presents the most common classification of these pigments.

TABLE 3   Inorganic Pigments Suitable for Vinyl Compounding

	White						Green			Yellow			
	Titanium dioxide, rutile	Titanium dioxide, anatase	Zinc oxide	Antimony oxide	Zinc sulfide	Chromium oxide	Hydrated chrome oxide	Yellow iron oxide, natural	Yellow iron oxide, synthetic	Natural siennas, ochres	Chrome yellows	Cadmium sulfides	Nickel titanate
Chemical class	OX	OX	OX	OX	MS	OX	OX	OX	OX	OX	MS	MS	MS
Brightness	H	H	VH	H	VH	VL	M	VL	VL	VL	H	H	H
Opacity	VH	H	L	VL	M	L	T	VH	VH	VH	H	H	VH
Lightfastness													
Masstone	VH	VH	VH	VH	VH	VH	VH	VH	VH	VH	MH	H	VH
Tint tone	VH	VH	VH	VH	VH	VH	VH	VH	VH	VH	M	L	VH
Heat stability	VH	VH	VH	VH	VH	VH	VH	L-M	H	L	H	H	VH
Resistance to reactivity with compound variables	H-VH	VH	VL	VH	VL	VH	VH	VL	H	VL	H	H	VH
Migration resistance	VH	VH	VH	VH	VH	VH	VH	VH	VH	VH	VH	VH	VH
Acid resistance	VH	VH	VL	H	L	VH	VH	L-M	VH	L-M	L	L	VH
Alkali resistance	VH	VH	M	H	M	VH	VH	VH	VH	VH	L	H	VH
Resistance to sulfide stain	VH	VH	VH	VH	VH	VH	VH	VH	VH	VH	VL	VH	VH
Weathering resistance	VH	M	M	H	L	VH	VH	L	VH	VL	M	VH	VH
Tint strength	VH	H	L	VL	L	VL	VL	VL	L	VL	L	L	VL
Economics of use	VH	H	L	VL	L	L	L	H	H	H	H	L	VL

Note: H, high. L, low. V, very. M, median; moderate. T, transparent. R, reactive in vinyls. Chemical class key: OX, oxide; MS, metallic salt; E, elemental.

	Oranges			Reds, maroons, violet				Blue		Brown		Black	
Molybdenum orange	Cadmium sulfo-selenide	Mercury cadmiums	Red iron oxide, synthetic	Cadmium sulfoselenides and selenides	Mercury sulfoselenides and selenides	Mineral violet	Ultramarine blue	Cobalt blue	Brown iron oxide	Burnt sienna	Carbon black, channel, furnace, lamp	Bone black	Black iron oxide
MS	MS	MS	OX	MS	MS	MS	MS	MS	OX	OX	E	E	OX
VH	VH	H	VL	H	H	MH	VH	M	VL	L	VL	VL	VL
VH	H	H	VH	H	H	M	M	L	H	T	VH	L	VL
H	VH	VH	VH	VH	H	H	H	VH	VH	VH	VH	R	VH
H	M	M	VH	M	1-M	M	R	VH	VH	VH	VH	R	VH
VH	VH	VH	VH	VH	VH	VH	VH	VH	H	H	VH	R	H
H	VH	VH	VH	VH	VH	VH	VL	VH	H	H	VH	VL	H
VH	VH	VH	VH	VH	VH	VH	VH	VH	VH	VH	VH	VH	VH
L	L	L	VH	L	L	H	VL	VH	H	VH	VH	VH	H
L	H	H	VH	H	H	L	VH	VH	VH	VH	VH	VH	VH
VL	VH	VH	VH	VH	VH	VH	VH	VH	VH	VH	VH	VH	VH
M	H	L-M	VH	VH	MH	L	VL	VH	VH	VH	VH	L	H
L	L	L	L	L	L	VL	VL	VL	L	L	VH	L	VL
H	L	M	H	L	M	VL	L	VL	VH	VH	VH	L	H

TABLE 4 Classification of Organic Pigments

Class	Typical structure	Typical common names	General properties
Basic	Precipitated basic dyes with tannic acid, phosphotungstic or phosphomolybdic acid	Auramines, Methyl Violet, Rhodamines, Victoria Blue, Methylene Blue	Not recommended: shows poor heat stability and lightfastness.
Insoluble azo	 TOLUIDINE RED HANSA G	Chlorinated paras, Fire Red, Para Red, Permanent Orange Toluidines, Arylamide Reds, Naphthols, Diansidine Orange, Nitraniline Oranges Hansa	Not recommended: shows bleeding and migration, crocking, poor heat stability, and plateout. Not recommended: shows bleeding, migration, crocking, plateout, sublimation, poor lightfastness [35].
Dis-azo	 DIARYLIDE YELLOW	Diarylide yellows	Most members of this class bleed, migrate, crock, and plate out. The AAOT type is used for flooring. The AAOA type is used for transparent yellows. All have poor lightfastness in tint tone and fair lightfastness in masstone.

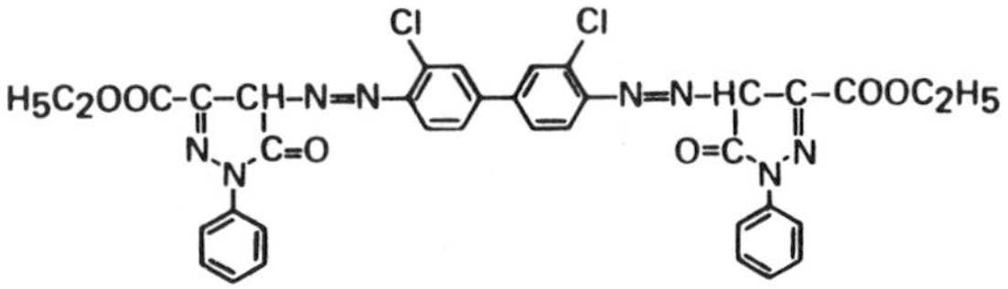

Soluble (precipitable) azo (pigments are formed by precipitation with calcium, barium, or manganese ion; all are nonbleeding and nonmigrating)

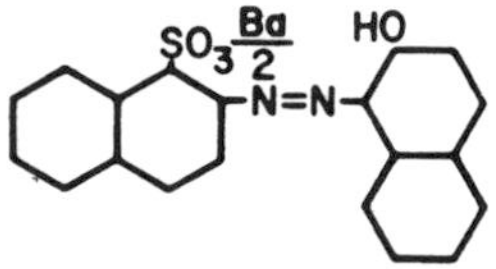

PYRAZALONE RED

Diarylide Orange, Pyrazalone Reds

Poor heat stability for long-term processing. Slight migration at room temperature; noticeable migration at high temperatures. Poor lightfastness in tint tone; fair to good lightfastness in masstone.

BARIUM LITHOL

Lithols, Persian Orange, Red Lake C

Sensitive to acids and alkalis. Poor heat stability and lightfastness in masstone. Poor lightfastness in tint tone.

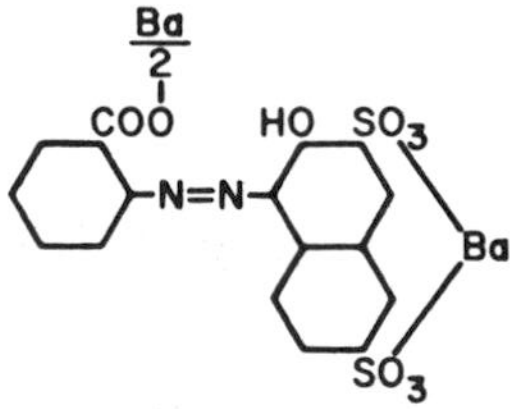

PIGMENT SCARLET

Pigment Scarlet

Sensitive to acids and alkalis. Good short-term heat stability. Poor long-term stability. Fair to good masstone lightfastness. Poor tint tone lightfastness. Permanent red 2B is subject to plateout.

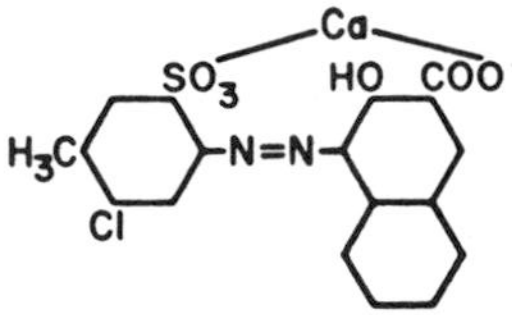

PERMANENT RED 2B

Permanent Red 2B, BON Reds and Maroons, Lithol Rubine

TABLE 4 (Continued)

Class	Typical structure	Typical common names	General properties
Miscellaneous	NICKEL AZO YELLOW	Metallized chelated azos	Excellent in most vinyl criteria, except for possible migration and bleeding.
	DISAZO	Dis-azo, condensation [39]	Excellent in all vinyl criteria, except for poor lightfastness. Many are excellent, some are good to light and exterior exposure. Available in yellow to deep red and maroon shades.
Condensation acid class		Alkali Blue, Acid Violet, Peacock Blue, Eosine, Quinoline Yellow	Not recommended: shows poor heat stability and lightfastness.
Vat pigments	ALIZARINE MAROON	Alizarin, Madder Lake	Good masstone lightfastness, poor tint tone lightfastness.

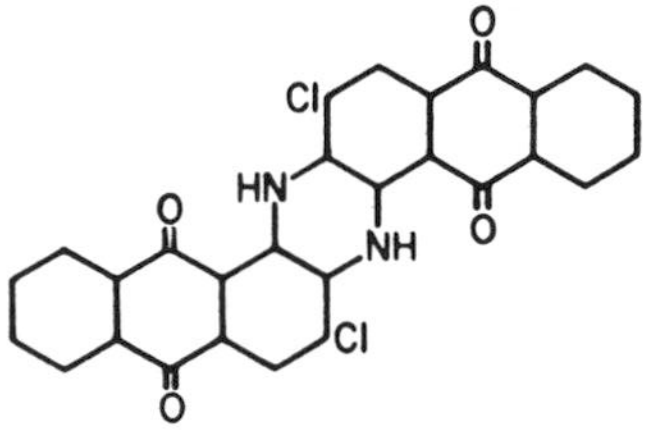

Those offered for vinyl formulation are nonbleeding, and nonmigrating, have excellent heat stability and lightfastness in masstone and tint tone. Stable to all types of chemical agents except reducing agents. Rated excellent in brilliance and transparency.

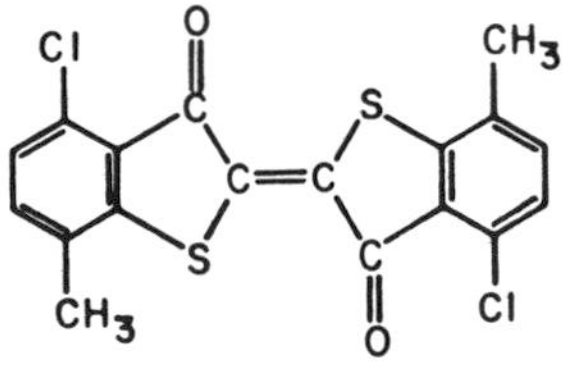

Anthrapyrimidine, Pyranthrones, Perylenes, Anthrimides, Indanthrone, Flavanthrone, Isoviolanthrone

Generally excellent lightfastness. Some numbers show slight bleeding. Used primarily for tinting.

Indigos, Thioindigo Reds, Thioindigo Maroons

Beta crystal form and alpha crystal forms with Cl or long-chain alkyl substituents on the benzene rings are stable to heat and solvents. Unmodified alpha crystal form is unstable. Heat- and solvent-stable forms meet all the criteria for excellent performance.

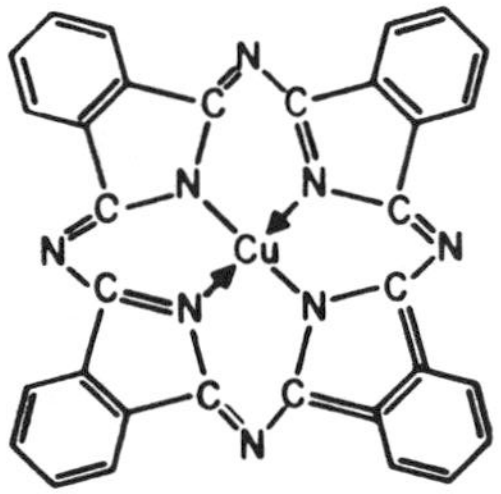

Phthalocyanine blues, vary in shade from very red undertone to very green undertone.

INDANTHRENE BLUE

THIOINDIGO

Phthalocyanines [36,37]

TABLE 4 (Continued)

Class	Typical structure	Typical common names	General properties
	PHTHALO GREEN	Phthalocyanine greens, vary in shade from blue undertone to yellow undertone with the number of chlorine atoms present. When bromine is used, the shade ia a very bright yellow green.	Excellent performance in all the criteria for vinyl requirements.
Miscellaneous class	**CARBAZOLE VIOLET**	Dioxazines [38]	Stable in all criteria for vinyl requirements.

Quinacridones

Stable in all criteria for vinyl requirements. Available in orange, red, maroon, violet shades.

QUINACRIDONE VIOLET

Isoindolinones

Excellent stability in all vinyl criteria.

ISOINDOLINONE

A. Basic Colorants

Basic colorants consist of basic dyestuffs precipitated with heteropoly acids. Even the "permanent" types are so poor in resistance to heat, light, and solvents that they are not recommended for use in vinyls.

B. Insoluble Azo Pigments

Colorants that are made by coupling various aromatic compounds to form monoazo compounds characterized by the—N≡N—linkage make up the insoluble azo pigments. They have no groups capable of forming salts with metals. As a class, they bleed, migrate, and crock in most vinyl formulations because of their solubility in plasticizers. They have poor heat stability and may sublime out of the compound during processing.

C. Disazo Pigments

Disazo pigments are formed from *o*-dichlorobenzidine, which can be diazotized and coupled at both ends of the molecule with other intermediates. They do not have as great a tendency to bleed as the monoazo class, and their heat stability and lightfastness is better.

D. Soluble (Precipitable) Azos

If azo pigments are formed from intermediates containing free phenolic, carboxylic, or sulfonic acid groups, they will be soluble in most solvents, including water. Adding calcium, barium, or manganese ions will precipitate these compounds to form nonbleeding pigments with good heat stability, fair to good lightfastness in masstone, and poor lightfastness in tint tone. Their resistance to alkali and to acid is poor because of the presence of the metal ion. This group of pigments in the orange to red and maroon shades are among the most popular pigments for use in vinyls because of their relatively low cost and excellent tinting strength.

E. Miscellaneous Azos

The two main members of this class are the chelated azos and the condensed azos. Azo pigments, chelated with a metal atom, have excellent characteristics for use in vinyls. Their increasing tendency to bleed and migrate, however, with increasing plasticizer concentration limits their use. Condensed azo pigments are formed by condensing two molecules of pigment from the insoluble azo group into a single molecule. This increase in size and molecular weight results in a series of brilliant red, yellow, and orange pigments with excellent properties for use in vinyls.

F. Condensation Acid Pigments

Dyestuffs that can be precipitated onto mordanting agent suitable for pigment use, such as aluminum hydrate, form this class of pigments. They are too water-soluble and poor in heat stability for vinyl use.

G.   Anthraquinone and Vat Pigments

Vat dyestuffs are colorless in their reduced or hydroquinone form

**HYDROQUINOID
(UNCOLORED)**

and colored in their oxidized or quinone form

**QUINOID
(COLORED)**

More than 400 such compounds are listed in the Color Index, and they have all been investigated for use as pigments. Those selected are prepared in their quinoid form, surface-treated for proper texture, and washed free of impurities. They have been chosen for their brilliance of color and for their ability to meet vinyl criteria. They are the most expensive of the pigments available and are used because of their excellent lightfastness.

H.   Phthalocyanine Pigments

The phthalocyanines are a series of blue and green pigments that have excellent stability to light, heat, and chemical reagents. They are insoluble in water and in all the usual ingredients of vinyl formulations. The tendency of the early phthalocyanines to crystallize under heat with a change in hue and in tinting strength has been overcome by adding substituents to the benzene ring and by careful processing in manufacture. Very red-shade blue pigments are available from the unsubstituted phthalocyanines, to blue-shade greens when 14 to 16 chlorine atoms are added to the possible 16 positions on the rings. Yellow-shade greens are made when bromine is substituted for part of the chlorine in the latter case. The phthalocyanines, as a class of pigments, appear to have most, if not all, of the properties considered desirable for use in vinyls.

I.   Miscellaneous Organic Pigments

The technique of tailoring an organic molecule to provide the color and stability characteristics required for vinyl processing has progressed to a fine
science.   The newer pigments such as the isoindolinones and quinacridones
fall into this class.   The specific capabilities and uses are best described
by the individual manufacturers, who continually modify and improve their
products.   Specific compositions are listed in the Color Index [2] and in
the references shown for each group.   Detailed recommendations for use
are constantly being revised because of these improvements.   Basic lists
for evaluation of pigments for use in vinyls appear in the technical literature on a regular basis [40–42] (see Table 5).

XIII.   PEARLESCENT AND OTHER INTERFERENCE PIGMENTS

A series of pigments is available in which the pigments exist as thin, transparent flakes and yield color effects due to reflection from both surfaces of
the particle and the consequent reinforcement or interference of these light
rays [14,43].   The effect is a clean, silvery multicolored appearance, which
can be modified with small amounts of durable colors.   Natural pearlescence
is too fragile and costly for use in most hot shear-processed vinyls, although it is occasionally used in plastisols and solution vinyls.   The same
visual effect is obtained through pigments made from lead carbonate, lead
phosphate, lead arsenate, and bismuth oxychloride.
    A new series of interference pigments made from titanium-coated mica
has also been introduced.   These pigments are easier to stabilize for heat
and light than some of the older forms.
    All these pigment particles are subject to fracture, with consequent
loss of "pearlescence" when subjected to too much shear.   For the best
appearance, they should be processed with a minimum of intensive shear
and in such a way that the particles orient in a plane parallel to the plastic
surface.

XIV.   FLUORESCENT AND OTHER OPTICALLY ACTIVE PIGMENTS

A.   Fluorescent Pigments

Fluorescent pigments [14,44] are based on fluorescent dyes that have been
incorporated into a thermosetting resin matrix, cured, and ground to a
suitable fine powder.   When the dye and resin are carefully chosen, pigments with excellent heat stability and resistance to bleeding and migration
result.   The original fluorescent pigments were deficient in lightfastness.
Newer pigments are reported to withstand 12 months or more of exterior
exposure, however.   Their fluorescence derives from their ability to absorb ultraviolet or visible light and reemit it at other visual wavelengths.
This has the effect of a brighter color being reflected from the surface
than would be expected from light incident on it.   Fluorescent pigments
are only of value when they are observed under a strong source of daylight or incandescent illumination reinforced with ultraviolet light.   Although
fluorescent light bulbs can be used, optimum color value is best manifested
under natural daylight.   Individual fluorescent pigments can be blended
for variation in color effects.   When they are combined with standard color
pigments, UV absorption of these latter pigments severely inhibits the
efficiency of fluorescence.

B.   Luminescent and Phosphorescent Pigments

The series of luminescent and phosphorescent pigments is based on the
variations possible in ZnS, CdS, and ZnO crystals, which can store UV
or visible light and reemit visible light when all sources of light are removed.   These pigments are highly fragile and are liable to fracture in
vinyl processing operations.   At best, the reemitted light is very weak in
intensity and is of relatively short duration.   These pigments are used in
acrylic systems or in vinyl-acrylic top coats as novelty coatings for various
commercial products [14].

C.   Emission Pigments

Pigments that emit light without needing any external light source are available from the vast source of radioactive materials now available.   Their price
is prohibitive and their use regulated by the Atomic Energy Commission.

D.   Optical Brighteners

The unusual group of colorless, transparent pigments composed of the optical brighteners can absorb ultraviolet light and reemit it in the visible
wavelengths without affecting transparency of the film.   They usually reemit light in the blue end of the spectrum and effectively make transparent
and white vinyls appear brighter and whiter (more blue white).   Although
the brighteners are relatively expensive, amounts of the order of 0.001 to
0.01 phr are usually sufficient.   Unfortunately, their lightfastness is limited and short periods of exterior exposure destroy their utility [57].

XV.   SPECIAL PIGMENT PROBLEMS

A.   Heat Sealing and Electrical Resistance

When heat sealing of vinyls is to be accomplished by directly applying heat
and pressure, the pigment will have no effect on the process so long as it
does not migrate or bloom.   When dielectric heat sealing is used, pigments
of minimum conductance are required.   Pigments that will pass the NEMA
requirements [11] are the most desirable.

B.   Toxicity of Pigments in Vinyl Formulations

There are basically four categories of use conditions of interest to vinyl
formulators:

1.   Products designated as toys, or other objects that children might
     put in their mouths
2.   Products to come in contact with food, such as food packaging,
     bread wrappers, and candy box separators
3.   Products to be used in packaging cosmetic products, like soaps or
     shampoos
4.   Products for which the ultimate in nontoxicity is required, such as
     bottles for salad oil, beer, other liquid food products and for med-
     ical prosthetic devices

TABLE 5   Organic Pigments Suitable for Vinyl Compounding

Group	Common pigment name	Chemical class	Brightness	Transparency	Lightfastness-Masstone	Lightfastness-Tint tone	Heat stability	Migration resistance	Acid resistance	Alkali resistance	Weathering resistance	Tint strength	For use in plasticized vinyls	For use in rigid vinyls	Economics of use
Green	Phthalocyanine green, all types	PCN	VH	VH	VH	VH	VH	VH	VH	VH	VH	VH	VH	VH	VH
Yellows	Flavanthrone	AN	VH	VH	VH	VH	VH	VH	VH	VH	VH	VH	VH	VH	L
	Diarylide yellow OT	DIS	VH	M	M	VL	VH	L	VH	VH	VL	VH	VL	L	VH
	Diarylide yellow AAOA	DIS	VH	VH	M	VL	VH	L	VH	VH	VL	VH	VL	L	VH
	Pigment yellow 83	DIS	VH	VH	M	M	H	VH	VH	VH	L	VH	VH	VH	VH
	Isoindolinones	MV	VH	VH	VH	VH	VH	VH	VH	VH	H	VH	VH	H	ML
	Nickel-azo	MA	H	M	VH	VH	M	M	L	M	M-H	VH	M	H	M
	Anthrapyrimidine	AN	VH	VH	VH	VH	H	H	VH	VH	H	VH	VH	H	L
	Dis-azo, condensation	MA	VH	VH	M-H	M-H	H	H	VH	VH	M-H	VH	H	H	M
Oranges	Diarylide orange	DIS	VH	VH	L	VL	VL	M	H	H	VL	H	VL	VL	H
	Nitraniline oranges	AZ	VH	VH	VL	VL	VL	VL	VH	VH	VL	VH	VL	VL	H
	Dianisidine orange	DIS	VH	H	L	L	H	L	VH	VH	VL	VH	VL	L	H
	Vat orange GR	AN	VH	VH	VH	H	H	VH	VH	VH	H	VH	VH	VH	VL
	Vat orange RK	AN	VH	VH	L-M	L-M	L-M	VH	VH	VH	M	VH	M	L	VL
	Isoindolinones	MV	VH	VH	VH	VH	H	VH	VH	VH	VH	VH	VH	VH	L

	Class													
Reds and Maroons														
Lithols	SA	VH	L	VL	VL	VL	VH	VL	L	VL	VH	VL	VL	VH
Permanent red 2B	SA	VH	M-H	M	VL	M-H	VH	L	L	VL	VH	VH	VH	VH
Lithol rubine	SA	VH	VH	L	VL	L	VH	L	L	VL	VH	VH	VH	VH
Pigment scarlet	SA	VH	VH	L	VL	H	VH	L	L	M	VH	VH	VH	VH
Red lake C	SA	VH	VH	VL	VL	VL	VH	VL	VL	VL	VH	VL	VL	VH
Bon reds and maroons	SA	VH	VH	M	VL	M	VH	M	VL	VH	VH	VH	VL	M-H
Pyrazalone	DIS	H	VH	H	VL	M	L-M	VH	VH	VL	VH	VH	H	H
Quinacridones	MV	VH	H	VH	VH	VH	VH	VH	H	H	H	VH	VH	L
Alizarine maroon	AN	VH	VH	H	VL	H	VH	M	M	VL	M	M	M	M
Perylenes	AN	VH	VH	H	VH	VH	VH	VH	VH	H	VH	VH	VH	L
Isoindolinone	MV	VH	VH	H	VH	VH	VH	VH	VH	H	VH	VH	VH	L
Dis-azo, condensation	MA	VH	VH	H	VH	VH	VH	VH	VH	H	VH	VH	VH	L
Thioindigos	AN	HH	VH	H	M-H	M-H	VH	VH	VH	M	VH	M	M	L
Blues and Violets														
Phthalocyanine	PCN	VH	VH	H	VH	VH	VH	VH	VH	VH	VH	VH	VH	VH
Indanthrone	AN	VH	VH	H	H	VH	VH	H	VH	VH	VH	VH	VH	M
Quinacridone	MV	VH	VH	VH	VH	VH	VH	H	VH	VH	VH	VH	VH	L
Carbazole	MV	VH	VH	VH	H	VH	VH	H	VH	VH	VH	VH	VH	M
Thioindigos	AN	VH	VH	M	M-H	M-H	VH	VH	M	VH	VH	M	M	L

Note: Key to chemical class: PCN, phthalocyanine; AN, anthraquinone: DIS, dis-azo; MA, miscellaneous azo; MV, miscellaneous vat; AZ, azo; SA, soluble azo. Ratings: V, very; H, high; M, median. Pigments not readily identified by common names are listed by Color Index number where possible [2].

Each category requires a different set of pigments, and cost of coloring increases significantly with increasing rigidity of specification. If the manufacturer can prove that there is no possibility for the pigment to be ingested-either because of an impermeable barrier between the pigments and the material to be eaten, or because of the insolubility and inertness of the pigment-then any pigment can be used [25]. Since this is difficult to prove, vinyl formulators follow the individual requirements of the four groups outlined above as set down by the U.S. Food and Drug Administration or other regulatory agencies.

1. Toys: Although individual state laws have varying requirements related to toxicity, the U.S. Department of Commerce Commercial Standard for Artist Materials to be used in Schools (CS-136-61) and the ANSI Standard [29] provide general guidelines for choosing pigments. In general, they place stringent limits on the amount of lead, mercury, bismuth, soluble barium, cadmium, antimony, and arsenic that can be used. This specification effectively proscribes the use of chrome yellows, molybdate oranges, cadmium pigments, bismuth, and synthetic pearlescence.

2. Food Packaging: In 1958 an amended form of prior legislation known as the Food Additives Amendment was passed by Congress [30]. This statute covers materials that are likely to become part of food as a result of food processing or packaging. The Food and Drug Administration regularly publishes lists of pigments (as well as other materials) that may be used for this application. Suitably prepared titanium dioxide, iron oxides, and ultramarine blue can be used in this application, as well as in foods, without regard to certification [31]. Other pigments can be used provided that they will not become part of the food. A series of test solvents and test procedures has been prepared that simulate exposure to various types of foods [32]. Although the tests do not specifically refer to vinyl, compliance with them can indicate intent to meet the safety requirements of this law.

3. Colors for Cosmetics: Suitable lists of colorants that can be used in products that come in contact with the skin have been prepared by the FDA under the Food, Drug, and Cosmetic Act of 1938, as amended in 1960 [32,33]. Provisions for certification of purity have been made. These D&C colors are prepared as lakes on aluminum hydrate for incorporation into vinyls. Their use as colorants in cosmetic packages exempts the formulator from complying with "the proof of nontoxicity" when noncertified colors are used. Unfortunately, these colorants are seriously deficient in lightfastness and heat stability and require special processing conditions for satisfactory use.

4. Food Colors: A list of FD&C colorants suitable for use in coloring food has also been prepared by the FDA. This list is smaller than the D&C list. and the laked colors on this list are also deficient in heat and light stability. The FD&C certified colorants can be used without restriction in any application where toxicity is a problem.

C. Outdoor Durability

The use of vinyls for outdoor applications is constantly increasing. Colorants can be important in increasing this use. Carbon black pigments will enhance the durability of any plasticized vinyl formulation because of their absorption of ultraviolet light. These short wavelengths of light in the sunlight radiation spectrum are the ones that most seriously affect vinyls.

Pastel colors and white vinyls should be formulated with the nonchalking grades of titanium dioxide to provide for maximum life of the vinyl and prevention of the disfiguring white powder, or "chalk," that can appear with other grades. Since many pigments have poorer lightfastness in tint tone than in masstone, only pigments that have been thoroughly evaluated with titanium dioxide should be used. Stabilizers specifically designed for transparent vinyls may enhance the lightfastness of such formulations.

Outdoor durability refers to exposure to all of the deteriorating factors that may affect a vinyl formula. Light is only one such factor, with humidity or moisture, acid attack from industrial fumes, and alkali and soap from cleansing agents representing other factors that affect weatherability. Pigments that show suitable resistance to all of these factors should therefore be chosen. If a compromise must be made, those pigments that darken or change hue to a darker value are preferable to those that lighten or fade.

XVI. COLOR AND CHEMICAL COMPOSITION

Color is the term that we give to the human eye's reaction when observing certain wavelengths of light. Colorants have the ability to accept incident light, absorb portions of this light, and reflect or transmit other portions. For example, when a colorant absorbs all of the light except blue light and when it reflects or transmits this blue light, we have a blue colorant. When a colorant absorbs all (or most) of the light equally, we are dealing with a black colorant. White colorants reflect equally all wavelengths of the incident radiation.

The color that we see depends on both the absorption characteristics of the colorant and the medium that surrounds it. Chemical groups that cause a stress on the nucleus of the molecule, and consequent response of electrons to the energy of the incident light, will be colored [3, 14, 38, 45, 48]. Interference between incident and reflected light can occur when pigment particles have two dimensions significantly different from the third, as with pearlescent or other flake-type pigments. Dichroism occurs when pigment crystals have significantly different refractive indices along different axes [3, 14] and when these pigments can orient themselves in specific planes. Interference pigments and dichroic pigments are difficult to evaluate for color because of the difficulty in duplicating particle orientation.

A. Color and Color Measurement

Color results from the effect on the eye of light modified by colorants [46]. It depends on

1. A light source
2. A modifier for the light source
3. An observer to interpret the modified light

Visible light is a physical phenomemon defined as that portion of the electromagnetic spectrum between 4000 and 7000 angstroms (Fig. 5). The

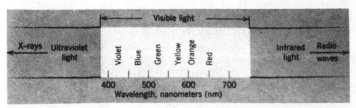

FIGURE 5   Visible light spectrum.

physicist can characterize the ability of a pigment to modify light by observing its spectrophotometric curve (Fig. 6).   The spectrophotometer measures the absorption of incident light on a colorant at each wavelength of visible light.

Spectrophotometric curves are not a measure of color because they do not provide a value for the type of light under which the colorant is viewed, and because they do not provide a value for the human eye-brain capability to see color.   The eye sees three attributes of color that Munsell has defined [9] (Fig. 7).

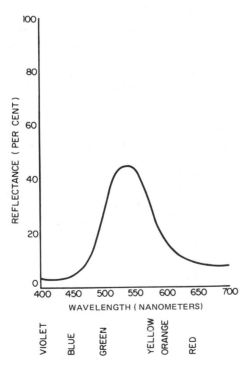

FIGURE   6   Spectrophotometric curve of a green pigment, by Davidson and Hemmendinger.   (Courtesy of *Color Engineering.*)

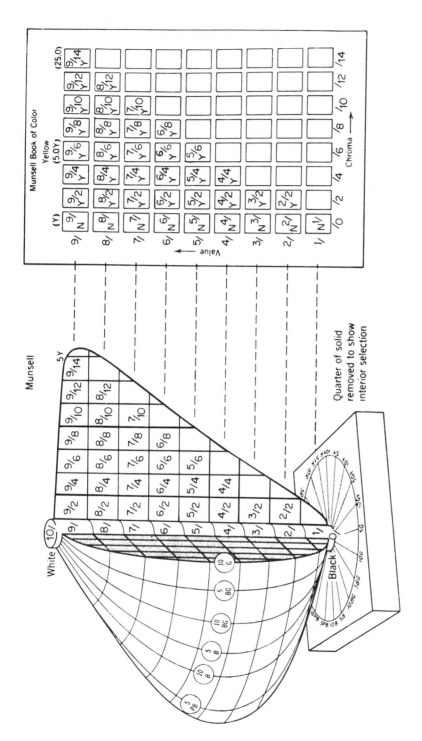

FIGURE 7 Munsell color system. A system of specifying objective colors on scales of hue, value, and chroma, exemplified by a collection of chips forming an atlas of charts that show scales for which two of the three variables are constant, the hue scales containing five principal and five intermediate hues, the value scale containing 10 steps from black to white, and the chroma scales showing up to 16 steps from the equivalent gray. All three scales are intended to represent equal visual (not physical) intervals for a normal observer and daylight viewing with gray to white surroundings. From ISCC *Newsletter*, no. 156 (Nov.-Dec. 1961). D. Nickerson.

1. Hue: the quality of color we describe as red, yellow, green, etc.
2. Value: the quality of color we describe as light or dark when
    compared with white (Munsell value of 10), through grays, to
    black (Munsell value of 0).
3. Chroma: the quality of color that distinguishes it from a gray of
    the same value. The terms *saturation* and *purity* are sometimes
    used to describe this quality.

In defining color, one must consider the characteristics of the light
under which the colored object is viewed. A red colorant (Fig. 8), il-
luminated by a blue light (curve C, Fig. 9), will appear quite dark and
different from its appearance under a red light (curve A, Fig. 9). All
these variables have been combined by the International Commission on
Illumination into the CIE system (Commission International de l'Eclairage)
of colorimetry [46,50]. This system develops its three-dimensional characteristics from the three lights (Fig. 10), which when properly combined
will produce white light; the definition of *white light* is "a bright light
containing all the wavelengths of visible light at equal intensity."

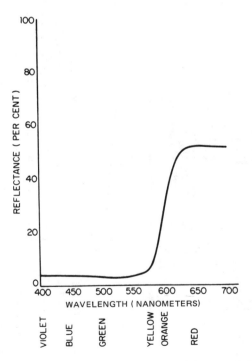

FIGURE 8 Spectrophotometric curve of a red pigment, by Davidson and
Hemmendinger. (Courtesy of *Color Engineering.*)

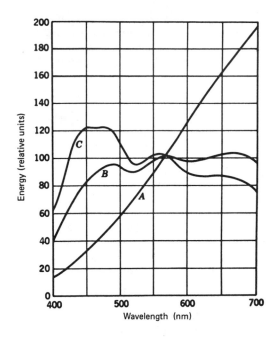

FIGURE 9 Illuminants A, B, C [58]. Reprinted by permission of John
Wiley & Sons, Inc., New York.

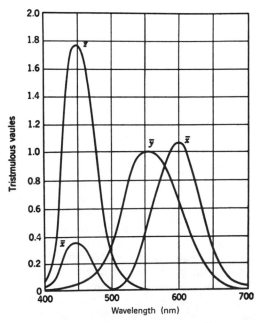

FIGURE 10 Characteristics of $\bar{x}$ (red light), $\bar{y}$ (green light), and $\bar{z}$ (blue
light) for the spectrum colors [58]. Reprinted by permission of John
Wiley & Sons, Inc., New York.

CIE measurements can be made with three-filter colorimeters that measure the light reflected or transmitted by a colored object under one of the standard sources of light (Fig. 9). These measurements are designated the tristimulus values, X, Y, Z. They are converted into tristimulus coordinates or trichromatic coefficients, x, y, z, by the equations

$$x = \frac{X}{X + Y + Z}$$

$$y = \frac{Y}{X + Y + Z}$$

$$z = \frac{Z}{X + Y + Z}$$

Tristimulus values are defined as the product of the relative energy of the light source used, the reflectance or transmittance of the object, and the tristimulus value of the equal-energy spectrum colors (Fig. 10). The tristimulus coordinates are the percentages of each component for the particular color observed [46]. These data can then be plotted on the CIE chromaticity diagram (Fig. 11).

Hue can be estimated from "dominant wavelength." Chroma is estimated from the value for "purity." The third dimension of value can be estimated directly from the Y reading determined on the colorimeter. The value obtained for Y has been determined to represent closely the brightness of a color as viewed by the human eye. [46,50].

The significance of the CIE system is that it permits the establishment of numerical definitions for color and for tolerance variations from color standards.

Computer programs (analog and digital) based on the measurements obtained from a spectrophotometric curve, the definitions of the standard light sources, and the definition of the CIE equal-energy colors have been devised [46,51,52,56].

B. Metamerism

Metamerism is a defect in a color match that is defined as the change in color relation between two colored objects when the conditions of viewing are changed. The phenomenon was first noted when colors were matched in artificial light (curve A, Fig. 9) and then observed to be different under daylight (curve C, Fig, 9). Spectrophotometric curves of two gray color chips (Fig. 12) illustrate two colors that are close matches under tungsten light but mismatches under fluorescent light and daylight. The "simple gray" of this set of colors becomes redder in contrast to the "complex gray" as the incident light becomes bluer. When this phenomenon occurs during routine color matching procedures, it invariably indicates that there is a difference in the type of colorants used in the two samples. This procedure has often been used for quality control analysis and pigment identification. Metamerism can be a serious problem to a vinyl formulator attempting to match a colored object (paint, paper, textile, ceramic) that has been made with colorants unsuitable for use in vinyl formulations. Adequate solutions to these problems can be arrived at by analyzing the spectrophotometric curves of colorants that are suitable in vinyl and combining them to eliminate most of the metamerism [54].

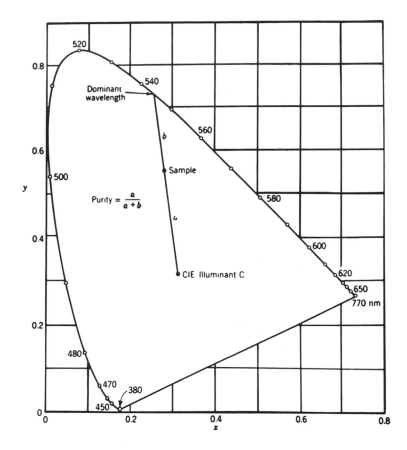

FIGURE 11 A typical chromaticity diagram [59]. Reprinted by permission of Interscience Division, John Wiley & Sons, Inc., New York.

Subsequent studies indicate that metamerism may also occur with differences in angle of light incidence or light reflection, with differences in observers, or with any change in conditions of color observation [46,53].

C. Visual Color Matching

The ultimate end of the coloration of vinyl compounds and the extensive color characterization procedures is the preparation of "suitable colors" for the purpose desired and the reproduction of these same colors. The original duplication of the desired color in vinyl formulations and its continued production is an engineering problem. Our engineering limits for suitable duplication of color are those that the human eye can distinguish. Whereas the human eye sees color essentially in terms of

1. Hue: dominant wavelength
2. Value: brightness, lightfastness
3. Chroma: saturation, purity

other factors in the character of the light that is transmitted to the eye are

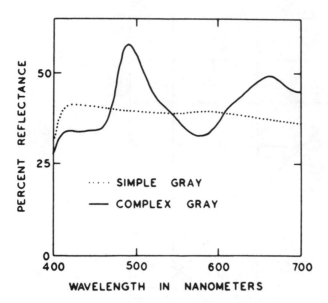

FIGURE 12 Spectrophotometric curves of a metameric pair of colors that will be similar in color in daylight but significantly different in blue (fluorescent tube) light and red (tungsten) light (By W. C. Granville. Courtesy of *Color Engineering.*)

important. Color can be affected by (1) a matte vs a press-polished surface, (2) the depth of embossed pattern, and (3) whether the colorants are at the surface (high concentration of pigment in a thin film) or submerged within the film, sheet, or slab (low concentration of pigment in a thick film or slab) [14,43,46,55]. *Constant care* must be taken, therefore, to ensure that colors are evaluated with all these other variables omitted. Color samples for comparison to the standard must be:

1. As close to each other in surface characteristics as possible.
2. Viewed with lights of the same characteristics as those under which they will be used.
3. Prepared on clean equipment and free of contamination from other colored objects.
4. Evaluated in terms of their end-use requirements.

This concept has been introduced as the Look-Think method of color evaluation and is valid regardless of whether the evaluation is made by eye or by instrument [46].

XVII. COLORANTS

PVC is used in a variety of applications and fabricated into finished materials on a broad range of equipment. Both the application and the fabrication technique can influence the choice of pigments. Not all pigments of the same chemical type perform in the same way. Variations in manufacture, surface treatments, and additives may, for example, make a particular

Pigment Blue 15 more desirable than another. A consistent, repeatable testing procedure is, therefore, useful in making proper pigment selections.

Colorants for PVC may be classified in three categories: (1) pigments, (2) dyes, and (3) special effect, that is metal flakes, mica, pearlescent. A pigment is an insoluble coloring material, whereas a dye is soluble in the polymer matrix. Although solubility of a colorant offers ease of incorporation (no shear forces needed to provide dispersion), in plasticized PVC, dyes are generally avoided because they tend to migrate. Migration is the movement of a colorant from one material into another material with which it is in contact. A classic example of migration is seen when a piece of stained wood is painted. In a short time, the dye in the stain will move into the paint, causing discoloration. Some pigments may also be partially soluble and exhibit some tendency to migrate. In rigid PVC, where no plasticizer is used, migration problems may be considerably reduced, but testing is advised.

Pigments are classified as either inorganic or organic. Both types are used in coloring PVC. The most commonly employed inorganic pigments are titanium dioxide, iron oxide reds and yellows, lead chromate yellows, and molybdate oranges. Also used, but less frequently, are cadmium reds, yellows, and oranges, selected mixed metal oxide pigments such as nickel titanate (greenish yellow) and chrome titanate (reddish yellow), and ultramarine blue.

Organic pigments are available in a wide range of properties, prices, colors, and chemical classes. With the exception of pigments having poor heat stability or tendencies to migrate, bloom, or plate out, most may be used in PVC. End-use properties such as lightfastness, weathering properties, electrical conductivity, or chemical resistance may be additional limiting factors. The pigments most commonly used in PVC are selected monoazo types, disazos, phthalocyanine blues and greens, quinacridone reds and violets, diarylide yellows, selected vat pigments, isoindolinone and isoindoline yellows, and quinophthalone yellows.

A. Pigment Incorporation in PVC

PVC, more than any other polymer, lends itself to a wide range of compounding and processing methods. The processes used may influence the choice of pigment incorporation technique. As with most polymers, there are three basic approaches to incorporating color; these involve:

1. Dry color
2. Color concentrates
3. Precolored compound

Although vinyl processors frequently purchase dry color, it is most likely put into a color concentrate which has a carrier, a plasticizer, or a vinyl compound. The vinyl industry saw its greatest development in the post-World War II years before the color concentrate industry became well established. Vinyl processors had no recourse but to develop their own dispersions. Today many processors rely on a combination of in-house concentrate preparation and outside purchases. Plasticizer concentrates have in part been displaced by homopolymer chip concentrates. Dry concentrates reduce waste, do not settle or separate, and are cleaner to handle than pastes. Pastes, on the other hand, can be readily mixed together, and even before a colored compound is processed, the color of the paste may give a reasonable indication of what to expect in the final product. Pastes are the preferred material for coloring plastisols and organosols. Although dry color may be used in plastisols, dispersion problems will probably arise.

Dry color may be used in compounding PVC when intensive mixers are used, but housekeeping and health considerations must be taken into account. The use of color concentrates eliminates the housekeeping problems associated with dry color and assures the processor that the color has been developed to optimize its properties.

Precolored compound finds relatively little use in the PVC industry. The rigid PVC industry, which up until the late 1970s used precolored compound extensively, now relies on in-house compounding or color concentrate.

B. Selecting Pigments for PVC

Because PVC is processed at a fairly low temperature and is easily compounded on a two-roll mill, it is relatively simple to evaluate pigment performance. A standardized program for pigment evaluation is valuable. Coupled with colorimetric data, such an evaluation will provide the user with colorant choices based on fastness properties, color properties, and economics.

The following factors may be considered in the course of evaluating pigments:

1. Heat stability
2. Lightfastness/weatherfastness
3. Migration resistance
4. Tinting strength
5. Money value
6. Chemical resistance
7. Blooming
8. Dispersibility/dispersion behavior
9. Behavior at low concentration
10. Plateout

Some aspects of pigment behavior are not constant and will vary depending on heat history, dispersion technique, concentration, and other additives. If feasible, an evaluation program will be designed to examine these variables.

C. Heat Stability

Since vinyl processing is done between 300°F (150°C) and 400°F (205°C), there are relatively few pigments which cannot be used on PVC solely because of poor heat stability. The monoazo pigments, in general, should be considered problematic with respect to heat stability. Included here are such products as toluidine red, pyrazolone red and orange, dinitraniline and dianisidine oranges, Red Lake C, and Permanent Red 2B. While some of these pigments have adequate heat stability at the low end of the temperature range used for PVC processing, they are marginal to unsatisfactory at the upper end. Residence time and concentration may be critical. In addition to questionable heat stability, most of the azo pigments exhibit other deficiencies: migration, blooming, plateout, and poor lightfastness. Among the products listed above, only the pyrazolones and the 2B Reds find fairly widespread application in PVC, but only in noncritical applications where migration, plateout, and limited lightfastness can be tolerated. Heat stability testing of pigments should be carried out at the highest temperature and the longest residence time that the compound may reasonably be expected to encounter. By also including intermediate points, a profile of the pigment's behavior can be derived. Several concentrations should also be tested. Many pigments have satisfactory heat stability at concentrations of 0.1% and higher, but are deficient below 0.1%.

Heat history is cumulative. Therefore, reworking of scrap or off standard material may cause a color failure. In reworking scrap, particular attention should be paid to pigment concentration. It may be inadvertently reduced below a critical minimum. This, coupled with additional heat history, may give rise to migration or reduced lightfastness with normally satisfactory pigments.

Depending on the equipment available, heat stability testing in PVC may be conducted in a variety of ways:

1. Prolonged two-roll milling with samples being taken at appropriate intervals
2. Static oven aging at varying times and/or temperatures
3. Compression molding at varying times and temperatures

The first method, while accurate, does not usually employ the highest temperature to which the compound may be subjected. The last two require care to ensure accuracy and reproducibility since ovens and presses are subject to rapid temperature changes or uneven heat distribution. A control in the form of an unpigmented sample should always be run.

Poor heat stability may appear as a darkening of the color (usually in masstones), a loss of color strength (tints or very low concentration masstones), or discoloration (usually yellowing or browning). If there is an apparent increase in color strength or a change in hue to a different but not undesirable color, it is likely that the pigment has dissolved and has, in effect, become a dye. A migration test on this heat-stressed material will confirm this.

D. Migration Resistance

Colorant migration may occur when the colorant is soluble in a polymer system. As long as the colored material does not come into contact with another material in which the colorant is also soluble, migration may not be a problem. Because PVC is used in a wide variety of end products, it is best to avoid colorants exhibiting severe migration tendencies. Migrating colors should not be used where printing is part of the process, for instance, in wall covering. A migrating color in a substrate can move into the ink, causing discoloration. Conversely, a migrating pigment in the ink would move into the substrate. Laminated or bonded constructions, particularly of different colors, should not employ migrating colors. It is common practice to color adhesives, either for identification or to aid in applying a uniform coat. If such adhesives are used with flexible PVC materials, migrating colors should not be used in the adhesive.

Depending on the degree of solubility and the severity of the conditions, migration may occur in a matter of minutes or it may take days. The most commonly used migration test is the "sandwich" test. A piece of colored vinyl is placed against a piece of white vinyl. Surfaces of both samples must be smooth to ensure contact. Placed between heavy metal plates, the sample is oven-aged at 80°C for 24 hr. On removal from the oven, the colored material is peeled away from the white. A stain on the white vinyl is evidence of migration. The intensity of the stain may be assessed by using a gray scale or other appropriate means. Because migration continues until equilbrium is reached (the white vinyl is evenly colored all the way through), assessment of migration should be done promptly. Otherwise, the stain which is found only on the surface of the white vinyl will appear to lessen in intensity over time.

It is recommended that several pigment concentrations, especially a very low one, be checked for migration. With some pigments, migration is a problem only at very low concentrations. In such a case it is thought that, at this low concentration, the pigment is fully dissolved. At a higher concentration, it is only partially dissolved and is continously dissolving and recrystallizing. In the former state migration may occur readily whereas in the latter case migration is virtually nonexistent.

In the preceding section on heat stability it was mentioned that prolonged heat history may cause a pigment to go into solution. This will usually show up as increased color intensity or transparency. Sometimes a shade shift may occur. A migration test on such material is in order.

Migration is not confined to flexible PVC. Testing may be devised which employs any substrate with which the material may come in contact.

E. Lightfastness/Weatherability

The ability of a colorant to hold its shade on exposure to light or weather is frequently critical. As with so many properties, lightfastness is often dependent on the pigment concentration and sometimes on heat history. A general rule of thumb is that, as the concentration of an organic pigment goes down, especially in tints, lightfastness goes down also. With inorganic pigments, the reverse is likely to occur. Masstones more often will darken, although tints are more stable.

Countless studies have been conducted which attempt to correlate natural weathering with accelerated or artificial weathering [60,61]. Because weather and climate are variable, correlations cannot be precise. The most commonly used tests for lightfastness employ either a carbon arc, a fluorescent sun lamp, or a xenon arc as a light source [62]. The carbon arc lamp emits some UV light energy at wavelengths which are not present at the earth's surface. Because of this changes due to exposure will occur two to three times faster under carbon arc illumination than under sunlight or xenon illumination, but it can be argued that the low wavelengths of the carbon arc provide an unrealistic environment. Many polymers are particularly sensitive to the UV wavelengths emitted by the carbon arc lamp. Although the carbon arc light source has a long history of use in the United States, there has been a movement toward the use of the xenon arc light source, notably in the textile industry. Newer xenon units offer precise temperature and humidity control and maintain more uniform light energy output over the life of the lamp than do older units.

Although lightfastness testing is useful for materials used indoors, it is imperative to test actual weathering characteristics if a product is to be used outdoors. Some pigments, both organic and inorganic, will exhibit excellent lightfastness, but a combination of light and moisture will cause rapid failure. Cadmium pigments, for example, exhibit very good lightfastness but poor weatherability. A commonly employed cycle in an accelerated weathering test is 102 minutes of light exposure at 145°F (63°C) followed by 18 minutes of darkness with water spray. Other cycles are available.

Even "accelerated" testing requires time. For products requiring good light or weatherfastness, a minimum exposure would be 500 to 1000 hr in a carbon arc unit and twice that in a xenon unit. In isolated cases, as for building materials, exposure periods of 5000 to 7000 hr may be appropriate.

Regardless of the exposure interval, an assessment of appearance change is required. Several methods are employed:

1. Description, for instance, slight fade, moderate darkening.
2. Color measurement data vs. an unexposed control.
3. Gray Scale rating from 1 through 5, where 5 means no change and 1 means severe change. Directionality of change may be indicated by an appropriate letter, such as d, darker; r, redder; or y, yellower.
4. Blue Wool Scale rating: This system is widely used in the textile industry in the United States. To a lesser extent, the plastics industry has adopted it. The European color industry, however, uses the Blue Wool Scale extensively. The American Association of Textile Chemists and Colorists' Blue Wool Scale, with ratings from 1 through 9, is slightly different from the European Blue Wool Scale, which has ratings from 1 through 8. Both rely on dyed fabric swatches of varying lightfastness. The fabrics are exposed along with a test sample. When the sample has shown a just discernible change, the numbered fabric which has also shown a just discernible change is taken as the rating. A higher number represents a better lightfastness rating. Each step on the AATCC Blue Wool Scale represents a doubling of the lightfastness:

Standard fading hours	AATCC rating
2.5	1
5	2
10	3
20	4
40	5
80	6
160	7
320	8
640	9

A standard fading hour is a unit of time exactly defined by the
U.S. National Bureau of Standards. It is determined with light-
sensitive paper available from the NBS.

Chemical resistance of pigments can be important both in process and
in end use. Some colorants, metal complex types, for example, may react with other additives (most frequently stabilizers) used in PVC compounds. This usually results in an unmistakable color change as a "new"
colorant is formed. When stabilizers and pigments contain barium, calcium,
tin, nickel, magnesium, zinc, lead, or sulfur, possibilities exist for reactions to occur. Careful testing is suggested if changes in additives are
contemplated.

Chemical resistance in end use may also be important. Sulfide staining
of lead-based pigments is well known. Alkali resistance is necessary for
materials subject to washing. Other end-use environments may require
specialized testing. Of course, pigments with acid sensitivity can be affected by the PVC itself, depending on processing conditions and stabilizer effectiveness.

F. Blooming

Blooming is the phenomenon of a pigment or other additive coming to the
surface of a vinyl or other plastic material. Depending on the pigment,
blooming may occur rapidly, that is, immediately after processing of the
vinyl, or slowly over a period of weeks or months. Whatever the case,
when the vinyl compound is wiped or rubbed, the color comes off. Blooming pigments include naphthol and toluidine reds, dianisidine and dinitraniline
oranges, Hansa yellows, and some diarylide yellows. Occasionally blooming
may occur only under certain conditions involving other additives, heat
history, or pigment concentration.

Testing for blooming should be done after an extended storage period.
Blooming may also appear after a relatively short exposure in an accelerated lightfastness test. If samples develop a blotchy or mottled appearance, test for blooming.

G. Plateout

Plateout refers to the tendency of certain pigments to form deposits on
metal parts (calender and mill rolls, embossing cylinders, extruder screws)
with which the colored vinyl comes in contact. In a continuous process,
this deposition can build up to a point where finished product quality is
affected. Embossing cylinders can become so caked that definition is lost.
The need for subsequent cleanup and the possibility of contaminating a
following run are problems created by the plateout phenomenon.

Additives such as TiO_2, fillers. lubricants, and stabilizers can reduce or eliminate plateout or, if not chosen carefully, may contribute to
an intensification of plateout. To test for plateout, incorporate the pigment into a standard, unfilled PVC compound on a two-roll mill. Following
the milling, run a standard compound containing a known quantity of
TiO_2 as a "cleaner" batch. The developed color intensity of the cleaner
batch is an indication of plateout. This test procedure is more fully
described in Chapter 13. If comparative tests are being run, exercise
care that the pigment or pigment dispersion does not cling to mill end
plates. It may subsequently wind up in the "cleaner batch" and distort
the test results.

As a rule, pigments which exhibit severe plateout tendencies should be avoided. However, slight plateout may often be tolerated. Many classical azo pigments, Permanent Red 2B, Hansa Yellow, some diarylide yellows, toluidine red, and dianisidine orange will plate out.

H. Low-Concentration Behavior

Under the sections on heat stability, lightfastness, and migration, the low concentration behavior of pigments was mentioned. Too often, pigment evaluations are carried out on the basis of a single masstone and tint. Many pigments, the organics especially, may exhibit anomalous behavior at very low concentrations. Heat stability and lightfastness frequently decrease as pigment concentration goes down. Less common, and therefore often overlooked, is the tendency of a pigment to go into solution or behave as a dye at a very low concentration. Low concentration can be defined as less than 0.1%, although with some pigments the anomalous behavior at low concentrations may not appear until the pigment loading is much lower. There are a number of instances where pigments can be employed inadvertently at levels that may give rise to problems:

1. Reworking scrap of unknown composition.
2. Reducing the total pigment loading of a formulation because there appears to be no justification for using so much expensive pigment. The reason may have been to keep one of the components above a critical minimum level, but three color formulators later, the original reason may have been forgotten!
3. Increasing a part thickness, which may allow a desired color to be achieved with less pigment.
4. Eliminating or reducing filler or flame retardant in a formulation, which may make it possible to reduce pigment loading with no color change.
5. Simply creating a new color match without regard to loading of individual pigments.
6. Using a contaminated pigment or pigment concentrate. As a result of careless housekeeping or poor cleaning of dispersion equipment, pigments which were not intended may wind up in a finished product.

It is recommended that evaluation of a pigment include a tint containing 0.01% pigment plus 1.00 or 2.00% titanium dioxide. Although testing such a concentration will not pinpoint the critical minimum concentration, simply knowing in advance what performance to expect can alert the formulator to excercise care.

I. Pigment Dispersions

Pure pigments are available as either a dry powder or presscake. The latter form, which normally contains about 70 to 75% water, is commonly used in preparing some types of printing inks or in making high-quality polyolefin color concentrates. The water in the presscake is displaced by an oil or carrier resin. Since the pigment particles are not dried and agglomerated, dispersions of very high quality can be achieved. However, for most vinyl applications this quality is not necessary and the cost of the

process is not justified.   Therefore, dry pigment is used in making
dispersions in the PVC industry.

   Dispersion of a pigment involves incorporating the pigment into a
plasticizer or resinous carrier and then grinding or shearing the mixture
in such a way that pigment aggregates or agglomerates are broken down
and the primary pigment particles are encapsulated or wetted by the carrier.

   In preparing dispersions for PVC, two methods are common:

   1. Dispersing into suitable plasticizer (s) on a three-roll mill or sand
       mill
   2. Dispersing into a flexible or semirigid PVC compound on a two-roll
       mill or intensive kneader such as a Banbury mixer

Historically, the former method was preferred, and it is still widely employed.   It is the only practical method of a PVC plastisol is to be colored.
In the past several years, the use of colored vinyl homopolymer chips
has become widespread.   Such chips are either soft or semirigid, depending on the plasticizer level employed.   Pigment content, especially with organic pigments, is higher than that found in plasticizer pastes, where viscosity requirements limit pigment loading.   Chips are nondusting compared
to dry color, easier to handle than either dry color or a liquid paste, and
minimize housekeeping problems.   Chips also do not settle.   The use of
chips may present a problem if the mixing efficiency of the equipment used
to incorporate them into a natural compound is poor.   If uneven color distribution is noticed, there are several possible remedies:

   1. Increase mixing time of fluxed material.
   2. Use a more efficient mixer.
   3. Reduce the chip size.
   4. Use chips with a lower pigment content.

The last two suggestions require working with the concentrate manufacturer.   The first two suggestions are process changes.   As with all
process changes, they should be evaluated for effects not only on color
but also on finished product properties.

J.   Color Measurement and Computer Color Matching

Computer color matching and auxiliary programs have taken much of the
guesswork and drudgery out of the color matching process.   They have
not, and probably never will, completely replace visual color matching.
In the imperfect world of color theory and perception, it is suggested
that a computer color matching system operated by a skilled visual color
matcher will give the best results.   A computer programmer/operator with
no knowledge of pigments may arrive at seemingly satisfactory answers
but technical and practical considerations related to the process will be
overlooked.   For example, a color match containing three different yellow
pigments might offer low cost and adequate fastness properties, but production color control would be extremely difficult, if not impossible, to
achieve.

   The computer color matching library is the key to success in establishing a reliable system.   The best hardware and software available cannot compensate for careless sample preparation.   It is tempting when
building a computer color matching library to prepare samples from material

at hand. The following precautions will save time and frustration:

1. Select a resin which is free of gels. Although not a problem in visual color matching, gels affect the optics of a system and may provide erratic data.

2. If a filler is to be used, treat it as a pigment. The color of some fillers is variable. Reserve enough material to eliminate this variable.

3. Do not take pigments from production stock. Ask your pigment suppliers for standards. Record batch numbers of all pigments being entered in the computer.

4. Use a dispersion technique which represents your production method; it should be repeatable.

5. Learn how to make black, white, and gray before working with chromatic colors. Time spent in this exercise will pay dividends later.

6. When colors are added, start with the commonly used inorganics such as chrome yellow, molybdate orange, and iron oxide. Since they are opaque and relatively free of dispersion problems, success with them is easier than with a shear-sensitive organic pigment. The inorganics mentioned above are extensively used in brown and beige shades in PVC. A small library containing two or three chrome yellows, two molybdate oranges, and three or four iron oxide yellows and reds is easily built and provides the opportunity to test the system before committing resources to a more extensive library.

7. Housekeeping and sample preparation procedures must be critically reviewed. Sample contamination is often the cause when satisfactory calibration data cannot be obtained. Sources of contamination range from the obvious, dirty mills or mixer, to the less obvious, a dirty knife or spatula, a dirty filter in an exhaust hood, or careless handling of a finished sample. One should assume that everything is a potential source of contamination. Train operators to keep raw materials covered, equipment clean, and samples protected.

A question frequently asked with regard to computer color matching libraries is, "Will one library serve all my color matching needs?" Unfortunately, the answer is no. If several types of processing or different dispersion techniques are employed, it may be necessary to enter colorimetric data for a single pigment based on the different processes or dispersions.

A somewhat time-consuming but useful method for determining pigment variation as a function of processing technique involves generating chromaticity diagrams of pigments which have been subjected to different dispersion methods.

K. Effect of Dispersion Technique on Pigment Behavior

In Figure 13, the chromaticity coordinates (x and y) are plotted for a series of tints of a red pigment. Two plots are illustrated: one for a chip dispersion made on a two-roll mill, the other for a plasticizer paste dispersion made on a three-roll mill. Both dispersions are "good" ones by the benchmarks customarily used to assess dispersion quality. It

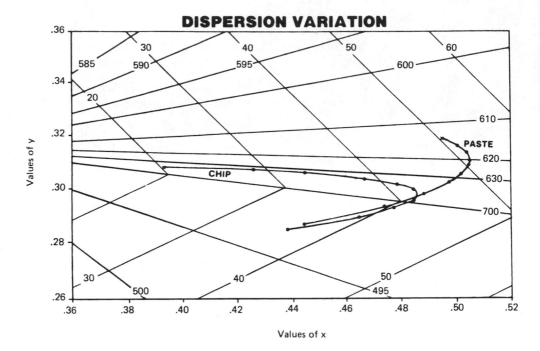

FIGURE 13 Chromaticity coordinates for dispersions made by different processes.

is obvious, however, that the chromaticity diagrams are significantly different. The masstone of the chip dispersion is much dirtier or lower in chroma than that of the paste dispersion (20% purity vs. 50%). As titanium dioxide is added to the chip dispersion, the shade maintains its hue (dominant wavelength) in the stronger tints, whereas the paste dispersion shows a pronounced variation in hue, becoming bluer as white is added. Because the third dimension of color space, value or lightness-darkness, is not shown in this illustration, even the point where the two diagrams cross (x = 0.485 y = 0.297) may not represent a match.

Although this illustration represents a fairly extreme example of variations as a result of dispersion technique, many organic pigments will exhibit variations as a result of dispersion differences. This, and not lot-to-lot variability in the pigment, explains why color shifts may occur when seemingly innocuous process changes are made. Changes in process machinery, temperatures, or raw materials other than pigment can affect, sometimes subtly but occasionally profoundly, the dispersion characteristics of a pigment.

L. Coloring Rigid PVC

The pigmentation of rigid PVC bears little relation to that of plasticized or flexible PVC. Rigid PVC is generally processed at somewhat higher temperatures than flexible PVC. Stabilizers and other additives in a rigid compound are not the same as those found in flexible PVC. End uses, particularly in such items as vinyl house siding and windows, require long-term weatherability not ordinarily expected of flexible vinyl. Virtually all rigid PVC compounding is done in a high-speed mixer. For these reasons, many of the pigments commonly used in flexible PVC are unsuitable for rigid PVC. If long-term weatherability is critical, very few pigments can be employed. The most weatherable pigments are nickel-antimony titanate, chromium-antimony titanate, and chromium oxide green. Somewhat less weatherable, but still widely used, are selected calcined inorganic brown pigments of either rutile or spinel crystal structure, phthalocyanine and indanthrone blue pigments, and Pigment Brown 23 in specialty areas.

For material not requiring long-term exterior durability, virtually any of the pigments used in flexible PVC may be used in rigid PVC. In the case of metal complex colorants, it is possible for colorant-stabilizer reactions to occur. This usually results in a metal ion exchange between the colorant and the stabilizer, thus creating a new colorant. Although the new colorant may be attractive in its own right, it is not the color originally specified, nor are its other fastness properties known.

Because rigid PVC compounds are usually prepared by powder blending in a high-intensity mixer, there are special precautions to be observed when pigments are added. High-intensity mixer is a misnomer. Such a "mixer" is most effective in dispersing pigments. At normal end-use concentrations, 1 or 2%, a high-intensity mixer can quickly and efficiently disperse most organic pigments in a powder blend compound. When only inorganic pigments are used, loadings may be even higher.

For compounds intended for outdoor use, inorganic pigment loadings of 10 to 12 phr are normal. Most of this is titanium dioxide. Because the inorganic pigments are abrasive, minimal mixing time is suggested. Otherwise, severe discoloration may result as abraded metal dust from the mixer blade and bowl contaminates the batch. In addition to discoloration (graying), the increased iron content may lead to compound instability problems during subsequent processing or on outdoor exposure.

Organic pigments, with their small particle size relative to inorganic pigments, vary in dispersion characteristics and in crystal structure. Pigments that are easy to disperse, especially at high loadings (above 5%), may be compacted in a high-intensity mixer, and aggregates can form with attendant loss of tinting strength. Acicular (needlelike) crystals can be fractured, altering both color and fastness properties.

There are several approaches to resolving pigment incorporation problems when using a high-intensity mixer:

1. Never add pigment at the beginning of the compounding cycle.
2. Preblend the pigment is a low-speed mixer with a portion of the compound. Depending on the amount required, 10 to 50% pigment may be used. Add the preblend to the high-speed mixer. Run the mixer at the lowest possible speed for the shortest time to achieve uniform distribution.
3. For easy-to-disperse pigments, do not add the pigment to the high-speed mixer, but add after the batch is dropped to the cooling mixer.
4. Use color concentrates specifically formulated for rigid PVC. These are available from several color concentrate manufacturers. Color concentrates offer pigment already dispersed, and the housekeeping problems associated with pure pigment are eliminated.

In unplasticized PVC, solvent dyes occasionally may be used for coloration. Dyes, by definition, are polymer-soluble and require only mixing to incorporate/dissolve them in the polymer melt. Dispersion or shear is not needed. The ease of incorporation, high tinting strength, extreme transparency, and good to excellent masstone lightfastness make dyes suitable for use in thin films of rigid PVC and a variety of packaging applications. Blue and violet anthraquinone dyes (Solvent Blue 97, Solvent Violet 13) are frequently used to impart a blue cast to otherwise yellowish resins.

Dyes may be extractable from rigid PVC. If nonextractability is required, testing under simulated use conditions is suggested.

REFERENCES

1. Hopkins, B. S., and Bailer, J. C., *Essentials of Chemistry*, Heath, Boston, 1946.
2. American Association of Textile Colorists and Chemists, *Color Index*, 1971.
3. Patterson, T., *Pigments, An Introduction to their Physical Chemistry*, Elsevier, New York, 1967.
4. Patton, T. C., *Paint Flow and Pigment Dispersion*, Interscience, New York, 1964.
5. Mattiello, J. J., *Protective and Decorative Coatings*, Vol. 4, Wiley, New York, 1946.
6. Mitton, P. B., and White, L. S., "Pigment Optical Behaviour-Evaluation on a Physical Basis, "*Official Digest, 30,* 1259 (Nov. 1958).
7. Titanium Pigment Corporation, *Handbook*, 1965.
8. Mie, Gustave, *Ann. Phys., 25,* 377 (1908).
9. American Association of Textile Colorists and Chemists, Research Triangle Park, Box 686, Durham, NC 27702, Standard Test Method 8-61.
10. Carr, D. S., Baum, B., Margosiak, A, and Llompart, A., "UV Stabilization of Thermoplastics with Zinc Oxide," *Mod. Plast., 48 (10),* 160 (1971).
11. National Electrical Manufacturers Association, 155 East 44 Street, New York 10017, Standards for Electrical Resistivity.
12. Florida East Coast Exposure: Subtropical Testing Service, Inc., 8920 S.W. 120 Street, Miami, FL 33143.
    Florida West Coast Exposure: Suncoast Testing Service, Inc., P.O. Box 2347, Sarasota, FL 33578.
    Arizona Exposure: Desert Sunshine Exposure Tests, 7740 N. 15 Avenue, Phoenix, AZ 85041.
13. American Society for Testing and Materials, Method D281-31, ASTM, 1916 Race St., Philadelphia, PA 19103.
14. Papillo, P., "Mass Coloring of Plastics," *Mod. Plast., 44 (12),* 131 (1967).
15. Devore, J. R., "Pigments Optics," American Chemical Society, Division of Organic Coatings and Plastics Chemistry, preprints of papers presented at New York City meeting, September 1963 (vol. 23, no. 2, p. 399).
16. Mitton, P. B., Vehnoska, L. W., and Frederick, M., "Iliding Power of White Pigments: Theory and Measurement," *Official Digest, 33,* 1264 (October 1961); *34,* 73 (June 1962).
17. Taylor, C. J., "Surface Treatment and Optical Properties of Titanium Dioxide," *J. Oil Colour Chem. Assoc. 49,* 1063 (Dec. 1966).
18. De Coste, J. B., and Hansen, R. H., "Colored Poly (Vinyl Chloride) Plastics for Outdoor Applications," *SPE J., 18,* 431 (April 1962).

19. Felsher, H. C., and Hanau, W. J., "Pigmented Plastics for Outdoor Exposure," preprints for Society of Plastics Engineers Retec, *Coloring of Plastics III*, June 10, 1966. Contact SPE, 14 Fairfield Dr., Brookfield, Ct 06805.

20. Mattielo, J. J., *Protective and Decorative Coatings*, Vol. 2, Wiley, New York, 1946.

21. Rober, Jr., S. G., "Carbon Black Pigments in Plastics," preprints for Society of Plastics Engineers Retec, *Coloring of Plastics III*, June 10, 1966.

22. Drogin, I., "The Role of Carbon Black as a Colorant", *Color Eng.*, *5*, 4 (July-Aug. 1967).

23. Drogin, I., "Carbon Black as a Pigment", *Color Eng.*, *2*, 3 (March 1964).

24. Venuto, L. J., and Hess, W. M., "A New Look at Carbon Blacks," *American Ink Maker* (Oct.-Nov. 1967).

25. Brody, D. E., "Pigment Technology", *Paint and Varnish Production* (July 1966).

26. Felsher, H. C., and Hanau, W. J., "Brilliance for Plastics Through Metallics", *Mod. Plast.*, *41*, 9 (1963).

27. Huckle, W. G., Swigert, G. F., and Wiberly, S. E., "Cadmium Pigments," *Ind. Eng. Chem. Prod. Res. Dev.*, *5*, 362-366 (Dec. 1966).

28. U.S. Congress (Public Law 86-618), Color Additive Amendment to the Food, Drug, and Cosmetic Act of 1938.

29. American National Standards Association, Inc., 1430 Broadway, New York, NY 10018.

30. U.S. Congress Public Law 85-929.

31. Dunn, M. J., "Colors for Food Packaging," *SPE Tech. Pap. 14* (1968).

32. "Part 121 - Food Additives - Title 21 - Food and Drugs," Federal Food and Cosmetic Act, Section 121.2514, Paragraph 34, Subpart F— Food Additives, 13.3-13.8.

33. Copies of the preceding reference and its amendments, and copies of the food additive amendments of the periodic changes in their regulations, can be obtained by writing to the Food and Drug Administration, Washington, DC.

34. Vesce, V. C., "Exposure Studies of Organic Pigments in Paint Systems," *Official Digest*, *31*, 419, part 2 (Dec. 1959).

35. Yao, H. C., and Resnick, P., "Degradation Reactions of Azo Pigments," American Chemical Society, Division of Organic Coatings and Plastics Chemistry, preprints of papers presented at New York meeting, *23*, 486 (Sept. 1963).

36. Hill, R. F., "Phthalocyanine Pigments and Their Application," *J. Oil Colour Chem. Assoc.*, *48*, 603 (July 1965).

37. Smith, F. M., and Easton, J. D., "Phthalocyanine Pigments, Their Form and Performance," *J. Oil Colour Chem. Assoc.*, *49*, 614 (Aug. 1966).

38. Pugin, A., "The Influence of Chemical Structure on the Color and Properties of Dioxazine Pigments," *Official Digest*, *37*, 782 (July 1965).

39. Gaertner, H., "Modern Chemistry of Organic Pigments," *J. Oil Colour Chem. Assoc.*, *46*, 13 (Jan. 1963).

40. Reeve, T. B., "Pigment Colors for Vinyl Coatings," *Color Eng.*, 19 (July-Aug. 1966).

41. Simpson, J. E., "How to Choose the Right Colorant for Plastics," *Mod. Plast.*, *40* (*12*), 90 (1962).

42. Simpson, J. E., "Coloring Plastics," series of articles in *Mod. Plast.*, *44* (1965).

43. Greenstein, L. M., and Miller, H. A., "The Properties of Nacreous Pigments," *SPE Tech. Pap.*, *13*, 1021 (1967).
44. Kazenas, Z., "Daylight Fluorescent Pigments," *Paint Industry Mag.* Feb. 1960).
45. Dunn, M. J., "Color and Constitution," *SPE Retec* Preprint, *Coloring of Plastics II*, Rochester, NY, May 12, 1965. Contact SPE, 14 Fairfield Drive, Brookfield, CT 06805.
46. Billmeyer, Jr., F. W., and Saltzman, M., "*Principles of Color Technology*, Interscience Div., Wiley, New York, 1966.
47. Orchard, S. E., "A New Look at Pigment Optics," *J. Oil Colour Chem. Assoc.*, *51*, 44 (Jan. 1968).
48. Pratt, L. S., *Chemistry and Physics of Organic Pigments*, Wiley, New York, 1947.
49. Munsell, A.E.O., *A Color Notation*, Munsell Color Co., Baltimore, 1963.
50. Judd, D. B., and Wyszecki, G., *Color in Business, Science, and Industry*, Wiley, New York, 1963.
51. Leete, C. G., and Lythe, J. R., "A Digital Computer Technique for Calculation of Dominant Wavelength, *Color Eng.*, 27 (Jan.-Feb. 1966).
52. Allen, E., "Analytical Color Matching," *J. Paint Technol*, *39*, 368 (June 1967).
53. Special Metamerism Issue, *Color Eng.* (May-June 1967).
54. Winey, R. K., and Longley, W. V., "Practical Solutions to Problems of Metamerism," *SPE Tech. Pap.*, *13*, 1110 (1967).
55. Hunter, R. S., "Measurements of Color and Other Appearance Attributes in the Plastics Industry," *SPE J.*, *23* (2), 51 (1967).
56. Blackwood, N. K., and Billmeyer, Jr., F. W., "A Computer Program for MacAdam PQS Color Difference," *Color Eng.*, 24 (March-April).
57. Zussman, H., "Optical Brighteners," *Mod. Plast. Encyc.*, *43* (*10-a*) 490, 495 (1966).
58. Burnham, R. W., et al., *Color: A Guide to Basic Facts and Concepts*, Wiley-Interscience, New York, 1963.
59. Billmeyer, Jr., F. W., and Saltzman, M., *Principles of Color Technology*, Wiley-Interscience, New York, 1966.
60. Stoloff, A., Interim Report of the Vinyl Siding Division Weathering Committee," Society of the Plastics Industry (Jan. 1980).
61. Stoloff, A., "Weathering Committee Report," Vinyl Siding Institute, Society of the Plastics Industry (Sept. 1980).
62. ASTM D1499, D1501, G53, G23, American Society for Testing and Materials, 1916 Race St., Philadelphia, PA.

8

Miscellaneous Modifying Agents

SAUL GOBSTEIN

Sa-Go Associates, Inc.
Shaker Heights, Ohio

I. COMPLEXITY OF ADDITIVE TERMINOLOGY

A. Additives Defined

It is known and documented that in the early development of polyvinyl chloride, useful products could not be manufactured unless certain additives were incorporated in the polymer matrix. The incorporation of these additives, forming a compounded plastic, overcame the limitations of the raw polymer. For example, the use of heat stabilizers allowed the polymer to be processed and gave the plastic a useful service life while plasticizers also assisted in processing and gave the product elasticity.

Mascia [1] defines plastic additives as "those materials that are physically dispersed in a polymer matrix without affecting significantly the molecular structure of the polymer." The definition should be expanded to include "but that do affect the processing of the polymer, as well as the performance in its service life." Based on this definition, all materials added to a polyvinyl chloride polymer are additives.

B. Importance of Additives

In the plastics industry, the words resin, polymer, and plastic are used interchangeably. They are nto synonomous. The American Heritage Collegiate Dictionary [2] defines the term resin as "any of numerous physically similar polymerized synthetic or chemical modified natural resins including thermoplastic materials such as polyvinyl chloride, polystyrene, and polyethylene; and thermosetting materials such as polyesters, epoxies, and silicones that are used with fillers, stabilizers, pigments and other components to form a plastic." Plastic [3] is defined as "capable of being shaped or formed" and comes from the Greek word "plastikos"—fit for molding." Therefore, it is the additives which transforms the resin or polymer into a useful plastic.

The first generation of additives was mainly concerned with overcoming the limitations of the polymers. The second generation of additive technology is directed to the building of desired properties into a plastic part by the use of additives.

C. Classification of Additives

There is no clear-cut classification system for plastics additives. Mascia [4] classifies additives according to function. This system is not adequate since an additive may perform more that one function. Another problem is that one additive with a specific function may interfere with another additive's function with regard to compatibility, interaction, dispersion, and/or distribution. Therefore, the final choice is a compromise based on the need for and the requirements of a specific function at the lowest cost.

II. ANTIBLOCKING AGENTS AND SLIP ADDITIVES

Antiblocking and slip agents have been used extensively in packaging film since the beginning of the cellophane era. The major market is in polyolefins [5]. However, there is also a market in polyvinyl chloride film and plastisols used in bottle caps.

The problem occurs between adjacent surfaces which are very smooth and have a high gloss. The contact between these types of surfaces is nearly perfect, eliminating air between the surfaces, and therefore the surfaces adhere to one another. Because there is no air layer between the surfaces, it is difficult to separate them. Blocking can also occur due to pressure or temperature effects that induce fusion of the surfaces in contact.

A. Blocking and Slip Defined

The term "block"[6] is defined as the ability of a layer of film to adhere to another layer of film. In many cases, this phenomenon may be minimized by processing conditions (line speed) and by avoiding high storage temperatures. It also can be controlled by antiblocking additives.

The term "slip" means the ability of a layer of film to slide past another layer of film. A high coefficient of friction [7] denotes low slip and a low coefficient of friction denotes high slip.

B. How They Function

Slip and antiblocking agents are microscopic-sized particles that are incorporated into a plastic formulation. These particles are completely or partially incompatible in the polymer matrix. The additive will exude or bloom to the surface, forming a microscopic bump, thus forming a surface irregularity that will give slip and antiblocking characteristics.

It is important to maintain a balance between slip and antiblocking agents. Blocking will usually decrease with the addition of an antiblocking agent up to a maximum point. Beyond that point, there is no further blocking. Excessive use of the antiblocking agent therefore only decreases the optical properties. The same is true for slip additives.

Antiblocking agents and slip additives do interact with one another as well as exerting an effect on the lubrication system, the film gauge, line speed, type of film, type of resin, and other additives in the formulation. It is possible to get "wet blocking" when a slip additive is used alone or in excess [8]. It is also possible to get "heavy blocking" when antiblocking agents are used alone or in excess due to the pressure, which flattens out the microscopic bumpy surface.

These additives have one characteristic in common; they are incompatible in vinyl. In the case of slip additives, they also migrate to the surface during or after processing and exhibit lubricating features. Increasing the slip additive concentration will decrease the coefficient of friction very quickly until it reaches the critical level. An increase in concentration after this point will not decrease the coefficient further.

C. Types of Materials Used

In PVC, the most popular antiblocking agents are diatomaceous earth or synthetic silicon dioxide. These products have the least interference with clarity and are usually approved by the U.S. Food and Drug Administration (FDA). Fatty acid amides are also used, in very small quantities—usually less than 0.25 phr. When the amides are used, the lubrication system should be checked to make sure it also is balanced.

The types of materials that are used as slip or antiblocking agents are given in Table 1. A partial list of manufacturers of antiblocking agents and slip additives is presented in Table 2.

III. ANTISTATIC AGENTS

A. Need for Antistatic Agents

Static electricity can cause several problems in the manufacture and fabrication of plastic parts as well as in the service life of the plastic. The charge on the plastic part or plastic particle, while in process, is unpredictable. A static electric discharge in a dust-filled or flammable atmosphere can cause an explosion. A static charge on a PVC resin particle in a low-humidity atmosphere can cause balling and agglomeration. Static charges on videotapes or disks can cause audio or video distortion. Static charges on computer components can cause loss of memory. The buildup of a static charge on a PVC film or sheet may cause difficulty in separation.

B. Theoretical Considerations

A static-charged particle or a static-charged plastic surface occurs when two surfaces are parted after close contact. The charge build-up is determined by the rate of generation and, simultaneously, by the rate of decay. Frictional forces on the polymer particles while in processing can build up a static charge on the particles.

The static electricity build-up on plastic parts and polymer particles is complex and not well understood. ASTM D257 states that "surface resistance or conductance cannot be measured accurately, only approximated, because more or less volume resistance or conductance is nearly always involved in the measurement. The measurement value is largely a property of the contamination that happens to be on the specimen at the time." This explains why DC resistance and conductance test methods may not be truly correlated with actual service life conditions.

Static electricity is a surface phenomenon. It is either a buildup or a deficiency of electrons. Electrostic charges on poor conductive surfaces such as plastics can be generated by frictional forces or ionized air [9].

To dissipate the charges, Mascia [10] teaches us that the rate of surface generation can be decreased to some extents by reducing the intimacy of contact, whereas the rate of charge decay can be increased substantially by rendering the surface conductive by the formation of a conductive layer. This conductive layer can be achieved by the addition of an antistatic agent which migrates to the surface and forms conductive paths through absorption of atmospheric moisture.

Schmidt [11] claims that charges can be dissipated into the air by use of an antistatic agent. The mechanism involves water molecules in the air becoming attracted to the charged end of the antistatic molecule. The charge is transferred from the surface of the plastic to the water molecules which then

TABLE 1 Products Used as Slip and Antiblocking Agents

	Antiblock	Slip
Saturated fatty acid amides	X	X
Unsaturated fatty acid amides	X	X
Metal salts of fatty acids	X	X
Diatomaceous earth	X	
Synthetic silica	X	
Waxes	X	X
Calcium carbonate	X	
Talc	X	

can diffuse the charge into the air environment. The antistatic agent is then free to pick up another charge from the plastic and transfer it to the water molecules in the air. This can go on and on unless the antistatic agent is removed from the surface of the plastic.

C. Practical Aspects

The buildup of electrical charges may be dissipated in several ways.

1. Grounding: If the problem occurs in processing, ground the equipment. This will undoubtedly solve the process problem but may not solve the problem of static buildup of the plastic part after it is manufactured and in service.

2. Increasing Moisture Content: Increasing the moisture content of the air can dissipate the static charge in processing. In the early days of dry blending rigid PVC, there were minor problems of balling and particle agglomeration while blending. With the new regulations for VCM, starting in 1974, blending problems became almost catastrophic, especially when humidity was

TABLE 2 U.S. Manufacturers of Antiblock and Slip Additive

Company	City, State
American Hoechst Corp., Ind. Chem. Div.	Somerville, NJ
Armak Co., Div. of Akzona	Chicago, IL
Cab-O-Sil Div., Cabot Corp.	Tuscola, IL
Davidson Chem. Div., W.R. Grace & Co.	Baltimore, MD
Degussa Corp.	Teterboro, NJ
Humko Chem. Div., Witco Chem. Corp.	Memphis, TN
Petrochemicals, Inc., Div. of De Soto Chemical	Forth Worth, TX

low. By introducing water while the resin was being charged into the blender, the problem was overcome. This required higher discharge temperatures from the blender (above 212°F), which in turn resulted in better dispersion and distribution of additives.

3. *Ionization*: The rate of static dissipation in a plastic part while in process can be increased by increasing the electrical conductance of the surrounding air by ionization. This method usually charges and discharges surface static electricity. In practice, this method has been found lacking as there may be static buildup farther down the process line or in the finished part.

4. *External Antistatic Agents*: An antistatic agent is applied to a plastic part, film, sheet, or monofilment surface in solution form. The solvent evaporates leaving the antistatic agent on the plastic surface. As long as the antistatic agent remains on the surface, protection is provided. Once the agent is removed by washing, rubbing, or solvent, the part will no longer be protected.

5. *Internal Antistatic Agents*: Internal antistatic agents are added to the polymer matrix. After processing, these additives exude to the surface, forming a conductive layer and thereby making the plastic itself more conductive. If the surface is washed, rubbed, or attacked by solvent, exudation will occur again when the surface is dried and will reestablish the conductive layer.

Care must be excercised in incorporating internal antistatic agents into a PVC formulation. The lubrication system should be checked to determine whether it is out of balance (internal vs. external lubrication). The performance of the stabilization system as well as other ingredients in the formulation should be checked for possible interaction with the antistatic agent.

D. Testing

Ash Pickup: Testing antistatic properties of a plastic material is done by many standard quantitative testing procedures as well as homegrown tests. A very popular homegrown test for rigid PVC phonograph records is the dust or ash pickup test. This is a simple test in which cigarette ashes are dumped on a phonograph record and the record is checked to see if the ashes are rejected from its surface.

2. *Corona discharge-Electrometer* [12]: In this test, a 5.5 in. × 5.5 in. × 103.0 in. pressed polished specimen is placed on a grounded aluminum plate for 5 sec. The corona discharge is 1 in. from the specimen. A charge of 10 kV ionizes the air without arcing, charges the panel, and completes the circuit through the aluminum plate. The charged panel is then placed under a 3 in. static detector. The detector is in an insulated glove box 36 in. × 16 in. × 24 in. The humidity is controlled by specific saturated salt solutions. Readings for the voltage decay can be taken from 1 to 5 minutes or up to 60 minutes, depending on the type of material used. With flexible PVC, which has relatively good antistatic properties (static charge dissipates quickly), readings may be taken every 15 sec. Rigid PVC (which holds its charge) can be tested for 1 hr.

3. *Surface Resistivity Measurements*: These measurements have been standardized under ASTM D257-66. As pointed out in this procedure and above, measuring surface resistivity has many drawbacks. Surface resistiv- is a measure of charge lost from the surface and not lost into the air.

It also does not take into account contamination or impurities that may improve or retard static dissipation. Therefore, it is possible to have a material with poor surface resistivity and good antistatic properties.

This test is an excellent one for measuring conductivity, but it does not actually measure antistatic behavior. Equipment for this test is available from Kiethley Instrument Co. of Cleveland, Ohio.

4. *Soot Chamber Test* [13]: This is a subjective test which attempts to measure dust buildup on a plastic part. A charge is induced on a plastic part, which is then placed in a chamber. The chamber contains a filter paper soaked with an aromatic solvent. The paper is ignited and allowed to burn completely. The chamber is then opened and the part examined for soot buildup. The degree of soot buildup is rated from 1 to 10.

5. *Charge Induction and Decay Between Electrodes*: This method is used to test plastic sheet for antistatic properties. The method which is adapted from method 4046 of Federal Test Method Standard 101B [14], essentially tests for sparking in an explosive atmosphere. It measures the time required to charge a plastic surface and the time it takes to dissipate that charge.

E. Materials Used

Materials used to impart antistatic properties to a plastic are ionic materials and hydrophilic materials which will attract moisture to the surface of the plastic [15]. The ionic group consists of quaternary ammonium compounds or amines. The hydrophilic group consists of polyglycols (derivatives of ethylene oxide). Derivatives of sulfonic acids, phosphoric acid, and polyhydric alcohols are also used as antistatic agents [15]. Most products are proprietary. A partial list of manufacturers is presented in Table 3.

F. Selection of an Antistat

Antistat consumption for all plastics grew 45% in a 5 year period from 2150 metric tons in 1980 to 3125 metric tons in 1985 [16]. Most of the growth was due to packaging of food and electronic parts. However, acceptance of antistats in PVC is still limited because, in most cases, they interfere with the function of other ingredients in the formulation.

For example, quaternary ammonium compounds give a rigid PVC compound good antistatic properties but hurt the heat stability. Polyether-type antistats have a smaller effect on heat stability but only render marginal antistatic properties.

In flexible PVC compounds the type and amount of plasticizers that are used play an important role in antistatic properties [17].

IV. FLAME RETARDANTS

The subject of flame retardancy is neither completely understood nor well defined. The literature contains many terms that are ambiguous and are interpreted emotionally. Because of this, the Federal Trade Commission no longer allows terms such as "flame resistant," "slow burning," or "self-extinguishing" to be used to describe a plastic product. These terms can be used only if they are meaningful in the sense of having a specific value in a standard test method. The industry is also trying to standardize other terms such as

TABLE 3 U.S. Manufacturers of Antistats for Vinyls

Manufacturer	Trade Name	City, State
Alframine	Electrosol 325	Paterson, NJ
American Cyanamid Co.	Cyastat	Wayne, NJ
American Hoechst	Hostastat	Somerville, NJ
Argus Chemical, Div. of Witco	Markstat	Brooklyn, NY
Buckman Laboratories	Bubond	Memphis, TN
Glyco, Inc.	Aldo	Greenwich, CT
Michel	Michel	New York, NY

"ease of ignition," "flame spread," "fuel contribution," "smoke production," "fire gases" and "fire endurance." The Society of the Plastics Industry's (SPI) Vinyl Institute in conjunction with the International Society of Fire Service Instructors has developed a film slide presentation designed to counteract misconceptions about vinyl and other plastics that exist within the fire community [18].

Polymers vary in their inherent resistance to fire and flame propagation. Mechanisms proposed to describe flame propagation, ignition, or flame retardancy are complex and controversial [19]. Part of the problem is that many researchers have studied the flammability of the PVC polymer and not of the plastic. Another view is that the results and conclusions drawn from test data are often interpreted emotionally and not objectively and therefore cannot be correlated with service life conditions. Another consideration is that flame retardancy test data are obtained under controlled laboratory conditions. In actual service life a fire is not controlled, and usually no two fires burn in the same way.

There is a terminology dilemma in the flame retardant area, as stated above. This dilemma is compounded in the PVC industry. The PVC polymer is inherently flame retardant since it contains 56% chlorine. The polymer is occasionally used as an additive to make other plastics (ABS) flame retardant. Rigid PVC plastic products are ordinarily flame retarded. Flexible PVC products usually are not flame retarded unless additional additives are incorated. The problems which lead to the confusion between rigid and flexible PVC are well illustrated in the proceedings of the 1983 SPE ANTEC Vinyl Forum [20].

At this forum, a panelist presented test data comparing burning rates of PVC and polyethylene. The test showed that when a Bunsen burner was removed, the flame of the PVC test bar extinguished itself while the polyethylene continued to burn. Not once did the speaker refer to the PVC as rigid (which it was). If the PVC test bar had been a 60 Durometer flexible compound, it would have continued to burn unless the compound had been formulated to be flame retarded.

For a more detailed review of the state of the art, the reader is referred to Hendersen and Wagner [21], O'Mara et al. [22], and the *Flammability Handbook*[23]. O'Mara et al. give a comprehensive review to PVC fire retardancy.

A. Burning Mechanism of Plastics

The chemistry of flame reactions in burning plastics is complex. The velocity of the flame reaction in one plastic can be slow, whereas in another plastic the reaction can be explosive. Sucessful use of a plastic in one application does not mean it can be used successfully in another with regard to fire hazard. For this reason, the SPI Committee on Consumer Safety advocates hazard analysis [24] for each application.

The heat generated by combustion may sustain the ignition by continually providing the necessary thermal energy to cause the material to burn and produce more volatiles. If non combustible products such as halides, amines, carbon dioxide, or fluorocarbons are formed, the ignition temperatures rise, thus increasing the oxygen demand for sustained ignition, and this may cause self-extinction of the flame.

When PVC plastic burns, the mechanism described above occurs, except that large amounts of HCl start to come off at relatively low temperatures (around 300°). After dehydrochlorination, black smoke is produced. The PVC has a self-ignition temperature above 500°C. The structure of the polymer causes self-extinction of a flame [26].

There are plastics that drip when heated to a point after ignition starts. Close examination of this drip shows that the flame drips away from the plastic; thus cooling the plastic. In some cases, the cooling can be below the point of ignition. Therefore, with respect to propagation of a flame, there may be applications where a plastic drip is desirable.

Interestingly, many authorities point out that a good part of the mechanism thesis must be taken on faith as there is no definitive reference that can be cited to correlate the theory with an actual fire.

The chemistry for the above thesis is demonstrated by the simple example of burning ethane [27]. The volatiles formed from thermal degradation of polymers can form sites for free-radical formation as well as oxygen attachment. Figure 1 shows the flame reaction. These reactions show the formation of the highly active HO· radical. This radical has a high energy level and can give a high velocity to a flame front.

The reaction leading to the formation of H_2O is exothermic. This excess heat provides additional acceleration to the oxidation reaction. It also contributes to the rate of combustion. The wall effect reaction shown in Figure 1 represents the barrier of a solid interface in the gas/vapor mixture acting to dissipate energy and promote the formation of the ·OOH radical instead of the HO· radical. The ·OOH radical has a much lower level of excitation than the HO· radical. Therefore, fire-dispersed solids, such as antimony compounds in the flame, will tend to reduce the violence of the reaction and promote the wall effect.

B. Promoting Flame Retardancy

There is general agreement that plastics cannot be made fireproof. Plastics are organic-based materials and will burn under the proper conditions. However, the susceptibility to fire can be reduced, and the spread of flame can be decreased by the incorporation of additives. Additives called "flame retardants" help reduce the ignition of the plastic and the rate of flame propagation.

In addition to the many individual research programs on plastic flammability under way in private industry, the Center for Fire Research (CFR) of the National Bureau of Standards conducts major research on plastic combusti-

$$CH_3-CH_2- \xrightarrow{\text{HEAT}} -CH_2-CH_2^{\bullet} + H^{\bullet}$$

$$-CH_2-CH_2^{\bullet} + O_2 \longrightarrow -CH_2-COO^{\bullet} + 2H^{\bullet}$$

$$-CH_2-COO^{\bullet} + 2H^{\bullet} \longrightarrow -CH_2CHO^{\bullet} + HO^{\bullet}$$

$$-CH_2-CH_2^{\bullet} + -CH_2-COO^{\bullet} + H^{\bullet} \longrightarrow 2(-CH=CH^{\bullet}) + H_2O + HO^{\bullet}$$

$$O^{\bullet} + H_2 \longrightarrow HO^{\bullet} + H^{\bullet}$$

$$H^{\bullet} + O_2 \longrightarrow HOO^{\bullet} \quad \text{(wall effect)}$$

$$HO^{\bullet} + H_2 \longrightarrow H_2O + H^{\bullet}$$

FIGURE 1 Flame reaction mechanism.

bility. Their approach in developing more effective fire retardants is to establish a theoretical basis for the performance of retardants and to evaluate
their toxicity and carcinogenicity. An important input is their hazard analysis. They are attempting to obtain correlations between the flame retardants
used and service life conditions [28].

C. Methods for Flameproofing Plastics

1. Protective and Intumescent Coatings: Coating the plastic part, usually with a paint that will provide an effective thermal insulation barrier, is one
method. The coating layer reduces the formation of volatiles, oxygen perme-
ation, and oxidative reactions. An intumescent coating for combustible material such as wallboard has been described which consists of PVC, diammon-
ium phosphate, pentaerythritol, and dicyandiamide [29].

2. Interfering with the Combustion Reaction: Some organic bromides and
chlorides function as free-radical trappers at high temperatures. These compounds decompose to produce a hydrogen halide. The halides react preferentially with the high-energy HO• radicals and thereby deactivate them. Most
rigid PVC formulations are inherently flame retardant. However, flexible
formulations containing plasticizers dilute the chlorine level. When the chlorine level falls below 35%, additional chlorine in the form of chlorinated paraffin, for example, must be added to restore flame retardancy.

3. Making the Products of Decomposition Less Flammable: Phosphates,
borates, silicates, and other noncombustible inorganic compounds in a flexible PVC formulation form a noncombustible barrier that excludes air from the
charred surface of the plastic. There are claims that the phosphorus compounds contribute to promotion of the "wall effect."

4. Forming NonCombustible Gases: Antimony trioxide in the presence of
halides at elevated temperatures forms antimony chlorides, oxychlorides, and
antimony pentoxide, all of which are gaseous at ignition temperatures. These
gases dilute the supply of oxygen and thereby reduce the rate of combustion.
Should the temperature fall below the ignition temperature, the flame would
self-extinguish.

D. Flame Retardant Chemicals

There are two broad groups of flame retardant chemicals for plastics. The additive flame retardant chemicals are used primarily in thermoplastics. They are chemicals that change the flammability characteristics of the polymer but do not react with the polymer.

The reactive flame retardants are used in polyesters, epoxides, polyurethanes, and other thermosets. They react with the polymer and are part of its backbone. They also change the flame characteristics of the polymer.

In PVC technology, the most common flame retardant chemicals are antimony trioxide, aluminum trihydrate, phosphate plasticizers, and chlorinated hydrocarbons. A partial list of producers of these chemicals is presented in Table 4.

1. Alumina Trihydrate: Alumina trihydrate (ATH) has been found to be a flame retardant chemical in flexible PVC Formulations as measured by the oxygen index test [30]. It is most effective in formulations with lower levels of plasticizer. Kaufman and Yocum [31] reported that ATH with a particle size of 1μm is more effective than coarser grades. ATH has a synergistic effect when used in conjunction with antimony trioxide and with phosphate plasticizers [32]. Sprague [33] reported the synergistic effect of Firebrake ZB with ATH, as measured by oxygen index.

Use of ATH has the advantage of replacing part of the antimony trioxide, thereby lowering formulation costs. It also can replace all or part of the calcium carbonate. This will increase cost as well as sacrifice some physical properties. The formulator should determine where a compromise can be made to obtain desired results.

2. Antimony Trioxide: When used by itself in plastics, antimony trioxide has very little flame retardant activity. However, when used in conjunction with a halide, there is a synergistic effect [34]. For that reason, antimony trioxide can be used in PVC by itself; the PVC supplies the halogen. One theory claims that the combination forms antimony halides and oxyhalides [35] during burning which are active free radical terminators. Another theory is that the combination promotes the wall effect.

Antimony trioxide is used in PVC formulations in concentrations from 1 to 5 phr. The most common range is 2 to 3 phr. It is available in several grades and varies in degree of tinctorial strength [36,37]. There is a grade that is coated in a silica base and used at the 4 to 8 phr level [38] and a collodial grade that can be used in special coating grades [39].

The ultrafine grade tint gives a slightly better oxygen index reading than the "lo tint." However, in white or pastel-colored plastics, the "hi tint" grade can reduce the amount of TiO_2 that is required. In formulations with no white, the "lo tint" should be used so that the cost of coloring is not significantly increased when attempting to overcome the white pigmentation contribution of the antimony trioxide [40].

3. Borates: It has been reported by Shen and Sprague [41] that zinc borate is an effective flame retardant additive in rigid PVC. In flexible PVC formulations, the synergism of antimony trioxide and zinc borate is recommended. Bonsignore [42] claims that there is a synergistic effect between zinc borate and alumina trihydrate.

Barium metaborate [43] can also be used as a partial replacement for antimony trioxide. An additional benefit of this additive is its antimicrobial activity and retardation of afterglow.

TABLE 4 U.S. Producers of Flame Retardant
Additives for PVC

Aluminum trihydrate

    Aluchem Inc., Reading, OH

    Aluminum Corp. of America, Pittsburgh, PA

    Custom Grinders, Chatworth, GA

    H & S Industries, Dalton, GA

    Reynolds Metals, Richmond, VA

    Solem Industries, Norcross, GA

Antimony trioxide

    Afrimet-Indussa, New York, NY

    Amspec Chemich., Glouster City, NJ

    Anzon America, Freehold, NJ

    Asarco Inc., New York, NY

    CoMetals Inc., New York, NY

    Laurel Industries, Pepper Pike, OH

    M & T Chemicals, Rahway, NJ

    McGean-Rohco, Cleveland, OH

    NYACOL Products, Ashland, MA

    Samin Corp., New York, NY

Borates

    Buckman Laboratories, Memphis, TN

    Humphrey Chemical, Edgewood Arsenal, MD

    U.S. Borax, Los Angeles, CA

Chlorinated Hydrocarbons

    Diamond Shamrock, Cleveland, OH

    Dover Chemical, Dover, OH

    Dow Chemical, Midland, MI

    Fire Tect, Los Angeles, CA

    Keil Chemical, Hammond, IN

    Occidental Chemical, Buffalo, NY

    Monsanto, St. Louis, MO

    Neville, Pittsburgh, PA

    Pearsall, Houston, TX

TABLE 4 (Continued)

Phosphate Esters

  FMC, New York, NY

  Monsanto, St. Louis, MO

  USX, Chemical Division, Pittsburgh, PA

  Stauffer Chem, Greenwich, CT

  Union Carbide, Danbury, CT

Chlorinated phosphate esters

  Stauffer Chem., Greenwich, CT

Brominated phosphate esters

  Michigan Chemical. St. Louis, MI

  Nuodex, Piscataway, NJ

---

    4. Phosphates: When flexible PVC compounds containing phosphates burn, a smoke is generated that deactivates the free radical via the wall effect. This in effect dilutes the combustibles and expels oxygen from the surface of the burning plastic by forming a glassy layer. The phosphates are also thought to promote the char mechanism [44].

    Transparent flame retardant flexible PVC compounds can be made by partial replacement of the primary plasticizer with an organic phosphate. Those most commonly used are tricresyl phosphate, cresyl diphenyl phosphate, 2-ethylhexyl diphenyl phosphate, decyldiphenyl phosphate, and tris(β-chloroethyl) phosphate. Others are available for specialty uses. For these, the reader is referred to the producers listed in Table 4. Useful levels of most phosphates range from 5 to 15 phr, depending on the degree of flameproofing required as well as the effects these types of plasticizers have on other properties.

    The phosphate plasticizers have adverse effects on heat stability and, in some cases, light stability (the latter point being extremely formulation-dependent is controversial). Most of these plasticizers also affect low-temperature flexibility. For these reasons, they are rarely used above 15 phr, even though many of the phosphates present no compatibility problems. Poor heat stability can be minimized by employing a balanced stabilization system or by increasing the one in use when the phosphates are being substituted.

    5. Miscellaneous Flame Retardant Systems for PVC: Other chemicals that have been suggested as PVC flame retardants include boranes [45], chlorophosphonates [46], bromophosphonates [46], and triphenyl stibene [47], also known as triphenyl antimony. A series of organophosphorus compounds containing halogens have been claimed to confer flame retardance on a wide variety of polymers [48]. Some members of the series are said to be applicable to PVC, although they have limited compatibility with the resin.

    Several patents claim flame retardant coatings, using PVC homopolymer and copolymers as binders for paper, fibers, and textiles. These coatings contain flame retardant systems such as zinc carbonate plus tricresyl phosphate [49–51]; diguanylurea pyrophosphates plus oxides of antimony, bismuth,

tin, or titanium [52]; antimony phosphate or antimony phosphate plus tricresyl phosphate [53]; chlorinated paraffin plus antimony trioxide [54]; and
chlorinated paraffin plus octadecylamine phosphate, a metal salt or organometallic salt PVC stabilizer, and an oxide of zinc, tin, lead, magnesium, iron,
manganese, molybdenum, tellurium, titanium, copper, chromium, aluminum,
vanadium, or tungsten [55].

The reader is cautioned to look at all the properties, not only the flame
retardancy property, when evaluating any of the above systems.

6. PVC as a Flame Retardant: Numerous patents have disclosed some interesting applications that take advantage of the inherent flame resistance of
PVC and of the ease with which this property can be enhanced to flameproof
other materials. A flame retardant electrical insulator is made by combining
epoxy resin with essentially equal amounts of PVC resin, adding smaller
amounts of plasticizer and a small amount of antimony trioxide [56]. The burning characteristics of fabrics made from acrylonitrile copolymers are improved
by adding vinyl chloride copolymers to the spinning dopes used to prepare
fibers. The rate of burning of these fabrics is reduced, and some of the
fabrics are self-extinguishing [57]. PVC resin is also used as an additive to
flame-retard ABS for injection molding and extrusion.

E. Method of Incorporation

Flame retardant additives are added to the resin in the blending operation.
Care should be taken to add these additives in the proper order. A plasticizer such as triaryl phosphate should be added when high solvating plasticizers are added. If there are no solvating plasticizers in the formulation,
it would be added at the end of the liquid addition. Also, it should be taken
into consideration that this type of plasticizer will decrease the time for fusion or lower the fusion temperature. More specific data on blending and mixing are presented in Chapters 12 and 20 through 23. Flammability test
methods are listed in Table 5.

F. Effect on Physical Properties

Every formulator knows that the addition of another chemical to the PVC
formulation will affect the physical properties of the PVC. The flame retardant may be selected on the basis of the system that changes the physical
properties least. The properties of the flame retardant system that can affect the physical properties are:

> The physical state: solid or liquid
> If liquid, compatibility with resin and other
>   ingredients in the recipe
> If solid, particle size, particle size distribution,
>   and melting point

The only way to tell what effect the flame retardant system has on the physical
properties is by an actual evaluation.

G. Smoke Suppressants

In recent years, due to increased losses of life and property as a result of
fires, much attention has been focused on the mechanism of such disasters.
The SPI has been playing an active role in conjunction with groups such as

TABLE 5 The More Popular Flammability Test Methods

ASTM D635 (Horizontal Burn Test)

UL 94

  a. Horizontal Burn

  b. Vertical Burn

ASTM D2863 (Oxygen Index Test)

DOT MUSS 302 (Horizontal Burn Test)

ASTM E84 (Tunnel Test)

Smoke Density

ASTM E162 (Flame Spread Radiant Heat)

ASTM D1433 (Flexible Thin Plastic Sheet)

ASTM D1692 (Plastic Foam and Plastic Sheet)

ASTM D229 (Rigid PVC)

UL tests:

  1. Hot Wire Ignition Test

  2. High Current A-C Ignition Test

  3. High Voltage A-C Ignition Test

  4. High Voltage A-C Tracking Test

Source: Sa-Go Associates, Inc.

the National Fire Protection Association and the National Fire Prevention and Control Administration in developing fire data [58]. Of particular importance is data on the actual cause of deaths associated with fires.

The major approach to combustibility research by the plastics industry is hazard analysis. This analysis attempts to determine all possible fire hazards, evaluate these hazards, and develop effective measurement and control methods for them [58]. The entire problem is very political and confusing. For example, although relatively little or no action is being taken against known hazards such as ignition sources (cigarettes, matches, lighters), legislation is being drafted to indiscriminately ban the use of rigid PVC electrical conduit due to hydrogen chloride toxicity. A benefits analysis of PVC electric conduit has been presented by Carlon in their disclosure papers CF-3 [59]. Flax, in his article entitled "The Dubious War On Plastic Pipe [59], discusses how PVC is being blamed unfairly for the cause of deaths in fires. In addition, results of the many studies made show that smoke generation is a significant factor in loss of life. Evidence from recent and continuing studies suggests that half of the fatalities from accidental fires may be caused by smoke rather than heat or burning. The severe human problems produced by smoke include inhalation of toxic particulates, disorientation and panic, eye irritation, and obstruction of vision. For example, investigations of airplane crash fatalities have led to the conclusion that some passengers survived the impact of the crash but that the smoke generated during ensuing fires was so dense that they were unable to see the escape exits.

There are two stages in the burning of the PVC polymer. In the first stage, HCl is evolved at a temperature usually below 300°C. This reaction has been studied and found to be endothermic. In the second stage, black smoke is produced. This stage takes place at 350°C. Smoke is also produced when the PVC polymer is smoldering. The smoke has been analyzed and shown to contain H_2, methane, ethylene, propylene, butene, benzene, toluene, and other alkylaromatics [60].

As a result of the awareness of the danger of smoke generation from the PVC polymer, research efforts are being directed toward discovering means of overcoming this problem. Some of the solutions that have been proposed in the patent literature involve mixtures of iron powder and molybdenum oxide or iron powder, molybdenum oxide, and copper oxide [61], alkali metal zinc ferrocyanides [62], vanadium oxides or vanadium acetonylacetonate [63], potassium zinc cuprocyanide [64], oxides, hydroxides, salts, or alkanecarboxylates of lithium, sodium, potassium, magnesium, calcium, barium, zirconium, manganese, or iron, in combination with a phosphate triester [65], iron ethyl dicyclopentadienyl [66], and AlB_2, Cr_2S_3N, Cu_2S, CuS, FeS, MoB_2, SnS_2, TiB_2, and mixtures thereof. [67,68].

A proprietary product DFR-121 identified as a secondary plasticizer with flame retardant and smoke suppressant properties has been offered for use in flexible PVC compositions [69]. Other products that are claimed to decrease smoke in PVC compounds are alumina trihydrate [70] zinc borate [71], molybdenum oxide [72], and metallic compounds of iron, zinc, and aluminum [73].

H. Test Methods

There has been limited success to date in research programs attempting to develop test procedures that can be correlated with actual fire conditions. The problem is that there are variables present during an actual fire which cannot be foreseen or evaluated such as flying sparks or a large quantity of adjacent combustibles caused by a sudden draft. It should be recognized that PVC plastic is a hydrocarbon, and burning a hydrocarbon under controlled conditions, as under test procedures, may not be the same as burning it under uncontrolled actual fire conditions.

The Fire Prevention and Control Act of 1974 has brought about standards for materials that should have better fire prevention performance. These standards recognize a comparison of flame properties between materials.

There are many test methods originating from government agencies, Underwriters Laboratories, ASTM, Factory Mutual, and individual organizations. Examples are the Monsanto Tunnel Test and the Arapahoe Smoke Chamber Test. The difficulty is in choosing the right tests. It is strongly recommemded that more than one test be used in flammability evaluation, since the tests have varying degrees of reproducibility. Also, as pointed out by Keeney [75], a material that may be marginal in one test may prove to be outstanding in other tests. Multiple testing helps to ensure that a good flame retardant system is not overlooked.

V. FOAMING AGENTS, FOAM PROMOTERS, AND FOAM ACCELERATORS

A. Introduction to Cellular Vinyls

Cellular PVC technology has had an impressive growth rate and continues to have commercial importance. Cellular vinyls have many desirable properties that cannot be achieved any other way. Products made of cellular PVC may

have improved thermal, insulating, and electrical properties. They are characterized by better resilience, cushioning, and a high degree of flexibility.
Cellular vinyl compositions impart superb hand and drape characteristics to
coated fabrics. Finally, a cellular vinyl product can be less expensive than
alternative products because of the lower weight per equal volume.

There are two types of cellular foamed vinyls: open cell and closed cell.
In the closed cell type, each cell is individual, usually spherical in shape, and
completely enclosed by plastic walls. This type of cell structure has good
insulating properties as well as a high degree of buoyancy. In the open cell
type, all the cells are interconnected. This type of cell structure is known
for its absorbency and capillary action.

A cellular vinyl is a solid-gas composition consisting of a continuous vinyl
phase and a gas phase, either continuous or discrete, created by a foaming
agent. The cellular structure depends on the nature of the cell-forming process, that is, whether it is a physical change of state, a chemical decomposition, or some other chemical reaction [76].

Cellular vinyl may be classified according to cell structure, density,
whether it is rigid or flexible, and the degree of flexibility. Further discussion of cellular vinyls may be found in Chapter 34 (Vol. 4) of this series.

Vinyls can be expanded in one of the following ways:

1. Whipping air into a plastisol, followed by rapid curing
2. Mixing in a liquid component which vaporizes when heated
3. Dissolving a gas into the vinyl compound when it is
   liquid or in the plastic state
4. Adding components that chemically react and generate
   a gas within the plastic
5. Incorporation of hollow glass beads
6. Thermal decomposition of a foaming agent that liberates a gas

B. Promoters for Mechanical Foaming of Vinyls

Foaming promoters are adjuncts that facilitate the formation of uniform cells
and can increase the stability of the foam. For example, they can change the
surface tension of a foamable plastisol, modify the melt viscosity of an expandable compound, or provide nucleation sites for cell formation in gas-saturated
vinyl melts. Adjuncts that can decrease the decomposition of the blowing
agent are called foaming agent (or blowing agent) promoters.

Vinyl foam can be prepared by vigorous mechanical agitation of plastisols.
In a typical mechanical foaming process, a compounded plastisol that contains
a suitable foam promoter and compressed air is fed into a mixing head of a
continuous mixer [77], where a flowable froth is produced. The froth can be
continuously cast on fabrics or release paper by a coating knife [78] or, if
desired, it can be discharged into a mold. The foam is gelled and fused in a
hot-air oven with radiant or high-frequency heating. Unless properly stabilized, the entrapped air diffuses rapidly and the froth collapses before or
during the fusion. Foam promoters are used to prevent this occurrence.

Foam promoters are surface-active agents of the nonionic or anionic type.
Usually, a combination of several oil-soluble surfactants affords the best balance of desired foaming properties[79]. Aqueous solutions of surfactants have
also been suggested for this purpose [80], but the most successful process
[81] uses a water-in-oil (plasticizer) emulsion prepared with anionic wetting
agents [82].

The currently marketed foam promoters [83] are blends of surface-active materials such as potassium oleate or morpholine linoleate [84] in a water-in-oil emulsion, combined with vinyl heat stabilizers and dissolved in a phthalate plasticizer.

Satisfactory foaming in the mixing device and trouble-free conveying of the froth require plastisols with either Newtonian or thixotropic flow properties. For good air retention properties, the foamable plastisol should be designed around highly solvating plasticizers such as butyl benzyl phthalate or dipropylene glycol dibenzoate.

Vinyl foam obtained by this process displays an open cell structure, and it can be made in a range of apparent densities from 12 to 60 lb/ft^3.

C. Physical Blowing Agents

Physical blowing agents undergo a phase change during foaming. For example, compressed gases dissolved under pressure in a plastisol can develop a cellular structure on release of the external pressure. Volatile liquids incorporated in a vinyl mix are capable of producing a foam by passing from a liquid to a gaseous state.

The preferred agents for physical foaming processes are usually odorless, nontoxic, chemically stable compounds that will not adversely effect the physical properties or the thermal stability of vinyl polymers. Inert gases and volatile liquids are the most commonly used agents of this class. Occasionally, leachable solids have also been used in preparing microporous vinyl compositions [85].

Since physical foaming agents leave no solid residues, nucleating additives are frequently needed to create sites for cell formation [86]. Although physical foaming agents are generally inexpensive, their efficient use ordinarily requires specialized equipment designed to produce one specific cellular product.

1. Gaseous Blowing Agents: Nitrogen, air, and carbon dioxide are prime examples of gaseous blowing agents used in preparing cellular vinyls. The first two are used in the high-pressure "gassing" process, while the latter lends itself to low-pressure adsorption methods.

The high-pressure gas expansion method was commercially developed in the mid-1930s for cellular materials based on cross-linkable elastomers [87]. Later, several modifications of the original process permitted the manufacture of expanded thermoplastics, primarily PVC [88]. The basic process operates as follows: an organosol (for rigid structures) or a plastisol (for flexibles) is placed in a mold or autoclave and is saturated under high pressure with nitrogen or air while heat is applied simultaneously to fuse the resin. Following the fusion step, the gas-saturated melt is cooled under pressure and, when cold, removed from the pressurized vessel. At this point, the uniformly distributed gas bubbles contain gas at higher than atmospheric pressure, but essentially no expansion occurs since the gas bubbles are encapsulated within the fused resin. When the preexpanded body is heated to temperatures above the heat distortion point of PVC, the gas in the cells expands the soft vinyl matrix. At the same time, a pressure equilibrium of the gas in the cells and the surrounding environment is brought about.

Since the solubility of nitrogen in PVC is rather low, extremely high pressures, on the order of 3,000 to 10,000 psi, are required to saturate the melt. An improved absorption device facilitates the gassing step [89]. Other improvements of the process include saturating plasticizers with gas prior to blending with PVC resins [90] and using special plasticizers that can readily be saturated with nitrogen [91]. In spite of all these improvements, gas expansion requires heavy-duty equipment capable of handling gas-saturated melts at high temperatures. Several overseas companies currently operate the high-pressure expansion process to produce closed-cell cellular vinyls [92]. Both rigid and flexible materials with densities from 0.05 g/cm^3 (3 to 8 lb/ft^3) are being offered [93].

Unlike the high-pressure process, the carbon dioxide foaming techniques permit the production of cellular vinyls on a continuous basis [94]. Two absorption processes were developed in early 1950; both operate with carbon dioxide as the blowing agent but differ in the gas absorption technique.

In the Elastomer process [95], a vinyl plastisol is mixed with carbon dioxide at moderate pressure (less than 100 psi) and simultaneously cooled in the Girdler Votator, a pressurized mixing chamber. The chilled wet foam is continuously discharged onto a conveyer belt. After leveling by a doctor blade, the foam slab, usually 2 in. thick, is fused by high-frequency heating supplemented by infrared or hot-air heaters for fusing the surfaces. The cooled foam can be easily stripped from the conveyer and, following conditioning, is ready for cutting, thermoforming, lamination, etc.

To meet a wide range of end-use requirements, plastisols must be tailored accordingly, but all formulations for the Elastomer process contain viscosity modifiers, such as aluminum stearate, to allow formation of foams with a homogeneous cell structure. Because the plastisol is refrigerated during foaming, careful control of the moisture content of all ingredients is mandatory [96]. In some cases, silicone oil is used to control the moisture content and cell-structure. The Elastomer process turns out open-cell vinyl foam at densities of 5 to 6 lb/ft^3; it is difficult to achieve lower densities. Elastomer-produced wet foam [97] can also be discharged into cored molds to obtain molded articles, such as pillows, in the density range 6 to 10 lb/ft^3.

The Dennis process [98] operates according to the same principles as the Elastomer technique but differs in the method of carbon dioxide absorption. The plastisol is saturated with the gas in a countercurrent absorption tower (Fig. 2) under moderate pressure (100 psig) and without refrigeration. Efficient gas saturation is acheived because the absorption tower permits intimate contact between the plastisol and the gaseous blowing agent. A wide range of densities can be obtained by varying the CO_2 pressure. Fusion of the foamed plastisol, whether cast or molded, is carried out by high-frequency heating.

The need for expensive plastisol grade resins, the inability to produce foams with densities below 3 lb/ft^3, and the availability of heat-sealable inexpensive polyurethane foams have contributed to the obsolescence of the CO_2 foaming processes in the United States. In many countries where economic conditions are different, the Elastomer process is still being operated successfully.

2. *Liquid Blowing Agents*: Many useful physical blowing agents are found among volatile liquids with boiling points not exceeding 110°C. These compounds are usually selected from odorless, nontoxic, noncorrosive, and nonflammable liquids having good thermal stability in the gaseous state.

Since the efficiency of liquid blowing agents is directly related to the ratio of the specific volume of vapor to the volume of liquid, products with high specific gravity combined with low molecular weight are most effective. Fluorinated aliphatic hydrocarbons have all these desirable properties and therefore are ideal physical blowing agents for vinyls [100] (Table 6). The less expensive chlorinated hydrocarbons, such as methylene chloride, also have been suggested for making cellular vinyl compositions [101].

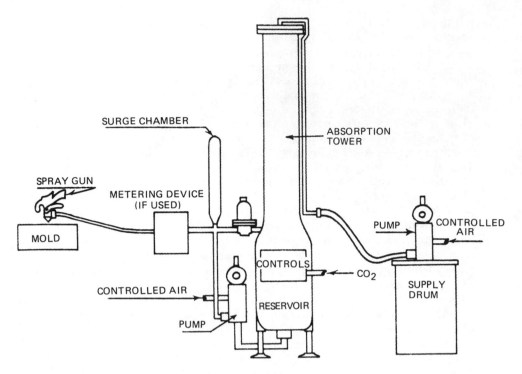

FIGURE 2 Flowchart for production of vinyl foam by the Dennis process.
(From *Modern Plastics*, July 1957, P. 117).

TABLE 6 Properties of Physical Blowing Agents for Vinyls[a]

Blowing agent	Molecular weight	Density (g/cm^3 at 25°C)	Boling point or range (°C)
Cyclohexane	84.17	0.774	80.8
Toluene	92.13	0.862	110.6
Trichloroethylene	131.40	1.466	87.2
1,2-Dichloroethane	98.97	1.245	83.5
Trichlorofluoromethane	137.88	1.476	23.8
1,1,2-Trichlorotrifluoroethane	187.39	1.565	47.6
Acetone	58.08	0.810	56.2
Methyl ethyl ketone	72.10	0.810	79.6

[a]From Ref. 99.

One of the main advantages of liquid blowing agents is the endotherm that accompanies the change from the liquid to the gaseous state. Hence, low-boiling liquids can be used in conjunction with chemical blowing agents to lower the decomposition exotherm of these agents. Tetrachloroethanes [102], methyl chloride, ketones, and aromatic hydrocarbons [103] are most widely used for this purpose.

An application using this technique is the production of ink-filled rolls for computers. A proprietary ink formulation is blended into a plastisol that contains a solvent. As the plastisol goes through the stage of gelation, the solvent evaporates, leaving a cellular structure. The ink fills the empty cells left by the solvent as it cannot fuse and it is incompatible with the vinyl. The resultant product is a plastisol roll which can ink characters on pressure.

D. Chemical Blowing Agents

Inorganic and organic compounds that liberate large volumes of gas, as a result of thermal decomposition at elevated temperatures constitute a group of additives known as chemical blowing agents. These agents are usually solid products with good thermal stability at ambient temperatures. At elevated temperatures, the agents undergo rapid decomposition in well-defined temperature intervals which are determined by the chemical nature of the agents and the environment in which they are being decomposed. The most commonly used foaming agents liberate, in addition to nitrogen, other noncondensable gases such as carbon dioxide, carbon monoxide, and hydrogen.

The ideal blowing agent would:

Be easily dispersed in a vinyl compound
Not have an adverse effect on process properties
Not plate out
Not have a deleterious effect on the heat or light stability of the vinyl compound
Leave and odorless, colorless, nonstaining, nonmigrating, nontoxic residue
Not leave a corrosive decomposition product
Release gas over a short, well-defined temperature range
Release gas at a controllable rate
Not affect the gelation or fusion rate of a vinyl plastisol
Function equally well in open or closed cellular production conditions
Be stable in storage
Leave a residue that could form nucleation sites for cells
Leave a residue that would not detract from electrical properties
Not have detrimental dermatological properties in applications requiring contact with skin
Allow the formulator to develop vinyl foams for a wide range of applications
Not be expensive

The major advantage of using a chemical blowing agent is that these additives can produce a cellular vinyl structure from a vinyl compound which is being processed in conventional vinyl processing equipment.

1. Inorganic Blowing Agents: Inorganic blowing agents, mostly alkali salts of weak acids, have been of limited use in the vinyl industry. The most important of these agents are listed in Table 7. They can liberate gas either by thermal dissociation or in the presence of activators by chemical decompo-

TABLE 7 Inorganic Blowing Agents for Vinyls

Chemical description[a]	Supplier	Decomposition temp. [°C (°F)]	Gas yield [ml (STP)/g]	Main decomposition products
Ammonium bi- carbonate	Many	60°C	850	NH_3, CO_2, H_2O
		(140°F)		
Sodium bicar- bonate	Many	100-140°C	267	CO_2, H_2O
$NaHCO_3$		(212-284°F)		
A		RT 100°C	534	CO_2, H_2O
B		(RT 212°F)		
Sodium boro- hydride	Ventron Div.	300°C		
$NaBH_4$	Morton- Thiokol Co.	(572°F)		
A		RT 100°C	2370	H_2
B		(RT 212°F)		

[a]Key: A, thermal dissociation; B, chemical decomposition in acidic medium.

sition. The thermal dissociation of an inorganic salt is an endothermic reversible reaction, and the reaction rate and equilibrium point depend on external
pressure. For this reason, the use of inorganic blowing agents is limited to
foaming at atmospheric conditions.

As foaming agents for vinyls, inorganic salts have the advantage of low
cost, but this is counterbalanced by poor storage stability and the difficult
dispersion properties of these products. Since gases generated by inorganic
foaming agents either are readily condensable (e.g., water vapor) or have
high diffusion rates (e.g., hydrogen), vinyl foams made with these agents
are dimensionally unstable and require long annealing periods.

Among the inorganic salts, ammonium bicarbonate is of some interest because it leaves no residue on decomposing:

$$(NH_4)HCO_3 \rightarrow NH_3 + CO_2 + H_2O$$

Assuming water vapor as the expandable gas, the gas yield of ammonium bicarbonate [850 ml (STP/g] is one of the highest among all chemical blowing
agents. The salt has been suggested as an additive to dinitrosopentamethyl-
enetetramine in the expansion of vinyl plastisol [104].

Sodium bicarbonate liberates a smaller volume of gas than the ammonium
salt but its thermal stability is better. In storage, it has a tendency to pick
up moisture and to cake. For use in vinyl compounds, sodium bicarbonate
must be carefully ground in plasticizer by repeated passing through a colloid
mill, three-roll mill, or similar high-shear device.

The thermal decomposition of sodium bicarbonate takes place in a wide temperature range (100° to 140°C) with the attendant evolution of carbon dioxide and water vapor:

$$2NaHCO_3 \rightarrow Na_2CO_3 + CO_2 + H_2O$$

The alkaline residue of sodium carbonate is objectionable in any application where it can damage coatings and substrates. Since the thermal dissociation of sodium bicarbonate decreases rapidly with increased pressure, application of the salt is limited to expansion of vinyl under atmospheric pressure [105].

To increase the efficiency of sodium bicarbonate, acidic additives such as stearic acid are used [106]. Total decomposition of the bicarbonate can be brought about by chemical reaction of the salt with water-soluble acids:

$$2NaHCO_3 + 2H^+ \rightarrow 2Na^+ + 2H_2O + 2CO_2$$

This system has been suggested for frothing of plastisols. The reaction may proceed at room temperature, and therefore the reactive ingredients must be predispersed in plastisol and stored separately. When the two components are blended, carbon dioxide foams the compound. So that the evolved gas in the plastisol can be entrapped and breakdown of the foam prevented, the compound should contain a fatty acid soap as a thickening agent.

A similar technique can be applied to sodium borohydride, which generates a much larger volume of gas than sodium bicarbonate:

$$NaBH_4 + 2H_2O \rightarrow 4H_2 + NaBO_2$$

A typical plastisol foaming system consists of a plastisol containing a buffered solution of sodium borohydride and a second plastisol with an acidic activator [107].

With strong organic acids, such as acetic and oxalic, the gas generation is extremely rapid, and it becomes difficult to control the frothing of plastisol. Using buffered acid solutions offers a convenient means of controlling the reaction. Best results have been reported with glycine [108], glycerol [109], and isocyanates [110].

2. *Organic Blowing Agents* [111] : Organic compounds that release nitrogen as the main component of the liberated gas are the most important foaming agents for cellular PVC. The thermal decomposition of organic blowing agents is an irreversible exothermic reaction independent of external pressure. The decomposition rates of these products are governed by temperature or time or both and are independent of concentration. Interaction with other ingredients in the formulation or impurities may change the course and the rate of the blowing agent decomposition.

Chemically, organic foaming agents are characterized by the functional groups shown in Table 8. Blowing agents with the same functional groups may display widely different decomposition temperatures, depending on their molecular structure, but their decomposition exotherms are essentially equal. For example, azides generate very high exotherms, whereas sulfonhydrazides have relatively low caloric effects. A large exotherm generated during rapid decomposition of the blowing agent may raise the internal temperature of the foaming system beyond the degradation point of vinyl polymers. Charring of the compound will occur if the heat is not dissipated quickly. Volatile solvents

TABLE 8 Characteristic Funtional Groups of Organic Blowing Agents

$-N{=}N-$	Azo
$>N{-}NO$	*N*-Nitroso
$-SO_2{-}NH{-}NH-$	Sulfohydrazo
$-N\begin{smallmatrix}N\\ \parallel\\ N\end{smallmatrix}$	Azido

having a high heat of evaporation can be used as a heat sink in foamable PVC compositions to control the decomposition exotherm of organic foaming agents.

With the exception of azides, all functional types of organic blowing agents are represented among commercial products. Table 9 gives the properties of commercially available blowing agents for vinyls [111].

a. Azo Compounds: Aromatic azo compounds, such as diazoaminobenzene, were the first useful nitrogen-releasing blowing agents for high polymers. Because of their toxic and staining properties, they became obsolete when aliphatic azo derivatives were made commercially available. Of these, azobisisobutyronitrile and azobisformamide are particularly significant in cellular vinyl technology.

b. 2,2'-Azobisisobutyronitrile (AZDN): This compound has found utility in the production of white, nonstaining, ordorless cellular vinyl products by the plastisol casting method. The low decomposition temperature of the blowing agent, combined with a moderate exotherm, permits the production of large cellular bodies without the danger of polymer degradation. Also helpful in this respect is the fact that the decomposition residue, tetramethylsuccinonitrile (TMSN), is a hydrogen chloride acceptor, which facilitates the stabilization of PVC compounds.

Unlike many other organic blowing agents, AZDN is insensitive to activation by vinyl compounding ingredients. The decomposition rate of AZDN depends only on the temperature of the environment (Table 9). At 30°C and below, the storage stability of the compound is excellent. AZDN is a flammable solid that ignites easily from an open flame and continues to burn when the flame is removed.

Simultaneously with the evolution of nitrogen, AZDN generates free radicals useful in the polymerization initiation or grafting of monomers with ethylenic linkages ($-C{=}C-$) [112]: Because of this dual functionality AZDN has found application in a novel process [113] for cellular cross-linked PVC having excellent dimensional and thermal stability. In the process, maleic anhydride and styrene (or acrylonitrile) are copolymerized and grafted onto the PVC chain. Maleic anhydride is then hydrolyzed to the acid, which is later caused to react with a diisocyanate to cross-link the polymer. As a free-radical donor, AZDN induces the copolymerization and grafting. At the same time, it forms minute gas cells that act as nucleation sites for the formation of gas cells when the isocyanate-terminated polymer is caused to react with water. This technique and the resultant cellular products are discussed further in Chapter 34 (Vol. 4).

The main disadvantage of AZDN as a blowing agent is the toxicity of the

TABLE 9 Commercial Blowing Agents for Vinyls

Chemical composition	Melting point, or decomp position point in air	Decomposition range in vinyls	Gas yield [ml (STP)/g]	Manufacturer
Azobisisobutyronitrile (AZDN) $(CH_3)_2(CN)C-N=N-C(CN)(CH_3)_2$	105°C (221°F)	90-105°C (194-221°F)	136	2
4,4'-Oxybis (benzenesulfonhydrazide) (OBSH) $O(C_6H_4-SO_2-NH-NH_2)_2$	164°C (327°F)	130-160°C (266-320°F)	125	4,5
1,1-Azobisformamide (ABFA) (azodicarbonamide) $H_2N-CO-N=N-CO-NH_2$	195-200°C (383-392°F)	150-200°C (302-392°F)	220	1,3,4,5
p-Toluenesulfonyl semicarbazide (TSC) $C_6H_4(CH_3)SO_2NNNHSONH_2$	227°C (441°F)	190-200°C (374-392°F)		5

Manufacturer	Address	Trade name
1. B.F.C. Chemicals	Wilmington, DE	Ficel
2. Dupont	Wilmington, DE	Vaso
3. Mobay Chemicals	Pittsburgh, PA	Porofor
4. Olin Chemicals	Stamford, CT	Nitropore, Kempore
5. Uniroyal	Naugatuck, CT	Celogen
6. Fairmont Chemical	Newark, NJ	Azocel

residual tetramethylsuccinonitrile. Since the residue is compatible with PVC, it can be driven off in the annealing ovens or removed from the surface of the expanded PVC by washing with warm water. The health hazards connected with handling and processing AZDN therefore require special safety measures. This limits the utility of AZDN as a foaming agent for vinyls. Today, free-radical polymerization of various vinyl monomers accounts for most of the AZDN consumption.

c. Sulfonyl Hydrazides: Monosubstituted sulfonhydrazides, such as benzene or toluenesulfonylhydrazide, have been of limited use in cellular vinyl technology [114]. They leave thio compounds as residues:

$$4R-SO_2-NH-NH_2 \rightarrow 4N_2 + 6H_2 + R-S-S-R + R-S-SO_2-R$$

Symmetric, disubstituted sulfonhydrazides offer a significant improvement in this respect because of the polymeric nature of the residue [115]. Decomposition of the sulfonhydrazides follows an internal redox reaction with a low level of heat evolution; the exotherm generated during the oxidation of the hydrazide groups is partially compensated by the endotherm of the sulfonyl reduction.

Among the symmetric sulfonhydrazides, 4,4'-oxybis (benzenesulfonyl-hydrazide) (OBSH) [116] is the most widely used in the United States, whereas in Europe, 3,3'-sulfonbis (benzenesulfonylhydrazide) [117] is more popular. Commercial OBSH is a finely ground powder that decomposes slowly at 130°C and more rapidly at 150°C. The generated noncondensable gas consists predominantly of nitrogen (98%) and the residue is a white, nontoxic solid. The blowing agent ignites easily from a spark or flame and continues to decompose with the evolution of copious smoke, even after the source of ignition has been removed. Under normal storage conditions, however, the compound is stable for a prolonged period. Like other sulfonhydrazides, OBSH can be activated to decompose at lower temperatures in the presence of oxidizing agents such as ferric chloride, peroxides, and organic bases such as triethanolamine.

Cellular vinyl polymers produced with OBSH display a uniform, medium to fine pore structure and are nontoxic, odorless, and nonstaining, although occasionally they discolor slightly under UV exposure. Because the decomposition temperature of OBSH is too high for chemical frothing of most plastisols and too low for processing in calendered vinyl compositions, this substance finds use only in expansion of plastisols based on low-molecular-weight PVC homopolymers or copolymers [118]. Another application of OBSH makes use of its bifunctionality, which is capable of cross-linking elastomeric blends containing diene rubbers, such as NBR, with PVC. In these compounds (useful as pipe insulation), OBSH acts as a blowing and vulcanizing agent [119].

d. Sulfonyl Semicarbazides: These compounds represent a novel group of chemical blowing agents [120] with higher decomposition temperatures than the hydrazides from which they have been derived. Of particular interest is tolune sulfonsemicarbazide (TSC) [121], a compound that has a decomposition range of 210° to 225°C and a gas yield of 143 ml (STP)/g. Because of its high decomposition point, TSC has been suggested as a foaming agent for rigid vinyl compositions.

e. 1,1'-Azobisformamide (ABFA): Frequently called azodicarbonamide, ABFA is the most versatile blowing agent for vinyls. It has a unique combination of valuable properties that fulfil the requirements of an ideal chemical foaming agent [122].

The commercially available azobisformamide is an orange yellow to pale yellow powder which decomposes in air above 195°C (the color depends on the particle size distribution). The agent and its residue are odorless, nonstaining, and, when properly compounded, nondiscoloring. Because ABFA and its residue are nontoxic, use of the blowing agent in nontoxic applications has been sanctioned under several FDA regulations [123].

Although ABFA is insoluble in common solvents and plasticizers, it disperses readily in vinyl compounds. Plasticizer dispersions of the foaming agent that can be blended directly with plastisols, dry blends, and other vinyl compounds are commerically available. The storage stability of the agent, both dry and in a dispersed form, is excellent. Significantly, ABFA is the only organic foaming agent that does not support combustion and is self-extinguishing.

When azobisformamide is heated in an ester plasticizer above its decomposition point, it delivers gas at 220 ml (STP)/g, corresponding to one-third of its molecular weight. The gas phase consists of 65% N_2, 32% CO, and 3% CO_2. Under alkaline conditions, ammonia is also generated. The composition of the residue depends on the medium in which the blowing agent is decomposed. The residue is usually white and consists mainly of hydrazobisformamide (biurea), cyanuric acid, and urazole. The ABFA residue may enhance the heat stability of vinyl compositions, particularly those stabilized with zinc salts. In the presence of alkalis, ABFA hydrolyzes rapidly at 100°C with the liberation of nitrogen, carbon dioxide, and ammonia. Polyhydric alcohols [124] and alcoholamines are equally capable of hydrolyzing ABFA at elevated temperatures.

Because of its high decomposition temperature in air, ABFA was originally considered unsuitable as a foaming agent for vinyls [125]. However, a later observation that many additives, including vinyl stabilizers, can lower the decomposition temperature of ABFA led to successful commercial application of the agent in the cellular vinyl industry [126].

E. Chemical Blowing Agent Activators

Vinyl stabilizers containing lead, cadmium, and zinc are the most efficient activators (Figure 3). These additives depress the decomposition temperature of ABFA and increase its rate of decomposition. Liquid stabilizers containing the same cations show increased activation due to their solubility in plasticizers [127]. The activating effect of the basic metal salts is derived from the instability of the corresponding azodicarboxylates that are formed as the first step in the decomposition of ABFA. Since the azodicarboxylates of lead, cadmium, and zinc decompose at lower temperatures than the diamide, they initiate the thermal breakdown of ABFA [128]. The activation of ABFA by metal salts is concentration-dependent [129], and the rate of ABFA decomposition increases as the amount of metal salt in the compound increases. Some proprietary compounds offered in the trade are said to activate the ABFA catalytically [130].

Figure 4, from a study by Union Carbide, shows how five proprietary stabilizers activate the blow in a vinyl compound [131].

Because of the high-temperature stability of barium azodicarboxylate, barium salts exert no activating effect on ABFA, and in their presence the decomposition rate of ABFA is as low as it is in air. However, barium compounds show synergistic activity with lead and zinc in reducing the de-

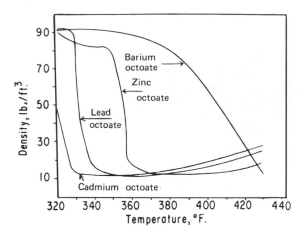

FIGURE 3 Activation of ABFA. (*Source*: Ref. 127)

composition temperature of ABFA [132]. The activation of ABFA is promoted
by alkaline conditions and inhibited in acidic media. For example, adding
fumaric acid to lead-stabilized, ABFA-containing plastisol increases the decomposition temperature of the blowing agent to 190°C, compared with 165°C
in the absence of the acid. This mechanism is used in one of the chemical embossing processes [133].

The particle size of ABFA is another factor that governs degree of activation by vinyl stabilizers [134]. Since ABFA is insoluble in vinyl plasticizers,
activation of the blowing agent takes place in a heterogeneous system and
therefore is directly related to the active surface area. ABFA of large particle
size and small surface area is more difficult to activate than material of fine
particle size having a large area of activation. Recognition of this fact has
led to the commercial acceptance of an azobisformamide series of blowing
agents with particle-size distribution tailored to specific applications [135].

ABFA lends itself to being compounded with dry blends [136] as well as
plastisols. Successful processing techniques applicable to these foamable
compounds have encompassed the entire spectrum of vinyl technology. The
chemical has been the preferred foaming agent in the manufacture of expanded
vinyl fabrics [137] and floor coverings [138]. Other uses include low-density
open-cell foam [139], extruded low-density gaskets [140], high-density profiles [141], expanded jackets for electrical wires and cables, slush-molded products [142], crown cap liners [143], injection-molded soles, and expandable
inks [144].

The barium salt of azodicarboxylic acid decomposes at 245°C [145] with
evolution of nitrogen and carbon monoxide, and it has been test-marketed as
a high-temperature blowing agent for rigid and semirigid vinyl compositions
[146]. Esters of azodicarboxylic acid are thermally unstable liquids, and it
has been proposed that they would be useful blowing agents for vinyls [147].

F. Methods of Incorporation

In the compounding operation, chemical blowing agents and promoters are
handled like many other vinyl additives. Because of the heat sensitivity
of the foaming agents, careful temperature control is essential to avoid

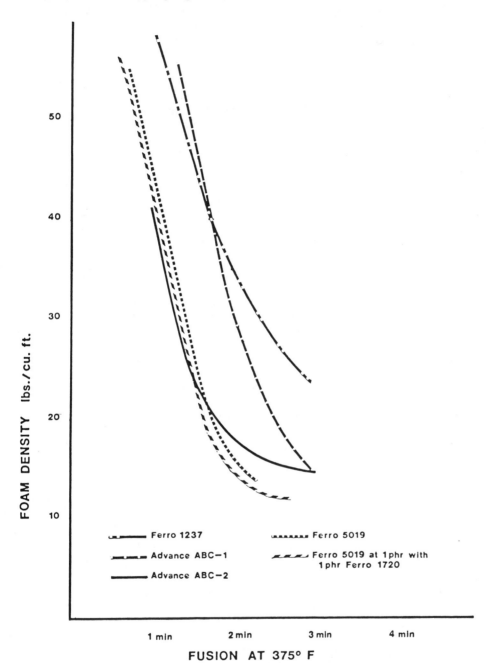

FIGURE 4 Plastisol blown with AZO and various activators. Formulation: 100 parts medium weight resin; 50 parts DOP; 5 parts ESO; 2.5 parts AZO; 2 parts activator.

premature gas generation.

For homogeneous distribution of the blowing agent in the vinyl compound, the agents are predispersed in liquid vehicles such as DOP, epoxidized soybean oil, or polymeric plasticizers. High-shear equipment, for example, three-roll mills and high-speed dissolvers, is usually used in this operation. Occasionally, plasticizer-soluble wetting agents are added to facilitate the dispersion of the solids. One of the most useful vehicles is DOP that contains 10 to 25% epoxidized soybean oil. For the convenience of processors, ready-to-use blowing agent dispersions are commercially available.

Blending of the blowing agent dispersion with the vinyl mix varies, depending on the type of vinyl compound involved in the processing. Low-shear churns and blenders are used for adding blowing agent dispersions to plastisols, and ribbon blenders or high-intensity mixers are used in processing dry blends and calendering compounds. When only small quantities of blowing agent are to be added to an extrusion or injection molding compound, resin pellets are coated with the blowing agent by tumbling.

G. Test Methods

The apparent density of a cellular vinyl product is obtained through the trial-and-error approach at tailoring the formulation to blow under optimum conditions. Unfortunately, there is no single definition of what constitutes an optimum condition. The optimum condition will depend on the type of process, the type and amount of blowing agent, and the formulation. The factors of chemical composition of the blowing agent that delineate its efficiency are its molecular weight, number of moles of gas that can be split off, and decomposition mechanism. The efficiency of chemical blowing agents is measured by the quantity of gas given off on decomposition. This quantity is called gas yield and is defined as the gas volume at normal temperature ($0°C$) and pressure (760 mm Hg) per unit of weight. The actual gas yield of a blowing agent in a vinyl composition may differ appreciably from the theoretical or stoichiometric gas yield because of side reactions induced by other additives present, for example, stabilizers, pigments, or resin impurities (e.g., residual emulsifiers). The gas yield values of most of the common chemical blowing agents are given in Tables 7 and 8.

Determination of the actual gas yield from a chemical blowing agent heated beyond its decomposition point is the subject of ASTM method D1715 [148], issued jointly by ASTM and SPI. In this test, since the blowing agent undergoes decomposition in an inert environment and the heating rate is low, the test has little relevance to blowing agent effectiveness in vinyls, nor is this method applicable to measurements of the rate of decomposition—for example, the gas yield as a function of temperature or time or both.

Differential thermal analysis (DTA) combined with thermogravimetric analysis (TGA) can be used advantageously to follow the decomposition behavior of chemical blowing agents. For example, an organic foaming agent generates an exotherm during the decomposition, and the evolved heat as well as the volume of the liberated gas can be plotted as a function of temperature. The interval between the onset and the peak of the exotherm is considered the decomposition temperature. From the corresponding TGA graph, the gas yield in this temperature range can be read off [149].

The Plastograph test method [150] allows the determination of blowing agent stability and gas evolution rate under conditions bearing a close relation to actual processing. Significant information on the thermal behavior of blowing agents in vinyl chloride polymers also can be obtained when their actual performance, that is, their efficiency in lowering the apparent density of the polymer, is determined as a function of time or temperature or both [151].

VI. FUNGICIDES, BACTERICIDES, AND OTHER BIOCIDES

Polymers and copolymers of vinyl chloride have excellent resistance to destruction by biological agents. When these polymers and copolymers are compounded with additives into a plastic to produce useful articles, the resultant plastic compositions are subject to biological attack of various types and degrees.

Biodeterioration of plastics can occur due to microorganisms and macroorganisms. Deterioration by microorganisms, fungi, and bacteria is probably the most commonly recognized form and is primarily a chemical destruction.

Damage by macroorganisms is chiefly physical rather than chemical. It involves gross destruction, such as the gnawing or boring of holes, instead of molecular changes and the rupturing of chemical bonds. Birds, insects, and rodents can penetrate PVC when it represents a barrier to a source of food or water. Among the macroorganisms, those broadly classified as foulers, such as barnacles, do no particular damage to PVC coatings. Those known as borers, including mollusks and crustaceans, are capable of penetrating the jackets of underwater cables.

End products known to have been damaged by microbes include upholstery, wall covering, floor covering, wire and cable, tarpaulins, tents, baby pants, outerwear, and shower curtains.

A problem caused by the indirect action of microorganisms is the phenomenon called pink staining of PVC [152-154]. Certain organisms produce colored metabolic by-products that are soluble in plasticized PVC and readily migrate into and through it. It is not necessary that they grow on the plastic itself; they can cause the staining by growing on some other substrate in contact with it. A typical example is vinyl-coated fabric in which the vinyl is frequently stained due to microorganisms that are flourishing in the supporting textile, even though the vinyl layer shows no evidence of growth. The name pink staining derives from the color of the stain and represents the first case in which a microbiological cause was proved. The pink stain-producing organism is *Streptoverticillum rubrireticule* [155]. Stains of different colors have been found to have a similar origin but to involve other microorganisms.

Biodeterioration of plastics, including PVC, has been comprehensively reviewed by Wessel [156] and the use of biocides for its control by Scullin et al. [157]. A textbook by Greathouse and Wessel [158] is a good source of general information on deterioration of materials.

Additive chemicals used to prevent degradation by microorganisms are collectively called microbicides; they included fungicides, which kill fungi, and bactericides, which kill bacteria. Although not scientifically correct, the term *microbicide* is often used in industry to include chemicals that inhibit growth but do not kill. The proper term for these kinds of agent are *fungistat* and *bacteristat*. Table 10 contains a partial listing of producers of microbicides for PVC. The chemicals used to prevent destruction by the higher forms of life, chiefly members of the animal kingdom, include both lethal agents, or poisons, and repellents.

Although PVC itself is not susceptible to microbial attack, as far as is

TABLE 10 U.S. Producers of Microbicides

M & T Chemicals. Inc	Rahway, NJ
Nuodex Chemicals, Inc.	Piscataway, NJ
Ferro Chemical	Toledo, OH
Ventron Corporation	Beverly, MA
R. T. Vanderbilt Company	East Norwalk, CT
Akzo America/Interstab Chemicals	New Brunswick, NJ

known, some of the modifying agents used with it provide a nutrient source
for both bacteria and fungi. The modifiers known to be microbial nutrients
include some lubricants, and plasticizers in particular. Even when only non-
susceptible ingredients are used to make a PVC compound, microbial growth
(commonly called mildew) may occur on the surface due to contamination with
nutrients from an external source.

A. Mechanism of Biodeterioration

Deterioration by fungi and bacteria is chemical, involving changes in composition and breaking of chemical bonds. It appears to be caused by the action
of enzymes, which the microorganisms produce, on additives in the formulation, expecially the plasticizer. Of the two, fungal deterioration seems to be
the most prevalent and has received the most attention, This may be because
fungi frequently cause discoloration, while bacterial damage is less readily
apparent. Other manifestations of microbial degradation include changes in
odor or weight, embrittlement, exudation, loss of tensile strength, loss of
elongation, and changes in electrical properties.

The enzymes furnish energy, provide a mechanism for moving nutrients
through the cell wall, allow the organism to extract nutrients from the substrate, and stabilize the internal chemistry of the cell to variable external
environments [159].

The large amount of investigation of fungal and bacterial deterioration
has been largely empirical [160]. The little that is known is not well understood and is controversial [161]. A major concern has been in the field of
electronics. It was thought that solving the moisture problem would also solve
the biodeterioration problem, based on the idea that the fungi and bacteria cannot live without water. In many cases this has solved the deterioration problem; however, there have been exceptions.

The adverse effects of fungi in flexible vinyl formulations are detracting
from the esthetic characteristics by discoloration and spotting as destruction
of the adjacent cellulasic components. In electrical or electronic applications,
fungal organisms may destroy the dielectric properties of the system.

Bacterial deterioration of vinyl compounds is more insidious than fungal
deterioration. The bacteria usually attack the plasticizer in the compound,
causing stiffening as well as destroying the physical properties and/or creating foul odors.

Brown [162] was among the first to present a systematic review of the susceptibility of plasticizers, and he drew some useful conclusions based on the
growth of a few organisms in a variety of plasticizers that were in use in 1946.

Stahl and Pessen [163], in 1953, studied the growth of one fungus and one bacterium on 47 plasticizers, including an 18-member homologous series of sebacates. In 1957 this work was expanded and extended by Berk et. al. [164], who presented results of an in vitro study of the ability of 24 species of fungi to use 127 plasticizers as the sole source of carbon. Hueck and van der Plas [165], in 1960, made a further contribution to the understanding of the biological deterioration of plasticizers.

Table 11 summarizes the relative microbial susceptibility of some important types of plasticizers. It is essential to realize that these are only generalizations; they are useful guidelines but should not be interpreted narrowly. For example, although phthalates and phosphates tend to be resistant and to be superior to saturated diesters of aliphatic dibasic acids, and branched-chain esters tend to be more resistant than straight chains, there are exceptions as well as varying degrees of microbial degradation. Resistance also varies according to the organism present, which ordinarily cannot be predicted or controlled during the service life of a plastic article, and according to the environmental conditions prevailing during exposure. Even for studies under carefully controlled laboratory test conditions and with pure cultures of microorganisms, the literature contains conflicting data reported by different investigators on a given plasticizer. Table 12 lists the microorganisms that are used to study the deterioration and degradation of PVC compounds.

Where resistance to microbiological degradation is important for the successful application of a PVC article, the compounder should determine its suitability by in vitro testing with the appropriate organism or organisms, or preferably by exposure to conditions which duplicate or closely simulate those of actual use, for instance, soil burial, tropical climate, or submersion under water.

B. Biocidal Chemicals

Many thousands of chemicals of widely differing structures are known to have biocidal activity. Of these, several hundred are used commercially as economic poisons for protection against insects, birds, rodents, marine organisms, weeds, algae, and microbes [166]. These have been chosen on the basis of a suitable combination of efficiency, cost, lack of danger to humans, and lack of side effects on the material being protected. The number of pesticidal chemicals suitable for polymers and plastics is still smaller, due to limitations of solubility and compatibility, among other factors [167]. The relatively few microbicides from this group that are useful in PVC compounds are listed in Table 13.

Three things are chiefly responsible for the small number of acceptable compounds: compatibility, heat stability, and processing temperature. Compatibility is inherent in the chemical structure of the biocide, but problems can sometimes be overcome by formulation with other additives, as in the case of copper 8-quinolinolate. Many potentially useful biocides are eliminated because of adverse effects on the heat stability of PVC. These effects can sometimes be alleviated by using other additives or by developing a carefully balanced stabilizer system. The high-temperature processing cycles used with PVC rule out other biocides. Some may decompose, others may volatilize completely from the compound, and still others may volatilize sufficiently to present a health hazard to workers during processing.

Of the microbicides shown in Table 13, those most widely used are

TABLE 11 Microbial Susceptibility
of Classes of Plasticizers

Susceptible

  Sebacates

  Epoxidized oils

  Epoxidized tallate esters

  Polyesters

  Glycolates

Moderately susceptible

  Adipates

  Azelates

  Pentaerythritol esters

Resistant

  Phthalates

  Phosphates

  Chlorinated hydrocarbons

  Citrates

N-(trichloromethylthio)phthalimide and N-(trichloromethylthio)-4-cyclohexene-
1,2-dicarboximide. They have high activity against a spectrum of organisms,
can be used to make transparent as well as opaque compounds, and are useful in calendering and extruding as well as in plastisols. The various 10,10'-
oxybisphenoxarsine derivatives are also widely employed [173], as is 2-n-
octyl-4-isothiazoline. The quaternary ammonium carboxylates are chiefly
limited to plastisols because of an adverse effect on PVC heat stability that
precludes use in the more rigorous conditions found in calendering and extrusion. They can be used in both clear and pigmented compounds.

Bis(8-quinolinolato)copper, commonly called copper-8-quinolinolate, is a
highly effective fungicide, but its use is primarily restricted to military applications because of the opacity and color it imparts to PVC. By itself, this
chemical has difficult and limited compatibility with plasticized compositions,
but it can be compounded to overcome this problem. Means for doing this
have been described in several patents [168-172].

Other materials that are occasionally used in special applications are
zinc borate and modified barium metaborate. These two chemicals can also be
used as part of a flame retardant system. Tri-n-butyltin compounds are
mainly used in marine applications such as boat bumpers, life preservers,
marine antifouling coatings, etc.

The usual concentration of microbicide ranges from 0.1 to 5%, based on
weight of plasticizer. The optimum level varies with the specific microbicide
used, the susceptibility of the plasticizer or other biodegradable additive, and
the degree of protection required by the end use. The recommendations of
manufacturers of microbicides should be followed in this regard.

TABLE 12 Microorganisms Used to Study Degradation

Fungi	*Mucor* sp.
Alternaria tenuis	*Curvularia geniculata*
Aspergillus flavus	*Stemphylium consortiale*
Cladosporium herbarum	*Glomerella cingulata*
Paecilomyces varioti	*Myrothecium verrucaria*
Penicillium funiculosum	*Stachybotrys atra*
Trichoderma viride	Bacteria
Pullulavia pullulans	*Pseudomonas aeruginosa*
Aspergillus niger	*Serratia marcescens*
Aspergillus versicolor	*Bacillus subtilis*
Penicillium piscarium	*Escherichia coli*
Penicillium luteum	*Staphylococcus aureus*
Aspergillus oryzae	*Streptomyces rubrireticuli*
Fusarium sp.	

The ideal biocide would:

Be highly effective as both a fungicide and a biocide
Be effective against a wide range of organisms
Not interact with other ingredients in the formulation
Easily be dispersed and distributed
Be essentially nontoxic
Be safe to handle
Not create any environmental problems
Not contribute color or odor to the PVC formulation
Not detract from the processing properties of the compound

TABLE 13 Microbicides for PVC Compounds

Quaternary ammonium carboxylates

N-(Trichloromethylthio) phthalimide

Bis(tri-*n*-butyltin) oxide

N-(Trichloromethylthio)-4-cyclohexene-1,2-dicarboximide

Bis(8-quinolinolato) copper

p-Hydroxybenzoic acid esters

Condensate of 10,10'-oxybisphenoxarsine and epoxidized soybean oil

The previous discussion has been concerned with the use of biocides to prevent microbial attack and degradation of PVC compounds themselves. There has been some interest in incorporating microbicides into plastics to keep the finished articles free from microbes in order to protect the user. Examples of such articles are toys, toothbrushes, hairbrushes, telephones, and dishes. Two of the chemicals claimed to be useful for this purpose are 2-hydroxy-5-chlorobenzoic-3', 4'-dichloroanilide and 3,5,3'4'-tetrachlorosalicylanilide [174].

C. Insecticides

Published information on the attack of insects on PVC is scanty and conflicting. Gay and Wetherly [175] reported that semirigid PVC and rigid PVC are virtually resistant to attack by termites. However, plasticized PVC used in wire and cable is very susceptible to attack. The degree of termite penetration varied with plasticizer concentration and structure. Susceptibility increased with increasing level of plasticizer and decreased when tricresyl phosphate was used. Other effective means of control were incorporation of the insecticides aldrin and dieldrin and incorporation of certain inorganic fillers [176].

Naphthalene has been incorporated into plasticized PVC, not to prevent degradation of the plastic but to make film for bags for protecting clothing against moths and other parasites [177].

D. Rodent Repellants

Polyvinyl chloride compounds such as cable jackets and food packages have been attacked by rodents, especially rats and mice. Damage to cables is apt to occur if they are placed in locations that block the passage of rodents to food sources. Welch and Duggan [178] studied effects of adding rodent-repellent chemicals to calendered film and found a trinitrobenzene complex at a concentration of 0.025 g/in.^{2} to be effective against roof rats, Norway rats, and house mice. Two others had some activity but were less efficient: tetramethylthiuram disulfide, and a zinc dimethyldithiocarbamate/cyclohexylamine complex.

E. Marine Microbiocides

The literature on damage to PVC compounds by marine organisms is scanty, and that on its control by chemical agents is even more sparse. Connolly [179] found that during a 7-year exposure in the sea, there was little damage by microorganisms, but extensive penetration by pholads. Carlston and Whiting [180] found a copolymer of vinyl chloride and vinyl acetate to be a suitable binder for a ship bottom paint to prevent the attachment of fouling organisms such as barnacles. This polymer provided controlled leaching of toxicants to inhibit growth of foulers but was not itself degraded. Suitable toxicants for the coating were copper, cuprous oxide, and mercuric oxide. As stated previously, various tri-*n*-butyltin derivatives have more recently been found useful in such coatings.

F. Test Methods

Many different approaches have been made toward evaluating resistance to microbial degradation and the effectiveness of microbicides. These range from laboratory Petri dish tests using pure cultures of organisms to actual environmental exposure under use conditions, and from a study of individual compounding ingredients to the testing of complete PVC compositions.

Because plasticizers are known to be used as a carbon source by microbes, much work has been done on plasticizers alone. A shake flask technique, such as Stahl and Pessen [163] used, has been frequently employed. In this method, plasticizer or plasticizer plus microbicide is added to Erlenmeyer flasks containing an aqueous solution of mineral salts and yeast extract. The flasks are then sterilized, cooled, inoculated with the desired organism, and placed in a horizontal reciprocating shaker in an incubator. They are shaken and incubated for 7 days at 30°C or other suitable combination of time and temperature. Because emulsions are formed, visual determination of the extent of growth cannot be made.

Fungal growth is determined gravimetrically by filtering out the mycelia, then washing, drying, and weighing the filtrate. Bacterial growth is determined turbidimetrically, and its weight estimated by comparison with standard curves.

Another technique, used by Berk et al. [164], is to add plasticizer or plasticizer plus biocide to sterile melted mineral salts agar, pour the whole into sterile Petri dishes, and allow the solution to solidify at room temperature. The solid medium is then inoculated with spore suspensions of the desired organisms; these are incubated and the dishes periodically examined visually for growth.

These two methods have the inherent disadvantage that they consider only the plasticizer as the nutrient source, whereas other additives may also contribute. In addition, the severity of microbial growth may be lessened in the presence of other compounding ingredients. This may be due to a simple dilution effect or to some biocidal activity of the other additives, especially the stabilizers.

A more realistic approach is to prepare samples of the total PVC composition, with and without microbicide, in some convenient form such as film, sheeting, or coated fabric. These can then be cut into small squares and placed in Petri dishes containing a medium of mineral salts agar or malt agar; the medium is inoculated with spore suspensions as before, and the dishes are incubated and observed for growth on the specimens as a function of time. A disadvantage of this procedure is that it gives only a qualitative picture, the visual observation of growth, but it is valuable as a rapid screening method. To make it more informative, larger containers for the agar medium, such as Pyrex baking dishes, can be used so that larger specimens can be exposed. After inoculation and incubation, the samples can be removed and subjected to some quantitative measurements. Such properties as stiffness, tensile strength, ultimate elongation, electrical resistance, color, surface tack, or change in weight can be measured by standard procedures.

An interesting technique is the respirometric method that Siu and Mandels [181,182] developed and Burgess and Darby [183,184] modified. The latter correlated results with a direct weight loss method. This procedure uses differential manometers to measure oxygen absorbed by micoorganisms growing on PVC specimens in inoculated nutrient agar. Quantitative estimations of microbial susceptibility in as little as 4 days are indicated.

The microorganisms associated with degradation of PVC compounds belong to different groups of fungi and bacteria. Although tens of thousands of microorganisms are known, relatively few have been isolated from PVC compositions that have deteriorated in actual service. There is little agreement regarding the organisms to use in laboratory testing. Table 12 presents some that have been used by investigators in this field. Both pure cultures and mixed cultures are used.

Several test methods are used: exposing vinyl compounds in such various finished forms as film, coated wire and cable, coated fabrics, and fibers in a tropical room, on an outdoor test fence, immersed in the sea, or buried in soil beds. These procedures give valuable information on the fate of PVC compounds under actual use conditions. In addition, they readily permit the exposure of entire finished products or of samples large enough to permit quantitative measurement of a variety of chemical and physical properties.

Along with visual observations, such data are helpful in characterizing a PVC formulation, estimating its service life, and studying the mechanism of degradation. As an example, Baskin and Kaplan [185] used a 14-day soil burial test to study the mildew resistance of vinyl-coated fabrics. They used changes in breaking (tensile) strength and stiffness, measured with the Clark flexibility tester, as criteria of resistance. The stiffness test gave the more reliable results. A typical soil burial procedure is method 5762, U.S. Federal Specification CCC-T-191b.

VII. REODORANTS AND ODOR PREVENTION

A frequent problem with PVC compounds is the presence of undesirable odors. These vary in intensity and in type, depending on many factors, including quality and composition of raw materials, thermal history during processing, and conditions of use. The importance of the problem is largely a function of the nature of the end product; a malodor might be tolerated in an industrial or agricultural tarpaulin but would be quite unacceptable in upholstery or clothing. Among the odors detectable in various PVC compounds are those that have been described as musty, rancid, butyric, phenolic, and mercaptan-like.

Although plasticizers or their impurities or degradation products are frequently responsible for odor formation, other compounding ingredients may also contribute and should be considered when a solution to an odor problem is sought. These other ingredients include stabilizers, lubricants, fillers, and colorants. Trace residues of solvents from inks used to print and decorate PVC films and sheeting have also been a source of odor. Finally, microbial degradation may cause malodor formation.

A. Chemicals for Odor Control

Products used to neutralize a malodor or to replace it with a pleasant odor are known variously as odorants, deodorants, reodorants, masking agents, perfumes, and fragrances. The development of a suitable material is an art, and the specific chemicals employed are closely guarded proprietary secrets. Blends of components are frequently employed from a variety of chemical classes including esters, ethers, alcohols, aldehydes, ketones, acids, terpenes, and phenols.

Factors to be considered in compounding a suitable product include (1) the nature of the malodor and its intensity, (2) the requisite lasting power of the deodorant or odorant, which is dependent on its evaporation rate, which may in turn be affected by the chemical nature of plasticizers, (3) any special requirements, such as FDA sanction if the end use is in food packaging, and (4) whether the objective is merely to neutralize an unpleasant odor or to replace it with a detectable, pleasant scent. Pantaleoni [186] has presented an excellent review of the subject of industrial perfumery.

A reodorant for a specific PVC compound is best selected through cooperation with a producer of fragrances, due to the sophisticated nature of the art. A partial list of these manufacturers can be found in Table 14. To achieve the best results, the reodorant specialist requires information on the identity of all compounding ingredients, the time and temperature of processing, the nature of the finished product, and its intended use and life expectancy. To supply a detectable odor when one is desired in the finished product for increased sales appeal or special effect, a wide variety of scents are available. These include leather, cedar, citrus, floral, mint, and talcum powder. It is also possible merely to neutralize an unwanted odor. Concentrations of reodorants range from 0.01 to as much as 1%.

Instead of using deodorants or perfumes, one can approach the problem in another way, preventing the occurrence of the odor by chemical means.

One source of odor in ester plasticizers involves trace quantities of aldehydes and ketones in the alcohols used in preparation of the plasticizers. It is possible to minimize or eliminate this source by chemically treating the alcohols. Sodium borohydride has been used for this purpose, to reduce the carbonyl compounds to alcohols [187,188]. Lithium borohydride has also been suggested [189].

Sears and Darby [190] showed that unpleasant odors can develop due to oxidation of plasticizers during processing of PVC compounds (see Chapter 9, Vol 1). The type and severity of the odors were influenced by plasticizer structure. They were influenced in an unpredictable manner when mixtures of plasticizers were used in varying proportions. Plasticizers studied included phthalates, adipates, phthaloyl alkyl glycolates, hydrogenated terphenyls, alkylated aromatic hydrocarbons, and epoxidized soybean oil. The last, although functioning as a color stabilizer, caused odor formation in some formulations. Other additives that caused an increase in odor included some lead stabilizers and some pigments, especially those containing manganese, iron, and chromium. Inhibition of odor was accomplished by using two antioxidants, bisphenol-A and 2,6-di-*tert*-butyl-*p*-cresol. Small concentrations of partially hydrogenated terphenyls, alkylated aromatic hydrocarbons, and carbon black were also beneficial. The inhibitors were added to the PVC compound before processing.

Yngve [191] found that odor in vinyl resins, including homopolymers and copolymers of vinyl chloride, plasticized with esters, was prevented by incorporating carboxylic acid anhydrides into the compound. An example of a preferred substance was phthalic anhydride, used at 0.25 to 5% of the total weight of resin and plasticizer. A similar problem, for development in vinyl compounds containing halogenated plasticizers and pigments containing sulfur or selenium, was solved by Rossig [192] through the inclusion of 1 to 5% of a lactam, such as caprolactam.

Organotin stabilizers that contain sulfur, such as dibutyltin bis(lauryl mercaptide), are known to impart disagreeable mercaptan odors to PVC compositions. Leistner and Knoepke [193] have used peroxygen compounds to overcome this problem. Effective representatives of this class are sodium

TABLE 14 U.S. Producers of Reodorants

Alpine Aromatics, Inc., Metuchen, NJ

Dodge & Olcott, Inc., New York, NY

Fritzsche Brothers, Inc., New York, NY

Givaudan Corporation, New York, NY

Harwick Standard Chemical Company, Akron, OH

Mermix Chemical Company, Chicago, IL

Noville Essential Oil Company, Inc., North Bergen, NJ

Polak's Frutal Works, Inc., Middletown, NY

---

perborate, potassium persulfate, dicumyl peroxide, and benzoyl peroxide. They were used at levels ranging from 0.05 to 0.5%, based on the weight of resin, and were added to the PVC compound before processing.

Attack by microorganisms is another cause of odor in PVC compounds during service. The use of chemicals for controlling microbial growth was discussed in a previous section of this chapter.

B. Test Methods

No satisfactory instrumental method has been devised for measuring and evaluating odor. It is usually judged by the human olfactory sense, by an organoleptic panel using an arbitrary scale previously agreed on. Such a panel generally consists of five or more persons, preferably trained in odor detection and evaluation.

It is interesting that there are people, known as anosmics, whose sense of smell is seriously impaired or totally lacking. It is also important to recognize that odor is quite subjective and that a given scent may be regarded as acceptable or even pleasant by one individual and unpleasant by another.

A common technique in odor evaluation is to accelerate or increase formation of the odor by exposing the test specimens in closed containers to an elevated temperature or a high relative humidity or both before the organoleptic panel begins the judging.

A problem related to odor is the taste imparted to food in contact with a PVC article or very near it. An example of such an article is a refrigerator door gasket. One test method is to place a PVC specimen in a closed jar containing butter, which does not touch the specimen. After a suitable aging period, an organoleptic panel tastes the butter and evaluates it for any off-flavor.

VIII. MISCELLANEOUS ADDITIVES

A variety of chemical modifiers for altering specific properties or producing special effects has been disclosed in the patent literature. Included here are modifications of the moisture sensitivity and of optical, electrical, and rheological properties of PVC.

The moisture vapor permeability was reduced by adding 1.5% dicetyl ether [194]. Calcium oxide was added to a plastisol to absorb traces of water and prevent bubble formation during fusion [195].

An electrically conductive composition was produced by incorporating carbon and a powered metal, such as silver [196].

A reaction product of cupric chloride and an amino hydroquinone dialkyl ether was used to prepare a material that is transparent to visible light but absorbs in the near-infrared and ultraviolet parts of the spectrum [197]. Salts of 9-phenylfluoren-9-ols, at about 0.01%, absorbed infrared but transmitted visible light [198]. Both these compositions were suggested as television filters and eye-protective lenses.

Fluorescent polyvinyl chloride screens were made by incorporating calcium and bismuth sulfides [199].

A composition opaque to x-rays resulted from adding bismuth carbonate [200].

A light-sensitive composition, suitable for making photographic images, was obtained by incorporating zinc oxide in PVC [201]. The sensitivity of the composition was increased by including a silver compound, such as silver naphthenate [202]. A modification of this was a compound consisting of PVC and a minor amount of a Friedel-Crafts catalyst or a Friedel-Crafts anion progenitor such as hexachloroethane [203]. A different approach used a composition comprising a vinyl chloride polymer, an aluminum compound such as aluminum oxide, and a vinyl chloride telomer to increase the photosensitivity [204].

Gelled plastisols (plastigels) were prepared by incorporating an amine adduct of bentonite, a silica aerogel, or aluminum distearate [205] and an alkaline earth salt of a hydroxy fatty acid [206].

The flow properties of plastisols were modified to make them suitable for reverse roll coating by the addition of an aqueous paste containing a pentaalkalimetal pentaalkyl tripolyphosphate [207].

A thermosetting molding compound resulted from the addition of 1 to 15% of a polyamide to a polyvinyl chloride resin containing plasticizer and stabilizer [208].

The air-release and viscosity characteristics of plastisols were improved by including boron esters [209] and zirconium alcoholates [210].

Viscosity of plastisols was reduced by adding liquid aliphatic amines [211], nonionic surfactants [212], or amine salts [213].

Methacrylate diesters of polyethylene glycols were used as partial replacements for conventional plasticizers to produce fluid plastisols that convert to semirigid solids during fusion [214]. Fluid plastisols containing very high concentrations, up to 82%, of PVC resin were prepared by using triethylene glycol dimethacrylate at 3 to 12 phr to replace a portion of conventional plasticizer [215].

Materials resistant to ionizing radiation and suitable for wire insulation and gaskets were prepared for PVC, bis(2-ethylhexyl) maleate, and styrene. They were cross-linked by radiation to improve hardness, tensile strength, and solvent resistance [216].

A "hammer finish" decorative surface coating composition was produced from an organic film-forming coating resin, a metallic pigment, and an organopolysiloxane. Polyvinyl chloride was disclosed as one of the film formers [217].

REFERENCES

1. Mascia, L., *The Role of Additives in Plastics*, Edward Arnold, London, 1974, p. 1.
2. *American Heritage Collegiate Dictionary*, American Heritage Publishing Co., New York, 1969, p. 1106.
3. *American Heritage Collegiate Dictionary*, American Heritage Publishing Co., New York, 1969, p. 1003.
4. Mascia, L., *The Role of Additives in Plastics*, Edward Arnold, London, 1974, p. 2.
5. Molnar, N.M., *J. Am. Oil Chem. Soc.*, *51*, 84–87, 1974.
6. ASTM D 1893. Method of Test for Blocking of Plastic Film. 1972.
7. ASTM D 1894. Method of Test for Coefficients of Friction of Plastic Film. 1975.
8. Birks, A. M., *Plast. Technol.*, 131 (July 1977).
9. Rogers, J. K., *SPE J.*, *29*, 28–34 (Jan. 1973).
10. Mascia, L., *The Role of Additives in Plastics*, Edward Arnold, London, 1974, p. 106.
11. Schmidt, W. I., *Modern Plastics Encyclopedia*, McGraw-Hill, New York, 1984, p. 103.
12. Roger, J. L., *SPE J.*, *29*, 28–34 (Jan. 1973).
13. Cubera, M., *Plast. Compounding*, *6(2)*, 29 (1983).
14. Federal Test Method Standard No. 101B, Method 4046, Electrostatic Properties of Material (1969).
15. Roger, J. L., *SPE J.*, *29*, 28–34 (Jan. 1973); Mascia, L., *The Role of Plastics in Additives*, Edward Arnold, London, 1974, p. 107.
16. Chemicals and Additives Special Report '83, *Mod. Plast.*, *60*, 75 (Sept.1983); Sa-Go Associates Estimates, private communications, Jan. 1986.
17. Rogers, J. L., *Plast. Eng.*, *29*, 52–56 (Feb. 1973).
18. *Plast. News Briefs*, *3(21)* (Oct. 21, 1983).
19. Guyout, A., Michel, A., Michel, B., and Van Hoang, T., Vinyl Technol., *3*, 189 (Sept. 1981).
20. Gobstein, S., *Plast. Compounding*, 12 (Oct. 1983).
21. For a comprehensive general review of flame retardancy: Hendersen, R. R., and Wagner, G. M., in *Encyclopedia of Polymer Science and Technology* I, Interscience, Wiley, New York, 1967; Boyer, N. E., and Vazda, A. E. *SPE Trans.* *4(1)*, 45 (1964).
22. For a comprehensive review of PVC flame retardancy: O'Mara, N. M., Ward, W., Knechtges, D.P., and Meyers, R. J., *Flame Retardancy of Polymeric Materials*, Dekker, New York, 1973, chapter 3; Hamilton, J. P., Flame Retardants for Thermoplastics, Part III, Polyvinal Chloride, *Plast. Compounding*, 54 (Oct. 1978).
23. Hilado, C. J., *Flammability Handbook for Plastics*, Technomics, Stamford, Conn., 1969, p. 82–83.
24. PVC Flammability, *Minitec Reprint* (Jan. 19, 1977), p. 15.
25. *Flame Retardant Mechanisms Involving Nitrogen, Phosphorus, Chlorine, Bromine, and Antimony*, Brochure, Monsanto, St. Louis, 1970.
26. Michel, A., Sainrat, A., and Michel, B., *J. Vinyl Technol.*, *3*, 182 (Sept. 1981).
27. Mascia, L., *The Role of Additives in Plastics*, Edward Arnold, London, 1974, p. 159.
28. Combustibility, *Plast. World*, 42 (Nov. 1977).

29. Stilbert, E. K., Jr., Cummings, I. J., and Talley, S. P. (to Dow Chemical Co.), U.S. Patent 2,755,260 (July 1956).
30. Bonsignore, P. V., and Claassen, P. L., *Vinyl Technol.*, *2(2)*, 114 (1980).
31. Kaufman, S., and Yocum, N. M., Balancing Flame Retardancy and Low Temperature Brittleness Properties in PVC in Fire Retardants, *Proceedings of the 1975 International Symposium on Flammability and Fire Retardancy.*
32. Bonsignore, P. V., Fire Retardant Inorganic Additives and Fillers for Flexible PVC, *Minitec Proc.*, 32 (Jan. 19, 1977).
33. Sprague, R. W., *Systematic Study of Firebrake ZB as a Fire Retardant in PVC*, Part IV, *Alumina Trihydrate as a Synergist*, U.S. Borax Research Corp., Los Angeles, Jan. 1975.
34. *Kirk-Othmer Encyclopedia of Chemical Technology*, 3rd ed., Vol. 10, Wiley, New York, 1980, p. 357.
35. *Flame Retardant Mechanisms Involving Nitrogen, Phosphorus, Chlorine, Bromine, and Antimony*, Brochure, Monsato, St. Louis, 1970.
36. M & T Chemicals Tech. Bull. 177, 178, 1976.
37. M & T Chemicals Tech. Bull. 239, 1976.
38. Anzon America, Technical Data Sheet, ONCOR 23A, 1984.
39. Colloidal Antimony Oxide. Nyacol Products Tech. Bull, 1977.
40. Abrams, R. L., *Vinyl Technol.*, *3(4)*, 205 (1981).
41. Shen, K. K., and Sprague, R. W., *J. Vinyl Technol.*, *4(3)*, 120 (1982).
42. Bonsignore, P. V., Fire Retardant Inorganic Additives and Fillers for Flexible PVC, *Minitec Proc.*, 32 (Jan. 19, 1977).
43. Baseman, A. L., *Plast. Technol.*, *12(6)*, 37 (1966).
44. *Flame Retardant Mechanisms Involving Nitrogen, Phosphorus, Chlorine, Bromine, and Antimony*, Brochure, Monsato, St. Louis, 1970.
45. Baseman, A. L., *Plast. Technol.*, *12(6)*, 37 (1966).
46. *Modern Plastics Encyclopedia*, McGraw-Hill, New York, (1966), p. 472.
47. Cooper, R. S. (to Diamond Alkali Co.), U.S. Patent 2,664,441 (Dec. 29, 1953).
48. Birum, G. H. (to Monsanto Chemical Co.), U.S. Patent 3,058,941 (Oct. 16, 1962).
49. Leatherman, M., U.S. Patent 2,407,668 (Sept. 17, 1946).
50. Leatherman, M., U.S. Patent 2,439,395 (April 13, 1948).
51. Leatherman, M., U.S. Patent 2,439,396 (April 13, 1948).
52. Creely, J. W., and Cooke, T. F. (to America Cyanamide), U.S. Patent 2,549,059 (April 17, 1951).
53. Broatch, J. D. (to British Jute Trade Research Association), U.S. Patent 2,852,414 (Sept. 16, 1958).
54. Read, N. J. (to Associated Lead Manufacturers), U.S. Patent 3,014,000 (Dec. 19, 1961).
55. Hopkinson, H., U.S. Patent 2,610,920 (Sept. 16, 1952).
56. Arone, N. F. (to General Electric Co.), U.S. Patent 2,717,216 (Sept. 6, 1955).
57. Hobson, P. H. (to Chemstrand Corp.), U.S. Patent 2,949,437 (Aug. 16, 1960).
58. *America Burning*, National Commission on Fire Prevention and Control, June 1973. See also Hilado [23, p. 15].
59. Carlon Electrical Sciences Inc., Disclosure Papers CF-3, "The Case For Plastic Conduit versus Steel conduit in Construction," March 1985; Flax, S. "The Dubious War on Plastic Pipe," *Fortune*, February 7, 1983.

60. Guyout, A., Michel, A., Michel, B., and Van Hoang, T., *J. Vinyl Technol.*, 3, 189 (Sept. 1981).

61. Mitchell, L. C. (to Ethyl Corp.), U.S. Patent 3,821,151 (June 28, 1974).

62. McRowe, A. W. (to B. F. Goodrich Co.), U.S. Patent 3,822,234 (July 2, 1974).

63. Mitchell, L. C. (to Ethyl Corp.), U.S. Patent 3,846,372 (Nov. 5, 1974).

64. McRowe, A. W. (to B. F. Goodrich), U.S. Patent 3,862,086

65. Mathis, T. C., and Morgan, A. W. (to Monsanto), Bel. Patent 808,824, (Dec. 19, 1973).

66. Syntex Corp., German Patent 2,307,387 (Feb. 15, 1973).

67. Kroenke, W.J. (to B. F. Goodrich), U.S. Patent 3,883,480 (May 13, 1975).

68. Kroenke, W. J. (to B. F. Goodrich, U.S. Patent 3,883,482 (May 13, 1975).

69. *Flame Retardancy Smoke Reduction—for the First Time You Can Have Both—with Arapahoe DFR-121*, Tech., Bull., Arapahoe Chemicals Inc., Boulder Colo.

70. Bonsignore, P. V., and Claassen P. L., *J. Vinyl Technol.*, 2(2), 114 (1980).

71. Shen, K. K., and Sprague, R. W., *J. Vinyl Technol.*, 4(3), 120 (1982).

72. Moore, F. W., and Church, D. A., Molybdenum Compounds as Flame Retardants and Smoke Suppressants for Polymers, presented at the International Symposium on Flammability and Flame Retardants, Toronto, May 7, 1976.

73. Michel, A., Sainrat A., and Bert, M., *J. Vinyl Technol.*, 3 (3), 182 (1981).

74. Combustibility, *Plast. World*, 42 (Nov. 1977).

75. Keeney, C. N., *Vinyl Technol.*, 3, 172 (1949).

76. ASTM Standards, part 35, D-883-75a, Nomenclature Relating to Plastics.

77. Oakes Mixer, E. T. Oakes Corp., Islip, N. Y.; Euromatic, Stork America Corp., New Canaan, Conn.

78. Schmidt, P., and Polte, A., *Kunststoffe 57*, 25 (1967).

79. Batsch, P. V. (to U.S. Rubber Co.), U.S. Patent 2,861,963 (Nov. 25, 1958).

80. Maltenfort, M. S. (to Chemical Research Association), U.S. Patent 2,966,470 (Dec. 27, 1960).

81. Deal, K. M., Morris, D. C., and Waterman, R. R., *Ind. Eng. Chem, Prod. Res. Dev.*, 3, 209 (1964); *Chem. Eng. News*, 63 (Sept. 23, 1963).

82. Waterman, R. R., Deal, K. M., and Whitman, P. A. (to R. T. Vanderbilt Co.), U.S. Patent 3,288,729 (Nov. 29, 1966).

83. Fomade, R. T., Vanderbilt Co., Inc, Norwalk, Conn. 06856.

84. Waterman, R. R., and Morris, D. C. (to R. T. Vanderbilt Co.), U.S. Patent 3,301,798 (Jan. 31, 1967).

85. *Chem. Eng. News*, 37(36), 42 (1959).

86. Hansen, R. H., *SPE J.*, 18, 77 (1962).

87. Osberg, E. V., *India Rubber World*, 97(37—39), 48 (1937).

88. Lindemann, H. (to Lonza Elect. & Chem. Works), U.S. Patent 2,751,627 (June 26, 1956).

89. Lindemann, H. (to Lonza Elect. & Chem. Works), U.S. Patent 2,829,117 (April 1, 1958).

90. Fuchs, O. (to Dynamit A. G.), German Patent 1,065,169 (Sept. 10, 1959).

91. Fuch, O. (to Dynamit A. G.), German Patent 1,060,508 (July 16, 1959).

92. Lindemnn, H., *Kunststoffe, 48,* 194 (1958).

93. Brunner, E., *Tech. Rundsch., 44,* 21 (1962).

94. Continuous Production of Vinyl Foam, *Br. Plast., 29,* 86 (1956); Vinyl Foam Process Finds Many Takers, *Chem. Eng.,* 122-126 (July 1956); Crowdes, G. J., Jr. *Mod. Plast. 34,* 117, 188, 212 (July 1957).

95. Schwenke, E. H. (to Elastomer Chemical Corp.), U.S. Patent 2,666,036 (January 12, 1954).

96. Smythe, W. J., and Mueller, E. (to Union Carbide Corp.), U.S. Patent 3,113,116 (Dec. 3, 1963).

97. Smythe, W. J. (to Union Carbide Corp.), U.S. Patent 2,881,141 (April 7, 1959).

98. Dennis, I., U.S. Patent 2,763,475 (Sept. 18, 1956).

99. *Encyclopedia of Polymer Science and Technology,* Vol. 2, Interscience, Wiley, New York, 1965, p. 534.

100. Lineberry, D. D. (to Union Carbide Corp.), U.S. Patent 3,052,643 (Sept. 4, 1962); Farbwerke Hoechst A. G., British Patent 890,398 (Feb. 28, 1962).

101. Peterson, R. H., and Asbeck, W. K. (to Union Carbide Corp.), U.S. Patent 3,122,515 (Feb. 25, 1964).

102. Hawkins, J. G. (to Whiffen & Sons), British Patent 818,224 (Aug. 12, 1959).

103. Sprague, G. R., and Scantlebury, F. M. (to B.F. Goodrich Co.), U.S. Patent 2,737,503 (March 6, 1956).

104. Sorbo Ltd., British Patent 728,666 (April 27, 1955).

105. Sarge, T. W., and Justin, F. H. (to Dow Chemical Co.), U.S. Patent 2,695,427 (Nov. 30, 1954).

106. Ten Broech, W. T., Jr. (to Wingfoot Corp.), U.S. Patent 2,478,879 (Aug. 9, 1949).

107. Bush, T. F. (to B.F. Goodrich Co.), U.S. Patent 2,909,493 (Oct. 20, 1959).

108. Jones, D. R., *Br. Plast.,* 248–250 (May 1962).

109. Farbenfabriken Bayer, A. G., German Patent 1,184,951 (July 28, 1962).

110. Vokousky, W. J. (to B.F. Goodrich Co.) U.S. Patent 3,084,127 (April 2, 1963); see also Tech. Service Bull. FF1202F/TF 915, B.F. Goodrich Chemical Co., Cleveland, Ohio.

111. The following papers deal in comprehensive manner with the chemistry and technology of organic blowing agents: Reed, R. A., *Plast. Progr. 1955,* 51 (1956); Reed, R. A., *Br. Plast. 33(10),* 468 (1960); Scheurlen, H. A., *Kunststoffe, 47,* 446 (1957); Lasman, H. R., *Encyc. Polym. Sci. Technol. 2,* 532 (1965); Heck R. L., III, *Plast. Compounding, 1(4),* 52 (Dec. 1978); *Plast. Compounding, 3(6),* 64 (1980).

112. Hunt, M. (to E.I. duPont de Nemours & Co.), U.S. Patent 2,471,959 (May 31, 1949); Burk, R. E. (to E.I. duPont de Nemours & Co.), U.S. Patent 2,500,023 (March 7, 1950).

113. Landler, Y. (to Kleber Colombes), U.S. Patent 3,200,089

(Aug. 10, 1965); Landler, Y., *J. Cell. Plast.* 3(9), 404 (1967);
Mod. Plast. 45(2), 94 (1967).

114. Lober, F., Bogemann, M. and Wegler, R. (to Farbenfabriken
Bayer, A. G.), U.S. Patent 2,626,933 (Jan. 27, 1953).

115. Hunter, B. A., and Schoene, D. L., *Ind. Eng. Chem.*, 44, 119
(1952).

116. Schoene, D. L. (to U.S. Rubber Co.), U.S. Patent 2,552,065
(May 8, 1951).

117. Wick, G., Homann, D., and Schmidt, P., *Kunststoffe*, 49
383 (1959).

118. Celogen OT, Bull. 200-B90, Uniroyal Chemical, Naugatuck, Conn;
Nitropore OVSH, Tech. Bull. ONE-01, National Polychemicals,
Inc., Wilmington, Mass.

119. Clark, L., Grabowsky, T., and Poshkus, A. (to Armstrong Cork Co.)
U.S. Patent 2,849,028 (Aug. 26, 1958); Clark, L. (to Armstrong
Cork Co.), U.S. Patent 2,873,259 (Feb. 10, 1959).

120. Hunter, B. A. (to U.S. Rubber Co.), U.S. Patent 3,152,176
(Oct. 6, 1964).

121. Hunter, B. A. (to U.S. Rubber Co.), U.S. Patent 3,235,519
(Feb. 15, 1966); Celogen RA, Bull. 200-B67, Uniroyal
Chemical, Naugatuck, Conn.

122. Reed, R. A., *Plast. Progr. 1955*, 51 (1956); Lasman, H. R.,
Mod. Plast. Encyc., 41-1A, 369 (1967).

123. Subpart F, 121.2550, 121.1085, 121.2562.

124. Curtis, W. B., and Hunter, B. A. (to U.S. Rubber Co.)
Patent 2,806,073 (Sept. 10, 1957).

125. BIOS Report 1150, No. 22, 21–23.

126. National Polychemicals, Inc., Wilmington, Mass., Bull. PKB-2,
Nov. 6, 1958; Barnhart, R. R., *Comp. Res. Rep. 38*, Celogen
AZ, 2-3 (1958); Reed, R. A., *Br. Plast.*, 33(10), 471 (1960).

127. Nass, L. I., *Mod. Plast.* 40(7), 151; 40(8), 127 (1963).

128. Genitron Blowing Agents, *Tech. News*, Whiffen & Sons,
Loughborough 1963, p. 12.

129. National Polychemicals, Inc., Wilmington, Mass., Bull. OKE-44;
Harris, E. B., 1(2), 296 (1965).

130. Advance Div., Carlisle Chem. Works, Inc. (now Akzo/Interstab
Chemicals), New Brunswick, N. J., ABC Blowing Agent Catalyst.

131. Union Carbide, Bound Brook, N. J.

132. Lally, R. E., and Alter, L. M., *SPE J.*, 23(11), 69 (1967).

133. Nairn, R. F., Harkins, J. C., Ehrenfeld, F. E., and Tarlow, H.
(to Congoleum-Nairn), U.S. Patent 3,293,094 (Dec. 20, 1966).

134. Lasman, H. R., and Blackwood, J. C., *Plast. Technol.*, 9(9),
37 (1963).

135. National Polychemicals, Inc., Wilmington, Mass., Bull. OKE-42.

136. Hartmen, T. V., Kozlowski, R. R., and Podnar, T., *Foams from
Vinyl Dryblend Powders*, SPE Retec, Akron, Ohio, 1965.

137. Cram, D., Lavender, M., Reed, R. A., and Schoffield, A., *Br.
Plast. 24* (Jan. 1961); National Polychemicals, Inc., Wilmington,
Mass., Bulletins OKE-02, OKE-46, OKE-47, and OKE-48; Werner, A.,
Mod. Plast. 39(2), 135 (1961); Hackert, A., *Kunststoffe*, 52(10), 624
(1962); Imperial Chemical Industries, Ltd., Plastics Div., IS 881;
Lavender, M., *Plastics*, 27(299), 65 (1962); Meyer, R. J., and Esarove,
D. I., *Plast. Technol*, 5(3), 27 (1959); General Tire & Rubber Co.,
British Patent 833,416 (April 21, 1960); Roggi, P. E., and Chartier,

138. R. A. (to U.S. Rubber Co.), U.S. Patent 2, 964,799 (Dec. 20, 1960);
      Smith P., *Melliand Textile Rep., Int. Ed.*, (*3*), (1963).
138. Conger, R. P., *SPE J.*, *24(3)*, 43 (1968).
139. Lanthier, P. V., and Lasman, H. R., Preparation of Low Density
      Cellular Thermoplastics, SPE Retec, Palisades Section, Nov. 1964.
140. Esarove, D., and Meyer, R. J., *Plast. Technol. 5(3)*, 27 (1959).
141. Meyer, R. J., and Esarove, D., *Plast. Technol.*, *5(4)*, 32 (1959);
      Dilley, E. R., *Plast. Inst. Trans.*, *34(109)*, 17 (1966).
142. Rhodes, T. J. (to U. S. Rubber Co.), U.S. Patent 2,907,074
      (Oct. 6, 1959); Streed, D. D. M., and Luxenberger, E. A. (to
      U. S. Rubber Co.), U.S. Patent 2,974,373 (Mar. 14, 1961);
      Hickler, W. R., and Powell, J. L. (to B. F. Goodrich Co.), U.S.
      Patent 2,939,180 (June 7, 1960); Strickhouser, S. L., and Van
      Twisk, R. J. (to U.S, Rubber Co.), U.S. Patent 2,917,749
      (Dec. 22, 1959).
143. Brillinger, T. H. (to W. R. Grace & Co.), U.S. Patent 3,032,826
      (May 8, 1962).
144. Nairn, R. F. (to Congoleum-Nairn), U.S. Patent 2,961,332
      (Nov. 22, 1960).
145. Hill, H. A. (to National Polychemicals), U.S. Patent 3,141,002
      (July 14, 1964).
146. Expandex 177, Tech. Bull. OOR-01, National Polychemicals, Inc.,
      Wilmington, Mass.
147. Shepard, C. S., Schack, N. H., and Mageli, O. L., *J. Cell. Plast.*,
      2(1), 97 (1966).
148. ASTM D1715. Method of Test for Gas Evolved from Chemical Blowing
      Agents for Cellular Plastics.
149. Hansen, R. H., *SPE J.*, *181*, 80–82 (1962).
150. Benning, C., *Plast. Technol.*, *13(4)*, 56 (1967).
151. Lasman, H. r., and Blackwood, J. C., *Plast. Technol.*, *9(9)*, 37
      (1963).
152. Girard, T. A., and Koda, C. F., *Mod. Plast. 36(10)*, 148 (1959).
153. Yeager, C. C., *Plast. World*, *20(12)*, 14 (1962).
154. Scullin, J. P., Girard, T. A., and Koda, C. F., *Rubber Plast. Age*,
      46(3), 267 (1965).
155. Yeager, C. C., Environmental Effects on Flexible Vinyl Systems,
      SPI Vinyl Film & Sheeting Meeting, Nov. 7, 1969.
156. Wessel, C. J., *SPE Trans.*, *4(3)*, 193 (1964).
157. Scullin, J. P., Dudarevitch, M. D., and Lowell, A. I., *in Encyclopedia
      of Polymer Science and Technology*, Vol.2, Interscience, Wiley, New
      York, 1965, p. 379.
158. Greathouse, G. A., and Wessel, C. J., eds., *Deterioration of Materials:
      Causes and Preventative Techniques*, Reinhold, New York, 1954.
159. Cadmus, E. L., *Additives For Plastics*, Vol. 1, Academic Press,
      New York, 1978, p. 219.
160. Rosato, D. V., and Schwartz, R. T., Environmental Effects on Poly-
      meric Materials, Vol I, Wiley & Sons Inc., New York, p.990, 1968.
161. Wessel, C. J., *Time Dependent Effect in Plastics Materials: Biode-
      terioration*, Plastics Institute of America, Dec. 3, 1963.
162. Brown, A. E., *Mod. Plast. 23(8)*, 189 (1946).
163. Stahl, W. H., and Pessen, H., *Appl. Microbiol.,1(1)*, 30 (1953).
164. Berk, S., Ebert, H., and Teitell, L., *Ind. Eng. Chem.*, *49(7)*,
      1115 (1957).
165. Hueck-van der Plas, E. H., *Plastica*, *13*, 1216 (1960).

166. Johnson, O., Krog, N., and Poland, J. L., *Chem. Week, 92(21)*, 117; *92(22)*, 55 (1963).

167. Scullin, J. P., Dudarevitch, M. D., and Lowell, A. I., in *Encyclopedia of Polymer Science and Technology*, Vol. 2, Interscience, Wiley, New York, 1965, p. 379.

168. Field, W. E. (to Monsanto Chemical Co.), U.S. Patent 2,567,905 (Sept. 11, 1951).

169. Malone, R. W. (to Monsanto Chemical Co.), U.S. Patent 2,567,910

170. Darby, J. R. (to Monsanto Chemical Co.), U.S. Patent 2,632,746 (Mar. 24, 1953).

171. Darby, J. R. (to Monsanto Chemical Co.), U.S. Patent 2,632,747 (Mar. 24, 1953).

172. Darby, J. R. (to Monsanto Chemical Co.), U.S. Patent 2,689,837 (Sept. 21, 1954).

173. Baker, P., *Plast. Compounding*, 35 (Oct. 1978).

174. Teller, W. K. (to Weco Products Co.), U.S. Patent 3,005,720 (Oct. 24, 1961).

175. Gay, F. J., and Wetherly, A. H., *Laboratory Studies of Termite Resistance*, IV. *The Termite Resistance of Plastics*, Australia Commonwealth Scientific and Industrial Research Organization, Div. of Entomology, Tech. Pap. No. 5, 1962, p.31.

176. Wessel, C. J., *SPE Trans.*, *4(3)*, 193 (1964).

177. Roncoroni, A. (to S.p.A. Fibre Tessili Artificiali), U.S. Patent 2,861,965 (Nov. 25, 1958).

178. Welch, J. F., and Duggan, E. W., *Mod. Packaging, 25(6)*, 130 (1952).

179. Connolly, R. A., *Mater. Res. Stand.*, *3*, 193 (1963).

180. Carlston, E. F., and Whiting, L. R. (to the United States of America), U.S. Patent 2,592,655 (April 15, 1952).

181. Siu, R. G. H., and Mandels, G. R., *Text. Res. J. 20*, 516 (1950).

182. Mandels, G. R., and Siu, R. G. H., *J. Bacteriol. 60*, 249 (1950).

183. Burgess, R., and Darby, A. E., *Br. Plast. 37(1)*, 32 (1964).

184. Burgess, R., and Darby, A. E., *Br. Plast. 38(3)*, 165 (1965).

185. Baskin, A. D., and Kaplan, A. M., *Appl. Microbiol.*, *4*, 288 (1956).

186. Pantaleoni, R., *Chem. Eng. News, 31(17)*, 1730 (1953).

187. Ventron Corporation, Metal Hydrides Division, Technical Booklet, *Process Stream Purification through Hydride Chemistry*.

188. Wise, R. H. (to Standard Oil Co.), U.S. Patent 2,867,651 (Jan. 6, 1959).

189. Dimler, W. A., and Schetelich, A. A. (to Esso Research and Engineering Co.), U.S. Patent 2,957,023

190. Sears, K., and Darby, J. F., *SPE J. 18(6)*, 671 (1962).

191. Yngve, V. (to Bakelite Corp.), U.S. Patent 2,394,417 (Feb. 5, 1946).

192. Rossig, L. (to Farbenfabriken Bayer A. G.), German Patent 1,035,895 (Aug. 7, 1958).

193. Leistner, W. E., and Knoepke, O. H. (to Argus Chemical Corp.), U.S. Patent 3,037,961 (June 5, 1962).

194. Radcliffe, M. R. (to the Firestone Tire and Rubber Co.), U.S. Patent 2,435,464 (Feb. 3, 1948).

195. Peciura, L. (to Stubnitz Greene Corp.), U.S. Patent 2,956,976 (Oct. 18, 1960).

196. Louis, A. S. (to Myron A. Coler), U.S. Patent 3,003,975 (Oct. 10, 1961).

197. Mahler, J. (to American Optical Co.), U.S. Patent 2,816,047 (Dec. 10, 1957).
198. Coleman, R. A., and Susi, P. V. (to American Cyanamid Co.), U.S. Patent 3,000,833 (Sept. 19, 1961).
199. Nanbu, H. (to Nagahama Resins Co.), Japanese Patent 11,535 ('60) (Aug. 19, 1960).
200. Telegraph Construction and Maintenance Co., British Patent 686,445 (Jan. 28, 1953).
201. Elliot, S. B. (to Ferro Corp.), U.S. Patent 2,772,158 (Nov. 27, 1956).
202. Elliot, S. B. (to Ferro Corp.), U.S. Patent 2,772,159 (Nov. 27, 1956).
203. Elliot, S. B. (to Ferro Corp.), U.S. Patent 2,789,052 (April 16, 1957).
204. Ogden, F. F. (to Monsanto Chemical Co.), U.S. Patent 2,046,137 (July 24, 1962).
205. Severs, E. T., and Frechtling, A. C. (to Union Carbide and Carbon Corp.), U.S. Patent 2,753,314 (July 3, 1956).
206. Patton, T. C., and Hall, F. M. (to the Baker Castor Oil Co.), U.S. Patent 2,794,791 (June 4, 1957).
207. White, W. H. (to United States Rubber Co.), U.S. Patent 2,831,824 (April 22, 1958).
208. Hogg, W. H., Coldfield, S., and Mobberley, W. (to Dunlop Rubber Co.), U.S. Patent 2,851,735 (Sept. 16, 1958).
209. Olson, H. M., U.S. Patent 2,912,400 (Nov. 10, 1959).
210. Olson, H. M. (to the Harshaw Chemical Co.), U.S. Patent 2,932,624 (April 12, 1960).
211. Klein, D. X., and Curgan, M. N. (to Heyden Chemical Corp.), U.S. Patent 2,548,433 (April 10, 1951).
212. Klein, D. X., and Curgan, M. N. (to Heyden Chemical Corp.), U.S. Patent 2,657,186 (Oct. 27, 1953).
213. Sims, H. J. (to Rohm & Haas Co.), U.S. Patent 2,810,703 (Oct. 22, 1957).
214. Burt, S. L. (to Union Carbide and Carbon Corp.), U.S. Patent 2,618,621 (Nov. 18, 1952).
215. Cornell, J. A. (to Sartomer Resins), U.S. Patent 3,066,110 (Nov. 27, 1962).
216. United States Rubber Company, British Patent 833,610 (April 27, 1960).
217. Hedlund, R. C. (to Dow Corning Corp.), U.S, Patent 2,884,388 (April 28, 1959).

Author Index

Italic numbers give the page on which the complete reference is cited.

Abrams, R. L., 611[40], *643*
Ackerson, E. R., 234[285], *260*
Alexander, C. H., 157[63], 165
    [94], *253, 254*
Ali, M. D., 148[25], *252*
Allen, E., 584[52], *600*
Allen, T., 524[45], *543*
Alter, L. M., 628[132], *646*
Amagi, Y., 448[97], *498*
Anagnostopoulos, C. E., 173[110],
    174[115], *254*
Anastasiu, S., 217[239], 237[239],
    258
Anderson, D. F., 70[40], 80[51],
    139, 140, 186[154], 215
    [154], *255*
Andrasik, I. J., 177[138], *255*
Andreasen, A. H., 525[53], *543*
Andrews, C. W., 231[278], *260*
Andrews, K. E., 370[25], *390*
Andrews, L., 525[52], *543*
Arone, N. F., 614[56], *643*
Artemova, T. G., 532[83], *544*
Asbeck, W. K., 619[101], *645*
Aston, J. E., 536[93], *544*
Atkins, J. T., 522[37], *543*
Aubrey, N. E., 419[42], *497*
Augustyn, E. J., 517[14], *542*

Baer, E., 436[80], *498*
Baer, M., 430[58], *497*
Bailer, J. C., 547[1], *598*

Baker, P., 634[173], *648*
Bakker, J., 145[16], 146[16], 147
    [16], *251*
Balinth, I. J., 159[84], *253*
Bargellini, F., 147[19], *251*
Barnhart, R. R., 627[126], *646*
Baronin, G. S., 532[83], *544*
Barrer, R. M., 535[92], *544*
Baseman, A. L., 611[43], 613[45],
    643
Baskin, A. D., 638[185], *648*
Batsch, P. V., 617[79], *644*
Bauer, W. H., 155[53], 217[247],
    252, 259
Baum, B., 554[10], *598*
Beach, L. K., 180[145], *255*
Beati, E., 467[132,133], 468[133],
    469[133], *499*
Benning, C., 630[150], *647*
Berens, A. R., 392[2], 438[83],
    496, 498
Berg, R. M., 160[311], *261*
Berg, S., 525[56], *543*
Bergen, H. S., 147[20], 153[20],
    251
Berk, S., 190[174,175], *256,* 633
    [164], 637[164], *647*
Bernardo, J. J., 150[39], *252*
Berndsten, N., 392[4], 393[4], *496*
Berndtsen, N., 442[89], 467[89],
    498
Bert, M., 49[11], *138,* 616[73], *644*
Best, C. E., 60[22], *139*
Bhatnagar, V. M., 517[18], *542*

653

Schmidt, P., 617[78], 626[117], *644, 646*

Schmidt, W. I., 604[11], *642*

Schoene, D. L., 626[115,116], *646*

Schoffield, A., 628[137], *646, 647*

Schreiber, H. P., 147[21,22], 153 [21], *251*

Schreider, H. P., 173[111], *254*

Schwarcz, J. M., 517[14], *542*

Schwartz, P. S., 177[128], 196 [128], *255*

Schwenke, E. H., 619[95], *645*

Scott, J. L., 121[73], *141*

Scott, R. L., 298[11], *390*

Scullin, J. P., 190[176], 246[313], *256, 260*, 631[154,157], 633 [167], *647, 648*

Searl, A. H., 230[275], *260*

Sears, J. K., 144[1], 148[1], 150 [1,40], 153[1], 155[1], 157 [1], 160[1], 161[1], 163[1], 170[1], 173[1], 175[1], 188 [160], 208[1], 210[1], 212 [1], 218[1], 220[1], 222[1], 226[1], 228[1], 231[1], 232 [1], *251, 252, 256*

Sears, K., 173[112], 174[112], *254*, 639[190], *648*

Semon, W. L., 46[8], *138*, 144 [4], 208[4], 210[4], 219 [4], 220[4], 223[4], 227 [4], 229[4], 233[4], *251*

Serdynsky, E. D., 461[122], *499*

Severs, E. T., 641[205], *649*

Seymour, R. B., 203[223], *258*

Shah, A. C., 195[204], *257*

Shah, M. R., 462[124], *499*

Shah, N. A., 465[127], *499*

Shah, P. L., 324[16], *390*

Shakirzyanova, S. S., 148[38], *252*

Shanks, H. G., 523[42], *543*

Sharatt, E., 525[57], *543*

Shen, K. K., 611[41], 616[71], *643, 644*

Shepard, C. S., 628[147], *647*

Sheth, K. B., 468[145], *500*

Shikata, T., 527[76], *544*

Shoets, L. V., 521[34], *543*

Shoulberg, R. H., 424[51], 425 [51], 428[51], *497*

Sidman, K. R., 177[128], 196 [128], *255*

Siegmann, A., 436[80], *498*

Silverman, A., 431[67], 432[67], 433[67], *497*

Silverman, B. H., 246[313], *260*

Simek, J., 470[159], *500*

Simha, R., *544*

Simpson, J. E., 574[41,42], *599*

Sims, H. J., 641[213], *649*

Sinclair, J., 524[47], *543*

Sing, K. S. W., 526[69], *544*

Siu, R. G. H., 637[181,182], *648*

Skrada, K. A., 470[163], 485[163], *500*

Small, 300[9], 301[9], 302[9], *390*

Smallwood, H., 535[91], *544*

Smietanska, G., 188[163], *256*

Smith, F. M., 569[37], *599*

Smith, H. V., 60[19], *139*

Smith, P., 628[137], *646, 647*

Smith, P. J., 61[27], 67[27], *139*

Smith, P. V., Jr., 225[257], *259*

Smith, R. R., 431[59], *497*

Smith, W. P., 158[80], 192[80], *253*

Smyth, H. F., Jr., 196[208], *257*

Smythe, W. J., 619[96,97], *645*

Sprague, G. R., 621[103], *645*

Sprague, R. W., 611[33,41], 616 [71], *643, 644*

Spreafico, C., 448[96], *498*

Stabenow, J., 436[78,79], *498*

Staff, C. E., 190[173], 217[232], *256, 258*

Stahl, W. H., 633[163], 637[163], *647*

Standen, A., 510[3], 519[28], *542*

Stapfer, C. H., 370[24], *390*

Steinkkopff, D., 237[286], *260*

Stepak, J., 470[159], *500*

Stepek, J., 185[153], *255*

Sternstein, S. S., 431[65,67], 432 [67], 433[67], *497*

Stilbert, E. K., Jr., 610[29], *643*

Stirton, R. I., 217[233], 242[233], 247[306], *258, 261*

Stoloff, A., 121[74,75], *141*, 460 [116], 470[163], 485[163], *499, 500*, 590[60,61], *600*

St. Pierre, L. E., 161[91], *254*

Streed, D. D. M., 628[142], *647*

Strella, S., 431[64], *497*

Stresinka, J., 469[153], *500*

Strickhouser, S. L., 628[142], *647*

Subject Index

674   *Subject Index*

Rodent repellants, 636
Roll release test, 323—324
Rubber toughening in PVC, 430—448

Secondary plasticizers, 6—7
Secondary stabilizers, 47
Shapes of fillers, 523—524
Shear burn test, 341
Sheets:
  calendered, 488—489
  exterior, 481—486
  extruded, 488—489
  lubricant selection for, 387—390
  stabilizers for, 95—96
Siding:
  stabilizers for, 95
  vinyl, 481—483
Silica gels, precipitated, 513—514
Silicas:
  precipitated, 513—514
  pyrogenic, 516—517
Silicates, colored, 563
Silicon oxide pigments, 554
Size distribution of fillers, 523—524
Slip additives, 603—604
Smoke suppressants, 2, 614—616
Solubility of lubricants, 297—298
  parameters for, 298—301
Specialty plasticizers, 7
Spherical fillers, composite
  moduli for, 536—537
Spiral flow test, 337—340
Stabilizer lubricant one-packs, 76
Stabilizers, 2, 8, 45—141
  dominant generic heat stabilizer
    classes, 55—73
    lead, 8, 55—60
    mixed metals, 67—73
    organotins, 60—67, 68, 70
  emerging stabilizer classes,
    73—76
    antimony mercaptides, 73—76
    stabilizer lubricant one-packs,
      76
  historical development of, 51—55
  individual applications, 94—96
  interactions between lubricants
    and, 368—370
  light stabilizers, 81—84

[Stabilizers]
  necessity of stabilization, 47
  organic synergists in stabilizer
    compounds, 76—81
  for plasticized vinyl compounds,
    4, 7—8
  properties of ideal stabilizer,
    49—51
  for rigid vinyl compounds, 31,
    32—33
  selection criteria for, 85—91
  selection parameters for, 52—53
  selection techniques for, 91—94
  stabilizer classification and
    world markets, 47—49, 50
  stabilizer optimization, 51
  testing and evaluation, 97—138
    chemical resistance testing,
      133—137
    color development and measure-
      ment, 97—109
    compatibility testing, 128—130
    exudation and migration, 130—132
    fogging and volatility testing,
      132—133
    heat stability, 109—115
    light stability and weather
      resistance, 115—128
    quality control testing, 137—138
  worldwide consumption of (1983),
    49, 50
  worldwide production of (1983),
    48
Stain resistance, 9, 24—25
  plasticizer effect on, 194—196
Stiffness of rigid vinyl compounds,
  39
Stress-strain properties of fillers,
  538—539
Strontium chromate pigments, 563
Structure of lubricants, 294—297
Surface treatment of fillers, 526—527

Talc, 9, 508—510
Tensile strength, 9, 12
Titanium dioxide:
  for fillers, 518
  for pigments, 554
Toxicity, 9, 22—23

APPLIED
FINITE
MATHEMATICS

APPLIED FINITE MATHEMATICS

Second Edition

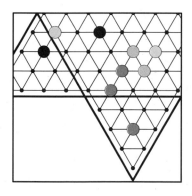

Alan Hoenig

John Jay College, City University of New York

HOUGHTON MIFFLIN COMPANY   Boston   Toronto
Geneva, Illinois   Palo Alto   Princeton, New Jersey

Sponsoring Editor: *Maureen O'Connor*
Senior Associate Editor: *Robert Hupp*
Senior Project Editor: *Cathy Labresh Brooks*
Production/Design Coordinator: *Sarah Ambrose*
Senior Manufacturing Coordinator: *Priscilla Bailey*
Marketing Manager: *Michael Ginley*

Cover designer: *Dick Hannus*
Cover image: *William Whitehurst*/The Stock Market

Alan Hoenig has been a Professor of Mathematics within the City University of New York since 1979. He earned his B.S. degree from Yale University and his Ph.D. from Harvard University. He has published many papers in Applied Mathematics, and seven books on computers and Finite Mathematics. He lives near New York City with his wife, who is a refugee from the former Soviet Union, and his two children.

Printed in the U.S.A.

Library of Congress Catalog Card Number: 93-78672

ISBN Numbers:
Text: 0-395-63778-3
Exam Copy: 0-395-69298-9
Printed Test Bank: 0-395-63781-3
Solutions Manual: 0-395-63779-1
Student's Solutions Manual: 0-395-63780-5

3456789-DH   00 99 98

CONTENTS

Part IV   PREDICTION AND PLANNING

PREFACE

Applied Finite Mathematics provides a rich and varied survey of topics generally taught in an introductory course taken by students majoring in management, life sciences, and social sciences. The only prerequisite is one and a half to two years of high school algebra or its equivalent. The presentation of material emphasizes a balance of methods and concepts, and features a clear, readable writing style that appeals, above all, to a student's intuition.

Finite mathematics courses present an enormous diversity with respect to content, coverage, and emphasis. Because of this, Houghton Mifflin and I selected a wide range of topics on the basis of my own classroom experience and extensive curriculum surveys done by the publisher. However, the final selection of topics was done with respect to the meaning of finite mathematics: *the application of these techniques allows managers and supervisors to make the best use of their valuable resources*, be these resources measured as time, money, information, or material goods.

- *Linear Programming*   With linear programming, we optimize our use of material resources.

- *Statistics*   Using statistical methods, we sift through large quantities of information to make decisions based on underlying trends.

- *Probability Theory*   We use probability in the opposite extreme—to help us make decisions when we have too little information.

- *Financial Mathematics* enables us to plan and use money in an optimal fashion.

- *Network Methods* (in a chapter new to this edition) help us run projects to use available time in the best possible way.

Goals of the Text

Mastery of these and other topics should serve as goals for a foundation upon which to build. Students should be able to absorb some of the excitement of discovering patterns and of experiencing the intellectual energy that motivates quality mathematics. So much of finite mathematics is genuinely fascinating; students should have the opportunity to experience this for themselves. This text provides such opportunities in a variety of ways, using proven and successful pedagogical techniques.

Pedagogical Features

- *Flexibility*   The text has been carefully designed to accommodate a variety of course needs and uses. Each chapter has been divided into sections of approximately the same length, and section subheads are used frequently to form natural blocks of teaching material.

- *Writing Style and Emphasis*   Finite mathematics is a unique course in that students can understand essentially all of the important ideas intuitively, bypassing the need for elaborate theoretical discussions. I have gone to great lengths to insure that the ideas are accessible. The writing style is clear, uncluttered, and direct. Precise mathematical language is used, though without excessive formalism. Key terms appear in the margin at their initial introduction to make reference by the student as straightforward as possible.

- *Motivation*   Each topic is discussed and reinforced to motivate the student in one of a variety of ways: appealing to intuition, utilizing interesting introductions, presenting graphical interpretations, and employing realistic and plausible applied problems to develop topics. I explain the ideas underlying a computation, and I encourage students to explore an idea.

- *Format*   The text layout, design, and use of second color are designed to provide a highly readable (and attractive) textbook for both student and instructor.

- *Graphics*   There are over 450 graphs, tables, and figures in the text to enhance discussions pedagogically and to help students visualize a problem for solution.

- *Examples*   Over 700 carefully chosen worked examples and skill enhancers are used to motivate and illustrate each topic or problem-solving strategy.

- *Exercises*   All exercise sets are carefully graded; interesting applications and challenging problems are included. Over 3000 exercises have been provided, all designed to either build competence or enhance understanding. The end-of-chapter exercises thoroughly reinforce a comprehensive understanding of the material presented.

- *Interesting Applications*   Finite math is without peer as a course that offers a vast selection of interesting problems. I have selected over 450 applied problems, both for the exercise sets and the textual presentations. All were selected for their pedagogical value and motivational appeal. In addition to many standard applications, I include problems dealing with epic poetry, history, perception, weather prediction, pollution, economic inflation, and other unusual and interesting applications for greater variety.

- *Historical Boxes*   Where appropriate and helpful, I have included brief historical discussions of important people and ideas in mathematics.

- *End-of-Chapter Reviews*   Each chapter concludes with a list of important terms, a prose summary of key concepts, and a rich and varied set of review exercises.

- *Accuracy*   Special care has been taken to insure that the mathematical presentations and answers to problems are free from mistakes and typographical errors.

New Features of the Second Edition

The new features and new material make this edition among the most complete texts available.

- The text now includes *many more examples and exercises*. There are over 3000 exercises and over 700 examples.

- Two new categories of exercises have been provided, identified by special icons. A few questions in each section will exercise skills that pertain to *writing across the curriculum*; simple numerical results will not suffice for their answer. It would have been easy (and it was tempting) to make such exercises occasions to recapitulate portions of the text's narrative. Instead, I have created questions that would exercise thinking skills as well as language skills.

- *Calculator problems* require the use of a scientific calculator. Problems in real life seldom confine themselves to whole number coefficients with solutions we can express in terms of simple fractions, so these problems add yet another dimension of meaningfulness to the text.

- Each example in the text is accompanied by a matching *skill enhancer*. These are similar to the example so the student can immediately practice the skills of the example. Answers to all the skill enhancer problems appear in a special section at the end of the text.

- Material pertaining to financial mathematics has been expanded and divided into two chapters, one on financial mathematics proper and one on *difference equations*.

- There is a new chapter on *graph theory* which presents a discussion of the *critical path method (CPM)* and *PERT* techniques for managing large scale projects.

- A new appendix presents a *review of algebraic topics* important for the material in this course.

- The material on the *simplex method* has been extensively revised to make the discussion more accessible.

- Each chapter begins with a short introduction and a brief chapter outline, a description of the material in the chapter, and reasons why the chapter is important.

- *Boxed Procedures* Procedures, definitions, and theorems are boxed and visually enhanced so students will find them easier to use.

Organization and Content

- *Linear Equations and Matrices:* Chapter 1 includes a review of graphing and a thorough discussion of linear equations, systems of linear equations, and applications. In addition to introductory matrix theory, Chapter 2 contains thorough discussions of the Gauss-Jordan technique, matrix inverses, and input–output analysis.

- *Linear Programming:* Chapters 3 and 4 present linear programming from several viewpoints. Chapter 3 covers the geometric approach, and the strengths, weaknesses, and limitations of this method are explored. Chapter 4 contains a complete discussion of the simplex method in a flexible manner, that is, in a series of sections. Instructors can adapt their presentations by including or omitting material, depending on their students and the depth of their course.

- *Probability and Its Applications:* Sets and counting topics are presented and carefully explained in Chapter 5, so that students will fully understand how to apply this material to probability in Chapter 6. Probability applications to Markov chains and game theory form the centerpieces of Chapters 7 and 8. The treatment of game theory includes games of mixed strategy and optimal mixed strategies in two-person games.

■ *Statistics and Financial Mathematics:* Chapter 9 surveys descriptive statistics and measures of central tendency, concluding with two widely used statistical distributions, the normal and binomial distributions. Chapter 10 presents the traditional topics in financial mathematics, and also the interesting (optional) topic of financial planning in an inflationary economy.

■ *Additional Topics (Difference Equations, Graphs and Network):* Difference equations provides a way for unifying the treatment of financial mathematics, and although the key ideas are deceptively simple, they do require a greater degree of mathematical sophistication than do other, more conventional treatments. Because of the inherent interest in this treatment, a new Chapter 11 on *difference equations* appears in this edition.

■ Another new chapter, Chapter 12, describes the *critical path* (CPM) and *PERT* methods for managing and scheduling extensive projects. The underlying mathematics, relying on *graphs* to realize many results, is quite simple, and the theme is consistent with the theme of the book—maximizing use of resources in various settings. In this case, the resource is that of available time.

Ancillary Materials

■ *Student Solutions Guide* This guide contains detailed solutions to selected problems in the exercise sets. It also includes brief discussions, hints, and explanations about how and why a given solution method was used. Discussion of special topics, such as the Karmarkar algorithm of linear programming, logic, and others have been included here. I hope that the inclusion of these extra discussions will spark students' enthusiasm.

■ *Instructor's Resource Manual*, including printed test bank. Available to adopters, this very helpful instructor's resource manual contains detailed solutions to even-numbered exercises, extensive teaching hints and suggestions, and transparency masters for classroom use and a variety of testing materials. In addition to a printout of the computerized Test Bank (approximately 1200 problems), it contains two chapter tests per chapter. Several "Interlude" discussions are included; these are a series of extended in-depth discussions on topics, including areas of technology and computer simulation, which instructors may offer as enrichment exercises.

■ *Computer Test Generator* The computerized test generator program is available to adopters for the IBM PC or compatible microcomputer. The program is very flexible and contains full question editing features. The accompanying test bank contains over 1200 problems, which may also be used to supplement the exercise sets in the text.

■ *PC-81 Emulation Software* Available for IBM PC (or compatible), this powerful and compact package completely emulates the functionality of the popular TI-81 graphing calculator, including operations involving matrices, statistical functions, factorials and other counting exercises, and graphing and shading. It is offered in cooperation with the Texas Instruments Corporation.

- *Math Assistant software* for IBM type or Macintosh computers. This is a disk containing generic software (aimed at both student and instructor) appropriate for common mathematical computation, including the graphing of functions, manipulation of matrices, and the finding of roots of equations.

- *Stat Plus* and *Data Plus* programs (IBM). With these programs, students can implement several important statistical procedures.

Acknowledgements

The creation of any textbook is a team effort, and I am pleased to give public thanks to the many members of that team and acknowledge their contributions.

First, it is with pleasure that I acknowledge all the manuscript reviewers who closely monitored my progress. They were always conscientious, and attacked my fledgling manuscript with gusto; their help and support has been invaluable:

Carol Achs
Mesa Community College, TX

Charles Burnap
*University of North Carolina—
Charlotte, NC*

Barbara Conway
Berkshire Community College, MA

Laura A. Dyer
Belleville Area College, IL

Mary S. Elick
Missouri Southern State College, MO

Phil Green
Tacoma Community College, WA

James Griesmer

Jeff Hoag
Providence College, RI

David Hoff
Indiana University, IN

Alec Ingraham
New Hampshire College, NH

Barry L. Jones
SUNY College at Brockport, NY

George Kolettis
University of Notre Dame, IN

E. L. Marsden
Norwich University, VT

Kenneth W. Meerdink
University of Idaho, ID

Robert J. Mergener
Moraine Valley College

Edward S. Miller
Lewis-Clark State College, ID

Professor Moshgi
Richland Community College, IL

Lawrence S. Moss
Indiana University, IN

Paul O'Heron
Broome Community College, NY

Eugene Robkin
University of Wisconsin—Baraboo, WI

Ray Rosentrater
Westmont College, CA

Joan Smith
Vincennes University, IN

Bryan Stewart
Tarrant County Junior College, TX

Lowell Stultz
Kalamazoo Valley Community College, MI

Peter Waterman
Northern Illinois University, IL

Joan Wyzkoski Weiss
Fairfield University, CT

H. Gordon Williams
Virginia Military Institute, VA

Deborah Woods
University of Cincinnati, OH

Earl Zwick
Indiana State University—Terre Haute, IN

I thank James Griesmer, Eugene Robkin, Bryan Stewart, and Earl Zwick for independently verifying the accuracy of the text. Their work was very thorough and I appreciate it sincerely.

This book would not be before you in its present form if Donald E. Knuth had not developed the T_EX typesetting program for the scientific publishing community.

The Houghton Mifflin production team was completely professional and supportive in every step of the process with this new computer technology. I thank Toni Haluga, Senior Project Editor, Cathy Brooks, Senior Project Editor, and Bailey Siletchnik, Art Editor, for their valuable contributions. Without the editorial guidance and commitment of my Sponsoring Editor Maureen O'Connor, this text would have been a shadow of its present self. At every step, Maureen and I were both thankful that we had Robert Hupp, whose title is Associate Editor but who is actually something much more, working so hard and caring so much for the quality of the project.

Finally, I commend my family for their steady support, from among whom I single out my wife Jozefa for her encouragement and my children Sam and Hannah. That they tolerated the presence of this project, an additional family member among them, for so long was an unexpected and unwarranted blessing.

Alan Hoenig
Huntington, Long Island
New York
Summer, 1994

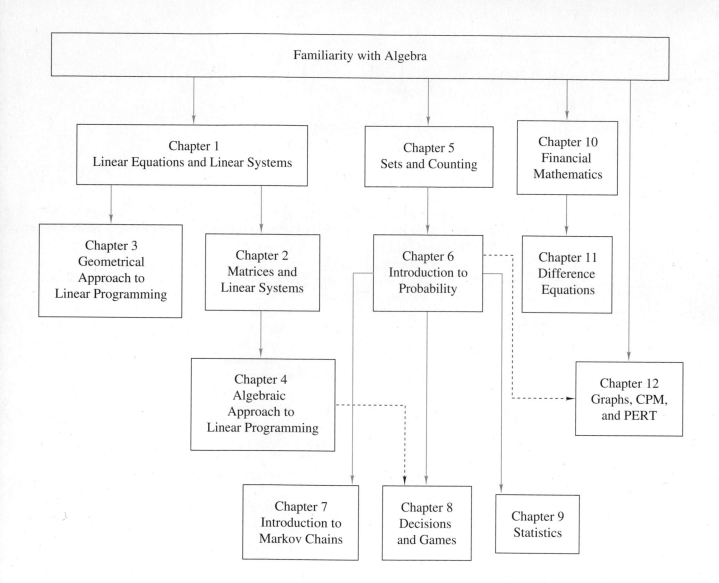

Familiarity with Algebra

Chapter 1
Linear Equations and Linear Systems

Chapter 5
Sets and Counting

Chapter 10
Financial
Mathematics

Chapter 3
Geometrical
Approach to
Linear Programming

Chapter 2
Matrices and
Linear Systems

Chapter 6
Introduction to
Probability

Chapter 11
Difference
Equations

Chapter 4
Algebraic
Approach to
Linear Programming

Chapter 12
Graphs, CPM,
and PERT

Chapter 7
Introduction to
Markov Chains

Chapter 8
Decisions
and Games

Chapter 9
Statistics

1

LINEAR EQUATIONS AND LINEAR SYSTEMS

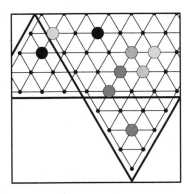

INTRODUCTION

In this chapter, we establish the basic foundations supporting linear mathematics. Much of finite mathematics involves linear mathematics, which accounts for the importance of this material.

Following a discussion of basic ideas, including the solution of linear equations containing a single variable, we explore *the graphing of linear equations* in greater detail, and take a brief look at the ways in which linear equations are useful in *real-world applications*. The final portion of the chapter emphasizes *systems* of linear equations.

Linear equations are basic to all work in mathematics. Despite their basic nature, they radiate great power. Can we estimate the effects of inflation or predict the future? Can we advise nations on their economic strategy or companies on cost-cutting measures? Can we help urban planners assess the flow of people moving to and from suburbs, or make an educated guess about weather during that next vacation? The answer is "yes" to all these questions, provided we have a firm understanding of linear equations. We touch on all these issues in this and later chapters.

Section 1.1 EQUATIONS AND GRAPHING

linear equation   A *linear equation* is one that contains only two types of terms:

1. Constants

2. Variables either appearing by themselves (raised to the first power) or multiplied only by a constant coefficient

Nothing else is allowed in a linear equation. Therefore,

$$x + 2y + 3z + 19 = 0$$

is linear (because each term is a constant or a constant times one of the variables x, y, or z), whereas

$$xy + z = 1$$

is *not* (because two variables are multiplied together in the first term).

Example 1   Which of the following are linear equations?

(a) $z = 144x - 81y + 64z - 49u + v - 16$ (b) $q = 4p - 3$ (c) $y = 8x - 3 + \dfrac{1}{x}$

(d) $\dfrac{1}{y} = x + 32$

Solution

The first two equations are linear because no variable is raised to any power other than 1. Both of the final two parts contain terms like $\frac{1}{x}$ or $\frac{1}{y}$, which are equivalent to x^{-1} and y^{-1}, terms with negative exponents. Therefore, these two equations are nonlinear.

Skill Enhancer 1   Which of the following are linear equations?

(a) $3x - 2y = 4z$ (b) $3^x + x^3 = 4$ (c) $\frac{1}{x} + \frac{1}{y} = \frac{1}{z}$ (d) $2u - 3t = 0$

Answer in Appendix E.

Subscripted Variables

Usually, we let letters like x or y represent variables. When a problem involves many variables, or when we need to translate a problem into a form that a computer can understand, we can replace a collection of *individual* variable names by a *family* of *subscripted* variable names. Whereas "conventional" variables use different letters of the alphabet (x, y, u, t, and so on), a family of subscripted variables has a single letter

subscript   name, to which we append numbers written "below the line," the *subscript*s. Different subscripts represent different variables. There is no limit to the number of variable names we can systematically assign, because we just keep increasing the subscript for new variables. Subscripted variables such as x_1, x_2, or y_6 are read "x-sub-1," "x-sub-2," "y-sub-6," and so on.

Example 2   Rewrite these equations using subscripted variables.
(a) $3x - 2y = 12$ (b) $9u - 2v + \frac{1}{2}w = 0$.

Solution

(a) We replace x and y by x_1 and x_2 to obtain the equivalent equation $3x_1 - 2x_2 = 12$.
(b) We may replace u, v, and w by x_1, x_2, and x_3. $9x_1 - 2x_2 + \frac{1}{2}x_3 = 0$ is the equivalent equation.

Skill Enhancer 2   Use subscripted variables to rewrite $x + y = 2$ and $\frac{1}{3}x - \frac{1}{4}y + \frac{1}{5}z + 10 = 0$.

Answer in Appendix E.

Example 3   The Seedy Compact Disc Company has a system that tracks the performance of its sales personnel. The company accountants want to use variables to record the monthly sales of all salespeople. As the company grows, the number of salespeople increases. Last year, for example, they had 10 salespeople, and this year they have 17. Explain how subscripted variables help set up this model.

Solution

The best idea is to use a family of subscripted variables x_i to keep track of the sales of the ith salesperson. For example, x_7 records the sales of the seventh saleswoman. When there are 10 salespeople, the model uses variables x_1 through x_{10}. When the sales team jumps to 17, the company include variables x_{11} through x_{17} in the model.

Skill Enhancer 3   How would subscripted variables be used to keep track of the hits scored by a team of little leaguers? As the season progresses, more and more children join the league.

Answer in Appendix E.

Example 4   As a small child, Sam operated a sidewalk lemonade stand, and sold his drinks for $0.25 per cup. He learned to express his daily profit p as the difference between his sales, which is the number of cups n sold times the price, and his expenses E:

$$p = 0.25n - E$$

The next summer, Sam "diversified" into fruit punch. If m represents the number of cups of punch sold, then his profit now becomes

$$p = 0.25(m + n) - E$$

Year after year, Sam continued to offer new flavors of drinks, and it became impractical to use different letters for the new variables for cups sold in each flavor. Assume there are now 31 flavors of drinks. Use subscripted variables to represent his daily profit.

Solution

Set up a family of subscripted variables n_1 through n_{31} so that n_j represents the number of cups sold of the jth flavor. Then the daily profit is

$$p = 0.25(n_1 + n_2 + \cdots + n_{30} + n_{31}) - E.$$

(Here the subscripts may be double-digit numbers.) Sam can adjust his mathematics for additional flavors at any time by creating additional variables of the form n_j.

Skill Enhancer 4   A student earns extra money selling magazine subscriptions door to door. Her commission is 15 percent of each subscription. Initially, she sells only five different magazines, but there is a possibility for expansion. Suggest variables she may use to keep track of her sales income. (Assume all subscriptions are for one year only.)

Answer in Appendix E.

Solving Linear Equations

solution set
The collection of all solutions to some given equation or some equation system is called the *solution set*. For a linear equation in one variable, the solution process involves implementing four well-defined steps, although not all steps will be needed for all exercises.

Solving Linear Equations in One Variable
Follow these steps to solve linear equations containing just one variable.

1. *Simplify* the equation by expanding all parentheses and combining like terms.

addition principle
2. The *addition principle* allows us to add identical quantities to (or subtract identical quantities from) both sides of an equation. Place the *constant terms and variable terms on opposite sides of the equal sign* by using the addition principle. Simplify again if necessary.

multiplication principle
3. The *multiplication principle* allows us to multiply (or divide) both sides of an equation by an identical, nonzero quantity. Use this principle to *solve for a single value of the variable* by dividing both sides of the equation by the coefficient of the variable.

4. *Check* the results.

Example 5   Solve $9(y+1) - 2y = 37$.

Solution

Step 1. $9y + 9 - 2y = 37$ *Simplify; expand parentheses*

                      $7y + 9 = 37$ *Combine like terms*

Step 2. $7y = 28$ *Add -9 to both sides*

Step 3. $y = 4$ *Multiply both sides by $\frac{1}{7}$*

Step 4. $9(4+1) - 2(4) \ ? \ 37$ *Check the results*

                  $45 - 8 \ ? \ 37$

                      $37 = 37$

The solution is $y = 4$.

Skill Enhancer 5   Solve $3y - 2(y - 1) = 2(y + 2)$.

Answer in Appendix E.

Not all equations have a solution.

Example 6   Solve

$$3(x_1 - 4) - x_1 = 2x_1$$

Solution

First expand parentheses and combine like terms. The equation becomes

$$2x_1 - 12 = 2x_1$$

Upon adding $-2x_1$ to both sides, we find that $12 = 0$, *a contradictory statement.* Since we applied valid operations throughout, and since we made no careless errors, we conclude that this equation has no solution.

Skill Enhancer 6   Solve $10(z - 2) - 8z = 2(z + 1)$.

Answer in Appendix E.

Example 7   Solve

$$4x - 2 = 2(2x - 2) + 2$$

Solution

Expand parentheses and combine like terms; this equation then becomes

$$4x - 2 = 4x - 2$$

Add $+2$ to both sides:

$$4x = 4x$$

This equation is *always* true, regardless of the value of x. Therefore, this equation has an infinite number of solutions, and no unique solution at all.

Skill Enhancer 7   Solve the equation $3u + 3 = 3(u + 1)$.

Answer in Appendix E.

Linear equations of one variable may have one unique solution, no solution, or an infinite number of solutions.

Graphing on Rectangular Coordinate Systems

There is a convenient technique for graphically representing equations and inequalities. With modest practice, we can create graphs that illuminate the relationships between variables within an equation.

Most graphing involves equations with two variables, such as $y = x^2$ or $x_1 + x_2 = 10$. The results of our graphing process will be a line or curve, the position and shape of which uniquely correspond to the given equation. The simplest and most widely used system is *Cartesian* or *rectangular coordinates*.

rectangular coordinates

Fundamentals of Cartesian Graphing

axis

1. Choose two perpendicular lines to serve as the *axes* (the singular is *axis*) of the graph. The lines cross at the *origin* and generate a rectangular grid as in Figure 1.1.

abscissa

2. The horizontal axis represents values of the first variable, often called x. The x value (or *abscissa*) of the origin is 0; horizontal distances to the right represent positive values, whereas horizontal distances to the left represent negative values.

ordinate

3. The vertical axis represents values of the second variable, often called y. The y value (or *ordinate*) of the origin is 0; distances above the origin stand for positive values, and distances below it correspond to negative values.

ordered pair

4. The position of any point is unambiguously determined by its x value and y value (its abscissa and ordinate), which are written as a pair of numbers separated by a comma, and surrounded by parentheses. The first value is always the x value, and this pair is called an *ordered pair* or the *coordinates of the point*.

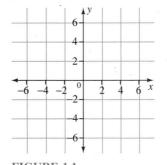

FIGURE 1.1
A coordinate axis system.

Historical Note

René Descartes   *The French mathematician and philosopher René Descartes (1596–1650) lends his name to the graphing technique he developed. Since he was French, his name is pronounced* day-CART, *but the technique is* car-TEEZH-ian. *His achievements in mathematics, optics (including the first satisfactory explanation for the appearance of rainbows), and philosophy made him preeminent in the learned society of his time. His work on coordinate systems appeared almost as an afterthought, in an appendix to his greatest work,* Discours de la méthode pour bien conduire sa raison et chercher la verité dans les sciences *("Discourse on a Method for Reasoning Well and Finding Truth in Science").*

Example 8   Graph these points:

(a) $(2, -5)$ (b) $(0, -1)$ (c) $(-1.5, 0)$ (d) $(-1, -1)$

Solution

The graphed points are shown in Figure 1.2.

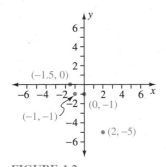

FIGURE 1.2
Graphs of the points $(2, -5)$, $(0, -1)$, $(-1.5, 0)$, and $(-1, -1)$.

Skill Enhancer 8   Graph the points $(3, 2)$, $(2, -1)$, $(-1, 0)$, and $(0, -2)$.

Answer in Appendix E.

General Remarks on Graphing

Any equation such as $y = 3x - 14$, $y = -x^3$, and so on, is a shorthand notation for an infinite collection of ordered pairs. Each possible value of x exists in exactly one pair, and we determine the corresponding y value according to the rule expressed by the equation.

Our concern in this text will be with *linear equations*. To graph an equation, we determine several representative points whose coordinates satisfy this equation, plot these points, and connect them with a smooth curve.

Two-dimensional linear equations contain only *two* variables and can always be rearranged to the form

$$ax + by = c$$

or

$$ax_1 + bx_2 = c$$

general form   for some constants a, b, and c. This is the *general form* of a linear equation.

We assert without proof this important theorem.

Theorem 1.1   Graph of a Linear Equation The graph of a two-dimensional linear equation will always be a straight line.

We will discuss two methods for graphing straight lines. Strictly speaking, we need to find only two points to fix the line. In the first method, we find three points. The third serves as an error-checking device.

Graphing a Straight Line: First Method

See Figure 1.3 on page 8 for this method.
1. Find the coordinates of three distinct points lying on the line.

2. Plot these points.

3. Connect the points with a straight line. If there is an error in computing or plotting the points, a single straight line will probably not connect all three points.

We need practice in being able to find the coordinates of points satisfying an equation.

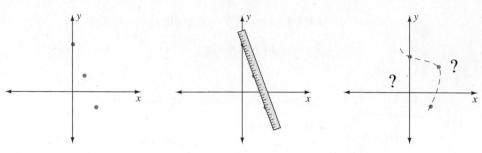

Plot the points.

Lay a straightedge along them and connect with a straight line.

If there is an error in determining and plotting the points, a straight line will probably not connect them.

FIGURE 1.3
Graphing a straight line using any three points on the line.

Example 9   Find three points satisfying the equation $2x - y = 4$.

Solution

We are always allowed to specify *one* of the two coordinates in any point. Therefore, we choose the x or y values that look the easiest to work with. Once we have selected one coordinate, we solve for the remaining one.

Here, we might choose $x = 0$ for one of the points. To find the corresponding y value, use this value in the given equation and solve for y.

$$2x - y = 4$$
$$2 \cdot 0 - y = 4$$
$$-y = 4$$
$$y = -4$$

$y = 0$ is another good value to choose, and we choose some arbitrary value for the third point. For this equation, then, we might start with the chart

$$
\begin{array}{c|c|c|c}
x & 0 & & 1 \\
\hline
y & & 0 &
\end{array}
$$

We already know how to fill in the first column. Work the same way for the remaining columns. The completed chart is

$$
\begin{array}{c|c|c|c}
x & 0 & 2 & 1 \\
\hline
y & -4 & 0 & -2
\end{array}
$$

The columns of this chart correspond to the three points $(0, -4)$, $(2, 0)$, and $(1, -2)$. These are three of the infinitely many points that satisfy this equation.

Skill Enhancer 9   Determine three points satisfying $3x + y = 6$.

Answer in Appendix E.

Example 10   Graph $x + 2y = 2$.

Solution

Let $x = 0$, 1, and 2. We will use these values to construct our chart:

x	0	1	2
y	1	$\frac{1}{2}$	0

Figure 1.4 shows this straight line. Do you see why linear equations deserve the name *linear*?

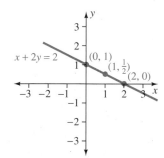

FIGURE 1.4

Graphing the straight line $x + 2y = 2$.

Skill Enhancer 10   Graph $2x + y = 4$.

Answer in Appendix E.

Example 11   Graph these linear equations:

(1.1) $y = x$

(1.2) $y = x + 2$

(1.3) $y = 2x$

Solution

For the three equations, let x have the values 0, 1, and 2. For each equation, begin by creating the appropriate chart of values.
(a) One chart for $y = x$ is

x	0	1	2
y	0	1	2

Figure 1.5*a* shows these points and the line connecting them.
(b) Here is the table, for $y = x + 2$:

x	0	1	2
y	2	3	4

These points, and the corresponding line, are plotted in Figure 1.5*b*.
(c) Here is the final table for $y = 2x$:

x	0	1	2
y	0	2	4

These points and the line they define are shown in Figure 1.5*c*.

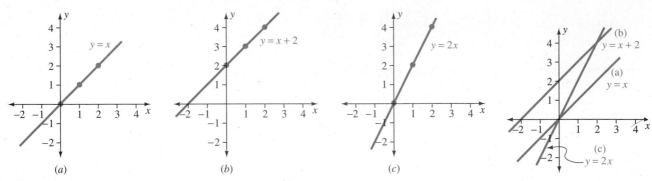

FIGURE 1.5
Graphing lines. (*a*) $y = x$; (*b*) $y = x + 2$; and (*c*) $y = 2x$.

FIGURE 1.6
Graphing $y = x$, $y = x + 2$, and $y = 2x$ on the same axes.

| Figure 1.6 displays the three lines on a single set of axes.

Skill Enhancer 11   Graph $y = 2x$, $y = 2x - 1$, and $y = -2x$ on a single set of coordinate axes.

Answer in Appendix E.

(Unless otherwise specified, we use the terms *line* and *straight line* interchangeably.)

The three equations of Example 11 are similar to one another. Examine Figure 1.6 carefully to see how changes to an equation can alter its graph.

The difference between equations (1.1) and (1.2) is the presence of the constant term +2 in Equation (1.2). The graphs of Equations (1.1) and (1.2) have the same steepness, but they intersect (cut) the coordinate axes in different places. Equations (1.1) and (1.3) are the same except for the coefficient of x, and their graphs intersect the axes in the same location, but have different slants. We surmise that somehow the presence of a constant term in the equation affects the intersection of the line with the axes, whereas the coefficient of x controls the steepness. The next section explores this correlation.

Intercepts

intercepts   Unless a line is horizontal or vertical, it will cut both of the axes. These points of intersection are the *intercepts*. (How many intercepts will a vertical or horizontal line have? Can a straight line ever have *no* intercepts at all?) Suppose a line intersects the x axis a units from the origin; the full coordinate representation of this point will be $(a, 0)$. (Do you see why?) Similarly, the y intercept of a line can be represented by a point $(0, b)$, where the intercept is b units above or below the origin, depending on the sign of b. (Why?) Because it is often easy to determine a line's intercepts, there is a second method we can use to graph a straight line. See Figure 1.7 for examples of lines possessing a single intercept.

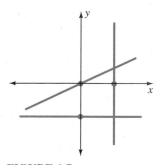

FIGURE 1.7
Three types of lines that have only a single intercept.

Graphing a Straight Line: Second (Intercept) Method

1. Find the intercepts of the line.

2. Plot these intercepts.

3. Connect the intercepts to determine the straight line.

   This method will work only for lines that have two distinct intercepts; see Figure 1.8. It will *fail* for lines that are horizontal or vertical, or that pass through the origin, as in Figure 1.7.

Example 12   Determine the intercepts of the equation $2x - 3y = 6$.

Solution

We find the two intercepts by first setting $x = 0$ and solving for y, then setting $y = 0$ and solving for x.

   When $x = 0$, we have $0 - 3y = 6$, so $y = -2$. When $y = 0$, we have $2x - 0 = 6$ and $x = 3$. Therefore, the two intercepts are $(0, -2)$ and $(3, 0)$.

Skill Enhancer 12   What are the intercepts to $x + 5y = 10$?

Answer in Appendix E.

Example 13   Determine the intercepts to $x = 3y$.

Solution

The only intercept is $(0, 0)$, which means the equation will pass through the origin.

Skill Enhancer 13   What are the intercepts to $2x = 5y$?

Answer in Appendix E.

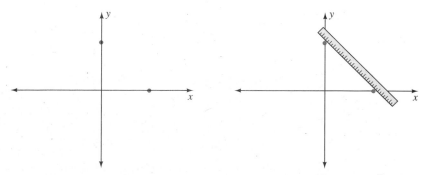

Find and plot the intercepts.   Use a straightedge to connect them.

FIGURE 1.8
Graphing lines using the intercept method.

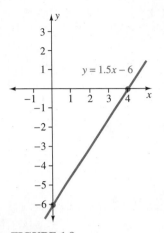

FIGURE 1.9
Graphing $y = 1.5x - 6$ using intercepts.

Example 14   Use the intercept method to graph $y = 1.5x - 6$.

Solution

To find the x intercept, let $y = 0$ and solve for x:

$$0 = 1.5x - 6$$
$$1.5x = 6$$
$$x = 4$$

Therefore, the x intercept is (4, 0).

To find the y intercept, set $x = 0$ and solve for y:

$$y = 0 - 6$$
$$y = -6$$

The y intercept is $(0, -6)$. These two intercepts and the equation are plotted in Figure 1.9.

Skill Enhancer 14   Use the intercept method to graph $2x + 3y = 6$.

Answer in Appendix E.

Horizontal and Vertical Lines

Some linear equations may contain only one variable. Any such equation can be transformed into either of the two forms

$$x = a$$

or

$$y = b$$

vertical lines   for some constants a and b. Equations like $x = a$ always graph as *vertical lines*, and
horizontal lines   equations like $y = b$ graph as *horizontal lines*.

Example 15
(a) Graph $x = 3$.
(b) Graph $y = -1$.

Solution

(a) $x = 3$ is a *vertical* line passing through the x axis at $(3, 0)$. The single requirement for a point to be on this line is that the x value must equal 3. Any y value is acceptable.
(b) $y = -1$ is a *horizontal* line passing through the y axis at $(0, -1)$. Any point whose y value is -1 will lie on this line, regardless of the x value.

Both of these lines are shown in Figure 1.10.

Skill Enhancer 15   Graph $y = 0$, $x = -1$.

Answer in Appendix E.

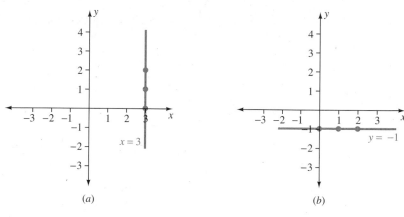

FIGURE 1.10
Vertical and horizontal lines. (a) $x = 3$; **(b)** $y = -1$.

Example 16   Write the equation of a line parallel to the x axis passing through the point $(-1, 3)$. What is the equation of a line parallel to the y axis passing through this point?

Solution

Any line parallel to the x axis is horizontal, and its equation is of the form $y = b$. Since it passes through $(-1, 3)$, the equation must be $y = 3$.

In the same way, a line parallel to the y axis has an equation like $x = a$. Hence, the equation we seek must be $x = -1$.

Skill Enhancer 16   What are the equations of the horizontal and vertical lines passing through the point $(-1, -4)$?

Answer in Appendix E.

Exercises 1.1

W 1. In deciding whether to place a traffic light at a busy intersection, a county official keeps track of the number and types of vehicles traveling on this street. Currently, the official notes whether each vehicle is a compact car, a full-size car, a light truck, or a heavy truck. Suggest a way in which subscripted variables can play a rôle in this record keeping.

W 2. A biologist keeps track of the varieties and genders of a particular type of beetle. Suggest how subscripted variables can aid this task.

3. Work the first part of Example 4 again. This time, break the expenses E into separate categories for the different flavors.

4. Work the *second* part of Example 4, supposing now that

there are only 5 flavors and that there are distinct fixed costs associated with each flavor. That is, there is a fixed cost f_1 to advertise and carry flavor 1 even if there are no expenses for buying ingredients for this flavor. (These quantities f_i are the same from day to day.) Write expressions for the daily profit.

5. Which of the following are linear equations?
   (a) $x + 3x^2 + 3 = 0$
   (b) $9y - 4 = 5.5y$
   (c) $x_1 + x_2 + x_3 = 7$

6. Which of the following are linear equations?
   (a) $x_1^2 + x_2^2 = 4$
   (b) $\frac{1}{x} + \frac{1}{y} = \frac{1}{16}$
   (c) $49y_1 - 64y_2 + 81y_3 - 100 = 0$

Solve the equations in Exercises 7–22.

7. $3x - 1 = 11$ **8.** $9(4z + 11) + 9 = 0$

9. $9y - 3 = 9y$ **10.** $2(x_2 - 5) + 3 = 4x_2 - 17$

11. $4(t_2 - 3) + 7 = 3t_2 - 1$

12. $7(u_1 + 3) - u_1 = 2u_1 + 22$

13. $8(x + 3) - 2x = 4 + 2(x - 2)$

14. $9z + 2(8 - z) = 10 + 2z$

15. $3(t_1 - 1) + 3 = 0$ **16.** $9(u_1 - 2) = u_1 - 18$

17. $\frac{1}{3}z - \frac{1}{6}z + \frac{1}{2}z = \frac{1}{10}$ **18.** $\frac{1}{4}q - \frac{2}{3} = q + \frac{1}{2}$

19. $\frac{1}{2}z - 6 = \frac{1}{4}z + 12$ **20.** $2y_1 - 1 = \frac{1}{2}y_1 + \frac{3}{2}$

21. $5(x + 3) - 2(x + 7) = 8x - 4$

22. $3(u + 2) - 2(u - 2) = 2(u + 2) - (u - 2)$

In Exercises 23–30, solve for x in each equation by assuming that all other quantities and variables are known.

23. $ax + by = c$ **24.** $7t - 3x = 1$

25. $3x + 9(y - x + 1) = 2x$ **26.** $19y - 3(x - 7) = 6x + 1$

27. $xy = 10$ **28.** $2y = \frac{1}{x}$

29. $xy^2 + 2y = 3$ **30.** $ay^2 + by + c = 2x$

31. Graph these points.
   (a) $(0, 1)$
   (b) $(1, 0)$
   (c) $(-1, -4.5)$

32. Graph these points.
   (a) $(3, 2)$
   (b) $(5, -1)$
   (c) $(-4, 3)$

33. Graph $(\frac{1}{2}, \frac{1}{2})$, $(-\frac{3}{4}, 2)$, and $(\frac{1}{4}, -\frac{5}{4})$.

34. Graph $(\frac{2}{3}, -\frac{2}{3})$, $(-1, -1\frac{1}{2})$, and $(0, 0)$.

35. Complete the table for $3x - y = 1$.

x	1	2	3
y	?	?	?

36. Complete the table for $x + y = 3$.

x	1	?	?
y	?	3	1

37. Complete the table for $2x - y + 2 = 0$.

x	?	?	?
y	4	6	8

38. Complete the table for $3x - 2y = 12$.

x	?	?	?
y	0	4	8

39. Complete the table for $y = 2x - 4$.

x	1	2	3
y	?	?	?

40. Complete the table for $y = \frac{1}{2}x + 1$.

x	2	4	?
y	?	?	-1

41. In graphing $2x - 4y = 10$, use the following table. Complete the missing values, graph the points, and draw the line.

x	0	2	?
y	?	?	-2

42. In graphing $x - 4y = 12$, use the following table. Complete the missing values, graph the points, and draw the line

x	0	?	?
y	?	-1	-2

43. Complete the table for the line $0.5x + 0.75y = 3$, graph the points, and draw the line.

x	?	0	-1
y	0	?	?

44. Complete the table for the line $0.5x + y = 2$, graph the points, and draw the line.

x	?	?	?
y	0	1	2

45. Complete the table for the line $\frac{1}{3}x - \frac{1}{6}y = 12$, graph the points, and draw the line.

x	?	18	0
y	0	?	?

46. Complete the table for the line $\frac{1}{4}x - \frac{1}{8}y = 16$, graph the points, and draw the line.

x	0	?	64
y	?	-64	?

Graph the equations in Exercises 47–62 by finding and plotting three points. (Use "the first method" of the text.)

47. $x + 5y = 6$ **48.** $y + 5x = 6$

49. $3x - 2y + 12 = 0$ **50.** $y = -x$

51. $x + 5y = 10$ **52.** $x - y = 0$

53. $x = 3$ **54.** $x = 0$

55. $y = -1.5$ **56.** $y = \sqrt{3}$

57. $\frac{1}{3}x - 2y = 2$ **58.** $x - 10y = 10$

59. $5x + 8y = 40$ **60.** $3y - 8x = 24$

61. $8x - 2y = 8$ **62.** $\frac{2}{3}x - \frac{1}{3}y = 3$

Graph the equations in Exercises 63–78 by finding and plotting the intercepts.

63. $y + 2x = -1$ **64.** $y = -4x + 12$

65. $y = -2.5x - 5$ **66.** $3x - 1.5y + 6 = 0$

67. $2x + 3y = 12$ **68.** $2x - y + 6 = 0$

69. $2y = 18 - 9x$ **70.** $y = -4 + 4x$

71. $8y - 7x = 56$ **72.** $5x + 2y = 20$

73. $6x - 7y = 42$ **74.** $x = \sqrt{8}$

75. $x = -2$ **76.** $y = 14$

77. $y = -0.5$ **78.** $7y - 3x = 21$

W 79. Graph these next equations on a single set of coordinate axes.

$$y = -2x \qquad y = -2x - 4 \qquad y = -2x + 5$$

These equations have the same coefficients for the x and y terms. Only the constant terms differ. What effect does this have on their graphs?

W 80. Graph the equations on a single set of coordinate axes.

$$y = x + 1 \qquad y = 2x + 1 \qquad y = -x + 1$$

These equations are the same except for the coefficient of the x term. What effect does this have on the graphs?

W 81. Why does the intercept method for graphing *fail* when the equation represents a horizontal or vertical line?

W 82. Why will the intercept method of graphing *fail* when the equation represents a straight line passing through the origin?

Applications

83. *(Athletics)* Miles N. Miles, varsity track star, first ran the mile in 5.5 minutes (min). After 1 week of practice, he ran it in 4.8 min. Ten days later, he ran a mile in 3.8 min. Display these data graphically.

84. Refer again to the previous problem. If Miles's improvement is linear, use this information to deduce what his performance would have been on the tenth day of practice. On what day would he have run the mile in exactly 4 min?

85. *(The Spread of Disease)* One person in the typing pool at Torus Tire Company had a cold on December 1. There are 45 typists in this group. Four days later, a total of 9 people had colds. Seven days later, 23 people had colds. Display these data graphically.

86. Refer to the previous problem. If you connect the data points with a smooth curve, what would the curve be? Use this curve to estimate the time at which all people in the typing pool have colds.

87. *(Fads)* Hula hoops were a popular fad in the 1950s. Assume that their usage peaked on July 1, at which time 10,000,000 hoops were in active use. On August 1, only 8,500,000 hoops were being used. On October 1, only 5,500,000 were in use.
(a) Plot these data.
(b) If the usage of hoops is linear in the time since July 1, estimate when the fad will have completely disappeared.

88. *(Weather)* By the 10:00 coffee break, 1 inch (in) of rain had fallen. By 11:00, a total of $1\frac{3}{4}$ in of rain had fallen, and by 1:00, when people were getting ready to leave for lunch, a total of $3\frac{1}{4}$ in of rain had come down. Plot this information on a graph. Connect the points with a smooth curve. Use this information to estimate the time at which the downpour began.

Calculator Exercises

Use a calculator to help solve the equations in Exercises 89–92. Express the solutions to two decimal place accuracy.

89. $5.6(-0.2y + 13) = 0$ **90.** $7.2y + 12 = 3.6y$

91. $5.9x - 3.7 = 6.8x$ **92.** $7.2(2t + 3) = 1.7t$

Section 1.2   *LINEAR EQUATIONS: MORE ON GRAPHING*

slope   The degree of *steepness* or *inclination* of a line is measured by a quantity called *slope*. Figure 1.11 shows a line graphed on a set of coordinate axes. We represent the coordinates of two distinct points using the notation (x_1, y_1) and (x_2, y_2); the subscripts here do *not* label unknown variables. A numerical measure for the slope involves these coordinates.

Slope of a Line

$$m = \frac{y_2 - y_1}{x_2 - x_1}$$

Given two points, use this formula to computer the *slope* of the line passing through them.

m is the slope.
(x_1, y_1) are the known coordinates of some arbitrary point on the line.
(x_2, y_2) are the known coordinates of a second arbitrary point on the line. It *must* be distinct from the first point (x_1, y_1).

If the two distinct points have the same x coordinate, the slope is *undefined*.

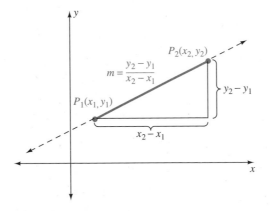

FIGURE 1.11
The geometric meaning of slope.

Example 17   Line segments pass through the following pairs of points. Compute (if possible) the slopes of these segments.
(a) $(-1, -1)$, $(2, 3)$ (b) $(0.125, 3.4)$, $(0.125, -8.3)$ (c) $(3, 4)$, $(8, 6)$
(d) $(-1, 6)$, $(2, 6)$

Solution

(a) To use the formula for slope, we need the numerical values for the x's and y's. It does not matter which of the two known points is point 1 and which is point 2, *provided the notation is consistent.* For example, for this first pair of points, here are the calculations under one system of labeling:

$$P_1(-1, -1) \qquad P_2(2, 3)$$

$$x_1 = -1 \qquad x_2 = 2$$

$$y_1 = -1 \qquad y_2 = 3$$

$$m = \frac{3 - (-1)}{2 - (-1)} = \frac{3 + 1}{2 + 1} = \frac{4}{3}$$

(b) If we try to compute the slope, we find that

$$m = \frac{-8.3 - 3.4}{0.125 - 0.125}$$

$$= \frac{-11.7}{0}$$

Division by zero is never allowed. This arises because the x values are the same for both points. In this case, the slope is undefined.

(c) If $P_1 = (8, 6)$ and $P_2 = (3, 4)$, then

$$x_1 = 8 \qquad y_1 = 6$$

$$x_2 = 3 \qquad y_2 = 4$$

so

$$m = \frac{4 - 6}{3 - 8} = \frac{-2}{-5} = \frac{2}{5}$$

(d) No matter which point is P_1 and which is P_2, we will have $y_1 = y_2 = 6$, $x_1 \neq x_2$, so that

$$m = \frac{y_2 - y_1}{x_2 - x_1} = \frac{6 - 6}{x_2 - x_1} = 0$$

A slope of 0 is possible and means the line is horizontal.

Skill Enhancer 17   Compute the slopes of the line segments connecting these pairs of points: $(\frac{1}{2}, \frac{1}{2})$ to $(\frac{3}{4}, \frac{3}{4})$; $(3, 1)$ to $(3, -1)$; and $(3.5, 5.5)$ to $(4, -2)$.

Answer in Appendix E.

Figure 1.11 illustrates the geometric meaning of slope. Erect a right triangle so that the line segment coincides with the hypotenuse (the side opposite the right angle) and the two given points mark the ends of the hypotenuse. The length of the vertical leg is $y_2 - y_1$, and that of the horizontal leg is $x_2 - x_1$. The slope, then, is just the ratio of the vertical leg to the horizontal leg—the ratio of "rise" (the length of the vertical leg) to "run" (the length of the horizontal leg).

The slope is the same for any two points on a straight line, so the formula for the slope makes sense. Figures 1.12 and 1.13 illustrate this fact, and also the fact that if the

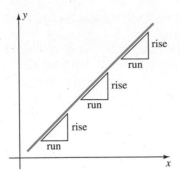

FIGURE 1.12
The slope is everywhere constant along a straight line.

slope is not constant, the curve is not a straight line. (The actual demonstrations are not included in this text.)

positive slope   A line with a *positive slope* inclines "up to the right." Conversely, we describe a
negative slope   line with a *negative slope* as leaning "down to the right." See Figure 1.14.

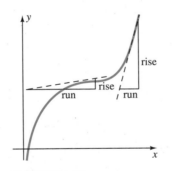

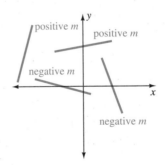

FIGURE 1.13
If the slope varies along a curve, then the curve *cannot* be a straight line.

FIGURE 1.14
Positive and negative slopes.

Slopes of Horizontal and Vertical Lines

What is the slope of a horizontal line? A horizontal line inclines neither up nor down, so its slope can be neither positive nor negative—it must be zero. *All horizontal lines have a slope of zero.*

Vertical lines have no run, so the definition of slope as the ratio of rise to run implies that *the slope is undefined* for any vertical line. (We can also use this concept to show that horizontal lines have zero slope. Horizontal lines have no rise. Therefore, the definition of slope as the ratio of rise to run implies that the slope is zero for horizontal lines.)

In general, the greater the absolute value of the slope, the more nearly vertical is the
absolute value   line. (There is a brief discussion of *absolute value* in the Algebra Review.) The closer

the absolute value of the slope is to zero, the more nearly horizontal is the line. See Figures 1.14 and 1.15.

parallel   Two lines are *parallel* if they never meet—like the rails of a long, straight railroad track. Two lines can be parallel only if they have the same steepness—the same slope. Lines with the same slope are always parallel, and parallel lines always have the same slope.

Determining the Equation of a Line

Suppose a line has some known slope m and passes through some point whose known coordinates are (x_1, y_1). What is an equation for that line?

An arbitrary point on that line has coordinates (x, y). *Determining an equation* for this line means discovering the relationship that exists between x and y so that (x, y) automatically lies *somewhere* on this line. Since it must be true that

$$m = \frac{y - y_1}{x - x_1}$$

point-slope equation   for points (x, y) on the line distinct from (x_1, y_1), we may multiply both sides by $(x - x_1)$ to get the *point-slope equation* for a straight line.

Point-Slope Form for Straight Lines

$$y - y_1 = m(x - x_1)$$

We use this formula to determine the equation for a straight line whenever we are given the slope of the line and the coordinates of one point on the line.

m is the known slope.
x_1 is the known x coordinate of a point on the line.
y_1 is the known y coordinate of the same point.

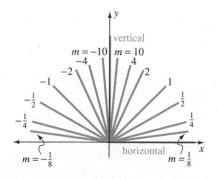

FIGURE 1.15
The greater the absolute value of m, the steeper the line.

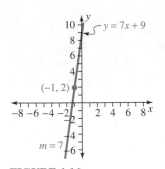

FIGURE 1.16
The line $y = 7x + 9$.

Example 18   What is an equation of a line with slope 7 passing through $(-1, 2)$?

Solution

Here, $m = 7$, $x_1 = -1$, and $y_1 = 2$. Therefore, the point-slope form $y - y_1 = m(x - x_1)$ becomes

$$y - 2 = 7[x - (-1)]$$

(watch out for the double negative!), which simplifies to

$$y - 2 = 7(x + 1) \qquad \text{or} \qquad y = 7x + 9$$

This equation is graphed in Figure 1.16. We usually solve for y to make clear the rôle of y as dependent variable in the equation.

Skill Enhancer 18   Determine an equation for the line of slope $m = -1$ passing through $(2, -3)$.

Answer in Appendix E.

Example 19   A line passes through $(-1, 2)$ and $(7, -4)$. What is an equation of such a line?

Solution

The information given to us in this problem is not quite in the form we need if we are to be able to use the point-slope form. An additional step is necessary.

First, we find the slope. From the definition of slope, we deduce that

$$m = \frac{2 - (-4)}{-1 - 7} = \frac{6}{-8} = -\frac{3}{4}$$

Now we use the point-slope form, with either one of the two given points.

Using the first point with the slope yields $y - 2 = -\frac{3}{4}[x - (-1)]$, which simplifies to

$$y = -\frac{3}{4}x + \frac{5}{4}$$

Using the second given point together with the slope gives

$$y - (-4) = -\frac{3}{4}(x - 7) \qquad \text{or} \qquad y = -\frac{3}{4}x + \frac{5}{4}$$

as before. It does not matter which known point we use; the mathematics is consistent! See Figure 1.17 for a graph of this equation.

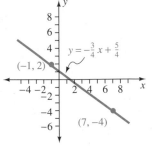

FIGURE 1.17
The line $y = -\frac{3}{4}x + \frac{5}{4}$.

Skill Enhancer 19   Find an equation for the line passing through the points $(2, -2)$ and $(0, -\frac{1}{2})$.

Answer in Appendix E.

Example 20   Village Catering was founded in 1957. In 1962, the firm did $300,000 worth of business. In 1992, it did $1.5 million worth of business. If a linear equation describes its growth since its founding, estimate the business it did during 1957. What is its growth per year?

Solution

Let x be the time elapsed since 1957, and let y be the volume of business transacted during the year x. Village Catering was founded when $x = 0$. The years 1962 and 1992 correspond to $x = 5$ and $x = 35$. When $x = 5$, $y = 300,000$. When $x = 35$, $y = 1,500,000$. A linear equation connects these points. Therefore, the slope m must be

$$m = \frac{1,500,000 - 300,000}{35 - 5} = \frac{1,200,000}{30} = 40,000$$

That is, its rate of growth is \$40,000 worth of business per year.

Using the first point $(5, 300,000)$ and the point-slope form, we find the equation governing this growth to be

or

$$y - 300,000 = 40,000(x - 5)$$
$$y = 40,000x + 100,000$$

This graph of this equation appears in Figure 1.18. When $x = 0$, $y = 100,000$. Thus, in its founding year, Village Catering did \$100,000 worth of business.

Skill Enhancer 20   In the previous example, suppose Village Catering did \$600,000 of business in 1992. If all the other facts remain the same, estimate the amount of business it did in its founding year (1957).

Answer in Appendix E.

There is another important special case. Suppose we know both the slope and the coordinates of a very special point on the line—the point of intersection with the y axis. Traditionally, the symbol b is used to represent the y intercept; its full coordinates are therefore $(0, b)$. Using this information in the point-slope form implies that

$$y - b = m(x - 0)$$

slope-intercept formula   We can do some further simplifying, and we arrive at the *slope-intercept formula* for a straight line.

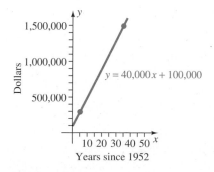

FIGURE 1.18
Growth of business: $y = 40,000x + 100,000$.

Slope-Intercept Form of a Straight Line

$$y = mx + b$$

Use this form in two situations.

1. If we know the slope of a line and its y intercept, we can find the equation for the line.

2. Suppose we need the slope of a line, and we are given the equation of the straight line. If we rewrite the given equation in the slope-intercept form, then the slope of the line is the coefficient of the x-term.

m is the known slope.
b is the y intercept.

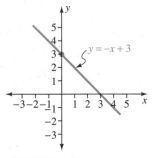

FIGURE 1.19
A line with $m = -1$ and $b = 3$.

Example 21   What is an equation of a line whose slope is -1 and whose y intercept is 3?

Solution

Here $m = -1$ and $b = 3$. Use this information directly in the slope intercept form:

$$y = (-1)x + 3$$

or

$$y = -x + 3$$

See Figure 1.19 for a graph of this equation.

Skill Enhancer 21   Write an equation of a line whose slope is $\frac{1}{2}$ which passes through the origin.

Answer in Appendix E.

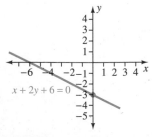

FIGURE 1.20
The line $x + 2y + 6 = 0$.

Example 22   A certain line has the equation $x + 2y + 6 = 0$. What is its slope?

Solution

Use algebra to manipulate the given equation into one that looks like the slope-intercept form. That is, solve for y on the left.

$$x + 2y + 6 = 0 \quad \text{becomes} \quad y = -\frac{1}{2}x - 3$$

The coefficient of x is the slope.

$$m = -\frac{1}{2}$$

It is also clear that $b = -3$. A graph of this equation appears in Figure 1.20.

Skill Enhancer 22 What is the slope of the line given by $3x - 2y = 17$?

Answer in Appendix E.

Example 23 Given the equations for three lines

$$2x + 3y = 4$$

$$y = -\frac{3}{2}x + 7$$

$$\frac{2}{3}x + y = 0$$

which of these three are parallel?

Solution

Graphs of these equations appear in Figure 1.21. Solve each of these equations for y:

$$y = -\frac{2}{3}x + \frac{4}{3}$$

$$y = -\frac{3}{2}x + 7$$

$$y = -\frac{2}{3}x$$

From our work with the slope-intercept form, we know that the coefficient of x in each yields the slope of each line. The first and third lines have the same slope, $m = -\frac{2}{3}$, and are parallel.

Skill Enhancer 23 Which are the equations for parallel lines? (a) $y = -\frac{5}{2}x - 3$; (b) $5x + 2y = 10$; and (c) $13 - 10x - 4y = 0$.

Answer in Appendix E.

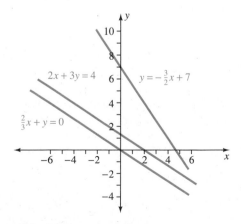

FIGURE 1.21
Which lines are parallel?

_____Exercises 1.2_____

In Exercises 1–16, compute the slope of a line which connects the given points:

1. $(1, 2)$, $(3, 4)$ **2.** $(7, -9)$, $(-1, 5)$

3. $(1, 0)$, $(1, 3)$ **4.** $(-7, -7)$, $(2, 2)$

5. $(3, 4)$, $(1, 4)$ **6.** $(\frac{1}{3}, \frac{2}{3})$, $(\frac{2}{3}, 1)$

7. $(0, -0.5)$, $(10, -0.5)$ **8.** $(a, \frac{2a}{3})$, $(a, \frac{a}{2})$, for $a \neq 0$

9. $(7, 3t)$, $(6, 2t)$, $t \neq 0$ **10.** $(-2q, 3)$, $(-4q, -4)$

11. $(-4, -12)$, $(-8, -3)$ **12.** $(-3.5, -7)$, $(-2, -5.5)$

13. $(\frac{1}{5}, \frac{3}{7})$, $(-\frac{3}{5}, -\frac{2}{7})$ **14.** $(\frac{1}{3}, \frac{1}{4})$, $(-\frac{1}{2}, \frac{2}{3})$

15. $(\frac{2}{5}, -9)$, $(\frac{3}{5}, -8)$ **16.** $(0.72, 0.38)$, $(0.44, 1)$

W 17. Compute in two ways the slope for the line connecting the points $(3, -4)$ and $(1, 2)$. First, let P_1 be $(3, -4)$. For the second way, let $P_1 = (1, 2)$. Does it matter which point you choose as P_2 and which as P_2? Why or why not?

18. Consider the line passing through the points $(9, 8)$ and $(4, 5)$. Compute the slope in two ways as in Exercise 17.

In Exercises 19–34, for each given pair of points find an equation of the line that passes through each of the given pairs. Then write each equation in the form $Ax + By = C$.

19. $(1, 1)$, $(2, 2)$ **20.** $(5, 3)$, $(3, 5)$

21. $(-2, 1)$, $(3, -2)$ **22.** $(1, 1)$, $(11, 1)$

23. $(-1, 12)$, $(-1, -1)$ **24.** $(3, 4)$, $(1, 4)$

25. $(\frac{1}{3}, \frac{2}{3})$, $(\frac{2}{3}, 1)$ **26.** $(0, -0.5)$, $(10, -0.5)$

27. $(99, 100)$, $(100, 101)$

28. $(-99, -100)$, $(-100, -101)$

29. $(-2, 3)$, $(-4, -4)$ **30.** $(-4, -12)$, $(-8, -3)$

31. $(-5, -7)$, $(-2, -5.5)$ **32.** $(\frac{1}{5}, \frac{3}{7})$, $(-\frac{3}{5}, -\frac{2}{7})$

33. $(\frac{1}{3}, \frac{1}{4})$, $(-\frac{1}{2}, \frac{2}{3})$ **34.** $(2, -9)$, $(\frac{3}{5}, -8)$

Find equations of the lines passing through the given points and having the specified slopes in Exercises 35–50. Give the answers in the form $y = mx + b$. Graph the lines.

35. $(3, 4)$; $m = -1$ **36.** $(2, -1)$; $m = 2$

37. $(4, 1)$; $m = 0$ **38.** $(4, 1)$; m undefined

39. $(-2, 3)$; $m = 0.5$ **40.** $(\frac{1}{3}, \frac{1}{3})$; $m = \frac{1}{3}$

41. $(-5, -7)$; $m = -\frac{4}{5}$ **42.** $(-4, -5)$; $m = 7$

43. $(18, 12)$; $m = -3$ **44.** $(0, 0)$; $m = 1$

45. $(-1, -1)$; $m = -10$ **46.** $(-19.87, 1)$; $m = 0$

47. $(-3, -\frac{1}{2})$; $m = \frac{1}{2}$ **48.** $(-\frac{1}{3}, -\frac{1}{4})$; $m = -\frac{1}{5}$

49. $(12, -0.78165)$; m undefined

50. $(8, 3)$; $m = 5$

51. Find the absolute value of x when x has the value **(a)** 3 **(b)** -12 **(c)** 0 **(d)** $-\frac{3}{2}$

W 52. Here is how B. C. Dull "proved" that the absolute value of -3 is -3. By definition, we find $|x|$ by dropping the sign. Therefore, $|x| = x$. Now let $x = -3$, and we have therefore "shown" that $|-3| = -3$. Where is the flaw in this argument?

Find equations of the lines passing through the given points and having the specified slopes in Exercises 53–68. Give the answers in the form $y = mx + b$.

53. $(0, 0)$; $m = -1$ **54.** $(0, -1)$; $m = 1$

55. $(0, \frac{1}{2})$; $m = -1$ **56.** $(0, 0)$; $m = \frac{5}{8}$

57. $(-1, -1)$; $m = -1$ **58.** $(2, 2)$; $m = -2$

59. $(1, -1)$; $m = 0$ **60.** $(\frac{1}{3}, -\frac{1}{3})$; m undefined

61. $(10, 4)$; m undefined   **62.** $(3, 8)$; $m = 0$

63. $(\frac{1}{2}, \frac{1}{2})$; $m = \frac{1}{2}$ **64.** $(-1, 3)$; $m = \frac{1}{10}$

65. $(2, 4)$; $m = 6$ **66.** $(6, 4)$; $m = -2$

67. $(1, 2)$; $m = -3$ **68.** $(3, 2)$; $m = -1$

Find the equation of the line passing through the pairs of points given in Exercises 69–80. Write each equation in the form $ax + by = c$.

69. $(0, 0)$, $(1, 1)$ **70.** $(-1, -1)$, $(2, 2)$

71. $(2, 2)$, $(-1, 2)$ **72.** $(-1, 4)$, $(-1, -1)$

73. $(2, 6)$, $(2, -3)$ **74.** $(5, -10)$, $(5.01, -10)$

75. $(2, 3)$, $(3, 2)$ **76.** $(5, 4)$, $(5, 5)$

77. $(\frac{1}{2}, \frac{3}{2})$, $(1\frac{1}{2}, 4)$ **78.** $(1\frac{1}{3}, -\frac{2}{3})$, $(0, -1)$

79. $(2, 1)$, $(\frac{1}{4}, -\frac{3}{4})$ **80.** $(0, \frac{2}{3})$, $(-\frac{2}{3}, 0)$

For each of Exercises 81–92, find an equation of the line having the given y intercept and slope. Then graph the line.

81. $b = 2$, $m = 2$ **82.** $b = -3$, $m = 0.5$

83. $b = 0$, $m = 1$ **84.** $b = m = -5$

85. $b = -\frac{7}{2}$, $m = -3$ **86.** $m = 2b$, $m = 0$

87. $b = 7$, $m + b = 10$ **88.** $b = m = 0$

89. $b = 3$, $m = -3$ **90.** $b = -7$, $m = \frac{1}{2}$

91. $b = -0.5$, $m = -7$ **92.** $b = 3m$, $b = -6$

Determine the slope of each line in Exercises 93–108.

93. $2x + y = 3$ **94.** $3x - 9y = 6$

95. $y - 20 = 0$ **96.** $x = 12$

97. $9y - 2x = 13$ **98.** $\frac{2}{3}x + \frac{3}{2}y = -1$

99. $2x - 3y = 12$ **100.** $100x - 0.5y = 2$

101. $52x = 13y - 4$ **102.** $2(x - y) + 2(x + y) = 4$

103. $2(x - y) - 3(x + y) - 12 = 0$

104. $ax + by + c = 0$, for $b \neq 0$

105. $3(x + 2y) + 3(y + 2x) = 7$

106. $4(x + 3y) + 4(x - 3y) = 7$

107. $2(y + 3x) + 2(y - 3x) = 10$

108. $37x + 43y = 1,000$

109. Let the point where any line intersects the x axis have coordinates represented by $(a, 0)$.
  (a) Show that the *slope–x intercept* form of a straight line is given by $x = \frac{y}{m} + a$.
  (b) Given the x and y intercepts, show that the *double-intercept* form of the line is $\frac{x}{a} + \frac{y}{b} = 1$, where $(0, b)$ is the y intercept as usual.

W **110.** *(Requires special thought)* In deriving the point-slope form for the equation of a straight line, why is it necessary to demand that the point (x, y) be distinct from the point (x_1, y_1)?

111. Find an equation of the line parallel to the line $2y - x = 4$ which passes through the point $(-1, 7)$.

112. Find an equation of the line parallel to the line $x + y = 3$ which passes through the point $(0, -1)$.

113. Find an equation of the line parallel to the line $3x + 2y + 1 = 0$ which passes through the point $(0, 0)$.

114. Find an equation of the line parallel to the line $2x - y = 9$ which contains the point $(-1, -2)$.

115. Two lines are perpendicular if the *product* of their slopes is -1.
  (a) What is the slope of the line perpendicular to $2y - x = 4$?
  (b) Determine an equation of the line perpendicular to $2y - x = 4$ which passes through the point $(-1, 7)$.

116. What is the slope of the line perpendicular to the equation $x - y = 0$? Determine the line perpendicular to this equation passing through $(0, -1)$. (Refer to Exercise 115.)

Applications

117. *(Restaurant Management)* A franchise of Big Boy Burgers reliably increases its business by about $60,000 of business per month. A new Big Boy Burger restaurant has opened up in Fort Wayne, IN. In its second month of operation, it did $500,000 worth of business. Estimate the business it will do during its sixth month of operation. What equation describes its growth?

118. *(Physics: Mechanics)* When a ball is dropped from a tower, its velocity v after t seconds is given by $v = gt$, where g is the so-called gravitational constant. Estimate the value of g if the ball is traveling with a velocity of 112 feet per second (ft/s) after 3.5 seconds (s) have elapsed.

119. *(Physics: Electricity)* In an electric circuit, the voltage V depends on the resistance R and current I according to the formula $V = IR$. Estimate the resistance of a particular circuit if a current of 3 amperes (A) produces 9.9 volts (V). Is this consistent with a current of 2.5 A producing a voltage of 8.25 V? According to this problem, what are the units in which resistance should be measured?

120. *(Library Acquisition)* The public library system of a certain large town has been ordered to purchase 150 books per month. The library opens with 155,000 volumes. What equation describes the library holdings as a function of the months since it opened? In what month will the library first own 200,000 books?

121. *(Forestry)* A team of foresters needs to prune 75 trees per week. Another team had pruned 160 at the time when this team took over. Write an equation relating the total trees pruned to the number of weeks that this team works. If this forest contains 450 trees, when will the team be finished with it?

W **122.** *(Buying Home Computers)* In 1982, Eric purchased the most powerful home computer system he could find. It cost him exactly $5,000 at that time. By 1985, the price for a system with *exactly* the same capabilities had declined to $2,300.
   (a) If the price decline was linear, estimate the price of the system in 1984.
   (b) Why do you think the price for an equivalent system could not continue to be linear much past 1986? What do you think might happen at that time to the price and performance of the system?

Calculator Exercises _____

Use a calculator to compute the answers to Exercises 123–

128. Express all answers and coefficients to two-decimal-place accuracy. What is the slope of the lines passing through the pairs of points in Exercises 123–124?

123. $(-3.7, -4.2)$, $(-6.8, 0.1)$

124. $(9.3, 6.3)$, $(-2.7, -4.4)$

What is the slope of the lines whose equations appear in Exercises 125 and 126?

125. $7.2x - 13.5y = -6$ **126.** $6.2(x - 3.1) = 5.4y$

Each of Exercises 127 and 128 lists the slope of a line and the coordinates of a point. Find an equation of the line passing through that point with the given slope. Express each equation as $ax + by = c$.

127. $m = -4.5$; $(3.4, 0.1)$ **128.** $m = 6.1$; $(12, -4.5)$

Section 1.3   APPLICATIONS OF LINEAR EQUATIONS

The successful application of linear equations is crucial to the solution of many real-world problems. In this section, we examine a few of these topics.

Simple Interest and Inflation

interest   A borrower pays *interest* for the use of someone else's money. The sum borrowed is the
principal   *principal*.

   Typically, the interest depends on three quantities:

   1. The *principal*.
   2. The *interest rate R,* which is a percentage denoting the fraction per year of the principal P that the borrower repays in the form of interest.
   3. The *time T* elapsed before the lender regains the principal. T is measured in years.

simple interest   The simplest interest is *simple interest,* for which the interest I is the product of the principal multiplied by the time and interest rate:

$$I = PRT$$

Example 24   Mitch wants to borrow $1,500 from Anita. Anita is happy to help out, provided Mitch will pay simple interest at 12 percent per year. In this problem,

$$I = 0.12 \times 1,500 \times T$$

or

(1.4) $$I = 180T$$

Simple interest (in units of dollars) is linearly related to the time the money is borrowed (in units of years).

(a) How much interest will accrue if Mitch keeps the borrowed money for 2 years?

(b) If he keeps it for 30 months?

Solution

(a) Let $T = 2$ in Equation (1.4).

$$I = 180(2) = \$360$$

(b) Since we have agreed to measure time in units of years, we convert 30 months to 2.5 years. Therefore,

$$I = 180(2.5) = \$450$$

A graph of Equation (1.4) appears in Figure 1.22.

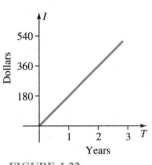

FIGURE 1.22
For the simple interest problem in the example, $I = 180T$**.**

Skill Enhancer 24   Compute the simple interest on $1,500 for 3 years and for 15 months. The rate is 8 percent.

Answer in Appendix E.

inflation   We use a similar calculation to compute the rise in prices of goods or services over time, which is called *inflation*. For short time periods, it is appropriate to use the simple interest formula to compute the effects of inflation.

Example 25   Economists predict a 7.5 percent inflation rate for the coming year. A car costs $9,500 today. Estimate its cost next year.

Solution

Let I stand for the estimated price increment. The period T is one year.

$$I = (0.075)(9,500)(T) = 712.50(1) = \$712.50$$

The new price equals the old price plus the inflationary increase.

$$\text{New price} = 9,500 + I = 9,500 + 712.50 = \$10,212.50$$

Inflation does not change all prices uniformly, so this is only an estimate.

Skill Enhancer 25   Redo this example if the inflation rate is 9.5 percent and the car costs $12,000 today.

Answer in Appendix E.

Example 26   A union has been unable to secure a salary cost of living increase for at least the next year and a half. Suppose a worker earns $23,500 per year now. Further assume that inflation persists at an annual rate of 8 percent for the next year and a half, and that there are no merit raises during that period. Estimate how much less the worker's salary will be worth at the end of the period.

Solution

The effect of inflation is the same as placing the salary in a bank granting a *negative* rate of interest equal to the inflation rate. Therefore, with $P = 23,500$, $T = 1.5$, and $R = -0.08$,

$$I = 23,500(-0.08)(1.5) = -\$2,820.00$$

The salary will be worth $2,820 less then than it is now.

Skill Enhancer 26   Another worker in the union currently makes $35,000 per year. How much less will her salary be worth at the end of the period?

Answer in Appendix E.

Interest and inflation are treated in greater detail in Chapter 10.

Depreciation

depreciation

All equipment and property wear out in time. In business, the amount by which something declines in value each year as a result of deterioration is a business "expense" to be deducted from the business's revenues. The amount by which an asset declines in value is the *depreciation*.

How long does it take for a 18-wheeler truck or an apartment house to fully depreciate? What portion wears out each year? No one can answer this exactly for all different assets, so there are rules that are assumed to hold for all depreciable assets. The simplest method for computing depreciation is a linear method, whereby the initial value is reduced by some equal amount each year, until at some future specified time the object retains no further economic worth.

Example 27   Ivor Gridley purchases a 40-unit apartment building in Dubuque, Iowa for $1.2 million. Of that sum, $300,000 represents the cost of the land on which the building sits. Land is assumed not to depreciate (after all, land never wears out), and for depreciation purposes, the building is said to be fully depreciated in 25 years. In doing the problem, we will assume that a linear depreciation scheme is appropriate. How much can be depreciated each year for income tax purposes? How much value will be assigned to the property in 15 years?

Solution

For purposes of depreciation, at the time of purchase the building is worth $900,000, the difference between the purchase price of $1.2 million and the presumed worth of the land. If we can depreciate the building over 25 years, then the amount we can depreciate

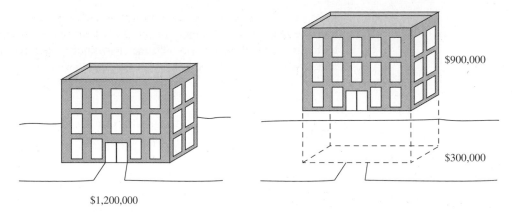

$900,000

$300,000

$1,200,000

FIGURE 1.23

each year is

$$\text{Depreciation amount} = \frac{\$900,000}{25} = \$36,000$$

Next, let x represent the time in years since the date of purchase, and let y represent the building's value at that time. Then, since the building's worth decreases by $36,000 each year, we have

(1.5) $$y = 900,000 - 36,000x$$

When $x = 15$, $y = 900,000 - 540,000 = 360,000$. Thus the building has a remaining economic value of $360,000, to which must be added the (nondepreciable) value of the land if we want to find the total value of the property at that time.

$$\text{Value} = 360,000 + 300,000$$
$$= \$660,000$$

Figure 1.24 displays the relation between y and x in Equation (1.5).

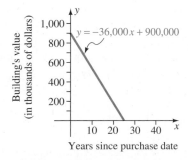

FIGURE 1.24
Depreciating an apartment house.

Skill Enhancer 27   Another investor pays $2 million for a small office building. Of that amount, $500,000 represents the cost of the land. The investor may depreciate the building over a 20-year period of time. How much depreciation is there each year? How much will the property be worth (for income tax purposes) after 10 years?

Answer in Appendix E.

Business and Economics: Break-Even Analysis; Supply and Demand

Any business fails or succeeds on the basis of the money that flows in and out. Money flows in as a result of selling products. Typically, the revenue R is simply equal to the selling price P of each item times the number x of items sold:

$$R = P \cdot x$$

At the same time, there are costs C associated with the production process, including a cost E for each item manufactured and some fixed cost F (which might include the cost of special molds or machinery that must be purchased before the manufacturing process can get under way). Therefore,

$$
\begin{array}{c}
\text{FIXED} \\
\text{COSTS} \\
\Downarrow \\
C = \underbrace{Ex}_{\substack{\Uparrow \\ \text{VARIABLE} \\ \text{COSTS}}} + F
\end{array}
$$

break-even point   When revenue exceeds costs, profit results. When costs exceed revenue, the business experiences a loss. The level of production at which revenues and costs exactly balance is called the *break-even point*. The importance of this value to the business planner lies in the fact that sales greater than this value result in profits.

Example 28   The fixed costs associated with the production of technical books, like this one, are quite high due to the cost of typesetting mathematics. The fixed costs (mostly for typesetting) for an upcoming edition of *Elements of Advanced Chromodynamics* are about $44,000. Each book has an additional cost of $8.50 associated with the actual printing and manufacturing of the book. The book should sell 15,000 copies at $19.50 apiece. (Assume all this revenue passes directly to the publisher.) Will the publisher make money on this venture? If so, how much?

Solution

Let y_1 and y_2 represent cost and revenue. If x represents the numbers of copies sold, then

$$\text{Cost} = y_1 = 44{,}000 + 8.5x$$
$$\text{Revenue} = y_2 = 19.5x$$

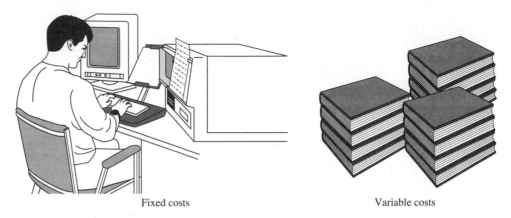

Fixed costs   Variable costs

FIGURE 1.25

The regions of success and failure are indicated in Figure 1.26 which shows graphs of both these equations. Break-even sales occur when $y_1 = y_2$ or

$$44,000 + 8.5x = 19.5x$$

The solution is $x = 4,000$. Since the publisher expects to sell 15,000 copies, this enterprise should be profitable. The cost of producing 15,000 copies is

$$y_1 = 44,000 + 8.5(15,000) = \$171,500$$

while the income from selling them will be

$$y_2 = 19.5(15,000) = \$292,500$$

The net profit should be $\$292,500 - \$171,500 = \$121,000$.

Skill Enhancer 28   In this example, suppose the book will be selling for \$44 a copy, but the fixed costs are also much higher—\$100,000. The publisher can only count on selling 9,000 copies. Variable costs are now \$10 per book. How much profit or loss will the publisher experience, based on these figures?

Answer in Appendix E.

demand   Linear equations may help economists analyze the eventual selling price of goods. In a market economy, consumers indicate a certain *demand* for an item. An equation attempting to relate this demand to the price that consumers are willing to pay would have to reflect an important fact: *as the available quantity of goods decreases, the price increases.* That is, the fewer there are of something, the more people will pay. And conversely, the price will decrease as the amount available increases.

supply   The producers have a certain *supply* of the item. An equation that relates the supplier's selling price to supply would show a price increase as the number produced increases. Otherwise, the suppliers have no incentive to produce more. (Think carefully about this. Some political observers feel that a contribution to the recent breakdown of the Soviet Union was precisely this lack of incentive.)

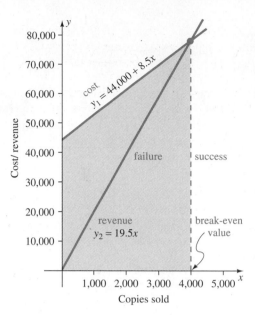

FIGURE 1.26
Break-even analysis in book publishing.

Often, supply and demand equations can be expressed using linear equations. We can use the equations to investigate the tug-of-war between the supply and demand forces.

Example 29   On one particular day, the price that investors are willing to pay for gold bullion depends on the available numbers of ounces:

   Demand equation: $p = -0.5x + 650$

where p is the price (in dollars) per ounce and x represents the numbers of ounces of gold in hundreds. The price the owners of gold are willing to accept for their gold is

   Supply equation: $p = 0.7x + 170$

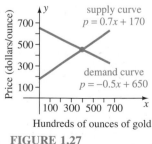

FIGURE 1.27

Supply and demand curves regulate the international gold market.

At what price will gold eventually sell on this day? Estimate the amount of gold that will change hands.

Solution

We have plotted the graphs of the demand and supply equations in Figure 1.27. Rational sellers observe that as they come forward with more gold, the price per ounce will increase. On the other hand, investors notice that as the supply increases, the price they are willing to pay decreases. These two market forces come into equilibrium at the point where the two graphs intersect. The point of intersection yields the actual market selling price and the amount of gold sold at that price.

   Setting the supply and demand equations equal yields

$$-0.5x + 650 = 0.7x + 170 \qquad \text{or} \qquad 1.2x = 480$$

The solution is $x = 400$, at which point the selling price of gold is $450 per ounce. Since x measures the amount sold in units of hundreds of ounces, the amount sold this day will be about $400 \times 100 = 40,000$ ounces of gold.

Skill Enhancer 29   Another economist determines that the demand curve for this problem really should be given by $p = 698 - 0.6x$. If the point at which the demand and supply curves intersect remains $x = 400$, and if the supply curve continues to have the same y intercept as that in the problem, determine the new supply curve for this problem.

Answer in Appendix E.

Prediction

Linear equations are often useful for predicting tomorrow's results from yesterday's data. Of course, these predictions are only estimates—no one can really predict the future with total accuracy. In the absence of additional information, it makes sense to use the simplest possible model: assume the process behaves in a linear fashion.

Example 30   The Akita is a large and handsome Japanese breed of dog, characterized by a prominent, bearlike head and a tail that curls over the animal's back. This dog has become quite fashionable in New York City. In 1989, a show-quality pup could be bought for $1,300. In 1993, a similar pup could cost $3,400. If Akitas maintain their trendiness, how much would you estimate a pup to cost in 1994?

Solution

Assume Akita prices increase linearly. Let x be the number of years since 1989, and let y be the price of an Akita in dollars that time. When $x = 0$ (1989 is 0 years from 1989), $y = 1,300$. When $x = 4$ in 1993, $y = 3,400$. The slope m is given from these two "points" by

$$m = \frac{3,400 - 1,300}{4 - 0} = 525$$

so Akita prices will have increased on average by $525 per year. This slope and the coordinates of the first point yield an equation relating price rises to years by

$$y - 1,300 = 525(x - 0)$$

or

$$y = 525x + 1,300$$

Figure 1.28 shows this relationship in graphic terms. In 1994, $x = 1994 - 1989 = 5$, so the price of the pup will be about

$$y = 525(5) + 1,300 = \$3,925$$

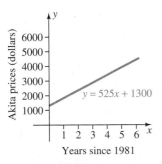

FIGURE 1.28
Akita prices depend on time.

Skill Enhancer 30   A show-quality pup in 1993 cost only $2,300. Estimate now the cost of a pup in 1994.

Answer in Appendix E.

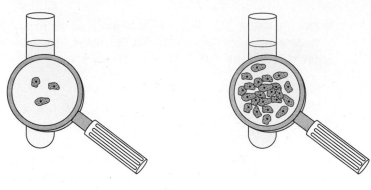

FIGURE 1.29

Example 31   (**Algal Blooms**)   *Algae* are single-celled plantlike organisms living in bodies of water. Certain species are harmful to people in high enough concentrations. One such species appeared in the waters off San Diego one spring. On March 1, marine biologists measure 32,000 algae cells per cubic centimeter (cc) of water. On March 21, they measure 62,000 algae per cc. These algae are toxic when the concentration reaches or exceeds 110,000 per cc. Assuming a linear growth rate, when will the authorities have to close the San Diego beaches?

Solution

Let x represent the number of days since March 1, and let y represent the concentration of algal organisms per cubic centimeter. On March 1, $x = 0$ and $y = 32,000$. On March 21, $x = 21 - 1 = 20$ and $y = 62,000$. These points lie on a straight line. Its slope m must be

$$m = \frac{62,000 - 32,000}{20 - 0} = 1,500$$

An equation governing this growth is

$$y - 32,000 = 1,500(x - 0)$$

or

(1.6) $$y = 1,500x + 32,000$$

where we have used the first point $(0, 32,000)$ in the point-slope formula. To determine the day of the beach closings, set $y = 110,000$ in this equation, and solve for x.

$$110,000 = 1,500x + 32,000$$
$$78,000 = 1,500x$$
$$52 = x$$

The authorities will close the beach 52 days after March 1. The graph of Equation 1.6 appears in Figure 1.30.

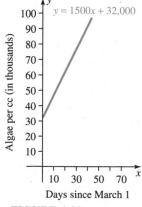

FIGURE 1.30

The concentration of algae in San Diego Bay changes over time.

Skill Enhancer 31 Redo the calculation if the danger level for algae is $106{,}000$ organisms per cc. Furthermore, although the reading on March 21 was correct, the initial reading on March 1 was found to be in error. The corrected reading is $y = 40{,}000$.

Answer in Appendix E.

Example 32 Oil Park, a suburb of Houston, has been experiencing the same boom as has that city. The Oil Park Police Department (OPPD) has been growing with the population. In 1987, there were 156 full-time police officers. In 1995, there will be 252. The town planner needs to know how many officers were on the force in 1992. Unfortunately, these records have been destroyed by fire. How can she reasonably estimate this figure?

Solution

We do not know that the OPPD has expanded linearly, but, in the absence of additional data, it is reasonable to assume that expansion has proceeded in a linear manner.

We may derive a relation by letting x represent the years since 1987 and y represent the level of employment of the OPPD at that time. When x is zero, y is 156. In 1995, $x = 8$ and $y = 252$. The slope is

$$m = \frac{252 - 156}{8 - 0} = 12$$

so that by the point-slope form

$$y - 156 = 12(x - 0)$$

and finally

$$y = 12x + 156$$

(we have used the coordinates of the first point together with the slope in the point-slope formula). We see this equation in Figure 1.31. In 1992, x is 5 ($= 1992 - 1987$), so the size of the OPPD at that time was probably close to

$$y = 12(5) + 156 = 216$$

men and women.

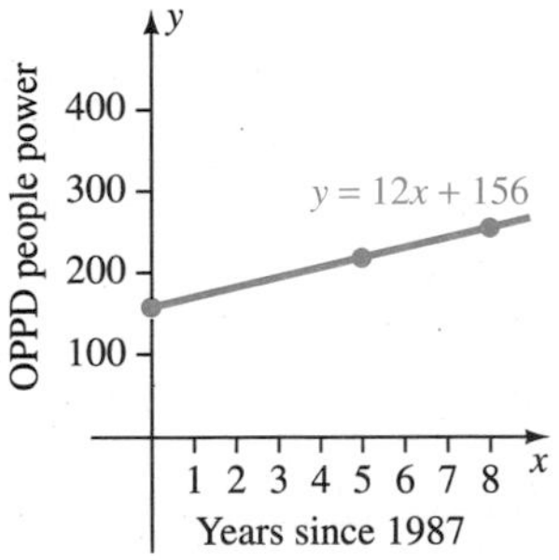

FIGURE 1.31
The size of a police force varies with time.

Skill Enhancer 32 Estimate the number of officers on the OPPD force in 1994.

Answer in Appendix E.

Exercises 1.3

Applications

1. *(Simple interest)* One of your classmates is in desperate need of cash and wants to borrow $100 from you today. She promises to return $115 to you in exactly 10 days. If you agree to this deal, at what rate of simple interest will your money be earning interest?

2. Redo Exercise 1 if your classmate returns $120 to you in exactly one month.

3. *(Inflation)* At the beginning of the school year, a round trip plane ticket to Fort Lauderdale costs $195. Nine months later, the same ticket costs $215. Estimate the annual rate of inflation from these data.

4. (Inflation) Inflation is currently running at 6.5 percent annually.
   (a) If a new sailboat costs $10,000 today, estimate its cost one year from today.
   (b) Estimate its cost 14 months from today.
   (c) Estimate its cost 18 months from today.

5. (Low-Temperature Physics) At a normal range of temperatures, let a volume of some gas be enclosed so that it is at constant pressure. This volume varies linearly with temperature. Suppose, for this gas, the volume at 0°C is 1 cc. At 100°C, its volume is 1.3663 cc. Physicists define *absolute zero* as the temperature below which all molecular motion ceases; this is the coldest temperature possible. In terms of the relationship between volume and temperature, this is the temperature at which the volume of the gas would decrease to 0 if the linear relation held good over this vast range of temperatures. Assume that it does, and use the above information about this given gas to estimate absolute zero on the Celsius scale.

6. (Entrepreneurship) Seven-year-old Sam sells lemonade for $0.25 per cup, and his fixed costs (sugar, drink mix, and so forth) are $4.35 per day. What is his break-even point?

7. (*Requires special thought*) Little Sam (see Exercise 6) sells lemonade, and his fixed costs have risen to $8.90 per day. He sells a regular cup of lemonade for $0.25, and a large-sized cup for $0.45. At the break-even point, what is the relationship between the number of large- and regular-sized cups that must hold?

8. (Gravity) There is a special relationship between the weight of an object and the mass of the planet on which the object is weighed. The weight is proportional to two factors: the mass of the planet and the quantity $1/R^2$, where R is the radius of the planet. How much would a 100-pound woman weigh on the planet Sram, which is one-sixteenth the mass of the earth but has the same radius?

9. (Biology) One part of a biology experiment involves measuring the population of one-celled protozoa (very tiny aquatic animals) in a test tube at various concentrations of nutrients. Here is a table summarizing some recorded data:

Protozoa	Nutrients
(No. per cc)	(mg/cc)
990	45
10,220	500

Estimate the population per cubic centimeter if the nutrient concentration is raised to 650 milligrams per cubic centimeter (mg/cc).

10. (Wine selling)
   (a) A simple analysis of the operation of Laredo Vineyards shows that fixed costs associated with this year's grape harvest are $1.2 million. The vineyard sells its wine directly to the public for $7.80 per bottle. What is the break-even point?
   (b) A more careful analysis finds that the fixed costs are only $750,000, but the bottling costs per bottle are $1.50. What now is the break-even point?

11. (Real estate) What is the annual depreciation on a small office building bought for $3.5 million? Land is assumed to account for one-seventh of the purchase price, and the building is depreciated over a 15-year period.

12. (Trucks) A medium-sized electronics company buys a truck for $75,000. If it is depreciated linearly over a 15-year period, what is the truck's remaining economic value at the end of 5 years? at the end of 12 years? at the end of 15 years? at the end of 16 years?

13. (Used cars) Harry just bought a new car for $13,000. He anticipates selling it in three years, at which point he will purchase a new car.
   (a) The economics of used cars are peculiar. The car declines in value by $3,000 the moment Harry drives it out of the showroom. Thereafter, it declines by 15 percent of its value annually. How much will it be worth in three years?
   (b) Assume that Harry wants to replace his car with a current model of the same car in three years. If inflation persists at a 5 percent annual rate between now and then, estimate the price of the new car in three years.
   (c) How much additional cash will Harry have to come up with in three years to buy the new car?

14. (Telephones) The phone company will sell a wall-unit telephone for $59.95. You may put $19.95 down and pay the balance in three monthly installments of $15.00. Suppose this is equivalent to borrowing the $49.95 purchase price for three months. This corresponds to simple interest at what percent?

15. (Economics) The bottle of wine which cost $10 last New Year's Eve costs $11.50 this year. Use these facts to estimate the rate of inflation for the intervening year.

16. (Civil Service) When the number of sanitation workers was 117, they were able to remove 130 tons of refuse weekly. When they numbered 169, they could remove 182 tons weekly. The town planners for this community anticipate that the community will be generating 286 tons per week by next year. Estimate how many sanitation people the community will require.

17. **(Temperature Physics)** Celsius and Fahrenheit are two different ways of measuring temperature. In the Fahrenheit scale, with which most of us are more familiar, the freezing point of water is 32 degrees and the boiling point is 212. In the Celsius scale, water's freezing and boiling points are defined to be 0 and 100 degrees. If the conversion between the two scales is linear (which it is), derive the relation that connects the two scales.

18. **(Gravity)** The speed of anything falling under the influence of the earth's gravity is given by the linear relation

$$v = gt + v_0$$

ignoring the resistance of air. Here, v is the velocity in meters per second (m/s) after t seconds have elapsed, when the object has an initial velocity of v_0 m/s. The constant g is the *gravitational constant,* roughly 9.8 m/s/s.

   (a) If a body falls from rest, its initial velocity is zero. If we drop an apple from the top of the Sears Tower in Chicago, how fast will it be traveling after 20 s? (Assume the Sears Tower is so tall that the apple takes longer than 20 s to reach the ground.)

   (b) If we throw it down, so that its initial velocity is 15 m/s, what then will its velocity be after 20 s?

19. **(Weapons Design)** The Prussian army contemplated using the following technique for estimating the muzzle velocity of a gun (the speed with which a bullet leaves a gun). They fired the gun straight upward, and noted how long it took before the bullet started on its downward descent. At that point, the bullet's velocity is precisely 0 m/s. Firing a bullet upward is tantamount to giving it a negative initial velocity. Use the formula given in the previous problem to estimate a bullet's muzzle velocity if it takes 40 s to reach the highest point of its path.

W 20. **(Weather)** On a bitter cold morning in Basalt, Colorado, Blossom notices that it is $-10°F$ at 8:00 in the morning. By 10:00, the temperature has risen to $+5°$. Assume that the temperature is rising linearly, and esti-

mate the temperature at 11:00. Why do you think you could not use this method to estimate the temperature at 8:00 that evening?

21. **(Restaurant Management)** Mario Cavaradossi has recently opened his own pizzeria. He quickly realizes that the bottom line is not how much he receives from selling pizza, but how much remains after paying expenses. The first month of business, he sold 1,000 pizzas, and was left with $1,200 after all expenses. The second month, after getting many of the new routines under his belt, he sold 1,500 pizzas, and found himself with $1,800 after expenses. In the third month, he is confident he can make and sell 2,000 pizzas. If his gross profit continues to rise in a linear manner, how much can he anticipate making in that month?

22. **(Employment Opportunities)** College students are tempted by the fabulous salaries available to computer science majors upon graduation. In fact, the demand curve relating the salaries S to the number of computer science majors in the country is

$$S = 37,000 - 2x$$

The supply equation, however, is

$$S = 6x + 1000$$

What is the equilibrium salary that these two curves determine, and how many computer science majors can the data processing industry expect to see?

23. **[Numismatics (Coin Collecting)]** Among the members of a club of coin collectors, an uncirculated 1955 penny minted in San Francisco is in great demand. The supply equation relating the price p in dollars to the number x of available coins is $p = 10 + 3x$. The demand equation is $p = 17 - 0.5x$. Estimate the number of such coins that will be traded. At what price will they be traded?

24. Refer to the previous exercise. Several months later, observers note that four of these pennies sold for $16 apiece. Coin collectors know that they won't pay more than $19 for one of the coins even when the available supply shrinks to zero. Assume that the demand equation is linear, and use this information to derive the demand equation for this new problem.

Section 1.4 *SOLVING LINEAR EQUATIONS WITH TWO VARIABLES*

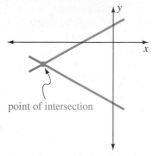

FIGURE 1.32
**The point of intersection
of two lines represents the
solution to a system of two
equations.**

When there are two variables in a linear equation, we will need the information contained in two equations to find a unique value for x and y (or for x_1 and x_2, if subscript notation is used). Even with two equations, there may not be a unique solution.

Two-Variable Systems: Classifying the Solutions

A single linear equation in two variables is equivalent to a single straight line graphed on a set of coordinate axes. Therefore, two equations correspond to a pair of lines.

What will a *unique solution* correspond to? A solution to both equations is a point lying on each line and yet satisfying both equations. When the lines are distinct, there is at most one such point—the point of intersection of the two lines, as lines can never intersect at more than one point. When we can solve two equations for a unique solution, the equations are *consistent equations*.

consistent equations

inconsistent equations

Pairs of straight lines need not be so obliging. Two equations will be *inconsistent equations* and have *no* solution if they represent two distinct parallel lines, as in Figure 1.33. To show that a pair of equations are inconsistent, it is sufficient to show they have the same slopes and different y intercepts. (Why is this condition sufficient?)

dependent equations

Two equations are *dependent equations* if they actually represent the same line. This is not trivial; equations may appear in such complicated disguises that their equivalence is obscured. In this case, an *infinite* number of points satisfy the equations, representing all points lying on the line, and so there are an infinite number of solutions to dependent equations. Dependent equations have the same slope-intercept form.

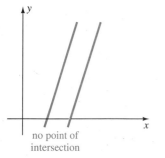

FIGURE 1.33
Inconsistent equations.

Systems of Two Linear Equations: Recognizing Types of Solutions.

1. *Consistent equations* yield unique values upon solution. The graph of two consistent linear equations is a pair of intersecting straight lines.

2. *Inconsistent equations* yield a contradictory result and have *no solution*. When graphed, inconsistent equations become a pair of parallel lines.

3. *Dependent equations* yield an equation that is satisfied by *many* values of the unknowns. These equations have an *infinite number* of solutions. The graph of two dependent equations appears to be a single straight line.

---

substitution method

elimination method

When two equations are consistent, our task will often be to solve these equations by finding the values of the variables that satisfy both (that is, finding the coordinates of the point of intersection of the two lines). Two widely used methods of solving pairs of linear equations are the substitution and elimination methods.

The Substitution Method.

1. *Select one variable and one of the equations.* Solve for this variable in terms of the other one.

2. *Substitute this expression* into the second equation. The second equation now contains only a single variable.

3. *Solve this equation* for the second variable.

4. *Return to the equation derived in step 1,* and solve for the first variable.

5. Always *check* the solution.

Item 1 suggests we solve for one of the variables in one of the equations, but it doesn't say which variable or which equation. Mathematically, it makes no difference. Practically it does, and we should try to solve for the variable that will involve the least amount of algebra. In particular, variables with coefficients of $+1$ or -1 are usually easiest to solve for.

Example 33   Use the substitution method to solve:

(1.7) $$x + y = 6$$
(1.8) $$3x - y = 6$$

Solution

Following step 1, choose to solve for x in Equation (1.7):

(1.9) $$x = -y + 6$$

Now substitute this expression for x in Equation (1.8):

(1.10) $$3(-y + 6) - y = 6$$

This is now an equation in a single variable, and can be readily solved. Here, without explanation, are the steps in the solution:

$$-3y + 18 - y = 6$$
$$-4y + 18 = 6$$
$$-4y = -12$$
$$y = 3$$

Use the value $y = 3$ in Equation (1.9):

$$x = -y + 6$$
$$= -3 + 6$$
$$x = 3$$

Finally, check these solutions.

$$x + y \; ? \; 6 \qquad 3x - y \; ? \; 6$$
$$3 + 3 \; ? \; 6 \qquad 3 \cdot 3 - 3 \; ? \; 6$$
$$6 = 6 \qquad\qquad 6 = 6$$

Figure 1.34 displays the solution graphically.

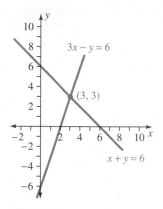

FIGURE 1.34
The solution to $x + y = 6$, $3x - y = 6$.

Skill Enhancer 33   Use the substitution method to solve the equations.

$$x - 3y = 12$$
$$2x + 5y = 2$$

Answer in Appendix E.

It is common to use *ordered pair notation* for the solution. In this form, the solution for the previous example is (3, 3).

Example 34   Last year, Sam was twice as old as Hannah. This year, Sam is one year older than Hannah. How old are they today?

Solution

Let *x* and *y* represent Sam's and Hannah's ages *last* year. This chart helps visualize how to represent their ages last year and this year.

	Last Year	This Year
SAM	x	$x + 1$
HANNAH	y	$y + 1$

FIGURE 1.35

Remember, we need their ages *this year,* although we will be solving for x and y, which represent their ages *last* year. The solution to the problem will be given by $x + 1$ and $y + 1$.

Last year,

(1.11) $$x = 2y$$

This year,

$$x + 1 = (y + 1) + 1$$

or

(1.12) $$x = y + 1$$

Solving for y in equations (11) and (12) yields

(1.13) $$y = \frac{1}{2}x$$

(1.14) $$y = x - 1$$

Figure 1.36 shows the graph of these equations.

Substituting the expression for x from Equation (1.11) into Equation (1.12) yields

$$2y = y + 1$$

or

$$y = 1$$

Use this information in Equation (1.11) to determine that

$$x = 2 \cdot 1 = 2$$

We express the solution concisely as $(2, 1)$. Last year, Sam was 2 years old and Hannah was just 1 year old. (Check this solution.) *This year,* Sam and Hannah are 3 and 2 years old.

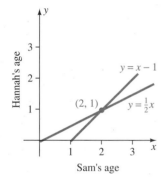

FIGURE 1.36
A graphic solution for the ages of Sam and Hannah.

Skill Enhancer 34   A year ago, Kenny had saved twice as much as Lenny. Kenny saved an additional $100 during the year. Lenny saved $200 in this time, and now he has $50 more than Kenny. What are the savings of each boy now?

Answer in Appendix E.

This example shows the importance of *answering the question* in a word problem. Don't focus too closely on simply solving the equations.

Example 35   Solve the system of equations

(1.15) $$x - y = 5$$
(1.16) $$y = x + 15$$

If no unique solution is possible, show why.

Solution

Using the value for y directly given by Equation (1.16), we find, from Equation (1.15), that

$$x - (x + 15) = 5$$

which seems to imply that $-15 = 5$, a clear contradiction. No unique solution is possible.

To see this, transform Equation (1.15) to the slope-intercept form.

(1.17) $$y = x - 5$$

[Equation (1.16) is already in that form.] From Equation (1.17), this equation represents a line with a slope of 1 and a y intercept of -5. The second equation represents a line with a slope of 1 and a y intercept of 15. Two lines with the same slope are parallel. They must be distinct, since they cut the y axis at different points. Therefore, this system is inconsistent. Figure 1.37 illustrates this.

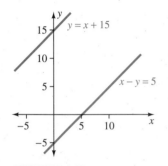

FIGURE 1.37
Inconsistent equations:
$x - y = 5$ **and** $y = x + 15.$

Skill Enhancer 35   Solve, if possible, these equations.

$$2s + 3t = 12$$
$$4s + 6t = 12$$

If no solution is possible, tell why.

Answer in Appendix E.

Before presenting a concise summary of the steps that make up the elimination method, let's examine a simple example that motivates this method.

Example 36   Solve

(1.18) $$x + 4y = 11$$
(1.19) $$x - 4y + 13 = 0$$

Solution

Multiplication of the equations is unnecessary. As given, the coefficients of the y terms in each equation are *negatives* of each other. As a result, we merely need to add the equations to each other, left side to left side and right side to right side.

$$
\begin{array}{r}
x + 4y \qquad\ = 11 \\
+(x - 4y + 13 = \ \ 0) \\
\hline
2x + 13 = 11
\end{array}
$$

so that $x = -1$. Use this value in either of the original equations to solve for y. Equation (1.18), for example, becomes

$$(-1) + 4y = 11$$

which implies that $y = 3$. The complete solution is thus $(-1, 3)$, which should be checked. Figure 1.38 displays a graphic solution to this problem.

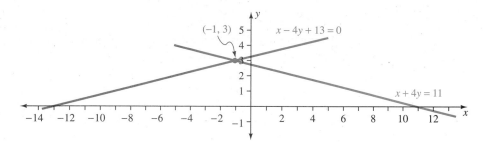

FIGURE 1.38
Solving $x + 4y = 11$, $x - 4y + 13 = 0$.

Skill Enhancer 36   Use this method to solve

$$2x - y = 4$$
$$-2x + 2y = 7$$

Answer in Appendix E.

Rarely are two equations as simple to solve as those in Example 36, but this method has so much appeal that mathematicians generalize this technique for use with any system of equations. Study this box and the following examples.

The Elimination Method.

1. *Choose multiplicative constants so that terms containing one of the variables vanish when the equations are added*; that is, when we multiply each equation by its constant, the coefficients of the variable we choose to eliminate will be additive inverses (additive opposites; terms which have opposite signs but are otherwise the same) of each other.

2. *Eliminate* one of the variables by multiplying one or both of the equations by an appropriate constant and *adding* the transformed equations together.

3. *Solve* for the variable that has not been eliminated. Substitute this value in either of the original equations and solve for the other variable.

4. Always *check* the results.

Both the elimination and substitution methods will work on any pair of linear equations, provided that the pair has a unique solution. Use the method which looks easier for the particular example. Either method must yield the same solution for the same pair of equations.

Example 37   Use the elimination method to solve

(1.20) $3x - 2y = 7$
(1.21) $x + 3y = 50$

by eliminating y.

Solution

Is it clear that if we multiply Equation (1.20) by 3 and Equation (1.21) by 2, the y terms will add to zero when we add the altered equations?

First multiply both sides of Equation (1.20) by 3:

$$3(3x - 2y) = 3(7) \quad \Rightarrow \quad 9x - 6y = 21$$

and multiply both sides of Equation (1.21) by 2:

$$2(x + 3y) = 2(50) \quad \Rightarrow \quad 2x + 6y = 100$$

Add the two equations:

$$
\begin{array}{r}
9x - 6y = 21 \\
+(2x + 6y = 100) \\
\hline
11x = 121
\end{array}
$$

so that

$$x = 11$$

To determine y, substitute $x = 11$ in Equation (1.21):

$$x + 3y = 50$$
$$11 + 3y = 50$$
$$3y = 39$$
$$y = 13$$

The solution is $(11, 13)$, which should be verified by checking.

Skill Enhancer 37   Use the elimination method to solve the system

$$2x - 3y = 3$$
$$3x + 2y = -2$$

Solve for x first.

Answer in Appendix E.

Example 38   Solve the system

(1.22) $$\qquad\qquad 2x + y = 3$$
(1.23) $$\qquad\qquad 2y = 6 - 4x$$

Use the elimination method. If no solution is possible, show why.

Solution

To use the elimination method, multiply Equation (1.22) by -2 and add the two equations. Equation (1.22) becomes

$$-2(2x + y) = -2(3)$$

or

$$-4x - 2y = -6$$

Adding, we obtain

$$-4x - 2y = -6$$
$$+(2y = +6 - 4x)$$
$$\overline{-4x \qquad = -6 + 6 - 4x}$$

or

$$-4x = -4x$$

This equation, while true, yields no information about a unique value of x or y. To see why, write both equations in the slope-intercept form. In this form, they both become

$$y = -2x + 3$$

The two equations therefore represent the same line (shown in Figure 1.39); this system is *dependent* and has no unique solution. [There *are* nonunique solutions, such as $(0, 3)$, $(-1, 5)$, etc.]

Skill Enhancer 38   Use any method to solve these equations. If no solution is possible, explain why.

$$x - 2y = 10$$
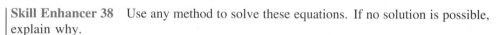
$$= \frac{1}{2}x - 5$$

Answer in Appendix E.

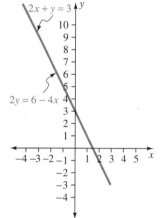

FIGURE 1.39
Dependent equations:
$2x + y = 3$, $2y = 6 - 4x$.

Example 39   A batch of Yuppie Delight gourmet ice cream calls for 200 gallons (gal) of a milk-cream mixture. Milk costs the factory $1.50 a gallon, and a gallon of cream costs $2.75. Profit studies seem to indicate that the company should spend $450 for the milk-cream mixture. How many gallons of milk should be mixed with how many gallons of cream to make a mixture that will cost $450?

Solution

First decide what the unknowns will be. In this problem, it will be gallons of milk m and gallons of cream c. Since there are two unknowns, there should be two equations.

The fluid volume of the milk products remains constant at 200 gal. Mix m gallons with c gallons to get the required 200 gal.

(1.24) $m + c = 200$

The company must spend $450, a portion for milk and a portion for cream:

Cost of milk + cost of cream = $450

Milk costs $1.50/gal, and the Yuppie Delight people are buying m gallons, so

$$\text{Cost of milk} = \text{price/gal} \times \text{no. of gallons}$$
$$= 1.5m$$

Similarly, Yuppie Delight will spend $2.75c$ for cream. Therefore, the total spent (in dollars) is

$$(1.25) \qquad\qquad 1.5m + 2.75c = 450$$

We will use the elimination method to solve Equations (1.24) and (1.25). Multiply Equation (1.24) by -1.5:

$$(1.26) \qquad -1.5(m + c) = -1.5(200) \quad\Rightarrow\quad -1.5m - 1.5c = -300$$

Add Equations (1.26) and (1.25) to determine that

$$1.25c = 150$$
$$c = 120 \text{ gal}$$

From Equation (1.24), if $c = 120$, then $m = 80$. (Be sure to check these answers.) Equations (1.24) and (1.25) appear in Figure 1.40. Since both $m \geq 0$ and $c \geq 0$, the graph is restricted to the first quadrant.

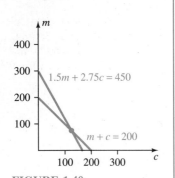

FIGURE 1.40
Determining the proper proportions of milk and cream.

Graph labels: $1.5m + 2.75c = 450$ and $m + c = 200$

Skill Enhancer 39 Several weeks later, the administrators redo the calculations in light of changes in the prices of milk and cream. Milk now costs \$2/gal, and cream is \$3.50/gal. As a result of these changes, the company should now spend \$550 for the mixture. How should the resulting mixture be prepared?

Answer in Appendix E.

___ *Exercises 1.4* _____

Solve the pairs of equations in Exercises 1–16 using the substitution method.

1. $x + y = 1$
    $2x - y = 8$

2. $u - 2t = 0$
    $2u + t = 5$

3. $3p + 4q = 14$
    $10p - 8q = 4$

4. $4x_2 - 3x_1 = 4$
       $2x_2 = 2x_1 + 1$

5. $2x_1 - x_2 = 2$
    $x_1 + 10x_2 = 1$

6. $3x_1 - x_2 = 4$
    $2x_2 - x_1 = -3$

7. $8y_1 + y_2 = 14$
    $2y_1 - 20y_2 = 10$

8. $2x - 30y = 15$
    $\dfrac{1}{2}x + 6y = 8$

9. $7u - 3t = 9$
    $u + 3t = 3$

10. $z_1 + 4z_2 = 0$
     $z_1 + z_2 = -0.75$

11. $3x + y = 28$
     $x - 2y = 14$

12. $x - \dfrac{1}{2}y = 0$
     $4x + y = 6$

13. $u + 7t = 175$
     $10u + 2t = 50$

14. $a - 4b + 9 = 0$
     $-3a + 2b - 27 = 0$

15. $4y_1 - 5y_2 = 0$
     $2y_1 + 3y_2 = 22$

16. $2x_1 + x_2 = 11$
     $2x_2 = 20 - 3x_1$

Use the elimination method to solve the sets of equations in Exercises 17–32.

17. $x + y = 1$
$2x - y = 8$

18. $3p - 5q = 1$
$5p + 3q = 13$

19. $t - 4u = 0$
$u + 2t = 9$

20. $8x - 7y = 56$
$x + 7y = 7$

21. $2u - v = 5$
$2v - u = -1$

22. $3s + t = 1$
$2s - 2t = 2$

23. $x - 2y = -4$
$x + y = 2$

24. $x_1 + 4x_2 = 1$
$2x_1 - 2x_2 = 12$

25. $x - 4y = 2$
$3x + y = 3$

26. $2x - y + 2 = 0$
$x + y - 2 = 0$

27. $4a - 3b = 1$
$5a + 6b = 11$

28. $10m + 10n = 0$
$-3m + 5n = 8$

29. $y_1 + y_2 = 1$
$3y_1 + 3y_2 = 1$

30. $x - 3y + 10 = 0$
$x = 3y - 10$

31. $4u + 3v = 24$
$-3u + 4v = 7$

32. $x_1 + x_2 = 1$
$-x_1 + 2x_2 = 11$

Use the method of your choice to solve for the variables in the sets of equations in Exercises 33–48. If the equations have no solutions, demonstrate this fact.

33. $x = 4y - 1$
$y = 2x + 3$

34. $y = 3x - 21$
$2y - 6x = 0$

35. $t + 7 = u - t$
$t - \dfrac{u}{2} + 3.5 = 0$

36. $10u - 3v = 29$
$8u + 3v = 7$

37. $-x + 2z = 3$
$2x + z = 9$

38. $3u - t = 8$
$t = -8 + 3u$

39. $x - y = -0.25$
$2x - y = -0.25$

40. $x_1 - 4x_2 = 5$
$x_2 - x_1 = 1$

41. $x - y = 2$
$y = x - 3$

42. $s + 2t = 4$
$t = 2 - \dfrac{1}{2}s$

43. $3u - 4t = 12$
$2t = \dfrac{3}{2}u - 6$

44. $2z_1 - 4z_2 = 12$
$2z_2 - z_1 = 13$

45. $2e - 3f = 5$
$3e + 2f = 1$

46. $3m - n = 9$
$m + 10n = 3$

47. $3x + 4y = 0$
$4x - 3y = 0$

48. $2u - 3v = 26$
$u + 2v = 6$

For the pairs of equations in Exercises 49–58, determine if there is a unique solution, no solution at all, or an infinite number of solutions by inspecting the slopes and the y intercepts of each equation. If there is a unique solution, find it (using any method).

49. $2a + 3b = 1$
$4a + 6b = 1$

50. $t_2 = 3t_1 - 1$
$6t_1 - 2t_2 = 2$

51. $2x_1 + 3x_2 = 6$
$3x_1 - 2x_2 = 9$

52. $2x - 9y = 10$
$y = \dfrac{2}{9}x - 1$

53. $3z_2 = 6z_1 + z_2 - 8$
$3z_1 - z_2 = 4$

54. $x_1 - x_2 = 0$
$x_1 + x_2 = 0$

55. $t_1 + 2t_2 = \dfrac{1}{2}$
$4t_2 - 2t_1 = 0$

56. $-(x_1 + x_2) = 5$
$2x_2 - x_1 = -4$

57. $2x_1 - x_2 = 1$
$x_1 + x_2 = -1$

58. $3x + y = 1$
$6(y - x) = 1$

59. Solve

$$3x - 2y = 7$$
$$x + 3y = 50$$

by eliminating x first.

60. Solve

$$2u - v = 5$$
$$2v - u = -1$$

by eliminating v first.

W **61.** Six subtracted from the product of 4 times a certain number is that number. What is the number? Is it faster to use algebra or guesswork to solve this problem? Despite the fact that guessing might indeed be faster, why is it "better" to use algebraic techniques?

62. The digits of a certain two-digit number add up to 6. Form a second number by reversing these digits. When the second is subtracted from the first, the result is 3 times the sum of the digits of either number. What is the first number? (Use any method but trial and error.)

Applications

63. The formula relating Celsius (C) and Fahrenheit (F) temperature is $F = \frac{9}{5}C + 32$. At what temperature is the reading in Fahrenheit numerically the same as the one in Celsius?

64. *(Medicine)* Use the formula in the previous problem to compute normal body temperature in the Celsius scale. (Normal temperature $= 98.6°$ F.)

W 65. *(Animal Food)* Puppy Pride dog food is a canned food containing a mixture of meat byproducts plus soybean. The food is made in 400-pound (lb) lots. Each pound of dog food must contain 50 percent protein by weight. Soybean is 80 percent protein, whereas the byproducts are 40 percent. How should the company mix the ingredients to satisfy exactly the minimum government requirement for protein content? What changes do you think might occur in the calculation if Puppy Pride had to equal *or exceed* these requirements?

66. *(Health Food)* "Here's to Your Health" Natural Food Stores carry sunflower seeds and raisins. The seeds sell for $1.50 a pound, and the raisins go for $2.50. One store can sell a total of 20 lb of the two foods together each business day. This store is thinking of mixing the two foods to get a gorp-like mix that can be sold for $2.00 per pound. How many pounds of seeds should be mixed with how many pounds of raisins if the income from the mixture should be the same as that from selling the foods separately?

67. *(Finance)* Rich N. Greety has come into a large inheritance totaling $750,000. When he was a minor, his guardian invested the money for him in bonds. Some of the riskier bonds have a per-annum yield of 12 percent, whereas the conservative remainder yield 7 percent. In the last six-month period, Rich received $35,000 from these investments. He would like to know how much is invested in each kind of bond. Can you help him? (Hint: $6\,\text{mo} = \frac{1}{2}\,\text{yr}$)

68. *(Handicrafts)* Seth Smith formed the Good Tree Furniture Company to produce handmade Shaker pieces of furniture. Originally, he was his own only employee, but business has been good, and he would like to hire additional craftspeople on a part-time basis. He keeps his workshop open 40 hours a week, and has space for two additional people to work. Furniture makers in his neighborhood group themselves as hewers and finishers, and command wages of $20 and $25 per hour, respectively. Seth can afford to pay $1,700 per week total for his helpers. For how many hours should he hire a hewer, and for how many a finisher?

W 69. *(Opera Administration)* The Fredonia Regional Opera Group (FROG) has a wealthy patron who will build them a new auditorium. Because of the shape of the land, a maximum of 2,500 seats can be fitted in the new hall. FROG tentatively plans two categories of tickets. The cheaper tickets, which will draw students and senior citizens, will be sold for $7.50 apiece. The other tickets will cost $20. Studies indicate that FROG will need to have ticket receipts of $37,500 each time they give a performance. Assuming that they have a full house at each event and that the architect has specified as many cheap seats as possible, how many seats should be allocated to each category? Why do you think it is necessary to specify that as many cheap seats as possible be present in the new auditorium?

W 70. *(Anthropology and Human Behavior)* Marcy is the victim of a strange superstition. Every evening she puts a coin, either a dime or a quarter, in a piggy bank to ward off bad luck and the evil eye. After 2 weeks, she knows she has $2.30 in the bank. Without breaking it open, how can she tell how many dimes and how many quarters are in there? How many of each coin are there?

71. *(Stamp Collecting)* A packet of 22 stamps from Sri Lanka and Burma costs $46. A stamp from Sri Lanka costs $1.75 singly, and a stamp from Burma costs $2.50. How many stamps from each country are there in this packet?

72. *(Gardening)* At one garden supply center, single begonia bulbs and single iris bulbs sell for $1.20 and $1.00, respectively. One mixture of begonia and iris bulbs containing 50 bulbs is priced by the managers as if each bulb in the mixture were sold for a 20 percent discount from the single-bulb prices. The mixture sells for $44.80. How many bulbs of each type are in the mixture?

🖩 Calculator Exercises

Solve the pairs of equations in Exercises 73 and 74 using any method. Express the solutions to two decimal place accuracy.

73. $2.1x - 3.4y = 6.6$

$\quad\ \ 2.3x + 6.8y = -6.7$

74. $\quad x - 6.7y = 12$

$\quad\ \ 8.2x + 12.4y = 9$

CHAPTER REVIEW

Terms

linear equation	subscript	solution set
addition principle	multiplication principle	rectangular coordinates
axis	abscissa	ordinate
ordered pair	general form	intercepts
vertical lines	horizontal lines	slope
positive slope	negative slope	absolute value
parallel	point-slope equation	slope-intercept formula
interest	principal	simple interest
inflation	depreciation	break-even point
demand	supply	consistent equations
consistent equations	inconsistent equations	dependent equations
substitution method	elimination method	

Key Concepts

- We sometimes use *subscripts* to distinguish one variable from another.

- *Cartesian graphing* techniques visually represent points, curves, and lines on a set of coordinate axes.

- Linear equations are a special class of equations in which terms containing variables have a special form. Any such variable term may contain only the variable multiplied by a numerical coefficient. The variable itself may be raised only to the first power. Terms like $3x$, $-4y$, x, and $\frac{1}{2}x$ are allowed in linear equations. Terms like $7xy$, $-\frac{8}{x}$, $5x^3$, and $-2\sqrt{x}$ are not. Linear equations with two variables always graph as straight lines.

- The slope m of a line connecting two points with coordinates (x_1, y_1) and (x_2, y_2) is

$$m = \frac{y_2 - y_1}{x_2 - x_1}$$

The slope of a line measures the amount of its slant. Horizontal lines have zero slope; the slopes of vertical lines are undefined. A negative slope indicates that the line slants *down* to the right. A line with positive slope slants *up* to the right. Two lines are *parallel* if they have the same slope.

- Given a point (x_1, y_1) lying on a line and the line's slope m, its equation is most easily written using the *point-slope form* as

$$y - y_1 = m(x - x_1)$$

Given the slope m and the y intercept b, the line's equation is most easily represented by the *slope-inter-cept form* as

$$y = mx + b$$

- Systems of two linear equations may determine a single point satisfying both equations in the system. This point represents the *solution* to the equations. Use the methods of *substitution* or *elimination* to solve them. Some pairs of equations may have no solution. (That is, the lines they represent may be parallel or may coincide.)

Review Exercises

Solve the equations in Exercises 1–6.

1. $2y - 1 = 5$ **2.** $\frac{1}{2}x = 2$

3. $3(x - 1) = 12$ **4.** $2(t + 1) = 3t - 4$

5. $-(x - 7) + 2x = 0$ **6.** $2x + 1 = 3x - \frac{1}{2}$

Graph the points in Exercises 7–12.

7. $(1, -1)$ **8.** $(-4, 3)$

9. $(1, 0)$ **10.** $(0, 1)$

11. $(7, 10)$ **12.** $(-7, -10)$

In Exercises 13–20, graph the linear equations.

13. $x + y = 1$ **14.** $p + q = 1$

15. $2x - 3y + 6 = 0$ **16.** $x_1 - 4x_2 = 10$

17. $7x - 3y = 0$ **18.** $y = 9$

19. $x = -\frac{5}{2}$ **20.** $3z_1 - 4z_2 = 12$

What are the slopes of the line segments connecting the pairs of points in Exercises 21–26?

21. $(1, 2), (3, 4)$ **22.** $(-1, -2), (-3, -4)$

23. $(6, 1), (-2, -3)$ **24.** $(a, 1), (a, 5)$ for any a

25. $(1, q), (100, q)$ for any q **26.** $(0, 0), (\pi, \frac{\pi}{2})$

Find an equation of the line connecting each pair of points in Exercises 27–32.

27. $(1, 2), (3, 4)$ **28.** $(-1, -2), (-3, -4)$

29. $(6, 1), (-2, -3)$ **30.** $(a, 1), (a, 5)$

31. $(1, q), (100, q)$ **32.** $(1 + \alpha, \alpha), (\alpha, 1 + \alpha)$, for any α

In Exercises 33–38, deduce equations of the lines possessing the given attributes. Here, m is the slope, b is the y intercept, and the coordinates of a point mean that the point lies on the line.

33. $(0, 0), m = 2$ **34.** $(-2, 3), (7, 6)$

35. $(2, 2), b = 3$ **36.** $m = -1, b = 7$

37. $(\frac{3}{2}, \frac{1}{2}), (-\frac{1}{2}, \frac{5}{2})$ **38.** $(3, 4), m$ undefined

What are the slopes of the lines in Exercises 39–42?

39. $2(y - 7x) + 12 = 3x$ **40.** $2(x - 3y) = -12$

41. $x + y = 7 + 2(3y - 4x)$

42. $4x + (7y - 2x + 1) = 3y - 2$

Use the substitution method to solve the pairs of equations in Exercises 43–46 (if possible). If no solutions are possible, say why.

43. $x - y = 5$ **44.** $x - y = 5$
      $2x + y = 17\frac{1}{2}$ $3x + 4y = 2$

45. $3p + 4q = 10$ **46.** $3p + 4q = 10$
      $8q = 20 - 6p$ $7p - q = 13$

Use the elimination method to solve the pairs of equations in Exercises 47–52 (if possible). If no solutions are possible, state why.

47. $2u + 2v = 9$ **48.** $6x + 2y = 9$
      $3u - 3v = 0$ $3x + 2y + 7 = 0$

49. $3p + 4q = 10$ **50.** $10x + y = 11$
      $7p - q = 13$ $x - 10y + 9 = 0$

51. $3y + 14z = 7$ **52.** $6x + 2y = 9$
      $-9y + 37z + 100 = 0$ $3x + 2y - 6 = 0$

Applications

W 53. *(Weather)* The mean temperature in February was 10 degrees. In March it was 17.5 degrees. What is a reasonable estimate for the mean April temperature? Why?

54. *(Fringe benefits)* The annual cost to a company of any employee includes the total annual wages paid plus the cost of fringe benefits. A pipe fitter costs the company $53,000 a year. If she works 2,000 hours a year at an hourly wage of $20, what is the cost to the company of her fringe benefits?

55. *(Politics)* In a certain municipality, voter turnout to elect the members of the school board, though always spotty, seems to be related to how long before the election the nominees' names are made public. In 1992, there was an 8-week lead time, and 14.5 percent of the eligible voters cast a ballot. In 1993, with a scant 2-week lead time, only 9.5 percent of the voters voted. In 1994, there will be a 12-week lead time. What proportion of the eligible voters would you estimate will show up at the polls?

56. *(Athletics)* The storage chest for the local bowling team contains 11 bowling balls. Some weigh 10 lb, and the remainder weigh the regulation 16 lb. The weight of the chest, with all the balls, weighs 152 lb. If the weight of the chest itself is negligible, how many of the balls are light and how many are regulation weight?

57. *(Physics)* The amount by which a spring can be stretched or compressed from its equilibrium length varies linearly with the tensile (stretching) or compressing force. The spring is unstretched when no forces are applied to it. A certain spring stretches 0.1 centimeters (cm) when a tensile force of 110 lb is applied. Estimate how much it *stretches* when a tensile force of 200 lb is applied. Estimate how much the spring *shrinks* when a compressive force of 110 lb is applied.

58. *(Botany)* The height of certain tropical plants is proportional to the time since the seed has germinated. Among a group of such plants, the average height is 12 cm. Seven days later, the average height is 19 cm. Estimate how many days before the first observation the plants germinated.

59. The *Cannonball Express* and *The Galloping Turtle* are two trains departing from Boston's South Station termi-

nal. The *Cannonball* travels twice as fast as the *Turtle*. If they both leave at the same time and travel parallel paths, they will be 25 miles (mi) apart after a half hour. What are their travel speeds?

60. *(Business)* Two business partners, Al and Ben, have to share the year's profits, a total (after taxes) of $75,000. According to the rules of the partnership, Al is to receive twice what Ben receives, plus additional compensation of 15 percent of the profits. How much does each partner receive?

61. *(Laboratory procedures)* A lab technician needs 50 ml (ml = milliliter, a metric unit of volume) of a solution which is 7 percent saline. She has unlimited quantities of 5 percent and 15 percent saline solution mixtures. How many milliliters of each should she combine to obtain the 7 percent solution?

W **62.** *(Economics and Advertising)* When economic conditions are poor, many stores start their Christmas advertising earlier and earlier. In 1988, as the economy was taking a severe downturn, Justin observed that Christmas advertising in Rochester, New York began right after Thanksgiving. In 1992, he noticed that this advertising began 3 weeks before Thanksgiving. He estimates that in 1994 (if there is no pickup in the economy), Christmas advertising will begin $4\frac{1}{2}$ weeks before the Thanksgiving holiday. Do you agree with this estimate? Why or why not?

2

MATRICES AND LINEAR SYSTEMS

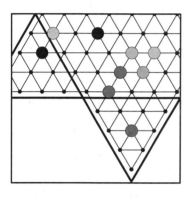

INTRODUCTION

A matrix is a shorthand way of representing a group of related quantities as a *table* or *chart*. The importance of matrices in so much of finite mathematics lies in the intimate connection between matrices and the solution of linear equation systems. This connection makes sense only if there are arithmetic rules for working with matrices. Answering simple questions—what are the total sales of two salespeople? how much better a salesperson is one than the other?—motivates most of these rules. The intimate and natural connection between matrices and systems of linear equations motivates the remaining rules. Using matrix techniques, it becomes natural and possible to solve linear equation systems with arbitrary numbers of variables and equations.

Section 2.1   *INTRODUCTION TO MATRICES*

matrix   Any rectangular array of numbers forms a *matrix*. *Charts* and *tables* are common examples of matrices (the plural of matrix). For example, these tables show how well two salespeople have done selling certain appliances in two consecutive months.

	MR. ANGST		MS. NOMER	
	Jan.	Feb.	Jan.	Feb.
Coffeepots	15	27	21	21
Toasters	8	17	14	19
Irons	21	42	35	37

row *i*

column *j*

FIGURE 2.1
Component notation.

If we agree that column headings refer to months and row labels to types of appliances, we can streamline this presentation by presenting the sales data as a rectangular array of numbers without the row and column headings. It is common to surround each array with large parentheses, in which form the arrays become *matrices*. We often use capital letters to represent arrays in the same way that lowercase letters often refer to variables. The above sales figures might have the following *matrix representation*:

$$
A = \begin{array}{c} \text{Row 1} \\ \text{Row 2} \\ \text{Row 3} \end{array}
\begin{matrix} \text{Col. 1} & \text{Col. 2} \end{matrix}
\begin{pmatrix} 15 & 27 \\ 8 & 17 \\ 21 & 42 \end{pmatrix}
\qquad
N = \begin{array}{c} \text{Row 1} \\ \text{Row 2} \\ \text{Row 3} \end{array}
\begin{matrix} \text{Col. 1} & \text{Col. 2} \end{matrix}
\begin{pmatrix} 21 & 21 \\ 14 & 19 \\ 35 & 37 \end{pmatrix}
$$

rows   *Rows* run horizontally in a matrix, and *columns* run vertically. Both A and N have 3
columns   rows and 2 columns. Generally, matrices may have any numbers of rows and columns.

Had we agreed to run the months along the sides and the item categories across the top, the corresponding matrices would have had 2 rows and 3 columns apiece. These alternative matrices, which we can call A^T and N^T, would be

$$
A^T = \begin{array}{c} \text{Jan.} \\ \text{Feb.} \end{array}
\begin{matrix} \text{Coffeepots} & \text{Toasters} & \text{Irons} \end{matrix}
\begin{pmatrix} 15 & 8 & 21 \\ 27 & 17 & 42 \end{pmatrix}
\qquad
N^T = \begin{pmatrix} 21 & 14 & 35 \\ 21 & 19 & 37 \end{pmatrix}
$$

(We use the superscript "T" because these alternative forms are called the *transpose* of the original matrix. Transposes will be important in later work with matrices.) The
matrix elements   individual entries within the matrix are called the *matrix elements*. Matrix elements are
matrix component   usually numbers or variables, but they may be any mathematical object at all. A *matrix component*, or just *component,* is another term for an element.

Historical Note

Matrices and Arthur Cayley   *The mathematical theory of matrices dates only from the middle of the nineteenth century, and was largely pioneered by the British mathematician Arthur Cayley (1821–1895), one of the most prolific mathematicians of all time.*

Cayley began by studying and practicing law, but became Sadlerian Professor of Mathematics at the University of Cambridge in 1863. He remained in this position for over 30 years.

If we represent a matrix by a particular uppercase letter, say B, for example, then we denote its elements by the corresponding lowercase letter with double subscripts to designate the rows and columns, such as b_{13}. The first subscript *always* labels the row, and the second labels the column. The matrix A has components

$$\begin{pmatrix} a_{11} = 15 & a_{12} = 27 \\ a_{21} = 8 & a_{22} = 17 \\ a_{31} = 21 & a_{32} = 42 \end{pmatrix}$$

whereas N has components

$$\begin{pmatrix} n_{11} = 21 & n_{12} = 21 \\ n_{21} = 14 & n_{22} = 19 \\ n_{31} = 35 & n_{32} = 37 \end{pmatrix}$$

More generally, we write

$$A = (a_{ij}) \qquad \text{and} \qquad N = (n_{ij})$$

where i and j stand for arbitrary subscripts, to denote that the matrices A and N have components a_{ij} and n_{ij}. See Figure 2.1.

dimension   The *dimension* of any matrix is the number of rows and columns. Rows come first. Both A and N are matrices with dimension 3 by 2, or 3×2.

Example 1   State the dimension for each of the following matrices.

$$A = \begin{pmatrix} 1 & 2 & 3 & 4 & 5 \\ 6 & 7 & 8 & 9 & -1 \end{pmatrix} \qquad B = \begin{pmatrix} \frac{1}{2} & \frac{2}{3} \\ \frac{3}{4} & \frac{4}{5} \\ 0 & \frac{5}{6} \end{pmatrix}$$

$$C = \begin{pmatrix} 1 \\ x \\ x^2 \\ x^3 \end{pmatrix} \qquad D = (-50 \quad +50) \qquad E = (7)$$

Solution

With 2 rows and 5 columns, A has dimension 2×5. B has dimension 3×2. C has only a single column—its dimension is 4×1. D has only a single row, so its dimension is 1×2. E has a dimension of 1×1.

Skill Enhancer 1 What is the dimension of each of these matrices?

$$\begin{pmatrix} 1 & 2 & -4 & 5 \\ 9 & 8 & 6 & -5 \end{pmatrix} \qquad \begin{pmatrix} \alpha & 0 \\ 0 & -2 \\ 1 & \beta \end{pmatrix}$$

Answer in Appendix E.

row matrix
column matrix
square matrix
matrix diagonal

Special matrices have particular names. A *row matrix* contains a single row, but an arbitrary number of columns; see Figure 2.3. A *column matrix* may have any number of rows, but will contain a single column; see Figure 2.2. A matrix is a *square matrix* if it has the same number of rows and of columns. The *matrix diagonal* of a square matrix is the collection of elements that lie atop an imaginary diagonal of the matrix, starting from the element in the first row and first column and sliding down to the last row and last column; see Figure 2.4. Only square matrices have diagonals.

$$\begin{pmatrix} 1 \\ 0 \end{pmatrix} \quad \begin{pmatrix} 1 \\ x \\ x^2 \\ x^3 \end{pmatrix} \quad \begin{pmatrix} -49 \\ +18 \\ +6.5 \end{pmatrix}$$

FIGURE 2.2
Column matrices.

$$(a \;\; b \;\; c) \qquad (-1 \;\; 1)$$
$$(.1 \;\; .2 \;\; .3 \;\; .4 \;\; 5) \qquad (1 \;\; 3 \;\; 3 \;\; 1)$$

FIGURE 2.3
Row matrices.

Example 2 Is

$$A = \begin{pmatrix} -0.5 & \pi & 0 \\ 14 & \frac{3}{2} & 0 \\ \sqrt{10} & -8 & 0 \end{pmatrix}$$

square? If so, what is its dimension? What are the elements of the diagonal?

Solution

Since there are 3 rows and 3 columns, it is square and of dimension 3×3. The diagonal elements are -0.5, $\frac{3}{2}$, and 0.

Skill Enhancer 2 What are the diagonal elements of $\begin{pmatrix} 0 & 1 & 0 \\ 1 & 0 & 1 \\ -1 & -1 & -2 \end{pmatrix}$?

Answer in Appendix E.

$$\begin{pmatrix} 3 & -1 \\ 4 & 0 \end{pmatrix} \quad \begin{pmatrix} 5 & -7 & 6 & \frac{1}{2} \\ \frac{3}{2} & \frac{1}{4} & 0 & -7 \\ 6 & -8 & -14 & -3 \\ 3 & 2 & \frac{3}{4} & \frac{7}{8} \end{pmatrix} \quad \begin{pmatrix} 1 & 3 & -.8 \\ 0 & 0 & -1.6 \\ 10 & 17 & 0 \end{pmatrix}$$

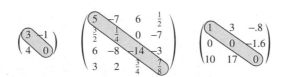

FIGURE 2.4
Square matrices and their diagonals.

matrix addition
matrix subtraction

Can matrices be added, subtracted, multiplied, and divided as other mathematical quantities can be? With several restrictions, the answer is "yes." The appliance sales problem above will suggest the means for *adding* and *subtracting* matrices.

Suppose we needed the total sales for both salespeople and the amount by which Nomer's sales exceeded Angst's. We can build two new matrices, called T for Total sales and E for excess sales, with components $n_{ij} + a_{ij}$ and $n_{ij} - a_{ij}$, respectively, that will display this information. We simply add or subtract the corresponding elements. The two new matrices are

$$T = \begin{pmatrix} 21+15 & 21+27 \\ 14+8 & 19+17 \\ 35+21 & 37+42 \end{pmatrix} = \begin{pmatrix} 36 & 48 \\ 22 & 36 \\ 56 & 79 \end{pmatrix}$$

$$E = \begin{pmatrix} 21-15 & 21-27 \\ 14-8 & 19-17 \\ 35-21 & 37-42 \end{pmatrix} = \begin{pmatrix} 6 & -6 \\ 6 & 2 \\ 14 & -5 \end{pmatrix}$$

We can generalize this operation to arbitrary matrices *of equal dimension.*

Matrix Addition and Subtraction

Two matrices can be added or subtracted *only* if they have the same dimension. Formally, we have

$$A + B = \begin{pmatrix} a_{11}+b_{11} & a_{12}+b_{12} & \cdots & a_{1n}+b_{1n} \\ \vdots & & & \vdots \\ a_{m1}+b_{m1} & a_{m2}+b_{m2} & \cdots & a_{mn}+b_{mn} \end{pmatrix}$$

$$A - B = \begin{pmatrix} a_{11}-b_{11} & a_{12}-b_{12} & \cdots & a_{1n}-b_{1n} \\ \vdots & & & \vdots \\ a_{m1}-b_{m1} & a_{m2}-b_{m2} & \cdots & a_{mn}-b_{mn} \end{pmatrix}$$

In words, we *add* two matrices by adding corresponding elements together. We *subtract* two matrices by subtracting corresponding elements from each other.

A, B are two matrices of equal dimension.
a_{ij} are the components of A.
b_{ij} are the components of B.

If A and B are *not* of the same dimension, their sum and difference are *undefined.*

Example 3   Suppose

$$A = \begin{pmatrix} 1 & -2 & 3 & -4 \\ -1 & 0 & -5 & 3 \end{pmatrix} \qquad B = \begin{pmatrix} 1 & 0 \\ 0 & 1 \end{pmatrix}$$

$$C = \begin{pmatrix} 1 & 0 & 1 & 1 \\ -1 & -2 & 1 & 0 \end{pmatrix}$$

Compute $A + B$, $A + C$, $B - C$, and $C - A$.

Solution

Neither $A + B$ nor $B - C$ can be computed, since A and B have different dimensions, as do B and C. The remaining expressions are easily evaluated:

$$A + C = \begin{pmatrix} 1 + 1 & -2 + 0 & 3 + 1 & -4 + 1 \\ -1 + (-1) & 0 + (-2) & -5 + 1 & 3 + 0 \end{pmatrix}$$

$$= \begin{pmatrix} 2 & -2 & 4 & -3 \\ -2 & -2 & -4 & 3 \end{pmatrix}$$

$$C - A = \begin{pmatrix} 1 - 1 & 0 - (-2) & 1 - 3 & 1 - (-4) \\ -1 - (-1) & -2 - 0 & 1 - (-5) & 0 - 3 \end{pmatrix}$$

$$= \begin{pmatrix} 0 & 2 & -2 & 5 \\ 0 & -2 & 6 & -3 \end{pmatrix}$$

Skill Enhancer 3   For

$$x = \begin{pmatrix} 2 \\ 3 \end{pmatrix} \quad y = (2 \quad 3) \quad z = \begin{pmatrix} -2 \\ -3 \end{pmatrix}, \quad w = (-2 \quad -3)$$

Compute

$$x + y, \quad z + x, \quad x - z, \quad \text{and } y + w.$$

Answer in Appendix E.

There are two kinds of matrix multiplication operations. Insight from the sales problem will suggest one type of multiplication operation.

The sales manager expects Angst's sales performance to more closely match Nomer's. Angst receives an ultimatum—his March and April sales for each item must at least be 50 percent greater than the respective figures for January and February. What are the sales goals? We multiply each element in A by 1.5 (which is 150 percent). That is,

$$\text{Angst's sales targets} = 1.5 \times \begin{pmatrix} 15 & 27 \\ 8 & 17 \\ 21 & 42 \end{pmatrix} = \begin{pmatrix} 22.5 & 40.5 \\ 12 & 25.5 \\ 31.5 & 63 \end{pmatrix}.$$

scalar matrix multiplication   This suggests that we define *scalar matrix multiplication* in the following way. (A *scalar*
scalar   is another name for an ordinary real number.)

Scalar Matrix Multiplication

Scalar matrix multiplication is the process whereby we multiply any matrix by a real number (scalar).

$$kA = k(a_{ij}) = (ka_{ij})$$

where

A is any matrix.
k is any scalar quantity.
a_{ij} are the components of A.

Example 4   Compute $-A$ and $7C$ for

$$A = \begin{pmatrix} 1 & -2 & 4 \end{pmatrix} \qquad C = \begin{pmatrix} 0 & 1 & 1 \\ 1 & 0 & 1 \\ 1 & 1 & 1 \end{pmatrix}$$

Solution

$$-A = (-1) \times A = \begin{pmatrix} -1 \cdot 1 & -1 \cdot (-2) & -1 \cdot 4 \end{pmatrix} = \begin{pmatrix} -1 & 2 & -4 \end{pmatrix}$$

$$7C = \begin{pmatrix} 7 \cdot 0 & 7 \cdot 1 & 7 \cdot 1 \\ 7 \cdot 1 & 7 \cdot 0 & 7 \cdot 1 \\ 7 \cdot 1 & 7 \cdot 1 & 7 \cdot 1 \end{pmatrix} = \begin{pmatrix} 0 & 7 & 7 \\ 7 & 0 & 7 \\ 7 & 7 & 7 \end{pmatrix}$$

Skill Enhancer 4   Compute $10A$ and $-\frac{1}{3}C$ where A and C are defined as in the previous example.

Answer in Appendix E.

Example 5   Nora Barnackle observes 3 varieties of New England lobster for a research project. In June, she is able to observe 19, 21, and 7 of the 3 varieties, respectively. In July, she observes 21, 13, and 20 of the 3 varieties. She spends the next fall, winter, and spring analyzing her data. Her advisor explains that she needs more observational data; she should target making observations on four times as many animals in the coming summer. Use a matrix to summarize her observations last year. Use matrix methods to determine her target goal for the coming summer.

Solution

She might use a 3×2 matrix D (for data) to summarize the animals she observed. The column headings might be the months of June and July, and the row labels would be the 3 varieties. The entries in the matrix would be the quantity of each (row) animal observed during that (column) month:

$$D = \begin{matrix} \text{Var. A} \\ \text{Var. B} \\ \text{Var. C} \end{matrix} \overset{\text{June} \quad \text{July}}{\begin{pmatrix} 19 & 21 \\ 21 & 13 \\ 7 & 20 \end{pmatrix}}.$$

Her target for the next summer is simply $4D$:

$$4D = \begin{matrix} \text{Var. A} \\ \text{Var. B} \\ \text{Var. C} \end{matrix} \overset{\text{June} \quad \text{July}}{\begin{pmatrix} 76 & 84 \\ 84 & 52 \\ 28 & 80 \end{pmatrix}}.$$

Skill Enhancer 5   The matrix

$$\begin{pmatrix} 1 & 0.5 \\ 0.7 & 0.7 \\ 0.5 & 1.1 \\ 0.8 & 0.9 \end{pmatrix}$$

summarizes rainfall data in Portland, Oregon during July and August of 1992. The columns refer to July and August, and the rows to the weeks of each month. (We assume that each month contains precisely four weeks.) Each number gives the rainfall in inches. A graduate student in meteorology is struck by a remarkable coincidence. The rainfall in each week for the entire period is exactly twice what it was in 1990. Determine the matrix giving the rainfall data for 1990.

Answer in Appendix E.

_____ *Exercises 2.1* _____

For Exercises 1–8, supply the dimension of each matrix. Where possible, identify the elements of the diagonal.

1. $\begin{pmatrix} 1 & 2 & 3 \\ 4 & 5 & 6 \end{pmatrix}$

2. $\begin{pmatrix} A & B & c & d \\ \pi & xyz & 0 & 0 \\ -e & 2 & 3 & 1 \end{pmatrix}$

3. (3)

4. $(9 \quad 8 \quad 6 \quad -2 \quad 1)$

5. $\begin{pmatrix} 0.12 \\ 0.23 \\ 0.34 \\ 0.45 \\ 0.56 \end{pmatrix}$

6. $\begin{pmatrix} 0 & 1 \\ 1 & 0 \end{pmatrix}$

7. $\begin{pmatrix} 1 & 0 & 1 \\ 0 & 1 & 1 \\ 1 & 1 & 1 \end{pmatrix}$

8. $\begin{pmatrix} 0 & 0 & 0 \\ 1 & 1 & 1 \\ x & y & z \\ x^2 & y^2 & z^2 \end{pmatrix}$

9. Let

$$Q = \begin{pmatrix} 1 & \frac{1}{2} & -\frac{3}{4} & -0 \\ \frac{3}{8} & 0 & 10 & -11 \\ 0 & 0 & \frac{3}{2} & -\frac{4}{8} \\ 1 & -1 & -2 & 2 \\ 0 & 8 & -8 & 0 \end{pmatrix}.$$

Use this matrix to identify the values for the components.

(a) q_{11} (b) q_{25} (c) q_{52} (d) q_{23} (e) q_{32}
(f) q_{42} (g) q_{14} (h) q_{44} (i) q_{52}

10. Given the matrix

$$Q = \begin{pmatrix} -4 & 5 & 1 & 0 & 0 & 6 \\ 5 & 5 & -5 & 2 & 2 & 0 \\ 1 & -8 & 3 & 3 & 0 & 0 \\ 7 & 9 & 9 & 1 & -2 & 1 \\ 10 & 5 & 2 & 9 & 0 & 9 \\ -7 & 0 & 0 & 0 & 2 & 4 \end{pmatrix}$$

answer the following questions.
(a) What is its dimension?   **(b)** $q_{11} = ?$
(c) $q_{23} = ?$ **(d)** $q_{32} = ?$ **(e)** $q_{45} = ?$
(f) $q_{71} = ?$ **(g)** $q_{14} = ?$ **(h)** What is the sum of
the diagonal elements?

For the expressions in Exercises 11–20, use these matrices to perform the indicated computations. If computation is not possible, explain why.

$$A = \begin{pmatrix} 1 & 2 \\ -1 & 0 \\ 1 & 1 \end{pmatrix} \qquad B = \begin{pmatrix} 3 & 0 & 1 \end{pmatrix}$$

$$C = \begin{pmatrix} 4 & -2 \\ 0 & 1 \\ 0 & 2 \end{pmatrix} \qquad D = \begin{pmatrix} -\frac{3}{2} & 2 & \frac{1}{4} \end{pmatrix}$$

11. $A + B$ **12.** $A + C$

13. $7A$ **14.** $2C - 5A$

15. $-6D + \frac{1}{2}B$ **16.** xB; x is any scalar.

17. $(3 - 6)B$ **18.** $-8B$

19. πC **20.** yC; y is any scalar.

For the expressions in Exercises 21–30, use these matrices to perform the indicated computations. If computation is not possible, explain why.

$$A = \begin{pmatrix} 2 & -2 \\ 1 & 1 \\ 2 & 2 \end{pmatrix} \qquad B = \begin{pmatrix} 2 & -2 & 2 \end{pmatrix}$$

$$C = \begin{pmatrix} 4 & 0 \\ 1 & 0 \\ 1 & 0 \end{pmatrix} \qquad D = \begin{pmatrix} 2 & 2 & \frac{1}{4} \end{pmatrix}$$

21. $A + B$ **22.** $A + C$

23. $7A$ **24.** $2C - 5A$

25. $-6D + \frac{1}{2}B$ **26.** xB; x is any scalar.

27. $(3 - 6)B$ **28.** $-8B$

29. $8A$ **30.** yB; y is any scalar.

Evaluate the matrix expressions in Exercises 31–42.

31. $\begin{pmatrix} 2 & -3 \\ 2 & -1 \end{pmatrix} + \begin{pmatrix} -2 & -1 \\ 3 & 3 \end{pmatrix}$

32. $\begin{pmatrix} 2 & -3 \\ 2 & -1 \end{pmatrix} - \begin{pmatrix} -2 & -1 \\ 3 & 3 \end{pmatrix}$

33. $\begin{pmatrix} 3 & -7 \\ 4 & 4 \end{pmatrix} - \begin{pmatrix} 0 & 1 \\ 1 & 0 \end{pmatrix}$

34. $\begin{pmatrix} 4 & -4 \\ 5 & -5 \end{pmatrix} + \begin{pmatrix} 3 & 0 \\ 0 & -2 \end{pmatrix}$

35. $-5 \begin{pmatrix} 7 & -3 \\ 4 & 5 \end{pmatrix}$ **36.** $\frac{2}{3} \begin{pmatrix} 0 & \frac{3}{4} \\ -1 & -6 \end{pmatrix}$

37. $\begin{pmatrix} -5.25 & 4.25 \\ 1.00 & 2.25 \\ -0.25 & 0.25 \end{pmatrix} - \begin{pmatrix} 3.25 & -2.25 \\ 1.00 & 2.50 \\ -0.50 & 1.75 \end{pmatrix}$

38. $-1 \cdot \begin{pmatrix} 2 & -2 & 2 & -2 \end{pmatrix}$

39. $\begin{pmatrix} 1 & 0 \\ 0 & 1 \end{pmatrix} + 7 \cdot \begin{pmatrix} 0 & -2 \\ 1 & -\frac{1}{7} \end{pmatrix}$

40. $-2 \cdot \begin{pmatrix} 3 & 2 & 1 \\ 0 & -2 & 0.5 \\ 0 & 0 & 2 \end{pmatrix} - (-2) \cdot \begin{pmatrix} -1 & -2 & 1 \\ 0 & -0.5 & 3 \\ 1 & 1 & -1 \end{pmatrix}$

41. $\frac{1}{3} \begin{pmatrix} 9 & 12 & -3 & 3 \\ 1 & 0 & 0 & -6 \end{pmatrix} - \frac{2}{3} \cdot \begin{pmatrix} -9 & 6 & -6 & 12 \\ 0 & 0 & 3 & -6 \end{pmatrix}$

42. $850 \cdot \begin{pmatrix} 4 \\ -2 \end{pmatrix} - 625 \begin{pmatrix} 7 \\ 1 \end{pmatrix}$

43. Given a matrix A with components a_{ij}, an associated matrix is the transpose matrix A^T formed by interchanging the rows with the columns of A; formally, A^T has components a_{ji}. For example, the matrix $A = \begin{pmatrix} 1 & 3 \\ 2 & 4 \\ 3 & 6 \end{pmatrix}$ has the transpose $A^T = \begin{pmatrix} 1 & 2 & 3 \\ 3 & 4 & 6 \end{pmatrix}$. If A has dimension $m \times n$, what is the dimension of its alternative form A^T?

44. Show that $(A^T)^T = A$ for all matrices A. Here, A^T is the transpose of a matrix, defined in the previous problem.

In Exercises 45–49, you are presented with statements of various rules that matrix arithmetic satisfies. Uppercase letters refer to matrices, and lowercase letters to scalars. In each case, you are to create examples for each of the matrices, and demonstrate that for these choices at least, the statement holds.

45. Commutative law of addition: $A + B = B + A$

46. Associative law of addition: $A+(B+C) = (A+B)+C$

47. $(c + d)A = cA + dA$

48. $c(A + B) = cA + cB$

49. $c(dA) = (cd)A$

Applications

Use matrix methods *only* on the following problems.

50. *(Salary Analysis)* Two years ago, you earned $10,000. Last year, you earned $14,500. This year, you expect to earn $19,750. Devise two different matrices to contain this information.

51. *(Academic Performance)* Imagine that you take four courses this semester, and that you have taken three exams in each course from which your final grade will be computed. Using invented grades, construct a matrix to contain this information.

52. *(Forest Management)* Two varieties of pine tree, one new and one standard, suitable for Christmas tree harvesting, are being tested for speed of growth. The new tree is being compared with the old variety. Two 12-in seedlings are planted, and their growth is recorded every two months for one year. The new variety grows 4 in between successive measurements, except for the last measurement, when it had grown 3 in. The standard seedling showed no growth until the third measurement. From then on, it grew 3 in between successive measurements.
(a) Use a single matrix to represent the recorded heights for both trees.
(b) The *average tree growth rate* is the increment in height between measurements divided by the time between measurements. Represent the average growth rates for the two seedlings by a single matrix.

W **53.** *(Movie Theaters)* The View-Rite Theatre Company owns 4 movie theaters in Davenport, Iowa. Ticket receipts vary with the time of week, with the different categories being weekdays, weekend matinees, and weekend evenings. Each theater has 500 seats, and full-priced tickets are $4.00 each. Senior citizens are entitled to a $1.50 discount when they present a valid Golden Age

card. Here are the percentages of patrons who are seniors at typical performances.

	Weekdays	Weekend Matinees	Weekend Evenings
# 1	80%	50%	30%
# 2	80%	75%	10%
# 3	75%	75%	20%
# 4	60%	50%	20%

Here also are the percentages of all theater seats that, on the average, are occupied.

	Weekdays	Weekend Matinees	Weekend Evenings
# 1	80%	75%	80%
# 2	80%	75%	80%
# 3	80%	75%	80%
# 4	80%	75%	80%

Use matrices and matrix arithmetic to answer the following questions.
(a) Define matrices S and T that give the senior occupancy percentages and the percentages of all seats that are sold.
(b) Write a matrix C (for capacity) that gives the number of seats, and hence the number of tickets that can be sold, for each theater.
(c) Why is matrix representation particularly well suited for a problem like this? Or is it?

54. *(Psychological Testing)* Three groups of students are tested for their ability to memorize and retain long numbers. The first group of students has received a short memory course, while the second group has received a more intensive course. The third group has no special training. Here are tables showing the average seconds of retention for the groups.

# of digits	Group 1	Group 2	Group 3
0-9	26	29	17
10-14	21	25	13
15-19	11	14	5

(a) Write these data as a matrix.
(b) Construct a matrix showing the percentage by which the performance of the first two groups surpasses that of the third.

55. *(New Zealand Sheep Farming)* A certain sheep rancher owns the following numbers of a given breed of sheep.

Age	Males	Females
0–1 yr	300	450
1–2 yr	350	400
2 yr and up	500	500

A neighboring ranch contains the following numbers of sheep.

Age	Males	Females
0–1 yr	350	400
1–2 yr	450	400
2 yr and up	650	550

(a) Represent the sheep population of each ranch as a matrix.

(b) Use matrix arithmetic to determine by how many sheep in each category the second ranch exceeds the first ranch.

(c) The second rancher is thinking of buying the first ranch. If she does, what will her total sheep holdings be in each of the categories? Use matrix arithmetic.

W **56.** Why can only square matrices have well-defined diagonals?

Calculator Exercises

Compute the answers to Exercises 57–60.

57. $\begin{pmatrix} 1.23 & -9.87 \\ 2.30 & 2.30 \end{pmatrix} + \begin{pmatrix} -4.66 & 9.87 \\ -2.30 & -1.15 \end{pmatrix}$

58. $\begin{pmatrix} 5.3 & -4.2 \\ 3.3 & -9.4 \end{pmatrix} - \begin{pmatrix} -4.3 & 3.3 \\ 3.3 & -5.3 \end{pmatrix}$

59. $-4.7 \begin{pmatrix} 2.1 & -4.2 \\ 3.7 & 7.9 \end{pmatrix}$

60. $\frac{1}{2} \begin{pmatrix} 1.0 & -2.3 \\ 7.6 & 7.2 \end{pmatrix} + \begin{pmatrix} 3.9 & 2.8 \\ 3.7 & 4.6 \end{pmatrix}$

Section 2.2 *MATRIX MULTIPLICATION*

Can one *multiply* one matrix by another? Beginners in matrices often want to define the product of two matrices of equal dimension as that matrix whose components are products of the corresponding components. This is *not* the way we define matrix multiplication, only because the combined 20-20 hindsight of generations of mathematicians has focused on a less intuitive but far more useful technique, a technique making possible the connection between matrix multiplication and the solution of linear equation systems.

Linear equations can have more than the two unknowns we worked with in the first chapter. A typical equation in three variables might be

(2.1) $$3x_1 - 4x_2 + x_3 = -2$$

Let's agree to write all linear equations in a *standard form* whenever we want to make a connection between them and matrices.

Standard Form for Linear Equations

1. We place the constant term on the *right* of the $=$ sign.

2. We write all variables as *subscripted* variables.

3. We agree to write all terms on the left of the $=$ sign in *ascending order of the variable subscripts.*

Is Equation (2.1) in standard form? The constant term (-2) is on the right-hand side, and the variables are all subscripted. Finally, the terms *are* written in ascending subscript order. (That is, the x_1 term precedes the x_2 term, which comes before the x_3 term.)

Define a *coefficient matrix L* containing the coefficients of the variables. In Equation (2.1), the coefficient matrix is a row matrix:

$$L = \begin{pmatrix} 3 & -4 & 1 \end{pmatrix}$$

Finally, define a *variable matrix R* to be a column matrix containing the variables of the problem.

$$R = \begin{pmatrix} x_1 \\ x_2 \\ x_3 \end{pmatrix}$$

We use the matrix names L and R to remind us that these matrices are the *left* and *right* factors of a matrix product. *Order is important* in matrix multiplication, and it seems preferable to use the L-R notation at this point to underscore this importance. After the development of matrix multiplication is complete, we will revert to notation that may be more appropriate for the matrix. For example, we might more suggestively label $\begin{pmatrix} x_1 \\ x_2 \\ x_3 \end{pmatrix}$ as X at that time.

From Equation (2.1), we write a column matrix

$$R = \begin{pmatrix} x_1 \\ x_2 \\ x_3 \end{pmatrix}$$

Is it possible to define matrix multiplication so that

$$LR = -2$$

makes sense? We let the *matrix product* be the *sum* of the products of the corresponding elements of the row and column matrix. That is, *by definition,*

$$\begin{pmatrix} 3 & -4 & 1 \end{pmatrix} \begin{pmatrix} x_1 \\ x_2 \\ x_3 \end{pmatrix} = 3 \cdot x_1 + (-4) \cdot x_2 + 1 \cdot x_3$$

$$= 3x_1 - 4x_2 + x_3 = -2$$

which *is* equivalent to Equation (2.1). With this example under our belt, we may generalize to the product of any row matrix with any column matrix, provided that

1. Each has the same number of elements.

2. The row is multiplied on the left of the column (the order of matrices in a matrix product is very important).

We denote the product as $L \cdot R$ or simply LR. The matrix product of any row matrix and column matrix is *always* a scalar. See figure 2.5.

We summarize what we now know.

Row-Column Matrix Multiplication

$$LR = \begin{pmatrix} l_1 & l_2 & \cdots & l_n \end{pmatrix} \begin{pmatrix} r_1 \\ r_2 \\ \vdots \\ r_n \end{pmatrix} = l_1 r_1 + l_2 r_2 + \cdots + l_n r_n$$

In words, we multiply corresponding elements in the factor matrices, and add these products together.

Here are the meanings for the symbols in this equation.

L is the *left* factor matrix, a row matrix with exactly n components.
R is the *right* factor matrix, a column matrix with exactly n components.
l_i are the components of L.
r_i are the components of R.

If L and R have different numbers of components, the matrix product is *undefined*. The product depends on the *order* in which the matrices are multiplied.

---

dot product multiplication   We also call the multiplication of a row and column matrix *dot product multiplication.*

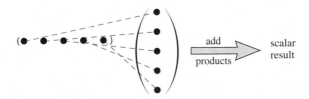

FIGURE 2.5
Row-column multiplication.

Matrix addition and subtraction operate on matrices of *equal* dimension to yield a new matrix of *equal* dimension. Row-column matrix multiplication takes two matrices of quite different dimension and creates a product matrix of yet another dimension. (We may regard a *scalar*, which is the product of row-column multiplication, as a matrix of dimension 1×1.) We will shortly construct the multiplication of more general matrices out of these basic row-column multiplications.

Example 6   Compute:

(a) $\begin{pmatrix} 3 & -2 \end{pmatrix} \begin{pmatrix} 2 \\ 2 \end{pmatrix}$

(b) $\begin{pmatrix} 1 & 1 & 4 \end{pmatrix} \cdot \begin{pmatrix} -1 \\ -1 \\ -7 \end{pmatrix}$

Solution

(a)

$$\begin{pmatrix} 3 & -2 \end{pmatrix} \begin{pmatrix} 2 \\ 2 \end{pmatrix} = 3 \cdot 2 + (-2) \cdot 2 = 2$$

(b)

$$\begin{pmatrix} 1 & 1 & 4 \end{pmatrix} \cdot \begin{pmatrix} -1 \\ -1 \\ -7 \end{pmatrix} = 1(-1) + 1(-1) + 4(-7) = -30$$

Skill Enhancer 6   Compute $\begin{pmatrix} 1 & -1 & 1 \end{pmatrix} \begin{pmatrix} 3 \\ 2 \\ -1 \end{pmatrix}$ and $\begin{pmatrix} 4 & -4 \end{pmatrix} \begin{pmatrix} 10 \\ 0 \end{pmatrix}$.

Answer in Appendix E.

Example 7   Let

$$A = \begin{pmatrix} -7 \\ 3 \\ 0 \\ 1 \end{pmatrix} \qquad B = \begin{pmatrix} 0 & 2 & 4 & 6 \end{pmatrix} \qquad C = \begin{pmatrix} 1 & 1 & 0 & -1 \end{pmatrix}$$

Compute $B \cdot A$, $(B \cdot A)A$, and $(B + C) \cdot A$.

Solution

$$B \cdot A = \begin{pmatrix} 0 & 2 & 4 & 6 \end{pmatrix} \cdot \begin{pmatrix} -7 \\ 3 \\ 0 \\ 1 \end{pmatrix} = 0 + 6 + 0 + 6 = 12$$

Since $B \cdot A$ is a scalar, we can use it to multiply the matrix A.

$$(B \cdot A)A = 12 \begin{pmatrix} -7 \\ 3 \\ 0 \\ 1 \end{pmatrix} = \begin{pmatrix} -84 \\ 36 \\ 0 \\ 12 \end{pmatrix}$$

For the last product, compute $B + C$ first.

$$B + C = \begin{pmatrix} 0+1 & 2+1 & 4+0 & 6-1 \end{pmatrix} = \begin{pmatrix} 1 & 3 & 4 & 5 \end{pmatrix}$$

so

$$(B+C) \cdot A = \begin{pmatrix} 1 & 3 & 4 & 5 \end{pmatrix} \cdot \begin{pmatrix} -7 \\ 3 \\ 0 \\ 1 \end{pmatrix} = \begin{pmatrix} -7+9+0+5 \end{pmatrix} = 7$$

Skill Enhancer 7   Use the matrices A, B, and C from this example to compute $C \cdot A$ and $(C \cdot A)B$.

Answer in Appendix E.

The generalization to arbitrary matrix products proceeds in two steps. First, suppose that the left matrix contains more than one row. Later, we will let the right matrix contain more than one column.

We embark on this first generalization by dealing with *systems* of linear equations. For a system such as

(2.2)
$$\begin{aligned} 3x_1 - 4x_2 + x_3 &= -2 \\ 2x_1 + x_2 - x_3 &= 0 \\ -x_1 + x_2 + x_3 &= -1 \end{aligned}$$

it is natural to retain the column matrix R with variable components and to create a column matrix of constants B as

$$B = \begin{pmatrix} -2 \\ 0 \\ -1 \end{pmatrix}$$

It is also natural to write the coefficient matrix L for this system as

$$L = \begin{pmatrix} 3 & -4 & 1 \\ 2 & 1 & -1 \\ -1 & 1 & 1 \end{pmatrix}$$

Is there a matrix multiplication operation such that the system [Equations (2.2)] can be written compactly in matrix notation as

$$LR = B$$

Is there a concept of matrix equality that will make this matrix equation even make sense?

matrix equality   Two conditions are necessary for *matrix equality* of two matrices:

1. The matrices must be of *identical dimension*.

2. Every component of the first *must equal the corresponding component* of the second.

Thus, if the product of L and R is a column matrix, then the matrix equation $LR = B$ makes sense under this definition, provided there is a way to extend the definition of matrix multiplication.

Several Rows in the Left Matrix

A natural way of extending multiplication suggests itself. The product of an $m \times n$ matrix A times a column matrix X of dimension $n \times 1$ shall be defined to be a column matrix of dimension $m \times 1$, where the ith component of the product is formed as the row-column product of the ith row of L with R. *Row-column multiplications are the building blocks of all matrix multiplication.*

Figure 2.6 displays the multiplication rule schematically. This rule reduces to row-column multiplication when L contains a single row. In words, we compute the component in the ith row of B by forming the row-column product of the ith row of L with the first (and only) column of R. We call this *matrix-column multiplication*.

Example 8   Compute $\begin{pmatrix} 2 & -4 & 3 \\ 0 & -1 & 1 \end{pmatrix} \begin{pmatrix} 2 \\ -1 \\ 3 \end{pmatrix}$.

Solution

$$\begin{pmatrix} 2 & -4 & 3 \\ 0 & -1 & 1 \end{pmatrix} \begin{pmatrix} 2 \\ -1 \\ 3 \end{pmatrix} = \begin{pmatrix} (2)(2) + (-4)(-1) + (3)(3) \\ (0)(2) + (-1)(-1) + (1)(3) \end{pmatrix}$$

$$= \begin{pmatrix} 17 \\ 4 \end{pmatrix}$$

Skill Enhancer 8   Compute $\begin{pmatrix} 6 & 5 & 4 \\ -1 & 0 & 0 \end{pmatrix} \begin{pmatrix} -3 \\ 2 \\ 4 \end{pmatrix}$.

Answer in Appendix E.

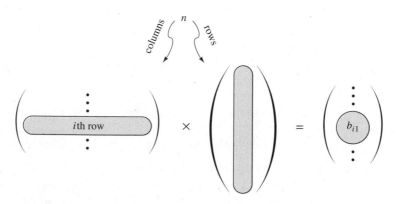

FIGURE 2.6
Multiplication of an $m \times n$ matrix times an $n \times 1$ matrix is a series of row-column multiplications.

Example 9 Consider these matrices:

$$A = \begin{pmatrix} 1 & 0 & 0 \\ 0 & 1 & 0 \\ 0 & 0 & 1 \end{pmatrix} \qquad B = \begin{pmatrix} -1 & 1 & -1 \\ 0 & 1 & 2 \end{pmatrix} \qquad C = \begin{pmatrix} 2 \\ 0 \\ 1 \end{pmatrix}$$

With reference *only* to the rules for matrix multiplication we have so far learned, which of the following matrix products can we compute? For the ones we can compute, find the product.
(a) CC;   (b) BC;   (c) CB;   (d) AC.

Solution

Neither CC nor CB is computable. In these cases, the dimensions of the matrix factors are incompatible with the rules for multiplication presented thus far.

(b)

$$BC = \begin{pmatrix} (-1 \quad 1 \quad -1) \cdot \begin{pmatrix} 2 \\ 0 \\ 1 \end{pmatrix} \\ (0 \quad 1 \quad 2) \cdot \begin{pmatrix} 2 \\ 0 \\ 1 \end{pmatrix} \end{pmatrix} = \begin{pmatrix} -3 \\ 2 \end{pmatrix}$$

(d)

$$AC = \begin{pmatrix} (1 \quad 0 \quad 0) \cdot \begin{pmatrix} 2 \\ 0 \\ 1 \end{pmatrix} \\ (0 \quad 1 \quad 0) \cdot \begin{pmatrix} 2 \\ 0 \\ 1 \end{pmatrix} \\ (0 \quad 0 \quad 1) \cdot \begin{pmatrix} 2 \\ 0 \\ 1 \end{pmatrix} \end{pmatrix} = \begin{pmatrix} 2 \\ 0 \\ 1 \end{pmatrix}$$

Skill Enhancer 9 If $D = \begin{pmatrix} 1 & 0 \\ 0 & 1 \end{pmatrix}$ and $E = \begin{pmatrix} 2 & -1 \\ 1 & 0 \end{pmatrix}$, compute (if possible) DE and ED.

Answer in Appendix E.

Several Columns in the Right Matrix

In what situations may one multiply a matrix L of arbitrary dimension by a matrix R, also of arbitrary dimension? When we added rows to L, we added rows to the product. Continuing the pattern, we expect to add *columns* to the product when we add *columns* to R. The jth column of the product will be the matrix-column product of L with the jth column of matrix R. It would follow from this that we can compute the component p_{ij} that is at the intersection of the ith row and the jth column of the product matrix P

by forming the row-column product of the ith row of L and the jth column of R. See Figure 2.7.

Example 10   Compute $\begin{pmatrix} 1 & 7 & 4 \end{pmatrix} \begin{pmatrix} 3 & 0 \\ 2 & -1 \\ 3 & -2 \end{pmatrix}$.

Solution

$$\begin{pmatrix} 1 & 7 & 4 \end{pmatrix} \begin{pmatrix} 3 & 0 \\ 2 & -1 \\ 3 & -2 \end{pmatrix} = \begin{pmatrix} 1(3) + 7(2) + 4(3) & 1(0) + 7(-1) + 4(-2) \end{pmatrix}$$

$$= \begin{pmatrix} 29 & -15 \end{pmatrix}$$

Skill Enhancer 10   What is the product of $\begin{pmatrix} -2 & 3 & 2 \end{pmatrix}$ and $\begin{pmatrix} 1 & -2 \\ 0 & -1 \\ 9 & 3 \end{pmatrix}$?

Answer in Appendix E.

General Matrix Multiplication

When is matrix multiplication possible? For all the row-column multiplications to be possible, there must be as many *columns* in the left matrix L as there are *rows* in the right matrix R. (Make sure you see why.) The product will contain as many *rows* as there are in L, and as many *columns* as there are in R. Schematically,

$$(2.3) \qquad \begin{pmatrix} m \text{ rows} \\ n \text{ columns} \end{pmatrix} \begin{pmatrix} n \text{ rows} \\ p \text{ columns} \end{pmatrix} = \begin{pmatrix} m \text{ rows} \\ p \text{ columns} \end{pmatrix}$$

A diagram makes this clear; see Figure 2.8.

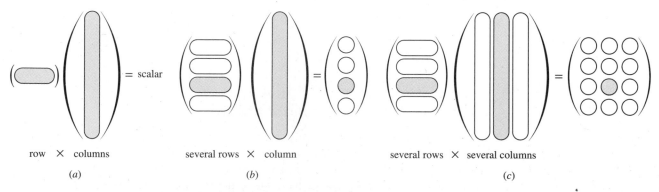

row × columns   several rows × column   several rows × several columns

(a)   (b)   (c)

FIGURE 2.7
Generalizing matrix multiplication.

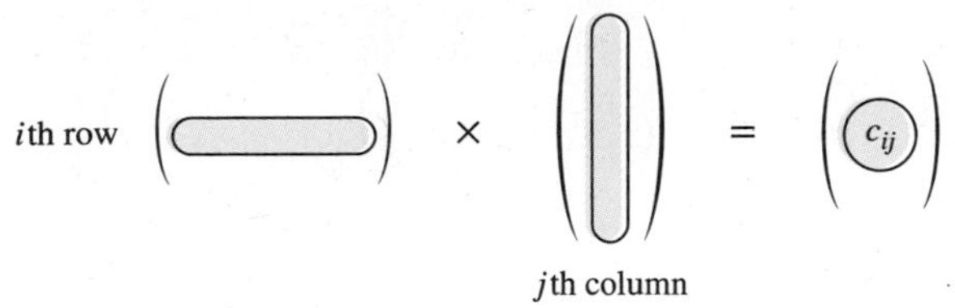

ith row

jth column

FIGURE 2.8
Matrix multiplication.

matrix multiplication

Matrix Multiplication

To form the product

$$LR = P$$

the dimensions of matrices L and R must match as in Equation (2.3). Any component p_{ij} of the product matrix P is the row-column product of the ith row of L and the jth column of R.

Example 11   Define matrices

$$L = \begin{pmatrix} 1 & 7 & 0 \\ 2 & 3 & 4 \end{pmatrix} \qquad R = \begin{pmatrix} 0 & 1 & 0 & 1 \\ -1 & 0 & -1 & 0 \\ 1 & 2 & 3 & 4 \end{pmatrix} \qquad P = LR$$

What is p_{13}, the component of P in the first row and third column? What is p_{22}?

Solution

From the definition of matrix multiplication,

$$p_{13} = \left(\text{row 1 of } L \right) \left(\text{column 3 of } R \right)$$

$$= \begin{pmatrix} 1 & 7 & 0 \end{pmatrix} \begin{pmatrix} 0 \\ -1 \\ 3 \end{pmatrix}$$

$$= -7$$

See Figure 2.9. In the same way,

$$p_{22} = \begin{pmatrix} 2 & 3 & 4 \end{pmatrix} \begin{pmatrix} 1 \\ 0 \\ 2 \end{pmatrix}$$

$$= 10$$

See Figure 2.10.

Skill Enhancer 11   What are p_{31}, p_{11}, and p_{24}?

Answer in Appendix E.

$$\begin{pmatrix} \boxed{1 \quad 7 \quad 0} \\ 2 \quad 3 \quad 4 \end{pmatrix} \quad \times \quad \begin{pmatrix} 0 & 1 & \boxed{0} & 1 \\ -1 & 0 & \boxed{-1} & 0 \\ 1 & 2 & \boxed{3} & 4 \end{pmatrix}$$

FIGURE 2.9
Computing p_{13}.

$$\begin{pmatrix} 1 \quad 7 \quad 0 \\ \boxed{2 \quad 3 \quad 4} \end{pmatrix} \quad \times \quad \begin{pmatrix} 0 & \boxed{1} & 0 & 1 \\ -1 & \boxed{0} & -1 & 0 \\ 1 & \boxed{2} & 3 & 4 \end{pmatrix}$$

FIGURE 2.10
Computing p_{22}.

Example 12   If

$$A = \begin{pmatrix} -1 & 1 \\ 0 & 1 \end{pmatrix} \qquad B = \begin{pmatrix} 2 & 0 \\ 1 & 2 \end{pmatrix}$$

then compute $C = AB$ and $D = BA$.

Solution

Since both A and B have two rows and two columns, so will C. Denote C's components by c_{ij}. Then

$$c_{11} = \begin{pmatrix} -1 & 1 \end{pmatrix} \cdot \begin{pmatrix} 2 \\ 1 \end{pmatrix} = -1 \qquad c_{12} = \begin{pmatrix} -1 & 1 \end{pmatrix} \cdot \begin{pmatrix} 0 \\ 2 \end{pmatrix} = 2$$

$$c_{21} = \begin{pmatrix} 0 & 1 \end{pmatrix} \cdot \begin{pmatrix} 2 \\ 1 \end{pmatrix} = 1 \qquad c_{22} = \begin{pmatrix} 0 & 1 \end{pmatrix} \cdot \begin{pmatrix} 0 \\ 2 \end{pmatrix} = 2$$

so that

$$C = \begin{pmatrix} -1 & 2 \\ 1 & 2 \end{pmatrix}$$

In the same way, we calculate the components of $D = BA$:

$$d_{11} = \begin{pmatrix} 2 & 0 \end{pmatrix} \cdot \begin{pmatrix} -1 \\ 0 \end{pmatrix} = -2 \qquad d_{12} = \begin{pmatrix} 2 & 0 \end{pmatrix} \cdot \begin{pmatrix} 1 \\ 1 \end{pmatrix} = 2$$

$$d_{21} = \begin{pmatrix} 1 & 2 \end{pmatrix} \cdot \begin{pmatrix} -1 \\ 0 \end{pmatrix} = -1 \qquad d_{22} = \begin{pmatrix} 1 & 2 \end{pmatrix} \cdot \begin{pmatrix} 1 \\ 1 \end{pmatrix} = 3$$

so that

$$D = \begin{pmatrix} -2 & 2 \\ -1 & 3 \end{pmatrix}$$

Skill Enhancer 12   If $X = \begin{pmatrix} 1 & -1 \\ 0 & 2 \end{pmatrix}$ and $Y = \begin{pmatrix} 0 & -2 \\ 2 & 1 \end{pmatrix}$, what is XY? What is YX?

Answer in Appendix E.

Like the previous example, the next example shows the importance of *order* in matrix multiplication. Although for any scalars a and b it is true that ab always equals ba, this will generally *not* be true for matrices. Be careful of the order of multiplication; matrix multiplication is *not commutative*. The products AB and BA in this example are quite different. In other examples, the matrix product AB may be legal, whereas BA may not even be computable.

Example 13   Let

$$A = \begin{pmatrix} -1 & 1 & -1 \\ 0 & 1 & 2 \end{pmatrix} \qquad B = \begin{pmatrix} 1 & 2 \\ 1 & 0 \\ 0 & 1 \end{pmatrix}$$

(a) Compute AB.
(b) Compute BA.

Solution

(a)

$$AB = \begin{pmatrix} -1 & 1 & -1 \\ 0 & 1 & 2 \end{pmatrix} \begin{pmatrix} 1 & 2 \\ 1 & 0 \\ 0 & 1 \end{pmatrix}$$

$$= \begin{pmatrix} (-1 \cdot 1 + 1 \cdot 1 - 1 \cdot 0) & (-1 \cdot 2 + 1 \cdot 0 - 1 \cdot 1) \\ (0 \cdot 1 + 1 \cdot 1 + 2 \cdot 0) & (0 \cdot 2 + 1 \cdot 0 + 2 \cdot 1) \end{pmatrix}$$

$$= \begin{pmatrix} 0 & -3 \\ 1 & 2 \end{pmatrix}$$

(b) In the same way,

$$BA = \begin{pmatrix} 1 & 2 \\ 1 & 0 \\ 0 & 1 \end{pmatrix} \begin{pmatrix} -1 & 1 & -1 \\ 0 & 1 & 2 \end{pmatrix}$$

$$= \begin{pmatrix} -1 & 3 & 3 \\ -1 & 1 & -1 \\ 0 & 1 & 2 \end{pmatrix}$$

Skill Enhancer 13   If $X = \begin{pmatrix} 1 & 2 \\ -1 & 1 \end{pmatrix}$ and $Y = \begin{pmatrix} 1 & 1 & 0 \\ -2 & 0 & -1 \end{pmatrix}$, what is XY? What is YX?

Answer in Appendix E.

Example 14   What is the dimension of the product

$$\begin{pmatrix} -6 & 5 & 0 \\ 1 & 3 & 2 \\ 0 & 1 & -1 \end{pmatrix} \begin{pmatrix} 0 & 1 \\ 2 & 2 \\ 1 & 0 \end{pmatrix}$$

What is the product?

Solution

Multiplication of a 3×3 matrix and a 3×2 matrix yields a product with dimension 3×2.

$$\begin{pmatrix} -6 & 5 & 0 \\ 1 & 3 & 2 \\ 0 & 1 & -1 \end{pmatrix} \begin{pmatrix} 0 & 1 \\ 2 & 2 \\ 1 & 0 \end{pmatrix} = \begin{pmatrix} 0+10+0 & -6+10+0 \\ 0+6+2 & 1+6+0 \\ 0+2-1 & 0+2+0 \end{pmatrix}$$

$$= \begin{pmatrix} 10 & 4 \\ 8 & 7 \\ 1 & 2 \end{pmatrix}$$

Skill Enhancer 14   Compute $\begin{pmatrix} 1 & -1 & 1 \\ 0 & 0 & -1 \\ 1 & 2 & 0 \end{pmatrix} \begin{pmatrix} 1 & -1 & 1 \\ 0 & 0 & -1 \\ 1 & 2 & 0 \end{pmatrix}$.

Answer in Appendix E.

Example 15   Compute

$$\begin{pmatrix} -1 & -\frac{3}{2} & 7 \\ 2 & 0 & 1 \end{pmatrix} \cdot \begin{pmatrix} 0 & -1 & 7 \\ 1 & 2 & 6 \\ 1 & -3 & -1 \end{pmatrix}$$

Solution

$$\begin{pmatrix} -1(0) - \frac{3}{2}(1) + 7(1) & -1(-1) - \frac{3}{2}(2) + 7(-3) & -1(7) - \frac{3}{2}(6) + 7(-1) \\ 2(0) + 0(1) + 1(1) & 2(-1) + 0(2) + 1(-3) & 2(7) + 0(6) + 1(-1) \end{pmatrix}$$

$$= \begin{pmatrix} 5\frac{1}{2} & -23 & -23 \\ 1 & -5 & 13 \end{pmatrix}$$

Skill Enhancer 15   Compute $\begin{pmatrix} 1 & 1 & \frac{1}{2} \\ 2 & -1 & 0 \end{pmatrix} \begin{pmatrix} -1 & 1 \\ 0 & \frac{1}{2} \\ \frac{3}{2} & -\frac{1}{2} \end{pmatrix}$.

Answer in Appendix E.

Matrix multiplication often represents a convenient shorthand for calculations in actual applications.

Example 16   A certain opera house can seat a total of 5,000 people. Tickets come in three categories: orchestra, with 2,000 seats, each selling for $35.00; the balcony, with 1,500 seats, each selling for $25.00; and 1,500 ring seats, at $15.00 per seat. In the first week of the season, there were two performances. For these performances, 1,800 and 1,500 orchestra seats were sold, and 1,400 and 1,200 balcony seats were sold. The ring

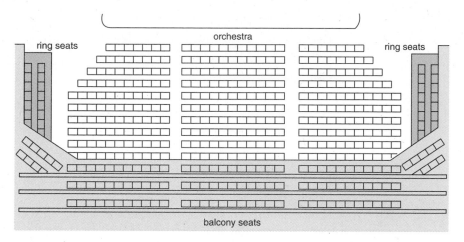

FIGURE 2.11

seats sold out for each performance. What were the total receipts for each performance? What was the total take for this week? Use matrix methods for the solution.

Solution

Define a sales matrix S whose columns give the sales for each performance and whose rows give the sales (number of tickets sold, that is) for each ticket category:

$$S = \begin{array}{c} \\ \text{orch.} \\ \text{balc.} \\ \text{ring} \end{array} \begin{pmatrix} \overset{\text{perf. 1}}{1,800} & \overset{\text{perf. 2}}{1,500} \\ 1,400 & 1,200 \\ 1,500 & 1,500 \end{pmatrix}$$

and a row matrix P containing the ticket prices:

$$P = \begin{pmatrix} 35 & 25 & 15 \end{pmatrix}$$

(Note that there are 1,500 ring tickets.) The total receipts are just the sum of the products of the number of tickets sold (in each category) times the price per ticket. For the first performance, total receipts are given by

$$1,800 \times 35 + 1,400 \times 25 + 1,500 \times 15$$

But this is just the first component in the matrix product PS. This suggests that the receipts for each performance are the components in the product PS. The receipts are therefore

$$PS = \begin{pmatrix} 120,500 & 105,000 \end{pmatrix}$$

The total receipts for the first week are the sum of these two components, $225,500.

Skill Enhancer 16   In the last week of the opera season, ticket sales were not so robust. There were three performances that week. The ticket sales for those performances were: 1,500, 1,100, and 1,000 for the orchestra; 1,200 for all balcony categories; and 1,100,

1,200, and 1,500 for the ring. What is the sales matrix S for this final week? Use matrix methods to compute the sales for each performance. What are the total ticket receipts for that week?

Answer in Appendix E.

Matrix equality is useful for solving some types of matrix problems.

Example 17   Solve for x_1, x_2, and x_3 if

$$A = \begin{pmatrix} 1 & 3 \\ 7 & 0 \end{pmatrix} \qquad X = \begin{pmatrix} x_1 & x_1 + x_2 \\ x_2 - x_3 & 0 \end{pmatrix} \qquad \text{and} \qquad A = X$$

Solution

Since the two matrices are equal only if all of their corresponding elements are equal, we must have the following scalar equations:

$$x_1 = 1$$
$$x_1 + x_2 = 3$$
$$x_2 - x_3 = 7$$

Of these, the first equation determines x_1 immediately. From the first pair of equations, we have

$$1 + x_2 = 3$$

so that $x_2 = 2$. Using this result in the last equation, we see that

$$2 - x_3 = 7$$

so that $x_3 = -5$.

Skill Enhancer 17   If X is as defined above, and $B = \begin{pmatrix} 1 & -1 \\ 2 & 0 \end{pmatrix}$, what are the values for x_1, x_2, and x_3 if $X = B$?

Answer in Appendix E.

Example 18   Define a 2×2 matrix

$$A = \begin{pmatrix} 2 & 1 \\ 1 & 2 \end{pmatrix}$$

Determine what restrictions there must be on the components of a 2×1 column matrix X such that

$$AX = X$$

Solution

Let $X = \begin{pmatrix} p \\ q \end{pmatrix}$. The corresponding matrix equation is

$$\begin{pmatrix} 2 & 1 \\ 1 & 2 \end{pmatrix} \begin{pmatrix} p \\ q \end{pmatrix} = \begin{pmatrix} p \\ q \end{pmatrix}$$

By the rules of matrix multiplication, the left side of this equation is

$$\begin{pmatrix} 2 & 1 \\ 1 & 2 \end{pmatrix} \begin{pmatrix} p \\ q \end{pmatrix} = \begin{pmatrix} 2p + q \\ p + 2q \end{pmatrix}.$$

Therefore,

$$\begin{pmatrix} p \\ q \end{pmatrix} = \begin{pmatrix} 2p + q \\ p + 2q \end{pmatrix}$$

By matrix equality, this last equation is true if and only if $p = 2p + q$ and $q = p + 2q$. Both of these reduce to the single equation $p + q = 0$. That is, the above matrix equation holds whenever the sum of the components of the matrix X is zero. Column matrices such as

$$\begin{pmatrix} -1 \\ 1 \end{pmatrix} \qquad \begin{pmatrix} 2 \\ -2 \end{pmatrix} \qquad \begin{pmatrix} \frac{17}{2} \\ -\frac{17}{2} \end{pmatrix} \quad \text{and} \quad \begin{pmatrix} 0 \\ 0 \end{pmatrix}$$

satisfy this property.

Skill Enhancer 18   Solve this example if now $S = \begin{pmatrix} 3 & 1 \\ 1 & 3 \end{pmatrix}$. As in the example, let $X = \begin{pmatrix} p \\ q \end{pmatrix}$.

Answer in Appendix E.

Matrix Identities

The number 0 is the *identity element of addition* because

$$0 + any\ number \qquad \text{or} \qquad any\ number + 0$$

is the original number (that is, if we add 0 to any number x, we still have x). In the same way, 1 is the *multiplicative identity element* because

$$1 \times any\ number \qquad \text{or} \qquad any\ number \times 1$$

leaves the original number unchanged. There are identity matrices for matrix addition and multiplication.

additive identity matrix   The *additive identity matrix* is straightforward. If A is some matrix, then let O be a matrix of the same dimension, each of whose components is 0: $o_{ij} = 0$. Then

$$A + O = O + A = A$$

multiplicative identity   Some care needs to be taken with the *multiplicative identity matrix*. The identity matrix
matrix I is one such that

$$AI = IA = A$$

Only square matrices A and I can satisfy this requirement, and only if A and I have the same dimension. To emphasize this connection, if A is square and of dimension $n \times n$, we label the identity I_n. (I_n is often written I if there is no possibility of confusion. Furthermore, we call it simply the identity element rather than the multiplicative identity;

usually there is no confusion.) A suitable definition for I_n is

$$I_n = \overbrace{\begin{pmatrix} 1 & 0 & \cdots & 0 \\ 0 & 1 & \cdots & 0 \\ \vdots & \vdots & \ddots & \vdots \\ 0 & 0 & \cdots & 1 \end{pmatrix}}^{n \text{ columns}} \left.\vphantom{\begin{pmatrix} 1 \\ 0 \\ \vdots \\ 0 \end{pmatrix}}\right\} n \text{ rows}$$

Under such a definition, any square $n \times n$ matrix A satisfies

$$AI_n = I_n A = A$$

Example 19   The definition for matrix identities implies that for two-dimensional matrices,

$$I_2 = \begin{pmatrix} 1 & 0 \\ 0 & 1 \end{pmatrix}$$

Show explicitly that I_2 is the multiplicative identity for all 2×2 matrices.

Solution

We may use the notation

$$A = \begin{pmatrix} a & b \\ c & d \end{pmatrix}$$

to represent an arbitrary 2×2 matrix A. I_2 will be the multiplicative identity if it satisfies the equations

$$AI_2 = I_2 A = A$$

It is sufficient to compute AI_2 and $I_2 A$. First,

$$AI_2 = \begin{pmatrix} a & b \\ c & d \end{pmatrix}\begin{pmatrix} 1 & 0 \\ 0 & 1 \end{pmatrix}$$
$$= \begin{pmatrix} a \cdot 1 + b \cdot 0 & a \cdot 0 + b \cdot 1 \\ c \cdot 1 + d \cdot 0 & c \cdot 0 + d \cdot 1 \end{pmatrix} = \begin{pmatrix} a & b \\ c & d \end{pmatrix}$$
$$= A$$

In the same way,

$$I_2 A = \begin{pmatrix} 1 & 0 \\ 0 & 1 \end{pmatrix}\begin{pmatrix} a & b \\ c & d \end{pmatrix} = \begin{pmatrix} a & b \\ c & d \end{pmatrix}$$
$$= A$$

so the assertion is proved.

Skill Enhancer 19   Show that

$$I_3 = \begin{pmatrix} 1 & 0 & 0 \\ 0 & 1 & 0 \\ 0 & 0 & 1 \end{pmatrix}$$

is the multiplicative identity for all 3×3 matrices.

Answer in Appendix E.

So far, we have explored the three matrix operations of addition, subtraction, and multiplication (two kinds). Matrix "division" is a stickier issue, and we defer it to the next two sections.

_____**Exercises 2.2** _____

Compute the dot products (row-column multiplications) in Exercises 1–14. If the multiplication is not possible, explain why.

1. $\begin{pmatrix} 1 & -1 & 0 \end{pmatrix} \cdot \begin{pmatrix} 3 \\ 2 \\ 3 \end{pmatrix}$

2. $\begin{pmatrix} 2 & 4 & 6 & 0 & -1 \end{pmatrix} \cdot \begin{pmatrix} \frac{1}{2} \\ -\frac{3}{2} \\ \frac{1}{2} \\ \frac{1}{3} \\ 0 \end{pmatrix}$

3. $\begin{pmatrix} 1 & 2 & -3 \end{pmatrix} \begin{pmatrix} 3 & -2 & -1 \end{pmatrix}$

4. $\begin{pmatrix} \frac{1}{2} & \frac{2}{3} & \frac{3}{4} \end{pmatrix} \begin{pmatrix} \frac{7}{8} \\ 1 \end{pmatrix}$

5. $\begin{pmatrix} 7 \\ -2 \\ 4 \end{pmatrix} \begin{pmatrix} 8 & 8 & -2 & 0 \end{pmatrix}$

6. $10.5 \begin{pmatrix} 7 & -2 & 6.5 \end{pmatrix} \begin{pmatrix} 13 \\ 10 \\ 3 \end{pmatrix}$

7. $\begin{pmatrix} 0 & -1 & 1 \end{pmatrix} \begin{pmatrix} 6 \\ 2 \\ 0 \end{pmatrix}$ **8.** $\begin{pmatrix} -1 & -1 & 1 \end{pmatrix} \begin{pmatrix} 2 \\ 2 \\ 1 \\ 0 \end{pmatrix}$

9. $\begin{pmatrix} 6 & 2 \end{pmatrix} \begin{pmatrix} 3 \\ 0 \end{pmatrix}$

10. $\begin{pmatrix} -1 & +1 & -1 & +1 \end{pmatrix} \cdot \begin{pmatrix} 1 \\ 1 \\ 1 \\ 1 \end{pmatrix}$

11. $\begin{pmatrix} \frac{1}{2} & 1 & \frac{1}{3} & 0 \end{pmatrix} \begin{pmatrix} 0 \\ \frac{1}{3} \\ 1 \\ 0 \end{pmatrix}$ **12.** $\begin{pmatrix} 3 & -3 \end{pmatrix} \begin{pmatrix} -1 \\ 0 \\ 1 \end{pmatrix}$

13. $\begin{pmatrix} -4 \\ 4 \\ 2 \\ 0 \end{pmatrix} \begin{pmatrix} 1 \\ -3 \\ 5 \\ 1 \end{pmatrix}$ **14.** $\begin{pmatrix} 2 & -3 & 1 & 0 \end{pmatrix} \begin{pmatrix} 4 \\ 3 \\ 2 \\ 1 \end{pmatrix}$

In Exercises 15–42, perform the indicated matrix arithmetic. If the operation is not possible, explain why.

15. $\begin{pmatrix} 1 & 0 \\ 0 & -1 \end{pmatrix} \begin{pmatrix} 2 \\ 2 \end{pmatrix}$ **16.** $\begin{pmatrix} 1 & 0 & 0 \\ 0 & 2 & 0 \\ 0 & 0 & 3 \end{pmatrix} \begin{pmatrix} \frac{1}{2} \\ \frac{1}{3} \\ \frac{1}{4} \end{pmatrix}$

17. $\begin{pmatrix} 3 & -1 & 0 \\ 2 & 0 & 1 \end{pmatrix} \begin{pmatrix} -1 \\ 0 \\ 1 \end{pmatrix}$ **18.** $\begin{pmatrix} -1 & 1 & -1 \\ 1 & -1 & 1 \\ -1 & 0 & -1 \end{pmatrix} \begin{pmatrix} 53 \\ 2 \end{pmatrix}$

19. $\begin{pmatrix} 8 & 2 & 3 \\ 1 & -1 & 0 \\ 7 & 0 & -5 \end{pmatrix} \begin{pmatrix} 0 \\ -2 \\ \frac{1}{3} \end{pmatrix}$

20. $\begin{pmatrix} 6 & -7 & 5 \\ 3 & -8 & 1 \\ 0 & 0 & 2 \end{pmatrix} \begin{pmatrix} -10 \\ 5 \\ 3 \end{pmatrix}$

21. $\begin{pmatrix} \frac{4}{5} & \frac{1}{5} \\ -10 & \frac{2}{5} \end{pmatrix} \begin{pmatrix} 0 & 25 \\ -15 & 30 \end{pmatrix}$

22. $\begin{pmatrix} 1 & 1 & 1 \\ -1 & -2 & -2 \end{pmatrix} \begin{pmatrix} 5 & 0 \\ -3 & 8 \end{pmatrix}$

23. $\begin{pmatrix} -2 & 31 \\ 4.5 & 7.2 \end{pmatrix} \begin{pmatrix} 10 & -12 \\ -12 & 0 \end{pmatrix}$

24. $\begin{pmatrix} 10 & -12 \\ -12 & 0 \end{pmatrix} \begin{pmatrix} -2 & 31 \\ 4.5 & 7.2 \end{pmatrix}$

25. $\begin{pmatrix} 1 & -3 & 9 \\ 2 & 4 & 8 \\ 6 & 0 & 0 \end{pmatrix} \begin{pmatrix} 2 & -1 & 3 \\ 0 & 2 & 0 \\ -4 & 0 & 3 \end{pmatrix}$

26. $\begin{pmatrix} 2 & 4 & -8 \\ 3 & 0 & -2 \end{pmatrix} \begin{pmatrix} 3 & 3 \\ 1 & 0 \\ 0 & -1 \end{pmatrix}$

27. $\begin{pmatrix} 6 & 9 & 0 \\ 0 & 0 & 2 \end{pmatrix} \begin{pmatrix} 1 & 2 & -1 \\ 3 & 0 & 4 \\ 4 & 0 & 1 \end{pmatrix}$

28. $\begin{pmatrix} 10 & 20 \\ -30 & 40 \end{pmatrix} \begin{pmatrix} 6 & 9 \\ -2 & -1 \end{pmatrix}$

29. $\left[\begin{pmatrix} 3 & -3 \\ 2 & 0 \end{pmatrix} \begin{pmatrix} -\frac{1}{2} & \frac{3}{2} \\ 1 & \frac{3}{2} \end{pmatrix} \right] \begin{pmatrix} 2 & -4 \\ 0 & 5 \end{pmatrix}$

30. $\left[\begin{pmatrix} 20 & 25 \end{pmatrix} + \begin{pmatrix} -35 & 10 \end{pmatrix} \right] \begin{pmatrix} 35 \\ 20 \end{pmatrix}$

31. $\begin{pmatrix} 2 & 2 \end{pmatrix} \cdot \begin{pmatrix} 2 & 2 \end{pmatrix}$

32. $\begin{pmatrix} 4,000 & 2,000 \\ 3,000 & 1,000 \\ -1,500 & 7,000 \end{pmatrix} \begin{pmatrix} 1 & 0 \\ 0 & 1 \end{pmatrix}$

33. $\begin{pmatrix} 1 & 0 & 0 \\ 0 & 1 & 0 \\ 0 & 0 & 1 \end{pmatrix} \begin{pmatrix} 0.027 & -0.331 \\ 0.956 & 0.002 \\ 0.007 & -0.598 \end{pmatrix}$

34. $\begin{pmatrix} -4 & 2 \\ 0 & 2 \end{pmatrix} \begin{pmatrix} 3 & 3 \\ -2 & 1 \end{pmatrix}$

35. $\begin{pmatrix} 0.254 & 0.004 & -0.127 \\ -0.993 & 0.385 & 0 \\ -0.004 & -0.999 & 0.925 \end{pmatrix} \begin{pmatrix} 0 & 1 & 0 \\ 1 & 0 & 0 \\ 0 & 0 & 1 \end{pmatrix}$

36. $\begin{pmatrix} \frac{1}{2} & -\frac{1}{2} & 1 \\ -3 & -3 & \frac{2}{3} \\ 0 & 0 & \frac{1}{2} \end{pmatrix} \left[\begin{pmatrix} 1 & 0 & 0 \\ 0 & 0 & 0 \\ 0 & 0 & 0 \end{pmatrix} + \begin{pmatrix} 0 & 0 & 0 \\ 0 & 1 & 0 \\ 0 & 0 & 1 \end{pmatrix} \right]$

37. $\begin{pmatrix} \frac{3}{4} & \frac{1}{3} \\ \frac{1}{4} & \frac{2}{3} \end{pmatrix} \begin{pmatrix} -1 & 7 \\ 3 & 0 \end{pmatrix}$

38. $\begin{pmatrix} 2 & -1 & 0 \\ 0 & 1 & 2 \\ 1 & 1 & -1 \end{pmatrix}^2$ (For a square matrix A, $A^2 = AA$.)

39. $\begin{pmatrix} 1 & 1 \\ 0 & 1 \end{pmatrix}^3$

40. $\begin{pmatrix} 1 & 0 & 0 \\ 0 & 1 & 0 \\ 0 & 0 & 1 \end{pmatrix}^{100}$

41. $\begin{pmatrix} -1 & 2 & 2 & 0 & 1 \\ 3 & 1 & 1 & -1 & 2 \\ 0 & -1 & 0 & 1 & 1 \\ -1 & 3 & -3 & -1 & 0 \\ -1 & 1 & -1 & 1 & -1 \end{pmatrix} \begin{pmatrix} 0 & -1 & 1 \\ 2 & 2 & -2 \\ 0 & 1 & 0 \\ 3 & 1 & -1 \\ 1 & 1 & 1 \end{pmatrix}$

42. $\begin{pmatrix} 1 & 1 & 1 \\ 2 & 2 & 2 \\ 3 & 3 & 3 \end{pmatrix} \begin{pmatrix} -1 & 3 & 0 \\ 0.5 & 0.75 & -0.25 \\ 0 & 0 & 1 \end{pmatrix}$

43. Let A and B be two arbitrary matrices with dimensions $m \times n$ and $p \times q$. What are the relationships that must hold between m, n, p, and q if it should be possible to compute both matrix products AB and BA? (In general, of course, $AB \neq BA$.)

44. Let A be some arbitrary matrix of dimension $m \times n$. Let I be a matrix of dimension $p \times q$ that should act as an identity; that is, we expect the equations

$$AI = IA = A$$

to be true. By considering only the dimensions of these products, show that this identity can be defined only when both A and I are square and of equal dimension.

45. We can represent the coordinates of a point as a 1×2 matrix. For example, the point $(2, 3)$ becomes the matrix

$(2 \quad 3)$. If P_1 and P_2 are the "matrices" for two points, define the matrix D as $D = P_2 - P_1$. Its transpose is D^T. Then the *distance* r between the two points is $r = \sqrt{D \cdot D^T}$. Use this formula to determine the distance between the following pairs of points:
   (a) $(2, 3)$, $(0, 0)$ (b) $(4, -1)$, $(1, 1)$ (c) $(1, 1)$, $(4, -1)$ (d) $(1, 1)$, $(4, 5)$ (e) $(-\frac{3}{2}, -\frac{1}{2})$, $(-1, -1)$ (f) $(-2, 3)$, $(-3, 2)$

46. Demonstrate that I_3 really is the multiplicative identity element for 3×3 matrices.

47. If $A = \begin{pmatrix} 2 & 1 \\ 1 & 2 \end{pmatrix}$, show that $Af = f$ for:
   (a) $f = \begin{pmatrix} -1 \\ 1 \end{pmatrix}$ (b) $f = \begin{pmatrix} 2 \\ -2 \end{pmatrix}$ (c) $f = \begin{pmatrix} -\frac{3}{2} \\ \frac{3}{2} \end{pmatrix}$

48. If A is the matrix defined in the previous problem, determine under what circumstances a matrix f will exist such that
   (a) $Af = f$ (b) $Af = 2f$ (c) $Af = nf$, for arbitrary scalar value n

49. **(Right-hand Matrix Identities)** Find a matrix R such that $AR = A$, even though RA may not be defined. Here,

$$A = \begin{pmatrix} 4 & 7 \\ 3 & -20 \\ 1 & 4 \end{pmatrix}$$

50. **(Left-hand Matrix Identities)** Using the matrix A defined in the previous exercise, determine a matrix L such that $LA = A$, even though the product AL will not be defined.

51. A square matrix is *symmetric* if $A = A^T$. It is *skew-symmetric* if $A^T = -A$. If $A = \begin{pmatrix} 1 & 2 \\ 2 & 0 \end{pmatrix}$, show that AA^T and $A + A^T$ are symmetric. Show that $A - A^T$ is skew-symmetric.

52. **(Requires special thought)** A matrix is *stochastic* if the sum of the elements in each *column* is 1. (All multiplicative identity matrices are stochastic, for example.) Let P and Q be two arbitrary 2×2 stochastic matrices. Show that both products PQ and QP are themselves stochastic.

Exercises 53–57 contains statements of various rules that matrix arithmetic satisfies. Uppercase letters refer to matrices, and lowercase letters to scalars. In each case, assume that the dimensions of the matrices are proper to permit the completion of the indicated operations. Verify each statement for 2×2 matrices.

53. If O is the zero matrix, then $A + O = O + A = A$.

54. *Associative law of multiplication:* $A(BC) = (AB)C$.

55. *Distributive law 1:* $A(B + C) = AB + AC$.

56. *Distributive law 2:* $(A + B)C = AC + AB$.

57. $A(cB) = (cA)B = c(AB)$.

Applications

W **58.** *(Academic Performance)* A professor's exams are classified as easy or hard. Seventy-five percent of the time, a hard test is followed by an easy one. Fifty percent of the time, an easy test is followed by a difficult one.

   (a) Construct a 2×2 matrix called P giving the percentages of the time that each type of exams will be succeeded by an easy and a difficult exam. The rows will be labeled "easy" and "hard," as will the columns. The cells of the matrix should contain the fraction of the time the "row" category of exam is followed by the "column" type.

   (b) (Challenging.) Everyone agrees that this professor's most recent exam was quite difficult. Another exam is scheduled for next week. Assess the possibility that it will be an easy one. Consider the matrix $T = \begin{pmatrix} 0 & 1 \end{pmatrix}$. Assign a meaning to TP.

59. *(Popular Fads)* Studies seem to indicate that preschool children are already quite susceptible to the influence of fads of popular culture (certain singers, types of clothing, cartoon characters, and so on). Two groups of 25 children are allowed to play with each other. In the first group, all the children seem to be fans of a certain popular singer. None of the children in the second group showed any interest in this singer. The interaction of the children produces a *bandwagon effect*. Seventy-five percent of the time, a child exposed to interest in the singer becomes a faddist by the end of the day. Twenty-five percent of the time, a child who starts the day as a supporter of the singer loses this interest by the end of the day.

   (a) Express the facts connecting the influence of children to the popular singer as a 2×2 matrix. The rows pertain to children who are under the influence of the fad and children who aren't. The columns contain the percentages switching to the cult state and those switching to the unenthused state.

   (b) Use matrix multiplication to estimate the number of children under the singer's influence on the second day of observation.

   (c) Estimate these numbers on the third and succeeding days.

W **60.** *(Efficient Shopping)* The following chart gives the prices of certain items that you need for a hiking trip you are planning.

	Backpack	Tent	Boots	Camping Utensils
Store 1	$140	$240	$69	$25
Store 2	$150	$200	$85	$29

There are many price differences.

   (a) Create a price matrix P whose rows contain the prices for the two stores. Let a third row contain the lowest price for each item. (That is, any element p_{3j} is the minimum value in the jth column.)

   (b) Show how to use P to decide on a best shopping strategy—making one trip and buying exclusively from one or the other of the stores, or a trip to each store to buy the cheapest items at each store. You will need a column matrix Q whose components are the quantities of each item you will need. (Assume you need 1 tent and 1 set of utensils, 2 backpacks, and 3 pair of boots.)

   (c) Do you think this is a reasonable application of matrix mathematics? Why or why not?

W **61.** *(Requires special thought)* *(Academic Performance)* Imagine that you are taking four courses this semester, and that you have taken three exams in each course from which your final grade will be computed. The third exam counts for 50 percent of the grade, and each of the first two count for 25 percent of your grade.

   (a) Using grades of your own invention, represent your work for this semester using a grade matrix G. The rows should contain the grade data for a particular course.

   (b) Construct a 3×1 column matrix W such that the sum of all three elements is 1. Discuss the significance of the product GW when all components of W are $\frac{1}{3}$.

   (c) Show how matrix operations can be used to compute your final averages. (*Hint*: Let $W = \begin{pmatrix} \frac{1}{4} \\ \frac{1}{4} \\ \frac{1}{2} \end{pmatrix}$. What is the significance of the product GW?)

   (d) In light of this and other exercises, please comment on the usefulness of matrix multiplication in applications.

W **62.** *(Research Methods)* B. C. Dull (a graduate student) knows that matrix multiplication is easy for computers, so it makes sense to organize large tables of data around the matrix concept whenever possible. For one research project, Dull needed to find the average number of children per family for 50 specially selected families in each of 10 small Indian villages. He decided to create two matrices, C and W, to help in this task. His first effort involved letting C be a matrix containing 10 rows and 50 columns and letting W be a row matrix with 50 components, each of which is $\frac{1}{50}$. His expectation was that the product CW would yield the average family number of children in each village. Why would this not work? What should he do to correct his error?

▨ Calculator Exercises

Compute the products in Exercises 63–66.

63. $\begin{pmatrix} 2.1 & -3.2 \\ 4.1 & 6.4 \end{pmatrix} \begin{pmatrix} 2.5 \\ 5.2 \end{pmatrix}$ 64. $\begin{pmatrix} 5.1 & 6.0 \end{pmatrix} \begin{pmatrix} 4.0 & 9.4 \\ 4.8 & 5.5 \end{pmatrix}$

65. $\begin{pmatrix} 1.4 & -1.2 \\ 3.1 & -3.4 \end{pmatrix} \begin{pmatrix} 6.9 & -2.7 \\ -4.2 & -1.8 \end{pmatrix}$

66. $\begin{pmatrix} -6.9 & -2.4 \\ 7.2 & 3.6 \end{pmatrix} \begin{pmatrix} -6.9 & -2.4 \\ 7.2 & 3.6 \end{pmatrix}$

Section 2.3   MATRIX METHODS FOR LINEAR EQUATION SYSTEMS

We may always write a general system of linear equations in this form:

$$a_{11}x_1 + a_{12}x_2 + \cdots + a_{1n}x_n = b_1$$
$$a_{21}x_1 + a_{22}x_2 + \cdots + a_{2n}x_n = b_2$$
$$\vdots$$
$$a_{m1}x_1 + a_{m2}x_2 + \cdots + a_{mn}x_n = b_m$$

and we can always write this more compactly using matrix notation. To do that, let X be a column matrix of subscripted variables:

$$X = \begin{pmatrix} x_1 \\ x_2 \\ \vdots \\ x_n \end{pmatrix}$$

and let A be a matrix of variable coefficients while B is the column matrix of constants b_1, b_2, and so on:

$$A = \begin{pmatrix} a_{11} & a_{12} & \cdots & a_{1n} \\ \vdots & \vdots & & \vdots \\ a_{m1} & a_{m2} & \cdots & a_{mn} \end{pmatrix} \qquad B = \begin{pmatrix} b_1 \\ b_2 \\ \vdots \\ b_m \end{pmatrix}$$

By matrix multiplication, this system can be written compactly using matrices as

$$AX = B \tag{2.4}$$

standard form for linear equations   The matrices A and B contain all the information necessary to determine a unique solution to the system, *if a unique solution is possible*. Equation (2.4) is the *standard form of a system of linear equations*.

augmented matrix   Rather than carry around two distinct matrices A and B, we define the *augmented matrix A'* consisting of the columns of A plus the single column of B on the right. Symbolically,

$$A' = \left(A | B \right)$$

where the vertical line serves as a reminder of the ancestry of A'; A' is the offspring of the marriage of the matrix on the left of the line to the matrix on the right.

Example 20   Represent

$$2x_1 - x_2 = 3$$
$$x_1 + x_2 = 3$$

by a single matrix equation. What is the augmented matrix for this system?

Solution

When $A = \begin{pmatrix} 2 & -1 \\ 1 & 1 \end{pmatrix}$, $B = \begin{pmatrix} 3 \\ 3 \end{pmatrix}$, and $X = \begin{pmatrix} x_1 \\ x_2 \end{pmatrix}$, this system is equivalent to $AX = B$. The augmented matrix A' is

$$A' = \left(A | B \right) = \left(\begin{array}{cc|c} 2 & -1 & 3 \\ 1 & 1 & 3 \end{array} \right)$$

Skill Enhancer 20   How would matrices be used to represent

$$u + 2v = -1$$
$$3u - 2v = 7$$

by a single matrix equation? What is the augmented matrix for this system?

Answer in Appendix E.

Example 21   Write the augmented matrix for the system

$$-x + 12y = 7$$
$$4x - 3y = -1$$

Solution

Mentally convert x and y to x_1 and x_2. After this conversion, these equations are in standard form, so

$$A' = \left(\begin{array}{cc|c} -1 & 12 & 7 \\ 4 & -3 & -1 \end{array} \right)$$

Skill Enhancer 21   Given the equations $2u - 3v = 5$ and $3u + 2v = -2$, what is the augmented matrix for this system?

Answer in Appendix E.

Matrix Row Operations

Addition, subtraction, and multiplication are examples of operations on entire matrices. Certain operations are permissible on elements *within* a given matrix. Because of the close connection between a matrix and a system of linear equations, we define operations that are permissible on equations to be permissible on individual rows of matrices.

Suppose we need to solve a system of linear equations. What operations do the rules of algebra permit? There are three basic operations.

1. Any two equations in the system may be *interchanged.*

2. Any equation may be *multiplied or divided* by a nonzero constant.

3. Any nonzero multiple of an equation may be added to or subtracted from any other equation.

Basic operations (1), (2), and (3) occurred frequently when we used elimination or substitution to solve systems of linear equations (refer again to Chapter 1). Any linear equation system corresponds to some augmented matrix, so there will be matrix operations equivalent to these three steps. Any matrix derived from an original matrix *row equivalent matrix* by means of these operations is said to be a *row equivalent matrix.* A straightforward translation of the above rules from the language of equations to the language of matrices is all that is needed to determine the permissible row operations.

---

matrix row operations

Matrix Row Operations

By virtue of the correspondence between systems of equations and matrices, we allow the following operations to be performed on matrices formed from equation systems.

1. Any two matrix rows may be *interchanged.*

2. Any row may be *multiplied or divided* by a nonzero constant.

3. A nonzero multiple of any row may be added to or subtracted from any other row.

Example 22   Given $S_0 = \begin{pmatrix} 1 & 2 \\ 0 & -1 \end{pmatrix}$, show that

$$S_1 = \begin{pmatrix} 0 & -1 \\ 1 & 2 \end{pmatrix} \qquad S_2 = \begin{pmatrix} \frac{1}{2} & 1 \\ 0 & -1 \end{pmatrix} \qquad S_3 = \begin{pmatrix} 1 & 2 \\ -1 & -3 \end{pmatrix}$$

are all row equivalent to it.

Solution

Interchanging the rows of S_0 yields S_1; by rule 1, this is a permissible row operation, so S_1 and S_0 are row equivalent.

S_2 derives from S_0 if we multiply the first row by $\frac{1}{2}$. By rule 2, S_2 and S_0 are row equivalent.

$$S_0 = \begin{pmatrix} 1 & 2 \\ 0 & -1 \end{pmatrix} \longrightarrow \begin{pmatrix} \frac{1}{2} \cdot 1 & \frac{1}{2} \cdot 2 \\ 0 & -1 \end{pmatrix} = \begin{pmatrix} \frac{1}{2} & 1 \\ 0 & -1 \end{pmatrix} = S_2$$

Rule 3 allows us to subtract row 1 from row 2 in S_0.

0	−1	original second row
−(1	2)	
−1	−3	new second row

minus original first row

Carrying out this procedure yields S_3. Therefore, this matrix is row equivalent to S_0.

Skill Enhancer 22   Let $T = \begin{pmatrix} 1 & 2 \\ 2 & 1 \end{pmatrix}$. How are

$$U = \begin{pmatrix} \frac{1}{2} & 1 \\ -1 & -\frac{1}{2} \end{pmatrix} \qquad V = \begin{pmatrix} 3 & 3 \\ 2 & 1 \end{pmatrix} \qquad W = \begin{pmatrix} 1 & 2 \\ 1 & -1 \end{pmatrix}$$

all row equivalent to T?

Answer in Appendix E.

Proper row operations on an augmented matrix often help us obtain the solutions to a system of linear equations, as an example illustrates. Suppose the ordered pair (a, b) represents the as-yet-unknown solution to the equation system

$$\begin{aligned} 2x - y &= 3 \\ x + y &= 3 \end{aligned} \tag{2.5}$$

Finding the solution is the same thing as transforming this system to the equivalent set of equations

$$\begin{aligned} 1 \cdot x + 0 \cdot y &= a \\ 0 \cdot x + 1 \cdot y &= b \end{aligned} \tag{2.6}$$

equivalent equations   (One set of equations is *equivalent* to another set if it has the same set of solutions as the other equations.) The augmented matrix for the original Equations (2.5) is

$$\begin{pmatrix} 2 & -1 & 3 \\ 1 & 1 & 3 \end{pmatrix}$$

and that for the final set [Equations (2.6)] is

$$\begin{pmatrix} 1 & 0 & a \\ 0 & 1 & b \end{pmatrix} \tag{2.7}$$

The left portion of the second augmented matrix [Equation (2.7)] is the identity matrix I_2 for 2×2 matrices.

These observations suggest a strategy: *Use row operations to transform the original augmented matrix to an augmented matrix with the multiplicative identity on the left.* The elements in the rightmost column of this matrix will be the solution. If a system of equations does not possess a unique solution, this strategy will fail. A diagram illustrating this strategy appears in Figure 2.12.

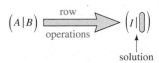

FIGURE 2.12
Solving equations using matrices.

Example 23   Use matrix methods to solve

$$2x - y = 3$$
$$x + y = 3$$

Solution

The augmented matrix for these equations is

$$\begin{pmatrix} 2 & -1 & | & 3 \\ 1 & 1 & | & 3 \end{pmatrix}$$

We multiply the first row by $\frac{1}{2}$ to place a 1 in the first row and column; this is a reasonable first step to generate the identity matrix. If we denote the first row of the current matrix as R_1, and the first row of the *new* row equivalent matrix as R_1', the transformation we carry out is $R_1' = \frac{1}{2}R_1$.

$$R_1' = \tfrac{1}{2}R_1 \qquad \begin{pmatrix} 1 & -\frac{1}{2} & | & \frac{3}{2} \\ 1 & 1 & | & 3 \end{pmatrix}$$

Next, subtract row 1 from row 2 to place a 0 below the 1 in the first column. (We are adopting a "seat of the pants" strategy. Starting at the upper left corner and proceeding down, we perform whatever row operations are necessary to generate the identity matrix. Shortly, we shall make this procedure more systematic.)

$$R_2' = R_2 - R_1 \qquad \begin{pmatrix} 1 & -\frac{1}{2} & | & \frac{3}{2} \\ 0 & \frac{3}{2} & | & \frac{3}{2} \end{pmatrix}$$

We focus on the second column now that the first column has taken such an encouraging form. Multiply the second row by $\frac{2}{3}$.

$$R_2' = \tfrac{2}{3}R_2 \qquad \begin{pmatrix} 1 & -\frac{1}{2} & | & \frac{3}{2} \\ 0 & 1 & | & 1 \end{pmatrix}$$

For the final operation, we appeal to rule 3, and add $\frac{1}{2}$ of the second row to the first row. (Make sure you see why—we need a 0 in that column.)

$$R_1' = R_1 + \tfrac{1}{2}R_2 \qquad \begin{pmatrix} 1 & 0 & | & 2 \\ 0 & 1 & | & 1 \end{pmatrix}$$

This is the last step because the identity matrix I_2 appears on the left of this final row-equivalent matrix. We read the solution $x = 2$, $y = 1$. (Check these results.)

Skill Enhancer 23   Use matrix methods to determine the solution to $u + v = -1$ and $u - v = 4$.

Answer in Appendix E.

Example 24   Use matrix methods to solve

$$x_1 + 4x_2 = 10$$
$$2x_1 - x_2 = -7$$

Solution

The augmented matrix is

$$A' = \begin{pmatrix} 1 & 4 & | & 10 \\ 2 & -1 & | & -7 \end{pmatrix}$$

Use row operations to find the solution. First,

$$R_2' = R_2 - 2R_1 \quad \begin{pmatrix} 1 & 4 & | & 10 \\ 0 & -9 & | & -27 \end{pmatrix}$$

Then

$$R_2' = -\tfrac{1}{9}R_2 \quad \begin{pmatrix} 1 & 4 & | & 10 \\ 0 & 1 & | & 3 \end{pmatrix}$$

Finally, we have

$$R_1' = R_1 - 4R_2 \quad \begin{pmatrix} 1 & 0 & | & -2 \\ 0 & 1 & | & 3 \end{pmatrix}$$

so that $x_1 = -2$ and $x_2 = 3$. Verify these results.

Skill Enhancer 24   Using matrix methods only, determine the solution to $x + 4y = -1$ and $x - y = 4$.

Answer in Appendix E.

Example 25   There is room for three major appliances in a warehouse cubicle containing 80 cubic meters (m^3) of space. One model refrigerator takes up $20\,\text{m}^3$, and another model requires $30\,\text{m}^3$. Use matrix methods to determine how many of each may be stored here.

FIGURE 2.13

Solution

Let x_1 be the number of the first model and x_2 be the number of the second. The first statement demands that

$$x_1 + x_2 = 3 \tag{2.8}$$

Now consider the space requirements of storage. Since each of the first model occupies $20\,\text{m}^3$ of space and since there are x_1 of them, these units use up $20x_1\,\text{m}^3$. Similarly, the units of the second model use up $30x_2\,\text{m}^3$. The total space occupied must equal $80\,\text{m}^3$:

$$20x_1 + 30x_2 = 80$$

or, after dividing both sides by 10,

$$2x_1 + 3x_2 = 8 \tag{2.9}$$

Make sure you see that Equations (2.8) and (2.9) give rise to the augmented matrix

$$\begin{pmatrix} 1 & 1 & | & 3 \\ 2 & 3 & | & 8 \end{pmatrix}$$

We can reduce this matrix in a few steps. First, subtract twice the first row from the second row:

$$R_2' = R_2 - 2R_1 \qquad \begin{pmatrix} 1 & 1 & | & 3 \\ 0 & 1 & | & 2 \end{pmatrix}$$

Next, using this matrix, subtract the second row from the first:

$$R_1' = R_1 - R_2 \qquad \begin{pmatrix} 1 & 0 & | & 1 \\ 0 & 1 & | & 2 \end{pmatrix}$$

This matrix is fully reduced, and we may read the solution from it: $x_1 = 1$, $x_2 = 2$. That is, there is room for one of the model 1 refrigerators and two of the second model.

Skill Enhancer 25   A larger cubicle contains $130\,\text{m}^3$ storage space, and there should be room for five appliances. Use matrix methods to determine how many of each can fit into the storage space.

Answer in Appendix E.

diagonalize   In the last three examples, we were able to *diagonalize* the augmented matrix—transform it to a row-equivalent matrix with the identity matrix on the left. Not every augmented matrix can be reduced to this convenient "diagonalized" form, since not every system of linear equations has a unique solution.

reduced form   For augmented matrices, the *Gauss-Jordan elimination method* is a set of well-defined procedures that will transform a matrix to its *reduced form*, its closest approach to complete diagonalization. (For systems with a unique solution, this technique *will* diagonalize the matrix.) A matrix is in reduced form when it satisfies four conditions.

> **Reduced Form of a Matrix**
>
> A matrix whose elements conform to the following requirements is said to be *reduced*.
>
> **1.** All rows *all* of whose components are zero are at the bottom of the matrix.
>
> **2.** Leading elements of all rows (except for those covered by condition 1) must equal 1. The *leading element* of any row is its leftmost nonzero element.
>
> **3.** The column containing any leading element has zeros above and below that element.
>
> **4.** The leading element of any row appears to the *right* of any leading element in any row *above* it. "The leading elements march down and to the right."

leading element

According to these rules,

$$A = \begin{pmatrix} 1 & 2 & 0 & 0 & 5 \\ 0 & 0 & 1 & 0 & 3 \\ 0 & 0 & 0 & 1 & 0 \\ 0 & 0 & 0 & 0 & 0 \end{pmatrix}$$

is in reduced form (see Figure 2.14), whereas

$$B = \begin{pmatrix} 1 & 2 & 0 & 0 & 5 \\ 0 & 0 & 0 & 0 & 0 \\ 0 & 0 & 0 & 1 & 0 \\ 0 & 0 & 2 & 3 & 3 \\ 0 & 0 & 0 & 0 & 0 \end{pmatrix}$$

is not (see Figure 2.15). Matrix B has been specially constructed to break *all* the rules. Row 2 contains only zero elements, and should be at the bottom of the matrix (by rule 1). The leading element of row 4 is not 1, violating rule 2. The leading 1 of row 3 has a nonzero element below it, breaking rule 3. The leading elements do not march down and to the right (rule 4), because the leading element of row 4 appears to the *left* of the leading element in the row above it.

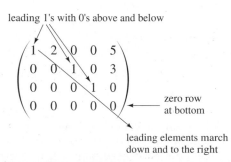

leading 1's with 0's above and below

zero row at bottom

leading elements march down and to the right

FIGURE 2.14
Why A is reduced.

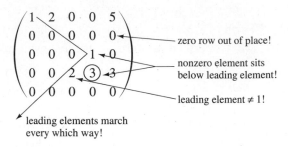

leading elements march
every which way!

FIGURE 2.15
How B breaks the rules.

Example 26   Examine the following matrices. Decide which are in reduced form. For those that are not, suggest row operations that could be performed to properly reduce the matrix.

(a) $\begin{pmatrix} 1 & 1 & 0 & 0 \\ 0 & 1 & 0 & 0 \\ 0 & 0 & 1 & 1 \end{pmatrix}$,   (b) $\begin{pmatrix} 1 & 2 & 3 & 0 \\ 0 & 1 & 0 & 1 \\ 0 & 0 & 1 & 0 \end{pmatrix}$,   (c) $\begin{pmatrix} 1 & 0 & 0 \\ 0 & 1 & 0 \\ 0 & 0 & 0 \end{pmatrix}$,

(d) $\begin{pmatrix} 0 & 0 & 0 \\ 0 & 1 & 0 \\ 1 & 0 & 0 \end{pmatrix}$.

Solution

Of these, only the third, (c), is in proper reduced form. (Make sure you see why.)

Matrix (a) violates rule 3, which demands that the columns containing the leading elements should have zeros above the leading ones. This can be easily corrected—simply subtract the second row from the first. The final reduced form of the matrix is

$$\begin{pmatrix} 1 & 0 & 0 & 0 \\ 0 & 1 & 0 & 0 \\ 0 & 0 & 1 & 1 \end{pmatrix}$$

Although the leading elements in (b) are ones, and these leading ones march down and to the right, one remaining rule is flouted. The second and third elements of the first row are nonzero, in conflict with the leading elements of the second and third rows appearing in those columns (violation of rule 3). A simple correction: subtract twice the second row and three times the third row from the first row to eliminate the unwanted elements. Symbolically, let $R_1' = R_1 - 2R_2 - 3R_3$. The result is this reduced matrix:

$$\begin{pmatrix} 1 & 0 & 0 & -2 \\ 0 & 1 & 0 & 1 \\ 0 & 0 & 1 & 0 \end{pmatrix}$$

The fourth matrix, (d), is upside down. Use one row interchange operation to transform this matrix to

$$\begin{pmatrix} 1 & 0 & 0 \\ 0 & 1 & 0 \\ 0 & 0 & 0 \end{pmatrix}$$

Skill Enhancer 26   Use row operations to reduce these matrices.

(a) $\begin{pmatrix} 0 & 0 & 1 \\ 0 & 1 & 0 \\ 1 & 0 & 0 \end{pmatrix}$;   (b) $\begin{pmatrix} 0 & 0 & 0 \\ 1 & 1 & 1 \\ 0 & 1 & 1 \end{pmatrix}$;   (c) $\begin{pmatrix} 1 & 2 \\ 1 & -1 \end{pmatrix}$.

Answer in Appendix E.

Exercises 2.3

Write the augmented matrix corresponding to the systems of equations in Exercises 1–8. Do *not* solve the equations.

1.
$$x_1 + x_2 = 3$$
$$-x_1 + x_2 = 2$$

2.
$$x - y + z = 0$$
$$-x + y + z = 1$$
$$z = 10$$

3.
$$u + v + w = 2$$
$$3u - 3v = 12$$
$$w = 7$$

4.
$$-s_1 + 4s_2 = -1$$
$$4s_1 - s_2 = 1$$

5.
$$x = 1$$
$$x + y = 2$$
$$x + y + z = 3$$
$$x + y + z + w = 4$$

6. $p + q = 4$

7.
$$0.03198y_1 - 2.2201y_2 = 6.33098$$
$$1.2983y_1 + 1.8000y_2 = 5.00159$$

8.
$$x = -1$$
$$y = 3$$
$$z = 2$$

Assume that the matrices in Exercises 9–16 are augmented and correspond to some system of equations. Using variables $x_1, x_2, x_3, \ldots$, write a system of equations corresponding to these matrices.

9. $\left(\begin{array}{ccc|c} 1 & 1 & 1 & 2 \\ 2 & 2 & 2 & 1 \end{array} \right)$

10. $\left(\begin{array}{cc|c} -25 & 30 & 125 \\ 40 & -50 & 100 \end{array} \right)$

11. $\left(\begin{array}{ccc|c} 0.93 & 0.77 & 0.00 & -4.22 \\ 0.00 & -0.34 & -0.43 & 6.01 \\ 0.01 & 0.04 & 0.93 & -0.92 \end{array} \right)$

12. $\left(\begin{array}{cccc|c} 1 & 2 & 3 & 4 & 5 \end{array} \right)$

13. $\left(\begin{array}{ccc|c} 1 & -1 & -1 & 3 \\ -1 & -2 & 0 & 1 \\ 2 & 0.5 & 0.5 & 1 \end{array} \right)$

14. $\left(\begin{array}{cc|c} 0.33 & 0.17 & 0.25 \\ 0.25 & 0.50 & 0.10 \end{array} \right)$

15. $\left(\begin{array}{ccc|c} 10 & 9 & 8 & 7 \\ 6 & 5 & 4 & 3 \\ 2 & 1 & 0 & 0 \end{array} \right)$

16. $\left(\begin{array}{ccc|c} 0.7 & 0.2 & 0 & 0.1 \\ 0.1 & 0.4 & 1 & -2 \end{array} \right)$

Using the matrix $M = \begin{pmatrix} 1 & 0 & 1 & 1 \\ 2 & 4 & 6 & 7 \\ 0 & 2 & -1 & 0 \end{pmatrix}$, specify the row operations necessary to transform M into the given matrices in Exercises 17–22. For example, given

$$\begin{pmatrix} -1 & 0 & -1 & -1 \\ 2 & 4 & 6 & 7 \\ 0 & 2 & -1 & 0 \end{pmatrix}$$

a correct response would be $R_1' = -R_1$.

17. $\begin{pmatrix} -2 & 0 & -2 & -2 \\ 2 & 4 & 6 & 7 \\ 0 & 2 & -1 & 0 \end{pmatrix}$

18. $\begin{pmatrix} 1 & 0 & 1 & 1 \\ 4 & 8 & 12 & 14 \\ 0 & 2 & -1 & 0 \end{pmatrix}$

19. $\begin{pmatrix} 1 & 0 & 1 & 1 \\ 3 & 4 & 7 & 8 \\ 0 & 2 & -1 & 0 \end{pmatrix}$

20. $\begin{pmatrix} 1 & 0 & 1 & 1 \\ 2 & 4 & 6 & 7 \\ 2 & 2 & 1 & 2 \end{pmatrix}$

21. $\begin{pmatrix} 1 & 0 & 1 & 1 \\ 0 & 2 & -1 & 0 \\ 2 & 4 & 6 & 7 \end{pmatrix}$

22. $\begin{pmatrix} 1 & 0 & 1 & 1 \\ 2 & 4 & 6 & 7 \\ 6 & 14 & 17 & 21 \end{pmatrix}$

Diagonalize the matrices in Exercises 23–36.

23. $\left(\begin{array}{cc|c} 1 & 0 & 3 \\ 1 & 1 & 5 \end{array} \right)$

24. $\left(\begin{array}{cc|c} 1 & 1 & 0 \\ 0 & 1 & -1 \end{array} \right)$

25. $\left(\begin{array}{cc|c} 2 & 2 & 2 \\ 1 & 1 & 1 \end{array} \right)$

26. $\left(\begin{array}{cc|c} 1 & -1 & 1 \\ -1 & 1 & 2 \end{array} \right)$

27. $\left(\begin{array}{cc|c} 1 & -1 & 1 \\ 0 & 1 & 0 \end{array} \right)$

28. $\left(\begin{array}{cc|c} 4 & 1 & 7 \\ 2 & 0 & 4 \end{array} \right)$

29. $\begin{pmatrix} -1 & -1 & | & -1 \\ 1 & 2 & | & 3 \end{pmatrix}$ **30.** $\begin{pmatrix} 1 & 2 & | & 5 \\ 3 & -1 & | & 8 \end{pmatrix}$

31. $\begin{pmatrix} 0 & 2 & | & 4 \\ 1 & 1 & | & 4 \end{pmatrix}$ **32.** $\begin{pmatrix} 1 & 1 & | & 0 \\ -1 & 1 & | & 4 \end{pmatrix}$

33. $\begin{pmatrix} 2 & 2 & | & 2 \\ 3 & 1 & | & 2 \end{pmatrix}$ **34.** $\begin{pmatrix} 2 & 4 & | & -6 \\ -2 & -2 & | & 4 \end{pmatrix}$

35. $\begin{pmatrix} 10 & 1 & | & 1 \\ 0 & 1 & | & 1 \end{pmatrix}$ **36.** $\begin{pmatrix} 1 & 2 & | & 6 \\ 2 & -4 & | & 4 \end{pmatrix}$

Use *matrix methods only* to solve the systems of equations in Exercises 37–48.

37. $x + 2y = 3$ **38.** $2x - y = 5$

$-x + y = 0$ $x + y = 1$

39. $u + v = 2$ **40.** $x + 3y = 6$

$u - v = 0$ $x - 2y = 1$

41. $2u_1 - 3u_2 = 2$ **42.** $2x_1 + 4x_2 = 5$

$u_1 + 6u_2 = 1$ $x_1 - x_2 = -2$

43. $-x + 3y = 10$ **44.** $x + 2y = 8$

$7x - y = -10$ $4x - 3y = 10$

45. $3s + 7t = -2$ **46.** $0.5z_1 + 0.5z_2 = 4$

$s + 2t = 0$ $1.5z_1 + 0.5z_2 = 9$

47. $-3x - 4y = 7$ **48.** $7.5z_1 + z_2 = 8$

$-4x + 2y = 2$ $2z_1 - 4z_2 = 0$

W 49. Remember that given a matrix $A = \left(a_{ij} \right)$, its transpose A^T is the matrix with components a_{ji}. Use the material in this chapter to deduce valid *column* operations for a matrix transpose.

50. Given a system of linear equations, we saw that this system can always be written compactly in matrix form as

$$AX = B$$

We then defined an augmented matrix $A' = \left(A|B \right)$. Show that if we define an *augmented variable matrix*

$$X' = \begin{pmatrix} x_1 \\ x_2 \\ \vdots \\ x_n \\ -1 \end{pmatrix}$$

then the matrix equation

$$A'X' = O$$

has the same solution set as the matrix equation $AX = B$. O is the zero matrix with as many rows as A' but a single column of its own. You may care to do a few examples of your own to convince yourself of the truth of this statement.

Applications

Use matrix methods only to solve these problems.

51. *(Chemistry)* A chemical storeroom in a certain factory contains unlimited amounts of one acid in 10 percent and 25 percent concentrations. A particular industrial application calls for 5 liters (l) of a 20 percent solution. How many liters of 10 percent acid must be mixed with how many liters of 25 percent acid to achieve the proper mix?

52. Refer to the previous exercise. If the application calls for 10 l of a 15 percent acid solution, how many liters of 10 percent acid must be mixed with how many liters of 25 percent acid to achieve the proper mix?

53. Once again, refer to Exercise 51 to determine how to mix unlimited supplies of solutions at 10 percent and 20 percent strengths to obtain 10 l of a 15 percent solution.

54. *(Performing Arts)* Magus the Magician has accepted an engagement. His performance will include 15 tricks, and will last 150 min. His tricks fall into two categories: illusions and sleight-of-hand performances. Illusions last 9 min, and sleight-of-hands last 12 min. How many of each should Magus program?

55. Consider again Exercise 54. Will it be possible for Magus to perform 8 tricks in 75 min?

56. Refer again to Exercise 54. After further practice, Magus finds he can perform an illusion in 5 min and a sleight-of-hand trick in 9 min. What kind of a program will include 10 tricks in 70 min?

57. *(Broadcasting)* Radio Station KTTT programs one hour of "easy listening" at 7:00 P.M. nightly. Short selections last for 4 min, and longer selections last for 6 min. (These timings include announcements and commercials that follow the music.) If the sponsors demand that there be 12 selections during this program, how many of each type shall there be?

58. Refer again to Exercise 57.
   (a) If the program is shortened to 45 min, suggest a programming strategy for the program manager. In 45 min, there should be 9 selections.
   (b) Will it be easier to program 9 selections in 50 min? If so, how?

Calculator Exercises _____

Solve Exercises 59–62.

59. Write the augmented matrix corresponding to these equa-

tions. Do not solve the equations.

$$4.6a - 3.7b = -3.1$$
$$-4.2a + 7.9b = -4.2$$

60. Write the system of equations corresponding to this augmented matrix. Do not solve the equations.

$$\begin{pmatrix} -12.3 & 4.6 & 13.4 & 10 \\ 44.4 & 9.2 & 19.5 & 15 \end{pmatrix}$$

61. Diagonalize $\begin{pmatrix} 1.2 & -1.1 & 1 \\ 0 & -1.2 & 0 \end{pmatrix}$.

62. Use matrix methods to solve

$$2.3x + 4.4y = 2.1$$
$$2.1x - 4.8y = -0.1$$

Section 2.4   GAUSS-JORDAN TECHNIQUES

Although the elimination and substitution methods for solving systems of linear equations *can* be extended to systems with more than two variables, it becomes tedious to do so. Using matrix notation whereby we reduce the augmented matrix is a superior approach, particularly when there are more than two variables or when the equations are to be solved by computer. The *Gauss-Jordan elimination method* is a systematic procedure, *algorithm*   or *algorithm*, for reducing an augmented matrix of arbitrary dimension.

Historical Note

Gauss and Jordan   *It is impossible to do justice to the career of the German mathematician Carl Friedrich Gauss (1777–1855) in a brief note such as this. He may be the greatest mathematician who ever lived, having made the first of his many significant contributions to mathematics and mathematical physics before he was nineteen years old. The mathematician Kronecker observed that his name is linked to almost everything that the mathematics of the nineteenth century brought forth in the way of original scientific ideas. It is much easier to summarize the career of the German surveyor Wilhelm Jordan (1842–1899), whose fame rests primarily on his Handbook of Geodesy (first published in 1873 and revised as recently as 1961). He presented this technique as an adjunct to performing certain complicated calculations in surveying. Curiously, Gauss was also motivated by problems in surveying (but also in astronomy) to develop this method. Do not confuse Wilhelm with the French mathematician Camille Jordan, to whom he is no (known) relation.*

pivoting   The operation within this Gauss-Jordan procedure is *pivoting* about a certain nonzero element of the matrix. As Figure 2.16 indicates, we have two reasons for wanting to pivot:

1. To use row operations to *replace this selected component with a value of 1*

2. To use row operations to create row-equivalent matrices with *zero elements above and below this pivot element*

The pivot operation is therefore a two-stage process. The prescription for pivoting will mention the *pivot element* (the element on which the pivot operation is being per-
pivot element
pivot position   formed), the *pivot position* (the column and row intersection marking the location of the
pivot column   pivot element), the *pivot column* (the column containing the pivot element), and the *pivot*
pivot row   *row* (the row containing the pivot element). These appear in Figure 2.17.

Pivoting

Follow these steps to *pivot* about a particular element of a matrix. (We have yet to learn how to choose the right pivot element.)

1. *Multiply* each element in the row by the *reciprocal of the pivot element*. The row-equivalent matrix obtained will now have a 1 in the pivot position.

2. *Add or subtract multiples of this pivot row to the rows above and below it* to ensure that the remaining entries in the pivot column are zero.

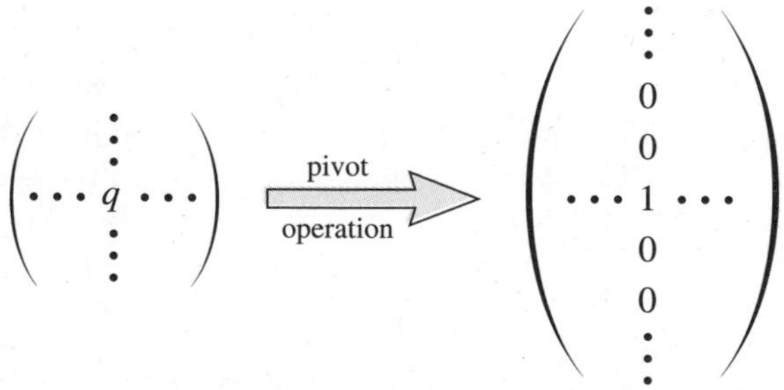

FIGURE 2.16
The goal of pivoting.

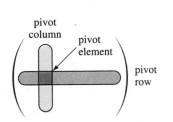

FIGURE 2.17
Pivot geography.

Example 27   Pivot about the boxed element of

$$\boxed{2} \begin{pmatrix} 2 & 4 & 10 \\ 1 & 3 & 3 & 5 \\ 1 & 2 & -1 & -2 \end{pmatrix} \tag{2.10}$$

Solution

The reciprocal of 2 is $\frac{1}{2}$; stage one requires the multiplication of row 1 by $\frac{1}{2}$.

$$R_1' = \tfrac{1}{2} R_1 \qquad \begin{pmatrix} 1 & 1 & 2 & 5 \\ 1 & 3 & 3 & 5 \\ 1 & 2 & -1 & -2 \end{pmatrix}$$

Generally, stage 1 is completed in a single step, but stage 2 may require several steps. We can transform the 1 in column 1, row 2 by subtracting row 1 from row 2:

$$R_2' = R_2 - R_1 \qquad \begin{pmatrix} 1 & 1 & 2 & 5 \\ 0 & 2 & 1 & 0 \\ 1 & 2 & -1 & -2 \end{pmatrix}$$

For the final step, subtract the first row from the third row (make sure you see why):

$$R_3' = R_3 - R_1 \qquad \begin{pmatrix} 1 & 1 & 2 & 5 \\ 0 & 2 & 1 & 0 \\ 0 & 1 & -3 & -7 \end{pmatrix}$$

This final matrix is row equivalent to the original matrix (2.10).

Skill Enhancer 27   Pivot about the boxed element.

$$\begin{pmatrix} \boxed{2} & 4 & 6 & 8 \\ 1 & 0 & 1 & 3 \\ -1 & 0 & 4 & 8 \end{pmatrix}$$

Answer in Appendix E.

Example 28   Pivot about the boxed element in

$$\begin{pmatrix} -1 & 2 & 3 \\ 4 & \boxed{3} & 9 \end{pmatrix}$$

Solution

First, multiply the second row by $\frac{1}{3}$:

$$R_2' = \tfrac{1}{3} R_2 \qquad \begin{pmatrix} -1 & 2 & 3 \\ \frac{4}{3} & 1 & 3 \end{pmatrix}$$

Next, form a new first row by subtracting twice the second row from the first row. In this way, the row-equivalent matrix created will have a zero above the pivot element.

$$R_1' = R_1 - 2R_2 \qquad \begin{pmatrix} -\frac{11}{3} & 0 & -3 \\ \frac{4}{3} & 1 & 3 \end{pmatrix}$$

This last matrix is the final pivoted matrix.

Skill Enhancer 28   Pivot about the boxed element in $\begin{pmatrix} 1 & \boxed{2} & 3 \\ 3 & -1 & 2 \end{pmatrix}$.

Answer in Appendix E.

Gauss-Jordan elimination   The *Gauss-Jordan elimination technique* is essentially a series of pivots.

Gauss-Jordan Elimination

These steps specify the row operations for reducing any matrix.

1. If any rows contain *all zeros*, transfer them to the bottom of the matrix.

2. If necessary, rearrange the remaining rows so that no leading element of any row appears to the *left* of the leading element of any row above it; this is a slight variation of the theme that *leading elements march down and to the right.*
  Begin by examining the element in the first row and first column.

3. *Pivot about this element.*

4. Mentally cover all rows containing all pivot elements. As the next candidate for pivot element, *choose the leading element of the top row in the uncovered matrix.*

5. *Return to step 3.* Continue until you have worked your way down to the bottom of the matrix, or until you have reached the first row with all zero components (whichever comes first).

Example 29   Use Gauss-Jordan methods to reduce

$$\begin{pmatrix} 2 & 2 & 4 & 10 \\ 1 & 3 & 3 & 5 \\ 1 & 2 & -1 & -2 \end{pmatrix}$$

Solution

There are no zero rows, and no rearrangement of the rows is called for. We begin by pivoting about the 2 in the first row and column, which we did in Example 27. This row-equivalent matrix is

$$\begin{pmatrix} 1 & 1 & 2 & 5 \\ 0 & 2 & 1 & 0 \\ 0 & 1 & -3 & -7 \end{pmatrix}$$

As in Figure 2.18, cover the top row. We need to pivot about the leading 2 in the second row. Here are the steps to take to pivot about this element:

$$R_2' = \tfrac{1}{2}R_2 \qquad \begin{pmatrix} 1 & 1 & 2 & 5 \\ 0 & 1 & \tfrac{1}{2} & 0 \\ 0 & 1 & -3 & -7 \end{pmatrix}$$

$$\begin{aligned} R_1' &= R_1 - R_2 \\ R_3' &= R_3 - R_2 \end{aligned} \qquad \begin{pmatrix} 1 & 0 & \tfrac{3}{2} & 5 \\ 0 & 1 & \tfrac{1}{2} & 0 \\ 0 & 0 & -\tfrac{7}{2} & -7 \end{pmatrix}$$

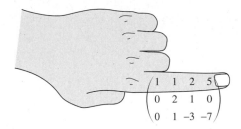

FIGURE 2.18
The "uncovered matrix" after the first pivot.

Mentally cover the top two rows (or use your fingers, as in Figure 2.19) to reveal that the final operation should be a pivot about the $-\frac{7}{2}$ in the third row.

$$R'_3 = -\tfrac{2}{7}R_3 \qquad \begin{pmatrix} 1 & 0 & \frac{3}{2} & 5 \\ 0 & 1 & \frac{1}{2} & 0 \\ 0 & 0 & 1 & 2 \end{pmatrix}$$

$$\begin{aligned} R'_1 &= R_1 - \tfrac{3}{2}R_3 \\ R'_2 &= R_2 - \tfrac{1}{2}R_3 \end{aligned} \qquad \begin{pmatrix} 1 & 0 & 0 & 2 \\ 0 & 1 & 0 & -1 \\ 0 & 0 & 1 & 2 \end{pmatrix}$$

The matrix $\begin{pmatrix} 1 & 0 & 0 & 2 \\ 0 & 1 & 0 & -1 \\ 0 & 0 & 1 & 2 \end{pmatrix}$ is the reduced matrix.

Skill Enhancer 29   Use Gauss-Jordan methods to reduce $\begin{pmatrix} 1 & 1 & 1 & 0 \\ 2 & 1 & 2 & 1 \\ 1 & 1 & -1 & 0 \end{pmatrix}$.

Answer in Appendix E.

Example 30   Use Gauss-Jordan methods to reduce

$$\begin{pmatrix} 1 & 2 & 3 & -2 \\ 0 & 0 & 1 & -1 \\ 2 & 0 & 4 & 2 \end{pmatrix}.$$

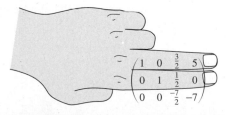

FIGURE 2.19
The "uncovered matrix" after two pivot operations.

Solution

The leading element of row 3 appears to the left of the leading element of row 2; these two rows must be rearranged by rule 2.

$$\begin{pmatrix} \boxed{1} & 2 & 3 & -2 \\ 2 & 0 & 4 & 2 \\ 0 & 0 & 1 & -1 \end{pmatrix}$$

First, pivot about the boxed 1.

$$R_2' = R_2 - 2R_1 \qquad \begin{pmatrix} 1 & 2 & 3 & -2 \\ 0 & \boxed{-4} & -2 & 6 \\ 0 & 0 & 1 & -1 \end{pmatrix}$$

Pivot about the boxed -4.

$$R_2' = -\tfrac{1}{4}R_2 \qquad \begin{pmatrix} 1 & 2 & 3 & -2 \\ 0 & 1 & \frac{1}{2} & -\frac{3}{2} \\ 0 & 0 & 1 & -1 \end{pmatrix}$$

$$R_1' = R_1 - 2R_2 \qquad \begin{pmatrix} 1 & 0 & 2 & 1 \\ 0 & 1 & \frac{1}{2} & -\frac{3}{2} \\ 0 & 0 & \boxed{1} & -1 \end{pmatrix}$$

The boxed 1 is the last element to pivot about.

$$\begin{array}{l} R_1' = R_1 - 2R_3 \\ R_2' = R_2 - \tfrac{1}{2}R_3 \end{array} \qquad \begin{pmatrix} 1 & 0 & 0 & 3 \\ 0 & 1 & 0 & -1 \\ 0 & 0 & 1 & -1 \end{pmatrix}$$

This final matrix $\begin{pmatrix} 1 & 0 & 0 & 3 \\ 0 & 1 & 0 & -1 \\ 0 & 0 & 1 & -1 \end{pmatrix}$ is the reduced form of the given matrix.

Skill Enhancer 30   Use the Gauss-Jordan technique to reduce $\begin{pmatrix} 1 & 1 & 1 & 2 \\ 2 & 1 & 0 & 0 \\ 1 & 0 & 1 & 0 \end{pmatrix}$.

Answer in Appendix E.

The Gauss-Jordan technique is a key element in the matrix process for solving systems of linear equations.

Using Matrix Methods to Solve Linear Equation Systems

1. Arrange the system in standard form, and form the equivalent *matrix equation*

$$AX = B$$

Form the augmented matrix

$$A' = \left(A | B \right)$$

2. Apply the steps of the Gauss-Jordan elimination procedure to A' to obtain the reduced matrix that is row equivalent to A'.

3. Read the solutions from the reduced matrix. If necessary, do this by recreating the system of equations from this reduced matrix.

   The solution will be *unique* only when the *matrix to the left of the vertical line in the reduced matrix is the identity matrix*. Otherwise, there may be an infinite set of solutions, or there may be no solution.

Example 31   Use matrix methods to solve the equations

$$2x + 2y + 4z = 10$$
$$x + 3y + 3z = 5$$
$$x + 2y - z = -2$$

Solution

The augmented matrix for these equations is

$$A' = \begin{pmatrix} 2 & 2 & 4 & | & 10 \\ 1 & 3 & 3 & | & 5 \\ 1 & 2 & -1 & | & -2 \end{pmatrix}$$

We reduced this matrix in Example 29 to

$$\begin{pmatrix} 1 & 0 & 0 & | & 2 \\ 0 & 1 & 0 & | & -1 \\ 0 & 0 & 1 & | & 2 \end{pmatrix}$$

This must represent a *unique* solution, because the left part of the reduced matrix (to the left of the vertical line) is the identity matrix. It is also clear that this represents a unique solution because it is the augmented matrix for the system

$$\begin{aligned} x & & = 2 \\ & y & = -1 \\ & & z = 2 \end{aligned}$$

which represents the solution.

Skill Enhancer 31   What are the solutions to

$$x + y +\ z = 3$$
$$x + y + 3z = 1$$
$$2x + y -\ z = 8$$

Answers in Appendix E.

Example 32   Use matrix methods to solve

$$x + y - 2z = 3$$
$$y\ + z = 4$$
$$2y + 2z = 8$$

Solution

The augmented matrix for this system is

$$\begin{pmatrix} 1 & 1 & -2 & | & 3 \\ 0 & 1 & 1 & | & 4 \\ 0 & 2 & 2 & | & 8 \end{pmatrix}$$

Normally, we need to begin pivoting about the 1 in the top left corner of this matrix. This matrix is already pivoted about that element, so we proceed, according to the rules of the Gauss-Jordan technique, to pivot about the leading 1 in the second row (make sure you see why). The steps to perform the pivot operation are

$$R_1' = R_1 - R_2$$
$$R_3' = R_3 - 2R_2$$
$$\begin{pmatrix} 1 & 0 & -3 & | & -1 \\ 0 & 1 & 1 & | & 4 \\ 0 & 0 & 0 & | & 0 \end{pmatrix}$$

In this form, the matrix has been reduced. To retrieve the solution, note that the equations that generate this augmented matrix are

$$x\qquad -\ 3z = -1$$
$$y +\ z = 4$$
$$0x + 0y + 0z = 0$$

There is no unique solution for x, y, and z; it is impossible to assign unique values to these variables using only information contained in these equations.

Skill Enhancer 32   Use matrix methods to solve (if possible)

$$2x + y = 1$$
$$x + y + z = 3$$
$$4x + 2y = 2$$

Answers in Appendix E.

The solution to Example 32 allows many sets of the variables x, y, and z to satisfy
parameter the equations. In *this* situation, it is usual to express one of the variables as a *parameter* t, and then to write all the variables in terms of this parameter. Here, for example, we

might let $z = t$. In terms of this parameter, the solution is

$$x = -1 + 3t$$
$$y = 4 - t$$
$$z = t$$

Now, x, y, and z are uniquely determined for any particular value of t. This table shows how x, y, and z assume different sets of values depending on values for t.

t	x	y	z
-5	-16	9	-5
-1	-4	5	-1
0	-1	4	0
7	20	-3	7

Many other values of t are possible.

Example 33   Use Gauss-Jordan methods to solve

$$x - y = -3$$
$$2y - 2x = -7$$

Solution

The augmented matrix

$$\begin{pmatrix} 1 & -1 & | & -3 \\ -2 & 2 & | & -7 \end{pmatrix}$$

is entirely reduced after the single pivot about the top left 1:

$$R_2' = R_2 + 2R_1 \qquad \begin{pmatrix} 1 & -1 & | & -3 \\ 0 & 0 & | & -13 \end{pmatrix}$$

(Can you see that no further reduction is possible?) This matrix does *not* admit any solution, since the last line corresponds to

$$0x + 0y = -13$$

an impossibility. This set of equations is *inconsistent* and has no solution. Since the left portion of the reduced augmented matrix is *not* an identity matrix, there can be no unique solution.

Skill Enhancer 33   Using Gauss-Jordan methods, solve (if possible) the equations

$$u + v = 2$$
$$2v + 2u = 6$$

Answers in Appendix E.

Example 34   Use matrix methods on

$$x_1 + 4x_2 - 7x_3 = 0$$
$$3x_1 - x_2 + 5x_3 = 0$$
$$x_1 + x_2 + x_3 = 2$$

Solution

The augmented matrix

$$\begin{pmatrix} 1 & 4 & -7 & | & 0 \\ 3 & -1 & 5 & | & 0 \\ 1 & 1 & 1 & | & 2 \end{pmatrix}$$

may be reduced according to the following steps:

$$\begin{pmatrix} \boxed{1} & 4 & -7 & | & 0 \\ 3 & -1 & 5 & | & 0 \\ 1 & 1 & 1 & | & 2 \end{pmatrix} \Rightarrow$$

Pivot about the boxed 1.
$$\begin{pmatrix} 1 & 4 & -7 & | & 0 \\ 0 & -13 & 26 & | & 0 \\ 0 & -3 & 8 & | & 2 \end{pmatrix} \Rightarrow$$

Pivot about the boxed element in row 2, column 2.
$$\begin{pmatrix} 1 & 4 & -7 & | & 0 \\ 0 & \boxed{1} & -2 & | & 0 \\ 0 & -3 & 8 & | & 2 \end{pmatrix} \Rightarrow$$

$$\begin{pmatrix} 1 & 0 & 1 & | & 0 \\ 0 & 1 & -2 & | & 0 \\ 0 & 0 & 2 & | & 2 \end{pmatrix} \Rightarrow$$

Finally, pivot about the third row, third column.
$$\begin{pmatrix} 1 & 0 & 1 & | & 0 \\ 0 & 1 & -2 & | & 0 \\ 0 & 0 & \boxed{1} & | & 1 \end{pmatrix} \Rightarrow$$

$$\begin{pmatrix} 1 & 0 & 0 & | & -1 \\ 0 & 1 & 0 & | & 2 \\ 0 & 0 & 1 & | & 1 \end{pmatrix}$$

From the last matrix, we read the solution:

$$x_1 = -1$$
$$x_2 = 2$$
$$x_3 = 1$$

Check this solution.

Skill Enhancer 34   Use matrix methods to solve

$$x_1 + 4x_2 + 2x_3 = 1$$
$$2x_2 + x_3 = 0$$
$$x_1 + x_2 = 0$$

Answer in Appendix E.

Example 35 What system of equations corresponds to the following augmented matrix, and what is its solution?

$$\begin{pmatrix} 1 & 0 & 1 & | & -1 \\ 1 & -1 & 0 & | & 1 \\ 0 & 3 & 2 & | & -3 \end{pmatrix}$$

Solution

The equations are

$$
\begin{array}{rcrcrcr}
x & & & + z & = & -1 \\
x & - & y & & = & 1 \\
& & 3y & + 2z & = & -3
\end{array}
$$

We can reduce the augmented matrix according to the following sequence of operations:

$$\begin{pmatrix} 1 & 0 & 1 & | & -1 \\ 1 & -1 & 0 & | & 1 \\ 0 & 3 & 2 & | & -3 \end{pmatrix}$$ The initial augmented matrix already has a leading 1 in the first row. Finish the pivot operation about this element.

⇓

$$\begin{pmatrix} 1 & 0 & 1 & | & -1 \\ 0 & 1 & 1 & | & -2 \\ 0 & 3 & 2 & | & -3 \end{pmatrix}$$ $R'_2 = -(R_2 - R_1)$

⇓

$$\begin{pmatrix} 1 & 0 & 1 & | & -1 \\ 0 & 1 & 1 & | & -2 \\ 0 & 0 & -1 & | & 3 \end{pmatrix}$$ $R'_3 = R_3 - 3R_2$

⇓

$$\begin{pmatrix} 1 & 0 & 1 & | & -1 \\ 0 & 1 & 1 & | & -2 \\ 0 & 0 & 1 & | & -3 \end{pmatrix}$$ $R'_3 = -R_3$

⇓

$$\begin{pmatrix} 1 & 0 & 0 & | & 2 \\ 0 & 1 & 0 & | & 1 \\ 0 & 0 & 1 & | & -3 \end{pmatrix}$$ $\begin{aligned} R'_1 &= R_1 - R_3 \\ R'_2 &= R_2 - R_3 \end{aligned}$

Read the solution from column 4. Thus, $x = 2$, $y = 1$, and $z = -3$. Check this solution.

Skill Enhancer 35 What set of equations corresponds to $\begin{pmatrix} 1 & 1 & 2 & 4 \\ 2 & -2 & 0 & 4 \\ 0 & 4 & 1 & 4 \end{pmatrix}$? What is the solution to those equations?

Answers in Appendix E.

Example 36   Use matrices to solve the system of equations

$$2x + y + z = 4$$
$$x - y + 2z = -1$$

Solution

The augmented matrix is

$$\begin{pmatrix} 2 & 1 & 1 & | & 4 \\ 1 & -1 & 2 & | & -1 \end{pmatrix}$$

The following sequence of row-equivalent matrices illustrates one way to reduce this matrix.

$$R_1' = \tfrac{1}{2}R_1 \qquad \begin{pmatrix} 1 & \tfrac{1}{2} & \tfrac{1}{2} & | & 2 \\ 1 & -1 & 2 & | & -1 \end{pmatrix}$$

$$\Downarrow$$

$$R_2' = R_2 - R_1 \qquad \begin{pmatrix} 1 & \tfrac{1}{2} & \tfrac{1}{2} & | & 2 \\ 0 & -\tfrac{3}{2} & \tfrac{3}{2} & | & -3 \end{pmatrix}$$

$$\Downarrow$$

$$R_2' = -\tfrac{2}{3}R_2 \qquad \begin{pmatrix} 1 & \tfrac{1}{2} & \tfrac{1}{2} & | & 2 \\ 0 & 1 & -1 & | & 2 \end{pmatrix}$$

$$\Downarrow$$

$$R_1' = R_1 - \tfrac{1}{2}R_2 \qquad \begin{pmatrix} 1 & 0 & 1 & | & 1 \\ 0 & 1 & -1 & | & 2 \end{pmatrix}$$

The final matrix is in reduced form. It will *not* permit us to deduce unique values for the variables x, y, and z because the left part of

$$\begin{pmatrix} 1 & 0 & 1 & | & 1 \\ 0 & 1 & -1 & | & 2 \end{pmatrix}$$

is *not* the identity matrix. The two original equations do not contain sufficient information to solve for unique values of the three variables. We may, though, express any two of the variables in terms of the remaining one. To emphasize this, we replace z by a variable t. In terms of this parameter, the solution is

$$x = 1 - t$$
$$y = t + 2$$
$$z = t$$

Given a particular value of this parameter t, it *is* possible to specify the variables uniquely. For example, you should verify that when $t = 0$, then $x = 1$, $y = 2$, and $z = 0$; when $t = 3$, then $x = -2$, $y = 5$, $z = 3$; and so on.

Skill Enhancer 36   Get the best possible solution for

$$x + y - z = 3$$
$$x - y + 2z = 1$$

Use matrix methods.

Answer in Appendix E.

Example 37   Farmers in Minnesota have a choice of growing 3 varieties of rye. One particular farmer wants to devote part of his acreage to each variety. The farm needs to harvest 9,000 bushels of grain at the end of the growing season to meet all expenses and make a reasonable profit. There are 45 acres of planting area on the farm. Each variety involves a different amount of labor. The first variety, let's call it variety A, requires 3 person-hours of labor per bushel during the growing season, and the farmer can expect to reap 400 bushels per acre. Variety B requires 2 person-hours of labor per bushel, but the farmer only gets 200 bushels per acre. Finally, variety C is the least labor intensive, requiring a single person-hour per bushel with a yield of only 100 bushels per acre. The farmer knows from past years that the farm work force can provide 20,000 person-hours of labor per growing season. How many bushels of each variety should the farmer plan to sow to allocate all resources? How many acres should the farmer devote to each variety?

Solution

Let a, b, and c represent the unknown bushels-worth of varieties A, B, and C that must be planted. The farm must yield $9,000$ bushels of rye:

$$a + b + c = 9,000 \qquad (2.11)$$

As far as labor goes, if a bushels of A are planted, at 3 person-hours per bushel, a total of $3a$ person-hours are used. In the same way, varieties B and C use $2b$ and $1 \cdot c = c$ person-hours, respectively. Their total must equal the total labor available:

$$3a + 2b + c = 20,000 \qquad (2.12)$$

If the farmer can squeeze 400 bushels of variety A from one acre, then each bushel requires 0.0025 acre ($1/400 = 0.0025$). Similarly, since one acre yields 200 and 100 bushels, respectively, of the other two varieties, one bushel of B and C require $\frac{1}{200} = 0.005$ and $\frac{1}{100} = 0.01$ acre. The total acreage for variety A is $0.0025a$, and so on for B and C. Therefore,

$$0.0025a + 0.005b + 0.01c = 45 \qquad (2.13)$$

where 45 is the total acreage of the farm. The augmented matrix corresponding to the system of Equations 2.11 through 2.13 is

$$\begin{pmatrix} 1 & 1 & 1 & | & 9,000 \\ 3 & 2 & 1 & | & 20,000 \\ 0.0025 & 0.005 & 0.01 & | & 45 \end{pmatrix}$$

Before beginning the reduction, it is a good idea to multiply the third row by 400 to eliminate some of the decimals.

$$\begin{pmatrix} 1 & 1 & 1 & | & 9{,}000 \\ 3 & 2 & 1 & | & 20{,}000 \\ 1 & 2 & 4 & | & 18{,}000 \end{pmatrix}$$

Since the first row starts with a leading 1, we can immediately add different multiples of the first row to the second and third rows to reduce to zeros the remaining elements in the first column.

$$\begin{pmatrix} 1 & 1 & 1 & | & 9{,}000 \\ 0 & -1 & -2 & | & -7{,}000 \\ 0 & 1 & 3 & | & 9{,}000 \end{pmatrix}$$

Begin the pivot operation by multiplying the second row by -1.

$$\begin{pmatrix} 1 & 1 & 1 & | & 9{,}000 \\ 0 & 1 & 2 & | & 7{,}000 \\ 0 & 1 & 3 & | & 9{,}000 \end{pmatrix}$$

The next matrix in the sequence is

$$\begin{pmatrix} 1 & 0 & -1 & | & 2{,}000 \\ 0 & 1 & 2 & | & 7{,}000 \\ 0 & 0 & 1 & | & 2{,}000 \end{pmatrix}$$

which completes the pivot. Make sure you see how to get it, and how to transform it into the final matrix:

$$\begin{pmatrix} 1 & 0 & 0 & | & 4{,}000 \\ 0 & 1 & 0 & | & 3{,}000 \\ 0 & 0 & 1 & | & 2{,}000 \end{pmatrix}$$

from which we can read the final answers. The farmer will plant 4,000 bushels of variety A, 3,000 of type B, and 2,000 of type C.

It is more useful to determine how many acres should be allotted to each variety. Four thousand bushels of type A, at 400 bushels to the acre, will need $4{,}000/400 = 10$ acres. In the same way, varieties B and C require 15 and 20 acres.

Skill Enhancer 37 A neighboring farm comprises 40 acres and needs to harvest 8,000 bushels. If the remaining facts are the same, comment on the strategy for this farm.

Answer in Appendix E.

_____ *Exercises 2.4* _____

Each of Exercises 1–20 consists of an augmented matrix. In each case, pivot about the boxed element.

1. $\begin{pmatrix} 1 & \boxed{0} & | & 2 \\ 1 & \boxed{-1} & | & 3 \end{pmatrix}$

2. $\begin{pmatrix} \boxed{1} & 0 & | & 2 \\ 1 & -1 & | & 3 \end{pmatrix}$

3. $\begin{pmatrix} 2 & 0 & | & 4 \\ \boxed{1} & 3 & | & 2 \end{pmatrix}$ **4.** $\begin{pmatrix} \boxed{2} & 0 & | & 4 \\ 1 & 3 & | & 2 \end{pmatrix}$

5. $\begin{pmatrix} \boxed{1} & 8 & | & 0 \\ 2 & 4 & | & 0 \end{pmatrix}$ **6.** $\begin{pmatrix} 3 & 0 & | & 2 \\ 1 & \boxed{2} & | & 2 \end{pmatrix}$

7. $\begin{pmatrix} \boxed{2} & 4 & 2 & | & 3 \\ 1 & 0 & 2 & | & 2 \end{pmatrix}$ **8.** $\begin{pmatrix} -2 & 1 & 0 & | & 3 \\ 3 & \boxed{4} & -1 & | & 0 \end{pmatrix}$

9. $\begin{pmatrix} \boxed{1} & 2 & 6 & | & 9 \\ 1 & 0 & 3 & | & 0 \\ 0 & 4 & 0 & | & 4 \end{pmatrix}$ **10.** $\begin{pmatrix} 1 & 2 & 6 & | & 9 \\ 1 & 0 & \boxed{3} & | & 0 \\ 0 & 4 & 0 & | & 4 \end{pmatrix}$

11. $\begin{pmatrix} 2 & 4 & \boxed{6} & | & 8 \\ 1 & 0 & 0 & | & 3 \\ 3 & 2 & 1 & | & 0 \end{pmatrix}$ **12.** $\begin{pmatrix} 2 & \boxed{4} & 6 & | & 8 \\ 1 & 0 & 0 & | & 3 \\ 3 & 2 & 1 & | & 0 \end{pmatrix}$

13. $\begin{pmatrix} \boxed{\frac{1}{2}} & 0 & 1 & | & -1 \\ -2 & 1 & 2 & | & 1 \end{pmatrix}$ **14.** $\begin{pmatrix} \frac{2}{3} & -\frac{1}{6} & | & 1 \\ \frac{1}{6} & \boxed{\frac{1}{2}} & | & 0 \end{pmatrix}$

15. $\begin{pmatrix} \frac{2}{3} & \boxed{-\frac{1}{6}} & | & 1 \\ \frac{1}{6} & \frac{1}{2} & | & 0 \end{pmatrix}$ **16.** $\begin{pmatrix} \frac{1}{2} & \boxed{-\frac{3}{4}} & | & \frac{1}{3} \\ \frac{1}{4} & 1 & | & 0 \end{pmatrix}$

17. $\begin{pmatrix} \frac{1}{3} & -\frac{1}{2} & | & \frac{1}{6} \\ 0 & \boxed{\frac{1}{4}} & | & 0 \end{pmatrix}$ **18.** $\begin{pmatrix} \frac{1}{3} & \boxed{-\frac{1}{2}} & | & \frac{1}{6} \\ 0 & \frac{1}{4} & | & 0 \end{pmatrix}$

19. $\begin{pmatrix} 1 & 0 & -1 & | & 2 \\ \boxed{2} & 1 & -3 & | & 0 \\ 0 & 0 & 2 & | & 4 \\ -1 & -1 & 0 & | & 3 \end{pmatrix}$ **20.** $\begin{pmatrix} 1 & 0 & -1 & | & 2 \\ 2 & 1 & -3 & | & 0 \\ 0 & 0 & \boxed{2} & | & 4 \\ -1 & -1 & 0 & | & 3 \end{pmatrix}$

In Exercises 21–40, use row operations to reduce each matrix.

21. $\begin{pmatrix} 1 & 3 & 8 \\ 3 & 1 & 0 \end{pmatrix}$ **22.** $\begin{pmatrix} 1 & 1 & 0 \\ 2 & 3 & -1 \end{pmatrix}$

23. $\begin{pmatrix} 2 & 4 & 6 \\ 1 & 2 & 3 \end{pmatrix}$ **24.** $\begin{pmatrix} 1 & 2 & 3 \\ 1 & 2 & 4 \end{pmatrix}$

25. $\begin{pmatrix} 0 & 0 & 1 \\ 0 & 1 & 0 \\ 1 & 0 & 0 \end{pmatrix}$ **26.** $\begin{pmatrix} 2 & 1 & 1 \\ 1 & 2 & 1 \end{pmatrix}$

27. $\begin{pmatrix} 1 & 0 & 0 & 0 & 1 \\ 1 & 1 & 0 & 0 & 2 \\ 1 & 1 & 1 & 0 & 3 \\ 1 & 1 & 1 & 1 & 4 \end{pmatrix}$ **28.** $\begin{pmatrix} -3 & 7 & 0 & -2.4 & 3 \end{pmatrix}$

29. $\begin{pmatrix} 0 & 0 \\ -0.2 & -4 \end{pmatrix}$ **30.** $\begin{pmatrix} 1 & 0 & 1 \\ 0 & 2 & 0 \\ 0 & 0 & 1 \end{pmatrix}$

31. $\begin{pmatrix} \frac{2}{3} & -\frac{1}{2} & 0 & \frac{1}{6} \\ \frac{1}{2} & 0 & -1 & -\frac{1}{6} \end{pmatrix}$ **32.** $\begin{pmatrix} 0.01 & 0 & 0.02 \\ 0.01 & 0.01 & 0.01 \end{pmatrix}$

33. $\begin{pmatrix} 1 & 0 & 1 & 1 \\ 2 & 4 & 6 & 7 \\ 0 & 2 & -1 & 0 \end{pmatrix}$

34. $\begin{pmatrix} 0.001 & 0 & 0.001 & 0.001 \\ 0.002 & 0.004 & 0.006 & 0.007 \\ 0 & 0.002 & -0.001 & 0 \end{pmatrix}$

35. $\begin{pmatrix} 1 & 2 & 3 \\ -1 & -1 & 2 \end{pmatrix}$ **36.** $\begin{pmatrix} 1 & 0 & -1 \\ 0 & 1 & -1 \end{pmatrix}$

37. $\begin{pmatrix} 1 & 2 & 4 & 8 & 0 \\ 0 & 1 & 2 & 4 & 0 \\ 0 & 0 & 0 & 0 & 0 \\ 0 & 0 & 1 & 2 & 0 \end{pmatrix}$

38. $\begin{pmatrix} 0 & 1 & 0 & 0 & 1 \\ 1 & 0 & -1 & 0 & 0 \\ 0 & 0 & -1 & 1 & -1 \\ -1 & 1 & 0 & -1 & 0 \\ 0 & 1 & 0 & 0 & 1 \end{pmatrix}$

39. $\begin{pmatrix} 1,000 & 0 & 1,000 & 1,000 \\ 2,000 & 4,000 & 6,000 & 7,000 \\ 0 & 2,000 & -1,000 & 0 \end{pmatrix}$

40. $\begin{pmatrix} 2 & 4 & 6 & 8 & 0 \\ 0 & 0 & 0 & 0 & 1 \\ 0 & 2 & 4 & 6 & 8 \\ 0 & 0 & 2 & 4 & 6 \\ 0 & 0 & 0 & 2 & 4 \end{pmatrix}$

In Exercises 41–54, use matrix methods to solve the systems of equations. In some cases, it will not be possible to obtain a unique solution. In that case, use matrix techniques to obtain the best possible solution and express each variable in terms of a parameter t. If there are *no* possible solutions, say so.

41. $x + y = 0$
$x - y = -2$

42. $3u - 2v = 8$
$2u + 3v = 1$

43. $x_1 - x_2 + x_3 = 3$
$x_1 + x_2 = 5$
$2x_2 - x_3 = 2$

44. $x + y + z = 1,000$
$y - z = 500$

45. $4a + 3b = 0$

$6a - 2b = 0$

46. $\dfrac{1}{2}x_1 + \dfrac{1}{3}x_2 = \dfrac{1}{6}$

$\dfrac{1}{3}x_1 + \dfrac{1}{4}x_2 = \dfrac{1}{12}$

47. $x_1 + 2x_2 = 3$

$4x_2 = 6 - 2x_1$

48. $2x_1 - 3x_2 = 5$

$2x_1 = 10 + 3x_2$

49. $\dfrac{1}{6}u_1 - \dfrac{1}{2}u_2 + \dfrac{1}{3}u_3 = 0$

$-u_1 + \dfrac{1}{4}u_2 + \dfrac{2}{3}u_3 = -1$

$\dfrac{3}{4}u_1 + \dfrac{1}{3}u_2 - \dfrac{1}{6}u_3 = 11$

50. $14y - 12z = 10$

$x = 3$

$7y - 6z = 5$

51. $x_1 = 1$

$x_1 + x_2 = 2$

$x_1 + x_2 + x_3 = 3$

$x_1 + x_2 + x_3 + x_4 = 4$

52. $p_1 + p_2 = 4$

$5p_1 = 20 - 5p_2$

53. $3a - b + c + \dfrac{2}{3}d = 0$

$a + b + d = -\dfrac{1}{3}$

$a = \dfrac{1}{3}$

$-a + b + 2c + 5d + 5 = 0$

54. $z_1 + z_4 = 0$

$z_2 + z_5 = 0$

$5z_1 - 2z_2 - z_3 + 3z_4 + z_5 = 3$

$z_1 + z_2 + z_3 + z_4 + z_5 = 1$

$z_1 + 2z_2 + 3z_3 + 4z_4 + z_5 = -3$

Applications

W55. **(Business Administration)** TrailBlazer Supermarkets is a chain in which all stores are open from 8:00 in the morning till midnight, 365 days a year. The TrailBlazer in Fox River, Illinois hires 60 workers, who must be apportioned in the 3 shifts—8 A.M. to 4 P.M., noon to 8 P.M., and 4 P.M. to midnight. The overlap between the first two shifts occurs between noon and 4 P.M., and the manager knows from long experience that 45 workers should be on hand at this time. The last two shifts overlap between 4 P.M. and 8 P.M., and 40 workers should be in the store in this period. Let x_1 be the number of workers needed in the first shift, and x_2 and x_3 the numbers for the second and third shifts, respectively.
(a) In terms of these variables, how many workers are present during the first overlap period? during the second overlap period?
(b) How should the 60 workers be allotted to the three shifts? Use matrix methods to solve.
(c) Why are matrix methods particularly useful for this problem?

56. There is another TrailBlazer store in South Bend, Indiana (see Exercise 55 for facts about these stores). This store hires 100 people, and the manager needs 60 people on hand during the period of the first overlap and 77 people during the second overlap. How many people should be allocated to each shift? Use matrix methods to solve.

57. Consider again the South Bend TrailBlazer store of Exercise 56. Due to expansion, the store needs 65 people to work during the first shift overlap, and 85 to work in the second. How large should the total work force be? There will be the same number of people during the second shift as in problem 56.

58. **(Employment Records)** Mr. Tuttle is trying to recreate certain records that were lost in a fire that destroyed the factory in Framingham, fortunately with no loss of life. He remembers that there were a total of 86 workers, categorized as either cutters or finishers. Someone else recalls that there were 50 more cutters than finishers. Use matrix methods to determine how many of each there were at the factory.

59. **(Library Acquisitions)** The Somerville Public Library buys books in three categories: fiction, nonfiction, and reference. Its charter specifies that it purchase 500 books each month, and further specifies additional restrictions on the book acquisitions. There should be 25 more fiction volumes than nonfiction, and 50 more nonfiction books than reference volumes. Use matrix methods to determine how many books in each category the library should buy.

60. **(Animal Husbandry)** Farmer Gray seeks to run his farm according to scientific principles. He has enough capital to purchase a total of 130 animals—pigs, cows, and horses. There should be twice as many cows as horses, and there should be 10 more pigs than there are horses. Determine how many of each animal should be on the farm, using only matrix methods.

boxed elements in the matrices of Exercises 61–62.

61. $\begin{pmatrix} \boxed{2.1} & 0.8 & 3.9 \\ \boxed{1} & 0 & -0.9 \end{pmatrix}$ **62.** $\begin{pmatrix} \boxed{2} & 0 & -2.4 \\ -1.1 & 4.5 & 6.3 \end{pmatrix}$

Calculator Exercises

Reduce the matrices in Exercises 63–64.

Use a calculator to solve Exercises 61–64. Pivot about the

63. $\begin{pmatrix} 1.2 & -12 & 15.36 \\ 9 & 11 & 21.40 \end{pmatrix}$ **64.** $\begin{pmatrix} 11 & 1.6 & 13.22 \\ 2 & 3 & 4.3 \end{pmatrix}$

Section 2.5   MATRIX INVERSES

The *inverse* of a matrix is another useful matrix tool for solving systems of equations. If some matrix A is square, and if there is another matrix A^{-1} (read "A inverse") such that

$$A A^{-1} = A^{-1} A = I_n$$

matrix inverse   (here A has dimension $n \times n$), then A^{-1} is the *matrix inverse* of A. **Note well** that the superscript -1 does not denote an exponent, but rather the inverse, in the same way that the multiplicative inverse to a nonzero scalar x is x^{-1}.

Example 38   If $A = \begin{pmatrix} 1 & 1 \\ \frac{1}{2} & \frac{3}{4} \end{pmatrix}$, show that

$$A^{-1} = \begin{pmatrix} 3 & -4 \\ -2 & 4 \end{pmatrix}$$

is the inverse of A.

Solution

Verifying that one matrix is the inverse of another is done by showing that the suspected inverse satisfies the equations $AA^{-1} = A^{-1}A = I$. In this case,

$$AA^{-1} = \begin{pmatrix} 1 & 1 \\ \frac{1}{2} & \frac{3}{4} \end{pmatrix} \begin{pmatrix} 3 & -4 \\ -2 & 4 \end{pmatrix} = \begin{pmatrix} 1 & 0 \\ 0 & 1 \end{pmatrix}$$

and

$$A^{-1} A = \begin{pmatrix} 3 & -4 \\ -2 & 4 \end{pmatrix} \begin{pmatrix} 1 & 1 \\ \frac{1}{2} & \frac{3}{4} \end{pmatrix} = \begin{pmatrix} 1 & 0 \\ 0 & 1 \end{pmatrix}$$

Therefore, A^{-1} is the inverse of A as claimed.

Skill Enhancer 38   If $B = \begin{pmatrix} 1 & 1 \\ \frac{1}{4} & \frac{1}{2} \end{pmatrix}$, show that $B^{-1} = \begin{pmatrix} 2 & -4 \\ -1 & 4 \end{pmatrix}$ is indeed the inverse of B.

Answers in Appendix E.

Not all square matrices have inverses. We will say more on this later. Matrix inverses are analogous to real number inverses; a real nonzero number a has inverse

$a^{-1} = \frac{1}{a}$ such that $a \cdot a^{-1} = a^{-1} \cdot a = 1$. We use this concept virtually without thinking to solve an equation such as

$$ax = b \tag{2.14}$$

given certain constants a and b where $a \neq 0$:

$$ax = b$$
$$a^{-1}(ax) = (a^{-1}a)x = a^{-1}b$$
$$1 \cdot x = x = \frac{1}{a} \cdot b$$
$$x = \frac{b}{a}$$

and we can run the analogy to solve a matrix equation the same way. Given a matrix equation $AX = B$, where A and B are known matrices, can we determine the matrix X? We can solve this equation by mirroring each step in the solution of Equation (2.14).

$$\begin{array}{ll} AX = B & a \cdot x = b \\ A^{-1}(AX) = A^{-1}B & a^{-1}(ax) = a^{-1}b \\ (A^{-1}A)X = A^{-1}B & (a^{-1}a)x = a^{-1}b \\ X = A^{-1}B & x = a^{-1}b = \frac{a}{b} \end{array}$$

We demanded $a \neq 0$ so that the inverse of a exists. The condition that A have an inverse is not so clear-cut; we will not, in fact, cover this topic in this text. For the time being, we will simply demand that A be a matrix with an inverse for this derivation to hold.

Solving a Matrix Equation

The solution to the matrix equation

$$AX = B$$

is

$$X = A^{-1}B$$

A and B are known matrices.
A^{-1} is the inverse to A.

X is the column matrix of variables $\begin{pmatrix} x_1 \\ x_2 \\ \vdots \\ x_n \end{pmatrix}$ (in any given problem, n is

the number of unknowns).

invertible matrix   This solution will not work unless A is an *invertible matrix*—that is, unless A^{-1} exists. It can be shown that if A is a square matrix and the solution to the system $AX = B$ is unique, then A^{-1} will exist. Matrices that are not square can never have an inverse.

Example 39   Use matrix methods to solve

$$x_1 + x_2 = 10$$
$$\frac{1}{2}x_1 + \frac{3}{4}x_2 = 6$$

Solution

The system is equivalent to

$$\begin{pmatrix} 1 & 1 \\ \frac{1}{2} & \frac{3}{4} \end{pmatrix} \begin{pmatrix} x_1 \\ x_2 \end{pmatrix} = \begin{pmatrix} 10 \\ 6 \end{pmatrix}$$

Identifying

$$A = \begin{pmatrix} 1 & 1 \\ \frac{1}{2} & \frac{3}{4} \end{pmatrix} \quad \text{and} \quad B = \begin{pmatrix} 10 \\ 6 \end{pmatrix}$$

leads us to expect the solution $X = A^{-1}B$. From Example 38,

$$A^{-1} = \begin{pmatrix} 3 & -4 \\ -2 & 4 \end{pmatrix}$$

Therefore,

$$X = \begin{pmatrix} 3 & -4 \\ -2 & 4 \end{pmatrix} \begin{pmatrix} 10 \\ 6 \end{pmatrix} = \begin{pmatrix} 6 \\ 4 \end{pmatrix}$$

The solution is $x_1 = 6$ and $x_2 = 4$. Check this solution.

Skill Enhancer 39   Using matrix methods, solve

$$x_1 + x_2 = 10$$
$$\frac{1}{4}x_1 + \frac{1}{2}x_2 = 4$$

Answers in Appendix E.

Solving equations via the matrix inverse has an advantage all its own. Inverses facilitate the solution of *several* sets of equations differing only in the values of the constants to the right of the equal signs. (This is another way of saying that the augmented matrices for each system are identical except for the rightmost column.) Essentially, the matrix inverse "saves" all the work of solution, so it is not necessary to begin at the very beginning for each related system of equations.

Example 40   One of the most popular products at Norman's Nut Hut is a mix of honey-roasted peanuts and cashews. Peanuts sell for 50 cents per pound, whereas cashews sell for 75 cents per pound.
(a) How many peanuts should be mixed with how many cashews to make 10 lb of a mixture that can be sold for 60 cents a pound?
(b) How many peanuts should be mixed with how many cashews to make 5 lb of a mixture that can be sold for 66 cents per pound?

Solution

Let x_1 and x_2 be the pounds of peanuts and cashews to be mixed together.
(a) Since Norman needs 10 lb of the mix,

$$x_1 + x_2 = 10 \tag{2.15}$$

At 60 cents per pound, 10 lb of the mix will cost \$6.00, and this value arises from its separate components—peanuts (\$0.50/lb times x pounds) and cashews (\$0.75/lb times y pounds). Remember, $\$0.50 = \frac{1}{2}$ dollar and $\$0.75 = \frac{3}{4}$ dollar, so

$$\frac{1}{2}x_1 + \frac{3}{4}x_2 = 6 \tag{2.16}$$

But Equations (2.15) and (2.16) are together equivalent to a *matrix* equation

$$\begin{pmatrix} 1 & 1 \\ \frac{1}{2} & \frac{3}{4} \end{pmatrix} \begin{pmatrix} x_1 \\ x_2 \end{pmatrix} = \begin{pmatrix} 10 \\ 6 \end{pmatrix}$$

whose solution (refer back to Example 39) is

$$X = \begin{pmatrix} 6 \\ 4 \end{pmatrix}$$

The proprietor should mix 6 lb of peanuts with 4 lb of cashews.
(b) Under these conditions, the matrix equation governing the unknowns is

$$\begin{pmatrix} 1 & 1 \\ \frac{1}{2} & \frac{3}{4} \end{pmatrix} \begin{pmatrix} x_1 \\ x_2 \end{pmatrix} = \begin{pmatrix} 5 \\ 3.3 \end{pmatrix}$$

(5 lb times \$0.66/lb = \$3.30). Rather than solve the equation anew, we may use the fact (verified in Example 38) that the inverse of the coefficient matrix

$$\begin{pmatrix} 1 & 1 \\ \frac{1}{2} & \frac{3}{4} \end{pmatrix}$$

is

$$\begin{pmatrix} 3 & -4 \\ -2 & 4 \end{pmatrix}$$

Matrix methods yield the solution as

$$X = \begin{pmatrix} 3 & -4 \\ -2 & 4 \end{pmatrix} \begin{pmatrix} 5 \\ 3.3 \end{pmatrix} = \begin{pmatrix} 1.8 \\ 3.2 \end{pmatrix}$$

The desired mix is a combination of 1.8 lb of peanuts and 3.2 lb of cashews.

Skill Enhancer 40   Referring again to this example, suppose peanuts now sell for \$0.25/lb, and cashews sell for \$0.50/lb. Now determine how many pounds of peanuts should be mixed with how many pounds of cashews to make 10 lb of a mixture selling for \$0.40/lb. How should the mixture be made to sell for \$0.30/lb.?

Answers in Appendix E.

Not all square matrices have inverses. To see this, appeal to the correspondence between matrices and linear equations. Whenever a square matrix can be inverted, we can uniquely solve a system of equations where the coefficients of the variables correspond to the elements of the matrix. But not every such system has a unique solution, and therefore not every square matrix is invertible.

Properties of Matrix Inverses

Readers should be thoroughly familiar with the following facts.

1. Only *square matrices* may have inverses, although not every square matrix possesses an inverse.

2. The inverse A^{-1} to a matrix A satisfies the equations

$$AA^{-1} = A^{-1}A = I_n$$

where I_n is a multiplicative identity matrix of the same dimension as A. If the inverse exists, it is *unique*.

3. If a system of linear equations (in any number of unknowns) corresponding to the matrix equation $AX = B$ has a unique solution, and if A is a square matrix, then the *solution* to this equation is

$$X = A^{-1}B$$

Multiplying by a matrix inverse is the closest we can come to actual matrix "division." Multiplication by the inverse of a matrix A resembles division by A in the same way that multiplication by the inverse of a real number a is division by the number a.

Finding a Matrix Inverse

To determine an inverse of a matrix, we rely again on the Gauss-Jordan methods for reducing matrices.

Example 41   Invert $A = \begin{pmatrix} 3 & 2 \\ 4 & 3 \end{pmatrix}$.

Solution

The inverse A^{-1} will be some matrix with unknown components:

$$A^{-1} = \begin{pmatrix} a & c \\ b & d \end{pmatrix}$$

We seek the four unknowns a, b, c, and d. The product AA^{-1} is

$$\begin{pmatrix} 3a + 2b & 3c + 2d \\ 4a + 3b & 4c + 3d \end{pmatrix}$$

which must also equal

$$\begin{pmatrix} 1 & 0 \\ 0 & 1 \end{pmatrix}$$

By the definition of matrix equality, we equate corresponding elements of these two matrices to derive four linear equations.

$$\begin{aligned} 3a + 2b &= 1 & 3c + 2d &= 0 \\ 4a + 3b &= 0 & 4c + 3d &= 1 \end{aligned}$$

Can you see that these two *pairs of equations* are equivalent to the following two *matrix equations*?

$$A\begin{pmatrix} a \\ b \end{pmatrix} = \begin{pmatrix} 1 \\ 0 \end{pmatrix} \qquad A\begin{pmatrix} c \\ d \end{pmatrix} = \begin{pmatrix} 0 \\ 1 \end{pmatrix}$$

The solution for the column matrices appears to be related to two augmented matrices

$$\left(A \middle| \begin{matrix} 1 \\ 0 \end{matrix} \right) \qquad \text{and} \qquad \left(A \middle| \begin{matrix} 0 \\ 1 \end{matrix} \right)$$

Instead of reducing *two* augmented matrices, it is possible to reduce the *single* augmented matrix

$$(A|I_2) = \begin{pmatrix} 3 & 2 & \vline & 1 & 0 \\ 4 & 3 & \vline & 0 & 1 \end{pmatrix}$$

where we allow A to be augmented by more than the single column that has so far been usual. If A^{-1} exists, then reducing the matrix $(A|I_2)$ to $(I_2|B)$ guarantees that B is the inverse we seek. In this example, then, we would reduce

$$\begin{pmatrix} 3 & 2 & \vline & 1 & 0 \\ 4 & 3 & \vline & 0 & 1 \end{pmatrix}$$

Here are the steps in the reduction process (as an exercise, you should verify that the indicated row operations transforming one intermediate matrix to the next are correct):

$$\begin{pmatrix} 3 & 2 & \vline & 1 & 0 \\ 4 & 3 & \vline & 0 & 1 \end{pmatrix} \rightarrow$$

$$R_1' = \tfrac{1}{3}R_1 \quad \begin{pmatrix} 1 & \tfrac{2}{3} & \vline & \tfrac{1}{3} & 0 \\ 4 & 3 & \vline & 0 & 1 \end{pmatrix} \rightarrow$$

$$R_2' = R_2 - 4R_1 \quad \begin{pmatrix} 1 & \tfrac{2}{3} & \vline & \tfrac{1}{3} & 0 \\ 0 & \tfrac{1}{3} & \vline & -\tfrac{4}{3} & 1 \end{pmatrix} \rightarrow$$

$$R_2' = 3R_2 \quad \begin{pmatrix} 1 & \tfrac{2}{3} & \vline & \tfrac{1}{3} & 0 \\ 0 & 1 & \vline & -4 & 3 \end{pmatrix} \rightarrow$$

$$R_1' = R_1 - \tfrac{2}{3}R_2 \quad \begin{pmatrix} 1 & 0 & \vline & 3 & -2 \\ 0 & 1 & \vline & -4 & 3 \end{pmatrix}$$

The matrix inverse is therefore

$$A^{-1} = \begin{pmatrix} 3 & -2 \\ -4 & 3 \end{pmatrix}$$

You should verify this independently by multiplying A by A^{-1}.

Skill Enhancer 41   Find the inverse to $B = \begin{pmatrix} 1 & 2 \\ 3 & 4 \end{pmatrix}$.

Answer in Appendix E.

This procedure works for inverting square matrices of *any dimension.*

Finding Matrix Inverses

Reduce the augmented matrix $\left(A | I \right)$ to the form $\left(I | A^{-1} \right)$. *If complete reduction to this form is not possible, then A has no inverse.*

Example 42   Find A^{-1} if $A = \begin{pmatrix} 1 & 2 & 1 \\ 2 & 1 & 0 \\ 1 & 0 & 1 \end{pmatrix}$.

Solution

Form the augmented matrix

$$A' = \begin{pmatrix} 1 & 2 & 1 & | & 1 & 0 & 0 \\ 2 & 1 & 0 & | & 0 & 1 & 0 \\ 1 & 0 & 1 & | & 0 & 0 & 1 \end{pmatrix}$$

and use the steps of the Gauss-Jordan process to reduce it.

$\begin{pmatrix} 1 & 2 & 1 & | & 1 & 0 & 0 \\ 2 & 1 & 0 & | & 0 & 1 & 0 \\ 1 & 0 & 1 & | & 0 & 0 & 1 \end{pmatrix}$ *The original augmented matrix already has a leading 1 in the first row.*

$\begin{pmatrix} 1 & 2 & 1 & | & 1 & 0 & 0 \\ 0 & -3 & -2 & | & -2 & 1 & 0 \\ 0 & -2 & 0 & | & -1 & 0 & 1 \end{pmatrix}$ *Use multiples of the first row to place 0s in the first column.*

$\begin{pmatrix} 1 & 2 & 1 & | & 1 & 0 & 0 \\ 0 & 1 & \frac{2}{3} & | & \frac{2}{3} & -\frac{1}{3} & 0 \\ 0 & -2 & 0 & | & -1 & 0 & 1 \end{pmatrix}$ *Leading 1 in the second row.*

$\begin{pmatrix} 1 & 0 & -\frac{1}{3} & | & -\frac{1}{3} & \frac{2}{3} & 0 \\ 0 & 1 & \frac{2}{3} & | & \frac{2}{3} & -\frac{1}{3} & 0 \\ 0 & 0 & \frac{4}{3} & | & \frac{1}{3} & -\frac{2}{3} & 1 \end{pmatrix}$ *Place 0s above and below the leading 1 in column 2.*

$\begin{pmatrix} 1 & 0 & -\frac{1}{3} & | & -\frac{1}{3} & \frac{2}{3} & 0 \\ 0 & 1 & \frac{2}{3} & | & \frac{2}{3} & -\frac{1}{3} & 0 \\ 0 & 0 & 1 & | & \frac{1}{4} & -\frac{1}{2} & \frac{3}{4} \end{pmatrix}$ *Multiply the third row by 3/4 to put a leading 1 in the third row.*

The final augmented matrix is

$$\left(\begin{array}{ccc|ccc} 1 & 0 & 0 & -\frac{1}{4} & \frac{1}{2} & \frac{1}{4} \\ 0 & 1 & 0 & \frac{1}{2} & 0 & -\frac{1}{2} \\ 0 & 0 & 1 & \frac{1}{4} & -\frac{1}{2} & \frac{3}{4} \end{array}\right)$$

(why?) and so

$$A^{-1} = \left(\begin{array}{ccc} -\frac{1}{4} & \frac{1}{2} & \frac{1}{4} \\ \frac{1}{2} & 0 & -\frac{1}{2} \\ \frac{1}{4} & -\frac{1}{2} & \frac{3}{4} \end{array}\right)$$

(Verify that $AA^{-1} = A^{-1}A = I_3$ as claimed.)

Skill Enhancer 42   Find the inverse to $B = \begin{pmatrix} 1 & 0 & 2 \\ 0 & 2 & 0 \\ 2 & 0 & 1 \end{pmatrix}$.

Answer in Appendix E.

Example 43   Invert $A = \begin{pmatrix} 2 & 0 \\ 1 & 0 \end{pmatrix}$.

Solution

The multiply augmented matrix is $\left(\begin{array}{cc|cc} 2 & 0 & 1 & 0 \\ 1 & 0 & 0 & 1 \end{array}\right)$.   Verify that the following two matrices are the intermediate row-equivalent matrices in the reduction:

$$\left(\begin{array}{cc|cc} 1 & 0 & \frac{1}{2} & 0 \\ 1 & 0 & 0 & 1 \end{array}\right) \Rightarrow \left(\begin{array}{cc|cc} 1 & 0 & \frac{1}{2} & 0 \\ 0 & 0 & -\frac{1}{2} & 1 \end{array}\right)$$

Since the part of the reduced matrix to the left of the vertical line is *not* the identity, A has no inverse.

Skill Enhancer 43   Invert $B = \begin{pmatrix} 2 & 1 \\ 4 & 2 \end{pmatrix}$. If no inverse exists, indicate how the reduced augmented matrix shows this.

Answer in Appendix E.

_____ *Exercises 2.5* _____

Find inverses, if possible, to the matrices in Exercises 1–20. If an inverse does not exist, state that the matrix is not invertible.

3. $\begin{pmatrix} \frac{1}{2} & \frac{1}{4} \\ 1 & 1 \end{pmatrix}$

4. $\begin{pmatrix} 1 & 1 & 0 \\ 0 & 1 & 1 \\ 1 & 0 & 1 \end{pmatrix}$

1. $\begin{pmatrix} 1 & 1 \\ 1 & 0 \end{pmatrix}$

2. $\begin{pmatrix} 1 & 2 \\ 0 & 2 \end{pmatrix}$

5. $\begin{pmatrix} 2 & -1 \\ 2 & -1 \end{pmatrix}$

6. $\begin{pmatrix} 1 & 2 \\ 3 & -4 \end{pmatrix}$

7. $\begin{pmatrix} -1 & -2 \\ -2 & -2 \end{pmatrix}$

8. $\begin{pmatrix} 8 & -2 \\ -4 & -3 \end{pmatrix}$

9. $\begin{pmatrix} 9 & -4 \\ 7 & 2 \end{pmatrix}$

10. $\begin{pmatrix} \frac{1}{2} & \frac{1}{3} \\ \frac{3}{4} & 0 \end{pmatrix}$

11. $\begin{pmatrix} 2 & 5 \\ 2 & -5 \end{pmatrix}$

12. $\begin{pmatrix} -2 & 3 \\ 2 & 3 \end{pmatrix}$

13. $\begin{pmatrix} 3 & 4 \\ -3 & 4 \end{pmatrix}$

14. $\begin{pmatrix} \frac{1}{5} & \frac{1}{5} \\ \frac{1}{4} & -\frac{1}{4} \end{pmatrix}$

15. $\begin{pmatrix} 2 & 1 & 2 \\ 1 & 1 & 1 \\ 0 & 2 & 0 \end{pmatrix}$

16. $\begin{pmatrix} -1 & -1 & -1 \\ 1 & 1 & 0 \\ 0 & 0 & 1 \end{pmatrix}$

17. $\begin{pmatrix} 2 & 3 & -1 \\ 1 & 0 & \frac{1}{2} \\ -1 & 0 & 0 \end{pmatrix}$

18. $\begin{pmatrix} 1 & 1 \\ 1 & \frac{1}{2} \end{pmatrix}$

19. $\begin{pmatrix} 5 & 10 & 0 \\ -10 & 5 & 0 \\ 0 & 0 & 1 \end{pmatrix}$

20. $\begin{pmatrix} \frac{1}{2} & \frac{1}{3} & \frac{1}{6} \\ 0 & \frac{1}{3} & -\frac{1}{2} \\ -\frac{2}{3} & \frac{1}{3} & \frac{1}{2} \end{pmatrix}$

21. Show that $\begin{pmatrix} 1 & 0 \\ 0 & 1 \end{pmatrix}$ is its own inverse.

W 22. (a) Show that $\begin{pmatrix} 1 & 1 \\ 1 & 1 \end{pmatrix}$ has no inverse.

(b) Explain why this matrix has no inverse by appealing to the close relationship between matrices and coefficients of systems of equations.

23. Find the inverse of $\begin{pmatrix} 1 & x \\ 1 & 1 \end{pmatrix}$ for arbitrary values of x. For what values of x will the inverse not exist?

24. Given an arbitrary matrix $\begin{pmatrix} a & b \\ c & d \end{pmatrix}$, show that the inverse, if it exists, is always given by

$$\frac{1}{ad - bc} \begin{pmatrix} d & -b \\ -c & a \end{pmatrix}$$

What relationship between a, b, c, and d guarantees that the inverse does *not* exist?

25. Use the results of Exercise 24 to show that $\begin{pmatrix} 2 & 1 \\ 0 & 0 \end{pmatrix}$ has no inverse.

26. Use the results of Exercise 24 to show that any 2×2 matrix with a row or column of zeros can never have an inverse.

27. Invert the matrix $A = \begin{pmatrix} 6 \end{pmatrix}$.
   (a) Use matrix methods to solve $6q = 18$.
   (b) Use the same matrix methods to solve $6q = 72$.

28. Invert the matrix $A = \begin{pmatrix} 4 \end{pmatrix}$.
   (a) Use matrix methods to solve $4x_1 = 22$.
   (b) Use the same matrix methods to solve $4t = 72$.

29. Invert $\begin{pmatrix} 3 & 4 \\ 1 & -2 \end{pmatrix}$.
   (a) Solve: $3u + 4v = 1$
   $$u - 2v = 0$$
   (b) Solve: $3x_1 + 4x_2 = 2$
   $$x_1 - 2x_2 = 0$$

30. Invert $\begin{pmatrix} 6 & 3 \\ 1 & -1 \end{pmatrix}$.
   (a) Solve: $6x_1 + 3x_2 = 3$
   $$x_1 - x_2 = -1$$
   (b) Solve: $6t_1 + 3t_2 = 0$
   $$t_1 - t_2 = -2$$

31. Invert $\begin{pmatrix} 3 & 3 \\ 1 & -2 \end{pmatrix}$.
   (a) Solve: $3z_1 + 3z_2 = 6$
   $$z_1 - 2z_2 = -1$$
   (b) Solve: $3z_1 + 3z_2 = -1$
   $$z_1 - 2z_2 = -1$$

32. Invert $\begin{pmatrix} 1 & 5 \\ 4 & -3 \end{pmatrix}$.
   (a) Solve: $t_1 + 5t_2 = -9$
   $$4t_1 - 3t_2 = 10$$
   (b) Solve: $t_1 + 5t_2 = 8$
   $$4t_1 - 3t_2 = 8$$

33. Invert $\begin{pmatrix} 1 & 1 \\ 1 & -1 \end{pmatrix}$.
   (a) Solve: $x + y = 2$
   $$x - y = -2$$
   (b) Solve: $x + y = 1$
   $$x - y = 0$$

34. Invert $\begin{pmatrix} 3 & 1 \\ 1 & 1 \end{pmatrix}$.

(a) Solve: $3x + y = 0$

$x + y = -2$

(b) Solve: $3x + y = 0$ (c) Solve: $3x + y = 3$

$x + y = -4$ $x + y = 0$

35. What is the inverse to $\begin{pmatrix} \frac{1}{4} & \frac{1}{4} \\ \frac{1}{10} & -\frac{1}{10} \end{pmatrix}$? Use this information to solve these systems of equations.

  (a) $\frac{1}{4}x + \frac{1}{4}y = 2$ (b) $\frac{1}{4}u + \frac{1}{4}v = 1$

       $\frac{1}{10}x - \frac{1}{10}y = 2$ $\frac{1}{10}u - \frac{1}{10}v = -1$

  (c) $\frac{1}{4}S + \frac{1}{4}T = 1$

       $\frac{1}{10}S - \frac{1}{10}T = 2$

36. What is the inverse to $\begin{pmatrix} -\frac{1}{4} & \frac{1}{4} \\ \frac{1}{6} & \frac{1}{6} \end{pmatrix}$? Solve:

  (a) $-\frac{1}{4}s + \frac{1}{4}t = 0$ (b) $-\frac{1}{4}s + \frac{1}{4}t = 1$

       $\frac{1}{6}s + \frac{1}{6}t = 1$ $\frac{1}{6}s + \frac{1}{6}t = -1$

37. Invert $\begin{pmatrix} 1 & -1 & 1 \\ 2 & 3 & -1 \\ -1 & 1 & -1 \end{pmatrix}$.

  (a) Solve: $x - y + z = 2$

        $2x + 3y - z = 3$

       $-x + y - z = 4$

  (b) Solve: $x - y + z = 2$

        $2x + 3y - z = 11$

       $-x + y - z = -2$

38. Invert $\begin{pmatrix} -1 & 1 & 0 \\ 1 & 2 & 1 \\ 0 & 1 & 0 \end{pmatrix}$.

  (a) Solve the system $-x + y = 1$

                $x + 2y + z = 0$

                     $y = \frac{1}{2}$

  (b) Solve the system $-x + y = 1$

                $x + 2y + z = 4$

                     $y = 1$

39. Invert $\begin{pmatrix} 1 & 2 & 3 \\ -1 & 0 & 1 \\ 0 & 4 & 0 \end{pmatrix}$.

(a) Solve the system $x + 2y + 3z = 10$

$-x + z = 4$

$4y = 4$

(b) Solve the system $x + 2y + 3z = -2$

$-x + z = 2$

$4y = 8$

40. Invert the matrix

$$A = \begin{pmatrix} 2 & 3 & -1 \\ 1 & 1 & 1 \\ 1 & 1 & 2 \end{pmatrix}$$

  (a) If $R = \begin{pmatrix} r_1 \\ r_2 \\ r_3 \end{pmatrix}$, solve the equation $AR = \begin{pmatrix} 5 \\ 0 \\ -1 \end{pmatrix}$.

  (b) Solve: $AR = \begin{pmatrix} -7 \\ 0 \\ 1 \end{pmatrix}$

41. Invert $\begin{pmatrix} 1 & 1 & 1 \\ 1 & 2 & 3 \\ 3 & 2 & 1 \end{pmatrix}$.

  (a) Solve: $x + y + z = 0$

         $x + 2y + 3z = -40$

        $3x + 2y + z = 40$

  (b) Solve: $x + y + z = 5$

         $x + 2y + 3z = 11$

        $3x + 2y + z = 9$

42. Invert $\begin{pmatrix} 1 & 1 & 1 & 1 \\ 1 & 2 & 0 & -1 \\ 2 & -1 & 2 & 0 \\ 0 & 1 & -1 & 1 \end{pmatrix}$.

  (a) Solve: $\begin{array}{rcrcrcrcr} x & + & y & + & z & + & u & = & 4 \\ x & + & 2y & & & - & u & = & 2 \\ 2x & - & y & + & 2z & & & = & 2 \\ & & y & - & z & + & u & = & 3 \end{array}$

  (b) Solve: $\begin{array}{rcrcrcrcr} x & + & y & + & z & + & u & = & 1 \\ x & + & 2y & & & - & u & = & 1 \\ 2x & - & y & + & 2z & & & = & 2 \\ & & y & - & z & + & u & = & 0 \end{array}$

43. Invert $\begin{pmatrix} 1 & -1 & 3 & -1 \\ 1 & 1 & 1 & 1 \\ 1 & -2 & -1 & 1 \\ 2 & 0 & 0 & 1 \end{pmatrix}$.

(a) Solve:
$$\begin{aligned} z_1 - z_2 + 3z_3 - z_4 &= 10 \\ z_1 + z_2 + z_3 + z_4 &= 0 \\ z_1 - 2z_2 - z_3 + z_4 &= 1 \\ 2z_1 \qquad\qquad\quad + z_4 &= 1 \end{aligned}$$

(b) Solve:
$$\begin{aligned} z_1 - z_2 + 3z_3 - z_4 &= -10 \\ z_1 + z_2 + z_3 + z_4 &= 3 \\ z_1 - 2z_2 - z_3 + z_4 &= 0 \\ 2z_1 \qquad\qquad\quad + z_4 &= 2 \end{aligned}$$

44. (*Requires special thought*) Show that the multiplicative identity matrix is unique. (Assume the contrary, and derive a contradiction.)

W **45.** What is there about the definition of matrix inverses that requires a matrix to be square in order to possess an inverse?

Applications

Use matrix methods to solve the following problems.

46. (*Real Estate Development*) Mansion Construction Company has acquired a tract of land for building single-family homes. It can build 28 homes on the site in either of two styles, a single-story ranch house or a two-story colonial. The more-popular ranches in this neighborhood sell for $150,000, whereas colonials sell for $200,000. The Mansion Co. feels that it needs to sell as many of the less-expensive ranches as possible but must also take in $5,000,000 from the sale of these homes to satisfy its board of directors.
 (a) How many of each style shall Mansion construct?
 (b) Because of a change in the zoning laws, the company may build only 20 houses. The company is willing to settle for $3,500,000 from the sale of them. How many of each style shall now be built?

47. (*Broadway Theater District*) Plays on New York's Broadway theater district are financed by *angels*, investors who hope to make handsome returns on their investments when and if their play is a success. One director has rounded up a cool million dollars from a band of angels, and this money must be budgeted among salaries, supplies, and advertising. As a rule of thumb, the amount spent on both salaries and supplies should equal the sum allocated to advertising. Furthermore, the difference between the advertising and salary budgets should be 10 percent of the total amount. How should the million dollars be split up?

48. Suppose angels contribute an additional $300,000 to the dramatic production. How should the money now be divided? Angels are defined in Exercise 47.

49. Again referring to Exercise 47, suppose one final angel adds enough money to make the total available money $1.75 million. How will the director divide her money?

50. Refer a final time to Exercise 47, and suppose the difference between staff salaries and the advertising budget is 20 percent of the angel's investment. How should the money be divided?

51. (*Tourist Trade*) Municipal authorities are anxious to best develop an island in their municipality for the tourist trade. The island attracts tourists interested in either its beautiful beaches or its numerous clay tennis courts; rarely are any tourists interested in both. During the last tourist season, a total of 300,000 visitors stayed on the island. Surveying the island innkeepers, the authorities determined that twice as many people are interested in the beach as in playing tennis. Use matrix methods to determine how many tourists are in each category.

52. The authorities referred to in Exercise 51 forecast that the tourist trade will grow to such an extent that in five years, 800,000 vacationers will be staying on the island during the year. How many of those are tennis enthusiasts, and how many are beach people? Use matrix methods.

53. (*Dinosaur Studies*) A graduate student in paleontology is analyzing prehistoric populations of three species of dinosaur. She was part of an expedition that discovered a mound of bones belonging to a total of 100 animals. There is some variation in all species, and she knows, for example, that long thigh bones are found in one-half of the animals in the first species, one-quarter of the animals in the second species, and all of the animals in the third species. Short forearm bones are found in three-quarters of the specimens of the first species, all of the second species, and one-half of the third species. In this mound, there are 60 long thigh bones and 75 short forearms.
 (a) How many animals of each species are in this mound?
 (b) A more careful count reveals that the mound is the grave of only 90 extinct beasts. Furthermore, only 50 of the thigh bones are long, and only 70 forearms are short. What now is the species count in this mound?

Calculator Exercises

Use a calculator in Exercises 54–57 to get numerical answers correct to two places to the right of the decimal point.

54. What is the inverse of $\begin{pmatrix} 1.1 & 1.9 \\ -0.1 & 2.2 \end{pmatrix}$?

55. Use the results of Exercise 54 to solve the equations

$$1.1x + 1.9y = 2$$
$$-0.1x + 2.2y = 0$$

56. Use the results of Exercise 54 to solve the equations

$$1.1X + 1.9Y = 1$$
$$-0.1X + 2.2Y = 1$$

W 57. Refer again to Exercise 56. Explore and explain how finding the components of the inverse to two-decimal-place accuracy affects the results by redoing the computation so that each component is accurate to three places to the right of the decimal point.

Section 2.6   INPUT-OUTPUT ANALYSIS

input-output analysis   *Input-output analysis* has become a basic mathematical tool for economists. Consider a community consisting of steelworkers and wheat farmers. The community itself has a certain need for steel and wheat, but beyond that, extra wheat and steel must be exported in return for goods and services that the members of the community cannot provide. The community would like to export, say, $100,000 worth of wheat. The trouble is, the steelworkers and farmers themselves need wheat to live, and the harder the farmers work to meet the export goal, the more wheat they eat. *Input-output analysis* helps the economic planners in this community. This concept involves harnessing common sense to our work on linear equations; it is a spectacular example of how, when this material is cast in matrix form, the resulting matrix equations reveal an unexpected clarity and simplicity. The American economist Wassily Leontieff is largely responsible for this important technique.

Historical Note

Wassily Leontieff   *Wassily Leontieff was born in Russia but emigrated to this country, where he joined the faculty of Harvard University in 1932. It was at Harvard that he developed his work on input-output analysis, for which he received the Nobel Prize in economics in 1973. Since 1975, he has been professor of economics at New York University, and has recently been active in affecting economic policies in Japan.*

     Input-output analysis begins by dividing the members of a community into two groups: producers and consumers. This division is not always clear-cut, because many goods (such as food) are consumed as well by farmers, butchers, and other food produc-

producers   ers. *Producers* generate surplus that is *output* from the community, whereas *consumers*

consumers   require *input* for their needs. Although this analysis received its name from the dual processes of input and output implicit in this model, we prefer to use the more familiar terms "consumers" and "producers"; terms like *input* and *output* smack too much of jargon.

In our community of steelworkers and wheat farmers, careful records are kept of production and consumption. A portion of all production must be used to sustain the population. Out of each dollars-worth of wheat grown by the farmer, the farmers use $0.15 for food and the steelworkers use $0.25. For each dollars-worth of steel, the farmers use $0.20 while the steelworkers themselves use $0.10 for replacing their tools and such. Figure 2.20 helps visualize these relations between the two industries. The community leaders need $10,000 and $25,000 of wheat and steel for export (in addition to the wheat and steel that the community consumes itself) and therefore ask: what is the *total* production of wheat and steel needed to maintain the population and yet produce the desired surplus?

input matrix   Placing the production data in matrix form helps in visualizing the processes involved. In this *input matrix*, each industry is allotted both a row and a column. The rows identify the industries as *consumers*, and the columns identify them as *producers*, always with respect to $1.00 worth of goods.

$$Q = \begin{array}{c} \\ \text{WHEAT} \\ \text{STEEL} \end{array} \begin{array}{cc} \text{WHEAT} & \text{STEEL} \\ \left(\begin{array}{cc} 0.15 & 0.20 \\ 0.25 & 0.10 \end{array} \right) \end{array}$$

The row marked STEEL summarizes the information about steelworkers as consumers: they consume $0.25 of wheat and $0.10 of steel for every $1.00 worth of steel they produce. Compare the matrix Q with Figure 2.20. In a more general treatment of an input-output model, the quantity q_{ij} at the intersection of row i and column j in the input matrix Q measures the amount consumed by the industry on row i of the output

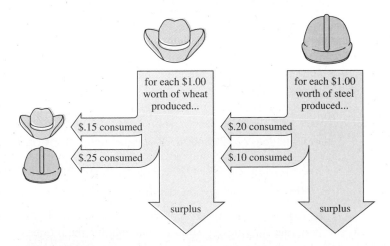

FIGURE 2.20
Here are the elements of wheat and steel production. Compare this diagram with the input matrix Q.

of the industry in column j in the production of a unit amount (usually $1.00 worth) of industry i.

In our example, $q_{21} = 0.25$, so that the amount consumed by industry 2 (the steel-workers) of industry 1 (wheat) in the production of a unit quantity of steel (industry 2) is $0.25.

demand The total production we seek is the sum of the goods consumed by the community plus the *demand*, the amount required for export. We can represent the unknown total *production* of wheat and steel by x_1 and x_2, respectively. The internal consumption depends on these variables.

The total wheat produced has three components. First, there is the demand, which we assume to be $10,000 in this example. Next, there is the wheat consumed by the farmers in the production of wheat. Since each $1.00 worth of wheat involves the consumption of $0.15 worth of wheat, the production of x_1 worth of wheat involves the consumption of $0.15x_1$. Finally, the steelworkers will consume their share of wheat. Since each $1.00 worth of steel production consumes $0.25 of wheat, steel production in the amount of x_2 will involve wheat consumption of $0.25x_2$.

Total wheat production	= Demand	+ Wheat used to produce wheat	+ Wheat used to produce steel	
x_1	$= 10,000$	$+ 0.15x_1$	$+ 0.25x_2$	(2.17)

In the same way, we can derive a second equation governing the production of steel.

Total steel production	= Demand	+ Steel required by wheat farmers	+ Steel required to produce steel	
x_2	$= 25,000$	$+ 0.20x_1$	$+ 0.10x_2$	(2.18)

We know several ways to solve pairs of simultaneous equations such as these, but the solution may be tedious. Furthermore, it is not clear from this analysis how to handle communities with more than two economic activities. The following observation motivates a better, matrix-oriented procedure.

If we define a row matrix $X = \begin{pmatrix} x_1 & x_2 \end{pmatrix}$, then the rightmost two terms in Equations (2.17) and (2.18) are the components of the matrix product $X \times Q$. If we define a

demand matrix *demand matrix*

$$D = \begin{pmatrix} 10,000 & 25,000 \end{pmatrix}$$

then Equations (2.17) and (2.18) can be compactly written as

$$X = D + XQ \qquad (2.19)$$

Equation (2.19) has several clear advantages:

1. This equation can be applied to communities involved in arbitrary numbers of economic activities by simply extending the definitions of the matrices X, D, and Q in a logical fashion.

2. It is easier and more insightful to solve the matrix Equation (2.19) rather than Equations (2.17) and (2.18).

Can we solve for X? Equation (2.19) implies

$$X - X \times Q = D$$
$$X \times I - X \times Q = D$$
$$X \times (I - Q) = D$$

For the last step, multiply both sides on the right by the inverse of the matrix $(I - Q)$:

$$X = D \times (I - Q)^{-1} \tag{2.20}$$

(The matrix $I - Q$ will always be square, so it is proper to speak of its inverse.) This result applies to any economy, no matter how many rows and columns Q has.

For this community of steelworkers and farmers,

$$I - Q = \begin{pmatrix} 0.85 & -0.20 \\ -0.25 & 0.90 \end{pmatrix}$$

and you should verify that

$$(I - Q)^{-1} = \begin{pmatrix} 1.26 & 0.28 \\ 0.35 & 1.19 \end{pmatrix}$$

To produce exportable wheat and steel in the amounts of \$10,000 and \$25,000, the total production should be

$$X = \begin{pmatrix} 10{,}000 & 25{,}000 \end{pmatrix} \times \begin{pmatrix} 1.26 & 0.28 \\ 0.35 & 1.19 \end{pmatrix}$$
$$= \begin{pmatrix} 21{,}340 & 32{,}525 \end{pmatrix}$$

That is, total wheat production should be worth \$21,340, and total steel production should be worth \$32,525.

Input-Output Analysis: Solving for Total Production

The solution of

$$X = D + X \times Q$$

is

$$X = D \times (I - Q)^{-1}$$

Here is the explanation for the quantities in this equation.

X is the row matrix with components x_i, the total production for industry i.

D is the demand row matrix.

Q is the input matrix.

I (really I_n, where n is number of industries) is the identity matrix for multiplication.

Example 44 The planning commissioners for the wheat-steel community mentioned above revise their export projections. They decide it would be better to export \$20,000 worth of both wheat and steel. What total production will produce these surpluses?

Solution
Use a revised demand matrix

$$D = (20{,}000 \quad 20{,}000)$$

in the solution for X. (The new requirements do not affect the matrix Q.) Therefore,

$$X = (20{,}000 \quad 20{,}000) \times \begin{pmatrix} 1.26 & 0.28 \\ 0.35 & 1.19 \end{pmatrix} = (32{,}200 \quad 29{,}400)$$

The total production for grain and steel should be \$32,200 and \$29,400, respectively.

Skill Enhancer 44   The community feels its demand for steel is \$30,000, although the demand for wheat remains \$20,000. What total production will permit this surplus to be exported?

Answer in Appendix E.

Example 45   Another community deals in three industries—food, shelter (housing), and clothing. The input matrix connecting the three industries is

$$\begin{array}{c} \\ Q = \begin{array}{c} \text{food} \\ \text{shelter} \\ \text{clothing} \end{array} \end{array} \begin{array}{ccc} \text{food} & \text{shelter} & \text{clothing} \\ \begin{pmatrix} 0.1 & 0 & 0.2 \\ 0.2 & 0.1 & 0.4 \\ 0 & 0 & 0.1 \end{pmatrix} \end{array}$$

What should the total production of the three industries be to generate a \$900,000 surplus of each? In another year, no shelter-related products can leave the community, but the community continues to meet the goal of \$900,000 of food and clothing. What should the total production be in that year?

Solution
We answer both questions via $X = D \times (I_3 - Q)^{-1}$. First, let

$$D = (900{,}000 \quad 900{,}000 \quad 900{,}000)$$

With the given definition of Q,

$$I - Q = \begin{pmatrix} 0.9 & 0 & -0.2 \\ -0.2 & 0.9 & -0.4 \\ 0 & 0 & 0.9 \end{pmatrix}$$

and the inverse is given *exactly* by

$$(I - Q)^{-1} = \begin{pmatrix} \frac{10}{9} & 0 & \frac{20}{9^2} \\ \frac{20}{9^2} & \frac{10}{9} & \frac{400}{9^3} \\ 0 & 0 & \frac{10}{9} \end{pmatrix}.$$

Therefore, find the total production goal by evaluating the matrix product $D \times (I - Q)^{-1}$:

$$X = \begin{pmatrix} 900,000 & 900,000 & 900,000 \end{pmatrix} \times \begin{pmatrix} \frac{10}{9} & 0 & \frac{20}{9^2} \\ \frac{20}{9^2} & \frac{10}{9} & \frac{400}{9^3} \\ 0 & 0 & \frac{10}{9} \end{pmatrix}$$

$$= \begin{pmatrix} 1,222,000 & 1,000,000 & 1,716,000 \end{pmatrix}$$

This little city-state will (to four significant digits) have to meet *total* production goals of \$1,222,000 for food, \$1,000,000 for shelter, and \$1,716,000 for clothing.

We use the demand matrix $\begin{pmatrix} 900,000 & 0 & 900,000 \end{pmatrix}$ for the second part of the problem. This demand matrix implies that the total production will be

$$X = \begin{pmatrix} 900,000 & 0 & 900,000 \end{pmatrix} \times \begin{pmatrix} \frac{10}{9} & 0 & \frac{20}{9^2} \\ \frac{20}{9^2} & \frac{10}{9} & \frac{400}{9^3} \\ 0 & 0 & \frac{10}{9} \end{pmatrix}$$

$$= \begin{pmatrix} 1,000,000 & 0 & 1,222,000 \end{pmatrix}$$

The community can count on producing no shelter, an even \$1 million worth of food, and \$1,222,000 worth of clothing.

Skill Enhancer 45   In yet a third year, the demand matrix is $D = \begin{pmatrix} \$1,800,000 & \$900,0000 \end{pmatrix}$. What total production will meet that demand?

Answer in Appendix E.

_____ *Exercises 2.6* _____

In Exercises 1–12, for each of the matrices Q, compute (if possible) both $I - Q$ and $(I - Q)^{-1}$. Use the identity matrix I_n appropriate for each exercise (for example, you would use I_2 for Exercise 1, but I_3 for Exercise 9).

1. $\begin{pmatrix} \frac{5}{4} & -\frac{1}{4} \\ -\frac{1}{6} & \frac{5}{6} \end{pmatrix}$

2. $\begin{pmatrix} \frac{5}{6} & \frac{1}{6} \\ -\frac{1}{4} & \frac{3}{4} \end{pmatrix}$

3. $\begin{pmatrix} \frac{1}{2} & \frac{1}{2} \\ \frac{1}{2} & \frac{1}{2} \end{pmatrix}$

4. $\begin{pmatrix} 0 & 1 \\ 1 & 0 \end{pmatrix}$

5. $\begin{pmatrix} (1-p) & p \\ p & (1-p) \end{pmatrix}$

6. $\begin{pmatrix} \frac{1}{2} & 1 \\ \frac{1}{2} & 0 \end{pmatrix}$

7. $\begin{pmatrix} \frac{1}{4} & \frac{1}{2} \\ \frac{1}{4} & 0 \end{pmatrix}$

8. $\begin{pmatrix} 0.1 & 0.1 \\ 0.1 & 0.1 \end{pmatrix}$

9. $\begin{pmatrix} 0.2 & 0.1 & 0 \\ 0.1 & 0.2 & 0 \\ 0 & 0 & 1 \end{pmatrix}$

10. $\begin{pmatrix} 0.1 & 0 & 0.1 \\ 0 & 0.1 & 0 \\ 0.1 & 0 & 0.1 \end{pmatrix}$

11. $\begin{pmatrix} 0.25 & 0.5 & 0 & 0 \\ 0.25 & 0 & 0 & 0 \\ 0 & 0 & 0.1 & 0.1 \\ 0 & 0 & 0.1 & 0.1 \end{pmatrix}$

12. $\begin{pmatrix} 0 & 0.1 & 0.2 & 0.3 \\ 0 & 0.1 & 0.2 & 0.3 \\ 0 & 0.1 & 0.2 & 0.3 \\ 0.5 & 0 & 0 & 0 \end{pmatrix}$

13. In the text, it was asserted that the expressions in parentheses in Equations (2.17) and (2.18) correspond to the components of the matrix product $X \times Q$. Show that this is in fact the case.

14. In this section, it was asserted that if $I - Q = \begin{pmatrix} 0.9 & 0 & -0.2 \\ -0.2 & 0.9 & -0.4 \\ 0 & 0 & 0.9 \end{pmatrix}$, then

$$(I - Q)^{-1} = \begin{pmatrix} \frac{10}{9} & 0 & \frac{20}{9^2} \\ \frac{20}{9^2} & \frac{10}{9} & \frac{400}{9^3} \\ 0 & 0 & \frac{10}{9} \end{pmatrix}$$

Verify this.

Applications

15. In the two-product community (wheat and steel) discussed in the text, suppose the external demand for the two products is $45,000 and $60,000. What should the total production be to support these goals?

16. In the same community, the goals for the next year are $50,000 for both wheat and steel. What should the total production be?

17. In one year in this community, the total production is $60,000 and $70,000 for the two kinds of goods. What external demands led to these total production goals?

W 18. (*Requires special thought*) A neighboring community produces guns and butter as its two industries. The total production for this community was $35,000 and $40,000 of guns and butter, respectively, and the external demand for these products was $30,000 and $35,000. Can you deduce from this information what are the components of the matrix Q? (You need not do the actual calculations.) Why or why not?

19. Focus on the three-industry state discussed in the text. Using the information there, deduce the total production needed to support a demand of $1,000,000 each for food and shelter, and $1,500,000 for clothing.

20. Suppose the figures in Exercise 19 describe the *total* production rather than the external demand. What external demand corresponds to those figures?

21. In the three-industry state discussed in the text, suppose the demand matrix is

$$\begin{pmatrix} 1,000,000 & 750,000 & 500,000 \end{pmatrix}$$

What will the total production be to meet this demand?

W 22. (*Requires special thought*) Imagine an arbitrary input matrix Q with components q_{ij}. The matrix has n rows and columns. Discuss the physical meaning of the statement that the *sum* of the components in any column j is the total fraction of $1.00 worth of goods in industry j that goes toward the manufacture of all other products in the economy.

W 23. Use the results of Exercise 22 to show that the sum of the components in any column of an input matrix must be less than or equal to 1 for any enterprise. Explain why the inequality must be *strict* for any *profit-making* enterprise.

24. *(Business Administration)* Large, modern corporations with many departments fit nicely into the input-output analysis model. The demand matrix represents either demand for the company's products or the sales goals decided upon by top management in the company. One large company has three departments—manufacturing, accounting, and advertising. An input matrix for this company might look like this:

$$Q = \begin{matrix} & \text{mfg.} & \text{acctg.} & \text{advtg.} \\ \text{manufacturing} & \\ \text{accounting} \\ \text{advertising} \end{matrix} \begin{pmatrix} 0.1 & 0.3 & 0.1 \\ 0 & 0.5 & 0 \\ 0 & 0.2 & 0 \end{pmatrix}$$

(a) Compute $Z = (I - Q)$.
(b) Compute Z^{-1}.
(c) What will the total manufacturing capacity be to support a sales goal of $1,000,000 of finished goods?
(d) What advertising effort will be needed for this goal?

▦ Calculator Exercises

Compute $I - Q$ and $(I - Q)^{-1}$ for the matrices Q in Exercises 25–26.

25. $\begin{pmatrix} -0.1 & -1.9 \\ 0.1 & -1.2 \end{pmatrix}$ **26.** $\begin{pmatrix} 0.3 & 0.4 \\ 0.4 & 0.2 \end{pmatrix}$

CHAPTER REVIEW

Terms

matrix
matrix elements
row matrix
matrix diagonal
scalar matrix multiplication
matrix equality
multiplicative identity matrix
row equivalent matrix
diagonalize
algorithm
pivot position
Gauss-Jordan elimination
invertible matrix
consumers
demand matrix

rows
matrix component
column matrix
matrix addition
scalar
matrix multiplication
standard form for linear equations
matrix row operations
reduced form
pivoting
pivot column
parameter
input-output analysis
input matrix
columns

dimension
square matrix
matrix subtraction
dot product multiplication
additive identity matrix
augmented matrix
equivalent equations
leading element
pivot element
pivot row
matrix inverse
producers
demand

Key Concepts

- A *matrix* is a rectangular array of numbers or other quantities.

- The individual components of any matrix are identified by *row* and *column*.

- One may *add* or *subtract* matrices of identical dimension by adding or subtracting corresponding elements.

- Any matrix can be multiplied by any scalar. Multiply each component of the given matrix by the scalar.

- *Matrix multiplication* is sensitive to the dimensions of the two factor matrices. General matrix multiplication is built up out of individual row-column multiplications. An $m \times n$ matrix can be multiplied by an $n \times p$ matrix to yield an $m \times p$ product matrix, where m, n, and p are any positive integers.

- Two matrices are *equal* if and only if they have identical dimension and all components of the first equal the corresponding components of the second.

- The *additive identity matrix* is a matrix all of whose components are zero; it is so called because adding it to any matrix leaves the value of the matrix unchanged. The *multiplicative identity matrix* is a square matrix all of whose components are zero except for the diagonal elements, which equal 1. Any matrix *times* the multiplicative identity matrix remains unchanged regardless of the order of multiplication.

- Any system of linear equations may be represented by a matrix equation of the form $A \times X = B$, for some coefficient matrix A, a column matrix B, and a column matrix X of the unknown variables of the problem.

- Such a system may be solved by forming the *augmented matrix* $(A|B)$ and performing suitable *row operations* on it to *reduce* it.

- Square matrices may have an *inverse*. When the matrix inverse is known, multiplication by that inverse is a quick way to solve a set of simultaneous equations that have been written in matrix form. If a matrix has an inverse, it can be found by reducing a matrix augmented by several columns.

- Matrix methods are important in *input-output analysis*, which attempts to relate the total output of an economic community to its internal needs and external export needs.

Review Exercises

In Exercises 1–6, give the dimensions of the following matrices.

1. $\begin{pmatrix} 1 & -0.0003 \\ 14 & 99.44 \end{pmatrix}$ **2.** $\begin{pmatrix} 1 & -1 & 1 \end{pmatrix}$

3. $\begin{pmatrix} 5 \\ 6 \\ -7 \end{pmatrix}$ **4.** $\begin{pmatrix} 7,624,912.01 \end{pmatrix}$

5. $\begin{pmatrix} 0 \\ 0 \\ 0 \end{pmatrix}$ **6.** $\begin{pmatrix} 0.4 & 0.3 & -0.3 & -0.6 \\ 0 & 0.2 & -1 & 0.1 \end{pmatrix}$

Perform the indicated matrix arithmetic in Exercises 7–18. If not possible, state why.

7. $\begin{pmatrix} 1 & 1 \\ 1 & 1 \end{pmatrix} + \begin{pmatrix} -1 & 0 \\ 0 & -1 \end{pmatrix}$

8. $-5\begin{pmatrix} -4 & -4 & 0 & -3 \end{pmatrix}$

9. $3\begin{pmatrix} 1 \\ 3 \\ \frac{3}{2} \end{pmatrix} - (-3)\begin{pmatrix} \frac{4}{3} & -2 \end{pmatrix}$

10. $-4\begin{pmatrix} 4.01 \end{pmatrix} + 2\begin{pmatrix} 4 & 0.01 \end{pmatrix}$

11. $-\frac{3}{2}\begin{pmatrix} 3 & 4 & 2 \\ 0 & \frac{2}{3} & 2 \end{pmatrix} - 2\begin{pmatrix} \frac{1}{2} & \frac{3}{2} & 0 \\ -5 & 6 & 1 \end{pmatrix}$

12. $\begin{pmatrix} -1 & \frac{2}{3} & 0 & \frac{1}{3} \end{pmatrix}\begin{pmatrix} 4 \\ -3 \\ -3 \\ 3 \end{pmatrix}$

13. $\begin{pmatrix} 3 \end{pmatrix}\begin{pmatrix} -\frac{1}{9} \end{pmatrix}$

14. $\begin{pmatrix} 1 & 1 & 0 & 1 \\ \frac{1}{4} & \frac{1}{2} & -\frac{1}{4} & -1 \end{pmatrix} \times \begin{pmatrix} 0 & 1 \\ -4 & 4 \\ 6 & \frac{1}{2} \\ 0 & 12 \end{pmatrix}$

15. $\begin{pmatrix} 0 & 1 \\ -4 & 4 \\ 6 & \frac{1}{2} \\ 0 & 12 \end{pmatrix} \times \begin{pmatrix} 1 & 1 & 0 & 1 \\ \frac{1}{4} & \frac{1}{2} & -\frac{1}{4} & -1 \end{pmatrix}$

16. $\begin{pmatrix} \frac{2}{3} & 2 & -3 & 1 \\ -2 & \frac{1}{6} & 0 & 0 \\ 0 & 1 & 1 & -\frac{1}{3} \end{pmatrix} \times \begin{pmatrix} 0 & 1 & 0 \\ 3 & -6 & 3 \\ 1 & 0 & \frac{1}{3} \\ 9 & -81 & 0 \end{pmatrix}$

17. $\left[\begin{pmatrix} 1 & 0 \\ 0 & 1 \end{pmatrix} + \begin{pmatrix} 1 & 1 \\ 1 & 0 \end{pmatrix}\right] \times \begin{pmatrix} 1 & \frac{1}{2} \\ -\frac{3}{4} & 1 \end{pmatrix}$

18. $\begin{pmatrix} 1 & 0 & 1 \\ -1 & 1 & 1 \\ 0 & -1 & -1 \end{pmatrix} \times \begin{pmatrix} \frac{1}{2} & 1 & 1 \\ 0 & \frac{3}{4} & 1 \\ 0 & -\frac{1}{2} & -\frac{1}{3} \end{pmatrix}$

Write the augmented matrix for each of the systems of equations in Exercises 19–22. Do *not* solve the equations.

19. $-2.304x + 9.88y = 12.034$

       $0.0003x - 0.004y = 7.645$

20. $3u - 2v = -4$

       $6u - 4v = -8$

       $12u - 8v = -16$

21. $10a - 7b = 12$

22. $29x_1 - 32x_2 - 17x_3 = 42$

       $-12x_1 + 11x_2 + 5.5x_3 = -21$

         $x_1 + 99x_2 - 87x_3 = 21$

In Exercises 23–32, reduce the following augmented matrices.

23. $\left(\begin{array}{ccc|c} 1 & 0 & 0 & 1 \\ 1 & 1 & 0 & 2 \\ 1 & 1 & 1 & 3 \end{array}\right)$ **24.** $\left(\begin{array}{ccc|c} 0 & 0 & 1 & \frac{2}{3} \\ 0 & 0 & 0 & 0 \\ 0 & 1 & 0 & -7 \\ 1 & 0 & 0 & 12.5 \end{array}\right)$

25. $\left(\begin{array}{cc|c} 1 & 1 & 3 \\ -1 & 1 & 2 \end{array}\right)$ **26.** $\left(\begin{array}{cc|c} -1 & 4 & -1 \\ 4 & -1 & 1 \end{array}\right)$

27. $\left(\begin{array}{cc|c} \frac{1}{6} & \frac{1}{4} & 5 \\ \frac{1}{3} & -\frac{1}{6} & 2 \end{array}\right)$ **28.** $\left(\begin{array}{ccc|c} 0.1 & 0.2 & 0.2 & 35 \\ 1 & 1 & -2 & 0 \\ 0 & 1 & 0 & 50 \end{array}\right)$

29. $\left(\begin{array}{ccc|c} 1 & 1 & 1 & 2 \\ -1 & 1 & 1 & -4 \\ 1 & 1 & -1 & 0 \end{array}\right)$ **30.** $\left(\begin{array}{cc|c} 0 & 0 & 0 \\ 2 & 1 & 20 \\ 1 & -2 & -15 \end{array}\right)$

31. $\left(\begin{array}{cc|c} 1 & 2 & 2 \\ 2 & 4 & 4 \\ -1 & -2 & -2 \end{array}\right)$ **32.** $\left(\begin{array}{cc|c} 1 & 0 & 1 \\ 2 & 0 & -2 \\ 3 & 0 & 4 \end{array}\right)$

Transform the systems of equations in Exercises 33–44 to a matrix equivalent and determine the best possible solution by reducing the augmented matrix formed from the coefficient and column matrices.

33. $\quad 0.4x - 0.2y = 2$

$\quad -0.1x + 0.5y = 4$

34. $\quad \dfrac{2}{3}x + \dfrac{3}{4}y = 12$

$\quad -x + 2y = 7$

35. $\quad x + y + z = 4$

$\quad x - y + z = 3$

36. $\quad x + y + z = 4$

$\quad x - y + z = 3$

$\quad x = 2$

37. $\quad x_1 - x_2 + x_3 = 4$

$\quad 2x_1 - x_2 - x_3 = 14$

$\quad x_1 + x_2 + 2x_3 = -5$

38. $\quad a + c - 4b = 0$

$\quad a + 2b + 2c = 2$

$\quad a + b - c = \dfrac{1}{4}$

39. $\quad x = 2$

$\quad x + 2y = 3z - 1$

40. $\quad x + y + 3z = 3$

$\quad x - 3z = 0$

$\quad 3y - 4z = \dfrac{5}{3}$

41. $\quad x + y - 3z = -4$

$\quad 2x + y + 2z = 7$

$\quad x + y = 3z - 4$

42. $\quad 2x - 3y + z = 1$

$\quad x + 4y - 2z = 1$

$\quad -x + 2y + z = 3$

43. $\quad x + y - z + 1 = 0$

$\quad 2x + 2y - z - 2 = 0$

$\quad x + 2y - 2z + 1 = 0$

44. $\quad z_1 + z_2 = 4$

$\quad z_2 + z_3 = -5$

$\quad z_3 + 6z_4 = 0$

$\quad 2z_4 + z_1 = 7$

Invert the matrices in Exercises 45–53 if possible.

45. $\begin{pmatrix} 1 & 0 \\ 0 & 1 \end{pmatrix}$

46. $\begin{pmatrix} 1 & 2 \\ 3 & 4 \end{pmatrix}$

47. $\begin{pmatrix} 0 & 1 \\ 1 & 1 \end{pmatrix}$

48. $\begin{pmatrix} \frac{2}{3} & -1 \\ \frac{1}{12} & \frac{1}{4} \end{pmatrix}$

49. $\begin{pmatrix} -0.3 & 1 \\ 0.003 & 0.02 \end{pmatrix}$

50. $\begin{pmatrix} 1 & -1 & 0 \\ 0 & 1 & -1 \\ -1 & 0 & 1 \end{pmatrix}$

51. $\begin{pmatrix} 1 & 1 & 1 \\ 2 & 4 & 0 \\ 0 & 1 & 3 \end{pmatrix}$

52. $\begin{pmatrix} -2 & -2 & 1 \\ 1 & 2 & 3 \\ 2 & 0 & 1 \end{pmatrix}$

53. $\begin{pmatrix} 2 & 2 & 1 \\ 0 & 1 & 1 \\ 1 & 0 & 2 \end{pmatrix}$

Use matrix inverses to solve the linear equation systems in Exercises 54–59.

54. (a) $x + 2y = 30$

$\quad x + y = 20$

(b) $x + 2y = 30$

$\quad x + y = 15$

55. (a) $2x - 3y = 1$

$\quad 3x + y = 7$

(b) $2a - 3b = 10$

$\quad 3a + b = 70$

56. (a) $\dfrac{2}{3}u + \dfrac{1}{3}v = 10$

$\quad \dfrac{1}{3}u - \dfrac{1}{2}v = -6$

(b) $\dfrac{2}{3}u + \dfrac{1}{3}v = -10$

$\quad \dfrac{1}{3}u - \dfrac{1}{2}v = 6$

57. (a) $0.8s + 0.4t = 10$

$\quad 0.3s - t = 2$

(b) $0.8s + 0.4t = 5$

$\quad 0.3s - t = 5$

58. (a) $x + 5y + 2z = 3$

$\quad -x + y - z = 0$

$\quad -x - 2z = 7$

(b) $x + 5y + 2y = 3$

$\quad -x + y - z = 1$

$\quad -x - 2z = 7$

59. (a) $4u + 3v - 4w = 9$

$\quad u + 3v - 2w = 5$

$\quad 2u + v + w = 15$

(b) $4u + 3v - 4w = 3$

$\quad u + 3v - 2w = 0$

$\quad 2u + v + w = 4$

Applications

60. (*Weather*) The average monthly daytime temperatures (in Celsius degrees) for New York City and Boston for June, July, and August are 28, 30, and 33 (NYC) and 29, 29, and 32 (Boston). A single matrix can display all these data. Discuss two different such matrices.

61. (*Agriculture*) Farmer Gray has four plots of land he uses for experimental agriculture. Two varieties of wheat are planted on each of the first pair of plots, and the following table gives the average heights of the wheat that he observes.

	Variety 1	Variety 2
Plot 1	48 in	51 in
Plot 2	52 in	50 in

Plots 3 and 4 have soil characteristics similar to the first two plots. A fertilizer "guaranteed" to increase plant height by 25 percent is spread over the second pair of plots, and wheat is planted.

(a) Represent the wheat data in the first pair of plots by a matrix.

(b) Use matrix arithmetic to anticipate the growth in the second pairs of fields.

(c) When the wheat was grown, the farmer carefully obtained the average height of the wheat fortified by the fertilizer. The actual recorded growths were these:

	Variety 1	Variety 2
Plot 3	48 in	50 in
Plot 4	58 in	59 in

Represent these data in matrix form.

(d) Use matrix arithmetic to determine the difference in growth between the anticipated and the actual wheat growth.

62. (Home Economy) Coffee, ground beef, and sugar cost $5.00, $1.80, and $1.00 per pound, respectively. You need 2, 3, and 5 lb of these products.

(a) Represent the commodity costs as a row matrix, and your commodity needs as a column matrix.

(b) Use matrix multiplication to determine your total cost.

63. (Chemistry) A certain industrial application calls for 10 tons of a mixture containing 30 percent of a certain metal. The supervisor has available mixtures containing 15 percent and 40 percent of the metal. How many tons of each should be mixed to yield the desired mix? Use matrix methods.

64. (Retailing) Heavenly Mash is a popular mixture of raisins and Hawaiian macadamia nuts selling for $3.00 per pound. Raisins sell for $2.00 per pound, and the nuts sell for $5.50 per pound.

(a) In order to prepare 50 lb of the mixture, how many pounds of raisins and nuts must a storekeeper combine? Use matrix methods.

(b) How might you adjust the components to prepare 100 lb of the Mash, this time to retail for $4.50 per pound? Use matrix methods.

65. (Television Polling) Television pollsters try to determine how many people are watching the competing programs in any given time slot. At 6:00 A.M., there are only two shows on in a particular midwest locality. The first show consistently has 25 percent more viewers than its competitor. Use matrix methods to answer the following questions.

(a) In one city, there are 200,000 viewers (on the average) for these two programs. How many watch the first show, and how many the second?

(b) In another city, there are 300,000 viewers. How many watch each of the programs?

66. (Biology) University researchers are interested in the effect of a certain nutriment on white rats. They are able to order unlimited quantities of feed in two grades. The first contains 20 percent of the nutriment, and the second contains 40 percent. Use matrix inverse methods to answer the following questions.

(a) How much of each should be mixed to obtain 100 kg of a mixture containing 25 percent of the nutriment?

(b) How much of each should be combined to get 150 kg of a mixture with 35 percent of the nutriment?

3

A GEOMETRICAL APPROACH TO LINEAR PROGRAMMING

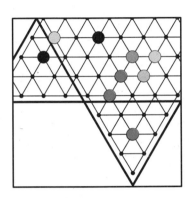

3.1 What Is Linear Programming?
3.2 Linear Inequalities and Their Graphs
3.3 Geometric Solutions to Linear Programming Problems
3.4 Applications
3.5 When the Geometric Method Fails
Chapter Review

INTRODUCTION

A group of craftspeople make and sell wooden tables and chairs. For this discussion, assume that each artisan is either a beginner or an "old hand." It takes a beginner 4 days of work to make a table and 5 days to make a chair. Experienced workers need 3 days for either a table or a chair. No one works more than 22 days per month. The profit is $150 for each table and $110 for each chair. There are many ways of deciding how many tables and chairs to make in any month. Of all these, which plan produces the most profit for the group? With the tools of *linear programming* at hand, it is possible to determine strategies for situations such as this where it is important to either *maximize the profit* or *minimize the costs or expenses*. (Do not confuse linear programming with computer programming, a different activity altogether.)

Linear programming saves American businesses and organizations hundreds of millions of dollars annually by teaching managers how best to allocate the resources and materials that are available.

After a brief introduction, we begin by extending our knowledge of linear mathematics to systems of linear *inequalities*. Our work with linear programming will draw heavily on this material. We will depend on sketches and geometric intuition to understand the major ideas behind linear programming.

Section 3.1   WHAT IS LINEAR PROGRAMMING?

• The proprietor of Nuts to You, a nut and appetizer retailer, wants to combine peanuts, sunflower kernels, and dried currants to form a quick-energy snack for hikers. Food is "quick energy" if it contains certain proportions of two slightly different kinds of sugars that the body can rapidly absorb into the bloodstream. Each of the three basic ingredients contains the two sugars in known proportions. In addition to being quick energy, the snack must possess "eye appeal," and the owner knows from long experience the maximum percentage of each ingredient allowable to maintain this important characteristic.

These rules still allow an immense latitude in the range of mixtures. Each of the component foods has a known cost per kilogram. How shall the dealer seek that mixture adhering to requirements for quick energy and for eye appeal *that has the lowest cost?*

• Grazioso Foods has long been famous for its fine Italian sausage. Sausage making is an art, involving the careful preparation of the stuffing and the insertion of this mix into the casing. A new factory employs four people to make two sausage varieties. Each variety requires different proportions of the mixing and stuffing operations. Suppose two of the workers are "mixers" and two are "stuffers." Mixers are paid much more than stuffers, because of the high technical competence that preparation of the sausage meat requires. The plant manager knows to a high degree of precision the relative proportions of mixing and stuffing appropriate for each sausage type. The first type of sausage needs more mixing than stuffing, but this type commands a higher price and generates more profits.

The four workers work 40 hours per week, no overtime. How much of each sausage type should be made, under the assumption that all sausage made will be sold? It is tempting to conclude that the best strategy is to manufacture only the costlier variety, since that type is the more profitable. However, it is doubtful that this is the most profitable business strategy. While the mixers are busy, the stuffers may stand idle, waiting for a new batch of mix, but continuing to draw wages and therefore detracting from the overall profit.

Problems like these in subjects as diverse as manufacturing, weapons deployment, science, finance, and transportation in which a best strategy is called for in response to various constraints can be solved by linear programming. *The best strategy is the one that will maximize profits or other quantities, or will minimize costs, time, or other quantities.*

Historical Note

Linear Programming   *Linear programming is the "new guy on the block," having been developed shortly after World War II by George Dantzig, now Professor Emeritus at Stanford University. During that conflict, military planners desperately needed to know the best way of allocating the depleted resources of a wartime economy. The "linear" part of linear programming refers to the fact that many constraints of the planning process involved linear expressions of the resources. The word "programming" refers* *not to computer programming, which linear programming (just barely) predates, but to a program as a procedure for accomplishing some goal.*

Elements of Any Linear Programming Problem

The above examples are useful for revealing the general structure of any situation for which linear programming is appropriate. In all examples, someone has to make a policy decision on the basis of constraints. Available materials must be *greater than* some given amount, or must be *not more than*, or *at least as much as*, or *less than* something else. This is the language of inequalities, and when the available resources of a problem are represented by variables, these constraints often take the form of linear inequalities. A

constraint   *constraint* is some restriction of a resource.

optimization   What is to be maximized or minimized? We shall frequently use the term *optimization* to refer to either minimization or maximization, depending on the details of the problem.

objective function   Generally, there will be an *objective function* to be optimized, which will be some linear expression representing either profit or costs. The objective function must be expressible in terms of the same variables that appear in the constraints.

Components of a Linear Programming Problem

The following will appear in any linear programming problem:

1. A set of *constraints*, which take the form of linear inequalities.

2. An *objective function* to be optimized. An objective function will take the form of a linear expression.

The techniques of this chapter draw heavily on the graphing of linear inequalities, to which we now turn.

Section 3.2   *LINEAR INEQUALITIES AND THEIR GRAPHS*

Often, knowledge is not precise. Perhaps a budget allows a department to spend *not more than* a certain amount for raw materials; maybe the vitamin content of a cereal mixture must be *greater than* some fixed amount; or suppose the available labor-hours to construct products must be *bounded* by some fixed, union-determined figure. Linear programming is the art of extracting the most precise information possible from sets of constraints such as these.

inequalities   *Inequalities* have the same form as equations except that the equal sign is replaced by any one of five inequality symbols:

$<$	strictly less than
$>$	strictly greater than
$\leq$	less than or equal to
$\geq$	greater than or equal to
$\neq$	not equal to

strict inequalities   Inequalities with $<$ or $>$ or $\neq$ are called *strict inequalities*.

Example 1   Which of the following are strict inequalities?
(a) $3x - 4y = 1$
(b) $0.2x_1 + 0.9x_2 \leq -0.003$
(c) $83z_1 + 44z_2 - 67z_3 < 1{,}024$

Solution

(a) is not an inequality.
(b) is not strict because of the $\leq$ sign.
(c) is strict.

Skill Enhancer 1   Which are strict inequalities?
(a) $x \geq 0$
(b) $-2 \leq y < 1$
(c) $3 > t$

Answer in Appendix E.

Can we *graph* inequalities as we can equations? Yes, but the graph (usually) covers a larger region than the graph of an equation. Instead of lines or curves, graphs of inequalities encompass entire regions.

corresponding equation   The *corresponding equation* to any inequality is that equation formed by replacing the inequality symbol with an equal sign. For example, the corresponding equation to the inequality $9x - 15y > 22.2$ is the equation $9x - 15y = 22.2$.

$$9x - 15y = 22.2 \quad \text{corresponds to} \quad 9x - 15y > 22.2$$

The corresponding equality is useful for graphing inequalities, because this equation forms the *boundary* to the graph of the inequality. To see how to use the corresponding equation, follow these steps to graph

$$y \geq x + 2$$

Because the inequality is not strict, the graph will at least consist of the graph of the corresponding equation, shown in Figure 3.1. (Make sure you understand why this statement is true.)

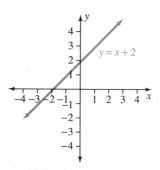

FIGURE 3.1
The graph of $y = x + 2$.

The final graph must surely include more than this line. For example, simple substitution shows that the points $(-2, 2)$, $(1, 4)$, and $(4, 1{,}000)$ are among the many additional points satisfying the inequality, and must therefore be part of the graph. All three of these points are above the line $y = x + 2$. In fact, *all points above the line also belong to the graph of the inequality.* A common way of showing this is by shading the region

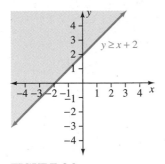

FIGURE 3.2
The graph of $y \geq x + 2$.

half planes

above the corresponding line, as in Figure 3.2. (The *corresponding line* is simply the graph of the corresponding equation.)

Graphing a Single Linear Inequality

Follow these steps when graphing a single linear inequality.

1. Begin by *graphing the corresponding line*, which divides the coordinate plane into two parts, or *half planes*.

2. For a *strict inequality*, the graph consists of exactly one of these two half planes, but *not* the boundary of this half plane.

3. If the inequality is *not strict*, the graph consists of exactly one of these half planes *together with* the graph of the corresponding equation.

---

test point

This list fails to provide a method for identifying the half plane. A simple technique, using a *test point*, identifies the proper half plane.

Select as a test point any point *not* lying on the corresponding line. If possible, the origin $(0, 0)$ is a convenient point to choose because working with two zeros keeps arithmetic calculation to a minimum and lessens the chance of careless error.

Next, *substitute* the coordinates of the test point in the inequality. If this point *satisfies* the inequality, then *every* point in the same half plane as the test point satisfies the inequality, and *that* is the side to shade. If the test point *fails* to satisfy the inequality, then *no* point in the same half plane will satisfy it, and we shade the *opposite* half plane. Refer to Figures 3.3 and 3.4 to see how to use these comments.

Example 2   Graph $x + y < 0$.

Solution

Because the inequality is strict, the points that satisfy the equation $x + y = 0$ do not satisfy the inequality. This line does, though, form the boundary between the region that

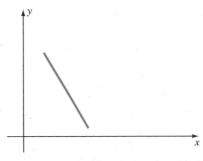
Graph the corresponding equation.

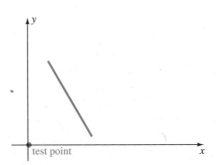
Select a test point (the origin if possible).

FIGURE 3.3
Graphing an inequality.

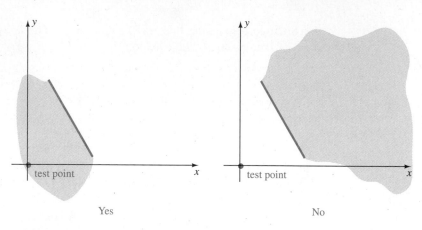

FIGURE 3.4
Does the test point satisfy the original inequality?

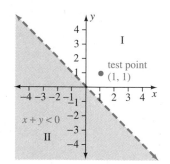

FIGURE 3.5
The graph of $x + y < 0$.

belongs to the graph and the region that does not, so it is helpful to graph it. Because it forms a boundary of the region, but is not itself part of the region, we graph it using a dashed line, as in Figure 3.5.

Choose a test point to decide which half of the plane, region I or II, to shade. We would like to choose the origin, but unfortunately it lies on the boundary line, and so cannot help us.

Choose the point $(1, 1)$ as test point. At this point, both x and y equal 1. When these values are substituted into the inequality, will the inequality be satisfied?

$$x + y < 0$$
$$1 + 1 < 0 \qquad ?$$
$$2 < 0 \qquad \text{Contradiction!}$$

Because 2 is not a negative number, this test point *fails* the test. Consequently, we shade the half plane on the *other* side of the line $x + y = 0$. The final graph is shown in Figure 3.5. The boundary is drawn with a dashed line to indicate that it does not belong to the graph.

Skill Enhancer 2   Graph $x - y > 0$.

Answer in Appendix E.

inequality systems   ## Systems of Inequalities

Example 3   Graphically solve the system

$$x + y \le 0$$
$$y \ge x + 2$$

Solution

First, graph both inequalities on the same set of coordinate axes. A strict version of the first inequality was graphed in Example 2. The second inequality was graphed at the outset of this section. Both graphs are redrawn on the same set of axes in Figure 3.6.

region of overlap   *Solving a system of inequalities* means identifying the region that satisfies *all* inequalities in the system. This area is given by the *region of overlap*, the area where the individual graphs overlap. Any point lying in this region, and only those points, satisfies the system of inequalities. In Figure 3.6, this solution is the wedge-shaped area opening to the left whose corner point lies at $(-1, 1)$. We determine this corner point by solving the corresponding equations simultaneously. In this case, we can solve

$$x + y = 0 \qquad \text{and} \qquad y = x + 2$$

using the substitution method in these steps. We use the expression for y from the second equation in the first equation.

$$x + x + 2 = 0$$
$$2x + 2 = 0$$
$$2x = -2$$
$$x = -1$$

Use this information in the second equation:

$$y = x + 2$$
$$= -1 + 2$$
$$= 1$$

This yields the coordinates of the corner point $(-1, 1)$ and so completes the solution.

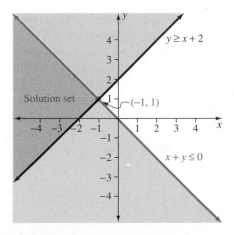

FIGURE 3.6
$x + y \leq 0,\ y \geq x + 2.$

Skill Enhancer 3   Use graphing methods to solve this system.

$$x - y > 0$$
$$y > -x$$

Answer in Appendix E.

feasible region   More formally, this overlap is called the *feasible region*. Only points within this
                          area have coordinates that can feasibly satisfy all the given inequalities. The feasible
allowable region   region may also be called the *allowable region*.

Graphing Linear Inequality Systems in Two Variables

Follow these steps to graphically solve systems of linear inequalities with two
variables (like x and y).

1. Graph each individual inequality on the same set of axes.

2. The region of overlap forms the graph of this system.

Example 4   Graphically determine the solution to

$$x \geq 0$$
$$y \geq 0$$

Solution

To graph $x \geq 0$, consider the equality $x = 0$. Its graph is the vertical y axis. The right
side of this line will contain points whose x values are positive.

    To graph $y \geq 0$, examine the equality $y = 0$. The horizontal x axis is its graphical
representation. Only those points lying above this axis will have positive y values.

    These two graphs are shown on the single set of axes in Figure 3.7. The region of
overlap is the top right quadrant of the plane. Since neither inequality is strict, the upper
part of the y axis, and the right half of the x axis are both included in the solution.

Skill Enhancer 4   Use graphing methods to solve this system of inequalities: $x \leq 0$
and $y \leq 0$.

Answer in Appendix E.

    The next example shows how important the direction of the inequality signs is in a
problem.

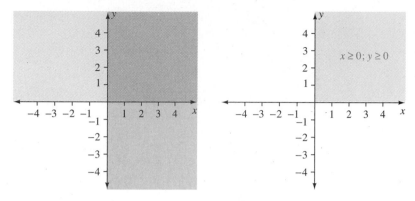

FIGURE 3.7
$x \geq 0$, $y \geq 0$.

Example 5   Solve graphically the following two pairs of inequalities:

(3.1) $\qquad\qquad\qquad\qquad x + 2y \leq 12$

(3.2) $\qquad\qquad\qquad\qquad x + 2y \geq 8$

(3.3) $\qquad\qquad\qquad\qquad x + 2y \geq 12$

(3.4) $\qquad\qquad\qquad\qquad x + 2y \leq 8$

Solution

The two pairs of inequalities differ only in the directions of the inequality signs. Respective inequalities in each pair will have the same corresponding equations.

The equation corresponding to Inequality (3.1) is

$$x + 2y = 12$$

whose intercepts are $(0, 6)$ and $(12, 0)$.

The points and the resulting line are plotted in Figure 3.8*a*. Which side of this line belongs to the inequality? Choose test point $(0, 0)$. At that point,

$$x + 2y = 0 + 2(0) \leq 12$$

so the "origin" side of the line should be shaded.

The intercepts for $x + 2y = 8$ are $(0, 4)$ and $(8, 0)$. Figure 3.8*a* shows this line. Both corresponding lines are parallel. (How can you demonstrate this mathematically?) For this line, test point $(0, 0)$ into Equation (3.2) yields

$$x + 2y = 0 + 2(0) = 0$$

which is definitely not greater than (or even equal to) 8. We shade *away* from the origin.

The region of overlap is the strip between the two lines, as in Figure 3.8*a*.

Now let's examine the second pair of inequalities. The same corresponding lines apply to this set. We see them again in Figure 3.8*b*. We have to retest each inequality, using the same test point, because of the change in inequality sign.

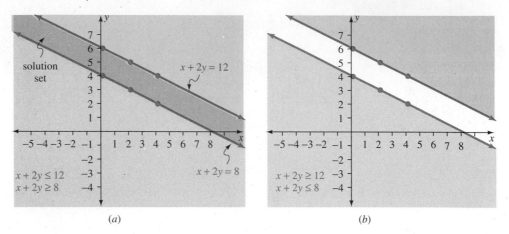

FIGURE 3.8
Two inequality systems.

Upon substitution of $(0, 0)$ into Inequality (3.3), we have

$$0 + 2(0) \text{ not} \geq 12$$

so we shade *away* from the origin. As for Equation (3.4),

$$0 + 2(0) < 8$$

so we would shade *toward* the origin. This shading is shown on Figure 3.8*b*. In this case, there is no overlap, so we conclude that no values for x and y simultaneously satisfy both inequalities; the solution set for this system of inequalities is *empty*.

Skill Enhancer 5   Use graphic methods to solve this system.

$$y - x \leq 3$$
$$y - x \geq 1$$

Answers in Appendix E.

Example 6   Solve this system graphically:

(3.5) $$y \leq 4$$
(3.6) $$x \geq 0$$
(3.7) $$x + y \geq 2$$
(3.8) $$2y \geq 3x - 1$$

Solution

We will graph the four corresponding equalities first. That corresponding to Inequality (3.5) is a horizontal line, intersecting the y axis at $(0, 4)$. Inequality (3.6) is even more straightforward. The equation $x = 0$ is a vertical line that coincides with the y axis.

The corresponding equation for Inequality (3.7) is

$$x + y = 2$$

whose intercepts are $(2, 0)$ and $(0, 2)$. We plot these points and connect them to show this line. Similarly, Inequality (3.8) yields the corresponding equation

$$2y = 3x - 1$$

which we can graph from intercepts $(0, -\frac{1}{2})$ and $(\frac{1}{3}, 0)$.

All four lines are shown in Figure 3.9*a*.

A test point of $(0, 0)$ will work for all inequalities except Inequality (3.6). (Why?) However, we need not be so formal for the first two inequalities [(3.5) and (3.6)]. It should be clear that the region below $y = 4$ and that to the right of $x = 0$ are the allowable regions for those inequalities.

For Inequality (3.7), substitute $(0, 0)$ into the inequality to obtain

$$0 + 0 \text{ not} \geq 2 \quad \text{or} \quad 0 \not\geq 2$$

The region above the line should be shaded. As far as Inequality (3.8) is concerned, we use the same test point:

$$2(0) > 0 - 1$$

is a true statement, so we will include the region containing the origin, the half of the plane lying above the line.

Four shaded regions on one plane may appear confusing, so it is possible to use an alternative convention and represent all the shading for a single inequality by a single arrow pointing to the allowable side. This alternative is followed in Figure 3.9*b*. The final solution is the interior of the quadrilateral with corners A, B, C, and D. Since none of the inequalities were strict, the sides of the quadrilateral themselves are included in the allowable region.

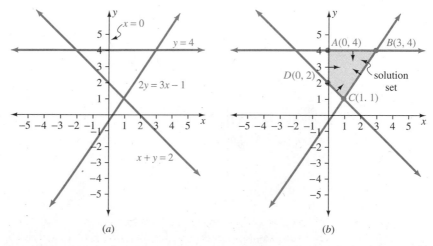

(a)   (b)

FIGURE 3.9

A system with four inequalities, $y \leq 4$, $x \geq 0$, $x + y \geq 2$, and $2y \geq 3x - 1$.

Skill Enhancer 6   Use graphical methods to solve this system.

$$x \geq 0$$
$$y \geq 0$$
$$y \leq 3$$
$$x + y \leq 4$$

Answer in Appendix E.

vertex   Problems in linear programming depend heavily on solving systems of inequalities. Important information lies in the coordinates of each corner, or *vertex*, of the allowable region. How can we determine the coordinates of vertices A, B, C, and D in the previous example? Any corner is the intersection of two of the corresponding lines of the system.

The Corners of the Feasible Region

To find a corner of the feasible region, solve simultaneously the corresponding equations for the two sides whose intersection forms the corner.

Often, a corner lies on a coordinate axis. Such corners may frequently be determined by inspection.

Corners A, B, and D in Figure 3.9 are found easily enough. A is the intersection of $x = 0$ with $y = 4$; this is the point $(0, 4)$. Point B is the intersection of $y = 4$ with $2y = 3x - 1$. Letting $y = 4$ in the second equation implies that $x = 3$, so the coordinates of B are $(3, 4)$. The corner at D is the y intercept of $x + y = 2$, or (equivalently) $y = -x + 2$. The y intercept has coordinates $(0, 2)$. (Why?)

The remaining corner C is the solution to

$$x + y = 2$$
$$2y = 3x - 1$$

You should use the substitution method to show that the solution is the point $(1, 1)$.

At Least 10 Sets
Per Month

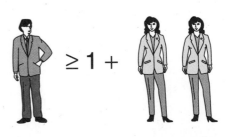

FIGURE 3.10

Example 7   Sam and Tamara sell sets of encyclopedias door to door. Their joint sales must total 10 sets per month. Because Sam has more experience, he is expected to sell at least 1 more than twice the number of sets that Tamara sells. These conditions impose constraints on their sales. Solve the constraints graphically. What is the least number of sets that Sam must sell in any month?

Solution

Represent the sets sold by Sam and Tamara by x_1 and x_2, respectively. Their total sales are $x_1 + x_2$, and since there must be at least 10 sets sold per month, we have

$$(3.9) \qquad\qquad x_1 + x_2 \geq 10$$

"One more than twice Tamara's sales" is equivalent to $1 + 2x_2$, and since Sam's sales must be at least this much, we have $x_1 \geq 1 + 2x_2$, or

$$(3.10) \qquad\qquad x_1 - 2x_2 \geq 1$$

Also, since it is impossible to sell fewer than zero sets, we demand that

$$(3.11) \qquad\qquad x_1 \geq 0 \qquad x_2 \geq 0$$

The four inequalities (3.9) to (3.11) for this situation are graphed in Figure 3.11. (Make sure you understand how we obtain this graph.)

The least amount of sales that Sam can ever get away with in any month is the x_1 coordinate of the "tip" of the wedge in Figure 3.11. The coordinates of this point are the simultaneous solution to the pair of equations

$$(3.12) \qquad\qquad x_1 + x_2 = 10$$
$$(3.13) \qquad\qquad x_1 - 2x_2 = 1$$

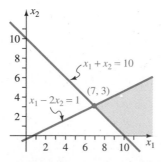

FIGURE 3.11

Selling encyclopedias: graphing the system $x_1 \geq 0$, $x_2 \geq 0$, $x_1 + x_2 \geq 10$, and $x_1 - 2x_2 \geq 1$.

(Why?) From Equation (3.13), we know that $x_1 = 1 + 2x_2$, and we can substitute this into Equation (3.12).

$$x_1 + x_2 = 10$$
$$(1 + 2x_2) + x_2 = 10$$
$$1 + 3x_2 = 10$$
$$3x_2 = 9$$
$$x_2 = 3$$

If $x_2 = 3$, then $x_1 = 7$ (why?). Therefore, Sam must always sell at least 7 sets of encyclopedias in any month.

Skill Enhancer 7   Suppose now that Sam must sell four sets more than Tamara sells. Does this change Sam's selling strategy? If so, how?

Answer in Appendix E.

Exercises 3.2

Graph the inequalities in Exercises 1–22.

1. $x > 3$

2. $x + y - 2 < 0$

3. $y \geq 2x - y + 3$

4. $9x - y \leq 18$

5. $y < -1.5$

6. $2x + 4y \leq 12$

7. $9y - x \leq 18$

8. $0.5x - 0.75y < 3$

9. $2x + 3y \geq 12$

10. $2x + 3y < 12$

11. $x - 4y \geq 8$

12. $x - 4y \leq 8$

13. $x + 2y \leq 3$

14. $x - 4y \leq 8$

15. $3x + 4y < 12$

16. $2x + 3y > 6$

17. $4x > -8$

18. $y - 3 \leq 0$

19. $y > 7$

20. $x - 2 < 0$

21. $x + y \leq \frac{1}{3}$

22. $\frac{1}{3}x + \frac{1}{2}y \leq \frac{3}{2}$

Solve graphically the systems of inequalities in Exercises 23–46. (Hint: Use the results of the above exercises wherever possible.) In all cases, identify the coordinates of the corners of the feasible region.

23.
$$x > 3$$
$$x + y - 2 < 0$$

24.
$$y > 2x - y + 3$$
$$9x - y < 18$$

25.
$$2x + 3y \leq 12$$
$$x - 4y \geq 8$$

26.
$$x \geq 0; y \geq 0$$
$$x + y \geq 2$$
$$x + y \leq 5$$

27.
$$x \geq 0; y \geq 0$$
$$x + y \geq 3$$
$$5y + 3x \leq 15$$
$$x + y \geq 1$$

28.
$$x \geq 0; y \geq 0$$
$$2x + 3y \geq 6$$
$$x + y \leq 6$$
$$3x + y \leq 12$$

29.
$$x \geq 0; y \geq 0$$
$$y - x \leq 0$$
$$2y - x \geq 0$$
$$x + y \leq 2$$

30.
$$x \geq 0; y \geq 0$$
$$y - x \leq 1$$
$$2y - x + 1 \geq 0$$
$$x \leq 4$$
$$x + y \leq 8$$

31.
$$9y - x \leq 18$$
$$9y - x \geq 10$$

32.
$$2x + 4y \leq 12$$
$$x + 2y \geq 4$$

33.
$$2x + 4y \geq 12$$
$$x + 2y \leq 4$$

34.
$$9y - x > 18$$
$$9y - x < 10$$

35.
$$2x + 3y \leq 12$$
$$2x + 3y \geq 13$$

36.
$$x + y < 2$$
$$x + y > -1$$

37.
$$x + y > 2$$
$$x + y < -1$$

38.
$$2x + 3y > 12$$
$$2x + 3y < 13$$

39.
$$x \geq 0; y \geq 0$$
$$x - 4y \leq 8$$
$$x - 4y \geq 4$$

40.
$$x \geq 0; y \geq 0$$
$$x + 2y \leq 3$$
$$x + 2y \geq 1$$

41.
$$x \geq 0; y \geq 0$$
$$3x + 4y \leq 12$$
$$3x + 4y \geq 6$$

42.
$$x \geq 0; y \geq 0$$
$$2x + 3y \geq 5.9$$
$$2x + 3y \leq 6.0$$

43. $x \geq 0;\ y \geq 0$
   $x \leq 2$
   $y \leq 3$

44. $x \geq 0;\ y \geq 0$
   $x \leq 3$
   $y \leq 2$

45. $x \geq 0;\ y \geq 0$
   $x + y \leq \frac{1}{3}$

46. $x \geq 0;\ y \geq 0$
   $\frac{1}{3}x + \frac{1}{2}y \leq \frac{3}{2}$

Applications

47. (Retailing) Computers 'R Us is a chain of computer stores in the southwest. The parent company expects each store to sell at least 15 computer systems per week. There are two types of systems. The Gold system is more profitable than the Silver system, and so the number of Gold systems each store is to have sold is at least 3 more than 3 times the number of Silver systems sold. Solve these constraints graphically. What is the least number of Gold systems that a store can sell any month?

48. (Diet and Nutrition) A certain diet requires a person to eat at least 100 grams (g) of two special types of food. For other nutritional reasons, a dieter must eat at least 10 g more of the second than of the first. Solve these constraints graphically. Since the second food is not very tasty, what is the minimum amount that must be consumed each day?

W 49. Monica and Veronica are discussing the best way to divide a group of beads, each of which is one of two colors. In symbols, Monica says that $x + y \geq 10$ and $y \leq 6$, where x and y represent the number of beads of each color that she will get. Translate Monica's demands back into English.

W 50. Barry is clearing out the basement, throwing out lots of stuff but saving various magazines and comic books. In symbols, Barry realizes that the material he has saved satisfies $x \geq 0$, $y \geq 0$, $x + y \leq 75$, and $y \leq 50$. Here, x and y are the numbers of books and comics he has retained. Translate this system of inequalities back into English.

Calculator Exercises

Graph the systems of inequalities in Exercises 51–52. Give the coordinates of all corners of the feasible region.

51. $x \geq 0;\ y \geq 0$
   $1.1x + 3y \geq 4.6$

52. $x \geq 0;\ y \geq 0$
   $x + y \leq 2$
   $1.1x + 0.8y \leq 2$

Section 3.3 GEOMETRIC SOLUTIONS TO LINEAR PROGRAMMING PROBLEMS

Follow and study Example 8 carefully; it displays the steps and motivation for a general solution scheme to solve linear programming problems. We will use this example to abstract a streamlined set of procedures that make the succeeding examples straightforward.

Example 8 Find values of x_1 and x_2 that make P, where

$$P = 2x_1 + x_2$$

as large as possible, given that x_1 and x_2 are constrained by the inequalities

(3.14) $x_1 + x_2 \leq 2$

(3.15) $x_1 \leq 1$

(3.16) $x_2 + 3x_1 \geq 2$

Solution

Determining the solution to the set of constraints—to the system of inequalities—is our first task. We discussed this technique in the previous section. The feasible region for the system of Inequalities (3.14) through (3.16) is displayed in Figure 3.12. Corners are found by solving the system formed by corresponding equations. In this case, we solve these pairs of equation systems:

$$\left.\begin{array}{r} x_1 + x_2 = 2 \\ x_1 = 1 \end{array}\right\} \text{ to find } A$$

$$\left.\begin{array}{r} x_1 = 1 \\ x_2 + 3x_1 = 2 \end{array}\right\} \text{ to find } B$$

$$\left.\begin{array}{r} x_2 + 3x_1 = 2 \\ x_1 + x_2 = 2 \end{array}\right\} \text{ to find } C$$

to determine that the corners of the feasible region are $A(1, 1)$, $B(1, -1)$, and $C(0, 2)$. (Make sure you agree with this statement.)

The expression for P takes a familiar and suggestive form if we rewrite in the *slope-intercept form* of a straight line, with P playing the role of x_2 intercept:

$$x_2 = -2x_1 + P$$

Imagine that P is a parameter that can assume any value. Then, this single equation is a shorthand representation for an entire *family* of straight lines, all with slope -2, where *each different value of P gives rise to a straight line with that value of P as its x_2 intercept.* That is, each line in this family intersects the x_2 axis at a different value of P. Figure 3.13 shows the feasible region, on top of which are drawn several members from this infinite family of straight lines.

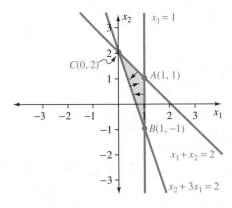

FIGURE 3.12
The inequality system $x_1 + x_2 \leq 2$, $x_1 \leq 1$, $x_2 + 3x_1 \geq 2$.

Under what circumstances will P be maximized? The question of finding values of x_1 and x_2 yielding the largest value of P is equivalent to identifying that straight line passing through the feasible region at at least *one* point with the greatest x_2 intercept. (This argument needs slight revision if the coefficient of x_2 in the given *objective function* is some number other than 1. Nevertheless, it will be true that the optimum will be proportional to the x_2 intercept. Revision is also needed if we need the smallest value of P instead of the largest. With these slight generalizations, this statement always applies to linear programming problems in two variables.)

Which x_1 and x_2 values yield the largest value of P?	$\Longleftrightarrow$	Which line crossing the feasible region has the greatest x_2 intercept?

In Figure 3.13, the line corresponding to $P = 3$ meets this criterion. Parallel lines above it give rise to a larger value of P, to be sure, but they do not pass through the feasible region. This line barely passes through the feasible region, as it only touches the region at the single corner point $(1, 1)$, but a single point is enough to make a line eligible for consideration.

For this problem, then, values of x_1 and x_2 given in coordinate form by $(1, 1)$ yield the largest value of $P = 3$.

Skill Enhancer 8   Using the same constraints, find the values of x_1 and x_2 that make $Q = 4x_1 + x_2$ as large as possible. What is the largest such value of Q?

Answer in Appendix E.

The reasoning we used in this example will work for almost *any* linear programming problem with two variables. (Certain exceptional situations are covered in the final section of this chapter.) Similar reasoning will also allow us to determine the *smallest* value of P possible permitted by the same constraints. To see how, refer back to

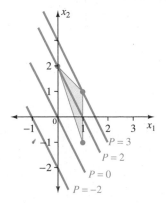

FIGURE 3.13

Several members of the family of straight lines $x_2 = -2x_1 + P$ superimposed on the feasible region.

Figure 3.13. The smallest value of P corresponds to the line crossing the feasible region at at least one point with the smallest x_2 intercept. The values of x_1 and x_2 yielding this value will typically correspond to the coordinates of one of the corners, although a different corner. (Again, certain exceptional cases are covered in the final section of this chapter.)

Restrictions on the Number of Variables

We solved this problem primarily because of the correspondence between linear expressions containing two distinct variables and graphs of these expressions on a set of coordinate axes. If there are more than two variables, we cannot graph the constraints on a two-dimensional set of axes. (There *are* algebraic techniques for solving such problems, and we shall examine a frequently used technique in the following chapter.)

With the appropriate tools, this method can be used for problems involving *three* variables. Graphs of linear equations in three variables correspond to *planes* in a coordinate space, and can be diagramed as in Figure 3.14. Planes also form the boundaries of linear inequalities in three dimensions. It is difficult, but not impossible, to graph the feasible region and determine its *vertices* (the plural of *vertex*, the three-dimensional analog of a two-dimensional corner).

Restrictions on the Feasible Region

simply connected
convex

The argument we used in the first example also relied on several further assumptions about the feasible region. First of all, it must always be *simply connected*, which means there are no holes inside the region. Second, it must also be *convex*—we allow no dips in the boundary of the region. Figure 3.15 shows legal and illegal feasible regions.

The essence of the approach we used to solve the problem of the first example is summarized by the following theorem, which we state without further proof.

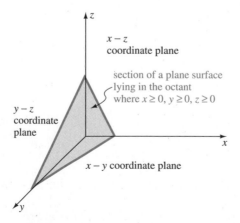

FIGURE 3.14
A three-dimensional graph.

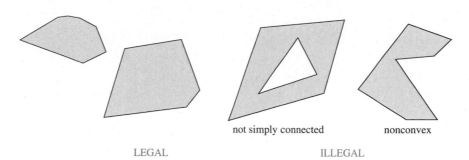

LEGAL   ILLEGAL

not simply connected   nonconvex

FIGURE 3.15
Feasible regions, good and bad.

Theorem of Linear
Programming

optimize

> **Fundamental Theorem of Linear Programming**
>
> In a linear programming problem, where the object is to (minimize or maximize) a linear objective function P subject to the constraints of a system of linear inequalities, *optimize* these inequalities will define a feasible region. If this region is simply connected and convex, and if P has the desired optimum, then that value is attained when the coordinates of *one* of the corners of the feasible region are substituted into the expression for P.

Read the theorem carefully as much for what it *does not* say as for what it does say:

1. *Must a solution exist?* The theorem does not guarantee a solution. (We will later examine situations in which linear programming problems have no solution.)

2. *Must the solution be unique?* By the very nature of an optimal solution, there will only be one value of the objective function P that is maximum or minimum. In certain situations, it is possible that more than one set of coordinates in the feasible region yield this optimum value, but *at least one such pair of coordinates* will correspond to a corner of the feasible region.

3. *How may we find the solution?* The theorem stands mute on the explicit procedure for determining the solution, assuming that it exists. Nevertheless, we *can* extract a workable strategy from the theorem that works particularly well when there are only two variables in the problem.

Finding Solutions to Linear Programming Problems

The solution, if it exists, must lie at a corner of the feasible region. That fact suggests that we construct a small table or chart listing all the corner points and the associated values of P. After that, it is a simple matter to examine the chart for the value of P that is largest or smallest.

We draw on the results of our experience with the first example and the recent theorem to summarize the steps we take to solve a linear programming problem geometrically. Figure 3.16 shows these steps in action.

> **Solving Linear Programming Problems Geometrically**
>
> 1. Graph the linear inequalities to determine the *feasible region*.
> 2. Determine the coordinates of each of the *corners of the region*. Do this by solving pairs of equations corresponding to adjacent sides of the region.
> 3. Prepare a two-column chart. In the first column, place each corner coordinate. In the second, compute the corresponding value of P, the objective function. Examine this column and select the optimum value.

Example 9   Minimize $p = 2x + 3y$ subject to

$$(3.17) \qquad\qquad x + y \geq 1$$

$$(3.18) \qquad\qquad x + y \leq 3$$

$$(3.19) \qquad\qquad x \geq \frac{1}{2}$$

$$(3.20) \qquad\qquad y \geq 0$$

Solution

We carefully follow the above sequence of steps.

1. We graph the inequalities in Figure 3.17.

2. We compute the coordinates of the corners. The point $(1, 0)$ is the x intercept of Inequality (3.17). (Actually, it is the x intercept of the line corresponding to that inequality.) Point $(3, 0)$ is the x intercept of Inequality (3.18). The remaining corners, $(\frac{1}{2}, 2\frac{1}{2})$ and $(\frac{1}{2}, \frac{1}{2})$, are the simultaneous solution of Inequality (3.19) with (3.18) and (3.17).

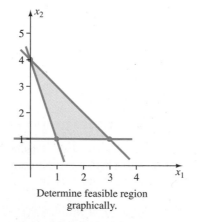

Determine feasible region
graphically.

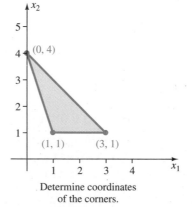

Determine coordinates
of the corners.

corners	objective function
$(0, 4)$	• • •
$(3, 1)$	• • •
$(1, 1)$	• • •

Prepare a chart.

FIGURE 3.16
Solving linear programming problems geometrically.

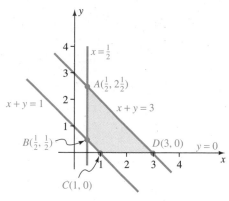

FIGURE 3.17
The feasible region for Example 9.

3. Finally, prepare a small chart for the values of p at each corner. For example, the corner $(\frac{1}{2}, 2\frac{1}{2})$ means $x = \frac{1}{2}$ and $y = 2\frac{1}{2}$, so at that point,

$$p = 2x + 3y = 2 \times \frac{1}{2} + 3 \times 2\frac{1}{2} = 8\frac{1}{2}$$

Here is the completed chart:

Corner	$p = 2x + 3y$
$(\frac{1}{2}, 2\frac{1}{2})$	$8\frac{1}{2}$
$(\frac{1}{2}, \frac{1}{2})$	$2\frac{1}{2}$
$(1, 0)$	2
$(3, 0)$	6

The minimum for p is 2, obtained when $x = 1$ and $y = 0$.

Skill Enhancer 9   Minimize $p = x + 4y$ subject to $x \geq 0$, $y \geq 0$, $x + y \geq 1$, and $3x + y \leq 3$.

Answer in Appendix E.

Could we have used this analysis to *maximize p*? Yes. Just look, in the above chart, for the largest value of p, which is $8\frac{1}{2}$ when $x = \frac{1}{2}$ and $y = 2\frac{1}{2}$.

Keep in mind that the solution to a linear programming problem contains the answer to two questions:

1. *What is the optimum value of P subject to the constraints?*

2. *How do we obtain this value of P?* That is, at what values of the variables is this optimum value achieved?

Example 10   Maximize $P = 2x + y$ subject to

(3.21)	$2y + x \leq 6$
(3.22)	$x + y \leq 4$
(3.23)	$y - x + 2 \geq 0$

and to

(3.24)	$x \geq 0$
(3.25)	$y \geq 0$

Solution

The feasible region defined by these five inequalities is sketched in Figure 3.18. The corners and their coordinates are labeled A through E.

This part of the procedure, the determination of the feasible region and the coordinates of the corners, constitutes the bulk of the work in any linear programming problem. The remaining work, creating and examining the chart containing the corners with their associated P values, is the only occasion on which the objective function P enters the problem.

Here is the table of corners with the associated values of the objective function.

Corner	$P = 2x + y$
$A(0, 3)$	3
$B(2, 2)$	6
$C(3, 1)$	7 $\Leftrightarrow$ MAX
$D(2, 0)$	4
$E(0, 0)$	0

Do not forget that there are *two* parts to the solution of this problem. The maximum value of P is 7. P reaches that value when $x = 3$ and $y = 1$.

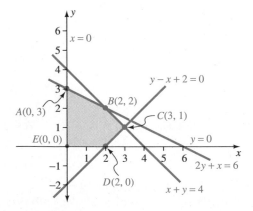

FIGURE 3.18
The feasible region for Example 10.

Skill Enhancer 10 Subject to $x \geq 0$, $y \geq 0$, $x + y \geq 1$, $y - x \leq 1$, and $5x + 2y \leq 10$, what is the largest value of $p = 3x + 4y$? How do we get that maximum value?

Answer in Appendix E.

_____ *Exercises 3.3* _____

In Exercises 1–12, find the maximum of the expression $M = 9x + 2y$ subject to the allowable region whose corner coordinates are given in the following exercises. (Assume that all sides of these regions are straight.) Graph the region.

1. $(0, 0)$, $(2, 0)$, $(1, 1)$ **2.** $(0, 0)$, $(2, 1)$, $(1, 2)$

3. $(0, 0)$, $(\frac{1}{2}, 0)$, $(1, \frac{1}{4})$, $(2, \frac{2}{3})$, $(1, 1)$, $(0, \frac{1}{2})$

4. $(1, 1)$, $(\frac{3}{2}, 1)$, $(2, \frac{5}{4})$, $(3, \frac{3}{2})$, $(2, 2)$, $(1, \frac{3}{2})$

5. $(0, 0)$, $(0, 4.5)$, $(1, 0)$ **6.** $(0, 0)$, $(1, -4.5)$, $(-1, -1)$

7. $(2, 0)$, $(3, 0)$, $(3, 1)$, $(2, 4)$, $(0, 4)$, $(0, 2)$

8. $(0, 0)$, $(1, 0)$, $(1\frac{1}{4}, \frac{1}{4})$, $(1\frac{1}{2}, \frac{3}{8})$, $(1\frac{3}{4}, 2\frac{1}{2})$

9. $(0, 1)$, $(1, 2)$, $(2, 4)$, $(3, 8)$

10. $(1, 1)$, $(2, 1)$, $(2, 2)$, $(1, 2)$

11. $(1, 0)$, $(0, 1)$, $(1, 2)$, $(2, 1)$

12. $(2, 0)$, $(0, 2)$, $(2, 4)$, $(4, 2)$

In Exercises 13–24, find the minimum value of the expression $m = \frac{1}{2}x + 5y$ subject to the feasible region whose corner coordinates are given in the following exercises. Graph the region.

13. $(1, 0)$, $(1, 2)$, $(2, 1)$, $(1, \frac{1}{2})$

14. $(\frac{1}{2}, -\frac{1}{2})$, $(\frac{1}{2}, \frac{3}{2})$, $(\frac{3}{2}, \frac{1}{2})$, $(\frac{1}{2}, 0)$

15. $(0, 2)$, $(1, 3)$, $(2, 3\frac{1}{2})$, $(3, \frac{7}{3})$, $(2, \frac{9}{8})$, $(1, 1)$

16. $(1, 2\frac{1}{2})$, $(2, 3\frac{1}{2})$, $(3, 4)$, $(4, 2\frac{5}{6})$, $(3, 1\frac{5}{8})$, $(2, \frac{3}{2})$

17. $(0, 0)$, $(10, -1)$, $(10, 0)$

18. $(0, \frac{1}{10})$, $(1, 0)$, $(\frac{3}{4}, \frac{7}{2})$

19. $(2, 0)$, $(3, 0)$, $(3, 1)$, $(2, 4)$, $(0, 4)$, $(0, 2)$

20. $(1, 0)$, $(1\frac{1}{4}, \frac{1}{4})$, $(1\frac{1}{2}, \frac{3}{8})$, $(1\frac{3}{4}, 2\frac{1}{2})$

21. $(0, 1)$, $(1, 2)$, $(2, 4)$, $(3, 8)$

22. $(1, 1)$, $(2, 1)$, $(2, 2)$, $(1, 2)$

23. $(1, 0)$, $(0, 1)$, $(1, 2)$, $(2, 1)$

24. $(2, 0)$, $(0, 2)$, $(2, 4)$, $(4, 2)$

In Exercises 25–42, find the indicated optimum of the given objective function subject to the set of constraints.

25. Maximize $M = x_1 + 3x_2$ subject to $x_1 \geq 0$, $x_2 \geq 0$, and $x_1 + x_2 \leq 4$.

26. Minimize $m = x_1 + x_2$ if it is true that $x_1 \geq 0$, $x_2 \geq 0$, $x_1 + 2x_2 \geq 2$, and $x_1 + 2x_2 \leq 10$.

27. Minimize $m = 2x_1 + 3x_2$ if it is true that $x_1 \geq 0$, $1 \leq x_2 \leq 4$, and $x_2 \geq x_1$.

28. Maximize $M = 2x_1 + 3x_2$ subject to $x_1 \geq 0$, $x_2 \geq 0$, $x_2 \leq 6$, and $x_2 \geq x_1$.

29. Maximize $M = -3x_1 + x_2$ subject to $1 \leq x_1 \leq 3$ and $2 \leq x_2 \leq 6$.

30. Minimize $m = 4x_1 - x_2$ subject to $-1 \leq x_1 \leq 3$ and $-1 \leq x_2 \leq 1$.

31. Minimize $m = -x + 4y$ subject to $1 \leq x + y \leq 3$ and $-1 \leq y - x \leq 1$.

32. Maximize $M = 3x - 2y$ subject to $1 \leq x + y \leq 3$ and $-1 \leq y - x \leq 1$.

33. Maximize $M = -x_1 + 2x_2$ subject to $x_1 + x_2 \geq 1$, $x_1 + x_2 \leq 3$, $x_2 - x_1 - 1 \leq 0$, and $x_2 - x_1 + 1 \geq 0$.

34. Minimize $m = 3x_1 - x_2$ subject to $x_1 \geq 0$, $x_2 \geq 0$, $x_1 + 2x_2 \leq 4$, and $2x_1 + x_2 \leq 4$.

35. Minimize $m = \frac{1}{2}x - y$ subject to $x \geq 0$, $y \geq 0$, $y \leq 2$, $x + y \leq 3$, and $x \leq 2$.

36. Maximize $M = 7x - 2y$ subject to $x + 3y \leq 2$ and $x + 3y \geq 8$.

37. Maximize $M = 3x_1 + 3x_2$ subject to $x_1 \geq 0$, $x_2 \geq 0$, $x_1 + \frac{1}{2}x_2 \geq 1$, and $x_1 + \frac{1}{2}x_2 \leq 4$.

38. Minimize $m = 2x_1 + 0.1x_2$ subject to $x_1 \geq 0$, $x_2 \geq 0$, $4x_1 + 3x_2 \leq 8$, and $4x_1 + 3x_2 \geq 1$.

39. Minimize $m = x + y$ subject to $x \geq 0$ and $y \geq 0$.

40. Maximize $M = 3x + \frac{3}{4}y$ subject to $x \geq 0$, $y \geq 0$, $y - 2x \leq 2$, $y + x \leq 4$, and $x \leq 4$.

41. Maximize $M = x_1 + 2x_2$ subject to $x_1 \geq 0$, $x_2 \geq 0$, $x_2 \leq 3$, $x_1 + 3x_2 \leq 15$, $x_1 + 3x_2 \geq 6$, and $x_1 \leq 6$.

42. Minimize $m = 2x_1 + 5x_2$ subject to $x_1 \geq 0$, $x_2 \geq 0$, $2x_2 + x_1 \leq 10$, $x_2 + x_1 \leq 5\frac{1}{2}$, $2x_1 + x_2 \leq 9\frac{1}{2}$, and $x_1 \leq 4\frac{1}{2}$.

43. *(A bug's-eye view of feasible regions)* A less formal approach to simply connected, convex regions involves imagining a tiny bug driving along the border of a feasible region. If the bug can traverse the entire boundary of the region *without* driving off the boundary, then the region is simply connected. If the bug is driving in a *clockwise direction* along the boundary and never turns the steering wheel to the left, then the region is convex; see Figure 3.19 below. Use this definition to tell which of the following regions are allowable in linear programming problems.
   (a) []   (b) []   (c) []   (d) []

W 44. In Exercise 43, a convex set was defined in terms of an insect driving along the boundary in a clockwise manner. How would the definition read if the bug was driving along the boundary in a counterclockwise manner?

W 45. At the beginning of the chapter, we showed how to maximize a quantity P by treating it as if it stood for a family of straight lines, each member of the family corresponding to a different value of p. Use the same reasoning to show that the minimum value of

$$p = 9x_1 + y_2$$

must also occur at one of the corners of the feasible region defined by Inequalities (3.14) through (3.16).

46. Earlier in this chapter, we showed that $P = 2x + y$, because it was equivalent to $y = -2x + P$, achieved its optimum at a corner of the feasible region. The text asserted that this would hold even when the coefficient of y was different from 1, and for an arbitrary coefficient of x. Using the method of the text, show that this more general assertion must be true.

Calculator Exercises

A feasible region has coordinates $(1.3, 2.1)$, $(1.5, 2.9)$, $(0.9, 4.1)$, and $(0, 1.2)$. Use this information in Exercises 47–48.

47. What point maximizes $M = 3.5x - 4.4y$, and what is the maximum value of M?

48. What point minimizes $m = 0.5x + 2.5y$? What is the minimum value of m?

A feasible region has coordinates $(-0.7, 0.1)$, $(-0.5, 0.9)$, $(-1.1, 2.1)$, and $(-2, -0.8)$. Use this information in Exercises 49–50.

49. Given $m = -3.4x + 6.2y$, what is the minimum value of this function within this feasible region? What are the coordinates of the point at which the minimum occurs?

50. Given $M = 0.001x + 0.001y$, what is the maximum value of M within the feasible region? Where does this maximum occur?

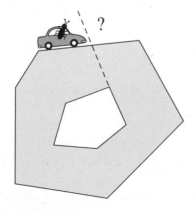

FIGURE 3.19
A bug's-eye view: non-simply connected regions, convex regions.

Section 3.4   APPLICATIONS

When solving an application problem, we have to perform an additional preliminary step—that of translating the conditions of the problem into mathematical form. We examine a production problem first.

Example 11   Extended example: Garment Manufacture. An elderly couple supplement their retirement income by manufacturing shirts and pants in their home. The husband is the cutter, and the wife sews the cut pieces together. A typical shirt requires 1 hour (h) of cutting and 2 h of sewing, and a pair of pants requires 2 h of cutting and 1 h of sewing. Because of their age, both partners in this enterprise rigidly restrict themselves to *not more than* 21 h of work per week. The profit on shirts is $2 per garment, whereas pants generate $3 profit per garment. How many pants and shirts should the couple assemble in a week to maximize their profits?

Solution

It is worth spending extra time on this first problem, since this is the first time we are applying linear programming techniques.

Look for Key Words

First of all, what is there about the problem that suggests the use of linear programming? There are two clues:

- The presence of constraints
- The need to find a best strategy

In this problem, the italicized phrase *not more than* is the flag that one or more conditions constrain the solution. The final question suggests that a strategy is called for—while many combinations of shirts and pants are feasible, what is the specific combination that makes the profit as large as possible?

Having decided that linear programming is appropriate for this problem, what next? The best procedure is to translate the details of the problem into a mathematical format, after which the solution should be straightforward. The elements of any linear programming problem are the presence of an objective function and constraints, all of which are expressed in terms of the same variables. The first step, therefore, should be to decide on the unknowns of the problem.

Identify the Unknowns

The problem itself will suggest the quantities that we will represent by unknowns. In this case, the problem asks for an optimal number of shirts and pairs of pants; these quantities are the unknowns. Let

$$x = \text{the number of shirts per week}$$
$$y = \text{the number of pants per week}$$

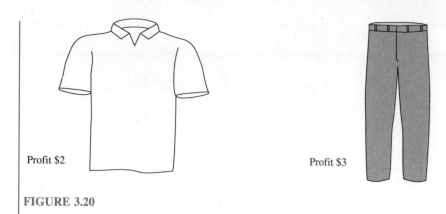

Profit $2

Profit $3

FIGURE 3.20

Write the Objective Function in Terms of the Unknowns

Next, how may we *represent the objective function* in terms of these variables? Look for a quantity that should be maximized or minimized. In this case, the couple will sensibly try to maximize the profit, and so the profit P will be the objective function. Part of the profit arises from the production of shirts, and the remainder from pants. For each shirt, the couple makes $2; if they make x shirts, the shirt profit will be $2x$. In the same way, the pants profit is $3y$, so the total profit, the sum of these, is

(3.26)
$$P = 2x + 3y$$

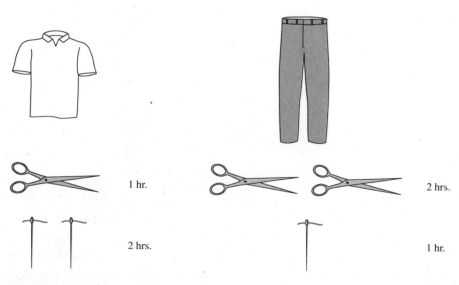

1 hr.

2 hrs.

2 hrs.

1 hr.

FIGURE 3.21

Write the Constraints in Terms of the Unknowns

The final stage in the setup process involves *identifying the constraints*. The couple's production is constrained entirely by the limits on their work time. The following table summarizes the needs of the garment production.

	Cutting	Sewing
Shirts	1 h	2 h
Pants	2 h	1 h

Only 21 h are available for cutting. Each shirt uses 1 h of that time; x shirts use $1 \cdot x = x$ hours of time. Each pair of pants consumes 2 h of cutting time, so y pairs of pants demand $2y$ hours. The total cutting time cannot be greater than 21 h:

$$(3.27) \qquad\qquad x + 2y \leq 21$$

In the same way, the total sewing time cannot be greater than 21 h, which implies that

$$(3.28) \qquad\qquad 2x + y \leq 21$$

It is clearly impossible to manufacture a negative number of shirts and pants, so we demand that

$$(3.29) \qquad\qquad \begin{aligned} x &\geq 0 \\ y &\geq 0 \end{aligned}$$

The objective function, Equation (3.26), and Inequalities (3.27) through (3.29) define a linear programming problem that we solve according to the procedure detailed in the previous section.

Solve the Mathematical Problem

The feasible region defined by these four inequalities appears as Figure 3.22. Corner D, for example, is the intersection of Equations (3.27) and (3.28) (actually, of the corresponding equations of these inequalities). The steps by which we solve these equations simultaneously to find D follow.

First subtract $2y$ from both sides of Equation (3.27) to get

$$(3.30) \qquad\qquad x = 21 - 2y$$

and substitute this expression for x in Equation (3.28):

$$\begin{aligned} 2(21 - 2y) + y &= 21 \\ 42 - 4y + y &= 21 \\ 42 - 3y &= 21 \\ -3y &= -21 \\ y &= 7 \end{aligned}$$

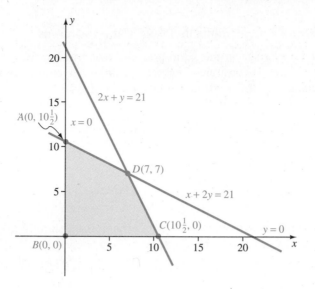

FIGURE 3.22
The feasible region for the clothing manufacture problem.

Use this value in Equation (3.30):

$$x = 21 - 2(7) = 7$$

The coordinates are therefore $(7, 7)$.
    Since $P = 2x + 3y$, at D,

$$P_D = 2x + 3y$$
$$= 2(7) + 3(7)$$
$$= 35$$

(The notation P_D refers to the value of the objective function P evaluated at the coordinates of point D.) Here is the chart relating each corner to its associated profit for all corners:

Corner	$P = 2x + 3y$
$A(0, \frac{21}{2})$	31.5
$B(0, 0)$	0
$C(\frac{21}{2}, 0)$	21
$D(7, 7)$	35 ⟺ MAX

    The mathematical solution yields a maximum value of 35, attained when x and y are both 7. Translating back into the language of the application, the couple can make a maximum of $35 per week. They do that by making 7 shirts and 7 pairs of pants.

Skill Enhancer 11   Fashions change, and so does the health of our couple. Profits remain the same on pants and shirts, and shirts require the same amounts of effort as before. Pants now need 3 h of cutting and 2 h of sewing. Furthermore, the woman can now only work 20 h per week, although the man can work 22 h. What now is the optimal strategy?

Answer in Appendix E.

We can use these steps to solve any linear programming application problem.

Formulating a Linear Programming Problem

1. *Look for key words* identifying constraints and seeking an optimum value of some expression.
2. *Identify the unknowns* of the problem.
3. *Write the objective function* in terms of the unknowns.
4. *Write the constraints of the problem* in terms of the unknowns.
5. *Solve* the resulting mathematical problem.

Example 12 A Nutrition Problem. A small local dairy company has developed a new dairy dessert that is sold in standard and premium grades, depending on the relative proportions of milk and cream present. Currently, they can process 310 gal milk and 300 gal cream per day. One gallon of the standard dessert is 20 percent milk and 15 percent cream by volume, whereas the premium grade contains 20 percent milk and 30 percent cream. Furthermore, the profits are $3 per gallon of the standard dessert, and $5 per gallon of the premium dessert. On a "gut" level, the foreperson knows that the company

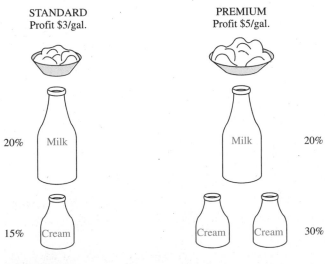

FIGURE 3.23

will have to manufacture at least 1,000 gal total of the two desserts in order to keep this branch of the company financially solvent. Furthermore, until the demand for this new product has proved itself, the amount of the premium-grade dessert should never exceed the quantity of the standard grade produced in a day. On the basis of these facts, what is the optimal daily production combination of premium and standard grades?

	Milk	Cream	Profit
Standard	20%	15%	$3/gal
Premium	20%	30%	$5/gal

Solution

(Look for key words.) This problem is a linear programming problem because of the constraints on the milk and cream and because of the need to know the maximum profit.

   (Identify the unknowns.) It makes sense to use two variables to represent the standard and premium grades:

$$x = \text{the number of gallons of } standard \text{ dessert produced per day}$$
$$y = \text{the number of gallons of } premium \text{ dessert produced per day}$$

   (Units are important in defining variables!)

   (Write the objective function.) We will want to maximize the manufacturing *profits*. For each gallon of standard dessert, the company makes $3; x gallons therefore makes $3x$ dollars. In the same way, y gallons of premium-grade product increase profits by $5y$ dollars. The total profit P arises from money earned from the standard grade plus money earned from the premium grade:

$$P = 3x + 5y$$

A rational businessperson will *maximize* this quantity, subject to the constraints of the problem.

   (Write the constraints.) And what are these constraints? The manufacturing strategy is limited by the raw materials available, in this case, milk and cream: 20 percent of both products, $0.2x + 0.2y$, cannot exceed the capacity of 310 gal. That is,

$$0.2x + 0.2y \leq 310$$

Similarly, since 15 percent of the standard product and 30 percent of the premium is cream,

$$0.15x + 0.3y \leq 300$$

The foreperson's "gut" feeling translates into the statement

$$x + y \geq 1,000$$

The constraint on the *relative* amounts of dessert produced translates into

$$x \geq y$$

Finally,

$$x \geq 0 \quad \text{and} \quad y \geq 0$$

(Why?) Here is a summary of these six linear inequalities:

(3.31) $$0.2x + 0.2y \le 310$$
(3.32) $$0.15x + 0.3y \le 300$$
(3.33) $$x + y \ge 1,000$$
(3.34) $$x - y \ge 0$$
(3.35) $$x \ge 0 \quad y \ge 0$$

(Solve.) Figure 3.24 displays the feasible region defined by these six inequalities together with the corner points and their coordinates. Once we have determined the coordinates, we can complete the problem by preparing the table listing the profit at each corner.

Corner	$P = 3x + 5y$
$A(500, 500)$	4,000
$B(1,000, 0)$	3,000
$C(1,550, 0)$	4,650
$D(1,100, 450)$	5,550 ⇔ OPTIMUM
$E(666\frac{2}{3}, 666\frac{2}{3})$	$5,333\frac{1}{3}$

The best production strategy calls for $1,100$ gal of the standard-grade product and 450 gal of the premium. In this way, the company earns the maximum daily profit of $5,550.

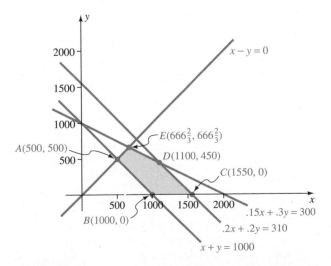

FIGURE 3.24
The feasible region defined by Inequalities (3.31) through (3.35) (the dairy dessert problem).

Skill Enhancer 12   A paint company wants to introduce two new pastel shades. The important ingredients of both are cadmium yellow and chromium white. We need to determine the most profitable daily manufacturing schedule. Doing that requires knowing that the profit per gallon of the shades are $3 and $5. Production is limited since the company can reliably get only 10 gal of the white pigment and 12 gal of the yellow pigment each day. Here is a chart showing the percentages of each pigment needed to create each new shade. (The percentages don't sum to 100 percent since additional ingredients are part of the paint. These are not critical.)

	White	Yellow
Pastel 1	50%	30%
Pastel 2	20%	60%

In solving the problem, let x and y be the optimal number of gallons of each shade.
(a) What is the objective function for this problem?
(b) What are the constraints of this problem?
(c) What are the corners of the feasible region?
(d) What is the optimal strategy? What is the maximum profit?

Answers in Appendix E.

What happens to the solution if the constraints are changed? Changing the constraints may disturb this solution, or leave it unchanged. Some of the exercises at the end of the section explore these possibilities.

Example 13   A Transportation Problem. Best Brands Appliance Mart is getting ready for its annual Labor Day sale. There are two Best Brands stores, one in midtown Manhattan (New York City) and another in Amityville, on nearby Long Island. Merchandise is stored in two warehouses, one in Brooklyn and one in Baldwin. From experience in past years, the owners know that the most popular product during the sale will be video-cassette recorders. The Manhattan store needs 500, and the Amityville store will require 400. The warehouses have 600 VCRs apiece in stock. It costs $1 and $2, respectively, to ship a VCR from Brooklyn to Manhattan and to Amityville, and $2 and $4 to ship one from Baldwin to Manhattan and to Amityville.

What is the best shipping strategy for getting the appliances from the warehouses into the stores?

Solution

(Keywords.) Since we need a strategy and since there are bounds on the VCRs available, we decide this is a linear programming problem.

(Unknowns.) The solution must reflect the various shipping costs involved: $1 per VCR from Brooklyn to Manhattan, and $2 per VCR from Brooklyn to Amityville; $2 per VCR from Baldwin to New York, and $4 per VCR from Baldwin to Amityville.

Shipping Costs per VCR

From	NYC	To Amityville
Brooklyn	$1	$2
Baldwin	$2	$4

At first glance, it might appear that four variables are involved in the resolution of the problem, each of which describes the movement of VCRs from one of the warehouses to one of the stores. (See Figure 3.25a.) Fortunately, the constraints of this problem reduce the mathematical description to two variables.

Figure 3.25b summarizes the stock information of the problem and shows why only two variables are necessary. We'll use these variables:

1. x = the quantity that we bring from Brooklyn to Manhattan

2. y = the number brought from Brooklyn to Amityville

(Write the objective function.) We have to determine the objective function—shipping costs—to *minimize*. This cost E (Expenses) will arise from the number shipped from Brooklyn to Manhattan and to Amityville, $1 \cdot x + 2 \cdot y$, plus the amount associated with the number shipped from Baldwin to the two stores, $2 \cdot (500 - x) + 4 \cdot (400 - y)$. Therefore

$$E = (x + 2y) + [2(500 - x) + 4(400 - y)]$$

or

$$E = 2{,}600 - x - 2y$$

(Write the constraints.) The number we will need from Baldwin is not an independent variable—since we need only 500 in total, of which x are coming from Brooklyn, we will need only $500 - x$ from Baldwin. Similarly, we will need only $400 - y$ to come from Baldwin to Amityville.

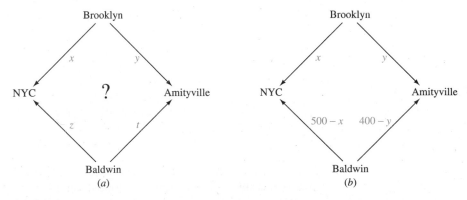

FIGURE 3.25
The first frame shows a "first approach" solution to this transport problem. The second frame suggests a better way to set the problem up.

The constraints on these quantities are two in nature. First, we cannot expect either warehouse to ship more VCRs than are currently in stock. From Brooklyn, a total of $x + y$ appliances will be shipped; that total may not exceed 600, the number in stock at the start of this exercise. Therefore,

(3.36)
$$x + y \leq 600$$

In the same way, the Baldwin shipments to each store may not exceed 600: $(500 - x) + (400 - y) \leq 600$, which simplifies to

(3.37)
$$x + y \geq 300$$

Since we cannot expect negative shipments, we demand that both

(3.38)
$$x \geq 0$$
$$y \geq 0$$

In addition, *neither of the Baldwin shipments may be negative*, so $500 - x \geq 0$ and $400 - y \geq 0$, or

(3.39)
$$x \leq 500$$
$$y \leq 400$$

(Is this clear? For example, if $500 - x \geq 0$, then add x to both sides to get $500 \geq x$, which is the same as $x \leq 500$.)

(Solve.) The six inequalities that define the feasible region appear in Figure 3.26.

The table listing the value of the expenses at each corner of the feasible region follows:

Corner	$E = 2{,}600 - x - 2y$	
$A(0, 300)$	2,000	
$B(0, 400)$	1,800	
$C(200, 400)$	1,600	⇔ MIN
$D(500, 100)$	1,900	
$E(500, 0)$	2,100	
$F(300, 0)$	2,300	

Point C yields the least shipping costs ($1,600), and dictates that we ship from the Brooklyn warehouse 200 VCRs to Manhattan and 400 to Long Island, while from the Baldwin spot, we should ship 300 ($= 500 - x$) to Manhattan and none ($400 - y$) to Amityville.

Skill Enhancer 13　The next year, for the same sale, some facts have changed. Business has been good. The warehouses each stock 1,000 VCRs, and the Manhattan store now needs 700 units, whereas the Amityville store needs 500. But the shipping costs have gone up; they are as in the following table.

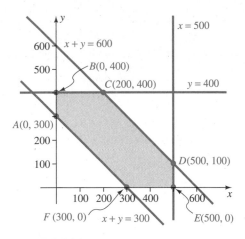

FIGURE 3.26
The feasible region for the transport-problem.

| | *Shipping Costs* | |
| | To | |
	NYC	Amityville
From Brooklyn	$2	$3
From Baldwin	$3	$5

Use this information to determine the optimal shipping strategy and the minimal shipping costs.

Answer in Appendix E.

Example 14 Sam and Tamara are a husband-and-wife team who sell sets of encyclopedias door to door. Together they must sell at least 10 sets per month. Sam, the more experienced salesperson, must sell at least 1 set more than twice the number Tamara sells. The company gives Sam a $200 commission for each set he sells, but gives Tamara only $150 for each set sold, since she has less experience. What is the least monthly income they can count on?

Solution

Since the team's sales are subject to certain constraints, and since we need to find a minimum commission, this is a linear programming problem. The unknowns are the numbers of sets of encyclopedias that each person sells; let x_1 equal Sam's sales, and let x_2 be Tamara's.

The *objective function* to be minimized is the commissions. Sam's sales generate $200x_1$ in commissions, whereas Tamara's generate $150x_2$. Therefore, the objective

function is

$$p = 200x_1 + 150x_2$$

The *constraints* of the problem arise from the numbers of sets the team must sell. Their total sales will be $x_1 + x_2$; this must equal or exceed 10 in any month, so the first constraint is

(3.40) $x_1 + x_2 \geq 10$

One more than twice Tamara's sales is $1 + 2x_2$. Since Sam's sales must exceed this, we demand that

(3.41) $x_1 \geq 1 + 2x_2$

Finally, we must have

(3.42) $x_1 \geq 0 \qquad x_2 \geq 0$

The feasible region defined by Inequalities (3.40) through (3.42) appears in Figure 3.14. Make sure you see how to get it, and how to get the coordinates of the three corners. It remains to prepare a chart showing the commissions generated from the corner values.

Corner	$p = 200x_1 + 150x_2$
(10,0)	2,000
(7,3)	1,850

The minimum commission is \$1,850, which happens when Tamara sells 3 sets and Sam sells 7.

Skill Enhancer 14   Sherman works in a showroom and sells large appliances, specifically refrigerators and stoves. His commission on each sale is \$50 and \$75, respectively. He must sell a total of 20 appliances every two weeks. To make sure he doesn't sell stoves exclusively, management decrees that the number of refrigerators he sells must

At Least 10 Sets
Per Month

FIGURE 3.27

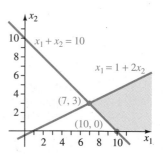

FIGURE 3.28
Graphing
Inequalities (3.40)
through (3.42).

exceed by at least 6 the number of stoves he sells. If he follows this strategy faithfully, what is the minimum commission he will make in each two-week period? What is the sales strategy to achieve this figure?

Answer in Appendix E.

_____ *Exercises 3.4*_____

The solution to linear programming problems depends heavily on the constants appearing in the constraints, as the first group of exercises, Exercises 1–4, demonstrate. You are to use these facts in the context of Example 12 in the text. Find the solution to this problem subject to these new facts.

W **1.** If the new dessert proves itself, so that the amount of standard dessert need not equal or exceed the premium dessert, will this change affect the optimal strategy? Why or why not? What is the objective function for this problem? What are the constraints?

2. Suppose the available milk (per day) is now 350 gal. What now is the best strategy? What is the objective function for this problem? What are the constraints?

3. What if the daily allotment of cream is upped to 350 gal? How will this change the strategy? What are the constraints of the problem?

4. Suppose now that the amount of standard dessert produced must equal or exceed *six times* the amount of premium dessert. What changes will this make in the optimal strategy?

Applications

5. *(Enterprise)* Two college students, Juana and Tran, wash cars and light trucks during weekends for extra money. They charge $10 for each car and $12.50 for each truck they wash. They know from experience that they get at least two cars, but never more than eight. Similarly, they wash between one and five trucks. Because of the time it takes to wash each vehicle, they can't wash more than ten vehicles between them. What is the largest amount they can make in a weekend? How many of each kind of vehicle must they wash to earn this amount?

6. *(Investment Strategy)* A pair of proud parents have up to $20,000 to invest for their newborn son. They decide to invest in bonds. Bonds return interest annually for a certain number of years, after which the bondholder receives the face value of the bond. The couple can select any of a number of bonds with widely varying interest rates. Although the higher the rate, the more attractive the bond, it is also a general rule of thumb that the higher the rate, the riskier the bond (the less chance that the issuer of the bond will be able to pay off the bond at the end of the interest-bearing period). After much thought, the couple decide to invest in two types of bonds, some of which yield $7\frac{1}{2}$ percent and others 15 percent. They also decide to adhere to an investment rule that the amount invested in the riskier but higher-yielding bonds shall not exceed the amount invested in the lower-yielding bonds. What is the parents' best investment strategy? What annual yield will their son see?

7. Suppose the couple in Exercise 6 had $30,000 to invest. Would the relative amounts of their investments change?

8. Another couple with $20,000 to invest for their newborn daughter also decides on bonds, and restricts their choices to two types of bonds—those yielding 5 percent and those yielding 12 percent. In addition to the investment rule followed by the couple in Exercise 7, this couple also decides that three times the low-risk investment plus the high-risk investment should not exceed $50,000. What, for this couple, should be the optimal investment strategy, and what is the maximum annual return on this investment?

9. If the couple in Exercise 8 decided to go with bonds yielding $7\frac{1}{2}$ percent and 15 percent, what then would be their best investment strategy? What now would be the best yield?

10. *(Construction)* Five Brothers Construction Company has acquired a large tract of land on which to build homes. Two types of homes can be built, ranch and colonial. The profit on each ranch house is $5,000, whereas profit on a colonial is only $4,000. Loosely speaking, work in housing construction is either "exterior" (digging the foundation, building the walls, erecting the chimney, and so on) or "interior" (plumbing, plastering, wiring, and so on). A ranch house requires 250 exterior hours of construction and 250 interior hours. A colonial house uses 300 exterior hours and 200 interior hours. There are currently 10 exterior workers and 8 interior workers on the payroll. All Five Brothers employees work 40 h a week. In any week, what is the best construction strategy to follow for maximum profit? What profit will that generate for the company?

11. Refer to Exercise 10. Because of improved construction techniques, the ranch and colonial houses now require only 200 and 250 exterior hours in their construction. If all other facts remain the same, what is the best strategy now? What is the profit?

12. *(Personal Computing)* The personal computers at a Seattle accounting firm can take either $5\frac{1}{4}''$ or $3\frac{1}{2}''$ floppy disks. The small disks cost $0.55 apiece, whereas the big ones cost only $0.30. The firm needs at least 1,500 new floppies, and the number of small disks must equal or exceed the number of large disks. (This is so because of the higher storage capacity of the smaller disks.) How shall the order be made to minimize costs for the company? What is the minimum cost?

13. *(Wildlife Preservation)* Animals destined for zoos are stored in either Tangiers or Capetown for eventual transport to either Liberia or Nairobi, from where they will travel to either Europe or America. Associated with the transportation to Liberia and Nairobi are certain transportation costs, given by this table.

	To ...	
	Nairobi	Liberia
From Capetown	100	150
From Tangiers	150	180

(Numbers are dollar amounts.) There are 200 animals at Capetown and 150 at Tangiers. We need 125 animals each at Liberia and Nairobi. How should animals be shipped from Tangiers and Capetown to minimize the cost? What is the minimal shipping cost?

14. *(Construction)* The Simpson Construction Company has been retained to build a brick wall using at least 12,000 bricks. The architect specifies a mixture of old brick and new brick, old brick being chosen for its picturesque effects while new brick provides the required strength. There have to be at least as many old bricks as new bricks, or the wall won't look nice. On the other hand, the local building code demands that the number of old bricks not exceed twice the number of new bricks, or the wall will be too weak. Old and new bricks cost $0.40 and $0.30 per brick. Use these facts to determine

the least-cost way to build the wall. What is the least cost?

15. *(Retailing)* A car dealership sells cars and trucks. In order to meet the manufacturer's quota, it must sell at least 100 vehicles per month. In order to cover its expenses, it must sell at least 1.5 times as many trucks as cars. The average truck commission is $2,000, and the average car commission is $1,350. What is the smallest amount of commissions the dealership generates during any single month? How does it obtain this minimum?

16. *(Psychology)* The department of psychology at a local university is preparing a study of male lifestyles. The research will involve interviewing men who recently graduated from college (category *A*) and men 20 years after their college graduation (category *B*). Professors and graduate students will participate in the gathering of information. Generally, professors are more adept at the interviewing process. Judging by past experience, here is how long a typical interview takes for students and professors:

	A	*B*
Professors	15	20
Graduate students	60	30

(Numbers are minutes per interview.) In a semester, the professors have a total of 1,000 minutes for the interviews, and the graduate students 2,400. How many men in each category should be interviewed to maximize the *total* number of men interviewed?

17. Refer to Exercise 16. If graduate students have 3,000 minutes available for interviewing, what is the best strategy?

18. Refer again to Exercise 16. Under the additional constraint that the number of recent college graduates participating in the study should not be less than the number of older men, what is the optimal strategy?

19. *(Car Sales)* Happy Harry's Used Cars employs two salespeople. Carrie is the newer salesperson. Barry, with more experience, is expected to sell at least 2 more than 5 times the cars that Carrie sells. Management (Harry) expects their joint sales to be at least 32 per month. No single salesperson in the history of Harry's Used Cars has ever sold more than 47 cars in any month. If Barry's commission is $300 per car sold and Carrie's is $200, what is the minimum amount of commissions that Harry has to pay out in any month? What is the maximum?

20. (Bicycles) The Wheeler Bicycle Corporation has two factories, at the north and south ends of a certain city. Their two showrooms are at the east and west ends of the same city. At the moment, the north factory has 300 bikes and the south factory 400 bikes to deliver to the two showrooms. The west showroom needs 200, and the east showroom needs 400. Bicycles are simultaneously road-tested and delivered by pedaling them from the factories to the showrooms. Delivery costs from the factories to the showrooms are therefore proportional to the distances. How should the deliveries be arranged in order to keep transportation costs at a minimum? The north factory is 10 kilometers (km) from the west showroom and 15 km from the east. The south factory is 20 km from both showrooms.

21. Wheeler (see Exercise 20) is thinking of relocating the east showroom. The projected new location would be 25 km from the north factory and still be 20 km from the south factory. Recompute the optimum transportation strategy for the projected new setup.

22. (Public Relations) Show how you might set this problem up *only*. (Do not solve.) The City Fathers of New Orleans are planning this year's Mardi Gras parade. The parade is made up of a succession of floats, which come in three levels of extravagance: *A*, *B*, and *C*, where *A* is the most ornate and impressive and *C* is the most restrained and sedate. Each type of float has a certain operating cost associated with it, and also a certain crowd-drawing power. There are only a certain number of floats of each type available. Here is all this information in chart form

	Cost	Drawing power	No. available
A	$100,000	100,000	3
B	$40,000	5,000	8
C	$20,000	3,000	24

Last year, 400,000 people attended the parade, which cost $500,000. How should the parade be organized in order to minimize expenses while at least equaling last year's spectator turnout? Set up only; do not solve.

23. Redo Exercise 22. This time, set up the problem so as to maximize spectator turnout while holding expenses down to the level of last year's parade. Set up only; do not solve.

W 24. Speculate as to how you might solve the preceding two problems. (Do not solve them.)

W 25. (Computer Solution) What problems (if any) do you foresee in attempting to use a computer to solve linear programming problems using the geometric approach of this chapter?

26. (Animal Husbandry) One type of chicken feed is a mixture of cracked corn, soybeans, and a granular artificial supplement. You can mix these components together to get a feed, but the mixture must contain minimum quantities of various nutrients, as specified by the U.S. Department of Agriculture (USDA). The quantities of three particular nutrients in the three components are these:

	No. 1	No. 2	No. 3	cost ($/lb)
Cracked corn	30%	20%	40%	0.20
Soybeans	30%	30%	10%	0.30
Artificial supp.	20%	40%	20%	0.10

Thus, the first nutrient makes up 30 percent of cracked corn by weight. The last column of this chart gives the cost per pound of each of these ingredients. Here are the USDA minimum requirements for the three nutrients per pound of mix:

	Nutrients	
No. 1	No. 2	No. 3
25%	30%	20%

What is the cheapest way to prepare 100 lb of this mixture subject to the USDA requirements? (Hint: Let x, y, and z be the fractions of the final mix, so that $x + y + z = 1$. Therefore, we can determine that $z = 1 - x - y$. Use this fact to transform this to a problem with two variables.)

27. Redo Exercise 26, ignoring the existence of artificial supplements—this is to be an all-natural chicken feed! How now should 10 lb be prepared?

28. Redo Exercise 26, ignoring for the moment any cost considerations because the USDA has issued an important new ruling. This agency has discovered that a fourth chemical present in these three ingredients can cause the chicken meat to take on a fishy taste, although the chemical is otherwise quite harmless. The object now is to create a chicken feed that satisfies USDA nutrition requirements but minimizes the presence of this fourth substance. Both corn and the artificial supplement contain 20 percent of this undesirable substance by weight.

Soybeans contain 30 percent. How now would you prepare 100 lb of chicken feed? How would you mix 500
lb?

and pair of pants is \$2.65 and \$3.77.

W **30.** Redo Example 11 again, using the new figures for profits
(see Exercise 29) and the fact that although the husband
can work for 20.5 h per week, his wife can work for
only 18.5 h. What are the average weekly profit and the
average weekly output?

▣ Calculator Exercises _____

29. Redo Example 11 from the text if the profit on each shirt

Section 3.5   *WHEN THE GEOMETRIC METHOD FAILS*

Sometimes linear programming techniques will not generate a solution, either because a
problem has no solution or because there is not enough information within the problem
to determine a solution. In this section, we examine these special cases.

The Constraints of the Problem May Define an Empty Feasible Region

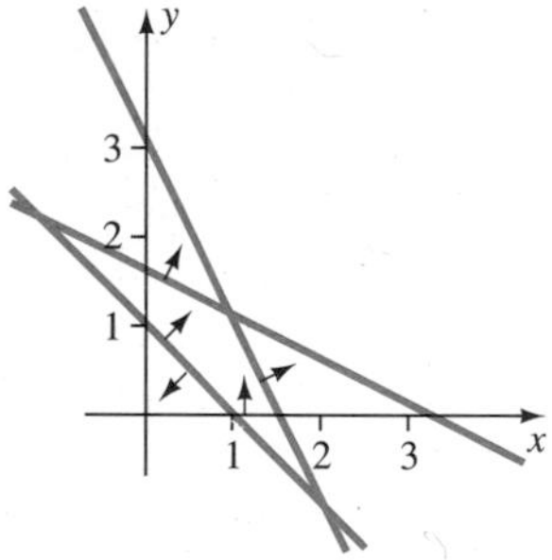

FIGURE 3.29
No feasible region.

Example 15   Maximize $P = 9x - \frac{1}{2}y$ subject to

(3.43)
$$x + y \leq 1$$
$$x + 2y \geq 3$$
$$2x + y \geq 3$$
$$x \geq 0 \qquad y \geq 0$$

Solution

The Inequalities (3.43) are graphed in Figure 3.29. There is no region of overlap of the
five inequalities, and hence no feasible region. Under these circumstances, it makes no
sense to speak of the optimization of P.

Skill Enhancer 15   We need to minimize $p = 5x + 7y$ subject to

$$x \geq 0 \qquad y \geq 0$$
$$2x + 3y \geq 6.5$$
$$2x + 3y \leq 6$$
$$y \leq 4$$
$$x \leq 3$$

What is the minimum and how do we obtain it?

Answer in Appendix E.

If the Feasible Region Is Unbounded, There May Not Be a Well-Defined Maximum or Minimum to the Problem

bounded   A feasible region is *bounded* as long as it does not contain arbitrarily large positive or
negative values of x or y. Practically speaking, if it is possible to draw the entire feasible

region on a graph, then the region is bounded. Unbounded feasible regions may or may not offer a solution.

Example 16   Maximize $P = x + y$ subject to

(3.44)
$$2x + y \geq 2$$
$$x + 4y \geq 2$$

Solution

unbounded   Inequalities (3.44) define the feasible region in Figure 3.30. This feasible region is *unbounded*, since it is impossible to draw the entire feasible region no matter how large a sheet of paper we use. It is not possible to specify the absolutely largest value of P. No matter what values of x and y we choose for solutions, we could choose larger values of both x and y, rendering obsolete any value of P that was our choice for maximum. There is no solution to this linear programming problem.

Skill Enhancer 16   Can we maximize $P = 4x + 9y$ subject to

$$x \geq 0 \qquad y \geq 0$$
$$x + y \geq 2$$
$$x \geq \frac{1}{2}$$
$$y \geq \frac{1}{3}$$

Why or why not?

Answer in Appendix E.

A slightly different problem would have a solution. Had we been asked to *minimize* P with respect to these constraints, there would have been no problem. In this case, the corner $(\frac{6}{7}, \frac{2}{7})$ yields the minimum strategy for the absolute minimum of $\frac{8}{7}$. Carefully

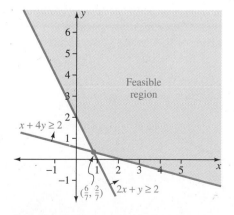

FIGURE 3.30
An unbounded feasible region.

check in cases where the feasible region is unbounded that the geometric method provides a valid solution.

Example 17   Consider again Example 14, page 165. There we wanted to determine the minimum commission that Sam and Tamara earn through the selling of sets of encyclopedias. Mathematically, the problem is equivalent to minimizing

(3.45) $$p = 200x_1 + 150x_2$$

subject to

$$x_1 + x_2 \geq 10$$
(3.46) $$x_1 \geq 1 + 2x_2$$
$$x_1 \geq 0 \qquad x_2 \geq 0$$

We show the feasible region again in Figure 3.31.

Could we determine the *maximum* commission per month? The feasible region is unbounded and has no absolute maximum (although the minimum does exist). (This makes sense from consideration of the problem. There is no constraint on selling ever more encyclopedias, so there is no limit on how large the commission could grow.)

Skill Enhancer 17   In this example, would we be able to find a maximum commission if we introduced the additional constraint

$$x_1 \leq 10 + 2x_2$$

Why or why not?

Answer in Appendix E.

The Optimum Solution Exists, but Is Not Unique

Earlier in this chapter, we justified the Fundamental Theorem by appealing to an analogy between the objective function and a family of parallel straight lines, where differing values of P (the objective function) corresponded to various y intercepts of the members of this family. By virtue of this analogy, an optimum value of P determines a line in

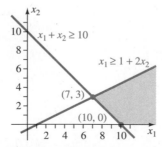

FIGURE 3.31
An unbounded feasible region.

this family that intersects the feasible region through one of its corners. Only this line has the smallest or largest intercept, and that is how we find the maximum or minimum.

Implicit in this derivation was a crucial fact: *the line related to the objective function was not parallel to any of the sides of the feasible region.* If it were parallel, then it would intersect, not a single corner, but the entire side. What does that mean from the standpoint of determining the solution? The following example helps answer this question.

Example 18   Maximize $P = x + 3y$ subject to the constraints

$$x + y \geq 1$$
$$x + 3y \leq 4$$
$$x \leq 2$$
$$x \geq 0$$
$$y \geq 0$$

The graph of these inequalities and their feasible region is the subject of Figure 3.32. Here is the usual chart for each corner and its objective value.

Corner	$P = x + 3y$
$A(0, 1)$	3
$B(1, 0)$	1
$C(2, 0)$	2
$D(2, \frac{2}{3})$	4 ⟺ MAX
$E(0, \frac{4}{3})$	4 ⟺ MAX

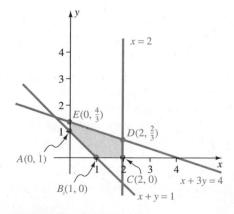

FIGURE 3.32
The strategy yielding the maximum need not be unique.

In the usual way, we can identify the maximum P value as 4, but there is some confusion as to the strategy to get it. Should we use the coordinates of corner D or corner E?

The line $x + 3y = 4$ has a slope of $-\frac{1}{3}$ (show this). Furthermore, the objective function $P = x + 3y$, when rewritten in the equivalent form $y = -\frac{1}{3}x + \frac{1}{3}P$, generates a family of lines, each of which also has a slope of $-\frac{1}{3}$. Each member of this family is a line with a different value of P.

Thus, the member of this family with the largest y intercept lies on an entire side of the feasible region, and it would be unreasonable of us not to accept any point on this side as corresponding to an acceptable *mathematical* strategy for the optimum value of P. That is, *any* point, not only the corners, lying on the line connecting corners D and E represents an acceptable strategy for generating the desired maximum.

Skill Enhancer 18   What is the minimum value of $p = 2x + 3y$ subject to the constraints

$$2x + 3y \geq 6$$
$$x + y \leq 5$$
$$y \leq x + 2$$
$$x \geq 0 \qquad y \geq 0$$

Discuss ways to achieve that minimum.

Answer in Appendix E.

The emphasis in the last paragraph in the solution is on the word "mathematical," because practical considerations often dictate the right choice of x and y values. Suppose that the variables in this example stood for pieces of two styles of furniture produced by a certain factory. Clearly, fractional values for x and y would be useless. In that case, we need a point with integral (whole-number) coordinates. The only such point on the line segment is $(1, 1)$.

When the Geometric Method Is Inappropriate

The linear programming problems of this chapter are interesting, but are contrived. No problem asked us to work with more than two or three unknown quantities. Real-life situations are rarely so considerate. A transportation problem, for example, is apt to involve the delivery of several products from a few factories to *many* stores. Preparation of food mixes usually calls for the combining of *many* components, all subject to many USDA constraints. Production processes have *many* substeps involved in the making of any number of different products. Can the method of this chapter be generalized in any way?

When a problem involves two variables, we use a two-dimensional plane on which the boundaries of the feasible region are linear equations of two variables, which happen to be straight lines.

Imagine three variables in a problem. The sides of the feasible region are formed from linear equations of *three* variables, which are three-dimensional *plane* surfaces.

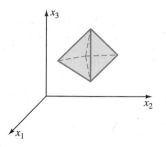

FIGURE 3.33
A three-dimensional feasible region.

Although three-dimensional feasible regions are hard to draw, they can sometimes be drawn, as in Figure 3.33. But what happens if the problem involves four variables, five variables, or more? How can we visualize linear equations of that many variables? It is impossible, and for this and other reasons, the geometric method breaks down. A different method, emphasizing algebraic manipulation instead of geometric visualization, forms the subject of the next chapter.

Exceptional Situations

1. The feasible region may be *empty*.

2. An *unbounded feasible region* may not permit a well-defined optimum solution to exist.

3. The *solution may not be unique*.

4. There may be *too many variables* for the geometric method to work. This geometric method *may* work when there are three variables in a problem, but it will *never* work for four or more variables.

Some visualizations of these items appears in Figure 3.34.

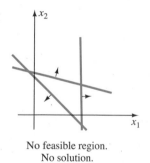

No feasible region.
No solution.

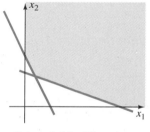

Unbounded feasible region.
Perhaps no solution.

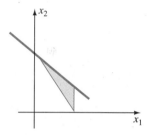

Objective function represents a straight line parallel to one side of feasible region.
Nonunique solution.

FIGURE 3.34
Exceptional situations.

_____ *Exercises 3.5* _____

Exercises 1–10 have no solution. Why not?

1. Maximize $M = 10x - 34$ when $x \geq 0$, $y \geq 0$, and $x + y + 1 \leq 0$.

2. Minimize $m = 32x + 14y$ when $x + 3y \leq 4$ and $x \leq 3$.

3. Minimize $m = 0.5x - 1.2y$ if $y \leq 1$, $x + y \leq 0.75$, and $x \leq 1$.

4. Maximize $M = x$ subject to $y \leq 10$, $x - 2y \geq 5$, and $2y \geq x + 2$.

5. Maximize $M = 2y + 1$ subject to $2y + 6x + 1 \leq 0$, $x \geq 0$, $y \geq 0$, and $3x + 2 \leq y$.

6. Minimize $m = -5y$ subject to $x + y \leq 5$, $x + y \geq 4$, $y \geq -2$, and $y + 3 \leq 0$.

7. Minimize $p = 3$ subject to $x \geq 0$, $y \geq 0$, and $x + y \leq 11$.

8. Minimize $p = 3x + 2y$ when $x \leq 3$, $y \leq 3$, and $x + 2y \leq 6$.

9. Maximize $P = x + \frac{1}{2}y$ when $x \geq 2$, $y \geq -1$, and $x + y \geq 55$.

10. Maximize $P = 2x + 3y$ if $x + y \geq 1$ and $-1 \leq y - x \leq 1$.

Exercises 11–20 are linear programming problems that can be solved. Solve them, and point out any interesting aspects of the solution. Beware especially of nonunique solutions.

11. Maximize $P = x + y$ if $x \geq 0$, $y \geq 0$, $x + 2y \geq 6$, and $x + 2y \leq 6$.

12. Minimize $p = x - y$ if $x \geq 0$, $y \geq 0$, $2x + y \leq 4$, and $y \geq 4 - 2x$.

13. Maximize $P = x + y$ subject to $x \geq 0$, $y \geq 0$, $y \leq 4$, and $2x + 2y \leq 12$.

14. Minimize $p = x + y$ subject to $x \geq 0$, $y \geq 0$, and $y \geq 5 - x$.

15. Maximize $P = 2.3x - 5.4y$ subject to $x + 3y \leq 10$, $7x + 2y \leq 14$, $x + y \leq 1$, and $2x + 2y \geq 2$.

16. Minimize $p = 9.1x + 8.2y$ if $x + y \leq 5$, $3x + y \leq 7$, $x \leq 2$, $x + 4y \leq 4$, and $x + 4y \geq 4$.

17. Maximize $P = 2x + 2y - 1$ if $x \geq 0$, $y \geq 0$, $y \leq 1$, $x + y \geq 1$, and $x + y \leq 2$.

18. Maximize $P = x + 2y$ subject to $x \geq 0$, $y \geq 0$, $x + y \geq 1$, $x + 2y \leq 4$, and $2x + y \leq 6$.

19. Minimize $p = \frac{1}{2}x + \frac{1}{2}y$ when $4x + y \geq 4$, $x + y \geq 2$, and $x + 6y \geq 3$.

20. Minimize $p = 2x + \frac{1}{2}y$ when $x \geq 0$, $y \geq 0$, $4x + y \geq 4$, $x + y \geq 2$, and $x + 6y \geq 3$.

W 21. (*Requires special thought*) What is the geometric interpretation of the feasible region for a linear programming problem with *three* variables?

W 22. (*Requires special thought*) Describe (in words) the geometric solution to the system of inequalities

$$x_1 \geq 0$$
$$x_2 \geq 0$$
$$x_3 \geq 0$$

Applications

23. (*Transportation*) Best Brands Appliance Mart has to transfer tape decks from its warehouses in Brooklyn and Baldwin to its two stores in Manhattan and Amityville. Each warehouse has 600 tape decks in stock. The Manhattan store needs 500 decks, and the Amityville store needs 400. The shipping costs per unit are as follows: Brooklyn to Manhattan, \$1; Brooklyn to Amityville, \$2; Baldwin to Manhattan, \$3; and Baldwin to Amityville, \$4. What shipping strategy or strategies will minimize the shipping costs? What is the minimum shipping cost?

24. (*Editorial Services*) A retired couple supplements their income by editing and typing. The wife is the editor, and the husband is the typist. A typical letter requires 1 h of editing and 2 h of typing, whereas a résumé requires 2 h of editing and 1 h of typing. Both people restrict their operation to 21 h per week each. The profit on a letter is \$20, and that on a résumé is \$10. They always have more work than they can handle, so what strategy or strategies will maximize their weekly profit? What is the maximum profit?

CHAPTER REVIEW

Terms

constraint	optimization	objective function
inequalities	strict inequalities	corresponding equation
half planes	test point	inequality systems

region of overlap	feasible region	allowable region
vertex	simply connected	convex
Theorem of Linear Programming	bounded	unbounded

Key Concepts

- Linear programming techniques help in discovering strategies for either maximizing gain or minimizing loss in problems where the gain or loss (the *objective function*) is described by a linear expression of the variables of the problem. At the same time, the constraints on this gain or loss must be expressible as linear inequalities in the variables of the problem.

- The graph of a single linear inequality consists of one of the half planes bounded by the corresponding equation. The graphic solution of several linear inequalities is generally a many-sided region in the plane.

- *A geometric approach to linear programming.* First, graph the constraints of the problem to determine the feasible region. Then, if there is a solution, it must lie at one of the corners of the feasible region. A *solution* is the pair of x and y values that yields the *optimum* value—either the maximum or the minimum—of the objective function. Substitute these coordinate values into the expression governing the objective function to obtain the optimum.

- Linear programming techniques can solve nutrition, production, and transportation problems.

- The geometric method may be applicable to a problem with three variables, but it can never be applied to problems with four or more variables.

- In certain special cases, a linear programming problem may have no solution. This could happen if the constraints do not define a feasible region or if the feasible region is unbounded. (Although a problem with an unbounded feasible region *may* have a solution. Check carefully when the region is not bounded.)

- In other special cases, the optimal strategy may not be unique. This occurs when the straight-line family corresponding to the objective function is parallel to one of the sides of the feasible region. In this case, the coordinates of any point on that side will yield the optimum value of the objective function.

Review Exercises

Graph the inequalities in Exercises 1–6.

1. $x + 2y \geq 0$

2. $x + 2y > 0$

3. $3x - 2y \leq 6$

4. $7x + 2y \geq 1$

5. $3u + 6v \geq 12$

6. $5u_1 - 8u_2 \leq 40$

Solve the systems of inequalities in Exercises 7–12 graphically, if possible. In all cases, identify the corners of the feasible (allowable) region (if possible).

7. $x + 2y > 0$
$x + 2y < 4$

8. $x \geq 10$
$x \leq 5$

9. $x \geq 0; y \geq 0$
$y - x \leq 1$
$y \leq 2$

10. $x \geq 0; y \geq 0$
$x + y \geq 1$
$y - x \leq 1$
$x \leq 3$

11. $x \geq 0; y \geq 0$
$y - 2x + 1 \leq 0$
$2y - x \geq 0$
$x + y \leq 10$

12. $x \geq 0; y \geq 0$
$y \geq x$
$x \geq 0.5$
$y - x \geq 1$
$y \leq 4$

Exercises 13–18 present the coordinates for the corners of a feasible region. For each region, determine the maximum *and* minimum values that P can have, where

$$P = 3x + \frac{1}{2}y$$

Indicate the strategy for obtaining the optimum. If the maximum or minimum of P can be obtained by using more than one strategy, state this, and the reason for it.

13. $(1, 1)$, $(2, 3)$, $(3, 2)$, $(4, 0)$

14. $(0, 1)$, $(0, 2)$, $(1, 2\frac{1}{2})$, $(1, 0)$

15. $(1, 0)$, $(1, 1)$, $(2, 0)$

16. $(1\frac{1}{2}, \frac{1}{3})$, $(1\frac{1}{2}, 1\frac{1}{3})$, $(2\frac{1}{2}, \frac{1}{3})$

17. $(1, 0)$, $(3, 1)$, $(4, 7)$, $(1, 7)$, $(0, 6)$

18. $(0, 5)$, $(2, 6)$, $(3, 6)$, $(4, 0)$, $(0, 0)$

In Exercises 19–24, find the indicated maximum or minimum subject to the listed constraints.

19. Maximize $P = 4x + \frac{1}{2}y$ if $x + y \geq 1$, $x - y \leq 3$, $x + y \leq 4$, $y \leq 2$, $x \geq 0$, and $y \geq 0$.

20. Minimize $p = 9x - 7y$ if $x \geq 0$, $y \geq 0$, $x - 4y \leq 24$, $x \leq 7$, $x - y \leq 5$, and $4x + y \geq 4$.

21. Maximize $P = 3x + 2y$ subject to $x \geq 0$, $y \geq 0$, $3x + y \leq 3$, $x + 4y \geq 20$, and $x \leq 4$.

22. Minimize $p = \frac{1}{2}x + \frac{2}{3}y$ subject to $x \geq 0$, $y \geq 0$, $x + y \leq 1$, $x \geq 1$, and $y \geq 1$.

23. Maximize $P = 3x - \frac{1}{4}y$ if $x \geq 0$, $x \leq 5$, and $x + y \geq 2$.

24. Minimize $p = x + \frac{1}{10}y$ subject to $y \leq 2$, $x + y \leq 2$, and $2x - y \leq 1$.

Applications

25. *(Manufacturing)* A hardware company makes two grades of aluminum ladders. Each grade requires two sets of procedures, construction and finishing. The first type calls for 2 h of construction and 1 h of finishing for each ladder. Each ladder of the second type requires a single hour of construction, but 3 h of finishing. The profit per ladder is $30 for the first type and $20 for the second. The company has 5 employees in its ladder division, each of whom works for 40 h per week. Of the 5, 3 are skilled constructors and 2 are finishers. How many ladders of each type should the company seek to manufacture per week in order to maximize its profit? What is this maximum profit?

26. Refer back to Exercise 25, but assume that construction requirements for the ladders are 3 h per ladder for the first type and 2 h per ladder for the second type. How will this change affect the maximum weekly profit and the strategy for obtaining that profit?

27. Refer again to Exercise 25 assuming that each ladder

calls for $3\frac{1}{2}$ h of finishing. Determine the optimal strategy. Are there any problems with this answer?

28. *(Office Management)* Five hundred reams of office copy paper are stored in each of two locations in the seven-story Greybar Building, the basement and the seventh floor. Office copiers are located on the second and fifth floors. Currently, each location is devoid of paper, and paper must be transported to each location quickly. The second floor needs 600 reams, whereas the fifth floor only needs 300 reams. Paper is heavy, and it costs the Greybar Company 2 units and 4 units of labor to transport a ream from the seventh floor to the fifth floor and the second floor, respectively. It costs 3 units and 5 units to transport a ream from the basement to the second and fifth floors. How shall the deliveries be made? (By the way, a *ream* of paper is a pack of 500 sheets.)

29. *(Chemistry)* Residential swimming pools need special mixes of chemicals to keep the water fresh and sweet. Two homeowners with identically sized pools decide to mix their own additives, using two industrial-strength fresheners. The first contains 5 and 10 ml of two substances, A and B, per liter of solution, and the second contains 20 and 5 ml/l of these same substances. Local health ordinances mandate that each pool have at least 40 ml of A and 30 ml of B within the pool. The first freshener costs $1 per liter, and the second costs $2 per liter. Unfortunately, each freshener also contains an ingredient that irritates swimmers' eyes in too great concentrations. Studies indicate that a pool should not have more than 40 ml of this noxious chemical C at any time. Each liter of the first freshener contains 10 ml of C, whereas a liter of the second contains 5 ml of this substance.
(a) How should the first homeowner mix the two industrial fresheners to satisfy all these constraints and yet pay the least for the chemicals?
(b) The second homeowner is not convinced that the studies about chemical C are conclusive by any means. He is willing to add as little of this chemical as possible to his pool, but his concern is the cost of the chemical additives. He is not willing to spend more than $8 for pool fresheners. What is the best strategy to take to satisfy these requirements?

4

AN ALGEBRAIC APPROACH TO LINEAR PROGRAMMING

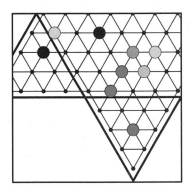

INTRODUCTION

The *simplex method* is an algebraic approach to linear programming that is especially appropriate for computer implementation or when there are more than three variables in the problem. Although all the theoretical underpinnings of this method are beyond this book's scope, the actual components of the approach are not, and we will learn in this chapter how to use this alternative technique.

The geometric method for linear programming is of limited use, since it mainly applies to problems described by two variables. Linear programming problems that occur in actual industrial and management situations are much more complex—we may need tens of thousands of variables to set them up these problems, and how could we ever adapt the geometric method to problems like these? The simplex method is purely algebraic and is especially appropriate for computer implementation.

Section 4.1 *INTRODUCING THE ALGEBRAIC APPROACH*

Can we free the linear programming concept from the geometric solution technique of the previous chapter? This question was first answered in the affirmative in the late 1940s by the contemporary mathematician George Dantzig, and we devote this chapter to his ideas. Dantzig's concepts evolved into the so-called simplex method for solving *simplex method* linear programming problems. The *simplex method* is an algebraic analysis of the system formed by the constraints together with the objective function. A complete justification for all the steps of this method is quite beyond the scope of this text, but by taking some key results on faith and adding a twist of common sense, we can become comfortable with the method. In this section, we restrict our attention to *maximum* linear programming *standard form* problems that can be put into a particular *standard form*.

Plan for This Section

Here is the plan for this section.

1. First, we explain what it is we mean by the standard form for a problem.
2. Then, we explore a new concept—that of slack variables.
3. At this point, we will find that our linear programming problem is described by several more variables than when we started, and it will be convenient to group these variables into two categories, basic and nonbasic.
4. Finally, a discussion of the Fundamental Theorem for Linear Programming will make clear why this grouping is important.

Standard Form

In words, a maximum problem is in standard form if all variables are nonnegative and all constraints are "less than" constraints.

Standard Form for a Maximum Linear Programming Problem

These items apply to the constraints and the objective function.

1. All variables $x_1, x_2, \ldots, x_u$ are *nonnegative*; that is, $x_1 \geq 0$, $x_2 \geq 0$, and so on.
2. All inequality constraints are in the form

$$\text{Linear expression} \leq \text{positive constant}$$

3. We seek to *maximize* the (linear) objective function P.

Example 1 Identify which of these collections of statements represents a linear programming problem in standard form.
(a) Minimize $p = 9(x_1 + x_2 - x_3)$ if $x_1 + 2x_2 + 3x_3 \geq 4$ and $3x_1 + 2x_2 + x_3 \leq 9$.

(b) Maximize $P = x_1 - x_2$ if $x_1 + x_2 - 3x_3 \leq 4$ and $5x_1 - 3x_2 + x_3 \leq -1$.

(c) Maximize $P = 2x_1 - 3x_2 - 2x_3$ for nonnegative x_1, x_2, x_3, $x_1 \leq 4$, $x_3 \leq 10$, $2x_3 - x_1 - x_2 \leq 7$, and $5x_1 + 3x_2 - 4x_3 \leq 8$.

Solution

Only (c) is in standard form; all three criteria are satisfied. Part (a) violates the definition three times. It asks for a minimum of the objective function and does not require that the variables be restricted to nonnegative values. Furthermore, the first inequality is $\geq$ rather than $\leq$. In (b), although we are to maximize P, we are *not* restricted to nonnegative variable values. Although the first constraint is proper, the second requires the expression to be less than a *negative* value, which is forbidden.

Skill Enhancer 1 Which (if any) are in standard form?

Minimize P subject to	Maximize P subject to
(a) $x_1 + 2x_2 \geq 4$	(b) $x_1 + 3x_2 \leq 4$
$x_1 \geq 0 \qquad x_2 \geq 0$	$2x_1 + x_2 \leq 5$

Maximize P subject to	Maximize P subject to
(c) $x_1 \geq 0 \qquad x_2 \geq 0$	(d) $x_1 \geq 0 \qquad x_2 \geq 0$
$10x_2 - 3x_1 \leq 12$	$4x_1 + 9x_2 \leq 6$
	$2x_1 + 3x_2 \geq \dfrac{1}{2}$

Answer in Appendix E.

It is natural to use subscript notation to describe variables used in a simplex problem. Subscript notation is always preferable in cases where the *number* of unknown variables varies from problem to problem, which may be the case with simplex problems. (Subscript notation also makes the transition to computer solution much easier.) For the remainder of this chapter, we will frequently use subscript notation to emphasize its importance for the simplex method.

Slack Variables

slack variables   Our first step shall be to convert all inequalities to equations by introducing additional *slack variables*. Slack variables represent the *nonnegative* difference (the "slack") of the two sides of an inequality. In any inequality, it is always possible to introduce such a quantity. For example,

$$3 < 7$$

and it is possible to introduce a slack variable s such that

$$3 + s = 7$$

In this trivial example, it is clear not only that s exists but that $s = 4$. We demand that any slack variable s introduced into a problem *must* satisfy $s \geq 0$.

As we see in Figure 4.1, it may help to think of a slack variable as a weight we add to the lighter side (the "less than" side) so that it weighs the same as—so that it equals—the heavier side.

Example 2   Suppose we wish to maximize

$$P = 2x_1 + x_2$$

subject to $x_1 \geq 0$, $x_2 \geq 0$ and to

(4.1) $$x_1 + 2x_2 \leq 6$$
(4.2) $$x_1 + x_2 \leq 4$$
(4.3) $$x_1 - x_2 \leq 2$$

Show how to introduce slack variables to transform the inequalities into a system of equations.

Solution

For Inequality (4.1), there is a variable s_1 such that

$$x_1 + 2x_2 + s_1 = 6$$

where $s_1 \geq 0$. Refer again to Figure 4.1.

In the same way, we may introduce slack variables s_2 and s_3 so that the three inequalities become the set of equations

$$
\begin{aligned}
x_1 + 2x_2 + s_1 &= 6 \\
x_1 + x_2 + s_2 &= 4 \\
x_1 - x_2 + s_3 &= 2
\end{aligned}
$$

In simplex method problems we also rewrite the objective function and include it in the system of equations that defines a problem. We do this by considering the function P to be yet another variable, then writing the objective function with all variables, now including P, on the left side. The coefficient of P should be $+1$. In this example, the objective function becomes

$$-2x_1 - x_2 + P = 0$$

(The equivalent equation $2x_1 + x_2 - P = 0$ is not acceptable since the coefficient of P would be -1.)

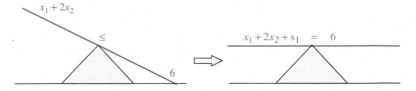

FIGURE 4.1
Slack variables make an inequality into a "balanced" equation.

In summary, this entire problem is now defined by the following system of *equations*:

$$
\begin{aligned}
x_1 + 2x_2 + s_1 \qquad\qquad\qquad &= 6 \\
x_1 + x_2 \quad + s_2 \qquad\qquad &= 4 \\
x_1 - x_2 \qquad\quad + s_3 \qquad &= 2 \\
-2x_1 - x_2 \qquad\qquad\quad + P &= 0
\end{aligned}
$$

This set of equations is described by six variables—the two original variables, three new slack variables (one for each constraint), and an additional variable corresponding to the objective function P. Remember, none of the slack variables (nor any of the original x_i variables) may ever take on *negative* values.

Skill Enhancer 2   Suppose we need to maximize $P = 3x_1 + x_2$ subject to $x_1 \geq 0$, $x_2 \geq 0$ and to

$$
\begin{aligned}
x_1 + x_2 &\leq 2 \\
2x_1 + x_2 &\leq 3
\end{aligned}
$$

Show how to introduce slack variables and how thereby to transform the system of inequalities into a system of equations.

Answer in Appendix E.

Maximum Problems: Getting Ready for an Algebraic Solution

1. Make sure the problem is in standard form.

2. Introduce nonnegative slack variables s_i.

3. Regard the objective function P as a variable. Rewrite the equation for P so that only constants appear to the right of the $=$, and so that the coefficient of P is $+1$.

Basic and Nonbasic Variables

In Example 2, there are four equations but *six* variables—too many variables or too few equations to determine all variables uniquely. We cannot hope to solve systems of equations *uniquely* unless there are *exactly* as many variables as equations. (Even then, unique solutions may not exist.) In the above equation system, two of the variables are somehow "in excess." Perhaps if we could somehow fix the values of two of them, it would be possible to discover unique values for the remaining four variables. It will be possible to do this.

nonbasic variables   The excess variables are termed *nonbasic variables*, whereas the remaining variables
basic variables   are *basic variables*.

Basic variables $\Leftrightarrow$ Variables we keep

Nonbasic variables $\Leftrightarrow$ Excess variables

Consider the equations in Example 2 that define the problem. If it were possible to declare x_1 and x_2 to be nonbasic, and further to set their values to 0, it would be possible to determine values for the basic variables s_1, s_2, s_3, and P. These solutions are particularly easy, because each of the basic variables occurs in a single equation and has a coefficient of $+1$.

$$
\begin{aligned}
0 + 2 \cdot 0 + s_1 &&&&&= 6 \\
0 + 0 &+ s_2 &&&&= 4 \\
0 - 0 && + s_3 &&&= 2 \\
-2 \cdot 0 - 0 &&&&+ P &= 0
\end{aligned}
$$

Make sure you see that the solutions under these circumstances for the basic variables are

$$s_1 = 6 \qquad s_2 = 4 \qquad s_3 = 2 \qquad P = 0$$

These comments will soon become particularly relevant in light of a restatement of the Fundamental Theorem of Linear Programming and the details of the simplex method. Under this restatement, it *will* be possible

- To *select nonbasic variables* that we may set to zero
- To ensure that *all basic variables occur in a single equation with a coefficient of $+1$* so that their values are easy to determine

The Fundamental Theorem: Algebraic Form

Example 2 provides a useful model for the kinds of things that happen in any simplex problem. To transform the three original inequalities to equations, we had to add the same number—three—of new slack variables.

$$3 \text{ inequalities} \Rightarrow 3 \text{ new slack variables}$$

There were two original variables in the problem, namely x_1 and x_2. Furthermore, we agreed to consider P to be another variable. We have in total

$$3 + 2 + 1$$

variables in the problem.

The Fundamental Theorem states that *if* there is a solution to this problem, then it will be possible to find a value for each of these $3 + 2 + 1 = 6$ variables, and this solution will be very special: in addition to P which will always remain basic, since there were three original inequalities, there will be three additional basic variables (a combination of the original variables or the slack variables). The two remaining variables (that is how many x_i variables there were) will be nonbasic, and can further be set equal to 0.

But what about a more general problem? We may not have just three inequalities, we may have c inequalities. (The parameter c should make you think of constraints.) And we may not just have two variables, we may have u distinct variables (here, u stands for unknown).

	c	u
	Constraints	Original variables
	Basic variables	Nonbasic variables
	Slack variables	0 values

(Remember, we use the term *constraint* as another word for *inequality* in this context.) That is, we have slack variables labeled

$$s_1, s_2, \ldots, s_c$$

which we introduced to transform the inequality constraints into equations. We also have original variables

$$x_1, x_2, \ldots, x_u$$

together with the objective function "variable" P for a total of

$$c + u + 1$$

distinct variables. The Fundamental Theorem states certain crucial information about the values of these variables at the point where P achieves its largest value (assuming an optimum value exists).

The Fundamental Theorem.
The $c + u + 1$ values of

$$x_1, \ x_2, \ \ldots, \ x_u, \ s_1, \ \ldots, \ s_c, \ P$$

that yield the largest value of P while satisfying the constraints of the problem include at least u zeros. That is, at least as many variables (in the augmented system) as there are original variables are excess (nonbasic) variables and have values of 0. The variable P is always basic.

It may be helpful to glance back at this chart as we proceed.

u original variables $\Leftrightarrow$ u nonbasic variables
c inequality constraints $\Leftrightarrow$ c basic variables
1 objective function $\Leftrightarrow$ 1 additional basic variable

As before, the Fundamental Theorem is silent about certain critical issues.

- Does a solution even exist?
- Which of the variables vanish at the solution point?
- And, most important, how do we identify the solution and the nonbasic variables of the solution?

Yet, as in the geometric approach, the theorem provides enough information to yield a strategy, as Example 3 shows.

Example 3   In Example 2 we examined the problem of maximizing

$$P = 2x_1 + x_2$$

subject to $x_1 \geq 0$, $x_2 \geq 0$ and to

$$x_1 + 2x_2 \leq 6$$
$$x_1 + x_2 \leq 4$$
$$x_1 - x_2 \leq 2$$

which gave rise to the system of equations

$$
\begin{aligned}
x_1 + 2x_2 + s_1 &= 6 \\
x_1 + x_2 \quad + s_2 &= 4 \\
x_1 - x_2 \quad\quad + s_3 &= 2 \\
-2x_1 - x_2 \quad\quad\quad + P &= 0
\end{aligned}
$$

Use the Fundamental Theorem to solve this linear programming problem.

Solution

There are 3 inequalities in the original formulation, and so there are 3 slack variables; therefore, $c = 3$ in the language of the theorem. There are 2 original variables, so $u = 2$. At the solution, there will be at least 2 zero values among the $5 = 3 + 2$ values of the x and s variables. The sixth variable P is always basic.

By experimentation, the two zeros can be distributed among the 5 variables in exactly 10 ways, and Table 4.1 displays them. (The asterisks in this table represent unknown values of the variables.) Since P is always basic, it will never take a value of 0, and we do not include it in this table.

TABLE 4.1 Distributing two zeros among five variables

	x_1	x_2	s_1	s_2	s_3
	0	0	*	*	*
	0	*	0	*	*
	0	*	*	0	*
	0	*	*	*	0
$\Rightarrow$	*	0	0	*	*
	*	0	*	0	*
	*	0	*	*	0
	*	*	0	0	*
	*	*	0	*	0
	*	*	*	0	0

When any 2 of the 5 values have been specified, the remaining 3 can be solved from the given equation system. For example, we will work out the case corresponding to the line tagged in Table 4.1 by the arrow, when $x_2 = s_1 = 0$. Then, the system

$$
\begin{aligned}
x_1 + 2x_2 + s_1 &= 6 \\
x_1 + x_2 \quad + s_2 &= 4 \\
x_1 - x_2 \quad\quad + s_3 &= 2
\end{aligned}
$$

becomes

(4.4) $$x_1 + 0 + 0 \qquad\qquad = 6$$
(4.5) $$x_1 + 0 \qquad + s_2 \qquad = 4$$
(4.6) $$x_1 - 0 \qquad\qquad + s_3 = 2$$

From Equation (4.4), we immediately conclude that $x_1 = 6$. Using this information in Equation (4.5), $s_2 = -2$. In the same way, using this fact in Equation (4.6) demands that $s_3 = -4$. We reject this solution right away, though, because no negative values for any variables are allowed; we say this potential solution is *not feasible*.

We follow this procedure—identify possible nonbasic variables, then solve for the values of the basic variables—to complete Table 4.2. The final column of this table contains additional information—the value of P. From this final column, we flag the largest value of P, and read off the values of the variables that lead to this maximum value.

TABLE 4.2 Solving for potential nonbasic variables

x_1	x_2	s_1	s_2	s_3	$P = 2x_1 + x_2$	
0	0	6	4	2	0	
0	3	0	1	5	3	
0	4	-2	0	6	nonfeasible	
0	-2	10	6	0	nonfeasible	
6	0	0	-2	-4	nonfeasible	
4	0	2	0	-2	nonfeasible	
2	0	4	2	0	4	
2	2	0	0	2	6	
$\frac{10}{3}$	$\frac{4}{3}$	0	$-\frac{2}{3}$	0	nonfeasible	
3	1	1	0	0	7	$\Leftrightarrow$ **max**

We have done this problem once before; using geometric methods in Example 10 of Chapter 3, page 152, we found that the maximum value is 7, occurring when $(x_1, x_2) = (3, 1)$. (There, we used the notation x and y.) The additional basic variable is $s_1 = 1$, and the nonbasic variables are s_2 and s_3. The solutions using either technique do agree, as they must.

Skill Enhancer 3 The previous Skill Enhancer showed a problem equivalent to the system of equations

$$x_1 + x_2 + s_1 \qquad\qquad = 2$$
$$2x_1 + x_2 \qquad + s_2 \qquad = 3$$
$$-3x_1 - x_2 \qquad\qquad + P = 0$$

(a) How many nonbasic variables are there? How many basic variables? If each nonbasic variable is equal to zero, how many ways are there of distributing these zero values among the variables of the problem?

(b) Complete a chart similar to that of Table 4.2 for this problem.

(c) Use the Fundamental Theorem to solve this problem.

Answer in Appendix E.

One strategy is clear.

1. Prepare a table like Table 4.1 (showing the ways values of 0 can be distributed among the variables of the problem).

2. Solve for the nonzero quantities, and reject any solutions where any of the x's or s's are negative.

3. Finally, compute P for each *feasible* solution. Choose the greatest value of P, noting the values of the basic variables that yield it.

Since P is always a basic variable of the problem, we frequently omit it from consideration in the process of setting up the problem for solution.

Example 4   Use an algebraic approach to find the largest value for

$$P = x - 2y$$

when $y \leq 3$ and $3x + 2y \leq 9$. Neither x nor y is permitted negative values.

Solution

The problem is in standard form. Upon switching notation so that

$$x_1 = x \qquad \text{and} \qquad x_2 = y$$

and introducing two slack variables, the constraints become

$$3x_1 + 2x_2 + s_1 \qquad = 9$$
$$x_2 \qquad + s_2 = 3$$

There are now a total of 4 variables but only 2 equations, so $4 - 2 = 2$ of them must assume zero values at the maximal point. Table 4.3 shows how these zeros can be distributed among 4 variables, and the computations are completed in Table 4.4.

TABLE 4.3   Distributing two zeros among four variables

x_1	x_2	s_1	s_2
0	0	*	*
0	*	0	*
0	*	*	0
*	0	0	*
*	0	*	0
*	*	0	0

The maximum value of $x_1 - 2x_2$ is 3, and this occurs when $x_1 = 3$ and $x_2 = 0$. (Make sure you agree with the results of Table 4.4. In particular, why does $x_2 = s_2 = 0$

TABLE 4.4 Completing the computations

x_1	x_2	s_1	s_2	$P = x_1 - 2x_2$	
0	0	9	3	0	
0	$4\frac{1}{2}$	0	$-1\frac{1}{2}$	nonfeasible	
0	3	3	0	−6	
3	**0**	**0**	**3**	**3**	$\Leftrightarrow$ **max**
	0		0	inconsistent	
1	3	0	0	−5	

lead to inconsistent results? Hint: Examine the constraints when these values are in force.)

Skill Enhancer 4   In some problem, we know that both x and y are nonnegative and that $x \le 4$ and $x + y \le 5$. We have to maximize the quantity $3x - y$ subject to these constraints.
(a) Introduce slack variables and write the system of equations equivalent to this linear programming problem.
(b) Construct a table similar to Table 4.4 for this problem.
(c) Use algebraic methods to find this maximum and the strategy for obtaining it.

Answer in Appendix E.

Example 5   A new development of homes will contain two types of houses. The profit from the sale of the Belvedere model is $20,000 per home, and that from the Huntington model is $30,000. The developer cannot build more than 10 homes per year. Each Huntington home uses twice as much electrical wire as the Belvedere, but because of industrywide shortages, the builder can obtain only as much wire as he would need for 12 Belvedere homes. What is his optimal building strategy? Use algebraic methods.

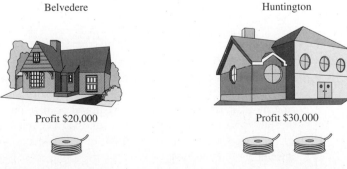

Belvedere

Profit $20,000

Huntington

Profit $30,000

FIGURE 4.2

Solution

This is a linear programming problem because we need to determine an optimal strategy subject to linear constraints which optimizes a linear objective function. Let x_1 and x_2 represent the number of Belvedere and Huntington homes to be built this year. Let P be the builder's profit.

$$P = 20,000x_1 + 30,000x_2 = 10,000(2x_1 + 3x_2)$$

The total that the builder will construct is $x_1 + x_2$; this quantity must be less than or equal to 10:

(4.7) $x_1 + x_2 \leq 10$

If q is the quantity of wire used in a single Belvedere home, then the builder knows that he can only obtain $12q$ wire in any year. Each Huntington home uses $2q$ wire. Therefore

$$qx_1 + 2qx_2 \leq 12q$$

Since $q \neq 0$, we may divide through to obtain

(4.8) $x_1 + 2x_2 \leq 12$

We transform Inequalities (4.7) and (4.8) by using slack variables s_1 and s_2:

$$x_1 + x_2 + s_1 \quad\quad = 10$$
$$x_1 + 2x_2 \quad\quad + s_2 = 12$$

These two equations contain 4 variables; this suggests that $2 = 4 - 2$ of the variables must be 0 at the optimal point. Refer again to Table 4.3 to see how the 0s can be distributed among 4 variables x_1, x_2, s_1, and s_2. In Table 4.5, the values of the nonzero variables are shown, along with the corresponding values of P. It is clear that the maximum profit is \$220,000, and the builder will earn this by building 8 Belvedere homes and 2 Huntington homes.

TABLE 4.5   Construction strategy

x_1	x_2	s_1	s_2	$P = 10,000(2x_1 + 3x_2)$
0	0	10	12	0
0	10	0	-8	nonfeasible
0	6	4	0	180,000
10	0	0	2	200,000
12	0	-2	0	nonfeasible
8	2	0	0	220,000

Skill Enhancer 5   Two types of bulb will be used in a small flower bed. We want to plant as many as possible to get the most flowers. However, more than 40 of the first kind or 30 of the second type will not thrive, as they will exhaust the soil's nutriments. Furthermore, studies at the local agricultural college indicate that three times the total of both bulbs plus the number of the second variety must not exceed 180 in a plot of this size. Let x and y be the number of the first and second varieties. Use algebraic methods to answer these questions.

(a) Introduce slack variables and write the system of equations for this problem.

(b) Complete the usual table like that of Table 4.2 for this problem.

(c) How many bulbs of each variety shall we plant?

Answer in Appendix E.

   Have we hit upon a workable strategy for solving linear programming problems algebraically? To be sure, we have freed ourselves from the geometric approach—this algebraic method will work for any *arbitrary* number of variables. The tradeoff appears to be an increase in calculation. When we solved Example 3 via geometry, we computed possible P values at 5 corners of a feasible region. With the algebraic approach, we computed these values at 10 points.

   Although any linear programming problem can be solved by this algebraic approach in theory, this method will not prove to be workable in practice. A relatively simple linear programming problem involving 20 variables and 10 constraints would require the computation of several hundred thousand possible P values, and a more realistic problem with perhaps 100 variables and 50 constraints might require something of the order of magnitude of 2^{100} P values to be computed, a task that would strain even the mightiest computers. The simplex method is a system for finding the optimal P that avoids testing the vast majority of potential P values. We will begin our exploration of the simplex method in the next section.

More About Basic Variables

algorithm   The heart of the simplex *algorithm*—steps for finding a solution—is a series of manipulations for identifying the basic variables and solving for their values. Remember, basic variables are easy to identify, because we will insist that *each basic variable occur in a single equation and have a coefficient of +1.* At the conclusion of the simplex method, we set the nonbasic variables to zero.

Example 6   Decide which variables in this system are basic, and which are nonbasic.

$$3x_1 + x_2 \qquad + 2s_1 + s_2 \qquad = 4$$
$$-x_1 \qquad + x_3 - 4s_1 + 3s_2 \qquad = 12$$
$$2x_1 \qquad \qquad + 8s_2 + P = \tfrac{7}{2}$$

Solution

Only P, x_2, and x_3 appear only once in the set of equations, and so are basic. The nonbasic variables are x_1, s_1, and s_2.

Skill Enhancer 6   In the system below, which variables are basic? Which are nonbasic?

$$2x_1 \qquad + 2s_2 + s_3 \qquad = 9$$
$$3x_1 + x_2 \qquad + 4s_2 \qquad = 9$$
$$3x_1 \qquad + s_1 - s_2 \qquad = 8$$
$$2x_1 \qquad \qquad + s_2 \qquad + P = 3$$

Answer in Appendix E.

As a general rule of thumb, *the number of nonbasic variables equals the number of original variables. There is one basic variable for each of the original inequality constraints, plus there is the additional basic variable P:*

$$\text{Number of } \textit{nonbasic} \text{ variables} = \text{Number of } \textit{original} \text{ variables}$$

$$\text{Number of } \textit{basic} \text{ variables} = \text{Number of } \textit{inequality constraints} + 1$$

Example 7   A certain problem may be put in the form of a standard linear programming problem in which there are four original variables constrained by six inequality constraints. When the problem is couched in algebraic language through the introduction of slack variables, how many basic and nonbasic variables will there be?

Solution

In the initial statement of the problem, the four original variables will appear as the nonbasic variables. The basic variables include the six slack variables (one for each constraint) plus one for the objective function "variable" P. Thus, there are 4 nonbasic variables and 7 basic variables.

Skill Enhancer 7   In another standard linear programming problem, five inequality constraints involve three variables x_1, x_2, and x_3. How many basic and nonbasic variables are there?

Answer in Appendix E.

Preparing for the Simplex Solution to Linear Programming Problems

1. Add one slack variable to convert each constraint inequality to an equation.

2. Regard the objective function P as another basic variable.

3. The problem is now described by $u + c + 1$ variables—u original variables, c slack variables, and the single objective function variable P. These variables occur in $c + 1$ equations—c from the constraints and 1 from the objective function.

_____ *Exercises 4.1* _____

In Exercises 1–6, determine which sets of inequality constraints are in standard form.

1. $x_1 \geq 1, x_2 \geq 1, x_1 + x_2 \leq 4$

2. $x_1 \geq 0, x_2 \geq 0, x_1 \leq 1, x_2 \leq 2, x_3 \leq 4$

3. $x_1 \geq 0, x_2 \geq 0, x_1 + x_2 + x_3 \leq 4.5$

4. $x_1 \geq 0, x_2 \geq 0, x_3 \geq 0, x_1 + x_2 + x_3 \leq 4.5, x_2 + x_3 \leq 3$

5. $x_1 \geq 0, x_2 \geq 0.001, x_1 + x_2 + 2x_3 \leq 4$

6. $x_1 \geq 0, x_2 \geq 0, x_3 \geq 0, x_4 \geq 0, x_1 + x_2 + 2x_3 \leq 4$

In Exercises 7–16, show how to transform the given group of inequality constraints into an equivalent group of equations via the addition of suitable slack variables.

7.
$$x_1 \geq 0$$
$$x_2 \geq 0$$
$$x_1 + 2x_2 \leq 3$$
$$2x_1 + x_2 \leq 2$$

8.
$$x_1 \geq 0$$
$$x_2 \geq 0$$
$$4x_1 + 3x_2 \leq 5$$
$$3x_1 + 2x_2 \leq 4$$

9.
$$x_1 \geq 0$$
$$x_2 \geq 0$$
$$3x_1 \leq 4$$
$$2x_1 + 2x_2 \leq 9$$

10.
$$x_1 \geq 0$$
$$x_2 \geq 0$$
$$x_1 \leq 3$$
$$x_2 \leq 3$$
$$x_1 + x_2 \leq 5$$

11.
$$x_1 \geq 0$$
$$x_2 \geq 0$$
$$x_1 + 2x_2 \leq 1$$
$$x_1 \leq \frac{3}{4}$$

12.
$$x_1 \geq 0$$
$$x_2 \geq 0$$
$$x_1 + x_2 \leq 3$$
$$x_2 \leq 2$$

13.
$$x_1 \geq 0$$
$$x_2 \geq 0$$
$$3x_1 + 4x_2 \leq 4$$
$$3x_1 + 3x_2 \leq 3.25$$

14.
$$x_1 \geq 0$$
$$x_2 \geq 0$$
$$x_1 \leq 3$$
$$x_2 \leq 4$$
$$x_1 + x_2 \leq 4$$

15.
$$x_1 \geq 0$$
$$x_2 \geq 0$$
$$x_3 \geq 0$$
$$x_1 + x_2 \leq 3$$
$$x_2 + x_3 \leq 4$$

16.
$$x_1 \geq 0$$
$$x_2 \geq 0$$
$$x_3 \geq 0$$
$$x_4 \geq 0$$
$$x_1 + 2x_2 + x_3 + x_4 \leq 5$$
$$x_1 + x_2 + 2x_3 + x_4 \leq 4$$
$$x_4 \leq 3$$
$$x_1 + x_2 + x_3 + 2x_4 \leq 4$$

Exercises 17–18 give practice in setting up problems. Do not solve them! Answer each of the questions about the setup instead.

17. We need to maximize $P = x_1 + x_2$ subject to the constraints
$$x_1 \geq 0 \qquad x_2 \geq 0$$
$$x_1 + 2x_2 \leq 6$$
$$2x_1 + x_2 \leq 6$$

(a) Use slack variables to transform this problem into a system of equations.
(b) How many basic variables are there in this problem?
(c) How many nonbasic variables are there?
(d) In solving this problem, we will (as in the text) try all the different ways of setting the nonbasic variables equal to zero. By experimentation, say how many ways there are of making this distribution.
(e) Setting s_1 and s_2 equal to zero is one of the ways of creating a feasible solution. What are the corresponding values of x_1, x_2, and P for this solution?
(f) Setting $x_1 = s_2 = 0$ gives rise to a nonfeasible solution. Why?

18. We need to maximize $P = x_1 + 2x_2$ subject to the constraints
$$x_1 \geq 0 \qquad x_2 \geq 0$$
$$2x_1 + 3x_2 \leq 10$$
$$3x_1 + 2x_2 \leq 10$$

(a) Use slack variables to transform this problem into a system of equations.
(b) How many basic variables are there in this problem?
(c) How many nonbasic variables are there?
(d) In solving this problem, we will (as in the text) try all the different ways of setting the nonbasic variables equal to zero. By experimentation, say how many ways there are of making this distribution.
(e) Setting s_1 and s_2 equal to zero is one of the ways of creating a feasible solution. What are the corresponding values of x_1, x_2, and P for this solution?
(f) Setting $x_1 = s_2 = 0$ gives rise to a nonfeasible solution. Why?

In each of Exercises 19–30, assume that all variables are prohibited from holding negative values (that is, $x_1 \geq 0$, $x_2 \geq 0$, and so on). Use the algebraic method outlined in this chapter to determine the maximum value of the objective function P and the proper values of the variables of the problem that achieve that solution.

19. $P = 2x_1 + 3x_2$
$$x_1 + x_2 \leq 2$$

20. $P = 3x_1 + 4x_2$
$$2x_1 + 3x_2 \leq 6$$

21. $P = x_1 + x_2$ **22.** $P = x_1 + 2x_2$

$x_1 + 2x_2 \leq 6$ $2x_1 + 3x_2 \leq 10$

$2x_1 + x_2 \leq 6$ $3x_1 + 2x_2 \leq 10$

23. $P = 2x_1 + 3x_2$ **24.** $P = 3x_1 + x_2$

$x_1 \leq 2$ $x_1 \leq 3$

$x_2 \leq 2$ $x_2 \leq 2$

$x_1 + x_2 \leq 3$ $x_1 + x_2 \leq 4$

25. $P = x_1 + x_2$ **26.** $P = 9x_1 - 3x_2$

$x_1 \leq 3$ $x_2 \leq 3$

$2x_1 + x_2 \leq 3$ $2x_1 + 3x_2 \leq 6$

27. $P = 5x_1 - x_2$ **28.** $P = -x_1 + 2x_2$

$2x_1 + x_2 \leq 3$ $x_1 \leq 3$

$x_1 + 2x_2 \leq 3$ $x_1 + x_2 \leq 4$

$x_1 + x_2 \leq \dfrac{3}{2}$ $x_1 + 3x_2 \leq 6$

29. $P = 4x_1 - 2x_2 + 3x_3$

$x_1 + x_2 \leq 4$

$x_1 + 3x_3 \leq 6$

30. $P = x_1 + x_2 + x_3$

$x_1 + 2x_2 + x_3 \leq 4$

$x_1 \leq 2$

$x_2 \leq 2$

$x_3 \leq 2$

Exercises 31–38 present a set of equations that describe linear programming problems. For each set, decide which variables are basic and which are nonbasic.

31. $x_1 + 2x_2 + s_1 = 3$ **32.** $x_1 + 2x_2 + s_1 = 6$

$-x_1 + s_1 + s_2 = 4$ $2x_1 - 3x_2 + s_2 = 8$

$-2x_1 + 3s_1 + P = 0$ $-2x_1 - 9x_2 + P = 0$

33. $x_1 + 2x_2 + s_1 = 4$ **34.** $2x_1 + x_2 + 3s_3 = 4$

$2x_1 + x_2 + s_2 = 0.5$ $-2x_1 + s_1 - s_3 = 2$

$3x_1 - 2x_2 + s_3 = 2$ $3x_1 + s_2 + 2s_3 = 2$

$-x_1 - 3x_2 + P = 0$ $-x_1 + 3s_3 + P = 5$

35. $x_1 + 2s_1 + s_2 = 1$ **36.** $x_1 + 2x_2 - s_1 = 4.5$

$x_2 + s_1 + 2s_2 = 7$ $3x_2 - s_1 + s_2 = 6.2$

$2s_1 + 3s_2 + s_3 = 5$ $2x_2 + s_1 + s_3 = 7.1$

$2s_1 + 4s_2 + P = 7$ $-3x_2 + 4s_1 + P = 2.1$

37. $2x_1 + x_2 + x_3 = 12$

$x_1 + 3x_2 + s_1 + 2s_2 = 17$

$-x_1 - 4x_2 + 2s_2 = -6$

38. $0.5x_1 - 0.2x_2 + 0.3x_3 + s_1 = 2$

$0.7x_1 + 0.1x_2 + 0.3x_3 + s_2 = 1$

$-0.8x_1 - 0.2x_2 - x_3 = 0$

Exercises 39–44 relate to linear programming problems that have been expressed in algebraic notation using the notation developed in this section. The number of original variables and inequality constraints is given for each such exercise. State how many basic and nonbasic variables there will be for each. (Do not forget to include the objective function P as a variable of the problem.)

39. three variables, three constraints

40. two variables, three constraints

41. four variables, three constraints

42. three variables, four constraints

43. three variables, five constraints

44. four variables, two constraints

45. one variable, two constraints

46. two variables, four constraints

Applications

47. *(Manufacturing)* A retired couple supplements their income by cleaning and repairing watches and old jewelry. The husband sees to the repairs, and the wife does the cleaning. Each works no more than 21 h per week. A watch needs 2 h of repairing and 1 h of cleaning, whereas a piece of old jewelry generally needs 1 h of repairing and 2 h of cleaning. They charge $20 to work on a watch, and $30 to work on a piece of jewelry. Use algebraic methods to determine the optimal strategy for the couple if they want to maximize their supplemental income.

48. *(Transportation)* A bicycle company has two factories, at the east and west ends of a certain city. Their two showrooms are in the north and south of the same city. At the moment, the east factory has 300 bikes and the west factory 400 bikes to deliver to the two showrooms. The south needs 200 and the north 400. The delivery charges from the east factory to the north and south are $10 and $15 for each bike. The charges from the west factory to the north and south are $20 and $20 per bike. Use algebraic methods to determine the shipping strategy that minimizes shipping costs.

 **Calculator Exercises**

Decide which of the variables in Exercises 49–50 are basic and which are nonbasic.

49. $2.634x_1 + x_2 + 4.777s_3 = 3.256$

$-2.22x_1 + s_1 - 7.234s_3 = 2.285$

$3.835x_1 + s_2 + 2.745s_3 = 3.374$

$-x_1 + 3s_3 + P = 5$

50. $x_1 + 1.96x_2 - 3.777s_1 = 5.182$

$3.295x_2 - 0.914s_1 + s_2 = 8.146$

$2.814x_2 + 0.714s_1 + s_3 = 1.621$

$-3x_2 + 4s_1 + P = 2.1$

Section 4.2   *MAXIMIZATION USING THE SIMPLEX METHOD*

The *algebraic* determination of a linear programming solution means finding the maximum value of P plus the values of the unknowns x_i and the values of the slack variables s_i that yield that maximum value. We often disregard the values of the slack variables when reporting the solution (but the values of s_i *may* be important, since each s_i gives the value of the unused resources in the ith constraint).

iterative method   The simplex method is an *iterative method* for determining this solution. That is, we begin with a guess for the x_i, a guess that is rarely correct, but that serves as a starting point. This initial guess must be feasible. On successive iterations, we refine the previous guess in a way that *guarantees* that we will sooner or later encounter the actual solution. Such a procedure—refining a guess based on the guess itself—may sound familiar. It is similar to any iterative procedure for solving problems on a computer (see Interlude 2), and the simplex method is well suited for computer implementation.

The Plan for This Section

Here is an outline of our tasks for this section.

1. We must first consider what goes into any iterative method. Any iterative method will consist of three steps.

2. We need to know how to make each of these steps apply to the simplex method.

3. Finally, we will examine an extended example to see how to solve maximum linear programming problems via the simplex method.

Designing an Iterative Method for the Simplex Method

Any iterative method should consist of three components:

1. A way to *make a first guess.*
2. A straightforward way to *refine the current guess.*
3. A way to test *whether the refined guess is the solution* or is acceptably close to the solution. If it is, stop. Otherwise, return to step 2.

Each of these steps has its counterpart in the simplex method.

Simplex Method: The First Guess

Make a *first guess* by letting the nonbasic variables be 0 and then solving for the remaining basic variables. At the outset, the nonbasic variables are the original variables, and the basic variables are the slack variables. This solution will never be correct, since it corresponds to the case where all original variables are 0. However, it is an easy first guess to make, it always leads to a feasible solution, and it gets the simplex method off to a smooth start.

Simplex Method: Refining the Guess

At the end of an iteration, we have a possible solution of the problem that is feasible. We know the basic variables and their values. Other variables are nonbasic, and have a value of zero. The expression for P will be given in terms of the nonbasic variables.

We will refine this guess (if possible) by rewriting the equations in a way that accomplishes two goals.

1. Exactly one nonbasic variable becomes basic (it will only appear in a single equation, and its coefficient is $+1$ in that equation).
2. Exactly one of the current basic variables becomes nonbasic.

It makes sense that the variable that contributes to the fastest growth of P should become basic (after all, we want to make P as big as possible). At the same time, one basic variable needs to become nonbasic. We choose that variable that limits the growth of the newly made basic variable.

Simplex Method: When to Stop

If the manipulations we have just described are not possible—that is, if none of the nonbasic variables will contribute to the growth of P—then the current guess is the solution. Stop at this point.

The flowchart of Figure 4.3 illustrates this strategy.

Although the simplex method itself is not geometric, a geometric interpretation of it is possible, as in Figure 4.4. Each iteration in the algorithm corresponds to a tracing from one corner to one of the adjacent ones. The path starts at some arbitrary corner

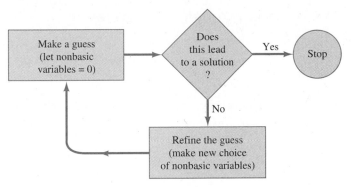

FIGURE 4.3
A flowchart for the simplex method.

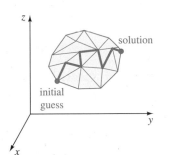

FIGURE 4.4
The "path" to a simplex solution.

and terminates at the corner yielding the optimum values of the original variables. The theory behind the simplex method guarantees that we will reach the optimal point. Computationally, this is an efficient method because of the many feasible corners that the method ignores on its way to the optimum.

An Extended Example

We will use the simplex method to solve the problem of finding the maximum value of

$$P = 3x + y$$

subject to nonnegative values for x and y and to

$$2x + 3y \leq 6$$
$$3x + 2y \leq 6$$
$$3x \leq 5$$

Since there are three nontrivial constraints, we need three slack variables. As a system of equations, this problem becomes

(4.9)
$$
\begin{aligned}
2x + 3y + s_1 &= 6 \\
3x + 2y \quad\;\; + s_2 \quad\quad &= 6 \\
3x \quad\quad\quad\;\; + s_3 &= 5 \\
-3x - y \quad\quad\quad\quad + P &= 0
\end{aligned}
$$

We choose the obvious *first guess*. Let x and y be nonbasic variables, with values of zero. Then

$$s_1 = 6 \qquad s_2 = 6 \qquad s_3 = 5 \qquad P = 0$$

a result that is neither surprising nor interesting. It is a feasible solution, however, and it enables us to begin the simplex method.

Extended Example: First Iteration

One of either x or y should become basic. Since

(4.10) $$P = 3x + y$$

it seems more advantageous to make x basic. Unit for unit, an increase in x contributes more to the growth of P than an increase of y, since the coefficient of x is greater than that of y. Referring back to Equations (4.9), the largest coefficient of Equation (4.10) appears as the variable with the most negative coefficient in the last equation of that system. Let's agree to *choose the variable with the most negative coefficient* in the last

pivot variable   line. We will call that variable the *pivot variable* of this iteration for reasons that will become clear when we use matrices to solve these problems. In this iteration, x is the pivot variable.

The larger x becomes, the larger P is, but the constraints limit the growth of x. What is this limit? To determine that, refer back to Equations (4.9). Remembering that y is nonbasic and has a value of 0, we can solve each of the first three equations for x.

$$x = 3 - \frac{s_1}{2} \quad \Rightarrow x_M = 3$$

$$x = 2 - \frac{s_2}{3} \quad \Rightarrow x_M = 2$$

$$x = \frac{5}{3} - \frac{s_3}{3} \quad \Rightarrow x_M = \frac{5}{3}$$

Since no slack variable can take a negative value, the maximum values for x occur when each of the s_i is zero. (The notation x_M suggests the maximum value for x.) For example, in the equation

$$x = 3 - \frac{s_1}{3}$$

x *decreases* as s_1 *increases*. The maximum value that x can achieve happens when s_1 assumes its minimum value, which is zero.

Of these three maximum values, only $x = \frac{5}{3}$ ensures that *all* the basic variables s_i are greater than or equal to zero. (Think about it. Should $x = 2$, for example, then $s_1 = 2$ but $s_3 = -1$.) In Equations (4.9), this constraint on x is the third equation. Referring back to this system, we see that if we compute the ratio of the constant at the right of each equation with the coefficient of the pivot variable, then this equation is the

pivot ratio   one with the lowest such ratio, or *pivot ratio*.

If we make s_3 nonbasic in the next iteration, then it will take a zero value and so guarantee that x assumes its maximum value. We make x basic and s_3 nonbasic by solving this equation for x and substituting the resulting expression in the remaining equations. From the third equation, we have

$$x = \frac{5}{3} - \frac{s_3}{3}$$

The first equation, $2x + 3y + s_1 = 6$, becomes, for example,

$$2x + 3y + s_1 = 6$$

$$2\left(\frac{5}{3} - \frac{s_3}{3}\right) + 3y + s_1 = 6$$

$$3y + s_1 - \frac{2}{3}s_3 = \frac{8}{3}$$

After performing the same substitution on the rest of Equations (4.9), they become

(4.11)
$$
\begin{aligned}
3y + s_1 & & - \tfrac{2}{3}s_3 & & = \tfrac{8}{3} \\
2y & + s_2 & - s_3 & & = 1 \\
x & & + \tfrac{1}{3}s_3 & & = \tfrac{5}{3} \\
-y & & + s_3 & + P & = 5
\end{aligned}
$$

This suggests that the basic variables are now x, s_1, and s_2 (and P), whereas y and s_3 are nonbasic.

Extended Example: Second Iteration

The last of Equations (4.11) implies that

$$P = 5 + y - s_3$$

and so we can increase P by increasing y. (If we tried to increase s_3, P would decrease, so s_3 cannot be the pivot variable.) We select y as the pivot variable in this iteration. This is consistent with our previous observation that the pivot variable is the one with the most negative coefficient in a system written like Equations (4.11).

Therefore, we seek to make y a basic variable in this iteration. Which variable will become nonbasic? We examine Equations (4.11) to see how they constrain y. As in the first iteration, we seek to find the bound on the increase in y.

After setting $s_3 = 0$ (since it is a nonbasic variable), the first and second equations become

$$3y + s_1 = \frac{8}{3} \quad \Rightarrow y_M = \frac{8}{9}$$

$$2y + s_2 = 1 \quad \Rightarrow y_M = \frac{1}{2}$$

The third equation contributes nothing to this discussion since it does not contain y, so we ignore it. We conclude that the largest value for y is $\frac{1}{2}$; this is the largest value that violates no constraints on any variables. The second equation determines this constraint. If we compute the pivot ratios—the ratios of the constant term on the right side of each equation with the coefficient of the current pivot variable—it is this second equation that has the lowest such ratio, consistent with our observation in the previous iteration.

To conclude this iteration, we make y basic by solving for y in the second equation and using this expression to eliminate y in the other equations. As an exercise, the reader should show that (4.11) becomes

$$
\begin{aligned}
s_1 - \tfrac{3}{2}s_2 + \tfrac{5}{6}s_3 \quad\quad &= \tfrac{7}{6} \\
y \;+\; \tfrac{1}{2}s_2 - \tfrac{1}{3}s_3 \quad\quad &= \tfrac{1}{2} \\
x \quad\quad\quad + \tfrac{1}{3}s_3 \quad\quad &= \tfrac{5}{3} \\
\tfrac{1}{2}s_2 + \tfrac{1}{2}s_3 + P &= 5\tfrac{1}{2}
\end{aligned}
$$

(4.12)

Extended Example: Done?

The final equation of (4.12) states that

$$
P = 5\frac{1}{2} - \frac{1}{2}(s_2 + s_3)
$$

Any nonzero value for s_2 or s_3 will reduce P, so *it is not possible to maximize P any further without violating constraints.* We agree therefore to stop the simplex method here. This corresponds to there being *no* negative coefficients in the last line of Equations (4.12).

And the solution? The maximum value of P is $5\frac{1}{2}$ when $x = \frac{5}{3}$ and $y = \frac{1}{2}$ (and $S_2 = S_3 = 0$).

With this example under our belts, let us summarize the steps involved in the simplex process.

The Simplex Algorithm

1. *Introduce slack variables* to the problem to transform it to a system of equations. Write each equation so it is set equal to a constant on the right side of the $=$ sign. The last equation of this system should contain P, the objective function. P should appear as a basic variable.

2. *Begin* the algorithm with an obvious first guess—only the slack variables (and P) are basic, and the original variables of the problem are nonbasic.

3. *Prepare for the next iteration.*
   - At the start of each iteration, *select as the pivot variable* the variable in the bottom row with the largest negative coefficient.
   - For each other equation, *compute the pivot ratio* and select as the *pivot constraint* the equation with the smallest positive ratio.
   - Use this equation to solve for the pivot variable and then substitute this equation for the pivot variable into the remaining equations of the iteration to make this variable basic.

4. After completing the calculations involved in the iteration, *examine the bottom equation,* the one containing P. Are there any negative coefficients? If not, stop and read the solution from the new system. If so, return to step 3, and perform another iteration.

pivot constraint

This extended example can also be solved using the geometric techniques of the previous chapter. The feasible region of the problem appears in Figure 4.5. There are

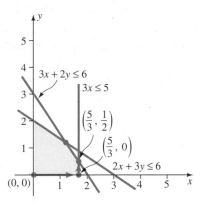

FIGURE 4.5
The feasible region for the extended example. The arrows indicate the path by which the simplex method approached the true solution.

five corners to the feasible region. When using the simplex method, our initial guess corresponds to the origin. The solution given by the first iteration corresponds to the corner lying on the x axis, and the second iteration takes us to the true solution.

_____Exercises 4.2_____

In Exercises 1–6, you are given a linear programming problem expressed "traditionally" in terms of inequalities. Show how to add slack variables to convert the set of inequalities to a system of equations.

1. Maximize $P = 10x_1 + 12x_2$ subject to $x_1 \geq 0$, $x_2 \geq 0$, $x_1 + 2x_2 \leq 4$, $4x_1 + 3x_2 \leq 4$, and $3x_1 + x_2 \leq 9$.

2. Maximize $P = 0.25x + 1.125y$ subject to $x \geq 0$, $y \geq 0$, $6.1x + 0.1y \leq 4.5$, $5.5x - 4.4y \leq 3.3$, and $x \leq 3$.

3. Maximize $P = \frac{3}{4}x_1 + \frac{3}{4}x_2$ when $x_1 \geq 0$, $x_2 \geq 0$, and $\frac{1}{2}x_1 + \frac{2}{3}x_2 \leq \frac{7}{8}$, $x_1 + 1\frac{1}{4}x_2 \leq 1\frac{1}{3}$, and $\frac{3}{4}x_1 + x_2 \leq 1$.

4. Maximize $P = \frac{1}{5}x_1 + \frac{2}{5}x_2$ subject to $x_1 \geq 0$, $x_2 \geq 0$, $x_1 + 2\frac{1}{5}x_2 \leq 2$, $\frac{4}{5}x_1 + 2x_2 \leq 2$, and $\frac{3}{5}x_1 + \frac{7}{10}x_2 \leq 3\frac{1}{5}$.

5. Maximize $P = x_1 - x_2 + 2x_3$ subject to $x_1 \geq 0$, $x_2 \geq 0$, $x_3 \geq 0$, and to $2x_1 + 3x_2 + x_3 \leq 7$, $x_1 + 3x_2 + 2x_3 \leq 6$, and $5x_1 + 3x_3 \leq 4$.

6. Maximize $P = 2x_1 + 3x_2$ subject to $x_1 \geq 0$, $x_2 \geq 0$, $x_3 \geq 0$, $x_1 - 3x_3 \leq 4$, $2x_2 + 3x_3 \leq 5$, and $x_1 + x_2 + 4x_3 \leq 4$.

Use the simplex method to solve the maximization problems

in Exercises 7–14. In all problems, assume that x_1 and x_2 are nonnegative.

7. Maximize $P = 2x_1 + 2x_2$ if $x_1 \leq 1$ and $x_2 \leq 2$.

8. Maximize $P = 4x_1 + x_2$ if $x_1 \leq \frac{3}{2}$ and $x_2 \leq 2$.

9. Maximize $P = 2x_1 + 3x_2$ if $x_1 \leq 2$ and $x_1 + x_2 \leq 2$.

10. Maximize $p = 8x_1 + 3x_2$ when $x_1 + 2x_2 \leq 5$ and $x_2 \leq 2$.

11. Maximize $P = x_1 + 3x_2$ if $2x_1 + 3x_2 \leq 6$ and $3x_1 + 2x_2 \leq 6$.

12. Maximize $P = \frac{1}{2}x_1 + x_2$ if $2x_1 + 3x_2 \leq 6$, $2x_1 + 2x_2 \leq 5$.

13. Maximize $P = x_1 + 2x_2$ if $x_1 \leq 2$, $x_2 \leq 2$, and $x_1 + x_2 \leq 3$.

14. Maximize $P = x_1 + 2x_2$ if $2x_1 + 3x_2 \leq 6$, $2x_1 + 2x_2 \leq 5$, and $x_2 \leq \frac{3}{2}$.

W 15. In any given problem, will the number of basic variables or the number of nonbasic variables ever change? Why or why not?

W 16. In any problem, will the variable P ever be a nonbasic variable? Why or why not?

Section 4.3   MATRICES AND THE SIMPLEX METHOD

The use of *augmented matrices* on the system of equations describing any linear programming problem streamlines the simplex method. The simplex method reduces to a sequence of matrix row transformations.

Why should this be so? Glance again at the summary of the simplex algorithm that appears on page 200. After we introduce slack variables, we know we can form a matrix of coefficients from these equations; this would be the augmented matrix. At any step in the procedure we can read off the current values for the variables by setting the nonbasic variables to zero and picking off the values for the basic variables.

Beginning an iteration by selecting a pivot variable means selecting the column with the most negative element in the bottom row. After computing the pivot ratios, the pivot constraint becomes the row with the smallest such ratio.

Making the pivot variable basic means manipulating the equations so that its coefficient in the pivot row is 1, while it disappears in the rows above and below. That is, we *pivot* the augmented matrix about this element. (Now the source of the term *pivot variable* becomes clear.) Upon completion of the pivot, we examine the elements of the bottom row to see if we are done or not.

We will begin by practicing the individual skills necessary to solve a simplex problem using matrices.

In any iteration, it is necessary to know which variables are the basic variables.

Identifying Basic Variables

In an augmented matrix from a linear programming problem, the basic variables have *columns* containing all zeros except for a single element whose value is 1.

Example 8   The following matrix represents the simplex tableau for some linear programming problem after several cycles in the simplex algorithm. Identify the basic variables.

$$
\begin{array}{cccccc}
x_1 & x_2 & s_1 & s_2 & s_3 & P \\
\end{array}
$$

$$
\left(
\begin{array}{cccccc|c}
3 & 0 & 1 & -1 & 0 & 0 & 2 \\
1 & 1 & 0 & 1 & 0 & 0 & 2 \\
1 & 0 & 0 & 0 & 1 & 0 & \frac{3}{4} \\
\hline
-1 & 0 & 0 & -3 & 0 & 1 & 6 \\
\end{array}
\right)
$$

Solution

Columns 1 and 4 contain several nonzero elements, so x_1 and s_2 are nonbasic. The remaining columns are all zeros except for a single component of value 1, so the variables x_2, s_1, s_3, and P are basic.

Skill Enhancer 8   In this simplex tableau, which variables are basic? Here, $*$ refers to some nonzero value in the matrix.

$$
\begin{array}{ccccccc}
x_1 & x_2 & x_3 & s_1 & s_2 & P & \\
\left(\begin{array}{cccccc|c}
0 & 0 & * & * & 0 & 0 & * \\
1 & * & * & 0 & 0 & 0 & 0 \\
0 & 0 & * & * & 1 & 0 & * \\
\hline
0 & * & * & 0 & 0 & 1 & *
\end{array}\right)
\end{array}
$$

Answer in Appendix E.

pivot column   Once we know the *pivot column*—the column corresponding to the pivot variable—we compute the pivot ratio for each row (except the bottom row).

Constructing the Pivot Ratio

$$\text{Pivot ratio} = \frac{\text{entry in rightmost column}}{\text{entry in pivot column}}$$

1. Compute pivot ratios for each row of the matrix except for the bottom one.

2. Ignore any pivot ratios that are negative or undefined.

Example 9   Calculate the pivot ratios for this simplex tableau.

$$
\begin{array}{cccccc}
x_1 & x_2 & s_1 & s_2 & s_3 & P \\
\left(\begin{array}{cccccc|c}
3 & 0 & 1 & -1 & 0 & 0 & 2 \\
1 & 1 & 0 & 1 & 0 & 0 & 2 \\
1 & 0 & 0 & 0 & 1 & 0 & \frac{3}{4} \\
\hline
-1 & 0 & 0 & -3 & 0 & 1 & 6
\end{array}\right)
\end{array}
$$

Solution

The pivot variable is s_2. The pivot ratio for the first row is therefore 2 divided by -1, the entry in the s_2 column in the first row. Here is the matrix with the pivot ratios displayed at the left.

$$
\begin{array}{r}
\frac{2}{-1} = -2 \\
\frac{2}{1} = 2 \\
\frac{(3/4)}{0} \Rightarrow \text{undefined}
\end{array}
\quad
\begin{array}{cccccc}
x_1 & x_2 & s_1 & s_2 & s_3 & P \\
\left(\begin{array}{cccccc|c}
3 & 0 & 1 & -1 & 0 & 0 & 2 \\
1 & 1 & 0 & 1 & 0 & 0 & 2 \\
1 & 0 & 0 & 0 & 1 & 0 & \frac{3}{4} \\
\hline
-1 & 0 & 0 & -3 & 0 & 1 & 6
\end{array}\right)
\end{array}
$$

We never compute the pivot ratio of the last row in any simplex tableau.

Skill Enhancer 9   What are the pivot ratios for this simplex tableau?

$$
\begin{array}{ccccc}
x_1 & x_2 & s_1 & s_2 & P \\
\left(\begin{array}{ccccc|c}
2 & 2 & 2 & 0 & 0 & 2 \\
1 & 0 & 3 & 1 & 0 & 6 \\
\hline
-1 & 0 & -10 & 0 & 1 & 11
\end{array}\right)
\end{array}
$$

Answer in Appendix E.

pivot element

The pivot row is the row with the *smallest* pivot ratio. The *pivot element* stands at the intersection of the pivot row with the pivot column. We recall that pivoting involves nothing more than the use of row reduction operations, and thus each successive matrix (in a series of pivots) is equivalent to the preceding one. That is, the set of solutions for each matrix remains the same.

Example 10   Given the matrix in Example 9, identify the pivot element and pivot about that element. For easy reference, this matrix is

$$
\begin{array}{cccccc}
x_1 & x_2 & s_1 & s_2 & s_3 & P \\
\end{array}
$$
$$
\left(
\begin{array}{cccccc|c}
3 & 0 & 1 & -1 & 0 & 0 & 2 \\
1 & 1 & 0 & 1 & 0 & 0 & 2 \\
1 & 0 & 0 & 0 & 1 & 0 & \frac{3}{4} \\
\hline
-1 & 0 & 0 & -3 & 0 & 1 & 6 \\
\end{array}
\right)
$$

Solution

In Example 9, we noted that the pivot variable is s_2 and that, among the permissible pivot ratios, the second row has the smallest. Therefore, the component at the second row and fourth column is the pivot element.

$$
\begin{array}{cccccc}
x_1 & x_2 & s_1 & s_2 & s_3 & P \\
\end{array}
$$
$$
\left(
\begin{array}{cccccc|c}
3 & 0 & 1 & -1 & 0 & 0 & 2 \\
1 & 1 & 0 & \boxed{1} & 0 & 0 & 2 \\
1 & 0 & 0 & 0 & 1 & 0 & \frac{3}{4} \\
\hline
-1 & 0 & 0 & -3 & 0 & 1 & 6 \\
\end{array}
\right)
$$

Pivoting about this boxed element using row reduction operations yields

$$
\left(
\begin{array}{cccccc|c}
4 & 1 & 1 & 0 & 0 & 0 & 4 \\
1 & 1 & 0 & 1 & 0 & 0 & 2 \\
1 & 0 & 0 & 0 & 1 & 0 & \frac{3}{4} \\
\hline
2 & 3 & 0 & 0 & 0 & 1 & 12 \\
\end{array}
\right)
$$

Skill Enhancer 10 Identify the pivot element and pivot about that element.

$$
\begin{array}{ccccc}
x_1 & x_2 & s_1 & s_2 & P \\
\end{array}
$$
$$
\left(
\begin{array}{ccccc|c}
2 & 2 & 2 & 0 & 0 & 2 \\
1 & 0 & 3 & 1 & 0 & 6 \\
\hline
-1 & 0 & -10 & 0 & 1 & 11 \\
\end{array}
\right)
$$

Answer in Appendix E.

Once the simplex algorithm is complete, it is possible to read off the answers from the final matrix.

Example 11   The augmented matrix

$$\begin{array}{cccccc} x_1 & x_2 & s_1 & s_2 & s_3 & P \\ \end{array}$$
$$\left(\begin{array}{cccccc|c} 4 & 1 & 1 & 0 & 0 & 0 & 4 \\ 1 & 0 & -2 & 1 & 0 & 0 & 2 \\ 1 & 0 & 0 & 0 & 1 & 0 & 2\frac{1}{2} \\ \hline 2 & 0 & 7 & 0 & 0 & 1 & 12 \end{array}\right)$$

appears in the final iteration in a simplex solution. Why? What is the solution to that problem?

Solution

Since there are no negative elements in the bottom row, this matrix is the final one. The maximum value of P is 12, since this is the element in the lower right corner of the matrix. The variables x_1 and s_1 are nonbasic, since there is more than one nonzero element in their columns; we conclude that $x_1 = s_1 = 0$ in this problem. Eliminating these columns and the column for P leaves

$$\begin{array}{ccc} x_2 & s_2 & s_3 \\ \end{array}$$
$$\left(\begin{array}{ccc|c} 1 & 0 & 0 & 4 \\ 0 & 1 & 0 & 2 \\ \hline 0 & 0 & 1 & 2\frac{1}{2} \end{array}\right)$$

from which $x_2 = 4$, $s_2 = 2$, and $s_3 = 2\frac{1}{2}$.

Skill Enhancer 11   Suppose

$$\begin{array}{ccccc} x_1 & x_2 & s_1 & s_2 & P \\ \end{array}$$
$$\left(\begin{array}{ccccc|c} 1 & 0 & 2 & 4 & 0 & 6 \\ 0 & 1 & -3 & 5 & 0 & 4 \\ \hline 0 & 0 & 9 & 2 & 1 & 14 \end{array}\right)$$

represents the final matrix in a simplex problem.
(a) What are the nonbasic variables and their values?
(b) What are the basic variables and their values?
(c) What is the maximum value of P?

Answer in Appendix E.

Simplex Algorithm: Augmented Matrix Form

1. *Add slack variables* to convert the constraints to a set of equalities. Include the objective function in this system.
2. Create the *augmented matrix* for this system.
3. *Select the pivot element* and pivot the matrix.
4. *Are all entries of the bottom row to the left of the P column nonnegative?* If so, stop and read off the answers. Otherwise, return to step 3.

Selection of the pivot element itself involves several steps.

> **The Simplex Algorithm: Selecting the Pivot Element**
>
> **1.** Examine the entries in the bottom row of the matrix. The *pivot column* contains the most negative of these entries.
>
> **2.** Compute the *pivot ratio* for each row except the bottom.
>
> **3.** Ignore negative or undefined pivot ratios.
>
> **4.** The *pivot row* is the row with the smallest positive pivot ratio. The *pivot element* is the element at the intersection of the pivot row and pivot column.
>
> **5.** P is always a basic variable, so neither of the rightmost two columns can ever be considered the pivot column.
>
> **6.** The bottom row can *never* be the pivot row.

Here's one additional comment. Every so often, the two smallest ratios will be equal. In that case, we may choose either ratio as the pivot ratio. In other words, choose either of the corresponding pivot elements when the two smallest pivot ratios are the same.

To see all the steps of the simplex method at work, we shall solve again the extended example which begins on page 197. This time, we use formal subscript notation, referring to x_1 and x_2 instead of x and y.

The problem is to maximize $P = 2x_1 + x_2$ subject to $x_1 \geq 0$, $x_2 \geq 0$ and to

$$x_1 + 2x_2 \leq 6$$
$$x_1 + x_2 \leq 4$$
$$x_1 - x_2 \leq 2$$

The equations

$$
\begin{aligned}
x_1 + 2x_2 + s_1 &= 6 \\
x_1 + x_2 \quad\quad + s_2 &= 4 \\
x_1 - x_2 \quad\quad\quad\quad + s_3 &= 2 \\
-2x_1 - x_2 \quad\quad\quad\quad\quad\quad + P &= 0
\end{aligned}
$$

initial simplex tableau   generate an augmented matrix called the *initial simplex tableau*. The columns contain the coefficients for all variables of the problem—original, slack, and objective function. The final *row* of the matrix *always* represents the equation for the objective function, and the horizontal line above it helps emphasize this.

(4.13)

$$
\begin{array}{cccccc|c}
x_1 & x_2 & s_1 & s_2 & s_3 & P & \\
1 & 2 & 1 & 0 & 0 & 0 & 6 \\
1 & 1 & 0 & 1 & 0 & 0 & 4 \\
1 & -1 & 0 & 0 & 1 & 0 & 2 \\
\hline
-2 & -1 & 0 & 0 & 0 & 1 & 0
\end{array}
$$

The basic variables at this point in the solution are s_1, s_2, s_3, and P. (Why?)

The first step in any cycle (iteration) of the simplex method is to identify the pivot column. In Equation (4.13), the most negative entry of the last row is -2, so x_1 is the pivot variable and the first column is the pivot column.

We need the pivot ratio for each row. Here is matrix (4.13) with the pivot ratios displayed at the left. The *boxed entry* marks the intersection of the pivot variable column with the row possessing the smallest pivot ratio.

$$
(4.14) \qquad
\begin{array}{c}
\frac{6}{1}=6 \\[2pt]
\frac{4}{1}=4 \\[2pt]
\frac{2}{1}=2 \\[2pt]
\\
\end{array}
\begin{array}{cccccc}
x_1 & x_2 & s_1 & s_2 & s_3 & P \\
\end{array}
\left(
\begin{array}{cccccc|c}
1 & 2 & 1 & 0 & 0 & 0 & 6 \\
1 & 1 & 0 & 1 & 0 & 0 & 4 \\
\boxed{1} & -1 & 0 & 0 & 1 & 0 & 2 \\
\hline
-2 & -1 & 0 & 0 & 0 & 1 & 0 \\
\end{array}
\right)
$$

Upon pivoting about the boxed entry in matrix (4.14), we obtain the following matrix.

$$
(4.15) \qquad
\begin{array}{cccccc}
x_1 & x_2 & s_1 & s_2 & s_3 & P \\
\end{array}
\left(
\begin{array}{cccccc|c}
0 & 3 & 1 & 0 & -1 & 0 & 4 \\
0 & 2 & 0 & 1 & -1 & 0 & 2 \\
1 & -1 & 0 & 0 & 1 & 0 & 2 \\
\hline
0 & -3 & 0 & 0 & 2 & 1 & 4 \\
\end{array}
\right)
$$

There is still one negative element in the bottom row, so we proceed.

To select the new pivot element, note first that -3 is the largest negative number (and the only negative number) that appears in the bottom row. The second column (the x_2 column) is the pivot column. The pivot ratios are

$$
\frac{4}{3} \qquad \frac{2}{2}=1 \qquad \text{and} \qquad \frac{2}{-1}=-2 \ \ (ignore)
$$

The second ratio is the smallest positive ratio, so the pivot element is the element in the second row and second column. Here is this matrix with the pivot element boxed, and the resulting matrix when the pivot operation is completed.

$$
\left(
\begin{array}{cccccc|c}
0 & 3 & 1 & 0 & -1 & 0 & 4 \\
0 & \boxed{2} & 0 & 1 & -1 & 0 & 2 \\
1 & -1 & 0 & 0 & 1 & 0 & 2 \\
\hline
0 & -3 & 0 & 0 & 2 & 1 & 4 \\
\end{array}
\right)
\Rightarrow
\begin{array}{cccccc}
x_1 & x_2 & s_1 & s_2 & s_3 & P \\
\end{array}
\left(
\begin{array}{cccccc|c}
0 & 0 & 1 & -\frac{3}{2} & \frac{1}{2} & 0 & 1 \\
0 & 1 & 0 & \frac{1}{2} & -\frac{1}{2} & 0 & 1 \\
1 & 0 & 0 & \frac{1}{2} & \frac{1}{2} & 0 & 3 \\
\hline
0 & 0 & 0 & \frac{3}{2} & \frac{1}{2} & 1 & 7 \\
\end{array}
\right)
$$

There are no negative elements in the bottom row, so this has been the final cycle. Note that the *nonbasic variables* are s_2 and s_3, so we set these values to 0. Mentally delete these columns from the final matrix to give

$$
\begin{array}{cccc}
x_1 & x_2 & s_1 & P \\
\end{array}
\left(
\begin{array}{ccc|c}
0 & 0 & 1 & 0 & 1 \\
0 & 1 & 0 & 0 & 1 \\
1 & 0 & 0 & 0 & 3 \\
0 & 0 & 0 & 1 & 7 \\
\end{array}
\right)
$$

To read the values of the basic variables, convert this *back* to a system of equations:

$$s_1 = 1 \qquad x_2 = 1 \qquad x_1 = 3 \qquad P = 7$$

Once again, we reproduce the actual solution. In Figure 4.6, we see the solution as the geometric method reveals it and the path traced out by the simplex method.

The next few examples provide practice in using matrix methods to solve simplex problems.

Example 12   Maximize $P = 2x_1 + 3x_2$ subject to $x_1 \geq 0$, $x_2 \geq 0$ and to

$$
\begin{aligned}
4x_1 + x_2 &\leq 4 \\
x_1 + x_2 &\leq 2 \\
x_1 &\leq \tfrac{3}{4}
\end{aligned}
$$

Use the simplex method.

Solution

Begin by introducing slack variables s_1, s_2, and s_3. Write the information of the problem in the usual way:

$$
\begin{aligned}
4x_1 + x_2 + s_1 &\qquad\qquad\qquad = 4 \\
x_1 + x_2 \qquad + s_2 &\qquad\qquad = 2 \\
x_1 \qquad\qquad + s_3 &\qquad = \tfrac{3}{4} \\
-2x_1 - 3x_2 \qquad\qquad\qquad &+ P = 0
\end{aligned}
$$

to which corresponds the augmented matrix

$$
\Rightarrow
\begin{array}{cccccc}
x_1 & x_2 & s_1 & s_2 & s_3 & P \\
\end{array}
\left(
\begin{array}{cccccc|c}
4 & 1 & 1 & 0 & 0 & 0 & 4 \\
1 & \boxed{1} & 0 & 1 & 0 & 0 & 2 \\
1 & 0 & 0 & 0 & 1 & 0 & \tfrac{3}{4} \\
\hline
-2 & -3 & 0 & 0 & 0 & 1 & 0
\end{array}
\right)
$$

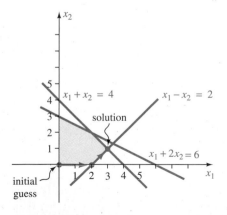

FIGURE 4.6
The solution to a linear programming problem—geometric and simplex methods.

In the bottom row, -3 is the most negative entry, so the second column is the pivot column. The pivot ratios with respect to this column are

$$\frac{4}{1} = 4 \qquad \frac{2}{1} = 2 \quad \text{and} \quad \frac{3/4}{0} \Rightarrow \text{undefined}$$

The smallest legitimate value is 2, so the second row is the pivot row. We pivot about the pivot element to get the following:

$$
\begin{array}{c}
4 \\ 1 \\ 1 \\ \\ -2
\end{array}
\left(
\begin{array}{ccccc|c}
1 & 1 & 0 & 0 & 0 & 4 \\
\boxed{1} & 0 & 1 & 0 & 0 & 2 \\
0 & 0 & 0 & 1 & 0 & \frac{3}{4} \\
\hline
-3 & 0 & 0 & 0 & 1 & 0
\end{array}
\right)
\Rightarrow
\begin{array}{ccccccc}
x_1 & x_2 & s_1 & s_2 & s_3 & P & \\
\end{array}
\left(
\begin{array}{ccccc|c}
3 & 0 & 1 & -1 & 0 & 0 & 2 \\
1 & 1 & 0 & 1 & 0 & 0 & 2 \\
1 & 0 & 0 & 0 & 1 & 0 & \frac{3}{4} \\
\hline
1 & 0 & 0 & 3 & 0 & 1 & 6
\end{array}
\right)
$$

The bottom row contains only positive or zero elements, so no more iterations are necessary (or possible).

    To read the solution, translate the matrix back into equation form. First, set to zero the nonbasic variables in this iteration. These are x_1 and s_2; ignore their columns. (Do you see how to identify these as nonbasic? These variables correspond to the variable columns with more than one nonzero element.) Now translate each row back into equation form to determine the values of the basic variables.

$$x_2 = 2 \qquad s_1 = 2 \qquad s_3 = \frac{3}{4} \qquad P = 6$$

In terms of the variables of the original problem, the maximum is $P = 6$, which occurs when $x_1 = 0$ and $x_2 = 2$.

Skill Enhancer 12   Find the largest value possible for $P = 3x_1 + 4x_2$ when we know that $x_1 \geq 0$, $x_2 \geq 0$, and that

$$x_1 + x_2 \leq 3$$
$$2x_1 + 3x_2 \leq 6$$

Answer in Appendix E.

Example 13   A certain linear programming problem in standard form can be reduced to the following augmented matrix. Deduce the maximum value of P. Is there enough information to decide how many variables there were in the original problem?

$$
\left(
\begin{array}{cccccc|c}
1 & 1 & 1 & 1 & 0 & 0 & 6 \\
2 & 2 & 1 & 0 & 1 & 0 & 4 \\
\hline
-2 & -1 & -1 & 0 & 0 & 1 & 0
\end{array}
\right)
$$

Solution

First, we will use the simplex algorithm to transform this matrix. The original matrix with the boxed pivot element is shown below, along with the calculations we needed to choose it. It should be clear that the pivot column is the first one; it stands atop the most negative element in the bottom row.

$$\begin{array}{c} \frac{6}{1} = 6 \\ \frac{4}{2} = 2 \end{array} \left(\begin{array}{cccccc|c} 1 & 1 & 1 & 1 & 0 & 0 & 6 \\ \boxed{2} & 2 & 1 & 0 & 1 & 0 & 4 \\ \hline -2 & -1 & -1 & 0 & 0 & 1 & 0 \end{array} \right) \Rightarrow \left(\begin{array}{cccccc|c} 0 & 0 & \frac{1}{2} & 1 & -\frac{1}{2} & 0 & 4 \\ 1 & 1 & \frac{1}{2} & 0 & \frac{1}{2} & 0 & 2 \\ \hline 0 & 1 & 0 & 0 & 1 & 1 & 4 \end{array} \right)$$

The transformed matrix contains no negative elements in the bottom row, and so becomes the final matrix in the simplex algorithm.

The rightmost element in the bottom row yields the maximum value of the objective function: $P = 4$.

Each row except the last corresponds to an original inequality. Here, two rows imply two original inequalities, and hence two slack variables. In order to fill up the remainder of the columns, there will have to be three original variables, x_1, x_2, and x_3. Three of the variables at any stage in the algorithm will therefore be nonbasic.

In the final iteration, the three nonbasic variables are x_2, x_3, and s_2, and these have zero value. The remaining variables are basic, and the matrix gives their values: $x_1 = 2$, $s_1 = 4$, $P = 4$.

The solution of the original problem will not involve any slack variables. In terms of the original variables, the solution is $P = 4$ when $x_1 = 2$ and $x_2 = x_3 = 0$.

Skill Enhancer 13   A certain maximization problem in linear programming leads to the initial simplex tableau

$$\left(\begin{array}{ccccc|c} 1 & 2 & 1 & 0 & 0 & 3 \\ 2 & 1 & 0 & 1 & 0 & 3 \\ \hline -3 & -4 & 0 & 0 & 1 & 0 \end{array} \right)$$

What is the maximum value of P for this problem? How many variables are there in the original problem, and what shall their values be to lead to the largest value of P?

Answer in Appendix E.

It is worth repeating a point from the solution to Example 13—the solution to the problem must *not* involve any of the slack variables.

Example 14   (Investment Strategy) An investment manager has no more than $50,000 to invest for a client. The manager will use it to buy bonds (which are sold in units of $1,000). One type of bond yields 8 percent per year, and another yields 12 percent.

FIGURE 4.7

The rules of the investment firm dictate that the manager shall purchase no more of the 12 percent bonds than 2 more than twice the number of the 8 percent type. Under these constraints, what strategy will bring in the largest annual income? Use algebraic methods.

Solution

We know this is a linear programming problem, because we need to determine an optimal strategy and an optimal income.

The unknowns of the problem are x_1, the number of 8 percent bonds, and x_2, the number of 12 percent bonds. The objective function we want to maximize is the annual income I of the bonds. If a bond has a *face value* of $1,000 and has an interest rate of 8 percent per year, then the owner of the bond will receive 8 percent of $1,000, or $80, each year. Since there are x_1 of these bonds, the income from them is $80x_1$. Similarly, the income from x_2 of the 12 percent bonds is $120x_2$, and the total income to be maximized is

$$I = 80x_1 + 120x_2 = 40(2x_1 + 3x_2)$$

Let $I' = 2x_1 + 3x_2$. I and I' differ by a factor of 40, and the values of x_1 and x_2 that make I maximum will also make I' maximum. Since I' has smaller coefficients, we prefer to work with it.

(4.16) $$I' = 2x_1 + 3x_2$$

Next, we need the constraints on the variables. The total number of bonds purchased is $x_1 + x_2$. Since at most 50 bonds can be purchased (remember, each bond costs $1,000, no fractions of bonds can be purchased, and there is a maximum of $50,000 at the investment manager's disposal), we demand that

(4.17) $$x_1 + x_2 \leq 50$$

The house rules with regard to investment strategy require that $x_2 \leq 2 + 2x_1$, or

(4.18) $$-2x_1 + x_2 \leq 2$$

Of course, it must also be true that

(4.19) $$x_1 \geq 0 \qquad x_2 \geq 0$$

Slack variables s_1 and s_2 transform the inequalities to equations, and so Inequalities (4.16) through (4.19) become this equation system:

(4.20)
$$
\begin{aligned}
x_1 + x_2 + s_1 \qquad\qquad &= 50 \\
-2x_1 + x_2 \qquad + s_2 \quad &= 2 \\
-2x_1 - 3x_2 \qquad\qquad + I' &= 0
\end{aligned}
$$

and the associated initial simplex tableau is

$$
\left(
\begin{array}{ccccc|c}
1 & 1 & 1 & 0 & 0 & 50 \\
-2 & \boxed{1} & 0 & 1 & 0 & 2 \\
\hline
-2 & -3 & 0 & 0 & 1 & 0
\end{array}
\right)
$$

The boxed 1 is the pivot element; make sure you see why. After pivoting about this element, the resulting matrix is

$$\left(\begin{array}{ccccc|c} \boxed{3} & 0 & 1 & -1 & 0 & 48 \\ -2 & 1 & 0 & 1 & 0 & 2 \\ \hline -8 & 0 & 0 & 3 & 1 & 6 \end{array}\right)$$

The boxed 3 is the pivot element of this matrix; again, make sure you see why. Upon carrying out this pivot, the next matrix in this sequence is

$$\left(\begin{array}{ccccc|c} 1 & 0 & \frac{1}{3} & -\frac{1}{3} & 0 & 16 \\ 0 & 1 & \frac{2}{3} & \frac{1}{3} & 0 & 34 \\ \hline 0 & 0 & \frac{8}{3} & \frac{1}{3} & 1 & 134 \end{array}\right)$$

We can read off the values of all variables from this matrix:

$$x_1 = 16 \qquad x_2 = 34 \qquad I' = 134$$
$$s_1 = s_2 = 0$$

That is, the investment manager should buy 16 of the 8 percent bonds and 34 of the 12 percent bonds to generate an annual income of $I = 40I' = 40 \times 134 = \$5,360$.

Skill Enhancer 14   Several years later, the investment climate has changed. The yields of the two kinds of bonds remain the same, but now the client has $100,000 to invest. The house rule relating the two bond types is somewhat different. It now states that the difference between four times the number of high-yield bonds and three times the number of low-yield bonds must be no more than one. What are the constraints for this problem? Now what is the optimal investment strategy? Use simplex methods.

Answer in Appendix E.

Example 15   (Retailing) The store Computers 'R Us assembles and sells two computer systems. The first uses a single floppy disk drive, and the second uses dual drives. The

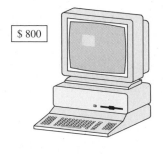

FIGURE 4.8

store can purchase only 100 of these drives per month. (There is a shortage of them.) Because of another shortage, of memory chips, the store cannot assemble more than 80 systems per month. The first system generates \$800 in profit; the second, \$1,200. What is the optimal assembly strategy to obtain the greatest profit? Use simplex methods.

Solution

We know this is a linear programming problem because of the need to determine a best strategy, one that will generate the largest monthly profit for the store. If P is this profit, and x_1 and x_2 represent the number of the first and second types of systems, then

$$P = 800x_1 + 1,200x_2 = 400(2x_1 + 3x_2)$$

If we let

(4.21) $$P' = 2x_1 + 3x_2$$

then we see that $P = 400P'$; both P and P' will achieve their maximum values at the same value of x_1 and x_2, and we prefer to work with P' since it has smaller coefficients.

The shortages of drives and memory chips constrain the values of x_1 and x_2. The total number of drives used is $x_1 + 2x_2$; this cannot exceed 100.

(4.22) $$x_1 + 2x_2 \leq 100$$

In the same way, the constraint on the number of systems built is

(4.23) $$x_1 + x_2 \leq 80$$

Of course, we also demand that $x_1 \geq 0$, $x_2 \geq 0$.

The system (4.21) through (4.23) becomes the following system of equations when we insert slack variables.

$$
\begin{aligned}
x_1 + 2x_2 + s_1 & = 100 \\
x_1 + x_2 + s_2 & = 80 \\
-2x_1 - 3x_2 + P' &= 0
\end{aligned}
$$

The initial simplex tableau is

$$
\left(
\begin{array}{ccccc|c}
1 & \boxed{2} & 1 & 0 & 0 & 100 \\
1 & 1 & 0 & 1 & 0 & 80 \\
\hline
-2 & -3 & 0 & 0 & 1 & 0
\end{array}
\right)
$$

(Make sure you see why the boxed element is the pivot element.) After carrying out the pivot, one obtains

$$
\left(
\begin{array}{ccccc|c}
\frac{1}{2} & 1 & \frac{1}{2} & 0 & 0 & 50 \\
\boxed{\frac{1}{2}} & 0 & -\frac{1}{2} & 1 & 0 & 30 \\
\hline
-\frac{1}{2} & 0 & \frac{3}{2} & 0 & 1 & 150
\end{array}
\right)
$$

We can carry out this pivot to obtain the reduced matrix

$$\left(\begin{array}{ccccc|c} 0 & 1 & 1 & -1 & 0 & 20 \\ 1 & 0 & -1 & 2 & 0 & 60 \\ \hline 0 & 0 & 1 & 1 & 1 & 180 \end{array}\right)$$

from which we read off the solution to the problem:

$$x_1 = 60 \qquad x_2 = 20 \qquad P' = 180$$
$$s_1 = s_2 = 0$$

That is, construct 60 of the single-drive systems and 20 of the dual-drive systems. For this configuration, $P' = 180$, and the actual monthly profit is 400 times this: $P = 400P' = 400 \times 180 = \$72,000$.

Skill Enhancer 15 Several years later, the details of computer construction have changed. Two hundred disk drives are now available per month, and the limit on systems sold is now 150. What now is the best strategy to maximize profits? Use simplex methods.

Answer in Appendix E.

Sizes of Augmented Matrices

How large will the augmented matrix be for any particular simplex problem? To answer this question, let there be c nontrivial constraints to the problem (those in addition to the usual $x_i \geq 0$) and u unknowns to the problem.

For each constraint, we will have another row in the augmented matrix. In addition, we have one row corresponding to the equation for the objective function P. All in all, there will be $c + 1$ rows to the matrix.

There will be one column for each original variable and one for each slack variable. Since there are as many slack variables as there are constraints, that is a total of $c + u$ so far. But we also have one column for P and one for the constants on the right side of each constraint. The grand total: $c + u + 2$ columns.

Dimensions of a Simplex Augmented Matrix

Each such matrix will have a dimension of $(c+1) \times (c+u+2)$, where the problem contains c constraints and u unknowns.

Example 16 An augmented matrix has dimension 4×8. Does this give enough information to deduce the number of constraints and unknowns in the original problem? What are these numbers?

Solution

Let c and u be the number of constraints and unknowns. We know that $c + 1$ is the number of rows, so

$$c + 1 = 4$$

from which $c = 3$. In the same way, $c + u + 2$ is the number of rows, so

$$c + u + 2 = 8$$

or $u + 5 = 8$ (using the known value of c) so that $u = 3$. There, we conclude that we have enough information. In the original problem, there were three constraints applied to three unknowns.

Skill Enhancer 16   If an augmented matrix with dimension 23×47 corresponds to a simplex problem, how many constraints and unknowns are in this problem?

Answer in Appendix E.

——— *Exercises 4.3* ———————————————————————————

Exercises 1–6 present augmented matrices that have been constructed from various linear programming problems. In each case, deduce the maximum value of P and indicate how many variables there were in the linear programming problem.

1. $\begin{pmatrix} 1 & 1 & 1 & 0 & | & 5 \\ -4 & -5 & 0 & 1 & | & 0 \end{pmatrix}$

2. $\begin{pmatrix} 1 & 2 & 1 & 0 & | & 3 \\ -3 & -2 & 0 & 1 & | & 0 \end{pmatrix}$

3. $\begin{pmatrix} 2 & 1 & 1 & 0 & 0 & | & 4 \\ 0 & 1 & 0 & 1 & 0 & | & 2 \\ -1 & -1 & 0 & 0 & 1 & | & 0 \end{pmatrix}$

4. $\begin{pmatrix} 1 & 1 & 1 & 0 & 0 & | & 3 \\ 1 & 0 & 0 & 1 & 0 & | & 2 \\ -2 & -1 & 0 & 0 & 1 & | & 0 \end{pmatrix}$

5. $\begin{pmatrix} 1 & 1 & 1 & 0 & 0 & 0 & 0 & | & 4 \\ 2 & 1 & 0 & 1 & 0 & 0 & 0 & | & 6 \\ 0 & 1 & 0 & 0 & 1 & 0 & 0 & | & 5 \\ 0 & 1 & 0 & 0 & 0 & 1 & 0 & | & 10 \\ -3 & -2 & 0 & 0 & 0 & 0 & 1 & | & 0 \end{pmatrix}$

6. $\begin{pmatrix} 2 & 3 & 1 & 0 & 0 & 0 & | & 12 \\ 3 & 2 & 0 & 1 & 0 & 0 & | & 12 \\ 1 & 0 & 0 & 0 & 1 & 0 & | & 3 \\ -1 & -1 & 0 & 0 & 0 & 1 & | & 0 \end{pmatrix}$

Write down the initial simplex tableau for the systems in Exercises 7–17, and circle the initial pivot element. You need not solve the problem further.

7. Maximize $P = 5x_1 + 6x_2$ subject to $x_1 \geq 0$, $x_2 \geq 0$ and $x_1 + x_2 \leq 4$.

8. Maximize $P = 3x_1 + 3x_2$ subject to $x_1 \geq 0$, $x_2 \geq 0$ and $4x_1 + 5x_2 \leq 20$.

9. Maximize $P = 2x_1 + 5x_2$ subject to $x_1 \geq 0$, $x_2 \geq 0$, $x_1 + x_2 \leq 4$, and $x_2 \leq 2$.

10. Maximize $P = 9x_1 + 11x_2$ subject to $x_1 \geq 0$, $x_2 \geq 0$, $x_1 \leq 2$, and $x_2 \leq 3$.

11. Maximize $P = 6x_1 + x_2$ subject to $x_1 \geq 0$, $x_2 \geq 0$, $x_1 \leq 3$, and $x_1 + 2x_2 \leq 4$.

12. Maximize $P = 10x_1 + 12x_2$ subject to $x_1 \geq 0$, $x_2 \geq 0$, $x_1 + 2x_2 \leq 4$, $4x_1 + 3x_2 \leq 4$, and $3x_1 + x_2 \leq 9$.

13. Maximize $P = 0.25x + 1.125y$ subject to $x \geq 0$, $y \geq 0$, $6.1x + 0.1y \leq 4.5$, $5.5x - 4.4y \leq 3.3$, and $x \leq 3$.

14. Maximize $P = \frac{3}{4}x_1 + \frac{3}{4}x_2$ when $x_1 \geq 0$, $x_2 \geq 0$, and $\frac{1}{2}x_1 + \frac{2}{3}x_2 \leq \frac{7}{8}$, $x_1 + 1\frac{1}{4}x_2 \leq 1\frac{1}{3}$, and $\frac{3}{4}x_1 + x_2 \leq 1$.

15. Maximize $P = \frac{1}{5}x_1 + \frac{2}{5}x_2$ subject to $x_1 \geq 0$, $x_2 \geq 0$, $x_1 + 2\frac{1}{5}x_2 \leq 2$, $\frac{4}{5}x_1 + 2x_2 \leq 2$, and $\frac{3}{5}x_1 + \frac{7}{10}x_2 \leq 3\frac{1}{5}$.

16. Maximize $P = x_1 - x_2 + 2x_3$ subject to $x_1 \geq 0$, $x_2 \geq 0$, $x_3 \geq 0$, and to $2x_1 + 3x_2 + x_3 \leq 7$, $x_1 + 3x_2 + 2x_3 \leq 6$, and $5x_1 + 3x_3 \leq 4$.

17. Maximize $P = 2x_1 + 3x_2$ subject to $x_1 \geq 0$, $x_2 \geq 0$, $x_3 \geq 0$, $x_1 - 3x_3 \leq 4$, $2x_2 + 3x_3 \leq 5$, $x_1 + x_2 + 4x_3 \leq 4$.

Use the simplex method to solve the linear programming problems in Exercises 18–33.

18. Maximize $P = 9x_1 + 11x_2$ subject to $x_1 \geq 0$, $x_2 \geq 0$, $x_1 \leq 2$, and $x_2 \leq 3$.

19. Maximize $P = 6x_1 + x_2$ subject to $x_1 \geq 0$, $x_2 \geq 0$, $x_1 \leq 3$, and to $x_1 + 2x_2 \leq 4$.

20. Maximize $P = x + 2y$ subject to $x \geq 0$, $y \geq 0$, $x + y \leq 3$, $x \leq 2$, $y \leq 2$.

21. Maximize $P = 2x_1 + x_2$ subject to $x_1 \geq 0$, $x_2 \geq 0$, $x_2 \leq 2$, $x_1 + x_2 \leq 4$.

22. Maximize $P = 5x_1 + 5x_2$ subject to $x_1 \geq 0$, $x_2 \geq 0$, $2x_1 + 3x_2 \leq 6$, and $x_1 \leq 4$.

23. Maximize $P = 4x_1 + 2x_2$ subject to $x_1 \geq 0$, $x_2 \geq 0$, $4x_1 + 3x_2 \leq 16$.

24. Maximize $P = 4x_1 + 2x_2$ subject to $x_1 \geq 0$, $x_2 \geq 0$, $4x_1 + 3x_2 \leq 16$, and $x_2 \leq 3$.

25. Maximize $P = x_1 + 2x_2$ when $x_1 \geq 0$, $x_2 \geq 0$, $3x_1 + 4x_2 \leq 12$, $x_1 \leq 1$, and $x_1 \leq 3$.

26. Maximize $P = 2x_1 + x_2 + x_3$ subject to $x_1 \geq 0$, $x_2 \geq 0$, $x_3 \geq 0$, $x_1 + x_2 \leq 2$, $x_2 + x_3 \leq 2$, and $x_1 + x_3 \leq 2$.

27. Maximize $P = 2x_1 + x_2 + x_3$ subject to $x_1 \geq 0$, $x_2 \geq 0$, $x_3 \geq 0$, $x_1 \leq 1$, $x_2 \leq 2$, $x_1 + x_2 + x_3 \leq 4$.

28. Maximize $P = x_1 + 2x_2$ subject to $x_1 \geq 0$, $x_2 \geq 0$, $x_3 \geq 0$, $x_1 + x_2 \leq 4$, and $x_2 + x_3 \leq 2$.

29. Redo Exercise 28 if $P = x_1 + x_2 + x_3$.

30. Maximize $P = x_1 + x_2 + 2x_3$ if $x_1 \geq 0$, $x_2 \geq 0$, $x_3 \geq 0$ and if $x_i \leq 2$, $i = 1, 2, 3$, and $x_1 + x_2 + x_3 \leq 3$.

31. Redo Exercise 30 if $P = x_1 + x_2 + x_3$.

32. Maximize $P = 3x_1 + 2x_2 + x_3$ subject to $x_1 \geq 0$, $x_2 \geq 0$, $x_3 \geq 0$, $x_1 + 2x_2 \leq 3$, $x_2 + 2x_3 \leq 3$.

33. Maximize $P = x_1 - x_2 + 2x_3 - x_4$ subject to $x_i \geq 0$, $i = 1, \ldots, 4$, $x_1 + 2x_2 + x_3 + x_4 \leq 4$, $2x_2 + x_4 \leq 4$, and $x_1 \leq 3$.

34. Show how to adapt the simplex method to solve this linear programming problem. *Minimize* $P = -2x - 3y$ subject to $x \geq 0$, $y \geq 0$, $x + 3y \leq 7$, $2x + y \leq 5$. Solve the problem using this method.

35. Show how to adapt the simplex method to solve the following linear programming problem, and then use this extension to solve the problem. Maximize $P = 4x + 6y$ subject to $x \geq 1$, $y \geq 2$, $x + 3y \leq 11$, $2x + y \leq 9$. Do not use the results of any future sections of this chapter. (Hint: Use a new variable x' where $x = x' + 1$. You will also need a new variable y'; what should its definition be?)

Applications

Use simplex techniques.

36. *(Investment Strategy)* An investor has up to $150,000 to invest in three types of mutual funds—stocks, municipal, and money market. The average rates of return for these are 10 percent, 8 percent, and 14 percent. Maximize the annual return subject to the fact that investments in the municipal fund should be at least as great as the investments in the two remaining funds.

37. *(Advertising Strategy)* A national chain of fast-food outlets has opened a new franchise in a medium-sized town. They have budgeted $10,000 for an initial advertising campaign to take place during one week. Ads can be placed with any of three media—in the local newspaper (it comes out daily excluding weekends), on radio (the leading local program is broadcast every day of the week), and by batches of fliers. A radio "spot" costs $500 per broadcast. Each newspaper ad costs $1,000. A batch of fliers costs $100 to distribute. The chain's advertising agency has found that each type of advertisement can be assigned an "effectiveness index." For radio spots, the index is 3 units per ad. For the newspaper advertisement, it is 8 units per ad, and for the fliers, it's 1 unit per batch of fliers. The most successful campaigns are the ones with the largest total effectiveness index values. What is the best way to plan the ad campaign for this town?

38. *(Business Administration)* The Morrisroe Construction Company is famous for its developments of low-cost, high-quality housing. Each development consists of three kinds of houses, "ranch," "colonial," and "manor." Each house calls for different amounts of rough work and finishing work, tasks performed by different categories of workers. Each house has a different profit associated with it. Here are the facts for these house models

expressed as a table:

	Rough	Finish	Profit
Ranch	500	400	$10,000
Colonial	500	400	$15,000
Manor	400	500	$12,500

All of Morrisroe's construction crew work 40-h weeks with no overtime. Currently, the company has 50 "roughers" and 60 finishers on its staff. How many houses of each type should the company build?

39. (Fundraising) The Fox Hollow Civic Association, a volunteer group, is planning a bake sale to raise much-needed funds. They plan to bake three different kinds of cakes, using eggs, flour, and sugar that have been donated to them. They have 8 dozen eggs, 25 lb flour, and 20 lb sugar, and the cake recipes basically involve mixing different proportions of these ingredients together (plus appropriate flavoring agents, which the club has in unlimited supply). The three recipes the club has decided upon call for these ingredients in each cake:

	Eggs	Flour (oz)	Sugar (oz)
Cake A	1	4	3
Cake B	2	4	4
Cake C	1	3	4

The club planning committee expects to make $1 on each cake of type A or B, and $1.25 on each cake of type C. How many of each cake should they bake?

40. (Fundraising) The Alumni Association is planning a fund-raising concert. The planning committee will book as many celebrities as they can, and hopefully will make as much money as possible through the sale of tickets. The Association has a budget of $20,000 for celebrity fees. When they approach an agent, they find that celebrities can be categorized into three groups, depending on the size of the fee they demand. Real stars command fees of $4,000 per performance, but this agent only has three individuals in this group. The second group consists of those stars whose luster has perhaps tarnished a bit—they only charge $2,500 per appearance, and the agent has eight of them from which to choose. Finally, there are those performers drawn from the resident population whose fame is purely local. Their fee is a mere $1,000, and there are 10 of them from which to choose. The choice should be based on the fact that each category of star has a different drawing power. Each celebrity in the top class will draw 6,000 spectators. Each of those in the intermediate category draws 3,500 spectators. Finally, each of the local stars draws 1,100 spectators. How should the program for the concert be structured to ensure the maximum attendance possible to the concert?

41. (Astronomy) A famous astronomical observatory is studying two types of galaxies. The goal is to maximize the number of galaxies studied. Each galaxy of the first type requires 2 h of observation, whereas each galaxy of the second type requires 3 h of observation. During the course of the study, the scientists assume they will have 200 h of observation time available to them. Each galaxy of the first type requires 3 min of computer analysis, whereas each of the second type needs only 2.5 min of analysis. The scientists expect to have 3 h of computer time available to them during the study. How many galaxies of each type should the study consist of?

42. (Ecology) Three different prairie environments are under the jurisdiction of the Department of Ecological Protection of some Midwestern state. The goal of the department is to maximize the total number of acres actively maintained by its staff. The maintenance costs for each type of environment per acre are $100, $100, and $150, respectively. The Department's annual budget is $1,000,000. The annual person-month requirement for each acre is 3, 2, and 2 for the three types; the Department has 300,000 person-months available to allocate during the year to prairie maintenance. How many acres of each type should the Department seek to maintain per year?

43. (Investment Strategy) An investment house has to invest money in bonds, which are sold in units of $1,000. In this situation, let x and y be the amounts of the two types of bonds, which have to satisfy certain house rules together with the needs of the client. Setting up the problem and solving the problem (many of the details of which are here omitted) indicates that $x = 59\frac{3}{5}$ and $y = 40\frac{2}{5}$ are the optimal values for the amounts of bonds. But the purchasing of fractional parts of bonds is not allowed! Discuss the ways in which these results should be interpreted.

▦ Calculator Exercises

The matrices in Exercises 44–49 represent augmented matrices that correspond to a linear programming program. Determine the pivot element for each matrix.

44. $\begin{pmatrix} 1 & 6.9 & 0 & 3.1 & 0 & 5.3 \\ 0 & 8.2 & 1 & 4.9 & 0 & 6.2 \\ \hline 0 & -7.1 & 0 & -2.3 & 1 & 0 \end{pmatrix}$

45. $\begin{pmatrix} 6.8 & 0 & 1 & 7.7 & 0 & 3.9 \\ -1.5 & 1 & 0 & 0 & 0 & 9.3 \\ \hline -6.9 & 0 & 0 & -5.4 & 1 & 0 \end{pmatrix}$

46. $\begin{pmatrix} 0 & 3.2 & 1 & -6.9 & 0 & 4.7 \\ 1 & 0 & 0 & 3.1 & 0 & 6.5 \\ \hline 0 & -7.8 & 0 & -5.4 & 1 & 0 \end{pmatrix}$

47. $\begin{pmatrix} 0 & 1 & 12.2 & 0 & 0 & 4.5 \\ 6.7 & 0 & 0 & 1 & 0 & 4.7 \\ \hline -3.1 & 0 & -3.2 & 0 & 1 & 0 \end{pmatrix}$

48. $\begin{pmatrix} 1 & 6.02 & 13.94 & 0 & 0 & 41.31 \\ 0 & 10.11 & 13.11 & 0 & 0 & 32.11 \\ 0 & 21.02 & 7.02 & 0 & 0 & 89.02 \\ 0 & 0 & 9.98 & 1 & 0 & 49.48 \\ \hline 0 & -6.82 & -6.02 & 0 & 1 & 0 \end{pmatrix}$

49. $\begin{pmatrix} 0 & 0 & 0 & 6.99 & 9.91 & 0 & 9.81 \\ 6.98 & 1 & 0 & 12.09 & 4.91 & 0 & 1.21 \\ -1.31 & 0 & 1 & 6.11 & 5.08 & 0 & 1.44 \\ 3.49 & 0 & 0 & 6.71 & 0 & 0 & 3.02 \\ \hline -12.1 & 0 & 0 & -8.41 & -6.02 & 1 & 0 \end{pmatrix}$

Section 4.4 DUALITY AND MINIMUM PROBLEMS

A slight variation of the simplex method allows us to solve linear programming *minimum* problems—problems in which the goal is the minimization of an objective function p subject to certain constraints. (We previously used an uppercase P to suggest quantities to be maximized; we use a lowercase p for objective functions that we want to *minimize*.) There is a *standard form* for minimization problems.

> **Standard Form: Minimization Problems**
>
> **1.** All variables must be *nonnegative*.
>
> **2.** All constraints must be *linear inequalities* in the special form
>
> $$\text{Linear expression} \geq \text{positive constant}$$
>
> **3.** We seek to *minimize* an objective function p, a linear expression of the variables of the problem.

Just as maximum problems had to have $\leq$ inequalities, so must *minimum* problems have $\geq$ inequalities. In both problem types, all variables must be ≥ 0.

Example 17   A certain quantity

$$p = 3x_1 + 7x_2 + x_3$$

needs to be minimized subject to several constraints. Which sets of constraints correspond to linear programming problems expressed in standard form?

(a) $x_i \geq 0$, $(i = 1, 2, 3)$, $x_1 \leq 4$, $x_1 + 3x_2 + 2x_3 \geq 3$, $x_1 + x_2 + x_3 \geq 7$.

(b) $x_1 \geq 4$, $x_1 + 3x_2 + 2x_3 \geq 3$, $x_1 + x_2 + x_3 \geq 7$.
(c) $x_i \geq 0$, $(i = 1, 2, 3)$, $x_1 \geq 4$, $x_1 + 3x_2 + 2x_3 \geq 3$, $x_1 + x_2 + x_3 \geq 7$.

Solution

In (a), one inequality goes the "wrong way;" $x_1 \leq 4$ is not in the standard form. The inequalities in (b) go the right way, but the variables are not constrained to be nonnegative. All parts of the definition are met in (c), so this is the only set in standard form.

Skill Enhancer 17   In one minimum linear programming problem, students have been asked to minimize $p = x + 2y + 3z$ subject to certain sets of constraints. Which of these constraints are in standard form?
(a) $x \geq 0$, $y \geq 0$, $z \geq 0$, $3x - 2y + 3z \geq 5$, $4x + 4y + 4z \geq 3$
(b) $x \geq 0$, $y \geq 0$, $z \geq 0$, $3x - 2y + 3z \geq 5$, $4x + 4y + 4z \geq 3$, $x = y \leq 10$
(c) $3x - 2y + 3z \geq 5$, $4x + 4y + 4z \geq 3$, $5x - 4y + 3z \geq -2$

Answer in Appendix E.

primary problem   The solution strategy for minimum problems involves associating the   *primary*
dual problem   *problem*—that is, the original minimization problem—with a corresponding *dual problem* —that is, a secondary problem—which will be a *maximization problem* in standard form. We can solve this dual problem using the procedures of the previous section, and determine the solutions for the original, primary problem from it.

Dual problems depend heavily on matrix manipulation. One new matrix operation
transpose   is that of the matrix *transpose*. Given a matrix M, its transpose M^T is formed by exchanging the rows and columns of M. An alternative (but equivalent) interpretation pictures the transpose as a *reflection* of M about an imaginary diagonal which starts at row 1 and column 1 of M. Figure 4.9 illustrates the procedure with several examples.

$$\begin{pmatrix} 1 & -1 & 3 \\ 2 & 0 & 1 \end{pmatrix} \xrightarrow{\text{transpose}} \begin{pmatrix} 1 & 2 \\ -1 & 0 \\ 3 & 1 \end{pmatrix}$$

$$\begin{pmatrix} 7 \\ -8 \\ \frac{1}{2} \end{pmatrix} \xrightarrow{\text{transpose}} \begin{pmatrix} 7 & -8 & \frac{1}{2} \end{pmatrix}$$

$$\begin{pmatrix} 1 & 9 & 8 \\ 7 & -3 & 0 \\ 5 & 5 & 2 \end{pmatrix} \xrightarrow{\text{transpose}} \begin{pmatrix} 1 & 7 & 5 \\ 9 & -3 & 5 \\ 8 & 0 & 2 \end{pmatrix}$$

FIGURE 4.9
Matrix transposes.

Example 18   Give the transpose matrices for these matrices.

$$A = \begin{pmatrix} 1 & 3 & -9 \\ 3 & 4 & 0 \end{pmatrix} \qquad B = \begin{pmatrix} x & y \\ y^2 & -x \end{pmatrix} \qquad C = \begin{pmatrix} \alpha & \beta \\ \gamma & \delta \\ \epsilon & \omega \end{pmatrix}$$

Solution

We have

$$A^T = \begin{pmatrix} 1 & 3 \\ 3 & 4 \\ -9 & 0 \end{pmatrix} \qquad B^T = \begin{pmatrix} x & y^2 \\ y & -x \end{pmatrix} \qquad C^T = \begin{pmatrix} \alpha & \gamma & \epsilon \\ \beta & \delta & \omega \end{pmatrix}$$

Skill Enhancer 18   What are the transposes to $M = \left(\frac{1}{2} \right)$, $N = \begin{pmatrix} 1 & x & x^2 \\ 1 & y & y^2 \end{pmatrix}$, and $P = \begin{pmatrix} 1 & 0 \\ 0 & 1 \end{pmatrix}$?

Answer in Appendix E.

Forming the Dual Problem

There are three large steps in connection with setting up and solving a minimum problem using the simplex method.

1. Create a matrix M that summarizes the coefficients of the original problem.

2. Use the information to create a dual problem in a way we will describe.

3. Solve the new, dual problem. The Von Neumann Duality Principle, which we are also about to discuss, allows us to read the answers to our *original* problem from this secondary, dual problem. (We will not show why the duality principle works.)

Let's see how to summarize the information in any minimization problem in a useful way. Suppose the constraints to some problem are

$$3x_1 + 2x_2 + 4x_3 \geq 10$$
$$2x_1 + x_2 + 2x_3 \geq 9$$

and the objective function is

$$p = 4x_1 + 3x_2 + 2x_3$$

Then we form a matrix M which has one more row than the number of constraints and one more column than the number of unknowns. The elements of this matrix are the coefficients from these relations, written down in an obvious way.

$$M = \begin{bmatrix} 3 & 2 & 4 & 10 \\ 2 & 1 & 2 & 9 \\ 4 & 3 & 2 & 0 \end{bmatrix}$$

This is *not* a simplex tableau, and the square brackets help emphasize this.

Next, form the transpose matrix M^T and reverse the process. That is, we use the rules to create M in the reverse order to write down a problem that corresponds to M^T. This shall be a maximum problem. In this case, for example,

$$M^T = \begin{bmatrix} 3 & 2 & 4 \\ 2 & 1 & 3 \\ 4 & 2 & 2 \\ 10 & 9 & 0 \end{bmatrix}.$$

We interpret this as a problem with the contraints

$$3y_1 + 2y_2 \leq 4$$
$$2y_1 + y_2 \leq 3$$
$$4y_1 + 2y_2 \leq 2$$

subject to which we need to maximize

$$P = 10y_1 + 9y_2$$

Example 19 illustrates this procedure again.

Example 19   In a certain problem, we are given the constraints $x_1 \geq 0$, $x_2 \geq 0$, and

$$x_1 + x_2 \geq 2$$
$$2x_1 + x_2 \geq 3$$
$$x_2 \geq 2$$

and we need to minimize

$$p = 4x_1 + 3x_2$$

What is the dual to this problem?

Solution

The matrix M is a matrix of the coefficients of the constraints and of p.

$$M = \begin{bmatrix} 1 & 1 & 2 \\ 2 & 1 & 3 \\ 0 & 1 & 2 \\ 4 & 3 & 0 \end{bmatrix}$$

Since there are three constraints, M will have $3 + 1 = 4$ rows. Since there are two unknowns, there will be $2 + 1 = 3$ columns.

We formed M without adding any slack variables, and p is not treated as a basic variable. The bottom right element of M will *always* be 0. M can *never* be the initial simplex tableau.

dual matrix   Next, form the transpose M^T, the *dual matrix*.

$$M^T = \begin{bmatrix} 1 & 2 & 0 & 4 \\ 1 & 1 & 1 & 3 \\ 2 & 3 & 2 & 0 \end{bmatrix}$$

Just as the original minimum problem corresponds to M, so will the dual *maximum* problem correspond to M^T. Create from M^T a set of inequalities and an objective function that form the *dual problem*.

$$\begin{bmatrix} 1 & 2 & 0 & 4 \\ 1 & 1 & 1 & 3 \\ 2 & 3 & 2 & 0 \end{bmatrix} \implies \begin{array}{l} y_1 + 2y_2 \qquad\;\; \leq 4 \\ y_1 + \;\; y_2 \;\; + y_3 \leq 3 \\ 2y_1 + 3y_2 + 2y_3 = \qquad P \end{array}$$

In the dual problem, we use notation y_i for the variables to emphasize the separate identity of this problem from the original problem. The dual problem is to *maximize* $P =$

$2y_1 + 3y_2 + 2y_3$ subject to $y_1 \geq 0$, $y_2 \geq 0$, $y_3 \geq 0$, and the remaining two inequalities. The dual problem is different from the original problem.

Skill Enhancer 19   Suppose we know that $x_1 \geq 0$ and $x_2 \geq 0$ and that

$$x_1 + 2x_2 \geq 3$$
$$2x_1 + x_2 \geq 3$$

We need to find the minimum value of $p = x_1 + x_2$.
(a) How many rows and columns will M have for this problem?
(b) What is M? What is M^T?
(c) What is the dual to this problem?

Answer in Appendix E.

The Von Neumann Duality Principle

Von Neumann duality principle

The *Von Neumann duality principle* accounts for the significance of the dual problem.

The Von Neumann Duality Principle

If some linear programming problem in standard form has a solution, then so does its dual problem.

- If the original, primary problem is a minimum problem, then its dual problem is a maximum problem.

- The value for P that *maximizes* the objective function of the dual (maximum) problem will *minimize* the objective function of the primary problem.

   We solve the dual problem using the simplex algorithm. The *final simplex tableau* will have the appearance shown in Figure 4.10. The solutions to the original, minimum problem appear on the bottom row.

- The rightmost component of the bottom row is the minimum value of p.

- To the left of that is a 1. To the left of this 1 appear a series of numbers in the columns corresponding to the slack variables of the dual problem, which we label u_1 through u_c (where c is the number of constraints in the dual problem). These components on the bottom, labeled a_1 through a_c in Figure 4.10, yield the values of x_1 through x_c that give rise to the minimum value of p.

Example 20   A dual maximum linear programming problem has three variables and two slack variables. The final tableau for the problem is

$$
\begin{array}{ccccccc}
y_1 & \cdots & y_n & u_1 & u_2 & \cdots & u_c & p \\
\end{array}
$$

$$
\begin{pmatrix}
* & \cdots & * & * & * & \cdots & * & * & * \\
\vdots & & & & & & & & \vdots \\
* & \cdots & * & a_1 & a_2 & \cdots & a_c & 1 & p
\end{pmatrix}
$$

FIGURE 4.10

The final tableau: the dual problem. Here, the value of P gives the maximum to the dual problem and the minimum to the primary problem. The components labeled a_1, a_2, and so on are the values of the variables of the *original*, primary problem which achieve this value. There are c original variables x_1 through x_c in the primary problem. We read these values—shown here as a_1 through a_c—from the bottom row. The asterisks $*$ indicate other components of the tableau.

$$
\left(
\begin{array}{cccccc|c}
\frac{1}{2} & 1 & 0 & \frac{1}{2} & 0 & 0 & 2 \\
\frac{1}{2} & 0 & 1 & -\frac{1}{2} & 1 & 0 & 1 \\
\hline
\frac{1}{2} & 0 & 0 & \frac{1}{2} & 2 & 1 & 8
\end{array}
\right)
$$

What is the solution to the original minimum problem?

Solution

Since the bottom right element is 8, $p = 8$. Since there are two slack variables in the dual problem, there must be two variables in the original problem. The optimal values for these lie to the left of the 1 in the bottom row. That is, $x_1 = \frac{1}{2}$ and $x_2 = 2$.

Skill Enhancer 20   The augmented matrix giving the solution for a dual maximum problem is

$$
\begin{pmatrix}
1 & 4 & 5 & 0 & 0 & 7 \\
0 & 9 & 6 & 1 & 0 & 8 \\
0 & 7 & 3 & 0 & 1 & 11
\end{pmatrix}
$$

What is the solution to the original minimum problem?

Answer in Appendix E.

We are ready to apply this new method to solve minimum problems using the simplex method.

Example 21   Minimize

(4.24) $$p = 4x_1 + 3x_2$$

subject to $x_1 \geq 0$, $x_2 \geq 0$, and to

$$x_1 + x_2 \geq 2$$

(4.25)
$$2x_1 + x_2 \geq 3$$

$$x_2 \geq 2$$

(Make sure you see that this problem is in standard form.)

Solution

This example continues the problem presented in Example 19. There we learned that the dual problem is to *maximize*

$$P = 2y_1 + 3y_2 + 2y_3$$

subject to $y_1 \geq 0$, $y_2 \geq 0$, $y_3 \geq 0$ and to

$$y_1 + 2y_2 \leq 4$$
$$y_1 + y_2 + y_3 \leq 3$$

The next step is to use the simplex method on this dual problem. The two constraints of the dual problem require two slack variables u_1 and u_2. We use the notation u_i for slack variables (rather than s_i) to emphasize again its distinction from the original problem. These equations give rise to the following initial simplex tableau.

$$
\begin{array}{c}
\frac{4}{2} = 2 \\
\frac{3}{1} = 3
\end{array}
\left(
\begin{array}{cccccc|c}
y_1 & y_2 & y_3 & u_1 & u_2 & p & \\
1 & \boxed{2} & 0 & 1 & 0 & 0 & 4 \\
1 & 1 & 1 & 0 & 1 & 0 & 3 \\
\hline
-2 & -3 & -2 & 0 & 0 & 1 & 0
\end{array}
\right)
$$

The pivot variable is y_2, and the pivot ratios are computed at the left. (Make sure you understand why the boxed element is the pivot element.) After performing this pivot operation, we obtain

$$
\frac{1}{1} = 1
\left(
\begin{array}{cccccc|c}
y_1 & y_2 & y_3 & u_1 & u_2 & p & \\
\frac{1}{2} & 1 & 0 & \frac{1}{2} & 0 & 0 & 2 \\
\frac{1}{2} & 0 & \boxed{1} & -\frac{1}{2} & 1 & 0 & 1 \\
\hline
-\frac{1}{2} & 0 & -2 & \frac{3}{2} & 0 & 1 & 6
\end{array}
\right)
$$

Because y_3 is the pivot variable in this iteration, and because 1 is the smallest pivot ratio (it's the only legitimate pivot ratio), the element in row 2 and column 3 is the pivot element. No further pivot is called for:

(4.26)
$$
\left(
\begin{array}{cccccc|c}
 & & & u_1 & u_2 & p & \\
\frac{1}{2} & 1 & 0 & \frac{1}{2} & 0 & 0 & 2 \\
\frac{1}{2} & 0 & 1 & -\frac{1}{2} & 1 & 0 & 1 \\
\hline
\frac{1}{2} & 0 & 0 & \frac{1}{2} & 2 & 1 & 8
\end{array}
\right) ;
$$

make sure you understand *why* this is the final pivot.

We can read off the solution to the primary problem from Matrix (4.26); see Example 20. The bottom element in the rightmost column—8—is the minimum value for p. (To the left of it is a 1, which should always occur.) Since there are two slack variables in the dual problem and two original variables in the original problem (x_1 and x_2), we conclude that the values for x_1 and x_2 that provide the minimum of p are the underlined components in Matrix (4.26), namely $\frac{1}{2}$ and 2.

Skill Enhancer 21   Suppose we know that $x_1 \geq 0$ and $x_2 \geq 0$ and that

$$x_1 + 2x_2 \geq 3$$
$$2x_1 + x_2 \geq 3$$

We need to find the minimum value of $p = x_1 + x_2$. We learned in the problem following the previous example that the dual to this problem is this: Subject to $y_1 \geq 0$ and $y_2 \geq 0$ and to

$$y_1 + 2y_2 \leq 1$$
$$2y_1 + y_2 \leq 1$$

what is the maximum value of $P = 3y_1 + 3y_2$?
(a) Introduce slack variables and write the system of equations equivalent to this problem.
(b) What is the initial simplex tableau for this problem?
(c) Pivot as many times as necessary to find the final simplex tableau for this problem. Show this final tableau.
(d) What is the solution to the *dual* problem?
(e) What is the solution to the original minimization problem?

Answer in Appendix E.

Creating the Dual Problem

1. *Write the coefficient matrix* for the original problem.

2. Take the *transpose of this matrix*.

3. *Construct a set of inequalities* using the coefficients of the transpose matrix. These form the dual problem.

We can summarize the entire procedure.

Solving Minimum Linear Programming Problems

When a minimum linear programming problem is in standard form, use this procedure.

1. Write the *dual problem,* which will be a maximum problem.

2. *Use the simplex method* to solve this dual problem.

3. *Read the answers to the primary problem* from the bottom row of the final simplex tableau matrix.

Make sure you understand the following rule of thumb. If the original minimum problem contains u unknowns bound by c constraints (not counting the trivial constraints on the variables to be nonnegative), then the dual problem will contain c variables bound by u constraints. Therefore, *the number of slack variables in the dual problem equals the number of original variables in the primary problem.*

Example 22   Minimize

$$p = 4x_1 + x_2$$

subject to

$$x_1 + 3x_2 \geq 4$$
$$x_1 + x_2 \geq 2$$

and to $x_1 \geq 0$, $x_2 \geq 0$.

Solution

This primary problem is in standard form, so we find its dual problem, and solve it by the simplex method. To find the dual, write the coefficients of the constraints and of the objective function as a matrix.

$$M = \begin{bmatrix} 1 & 3 & 4 \\ 1 & 1 & 2 \\ 4 & 1 & 0 \end{bmatrix} \quad \Rightarrow \quad M^T = \begin{bmatrix} 1 & 1 & 4 \\ 3 & 1 & 1 \\ 4 & 2 & 0 \end{bmatrix}$$

The dual problem seeks a maximization of $P = 4y_1 + 2y_2$ subject to

$$y_1 + y_2 \leq 4$$
$$3y_1 + y_2 \leq 1$$

(and subject to all $y_i \geq 0$). Introduce the two slack variables u_1 and u_2 to transform the system of inequalities to a system of equations. The initial simplex tableau becomes

$$\begin{array}{ccccc} y_1 & y_2 & u_1 & u_2 & p \\ \end{array}$$
$$\left(\begin{array}{ccccc|c} 1 & 1 & 1 & 0 & 0 & 4 \\ \boxed{3} & 1 & 0 & 1 & 0 & 1 \\ \hline -4 & -2 & 0 & 0 & 1 & 0 \end{array} \right)$$

Pivot about the boxed element (make sure you see why *this* element is boxed) to obtain

$$\begin{array}{ccccc} y_1 & y_2 & u_1 & u_2 & p \\ \end{array}$$
$$\left(\begin{array}{ccccc|c} 0 & \frac{2}{3} & 1 & -\frac{1}{3} & 0 & \frac{11}{3} \\ 1 & \boxed{\frac{1}{3}} & 0 & \frac{1}{3} & 0 & \frac{1}{3} \\ \hline 0 & -\frac{2}{3} & 0 & \frac{4}{3} & 1 & \frac{4}{3} \end{array} \right)$$

The boxed element is the pivot element (again, why?). Pivot about it for the final matrix in this sequence.

$$\begin{array}{ccc} & u_1 & u_2 & p \\ \end{array}$$
$$\left(\begin{array}{ccccc|c} -2 & 0 & 1 & -1 & 0 & 3 \\ 3 & 1 & 0 & 1 & 0 & 1 \\ \hline 2 & 0 & \underline{0} & \underline{2} & 1 & 2 \end{array} \right)$$

Read the answer to the primary problem from the bottom line. The minimum value of p is 2, attained when $x_1 = 0$ and $x_2 = 2$.

Skill Enhancer 22   We want to minimize $p = 2x_1 + 3x_2$ if $x_1 \geq 0$, $x_2 \geq 0$, $x_1 + x_2 \geq 2$, and $x_1 + 2x_2 \geq 3$.
(a) What is the transpose of the coefficient matrix?
(b) In the dual problem, what function P must we maximize? Besides $y_1 \geq 0$ and $y_2 \geq 0$, what are the constraints on P?
(c) What is the initial simplex tableau for the dual problem?
(d) What is the final simplex tableau for the dual problem?
(e) What is the solution to the original, primary problem?

Answer in Appendix E.

The optimal strategy to attain the minimum may not be unique. Study the next example.

Example 23   Minimize $p = x_1 + x_2 + x_3$ if all variables must be greater than or equal to zero and if $2x_1 + 2x_2 + x_3 \geq 4$ and $x_1 + 2x_2 + 2x_3 \geq 5$.

Solution

The problem is in standard form, so we are free to use the simplex method to solve the dual problem, from which we fully expect to read the solution for the original, primary problem. Here are M and M^T:

$$M = \begin{bmatrix} 2 & 2 & 1 & 4 \\ 1 & 2 & 2 & 5 \\ 1 & 1 & 1 & 0 \end{bmatrix} \qquad M^T = \begin{bmatrix} 2 & 1 & 1 \\ 2 & 2 & 1 \\ 1 & 2 & 1 \\ 4 & 5 & 0 \end{bmatrix}$$

Construct the dual problem from M^T: Maximize $P = 4y_1 + 5y_2$ subject to

$$\begin{aligned} 2y_1 + y_2 &\leq 1 \\ 2y_1 + 2y_2 &\leq 1 \\ y_1 + 2y_2 &\leq 1 \end{aligned}$$

and to nonnegative y_i.

We introduce three slack variables u_1, u_2, and u_3 to form this initial simplex tableau:

y_1	y_2	u_1	u_2	u_3	p	
2	1	1	0	0	0	1
2	2	0	1	0	0	1
1	2	0	0	1	0	1
−4	−5	0	0	0	1	0

The pivot column is the *second* column, and the pivot ratios for the top three rows are 1, $\frac{1}{2}$, and $\frac{1}{2}$, respectively. The two smallest ratios are equal. Which one shall we choose? The existence of *several ratios*, all of which are equal, is a signal that *there may be a range of variables* that yield the unique minimum value of p.

This situation is analogous to the two-dimensional case. In certain cases, all pairs of x and y lying on an edge of the feasible region yielded the optimum. Here, too, in the multidimensional case, sometimes a range of variable values corresponds to those values lying on an edge of the feasible region that give rise to the same minimum. To verify that this is so, we will solve this problem *twice*, pivoting about each of the feasible pivot elements.

In the first case, when we pivot about the second element in the pivot column, a single transformation is sufficient.

$$
\begin{pmatrix}
2 & 1 & 1 & 0 & 0 & 0 & | & 1 \\
2 & \boxed{2} & 0 & 1 & 0 & 0 & | & 1 \\
1 & 2 & 0 & 0 & 1 & 0 & | & 1 \\
\hline
-4 & -5 & 0 & 0 & 0 & 1 & | & 0
\end{pmatrix}
\Rightarrow
\begin{array}{cccccc}
 & & u_1 & u_2 & u_3 & \\
\end{array}
\begin{pmatrix}
1 & 0 & 1 & -\frac{1}{2} & 0 & 0 & | & \frac{1}{2} \\
1 & 1 & 0 & \frac{1}{2} & 0 & 0 & | & \frac{1}{2} \\
-1 & 0 & 0 & -1 & 1 & 0 & | & 0 \\
\hline
1 & 0 & 0 & \frac{5}{2} & 0 & 1 & | & \frac{5}{2}
\end{pmatrix}
$$

This matrix implies that $p = \frac{5}{2}$ is the minimum of the original problem, which happens when $x_1 = 0$, $x_2 = \frac{5}{2}$, and $x_3 = 0$. (Note that the solution of the *dual* linear programming problem is $P_{\max} = \frac{5}{2}$ when $y_1 = u_2 = u_3 = 0$, and $y_2 = u_1 = \frac{1}{2}$.) ($u_3 = 0$ not because it is nonbasic but because that is its value upon solution.)

On the other hand, when we pivot about the third element of the pivot column in the initial tableau, we will need two transformations; here is the sequence.

$$
\begin{pmatrix}
2 & 1 & 1 & 0 & 0 & 0 & | & 1 \\
2 & 2 & 0 & 1 & 0 & 0 & | & 1 \\
1 & \boxed{2} & 0 & 0 & 1 & 0 & | & 1 \\
\hline
-4 & -5 & 0 & 0 & 0 & 1 & | & 0
\end{pmatrix}
\Rightarrow
\begin{pmatrix}
\frac{3}{2} & 0 & 1 & 0 & -\frac{1}{2} & 0 & | & \frac{1}{2} \\
\boxed{1} & 0 & 0 & 1 & -1 & 0 & | & 0 \\
\frac{1}{2} & 1 & 0 & 0 & \frac{1}{2} & 0 & | & \frac{1}{2} \\
\hline
-\frac{3}{2} & 0 & 0 & 0 & \frac{5}{2} & 1 & | & \frac{5}{2}
\end{pmatrix}
$$

$$
\Rightarrow
\begin{array}{cccccc}
 & & u_1 & u_2 & u_3 & \\
\end{array}
\begin{pmatrix}
0 & 0 & 1 & -\frac{3}{2} & 1 & 0 & | & \frac{1}{2} \\
1 & 0 & 0 & 1 & -1 & 0 & | & 0 \\
0 & 1 & 0 & -\frac{1}{2} & 1 & 0 & | & \frac{1}{2} \\
\hline
0 & 0 & 0 & \frac{3}{2} & 1 & 1 & | & \frac{5}{2}
\end{pmatrix}
$$

Although this final tableau is different from the final tableau above, it still yields the *same* answer to the *dual* problem. (Make sure you are convinced.) However, this tableau implies that the solution to the primary problem is

$$
p = \frac{5}{2} \qquad \text{when } x_1 = 0,\ x_2 = \frac{3}{2},\ \text{and } x_3 = 1.
$$

If you can, you may care to verify that any points (x_1, x_2, x_3) satisfying $x_1 = 0$ and lying on the line in the x_2–x_3 plane connecting $(x_2, x_3) = (\frac{5}{2}, 0)$ and $(x_2, x_3) = (\frac{3}{2}, 1)$ yield $p_{\min} = \frac{5}{2}$.

Skill Enhancer 23   Minimize $p = x_1 + x_2 + x_3$ for $x_1 \geq 0$, $x_2 \geq 0$, $x_3 \geq 0$, $x_1 + 2x_2 + x_3 \geq 4$, and $x_1 + 2x_2 + 2x_3 \geq 5$.

Answer in Appendix E.

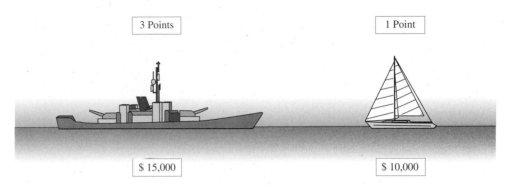

3 Points   1 Point

$ 15,000 $ 10,000

FIGURE 4.11

Example 24 A naval display is to involve two types of ships. An effective display should contain at least 10 vessels. Each of the larger ships is worth 3 points, and a small ship is worth 1 point. These "points" are public relations units to measure the effectiveness of an event. An effective display should be worth at least 16 points. There is a cost associated with the display. Each small ship costs $10,000 to use, and each large ship costs $15,000. How shall the navy arrange the display to satisfy the public relations constraints yet minimize the cost? Use simplex methods to determine the solution.

Solution

Make sure you see why this is a linear programming problem. If x_1 and x_2 stand for the number of small and large ships, respectively, then we need to minimize the cost

$$p = 10,000x_1 + 15,000x_2 = 5,000(2x_1 + 3x_2)$$

We will minimize

(4.27) $$p' = 2x_1 + 3x_2$$

both p and p' have minimum values at the same values of x_1 and x_2, but since the coefficients of p' are smaller, it is easier to work with.

That there shall be at least 10 ships translates to

(4.28) $$x_1 + x_2 \geq 10$$

The constraint on public relations "points" becomes

(4.29) $$x_1 + 3x_2 \geq 16$$

Writing (4.27) through (4.29) as

$$x_1 + x_2 \geq 10$$
$$x_1 + 3x_2 \geq 16$$
$$2x_1 + 3x_2 = p'$$

implies that the matrices M and M^T are

$$M = \begin{bmatrix} 1 & 1 & 10 \\ 1 & 3 & 16 \\ 2 & 3 & 0 \end{bmatrix} \qquad M^T = \begin{bmatrix} 1 & 1 & 2 \\ 1 & 3 & 3 \\ 10 & 16 & 0 \end{bmatrix}$$

M^T suggests that the dual problem is defined in terms of variables y_1, y_2 by

$$y_1 + y_2 \leq 2$$
$$y_1 + 3y_2 \leq 3$$
$$10y_1 + 16y_2 = P$$

where P is to be maximized. This system of inequalities becomes a system of equations when we include slack variables s_1 and s_2.

$$y_1 + y_2 + s_1 \qquad\qquad = 2$$
$$y_1 + 3y_2 \qquad + s_2 \qquad = 3$$
$$-10y_1 - 16y_2 \qquad\qquad + P = 0$$

The initial simplex tableau is

$$\left(\begin{array}{ccccc|c} 1 & 1 & 1 & 0 & 0 & 2 \\ 1 & \boxed{3} & 0 & 1 & 0 & 3 \\ \hline -10 & -16 & 0 & 0 & 1 & 0 \end{array}\right)$$

The reader should understand why the boxed 3 is the pivot element, and that when pivoting is done, we obtain

$$\left(\begin{array}{ccccc|c} \boxed{\frac{2}{3}} & 0 & 1 & -\frac{1}{3} & 0 & 1 \\ \frac{1}{3} & 1 & 0 & \frac{1}{3} & 0 & 1 \\ \hline -\frac{14}{3} & 0 & 0 & \frac{16}{3} & 1 & 16 \end{array}\right).$$

Upon carrying out this pivot operation (about the boxed $\frac{2}{3}$), we obtain the final reduced matrix

$$\begin{array}{ccccc} y_1 & y_2 & u_1 & u_2 & P \end{array}$$
$$\left(\begin{array}{ccccc|c} 1 & 0 & \frac{3}{2} & -\frac{1}{2} & 0 & \frac{3}{2} \\ 0 & 1 & -\frac{1}{2} & \frac{1}{2} & 0 & \frac{1}{2} \\ \hline 0 & 0 & 7 & 3 & 1 & 23 \end{array}\right).$$

This provides the answer to the *dual* problem. Since there were two variables in the original problem, there were two slack variables in the dual problem, and the entries at the bottom of the slack variable columns are the optimal values for the original x_1 and x_2. Furthermore, the entry in the lower right-hand position is the value of p' in the original problem. Therefore, we conclude that

$$x_1 = 7 \qquad x_2 = 3 \qquad p' = 23$$

That is, use 7 small ships and 3 large ships. The associated cost is $p = 5{,}000p' = 5{,}000 \times 23 = \$115{,}000$.

Skill Enhancer 24   A student will use linear programming techniques to decide on an optimum display of hot air balloons at the County Fair. The display should involve at least 7 balloons, of which there are two types (large and small). With regard to public relations "points" (as per the previous example), a display should involve at least

5 Points

$1200

2 Points

$800

FIGURE 4.12

29 points. A large balloon is worth 5 points, and a small balloon is worth 2 points. On the other hand, each small balloon costs $800 to launch, whereas a large balloon costs $1,200. How shall the display be arranged so as to satisfy the public relations constraints and yet minimize the cost of the display?

Answer in Appendix E.

_____ *Exercises 4.4* _____

In Exercises 1–14, give the transposes (duals) for the following matrices.

1. (3)

2. (-4)

3. $\begin{pmatrix} 1 & 2 \\ 3 & 4 \end{pmatrix}$

4. $\begin{pmatrix} 1 & 1 \\ 1 & 0 \end{pmatrix}$

5. $\begin{pmatrix} 1 & 0 \\ 1 & 1 \end{pmatrix}$

6. $\begin{pmatrix} 1 & -2 \\ -3 & 4 \end{pmatrix}$

7. $\begin{pmatrix} 0 & 0 \\ 0 & 0 \end{pmatrix}$

8. $\begin{pmatrix} -2 & -2 \\ -2 & -2 \end{pmatrix}$

9. $\begin{pmatrix} 1 & 2 & 3 \\ -3 & -2 & -1 \end{pmatrix}$

10. $\begin{pmatrix} 0.1 \\ 0.1 \\ 0 \end{pmatrix}$

11. $\begin{pmatrix} \frac{1}{2} & 2 \\ -1 & 1 \\ 0 & \frac{1}{2} \end{pmatrix}$

12. $(7 \quad -7)$

13. $(-1 \quad 1 \quad 0 \quad 2 \quad 4 \quad 6 \quad 8)$

14. $\begin{pmatrix} 1 & 0 \\ 0 & 0 \\ 0 & 0 \\ -1 & -1 \\ 0 & 0 \end{pmatrix}$

In Exercises 15–20, assume an objective function

$$p = 2x_1 + x_2 + 3x_3$$

to be minimized with respect to the individual sets of constraints given below. Also, assume in all cases that $x_i \geq 0$ for $i = 1, 2, 3$. Do not solve these problems but give the dual maximal linear programming problem.

15. $x_1 \geq 3$
$x_2 \geq 1$
$x_3 \geq 1$

16. $x_1 \geq 1$
$x_2 \geq 2$
$x_3 \geq 2$

17. $x_1 + x_2 + x_3 \geq 3$

18. $2x_1 + x_3 \geq 3$
$x_1 + 2x_2 + x_3 \geq 4$

19. $x_1 + 2x_2 + 3x_3 \geq 6$

$2x_1 + x_2 + x_3 \geq 4$

$x_1 + x_2 \geq 2$

20. $3x_1 + 9x_2 + x_3 \geq 10$

$4x_1 + 5x_2 - 3x_3 \geq 6$

$9x_1 + 10x_2 + x_3 \geq 10$

In Exercises 21–24, assume an objective function

$$p = 3x_1 + x_3$$

to be minimized with respect to the individual sets of constraints given below. Also, assume in all cases that $x_i \geq 0$ for $i = 1, 2, 3$. Do not solve these problems. Rather, give the dual maximal linear programming problem.

21. $x_1 \geq 2$, $x_2 \geq 2$, $x_3 \geq 3$ **22.** $x_1 \geq 3$

$x_1 + 2x_2 + x_3 \geq 7$

23. $x_1 + 2x_2 \geq 3$ **24.** $3x_1 + 6x_2 + x_3 \geq 6$

$2x_1 + x_3 \geq 3$

In Exercises 23–26, assume an objective function

$$P = 3y_1 + 4y_2$$

to be maximized subject to the sets of constraints given in each exercise. Assume that all variables y_i are greater than or equal to zero. Do not solve the problems, but give the dual minimal problem. (That is, give the objective function p that will be minimized subject to the constraints of the dual problem.)

25. $y_1 + y_2 \leq 3$ **26.** $y_1 \leq 1$

$y_2 \leq 2$

27. $y_1 + 2y_2 \leq 4$ **28.** $3.5y_1 + 7.2y_2 \leq 9.5$

$3y_1 + 2y_2 \leq 3$ $1.2y_1 + 3.3y_2 \leq 5.4$

29. Consider this problem. What is the minimum value for

$$p = x_1 + x_2$$

when

$$x_1 + 2x_2 \geq 3$$
$$2x_1 + 3x_2 \geq 5$$

and (of course) $x_1 \geq 0$, $x_2 \geq 0$. Answer these questions.
(a) What are the matrices M and M^T for this problem?
(b) State the dual maximum problem for this problem.
(c) What is the initial simplex tableau for the dual problem? What is the pivot element for this matrix?

(d) Pivot about this element. What is the resulting matrix?
(e) What is the solution to the *dual* problem?
(f) What is the solution to the original problem?

30. Subject to $x_1 \geq 0$ and $x_2 \geq 0$ and to

$$x_1 + 3x_2 \geq 4$$
$$3x_1 + x_2 \geq 4$$

we want the minimum value for

$$p = x_1 + 2x_2.$$

Answer these questions.
(a) What are the matrices M and M^T for this problem?
(b) State the dual maximum problem for this problem.
(c) Use slack variables to transform the dual problem into a system of equations.
(d) What is the initial simplex tableau for the dual problem? What is the pivot element for this matrix?
(e) What is the final matrix in the series of pivots?
(f) What is the solution to the *dual* problem?
(g) What is the solution to the original problem?

31. Redo Example 22 using geometric methods, and verify that the answer one obtains using the simplex method is identical to the solution obtained by means of these geometrical methods.

32. Consider the problem of minimizing $p = 3x + 4y$ if $x + 2y \geq 2$, $2x + 2y \geq 3$, and both x and y are nonnegative. Solve this problem by geometric methods and by the simplex method, and show that you get the same answer in both cases.

Use the simplex method to solve the linear programming problems in Exercises 33–38. Minimize each of the objective functions p and indicate the values of the variables that yield the minimum.

33. Minimize $p = 10x_1 + 4x_2$ subject to $x_1 \geq 3$, $x_2 \geq 4$.

34. Minimize $p = x_1 + 2x_2$ subject to $x_1 \geq 0$, $x_2 \geq 0$ and to $2x_1 + x_2 \geq 2$, $x_1 + 4x_2 \geq 4$.

35. Minimize $p = 3x_1 + x_2$ subject to $x_1 \geq 0$, $x_2 \geq 0$ and to $x_1 + x_2 \geq 2$ and $2x_1 + 2x_2 \geq 4$.

36. Minimize $p = x_1 + x_2$ subject to $x_1 \geq 0$, $x_2 \geq 0$, $2x_1 + 4x_2 \leq 4$, $x_1 + x_2 \geq 2$, and $3x_1 + 2x_2 \leq 6$.

37. Minimize $p = x_1 + x_2 + x_3$ subject to $x_1 \geq 0$, $x_2 \geq 0$ and to $x_1 + 2x_2 \geq 2$, $3x_1 + x_3 \geq 3$, and $2x_2 + x_3 \geq 4$.

38. Minimize $p = x_1 + x_2 + x_3$ subject to $x_1 \geq 0$, $x_2 \geq 0$ and to $x_1 + 2x_2 + 2x_3 \geq 4$ and $2x_1 + x_2 + 3x_3 \geq 4$.

Applications

Use the simplex method for these problems.

39. **(Banquet Planning)** The Pastry Club of America is getting ready for the annual pastry tasting banquet. Each member receives her or his own plate consisting of pastry drawn from each of three kinds of pastry. Call the varieties A, B, and C. Associated with each type of pastry is a cost and a number indicating the richness of the pastry. Here are the figures for the three pastry types:

	Cost	Richness
A	$5	1,100
B	$7	1,500
C	$9	2,000

Each plate must consist of at least one of pastry type A and B and at least two of C. Last year, each plate contained 6,000 richness units and $30 worth of the sweet confections.

(a) How should this year's plates be composed if the banquet chairperson wants to minimize costs while at least equaling last year's richness level?

(b) Some of the club members are getting on in years, and have to watch their intake of rich foods. The chairperson could cater to this faction by minimizing the richness level and yet seeing that each plate contains pastry costing at least as much as last year's plate. What strategy will accomplish this goal?

40. **(Chemistry)** Chemists working for the Fadeless Paint Company have discovered that adding quantities of three different chemicals to a gallon of paint will dramatically increase the paint's ability to cover existing coats and will decrease the drying time. To preserve Fadeless's trade secrets, call these chemicals α, β, and γ. These are expensive chemicals, costing $0.50, $0.50, and $0.75 per ounce, respectively. To acquire the desirable properties from these chemicals, a gallon of paint needs at least 2 oz of α and at least 1 oz of the two others. Furthermore, because of chemical reactions between the three, the total amount of chemicals α and β must at least equal the amount of γ. How much of each chemical should a gallon of paint contain in order to minimize the cost of the chemical additives?

41. Refer back to Exercise 40. Suppose now that the federal government has ascertained that these additives are highly toxic. The government has assigned units of toxicity to each compound. Per ounce, the chemicals α, β, and γ contain 1, 2, and 4 units of toxicity. If cost is no longer an object, how can you mix a gallon of paint subject to the above chemical requirements that possesses the minimum possible toxicity?

42. **(Shoe Manufacture)** The GoodEarth Footwear Corporation is a small operation producing a line of shoes and of sneakers. Each pair of shoes or sneakers requires a certain amount of cutting, sewing, and gluing in the manufacturing process given by this chart:

	Cutting	Sewing	Gluing
Shoes	2 h	3 h	1 h
Sneakers	1 h	2 h	2 h

There are 3 cutters, 2 seamstresses, and 1 gluer in the company. Normally, each of these individuals works a maximum of 40 h per week. Each pair of shoes uses $14 of materials, whereas a pair of sneakers uses only $10 worth. The company has just entered the busy end-of-year shopping season, and has plenty of orders to fill. All the workers have graciously agreed to work overtime if need be. How many pairs of shoes and sneakers should be made to minimize the material costs of manufacturing and yet ensure that *each* worker works *at least* the full 40 h per week? What will be the materials cost associated with this strategy? Redo this problem using the geometric method, and make sure you get the same answer by both methods!

43. **(Social Services)** Violent political unrest continues to shame the twentieth century. Imagine that the United States accepts 1,100 recently orphaned refugee children. Six hundred of these youngsters enter the country at Los Angeles, and 500 enter at New York. They remain temporarily at these locations before being assigned to three shelters located in Albuquerque, Buffalo, and Chicago. At the shelters, the children will be assigned to couples for adoption. The Albuquerque and Chicago shelters need a certain minimum number of children before they can be open for operation. Specifically, Albuquerque needs at least 400 children, and Chicago needs 200. The Buffalo shelter has been in operation for a while, and places a constraint on the maximum number of children it can contain, namely 300. Associated with a child is a travel cost from the point of embarkation to the shelter. Here are those costs.

	Albuquerque	Chicago	Buffalo
New York to...	$300	$250	$200
Los Angeles to...	$300	$100	$100

Making sure that all children leave New York and Los Angeles, how should the orphans be distributed to minimize their travel costs? (Hint: Use four variables. For example, let x_1 be the number traveling from New York to Albuquerque, let x_2 be the number traveling from New York to Chicago, and let $[600 - (x_1 + x_2)]$ be the number going from New York to Buffalo.)

W 44. Is it preferable to use the simplex method in the case where there are only two or three constraints and only two variables? (This includes most of the examples of this chapter.) Why or why not?

▦ Calculator Exercises_____

In Exercises 45–46, select the pivot element and pivot about that element. Express components to two decimal places.

45. $\begin{pmatrix} 1.1 & 1.9 & 1 & 0 & 0 & | & 1 \\ 2.2 & 2.1 & 0 & 1 & 0 & | & 1 \\ -4.2 & -3.1 & 0 & 0 & 1 & | & 0 \end{pmatrix}$

46. $\begin{pmatrix} 2.3 & 5.0 & 1 & 0 & 0 & | & 1 \\ 4.6 & 2.0 & 0 & 1 & 0 & | & 1 \\ -2.0 & -3.0 & 0 & 0 & 1 & | & 0 \end{pmatrix}$

Section 4.5   MAXIMIZATION WITH MIXED CONSTRAINTS

mixed constraints

Mathematicians have determined how to apply the simplex algorithm to other kinds of linear programming problems, especially to problems with *mixed constraints*; some constraints are $\geq$ and some are $\leq$. This extension involves the creation of two new types of *supplemental variables* in addition to slack variables. Once the problem has been set up with these new variables, we pivot the resulting matrix until we find a solution. As has been the case throughout this chapter, we omit the details that rigorously justify the steps we take.

So far, our only experience with supplemental variables has been with slack variables, which were forbidden to assume negative values and which were *basic* in the initial simplex tableau.

surplus variable

In addition to slack variables, which will continue to occur in $\leq$ constraints, we will need to add surplus and artificial variables to $\geq$ constraints. We add a *surplus variable* S_i to the right side of each $\geq$ constraint to transform the constraint into an equation. For example,

$$-x_1 + x_2 \geq 1$$

becomes

$$-x_1 + x_2 = 1 + S_1$$

or

(4.30) $$-x_1 + x_2 - S_1 = 1$$

artificial variable

as in Figure 4.13. Since the coefficient of $-S_1$ is -1 (and not $+1$), this variable is not basic. We agree, therefore, to add a purely *artificial variable* a_i to the left side of each equation containing a surplus variable. Equation (4.30) becomes

$$-x_1 + x_2 - S_1 + a_1 = 1$$

Note that each surplus variable is greater than or equal to zero, and we demand that each artificial variable satisfy $a_i \geq 0$ as well.

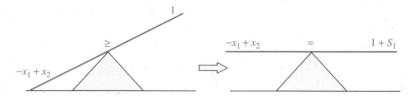

FIGURE 4.13
Surplus variables.

We introduce the artificial variables purely as a "trick" to make sure we have the right number of basic variables in the set of equations. Since it is artificial, how can we ensure that its final value is zero? For each artificial variable a_i we introduce in an equation, we add a term $-Ma_i$ to the right side of the objective function P:

$$P = \text{terms in } x_i - Ma_1 - Ma_2 - \cdots$$

cost penalty   We call M the *cost penalty* and imagine that M is a very large positive number, one that we can make to be as large as possible.

If the artificial variables take on values other than 0, then P cannot achieve its largest value; it pays a cost penalty proportional to M and to the artificial variables. If we set up a mixed constraint problem properly, solve it, and determine that some artificial variables have nonzero values, then the problem has no solution. For any problem with a valid solution, all the a_i must be 0 at the final iteration. For then, each product Ma_i is also 0, and P will suffer no cost penalty.

This brief discussion of artificial variables seems to imply that each a_i is not basic, for a_i occurs in two equations—initially, in the same equation as the surplus variable S_i and in the final equation as a product with the cost penalty M. For that reason, the first course of action shall be the proper transformations necessary to make these artificial variables basic.

Before examining several examples, let us summarize these observations.

Maximization with Mixed Constraints: Introducing Supplemental Variables

1. Write each inequality constraint in standard form—variable terms on the left and the purely constant term on the right. The objective function should be in the form $P = \cdots$.

2. To change each $\leq$ into an equation, introduce a slack variable term $+s_{i*}$ to the left of the inequality.

3. To change each $\geq$ into an equation, introduce two terms $-S_i + a_i$ to the left of the inequality.

4. For each artificial variable a_i, introduce a term $-Ma_i$ to the right of the equation for P.

5. Perform the necessary transformation to make all the artificial variables a_i basic.

Big-M method   The cost penalty M frequently gives its name to this technique—the *Big-M method*.

Example 25   Suppose we know that $x_1 \geq 0$, $x_2 \geq 0$, and

(4.31) $x_1 + x_2 \leq 2$
(4.32) $-x_1 + x_2 \geq 1$.

Eventually, we will need to find the maximum value of $P = 3x_1 + x_2$ subject to these constraints. How shall we use supplementary variables to set up a system of equations?

Solution

For Equation (4.31), we introduce a slack variable $s_1 \geq 0$ in the usual way.

(4.33) $x_1 + x_2 + s_1 = 2$

For Equation (4.32), we introduce the surplus variable $S_1 \geq 0$ that represents the amount by which the left-hand side exceeds the right:

$$-x_1 + x_2 = 1 + S_1$$

or

(4.34) $-x_1 + x_2 - S_1 = 1$

We also need an artificial variable a_1, so Equation (4.34) takes the form

(4.35) $-x_1 + x_2 - S_1 + a_1 = 1$ $S_1 \geq 0,\ a_1 \geq 0$

The objective function in our example now becomes

$$P = 3x_1 + x_2 - Ma_1$$

upon addition of a cost penalty term.

Following earlier patterns, we write the equations for this problem as

$$
\begin{aligned}
x_1 + x_2 + s_1 \qquad\qquad\qquad\qquad &= 2 \\
-x_1 + x_2 \qquad - S_1 + \quad a_1 \qquad\quad &= 1 \\
-3x_1 - x_2 \qquad\qquad\quad + Ma_1 + P\ &= 0
\end{aligned}
$$

Skill Enhancer 25   Suppose we know that $x_1 \geq 0$ and $x_2 \geq 0$ and that $x_1 + x_2 \leq 2$ and $x_1 - x_2 \geq 1$, and we will soon want to know the largest value that $P = 3x_1 + 2x_2$ can be. What supplementary variables do we need to transform this problem to a system of equations?

Answer in Appendix E.

In the usual way, we extract an initial simplex tableau from the equations that describe the mixed problem. One additional step is necessary to make sure that the artificial examples are basic. Study the following example.

Example 26   Deduce the maximum value for

$$P = 3x_1 + x_2$$

subject to $x_1,\ x_2 \geq 0$ and

$$x_1 + x_2 \leq 2$$
$$-x_1 + x_2 \geq 1$$

Solution

In Example 25 we introduced supplementary variables to transform this problem to a series of equations. From this system, we extract the augmented matrix:

(4.36)

$$
\begin{array}{cccccc}
x_1 & x_2 & s_1 & S_1 & a_1 & P \\
\end{array}
$$

$$
\left(
\begin{array}{cccccc|c}
1 & 1 & 1 & 0 & 0 & 0 & 2 \\
-1 & 1 & 0 & -1 & 1 & 0 & 1 \\
\hline
-3 & -1 & 0 & 0 & \boxed{M} & 1 & 0 \\
\end{array}
\right)
$$

As written, a_1 is not basic, since it appears in two equations. We can eliminate the bottom entry of this column by letting a new row 3 be equal to the old row 3 minus M times the second row. That is, we pivot about the boxed M in the bottom row.

$$
R_3' = R_3 - M R_2
\left(
\begin{array}{cccccc|c}
1 & 1 & 1 & 0 & 0 & 0 & 2 \\
-1 & 1 & 0 & -1 & 1 & 0 & 1 \\
\hline
M-3 & -M-1 & 0 & M & 0 & 1 & -M \\
\end{array}
\right) \cdot
$$
$$\Uparrow$$

Now we may begin the standard sequence of pivot operations.

Which variable is the pivot variable? Use the fact that M is as large as we care to make it. Therefore, since M is so large (it may help to imagine that it has a large definite value, like $M = 1,000,000$), x_2 is the pivot variable, and the second column is the pivot column. The following sequence of matrices are the matrices in each iteration. In each case, the up arrow in the bottom row identifies the pivot column. The pivot ratios are displayed at left.

$$
\begin{array}{l}
\frac{2}{1} = 2 \\
\frac{1}{1} = 1
\end{array}
\left(
\begin{array}{cccccc|c}
1 & 1 & 1 & 0 & 0 & 0 & 2 \\
-1 & \boxed{1} & 0 & -1 & 1 & 0 & 1 \\
\hline
M-3 & -M-1 & 0 & M & 0 & 1 & -M \\
\end{array}
\right) \Rightarrow
$$
$$\Uparrow$$

$$
\begin{array}{l}
\frac{1}{2} \\
-1
\end{array}
\left(
\begin{array}{cccccc|c}
\boxed{2} & 0 & 1 & 1 & -1 & 0 & 1 \\
-1 & 1 & 0 & -1 & 1 & 0 & 1 \\
\hline
-4 & 0 & 0 & -1 & M+1 & 1 & 1 \\
\end{array}
\right) \Rightarrow
$$
$$\Uparrow$$

$$
\left(
\begin{array}{cccccc|c}
1 & 0 & \frac{1}{2} & \frac{1}{2} & -\frac{1}{2} & 0 & \frac{1}{2} \\
0 & 1 & \frac{1}{2} & -\frac{1}{2} & \frac{1}{2} & 0 & \frac{3}{2} \\
\hline
0 & 0 & 2 & 1 & M-1 & 1 & 3 \\
\end{array}
\right)
$$

Pivoting stops here, because all the elements of the bottom row are positive. We read the answers in the usual way: Because the third, fourth, and fifth columns have more than one nonzero element, we conclude that s_1, S_1, and a_1 are nonbasic variables that we may set to zero. (The variable a_1 did turn out to be zero as required.) Then, we see $x_1 = \frac{1}{2}$, $x_2 = \frac{3}{2}$, and $P = 3$. (In the exercises, you are asked to do this problem using geometric methods to verify that you obtain the same solution using that method.)

Skill Enhancer 26   If we know that $x_1 \geq 0$ and $x_2 \geq 0$ and that $x_1 + x_2 \leq 2$ and $x_1 - x_2 \geq 1$, what is the largest value that $P = 3x_1 + 2x_2$ can be? How do we get that value? What is the sequence of augmented matrices that determines the solution?

Answer in Appendix E.

Maximization with Mixed Constraints: Solving Problems

1. Express the problem in *standard form*.
2. Convert the problem to a sequence of equations by adding *supplemental variables*.
3. *Extract the initial simplex tableau for this system of equations,* and use row operations to make sure that the a_i are basic.
4. *Obtain the solution by iteration.*

Example 27   Maximize

$$P = 2x_1 + 3x_2$$

subject to $x_1,\ x_2 \geq 0$ and to

$$x_1 + x_2 \leq 4$$
$$2x_1 + x_2 \geq 2$$
$$x_1 \geq \frac{1}{2}$$

Set up and solve.

Solution

Since there is one $\leq$ constraint, we need a single slack variable. There are two $\geq$ constraints as well, so the problem will require the addition of two surplus and two additional variables. These supplementary variables appear solely in the inequality constraints, which will now become equalities, but the two artificial variables plus a cost penalty M will appear in the equation for the objective function. Specifically, the constraints become

$$
\begin{aligned}
x_1 + \ x_2 + s_1 &&&&&&= 4 \\
2x_1 + \ x_2 && - S_1 && + \ a_1 &&&= 2 \\
x_1 &&& - S_2 && + \ a_2 &&= \tfrac{1}{2}
\end{aligned}
$$

and the objective function becomes

$$-2x_1 - 3x_2 \qquad\qquad + Ma_1 + Ma_2 + P = 0$$

Extracting the coefficients from this system yields this augmented matrix:

$$
\begin{array}{cccccccc}
x_1 & x_2 & s_1 & S_1 & S_2 & a_1 & a_2 & P \\
\end{array}
$$

$$
\left(
\begin{array}{cccccccc|c}
1 & 1 & 1 & 0 & 0 & 0 & 0 & 0 & 4 \\
2 & 1 & 0 & -1 & 0 & 1 & 0 & 0 & 2 \\
1 & 0 & 0 & 0 & -1 & 0 & 1 & 0 & \frac{1}{2} \\
\hline
-2 & -3 & 0 & 0 & 0 & M & M & 1 & 0 \\
\end{array}
\right)
$$

To make the artificial variables a_1 and a_2 basic in the first iteration, we begin by subtracting M copies of the second and third rows from the fourth row.

$$
\begin{array}{cccccccc}
x_1 & x_2 & s_1 & S_1 & S_2 & a_1 & a_2 & P \\
\end{array}
$$

$$
\left(
\begin{array}{cccccccc|c}
1 & 1 & 1 & 0 & 0 & 0 & 0 & 0 & 4 \\
2 & 1 & 0 & -1 & 0 & 1 & 0 & 0 & 2 \\
1 & 0 & 0 & 0 & -1 & 0 & 1 & 0 & \frac{1}{2} \\
\hline
-2 & -3 & 0 & 0 & 0 & M & M & 1 & 0 \\
\end{array}
\right) \Rightarrow
$$

$$
\begin{array}{cccccccc}
x_1 & x_2 & s_1 & S_1 & S_2 & a_1 & a_2 & P \\
\end{array}
$$

$$
R_4' = R_4 - MR_2 - MR_3 \left(
\begin{array}{cccccccc|c}
1 & 1 & 1 & 0 & 0 & 0 & 0 & 0 & 4 \\
2 & 1 & 0 & -1 & 0 & 1 & 0 & 0 & 2 \\
1 & 0 & 0 & 0 & -1 & 0 & 1 & 0 & \frac{1}{2} \\
\hline
-3M-2 & -M-3 & 0 & M & M & 0 & 0 & 1 & -\frac{5}{2}M \\
\end{array}
\right)
$$

At this point simply identify the pivot element and perform the usual series of pivot operations until all elements on the bottom row of the latest iteration are all nonnegative. Here are the successive matrices in that series (make sure you can obtain them yourself). In the usual way, the calculations to the left of each matrix yield the pivot ratios. The vertical arrow at the bottom identifies the pivot column. *Remember, M is always considered to be a very large number.* It may be helpful to regard it as a number like 1,000 or 1,000,000 (or even larger).

$$
\begin{array}{cccccccc}
 & x_1 & x_2 & s_1 & S_1 & S_2 & a_1 & a_2 & P \\
\end{array}
$$

$$
\begin{array}{c}
\frac{4}{1}=4 \\
\frac{2}{2}=1 \\
\frac{1/2}{1}=\frac{1}{2}
\end{array}
\left(
\begin{array}{cccccccc|c}
1 & 1 & 1 & 0 & 0 & 0 & 0 & 0 & 4 \\
2 & 1 & 0 & -1 & 0 & 1 & 0 & 0 & 2 \\
\boxed{1} & 0 & 0 & 0 & -1 & 0 & 1 & 0 & \frac{1}{2} \\
\hline
-3M-2 & -M-3 & 0 & M & M & 0 & 0 & 1 & -\frac{5}{2}M \\
\end{array}
\right) \Rightarrow
$$

$$\Uparrow$$

$$
\begin{array}{c}
3\frac{1}{2} \\
\frac{1}{2} \\
-\frac{1}{2}
\end{array}
\left(
\begin{array}{cccccccc|c}
0 & 1 & 1 & 0 & 1 & 0 & -1 & 0 & 3\frac{1}{2} \\
0 & 1 & 0 & -1 & \boxed{2} & 1 & -2 & 0 & 1 \\
1 & 0 & 0 & 0 & -1 & 0 & 1 & 0 & \frac{1}{2} \\
\hline
0 & -M-3 & 0 & M & -2M-2 & 0 & 3M+2 & 1 & -M+1 \\
\end{array}
\right) \Rightarrow
$$

$$\Uparrow$$

$$6 \begin{pmatrix} 0 & \frac{1}{2} & 1 & \frac{1}{2} & 0 & -\frac{1}{2} & 0 & 0 & 3 \\ 1 & 0 & \boxed{\frac{1}{2}} & 0 & -\frac{1}{2} & 1 & \frac{1}{2} & -1 & 0 & \frac{1}{2} \\ 2 & 1 & \frac{1}{2} & 0 & -\frac{1}{2} & 0 & \frac{1}{2} & 0 & 0 & 1 \\ \hline & 0 & -2 & 0 & -1 & 0 & M+1 & M & 1 & 2 \end{pmatrix} \Rightarrow$$

$$\begin{matrix} 2\frac{1}{2} \\ - \\ - \end{matrix} \begin{pmatrix} 0 & 0 & 1 & \boxed{1} & -1 & -1 & 1 & 0 & 2\frac{1}{2} \\ 0 & 1 & 0 & -1 & 2 & 1 & -2 & 0 & 1 \\ 1 & 0 & 0 & 0 & -1 & 0 & 1 & 0 & \frac{1}{2} \\ \hline 0 & 0 & 0 & -3 & 4 & M+3 & M-4 & 1 & 4 \end{pmatrix} \Rightarrow$$

$$\begin{matrix} x_1 & x_2 & s_1 & S_1 & S_2 & a_1 & a_2 & P \end{matrix}$$
$$\begin{pmatrix} 0 & 0 & 1 & 1 & -1 & -1 & 1 & 0 & 2\frac{1}{2} \\ 0 & 1 & 1 & 0 & 1 & 0 & -1 & 0 & 3\frac{1}{2} \\ 1 & 0 & 0 & 0 & -1 & 0 & 1 & 0 & \frac{1}{2} \\ \hline 0 & 0 & 3 & 0 & 1 & M & M-1 & 1 & 11\frac{1}{2} \end{pmatrix}$$

The answers can be read from this last matrix: $x_1 = \frac{1}{2}$, $x_2 = 3\frac{1}{2}$, and $P = 11\frac{1}{2}$. In addition, the values for the supplementary variables are $s_1 = S_2 = a_1 = a_2 = 0$ and $S_1 = 2\frac{1}{2}$.

Skill Enhancer 27   What is the maximum value of $P = 3x_1 + x_2$ if $x_1 + x_2 \leq 4$, $x_1 + x_2 \geq 2$, $x_2 \leq 2$, and (of course) $x_1 \geq 0$ and $x_2 \geq 0$? What values of x_1 and x_2 give that value?

Answer in Appendix E.

Supplementary Variables for Mixed Problems: How Many of Each?

A problem involving mixed constraints will require three different supplementary variables. We add one *slack* variable for each $\leq$ constraint, and one *surplus* variable for each $\geq$ constraint, so the total number of slack plus surplus variables gives the total number of inequalities, which is also the number of basic variables. Furthermore, there will be as many *artificial* variables as there are surplus variables.

Example 28   A factory in this country manufactures computer chips. One type sells for $2 apiece and a second for $3 apiece. The profit for these two chips is $1 and $2 per chip, respectively. The cost of the raw materials for each chip is $1 per chip no matter what variety it is. The monthly profit from the chip manufacture must be at least

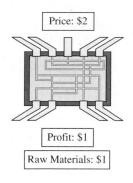

Price: $2

Profit: $1

Raw Materials: $1

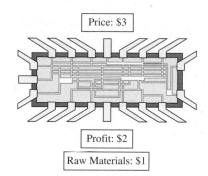

Price: $3

Profit: $2

Raw Materials: $1

FIGURE 4.14

$10,000, but the monthly cost of raw materials must not exceed $7,000. Furthermore, the marketplace will not absorb more than $3,000 worth of the first chip per month. What manufacturing strategy will maximize the *income*? What is the maximum income? Use simplex methods.

Solution

For this linear programming problem, let x_1 and x_2 be the amounts of the first and second chips, measured in units of thousands of dollars. Clearly, $x_1 \geq 0$ and $x_2 \geq 0$.

The two "less than" constraints are

$$(4.37) \qquad\qquad x_1 + x_2 \leq 7$$

arising from the constraint on monthly expenditures of raw materials, and

$$(4.38) \qquad\qquad x_1 \leq 3$$

because we may not produce more than $3,000 worth of the first chip.

There is a single "greater than" inequality

$$(4.39) \qquad\qquad x_1 + 2x_2 \geq 10$$

due to the demand that monthly profit must equal or exceed $10,000.

If I stands for income, then we must maximize the objective function

$$(4.40) \qquad\qquad I = 2x_1 + 3x_2$$

Upon the inclusion of slack variables s_1 and s_2, one surplus variable S_1, and one artificial variable a_1, the relations (4.37) through (4.39) become

$$
\begin{aligned}
x_1 &+ x_2 &+ s_1 && && && &= 7 \\
x_1 && && &+ s_2 && && &= 3 \\
x_1 &+ 2x_2 && && &- S_1 &+ a_1 && &= 10
\end{aligned}
$$

while the objective function (4.40) becomes

$$
-2x_1 \quad -3x_2 \qquad\qquad + Ma_1 \quad + I \ = 0
$$

(M is the cost penalty). The initial simplex tableau is

$$\begin{array}{ccccccc}
x_1 & x_2 & s_1 & s_2 & S_1 & a_1 & I \\
\end{array}$$
$$\left(\begin{array}{ccccccc|c}
1 & 1 & 1 & 0 & 0 & 0 & 0 & 7 \\
1 & 0 & 0 & 1 & 0 & 0 & 0 & 3 \\
1 & 2 & 0 & 0 & -1 & 1 & 0 & 10 \\
\hline
-2 & -3 & 0 & 0 & 0 & M & 1 & 0
\end{array}\right)$$

Before beginning the series of pivot operations, we make a_1 a basic variable by subtracting M multiples of the third row from the last row.

$$\begin{array}{ccccccc}
x_1 & x_2 & s_1 & s_2 & S_1 & a_1 & I \\
\end{array}$$
$$\left(\begin{array}{ccccccc|c}
1 & 1 & 1 & 0 & 0 & 0 & 0 & 7 \\
1 & 0 & 0 & 1 & 0 & 0 & 0 & 3 \\
1 & \boxed{2} & 0 & 0 & -1 & 1 & 0 & 10 \\
\hline
-M-2 & -2M-3 & 0 & 0 & M & 0 & 1 & -10M
\end{array}\right)$$

We choose the cost penalty to be so large that $-2M - 3$ will have a larger absolute value than $-M - 2$, so that the second column will be the pivot column. The boxed 2 is the pivot element. (Why?) The reader should show that there are two additional pivots before we reach the final reduced matrix, which is

$$\begin{array}{ccccccc}
x_1 & x_2 & s_1 & s_2 & S_1 & a_1 & I \\
\end{array}$$
$$\left(\begin{array}{ccccccc|c}
1 & 0 & 2 & 0 & 1 & -1 & 0 & 4 \\
1 & 0 & 0 & 1 & 0 & 0 & 0 & 3 \\
1 & 1 & 1 & 0 & 0 & 0 & 0 & 7 \\
\hline
1 & 0 & 3 & 0 & 0 & M & 1 & 21
\end{array}\right)$$

The nonbasic variables are x_1, s_1, and a_1; we set them to 0:

$$x_1 = s_1 = a_1 = 0$$

The remaining variables have the following values:

$$x_2 = 7 \qquad s_2 = 3 \qquad S_1 = 4 \qquad I = 21$$

That is, the optimal strategy consists of manufacturing none of the first chip and 7,000 of the second chip. (Remember, units are in thousands of chips.) The resulting income to the firm is $21,000 per month.

Skill Enhancer 28   Does the optimal strategy for this example change if the income function becomes $I = 3x_1 + 2x_2$? If so, what is the new optimal strategy? What is the new maximum value for I?

Answer in Appendix E.

_____ *Exercises 4.5* _____

For Exercises 1–4, introduce supplementary variables (slack, surplus, and artificial variables) to put the objective function and inequality constraints into the equation form that the simplex form requires. From these equations, write down the augmented matrix that will be used in the simplex algorithm. *Do not solve any further.* For these problems, assume all variables are nonnegative, and the objective function

$$P = 2x + 4y$$

is to be maximized.

1. $x \geq 1, \ x \leq 4$
$\quad y \geq 1, \ y \leq 3$

2. $x \geq \frac{1}{2}, \ x \leq 1$
$\quad y \geq 1, \ y \leq 2$

3. $x + y \geq 1$
$\quad y \leq 2, \ x \leq 2$

4. $2x + y \geq 4$
$\quad y \leq 4, \ x \leq 9$

Set up Exercises 5–8 assuming that the quantity

$$P = x + 2y + 3z$$

is the objective function to be maximized. Do not solve. Assume that all variables are nonnegative.

5. $x + y + z \geq 3$

$\quad x + y \leq 4$

$\quad y + z \leq 4$

$\quad z + x \leq 4$

6. $\qquad x + 2y + z \geq 4$

$\quad x \leq 10, \ y \leq 8, \ z \leq 8$

7. $\qquad x + 2y + 3z \geq 4$

$\quad x \leq 3, \ y \leq 2, \ z \leq 2$

8. $\qquad x + y \leq 3$

$\qquad y + z \leq 2$

$\qquad z + x \leq 2$

$\qquad x + y + z \geq 3$

9. Consider the problem where

$$x + y \leq 2$$
$$x + y \geq 1$$

and $x \geq 0, \ y \geq 0$. We need to maximize

$$P = x + 2y$$

subject to these constraints.
(a) Add supplementary variables to transform this problem to a set of equations.
(b) What is the initial simplex tableau for this problem?
(c) What is the final tableau in the series of pivots?
(d) What is the solution to this problem?

10. Consider the problem where

$$x + 2y \leq 4$$
$$x + 2y \geq 2$$

and $x \geq 0, \ y \geq 0$. We need to maximize

$$P = 3x + 2y$$

subject to these constraints.
(a) Add supplementary variables to transform this problem to a set of equations.
(b) What is the initial simplex tableau for this problem?

(c) What is the final tableau in the series of pivots?
(d) What is the solution to this problem?

11. Use simplex methods to solve Exercise 1 above.

12. Use simplex methods to solve Exercise 2 above.

13. Use simplex methods to solve Exercise 3 above.

14. Use simplex methods to solve Exercise 4 above.

15. Use simplex methods to solve Exercise 5 above.

16. Use simplex methods to solve Exercise 6 above.

17. Use simplex methods to solve Exercise 7 above.

18. Use simplex methods to solve Exercise 8 above.

19. Solve Example 26 using geometric methods to verify that that method and the simplex method yield the same answer.

20. Solve Example 27 in the text using geometric methods. Show that the answers obtained by both methods are identical.

Applications

Use simplex methods to solve these problems.

21. *(Manufacturing)* A factory manufactures tennis and golf balls. Tennis balls sell for \$3 apiece and golf balls for \$4 apiece. The profit from these two products is \$2 and \$3 per ball. The cost of the raw materials for each product is \$1 per ball. The monthly profit from the manufacture must be at least \$10,000, but the monthly cost of raw materials must not exceed \$7,000. Furthermore, the marketplace will not absorb more than \$4,000 worth of golf balls per month. What manufacturing strategy will maximize the *income*? What is the maximum income?

22. *(Education and Administration)* A college faculty member wants to maximize her chances for promotion from associate professor to full professor. Promotion depends heavily upon the quantity and quality of papers (journal articles) published by candidates for promotion. She feels that long papers are worth 5 points to the personnel committee, whereas short papers are worth only 3 points apiece. The personnel committee expect at least 3 of each. Furthermore, long papers require 150 h for research, whereas short papers only demand 100 h. She feels that she has only 1,000 h in the coming academic year to devote to research. What is her optimal strategy for maximizing the points she may earn?

23. Set up but *do not solve* the previous problem if, in addition, the personnel committee expects at least one more long paper than short paper.

24. **(Manufacturing)** Set up but *do not solve* Example 28 if the *profit* is to be as large as possible and the income must be at least $18,000 per month.

W 25. Some mixed constraint problems are difficult or tedious to solve by either method we have studied. In light of your experience, summarize the conditions under which

you might consider solving a problem using the geometric method and the conditions under which you might prefer using the simplex method.

W 26. Speculate on techniques that seem reasonable for solving *minimum* linear programming problems with mixed constraints.

CHAPTER REVIEW

Terms

simplex method	standard form	slack variables
nonbasic variables	basic variables	algorithm
iterative method	pivot variable	pivot ratio
pivot constraint	pivot column	pivot element
initial simplex tableau	primary problem	dual problem
transpose	dual matrix	Von Neumann duality principle
mixed constraints	surplus variable	artificial variable
cost penalty	Big-M method	

Key Concepts

- The geometric methods of linear programming are inadequate for large-scale, real-life problems, so mathematicians have developed an algebraic approach to linear programming. The *simplex method* is an iterative technique for determining the maximum or minimum value of an objective function subject to special classes of constraints.

- Linear programming problems must be posed in a special *standard form*. (The simplex method *can* be extended to other linear programming problems, but these methods are beyond the scope of the text.) In maximum problems, all inequality constraints take the form of "less than" inequalities. (That is, linear expressions of the variables must be less than a positive constant.) In contrast, the constraints must be "greater than" for minimum problems. In all cases, the variables of the problem may not be less than zero.

- Minimum problems are the *duals* of maximum problems. That is, if we reduce the minimum problem to a special matrix (*not* the simplex tableau), then the dual of this matrix

can be transformed to an equivalent maximum problem whose solution bears a close relationship to the original minimum problem. Therefore, it is sufficient to consider the maximum case in detail.

- Begin the simplex method by adding a series of *slack variables* to transform the original constraint inequalities into a series of equations. The Fundamental Theorem makes a precise statement about the nature of the optimal solution.

- The simplex method is iterative. That is, we begin with a guess, even though this guess may be demonstrably way off the mark. Then, refine this guess in a special way so that the next guess is closer to the optimum. If there is a solution, the simplex method guarantees that its implementation will yield the actual optimum after a finite number of steps.

- Solve maximum problems with mixed constraints by introducing slack, surplus, and artificial variables into the problem. A *cost penalty* ensures that the artificial variables are zero at the solution of the problem.

Review Exercises

Show how to add slack variables to the systems of inequalities in Exercises 1–4 to transform them to systems of equations.

1. $x_1 \leq 4$

$x_1 + x_2 \leq 5$

2. $3x_1 + 2x_3 \leq 3$

$2x_2 + 3x_3 \leq 3$

3. $x_1 + x_2 + x_3 \geq 4$

$3x_1 + 2x_2 + x_3 \geq 10$

4. $30x_1 + 65x_2 + 108x_3 \geq 144$

$0.01x_1 + 0.18x_2 + 0.33x_3 \geq 1.32$

Give the transposes for the matrices in Exercises 5–10.

5. $\begin{pmatrix} -1 & -1 & 0 & -1 \\ 1 & 1 & -2 & -1 \end{pmatrix}$

6. $\begin{pmatrix} 0.0012 \end{pmatrix}$

7. $\begin{pmatrix} 0.012 & -0.444 \\ -2.01 & -0.079 \end{pmatrix}$

8. $\begin{pmatrix} 0 & \frac{1}{2} \\ \frac{1}{2} & -\frac{3}{4} \\ -\frac{5}{6} & 0 \\ 0 & -\frac{3}{2} \end{pmatrix}$

9. $\begin{pmatrix} 1 & 0.1 \\ 0 & 1 \end{pmatrix}$

10. $\begin{pmatrix} 11 & 12 & -13 \end{pmatrix}$

In each of Exercises 11–16, is presented a linear programming problem in standard form. Give the dual problem for each. (Assume that all variables $x_i \geq 0$.) You need not solve the problems.

11. Maximize $P = 45x_1 + 32x_2 + 14x_3$ subject to $x_1 \leq 2$, $x_2 \leq 4$, and $x_3 \leq 2$.

12. Maximize $P = 7x_1 + 3x_2$ subject to $x_1 + x_2 \leq 10$ and $x_2 \leq 7$.

13. Maximize $P = x_1 + x_2 + x_3 + x_4$ subject to $2x_1 + x_3 \leq 4$, $3x_2 + x_4 \leq 1$, $7x_1 + x_2 \leq 5$, and $3x_3 + 4x_4 \leq 9$.

14. Minimize $p = x_1 + 4x_2$ subject to $x_1 \geq 2$ and $x_1 + 3x_2 \geq 4$.

15. Minimize $p = 2x_1 + x_2 + 3x_3$ subject to $x_1 + x_3 \geq 4$, $2x_1 + 2x_2 + 3x_3 \geq 7$, and to $4x_2 + x_3 \geq 3$.

16. Minimize $p = 4x_1 + 3x_2 + 2x_3$ subject to $x_1 \geq 1$, $2x_1 + x_2 + x_3 \geq 4$, $x_1 + x_2 + 3x_3 \geq 5$, and $4x_2 + 3x_3 \geq 4$.

Use the simplex method to solve Exercises 17–22. Determine both the optimum value of the objective function and the strategy for achieving this value.

17. Maximize $P = 2x + 3y$ subject to $y \leq 4$ and $x + y \leq 8$. Both x and y may not be less than 0.

18. Maximize $P = x_1 + x_2 + x_3$ subject to $x_1 \leq 3$, $x_2 \leq 3$, $x_3 \leq 3$, and $2x_1 + 2x_2 + x_3 \leq 15$. All the x_i are ≥ 0.

19. Maximize $P = 4x_1 + 5x_3$ subject to $x_2 \leq 4$, $x_1 + x_3 \leq 3$, and $x_i \geq 0$ for $i = 1, 2, 3$.

20. Minimize $p = 2x_1 + 2x_2$ for $x_1 + 3x_2 \geq 4$, $3x_1 + x_2 \geq 4$, and for $x_1 \geq 0$, $x_2 \geq 0$.

21. Minimize $p = 3x_1 + x_2 + 3x_3$ subject to no x_i being less than zero, and to $x_1 \geq 2$, $x_1 + x_2 + x_3 \geq 6$, and $2x_1 + 3x_2 + x_3 \geq 12$.

22. Minimize $p = x_1 + 4x_2 + x_3$ subject to $x_1 + x_2 + x_3 \geq 4$, $x_1 \geq 2$, $x_2 + 2x_3 \geq 4$, and to $x_i \geq 0$, $i = 1, 2, 3$.

Set up the mixed constraint problems in Exercises 23–24. Do not solve. For both exercises, maximize the quantity $P = 2x_1 + 3\frac{1}{2}x_2$. Assume $x_1 \geq 0$, $x_2 \geq 0$.

23. $x_1 \geq 1$

$x_1 + x_2 \leq 2$

24. $x_1 \geq 1$

$x_2 \geq 1$

$x_1 + x_2 \leq 4$

25. Use simplex techniques to solve Exercise 23.

26. Use simplex techniques to solve Exercise 24.

Applications

Use the simplex method in each problem.

27. (Manufacturing) The UFO Surfboard Company is a small surfboard manufacturer in southern California. Loosely speaking, any surfboard is manufactured in two steps: shaping and finishing. UFO manufactures three models of boards. Here's a chart detailing the hours and profit per surfboard in each model class.

	Shaping	Finishing	Profit
Model A	1	2	$10
Model B	2	2	$10
Model C	3	3	$15

The company employs 20 shapers and 10 finishers, each of whom works not more than 30 h per week. Determine the best weekly manufacturing strategy for the company to maximize its profit.

28. **(Biological Research)** The biology department of a great Midwestern university is engaged in significant genetic research. The department must conduct a series of genetic experiments on three species of fruit flies to gain data for the project. Each experiment involves several weeks of setting up and several weeks of observation. Laboratory technicians set up each experiment, and graduate students observe and record data. Here are the setup and observation requirements (in weeks) for each species.

	Setup	Observation
Species 1	1	2
Species 2	2	2
Species 3	3	3

The professor designing the experiment points to each of the three types of experiments indicating the relative worth of a single experiment on a species to the entire project. The three species have point values of 100, 100, and 150, respectively. If there are 20 lab technicians and 10 graduate students available to this project, how many experiments shall be conducted with each kind of fly to maximize the point value of the entire project? Assume that all personnel have 30 weeks to devote to the project.

29. Here are some additional data pertaining to the experiment described in Exercise 28. The cost of the materials for an experiment on each kind of fly is $20, $15, and $25 for each. How now should the project be conducted to *minimize* the cost of materials? Assume that each lab technician and graduate student will devote *at least* 30 weeks to the project.

30. **(City Planning)** The Planning Department of the City of New York is surveying two recent groups of immigrants, Taiwanese and Russians. Some immigrants are selected, and the survey consists of two interviews with that individual. Generally, for the Russian-speaking immigrants, the initial interview takes $2\frac{1}{2}$ h and the follow-up interview lasts $1\frac{1}{2}$ h. For the Taiwanese, the initial and follow-up interviews generally last 2 h each. Two groups of city employees take the data. The initial interviews are conducted by trainees, who are paid $7.50 per hour. Professional social workers, paid $10 per hour, conduct the follow-up interviews. If the City wants to minimize the interviewing costs, how many immigrants of each group shall be interviewed per week if both the trainees and social workers work at least 40 h per week?

31. Referring back to Exercise 30, what is the maximum number of immigrants that can be interviewed per week if the interviewers work *at most* 40 hours per week?

32. Juanita is a traveling sales representative for a publishing company that furnishes a certain encyclopedia. This product comes in two grades: deluxe (leather binding) and standard. Her commission varies with the grade of encyclopedia she sells. It is $200 for each deluxe set, but only $100 for each standard set. Historically, she has always met the company's quota of selling a total of 10 sets per month, and has on occasion sold as many as 20 sets per month, but never more. Under these conditions, what is the maximum commission she can hope to earn in any month?

5

SETS AND COUNTING

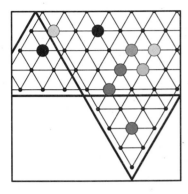

INTRODUCTION

Mathematicians use *sets* the way some people use oversized, floppy handbags—as convenient containers for just about anything. We will study relationships between these mathematical handbags and develop a method for visualizing these relationships. We will learn how to *count* the elements of sets, a surprisingly nontrivial topic, and we will use this skill to extract information about groups of things, some of which share attributes.

We may generalize these concepts of counting from sets to activities in which there are several ways of performing or choosing something. Discussions on *permutations* and *combinations* enable us to compute the number of ways.

Apart from the importance of the new counting techniques for their own sakes, this material is fundamental to the study of probability. It is difficult to state a useful collection of rules for computing probabilities without drawing upon set theory.

Section 5.1 INTRODUCTION TO SETS

set

element

We can group a collection of objects together into a mathematical "container" called a *set*. Sets may contain anything at all, but typically the members of a set are other mathematical objects—numbers, variables, or other sets. Any member of a set is referred to as an *element* of the set. Figure 5.1 shows a general set and part of its diversified contents.

One way to *define a set* is by listing the elements of the set within curly brackets:

$$\{\text{Thursday, Friday, Saturday}\} \qquad \{\text{s, o, f, t, e, n}\} \qquad \{1, 0, -1\}$$

The first and third sets of this series each contains three elements. The second set contains six elements.

Often, though, it is more convenient to define a set by stating the rule or pattern to which all elements conform:

$$\{\text{The days of the week containing the letter } r\}$$
$$\{\text{The initial letters for the English names}$$
$$\text{of the integers from one to ten inclusive}\}$$
$$\{\text{The natural numbers between } -1 \text{ and } +1 \text{ inclusive}\}$$

Sometimes we may describe a set yet be unable (or unwilling) to list all its members:

$$\{\text{All distinct hands in a game of bridge}\}$$
$$\{\text{Closing stock prices for all stocks on the}$$
$$\text{American Stock Exchange on October 19, 1987}\}$$
$$\{\text{All integers evenly divisible by 3}\}$$

Sets are usually named using capital letters:

$$A = \{\text{Thursday, Friday, Saturday}\}$$
$$B = \{\text{s, o, f, t, e, n}\}$$
$$C = \{1, 0, -1\}$$

set equality

Two sets are *equal* when they contain the same elements, regardless of the order in which the elements are listed. *The order of elements in a set is unimportant.* For example, if

$$D = \{\text{e, f, n, o, s, t}\}$$

we can say

$$B = D$$

When listing elements of a set, list only the *distinct* elements. That means that

$$E = \{\text{r, i, f, f, r, a, f, f}\}$$

is redundant; more concisely, we have

$$E = \{\text{a, f, i, r}\}$$

where the distinct elements are in alphabetical order. (We use alphabetical order because it is useful for a human reader, but it has no significance to the construction of the set E.)

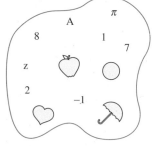

FIGURE 5.1
Set elements may be anything at all.

Relationships are important in mathematics. We can say that element e belongs to a set A by writing

$$e \in A$$

The $\in$ operator *only* connects an element to a set. If B is another set, it is *never* proper to state

$$B \in A$$

(unless A is a set whose elements are themselves sets). To indicate that some element f is *not* in A, we write

$$f \notin A$$

Example 1   Let

$$A = \{\text{place names beginning with the uppercase letter } A\}$$

$$B = \{\text{place names ending with the lowercase letter } a\}$$

Identify the following statements as true or false.
(a) Allentown $\in A$
(b) Allentown $\notin B$
(c) Botswana $\in A$
(d) Botswana $\in B$
(e) Arabia $\in A$ and Arabia $\in B$
(f) Peoria $\in A$ or Peoria $\in B$
(g) Let

$$C = \{\text{all place names both beginning and ending in } A\}$$

Then $C \in A$.

Solution

Statements (a), (b), (d), (e), and (f) are true; the remainder are false. Part (g) is false because the $\in$ operator connects elements and sets, not sets and sets.

Skill Enhancer 1   If $C = \{\text{years in the twentieth century}\}$ and $D = \{\text{years from 1850 to 1950 exclusive}\}$, then which of the following statements are true?
(a) $1899 \in C$
(b) $1899 \in D$
(c) $C \in D$
(d) $1901 \in C$ and $1901 \in D$
(e) $1950 \in D$

Answers in Appendix E.

Sets may have an *infinite* number of elements:

$$I = \{\text{all positive integers}\} = \{1, 2, 3, 4, \dots\}$$

$$R = \{\text{all rational numbers}\}$$

$$O = \{\text{all positive, odd integers}\} = \{1, 3, 5, 7, 9, \dots\}$$

$$M = \{\text{all positive integer multiples of 11}\} = \{11, 22, 33, 44, \dots\}$$

infinite set   Any set containing an infinite number of members is an *infinite set*. It is permissible to specify the elements of an infinite set by listing the first few elements, followed by an *ellipsis* (three dots, ...). List enough elements to make the pattern clear. We specified sets I, O, and M by describing the elements and listing them. R can only be described. (It is not possible to list the elements of R in numerical order. If you doubt this statement, try to answer this question: What is the numerically next rational number after $\frac{1}{2}$? All rational numbers *can* be ordered, but this ordering is not numerical.)

Subsets, Proper and Improper

The elements within a set completely determine that set. Therefore, we may compare two sets on the basis of elements they may have in common.

subset   If all elements of some set A are found in some set B, we call A a *subset* of B and write

$$A \subseteq B$$

Figure 5.2 illustrates the subset relationship.

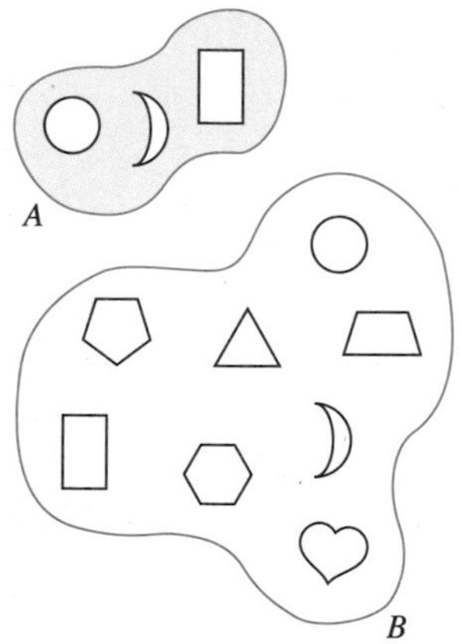

Example 2   Here are two sets:

$$I = \{\text{all integers between 1 and 10 inclusive}\}$$
$$O = \{\text{all odd integers } x \text{ satisfying } 1 < x < 10\}$$

Is $O \subseteq I$ a true statement?

Solution

Every odd number between 1 and 10 is also an integer between 1 and 10. Therefore, $O \subseteq I$.

Skill Enhancer 2   Let $E = \{1, 2, 3, 4, 5\}$ and $F = \{5, 4, 3, 2, 1\}$. Is it true that $E \subseteq F$? Is it true that $F \subseteq E$?

Answer in Appendix E.

FIGURE 5.2
Subsets. Since every element in the shaded set A occurs in the set B, $A \subseteq B$.

Suppose C and D are two *equal* sets. Is it true that $C \subseteq D$? Yes, because since C and D contain the same elements, every element of C is within D. By the same token, $D \subseteq C$. This observation leads to a more concise definition of set equality. Two sets A and B are *equal* if and only if it is true that

$$A \subseteq B$$

and

$$B \subseteq A$$

proper subset   We sharpen our focus with the concept of *proper subset*. A set A is a proper subset of B if $A \subseteq B$ and $A \neq B$. That is, A is entirely contained within B, and there is at least one element of B not in A. If A is a proper subset of B, we write

$$A \subset B$$

In Example 2, not only is $O \subseteq I$, but we can make the stronger statement

$$O \subset I$$

If e is some element in some set A, one may say

$$\{e\} \subset A$$

or

$$e \in A$$

empty set   The *empty set* or *null set* is the special set containing no elements. Designated by the
null set   symbol $\emptyset$, it is a subset of every set, including itself:

$$\emptyset \subseteq A$$

for any set A.

Example 3   How many subsets (proper and improper) are there in the set $Z = \{0, 2, 4\}$?

Solution

Write down all subsets, and then count them. The most obvious subsets are

$$\{0\}, \{0, 2\}, \{0, 4\}$$
$$\{2\}, \{2, 4\}$$
$$\{4\}$$

Where is the subset $\{2, 0\}$? Since the order in which the elements are listed does not
matter, this subset has already been listed—it is the second set in the first row.

What other subsets have we missed? Every set is a subset of itself (but not a proper
subset), so we must include Z itself in the list of subsets. Finally, we must include the
null set $\emptyset$ in the list.

There are a total of eight subsets of Z.

Skill Enhancer 3   How many *proper* non-empty subsets are there of M, where $M = \{\text{Abel, Baker, Charlie}\}$?

Answer in Appendix E.

Note that $\{0\}$ is *not* the same as $\emptyset$. The set $\{0\}$ contains one element—the number 0.
The set $\emptyset$ contains *no* elements, and could never equal a set containing one element.

Example 4   A committee will be formed from a group of three officers in the Air
Force. The three officers may not all be on the committee at the same time. How many
committees can be formed?

Solution

Let A be the set containing the three officers. Each possible committee corresponds to a
subset of A, except that no committee can correspond to subsets $\emptyset$ or A.

In Example 3, we learned that a set containing three elements has a total of eight
subsets. This conclusion holds no matter what the elements of the set are. Thus, A

has eight subsets, of which two cannot give rise to a committee. Therefore, $8 - 2 = 6$ committees are possible.

Skill Enhancer 4   A committee will be formed from a group of *four* officers in the Air Force. How many committees can be formed if the four officers cannot all be on it?

Answer in Appendix E.

Set Symbols

$\subseteq$	subset
$\subset$	proper subset
$\in$	an element belongs to a set
$\varnothing$	the empty set

_____Exercises 5.1_____

If $A = \{-1, 0, 2, 4, 7\}$, $B = \{1, 2, 3, 4\}$, $C = \{0, 2, 4\}$, and $D = \{$integers between 0 and 5 exclusive$\}$, state whether Exercises 1–16 are true or false.

1. $0 \in A$ **2.** $0 \subseteq B$

3. $\{0\} \in C$ **4.** $B \subset D$

5. $B \subseteq D$ **6.** $D \subseteq B$

7. $\varnothing \subset A$ **8.** $\varnothing \subseteq A$

9. $\{2, 4\} \in C$ **10.** $B = D$

11. $A = B$ **12.** $\varnothing = \{A, B\}$

13. $\{0\} \in \{A, B\}$ **14.** $\{1, 2, 3, 4\} \in \{A, B\}$

15. $\{1, 2, 3, 4\} \subset \{A, B\}$ **16.** $\{\varnothing\} \nsubseteq A$

17. Create separate sets containing the names of your best friends, the last ten days in which the weather was acceptable to you, and the integers less than 100 that are equal to the sum of all their factors.

18. List the elements of the set of whole numbers between 1 and 100 that are equal to the sum of all factors less than themselves. (For example, $6 = 1 \times 6 = 2 \times 3$; the factors less than 6 are 1, 2, 3, and $6 = 1 + 2 + 3$.) Numbers with this property are called *perfect*.

19. List the elements in the set {letters in "MISSISSIPPI"}.

20. List the elements in the set {letters in "ZANZIBAR"}.

Let $E = \{$colors of the rainbow$\}$, $F = \{$violet, indigo, blue, green$\}$, $G = \{$yellow, orange, red$\}$, $H = \{$red, yellow, blue$\}$, and $J = \{$brown$\}$. Discuss the truth of Exercises 21–26, given that the rainbow's colors are violet, indigo, blue, green, yellow, orange, and red.

21. $E = G$ **22.** $\{$violet$\} \in F$

23. green $\notin J$ **24.** $J \subseteq H$

25. $G \subseteq E$ **26.** $G \subset E$

27. There is a pair of dice, one of which is red and the other of which is white. (Dice are cubes of ivory or plastic, each of whose six faces contains an array of dots for the numbers from 1 to 6.) Let $A = \{$all possible outcomes of rolling this pair of dice$\}$. List the elements of A.

28. Suppose the dice in a pair are both white and identical. Let the set A be as defined in Exercise 27. List the elements of A.

29. Someone flips a coin three times. Let $B = \{$the outcomes of this "experiment"$\}$. List the elements of B.

30. The set B is as described in Exercise 29. List the subsets of B.

Let $C = \{$the letters in the word "Tennessee"$\}$.

31. List the proper subsets of C.

32. What subsets of C are not proper subsets?

33. Abel, Baker, Charlie, and David are four old army buddies who meet for the 30th anniversary of their demobilization. Let H be the set containing the different ways in which the quartet can shake hands. How many elements are in the set? How many subsets?

34. Kermit McHermit owns just five articles of clothing—two pairs of pants and three shirts. The pants are khaki and black, and the three shirts are mauve, fuchsia, and chartreuse. Let C = {all possible clothing ensembles}. An ensemble consists of one shirt and a pair of pants.
(a) List all elements of C.
(b) List all subsets of C.

35. How many ways are there of making a selection from among four books for a course reading list? No list can contain all four books, and each list must contain at least one title.

36. How many committees can be formed from five students?

37. How many ways are there of making course selections from among eight distinct courses? All students must select at least one course.

38. How many ways are there of selecting a meal from a menu of four dishes? At least one dish must be selected.

39. How many ways are there of planning a trip that can include as many as six different European cities?

40. A student sells magazine subscriptions, portions of the proceeds from which will go to charity. A customer can choose from among eight magazines. How many ways can a selection be made by a customer?

41. There are two indistinguishable Ping-Pong balls in a hat. How many different ways are there of drawing some of the balls from the hat? At least one ball must be drawn.

42. B. C. Dull, a student, has three coins in his pocket—two dimes and one nickel. He reaches in and pulls out some coins (a random number) at random. How many different ways are there of pulling coins out in this random manner?

43. On another day, B. C. Dull has four coins in his pocket— two dimes, one nickel, and one penny. As in Exercise 42, he reaches in and pulls out some coins (a random number) at random. How many different ways are there of pulling coins out in this random manner?

44. The next week, B. C. Dull has four coins in his pocket— two dimes and two nickels. As in Exercise 42, he reaches in and pulls out some coins (a random number) at random. How many different ways are there of pulling coins out in this random manner?

45. Let Z_i be the set containing all positive integers less than or equal to i. For example, $Z_2 = \{1, 2\}$ and $Z_{10} = \{1, 2, 3, \ldots, 8, 9, 10\}$.
(a) How many subsets are there in Z_1?
(b) How many subsets are there in Z_2?
(c) How many subsets are there in Z_3?
(d) How many subsets are there in Z_4?
(e) Examine your answers to the first four parts of this exercise. Generalize the results to give the number of subsets for any of the sets Z_i.
(f) Suppose an arbitrary set Q contains j elements. How many subsets are there of Q?

W 46. Describe in your own words the difference between the general concepts of *subset* and of *proper subset*.

W 47. The text asserted that if $e \in A$, $\{e\} \subset A$. When is this statement incorrect?

Calculator Exercises

48. A master computer file contains 33 names. How many ways are there of making subfiles containing names from this file? There must be at least one name in a subfile, but the subfile must be smaller than the master file. The order in which names appear in the subfile is not important.

Section 5.2   SET OPERATIONS

complement   The *complement* of a given set A is the set containing all elements *not* in A. We denote the complement of A as A', but other writers use $\overline{A}$, $-A$, or $\tilde{A}$ to mean the same thing.

Suppose

$$A = \{1, 3, 5, 7, 9\}$$

What are the elements of A'? Well, the numbers 0, 2, 4, 6, 8, … will be in A'. So will $-1, -2, -3, \ldots$, and so will all real numbers except for the elements of A. Furthermore, aardvarks, ants, antelopes, … , and zebras are also in A', and so on, and so on. Clearly, A' is a cluttered set; see Figure 5.3. Without some restrictions, it is too general to be of use.

universe   It is customary, therefore, to restrict the elements of all sets under discussion to those that occur in a "master set." This master set is the *universe* (or *universe of discourse*), denoted by U, and needs to be specified for each problem.

In the above example, if

$$U = \{\text{whole numbers between 0 and 10 inclusive}\}$$

then A' is a more compact object:

$$A' = \{0, 2, 4, 6, 8, 10\}$$

Once the universe U has been specified, the complement to any set A is uniquely determined.

Example 5   If

$$A = \{1, 3, 5, 7, 9\} \qquad \text{and} \qquad U = \{\text{odd integers between 0 and 20}\}$$

determine A'.

Solution

$A' = \{11, 13, 15, 17, 19\}$

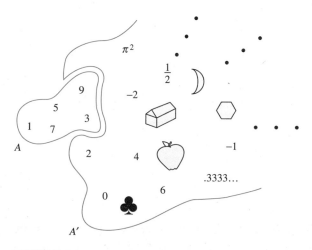

FIGURE 5.3
The set A and its cluttered complement.

Skill Enhancer 5   If U = {letters of the alphabet}, and

A = {letters in the sentence "The quick brown fox jumped over the dog."}

what is A'?

Answer in Appendix E.

Example 6   What is $\emptyset'$? What is U'?

Solution

It will always be true that $\emptyset' = U$ and that $U' = \emptyset$.

Skill Enhancer 6   What is $\emptyset''$? What is U''?

Answers in Appendix E.

Example 7   Imagine a family with three children. Let U be the set containing all descriptions of the birth order and sex of the children. (A typical element of U might be BBG, which indicates that the two eldest are boys and the youngest is a girl.) How many elements are there in U? Suppose A is the set of all ways of having three children, of which the eldest is a boy. What is A'?

Solution

The easiest way (at this point) to determine the number of elements in U is to write them down and count them.

$$BBB \quad GBB$$
$$BBG \quad GBG$$
$$BGB \quad GGB$$
$$BGG \quad GGG$$

These eight triples are the only elements of U; see Figure 5.4.

The elements of A are those triples lying in the left column. The elements of A', by default, are those in the right.

Skill Enhancer 7   Two pennies are tossed. Let U be the description of all possible ways the pennies are tossed. How many elements are there in U? What are these elements?

Answers in Appendix E.

Intersection and Union

intersection   The *intersection* of two sets A and B is a new set containing the elements (if any) that A and B have in common. We write this new set as

$$A \cap B$$

FIGURE 5.4

The mathematical relation of *intersection* is equivalent to the relationship implied by the word "and," because the intersection of A and B contains the elements that are both in A *and* in B.

Example 8 Suppose

$$P = \{\text{all even integers from 2 to 100 inclusive}\}$$
$$Q = \{\text{squares of all real numbers } x \text{ such that } x^2 < 100\}$$

What is $P \cap Q$?

Solution

P and Q are sets whose elements can be enumerated. The intersection is

$$P \cap Q = \{4, 16, 36, 64\}$$

Skill Enhancer 8 If $L = \{5, 6, \ldots, 14\}$ and $M = \{1, 2, 3, \ldots, 10\}$, what is $L \cap M$?

Answer in Appendix E.

union   The *union* of two sets is the set containing all elements that appear in *either* of the two given sets. The union of sets A and B is written

$$A \cup B$$

where the symbol $\cup$ suggests the U of *union*. (Do not let it suggest *universe*, another *u* word.) The union operation is akin to the word "or," because the union of sets A and B is the set containing all elements that are either in A *or* in B *or* in both sets.

Example 9 If

$$A = \{\triangle, \bigcirc, \heartsuit, \nabla\}$$

and

$$B = \{\clubsuit, \diamondsuit, \heartsuit, \spadesuit\}$$

what is $A \cup B$?

Solution

$$A \cup B = \{\triangle, \bigcirc, \nabla, \clubsuit, \diamondsuit, \heartsuit, \spadesuit\}$$

Skill Enhancer 9   For any set A, what is $A \cup A$?

Answer in Appendix E.

Set Relationships

A'	*Complement:* All elements in U but *not* in A.
$A \cap B$	*Intersection:* All elements in A and in B.
$A \cup B$	*Union:* All elements either in A *or* in B or in both.

$\cap$ is like *and*
$\cup$ is like *or*
A' is like *not*

Example 10   Let $A = \{$letters in "recover"$\}$. Let $B = \{$letters in "louvre"$\}$. When U is the set of lowercase letters of the English alphabet, write $A \cap B$, $A \cup B$, and $(A \cup B)'$.

Solution

$A \cap B = \{r, e, o, v\}$. $A \cup B = \{r, e, c, o, v, l, u\}$. Remember, in union, duplicate elements are not listed. $(A \cup B)' = \{$all lowercase letters except $r, e, c, o, v, l,$ and $u\}$.

Skill Enhancer 10   If $P = \{3, 1, 4, 1, 5, 9, 2, 6, 5\}$ and $E = \{2, 7, 1, 8, 2, 8\}$, what are $P \cap E$, $P \cup E$, and $(P \cap E)'$? Let U be the set of digits.

Answer in Appendix E.

Set notation is handy in representing involved relationships.

Example 11   Cantor Polls recently surveyed a broad group of people to gauge political opinions. In this particular poll, they are confining their inquiries to the people sailing on board the ship *Queen Elizabeth 2* from New York Harbor to England. For purposes of the survey, the passengers have been categorized as male, female, very rich, or merely comfortably well off.

We will categorize the passengers by placing passengers with similar attributes in the same set. We form these sets:

$$M = \{\text{Male passengers}\}$$
$$F = \{\text{Female passengers}\}$$
$$W = \{\text{Wealthy passengers}\}$$
$$C = \{\text{merely comfortably well off passengers}\}$$

The universe U is the set of all passengers on the cruise. Use only these sets to describe the classes of passengers who are
(a) Male or female
(b) Wealthy men or any woman
(c) Wealthy women and anyone who isn't comfortably well off

Solution

Translation depends heavily on the correspondence between the set operations "and", "or", and "not".
(a) Male *or* female passengers become the set $M \cup F = U$.
(b) "Wealthy men" are passengers who are both men *and* wealthy, $M \cap W$. The set containing wealthy men or any woman is $(M \cap W) \cup F$.
(c) "Wealthy women" consists of those passengers who are wealthy *and* who are women, $W \cap F$. The set of passengers who are *not* comfortably well off is C'. Therefore, the collection of wealthy women *and* anyone who isn't comfortably well off is $(W \cap F) \cap C'$.

Skill Enhancer 11   Volumes in one section of the university library may be fiction or nonfiction, and may be about crime. Using F as the set of fiction books, N as the nonfiction books, and C as the crime volumes, use set relationships to describe these groups:
(a) Nonfiction crime books
(b) Fiction not concerning crime
(c) All books about crime

Answers in Appendix E.

Visualizing Set Relationships: Venn Diagrams

Venn diagrams

Just as graphing helped us visualize linear equations, so do *Venn diagrams* help us visualize set relationships. We represent the universe U by a rectangle and its interior, as in Figure 5.5. Subsets of U take the form of circles (or other simple shapes) within U, and relationships between sets are displayed by overlapping or embedding these sets.

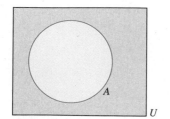

FIGURE 5.5
A set A and its universe.

Historical Note

John Venn   *The English logician John Venn (1834–1923) was a lecturer in the moral sciences at Cambridge University. By the late nineteenth century several methods for representing relationships between sets had been developed, including one technique due to the mathematician Leonard Euler. All have been supplanted by Venn's method,* *and these other techniques rate only footnotes in the history of mathematics. The set representations discussed in this text are also important in the formal study of logic. Up until a few years ago, this material was considered stodgy mathematics. But because it is important in the design of computers and computer circuits, there has been renewed interest in Venn diagrams and logic.*

Figure 5.6 shows how Venn diagrams display basic relationships between two sets. This system of intersecting coffee-cup rings helps put set concepts into perspective.

Example 12 Figure 5.7 shows the Venn diagram for two sets *A* and *B* whose intersection is nonempty. Use set notation to identify each of the distinct, numbered regions in this diagram.

Solution

The football-shaped region 2 is the intersection $A \cap B$ of the two sets. Region 1 contains those elements that are in *A* and *not* in *B*; its set representation is therefore $A \cap B'$. In the same way, the right crescent shape, region 3, is $B \cap A'$. Region 4 contains those elements not in either *A* or *B*, that is, $(A \cup B)'$. These notations appear in Figure 5.8.

Skill Enhancer 12 Refer to Figure 5.7. What set corresponds to the region formed by 1 and 2 together? to the region formed by 1 and 3 together?

Answers in Appendix E.

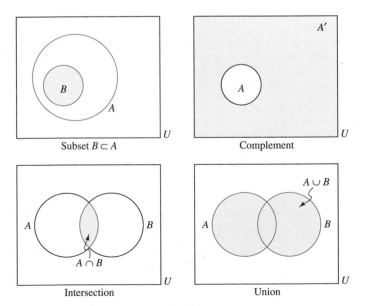

FIGURE 5.6
Subset, complement, intersection, and union via Venn diagrams.

Example 13   What is a Venn diagram for $(A \cap B)'$?

Solution

The solution is shown in Figure 5.9.

Skill Enhancer 13   What is the Venn diagram for $(A \cup B)'$?

Answer in Appendix E.

Example 14   What is the Venn diagram for $A' \cup B'$?

Solution

In Figure 5.10a, we see A' and B'. Their union is Figure 5.10b.

Skill Enhancer 14   What is the Venn diagram for $A' \cap B'$?

Answer in Appendix E.

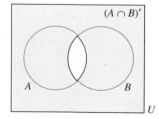

FIGURE 5.7
Four regions need labels.

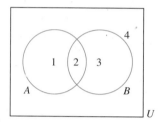

FIGURE 5.8
$(A \cup B)'$.

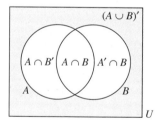

FIGURE 5.9
$(A \cap B)'$.

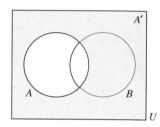

(a)

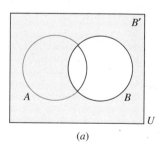

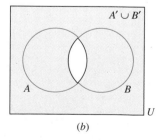

(b)

FIGURE 5.10
Determining $A' \cup B'$.

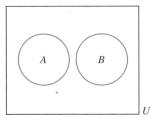

FIGURE 5.11
Venn diagram for two sets with no elements in common.

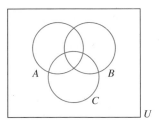

FIGURE 5.12
A Venn diagram for three sets.

Example 15   Use Venn diagrams to prove the theorem

$$(A \cap B)' = A' \cup B'$$

Solution

The right side of the theorem has the Venn diagram of Figure 5.10*b*. The left side has the diagram of Figure 5.9. The two diagrams are the same. Since the two expressions give rise to identical Venn diagrams, we conclude they are the same sets, and hence the assertion is proved. (Pure mathematicians might not be so easily convinced, but the theorem is true nevertheless.)

Skill Enhancer 15   Use Venn diagrams to prove the theorem $(A \cup B)' = A' \cap B'$.

Answer in Appendix E.

Example 16   Can Venn diagrams be used to illustrate two sets with *no* elements in common (that is, with an empty intersection)? If so, draw the Venn diagram.

Solution

Yes; see Figure 5.11.

Skill Enhancer 16   Describe the Venn diagram for two equal sets.

Answer in Appendix E.

We can use Venn diagrams to display the relationships among *three* sets. The basic display involves three interlocking rings, like the trademark of a well-known brand of beer; see Figure 5.12.

Example 17   What is the Venn diagram for $A \cap (B \cup C)$?

Solution

We proceed in stages, first considering the grouped expression $(B \cup C)$, as in Figure 5.13*a*. In Figure 5.13*b*, we form the intersection of this set and A, to yield the final picture.

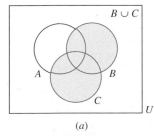
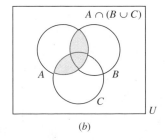

(a)   (b)

FIGURE 5.13
Determining the Venn diagram for $A \cap (B \cup C)$.

Skill Enhancer 17   Use Venn diagrams to illustrate the set $A \cap (B \cap C)$. Is this set different from $(A \cap B) \cap C$?

Answer in Appendix E.

The intersection of sets *always* corresponds to the overlap of these sets in a Venn diagram.

Example 18   Use Venn diagrams to prove the theorem

$$A \cap (B \cup C) = (A \cap B) \cup (A \cap C)$$

Solution

The diagram for the left side is the result of Example 17, and appears in Figure 5.13. We need only diagram the right side, and compare it with this figure.

Figures 5.14*a* and 5.14*b* show the sets $A \cap B$ and $A \cap C$. Their union appears in Figure 5.15. This figure is the same as Figure 5.13; the theorem is proved.

Skill Enhancer 18   What is the Venn diagram for the set $(A \cap B') \cup C$?

Answer in Appendix E.

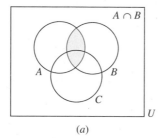 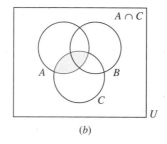

(a)   (b)

FIGURE 5.14
(a) $A \cap B$; (b) $A \cap C$.

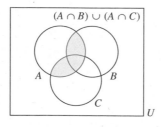

FIGURE 5.15
$(A \cap B) \cup (A \cap C)$.

_____ *Exercises 5.2* _____

In Exercises 1–6, let the universe U be the whole numbers from 0 to 9. Let E be the set containing the even numbers, and B be the set containing the numbers less than seven. List the elements for the given sets.

1. B **2.** E

3. E' **4.** $B \cap E$

5. B' **6.** $B \cup E$

Let F be the set of letters in the word "computer." Let G be the set of letters in the word "commuter." The universe is the entire alphabet. Use these sets to list the elements in the sets given in Exercises 7–12.

7. $F \cup G$ **8.** $F \cup F$

9. $F \cap G$ **10.** $G \cap G'$

11. $F \cup F'$ **12.** $F \cap G'$

Let $A = \{1, 2, 3, 4, 5, 6, 7, 8, 9, 10\}$, $B = \{2, 4, 6, 8, 10\}$, $C = \{1, 3, 5, 7, 9\}$. Answer Exercises 13–18 using these sets.

13. What is $B \cup C$?   **14.** What is $B \cap C$?

15. What is $A \cup C$?   **16.** What is $A \cap B$?

17. What is $(B \cup C) \cup A$?   **18.** Identify $(B \cup C)'$.

Use a Venn diagram for two sets A and B, and shade in the regions specified in Exercises 19–26.

19. $A \cap B'$ **20.** $A' \cup B$

21. $(A \cap B)'$ **22.** $A \cup B'$

23. $A' \cap (A \cup B)$ **24.** $((A \cap B)')'$

25. $A \cup A'$ **26.** $A \cap A'$

Let

$$U = \{\text{months of the year}\}$$
$$R = \{\text{months containing an } r\}$$
$$J = \{\text{months beginning with } J\}$$
$$A = \{\text{months beginning with } A\}$$
$$S = \{\text{the months with names of five letters or less}\}.$$

Use this information to describe the elements of the sets in Exercises 27–32.

27. A' **28.** $S \cup A$

29. $(U' \cap J)'$ **30.** $(A \cap J) \cap S'$

31. (*Requires special thought*) $(J' \cup A')' \cap S$

32. $A \cup (A \cap (S \cap J'))$

33. Suppose $A \cup B$ is $\{a, b, c, d, e\}$ and $A \cap B$ is $\{d\}$. What is the smallest universal set U containing both A and B?

34. Suppose $A \cup B = \{\diamond, \bigcirc, \clubsuit, \spadesuit, \heartsuit, \dagger\}$ and $A \cap B = \{\clubsuit, \dagger\}$. What is the smallest universal set U containing both A and B?

Use Venn diagrams for three sets A, B, and C to illustrate the sets in Exercises 35–40.

35. $A \cup B'$ **36.** $A' \cap B'$

37. $A \cup (B \cap C)$ **38.** $A \cup (B' \cap C)$

39. (*Requires special thought*) $(A' \cup (B \cup C'))'$

40. $(A \cap B)' \cup C$

41. Use Venn diagrams to prove the theorem

$$A \cup (B \cap C) = (A \cup B) \cap (A \cup C)$$

42. Use Venn diagrams to find and derive alternative expressions for $(A \cap B)'$ and $(A \cup B)'$.

43. Use Venn diagrams to show that

$$A \cap B \subseteq A \cup B$$

44. Use Venn diagrams to show that

$$A \cup B = (A' \cap B) \cup (A \cap B') \cup (A \cap B)$$

45. Figure 5.16 contains a typical setup for a Venn diagram of three sets. Use set notation to describe each of the adjacent regions of the figure. For example, the central portion is $A \cap B \cap C$.

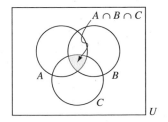

FIGURE 5.16
Venn diagram for three sets.

Applications

(Marketing) Mexicali Maid, a company that produces frozen orange juice, wants to make sure that its product appeals to the greatest majority of its customers. Periodically, the company engages a market research firm to survey communities throughout the United States. Most recently, the market research company performed an exhaustive survey of 80 families in the Pontchartrain suburb of New Orleans. These families were questioned as to their preferences for pulpy juice, juice that is red-orange in color versus yellow-orange, and sweet versus tart. Let P = {families preferring pulpy juice}, R = {families preferring the darker, reddish orange juice color}, and S = {families preferring sweet juice}. Use these sets and the appropriate set connecting symbols to represent the groups of families in Exercises 46–51.

46. Families who like tart juice

47. Families who like either tart or pulpy juice

48. Those preferring nonpulpy and yellow juice

49. Those who do not like either pulpy or dark juice but do like their juice slightly tart

50. Families preferring pulpy, dark, and tart juice

51. Families who specifically do not care for pulpy, dark, and tart juice

(Passengers on QE 2) Refer again to Example 11. Use those sets to represent the groups in Exercises 52–57.

52. Wealthy women

53. Men who are not wealthy

54. Passengers who are women and who are not merely comfortably well off

55. Passengers who are not merely comfortably well off men

56. Wealthy women and anyone who isn't a merely comfortably well off man

57. Men who are not wealthy or women who are merely comfortably well off

(College Education) An evening program at a local community college is attempting to analyze its student body. Each of the students may attend evening only or part-time, and the analysis focuses on these groups. Designate these categories of students with the sets E and P, and use set notation to describe the sets in Exercises 58–61.

58. Students who are not part-timers or are evening students

59. Students who are part-timers or attend in the evening

60. Students who are part-timers and who do not attend in the evening

61. Students who are not either part-timers or evening attendees

62. What is the connection between set intersection and the area of overlap that was so important when graphing inequalities and when solving linear programming problems using the geometric method?

63. Without looking ahead in the text, can you speculate on reasons why it would be advantageous to classify groups using the set notation and concepts of this section?

Section 5.3 *COUNTING ELEMENTS OF SETS*

one-to-one

What does it mean to *count* the elements of a set? When we count, we attempt to match the elements of a set on a *one-to-one* basis with the elements of a second set whose count is known. This second, known set is the set of positive integers. If a known set A contains the integers from 1 through 7, we say it has seven elements. We can (trivially) use this information to determine how many days there are in a week. Let the set W contain the names of the days of the week. How many elements are there in W? The answer is 7, because we can match each of the day names with one of the integers from 1 to 7. See Figure 5.17.

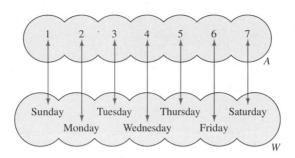

FIGURE 5.17
A one-to-one correspondence.

cardinality; count   We say the *cardinality* or *count* of W is 7, and denote this with the notation

$$c(W) = 7$$

Problems in counting most often arise in sets formed from intersections and unions. In these cases, we appeal to Venn diagrams for help, as in the following example.

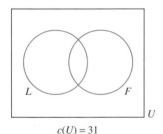

$c(U) = 31$

FIGURE 5.18
Student analysis using Venn diagrams.

Example 19   A class contains 31 students. Of these, 21 students have taken Latin and 17 have studied French; 11 have studied both. Studying a foreign language is one of the requirements for graduation, and both French and Latin qualify as foreign languages. How many students still need to satisfy the foreign language requirement?

Solution

We restate the problem in set notation, and then use a Venn diagram to organize our thinking.

There are two sets. The set L contains all students who have studied Latin, and the set F encompasses those with some French coursework. The universe consists of the 31 students in the class. The associated Venn diagram might look like Figure 5.18. This diagram helps us combine information about both foreign languages.

Begin in the middle of the diagram and work to the outside. The mid-portion is the pointed oval shape that represents $F \cap L$, the students who have taken both courses (French *and* Latin). Eleven such students fit that description, so we place the number 11 within the region. (Refer ahead to Figure 5.19.)

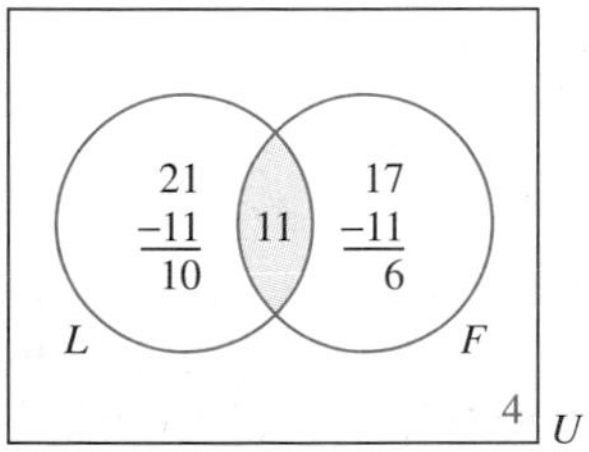

The next step is to determine how many students are contained in the left and right "crescent moons." The left crescent, the Latin circle with a bite out of it, contains those students who have taken Latin, but *excludes* those who have taken both languages. (Such students have already been accounted for; they are included in the central intersection.) A total of 21 individuals have taken Latin, but of that 21, 11 have been accounted for in the group taking both languages. Consequently, the number having taken Latin but not French is

$$c(L) - c(L \cap F)$$

FIGURE 5.19
Finishing the student analysis.

or

$$21 - 11 = 10$$

Place a 10 in the left area. By the same reasoning, the right crescent represents those students who have taken French, but excludes those who have taken Latin (because they have already been accounted for in the central region). Those students number

17 (the number taking French) $-$ 11 (the number for both languages) $= 6$

Now, how many students have taken either French or Latin (or both)? The Venn representation of this set, $F \cup L$, is in Figure 5.18. The number of students contained in this set is the sum of the students in the three adjacent regions:

$$10 + 11 + 6 = 27$$

How many students have yet to study a foreign language? Since there are 31 students in the class, and 27 of them have taken one language or the other, there remain 4 students who must yet satisfy their language requirement. (Where in the diagram might we place a numeral 4?)

Skill Enhancer 19   There are 45 students enrolled in one section of Advanced Swimming, of whom 20 have taken a lifeguard qualifying exam and the class on Intermediate Swimming, 25 have taken the qualifying exam, and 30 have taken the intermediate class. How many have taken neither?

Answer in Appendix E.

Example 20   The family records for the Duke of Strathcona are unusually complete, owing to his being a member of the British peerage. The Duke's ancestors include 89 men since the inception of family records in the late 1600s. Of these 89, 43 were below 5'4'' in height, and 14 were left-handed; 9 possessed both characteristics. These two traits were deemed particularly unlucky by the other members of the family. How many of the Duke's male ancestors were short (below 5'4'') but right-handed? How many were not short and left-handed? How many possessed neither of these unlucky traits?

Solution

The set representation involves the two intersecting sets L and S, containing the left-handed and short men, respectively, enclosed by the universe consisting of the 89 male ancestors, as in Figure 5.20. We work from the inside of this diagram to the outside. In the central oval, which contains those individuals who were both short and left-handed, we place the number 9. To answer the first question, simply subtract the number of short and left-handed members from the number who are merely short:

$$43 - 9 = 34$$

The answer to the second question consists of the total number who were left-handed minus the number who shared both characteristics:

$$14 - 9 = 5$$

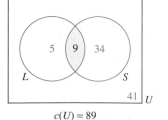

$c(U) = 89$

FIGURE 5.20
The Strathcona family.

These numbers are shown in Figure 5.20. The number of male ancestors who possessed either of the attributes (or both) is

$$34 + 9 + 5 = 48$$

There were a total of 89 ancestors, so that leaves 41 ($= 89 - 48$) who were troubled by neither of these so-called unlucky attributes.

Skill Enhancer 20   River City dog pound contains 43 dogs. Of these, 17 are short-haired and single-colored, 20 are short-haired, and 35 are single-colored. How many are neither short-haired nor single-colored?

Answer in Appendix E.

Given two sets E and F, what can we say about the count of their union $c(E \cup F)$? A common error is to assume that the answer is $c(E) + c(F)$, but this sum actually overstates the result. Since the set $E \cap F$ is a subset of *both* E and F, the elements represented in its count are included in both $c(E)$ and $c(F)$. We can correct this overstatement, therefore, by subtracting the extra amount. The correct formula is

(5.1) $$c(E \cup F) = c(E) + c(F) - c(E \cap F)$$

whose derivation is illustrated by Figure 5.21.

Example 21   A company hires consultants to analyze the effectiveness of a new drug targeted to cold and hay fever sufferers. There were 45 people participating in the survey, and the consultants reported that although 14 people experienced no relief with the new medicine, 20 people with colds experienced relief, 25 hay fever sufferers experienced relief, and 17 hay fever sufferers who also had colds experienced relief. Of the people who experienced relief, 6 had colds only, and 10 suffered only from hay fever. The consultants were discharged. Why?

Solution

From the information in the question, we know that 17 people who were helped by the drug both had a cold and suffered from hay fever. We also know that according to the consultants, only 6 people helped by the new drug had colds only. This implies that a total of $17 + 6 = 23$ people who responded to the drug had colds, and yet the consultants reported that only 20 people who responded to the drug had colds. The drug company could not tolerate this inconsistency, and fired the consultants. (There are other inconsistencies; can you find them?)

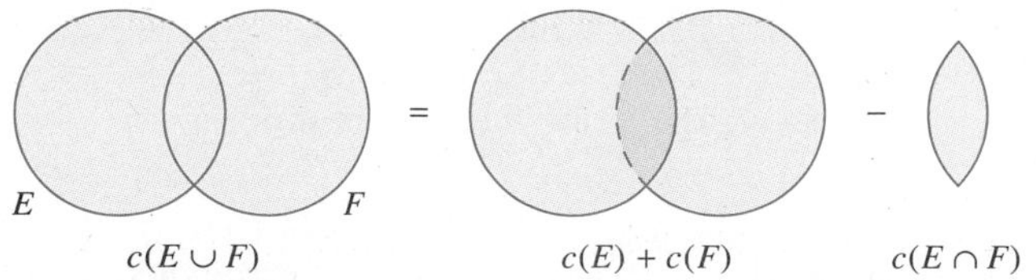

$$E \qquad\qquad F$$
$$c(E \cup F) \qquad = \qquad c(E) + c(F) \qquad - \qquad c(E \cap F)$$

FIGURE 5.21
The cardinality of a set union.

Skill Enhancer 21   Fire destroyed the contents of the Acme Warehouse, including 59 priceless works of art. In the fire, 30 German sculptures were completely ruined. A total of 40 German works were destroyed, and a total of 33 sculptures were burned. Of the art works, how many were neither German in origin nor sculptures?

Answer in Appendix E.

Example 22   In a sample of people who are proprietors of their own small businesses, there are 58 individuals who import goods or export goods as their principal businesses. Of these, 12 do both exporting and importing. There are an equal number of importers and exporters. How many exporters are there?

Solution

Let I and E be sets containing the importers and exporters. From the given facts, we know that

$$c(I) = c(E) = x \qquad c(E \cup I) = 58 \qquad \text{and} \qquad c(E \cap I) = 12$$

From Equation 5.1 above, we know that

$$c(E \cup I) = c(E) + c(I) - c(E \cap I)$$

must hold. This implies that

$$58 = 2x - 12$$

so that $x = 35$. Check this answer.

Skill Enhancer 22   The grocery order for the Kappa Eta Pi fraternity included 17 cases of beer, some of which was "lite" (reduced in calorie content) and some of which was imported—5 cases were neither, 8 cases were lite, and 8 were imported. How many were both lite beer and imported?

Answer in Appendix E.

Counting with Venn Diagrams

1. Create the initial Venn diagram.
2. Use the information in the problem to determine the cardinality of each distinct region in the diagram. (Work from the inside out.)
3. Use these cardinalities to solve the problem.

We can also use Venn diagrams in problems involving three sets.

Example 23   Parents for Better TV have made a preliminary survey of children's television on Saturday morning. They have sampled 100 hours worth of programs, and rated them as being cartoon or filmed with live actors, fantasy or realistic, violent or nonviolent:

cartoon

fantasy

violent

film still

realistic

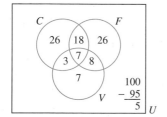
peaceful

FIGURE 5.22

1. There are 54 hours containing cartoons.

2. There are 59 hours that seem to be of a fantastic nature.

3. Only 25 hours of programs could be classified as violent.

4. Cartoon fantasies account for 25 hours.

5. Violent cartoons are shown for 10 hours of broadcasting.

6. Violent fantasies are shown for 15 hours.

7. The survey further indicates that 7 hours of programs are violent cartoons organized around fantasy themes.

On the basis of this information, answer the following questions.

(a) How many hours of nonviolent cartoons and of nonviolent fantasies are there?

(b) How many hours of fantasy programming are there that are neither cartoonlike nor violent?

(c) How many hours of nonviolent cartoon fantasies are there?

(d) Finally, how many hours of programming fit into none of the categories?

Solution

Set up the Venn diagram as in Figure 5.23, and use the given information to assign the numbers of hours for each region. Using set notation, the information is summarized as follows:

$$
\begin{array}{ll}
c(U) = 100 & c(C) = 54 \\
c(V) = 25 & c(F) = 59 \\
c(V \cap C) = 10 & c(C \cap F) = 25 \\
c(V \cap F) = 15 & c(C \cap V \cap F) = 7
\end{array}
$$

FIGURE 5.23
A television survey.

The three circular sets contain those programs that are cartoons, fantasies, and of violent nature. The innermost region corresponds to those programs that are simultaneously cartoons (C), fantasies (F), and violent (V). Seven hours of programming fit that bill, so place a 7 in the central portion of the diagram.

Move now to compute the hours in each of the three triangular areas surrounding the center. We start at the top, and work clockwise. The top region corresponds to those nonviolent programs that are fantasies and cartoons. A total of 25 hours are cartoon fantasies, but we must exclude the 7 hours that we have already counted in the cartoon-fantasy-violent category. Therefore, the top region contains $25 - 7 = 18$ hours. In the same way, the remaining triangles have $8 \, (= 15 - 7)$ and $3 \, (= 10 - 7)$ hours in them.

The remaining areas are the exterior regions of each set, each resembling a cookie with two bites out of it. First consider the outer part of the cartoon set, which corresponds to those cartoon programs that are neither fantastic nor violent. We were given that $c(C) = 54$, but that includes those cartoons that are fantastic and/or violent. Thus

$$c[C \cap (F \cup V)']$$

is 54 less the hours already accounted for:

$$c[C \cap (F \cup V)'] = 54 - 18 - 7 - 3$$
$$= 26$$

Moving clockwise, the number of fantasies that are not cartoons or violent is

$$59 - 18 - 7 - 8 = 26$$

Finally, the number of hours of violent programming that are not fantasies or cartoons is

$$25 - 3 - 7 - 8 = 7$$

The fully labeled Venn diagram is shown in Figure 5.23.

(a) The hours of nonviolent cartoons consist of all cartoons, less those in the violent "circle," or $54 - 3 - 7 = 44$ hours of programming. In the same way, the nonviolent fantasies are those in the fantasy set excluding those also in the violent set: $59 - 7 - 8 = 44$ hours of programming.

(b) We immediately read from the diagram that there are 26 hours of fantasy programs that are neither cartoons nor violent.

(c) There are a total of 18 hours of nonviolent cartoon fantasies.

(d) To determine the hours of programming in the "other" category, add all the hours in each of the separate regions. This number must be subtracted from the total number of hours sampled. A total of 95 hours fits into the three categories, and 100 hours were sampled. Therefore, 5 hours of programming could not be neatly pigeonholed.

Skill Enhancer 23   The nutritionist at the Huntington Elementary School examines 50 dishes and classifies them with respect to three categories: no cholesterol (NC), fat-free (FF), and low carbohydrate content (LC). Of these dishes, 8 are in all categories, 15 are NC and FF, 15 are FF and LC, and 9 are LC and NC. Also, 19 are NC, 32 are FF, and 21 are LC.

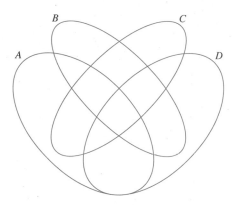 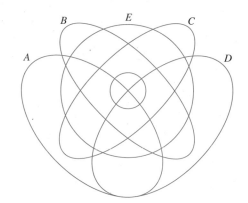

FIGURE 5.24
Venn diagrams for four and five sets.

(a) How many dishes are NC only?
(b) How many dishes are FF only?
(c) How many dishes are LC only?
(d) How many dishes are neither?

Answers in Appendix E.

What of situations where there are not two or three distinct sets, but four, five, or more? What would the Venn diagrams for these problems resemble? It is possible to create Venn diagrams for any number of terms, but beyond three, the resulting diagram loses the clarity that we have noticed in this chapter. Figure 5.24 displays the base diagrams for four and five sets. In neither case is it possible to use circles; we have had to resort to ellipses as shown. The diagram for five sets has an additional disadvantage. Set E is really in the form of a doughnut—the small circle in the "center" is *outside* E but *inside* B and C. It is possible to use convex closed curves for any number of sets. Venn diagrams originally arose in the analysis of problems of logic. For a clear exposition of these and other aids to logical analysis, see Martin Gardner's brief volume, *Logic Machines and Diagrams* (University of Chicago Press, 1982).

_____ *Exercises 5.3* _____

In Exercises 1–6, use the given information to determine the indicated cardinalities.

1. $c(A) = 12$, $c(B) = 10$, $c(A \cap B) = 8$, $c(A \cup B) = ?$

2. $c(A) = 31$, $c(B) = 5$, $c(A \cap B) = 23$, $c(A \cup B) = ?$

3. $c(B) = 14$, $c(A \cap B) = 1$, $c(A \cup B) = 20$, $c(A) = ?$

4. $c(A \cap B) = 75$, $c(A \cup B) = 80$, $c(A) = 50$, $c(B) = ?$

5. $c(A) = c(B)$, $c(A \cap B) = 0$, $c(A \cup B) = 100$, $c(A) = ?$

6. $c(A) = 0$, $c(A \cap B) = 17$, $c(B) = ?$

A particular universal set U consists of 20 elements and contains two nonempty sets A and B. The intersection of these two sets contains 7 elements. Set A contains a total of 12 elements. Set B consists of 5 members that are not in A, plus whatever is in the intersection. Use this information to compute the number of elements in the sets in Exercises 7–12.

7. B

8. $A' \cap B$

9. $A \cup B$

10. $A \cup (A' \cup B')$

11. $A' \cap B'$

12. $(A \cap B) \cap U$

Figure 5.25 is a Venn diagram involving three sets and their universe. The universe U contains 100 members, and sets A, B, and C have 25, 33, and 50 elements, respectively. Various intersection sets have the cardinalities indicated on the figure. Use this information to find the cardinalities of the sets in Exercises 13–18.

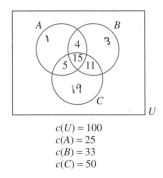

$c(U) = 100$
$c(A) = 25$
$c(B) = 33$
$c(C) = 50$

FIGURE 5.25
Venn diagram for three sets.

13. $A \cap B$

14. $B \cup C$

15. $U \cap (A \cup B \cup C)$

16. $A \cap B \cap C$

17. $B \cap (A \cup B)'$

18. $C \cap B'$

19. Use Venn diagrams to prove the formula

$$c(A \cup B) = c(A) + c(B) - c(A \cap B)$$

20. Use the results of Exercise 45 in the previous section to show that

$$c(A \cup B \cup C) = c(A) + c(B) + c(C) - c(A \cap B)$$
$$- c(B \cap C) - c(C \cap A) + c(A \cap B \cap C)$$

Each of Exercises 21–26 presents counting information about three sets, called A, B, and C in each case. Deduce the minimum number of elements in each set for each of these problems. You may find Figure 5.25 helpful in thinking about these problems.

21. $c(A \cap B) = 13$ **22.** $c(A \cap B) = 9$

$c(A \cap C) = 16$ $c(A \cap C) = 8$

$c(B \cap C) = 14$ $c(B \cap C) = 7$

$c(A \cap B \cap C) = 12$ $c(A \cap B \cap C) = 7$

23. $c(A \cap B) = 200$ **24.** $c(A \cap B) = 4$

$c(A \cap C) = 200$ $c(A \cap C) = 3$

$c(B \cap C) = 200$ $c(B \cap C) = 2$

$c(A \cap B \cap C) = 100$ $c(A \cap B \cap C) = 1$

25. $c(A \cap B) = 10$ **26.** $c(A \cap B) = 17$

$c(A \cap C) = 11$ $c(A \cap C) = 21$

$c(B \cap C) = 10$ $c(B \cap C) = 21$

$c(A \cap B \cap C) = 9$ $c(A \cap B \cap C) = 17$

Applications

(Retailing) Right-Bye Bargain Outlets has just received a shipment of boys' irregular thermal underwear tops. Ten dozen units are in the shipment. The irregularities can be in the color of the garment or in the sewing and stitching. Of the tops, 56 have bad color and 89 have defective stitching; 26 have both defects. Use this information in Exercises 27–30.

27. How many tops have just the dye irregularity?

28. How many have just the stitching irregularity?

29. By good luck, some of the garments aren't irregular at all. How many fit into this category?

30. How many garments have sewing or stitching irregularities, but not both?

(Kennel Care) K-9 Korral Kennels has just received a new bunch of cocker spaniel puppies. Five of the puppies have black in their coats, and four have golden-brown coats. Three have both, and six have neither. Use this information to solve Exercises 31–34.

31. How many puppies have black coats with no brown?

32. How many have brown coats with no black?

33. How many puppies have black coats or brown coats or both?

34. How many puppies were in the shipment?

(Anthropology) An isolated tribe in New Guinea has been the subject of intensive investigation by anthropologists. By virtue of their isolation, many members of the tribe display unusual congenital characteristics. Of 133 tribal elders, 74 are exceptionally tall (2 m or more), 87 have a rare form of anemia, and 66 have six fingers on one hand. Also, 45 have

six fingers and anemia, 51 tall elders have anemia, and 31 tall elders have six fingers; 17 of them have all three characteristics. Use this information to solve Exercises 35–38.

35. How many of these men were tall and showed no other of these traits?

36. How many were anemic only?

37. How many were six-fingered only?

38. How many of the elders possessed none of these characteristics?

(Imported Cars) A shipment of 100 imported cars has just been unloaded at the Port of Seattle. Of these, 33 have a four-cylinder engine, 45 have tilt steering wheels, and 27 have air conditioning already installed; 30 cars have both four-cylinder engines and tilt steering; 25 have tilt steering and air conditioning, and 20 have air conditioning and a four-cylinder engine; 18 cars have all three characteristics. Use this information in Exercises 39–44.

39. Draw a Venn diagram to make use of these data.

40. How many cars have none of these three characteristics?

41. What *fraction* of the auto shipment has tilt steering but not air conditioning?

42. How many cars have either air conditioning or a four-cylinder engine but not both?

43. How many cars have either tilt steering or a four-cylinder engine but not both?

44. How many cars have two of these characteristics but not all three? (A car in this subset may have any two characteristics.)

45. *(Child Care)* Wee Care Day Care Centers are naturally very careful about their wee charges. Dan DeLeon is a new counselor. Reporting on the afternoon activities of a group of 41 youngsters, he notes that 24 are both eating cookies and drawing, and 22 are playing with blocks. Also, 13 seem to be drawing and playing with blocks at the same time; 14 are drawing and eating simultaneously, and 10 are eating and playing with blocks. There are 6 precocious children who are engaged in all three activities at once. On the basis of this report, DeLeon was fired immediately. Why?

46. Refer again to Example 21. What other inconsistencies are there in the reported data?

W **47.** What do you think limits the usefulness of the Venn diagram counting techniques presented in this section?

W **48.** Suggest more practical ways for dealing with Venn diagram problems containing four or five sets.

Calculator Exercises

49. We need to know how many elements are in the intersection of two sets A and B. There are $40,890$ and $20,480$ elements in the sets, and their union contains $51,200$ elements.

50. Given three sets, suppose we know that
$$c(A \cap B) = 5,612$$
$$c(A \cap C) = 7,012$$
$$c(B \cap C) = 4,019$$
$$c(A \cap B \cap C) = 3,719.$$
What are the minimum numbers of elements contained in sets A, B, and C?

Section 5.4 COUNTING AND PERMUTATIONS

Different types of counting problems arise in other situations. Suppose, for example, a chemist mixes a solution containing two types of positive ions with one containing three types of negative ions. A positive and a negative ion combine in solution to produce a distinct chemical compound. How many types of compounds will be present in the mixture?

tree diagram   The easiest method of solution is to list all possible compounds, and then count them. A *tree diagram* provides a systematic way of doing this, as in Figure 5.26. This

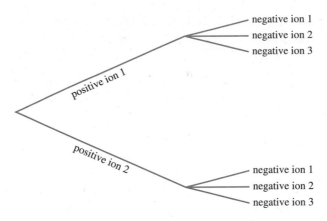

FIGURE 5.26
A tree diagram.

diagram is horizontal; imagine it in a vertical position to understand its name. The paths spread out like the branches of a tree.

Each path from the left beginning node to the end of any twig represents a valid chemical compound. The number of twigs at the right yields the number of compounds. There are six. Note that each of the two main trunks splits into the same three branches. A quick multiplication—2 × 3—yields the answer.

Principle of Counting   This demonstrates the useful *Principle of Counting*. Formally, suppose that an event consists of n separate stages, which we call S_1 through S_n. The stages occur one after the other. Each of these stages is itself an event that has any one of O_i outcomes. In Figure 5.27 there are two stages, S_1 and S_2. There are two ways S_1 can occur—representing each of two positive ions—so $O_1 = 2$. The second stage can occur in any of three ways, corresponding to any one of three negative ions. Therefore, $O_2 = 3$. (It is not necessary to know what an ion is in order to proceed with this analysis.)

How many different, distinct outcomes are there for the *entire n*-staged process? By the *Principle of Counting*, this is the product of the O_i's.

The Principle of Counting

In a multistage process, we often need to know the total number of ways the process can occur.

$$\text{Total number of ways} = O_1 \times O_2 \times \cdots \times O_n$$

Here,
n is the number of stages.
O_i is the number of outcomes of the ith stage.

For our chemistry problem, the final number of compounds is the number of outcomes of a two-stage problem—the mixing of the positive ions ($O_1 = 2$) with the negative ions ($O_2 = 3$). By the Principle of Counting, the final number is

$$2 \times 3 = 6$$

which agrees with our tree diagram analysis.

Example 24   The Village of Missasaucky will shortly elect new members to its governing board. There are 4 declared candidates for the position of village executive, 2 for village secretary, and 3 for treasurer. How many different governing boards are possible?

Solution

Three separate steps are needed to make up the slate for a governing board. The number of candidates for each position is the number of outcomes for each step. By the Principle of Counting, multiply the number of different candidates for each position to find the number of possible boards:

$$4 \times 2 \times 3 = 24$$

There are 24 different boards possible. The tree diagram for this problem appears in Figure 5.27. We prefer to use the Principle of Counting instead of tree diagrams, which quickly become unwieldy to create, whenever possible.

Skill Enhancer 24   To get to class, a student can take any of 3 bus lines to a subway station, from which she can take any of 3 subway lines to a particular stop. From there,

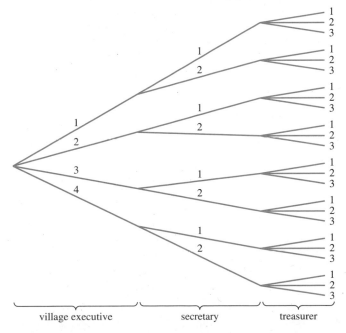

FIGURE 5.27
Missasaucky village elections.

the classroom building is several blocks away, and she can take either of 2 walking routes to get there. In how many ways can she get to class?

Answer in Appendix E.

Factorials

factorial   *Factorial* notation is shorthand for a useful product that arises often in counting problems. The quantity *n factorial*, written $n!$, is the product of the first n integers.

There is one special case that students should memorize.

$$0! = 1$$

Many calculators have factorial keys that approximate $n!$ for large n. They allow the approximation of $n!$ up to about $13!$.

Factorial

$$n! = 1 \times 2 \times \ldots \times n, \text{ for } n \geq 1$$
$$0! = 1$$

Example 25   Compute $1!$, $2!$, $3!$, $4!$, $6!$, and $10!$.

Solution

$$1! = 1$$
$$2! = 1 \times 2 = 2$$
$$3! = 1 \times 2 \times 3 = 6$$
$$4! = 1 \times 2 \times 3 \times 4 = 24$$
$$6! = 1 \times 2 \times 3 \times 4 \times 5 \times 6 = 720$$
$$10! = 1 \times 2 \times 3 \times 4 \times 5 \times 6 \times 7 \times 8 \times 9 \times 10 = 3,628,800$$

Factorials grow large quite rapidly.

Skill Enhancer 25   Compute $5!$ and $7!$.

Answer in Appendix E.

Example 26   There are 9 players on a baseball team. How many different batting orders are possible?

Solution

There are 9 positions in a batting order, or line-up. For this analysis, regard the batting order as a process in which each position in the order is a distinct step. Since there are 9 individuals on a team, any of the 9 can be chosen to bat first. Once that person has been chosen, 8 players remain. Consequently, the person batting second can be chosen from among 8 players. This leaves 7, from which the third batter is chosen. Figure 5.28 helps make this clear. Continue in the same way, and then apply the Principle of Counting.

FIGURE 5.28

The number of batting orders is

$$9 \times 8 \times 7 \times 6 \times 5 \times 4 \times 3 \times 2 \times 1$$

or 9!. Check that $9! = 362,880$.

Skill Enhancer 26   For a certain Little League baseball team, each team has only 6 players instead of the usual 9. How many different batting orders are possible for this team?

Answer in Appendix E.

Permutations

permutation   The determination of baseball line-ups is a special case of a wider class of *permutation* problems. In the general case, we have a collection of n objects, and suppose r is some number such that $0 \leq r \leq n$. Often, we need to choose r items out of the original n, and if this process of selection is subject to a few restrictions, an important question arises: How many ways are there of choosing this smaller group? When the order of the selection is important, this method of selection is a *permutation*.

Permutations

The choice of some objects out of a group of more objects is a permutation if it conforms to these attributes.

1. The *order* of choice is *important*.

2. An object once chosen is *not reused*.

3. All n items in the main collection are *different*.

4. The choice of one object depends *neither* on the items chosen before it nor on the items chosen after it.

It is also possible to consider a permutation to be an *arrangement* of a subgroup of the original collection, an arrangement in which the order of arrangement is important.

When r items are chosen from a total of n, subject to the above restrictions, we have a *permutation of n things taken r at a time*, written most frequently as $P(n, r)$. Other acceptable notations include $_nP_r$, P_{nr}, and P_r^n.

Why is the problem of baseball batting orders a problem in permutations?

1. Order *is* important. (Compare this with choosing players for a basketball team, where order is not important.)

2. Each player appears once in the batting order.

3. All players are clearly different people.

4. We have implicitly assumed that each player's slot in the line-up depends on nobody else's.

Batting orders are permutations of 9 things taken 9 at a time—$P(9, 9)$.

To develop a general formula for permutations, consider a slightly larger baseball team. That is, suppose teams in an amateur league have agreed to allow 14 players on each team, although only 9 of these are allowed at bat in any single game. It can be any nine, but batters must come from this line-up for the entire game. Under these circumstances, how many batting orders are possible? More formally, what is $P(14, 9)$?

Start by choosing any of the 14 players as a lead batter. Having chosen her, we choose one of the remaining 13 to bat second, and so on for a total of nine choices. By the fundamental principle of counting,

$$P(14, 9) = \underbrace{14 \times 13 \times 12 \times 11 \times 10 \times 9 \times 8 \times 7 \times 6}_{9 \text{ factors}}$$

Counting problems frequently generate large numbers, so where possible we express answers compactly using factorial notation. To do that here, multiply and divide this expression by 5!. (Remember, simultaneously multiplying and dividing by the same number will not change the value of a number.) We obtained the 5 as the difference $14 - 9$.

$$P(14, 9) = \frac{14 \times 13 \times 12 \times 11 \times 10 \times 9 \times 8 \times 7 \times 6 \times 5!}{5!} = \frac{14!}{5!}$$

Since this calculation would work for a "general" baseball team of n players from whom we can choose r batters, we conclude that the general formula is

$$P(n, r) = \frac{n!}{(n - r)!}$$

for all values of n and r, $n \geq r \geq 0$.

Example 27 Compute (a) $P(4, 2)$; (b) $P(4, 3)$; (c) $P(4, 4)$; (d) $P(4, 0)$.

Solution

(a)
$$P(4, 2) = \frac{4!}{(4 - 2)!} = \frac{4!}{2!} = \frac{1 \times 2 \times 3 \times 4}{1 \times 2} = 3 \times 4 = 12$$

(b)
$$P(4, 3) = \frac{4!}{(4 - 3)!} = \frac{4!}{1!} = 4! = 24$$

(c)
$$P(4, 4) = \frac{4!}{(4 - 4)!} = \frac{4!}{0!} = 4! = 24$$

(d)
$$P(4, 0) = \frac{4!}{(4 - 0)!} = \frac{4!}{4!} = 1$$

Skill Enhancer 27   Compute (a) $P(2, 2)$;   (b) $P(3, 2)$;   (c) $P(5, 2)$.

Answers in Appendix E.

Example 28   Mr. and Mrs. Flood plan a vacation in the American southwest. Of the 13 vacation spots that appeal to them, they will have time to visit eight only 8. How many different tour itineraries are possible?

Solution

A tour itinerary specifies thc places to be visited and the order in which they will be seen. The solution is therefore the number of permutations of 13 things taken 8 at time, $P(13, 8) = \frac{13}{(13-8)!} = \frac{13!}{5!}$.

Skill Enhancer 28   The next year, the Floods contemplate a vacation in which they can choose from 10 possible locations. They will have time to visit only 7 of these tourist spots. Now how many different tour itineraries are possible?

Answers in Appendix E.

Exercises 5.4

In Exercises 1–12, evaluate:

1. $5!$

2. $3!4!$

3. $(4!)^2$

4. $(3!)!$

5. $2!/4!$

6. $(0!)^{10!}$

7. $\frac{12!}{4!8!}$

8. $\frac{19!}{17!}$

9. $\frac{7!}{3!}$

10. $\frac{9!}{2!}$

11. $\frac{10!}{3!7!}$

12. $\frac{9!}{4!5!}$

Rearrange the expressions in Exercises 13–16 and write them solely in terms of factorials and powers of two (if necessary). For example, $3 \times 4 \times 5 = \frac{5!}{2!}$. Do not bother to evaluate them.

13. $5 \times 6 \times 7$

14. $2 \times 4 \times 6 \times 8$

15. $\dfrac{2 \times 4 \times 6 \times 8}{16}$

16. (*Requires special thought*)

$1 \times 3 \times 5 \times 7$

17. Show that for any whole value of n greater than 0,

$$n! = n \times (n - 1)!$$

18. Use the equation in Exercise 17 to justify $0! = 1$. (*Hint:* Let $n = 1$.)

W 19. You may be tempted to define factorials for negative integers the same way we defined a value for 0!. Use Exercise 17 yet again to suggest why no such definitions are possible.

Evaluate the expressions in Exercises 20–31.

20. $P(2, 2)$

21. $P(4, 0)$

22. $P(1, 3)$

23. $P(3, 1)$

24. $P(7, 4)$

25. $P(0, 0)$

26. $P(4, P(3, 1))$

27. $P(n + 1, 1)$

28. $P(2n, n)$

29. $P(3n, 2n)$

30. $P(3n, n)$

31. $P(8, 6)$

Applications

32. The new diet program Weight Waiters gives you a choice of 3 foods for an entrée and 4 foods for dessert. How many different meals are possible if each meal consists of one entrée and one dessert?

33. Free World Chinese Restaurant provides a family menu option. For example, a party of 2 can choose one of the 5 selections in column A and one of the 7 in column B. How many different meals are possible?

34. Polly Darton, a renowned country-western singer, is planning her next concert. Polly is a highly organized performer, and she has divided her repertoire into 10 categories, each of which contains 4 songs. The concert will include 10 songs, one from each of her categories. How many different programs are possible? (Use exponential notation.)

35. A pair of dice are rolled. One die is red, and the other is green. (*Die* is the singular of dice.) How many different rolls of the dice are possible?

36. An intelligence service is trying to make up a code. Because of the nature of the transmitting device, only 5 letters of the alphabet can be used. Information will be coded into special "words," each of which will contain 3 letters. No letter can be repeated within a word. How many different words are possible? Suppose each "word" contains 4 letters. How many are possible? Suppose each word can contain up to 4 letters. Now how many are possible?

37. For a project in a comparative literature class, a computer is fed 8 sentences, which it will arrange to form a paragraph. How many different paragraphs are possible?

38. There are not enough floor samples of the new spring fashion line by Toulouse Chartreuse to go to each store in the Glimmerglass department store chain. Toulouse has created 4 new swimsuits, 5 spring dresses, and 3 pant suits. Each store will receive one garment in each category. How many different sample selections are possible?

39. In a computer code, 26 letters are to be used to form words, each containing 8 letters. No single letter may occur more than once in a code word, and the order of spelling is important. How many such words are there?

40. There is room for 7 out of 15 students in a shuttle bus going from the north end to the south end of a college campus. How many different seating arrangements are there?

41. A concert will consist of 5 pieces of "new music." How many different presentations of these pieces are there?

42. A retired couple from Tucson, Arizona has the option of taking a tour of 4 out of 10 European capitals. How many tour itineraries are possible?

43. There are 7 members of the Chappauqua Debating Society who are eligible for election to any of 3 elected offices (president, secretary, and treasurer). How many different slates are possible?

44. There are 7 members of the Halesite Volunteer Fire Department who are to be awarded citations for valor in the performance of their duties. How many different ways of ordering the ceremonies are there?

45. The Russian pianist Dura Durakova is making her first tour of the United States and needs to organize her initial Carnegie Hall recital with some care. The order in which pieces are presented in a concert program is important because the order creates different moods in listeners and critics. She has 8 virtuoso pieces in her concert repertoire. How many different concert programs are possible? If there are 10 pieces in her repertoire, out of which she must create a program of 7 pieces, then how many different concerts are possible?

W 46. Explain why diagrams like Figure 5.27 or Figure 5.28 are called tree diagrams. Why might they also be called *root diagrams*?

Calculator Exercises

Evaluate the factorials in Exercises 47–49.

47. 14!   **48.** 12!   **49.** 11!

49. Experiment with your calculator to determine the largest value of x so that $x!$ is explicitly shown.

Section 5.5 COMBINATIONS

combination   Often the order of chosen objects is *not* important, and the arrangement is then called a *combination* of n things taken r at a time. Symbolically, we write $C(n, r)$. Other notations are C_{nr}, $_nC_r$, or $\binom{n}{r}$.

Combinations

Often, the selection of a group of objects out of a larger group meets these conditions.

1. Order is *not important.*

2. An object once chosen may *not be reused.*

3. All *n* items are *different.*

4. The choice of one object depends *neither on the items chosen before it nor on the items chosen after it.*

In that case, the selection is a *combination.*

Combinations and permutations differ *only* in the requirement that for permutations the order of the selection be considered.

How many ways are there of choosing combinations of r things from a group of n things? To derive a formula for combinations (in the same way, but different from, the permutation formula), let us examine the 14-member baseball team from the previous section once again, this time from a slightly different perspective.

We know that the number of batting orders of this team is the *permutation* $P(14, 9)$. Originally, we regarded a batting order as a nine-stage process. This time, let us regard it as a *two-stage* process.

As the first stage, we choose a subgroup of 9 players, *order being of no importance in this selection.* Note that the number of ways of choosing this subgroup is the as-yet-unknown number of combinations $C(14, 9)$.

In the second stage, we order this subgroup of 9 players. How many ways are there of performing this ordering? We have already computed this as 9!.

We use this information by appealing to the Principle of Counting,

$$\begin{array}{ccc} \textit{Number of} \\ \textit{batting orders} \end{array} = \begin{array}{c} \textit{ways of choosing 9} \\ \textit{players with no order} \end{array} \times \begin{array}{c} \textit{ways of ordering} \\ \textit{this group of 9} \end{array}$$

We have expressions for each of these factors. Therefore, we have the equation

$$P(14, 9) = C(14, 9) \times 9!$$

from which we can solve for the unknown quantity $C(14, 9)$:

$$C(14, 9) = \frac{P(14, 9)}{9!} = \frac{14!}{(14 - 9)!\, 9!}$$

which evaluates to 2,002. Again, we generalize to a baseball team of n players from whom r are chosen to deduce the general formula

$$C(n, r) = \frac{n!}{(n - r)!\, r!}$$

Example 29   Compute the following: (a) $C(4, 2)$ (b) $C(3, 0)$ (c) $C(2, 3)$
(d) $C(3, 2)$ and $P(3, 2)$

Solution

(a)

$$C(4, 2) = \frac{4!}{2!(4-2)!} = \frac{4!}{2!2!} = \frac{24}{(2 \times 2)} = 6$$

(b)

$$C(3, 0) = \frac{3!}{0!(3-0)!} = \frac{3!}{3!} = 1$$

(c) Not computable—n must be greater than or equal to r.
(d)

$$C(3, 2) = \frac{3!}{2!(3-2)!} = \frac{3!}{2!1!} = 3$$

$$P(3, 2) = \frac{3!}{(3-2)!} = \frac{3!}{1!} = 3! = 6$$

Comparing the two results of part (d) suggests that given n and r, $P(n, r)$ is greater than or equal to $C(n, r)$. (From the formulas below, do you see why this is always the case?)

Skill Enhancer 29   Compute
(a) $C(3, 3)$ (b) $C(4, 3)$ (c) $C(5, 3)$

Answers in Appendix E.

Permutations and Combinations

Permutations—order counts.

$$P(n, r) = \frac{n!}{(n-r)!}$$

Combinations—order is unimportant.

$$C(n, r) = \frac{n!}{r!(n-r)!}$$

Example 30   In how many ways can a committee of 5 people be chosen from a group of 11 students?

Solution

The characteristics of the problem all point to the use of the combinations formula—each student is unique, no student can serve more than once, and, most important, the order of the choosing is totally irrelevant. It makes no difference whether Anita is chosen before Bernardo, or vice versa. The answer is

$$C(11, 5) = \frac{11!}{5!6!} = 462$$

Skill Enhancer 30   If 11 students try out for the basketball team, which consists of 5 players, how many different teams are possible?

Answer in Appendix E.

- When order is important, use permutations.
- When order is irrelevant, use combinations.

Example 31   In the following situations, is order important or irrelevant?
(a) Choosing a subcommittee from a larger committee
(b) Finding words embedded in larger words
(c) Choosing two scoops of ice cream from among five different flavors
(d) Choosing an election ticket from among a group of prominent citizens

Solution

(a) irrelevant;   (b) important;   (c) irrelevant;   (d) important.

Skill Enhancer 31   In these situations, would you use combinations or permutations?
(a) Deciding how to display four posters from a larger collection
(b) Creating clothing outfits
(c) Seating family members in the family station wagon in every seat

Answers in Appendix E.

Example 32   Compute $C(n, n)$ in *two* ways, and use this result to suggest why we should define $0! = 1$. Here, n is any whole number greater than 0.

Solution

Let $x = 0!$. In a straightforward use of the formula for combinations,

$$C(n, n) = \frac{n!}{n!\,(n - n)!} = \frac{n!}{n!\,x}$$
$$= \frac{1}{x}$$

On the other hand, $C(n, n)$ asks us to compute the number of ways of choosing n things from a group of n things (order being unimportant). But there is only one way of making this selection! (Make sure you see why.) Therefore, we must have the equation

$$\frac{1}{x} = 1$$

so that

$$x = 0! = 1$$

Skill Enhancer 32   Verify that $C(n, n) = 1$ for all n.

Answer in Appendix E.

More Counting Problems

On occasion, the Principle of Counting and the combinations formula occur in a single problem.

Example 33   Suppose that two committees are to be formed from a group of 11 students. The first will contain 5 students, and the second 3. No student is to serve on more than one committee. How many ways are there to compose these two committees?

Solution

Composing the dual committees is a two-step process. First, choose the 5-member committee from the 11 students. Once the 5 are selected, there are $11 - 5 = 6$ students remaining. The remaining committee of 3 is chosen out of these 6. By the principle of counting, therefore, the number of ways is

$$C(11, 5) \times C(6, 3) = \frac{11!}{5!6!} \times \frac{6!}{3!3!} = \frac{11!}{5!3!3!} = 9,240$$

Skill Enhancer 33   Two basketball teams (each containing 5 students) are to be formed from a group of 11 students. In how many ways can this be done?

Answer in Appendix E.

Example 34   The New American Regional Encyclopedia is an 11-volume set of books. The publishers ship it in three unequally sized boxes. The first box holds 5 volumes, the second holds 4, and the remaining 2 are placed in a heavy mailing envelope. How many ways are there of shipping an encyclopedia set?

FIGURE 5.29

Solution

When the first box of 5 books is packed, there are 6 left. When the second is packed, 2 are left. Therefore, the number of ways is

$$C(11, 5) \times C(6, 4) \times C(2, 2) = \frac{11!}{5!6!} \times \frac{6!}{4!2!} \times \frac{2!}{2!0!} = \frac{11!}{5!4!2!} = 6,930$$

Skill Enhancer 34   Three parents have agreed to help drive a class on a field trip. The first two vans hold 7 students apiece, and the third holds 6. There are 20 students in the class. How many different seating arrangements are there?

Answer in Appendix E.

Multinomial Coefficients

Mathematicians like to make formal the computations we have been doing in Examples 33 and 34.

ordered partition   Suppose some set S contains n elements. We can divide these into several smaller subsets; suppose there are m of these subsets. Then these subsets form an *ordered partition* of S provided that these subsets contain all n elements of S and each pair of the m subsets have no elements in common. See Figure 5.30. Why are the subsets shown in Figure 5.31 not an ordered partition?

If the ordered partition of m subsets contain $n_1, n_2, \ldots, n_m$ elements, where

$$n_1 + n_2 + \cdots + n_m = n$$

we use the notation $(n_1, n_2, \ldots, n_m)$ for this partition.

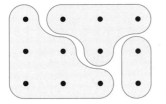

FIGURE 5.30
An ordered partition of a set.

Example 35   What are all ordered partitions of type $(2, 1, 1)$ for the set $S = \{\heartsuit, \diamondsuit, \clubsuit, \spadesuit\}$?

Solution

The ordered partition $(2, 1, 1)$ is valid, since $2 + 1 + 1 = 4$, the number of elements in S. There are to be two elements in the first subset, and a single element in the second and third subsets. Here is the list:

$(\{\clubsuit, \diamondsuit\}, \{\spadesuit\}, \{\heartsuit\})$	$(\{\diamondsuit, \spadesuit\}, \{\clubsuit\}, \{\heartsuit\})$
$(\{\clubsuit, \diamondsuit\}, \{\heartsuit\}, \{\spadesuit\})$	$(\{\diamondsuit, \spadesuit\}, \{\heartsuit\}, \{\clubsuit\})$
$(\{\clubsuit, \spadesuit\}, \{\diamondsuit\}, \{\heartsuit\})$	$(\{\diamondsuit, \heartsuit\}, \{\clubsuit\}, \{\spadesuit\})$
$(\{\clubsuit, \spadesuit\}, \{\heartsuit\}, \{\diamondsuit\})$	$(\{\diamondsuit, \heartsuit\}, \{\spadesuit\}, \{\clubsuit\})$
$(\{\clubsuit, \heartsuit\}, \{\diamondsuit\}, \{\spadesuit\})$	$(\{\spadesuit, \heartsuit\}, \{\clubsuit\}, \{\diamondsuit\})$
$(\{\clubsuit, \heartsuit\}, \{\spadesuit\}, \{\diamondsuit\})$	$(\{\spadesuit, \heartsuit\}, \{\diamondsuit\}, \{\clubsuit\})$

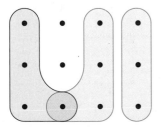

FIGURE 5.31
These subsets do not form an ordered partition.

Study it closely to see how important order is in the listing.

Skill Enhancer 35   If $T = \{i, j, k\}$, what are all ordered partitions of type $(1, 1, 1)$?

Answer in Appendix E.

multinomial coefficient   How many ordered partitions of type $(n_1, \ldots, n_m)$ are there? The answer is given by the *multinomial coefficient* $\binom{n}{n_1, n_2, \ldots, n_m}$, where

$$\binom{n}{n_1, n_2, \cdots, n_m} = \frac{n!}{n_1! n_2! \ldots n_m!}$$

and where $n_1 + \cdots + n_m = n$ (of course). The ambitious reader can justify this result by using the principle of multiplication (as we did in Examples 33 and 34).

Example 36   Refer to Example 34. How many ways are there of shipping an 11-volume set in cartons containing 5, 4, and 2 books? Use multinomial coefficients.

Solution

The number is given by the multinomial coefficient

$$\binom{11}{5, 4, 2} = \frac{11!}{5! 4! 2!} = 6,930$$

as we saw previously.

Skill Enhancer 36   Three parents have agreed to help drive a class on a field trip. The first two vans hold 7 students apiece, and the third holds 6. There are 20 students in the class. How many different seating arrangements are there? Use multinomial coefficients.

Answer in Appendix E.

The Binomial Theorem

binomials   One final application of combinations is to the expansion of *binomials* such as $(x + y)^4$. Although we can certainly multiply out this expression and simplify the resulting product, this is a tedious operation, especially when the exponent is larger than 4. It is easy to see that, upon expansion and simplification of the product $(x + y)^4$, there will be five terms, all in the form $x^i y^{n-i}$ for $0 \le i \le n$. That is, for this case ($n = 4$), the expanded product will contain terms in $x^0 y^4$, $x^1 y^3$, $x^2 y^2$, $x^3 y^1$, and $x^4 y^0$. (In the general case of the expansion of $(x + y)^n$, the simplified product will contain $n + 1$ terms, each of the form $x^i y^{n-i}$ for $0 \le i \le n$.) But what is the coefficient of each of these terms? Can we ascertain these coefficients without actually carrying out the multiplication?

To provide insight, we may expand $(x + y)^4$ *without* simplifying and *without* combining like terms.

$$(x + y)^4$$

$$= xxxx + xxyx + xyxx + xyyx + xxxy + xxyy + xyxy + xyyy$$

$$+ yxxx + yxyx + yyxx + yyyx + yxxy + yxyy + yyxy + yyyy$$

How many of these terms will simplify to $x^2 y^2$, for example? There will be as many terms as there are ways of distributing x and y in four positions. But this is simply $C(4, 2)$ because the order in which the two x's appear is unimportant. Therefore, the coefficient of $x^2 y^2$ will be $C(4, 2)$. We may generalize this conclusion to learn that the coefficient of any term $x^i y^{4-i}$ is $C(4, i)$ in this problem. For general n, the coefficient of $x^i y^{n-i}$ in the expansion of $(x + y)^n$ is $C(n, i)$. This holds for $0 \leq i \leq n$, and for all whole numbers $n \geq 1$. Mathematicians call this conclusion the *Binomial Theorem*.

The Binomial Theorem

$$(x + y)^n = x^n + \cdots + C(n, i)x^i y^{n-i} + \cdots + y^n$$

for $n \geq 1$, for n a whole number.

---

binomial coefficients   The combinations $C(n, r)$ are also called *binomial coefficients* because $C(n, r)$ is the coefficient of $x^i y^{n-i}$ in the expansion of $(x + y)^n$. In the same way, the multinomial coefficient $\binom{n}{n_1, n_2, \ldots, n_m}$ is the coefficient of

$$x_1^{n_1} x_2^{n_2} \ldots x_m^{n_m}$$

in the expansion of $(x_1 + x_2 + \cdots + x_m)^n$.

Example 37   Use the Binomial Theorem to determine the coefficient of $x^2 y^3$ in the expansion of $(x + y)^5$.

Solution

By the Binomial Theorem, this coefficient is $C(5, 2) = \frac{5!}{2!3!} = 10$.

Skill Enhancer 37   Without expanding, determine the coefficient of xy^3 in the expansion of $(x + y)^4$.

Answer in Appendix E.

Example 38   Use the Binomial Theorem to show that

$$C(n, 0) + C(n, 1) + C(n, 2) + \cdots + C(n, n - 1) + C(n, n) = 2^n$$

for all whole numbers $n \geq 0$.

Solution

Consider the expression $(1 + 1)^n$. We may evaluate this expression in two ways, and set the two expansions equal.

On the one hand, since $1 + 1 = 2$, we have

(5.2) $$(1 + 1)^n = 2^n$$

On the other hand, by the Binomial Theorem,

(5.3)
$$(1 + 1)^n = C(n, 0)1^0 1^n + C(n, 1)1^1 1^{n-1} + C(n, 2)1^2 1^{n-2} + \cdots$$
$$+ C(n, n)1^n 1^0$$

But $1^i 1^{n-i} = 1$ for all i between 0 and n inclusive. Equation (5.3) becomes simply

$$(5.4) \qquad (1+1)^n = C(n,0) + C(n,1) + C(n,2) + \cdots + C(n,n)$$

Setting equal the right-hand sides of Equations (5.2) and (5.4) proves the assertion.

Skill Enhancer 38 Why does the Binomial Theorem suggest that $C(n,i) = C(n, n-i)$?

Answer in Appendix E.

We will use the Binomial Theorem in our later work on statistics.

Exercises 5.5

Evaluate the expressions in Exercises 1–14.

1. $C(3,2)$

2. $C(3,1)$

3. $C(7,4)$

4. $C(5,7)$

5. $C(7,5)$

6. $C(9,2)$

7. $C(4,1)$

8. $C(C(4,1), C(4,4))$

9. $C(6,3)$

10. $C(n, n-r)$

11. $C(n+1, n)$

12. $C(6,5)$

13. $C(n,n)$

14. $C(100, 100)$

15. The text states that $P(n,r) \geq C(n,r)$. Under what circumstances is the equation $P(n,r) = C(n,r)$ true?

Consider the expression $(x+y)^7$. Without directly expanding this expression, give the coefficients of the terms indicated in Exercises 16–21.

16. x^7

17. y^7

18. $x^2 y^5$

19. xy^6

20. $x^3 y^4$

21. $x^4 y^3$

Evaluate the multinomial coefficients in Exercises 22–27.

22. $\binom{4}{2,1,1}$

23. $\binom{3}{2,0,1}$

24. $\binom{5}{2,1,2}$

25. $\binom{5}{1,3,1}$

26. $\binom{5}{3,0,2}$

27. $\binom{5}{1,2,2}$

W 28. (a) For what value (or values) of r is $C(4,r)$ a maximum (that is, as large as possible)?
(b) For what value of r is $C(5,r)$ a maximum?
(c) For what value of r is $C(6,r)$ a maximum?
(d) For what value of r is $C(7,r)$ a maximum?

(e) Based on your exploration here, can you draw a conclusion about when $C(n,r)$ will take on its largest value?

W 29. For what values of r does $C(n,r)$ take on its smallest value? If necessary, try to answer this question by reminding yourself of the "physical" meaning of $C(n,r)$. Otherwise, you may find it helpful to experiment with particular values of n and r, as in Exercise 28.

Applications

30. A branch music store can only carry 100 compact discs out of the 300 that the parent company maintains in stock. How many ways are there of choosing these 100?

31. For a psychology experiment, a graduate student needs to select 8 students out of 12 volunteers. In how many ways can this choice be made?

32. As a reward for conscientious and aggressive selling of encyclopedias door-to-door, the company will pay for 3 magazine subscriptions out of 10 possible titles for Malcolm. In how many ways can Malcolm choose these 3?

33. Certain high-quality computer monitors can display any of 4,096 different and distinct colors, but only 12 colors can be displayed at any one time. The 12 colors can be chosen by the computer programmer. In how many ways can this choice be made?

34. There are 8 students who are eligible for the varsity basketball team, consisting of 5 players and 1 backup. How many different basketball teams can be formed from the 8 students?

35. In another year, 14 students are eligible for the school's two basketball teams, each consisting of 6 players. How many ways are there of forming the two teams? No player is allowed on both teams.

36. A hand of poker consists of 5 cards dealt from a pack of 52, all of which are different. How many such hands are possible?

37. Suppose there are 5 players in the card game of Exercise 36. How many different deals are possible?

38. A committee at school consists of 10 students and 5 administrators. A special committee is to be composed of 7 students and 3 administrators. How many ways are there of forming this special committee?

39. The boss has drafted 10 middle-level managers to study ways to revamp the company. One committee of 3 will study the production line. A second committee of 4 will review the advertising policies of the company. The final committee of 3 will examine the accounting procedures. No manager may serve on more than one committee. In how many ways can the committees be formed?

40. Detroit's newest sales gimmick when you buy a new car is to give you a choice of "free" options. For a particular model, you can choose 5 out of 8 possible options. How many different variations of this model are possible?

41. A piggy bank contains 3 dimes, 4 nickels, 5 quarters, and 2 pennies. You shake the bank and manage to dislodge 5 coins. Assume that all coins are distinguishable.
   (a) How many different groups of 5 coins are possible?
   (b) In how many ways can you shake the coins so that 2 are nickels and 3 are quarters?
   (c) In how many ways can you shake the coins so that 4 are nickels and 1 is a dime?
   (d) In how many ways can you shake the coins so that 1 is a dime, 1 is a nickel, 1 is a quarter, and 2 are pennies?
   (e) What is the maximum amount of money you can get in 5 coins, and how many ways are there to get that amount?

42. A toy chest contains 4 rubber balls (different colors), 3 rubber animals (different species), and 3 different stuffed animals. A child runs over and chooses 3 toys at random.
   (a) How many different selections of toys are possible?
   (b) How many ways are there to choose 3 balls?
   (c) How many ways are there to choose 1 stuffed animal and 2 balls?
   (d) How many ways are there to choose one of each kind of toy?

43. The card game of bridge is played with four players each receiving 13 cards. How many different hands are possible? How many different deals are possible? Leave the answers in factored form.

W 44. Since $n = n_1 + \cdots + n_m$, we can write a multinomial coefficient as

$$\binom{n_1 + n_2 + \cdots + n_m}{n_1, n_2, \ldots, n_m}.$$

Expand this using the definitions, together with the laws of algebra, to suggest why the relation

$$\binom{n_1 + n_2 + \cdots + n_m}{n_1, n_2, \ldots, n_m} = \binom{n}{n_1}\binom{n_2 + \cdots + n_m}{n_2, \ldots, n_m}$$

must hold. Here, we use an alternative notation $\binom{n}{r}$ which means the same thing as $C(n, r)$.

W 45. It's easy to show from the formula for combinatorial coefficients that $C(n, n) = 1$. Appeal to the meaning of combinations to show why this must be so. (No formulas should be part of your answer.)

**Calculator Exercises** _____

Calculate the numbers of combinations in Exercises 46–47.

46. $C(13, 3)$ **47.** $C(12, 8)$

CHAPTER REVIEW _____

Terms _____

set	element	set equality
infinite set	subset	proper subset
empty set	null set	complement
universe	intersection	union

Venn diagrams   one-to-one   cardinality
count of a set   tree diagram   Principle of Counting
factorial   permutation   combination
ordered partition   multinomial coefficient   Binomial Theorem
binomials   binomial coefficients

Key Concepts

- A set is the container a mathematician uses to group a collection of objects together. Sets contain elements or members. Collections of objects wholly within a set are called subsets. A proper subset is contained in the set but does not include all elements of the set.

- The empty set or null set $\emptyset$ is the set containing precisely no elements. The universal set or universe U is the set of all elements under consideration.

- *Set operations.* The complement A' of a set A is the set containing those elements in the universe not in A. The intersection $A \cap B$ of two sets A and B is the set containing those elements both in A and in B. The set union $A \cup B$ is the set containing the elements that are in either A or B or both. Set relationships can be easily visualized with the aid of Venn diagrams, a system whereby circles are made to represent sets. Two or more circles will overlap to indicate a nonempty set intersection.

- The number of elements in a set is that set's cardinality.

- *Principal of Counting.* Use the Principle of Counting to find the number of ways a process composed of several subprocesses can be performed. If each of the n subprocesses can be carried out in O_i ways, then the total number of ways for the main process is the product $O_1 \times O_2 \times \cdots \times O_n$.

- *Factorial.* For positive integers n, $n! = 1 \times 2 \times \cdots \times n$. In particular, $1! = 1$, and there is a special case $0! = 1$.

- *Permutations and combinations.* To find the number of ways of arranging r items plucked from a group of n such items, use the formulas for permutations or combinations. All n items are assumed unique, and no item can appear in the subgroup more than once. The choice of each item depends neither on the elements chosen before it nor on those chosen after it. When the order in which the items are chosen is important, the number of permutations $P(n, r)$ is given by

$$P(n, r) = \frac{n!}{(n-r)!}$$

When order is unimportant and irrelevant, the number of combinations is given by

$$C(n, r) = \frac{n!}{r!(n-r)!}$$

Review Exercises

Let $A = \{a, c, e, g, h, i, j\}$, $B = \{a, b, c, d\}$, $C = \{e, f, g, h\}$. Indicate whether the statements in Exercises 1–4 are true or false.

1. $b \in B$ **2.** $b \subset B$

3. $B \subseteq A$ **4.** $a, d \subset A$

5. List the elements in the set $A = \{$letters in "Mississippi"$\}$

6. List all the subsets of $\{a, b, c, d\}$.

Let $A = \{a, c, e, g, h, i, j\}$, $B = \{a, b, c, d\}$, $C = \{e, f, g, h\}$, and the universe

$$U = \{a, b, c, d, e, f, g, h, i, j, z\}$$

Use this information in Exercises 7–12.

7. Describe $A \cap B$.   **8.** Describe U'.

9. Describe A'.   **10.** Describe $A' \cap B$.

11. Describe $C \cup B$.   **12.** Describe $(A \cup B') \cap C$.

Use a Venn diagram for two sets A and B to answer Exercises 13–14.

13. Shade $A \cap B$.   **14.** Shade $A \cup B'$.

Use a Venn diagram for three sets A, B, and C to answer Exercises 15–22.

15. Shade $A \cap C$.

16. Shade $A \cap (B \cup C)$.

17. Shade $(A \cup B') \cap C$.

18. Shade $A \cap (B \cup C)'$.

19. Shade $A \cap B \cap C$.

20. Shade $(B \cap C) \cup A$.

21. Shade $((B' \cap C) \cup A)'$.

22. Shade $(A' \cup B') \cap C$.

Consider two intersecting sets E and F for Exercises 23–26.

23. If $c(E) = c(F) = 30$, and $c(E \cup F) = 55$, then what is $c(E \cap F)$?

24. If $c(E) = 2c(F)$, $c(E \cup F) = 45$, and $c(E \cap F) = 10$, then what is $c(F)$?

25. If $c(E \cap F) = 0$ and $c(E) = c(F) = 7$, what is $c(E \cup F)$?

26. If the cardinalities of E and of F are restricted to be nonnegative integers, and if $c(E \cap F) = c(E \cup F)$, what can you conclude about E and F?

Evaluate the expressions in Exercises 27–34.

27. $4!$

28. $1!$

29. $0!$

30. $C(3, 2)$

31. $P(3, 2)$

32. $C(5, 3)$

33. $P(5, 3)$

34. $\frac{C(9,7)}{C(7,4)}$

Applications

(*Market Research*) A market research survey of 100 consumers determines that 48 buy fluoride toothpaste, 67 purchase gel pastes, and 27 purchase fluoride gels. Use this information in Exercises 35–36.

35. How many in the survey sample purchase some other type of dentifrice?

36. How many use a gel toothpaste with no fluoride?

(*Planned Communities*) The original Levittown community of tract houses built after World War II has long been of interest to sociologists searching for patterns in current American life styles. The original homes were simple. In the intervening years, many improvements have been added. In a group of 80 homes, 40 homeowners had added a new room, 52 had redone their kitchens, and 47 had finished their basements; 25 have both new kitchens and add-on rooms, 30 have new kitchens and finished basements, and 27 have finished basements and add-on rooms; and 20 homeowners have made all three improvements. Use this information to solve Exercises 37–40.

37. How many have made some alteration other than a new kitchen?

38. How many have added new rooms and not touched their kitchens or basements?

39. How many have new kitchens or finished basements but not any new rooms?

40. How many have not added any of these 3 improvements to their houses?

41. A new teenage fad involves wearing unmatched pairs of shoes. One young individual owns 13 pairs of shoes. How many different ways may the shoes be worn? How many if shoes are constrained not to match?

42. There are three positions in your local government—executive, secretary, and treasurer. At present, 5 persons are running for chief executive, 3 for secretary, and 6 for treasurer. Ignoring party politics, how many different governing bodies could result from the election?

43. (*Drug Certification*) The Food and Drug Administration has to approve 37 new medicines, classified as pain-killers or antibiotics. Of these, 15 are both, 8 are neither, and 10 are antibiotics *only*. How many are pain-killers only?

44. Compute the number of batting orders for a 9-person baseball team if the star hitter bats first and the pitcher bats fourth. Assume that the star hitter and the pitcher are distinct players.

45. A computer dating service has a pool consisting of 34 men and 48 women from which to draw. How many dates could be set up?

46. The Central Committee of a certain country contains 20 members, of which 11 are thought to be undercover agents of a foreign power. Agents in the country's intelligence service operate according to strict, bureaucratic rules. An agent is responsible for a particular group of 11, which she investigates secretly. How many groups of 11 must the country's intelligence service investigate to find the foreign agents?

47. For the upcoming semester, Sally Student has to take 5 courses from among 9 offered electives in order to graduate. In how many ways can she satisfy the graduation requirement?

48. A biologist is about to perform a crucial experiment. The subjects are a group of 15 labeled laboratory white mice, of which 6 take one drug, 6 take another, and the remaining 3 are given the control. How many different experimental protocols are possible?

6

INTRODUCTION TO PROBABILITY

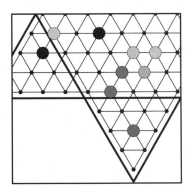

6.1 Sample Spaces; Basic Properties of Probability

6.2 Equally Likely Probabilities

6.3 Probability and Set Theory

6.4 Conditional Probability

6.5 Bayes' Theorem

6.6 Expectation

Chapter Review

INTRODUCTION

So far, all our problems have involved conditions of *perfect information*—the conditions of a problem were known perfectly, and it was up to us to use this information to find an accurate solution. But what if the initial conditions are *not* known perfectly? The best we can then do is assess the *likelihood* of various outcomes. This likelihood, or *probability*, is familiar from everyday life.

We use the techniques of probability to determine answers to familiar questions: Should we prepare for rain tomorrow?
How should insurance premiums be adjusted for a 30-year-old man? for a 40-year-old woman? How likely is it that one of us will win "big" in Atlantic City or Las Vegas? If there are several defective light bulbs in a batch of 100, is this a fluke or should the whole lot be discarded?

In this chapter, we will examine the basic underpinnings of a mathematical theory of probability. In the first section, we examine the basic attributes common to *any* theory of probability. (There is more than one way to assign probabilities to events.) The easiest probabilities to compute are those of equally likely events. We show how to compute these probabilities, and then show how to apply them to situations of more general interest. We then explore some theory and application of the powerful Bayes' Theorem.

The chapter concludes with a discussion of expectation, whereby we can make precise what we mean when we talk about the long-term gains or losses in games and experiments.

Section 6.1 SAMPLE SPACES; BASIC PROPERTIES OF PROBABILITY

probability A theory of *probability* attempts to attach a number to some event that has not yet occurred. We will use this number to judge how likely will be the occurrence of that event. We need this probability for two reasons.

1. Either no one knows enough to predict the outcome with a 100 percent certainty, or

2. It would not be practical to process all the facts to predict the outcome with complete assurance.

Here are some questions whose answers cannot be known (in advance) with total precision:

- A friend is about to flip a fair coin. Will it land heads or tails?

- A married couple plans to have two children. How likely is it that both children will be of the same sex? How likely is it that the eldest will be a boy?

- How likely is it to rain on a one-day trip to visit a friend?

- *The Birthday Problem.* Twenty-three people sit together in a classroom. What are the chances that *at least two people* will have the *same* birthday?

- *Chevalier de Méré's Problem.* A gambler chooses between two games. In the first, the croupier throws four dice simultaneously, and the gambler wagers that a 4 will appear on one of the dice. In the second, a pair of fair dice are thrown 24 times, and the gambler bets that a pair of 4s will turn up at least once in this sequence. Which game is better for the bettor?

Historical Note

Founding Figures of Probability *In 1654, the Chevalier de Méré posed his problem to the great mathematician Blaise Pascal (1623–1662). Pascal's interest was sufficiently piqued for him to discuss it with his friend, the equally eminent mathematician Pierre de Fermat (1601–1665). Their correspondence exerted a strong influence on the subsequent development of probability theory.*

Pascal's other achievements in mathematics centered around geometry and around the development of the first mechanical calculating machine. He is therefore honored by having a prominent computer programming language named after him (PASCAL). Fermat was a lawyer and politician who did mathematics—mostly geometry—as a hobby. His most famous achievement is a theorem he claimed to have proven. His claim appears as a handwritten marginal note in one of the books of his library, but he omitted the proof as it was too long to fit in the margin. No mathematician from then until now has succeeded in reproducing that proof (or any other proof of that theorem), although a lengthy (200-page+) proof presented in the summer of 1993 is now being evaluated by mathematicians.

Sample Spaces and Simple Events

experiment   We use probability to find the likelihood of the results of some *experiment*, where this word has a slightly different meaning from the usual scientific one. A probabilistic experiment is *any* situation about whose outcome we are uncertain. The results of tossing a coin, assessing tomorrow's weather, playing a hand of bridge, or predicting the way a rat will behave in a scientific experiment are all probabilistic experiments. For the remainder of this chapter, this word will only be used in its new, probabilistic sense.

sample space   The *sample space* is the set of all individual outcomes of a probabilistic experiment. A sample space for a probability problem plays the same role as the universe in a set theory problem. Although the elements of a universe may be any items at all, the elements of a sample space are restricted to outcomes of an experiment. We will pursue the analogy between concepts in set theory and concepts in probability theory. Many results of set theory becomes useful for probability.

event   An *event* is any outcome of an experiment, and is a subset of the sample space. A
simple event   *simple event* is a subset containing a single outcome; in other words, it is a subset of the sample space containing only one element. Figure 6.1 illustrates these concepts.

The next few examples provide practice in thinking of a situation as a probabilistic experiment.

Example 1   A magician asks us to guess a whole number at random between 1 and 10 inclusive. Discuss the experiment using probability theory.
Solution

The simplest sample space S is one in which each of the simple events is one of the possible numbers:

$$S = \{1, 2, 3, 4, 5, 6, 7, 8, 9, 10\}$$

Skill Enhancer 1   The colors of the rainbow are violet, indigo, blue, green, yellow, orange, and red. Someone asks us to choose one of these colors at random. What is the simplest sample space to describe this experiment?

Answer in Appendix E.

Example 2   We toss a coin, weighted so that heads comes up three times as often as tails. Describe this experiment in terms of probability theory.
Solution

The sample space S is

$$S = \{\text{heads, tails}\}$$

The weighting of the coin does not enter into this description. The events in the sample space are the subsets {heads}, {tails}, the empty set Ø, and the sample space S itself. Ø is an impossible event. (Why?) S is the event that upon tossing, the coin lands with either heads or tails showing. S is a certain event.

Skill Enhancer 2   Baby Hannah throws a piece of buttered bread on the floor. Describe this experiment in terms of probability theory.

Answer in Appendix E.

Some events are themselves processes consisting of several stages.

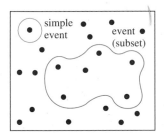

simple event   event (subset)

sample space (universe)

FIGURE 6.1
Concepts in probability theory are similar to those in set theory.

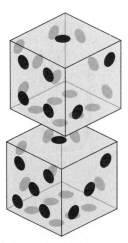

FIGURE 6.2

Example 3   Someone rolls a pair of dice. Describe one sample space in which each element stores information about both dice.

Solution

We let each element in S be an ordered pair, each number of which stands for a number on the first or second die. Thus,

$$S = \{(1, 1), (1, 2), (1, 3), (1, 4), (1, 5), (1, 6),$$
$$(2, 1), (2, 2), \ldots , (2, 6)$$
$$\vdots$$
$$(6, 1), (6, 2), \ldots , (6, 6)\}$$

Skill Enhancer 3   What is a sample space S to represent the tossing of two coins? Each element of S should say something about each of the coins.

Answer in Appendix E.

Example 4   Professor Stutter is famous for his exams, which tend to be easy or difficult. Seventy-five percent of the time, an easy test is followed by a difficult one. Fifty percent of the time, a difficult test is followed by an easy one. Set up a sample space S describing the first two tests of the term.

Solution

Let E and D represent easy and difficult tests, respectively. The sample space S is the union of the simple events $\{EE\}$, $\{ED\}$, $\{DE\}$, and $\{DD\}$.

Skill Enhancer 4   Weather on Saturday will be rainy or sunny, as will be the weather on Sunday. Give one sample space describing the weekend weather.

Answer in Appendix E.

For any experiment there may be many ways of constructing the sample space. At this point, we have no way of deciding which is the best, but this will change as our experience increases.

Example 5   A couple is planning to have three children. Construct several plausible sample spaces for this "experiment."

Solution

One way of describing the sample space is to consider the number of girls among the three children. This gives rise to one sample space

$$S_1 = \{0, 1, 2, 3\}$$

where each number refers to a possible number of girls among three children.

If B represents *boy* and G represents *girl*, a second sample space S_2 is

$$S_2 = \{BBB, BBG, BGB, BGG, GBB, GBG, GGB, GGG\}$$

where the position of the letter indicates the birth order of the three children. (Thus, for example, $\{GBG\}$ is the event that a son in this family has both an older and younger sister.) Both S_1 and S_2 are equally valid representations of the sample space, although (as we shall see) S_2 is usually more useful for calculating probabilities.

Skill Enhancer 5   A coin is tossed three times. Give two distinct sample spaces for this experiment.

Answer in Appendix E.

tree diagrams   *Tree diagrams* are as useful in summarizing the elements of sample spaces as they were in enumerating all possibilities in various set theory counting problems.

Example 6   A professor in Boston travels to Salt Lake City for a scientific meeting. He can take either of two subway lines from his house to the airport, and fly on any of three airlines from the Boston airport to Salt Lake City. (The meeting is held in a hotel located within the Utah airport.) Construct a sample space for this trip.

Solution

Denote the two subway lines by X and Y and the three airlines by 1, 2, and 3. We know by the Principle of Counting that there are $2 \times 3 = 6$ different ways to make the trip, and we shall set up the sample space so that each of these six feasible travel itineraries is an element, as the tree diagram in Figure 6.3 shows.

Using obvious notation, we can write the sample space as

$$S = \{X1, X2, X3, Y1, Y2, Y3\}$$

Skill Enhancer 6   Any of three candidates may be elected president of a club, and any of two candidates may be elected treasurer. A tree diagram suggests what sample space for this experiment?

Answer in Appendix E.

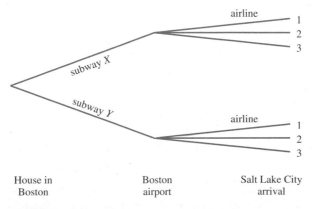

FIGURE 6.3
Boston to Salt Lake City: a tree diagram.

Later, we will not be able to compute numerical probabilities (although we don't yet know what this expression really means) without being able to make the connection between events described in words and events of a sample space. The next examples help show how to make an equivalence between any outcome and the simple events of a sample space.

Example 7   We roll a single die. A reasonable sample space is

$$S = \{1, 2, 3, 4, 5, 6\}$$

Describe the following outcomes as a subset of S.
(a) The number is greater than 2.
(b) The number is odd.
(c) Either the number is even or it is less than 3.

Solution

(a) $\{3, 4, 5, 6\}$.
(b) $\{1, 3, 5\}$.
(c) $\{1, 2, 4, 6\}$.

Skill Enhancer 7   One letter of the alphabet is chosen at random. A sample space for this experiment is $S = \{a, b, c, d, e, f, \ldots, x, y, z\}$. Describe these outcomes as subsets of S.
(a) The letter is part of the word "mathematics."
(b) The letter is a vowel or is in the word "lousy."
(c) The letter is a vowel and is in the exclamation "eureka."

Answer in Appendix E.

Example 8   Consider the following events in light of Example 5 above; use the second sample space S_2 from that example. If an event is simple, say so; otherwise, state which simple events the given event comprises.
(a) No boys are born.
(b) The oldest child is a boy.
(c) At least as many girls as boys are born.
(d) No male child has sisters older than himself.
(e) An oldest son has two younger sisters.
(f) Which of these events can be described in terms of the events of sample space S_1 described in Example 5?

Solution

Each element of S_2 is a simple event.
(a) This is $\{GGG\}$, a simple event.
(b) This event E_2 can occur in any of four ways. We may write $E_2 = \{BBB, BBG, BGB, BGG\}$.
(c) This event $E_3 = \{GGG, GGB, GBG, BGG\}$.
(d) Here, $E_4 = \{BBB, BBG, BGG, GGG\}$.
(e) This is the simple event $\{BGG\}$.
(f) Subsets of S_1 can be used to describe event (a) and event (c) only. (Do you see why?)

Skill Enhancer 8   Refer again to S_2 in Example 5. What simple events correspond to these descriptions?
(a) No girls are born.
(b) The gender of the middle child is unimportant.
(c) The youngest child is a girl.

Answer in Appendix E.

Mathematical Probability

probability function

A *probability function* is a way of assigning a number to an event E in a sample space so that $P(E)$ is the probability, or likelihood, that the event E happens. In seeking this function P, we have to satisfy good mathematical sense and then satisfy our physical intuition so that $P(E)$ will measure likelihood in ways we have come to expect. In particular, we expect the probability of an outcome to coincide with its relative frequency.

relative frequency

The *relative frequency* of any outcome is the fraction p of the time in the long run that we observe this outcome. This motivates the mathematical rules of probability that we present below.

impossible

If E is an *impossible* event, like tossing a coin and expecting it to land on its edge, then we assign a probability of 0 to that event; it has no likelihood of occurring. If some

certain

other event F is *certain* to occur, we assign a probability of 1 to it. For any arbitrary event E, its probability $P(E)$ must satisfy the inequality

(6.1) $$0 \leq P(E) \leq 1$$

Next, suppose there are m *simple* and *distinct* events E_1 through E_m that have probabilities $P(E_1)$ through $P(E_m)$. If E is the union of these simple events,

$$E = E_1 \cup E_2 \cup \cdots \cup E_m$$

then E represents the event that *either* E_1 or E_2 or ... or E_m will occur. (Remember, set *union* corresponds to *or*.) Then we say that the probability of E is given by

(6.2) $$P(E) = P(E_1) + P(E_2) + \cdots + P(E_m)$$

Equation (6.2) holds for simple, distinct events *only*.

Finally, suppose there are n simple and distinct events E_i in a sample space S. Then

(6.3) $$P(E_1) + P(E_2) + \cdots + P(E_n) = 1$$

These three equations summarize the three fundamental properties of any probability function.

Properties of Probability

1. $0 \leq P(E) \leq 1$ for any event E.

2. $P(E_1 \cup E_2 \cup \ldots \cup E_m) = P(E_1) + P(E_2) + \cdots + P(E_m)$ for any group of m simple and distinct events in the sample space.

3. If the sample space contains a total of n simple, distinct events E_i, then $P(E_1) + P(E_2) + \cdots + P(E_n) = 1$.

With these rules, we can create numeric probabilities such that the probability of an event that occurs over and over again is the same as that event's relative frequency. We assume that these rules will also hold for events that occur once or only a few times, for which the concept of relative frequency is not well defined.

Example 9   In the last month (30 days), it was sunny on 26 days in a resort town on the French Riviera. What is the relative frequency of sun over the last 30 days?

Solution

The relative frequency is $\frac{26}{30} = \frac{13}{15}$.

Skill Enhancer 9   Of 20 cars in an automobile showroom, 15 are station wagons. What is the relative frequency of station wagons in this showroom?

Answer in Appendix E.

Example 10   The 33 students in a math class drop 3 coins each. Of the 99 distinct outcomes, exactly 45 heads are observed. What is the relative frequency with which heads occurred in this experiment? What is the relative frequency with which tails occurred?

Solution

The relative frequency for heads is $\frac{45}{99}$. If 45 heads occurred, then $99 - 45 = 44$ tails occurred. The relative frequency for tails is therefore $\frac{44}{99}$.

Skill Enhancer 10   Baby Hannah drops a piece of buttered bread on the floor 12 times. Her distraught father sourly observes that the butter landed face down 9 times. What is the relative frequency with which the buttered side landed down in this experiment?

Answer in Appendix E.

probability distribution   We speak of a *probability distribution* as the rule (or collection of rules) that gives the probability for any simple event in a sample space.

Example 11   Perform the experiment of tossing two coins. The sample space consists of four simple events—*HH, HT, TH, TT*. Decide whether the following probability distributions are mathematically valid.
(a) $P(HH) = P(TT) = \frac{1}{2}$; $P(HT) = P(TH) = 0$.
(b) The probabilities of all four simple events are $\frac{1}{6}$.
(c) $P(HH) = P(HT) = \frac{3}{4}$, $P(TH) = 0$, $P(TT) = -\frac{1}{2}$
(d) $P(HH) = \frac{1}{2}$, $P(HT) = \frac{1}{3}$, $P(TH) = \frac{1}{4}$, $P(TT) = \frac{1}{5}$.
(e) The probabilities of all four simple events are each equal to $\frac{1}{4}$.

Solution

(a) This is a valid probability assignment, since all probabilities lie within the range of 0 to 1, and the sum of the four probabilities equals 1. (Although it is valid because it satisfies the rules, it is not a probability distribution that we would expect to encounter with any real pair of coins.)

(b) Invalid—the sum of the probabilities of all simple events is not 1.
(c) Invalid—negative probabilities are not allowed.
(d) Invalid—the sum of probabilities is not 1.
(e) Valid. (We expect that any pair of fair coins will illustrate probabilities according to these rules when we toss the coins.)

Skill Enhancer 11   A baby drops a piece of buttered bread on the floor. There are two simple events—D and U, for buttered side lands face down or face up. Which of the following are valid probability distributions?
(a) $P(D) = \frac{2}{3}$ and $P(U) = \frac{1}{3}$
(b) $P(D) = P(U) = \frac{3}{8}$
(c) $P(D) = P(U) = 23$
(d) $P(D) = -\frac{1}{8}$ and $P(U) = 0$

Answer in Appendix E.

Example 12   A game involves rolling a single die. Each of the six faces is marked with a number of dots from 1 to 6. For this die, the probabilities of rolling a 5 or a 6 are each b. The probabilities of rolling a 1 through a 4 are each a. Use these facts to answer these questions.
(a) Determine one relationship between a and b.
(b) What is the probability of rolling a 2?
(c) What is the probability of rolling a 6?
(d) What is the probability of rolling an even number?
(e) What is the probability of rolling a prime number?
(f) What is the probability of rolling a 3 or higher?

Solution

The sample space for this experiment consists of the six simple events "rolling a 1" through "rolling a 6." Use the notation $e_1, \ldots, e_6$ for these outcomes and let $E_1 = \{e_1\}$ and so on. The problem states that $P(E_1) = P(E_2) = P(E_3) = P(E_4) = a$ and $P(E_5) = P(E_6) = b$.
(a) The sum of all six probabilities must be 1. Therefore, $4a + 2b = 1$ is the desired relation.
(b) $P(E_2) = a$
(c) $P(E_6) = b$
(d) Let E be the event of rolling an even number, that is, a 2, 4, or 6. Since $E = E_2 \cup E_4 \cup E_6$, we have that $P(E) = P(E_2) + P(E_4) + P(E_6) = 2a + b$.
(e) The prime numbers below 6 are 2, 3, and 5. If F is the probability of rolling a prime number on a single die, then $F = E_2 \cup E_3 \cup E_5$, so that $P(F) = 2a + b$.
(f) Let G be the outcome of rolling a 3 or higher. Then, $G = E_3 \cup E_4 \cup E_5 \cup E_6$, so that $P(G) = a + a + b + b = 2a + 2b$.

Skill Enhancer 12   Play the same game with a slightly different die. This time, the probabilities of rolling a 1, 2, or 3 are each c and the probabilities of rolling a 4, 5, or 6 are each d.
(a) What is one relationship between c and d?
(b) What is the probability of rolling a 3?

(c) What is the probability of rolling 3 or higher?

(d) What is the probability of rolling an odd number?

Answer in Appendix E.

We can use this formalism to answer many probability questions, but we cannot use these laws to determine the probabilities of simple events in the vast majority of problems. For that, we have to work somewhat harder. In the next section, we will investigate a powerful concept that will enable us to actually assign numerical probabilities, but there are two additional methods for computing probabilities that we will look at now.

Empirical Probability

empirical probabilities   We may compute *empirical probabilities* by examining a record of past occurrences and using that record to estimate the likelihood of future occurrences. The empirical probability of any event is the *relative frequency* with which it occurs.

For example, a woman about to give birth wants to know the probability with which she will deliver a girl. Upon researching the hospital records for the past five years, she discovers that there were slightly more boys born than girls during that time. Specifically, out of 28,815 healthy deliveries, exactly 14,753 were boys and 14,062 were girls. The fraction

$$\frac{14,753}{28,815} \approx 0.5120$$

represents the empirical probability that a woman will deliver a boy at this hospital, whereas

$$\frac{14,062}{28,815} \approx 0.4880$$

is the empirical probability that a woman gives birth to a girl. The woman therefore estimates her likelihood of giving birth to a girl at 0.4880, or 48.80 percent.

Example 13   The fog rolled in from San Francisco Bay on 13 days in the last three weeks. Estimate the probability that it will roll in tomorrow.

Solution

The relative frequency of fog is $\frac{13}{21}$; this is the empirical probability that it will roll in tomorrow.

Skill Enhancer 13   Out of 100 tulip bulbs planted, only 74 germinated. There is one more bulb to plant. Estimate the probability that it will germinate. What numerical quantity provides this estimation?

Answer in Appendix E.

Example 14   A special "coin" is created by cementing a New York City subway token to a token for the Connecticut turnpike. Both tokens are approximately the same size. The coin is tossed a great many times, and the number of heads is counted. By definition, this coin shows "heads" when the subway token faces up. Because of the differing weights of the tokens, heads comes up 3,916 times out of a total of 5,161 tosses. What is a reasonable estimate for the probability that heads will come up on any future toss of this coin?

Solution

The empirical probability is

$$P(\text{heads}) = \frac{3,916}{5,161} \approx 0.759$$

Skill Enhancer 14   A compulsive compiler of useless information, Mr. Grundy notes that of the past 313 times his newspaper carrier has delivered the paper, it was torn 211 times. What is a reasonable estimate for the probability that the paper will be torn on any given morning in the future?

Answer in Appendix E.

Odds and Subjective Probability

subjective probability   Probability values can also be assigned *subjectively*—on the basis of opinion. Probabilities governing tomorrow's weather, the winner of the fifth race at Hialeah racetrack, the success of various candidates in an upcoming election, the winner of an athletic competition, and the behavior of the stock market are all examples of probabilities that are computed subjectively.

odds   We often express subjective probabilities in the form of *odds*. The home basketball team may be favored to win tomorrow's big game with odds of 7 to 5. That means that someone judges the probability of the home team's win to be

$$\frac{7}{7+5} = \frac{7}{12}$$

These odds can also be written $7 : 5$.

Odds

If event E is given odds of a to b of occurring, then

(6.4) $P(E) = \dfrac{a}{a+b}$

These odds may also be written $a : b$.

Example 15   If the odds of an event E are $4 : 5$, what is the probability that the event will occur?

Solution

To use Equation (6.4), we let $a = 4$ and $b = 5$, so $P(E) = \frac{4}{4+5} = \frac{4}{9}$.

Skill Enhancer 15   The odds of an event F are $3 : 6$. What values correspond to a and b in Equation (6.4)? What is $P(F)$?

Answer in Appendix E.

Example 16   It is "fifty-fifty" that the physics final will be easy. What is the probability that the exam will be easy?

Solution

$P(\text{easy test}) = \frac{50}{50+50} = \frac{1}{2}$.

Skill Enhancer 16   The odds that a certain painting in the Municipal Art Gallery was really painted by Rembrandt are given as 3 to 5. What is the probability that the portrait was really by Rembrandt?

Answer in Appendix E.

Example 17   The probability of an event E is $\frac{1}{3}$. What are the odds that the event occurs?

Solution

From Equation (6.4), we know that the odds are $a : b$, where $\frac{a}{a+b} = \frac{1}{3}$. There are many solutions to this equation. To get the simplest, let

$$a = 1 \qquad a + b = 3$$

(that is, we set numerators equal and set denominators equal). This gives $b = 2$, so the odds are $1 : 2$.

Skill Enhancer 17   For some event F, $P(F) = \frac{5}{8}$. We want the odds. What is one set of equations for a and b? What are the values for a and b? What are the odds?

Answer in Appendix E.

Example 18   The National Weather Service predicts a hailstorm tomorrow. The Service predicts that the storm will occur with probability 0.3. Express this as odds.

Solution

A given odds for an event determines a unique probability. Given the probability, the corresponding odds are not unique. If the odds that the storm occur are a to b, then the only equation that determines these two parameters is

$$\frac{a}{a+b} = 0.3$$

which simplifies to $7a = 3b$. (Make sure you see this.) There is an infinite set of pairs of numbers satisfying this equality, such as 3,7; 6,14; 15,35; 30,70; and so on. Most likely, though, the odds would be quoted as 3 to 7 that the storm will occur or 7 to 3 against. (Often, odds are quoted with the highest number first.)

Skill Enhancer 18   The probability over his career that a certain baseball player will get a hit when up at bat is $\frac{1}{4}$. In light of this information, what are the odds that he will get a hit when next up at bat?

Answer in Appendix E.

_____**Exercises 6.1**_____

Discuss the events in Exercises 1–6 in terms of probability theory. See Examples 1, 2, and 3.

1. A friend chooses one letter from the word "loquacious."

2. A comrade chooses one letter from the word "antidisestablishmentarianism."

3. I choose one letter from the word "banana."

4. You choose one letter from among the words "army," "yard," and "rays."

5. A team chooses two letters, one each from the words "apple" and "pear." The order in which the letters are chosen is important.

6. A team chooses two letters, one each from the words "apple" and "pear." The order in which the letters are chosen is irrelevant.

Specify sample spaces for outcomes of the experiments in Exercises 7–18. See Examples 1, 2, and 3.

7. Two tacks are tossed onto a table.

8. A pair of dice are rolled. One die is red, and the other is green.

9. Two books are selected randomly from a group of four books.

10. Two balls are selected from an urn containing three balls. Two of the balls are black, and the other is white.

11. A dog is about to deliver a litter of puppies. In this particular breed of dog, litters never contain more than four pups. Assume all pups are born alive and healthy.

12. A coin is tossed, and then a single die is rolled.

13. A card is drawn from a deck, and then another card is drawn.

14. A card is drawn from a deck and noted. It is returned to the deck, and then a second card is picked.

15. A _T-maze_ is a runway for laboratory animals constructed like the letter T; there is a straight runway, at the end of which is a crosspiece giving the animal the option of turning to the right or the left. Three rats are given the opportunity to run the maze.

16. If a hospital laboratory test comes back _negative_ (nothing abnormal), a patient will be discharged from the hospital. Otherwise, a second lab test will be done.

17. Four random passersby are asked to comment on which of two soft drinks they prefer.

18. In the greater Phoenix metropolitan region, there are four television channels that can be received. A viewer in this city watches one half-hour situation comedy and then another sitcom on the same or a different channel.

Two friends always do their supermarket shopping together. Here is a list of products and their prices from a recent trip.

Apples	$10.19
Bananas	0.98
Cherries	5.02
Dates	3.11
Eggplant	2.11
Figs	0.99

Describe the categories in Exercises 19–24 in terms of the elements of the sample space for this shopping trip. See Examples 1, 2, and 3.

19. Items beginning with e

20. Items not ending with s

21. Items costing less than one dollar

22. Items costing more than five dollars

23. Items costing less than two dollars and beginning with the letter c

24. Items costing less than two dollars or beginning with the letter c

Provide at least two different sample space descriptions for the experiments in Exercises 25–26.

25. Rolling a pair of dice

26. Tossing two coins

27. Imagine traveling from a house to downtown by bus by taking any of three buses to a midpoint location, and from there picking up either of two buses to the downtown section. Choose whichever bus stops first. Denote the bus lines for the first part of the trip as 1, 2, and 3, and the lines for the second part as A and B. See Example 6. Set up a sample space for this experiment.

Examine the following chart, which gives probability distributions for the simple events in the sample space, and decide which probability distributions in Exercises 28–33 are valid mathematically. If a distribution is invalid, state why. See Example 11.

	1A	1B	2A	2B	3A	3B
28.	1	1	1	1	1	1
29.	1	1	1	-1	-1	0
30.	$\frac{1}{4}$	$\frac{1}{4}$	$\frac{1}{4}$	$\frac{1}{4}$	$\frac{1}{4}$	$\frac{1}{4}$
31.	$\frac{1}{6}$	$\frac{1}{6}$	$\frac{1}{6}$	$\frac{1}{6}$	$\frac{1}{6}$	$\frac{1}{6}$
32.	$\frac{1}{7}$	$\frac{1}{7}$	$\frac{1}{7}$	$\frac{1}{7}$	$\frac{1}{7}$	$\frac{1}{7}$
33.	$\frac{1}{6}$	$\frac{1}{6}$	$\frac{1}{12}$	$\frac{1}{6}$	$\frac{1}{12}$	$\frac{1}{3}$

34. In an office building in your city, you can take one of two elevators in the lobby. You enter the first available one, and ascend to your floor, where you enter one of four offices. Set up a sample space.

Examine the probability distributions in Exercises 35–38, and comment on their validity purely from a mathematical point of view. See Example 11.

	11	12	12	14	21	22	23	24
35.	0	0	0	1	0	0	0	0
36.	0.1	0.2	0.05	0.05	0.1	0.4	0.05	0.05
37.	$\frac{1}{8}$	$\frac{1}{8}$	$\frac{1}{8}$	$\frac{1}{8}$	$\frac{1}{8}$	$\frac{1}{8}$	$\frac{1}{8}$	$\frac{1}{8}$
38.	$\frac{1}{8}$	$\frac{1}{8}$	$\frac{1}{8}$	$\frac{1}{8}$	$\frac{1}{8}$	$\frac{3}{8}$	$-\frac{1}{8}$	$\frac{1}{8}$

39. Imagine rolling a pair of dice. Set up a sample space for this experiment. The space should be able to help answer Exercises 40–51.

In Exercises 40–51, express the given events in terms of the simple events of the sample space in Exercise 39. See Examples 7 and 8.

40. The sum of the dice is 3.

41. The sum of the dice is 10.

42. The sum of the dice is 13.

43. One die shows a 1.

44. One of the dice shows an odd number.

45. The sum of the dice is odd, and both dice show an even number.

46. At least one die is odd, and the sum is an odd number strictly less than 7.

47. The sum of the dice is a perfect square.

48. The sum of the dice is a perfect cube.

49. The sum of the dice is 10 and both dice are odd.

50. The sum of the dice is odd, and both dice show an odd number of dots.

51. The sum of the dice is odd, and only one die shows an odd number.

52. A couple is planning to have three children. Express the possible family configurations as a sample space.

Express the family types in Exercises 53–60 in terms of the simple events of the sample space in Exercise 52. Then give the probability of each event, assuming that all simple events in the sample space have the same probability. See Examples 7 and 8.

53. A daughter is the eldest child.

54. Both the eldest and youngest children are daughters.

55. No boy is older than any girl.

56. No child is a girl.

57. At least two children are girls.

58. Exactly two children are girls.

59. Either all children are girls or there is at least one son among the three children.

60. A son is the eldest child and, at the same time, no daughter is younger than any son.

Dr. T. H. H. Horne (1780–1862) spent his whole life compiling a statistical description of the Bible. This table lists some of the things he discovered.

	Old Testament	New Testament	Total
Books	39	27	66
Chapters	929	260	1,189
Verses	23,214	7,959	31,173
Words	593,493	181,253	774,746
Letters	2,728,100	838,380	3,566,480

(It is not recorded which version he studied.) Note that the middle verse of the Bible is the eighth verse of Psalm 118. Use this information to compute empirical probabilities for the situations described in Exercises 61–68. See Examples 14 and 13.

61. The book of Psalms contains the midpoint of the Bible. If a book from the Old Testament is chosen at random, what is the probability of choosing this book? If the book is chosen from the entire Bible, what is the empirical probability?

62. The word "and" occurs 46,277 times in the entire Bible. What is the probability of choosing this word if a word is chosen at random?

63. A book is chosen at random from the Old Testament. Estimate the empirical probability of choosing any of the books of Genesis, Esther, or Isaiah.

64. Dr. Horne regarded Chapter 26 of Acts as the finest piece of reading in the entire Bible. If a chapter is chosen at random, what is the probability of choosing this chapter? If a chapter is chosen at random from the New Testament, what then is the probability?

65. If a verse is chosen at random from the entire Bible, what is the probability that it is the middle verse?

66. If a chapter is chosen at random from the entire Bible, what is the probability that it contains the middle verse?

67. If a book is chosen at random from the entire Bible, what is the probability that it contains the middle verse?

68. If a chapter is chosen at random from the entire Bible, what is the probability that it is Psalm 118 or Chapter 26 of Acts?

A typical year of 365 days contains 104 weekend days and (by one count) 44 holidays. In one particular year, 13 of those holidays occur on a weekend. Compute empirical probabilities for the situations described in Exercises 69–72. See Examples 14 and 13.

69. If a day is chosen at random, what is the probability that it is a weekend day?

70. A day is chosen at random. What is the probability that it is a holiday?

71. If a day of the year is chosen at random, what is the probability that it is a weekend day and a holiday?

72. What is the probability of choosing a weekend day or a holiday (or both) if one day of the year is chosen at random?

Let the notation $a{:}b$ represent odds of a to b. Convert the odds in Exercises 73–90 to probability. See Examples 15 and 16.

73. 3:4   74. 4:5

75. 5:6   76. 1:1

77. 2:2   78. a million to 1

79. 4:2   80. 0:1

81. 4:7   82. 7:4

83. 1:0   84. 7:1

85. 3:3   86. 9:10

87. 10:9   88. 1 to a million

89. 3:2   90. 2:3

Suggest odds that are equivalent to the probabilities in Exercises 91–108. See Examples 17 and 18.

91. 0.7   92. $\frac{1}{7}$

93. 0   94. $\frac{2}{3}$

95. $\frac{4}{5}$ 96. 1

97. $\frac{2}{7}$ 98. $\frac{3}{4}$

99. $\frac{3}{2}$ 100. $\frac{6}{11}$

101. $\frac{1}{9}$ 102. $\frac{8}{9}$

103. $\frac{4}{9}$ 104. $\frac{4}{4}$

105. $\frac{19}{20}$ 106. $\frac{75}{100}$

107. $\frac{5}{8}$ 108. $\frac{5}{7}$

W 109. Explain why the requirement $P(E_1) + \cdots + P(E_n) = 1$ for the n simple events of some sample space S is equivalent to the statement that some outcome of the experiment *must* occur.

110. Suppose that a sample space contains a finite number of elements. Let E_1 and E_2 be two events with *no* outcomes in common. E_1 and E_2 are not necessarily simple.
(a) Show that $P(E_1 \cup E_2) = P(E_1) + P(E_2)$.
(b) Generalize this result to apply to n events E_1, E_2, ..., E_n that have no outcomes in common.

111. Suppose that two events E and F have some events in common. We may write these common outcomes as $E \cap F$. In this problem we use the notation E' for the complement of event E, the event that E does *not* occur.
(a) Show that $E \cup F = E \cup (F \cap E')$. Use Venn diagrams.
(b) Use the results of Exercise 110 to show that $P(E \cup F) = P(E) + P(F \cap E')$.
(c) Now show that $F = (E' \cap F) \cup (E \cap F)$.
(d) Derive an expression for $P(F \cap E')$.
(e) Finally, show that, in general,

$$P(E \cup F) = P(E) + P(F) - P(E \cap F)$$

112. Use the results of Exercise 110 to show that for any event E, the probability that E does not occur is given by $P(E') = 1 - P(E)$. (*Hint*: Note that $E \cup E' = S$, the sample space, and that $E \cap E' = \emptyset$.)

Applications

Compute empirical probabilities for Exercises 113–122.

113. List all the ways I can put on any of 3 shirts and 2 pair of pants.

114. List all the ways a new car can be painted. Each car may be painted with up to 2 colors, and there are 4 colors from which to choose.

115. Student B. C. Dull will have to do one of 3 projects for his math class and one of 4 assignments for an English class. List all the ways this can happen.

116. In a restaurant, diners have the choice of among any of three entrées and two soups in a special family dinner menu. List all the dining choices for anybody who orders the special family dinner.

117. Among a group of 210 students, 45 are 6 feet tall or over. One student is randomly selected. What is the probability that this student is 6 feet tall or over?

118. Heralds Department Store samples 77 shoppers, and discovers that 11 of them are 65 years old or older. One shopper is chosen at random for a follow-up survey. With what probability will that shopper be 65 or older? With what probability will the shopper be younger than 65?

119. Over the years, out of 10,723 chicks born at the Everfresh Poultry Farm, 4,719 have been born with the right wing larger than the left. What probability would you assign to a newly hatched chick having its right wing larger?

120. Of 10,000 rural inhabitants who participate in a survey on the occurrence of lung cancer, 40 have a detected cancer, and of these 40, 8 have lung cancer.
(a) What is the probability that a person selected randomly from the 10,000 has lung cancer?
(b) What is the probability that a person selected randomly has any cancer?
(c) A researcher selects one of the individuals with cancer. What is the probability that that person has lung cancer?

121. Beneficial Electric Products Company has just begun the manufacture of light bulbs. In a preliminary batch of 1,000 bulbs, 469 were defective. In a second batch of 500 bulbs, 236 were defective. Finally, in a third trial batch of 1,500 bulbs, 699 were defective. A fourth batch has been prepared. If I select a bulb at random from this batch, what is the most reasonable empirical probability that it will be defective?

122. The East-West Trading Company has imported 500 Japanese bicycles. 325 are red; the rest are blue. Of the bikes, 200 are constructed out of a special titanium-molybdenum alloy, and the remainder are titanium-aluminum. There are 25 red, molybdenum bikes. In the shipment, the labels of all crates have been lost. Upon receipt of the shipment, the importer opens a bicycle box at random. What is the probability that the bike meets the following characteristics?
(a) The bike is red.
(b) The bike is blue.
(c) The bike is of the molybdenum alloy.
(d) It is red *and* molybdenum.
(e) The box has a small hole in it, and the importer could see that it was red when she chose it. What is the probability that it is of the stronger molybdenum alloy?

123. Fat Louie the Bookie gives odds of 3:6 that a team will win. Honest Harry, another bookie, gives odds of 4:7 for the same event. Harry therefore thinks it more probable than Louie does that the team will win. How much more probable does Harry think it is?

124. The home team won 14 of 21 games last year. Estimate the probability that the team will win its first game this year.

125. Ms. Nomer's mail last month included 16 personal letters, 38 bills, and 162 pieces of junk mail. Today, she counted 20 pieces of mail in her mailbox. Estimate how many of these are likely to be pieces of junk mail.

W **126.** Think about the properties of probability and explain

what is missing to enable us to calculate the numerical probabilities for any event.

**Calculator Exercises** _____

127. In one area of India, it has rained 6,471 days of the previous 10,946. Use this information to compute the empirical probability that it will rain today in that region. Express the result to four decimal places.

128. The probability of a certain event is 0.446. What are the odds of that event's happening?

Section 6.2 EQUALLY LIKELY PROBABILITIES

Unless we are satisfied with empirical or subjective probabilities, we have as yet no way to compute actual probability values. We will address this question in this section.

equiprobable
equally likely

One important situation arises when each simple event in the sample space is as likely to occur as any other. These events are *equally likely* or *equiprobable*. When all simple events are equiprobable, it is often easy to compute probabilities, as the following examples show.

Example 19 A ten-year-old boy dresses himself. He chooses a shirt by reaching into his dresser and pulling out a shirt at random. Only one of the five shirts is red. If he is as likely to pick any of the shirts, what is the probability of his putting on the red shirt?

Solution

The sample space contains five elements, one for the selection of each of the five shirts. Each event has the same probability

$$P(\text{any shirt}) = x \qquad \text{for some } x$$

and the sum of the probabilities must be 1.

$$P(\text{shirt 1}) + P(\text{shirt 2}) + P(\text{shirt 3}) + P(\text{shirt 4}) + P(\text{shirt 5}) = 1$$
$$x + x + x + x + x = 1$$
$$5x = 1$$
$$x = \frac{1}{5}$$

The probability is $\frac{1}{5}$.

Skill Enhancer 19 A jar holds seven jelly beans, all of which are different colors. Amos reaches in and chooses one piece of candy. Are all such events equally likely? How many such events are there? What is the probability of clasping the yellow jelly bean?

Answer in Appendix E.

urn   Many probability problems involve plucking objects out of urns. *Urn* is a special word for *vase* used almost exclusively in poetry and mathematics.

FIGURE 6.4

Example 20   Ten marbles in an urn are all white, except for one marble that is black. One marble is chosen at random. What is the probability of choosing the black marble? What is the probability if there are five marbles in the urn?

Solution

Since only one of ten marbles is black,

$$P(\text{black}) = \frac{1}{10}$$

Similarly, if there are only five marbles,

$$P(\text{black}) = \frac{1}{5}$$

Skill Enhancer 20   Widow Herman hid her will in one of the 22 books found by her bedside. If a book is chosen at random, what is the probability that it will contain the will?

Answer in Appendix E.

In the general case, there will be n simple events, each with the same probability x. We know that

$$P(E_1) + P(E_2) + \cdots + P(E_n) = \underbrace{x + x + \cdots + x}_{n \text{ terms}} = nx$$

On the other hand,

$$P(E_1) + P(E_2) + \cdots + P(E_n) = 1$$

(why?). Therefore,

$$nx = 1$$

so that

(6.5) $$x = P(\text{any simple event}) = \frac{1}{n}$$

Suppose further that some event E can occur in any of m different ways. Using set cardinalities, we may say that

$$c(E) = m \qquad c(S) = n$$

where S is the sample space. That is,

$$E = E_1 \cup E_2 \cup \ldots \cup E_m$$

for some simple events E_1 through E_m. Therefore,

(6.6) $$P(E) = P(E_1) + \cdots + P(E_m) = \underbrace{\frac{1}{n} + \cdots + \frac{1}{n}}_{m \text{ terms}} = \frac{m}{n}$$

Using set cardinalities, this becomes

$$P(E) = \frac{c(E)}{c(S)}$$

under the assumption that all simple events have the same probability.

Probability of Equally Likely Events

These equations hold true in any sample space for which all simple events are equally likely.

$$P(\text{any simple event}) = \frac{1}{n}$$

$$P(\text{any event}) = \frac{m}{n}$$

$$P(\text{any event}) = \frac{c(E)}{c(S)}$$

n is the number of simple events in the sample space.

m is the number of outcomes of any event.

$c(E)$ and $c(S)$ are the set theory cardinalities of E and S.

These observations suggest a guiding principle in the construction of a sample space.

Creating a Sample Space

Whenever possible, create a sample space such that each simple event is *equally likely*. That way, the probabilities of all simple events can be computed easily, and the way is clear to compute the numerical probabilities of any event in the sample space.

Example 21   Select the best sample space that applies to a married couple planning to have three children.

Solution

We earlier provided two solutions:

$$S_1 = \{0, 1, 2, 3\}$$

where the numbers refer to the possible number of daughters, and

$$S_2 = \{BBB, BBG, BGB, BGG, GBB, GBG, GGB, GGG\}$$

The elements of S_1 are not equally likely, whereas the elements of S_2 are. S_2 is the preferred sample space.

Skill Enhancer 21   Provide two possible sample spaces for an experiment in which a player tosses three coins. Which of these two is the preferred sample space?

Answer in Appendix E.

In any problem where the simple events of the sample space are *equally likely*, there is an important procedure for computing the probability of an arbitrary event.

Computing the Probability of an Event

Follow this procedure for computing the probability of *any* event whenever the n simple events of the sample space are equiprobable.

1. Write the event E as the union of simple events, each of whose probability is $\frac{1}{n}$.
2. Count the number of simple events in this union. Suppose this number is m.
3. The probability of the event is $m \times \frac{1}{n} = \frac{m}{n}$.

The following examples provide practice in computing probabilities by analyzing an outcome as the union of simple events, all of which are equiprobable.

Example 22   Given the family from Example 21, compute the probabilities of the following family configurations.
(a) F_1 = The eldest child is a boy.
(b) F_2 = The eldest and youngest are girls.
(c) F_3 = No boy is older than any girl.
(d) F_4 = All children are girls.
(e) F_5 = Either the eldest is a boy or the two youngest are girls.

Solution

We use S_2 from that example as a sample space. Since there are 8 equally likely simple events in this sample space, the probability of any single event—that is, the probability of any single family configuration—occurring is $\frac{1}{8}$.
(a) F_1 is the union of all simple events in which the eldest child is a boy. Specifically,

$$F_1 = \{BBB, BBG, BGB, BGG\}$$

Since the cardinality of F_1, $c(F_1)$, is 4, then $P(F_1) = 4 \times \frac{1}{8} = \frac{1}{2}$.
(b) Here, $F_2 = \{GBG, GGG\}$, so that $P(F_2) = 2 \times \frac{1}{8} = \frac{1}{4}$.
(c) $F_3 = \{BBB, GBB, GGB, GGG\}$ and $P(F_3) = \frac{1}{2}$.
(d) Since $F_4 = \{GGG\}$, $P(F_4) = \frac{1}{8}$.
(e) In this case, F_5 is the union of two subevents. We have seen that

$$\{\text{eldest is a boy}\} = \{BBB, BGB, BBG, BGG\}.$$

We know that

$$\{\text{the two youngest are girls}\} = \{BGG, GGG\}$$

FIGURE 6.5

so that

$$F_5 = \{BBB, BGB, BBG, BGG, GGG\}$$

and

$$P(F_5) = \frac{5}{8}$$

Skill Enhancer 22   Two coins are tossed. What is the probability of these events?
(a) Both coins are heads.
(b) A head is followed by a tail.
(c) Both coins are different.

Answer in Appendix E.

Example 23   Little Sam picks a number between 1 and 10. What is the probability of the following events?
(a) The number is even.
(b) The number is odd.
(c) The date today is 6/18/93. What is the probability that the number is one of the digits in today's date using this format?

Solution

(a) Since there are five even numbers, $P(\text{even}) = \frac{5}{10} = \frac{1}{2}$.
(b) There are five odd numbers in this range, so $P(\text{odd}) = \frac{5}{10} = \frac{1}{2}$.
(c) The digits in the current date are 1, 3, 6, 8, and 9. Since there are five of them, the probability is (once again) $\frac{5}{10} = \frac{1}{2}$.

Skill Enhancer 23   Redo this example if Sam picks a number between 1 and 15 inclusive.

Answer in Appendix E.

Example 24   A pair of fair dice are rolled. Compute the probabilities of the following events.
(a) The sum of the dice is 12.
(b) One die shows a 3, and the other shows a 4.
(c) The number on one die is twice the number on the other.

Solution

A useful sample space is one in which each simple event is an individual roll of the dice. Each such event is as probable as any other. There are 36 possible rolls of the dice (why?), so the probability of obtaining any roll is $\frac{1}{36}$. We will use an ordered pair notation to refer to individual rolls. Thus, $(1, 4)$ refers to a roll where the first die is 1 and the second is 4. [This roll is different from $(4, 1)$—why?]
(a) The only way the sum of the dice can be 12 is if both dice come up 6. There is only one way of achieving that, so the probability is $\frac{1}{36}$. This is a simple event.
(b) There are two rolls of the dice that produce this configuration, $(3, 4)$ *and* $(4, 3)$. These are not the same! Consequently, the probability is $2 \times \frac{1}{36} = \frac{1}{18}$.

(c) Here are all the rolls for which one die is twice the other.

$$(1, 2) \quad (2, 1)$$
$$(2, 4) \quad (4, 2)$$
$$(3, 6) \quad (6, 3)$$

The probability is $6 \times \frac{1}{36} = \frac{1}{6}$.

Skill Enhancer 24   Using the same pair of dice, compute these probabilities.
(a) The sum of the dice is 3.
(b) One die is three times the other.
(c) Neither of the dice shows 3, 4, 5, or 6.

Answer in Appendix E.

In dice problems, beginners often wonder why the event $(1, 4)$—that is, a one followed by a four—is distinct from $(4, 1)$. Although in practice the two rolls are indistinguishable, we create the sample space as if the two dice were distinct. This makes each simple event is equiprobable. It may help to regard one die as white and the other slightly off-white, so that distinction is possible.

All too often, it is impractical to enumerate the individual outcomes in a sample space. In such instances, we resort to *counting techniques* to count the simple events in the sample space and to count the individual outcomes of the event. The most common such problems involve decks of playing cards, where a deck consists of 52 pieces of light cardboard. Each card is marked in two ways—with one of four suits, and with a number from 1 to 13, where we assign the value 1 to an ace, 11 to a jack, 12 to a queen, and 13 to a king.

Example 25   A hand of poker consists of 5 cards dealt at random from a deck of 52 different playing cards. What is the probability of receiving a hand containing all 4 aces?

Solution

The sample space S consists of all distinct hands of cards—clearly, this is an impractical set to list. But we can compute the number of elements it contains, $c(S)$, because a hand of cards is just a way of choosing 5 cards from the deck when order is *not* important—

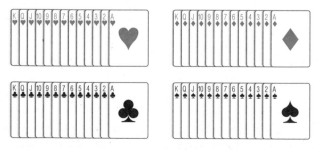

FIGURE 6.6

$C(52, 5)$. Order is unimportant because a player is free to rearrange the cards after the deal.

$$C(52, 5) = \frac{52!}{5!47!}$$

$$= \frac{52 \cdot 51 \cdot 50 \cdot 49 \cdot 48 \cdot \cancel{47} \cdot \cancel{46} \cdot \cdots}{5 \cdot 4 \cdot 3 \cdot 2 \cdot 1 \cdot \cancel{47} \cdot \cancel{46} \cdot \cdots}$$

$$= \frac{52 \cdot 51 \cdot \overset{10}{\cancel{50}} \cdot 49 \cdot \overset{2}{\cancel{48}}}{\cancel{5} \cdot \cancel{4} \cdot \cancel{3} \cdot \cancel{2} \cdot 1}$$

$$= 52 \cdot 51 \cdot 49 \cdot 20$$

Next, how many hands contain all 4 aces? Four of the cards must be aces, and the remaining (fifth) card may be any of $52 - 4 = 48$ distinct cards. The number of ways of choosing 4 aces out of 4 aces is

$$C(4, 4) = 1$$

By the Principal of Counting, the number of hands with 4 aces is the number of ways of choosing four aces multiplied by the number of ways of choosing the remaining card:

$$1 \times 48 = 48$$

The probability is

$$\frac{48}{C(52, 5)} = \frac{48}{52 \cdot 51 \cdot 49 \cdot 20} \approx 0.000018$$

Skill Enhancer 25   The Interplanetary Expeditionary Force discovers the remains of an ancient civilization on the planet Mars. One prized find is a deck of Martian cards. It consists of 40 cards, and there appear to be 6 aces in the deck. If the cards were dealt randomly (as in an earthly game of poker), what is the probability of receiving a hand containing all 6 aces? A hand of Martian poker consists of 7 cards.

Answer in Appendix E.

Example 26   An increasing number of state governments are holding *lotteries*—large-scale raffles—to raise money for worthy purposes. At one time in New York State, a person played a game in the lottery by choosing 6 different numbers in the range 1 to 44, and also choosing a seventh number, the so-called "supplementary number." The Lottery Commission chose 6 numbers randomly. A player won the *grand prize* if all 6 numbers matched the Commission's. Second prize was awarded to persons with a match of 5 of 6 numbers. For third prize, 4 numbers must have matched. Fourth prize was the only one involving the supplementary number, and went to any individual with 3 numbers plus the supplementary number matching any 4 of the winning numbers.
(a) What was the probability of winning the grand prize?
(b) What was the probability of winning second prize?

Solution

(a) The sample space consists of all possible 6-number combinations, and contains $C(44, 6) = 7,059,052$ different outcomes. There is only one way to select the 6 winning numbers, so the probability of winning is a tiny $\frac{1}{7,059,052} \approx 0.00000014$.

(b) How many ways can we choose 6 numbers so that 5 of them match the winning numbers? There are

$$C(6,5) = 6$$

ways to choose 5 out of 6 winning numbers. The remaining number can be any of $44 - 6 = 38$ choices, so the total number of ways (by the Principle of Counting) is

$$6 \times 38 = 228$$

Consequently the probability of winning second prize is

$$\frac{228}{7,059,052} \approx 0.0000323$$

Skill Enhancer 26   Which event has the greater probability—winning second prize in the New York State lottery described above, or winning first place in a state lottery in which the player needs to get 6 numbers out of 20?

Answer in Appendix E.

Example 27   An urn contains 5 blue marbles and 4 white ones. An experiment consists of picking 3 marbles randomly from the nine.
(a) What is the probability of picking out all blue marbles?
(b) What is the probability of choosing all white marbles?
(c) What is the probability of plucking 2 blue marbles and 1 white marble?

FIGURE 6.7

Solution

One sample space whose elements are all equally likely is the one whose simple events consist of all possible ways of choosing 3 marbles from the combined total of 9 in the urn. We won't list all these elements, but we know how many there are:

$$C(9,3) = \frac{9!}{3!6!} = 84$$

(a) In how many ways can a group of 3 blue marbles be chosen from a group of 5 blue marbles? Answer: $C(5,3) = \frac{5!}{2!3!} = 10$. Therefore, the probability of choosing three blue marbles out of the total of nine marbles is

$$\frac{C(5,3)}{C(9,3)} = \frac{10}{84} \approx 0.119$$

(b) In the same way, we can choose $C(4,3) = 4$ groups of 3 white marbles out of 4. The probability of choosing all white marbles out of the total is therefore

$$\frac{4}{84} = \frac{1}{21} \approx 0.048$$

(c) For the final question, imagine that the group of 3 consists of two subgroups—one with 2 blue marbles, and the other with 1 white marble. For the first subgroup, there are $C(5,2)$ ways for us to choose 2 blue marbles out of 5 blue ones. In the same way, there are $C(4,1)$ ways of choosing 1 white marble. By the Principle of Counting, there are

$C(5, 2) \cdot C(4, 1) = 10 \cdot 4 = 40$ ways of constructing a group consisting of 2 blue and 1 white marbles. Consequently, the probability of this event is

$$\frac{40}{84} \approx 0.476$$

Skill Enhancer 27   An urn contains 3 each of red and black marbles. A student reaches in and retrieves 3 from the collection. What is the probability that the following events occur?
(a) All marbles are red.
(b) All marbles are red or all are black.
(c) Exactly 2 marbles are red.
(d) At least 2 marbles are red.

Answer in Appendix E.

The Birthday Problem

There is a group of n people sitting in a class. What is the probability that at least two of them have the same birthday (the same month and day, but not necessarily the same year)? This problem intrigues generation after generation of students, as the answer for classroom-sized groups is so counterintuitive.

This problem requires considering a related question—what is the probability that *no two* people in a group of n people have the same birthday? This related problem is easier to answer, and we are able to relate the answer to this second problem to the original birthday problem.

There are many different ways that n people can have birthdays. For both problems, let the sample space S consist of all these ways. Each such way is equally likely. (We eliminate February 29 as a feasible birth date to avoid tricky complications.) How many such ways are there? The first person can have any of 365 birthdays, the second can have any of 365 birthdays, and so on. By the Principle of Counting, there are a total of

$$\underbrace{365 \cdot 365 \cdot \cdots \cdot 365}_{n \text{ factors}} = 365^n$$

such ways.

Now, how many ways can the n people *not* have common birthdays? If the first person has any of 365 birthdays, then the second may have any of $365 - 1 = 364$ birthdays—all but the first person's. The third person may have any of $365 - 2 = 363$, and so on, down to the nth person, who may have any of $365 - (n - 1)$. If K represents the total number of ways that n people *not* have birthdays in common, then the Principle of Counting tells us that

$$K = 365 \cdot 364 \cdot \cdots \cdot [365 - (n - 1)]$$

The number of ways that people *may* have birthdays in common is simply $365^n - K$. Let $P(n)$ represent the probability that among n people, *at least two* people have a common birthday. Then

$$P(n) = \frac{365^n - K}{365^n} = \frac{365^n - 365 \cdot 364 \cdot \cdots \cdot [365 - (n - 1)]}{365^n}$$

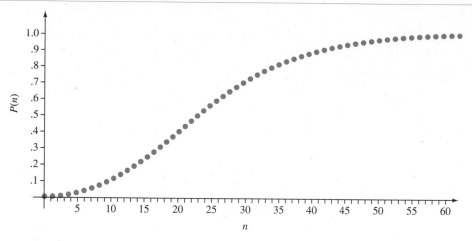

FIGURE 6.8

The function $P(n)$ graphed against n.

or

(6.7) $$P(n) = 1 - \frac{365 \cdot 364 \cdots\cdots [365 - (n-1)]}{365^n}$$

Most people would guess that $P(n)$ is a small number regardless of the value of n. It *is* small for small values of n, but rapidly increases as n increases. When there are 23 people in the class, $P(n)$ is about equal to one-half—roughly even odds. The following table gives the numerical values of the probabilities for various values of n.

n	5	10	15	20	21	22	23	24	25	30	40	50	60
$P(n)$	0.027	0.117	0.253	0.411	0.444	0.476	0.507	0.538	0.569	0.706	0.891	0.970	0.994

Figure 6.8 shows the graph of $P(n)$ versus n. For small n, the probability is small that a pair or more of people have the same birthday. As n increases, this probability increases rapidly, although it can never exceed a value of 1 (why?). For $n > 365$, $P(n) = 1$ identically. Figure 6.8 makes this point graphically. For modest values of n, say $n = 40$ or more, it is almost a sure thing that at least two people in the class have the same birthday. Test this theory in class.

_____ *Exercises 6.2* _____

Compute the probabilities of the events in Exercises 1–12. Use sample spaces in which all simple events are equally likely.

1. Guessing a person's birthday in a month containing 30 days.

2. Michiko plays 5 selections by heart on the violin. Her mother chooses 1 at random for her to play at a party.

3. A surprise quiz is given on a day of the week beginning with T. Quizzes are only given on Monday through Friday.

4. A surprise quiz is given on a day of the week beginning with T. Quizzes are given on any day of the week.

5. A letter chosen from the alphabet is a vowel. (The letter y is a vowel in this problem.)

6. A letter chosen from the alphabet is a vowel. (The letter y is *not* a vowel in this problem.)

7. Doris chooses one of her mink coats to wear. She has 7 coats, 2 of which are mink.

8. Rich N. Greedy's chauffeur chooses to drive a Rolls Royce on Monday. Of Greedy's 9 cars, 2 are Rolls Royces.

9. One of the white rats survived the scientific experiment; 3 of the 12 rats used were white.

10. One of the utility stocks watched by an investor survives bankruptcy; 4 out of 19 stocks declaring bankruptcy are utilities.

11. Albert, Bernard, and Corey ask Daisy to the prom. She chooses at random, and Corey is her choice.

12. On a quiz show, Bertram decides to spin the wheel of fortune in which there are 20 possible outcomes. The wheel lands on either "lose one turn" or "bankrupt."

A carnival game consists of flipping a fair coin followed by spinning a roulette wheel. The wheel is numbered 1 to 36 and has a ratchet mechanism so it will select a number unambiguously when the carnival proprietor spins it.

13. (a) Create a sample space for this experiment in which all elements of the space are equally likely.
    (b) Write another sample space for this game. (The events need not be equally likely in this space.)

Use the sample space of Exercise 13(a) to compute the probability of the events in Exercises 14–19.

14. The coin shows heads.

15. The roulette wheel shows a 14.

16. The roulette wheel shows a 14 and the coin shows heads.

17. A number 27 or higher shows on the wheel.

18. The coin shows tails and the wheel shows a number either higher than 20 or lower than 10.

19. The coin shows tails and the wheel shows 11 *or* the coin is heads and the roulette wheel shows a number strictly less than 10.

Three dice are rolled. Compute the probabilities of observing the events in Exercises 20–29.

20. Three 1s
21. The sum is 4.

22. The sum is 5.

23. No number is greater than 4.

24. A 1, 2, and 3 come up on the dice in any order.

25. A 1, 2, and 3 come up on the dice in this given order.

26. Exactly one die shows a 3.

27. One or more dice show a 3.

28. (a) No dice come up 3.
    (b) Can you see a relationship between this problem and the previous one?

29. All 3 dice come up showing prime numbers.

A coin is flipped 7 times.

30. Imagine, without listing the elements, a sample space in which all simple events are equiprobable. How many elements are in this sample space?

Using the sample space in Exercise 30, determine the probability of the events in Exercises 31–38.

31. The first coin comes up heads.

32. The seventh coin comes up tails.

33. The first, third, fifth, and seventh coins come up heads.

34. The first, third, fifth, and seventh coins come up the same.

35. A sequence of 5 consecutive heads appears somewhere in the 7 flips.

36. Exactly 5 heads appear in the sequence of 7 flips.

37. All flips come up the same.

38. The second and fourth flips are different.

A single card is drawn randomly from a standard deck. Compute the probabilities of the events in Exercises 39–46.

39. It will be a diamond.

40. It will be a 4.

41. It will be the 4 of diamonds.

42. It will be a picture card.

43. It will be the 6 of clubs.

44. It will be a 6 or a 4 in any suit.

45. It will be the 6 of clubs or 4 of diamonds.

46. It will be a red card or an even card. (Picture cards are neither odd nor even.)

The game of bridge is played by dealing all 52 cards in a standard deck into 4 hands, each containing 13 cards. There are $C(52, 13) \approx 6.35 \times 10^{11}$ different bridge hands possible. In answering Exercises 47–48, you may leave the answers in terms of factorials and in C and P notation.

47. What is the probability of being dealt cards of a single suit?

48. What is the probability of a hand containing a sequence of cards 2 through 10 in a single suit?

One letter is chosen at random from among the letters of the alphabet. Compute the probabilities of the events described in Exercises 49–54.

49. The letter comes before m.

50. The letter comes after a.

51. The letter is in the word "abacus."

52. The letter occurs in the word "banana."

53. The letter occurs in either "lead" or "free."

54. The letter occurs in "apple" and in "pear."

Two numbers are chosen from the whole numbers 1 through 10 inclusive. Compute the probabilities of the events in Exercises 56–64. Use a sample space in which all simple events are equally likely.

55. How many elements are in the sample space?

56. Both numbers equal 1.

57. Both numbers don't equal 1.

58. Both numbers are odd.

59. An even number is followed by an odd number.

60. An odd number is followed by an even number.

61. Both numbers are even.

62. The product of the numbers is ≥ 64.

63. The names of both numbers begin with t.

64. The names of both numbers end in a vowel.

An urn contains 2 black marbles and 8 clear ones; 2 marbles are plucked from it. Use this information to compute the probabilities of picking the arrangements described in Exercises 65–68 and 70–72.

65. 2 black marbles

66. 2 clear marbles

67. 1 black and 1 clear marble

68. No clear marbles

69. How many elements are there in the sample space?

70. At least 1 black marble

71. At least 1 clear marble

72. Zero or more clear marbles

The urn now contains 2 black marbles but only 3 clear ones. Again, 2 marbles are plucked from it. Use this information to compute the probabilities of picking the arrangements described in Exercises 74–80.

73. How many elements are there in the sample space?

74. 2 black marbles

75. 2 clear marbles

76. 1 black and 1 clear marble

77. No clear marbles

78. At least 1 black marble

79. At least 1 clear marble

80. Zero or more clear marbles

An urn contains 2 yellow marbles and 6 red ones. Choose 4 marbles at random. Use this information in answering Exercises 81–83.

81. What is the probability of picking only red marbles?

82. How probable is choosing 2 yellow and 2 red ones?

83. What is the probability of choosing a group containing more red marbles than yellow ones?

This time, the urn contains 3 yellow, 3 black, and 3 red marbles. Again, 4 marbles are chosen. Use this information in answering Exercises 84–87.

84. How probable is it to choose marbles so that none of them are black?

85. What is the probability of choosing 2 yellow, 1 black, and 1 red marble?

86. What is the probability of choosing a group containing all 3 yellow marbles plus a single red one?

87. What is the probability of choosing marbles so that 2 or less are red?

A gardener plants grass seed. Not all grass is alike. Individual grass plants are characterized as short or tall and as dark or light green. In a particular batch of seed, it is known that 75 percent grow tall and $33\frac{1}{3}$ percent are light. One seed is planted and successfully germinates. Use this information to answer Exercises 88–90.

88. Can you create a sample space for this experiment? How would you translate percentages into equally likely elements of a sample space?

89. How likely is it that the grass plant will grow tall? will grow dark?

90. The gardener would like to assess the probability that the plant will grow to be both tall *and* dark. Use any reasonable assumption you need and compute this probability.

Probability theory can be applied to problems of industrial quality control—sampling the products made in some industrial process, counting the rejects, and quantifying the quality of the production on the basis of this sample. An assembly line puts out television sets. In every finished group of 100, the foreperson expects to see 7 defective sets. Now 10 sets are sampled. For Exercises 91–93, assume that there are *exactly* 7 defective TVs in each batch of 100.

91. What is the probability of obtaining 7 defective sets in the group of 10?

92. What is the probability of finding no defective sets?

93. What is the probability of finding 1 defective set?

94. *(The New York State Lottery Revisited)* Refer again to Example 26 describing the NYS Lottery. What is the probability of winning third prize?

95. What is the probability of winning fourth prize in the New York State Lottery?

96. Refer back to Example 21. Use the sample space S_1 to compute the probabilities of the elements of the sample space S_1.

W 97. Suppose an experiment was done in which it was impossible to create a sample space containing equiprobable events. Speculate on the kind of information you would need to compute numerical probabilities for the outcomes of this experiment.

W 98. Which is more advantageous—a sample space with the least number of simple events in it, or a sample space all of whose simple events are equiprobable? Why?

Calculator Exercises _____

A bowl contains 7 black sunflower seeds and 6 white pumpkin seeds; 5 seeds are chosen at random. Use this information to solve Exercises 99–100.

99. How many ways are there of choosing the 5 seeds?

100. What is the probability of choosing 1 pumpkin seed and 4 sunflower seeds?

Section 6.3   PROBABILITY AND SET THEORY

We established a connection between probability and set theory early on by treating events as *subsets* of sample spaces. There is more to this correspondence, and we will exploit it in this section. With the help of set theory we can answer questions such as, if we know the probabilities of two events E and F, what is the probability that *either* E or F will occur?

Mutually Exclusive Events

mutually exclusive   Two events are *mutually exclusive* if they have no outcomes in common. Whether two events are mutually exclusive is important. The following example helps give practice in this.

Example 28   Suppose a sample space consists of all letters of the alphabet. Let the event E represent the letter we pick. Are the following pairs of events mutually exclusive or not? If not, why not?
(a) E is in the word "aardvark"; E is in "syzygy."
(b) E is in the word "banana"; E is in "eggplant."
Solution

Figure 6.9 helps illustrate the relationships between these events.
(a) "Aardvark" and "syzygy" have no letters in common. Therefore, these two events *are* mutually exclusive.
(b) The letters a and n are in both words. These two events, since they have events in common, are *not* mutually exclusive.

Skill Enhancer 28   Are these pairs mutually exclusive?
(a) E is in "vanity"; E is in "modesty."
(b) E is in "angelic"; E is in "devilish."

Answer in Appendix E.

Example 29   A pair of dice is rolled. Are the following pairs of events mutually exclusive?
(a) The sum of the dice is even; one die is twice the other.
(b) The sum of the dice is greater than 10; one die is twice the other.
Solution

(a) The sum is even if we roll any of these combinations:

$$
\begin{array}{ccc}
(1,1) & (1,3) & (1,5) \\
(2,2) & (2,4) & (2,6) \\
(3,1) & (3,3) & (3,5) \\
(4,2) & (4,4) & (4,6) \\
(5,1) & (5,3) & (5,5) \\
(6,2) & (6,4) & (6,6)
\end{array}
$$

If one die is twice the other, the dice will be one of the following rolls.

$$(1,2) \quad (2,4) \quad (3,6) \quad (2,1) \quad (4,2) \quad (6,3)$$

Carefully study both lists. Two rolls of the dice appear in both—$(2,4)$ and $(4,2)$. Therefore, the two events *do* have outcomes in common; they are *not* mutually exclusive.
(b) The sum of the dice can be greater than 10 only if we roll $(5,6)$, $(6,5)$, or $(6,6)$. In none of these outcomes is one die twice the other. Hence, these two given events *are* mutually exclusive.

```
                                                        E
                                              E         G
                                              G         G
              (a)                     (b)     G         P
                    ┌──────────┐              P         L
                    │ aardvark │              L         A
                    └──────────┘        B  A  N  A  N   A
                    ┌──────────┐              A         N
                    │  syzygy  │              N         T
                    └──────────┘              T
```

FIGURE 6.9

Are these events mutually exclusive?

Skill Enhancer 29   Little Louie throws a dart at a monthly calendar. His dart always lands on a number at random. Are the following pairs of events mutually exclusive or not? Why or why not?

(a) The dart lands on a day in one of the first two weeks; the dart lands on a even-numbered day of the month.

(b) The dart lands on a day that is Thursday through Saturday; the dart lands on a day of the week beginning with M.

Answer in Appendix E.

Venn diagrams are useful for making relationships between events visible. A sample space corresponds to some set universe. Since our sample spaces contain a finite number of simple events, we can represent the contents of a sample space by a pattern of dots, each of which represents one of the simple events. An event, which is a set of these dots, can be shown by circling some of the dots. Mutually exclusive events might appear as in Figure 6.10.

This sketch suggests a way of computing the probability that either of two mutually exclusive events will occur. In the language of set theory, the set relationship *or* translates into the *union* operator $\cup$, so E or F is the same as $E \cup F$. *No events in common* translates into $E \cap F = \emptyset$.

Because events E and F have an empty intersection set, Figure 6.10 suggests that the probability of either of two mutually exclusive events is the sum of the probabilities of the events, which we now show.

First, let the simple events that make up E and F be E_1 through E_p and F_1 through F_q. That is,

$$E = E_1 \cup E_2 \cup \ldots \cup E_p$$
$$F = F_1 \cup F_2 \cup \ldots \cup F_q$$

Since E and F are mutually exclusive, the simple events E_i and F_j must be distinct, so that

$$E \cup F = E_1 \cup \ldots \cup E_p \cup F_1 \cup \ldots \cup F_q$$

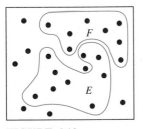

FIGURE 6.10

Venn diagrams can represent mutually exclusive events.

$$\begin{array}{ccc} 1 & 8 & 4 \\ 5 & 9 & 3 \\ & 7 & \\ 2 & & 6 \end{array}$$

FIGURE 6.11

(that is, the union of the given events E and F is the union of their component simple events) so that

$$P(E \cup F) = P(E_1) + P(E_2) + \cdots + P(F_q)$$
$$= P(E) + P(F)$$

(that is, the probability of the union is the sum of the probabilities of the component simple events; why?). Therefore, for *mutually exclusive events only*,

(6.8) $$P(E \text{ or } F) = P(E \cup F) = P(E) + P(F)$$

For mutually exclusive events, the probability that either of two events will occur is the sum of the probabilities of the two events.

The next few examples provide practice in using this rule, Equation (6.8).

Example 30   Marvin the Magician selects two single-digit whole numbers between 1 and 9 inclusive purely at random. What is the probability of selecting two even numbers or two odd numbers?

Solution

By virtue of the word "or" in the question, we turn our attention first to Equation (6.8). Before using it, we must satisfy ourselves that the two events E and F, defined by

$$E = \text{selecting two even numbers}$$

and

$$F = \text{selecting two odd numbers}$$

are mutually exclusive.

But they are. If they weren't, it would be possible to select numbers that are both even and odd, an impossibility. Therefore, we may compute

$$P(E \cup F) = P(E \text{ or } F)$$

as the sum .

$$P(E) + P(F)$$

Now, there are 81 ways of making the selection: that's nine ways to choose the first number, followed by nine ways to choose the second. Of those 81, there are 16 ways that E can occur: 22, 24, ..., 86, 88, so

$$P(E) = \frac{16}{81}$$

In the same way, since there are 25 ways that F may occur (that is, 11, 13, and so on, through to 97, 99), so

$$P(F) = \frac{25}{81}$$

Therefore,

$$P(E \cup F) = \frac{16}{81} + \frac{25}{81} = \frac{41}{81}$$

The probability is just slightly greater than 50 percent.

Skill Enhancer 30   Now let G be choosing an odd followed by an even and let H be the choice of an even followed by an odd.
(a) Are G and H mutually exclusive?
(b) What is $P(G)$?
(c) What is $P(H)$?
(d) What is the notation for event G or H occurring?
(e) What is the probability that either G or H will occur?

Answer in Appendix E.

Example 31   A coin is flipped three times in a row. What is the probability of obtaining either all heads or more tails than heads?

Solution

Since there are two possible results for each flip of a coin, there are a total of $2 \times 2 \times 2 = 8$ equally likely patterns for the coin flips. Let E be the event that all heads show, and let F represent the event that more tails than heads show up.

   E is the single simple event $\{HHH\}$, and $P(E) = \frac{1}{8}$. We have

$$F = \{TTT, TTH, THT, HTT\}$$

so that $P(F) = \frac{4}{8} = \frac{1}{2}$.

   By inspection, $E \cap F = \emptyset$, so E and F are mutually exclusive. Therefore,

(6.9) $$P(E \text{ or } F) = \frac{1}{8} + \frac{1}{2} = \frac{5}{8}$$

Skill Enhancer 31

(a) In this same experiment, what is the probability of getting all tails?
(b) Are the events "getting all heads" and "getting all tails" (in this experiment) mutually exclusive?
(c) What is the probability of getting all heads or all tails in three flips of a coin?

Answer in Appendix E.

The Addition Law

If two events are *not* mutually exclusive, what then is the probability that either event will occur? Venn diagrams help suggest the answer.

   Venn diagrams for two arbitrary events might look like Figure 6.12 on page 326. Although E and F might overlap, as in this figure, the figure does suggest how to use two nonoverlapping (and hence mutually exclusive) events to compute the probability that we want. First, choose as two mutually exclusive events E and the event represented by the right "crescent moon." This is the event of all outcomes in F and not in E, which is $E' \cap F$. Therefore,

(6.10) $$P(E \cup F) = P(E) + P(E' \cap F)$$

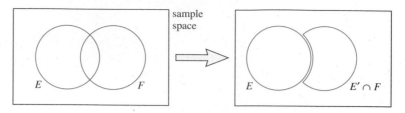

FIGURE 6.12
Two events, not necessarily mutually exclusive.

To derive an expression for $P(E' \cap F)$, take a closer look at F and the Venn diagram of Figure 6.13. Here, F itself is broken down in a special way into two component events. F consists of the events that are both in F and in E, and those that are in F and *not* in E:

$$F = (E \cap F) \cup (E' \cap F)$$

Figure 6.13 shows that these events are mutually exclusive, so that

$$P(F) = P(E \cap F) + P(E' \cap F)$$

or [solving for $P(E' \cap F)$]

(6.11) $$P(E' \cap F) = P(F) - P(E \cap F)$$

addition formula   We can substitute the results of Equation (6.11) into Equation (6.10) to obtain the *addition formula* of probability:

(6.12) $$P(E \cup F) = P(E) + P(F) - P(E \cap F)$$

This formula applies to *any* two events. If the two events are mutually exclusive, then $E \cap F = \emptyset$ and $P(E \cap F) = 0$, so the addition formula, Equation (6.12), reduces to Equation (6.8).

Example 32 A roulette wheel is marked with numbers from 1 to 36. Numbers 1 through 18 are colored red, while the remaining ones are black.
(a) What is the probability of a red, even number showing?
(b) What is the probability of a red number *or* an even number coming up?

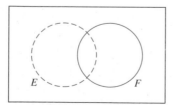
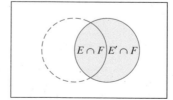

FIGURE 6.13
Deriving the addition law.

Solution

We will need to assume that all numbers are equally likely, and thus that there are 36 equally likely outcomes of a spin of the wheel. Let R represent a red number, and E be an even number.

(a) A number both red *and* even corresponds to $E \cap R$, comprising the 9 outcomes 2, 4, 6, 8, and so on, up to 18. Therefore,

$$P(E \cap R) = 9 \times \frac{1}{36} = \frac{1}{4}$$

(b) To answer this part, we need $P(E \cup R)$. First, note that both $P(E)$ and $P(R)$ equal $\frac{1}{2}$ (why?). From part (a) we know that $P(E \cap R) = \frac{1}{4}$. By the addition formula, then,

$$P(E \cup R) = P(E) + P(R) - P(E \cap R) = \frac{1}{2} + \frac{1}{2} - \frac{1}{4} = \frac{3}{4}$$

Skill Enhancer 32   A single die is colored so that the faces with 1, 2, and 3 are green; the remaining faces are red. This die is rolled.
(a) What is the probability of an even face showing? of a green face showing?
(b) What is the probability of an even, green face landing on top?
(c) What is the probability that the top face is green or even?

Answer in Appendix E.

Example 33   There are 29 high school students in a room. Of these, 19 are seniors, 11 take French, and 7 are seniors who take French. A student is selected at random.
(a) What is the probability that the student is a senior who takes French?
(b) What is the probability that the student either is a senior or takes French?

Solution

Let F stand for a student who takes French, and let S represent a senior. The Venn diagram of Figure 6.14 helps make clear the relationships between these attributes. It also demonstrates that F and S are *not* mutually exclusive, so we need to use Equation (6.12) to compute the second probability.

(a) A student who is a "senior who takes French" corresponds to $S \cap F$.

$$P(S \cap F) = \frac{7}{29}$$

(b) In preparation for using Equation (6.12), note that

$$P(S) = \frac{19}{29} \quad \text{and} \quad P(F) = \frac{11}{29}$$

In this case, using the results of (a) and Equation (6.12), we have

$$P(S \cup F) = P(S) + P(F) - P(S \cap F)$$
$$= \frac{19}{29} + \frac{11}{29} - \frac{7}{29}$$
$$= \frac{23}{29}$$

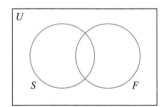

FIGURE 6.14

Some students are seniors, some take French, some are seniors who take French.

Skill Enhancer 33   In a group of 20 dogs, 15 are females and 7 have pedigrees; 5 females are pedigreed. One dog is chosen at random. Let F stand for a female dog, and let B stand for pedigree.
(a) Are F and B mutually exclusive events?
(b) What is $P(F)$? What is $P(B)$?
(c) What is $P(F \cap B)$?
(d) What is the probability that the chosen dog is pedigreed or female?

Answer in Appendix E.

Probability of Complementary Events

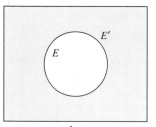

sample space

FIGURE 6.15
Set complement.

If we know the probability that an event E *will* occur, what is the probability of E', that the event will *not* occur?

First, $E \cap E' = \emptyset$ because E and E' are mutually exclusive, and $E \cup E' = S$ where S is the sample space. (Why? See Figure 6.15.) What is the probability of the event S, that is, that at least one outcome occurs?

Since S is the sample space, some outcome is certain.

$$P(S) = 1$$

On the other hand, since $S = E \cup E'$,

$$P(S) = P(E \cup E') = P(E) + P(E')$$

Finally, therefore,

$$P(E) + P(E') = 1$$

for any event E, so that

(6.13) $$P(E') = 1 - P(E)$$

Complementary Events
E' is the event that E will *not* occur.
$$P(E') = 1 - P(E)$$

Example 34   The probability of winning a raffle is $\frac{1}{1,000}$. What is the probability of losing?

Solution

If winning is the event W, then losing is W'. Therefore,

$$P(W') = 1 - \frac{1}{1,000} = \frac{999}{1,000}$$

Skill Enhancer 34   The probability that women 85 years old or older will contract a certain type of cancer is $\frac{1}{9}$. What is the probability that a woman in this category will not contract this cancer?

Answer in Appendix E.

Example 35   An experimenter rolls a pair of dice. What is the probability of *not* rolling a 7 or an 11?

Solution

Rolling a 7 and rolling an 11 are mutually exclusive events. We would achieve a 7 through any of the six combinations $(1, 6)$, $(2, 5)$, $(3, 4)$, $(4, 3)$, $(5, 2)$, or $(6, 1)$. We achieve an 11 through a $(5, 6)$ or $(6, 5)$. Therefore,

$$P(7) = 6 \times \frac{1}{36} = \frac{1}{6}$$

and

$$P(11) = 2 \times \frac{1}{36} = \frac{1}{18}$$

By the addition formula,

$$P(7 \text{ or } 11) = \frac{1}{6} + \frac{1}{18} = \frac{2}{9}$$

The probability of not rolling a 7 or an 11 is

$$1 - \frac{2}{9} = \frac{7}{9}$$

by Equation (6.13).

Skill Enhancer 35   Three coins are tossed. What is the probability of seeing three heads? What is the probability of *not* seeing three heads?

Answer in Appendix E.

We proved three main results in this section.

Further Properties of Probability

1. $P(E \cup F) = P(E) + P(F)$ for *mutually exclusive events* E and F.
2. $P(E \cup F) = P(E) + P(F) - P(E \cap F)$ for *arbitrary* events E and F.
3. $P(E') = 1 - P(E)$.

These laws apply to *any* probability distribution whatever, although in the examples we assumed only simple events that are equiprobable.

Example 36   Miracle Floor Coverings plans the introduction of two lines of new carpets. An advertising agency has prepared an expensive report declaring that a typical sales lead will choose the first type of rug with probability 0.4, the second type with probability 0.55, or either of the two with a probability of 0.6, and may even buy both with a probability of 0.5. The agency was fired. Why?

Solution

Any probability assignments must satisfy the laws of probability. Let R_1 and R_2 correspond to choosing the first or second line of rugs. The data in the report translate to

$$P(R_1) = 0.4 \qquad P(R_2) = 0.55 \qquad P(R_1 \cup R_2) = 0.6 \qquad \text{and} \qquad P(R_1 \cap R_2) = 0.5$$

The addition law demands that

$$P(R_1 \cup R_2) = P(R_1) + P(R_2) - P(R_1 \cap R_2)$$

But the above numbers do not satisfy this equation:

$$0.6 \neq 0.4 + 0.55 - 0.5 = 0.45$$

The agency's probabilities are inconsistent.

Skill Enhancer 36   The probability that tomorrow will be hot is 0.7. The probability that it will rain is 0.4. It will be hot and rainy with a probability of 0.3. (Let H stand for hot weather and R for rainy weather tomorrow.)
(a) What is the probability that it will be rainy or hot?
(b) What is the probability that it will be neither hot nor rainy?

Answer in Appendix E.

———— *Exercises 6.3* ————————————————————

Are the pairs of events in Exercises 1–12 mutually exclusive or not?

1. A letter is chosen. It is in the word "army." It is in "navy."

2. A letter is chosen. It is in "alpha." It is in "omega."

3. A letter is chosen. It is in "black." It is in "white."

4. A letter is chosen. It is in "even." It is in "odd."

5. Two letters are chosen. Both are in "heaven." Both are in "purgatory."

6. Two letters are chosen. Both are in "bread." Both are in "butter."

7. A whole number is chosen. It is even. It is odd.

8. A whole number is chosen. It is even. It is a square number.

9. A whole number is chosen. It is odd. It is prime.

10. A whole number is chosen. It is less than 7. It is divisible by 8.

11. A whole number less than 10 is chosen. It is a square. It is greater than 7.

12. A whole number less than 20 is chosen. It is greater than 16. It is a perfect square.

Compute the probabilities of the events in Exercises 13–24.

13. A letter is chosen. It is in the word "army" or it is in "navy."

14. A letter is chosen. It is in "alpha" or it is in "omega."

15. A letter is chosen. It is in "black" or it is in "white."

16. A letter is chosen. It is in "even" or it is in "odd."

17. A positive whole number is chosen. It is even or it is odd.

18. A positive whole number less than 11 is chosen. It is even or it is a square number.

19. A positive whole number less than 11 is chosen. It is odd or it is prime.

20. A positive whole number less than 11 is chosen. It is less than 7 or it is divisible by 8.

21. A positive whole number less than 10 is chosen. It is a square or it is greater than 7.

22. A positive whole number less than 20 is chosen. It is greater than 16 or it is a perfect square.

23. The probability of "bread" is 0.4. The probability of "butter" is 0.65. What is the smallest value that the probability "bread and butter" must have?

24. The probabilities of two events E and F are both equal to 0.4. The probability that either will occur is 0.7. What is the probability that both will occur? Can the two events be mutually exclusive?

In Exercises 25–30, compute the probabilities that the events do *not* occur.

25. The letter e is chosen from all letters.

26. The letter e or g is chosen from all letters.

27. A letter in the word "mathematics" is chosen from all letters.

28. A letter in the phrase "finite mathematics" is chosen from among all letters.

29. A perfect square is chosen from among all whole numbers less than 100.

30. (*Requires special thought*) A perfect square is *not* chosen from among all whole numbers less than 100.

A "triple" of dice are rolled—that is, three dice instead of two are played at once. Compute the probabilities of the events in Exercises 31–40.

31. Rolling three 1s or three 3s.

32. Rolling a pair of 2s or a triple of 2s.

33. The sum of the faces is 3 or 18.

34. The sum is greater than 15 or less than 5.

35. Rolling three 6s or a sum greater than 16.

36. Not rolling three 6s.

37. Rolling a 1, 2, and 3 or not rolling the sum of 7.

38. Rolling a 1, 2, and 3 or any sum but 18.

39. None of the dice come up 2.

40. It is not the case that either the sum is 18 or the dice show three 6s.

Using advanced sonographic techniques, Farmer Grey learns that his prize sow, the Duchess of Blandings, will shortly give birth to 4 piglets. The sonograms say nothing, though, about the sex of the piglets-to-be. Compute the probabilities of the events in Exercises 41–52, under the assumption that a piglet is as apt to be born male as female.

41. Either the eldest or the youngest is a male.

42. Both the eldest and the youngest are male.

43. Not all piglets are of the same sex.

44. All piglets are not females.

45. Neither the eldest 2 nor the youngest 2 piglets are males.

46. Either the eldest 3 piglets or the youngest 3 piglets are all females.

47. It is not the case that there are exactly 3 females born in the litter.

48. At least 3 females are born.

49. At most 1 female is born.

50. Either the eldest 2 piglets are males or it is not the case that the youngest 3 are males.

51. Either there are exactly 2 males and 2 females or the eldest and youngest piglets are both males.

52. Either the eldest is a male or the youngest 2 are males or not all piglets are females.

An urn contains 4 red marbles, 2 yellow ones, and 3 black ones; 3 marbles are picked. Compute the probabilities of the events in Exercises 53–58 using the three laws of probability covered in this section wherever possible.

53. Picking 1 marble of each color.

54. Not picking 3 black marbles.

55. Either picking 1 marble of each color or not picking 3 black marbles.

56. Either 2 of the marbles are red or 2 are yellow.

57. Either 2 of the marbles are yellow or exactly 1 is red.

58. It is not the case that either all marbles are red or all marbles are black.

59. What is the probability of choosing at random the green pair of pants out of 7 pair of pants?

60. What is the probability of selecting a penny out of a collection of change consisting of 4 nickels, 3 dimes, and 1 penny?

61. What is the probability of selecting a penny out of a collection of change consisting of 4 nickels, 2 dimes, and 2 pennies?

62. What is the probability of your choosing the whole number between 1 and 12 inclusive that I am thinking of?

63. What is the probability of drawing an ace from a deck of regulation playing cards?

64. What is the probability of drawing a red card from a deck of regulation playing cards?

65. What is the probability of drawing a picture card from a deck of regulation playing cards?

66. What is the probability of drawing a red picture card from a deck of regulation playing cards? What is the probability of drawing a red card or a picture card from the deck?

67. I have 3 each of pennies, nickels, dimes, and quarters in my pocket. If I choose a coin at random, what is the probability of choosing a nickel or a dime?

68. I have 3 each of pennies, nickels, dimes, and quarters in my pocket. If I choose two coins at random, what is the probability of choosing change worth 30 cents or more? How many elements are there in the sample space?

69. What is the probability of choosing a shirt with a collar out of 7, only 2 of which have collars?

70. What is the probability of choosing a shirt with or without a collar out of 7, only 2 of which have collars?

71. In one listening area it's possible to receive any of 11 radio stations. Of these, 4 play rock music, 3 play country, and 2 play a mix of the two *genres*. If a visitor chooses a station at random, what is the probability that it will play both rock and country music?

72. Refer again to Exercise 71. What is the probability that the selected station plays rock or country music?

73. Of 144 files on a computer, 12 are database files, 14 are spreadsheet, and 11 contain elements of both. If a file is chosen at random, what is the probability that it is one of the files containing both types of data? What is the probability of choosing a spreadsheet or database file?

74. Out of a total of 52 recent college graduates, 30 got a job in sales, 26 went on to pursue graduate studies, and 14 do both. One of these students recently won the state lottery. What is the probability that he or she was one of the ones pursuing graduate studies while working? What is the probability that this person pursued graduate studies or is working?

75. A box of books contains 33 novels, 14 cookbooks, and 12 college texts. How many elements are there in the sample space? What is the probability that one randomly selected book is a cookbook or a text? a novel or a text?

76. A certain computer magazine reviews 100 new software products each month. One month, 33 are word processing programs, 14 are database packages, and 10 are hybrid programs combining both. A reviewer selects one of the new products to examine. Are word processors and database programs mutually exclusive? What is the probability that the selected package is word processing or database?

Applications

77. *(Weather Prediction)* The local weatherperson predicts a 45 percent chance of rain tomorrow. What is the probability that it won't rain?

78. *(Political Polls)* A sample of 1,000 people indicates that 735 are satisfied with the current Presidential administration. If a person from this sample is selected at random, what is the probability that the person is *not* satisfied with the administration?

79. *(Psychology)* A particular scientific experiment involves working with a population of 150 laboratory rats, 75 of which are short-haired, 100 of which are red-eyed, and 45 of which possess both attributes.
    (a) If a rat is selected at random, what is the probability that it is both short-haired and red-eyed?
    (b) What is the probability that it is either short-haired or red-eyed?
    (c) What is the probability that it is not short-haired?

80. *(Weather Prediction)* The chance of snow tomorrow is 33 percent. The chance of overcast skies tomorrow is 45 percent. The chance that either of the two weather conditions will occur tomorrow is 50 percent. Deduce the probability that it will be both overcast and snowy tomorrow.

81. *(Manufacturing)* In a typical batch of 200 pairs of shoes made at the Lowell Shoe Mill and Factory, 25 are

defective. In that same batch, 120 pairs are brown and 80 are black.

(a) If a pair of shoes from this batch is selected at random, with what probability will it be free of defects?

(b) With what probability will a randomly selected pair of shoes be brown?

(c) Long experience shows that 15 out of 200 pairs of shoes are brown and defective. What is the probability that a randomly selected pair will be either *black* or defective?

82. *(Political Preferences)* The town of Red Gulch contains 438 registered voters, categorized by class and party affiliation. The breakdown of voters is as follows:

	Republican	Democratic
Upper-class	160	45
Lower-class	100	133

One of the voters is selected at random as he or she walks down the street.

(a) What is the probability that the person will be an upper-class Republican?

(b) What is the probability that the person will be upper-class *or* Republican?

(c) With what probability will the person be either a Democrat or not a Republican?

83. *(Retailing)* A direct-mail merchandiser (mail-order house) has ordered 10,000 catalogs for mailing. The catalogs are specially slanted to appeal to certain groups of people or geographic regions. Of these catalogs, 6,500 of the catalogs appeal to eastern tastes (as opposed to the western part of the country), 6,500 are slanted to men, and 6,000 are slanted towards Golden Seniors (citizens past the age of 65). Also, 3,500 appeal to eastern men, 3,500 to eastern seniors, 4,000 to male seniors, and 2,000 catalogs are aimed towards eastern, male, Golden Seniors. If a catalog is randomly selected, compute the probabilities that the catalog fits into the following categories.

(a) Male Golden Seniors

(b) Eastern, male Golden Seniors

(c) Eastern men below the Golden Senior age

(d) Golden Senior men not in the east

(e) Eastern men or Golden Senior men

(f) Eastern Golden Seniors or younger men not from the east

W 84. In your own words, express the difference between events $E \cap F$ and $E \cup F$.

W 85. What is the difference in the methods you might use for *calculating* $P(E \cap F)$ versus $P(E \cup F)$ for arbitrary events E, F? For the sake of simplicity, let's assume that the simple events in the sample space S are equiprobable.

Calculator Exercises

86. The probability of a certain event is 0.5748. What is the probability that the event does not occur?

87. Two events E and F have equal probability of 0.5. The product of the probabilities of the union and intersection of these events is $\frac{1}{5}$. What is $P(E \cup F)$?

Section 6.4   CONDITIONAL PROBABILITY

a priori probability   All our examples so far illustrate the computing of *a priori probability*. We computed the probabilities prior to (before) learning additional information that might affect the outcome. Sometimes, we will have access to information that, though not enough to let us determine the outcome uniquely, will alter the distribution of probability.

Example 37   Discuss qualitatively the probabilities of the following events.

(a) Two coins are tossed. What is the probability of obtaining two heads?

(b) A friend flips two coins in a darkened room. In the darkness, she can only say for sure that at least *one* of the coins is a head. Given this additional information, what now is the probability of obtaining two heads?

Solution

(a) The probability is $\frac{1}{4}$. (Why?)

(b) The probabilities are changed somewhat, because TT is now an impossible outcome. With this additional information, we expect that $P(HH)$ will be more likely than the a priori value of $\frac{1}{4}$ (but we are not yet in a position to say how much more likely).

Skill Enhancer 37　Three coins are tossed. In your opinion, which has greater probability: (1) all three coins land heads up; or (2) all three coins land face up if we observe the first two coins before we toss the third coin and, by chance, the first two coins land face up.

Answer in Appendix E.

conditional probability

posterior probability

The second part of this example is a problem in *conditional probability*, computing probability *conditional* upon additional information. Conditional probabilities are sometimes called *posterior* probabilities, because we calculate them *after* (posterior to) learning additional facts.

We are about to solve the problem of determining the probability of E given that some other event F has already taken place. We write this probability as

$$P(E|F)$$

the *probability of E given F*. Clues in any problem that suggest conditional probability are phrases such as "given," "if it is known that," "assuming that such-and-such has already occurred," and similar wordings.

Computing Conditional Probabilities

Since a conditional probability involves knowledge about two events E and F, it is tempting (but **wrong**, as we will see) to define conditional probability as the probability that *both* events occur: $P(E|F) = P(E \cap F)$. This reasonable first guess is wrong, primarily because it is *not a valid assignment of probability*.

To see why not, consider the special case that $E = F$. In this case, by our tentative definition, $P(F|F) = P(F)$. On the other hand, we expect it to always equal 1, for it is a certain thing that an event has happened if it has occurred. In general, though, $P(F)$ is not equal to 1.

This special case suggests a way to refine the definition, namely

$$P(E|F) = \frac{P(E \cap F)}{P(F)}$$

By this revised definition, the special case $P(F|F) = 1$ as we expect. This definition also satisfies all the axioms of probability (as you should show).

To see why this definition must hold, consider this example. A bowl contains five eggs, two of which are white and three of which are brown. One of the white eggs is cracked, as are two of the brown eggs, as in Figure 6.16. An egg is randomly chosen from the bowl. Let E be the event that the egg is cracked. Then,

$$P(E) = \frac{3}{5} = 0.60$$

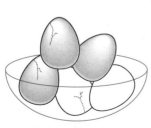

FIGURE 6.16
Five eggs in a bowl.

because a total of three out of the five eggs are cracked. But suppose we know that the egg we chose is brown. What is the probability that the egg is cracked *given that it is brown?* That is, letting the event F be the event that the egg is brown, what is $P(E|F)$?

When F has occurred, does S remain the same sample space? No! Any outcome that is not in F is no longer a permissible outcome, so knowing that F has occurred restricts the conditional outcome of E to outcomes that are in F. We might say that

$$S_{\text{new}} = F$$

That is, the permissible outcomes are only those found in F, as in Figure 6.17. Since the brown eggs are chosen equiprobably, we have that the conditional probability is the number of outcomes that are in both E and F divided by the number of outcomes in the sample space $S_{\text{new}} = F$:

$$P(E|F) = \frac{c(E \cap F)}{c(F)}$$
$$= \frac{c(E \cap F)/c(S_{\text{new}})}{c(F)/c(S_{\text{new}})}$$

Here, we divided the top and bottom by $c(S_{\text{new}})$ because we want to express $P(E|F)$ in terms of probabilities rather than cardinalities. We remember that

$$P(E) = \frac{c(E)}{c(S)}$$

for any event E in a sample space S. Therefore,

$$P(E|F) = \frac{P(E \cap F)}{P(F)}$$

For this example,

$$P(E \cap F) = P(\text{cracked, brown eggs}) = \frac{2}{5}$$
$$P(F) = \frac{3}{5}$$

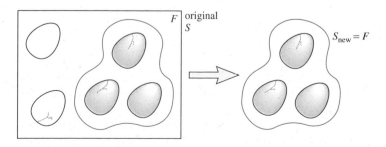

FIGURE 6.17
The new sample space.

so that

$$P(E|F) = \frac{\frac{2}{5}}{\frac{3}{5}} = \frac{2}{3}$$

Conditional Probability

For any two events E and F, we have

(6.14) $$P(E|F) = \frac{P(E \cap F)}{P(F)}$$

Values for conditional probabilities can be quite different from unconditional probabilities.

Example 38   If $P(E) = 0.3$, $P(F) = 0.5$, and $P(E \cap F) = 0.2$, what is $P(E|F)$?

Solution

From Equation (6.14), $P(E|F) = \frac{P(E \cap F)}{P(F)} = \frac{0.2}{0.5} = 0.4$. $P(E)$ plays no rôle in this computation.

Skill Enhancer 38   If $P(F) = \frac{2}{5}$ and $P(E \cap F) = \frac{1}{5}$, what is $P(E|F)$? ⁎

Answer in Appendix E.

Example 39   Two coins are flipped. What is the probability of two heads, if we know at least one of the flips is a head?

Solution

We will solve this problem twice. Our first approach will use our experience with probability, whereas the second will show how a more formal analysis leads to the same result.

First, the original sample space has four outcomes—HH, HT, TH, and TT—but the restricted sample space, after we know that one of the coins is a head, contains only the outcomes TH, HT, and HH. The event HH is only one of three equally likely outcomes, so the conditional probability is $\frac{1}{3}$.

Formally, let E be the event that there are two heads, and let F be the event that one of the coins comes up heads. Then

$$E \cap F = \{HH\}$$
$$P(E \cap F) = \frac{1}{4}$$

Also, $P(F) = \frac{3}{4}$ (why?), so by Equation (6.14)

$$P(E|F) = \frac{\frac{1}{4}}{\frac{3}{4}} = \frac{1}{3}$$

which agrees with the previous computation. Compare this with the *unconditional* probability $P(E) = \frac{1}{4}$.

Skill Enhancer 39   There are 4 socks, 2 right and 2 left, in a drawer in Missy's dresser. She picks two socks at random.
(a) What is the probability of picking a left and right sock?
(b) What is the probability of picking a left and right sock if we somehow know that one of the socks will be for the left foot?

Answer in Appendix E.

Example 40   Out of a litter of 10 newborn kittens, 4 are grey, 4 are female, and 2 are both. A neighbor adopts one of the grey kittens. What is the probability that it is female?

Solution

Let G and F stand for events that the chosen kitten is grey and female. The event $G \cap F$ stands for a grey female. Equation (6.14) becomes $P(F|G) = \frac{P(F \cap G)}{P(G)}$. Here, we have $P(G) = \frac{6}{10}$ and $P(G \cap F) = \frac{2}{10}$. Therefore, $P(F|G) = \frac{2/10}{6/10} = \frac{1}{3}$.

Skill Enhancer 40   The same neighbor changed her mind. Instead of choosing a grey cat, she chose a female from the start. Given that the kitten is female, we will want to know the probability that it is grey.
(a) What is the probability that the cat is a female?
(b) What form does Equation (6.14) take in this case?
(c) If it is known that the cat is female, what's the probability that it is grey?

Answer in Appendix E.

Example 41   A student rolls a pair of fair dice. If the sum is less than 7, what is the probability that one die will be twice another?

Solution

Let U represent the event that the sum is strictly less than 7, and let V be the event that one die is twice the other. By definition,

$$P(V|U) = P(V \cap U)/P(U)$$

The sum is less than 7 if it is among any of the rolls of the dice in the following box.

$$
\begin{array}{lllll}
(1,1) & & & & \\
(1,2) & (2,1) & & & \\
(1,3) & (2,2) & (3,1) & & \\
(1,4) & (2,3) & (3,2) & (4,1) & \\
(1,5) & (2,4) & (3,3) & (4,2) & (5,1)
\end{array}
$$

Since there are 15 outcomes shown, $P(U) = \frac{15}{36}$. $V \cap U$ consists of all outcomes such that the sum is less than 7 *and* one die is twice another. Referring to the box, the only outcomes with this relationship between the dice are $(1,2)$, $(2,1)$, $(2,4)$, and $(4,2)$, so

$P(V \cap U) = \frac{4}{36}$. Finally, therefore,

$$P(V|U) = \frac{\frac{4}{36}}{\frac{15}{36}} = \frac{4}{15}$$

Skill Enhancer 41   Natasha rolls a single die. What is the probability of rolling the 4? What is this probability if we know the die is "loaded" so that only even numbers show?

Answer in Appendix E.

Example 42   In a recent expedition to Brazil's Amazon Basin, Professor Maclaurin brought back specimens of 1,433 beetles. By some quirk, 83 percent of these were males, probably because females blended much better with the local environment and were more difficult to discover. Furthermore, many were of a single genus, call it A, and exactly 500 of these A beetles were males. A specimen is chosen at random—it is a male. With what probability will it be a member of genus A?

Solution

First, $P(\text{male}) = 0.83$. Next, since 500 beetles are male A beetles, then $P(A \cap \text{male}) = \frac{500}{1,433} \approx 0.35$. Finally

$$P(A|\text{male}) = \frac{P(A \cap \text{male})}{P(\text{male})} = \frac{0.35}{0.83} \approx 0.42$$

The required probability is about 42 percent.

Skill Enhancer 42   In 1949, of the first 500 children born in the Methodist Hospital in Brooklyn, New York, exactly 269 were boys. Of all 500 children, a total of 64 grew up left-handed, and of these 55 were boys. One day in 1992, one of these children (now middle-aged and fat) returns to the hospital. Let B and L be the events that the person was a boy and is left-handed.
(a) What is the probability that this individual was a boy?
(b) What is the probability that this person is left-handed?
(c) Give the expression (in terms of probabilities for B and L) that the person is a lefty given that this visitor is a man.
(d) What is the probability that the person is a lefty given that he is a "he"?

Answer in Appendix E.

Independent Events

independent   Two events are *independent* when the outcome of one does not affect the outcome of the other. If two events A and B are independent, then the probability of A is the same whether or not we know B has occurred. That is,

$$P(A|B) = P(A) \qquad \text{for } independent \text{ events.}$$

In the same way,

$$P(B|A) = P(B) \qquad \text{for } independent \text{ events.}$$

Appealing to Equation (6.14) for the definition of conditional probability and noting that $P(A \cap B) = P(B \cap A)$, we can derive the following equation between the probabilities of independent events.

Independent Events

$$P(A \cap B) = P(A) \cdot P(B) \qquad\qquad (6.15)$$

where

A, B are any two independent events.

The following examples show how knowledge of independence between events can help solve problems.

Example 43   Two events A and B have probabilities $P(A) = 0.3$ and $P(B) = 0.7$. It is known that $P(A \cap B) = 0.21$. Are A and B independent?

Solution

Since $0.3 \times 0.7 = 0.21$, the probabilities obey Equation (6.15). We conclude that A and B are independent.

Skill Enhancer 43   If C and D are two events such that $P(C) = 0.6$, $P(D) = 0.7$, and $P(C \cap D) = 0.41$, are C and D independent? Why or why not?

Answer in Appendix E.

Example 44   Out of 1,000 typical commuters in New York City, 550 take the IND subway line, 320 take the BMT line, and 150 take both. If a commuter is chosen at random, are the acts of taking these two subway lines independent events?

Solution

From the data, $P(\text{IND}) = 0.55$, $P(\text{BMT}) = 0.32$, and $P(\text{IND} \cap \text{BMT}) = 0.15$. But Equation (6.15) asserts independence only if $P(\text{IND} \cap \text{BMT}) = P(\text{IND}) \cdot P(\text{BMT}) = 0.55 \cdot 0.32 = 0.176 \neq 0.15$. These two events are not independent.

Skill Enhancer 44   In this example, how many more commuters would need to take both subway lines to make the events independent?

Answer in Appendix E.

Example 45   The St. Andrews Golf Club is testing new grass varieties for its putting greens. In one variety, grass plants, which are characterized as being short or tall versus dark or light green, will produce tall plants 75 percent of the time and light plants $33\frac{1}{3}$ percent of the time. A seed is planted, and successfully germinates. With what probability will it grow to be both tall *and* dark?

Solution

First, note that since a grass plant is either dark or light, then $P(\text{light}) = \frac{1}{3}$ implies that $P(\text{dark}) = \frac{2}{3}$.

Next, assume that the qualities of color and height are independent. Therefore,

$$P(\text{tall} \cap \text{dark}) = P(\text{tall}) \cdot P(\text{dark})$$
$$= \frac{3}{4} \times \frac{2}{3} = \frac{1}{2}$$

Skill Enhancer 45   The next year, a different grass mixture is tested. This time, tall plants occur two-thirds of the time, and light plants appear one-fourth of the time. Now with what probability will a single grass plant grow to be both tall *and* dark?

Answer in Appendix E.

Example 46   The management of a large investment banking firm is about to evaluate the recent performance of a trainee. She has unearthed 36 leads, of which 12 turned into deals. As of the evaluation, 25 percent of these deals are profitable. Throughout the history of this firm, leads become deals in 20 percent of cases, and deals possess the potential to become profitable deals 60 percent of the time. (Not every deal generates profit for the firm.) Assuming these are independent events, is her performance consistent with the firm's historical averages?

Solution

Let D be the event that a lead turns into a deal, and let P_D be the occurrence of a profitable deal. Within the firm, then, $P(D) = 0.2$ and $P(P_D) = 0.6$. Since D and P_D are independent, then $P(D \cap P_D) = 0.2 \times 0.6 = 0.12$. The trainee made a total of 12 deals, so for her

$$P(D) = \frac{12}{36} = \frac{1}{3}$$

Since her performance indicates that $P(P_D|D) = 0.25$, then $P(D \cap P_D) = 0.33 \times 0.25 \approx 0.08 = 8$ percent. Her performance, strictly based on these facts, is below the historical record.

Skill Enhancer 46   An additional five of her deals became profitable.
(a) How many profitable deals did she make in all, and now what is $P(P_D|D)$ for her?
(b) What is $P(D \cap P_D)$ for her and now how does she compare with historical performance figures of the firm?

Answer in Appendix E.

Mutually Exclusive Events versus Independent Events

Students commonly suffer confusion between the concepts of *mutually exclusive* and *independent events*. The source of the confusion probably lies thinking of both concepts as pertaining to events "that have nothing to do with each other." The following two discussions compare these two concepts.

FIGURE 6.18

Dependent vs. Mutually Exclusive

This example concerns a new radio with a spin-dial station selector. Suppose someone buys one such radio in an area where there are 10 distinct radio stations. If he twirls the station selection knob, he is equally likely to select any of these 10 stations. When the radio is brand-new, the station he selects *always* corresponds to the station he hears. As the radio ages, the selector mechanism starts slipping. After a few years, the owner can point the selector to some station, but end up hearing a different station because of this slippage.

Formally, let S_i be the event that he selects station i, and let H_i be the event that he hears station i. When the radio is new,

$$P(H_i|S_i) = 1$$

that is, the purchaser hears the station he selects. The dependence between the station heard and the station selected is as strong as possible. As the years pass, because of the mechanical deterioration of the selector device, the conditional probabilities $P(H_i|S_i)$ become less than 1. We see that the dependence between selecting and hearing a station becomes weaker, and we can imagine a time in the distant future when someone selects a station and winds up listening to one of the 10 stations completely at random. At such a point in time, the events S_i and H_i have become completely independent. On the other hand, they are by no means mutually exclusive, because it is always possible to select the ith station and actually end up listening to it.

Mutually Exclusive vs. Independent

The twins Castor and Pollux are enrolled in the same history course. The sole requirement for the course is a term paper written on any of n topics. Castor will select his topic randomly, as will Pollux, with the proviso that if Pollux chooses the same topic as Castor, he will choose again, and keep choosing as long as necessary to ensure that he has a different topic.

By the nature of this selection process, the two topics are mutually exclusive. Nevertheless, they *are* dependent. It is easiest to see this dependence when $n = 2$, for Pollux's term paper topic is completely determined by Castor's preliminary choice. As n increases, this dependence weakens, but it never completely disappears. Although these two events are never independent, they are always mutually exclusive.

_____ *Exercises 6.4* _____

In Exercises 1–2, simply set up the problems according to the questions asked. You need not solve the problems.

1. A pair of dice is rolled. We might wish to know the probability that double sixes show, given that the sum of the dice is greater than 7.
   (a) Describe appropriate events E and F for this problem.
   (b) What, *in words*, is the event $E \cap F$?

2. A roulette wheel is spun. Eventually, we will want to compute the probability that the number is odd, given that the number is red.
   (a) Describe appropriate events E and F for this problem.
   (b) What, *in words*, is the event $E \cap F$?

Given the numerical probabilities in Exercises 3–14, compute the indicated probabilities.

3. $P(E \cap F) = 0.3$, $P(F) = 0.4$, $P(E|F) = ?$

4. $P(A \cap B) = 0.1$, $P(B) = 0.1$, $P(A|B) = ?$

5. $P(X \cap Y) = \frac{1}{4}$, $P(Y) = \frac{3}{8}$, $P(X|Y) = ?$

6. $P(E \cap F) = \frac{1}{3}$, $P(F) = \frac{2}{3}$, $P(E|F) = ?$

7. $P(S \cap T) = 0.25$, $P(T) = 0.35$, $P(S|T) = ?$

8. $P(U \cap V) = \frac{5}{7}$, $P(V) = \frac{5}{6}$, $P(U|V) = ?$

9. $P(E|F) = 0.5$, $P(F) = 0.5$, $P(E \cap F) = ?$

10. $P(U|V) = 0.3$, $P(V) = 0.6$, $P(U \cap V) = ?$

11. $P(Q|R) = \frac{4}{5}$, $P(R) = 0.5$, $P(Q \cap R) = ?$

12. $P(E \cap F) = \frac{1}{2}$, $P(E|F) = \frac{3}{4}$, $P(F) = ?$

13. $P(X \cap Y) = 0.9$, $P(X|Y) = 0.95$, $P(Y) = ?$

14. $P(A \cap B) = \frac{2}{3}$, $P(A|B) = \frac{3}{4}$, $P(B) = ?$

For Exercises 15–20, imagine that someone has rolled a pair of dice. Compute the probabilities requested.

15. The sum is 2 given that one number is a 1.

16. The sum is 2 given that one number is a 3.

17. No number is odd given that the sum is 4.

18. The sum is 3 given that one number is even or odd.

19. The sum is 8 given that the dice are equal.

20. The dice are equal given that the sum is 8.

Three fair coins are tossed. Use this information in Exercises 21–26 to compute the probabilities of the events.

21. Three heads show given that the first two are heads.

22. Three heads show given two tails.

23. Getting one or more heads given that the first toss is a tail.

24. Observing one or more heads when the first toss is a head.

25. Observing no tails given that the first toss is a tail.

26. Seeing no tails if the first toss is heads.

A roulette wheel is marked with numbers 1 through 36. When spun, a special marker selects a single number with equal probability. Suppose all numbers from 1 through 18 inclusive are colored red, and the numbers 19 through 36 are black. The croupier spins the wheel; compute the probabilities of the events in Exercises 27–34.

27. An odd number given that the number is red.

28. A red number given that the number is odd.

29. A black number given that the number is prime.

30. An odd number given that the number is black.

31. A number between 10 and 20 inclusive given that it is both odd and black.

32. A perfect square given that the number is red.

33. A perfect square given that the number is black.

34. The number is odd given that it is a perfect square.

A soft-drink company is about to introduce a revolutionary new soft drink, for which there are five candidates for names. The five names are printed on slips of paper, which will be placed in a hat. One slip will be picked blindly. Compute the probabilities of the events in Exercises 35–40. The possible names are AZAZ, ZZAZZ, ZAZAZ, YAZAZ, and ZAZAY.

35. (a) YAZAZ
    (b) YAZAZ given that the choice contains a Y.

36. The name contains 5 letters given that it contains 2 A's.

37. The name contains exactly 2 A's given that it is five letters long.

38. AZAZ given that the name contains no Y's.

39. AZAZ given that the name contains 4 letters.

40. ZAZAZ given that the name does not contain two Z's in a row.

A pair of fair dice are rolled. One die is red, and the other is green. Decide whether the pairs of events in Exercises 41–46 are independent.

41. The sum is less than 10; both dice show a 3.

42. A red 5; a green 2.

43. A red 5; the green shows a greater number than the red die.

44. The sum is odd; the sum is prime.

45. The red die is odd; the green die is prime.

46. The sum of the dice is 11; the sum is odd.

Given the probabilities in Exercises 47–52, are the events independent?

47. $P(A) = 0.1$, $P(B) = 0.2$, and $P(A \cap B) = 0.3$.

48. $P(A) = 0.1$, $P(B) = 0.2$, and $P(A \cap B) = 0.03$.

49. $P(A) = 0.5$, $P(B) = 0.5$, and $P(A \cap B) = 0.5$.

50. $P(A) = 0.3$, $P(B) = 0.4$, and $P(A \cap B) = 0.12$.

51. $P(A) = 0.7$, $P(B) = 0.5$, and $P(A \cap B) = 0.35$.

52. $P(A) = 0.8$, $P(B) = 0.9$, and $P(A \cap B) = 1$.

In Exercises 53–64, assume that A and B are independent. Use the given information to deduce the indicated probability.

53. $P(A) = 0.1$, $P(B) = 0.7$, $P(A \cap B) = ?$

54. $P(A) = \frac{2}{3}$, $P(B) = \frac{1}{6}$, $P(A \cap B) = ?$

55. $P(U) = 0.5$, $P(V) = 0.5$, $P(U \cap V) = ?$

56. $P(X) = 0.3$, $P(Y) = 0.2$, $P(U \cap V) = ?$

57. $P(E) = 0.2$, $P(F) = 0.2$, $P(E|F) = ?$

58. $P(S) = \frac{1}{4}$, $P(T) = \frac{1}{3}$, $P(T|S) = ?$

59. $P(L) = 0.4$, $P(M|L) = 0.3$, $P(M) = ?$

60. $P(P|Q) = 0.4$, $P(Q|P) = 0.2$, $P(Q \cap P) = ?$

61. $P(A|B) = \frac{3}{4}$, $P(B|A) = \frac{1}{3}$, $P(A \cap B) = ?$

62. $P(C) = 0.5$, $P(C|D) = 0.5$, $P(D) = ?$

63. $P(G) = 0.7$, $P(H|G) = 0.44$, $P(H) = ?$

64. $P(E \cap F) = 0.5$, $P(E) = 0.75$, $P(F) = ?$

65. The text stated that for independent events, the definitions of conditional probability imply that $P(A \cap B) = P(A) \cdot P(B)$. Show this.

66. If E and F are two mutually exclusive events, then $E \cap F$ must be empty. Justify this statement, and use it to show that two mutually exclusive events can never be independent.

67. Refer again to the discussion of mutually exclusive versus independent events at the end of this section. Let C_1 be the event that Castor selects the first topic on the list, and let P_2 be the event that Pollux selects the second. Compute $P(P_2)$, $P(C_1)$, and $P(C_1 \cap P_2)$ as functions of n the number of term paper topics and use this calculation to show that P_2 and C_1 can never be independent.

Applications

(Horticulture) Compute probabilities for the events in Exercises 68–71 using information provided in Example 45 in the text.

68. The grass plant grows both tall and light.

69. It grows short.

70. It grows short and dark.

71. Not all grass seed will germinate. This grass seed is rated at an 80 percent germination rate. Suppose now that the gardener plants a seed and waters it, but before waiting to see if it germinates, needs to know the probability now that it grows to be both tall and dark. What is this probability?

72. *(Strategy for One-Armed Bandits)* As an experiment, imagine flipping a perfectly fair coin many times, and tabulating the results of the throws.
   (a) Is it possible, even if unlikely, that the coin might land "heads" 500 times in a row? If so, compute the probability of achieving 500 heads in 500 tosses of the coin.
   (b) If we know that the coin is fair, and that it has just gone through a run of 500 heads in a row, compute the probability of obtaining a *tails* in the next toss of the coin.
   (c) Slot machines in gambling casinos are often called one-armed bandits (for good reason). A popular strategy consists of standing on the sidelines to observe which machines have long runs of bad luck for the gamblers. If these machines become free before the run of no winners ends, popular feeling anticipates that the odds of winning are now better than average, to make up for the worse than average play that the machine just experienced. Therefore, it should now be possible to rack up impressive winnings from these machines. Comment on this strategy in the light of your results from the previous parts of this example.

73. *(Corporate Structure)* In a certain company, 3 assistant vice presidents (AVPs), whose identity is so far a secret, will be considered for promotion to full vice presidential status. Assume that all candidates are equally likely to be chosen. There are 25 AVPs at the moment.
   (a) What is the probability that any one AVP will be promoted?
   (b) With what probability will one of the favored candidates be promoted?
   (c) One AVP somehow determines that his chance to be one of the 3 candidates is 0.75. In light of this information, what is his chance of receiving the promotion?

74. *(Taxes)* In the vast majority of states, taxpayers owe tax to both federal and state governments. Suppose that the IRS, a federal agency, audits 5 percent of returns submitted to it, and that the state tax collection agency audits 3 percent of its returns. These two events are independent.

(a) What then is the probability that you will be audited by *either* of the agencies?

(b) What are your chances of being audited by *both* agencies?

75. *(Pop Culture: Elvis)* According to the Elvis Institute, 45 percent of all Elvis sightings are made west of the Mississippi, and 63 percent of the sightings are made after 2 P.M. Twenty-one percent are both. A report of yet another sighting is received by the Institute. What is the probability of spotting Elvis east of the Mississippi before 2 P.M.?

(Public Opinion) The *New Haven Register*, an important newspaper in Connecticut, recently conducted a survey among 850 of its adult readers to determine their opinion on armed retaliation against countries that have sponsored terrorist attacks against American civilians or bases anywhere in the world. All respondents felt strongly enough about the issue to answer "yes" or "no"; no one took a neutral stance. Of the 850, 400 were men and 450 were women. There were 290 men and 390 women who answered yes. One of the members of the survey pool is chosen at random. Compute the probabilities of the events in Exercises 76–77.

76. The person is a woman, given that the individual was not in favor of armed retaliation.

77. The respondent favors armed retaliation, given that the respondent is a man.

W 78. *(Economics and Cars)* Studies show that one-fourth of all new car buyers choose bright red cars. If R is the event that a red car is chosen, and C is the event that a car is bought, does the statistic we have quoted give the value of $P(R)$, $P(C)$, or $P(R|C)$? Why?

W 79. *(Forestry)* A forestry ranger examines a load of diseased trees cut from two different forests. He selects at random a diseased tree trunk from the first forest. We know only the proportions of sick trees in each forest. Would we examine this event by computing $P(D|F_1)$ or $P(D \cap F_1)$?

Calculator Exercises

80. A library contains 97,324 volumes containing less than 200 pages of text. Of its collection of books written in English, 91,277 books are shorter than 200 pages. A book chosen at random is found to be less than 200 pages long. Given that fact, what is the probability that the book is written in English? There are a total of 100,372 books in the library.

81. As of the most recent census, there were exactly 97,209 residents in the Town of Huntington. Of these, 43,219 were women or girls, and 49,111 were too young to vote. Exactly 30,529 females were too young to vote. A Huntington resident too young to vote stops by the town councilman's office. With what probability will that person be female?

Section 6.5   BAYES' THEOREM

Consider a laboratory that is studying Dutch elm disease, an incurable and untreatable disease that kills elm trees and related species. Sapling tree trunks from two experimental forests have been shipped to the lab; 2 percent of the trees from the first forest are diseased, as are 5 percent from the second. A shipment of 300 trunks consists of 100 trees from the first forest and 200 from the second. We have two questions.

1. One tree is pulled at random from this lot for testing. What is the probability that it will be diseased?

2. Suppose it is found to be diseased. What is the probability that it comes from the first forest?

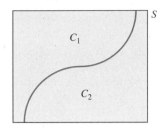

FIGURE 6.19
C_1 **and** C_2 **span the sample space.**

We can solve problems such as this by putting together in the proper way all the concepts we have developed in this chapter. We begin as always by creating a sample space. There are four outcomes in this space—a tree can be healthy or sick from either of two forests—but these outcomes are *not* equally likely. The likelihood of these outcomes is provided by the given facts.

$$P(F_1) = \frac{100}{300} \qquad P(F_2) = \frac{200}{300}$$

$$P(D|F_1) = \frac{2}{100} \qquad P(D|F_2) = \frac{5}{100}$$

The events F_1 and F_2 refer to trees taken from the first or second forest, and D to the discovery of a diseased tree. The problem asks for two quantities:

- An *unconditional* probability $P(D)$
- The *conditional* probability $P(F_1|D)$

The conditional probability $P(F_1|D)$ that a tree is from the first forest given that it is diseased is a new kind of conditional probability in which the arguments within the parentheses are "inverted." That is, we know $P(D|F_1)$; we need $P(F_1|D)$.

In problems such as this, we first split the sample space S into two (or more, if need be) subsets C_1 and C_2 that are mutually exclusive and yet whose union is identical with *span* S. Such sets are said to *span* the sample space, and any such collection of spanning *partition* subsets forms a *partition* of the sample space. For this example, C_1 consists of the outcomes for the first forest and C_2 contains the outcomes for the second forest. A Venn diagram for S appears in Figure 6.19.

Since all events in S occur in either C_1 or C_2 but not in both, all the outcomes in some event such as D are found in either C_1 or C_2. That is, all sick trees come from one of the two forests. It is always possible to split D into two mutually exclusive subsets, $D \cap C_1$ and $D \cap C_2$, such that

$$D = (D \cap C_1) \cup (D \cap C_2) \tag{6.16}$$

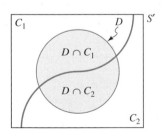

FIGURE 6.20
Any event D **can be decomposed into two mutually exclusive events** $D \cap C_1$ **and** $D \cap C_2$ **where** C_1 **and** C_2 **span the sample space.**

Figure 6.20 helps visualize this result. Because the events on the right side of Equation (6.16) are mutually exclusive, we may further conclude that

$$P(D) = P(D \cap C_1) + P(D \cap C_2) \tag{6.17}$$

Equation (6.17) appears daunting until we realize that each term on the right appears in a definition for conditional probability. For, since

$$P(D|C_1) = \frac{P(D \cap C_1)}{P(C_1)}$$

we know that

$$P(D \cap C_1) = P(C_1) \cdot P(D|C_1)$$

Similarly

$$P(D \cap C_2) = P(C_2) \cdot P(D|C_2)$$

Therefore, Equation (6.17) becomes

$$P(D) = P(C_1)P(D|C_1) + P(C_2)P(D|C_2) \tag{6.18}$$

in which all quantities on the right are known from the information supplied with the problem. The more general formula

$$P(D) = P(C_1)P(D|C_1) + P(C_2)P(D|C_2) + \cdots + P(C_n)P(D|C_n) \qquad (6.19)$$

works whenever events C_1 through C_n span the sample space. Although this formula appears overwhelming, it follows a simple pattern:

1. The right-hand side is a sum of terms.
2. Each term contains two factors. For an arbitrary, ith term, the first factor is $P(C_i)$, and the second is $P(D|C_i)$.

We can use Equation (6.18) to answer the first question of this section. The data given in this example together with Equation (6.18) imply that

$$P(D) = \frac{100}{300} \cdot \frac{2}{100} + \frac{200}{300} \cdot \frac{5}{100}$$
$$= \frac{1}{25}$$

or 4 percent. (We need to do additional work before answering the second question.)

Example 47   Suppose S_1 and S_2 are events that span a sample space S such that $P(S_1) = 0.4$ and $P(S_2) = 0.6$. Furthermore, suppose there is an event B such that $P(B|S_1) = 0.3$ and $P(B|S_2) = 0.5$. What is $P(B)$?

Solution

From Equation (6.18), we have

$$P(B) = P(S_1)P(B|S_1) + P(S_2)P(B|S_2)$$
$$= (0.4)(0.3) + (0.6)(0.5) = 0.12 + 0.3$$
$$= 0.42$$

Skill Enhancer 47 S_1 and S_2 are events that span a sample space S. Here, $P(S_1) = 0.7$ and $P(S_2) = 0.3$. Furthermore, suppose there is an event B such that $P(B|S_1) = 0.2$ and $P(B|S_2) = 0.9$. What is $P(B)$?

Answer in Appendix E.

Bayes' Theorem: Setting Up Problems

When solving a problem, first decide which parts of the problem are candidates for the spanning subsets C_i. *Use these three tips:*

1. Since the C_i's are mutually exclusive and completely span S, the sum of the probabilities of the events that we identify as the C_i's must add up to 1.
2. Check that the events called C_i are mutually exclusive.
3. If the problem seeks an *unconditional* probability of an event, this event should *not* be one of the partition subsets.

In the discussion of Dutch elm disease, it is certain that a tree comes from the first or the second forest, but not from both. We sought the unconditional probability that a tree is sick; the events *well* or *diseased* would have made an unsuitable choice for events to span the sample space.

FIGURE 6.21
Urns on a lazy Susan

Example 48   Suppose one urn contains 3 marbles, only 1 of which is black, while a second contains 9 marbles, exactly 2 of which are black. Place the two urns on a lazy Susan (a small, rotating platform) as in Figure 6.21 and spin it. Someone draws a single marble randomly from the urn that lands in front of that person. What is the probability that the marble is black?

Solution

How do we decide that Equation (6.18) is appropriate for this problem? The problem seeks an *unconditional* probability, in this case that a marble taken from the urn is black. The *given* information about colored marbles is conditional—conditional on knowing which urn contains the black marble. This pattern—given conditional information to find unconditional information—is typical of these types of problems.

Which events span the sample space? Likely choices are two:

1. That a marble comes from urn 1 or urn 2; or

2. That a marble is black or not.

Since the problem asks for a probability that the marble was black, choosing as events U_1 and U_2 that the marble comes from urn 1 or 2 seems a better choice for mutually exclusive events to span the sample space.

Now, it remains to set the event B as the outcome that a black marble is chosen, use the facts of the problem, and substitute them in Equation (6.18).

Set the problem up and solve. The problem implies that each urn is as likely to land in front of you when the turntable is spun. Therefore,

$$P(U_1) = \tfrac{1}{2} \qquad P(U_2) = \tfrac{1}{2}$$
$$P(B|U_1) = \tfrac{1}{3} \qquad P(B|U_2) = \tfrac{2}{9}$$

Use the proper form of Equation (6.18). For this problem, this equation becomes

$$P(B) = P(U_1)P(B|U_1) + P(U_2)P(B|U_2)$$
$$= \frac{1}{2} \cdot \frac{1}{3} + \frac{1}{2} \cdot \frac{2}{9}$$
$$= \frac{5}{18}$$

Skill Enhancer 48   Redo the problem, subject to these new facts. All 3 marbles in the first urn are black. The second urn now contains 3 black marbles out of 9. When a marble is drawn (under the same conditions as above), what now is the probability that it is black?

Answer in Appendix E.

Bayes' Theorem

Bayes' Theorem

Although Equation (6.18) is a powerful result in its own right, it has never been honored with a name of its own. The even more powerful *Bayes' Theorem*, of which Equation (6.18) is a step in the derivation, has always overshadowed it. Given $P(C_i)$ and $P(D|C_i)$, Bayes' Theorem provides information about $P(C_i|D)$.

Historical Note

Thomas Bayes   *Thomas Bayes, an eighteenth-century British clergyman (1702–1761), derived his results on "inverse probabilities" in a different way than we do in this text. His work on the "doctrine of chances"—probability—was merely a hobby. Published after his death, it attracted little attention. It fortunately came to the notice of the eminent mathematician Pierre Simon Laplace (1749–1827), who gave an account of it in his "Essai philosophique sur les probabilités," a still readable introduction to probability theory. Application of Bayes' work to statistics in our twentieth century has inspired the Bayesian school of statistical analysis.*

We have already set the stage for the derivation of Bayes' result. By definition,

$$P(C_1|D) = \frac{P(C_1 \cap D)}{P(D)} = \frac{P(D \cap C_1)}{P(D)}$$

As above, note that $P(D \cap C_1) = P(C_1) \cdot P(D|C_1)$, so that

$$P(C_1|D) = \frac{P(C_1)P(D|C_1)}{P(D)} \tag{6.20}$$

where $P(D)$ is given by Equation (6.18). Equation (6.20) forms the content to *Bayes' Theorem,* and similar results follow for C_2 and any other C_i into which the sample space is partitioned.

Bayes' Theorem

$$P(C_i|D) = \frac{P(C_i)P(D|C_i)}{P(D)}$$

$$P(D) = P(C_1)P(D|C_1) + \cdots + P(C_n)P(D|C_n)$$

where

$C_1, \ldots, C_n$ are mutually exclusive events spanning the sample space, and

D is any possible event (so that $P(D) \neq 0$).

An important special case is $n = 2$. Then the preceding equations simplify to

$$P(C_1|D) = \frac{P(C_1)P(D|C_1)}{P(D)}$$

$$P(C_2|D) = \frac{P(C_2)P(D|C_2)}{P(D)}$$

$$P(D) = P(C_1)P(D|C_1) + P(C_2)P(D|C_2)$$

We will use these special formulas to finally answer the questions posed at the outset of this section.

We still have to compute $P(C_1|D)$, the probability that a diseased tree comes from the first forest. We do this via Bayes' Theorem and quantities that we know or have computed. We found that $P(D) = \frac{1}{25}$, and the problem told us that $P(C_1) = \frac{1}{3}$ and that $P(D|C_1) = \frac{2}{100}$. Therefore, by Equation (6.20),

$$P(C_1|D) = \frac{\frac{1}{3} \times \frac{2}{100}}{\frac{1}{25}}$$

$$= \frac{1}{6}$$

Example 49   In Example 47, we considered the situation that S_1 and S_2 are events that span a sample space S such that $P(S_1) = 0.4$ and $P(S_2) = 0.6$. There is an event B such that $P(B|S_1) = 0.3$ and $P(B|S_2) = 0.5$. What is $P(S_1|B)$? What is $P(S_2|B)$?

Solution

In Example 47 we saw that $P(B) = 0.42$. By applying Equation (6.20), we learn that

$$P(S_1|B) = \frac{P(S_1)P(B|S_1)}{P(B)}$$

$$= \frac{0.4(0.3)}{0.42} = \frac{4}{14} = \frac{2}{7}$$

In the same way,

$$P(S_2|B) = \frac{P(S_2)P(B|S_2)}{P(B)}$$

$$= \frac{0.6(0.5)}{0.42} = \frac{5}{7}$$

Skill Enhancer 49 S_1 and S_2 are events that span a sample space S. Here, $P(S_1) = 0.7$ and $P(S_2) = 0.3$. Furthermore, suppose there is an event B such that $P(B|S_1) = 0.2$ and $P(B|S_2) = 0.9$. We have earlier seen that $P(B) = 0.41$. What is $P(S_1|B)$? What is $P(S_2|B)$?

Answer in Appendix E.

Example 50   A consignment of canned fruit from Taiwan has had its labels washed off in a sprinkler malfunction in the warehouse. This is part of a shipment of 3,000 cans,

FIGURE 6.22

1,200 of which contain pears whereas the remainder hold apricots. Because of imperfections in the canning process, natural chemicals in the pears cause metal corrosion in 1 percent of the cans, and certain agents in the apricots cause the same type of damage in 7 percent of cans. The line manager notices a corroded can and takes it back to his office. What is the likelihood that it contains pears? that it contains apricots?

Solution

Partition the sample space into two sets C_A and C_P, for cans that contain apricots and pears. D is the event that a can is corroded (defective). According to the facts,

$$P(C_P) = \frac{1,200}{3,000} = \frac{2}{5} \qquad P(C_A) = \frac{1,800}{3,000} = \frac{3}{5}$$
$$P(D|C_P) = 0.01 \qquad P(D|C_A) = 0.07$$

By Equation (6.18),

$$P(D) = \frac{2}{5} \times 0.01 + \frac{3}{5} \times 0.07 = 0.046 = 4.6 \text{ percent}$$

The problem asks us to determine $P(C_P|D)$ and $P(C_A|D)$. They are

$$P(C_P|D) = \frac{P(C_P)P(D|C_P)}{P(D)} = \frac{\frac{2}{5} \times 0.01}{0.046} \approx 0.087$$

$$P(C_A|D) = \frac{P(C_A)P(D|C_A)}{P(D)} = \frac{\frac{3}{5} \times 0.07}{0.046} \approx 0.913$$

The can is substantially more likely to contain apricots, with a probability in excess of 91 percent.

Skill Enhancer 50   If the shipment contained only 300 cans, 120 of which contain pears and the remainder hold apricots, and the corrosion rates remain the same, what effect would this have on the likelihoods we just computed?

Answer in Appendix E.

Example 51   *Computer viruses* are small programs that can infiltrate themselves into data files on a computer. The Reliant Data Processing Company examines files from two different clients. In disks from client A, 5 percent of the files are corrupted with a virus. Of the files from client B, 10 percent are corrupted. Overall, there are five times as many files in the collection from client A than client B.

(a) A file is selected at random. With what probability will it be infected by a virus?

(b) If it turns out to have been infected, with what probability will it have come from the first client?

Solution

We let A and B stand for the event that the file comes from the first or second client, and V be the event that a virus infects a selected file. If $P(B) = x$, then $P(A) = 5x$, and we solve for x by noting that $5x + x = 6x = 1$, so $x = \frac{1}{6}$. Therefore,

$$P(A) = \frac{5}{6} \qquad P(B) = \frac{1}{6}$$

$$P(V|A) = 0.05 \quad P(V|B) = 0.1 \tag{6.21}$$

(a) From Equation (6.18), we compute

$$P(V) = \frac{5}{6}(0.05) + \frac{1}{6}(0.1) \approx 0.058$$

(b) Therefore,

$$P(A|V) = \frac{\frac{5}{6}(0.05)}{0.068} \approx 0.718$$

Skill Enhancer 51   Recompute this example if there are equal numbers of files from the two clients.

(a) Set up a table like Equations (6.21).

(b) What is $P(V)$?

(c) What is $P(A|V)$?

Answer in Appendix E.

Example 52   In Ms. Sauer's first grade class, there are normally 15 children, but 3 are out with chicken pox. In Mrs. Lang's class, there are 25 children, but 5 are out with measles. The school principal will be selecting a group of student names for a survey. What is the probability that one of the chosen students will be sick? If the child is sick, what is the probability of her being a student in Ms. Sauer's class?

Solution

We have this chart:

$$P(S) = \frac{3}{8} \qquad P(L) = \frac{5}{8}$$

$$P(I|S) = \frac{1}{5} \qquad P(I|L) = \frac{1}{5}$$

where S and L refer to students of Ms. Sauer and Mrs. Lang, and I refers to an ill child. Since there are a total of $15 + 25 = 40$ first graders, $P(S) = \frac{15}{40} = \frac{3}{8}$ [and $P(L) = \frac{25}{40} = \frac{5}{8}$]. We have

$$P(I) = \frac{3}{8} \cdot \frac{1}{5} + \frac{5}{8} \cdot \frac{1}{5} = \frac{1}{5}$$

Therefore,

$$P(S|I) = \frac{\frac{3}{8} \cdot \frac{1}{5}}{\frac{1}{5}} = \frac{3}{8}$$

Skill Enhancer 52 There are 30 students in one class, and 40 in another. There are 20 young women in the first, but only 10 in the other. If a student is picked from the classes, what's the probability of the student being a woman? If it is a woman, what's the probability of her being in the first class? in the second class? Let $P(1)$ and $P(2)$ be the probabilities of being in the first and second classes, and W the choice of a woman.

Answer in Appendix E.

Example 53 The roster of workers at a shoe factory consists of cutters, lashers, and tanners. The following chart gives the number of employees in each category and the number of workers in that category who have passed the company's tests for superior workmanship.

	Cutters	Lashers	Tanners
Total number	150	180	90
Passed test	50	45	45

Management selects one of the workers at random from among those who passed the workmanship test. What is the probability that this worker is a tanner?

Solution

The solution requires setting up *three* subsets to span the sample space. Let W represent the selection of a worker passing the workmanship test, and let E_C, E_L, and E_T represent the groups of employees who are cutters, lashers, and tanners. According to the problem,

$$P(E_C) = \frac{150}{420} = \frac{5}{14} \qquad P(E_L) = \frac{180}{420} = \frac{3}{7} \qquad P(E_T) = \frac{90}{420} = \frac{3}{14}$$
$$P(W|E_C) = \frac{50}{150} = \frac{1}{3} \qquad P(W|E_L) = \frac{45}{180} = \frac{1}{4} \qquad P(W|E_T) = \frac{45}{90} = \frac{1}{2}$$

(Note that there are a total of $150 + 180 + 90 = 420$ workers in all.) First compute $P(W)$ by means of

$$P(W) = \frac{5}{14} \times \frac{1}{3} + \frac{3}{7} \times \frac{1}{4} + \frac{3}{14} \times \frac{1}{2} = \frac{1}{3}$$

and then $P(E_T|W)$ via

$$P(E_T|W) = \frac{P(E_T)P(W|E_T)}{P(W)} = \frac{\frac{3}{14} \times \frac{1}{2}}{\frac{1}{3}} = \frac{9}{28} \approx 0.321$$

Skill Enhancer 53 Suppose the chart for the workers looks like this.

	Cutters	Lashers	Tanners
Total number	150	180	90
Passed test	75	60	30

(a) Management selects one of the workers at random. What is the probability that she has passed the proficiency exam?
(b) By good fortune, this selected worker indeed has passed the proficiency exam. What is the probability that she is a tanner?
(c) What is the probability that she is a cutter?

Answer in Appendix E.

Example 54   B. C. Dull, a student, is looking for a book to read to prepare a book report. Of 15 novels, 10 were less than 150 pages long. Of 20 books on current events, only 5 were less than 150 pages long. Of five biographies, only one was less than 150 pages in length. If he selects a book at random, with what probability will it be short (less than 150 pages)? If the chosen book is short, with what probability will it be a biography, his favorite form of literature?

Solution

Let N, C, B, and S stand for books that are novels, current events, biographies, and short. There are 40 books in all. Therefore,

$$P(N) = \frac{3}{8} \qquad P(C) = \frac{1}{2} \qquad P(B) = \frac{1}{8}$$

$$P(S|N) = \frac{2}{3} \qquad P(S|C) = \frac{1}{4} \qquad P(S|B) = \frac{1}{5}$$

Therefore,

$$P(S) = \frac{3}{8} \times \frac{2}{3} + \frac{1}{2} \times \frac{1}{4} + \frac{1}{8} \times \frac{1}{5} = \frac{16}{40} = \frac{2}{5}$$

The probability that the book is a biography if it turns out to be short is

$$P(B|S) = \frac{\frac{1}{8} \cdot \frac{1}{5}}{\frac{16}{40}} = \frac{1}{16}$$

Skill Enhancer 54   Redo the above example. Suppose now there are five short books in each category.
(a) What is $P(S|N)$? $P(S|C)$? $P(S|B)$?
(b) What is $P(S)$?
(c) What is $P(B|S)$? What is $P(C|S)$? What is $P(N|S)$?

Answer in Appendix E.

Bayes' Theorem is useful in problems in which researchers or physicians are trying to determine a screening test, such as a test for tuberculosis, cancer, or college aptitude. Such tests are never perfect. A medical test result is said to be *positive* when it indicates the presence of the disease or condition. By chance, some one or some condition slips through the test, so that a healthy or qualified person will unaccountably fail the test—that *false positive*   is, it gives a positive result which is false (a *false positive*). Study this final example.

Example 55   In a remote African town, a full 45 percent of the inhabitants host germs that will lead to tuberculosis. The World Health Organization (WHO) enters the town with a new test for the disease. Individuals with the disease are correctly identified by the test 99 percent of the time, but 30 percent of the time, perfectly healthy people are falsely identified as hosting the disease organisms. The head of the town tests positive in this test—that means the test "thinks" he has the disease. What is the probability that he *really* has tuberculosis?

Solution

Let T and T' be the events that a person really does or does not have tuberculosis. (We are using *set complement notation* to express the negative of a condition.) Let $+$ be the event that the result of the tuberculosis test is positive, that is, that the test indicates a person with tuberculosis. $P(T)$ and $P(T')$ are 0.45 and $1 - 0.45 = 0.55$, respectively. $P(+|T) = 0.99$ and $P(+|T') = 0.30$. With these facts, we compute

$$P(+) = P(T)P(+|T) + P(T')P(+|T')$$
$$= 0.45 \times 0.99 + 0.55 \times 0.30$$
$$= 0.6105$$

Thus, 61 percent of the time, regardless of their actual health, people think they have the disease after learning the test results. For the chief,

$$P(T|+) = \frac{P(T)P(+|T)}{P(+)}$$
$$= \frac{0.45 \times 0.99}{0.6105} \approx 0.730$$

The chief, then, has a slightly less than 75 percent probability of having TB.

Skill Enhancer 55   In one Thai community, 33 percent of the children are infected with a troublesome parasitic condition (schistosomiasis). The government has a new test for the condition. Children with the condition are correctly identified by the test 95 percent of the time, but 20 percent of the time, perfectly healthy children are falsely identified as hosting the parasite. One child tests positive.
(a) If T is the event that a child hosts the parasite, what is $P(T)$ and $P(T')$?
(b) If $+$ represents a positive result on the test, what is $P(+)$?
(c) What is the probability that this small person really hosts the parasite? Use Bayes' Theorem.

Answer in Appendix E.

_____ *Exercises 6.5* _____

Suppose two sets A_1 and A_2 span a sample space S, and suppose B represents another event. If

$$P(A_1) = 0.5 \qquad P(B|A_1) = 0.2 \qquad P(B|A_2) = 0.4$$

then use this information to compute the probabilities of Exercises 1–4.

1. $P(B)$

2. $P(A_2)$

3. $P(A_1|B)$

4. $P(A_2|B)$

Suppose two sets A_1 and A_2 span a sample space S, and suppose B represents another event. If

$$P(A_1) = 0.3 \qquad P(B|A_1) = 0.5 \qquad P(B|A_2) = 0.5$$

then use this information to compute the probabilities of Exercises 5–8.

5. $P(A_2)$ **6.** $P(B)$

7. $P(A_1|B)$ **8.** $P(A_2|B)$

Suppose two sets A_1 and A_2 span a sample space S, and suppose B represents another event. If

$$P(A_1) = 0.35 \qquad P(B|A_1) = 0.1 \qquad P(B|A_2) = 0.15$$

then use this information to compute the probabilities of Exercises 9–12.

9. $P(B)$ **10.** $P(A_2)$

11. $P(A_1|B)$ **12.** $P(A_2|B)$

Suppose two sets A_1 and A_2 form a partition to a sample space S, and suppose B is some other event in S. If $P(A_1) = 0.3$, $P(B|A_1) = 0.1$, and $P(B|A_2) = 0.5$, then compute:

13. $P(A_2)$ **14.** $P(B)$

15. $P(A_1|B)$ **16.** $P(A_2|B)$

A certain sample space S is partitioned into three events X_1, X_2, and X_3, where

$$P(X_1) = P(X_2) = P(X_3)$$

For some event Y,

$$P(Y|X_1) = P(Y|X_2) = 2P(Y|X_3) = 0.5$$

Compute the probabilities in Exercises 17–20.

17. $P(Y)$ **18.** $P(X_2)$

19. $P(X_1|Y)$ **20.** $P(X_3|Y)$

A certain sample space S is partitioned into three events X_1, X_2, and X_3, where

$$P(X_1) = 0.2 \qquad\qquad P(X_3) = 0.4$$
$$P(Y|X_1) = 0.6 \quad P(Y|X_2) = 0.5 \quad P(Y|X_3) = 0.4$$

for some other event Y. Compute the probabilities in Exercises 21–24.

21. $P(X_2)$ **22.** $P(Y)$

23. $P(X_1|Y)$ **24.** $P(X_2|Y)$

A certain sample space S is partitioned into three events X_1, X_2, and X_3 where

$$P(X_1) = P(X_3) = 0.3$$

For some event Y,

$$P(Y|X_1) = P(Y|X_2) = 2P(Y|X_3) = 0.6.$$

Compute the probabilities in Exercises 25–30.

25. $P(Y')$ **26.** $P(X_2)$

27. $P(X_1|Y)$ **28.** $P(X_2|Y)$

29. (*Requires special thought*)
$P(X_3'|Y)$

30. (*Requires special thought*)
$P((X_1 \cup X_3)'|Y)$

A game consists of rolling a die, possibly followed by tossing a coin. There are two ways to win this game. One way is to roll an even number with the die. However, if the die is not even, the player tosses the coin, and will then win only if the coin comes up heads. Compute the probabilities of the plays described in Exercises 31–34 given these facts.

31. Getting tails.

32. Winning.

33. Given the final outcome of heads, that you rolled an even die.

34. Given the final outcome of tails, that you rolled an odd die.

In a certain game of chance, one player rolls any number of dice from 2 to 4 and then announces the *sum* of the dots. The rest of the players do not know how many dice have been rolled. Use this information to compute the probability of the events in Exercises 35–40. Assume that the player is equally likely to roll any number of the dice.

35. Rolling the sum of 3.

36. Rolling the sum of 4.

37. Rolling the sum of 24.

38. Rolling 4 dice given that the sum is 3.

39. Rolling 3 dice given that the sum is 3.

40. Rolling an even number of dice given that the sum of the dice is 6.

Applications

41. There are two classes, each containing 30 children. In the first, 5 have chicken pox, whereas 10 in the second class have it.

(a) What is the probability of choosing a child with the disease?

(b) Given that a child has chicken pox, what is the probability that he or she was in the first class?

42. Refer again to Exercise 41. Suppose this time that 10 children in each class have the disease.

(a) What is the probability of choosing a child with the disease?

(b) Given that a child has chicken pox, what is the probability he or she was in the first class?

43. *(Medical Testing)* In Example 55 in the text, what is the probability that a person falsely thinks he or she does not have tuberculosis?

44. *(Athletics)* Openings currently exist on the track and gymnastic teams in your university. Since gymnastics is more popular, a qualified athlete is twice as likely to be picked for that team as for the track team. Both coaches favor a 15-week Zen meditation course. For an athlete who has undergone this program, the chances of joining the squad rise to 50 percent for the track squad, and to $\frac{2}{3}$ for the gymnastic squad. After a recent triumph, a reporter interviews an athlete, but has trouble making sense of the answers (the athlete is completely out of breath). What are the chances that this athlete completed the meditation course? If this athlete did complete the course, what are the chances of her being a runner?

45. *(Clothing Manufacturing)* Baby Waves is a manufacturer of baby's clothing and consists of two factories. Currently, there is some trouble with the lines of toddler's shorts. The first factory has an acceptable 3 percent defect rate, while the second has an unacceptable 10 percent defect rate. Fortunately, the first factory manufactures three times as many garments as the second. A random pair of shorts is sampled and found to be defective. With what probability did it come from the second (troubled) factory?

46. *(Cancer Detection)* The government has sponsored the development of tests for the special types of cancers common around certain dumping grounds for toxic wastes. Typically, 10 percent of such a community will have such a cancer, which fortunately can be cured if caught in time. The test accurately detects the presence of disease in a person 95 percent of the time. In people demonstrably free of the disease, the test indicates this negative result in 60 percent of people taking the test. What is the probability of any person selected at random testing positive for the disease? Given that a person tests positively for cancer, with what probability will that person actually have the disease? What do you think is the

benefit in a test whose results are *false positives* so much of the time?

47. *(Cancer Detection)* The probability that women over the age of 85 will develop a certain kind of cancer is $\frac{1}{9}$. The probability that women below the age of 35 will develop the same kind of cancer is $\frac{1}{900}$. A test group consists of 900 women below the age of 35 and 100 women above the age of 85.

(a) If one is selected at random, what is the probability that that woman will develop this type of cancer?

(b) If that woman develops this cancer, what is the probability that she is 85 or older?

48. *(Urban Crime)* Studies of violent offenders in one prison relate the type of crime to the age of the offender. If the offender is below the age of 30, the probability that he or she has committed murder is one-third. If the offender is 30 or older, the probability drops to one-tenth. A certain prison population consists of 60 of the younger offenders and 100 offenders 30 or older. One of the offenders is chosen at random.

(a) What is the probability that this one has committed murder?

(b) If we know that he is a murderer (and murderers overwhelmingly tend to be male), what is the probability that he is below the age of 30?

49. *(Diplomacy and Foreign Policy)* One nineteenth-century German kingdom, call it Gerolstein, was noted for its belligerent foreign policy. In any altercation with a foreign power, this kingdom was inclined to appeal to diplomacy to resolve the issue in only 35 percent of the issues of which historians have record. Situations with diplomatic arbitration led to armed engagements only 15 percent of these times. When diplomacy was not used in an altercation, Gerolstein resorted to some kind of armed attack in fully 80 percent of recorded instances. In one famous incident, in which the Gerolstein army attacked and overcame an army three times its size, it is not known with precision whether diplomatic talks preceded the army action. Assess the likelihood that these talks did occur.

50. *(Effects of Exercise)* The results of a multidecade study of the effects of moderate exercise on the male alumni of a famous Ivy League college were recently made public. Throughout the course of the study, 35 percent of the alumni engaged in exercise defined by the study to be moderate, 20 percent engaged in heavy exercise, and the remainder led sedentary lives. The effect of exercise was found to be strongest for those who exercised moderately. For these individuals, the probability of living past the age of 80 was 35 percent. For the sedentary group, it was 10 percent, and for the heavy exercisers, it was 15 percent.

(a) Given a random male student currently enrolled at this university, what is the probability that he will make it to age 80?

(b) One particularly wealthy and generous alumnus recently died at age 85. What is the probability that he practiced moderate exercise throughout his life? that he practiced heavy exercise throughout his life?

51. **(Automobile Manufacturing)** A well-known luxury car manufacturer has three factories in the Detroit area. Factory 1 produces three times as many cars as factory 2, and factory 3 produces twice as many cars as factory 2. Factory 2, though, has the best track record in producing cars that need no major repairs in the first 10,000 miles of customer use—only 3 percent of its output falls into this category, as opposed to 7 percent of the output of factory 1 and 10 percent of cars produced by factory 3. Miles N. Miles purchases one of these luxury cars, but bitterly rues the transmission job that was needed before he had driven 7,000 miles.

(a) What is the probability that the car was produced by factory 3? by factory 1?

(b) Miles's brother-in-law bought a similar car, and, wouldn't you know it—the car did not need so much as an oil change for the first 10,000 miles. What is the probability that the car was manufactured in factory 1? in factory 2?

W 52. State in your own words why Bayes' theory is so important to medical testing. Why do you think most serious medical tests give rise to so many false positives?

Calculator Exercises

53. The population of Hatfield County in Kentucky is 786,123, of whom 500,692 are fair-haired. The population of neighboring McCoy County is 692,102, and 123,781 of its residents are fair-haired. A traveler is stopped at random, and lives in one of these two counties.

(a) With what probability is she fair-haired?

(b) Given that she is fair-haired, with what probability does she come from Hatfield County?

54. The Black and Brown Shoe Warehouse contains 10,623 pairs of brown shoes and 9,828 pairs of black shoes. Within these categories, 9,000 and 2,000 of the pairs of shoes are made with leather uppers. The remainder were entirely constructed out of artificial materials. A recent thunderstorm flooded the entire warehouse.

(a) What is the probability that a pair of shoes chosen at random is made of leather?

(b) If a pair of shoes is black, with what probability will it be leather?

(c) If a pair of waterlogged shoes is chosen at random, with what probability will it be black?

(d) The chosen waterlogged shoes are found to be leather. What is the probability that they will be black?

Section 6.6   *EXPECTATION*

Anita and Barbara each draw a card randomly from a standard deck of playing cards. If the card is a picture card, Anita gives Barbara $2. If the card is even, Barbara gives Anita $1.50. Any other card is ignored. They play this game over and over. In the long run, who comes out ahead, and by how much?

We can compute the probability of each outcome in this game. Associated with each outcome is a particular payoff. To answer the questions in this example, we need to determine the average payoff per play, where each distinct payoff is weighted by the probability of its occurrence. This weighted average is the *expectation* or *expected payoff*.

expectation

The payoff is determined by three possible outcomes: that the card is a picture card (p), that it is even (e), or that it is anything else (o for "other"). The probabilities for

Here Anita draws the King of Hearts from the deck which Barbara holds.

Here, Barbara chooses the 4 of Diamonds

FIGURE 6.23

these outcomes are

$$P(p) = \frac{3}{13} \qquad P(e) = \frac{5}{13} \qquad P(o) = \frac{5}{13}$$

Let us compute the payoffs from Anita's point of view. If the card is a picture card, her reward R_p is actually a loss of \$2, which we denote with a negative sign.

$$R_p = -2$$

If the card is even, her reward R_e is \$1.50.

$$R_e = 1.5$$

In any other situation, she neither wins nor loses any money.

$$R_o = 0$$

In the long run, she expects to pay out \$2 in each game $\frac{3}{13}$ of the time, gain \$1.50 in each game $\frac{5}{13}$ of the time, and have no loss or gain $\frac{5}{13}$ of the time. Her expected payoff E per game, or *expectation*, is therefore

$$E = -2 \cdot \frac{3}{13} + 1.5 \cdot \frac{5}{13} + 0 \cdot \frac{5}{13}$$

$$= \frac{1.5}{13} \approx \$0.12$$

In the long run and on the average, Anita expects to come out ahead about 12 cents per game. The expected value need not be one of the possible outcomes of the game.

Example 56   Three outcomes E_1, E_2, and E_3 are associated with some game. The probability for each E_i is $P(E_i)$ where

$$P(E_1) = 0.5 \qquad P(E_2) = 0.3 \qquad P(E_3) = 0.2$$

while the reward R_i for each outcome is

$$R_1 = -1 \qquad R_2 = +2 \qquad R_3 = 0$$

What is the expected value for this game?

Solution

We add together the products of the probabilities and the rewards for each outcome. Don't forget the sign of each reward.

$$\text{Expected value} = (0.5)(-1) + (0.3)2 + (0.2)0 = +0.1$$

Skill Enhancer 56   For some game with three outcomes, $P(E_1) = P(E_3) = 0.3$ and $P(E_2) = 0.4$ while $R_1 = 2$, $R_2 = -1$, and $R_3 = -1$. What is the expected value for this game?

Answer in Appendix E.

Example 57   Here are the rules for a game. It costs you \$1.00 to play the game once. A coin is tossed. If it comes up heads, you get \$1.50; otherwise you get nothing. What is the expected value of the game?

Solution

Clearly, $P(H) = P(T) = \frac{1}{2}$. Also,

$$R_H = 1.50 \qquad R_T = 0$$

so

$$\text{Expected value} = 1.5\left(\frac{1}{2}\right) + 0\left(\frac{1}{2}\right)$$
$$= 0.75$$

The expected value is \$0.75. (Since this is less than the entry fee, the "house" will make money in the long run.)

Skill Enhancer 57   Recompute the expected value if the payoff is \$2.00 for heads, but you pay an additional \$1.00 if you lose. What is the meaning of the expected value? If the entrance fee for this game is \$0.30, would you play?

Answer in Appendix E.

Example 58   Consider a coin weighted so that $P(H) = \frac{1}{3}$ and $P(T) = \frac{2}{3}$. A player gets \$1.00 on heads and loses \$1.00 on tails. What is the expected value?

Solution

From the player's point of view, $R_H = 1$ and $R_T = -1$. The expected value E is

$$E = \frac{1}{3}(+1) + \frac{2}{3}(-1) = -\frac{1}{3}$$

Skill Enhancer 58   A coin is weighted so that $P(H) = 0.6$. A player wins \$1.00 on H and loses on T. Over the long run, a mathematician observes that the expected value of the game is -0.2, which means that a player loses 20 cents on each play in the long run. Let x be the amount a player loses on tails.
(a) For this coin, what is $P(T)$?
(b) What is the defining equation for x?
(c) What is x?

Answer in Appendix E.

Example 59   Some years ago, the American medical establishment had to decide which of two polio vaccines to distribute. The vaccine made from *live virus* is the more effective, but because it uses live virus, 2 children out of 200,000 contracted the disease. On the other hand, vaccine made from *dead virus* never causes polio, but it fails to protect 1 child out of 100. In one substantial survey done before the introduction of either vaccine, 15 children out of the 1,000 studied contracted the disease. On the basis of these data, which type of vaccine should be used?

Solution

For both vaccines, we will compute the expectation of contracting polio. We prefer the vaccine with the *lowest* expectation of childhood disease. In each case, there are two probabilities—contracting the disease or not contracting the disease. We will assign a payoff of 1 for contracting the disease, and 0 otherwise.

The calculation is easiest for the live-virus vaccine. Here,

$$E_{\text{live}} = P(\text{protected}) \cdot 0 + P(\text{polio}) \cdot 1$$
$$= \frac{199,998}{200,000} \cdot 0 + \frac{2}{200,000} \cdot 1$$
$$= \frac{1}{100,000} = 0.00001 = 0.001 \text{ percent.}$$

For dead-virus vaccine, no child contracts the disease directly. However, some small fraction of children fail to be protected, and a small fraction of those *will* contract polio. First compute the probability that an injected child remains unprotected. Empirically, this probability is $P_u = 0.01$. Also empirically, the probability that an unprotected child will contract polio is $P_p = \frac{15}{1,000} = 0.015$. Because these events are independent

$$P(\text{polio}) = P_p \cdot P_u = (0.015)(0.01) = 0.00015$$

Therefore, the expectation is

$$E_{\text{dead}} = P(\text{protected}) \cdot 0 + P(\text{polio}) \cdot 1$$
$$= (0.00015)(1) = 0.00015 = 0.015 \text{ percent}$$

Since $E_{\text{live}} < E_{\text{dead}}$, the better choice is the live vaccine.

Skill Enhancer 59   Recompute E_{live} and E_{dead} in this problem if additional research casts new light on the statistics. Now it appears that only 1 child out of 200,000 contracts the disease when live vaccine is used. On the other hand, when vaccine from dead virus is used, 1 child out of 150 is not protected.

Answer in Appendix E.

Expectation (Expected Value)

$$E = P(E_1) \times R_1 + \cdots + P(E_n) \times R_n$$

where

E	is the expectation,
$E_1, E_2, \ldots, E_n$	are the events, and
R_i	is the reward (payoff) of event E_i.

Example 60   A single die is weighted so that the number n is n times as likely to appear as the number 1. (That is, a 3 is three times as likely as a 1, a 5 is five times as likely as a 1, and so on.) What is the expected value of the number that appears on a roll of this die?

Solution

For our first step, we must compute the probabilities of the six different outcomes. Let e_i be the event that i dots show on the die. Then

$$P(e_i) = ix$$

for some x, and also

$$P(e_1) + P(e_2) + P(e_3) + P(e_4) + P(e_5) + P(e_6) = 1$$
$$x + 2x + 3x + 4x + 5x + 6x = 1$$
$$21x = 1$$

so that $x = \frac{1}{21}$. Therefore, $P(e_i) = \frac{i}{21}$.

The expectation E is the sum of the products of each outcome with its probability:

$$E = 1 \cdot \frac{1}{21} + 2 \cdot \frac{2}{21} + 3 \cdot \frac{3}{21} + 4 \cdot \frac{4}{21} + 5 \cdot \frac{5}{21} + 6 \cdot \frac{6}{21}$$
$$= \frac{91}{21} = 4\frac{1}{3}$$

Skill Enhancer 60   A single die is weighted so that 1 and 2 never show, whereas the 3 and 4 are each twice as likely to appear as 5 and 6.
(a) What is the probability of rolling a 3?
(b) What is the probability of rolling a 6?
(c) What is the expected value of the die?

Answer in Appendix E.

Example 61   Professor Stutter has the annoying habit of giving frequent surprise quizzes to his classes. Each night, every student is faced with the same problem: to study or not to study. B. C. Dull (a student) tries to assign numerical payoffs to each of the four possible combinations of events. If he studies and there is a quiz, he awards himself 10 points. There is mild frustration if he studies for a quiz and there is none; in this case, he gives himself 4 points. The worst case occurs when a quiz is given, and he has not studied. In this case, he gives himself -15 points. The most satisfying situation from this student's point of view happens when he fails to study but there is no quiz. He then awards himself 12 points. If the professor is as likely to give a surprise quiz as not, what should Dull's study plan be if he wishes to maximize expectation?

Solution

We will compute expectation for two cases: (1) Dull studies, (2) he fails to study. In either case, there are two outcomes: quiz or no quiz. Each of these events is equally likely; the probability of each is $\frac{1}{2}$.

To compute E_1 for the first case, the outcome of a quiz yields a payoff of 10 points, while no quiz has a payoff of 4 points. Therefore,

$$E_1 = 10\left(\frac{1}{2}\right) + 4\left(\frac{1}{2}\right) = 7$$

In the second case, that Dull does not study, the payoff if there is a quiz is -15. If there is no quiz, the payoff is 12 points. Therefore,

$$E_2 = -15\left(\frac{1}{2}\right) + 12\left(\frac{1}{2}\right) = -\frac{3}{2}$$

The student should always study.

Skill Enhancer 61   Sam, a rambunctious six-year-old boy, sometimes gives his parents trouble when being put to bed at night. He goes to bed quietly one-third of the time. When he does not go quietly, his mother must take two aspirins and his father one. What is the daily expected number of aspirins consumed in this household? (Assume there is no other consumption of aspirin.)

Answer in Appendix E.

Expectation as a Decision-Making Tool

In case an event will be repeated over and over, it often makes sense to choose the alternative that optimizes the expectation. That is, we seek to maximize it when it represents money or profit but to minimize it when it represents expenses, costs, death, and so on. Even when some event *cannot* be repeated, optimization of the expectation may be the only reasonable way to make a decision.

Example 62   Management at one company has to decide between two alternatives. Associated with the first alternative is a definite cost of $150,000. Two things can happen in the second choice with equal probability: Management will have to spend $100,000, or management will have to spend $190,000. On the basis of expectation, which alternative should management prefer?

Solution

Let E_I and E_{II} represent the expected values for the two alternatives. For the first,

$$E_I = -150,000$$

for sure. For the second, we have

$$E_{II} = \frac{1}{2}(-100,000) + \frac{1}{2}(-190,000) = -145,000$$

we use negative signs to indicated that these "payoffs" are actually costs to the company. Since $E_{II} < E_I$, management may prefer the second alternative.

Skill Enhancer 62   Redo the calculation if the expenses for the second alternative are $130,000 and $170,000.

(a) What now is E_{II}?

(b) Comment on management's decision making.

Answer in Appendix E.

Example 63   The steel plant in Iron City can use either of two iron refining machines, costing $1 million and $1.6 million, respectively. They each do the job efficiently, but the first (cheaper) machine will certainly wear out within the first year of operation, whereas the second has a probability of only $\frac{1}{5}$ of wearing out. (These machines cannot be repaired. When they break down, they are replaced by new units of the same type.) The company is interested in minimizing refining expenses *for the first year only*. Which machine should be bought? (Assume that a second replacement in one year for the first machine would never be needed.)

Solution

The total refining expense consists of the original cost of the refining machine plus the expected cost of replacing it. For the first machine, the original cost is $1 million, as is the expected cost of replacing it, since it will break down for sure.

$$\text{Total cost of machine } 1 = \$1 \text{ million} + \$1 \text{ million} = \$2 \text{ million}$$

For the second machine, the expected replacement cost is $\frac{1}{5} \cdot \$1.6$ million $= \$320,000$. The total cost is therefore

$$\text{Total cost of machine } 2 = \$1.6 \text{ million} + \$0.32 \text{ million} = \$1.92 \text{ million}$$

The second machine will have a smaller expected total cost associated with it; management should choose this machine on the basis of its lower expected costs.

Skill Enhancer 63   Now, the second machine only costs $1.2 million, but has a 0.7 chance of breaking down.

(a) What is the expected cost associated with the second machine?

(b) On the basis of expected value, which machine should management purchase?

Answer in Appendix E.

_____ *Exercises 6.6* _____

1. Compute the expectation given the following probabilities $P(e_i)$ and payoffs (rewards) R_i.

	e_1	e_2
$P(e_i)$	$\frac{1}{3}$	$\frac{2}{3}$
R_i	1	-1

2. Compute the expectation given the following probabili-

ties $P(e_i)$ and payoffs (rewards) R_i.

	e_1	e_2
$P(e_i)$	0.2	0.8
R_i	$\frac{1}{2}$	$\frac{1}{2}$

3. Compute the expectation given the following probabilities $P(e_i)$ and payoffs (rewards) R_i.

	e_1	e_2
$P(e_i)$	$\frac{3}{4}$	$\frac{1}{4}$
R_i	$\frac{1}{4}$	$-\frac{3}{4}$

4. Compute the expectation given the following probabilities $P(e_i)$ and payoffs (rewards) R_i.

	e_1	e_2
$P(e_i)$	0.52	0.48
R_i	-1	$+1$

5. Compute the expectation given the following probabilities $P(e_i)$ and payoffs (rewards) R_i.

	e_1	e_2
$P(e_i)$	$\frac{1}{9}$	$\frac{8}{9}$
R_i	$\frac{3}{4}$	$\frac{3}{4}$

6. Compute the expectation given the following probabilities $P(e_i)$ and payoffs (rewards) R_i.

	e_1	e_2	e_3
$P(e_i)$	$\frac{1}{3}$	$\frac{1}{3}$	$\frac{1}{3}$
R_i	-2	0	1

7. Compute the expectation given the following probabilities $P(e_i)$ and payoffs (rewards) R_i.

	e_1	e_2	e_3
$P(e_i)$	$\frac{1}{3}$	$\frac{1}{3}$	$\frac{1}{3}$
R_i	1	-1	1

8. Compute the expectation given the following probabilities $P(e_i)$ and payoffs (rewards) R_i.

	e_1	e_2	e_3
$P(e_i)$	0.5	0.2	0.3
R_i	0.3	0.2	0.5

9. Compute the expectation given the following probabilities $P(e_i)$ and payoffs (rewards) R_i.

	e_1	e_2	e_3
$P(e_i)$	0.5	0.2	0.3
R_i	-0.5	0.1	0.2

10. Compute the expectation given the following probabilities $P(e_i)$ and payoffs (rewards) R_i.

	e_1	e_2	e_3
$P(e_i)$	$\frac{1}{6}$	$\frac{1}{2}$	$\frac{1}{3}$
R_i	1	1	-2

11. Compute the expectation given the following probabilities $P(e_i)$ and payoffs (rewards) R_i.

	e_1	e_2	e_3
$P(e_i)$	$\frac{1}{2}$	$\frac{1}{3}$	$\frac{1}{6}$
R_i	5	-7	11

12. Compute the expectation given the following probabilities $P(e_i)$ and payoffs (rewards) R_i.

	e_1	e_2	e_3
$P(e_i)$	$\frac{1}{8}$	$\frac{1}{3}$	$\frac{13}{24}$
R_i	5	-7	11

13. Compute the expectation given the following probabilities $P(e_i)$ and payoffs (rewards) R_i.

	e_1	e_2	e_3	e_4
$P(e_i)$	$\frac{1}{3}$	0	$\frac{1}{3}$	$\frac{1}{3}$
R_i	-9	$1,000,000$	6	3

14. Compute the expectation given the following probabilities $P(e_i)$ and payoffs (rewards) R_i.

	e_1	e_2	e_3	e_4
$P(e_i)$	$\frac{1}{4}$	$\frac{1}{4}$	$\frac{1}{4}$	$\frac{1}{4}$
R_i	-1	-1	4	-1

15. Compute the expectation given the following probabilities $P(e_i)$ and payoffs (rewards) R_i.

	e_1	e_2	e_3	e_4
$P(e_i)$	0.3	0.3	0.2	0.2
R_i	0.5	−0.5	−0.8	0.8

16. Compute the expectation given the following probabilities $P(e_i)$ and payoffs (rewards) R_i.

	e_1	e_2	e_3	e_4
$P(e_i)$	0.1	0.2	0.3	0.4
R_i	−0.4	−0.3	0.2	0.1

17. Compute the expectation given the following probabilities $P(e_i)$ and payoffs (rewards) R_i.

	e_1	e_2	e_3	e_4
$P(e_i)$	$\frac{1}{2}$	$\frac{1}{4}$	$\frac{1}{8}$	$\frac{1}{8}$
R_i	1	−2	3	−4

18. Compute the expectation given the following probabilities $P(e_i)$ and payoffs (rewards) R_i.

	e_1	e_2	e_3	e_4
$P(e_i)$	$\frac{1}{10}$	$\frac{1}{10}$	$\frac{3}{5}$	$\frac{1}{5}$
R_i	−2	2	1	−1

19. Compute the expectation for the following process. The probabilities and payoffs are given in the following table.

	e_1	e_2	e_3	e_4	e_5
$P(e_i)$	0.05	0.1	0.13	0.6	0.12
R_i	1	0	−1	1.5	1.5

20. Compute the expectation if the following table gives the probabilities and payoffs for events in some sample space.

	e_1	e_2	e_3	e_4	e_5	e_6
$P(e_i)$	0.1	0.2	0.07	0.03	0.24	0.36
R_i	0.1	10	11	−1.5	−1.5	−1.5

21. A coin is weighted so that it is twice as likely to come up heads as tails. If it comes up heads, you get 1 penny; otherwise, you give the house 2 pennies. What is the expected outcome of tossing this coin?

22. A coin is weighted so that it is 50 percent more likely to come up tails than heads. If it comes up tails, you win 25 cents; otherwise, you pay 15 cents. What is the expected outcome of playing this game?

23. A single die is rolled over and over again many times. What is the expected value of dots showing?

24. A single die is weighted so that the even numbers are *three times* as likely to appear as the odd numbers. What is the expected value of the number that appears on a roll of this die?

25. A game consists of rolling a single die. You get as many dollars as dots showing face up. What is the most you should be willing to pay to play this game?

26. Your employer, an individual with a strange sense of humor, gives you a monthly bonus of $250 at the end of every month whose name contains an *R*. What is your expected monthly bonus?

Applications

27. A game at a carnival pits you against the house, a roulette wheel containing the numbers 1 through 36, 0, and 00. The wheel is spun. If the number comes up even, you get one dollar. Otherwise, you pay one dollar. (Neither 0 nor 00 is an even number.)
   (a) What is the expectation for this game?
   (b) The carnival expects to play this game 700 times during the course of the carnival. What is the most likely estimate for the total amount of money the carnival will win or lose?

28. *(Carnival Fun)* A mathematician keeps careful records concerning a particular carnival game. There are five equiprobable outcomes in each outcome of the game. The mathematician knows that $R_1 = 0$, $R_5 = 1$, and the expected value for each play of the game is $0.50 for the carnival. What relationship can she derive for the remaining payoffs R_2, R_3, and R_4?

29. *(Insurance)* A certain church is planning a very important outdoor fair. Rainy weather would ruin their plans, so they plan to insure the event with the ABC Insurance Company for $50,000. The premium is $5000. That is, they pay the insurance company five thousand dollars. In the event that the fair is cancelled due to bad weather, the insurance company will then pay the church fifty thousand dollars. The insurance company does this because they calculate the probability of severe rainy weather to be only 0.05. What is the expected value of this policy to the company?

30. **(State Aid)** The state government will give $5 in aid for each day a pupil attends a certain local school. 500 pupils are enrolled, but there are some absences each day, and the number of absences seems to depend on the day of the week. During the last year, these were the average attendance figures for each day of the week.

Monday	Tuesday	Wednesday	Thursday	Friday
420	440	475	460	410

What is the expected daily state aid for this school?

31. **(New Products)** Lucy and Lacy do their food shopping together. Many new products appear on the shelves. Lucy wants to try a new breakfast cereal, which comes in two packages—a 4 oz. box for $1.99 or an 8 oz. box for $2.75. Lucy likes sampling new things, but the chances that she will like a new product are only $\frac{2}{3}$. The larger box is far more economical than the small box, but if she does not like the cereal, the waste will be larger. What box should she buy, assuming she tries the new cereal?

32. Refer to Exercise 31. Lacy is much less adventurous than Lucy, and the probability that she will like the new cereal is only $\frac{1}{4}$. If she is persuaded to try this new product, which box should she buy—the larger or the smaller?

33. **(Real Estate)** All other things being equal, the construction costs of a building vary with the geographic nature of the building site. A new library can be built at any of three locations. At a construction site, if the soil is rock, sand, or topsoil, the costs are $800,000, $1,000,000, or $500,000. At the first possible site, the probabilities that the site is rock, sand, or topsoil are $\frac{1}{2}$, $\frac{1}{6}$, and $\frac{1}{3}$. At the second site, the probabilities are 0.4, 0.3, and 0.3. At site 3, they are 0.27, 0.48, and $\frac{1}{4}$. Where should the city build the library?

34. **(Study Habits)** In Example 61, suppose the professor is three times as likely to give a quiz as not, what should Dull's study plan be if he acts to maximize the expectation?

35. In Example 61, what should Dull's study plan be if Professor Stutter is two-and-one-half times as likely *not* to

give a quiz as to give one. Assume once again that Dull acts to maximize the expectation.

(Income Tax Preparation) Many taxpayers are tempted to under-report their annual income on tax returns so as to pay less tax. The government tries to discourage this practice by auditing the returns of selected taxpayers and levying stiff penalties when there is evidence of deliberate under-reporting. One taxpayer attempts to use decision theory to decide whether to be truthful or not on her return. If she is accurate (truthful) on her return, then there is no payoff associated either with an audit or without one. If she under-reports income and is audited, then she will pay a penalty of $2,000 to the government. If she under-reports and is *not* audited, then her gain is $1,000, the tax she saves. She believes that there is no way to tell whether the government will audit her or not. Use this information to solve Exercises 36–38. In each of these questions, assume that the taxpayer acts in the way which maximizes her expectation.

36. How should the taxpayer behave if she believes that an audit is as likely as no audit?

37. According to one survey the taxpayer has seen, only 3 tax returns out of every 100 are audited. Compute the expectation in this case. How should the taxpayer prepare her return now?

38. This taxpayer continues to believe that only 3 out of every 100 tax returns are audited. Her accountant has recently told her that her penalty for under-reporting should she be caught in an audit would be $10,000. Will this have any effect on the way she fills out her return? Why or why not?

39. Solve again the game problem from the beginning of this section, this time from *Barbara's* point of view, and show that the answer under this computation is equivalent to the solution in the text.

40. **(Church Raffle)** A church raffle involves selling 800 tickets, the single prize of which is a video cassette recorder with a retail value of $250. What is the maximum you would pay for a ticket if you are solely interested in participating in a fair game? What is the minimum that the church should charge for each ticket?

41. **(Software Development)** A software developer has written a clever word processing program which she wants to license to a marketing company. She can either accept a grant of $100,000 for the program plus a royalty of $1 for each copy of the program sold, or she can accept $10,000 plus a royalty of $5 per copy sold. There is a 50 percent probability that the marketer will go out

of business in the first year of business, and a 25 percent probability it will go out of business in the second year. Independently of this, there is a 33 percent probability of selling 500,000 copies of the program, and a 20 percent probability of selling 1,000,000 copies. Use the expectation concept to determine which offer the software developer should accept.

42. Despite the high values of the grand prize, the fact that the chances that any individual ticket holder will win a state lottery is vanishingly small. Estimate the estimated value of the expected value of any state lottery ticket and comment on the reasons why so many lottery tickets are sold.

Calculator Exercises

43. A single die is weighted. Each face will show with the following probabilities.

Face	1	2	3	4	5	6
Probability	0.1	0.25	$\frac{1}{6}$	$\frac{1}{7}$	$\frac{1}{9}$	?

(a) What is the probability that a 6 will show?

(b) What is the expected value of the faces that show?

44. In the tiny country of Bohemia, there are four income tax brackets. A citizen owes the government the following amounts depending on the tax bracket.

Bracket	1	2	3	4
Amount	123	456	789	1,500

(The unit of currency is the Bohemian dollar.) In addition, if a taxpayer is caught cheating on his or her tax return, he is fined 5,000 Bohemian dollars instead of paying any tax. In a recent year, the government found that 5,238 taxpayers were cheaters. In addition there were 9,282, 10,295, 11,358, and 5,444 returns in the first, second, third, and fourth tax brackets. What is the average "worth" of a taxpayer to the government?

CHAPTER REVIEW

Terms

probability	experiment	sample space
event	simple event	tree diagrams
probability function	relative frequency	impossible
certain	probability distribution	empirical probabilities
subjective probability	odds	equiprobable
equally likely	urn	mutually exclusive
addition formula	a priori probability	conditional probability
posterior probability	independent	span
partition	Bayes' Theorem	false positive
expectation	expected payoff	

Key Concepts

- Begin probabilistic analysis of a problem by creating an appropriate sample space to describe the experiment.

- Probabilities may be assigned to the simple events of a sample space so long as they conform to three mathematical probability laws. (Of course, this assignment should also correspond to common sense.) Some probability as-

signments are made by appealing to empirical or subjective results, others to rules concerning equally likely probability assignments.

- For most problems, it is best to decide on a sample space that assigns equal probabilities to simple events. We build up complex events using set union operations on the events themselves, and use the additive probability property to determine the probability of the nonsimple events.

- If two events are mutually exclusive, the probability of their union is the sum of their probabilities. Otherwise, this probability obeys the addition formula.

- The sum of the probability of an event and the probability of its complement is one.

- The formula for conditional probability is used to compute probabilities for any event with some restrictive information on the outcome of the event.

- Independent events are events whose outcomes fail to influence each other. It is important to distinguish between the concepts of *mutually exclusive* and *independent*.

- Bayes' Theorem is used to "invert the arguments" in a situation involving conditional probability. This theorem states that for n sets C_i that span the sample space S, then

$$P(C_i|D) = \frac{P(C_i)P(D|C_i)}{P(D)}$$

where

$$P(D) = P(C_i)P(D|C_i) + \cdots + P(C_n)P(D|C_n)$$

- *Expectation* is an average suitably weighted by the probabilities of the outcomes that are part of this average. If each of the events E_i has reward R_i and probability $P(E_i)$ associated with it, the expectation is

$$E = P(E_1)R_1 + P(E_2)R_2 + \cdots + P(E_n)R_n$$

Review Exercises

Suggest sample spaces for the experiments described in Exercises 1–6.

1. Tomorrow's weather.

2. Three seeds of different species are planted in a pot and watered. If they are viable, they will germinate in seven days.

3. Michael had four different job interviews, and waits for their results.

4. An environmental chemist tests four different samples of water from a toxic waste site.

5. A single Korean die is rolled. Korean dice have only four sides, numbered 1 through 4.

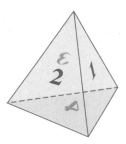

FIGURE 6.24
A Korean die.

6. Imagine rolling a pair of Korean dice. (See Figure 6.24 and Exercise 5 for a description.) Set up an appropriate sample space.

Now express the events of Exercises 7–10 in terms of simple events of the sample space of Exercise 6.

7. The sum of Korean dice is 8.

8. The sum of Korean dice is 3.

9. One Korean die shows an odd number, while the other is even.

10. The sum of the Korean dice is a prime number greater than 3.

11. Compute the probabilities of the event in Exercise 7 assuming that each roll of the dice is equally likely.

12. Compute the probabilities of the event in Exercise 8 assuming that each roll of the dice is equally likely.

13. Compute the probabilities of the event in Exercise 9 assuming that each roll of the dice is equally likely.

14. Compute the probabilities of the event in Exercise 10 assuming that each roll of the dice is equally likely.

15. Compute the probabilities of the event in Exercise 7 assuming that a 4 on any die is twice as likely to appear as any of the other numbers.

16. Compute the probabilities of the event in Exercise 8 assuming that a 4 on any die is twice as likely to appear as any of the other numbers.

17. Compute the probabilities of the event in Exercise 9 assuming that a 4 on any die is twice as likely to appear as any of the other numbers.

18. Compute the probabilities of the event in Exercise 10 assuming that a 4 on any die is twice as likely to appear as any of the other numbers.

19. Compute the probabilities of the event in Exercise 7 assuming that a 2 is twice as likely as a 1, that a 3 is three

times as likely as a 1, and that a 4 is twice as likely as a 2.

20. Compute the probabilities of the event in Exercise 8 assuming that a 2 is twice as likely as a 1, that a 3 is three times as likely as a 1, and that a 4 is twice as likely as a 2.

21. Compute the probabilities of the event in Exercise 9 assuming that a 2 is twice as likely as a 1, that a 3 is three times as likely as a 1, and that a 4 is twice as likely as a 2.

22. Compute the probabilities of the event in Exercise 10 assuming that a 2 is twice as likely as a 1, that a 3 is three times as likely as a 1, and that a 4 is twice as likely as a 2.

The odds in Exercises 23–26 correspond to what probabilities?

23. 2:3   **24.** 3:2

25. 10:9   **26.** 4:3

The probabilities in Exercises 27–30 correspond to what odds? Of course, there are an infinite number of pairs of odds for any single probability. Choose the pair $a : b$ with the smallest value of a and b.

27. $\frac{2}{3}$ **28.** 0.3

29. 0.44   **30.** $\frac{123}{247}$

31. The odds that it will rain tomorrow have been quoted as being 4:8. What is the probability that it will rain? that it will not rain?

32. Because she does not smoke, an insurance agent assured Brenda that the odds of her living past age 75 were better than 4:9. What is the minimum probability that the agent is speaking about?

33. The probability that Carl Klutz will pass his driving test is thought to be about $\frac{1}{12}$. What are the odds of his passing?

34. A friend drops three quarters on the ground. What is the probability that there will be more heads than tails showing?

35. Estimate the minimum number of people necessary to sit in a room so the probability that at least two of them were born in the same month is 50 percent or more.

36. Do you think that you can use the results of the birthday problem to estimate how many members of some historical group *died* on the same day of the year (same month and day, possibly different year)? Making whatever adjustments to the theory that you deem necessary, estimate how many of the first 30 Presidents of the United States will have died on the same day.

A single card is chosen from a standard deck. What is the probability of the events in Exercises 37–40?

37. Picking the ace of clubs.

38. Choosing a club or a spade.

39. Choosing any club or the ten of diamonds.

40. Choosing any heart or the ten of hearts.

An urn contains 2 black marbles and 3 white ones. An experiment consists of drawing 2 marbles from the urn. Determine the probabilities of the outcomes listed in Exercises 41–44.

41. How many elements are there in the usual sample space?

42. Choosing the 2 black marbles.

43. Choosing 2 white marbles.

44. Choosing 1 white and 1 black marble.

The numbers in Exercises 45–48 are the probabilities that some event E will occur. What is the probability that E will *not* occur?

45. 0.3   **46.** $\frac{3}{11}$

47. $\frac{44}{45}$ **48.** 0.123

Once again, a single card is drawn from a standard deck. Compute the probabilities of the events given in Exercises 49–54.

49. The card is red and a picture card.

50. The card is red or a picture card.

51. The card is a red, even card.

52. The card is an ace or black.

53. The card is a black ace.

54. The card is an ace, if we know it is black.

A box contains 11 lottery tickets numbered 5 through 15 inclusive. One ticket is chosen. Determine the probabilities of the events in Exercises 55–58.

55. Number 8 is a winner.

56. The winning number is odd.

57. The winning number contains two digits.

58. The sum of the digits in the winning number is greater than or equal to 5.

59. An urn contains 10 marbles—3 red, 3 blue, 3 green, and 1 black; 4 are selected at random. What is the probability that 2 are red and 2 blue? 2 are red and 2 are green? that the 4 marbles consist of two pairs of identically colored marbles?

60. Three cubical dice are rolled. What is the probability of rolling either the sum 7 or double 1s?

61. A woman learns with astonishment that she will soon give birth to triplets. What is the probability of her delivering either all boys or all girls? exactly 2 boys or 1 or more girls?

62. A pair of dice is rolled. What is the probability that the sum is greater than 10 given that each die shows a 4 or more? given that each die shows a 5 or more?

63. An airplane carrying 50 passengers is carrying your long-lost Uncle Harris, whom you go to pick up. All you know about him is that he is bald and wears glasses. Of the passengers, 20 are bald and 30 wear glasses; 15 are both bald and wear glasses. The passengers disembark and head for the luggage area.
   (a) If you approach a bald man at random, what are the chances that he is your uncle?
   (b) If you approach a man wearing glasses, what is the probability that he is your uncle?

Suppose two sets A and B form a partition to a sample space S, and suppose E is some other event. $P(B) = 0.75$ and $P(E|B) = 0.7$. Also, $P(E|A) = 0.35$. Use this information to compute the probabilities in Exercises 64–67.

64. $P(A)$ **65.** $P(E)$

66. $P(A|E)$ **67.** $P(B|E)$

Suppose sets A_1, A_2, and A_3 form a partition to a sample space, with the respective probabilities of the events being $\frac{1}{2}$, $\frac{1}{3}$, and $\frac{1}{6}$. For some event F, $P(F|A_i) = \frac{1}{i}$ for $i = 1, 2, 3$. Compute the probabilities in Exercises 68–73.

68. $P(F)$ **69.** $P(A_1|F)$

70. $P(A_1'|F)$ **71.** $P(A_2|F)$

72. $P(A_3|F)$ **73.** $P(A_1 \cup A_2|F)$

74. Compute the expectation given the following probabilities $P(e_i)$ and payoffs (rewards) R_i.

	e_1	e_2
$P(e_i)$	0.6	0.4
R_i	-7	$+8$

75. Compute the expectation given the following probabilities $P(e_i)$ and payoffs (rewards) R_i.

	e_1	e_2
$P(e_i)$	$\frac{3}{5}$	$\frac{2}{5}$
R_i	1	-1

76. Compute the expectation given the following probabilities $P(e_i)$ and payoffs (rewards) R_i.

	e_1	e_2	e_3
$P(e_i)$	$\frac{1}{2}$	$\frac{1}{3}$	$\frac{1}{6}$
R_i	1	-1	-1

77. Compute the expectation given the following probabilities $P(e_i)$ and payoffs (rewards) R_i.

	e_1	e_2	e_3
$P(e_i)$	$\frac{3}{4}$	$\frac{1}{4}$	0
R_i	-1	-1	$+1,000,000$

Applications

78. *(Traffic)* Baby Hannah counted 27 American cars out of the 34 that passed by her window since her mother put her to bed. Suddenly, she heard the sound of a crash! One of the cars she saw must have been involved. What is the probability that an American car crashed?

79. *(Zoology)* In the new vertebrate pavilion at the municipal zoo, 34 of the 50 new inhabitants are mammals; 6 are birds, and the remainder are reptiles. If one of the animals escapes, what is the probability that it is a bird? a reptile?

80. *(Manufacturing)* Thermometers are difficult to manufacture. In a batch of 999 thermometers, 333 were defective. In a second batch of 600, fully 200 were defective. Inadvertently, the two batches are mixed together. If a thermometer is selected at random, what are the odds that it will be defective? The vice president in charge of manufacturing is upset that the two batches have been consolidated. He wanted to prepare statistics on the likelihood of encountering defective instruments from different batches. Does it matter in this example? Why or why not?

81. *(Designing Gaming Dice)* Dice can only be manufactured in very special shapes, those of *regular polyhedra*, like tetrahedrons or cubes, which are many-sided bodies, each of whose faces is identical to the others. There are only a few such special shapes, of which the cube is probably the most familiar. Another less familiar one is the *dodecahedron*, which consists of 12 identical pentagonal (five-sided) faces.
  (a) If a pair of dodecahedral dice are fair, how many different rolls of the dice are there?
  (b) If each dodecahedral die is marked with numbers 1 through 12, what is the probability of rolling a pair of 1s? the sum 22? a die four times its mate? Assume the dice are fair.
  (c) Each dodecahedral die is marked with the numbers 1 through 6 *twice*. Compare the probability of rolling, say, a pair of 5s with the probability of rolling a pair of 5s on a conventional pair of cubical dice.

82. *(Weather Prediction)* The chance of 100° heat tomorrow is 50 percent. The chance of muggy humidity tomorrow is 33 percent. The chance that tomorrow will be hot and muggy is 20 percent. What is the probability that it will be either hot or very humid tomorrow?

83. *(Social Services Studies)* In a group of 200 residents of inner city neighborhoods, sociologists encounter 40 who are physically handicapped. In this group, 110 are female and 90 male; 30 individuals are handicapped men. One of these residents is selected at random.
  (a) What is the probability that this person will be female?
  (b) What is the probability that this person will be male or handicapped?
  (c) What is the probability that this person will be female or handicapped?

84. *(Retailing)* Shoppers in a major department store observe that of every 100 shoppers, 60 will be women. Of that same typical group, 40 will actually purchase something. If these are independent qualities, what is the probability that a man will buy something? that a woman will not buy anything?

85. *(Retailing)* A major conglomerate classifies potential store locations as upscale or medium-scale. In upscale locations, 75 percent of shoppers buy something during a typical shopping expedition. The percentage in medium-scale locations is typically more like 50 percent. Medium-scale locations are five times more common than upscale locations. An official from the conglomerate is scouting out another location. She focuses her attention on a particular shopper.
  (a) If the official knows nothing about the nature of the location, what is the probability that the shopper will buy something?
  (b) The shopper made a purchase! What now is the probability that the location is upscale? that the location is medium-scale?

86. *(The Melting Pot)* New York City is famous for its ethnic variety, and New York politicians for the way they play on the views of these groups. One typical Brooklyn neighborhood contains three dominant groups.

	Russian	*Haitian*	*Greek*
Percent of population	45	35	20
Percent Democratic	75	80	50

  (a) What is the probability that a voter selected at random from this neighborhood will be Democratic?
  (b) Given that the person is a Democrat, what is the probability that he or she will be Russian? Haitian?

87. *(Civil Engineering)* The San Antonio city council has retained two groups of consulting engineers to inspect an aging system of 10 footbridges throughout this city. The first group finds evidence of structural decay in 6 bridges. The second group finds that 7 bridges require extensive preventive maintenance work. Also, 4 bridges are structurally unsound but also need preventive maintenance.
  (a) The mayor chooses one bridge to personally inspect himself. What is the probability that it will be structurally unsound? that it will both be structurally unsound and require preventive maintenance?
  (b) Given that the bridge needs preventive maintenance, what is the probability that it will be structurally unsound?

88. *(Gambling: Pay to Play)* Some game involves drawing a card at random from a regulation deck of cards. If the card is an ace, you give up $1. If the card is a picture card, you give $2. For any other card, you get the number of dollars equal to the numeric value of the card. How much should you be willing to pay to play this game?

89. *(Games of Chance)* An otherwise fair die is weighted so that the 6 and the 4 come up twice as often as the other numbers.
    (a) What is the expected number on the die?
    (b) If you get a dollar whenever the die comes up odd and pay out a dollar if the die is even, what is your expected loss or gain each time you roll the die?

7

INTRODUCTION TO MARKOV CHAINS

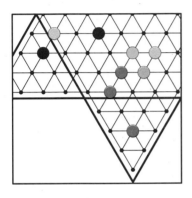

INTRODUCTION

Question: American consumers can buy American or Japanese cars. Studies show that the owner of an American car will purchase an American car as the next car with probability 0.4, whereas owners of a Japanese vehicle will "buy Japanese" for the next car with probability 0.8. Suppose a teenager receives a Japanese car from her parents as a graduation present. What is the probability that her next car will be Japanese? that the car after that will be Japanese? that she will always buy Japanese cars?

A special class of experiments with probabilistic outcomes consists of those whose outcomes depend on the results of a *previous* experiment. If the experiments in such a sequence meet certain additional requirements, this sequence is called a *Markov chain*. Markov chain analysis combines our work on matrices with that of probability.

It is often possible to express behavior that changes over time as a Markov chain. If so, we can use our knowledge of Markov chains to help predict what will happen, and to assess the probabilities of these results.

In the first section of this chapter, we explore the foundations of the Markov chain concept, and see the kinds of problems for which this material is important. In the second section, we encounter several instances where we can ascertain the "long-term" behavior of a Markov process with little difficulty.

Section 7.1   INTRODUCTION TO MARKOV CHAINS

Many real-life situations can be described by a *chain* of outcomes or sequence of events. Many times, the outcome of one link or *step* in this chain depends somehow on the outcome of the previous step and *only* on this previous link. At one time, weather prediction fell into this category. A highly simplified scheme for weather prediction might employ rules like the following.

- If it is sunny today, we expect sunshine tomorrow with 50 percent probability.

- If it rains today, the likelihood of rain tomorrow is 75 percent.

This simple model assumes only two states of weather—sunshine and rain.

Given this model, one might reasonably ask, what will the weather be tomorrow if it is sunny today? What will the weather be the next day, or a week from now? Can we say anything about the weather during the summer? (Of course, at best we can determine probabilistic answers to these questions.) Each day's weather in this simple

steps   example constitutes one of the links or *steps* in this chain of daily weather patterns.

Markov chain   Informally, a *Markov chain* is a sequence of events. The outcome of each event depends only on the outcome of the previous event. All the possible outcomes of each

states   event constitute the *states* of the Markov chain. Each state is therefore one simple event of the sample space of each event in the chain.

Formally, each step in a Markov chain must satisfy three conditions.

1. The sample space *remains the same* from step to step.

2. The state of any step depends only on the *state of the previous step.*

3. The *probabilities governing the transitions from one state (in one step) to another state (in the next step) remain constant* throughout the Markov chain.

In the weather example, we assume that only two kinds of weather are possible in any day. Furthermore, the weather on any day depends (by assumption) on the weather of the previous day, and the probabilities governing the weather are unvarying. For these reasons, this simple model of weather prediction is an example of a Markov chain.

Historical Note

A. A. Markov   *Markov chain analysis is one of several twentieth-century mathematical fields of investigation representing a radical departure from the mathematics of the past. The initial work was done in 1906–1907 by the Russian mathematician Andrei Andreyevich Markov (1856–1922). The subject received an important additional boost from work done by the contemporary Russian mathematician Andrei N. Kolmagoroff, whose work appeared first in 1931.*

We can use these rules for weather prediction to determine the probability that it will be sunny tomorrow. According to our model, weather occurs in two mutually exclusive states.

State 1   Sunshine
State 2   Rain

Since there are only two states in any step of the chain, tomorrow will be sunny as a result of either of *two* situations.

1. It is sunny today, and sunny weather today leads to sunshine tomorrow.

2. It is raining today, and rain today leads to sun tomorrow.

More formally, if $S1$ and $S2$ stand for states one (sunshine) and two (rain), then we can summarize these situations as

(7.1)

$$S1 \ and \ S1 \text{ leads to } S1$$
$$or$$
$$S2 \ and \ S2 \text{ leads to } S1$$

A tree diagram showing these relationships appears in Figure 7.1, and this figure helps us translate these relationships into an equation between probabilities. In this translation process, we make use of the correspondence between multiplication and the word *and* and between addition and *or*. (Why may we do this?) First, though, we need some new notation.

transition probability   Let p_{ij} represent the *transition probability* that state i in one step leads to state j in the next. Transition probabilities are *conditional probabilities*.

$$p_{ij} = P(\text{state } j \text{ in current step} \mid \text{state } i \text{ in the previous step})$$

It helps to think of the subscript notation as a reminder that the probability is for state i going to state j in one step.

$$\text{state } i \xrightarrow{\text{one step}} \text{state } j$$

transition matrix   This notation naturally suggests a *transition matrix* P whose components are p_{ij}.

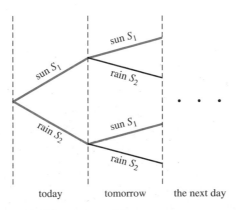

today   tomorrow   the next day

FIGURE 7.1
A tree diagram.

Let P_i be the probability of being in state i in the current step. The P_i's define a row matrix

$$A = \begin{pmatrix} P_1 & P_2 & \ldots & P_n \end{pmatrix}$$

for a general case. Since there are only two states in the weather example, then

$$A = \begin{pmatrix} P_1 & P_2 \end{pmatrix}$$

for this example, where P_1 and P_2 are the probabilities of sun and rain today.

Let the row matrix A' represent the probabilities for *tomorrow's* weather. Is there a relationship between A and A'? From Relation 7.1 we may conclude that

(7.2) $$P_1' = P_1 \cdot p_{11} + P_2 \cdot p_{21}$$

and

(7.3) $$P_2' = P_1 \cdot p_{12} + P_2 \cdot p_{22}$$

Equations (7.2) and 7.3 together form the single matrix equation

(7.4) $$\mathbf{A}' = \mathbf{A}\mathbf{P}$$

Although Equations (7.2) and 7.3 hold for a Markov chain of only two states, Equation (7.4) holds for Markov chains with *arbitrary* numbers of states per step. (In the general case where there are n states per step, the matrix P will be an $n \times n$ matrix, while the probability row matrices A and A' will be of dimension $1 \times n$.)

It is possible to generalize Equation (7.4) to apply to an *arbitrary* step within the chain, but first we must generalize our notation somewhat. Let A^i be the row matrix containing the state probabilities at step i. By convention, the initial step in any Markov chain is step 0, whose probability row matrix is therefore A^0.

The Defining Equations for Markov Chains

$$\mathbf{A}^{i+1} = \mathbf{A}^i \, \mathbf{P}$$

where
P is the $n \times n$ matrix of transition probabilities;
i is an arbitrary step in the chain; $i = 0, 1, 2, \ldots$; and
$\mathbf{A}^i$ is the probability row matrix for step i.

---

stochastic matrix

The transition matrix **P** is an example of a *stochastic matrix*, a matrix all of whose rows sum to 1. The matrices $\mathbf{A}^i$ are also stochastic. Another term for stochastic matrix

probability matrix is *probability matrix*.

Example 1   Use the rules given in the discussion at the beginning of this section.
(a) It is sunny on Monday. Comment on the likelihoods for Tuesday's weather.
(b) It is a rainy Friday. Assess the probabilities for the weekend weather.

Solution

Asserting that one day is sunny means that $S1$ is certain. In such a case,

$$P_1 = 1 \qquad P_2 = 0$$

According to the rules given at the beginning of the section,

$$p_{11} = 0.5 \qquad p_{22} = 0.75$$

which implies that

$$p_{12} = 0.5 \qquad p_{21} = 0.25$$

(why?) so the transition matrix is

$$\mathbf{P} = \begin{array}{c} \\ Sun \\ Rain \end{array} \begin{pmatrix} Sun & Rain \\ 0.5 & 0.5 \\ 0.25 & 0.75 \end{pmatrix}$$

(a) By virtue of these observations, and noting if Monday's weather constitutes step 0, then $\mathbf{A}^0 = \begin{pmatrix} 1 & 0 \end{pmatrix}$. We have

$$\mathbf{A}^1 = \mathbf{A}^0 \mathbf{P}$$

$$= \begin{pmatrix} 1 & 0 \end{pmatrix} \begin{pmatrix} 0.5 & 0.5 \\ 0.25 & 0.75 \end{pmatrix}$$

$$= \begin{pmatrix} 0.5 & 0.5 \end{pmatrix}$$

It is as likely to be sunny as rainy on Tuesday.

(b) Rainy Friday corresponds to an initial state matrix of $\begin{pmatrix} 0 & 1 \end{pmatrix}$. If Friday is step 0 of this chain, then Saturday and Sunday are steps 1 and 2. We can predict Saturday's weather via the matrix equation

$$\mathbf{A}^1 = \begin{pmatrix} 0 & 1 \end{pmatrix} \begin{pmatrix} 0.5 & 0.5 \\ 0.25 & 0.75 \end{pmatrix} = \begin{pmatrix} 0.25 & 0.75 \end{pmatrix}$$

We predict Sunday's weather (Sunday is step 2 of the chain beginning on Friday) by means of

$$\mathbf{A}^2 = \mathbf{A}^1 \mathbf{P}$$

so that

$$\mathbf{A}^2 = \begin{pmatrix} 0.25 & 0.75 \end{pmatrix} \begin{pmatrix} 0.5 & 0.5 \\ 0.25 & 0.75 \end{pmatrix} = \begin{pmatrix} 0.3125 & 0.6875 \end{pmatrix}$$

There is a 75 percent chance of rain Saturday, decreasing slightly to a 69 percent chance on Sunday.

Skill Enhancer 1   Redo this question if the transition matrix is given by

$$\begin{pmatrix} 0.75 & 0.25 \\ 0.25 & 0.75 \end{pmatrix}$$

Answer in Appendix E.

Starting with initial conditions and the transition matrix $\mathbf{P}$, we can say something about all successive steps in the chain. If we compute the matrix $\mathbf{A}^1$, then we can compute $\mathbf{A}^2$ as $\mathbf{A}^2 = \mathbf{A}^1\mathbf{P}$. But $\mathbf{A}^1 = \mathbf{A}^0\mathbf{P}$, so that

$$\mathbf{A}^2 = \mathbf{A}^1 \mathbf{P} = \left(\mathbf{A}^0 \mathbf{P} \right) \mathbf{P}$$

Matrix multiplication is associative, so finally

$$\mathbf{A}^2 = \mathbf{A}^0 \, (\mathbf{P}\mathbf{P}) = \mathbf{A}^0 \, \mathbf{P}^2$$

and we may generalize to this alternative form of the defining matrix equation.

Equations for Markov Chains: Alternative Form

$$\mathbf{A}^n = \mathbf{A}^0 \, \mathbf{P}^n \qquad\qquad n = 0, 1, \ldots$$

where
$$\mathbf{P}^2 = \mathbf{P}\mathbf{P}, \ \mathbf{P}^3 = \mathbf{P}\mathbf{P}\mathbf{P}, \quad \text{and so on; and}$$
$$\mathbf{P}^0 = I \qquad\qquad\qquad \text{is the identity matrix.}$$

Meaning of the components of P^n

Meaning of the Components of P^n

The component at the intersection of row i and column j of matrix $\mathbf{P}^n$ reveals the probability of leaving state i at some step and arriving, n steps later, at state j. We use the notation $(\mathbf{P}^n)_{ij}$ to represent component (ij) of matrix $\mathbf{P}^n$.

Example 2   A Markov chain with three states has the transition matrix

$$\mathbf{P} = \begin{pmatrix} 0.2 & 0.4 & 0.4 \\ 0.4 & 0 & 0.6 \\ 0.5 & 0.3 & 0.2 \end{pmatrix}$$

What is the probability of proceeding from $S2$ (state 2) to $S1$ in one step? from $S1$ to $S3$ in two steps? from $S2$ to $S2$ in exactly 3 steps?

Solution

We are given P, and we need P^2 and P^3.

$$\mathbf{P}^2 = \mathbf{P}\mathbf{P} = \begin{pmatrix} 0.2 & 0.4 & 0.4 \\ 0.4 & 0 & 0.6 \\ 0.5 & 0.3 & 0.2 \end{pmatrix} \begin{pmatrix} 0.2 & 0.4 & 0.4 \\ 0.4 & 0 & 0.6 \\ 0.5 & 0.3 & 0.2 \end{pmatrix}$$

$$= \begin{pmatrix} 0.4 & 0.2 & 0.4 \\ 0.38 & 0.34 & 0.28 \\ 0.32 & 0.26 & 0.42 \end{pmatrix}$$

$$\mathbf{P}^3 = \mathbf{P}^2\mathbf{P} = \begin{pmatrix} 0.36 & 0.28 & 0.36 \\ 0.352 & 0.236 & 0.412 \\ 0.378 & 0.254 & 0.368 \end{pmatrix}$$

The probability of the transition $S2 \Rightarrow S1$ is simply component p_{21} of matrix $\mathbf{P}$. This component, at the intersection of row 2 and column 1, is 0.4. The transition

$$S1 \Rightarrow \text{any state} \Rightarrow S3$$

which is what we mean by transition to $S3$ from $S1$ in two steps, is component $(\mathbf{P}^2)_{13}$ of matrix $\mathbf{P}^2$.   [Do *not* make the common mistake of assuming that this is simply $(p_{13})^2 = (0.4)^2$.]   Read this value as 0.4.   Finally,

$$S2 \stackrel{3 \text{ steps}}{\Longrightarrow} S2$$

is component $(\mathbf{P}^3)_{22} = 0.236$.   (Notice that in this problem it is impossible to go from $S2$ to $S2$ in one step since $p_{22} = 0$.)

Skill Enhancer 2   Using this same transition matrix, determine these probabilities: $S2$ to $S2$ in one step, $S2$ to $S2$ in two steps, $S2$ to $S2$ in three steps, and $S2$ to $S1$ in three steps.

Answer in Appendix E.

P, $\mathbf{P}^2$, and $\mathbf{P}^3$ are all stochastic matrices.

Example 3   A certain metropolitan region boasts of three major television networks; call them 1, 2, and 3.   Channel 1 has conducted a study to test viewer loyalty, and determined that at the end of any half-hour segment, any viewer is as likely to stay with the current channel as switch to some other.   A viewer is likely to switch from channel 1 to 3 or from 3 to 2 with 30 percent probability.   Viewers watching channel 2 never switch to channel 3.   Assume that week night prime-time viewing consists of the four

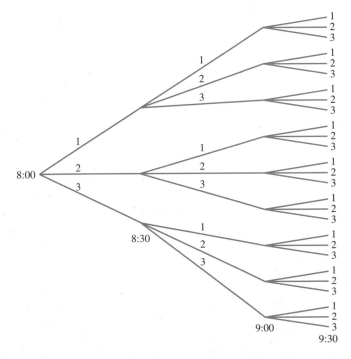

FIGURE 7.2

half-hour segments from 8:00 to 10:00 P.M. Set up the transition matrix for this problem. If a viewer starts with channel 2 at 8:00, may she ever watch channel 3 at 9:00? If a family starts watching channel 1 at 8:00, how likely is it that they will be watching channel 1 in the 9:30–10:00 segment? For the purposes of this problem, all television watchers view the entire prime time segment in its entirety—assume no one shuts off the television before 10:00, and no one turns on their TV after 8:00.

Solution

The steps in the Markov chain are the successive half-hour segments in TV prime time. From the problem,

$$p_{11} = p_{22} = p_{33} = 0.5$$
$$p_{13} = p_{32} = 0.3$$
$$p_{23} = 0$$

Because the sum of the row elements must be 1, we can deduce the remaining components of the transition matrix. This matrix becomes

$$
\mathbf{P} = \begin{array}{c} 1 \\ 2 \\ 3 \end{array}
\begin{pmatrix}
0.5 & 0.2 & 0.3 \\
0.5 & 0.5 & 0 \\
0.2 & 0.3 & 0.5
\end{pmatrix}
\quad \begin{array}{ccc} 1 & 2 & 3 \end{array}
$$

We will need the matrices $\mathbf{P}^2$ and $\mathbf{P}^3$.

$$
\mathbf{P}^2 = \mathbf{P}\,\mathbf{P} =
\begin{pmatrix}
0.5 & 0.2 & 0.3 \\
0.5 & 0.5 & 0 \\
0.2 & 0.3 & 0.5
\end{pmatrix}
\begin{pmatrix}
0.5 & 0.2 & 0.3 \\
0.5 & 0.5 & 0 \\
0.2 & 0.3 & 0.5
\end{pmatrix}
=
\begin{pmatrix}
0.41 & 0.29 & 0.3 \\
0.5 & 0.35 & 0.15 \\
0.35 & 0.34 & 0.31
\end{pmatrix}
$$

$$
\mathbf{P}^3 = \mathbf{P}^2\,\mathbf{P} =
\begin{pmatrix}
0.41 & 0.29 & 0.3 \\
0.5 & 0.35 & 0.15 \\
0.35 & 0.34 & 0.31
\end{pmatrix}
$$
$$
=
\begin{pmatrix}
0.410 & 0.317 & 0.273 \\
0.455 & 0.320 & 0.225 \\
0.407 & 0.333 & 0.260
\end{pmatrix}
$$

To see if someone watching channel 2 at 8:00 (step 0) may be watching channel 3 at 9:00 (step 2), we inspect component (23) of $\mathbf{P}^2$. If it is nonzero, this event may well happen. In this case, $(\mathbf{P}^2)_{23} = 0.15$.

The 9:30 program constitutes step 3 of the Markov chain. To determine the probability that a family viewing channel 1 at step 0 (8:00) will be viewing channel 1 regardless of the number of channel changes in between is given as $(\mathbf{P}^3)_{11}$, which is 0.410.

Skill Enhancer 3   A more extensive analysis of the data indicates now that a viewer is likely to switch from channel 1 to 3 or from 3 to 2 with a 50-percent probability.
(a) What is the transition matrix $\mathbf{P}$ now?
(b) Compute $\mathbf{P}^2$ and $\mathbf{P}^3$.
(c) If a viewer starts with channel 2 at 8:00, may she ever watch channel 3 at 9:00?

(d) If a family starts watching channel 1 at 8:00, how likely is it that they will be watching channel 1 in the 9:30–10:00 segment?

Answer in Appendix E.

_____ *Exercises 7.1* _____

1. Suppose there are two events A, B such that $P(A|A) = P(B|B) = p$ and $P(A|B) = P(B|A) = 1 - p$. What is the transition matrix for these events?

2. Suppose there are two events X, Y such that $P(X|X) = P(Y|Y) = 0.6$ and $P(X|Y) = P(Y|X) = 0.4$. What is the transition matrix for these events?

Which of the matrices in Exercises 3–8 are transition matrices? If a matrix is not a transition matrix, explain why not.

3. $\begin{pmatrix} 1 & 0 \\ 0 & 1 \end{pmatrix}$

4. $\begin{pmatrix} -0.5 & 1.5 \\ 0.2 & 0.8 \end{pmatrix}$

5. $\begin{pmatrix} 0.7 & 0 & 0.299 \\ 0.3 & 0.3 & 0.4 \end{pmatrix}$

6. $\begin{pmatrix} 0.2 & 0.2 & 0.3 \\ 1 & 0 & 0 \\ 0.8 & 0.15 & 0.05 \end{pmatrix}$

7. $\begin{pmatrix} 0.55 & 0.45 & 0 \\ 0 & 0 & 1 \\ 0 & 0.95 & 0.05 \end{pmatrix}$

8. $\begin{pmatrix} 0.3 & 0.3 & 0.3 \\ 0.3 & 0.3 & 0.3 \\ 0.3 & 0.3 & 0.4 \end{pmatrix}$

If a transition matrix $\mathbf{P}$ is given by $\begin{pmatrix} \frac{2}{3} & \frac{1}{3} \\ \frac{1}{2} & \frac{1}{2} \end{pmatrix}$, then compute $\mathbf{A}^1$ given the row matrix $\mathbf{A}^0$ in each of Exercises 9–20.

9. $\begin{pmatrix} 0 & 1 \end{pmatrix}$

10. $\begin{pmatrix} \frac{1}{3} & \frac{2}{3} \end{pmatrix}$

11. $\begin{pmatrix} \frac{4}{7} & \frac{3}{7} \end{pmatrix}$

12. $\begin{pmatrix} \frac{1}{2} & \frac{1}{2} \end{pmatrix}$

13. $\begin{pmatrix} 0.8 & 0.2 \end{pmatrix}$

14. $\begin{pmatrix} 0.2 & 0.8 \end{pmatrix}$

15. $\begin{pmatrix} 0.4 & 0.6 \end{pmatrix}$

16. $\begin{pmatrix} 0.7 & 0.3 \end{pmatrix}$

17. $\begin{pmatrix} p & 1 - p \end{pmatrix}$

18. $\begin{pmatrix} 1 - p & p \end{pmatrix}$

19. $\begin{pmatrix} 1 & 0 \end{pmatrix}$

20. $\begin{pmatrix} \frac{1}{6} & \frac{5}{6} \end{pmatrix}$

If a transition matrix $\mathbf{P}$ is given by $\begin{pmatrix} \frac{3}{4} & \frac{1}{4} \\ \frac{1}{3} & \frac{2}{3} \end{pmatrix}$, then compute $\mathbf{A}^1$ given the row matrix $\mathbf{A}^0$ in each of Exercises 21–26.

21. $\begin{pmatrix} 0 & 1 \end{pmatrix}$

22. $\begin{pmatrix} \frac{1}{3} & \frac{2}{3} \end{pmatrix}$

23. $\begin{pmatrix} \frac{4}{7} & \frac{3}{7} \end{pmatrix}$

24. $\begin{pmatrix} \frac{1}{2} & \frac{1}{2} \end{pmatrix}$

25. $\begin{pmatrix} 1 & 0 \end{pmatrix}$

26. $\begin{pmatrix} p & 1 - p \end{pmatrix}$

If a transition matrix $\mathbf{P}$ is given by $\begin{pmatrix} \frac{1}{3} & 0 & \frac{2}{3} \\ 0 & \frac{1}{3} & \frac{2}{3} \\ \frac{1}{3} & \frac{2}{3} & 0 \end{pmatrix}$, then compute $\mathbf{A}^1$ given the row matrix $\mathbf{A}^0$ in each of Exercises 27–32.

27. $\begin{pmatrix} 1 & 0 & 0 \end{pmatrix}$

28. $\begin{pmatrix} \frac{1}{4} & \frac{3}{4} & 0 \end{pmatrix}$

29. $\begin{pmatrix} \frac{1}{3} & \frac{1}{3} & \frac{1}{3} \end{pmatrix}$

30. $\begin{pmatrix} \frac{1}{2} & 0 & \frac{1}{2} \end{pmatrix}$

31. $\begin{pmatrix} \frac{1}{4} & \frac{1}{4} & \frac{1}{2} \end{pmatrix}$

32. $\begin{pmatrix} 0 & 0 & 1 \end{pmatrix}$

If a transition matrix $\mathbf{P}$ is given by $\begin{pmatrix} \frac{1}{3} & \frac{1}{3} & \frac{1}{3} \\ 0 & \frac{1}{2} & \frac{1}{2} \\ \frac{1}{2} & \frac{1}{4} & \frac{1}{4} \end{pmatrix}$, then compute $\mathbf{A}^1$ given the row matrix $\mathbf{A}^0$ in each of Exercises 33–38.

33. $\begin{pmatrix} \frac{1}{3} & \frac{1}{3} & \frac{1}{3} \end{pmatrix}$

34. $\begin{pmatrix} 0 & \frac{1}{2} & \frac{1}{2} \end{pmatrix}$

35. $\begin{pmatrix} 0.5 & 0.3 & 0.2 \end{pmatrix}$

36. $\begin{pmatrix} \frac{1}{2} & \frac{1}{3} & \frac{1}{6} \end{pmatrix}$

37. $\begin{pmatrix} 0.4 & 0.2 & 0.4 \end{pmatrix}$

38. $\begin{pmatrix} 0.1 & 0.1 & 0.8 \end{pmatrix}$

If a transition matrix $\mathbf{P}$ is given by $\begin{pmatrix} 0 & 1 & 0 \\ \frac{1}{2} & \frac{1}{4} & \frac{1}{4} \\ \frac{1}{8} & \frac{1}{2} & \frac{3}{8} \end{pmatrix}$ then compute $\mathbf{A}^1$ given the row matrix $\mathbf{A}^0$ in each of Exercises 39–44.

39. $\begin{pmatrix} \frac{1}{2} & \frac{1}{2} & 0 \end{pmatrix}$

40. $\begin{pmatrix} \frac{1}{3} & \frac{1}{3} & \frac{1}{3} \end{pmatrix}$

41. $\begin{pmatrix} 0.2 & 0.4 & 0.4 \end{pmatrix}$

42. $\begin{pmatrix} 0.45 & 0.1 & 0.45 \end{pmatrix}$

43. $\begin{pmatrix} 0 & 0.7 & 0.3 \end{pmatrix}$

44. $\begin{pmatrix} 0.3 & 0.5 & 0.2 \end{pmatrix}$

45. Let $\mathbf{P} = \begin{pmatrix} 0.6 & 0.4 \\ 0.4 & 0.6 \end{pmatrix}$ be a transition matrix for some Markov chain.

  (a) What is the probability that the chain will proceed from state 2 to state 1 in one step?

  (b) What is the probability of proceeding to state 2 from state 1 in a single step?

46. Suppose a transition matrix for a Markov chain has the form $\mathbf{P} = \begin{pmatrix} 1 & 0 \\ 0.7 & 0.3 \end{pmatrix}$.

  (a) Compute $\mathbf{P}^2$, $\mathbf{P}^3$.

  (b) What is the probability of proceeding from state 1 to state 2 in one step? 2 steps? 3 steps?

  (c) What is the probability of being in state 1 starting from state 2 in a single step? in 3 steps?

  (d) What is the probability of starting from state 1 and returning to state 1 in a single step? 2 steps? 3 steps?

  (e) As the number of steps gets larger and larger, speculate on the long-term behavior of the chain.

47. Suppose a Markov process has the transition matrix $\mathbf{P} = \begin{pmatrix} \frac{1}{8} & \frac{7}{8} \\ \frac{3}{4} & \frac{1}{4} \end{pmatrix}$.

  (a) Compute $\mathbf{P}^2$ and $\mathbf{P}^3$.

  (b) What is the probability of making the transition from state 1 to state 1 in one step? state 2 to state 1 in a single step?

  (c) Suppose that at the initial step, the process will be in either state with equal probability. What row matrix describes the system after the first step? after the second step?

  (d) Suppose that the system will be in state 2 with odds of 3 to 1. What row matrix describes the system after an additional two steps have transpired?

48. A transition matrix $\mathbf{P} = \begin{pmatrix} 0.5 & 0 & 0.5 \\ 0 & 0.5 & 0.5 \\ 0.5 & 0.5 & 0 \end{pmatrix}$ governs a three-state Markov process.

  (a) Compute $\mathbf{P}^2$ and $\mathbf{P}^3$.

  (b) What is the probability of going from state 1 to state 2 in a single step? in two steps? in three steps?

  (c) At some point, a row matrix describing the probabilities of being in the three states is $\begin{pmatrix} 0.4 & 0.5 & 0.1 \end{pmatrix}$. What row matrix describes the states of the system in one step? in two steps?

49. Suppose a transition matrix for a three-state Markov process has the form $\mathbf{P} = \begin{pmatrix} 1 & 0 & 0 \\ 0 & 1 & 0 \\ 0 & 0 & 1 \end{pmatrix}$. Given some initial state matrix $\mathbf{A}^0$, what can you say about the *long-term* behavior of this chain? That is, after a large number n of steps, what will the form of $\mathbf{A}^n$ be? What will the form of $\mathbf{A}^n$ be after a few steps?

50. Suppose a transition matrix for a three-state Markov chain has the form $\mathbf{P} = \begin{pmatrix} 0 & 1 & 0 \\ 0 & 0 & 1 \\ 1 & 0 & 0 \end{pmatrix}$. Let the initial probability row matrix be $\mathbf{A}^0 = \begin{pmatrix} a_1 & a_2 & a_3 \end{pmatrix}$.

  (a) What is $\mathbf{A}^1$? What is $\mathbf{A}^2$? What is $\mathbf{A}^3$? What is $\mathbf{A}^4$?

  (b) What is $\mathbf{A}^n$ for arbitrary n? (*Hint*: Try to determine and generalize a pattern that the answers to the first part of this question generate.)

51. Show why the sum of the elements in any *row* of any transition matrix must be 1.

52. Refer to the transition matrix in Example 2. Show that the three matrices $\mathbf{P}$, $\mathbf{P}^2$, and $\mathbf{P}^3$ are stochastic.

53. In the discussion at the beginning of this section (and preceding Example 1) of the text, we asserted that the probability of sun tomorrow is the first component in a matrix multiplication of a row matrix times the transition matrix. Show this, and show that the probability of rain tomorrow is the second component of this product.

54. **(a)** Why must the initial state row matrix for a Markov process be stochastic?

  (b) Must the row matrix describing successive states in a Markov chain be stochastic? Why or why not?

Applications

55. *(Higher Education)* A professor's exams are either easy or hard—there seems to be no middle ground. If the exam was easy last time, it will be easy this time with a 40 percent probability. If it was hard last time, it will be easy this time with a 50 percent probability.

  (a) The first exam of the semester was easy. What is the probability that the second will be hard?

  (b) The professor will give three exams throughout the semester. If the first was difficult, what can you say about the second and third exams?

56. *(Urban Demography)* Many American cities have in recent years been plagued by the flight of the middle class to the suburbs. Now, some of these people are returning to historic regions of the central cities. The Department of Urban Planning of one typical city reports that 5 percent of the middle-class population move from the city to one of its suburbs each year, but that 3 percent return from some suburb to the city proper.

(a) If the population movement during one year counts as a single step in a Markov process, set up the Markov process transition matrix.

(b) In the year the Department of Urban Planning started keeping records, 65 percent of the middle-class population under consideration lived in the central city. What will that statistic be for the next year, assuming the Markov process model applies?

(c) What population distribution is predicted for the second year of this study?

57. *(Public Transportation)* The streets of many major cities are congested with cars. The City of Missasaucki instituted an intensive campaign advocating public bus transportation. After the campaign was well underway, it was determined that a commuter who normally travels to work by car is 80 percent likely to switch to taking buses in the next month. On the other hand, because the bus system is so archaic, bus travelers are apt to switch back to cars in the following month with a probability of 60 percent.

(a) Set up a Markov process with two states, and write the appropriate stochastic matrix.

(b) At the commencement of the study, commuters are evenly divided between people who drive and those who take the bus. What will the distribution be in the second year of the study? In the third year?

58. *(Wine Retailing)* A small New York vineyard, producer of the highly regarded Mendicant Friars (MF) label of wines, has been acquired by an aggressive conglomerate that wants to break into the lucrative table wine market. The major competitors are the Happy Valley (HV) and Fuchsia Nun (FN) labels. Preliminary marketing studies indicate what the results of a multimillion-dollar advertising campaign might be. After one month, 40 percent and 50 percent of the HV and FN drinkers will switch to MF. Eighty percent of the MF drinkers continue to drink this wine after any given month. Twenty percent of HV and FN drinkers switch to each other's brands after any given month. Equal percentages of MF drinkers revert to HV and FN vintages after any given month.

(a) What transition matrix describes this situation?

(b) At the start of the campaign, equal 40 percent shares of the market consume HV and FN wines. What will the percentages describing market shares be after a single month? After two months?

(c) The administrators in this company would like to know the eventual market share they can count on. Without performing any calculations, suggest how you might determine whether or not there is a unique

answer to that question, and what the numerical percentage might be.

59. *(Tendencies Towards Criminal Behavior)* Studies of 18–30-year-old males indicate that in any given year 5 percent of them will commit some felonious act that leads to arrest and a short prison term. Among youthful offenders, 70 percent never again exhibit this criminal behavior.

(a) Model this description as a Markov process, and write the transition matrix for the process.

(b) In one graduating class of 18-year-old seniors, no one of them has shown any criminal behavior. What do you expect the percentages to be a year hence? two years hence?

(c) In another group, a full third of a group of 18-year-old graduating senior boys has an arrest record. What will the percentages be a year down the road? Two years down the road?

60. *(Weather Prediction Revisited)* A Markov chain description of weather prediction might be made more realistic by allowing for more states—weather conditions—on a given day. Suppose some resort has weather that can unambiguously be described as sunny, cloudy, or rainy. Sunny days are followed by sunny days in half the cases, and by rainy days in one-quarter of the cases. Cloudy days are followed by sun half the time, and by another cloudy day with probability $\frac{1}{3}$. Rainy days are always followed by sunny days.

(a) Set up a transition matrix.

(b) You arrive at this resort one rainy Friday. Assess the possible weather conditions for that weekend, including Monday, the day you depart.

61. *(Animal Training)* Killer whales are not that ferocious, and can be trained to do tricks much like a family dog, except that the whales are never 100 percent reliable. Suppose the sea animal is completely trained. Even then, if the animal has successfully performed one trick, there is only a 90 percent chance that it will perform the same trick the next day. On the other hand, if the whale has *not* performed the trick on command, the likelihood that it will do so the next day is 60 percent. Set up a transition matrix for this problem. Assess the probability that if the whale performs the trick on Friday, it will perform it on Sunday. (It attempts the trick once a day.)

62. *(Buying Cars)* American consumers can choose to buy an American car or a Japanese car. Studies show that the owner of an American car will purchase an American car as her next car with probability 0.4, whereas owners of a Japanese vehicle will "buy Japanese" for the next car with probability 0.8. Set up a transition matrix for this problem. What is the probability that if a teenager receives a Japanese car as a graduation present from her parents, her next car will be Japanese? that her third car will be Japanese?

63. **(Economics)** The fortunes of the computer industry in California's Silicon Valley are closely watched. One economist has devised a measure of this industry's performance that appears to exhibit Markov behavior. If this indicator rises one month, it will rise the next month with a 50 percent probability. If it falls one month, it will fall the next month with a 25 percent probability. If the indicator falls in January, what is the probability that it will fall in February? in March?

64. **(Animal Behavior Studies)** Rats will willingly learn to run a T maze. One experiment is so set up that if a rat runs to the right on one trial, it will run to the right on the next with a probability of $\frac{3}{4}$. If it runs to the left on one trial, it will run to the left on the next trial with probability $\frac{1}{3}$. Set up a transition matrix for this

problem. Suppose a rat is placed at the entrance to the maze. For its initial run, it will go to the left or right with equal probability. What is the probability that it will turn to the left on the second trial? What is the probability that it will turn to the left on the third trial?

W 65. We used a simple model of weather prediction as an introduction to Markov chains. What factors might make weather prediction an unsuitable subject for Markov chain analysis?

W 66. **(Advertising)** Why might advertisers testing product loyalty be interested in Markov chain analysis?

Calculator Exercises

67. For $\mathbf{A}^0 = (0.45 \quad 0.55)$ and $\mathbf{P} = \begin{pmatrix} 0.21 & 0.79 \\ 0.63 & 0.37 \end{pmatrix}$, compute $\mathbf{A}^1$.

68. For $\mathbf{A}^0 = (0.33 \quad 0.67)$ and $\mathbf{P} = \begin{pmatrix} 0.26 & 0.74 \\ 0.51 & 0.49 \end{pmatrix}$, compute $\mathbf{A}^1$ and $\mathbf{A}^2$.

Section 7.2   REGULAR MARKOV CHAINS

What can be said about the *long-term behavior* of Markov chains? An acceptable answer to this question will help us resolve two issues.

1. The components of the transition matrices $\mathbf{P}^n$ for a Markov chain can vary significantly as the number of steps n changes. Will these components ever "settle down" to values that vary little from one step to the next?

2. Given some initial probability row matrix $\mathbf{A}^0$ that describes the system at some particular step, we can construct a row matrix $\mathbf{A}^n$ whose components, being probabilities, describe the system at some future step n. As the number of steps n becomes large, how much will the components of $\mathbf{A}^n$ change from step to step?

As a first step towards answering these questions, let us choose some $\mathbf{P}$ and compute successive powers of $\mathbf{P}$ to see if any significant patterns emerge. In the first section, we used the transition matrix

$$\mathbf{P} = \begin{pmatrix} 0.5 & 0.5 \\ 0.25 & 0.75 \end{pmatrix}$$

to help us analyze weather. The table in Figure 7.3 displays the components of $\mathbf{P}^n$ for various values of n. We write all four components in a row for easier visualization.

Each of these components *does* appear to approach some limiting value. Let $\mathbf{W}$ be the matrix whose components are these limiting values. By inspection, $\mathbf{W}$ appears to be

$$\mathbf{W} = \begin{pmatrix} \frac{1}{3} & \frac{2}{3} \\ \frac{1}{3} & \frac{2}{3} \end{pmatrix}$$

(This demonstration is not at all rigorous.) At least for this Markov chain, the transition matrices $\mathbf{P}^n$ do reach some stable value for values of n above $n = 7$.

We can use this example to investigate the second of the two questions we posed at the beginning of the section. Although we know that

$$\mathbf{A}^n = \mathbf{A}^0 \, \mathbf{P}^n$$

exactly, it appears in this example that for large n, $\mathbf{P}^n$ is essentially the same as $\mathbf{W}$. In cases similar to this example, we therefore surmise that the equality

$$\mathbf{A}^{\text{limit}} = \mathbf{A}^0 \, \mathbf{W}$$

yields the limiting value for $\mathbf{A}^n$, which we denote by $\mathbf{A}^{\text{limit}}$.

We may apply these results to the weather problem. If it is sunny today, what is the weather likely to be several weeks from now? An answer is given by

$$\begin{aligned} \mathbf{A}^{\text{limit}} &= \begin{pmatrix} 1 & 0 \end{pmatrix} \times \mathbf{W} \\ &= \begin{pmatrix} 1 & 0 \end{pmatrix} \begin{pmatrix} \frac{1}{3} & \frac{2}{3} \\ \frac{1}{3} & \frac{2}{3} \end{pmatrix} \\ &= \begin{pmatrix} \frac{1}{3} & \frac{2}{3} \end{pmatrix} \end{aligned}$$

If it is sunny today, then several weeks from now, on any day, it will always be twice as likely to rain.

n	p_{11}	p_{12}	p_{21}	p_{22}
1	0.5000	0.5000	0.2500	0.7500
2	0.3750	0.6250	0.3125	0.6875
3	0.3438	0.6562	0.3281	0.6719
4	0.3360	0.6640	0.3320	0.6680
5	0.3340	0.6660	0.3330	0.6670
6	0.3335	0.6665	0.3332	0.6668
7	0.3334	0.6666	0.3333	0.6667
8	0.3333	0.6667	0.3333	0.6667
9	0.3333	0.6667	0.3333	0.6667

FIGURE 7.3

Successive components of P^n for various values of n.

Let us perform these calculations again, supposing now that it is raining today. Several weeks from now, the probabilities for the weather will be

$$\mathbf{A}^{\text{limit}} = \begin{pmatrix} 0 & 1 \end{pmatrix} \begin{pmatrix} \frac{1}{3} & \frac{2}{3} \\ \frac{1}{3} & \frac{2}{3} \end{pmatrix}$$

$$= \begin{pmatrix} \frac{1}{3} & \frac{2}{3} \end{pmatrix}$$

Again, rain will be twice as likely as sun. In this case, *the limit probabilities are the same regardless of the initial state of the system.* Is this coincidence, or should we expect this pattern? In fact, when the limiting transition matrix $\mathbf{W}$ is well defined, then $\mathbf{A}^{\text{limit}}$ will *not* depend on the initial state of the Markov chain, as we have here observed.

Regular Matrices

regular matrix   A transition matrix $\mathbf{P}$ is *regular* if all the elements of $\mathbf{P}^n$ are strictly greater than zero for at least one value of n. The importance of regular matrices lies in their connection with limit matrices.

Example 4   Which of the following are regular matrices?

(a) $\begin{pmatrix} 0.5 & 0.5 \\ 1 & 0 \end{pmatrix}$ (b) $\begin{pmatrix} \frac{1}{3} & \frac{1}{2} & \frac{1}{6} \\ \frac{1}{2} & \frac{1}{4} & \frac{1}{4} \\ \frac{1}{20} & \frac{1}{20} & \frac{9}{10} \end{pmatrix}$ (c) $\mathbf{Q} = \begin{pmatrix} 0 & 1 & 0 \\ 0.5 & 0 & 0.5 \\ 0 & 1 & 0 \end{pmatrix}$

Solution

(a) Squaring this matrix, we find that

$$\begin{pmatrix} 0.5 & 0.5 \\ 1 & 0 \end{pmatrix}^2 = \begin{pmatrix} 0.75 & 0.25 \\ 0.5 & 0.5 \end{pmatrix}$$

Since all components of this right-hand matrix are positive, the given matrix is regular.
(b) All components of this matrix as given are greater than zero, so this matrix is regular.
(c) Since several components of $\mathbf{Q}$ are zero, we need to check successive powers. Observe first that

$$\mathbf{Q}^2 = \begin{pmatrix} 0 & 1 & 0 \\ 0.5 & 0 & 0.5 \\ 0 & 1 & 0 \end{pmatrix}^2 = \begin{pmatrix} 0.5 & 0 & 0.5 \\ 0 & 1 & 0 \\ 0.5 & 0 & 0.5 \end{pmatrix}$$

Define a new matrix

$$\mathbf{R} = \mathbf{Q}^2 = \begin{pmatrix} 0.5 & 0 & 0.5 \\ 0 & 1 & 0 \\ 0.5 & 0 & 0.5 \end{pmatrix}$$

Next, we find that $\mathbf{Q}^3$ is

$$\mathbf{Q}^3 = \mathbf{R} \times \mathbf{Q} = \begin{pmatrix} 0.5 & 0 & 0.5 \\ 0 & 1 & 0 \\ 0.5 & 0 & 0.5 \end{pmatrix} \begin{pmatrix} 0 & 1 & 0 \\ 0.5 & 0 & 0.5 \\ 0 & 1 & 0 \end{pmatrix}$$

$$= \begin{pmatrix} 0 & 1 & 0 \\ 0.5 & 0 & 0.5 \\ 0 & 1 & 0 \end{pmatrix} = \mathbf{Q}$$

Continuing in this way, we can show that $\mathbf{Q}^n$ is either $\mathbf{Q}$ (if n is odd) or $\mathbf{R}$ (for even n). In other words, powers of $\mathbf{Q}$ alternate between $\mathbf{Q}$ and $\mathbf{R}$. Since both $\mathbf{Q}$ and $\mathbf{R}$ have several zero components, $\mathbf{Q}$ is not regular. (Notice that $\mathbf{Q}^n$ will therefore *not* tend to a limiting value for large value of n.)

Skill Enhancer 4   Which are regular? Why or why not? $\mathbf{L} = \begin{pmatrix} 0 & 1 \\ \frac{1}{3} & \frac{2}{3} \end{pmatrix}$; $\mathbf{M} = \begin{pmatrix} 0.001 & 0.999 \\ 0.999 & 0.001 \end{pmatrix}$; $\mathbf{N} = \begin{pmatrix} 0.5 & 0 & 0.5 \\ 0 & 1 & 0 \\ 0.5 & 0 & 0.5 \end{pmatrix}$.

Answer in Appendix E.

The following theorem establishes the relationship between regular transition matrices and the long-term behavior of Markov chains. We omit the proof.

> **Theorem 7.1   Regular Transition Matrices** If the transition matrix in any Markov process is *regular,* then:
>
> 1. The matrix product $\mathbf{P}^n$ approaches a limit matrix $\mathbf{W}$ as n gets larger and larger.
> 2. Each row of $\mathbf{W}$ is identical to every other row of $\mathbf{W}$.

How may we determine the components of $\mathbf{W}$? This is our next task.

Limit Values of $w = \mathbf{A}^{\text{limit}}$

Let us use $\mathbf{w}$ to denote the $\mathbf{A}^{\text{limit}}$ matrix.

What does it mean for a row vector $\mathbf{w}$ to be a limit? As the number of steps in the chain increases, the difference between consecutive matrices $\mathbf{A}^n$ should become closer and closer to the zero matrix. That is, for ever-increasing values of n, we expect that any differences between $\mathbf{A}^n$ and $\mathbf{A}^{n-1}$ will become less and less important.

Eventually this difference should become smaller than we can distinguish on any practical level; assume that at some step n we can no longer measure the difference between successive row matrices $\mathbf{A}^n$, $\mathbf{A}^{n+1}$, and so on. Since

$$\mathbf{A}^{n+1} = \mathbf{A}^n \mathbf{P}$$

in the limit we expect $\mathbf{A}^{\text{lim}} = \mathbf{w}$ to be unchanged when multiplied by $\mathbf{P}$. That is,

$$(7.5) \qquad\qquad \mathbf{w} = \mathbf{w}\mathbf{P}$$

which is the defining equation for $\mathbf{w}$. Since all row matrices $\mathbf{A}^n$ are probability row matrices, so is $\mathbf{w}$, and we must further have that

$$(7.6) \qquad\qquad w_1 + w_2 + \cdots + w_m = 1$$

fixed-point matrix

where the w_i are the components of **w** and there are m states in this Markov process. Equations (7.5) and (7.6) together determine **w**. Any matrix **w** satisfying an equation like Equation (7.5) is called a *fixed-point matrix*.

Example 5   In the weather problem with which the chapter began, the transition matrix is

$$\mathbf{P} = \begin{pmatrix} 0.5 & 0.5 \\ 0.25 & 0.75 \end{pmatrix}$$

What is the fixed-point matrix **w** satisfying $\mathbf{w} = \mathbf{wP}$, and what is its significance in Markov chain analysis?

Solution

Let $\mathbf{w} = \begin{pmatrix} u & v \end{pmatrix}$. Then, we seek solutions to

$$\begin{pmatrix} u & v \end{pmatrix} = \begin{pmatrix} u & v \end{pmatrix} \begin{pmatrix} 0.5 & 0.5 \\ 0.25 & 0.75 \end{pmatrix} \quad \text{and} \quad u + v = 1$$

Upon performing the matrix multiplication we learn that

$$\begin{pmatrix} u & v \end{pmatrix} = \begin{pmatrix} 0.5u + 0.25v & 0.5u + 0.75v \end{pmatrix}$$

Since two matrices can be equal only if each of their corresponding components are equal, we must have

$$u = 0.5u + 0.25v \qquad v = 0.5u + 0.75v$$

Both of these equations reduce to $2u = v$; this plus the requirement that $u + v = 1$ allows the conclusion that

$$u = \frac{1}{3}$$
$$v = \frac{2}{3}$$

The significance of **w** lies in the fact the two components give the probability for the weather many days from now.

Skill Enhancer 5   Suppose a transition probability matrix is given by

$$\begin{pmatrix} \frac{1}{3} & \frac{2}{3} \\ \frac{1}{2} & \frac{1}{2} \end{pmatrix}$$

What are conditions that a fixed-point matrix $\mathbf{w} = \begin{pmatrix} u & v \end{pmatrix}$ must satisfy? What is **w** for this transition matrix?

Answer in Appendix E.

Determining *W*

The next issue is the determination of **W** itself. Example 5 suggests how. We know that all the rows of **W** are identical. (Why?) Might there be a relationship between each row and **w**? For the weather problem, each row of **W** is precisely identical with **w**, and this is more than coincidence. In general, *each row of the matrix* **W** *is a copy of the row matrix* **w**. (We omit the proof.)

Properties of the Limit Matrix W

1. Each row of **W** is the row matrix **w**, where **w** is the limit value of $\mathbf{A}^n$ for large n.

2. Determine **w** by solving the equation

$$\mathbf{w} = \mathbf{w}\,\mathbf{P}$$

together with the condition $w_1 + w_2 + \cdots + w_m = 1$. Here, the w_i are the components of **W**. **P** must be regular.

3. For *any* probability row matrix **A**, then the matrix product $\mathbf{A}^n\mathbf{P}$ approaches **w** after a large number n of steps have occurred.

Example 6   Calculate the fixed-point matrix for the regular matrix

$$\mathbf{P} = \begin{pmatrix} \frac{1}{2} & \frac{1}{4} & \frac{1}{4} \\ \frac{1}{8} & \frac{3}{4} & \frac{1}{8} \\ \frac{1}{4} & \frac{1}{4} & \frac{1}{2} \end{pmatrix}$$

As n gets larger and larger, what matrix does $\mathbf{P}^n$ approach?

Solution

The fixed-point matrix $\mathbf{w} = \begin{pmatrix} w_1 & w_2 & w_3 \end{pmatrix}$ satisfies

$$\mathbf{w} = \mathbf{w}\,\mathbf{P}$$

and $w_1 + w_2 + w_3 = 1$. We can carry out the matrix multiplication, and equate the components in the matrix product. These equations, together with the condition that **w** be stochastic, will determine **w**. In this example, we will use the first two components of the product plus the stochastic condition $w_1 + w_2 + w_3 = 1$ to obtain

$$\frac{1}{2}w_1 + \frac{1}{8}w_2 + \frac{1}{4}w_3 = w_1$$
$$\frac{1}{4}w_1 + \frac{3}{4}w_2 + \frac{1}{4}w_3 = w_2$$
$$1 - w_1 - w_2 = w_3$$

Substituting the expression for w_3 from the third equation into the first two reduces the system of three equations to two:

$$6w_1 + w_2 = 2$$
$$w_2 = \frac{1}{2}$$

The final solution is then $w_1 = \frac{1}{4}$, $w_2 = \frac{1}{2}$, and $w_3 = \frac{1}{4}$. The limit form of **P** is a 3×3 matrix with identical rows:

$$\begin{pmatrix} \frac{1}{4} & \frac{1}{2} & \frac{1}{4} \\ \frac{1}{4} & \frac{1}{2} & \frac{1}{4} \\ \frac{1}{4} & \frac{1}{2} & \frac{1}{4} \end{pmatrix}$$

Skill Enhancer 6 We need the fixed-point matrix $\mathbf{w} = \begin{pmatrix} u & v & w \end{pmatrix}$ for

$$\mathbf{S} = \begin{pmatrix} \frac{1}{4} & \frac{1}{2} & \frac{1}{4} \\ \frac{1}{2} & \frac{1}{4} & \frac{1}{4} \\ \frac{1}{4} & \frac{1}{4} & \frac{1}{2} \end{pmatrix}$$

For large n, what matrix $\mathbf{W}$ does $\mathbf{P}^n$ approach?

Answer in Appendix E.

Example 7 The President of the United States has been in isolation, studying the issues, and deciding whether to veto the recent tax reform program that came before him. At last, his decision is made. Thoughtlessly, he whispers his decision—yea or nay—to one individual, who whispers it to one friend, and so on. Slowly but surely, many people later, the news reaches a reporter for the *Philadelphia Enquirer*, who determines to print the story as truth. What is the probability that the reporter prints a *true* story of the President's veto?

Solution

Assume that there is a strictly positive probability p that any person in the chain will incorrectly hear the rumor that his or her predecessor communicates. Any person in the chain of rumor-mongers thinks that the President will either support or veto the tax program, so he or she stands in one of the two states of a Markov process. The transition matrix for this chain is

$$\mathbf{P} = \begin{matrix} yea \\ nay \end{matrix} \begin{pmatrix} \overset{yea}{1-p} & \overset{nay}{p} \\ p & 1-p \end{pmatrix}$$

FIGURE 7.4

The fixed-point row matrix for **P** is $\mathbf{w} = \begin{pmatrix} u & v \end{pmatrix}$, which satisfies the matrix equation

$$\begin{pmatrix} u & v \end{pmatrix} = \begin{pmatrix} u & v \end{pmatrix} \begin{pmatrix} 1-p & p \\ p & 1-p \end{pmatrix}$$

and the additional equation $u + v = 1$. Upon expansion of the matrix equation, we have

$$u = (1-p)u + \quad p \quad v$$
$$v = \quad p \quad u + (1-p)v$$

both of which reduce to $pu = pv$, or

$$u = v$$

(Why may we cancel p in this equation?) This equation may be solved together with $u + v = 1$ to yield the solutions

$$u = \frac{1}{2}$$
$$v = \frac{1}{2}$$

which means that the long-term behavior of the spread of rumor (at least in this model) depends *in no way* on the value of the probability p. Thus, no matter how carefully people listen to their informer, provided there is some chance that they hear incorrectly, and provided that this probability is the same for each person in the chain, then far enough along the chain, any individual, including the *Enquirer* reporter, has only a 50 percent chance of hearing the correct story. Do not believe everything you hear or read!

Skill Enhancer 7 Compute $\mathbf{P}^2$ and $\mathbf{P}^3$ when $p = \frac{1}{2}$. Compute $\mathbf{P}^2$ and $\mathbf{P}^3$ when $p = \frac{1}{3}$.

Answer in Appendix E.

_____Exercises 7.2_____

Each of Exercises 1–4 displays a transition matrix **P** for some Markov process. Determine the limit matrix **W** for each by calculating successive powers of **P**. For each problem, prepare a table similar to the one in Figure 7.3 at the beginning of this section.

1. $\begin{pmatrix} \frac{1}{4} & \frac{3}{4} \\ \frac{3}{4} & \frac{1}{4} \end{pmatrix}$

2. $\begin{pmatrix} \frac{1}{3} & \frac{2}{3} \\ \frac{1}{2} & \frac{1}{2} \end{pmatrix}$

3. $\begin{pmatrix} \frac{2}{5} & \frac{3}{5} \\ 0 & 1 \end{pmatrix}$

4. $\begin{pmatrix} 0 & 1 & 0 \\ \frac{1}{3} & \frac{1}{3} & \frac{1}{3} \\ \frac{1}{2} & 0 & \frac{1}{2} \end{pmatrix}$

In Exercises 5–12, decide which of the given matrices is regular. If it is regular, determine the fixed-point matrix.

5. $\begin{pmatrix} 0.08995 & 0.91005 \\ 0.612 & 0.382 \end{pmatrix}$

6. $\begin{pmatrix} 0 & 0 & 1 \\ 0 & 0.7 & 0.3 \\ 0.2 & 0.6 & 0.2 \end{pmatrix}$

7. $\begin{pmatrix} 0 & 0 & 0 \\ 0.3 & 0.4 & 0.3 \\ 0.5 & 0 & 0.5 \end{pmatrix}$

8. $\begin{pmatrix} 0.4 & 0 & 0.6 \\ 0.8 & 0.1 & 0.1 \\ 0 & 1 & 0 \end{pmatrix}$

9. $\begin{pmatrix} 1 & 0 \\ \frac{1}{3} & \frac{2}{3} \end{pmatrix}$

10. $\begin{pmatrix} 0 & 1 \\ \frac{3}{4} & \frac{1}{4} \end{pmatrix}$

11. $\begin{pmatrix} 0.65 & 0.35 \\ 0 & 1 \end{pmatrix}$ **12.** $\begin{pmatrix} \frac{3}{8} & \frac{5}{8} \\ 1 & 0 \end{pmatrix}$

Determine the limit matrix **W** for each of the transition matrices in Exercises 13–24. (Use any method.) If limit matrices do not exist, state why.

13. $\begin{pmatrix} 1 & 0 \\ 0 & 1 \end{pmatrix}$ **14.** $\begin{pmatrix} 0 & 1 \\ 1 & 0 \end{pmatrix}$

15. $\begin{pmatrix} 0.5 & 0.5 \\ 0.5 & 0.5 \end{pmatrix}$ **16.** $\begin{pmatrix} 0.6 & 0.4 \\ 0.4 & 0.6 \end{pmatrix}$

17. $\begin{pmatrix} p & 1-p \\ 1-p & p \end{pmatrix}$ **18.** $\begin{pmatrix} \frac{1}{3} & \frac{1}{3} & \frac{1}{3} \\ \frac{1}{3} & \frac{1}{3} & \frac{1}{3} \\ \frac{1}{3} & \frac{1}{3} & \frac{1}{3} \end{pmatrix}$
$0 \le p \le 1$

19. $\begin{pmatrix} 0.4 & 0.3 & 0.3 \\ 0.4 & 0.3 & 0.3 \\ 0.4 & 0.3 & 0.3 \end{pmatrix}$ **20.** $\begin{pmatrix} \frac{1}{6} & \frac{1}{2} & \frac{1}{3} \\ \frac{1}{6} & \frac{1}{2} & \frac{1}{3} \\ \frac{1}{6} & \frac{1}{2} & \frac{1}{3} \end{pmatrix}$

21. $\begin{pmatrix} a & b & b \\ b & a & b \\ b & b & a \end{pmatrix}$, $a + 2b = 1$

22. $\begin{pmatrix} \frac{1}{2} & \frac{1}{2} \\ \frac{1}{3} & \frac{2}{3} \end{pmatrix}$ **23.** $\begin{pmatrix} \frac{1}{3} & \frac{2}{3} \\ \frac{1}{2} & \frac{1}{2} \end{pmatrix}$

24. $\begin{pmatrix} \frac{1}{2} & \frac{1}{2} \\ \frac{1}{4} & \frac{3}{4} \end{pmatrix}$

Calculate the fixed-point row matrices for Exercises 25–36. Each of these transition matrices is (clearly) regular.

25. $\begin{pmatrix} 0.5 & 0.5 \\ 0.4 & 0.6 \end{pmatrix}$ **26.** $\begin{pmatrix} 0.1 & 0.9 \\ 0.9 & 0.1 \end{pmatrix}$

27. $\begin{pmatrix} 0.2 & 0.8 \\ 0.3 & 0.7 \end{pmatrix}$ **28.** $\begin{pmatrix} \frac{3}{4} & \frac{1}{4} \\ \frac{9}{10} & \frac{1}{10} \end{pmatrix}$

29. $\begin{pmatrix} 0.6 & 0.4 \\ 0.55 & 0.45 \end{pmatrix}$ **30.** $\begin{pmatrix} \frac{1}{3} & \frac{1}{3} & \frac{1}{3} \\ \frac{1}{4} & \frac{1}{2} & \frac{1}{4} \\ \frac{1}{6} & \frac{1}{3} & \frac{1}{2} \end{pmatrix}$

31. $\begin{pmatrix} 0.3 & 0.7 \\ 0.3 & 0.7 \end{pmatrix}$ **32.** $\begin{pmatrix} 0.6 & 0.4 \\ 0.6 & 0.4 \end{pmatrix}$

33. $\begin{pmatrix} 0.8 & 0.2 \\ 0.2 & 0.8 \end{pmatrix}$ **34.** $\begin{pmatrix} 0.6 & 0.4 \\ 0.4 & 0.6 \end{pmatrix}$

35. $\begin{pmatrix} p & 1-p \\ 1-p & p \end{pmatrix}$, $0 \le p \le 1$

36. $\begin{pmatrix} 0.2 & 0.4 & 0.4 \\ 0.4 & 0.2 & 0.4 \\ 0.4 & 0.4 & 0.2 \end{pmatrix}$

W 37. A certain square matrix **P** is a regular, stochastic matrix all of whose elements are greater than zero. Why must the elements of all powers of **P** also be greater than zero?

W 38. What step in the solution of Example 7 depends on the inequality $p > 0$?

Applications

39. *(Genetics)* The trait of left-handedness seems to be prevalent in a particular British aristocratic family. If a man in this family is a lefty, his son will be lefty with a probability of $\frac{3}{4}$. If a man is right-handed, then his son will be a lefty with probability of $\frac{1}{3}$. Assume that each man in this family has at least one son as far back as records go, some several hundred years. With no knowledge about the handedness of the current lord, what do you predict will be the handedness of the newborn male heir to the title, and with what probability?

40. *(Genetics)* *Polycystic kidney disease* is a disease of the kidneys that usually leads to complete kidney failure in the victim by age 50. This disease occurs almost exclusively in men and must be inherited from one's father, but it poorly obeys the textbook laws of genetics. If one man has it, the chances are 60 percent that any of his sons will have it. If one man in this family does not have this condition, the chances are still 25 percent that any of his sons will have it. A boy is born into a family in which this disease occurs. What are his chances of being stricken with this disease?

41. *(Voting)* The voting patterns of several small New England hamlets seem to depend on past voting behavior rather than on the current candidates. If one such town voted Democratic in the last election, it will vote Republican, Independent, or Democratic with equal likelihood. If, on the other hand, the town voted Republican in that election, it will elect the Democratic candidate with a 50 percent likelihood; Republican and Independent victories are equally likely. Finally, if the town elected an Independent official in the last election, it will return that official with a probability of $\frac{3}{5}$. Again, the remaining two parties will win the election with equal likelihoods.
   (a) If a Republican official is elected in one election, what is the probability that this official will be reelected?

(b) A Democratic official was elected shortly after World War II. Knowing nothing else about the elections that have been held in this particular town, speculate on the probabilities that the current Republican, Democratic, and Independent official have of being elected. How do these probabilities change if you learn that a Republican official was elected before the outset of the Korean War?

42. *(Musical Ability)* In some famous families, such as that of Johann Sebastian Bach, musical talent seems to be an inherited characteristic because so many members of these families over long periods of time are musically gifted. Suppose this process forms a Markov process, such that a gifted member of this family is apt to have a gifted child with probability 75 percent. A member of this family with no special talents is still apt to have a musically gifted child with probability 40 percent. If a daughter is born into one such family, with what probability will she develop into a gifted musician?

43. *(Bad Habits)* Smoking is a good example of a bad habit that is difficult to shake. If a smoker smokes a cigarette one day, the chances that he will smoke the next day are 70 percent. If the smoker gets through the day without smoking, his chances of smoking the next day are 40 percent. What are this smoker's long-term chances of not smoking?

44. *(Consumer Shopping Patterns)* Family shoppers buy new packages of cold cereal once a week on the average. United Grains is interested in introducing a new bran cereal, Bran Clumps, and is positioning this product against the two biggest competing bran cereals, call them A and B. Market data give the probabilities that families will switch brands from one week to the next, and these data are summarized in the table below. The columns relate to brands a family is using in a given week, and the rows refer to the brands they might switch to in the coming week.

	A	B	Bran Clumps
A	0.3	0.4	0.3
B	0.2	0.7	0.1
Bran Clumps	0.5	0.3	0.2

United Grains begins the marketing campaign shortly after the New Year. How much of the market can the company expect to control by the next New Year? (As

the "market," consider only that group that regularly consumes one of the three bran cereals.)

45. *(Urban Dynamics)* Five percent of the middle class flees from the city to one of the suburbs each year, and 3 percent return in any given year from the suburbs to the city. Several years after these figures have been collected, what do you expect will be the percentage of the metropolitan middle class living in the city?

46. Calculate Exercise 45 if the two percentages governing urban movement are 5 percent and 10 percent, respectively.

47. *(Biology)* Breeders of laboratory animals are careful to provide animals and insects that are as genetically pure as possible, so that scientists performing experiments need not concern themselves with genetic effects influencing the experimental outcomes. Nevertheless, even after many generations of controlled breeding, it may not be possible to predict genetic traits exactly. For example, a type of fruit fly can possess long wings (L), stubby wings (S), or crinkly wings (C). The following table gives the probability that a pair of parent flies, possessing the characteristics given on the left column, will give rise to offspring possessing the wing type given in the column headings.

	L	S	C
L	$\frac{3}{4}$	$\frac{1}{8}$	$\frac{1}{8}$
S	$\frac{3}{16}$	$\frac{3}{4}$	$\frac{1}{16}$
C	$\frac{1}{8}$	$\frac{3}{8}$	$\frac{1}{2}$

Assuming that breeders are successful in restricting fly breeding so that parents always have identical wing characteristics, estimate that proportion of the fly population that will have long wings, stubby wings, and crinkled wings.

48. *(Social Mobility)* Social mobility between socioeconomic classes can be successfully modeled as a Markov process. Parents in one class produce a child who moves to another class. This type of data is relatively easy to come by in Europe, where countries have a long history of stratified social structure. One researcher has obtained the following data for one of the *arondissements* (administrative departments) of Paris. He assumed that there are three classes—gentry (G), tradespeople (T), and manual workers (W). His observations form the following table, which gives the proportions of children from one class that land in other classes.

	G	T	W
G	0.9	0.1	0
T	0.2	0.6	0.2
W	0	0.1	0.9

If these data hold true for many generations, predict the proportions of people who will be in each class.

49. Compute Exercise 48 if the intergeneration data are given by

	G	T	W
G	$\frac{3}{4}$	$\frac{1}{8}$	$\frac{1}{8}$
T	$\frac{3}{16}$	$\frac{3}{4}$	$\frac{1}{16}$
W	$\frac{1}{8}$	$\frac{3}{8}$	$\frac{1}{2}$

Hint: Refer to Exercise 47.

Calculator Exercises

50. Determine the limit matrix for $\mathbf{P} = \begin{pmatrix} 0.24 & 0.76 \\ 0.47 & 0.53 \end{pmatrix}$.

51. Determine the limit matrix for $\mathbf{P} = \begin{pmatrix} 0.31 & 0.40 & 0.29 \\ 0.50 & 0.37 & 0.13 \\ 0.11 & 0.22 & 0.67 \end{pmatrix}$.

Section 7.3   ABSORBING MARKOV CHAINS

absorbing state   An *absorbing state* is one which, once entered, there is no probability of leaving. Here are some examples of absorbing states.

- *Death.* If a Markov chain is describing states of illness, with one state being "death," then this is an absorbing state. Once an organism has died, there is no possibility of a return to health.

- *Loss of Money.* A Markov chain might be used to describe a business's economic health. Given some nonzero level of financial resources, the company can acquire some, lose some, or neither. Once it loses all its money, the game is over—bankruptcy ensues.

absorbing Markov chain   Any Markov chain containing one or more absorbing states is an *absorbing Markov chain*. It must also be possible to get from any nonabsorbing state to any absorbing state in one or more steps (but it must not take an infinite number of states to get to the absorbing state). Given a stochastic matrix for a Markov chain, it is easy to tell if the chain is absorbing. *Any row containing a 1 on the main diagonal is an absorbing state.* (Why must this be so?) Of course, this statement implies that all other elements on the row are zero. (Why?)

Example 8   The following matrices describe certain Markov chains. Which (if any) of them are absorbing matrices (that is, pertain to absorbing Markov chains)?

(a) $\begin{pmatrix} \frac{1}{2} & \frac{1}{2} & 0 \\ \frac{1}{3} & \frac{1}{3} & \frac{1}{3} \\ \frac{2}{5} & \frac{1}{5} & \frac{3}{5} \end{pmatrix}$

$$\text{(b)} \begin{pmatrix} 0.31 & 0.52 & 0.17 \\ 0 & 0 & 1 \\ 0.72 & 0.10 & 0.18 \end{pmatrix}$$

$$\text{(c)} \begin{pmatrix} 0.3 & 0.4 & 0.3 \\ 0 & 1 & 0 \\ 0.2 & 0.1 & 0.7 \end{pmatrix}$$

Solution

None of the rows of (a) resemble those necessary for an absorbing state. This matrix is not absorbing. In (b), one row does contain a 1 surrounded by 0s, but the 1 does *not* lie on the main diagonal. Finally, row 2 of (c) has a 1 on the main diagonal that is surrounded only by 0s. This matrix *is* an absorbing matrix.

Skill Enhancer 8   Consider these transition matrices.

$$\mathbf{L} = \begin{pmatrix} 1 & 0 \\ 0 & 1 \end{pmatrix} \qquad \mathbf{M} = \begin{pmatrix} 0 & 1 & 0 \\ 0 & 0 & 1 \\ 1 & 0 & 0 \end{pmatrix} \qquad \mathbf{N} = \begin{pmatrix} 1 & 0 & 0 \\ 0 & 1 & 0 \\ 0 & 0 & 1 \end{pmatrix}$$

Which, if any, are absorbing matrices?

Answer in Appendix E.

When dealing with an absorbing Markov chain, it is convenient to relabel the states so that all the absorbing states come first. (Numbers to the top and left of a matrix label these states.) Matrix (c) in Example 8 might become

$$\begin{array}{c c c c} & 1 & 2 & 3 \\ \begin{matrix} 1 \\ 2 \\ 3 \end{matrix} & \begin{pmatrix} 0.3 & 0.4 & 0.3 \\ 0 & 1 & 0 \\ 0.2 & 0.1 & 0.7 \end{pmatrix} \end{array} \Rightarrow \begin{array}{c c c c} & 2 & 1 & 3 \\ \begin{matrix} 2 \\ 1 \\ 3 \end{matrix} & \begin{pmatrix} 1 & 0 & 0 \\ 0.4 & 0.3 & 0.3 \\ 0.1 & 0.2 & 0.7 \end{pmatrix} \end{array}$$

in this reordering.

Examine these hypothetical absorbing Markov matrices.

$$\left(\begin{array}{cc|cc} 1 & 0 & 0 & 0 \\ 0 & 1 & 0 & 0 \\ \hline 0.25 & 0.25 & 0.3 & 0.2 \\ 0.3 & 0.2 & 0.1 & 0.4 \end{array} \right) \qquad \left(\begin{array}{cc|ccc} 1 & 0 & 0 & 0 & 0 \\ 0 & 1 & 0 & 0 & 0 \\ \hline 0.2 & 0.2 & 0.2 & 0.2 & 0.2 \\ 0.1 & 0.2 & 0.4 & 0.2 & 0.1 \\ 0.4 & 0.2 & 0.2 & 0.1 & 0.1 \end{array} \right)$$

$$\left(\begin{array}{ccc|cc} 1 & 0 & 0 & 0 & 0 \\ 0 & 1 & 0 & 0 & 0 \\ 0 & 0 & 1 & 0 & 0 \\ \hline 0.3 & 0.1 & 0.3 & 0.1 & 0.2 \\ 1 & 0 & 0 & 0 & 0 \end{array} \right)$$

Each matrix has a similar structure. For each matrix, we may divide it into four submatrices, the top left of which is an identity matrix, and the top right of which is some (square or nonsquare) zero matrix. That is, any absorbing matrix $\mathbf{P}$ may be placed in the standard form

$$\mathbf{P} = \begin{pmatrix} I & O \\ R & Q \end{pmatrix}$$

where I, O, R, and Q are themselves matrices. Q will always be square, whereas matrix R may or may not be. I is the (square) identity matrix with as many rows as there are absorbing states. O is a zero matrix. Often, O is nonsquare with as many rows as I but a different number of columns. It is possible to state results in the theory of absorbing Markov chains using these submatrices.

Example 9 Identify matrices R and Q for the absorbing matrices $\begin{pmatrix} 1 & 0 \\ \frac{2}{5} & \frac{3}{5} \end{pmatrix}$ and

$$\begin{pmatrix} 1 & 0 & 0 & 0 \\ 0 & 1 & 0 & 0 \\ 0.25 & 0.25 & 0.3 & 0.2 \\ 0.2 & 0.3 & 0.25 & 0.25 \end{pmatrix}$$

Solution

In the first case, R and Q are both 1×1 matrices: $R = (\frac{2}{5})$ and $Q = (\frac{3}{5})$. In the second case, we have

$$R = \begin{pmatrix} 0.25 & 0.25 \\ 0.2 & 0.3 \end{pmatrix} \qquad Q = \begin{pmatrix} 0.3 & 0.2 \\ 0.25 & 0.25 \end{pmatrix}$$

Skill Enhancer 9 What are I, O, R, and Q for the absorbing matrix

$$\begin{pmatrix} 1 & 0 & 0 & 0 \\ 0 & 1 & 0 & 0 \\ \frac{1}{3} & \frac{1}{6} & \frac{1}{6} & \frac{1}{3} \\ \frac{1}{4} & \frac{1}{4} & \frac{1}{4} & \frac{1}{4} \end{pmatrix}$$

How many absorbing states are there? How many nonabsorbing states?

Answer in Appendix E.

Example 10 If $\mathbf{P}$ is an absorbing Markov chain with k absorbing states, show that $\mathbf{P}^2$ is also absorbing with k states.

Solution

It is a fact that matrix multiplication using submatrices instead of components is a valid operation. That is,

$$\mathbf{P}^2 = \begin{pmatrix} I & O \\ R & Q \end{pmatrix} \begin{pmatrix} I & O \\ R & Q \end{pmatrix} = \begin{pmatrix} I^2 + OR & IO + OQ \\ RI + QR & RO + Q^2 \end{pmatrix}$$

Multiplying by a zero matrix yields another zero matrix. A basic property of I is $I = I^2$. Therefore, this product simplifies to

$$\begin{pmatrix} I & O \\ R + QR & Q^2 \end{pmatrix}$$

This is an absorbing Markov matrix, still containing k absorbing states. The matrices R' and Q' for $\mathbf{P}^2$ are given in terms of the original matrices R and Q by

$$R' = R + QR$$
$$Q' = Q^2$$

Skill Enhancer 10   Suppose there are m nonabsorbing states in the matrix $\mathbf{P}$. How many nonabsorbing states are there in $\mathbf{P}^2$? Why?

Answer in Appendix E.

Long-Term Behavior of Absorbing Markov Chains

In any absorbing Markov process, *all outcomes will eventually reach one of the absorbing states*. (We omit the proof in this text.) That is, the absorbing states act like magnets, inevitably drawing the Markov process to them, however slowly. Two questions we can investigate are:

1. What is the probability of ending up in any of the absorbing states?
2. How many steps, on the average, will it take for the process to be absorbed?

fundamental matrix   The answers to both these questions lie with the *fundamental matrix* of the absorbing matrix. The fundamental matrix F is

$$F = (I - Q)^{-1}$$

that is, it is the inverse to $I - Q$. It is understood that we choose the proper identity matrix I so that I and Q have the same dimension. (In general, this I will have different dimension than the top left submatrix of $\mathbf{P}$.)

Example 11   What is the fundamental matrix for $\begin{pmatrix} 1 & 0 & 0 & 0 \\ 0 & 1 & 0 & 0 \\ 0.25 & 0.25 & 0.3 & 0.2 \\ 0.2 & 0.3 & 0.25 & 0.25 \end{pmatrix}$?

Solution

For this matrix, we have $Q = \begin{pmatrix} 0.3 & 0.2 \\ 0.25 & 0.25 \end{pmatrix}$, so that

$$I - Q = \begin{pmatrix} 1 & 0 \\ 0 & 1 \end{pmatrix} - \begin{pmatrix} 0.3 & 0.2 \\ 0.25 & 0.25 \end{pmatrix} = \begin{pmatrix} 0.7 & -0.2 \\ -0.25 & 0.75 \end{pmatrix}$$

Finally,

$$F = (I - Q)^{-1} = \frac{1}{0.475} \begin{pmatrix} 0.75 & 0.2 \\ 0.25 & 0.7 \end{pmatrix} = \begin{pmatrix} 1.5789 & 0.4210 \\ 0.5263 & 1.4737 \end{pmatrix}$$

Skill Enhancer 11   What is $I - Q$ for $\begin{pmatrix} 1 & 0 & 0 & 0 \\ 0 & 1 & 0 & 0 \\ \frac{1}{3} & \frac{1}{6} & \frac{1}{6} & \frac{1}{3} \\ \frac{1}{4} & \frac{1}{4} & \frac{1}{4} & \frac{1}{4} \end{pmatrix}$? What is the fundamental matrix for this matrix?

Answer in Appendix E.

We assert the following facts without proof.

Probabilities of Ending Up in the Absorbing States

The probability of being absorbed in absorbing state j after starting out in nonabsorbing state i is given by the ij component of the matrix FR. Here

F is the fundamental matrix of the Markov chain, and
R is the lower left submatrix of the absorbing matrix.

Note well: We observe a slightly different subscript convention. The subscript i (row numbers) runs from $k + 1$ to n, the number of states, whereas j (column numbers) runs from 1 to k, the total number of absorbing states in the Markov chain.

Example 12   The matrix

$$\mathbf{P} = \begin{array}{c} \\ 1 \\ 2 \\ 3 \\ 4 \end{array} \begin{array}{cccc} 1 & 2 & 3 & 4 \\ \left(\begin{array}{cccc} 1 & 0 & 0 & 0 \\ 0 & 1 & 0 & 0 \\ 0.25 & 0.25 & 0.3 & 0.2 \\ 0.2 & 0.3 & 0.25 & 0.25 \end{array}\right) \end{array}$$

has as its fundamental matrix $F = \dfrac{1}{0.475}\begin{pmatrix} 0.75 & 0.2 \\ 0.25 & 0.7 \end{pmatrix}$. Assess the likelihood of ending up in states 1 or 2.

Solution

For this matrix, $R = \begin{pmatrix} 0.25 & 0.25 \\ 0.2 & 0.3 \end{pmatrix}$. The product

$$FR = \frac{1}{0.475}\begin{pmatrix} 0.2275 & 0.2475 \\ 0.2025 & 0.2725 \end{pmatrix} = \begin{array}{c} \\ 3 \\ 4 \end{array}\begin{array}{cc} 1 & 2 \\ \left(\begin{array}{cc} 0.4789 & 0.5211 \\ 0.4263 & 0.5737 \end{array}\right) \end{array}$$

The rows refer to the two nonabsorbing states, which are 3 and 4. The columns refer to the absorbing states 1 and 2. According to this result, the probability of proceeding from state 3 to state 1 is 0.4789, while the probability of proceeding from state 4 and ending up in state 1 is 0.4263. The probability of the transition from state 3 to 2 is 0.5211, and the transition probability for proceeding from state 4 to state 2 is 0.5737.

Skill Enhancer 12   Consider the matrix

$$\begin{array}{c} \\ 1 \\ 2 \\ 3 \\ 4 \end{array}\begin{array}{cccc} 1 & 2 & 3 & 4 \\ \left(\begin{array}{cccc} 1 & 0 & 0 & 0 \\ 0 & 1 & 0 & 0 \\ \frac{1}{3} & \frac{1}{6} & \frac{1}{6} & \frac{1}{3} \\ \frac{1}{4} & \frac{1}{4} & \frac{1}{4} & \frac{1}{4} \end{array}\right) \end{array}$$

(a) What is R? What is FR?
(b) What is the probability of ending up in absorbing states 1 or 2 if the process begins in nonabsorbing state 3?

(c) What is the probability of ending up in absorbing states 1 or 2 if the process begins in nonabsorbing state 4?

Answer in Appendix E.

We assert, again without proof, a fact pertaining to the expected number of steps the chain goes through before being trapped by one of the absorbing states.

Length of Stay in Nonabsorbing States

The components of F give the expected number of times that the Markov process remains in nonabsorbing state j if it starts in nonabsorbing state i.

Yet another slightly different subscript convention is in force. Both i and j run from values $k+1$ to n where there are k absorbing states and a total of number of states n.

The expected length of time the process remains in a nonabsorbing state if it starts in state i is the number of steps it stays in the first nonabsorbing state, plus the number of steps it remains in the second nonabsorbing state, $\dots$, plus the number of steps it remains in the last nonabsorbing state. Their total—the sum of all components in the ith row of F—represents the average number of steps the process enjoys before absorption, given that the process began in state i.

Example 13 The matrix $P = \begin{pmatrix} 1 & 0 & 0 & 0 \\ 0 & 1 & 0 & 0 \\ 0.25 & 0.25 & 0.3 & 0.2 \\ 0.2 & 0.3 & 0.25 & 0.25 \end{pmatrix}$ has as its fundamental matrix

$$F = \begin{array}{c} 3 \\ 4 \end{array}\begin{pmatrix} \overset{3}{1.5789} & \overset{4}{0.4210} \\ 0.5263 & 1.4737 \end{pmatrix}$$

How many times does the process remain in state 4 if it starts in state 3? How many times does the process remain in state 3 before being absorbed? in state 4?

Solution

On the average, if a process starts in state 3, it remains 0.3846 times in state 4. It remains a total number of $1.5789+0.4210 = 1.999$ steps in state 3 before absorption. In addition, if it starts in state 4, it remains in the nonabsorbing states for $0.5263 + 1.4737 = 2.000$ times before being absorbed.

Skill Enhancer 13 The matrix $\begin{array}{c} 1 \\ 2 \\ 3 \\ 4 \end{array}\begin{pmatrix} \overset{1}{1} & \overset{2}{0} & \overset{3}{0} & \overset{4}{0} \\ 0 & 1 & 0 & 0 \\ \frac{1}{3} & \frac{1}{6} & \frac{1}{6} & \frac{1}{3} \\ \frac{1}{4} & \frac{1}{4} & \frac{1}{4} & \frac{1}{4} \end{pmatrix}$ has fundamental matrix

$$F = \frac{1}{13} \begin{array}{c} \\ \begin{pmatrix} 18 & 8 \\ 6 & 20 \end{pmatrix} \end{array} \begin{array}{c} 3 \\ 4 \end{array}$$

(a) Assume the process begins in state 4. How many times on average will the process remain in state 3? in state 4? How many steps on average before the process is absorbed? (b) If the process begins in state 3, how many steps will it undergo on average before being absorbed?

Answer in Appendix E.

Additional properties of the absorbing stochastic matrix are explored in the exercises.

Gambler's Ruin

The classic application of absorbing Markov chains is to the problem of a gambler's ruin. A play of a game involves betting $1. If the game is won, the gambler receives winnings of $1. (In a winning play, the gambler receives the winnings in addition to the cost of the wager.) If the game is lost, the gambler loses the wager. The game stops when the gambler either has lost all the money or has acquired $4. In addition, the probability of winning is $\frac{1}{3}$, and that of losing is $\frac{2}{3}$. Use Markov chain theory to analyze the game.

The player can be in any of five states: owning 0, 1, 2, 3, or 4 dollars. Suppose, for example, that the gambler has 2 dollars. The probability of raising that to 3 dollars on the next play is $\frac{1}{3}$, and that of losing 1 dollar to have 1 dollar is $\frac{2}{3}$. If the gambler has 4 dollars or no dollars, the game stops; these are absorbing states. These considerations lead to this transition matrix.

$$\mathbf{P} = \begin{array}{c} \\ 0 \\ 1 \\ 2 \\ 3 \\ 4 \end{array} \begin{array}{ccccc} 0 & 1 & 2 & 3 & 4 \end{array} \\ \begin{pmatrix} 1 & 0 & 0 & 0 & 0 \\ \frac{2}{3} & 0 & \frac{1}{3} & 0 & 0 \\ 0 & \frac{2}{3} & 0 & \frac{1}{3} & 0 \\ 0 & 0 & \frac{2}{3} & 0 & \frac{1}{3} \\ 0 & 0 & 0 & 0 & 1 \end{pmatrix}$$

Next, we "shuffle" the order of the states so that the absorbing states are adjacent in this matrix.

$$\begin{array}{c} \\ 0 \\ 4 \\ 1 \\ 2 \\ 3 \end{array} \begin{array}{ccccc} 0 & 4 & 1 & 2 & 3 \end{array} \\ \begin{pmatrix} 1 & 0 & 0 & 0 & 0 \\ 0 & 1 & 0 & 0 & 0 \\ \frac{2}{3} & 0 & \frac{1}{3} & 0 & \frac{2}{3} \\ 0 & 0 & \frac{2}{3} & 0 & \frac{1}{3} \\ 0 & \frac{1}{3} & 0 & \frac{2}{3} & 0 \end{pmatrix}$$

In the context of this section, we identify the submatrices

$$R = \begin{pmatrix} \frac{2}{3} & 0 \\ 0 & 0 \\ 0 & \frac{1}{3} \end{pmatrix} \qquad Q = \begin{pmatrix} 0 & \frac{1}{3} & 0 \\ \frac{2}{3} & 0 & \frac{1}{3} \\ 0 & \frac{2}{3} & 0 \end{pmatrix}$$

Also,

$$I - Q = \begin{pmatrix} 1 & -\frac{1}{3} & 0 \\ -\frac{2}{3} & 1 & -\frac{1}{3} \\ 0 & -\frac{2}{3} & 1 \end{pmatrix}$$

so that the fundamental matrix F and product FR are

$$F = (I - Q)^{-1} = \frac{9}{5} \begin{pmatrix} \frac{7}{9} & \frac{1}{3} & \frac{1}{9} \\ \frac{2}{3} & 1 & \frac{1}{3} \\ \frac{4}{9} & \frac{2}{3} & \frac{7}{9} \end{pmatrix} \begin{matrix} 1 \\ 2 \\ 3 \end{matrix} = \begin{pmatrix} \frac{7}{5} & \frac{3}{5} & \frac{1}{5} \\ \frac{6}{5} & \frac{9}{5} & \frac{3}{5} \\ \frac{4}{5} & \frac{6}{5} & \frac{7}{5} \end{pmatrix}$$

$$FR = \begin{matrix} 1 \\ 2 \\ 3 \end{matrix} \begin{pmatrix} \frac{14}{15} & \frac{1}{15} \\ \frac{4}{5} & \frac{1}{5} \\ \frac{8}{15} & \frac{7}{15} \end{pmatrix}$$

From the product FR, it is clear that no matter how much the gambler starts out with, he is always more likely to lose than to win, as we would expect. FR quantifies this information. For example, if the gambler starts with 3 dollars, his probability of losing is $\frac{8}{15}$ versus a probability of $\frac{7}{15}$ of winning. If he starts with 1 dollar, he is fourteen times as likely to lose everything as to "break the bank."

From the fundamental matrix F, we learn (for example) that if the gambler starts with 1 dollar, he will have 1 dollar for $\frac{7}{5}$ of a move. The total number of moves before he enters an absorbing state is $\frac{7}{5} + \frac{3}{5} + \frac{1}{5} = 2\frac{1}{5}$ on the average if he started with one dollar.

_____ *Exercises 7.3* _____

In Exercises 1–12, decide which of the given matrices can represent absorbing Markov chains.

1. $\begin{pmatrix} 1 & 0 \\ 0.2 & 0.8 \end{pmatrix}$

2. $\begin{pmatrix} 0 & 1 \\ 0.3 & 0.7 \end{pmatrix}$

3. $\begin{pmatrix} 1 & 0 & 0 \\ 0 & 0 & 1 \\ \frac{1}{3} & \frac{2}{3} & 0 \end{pmatrix}$

4. $\begin{pmatrix} 0.2 & 0.7 & 0.1 \\ 0 & 1 & 0 \\ 0.3 & 0.3 & 0.4 \end{pmatrix}$

5. $\begin{pmatrix} 1 & 0 & 0 \\ 0 & 1 & 0 \\ 0 & 0 & 1 \end{pmatrix}$

6. $\begin{pmatrix} 0 & 0 & 1 \\ 0 & 1 & 0 \\ 1 & 0 & 0 \end{pmatrix}$

7. $\begin{pmatrix} \frac{1}{2} & \frac{1}{2} \\ \frac{1}{2} & \frac{1}{2} \end{pmatrix}$

8. $\begin{pmatrix} 1 & 0 \\ \frac{3}{4} & \frac{1}{3} \end{pmatrix}$

9. $\begin{pmatrix} 1 & 0 & 0 \\ 0 & 1 & 0 \\ \frac{1}{3} & \frac{2}{3} & \frac{1}{6} \end{pmatrix}$

10. $\begin{pmatrix} \frac{5}{4} & -\frac{1}{4} \\ 0 & 1 \end{pmatrix}$

11. $\begin{pmatrix} \frac{2}{7} & \frac{1}{7} & \frac{4}{7} \\ 0 & 1 & 0 \\ \frac{5}{9} & \frac{5}{9} & -\frac{1}{9} \end{pmatrix}$

12. $\begin{pmatrix} 1 & 0 & 0 & 0 \\ 0 & 1 & 0 & 0 \\ 0.25 & 0.25 & 0.3 & 0.2 \\ 0.4 & 0.4 & 0.1 & 0.1 \end{pmatrix}$

For each of the absorbing stochastic matrices in Exercises 13–18, identify the submatrices R and Q.

13. $\begin{pmatrix} 1 & 0 \\ \frac{1}{4} & \frac{3}{4} \end{pmatrix}$

14. $\begin{pmatrix} 1 & 0 \\ \frac{1}{2} & \frac{1}{2} \end{pmatrix}$

15. $\begin{pmatrix} \frac{1}{2} & \frac{1}{2} \\ 0 & 1 \end{pmatrix}$ **16.** $\begin{pmatrix} \frac{1}{4} & \frac{3}{4} \\ 0 & 1 \end{pmatrix}$

17. $\begin{pmatrix} \frac{2}{3} & \frac{1}{6} & \frac{1}{6} \\ 0 & 1 & 0 \\ \frac{1}{2} & \frac{1}{4} & \frac{1}{4} \end{pmatrix}$ **18.** $\begin{pmatrix} 0.2 & 0.3 & 0.5 \\ 0.6 & 0.3 & 0.1 \\ 0 & 0 & 1 \end{pmatrix}$

There is an absorbing stochastic matrix in each of Exercises 19–24. In each, identify the submatrices R and Q. Calculate the fundamental matrix F for each.

19. $\begin{pmatrix} 1 & 0 & 0 \\ 0 & 1 & 0 \\ 0.25 & 0.35 & 0.4 \end{pmatrix}$ **20.** $\begin{pmatrix} 1 & 0 & 0 \\ 0 & 1 & 0 \\ \frac{1}{3} & \frac{1}{3} & \frac{1}{3} \end{pmatrix}$

21. $\begin{pmatrix} 1 & 0 & 0 \\ \frac{1}{3} & \frac{1}{3} & \frac{1}{3} \\ \frac{1}{4} & \frac{1}{4} & \frac{1}{2} \end{pmatrix}$ **22.** $\begin{pmatrix} 1 & 0 & 0 \\ \frac{1}{3} & \frac{1}{3} & \frac{1}{3} \\ \frac{1}{3} & \frac{1}{3} & \frac{1}{3} \end{pmatrix}$

23. $\begin{pmatrix} 1 & 0 & 0 & 0 \\ 0 & 1 & 0 & 0 \\ \frac{1}{5} & \frac{1}{5} & \frac{3}{10} & \frac{3}{10} \\ \frac{1}{4} & \frac{1}{4} & \frac{1}{4} & \frac{1}{4} \end{pmatrix}$ **24.** $\begin{pmatrix} 1 & 0 & 0 & 0 \\ 0 & 1 & 0 & 0 \\ \frac{1}{4} & \frac{1}{4} & \frac{1}{4} & \frac{1}{4} \\ \frac{1}{8} & \frac{1}{4} & \frac{1}{4} & \frac{3}{8} \end{pmatrix}$

In each of Exercises 25–30, assume that the process begins in the first nonabsorbing state. Decide how many steps the process will take on the average to reach the second nonabsorbing state. For each nonabsorbing state, determine how many steps on the average the process will endure before absorption.

25. The matrix in Exercise 19.

26. The matrix in Exercise 20.

27. The matrix in Exercise 21.

28. The matrix in Exercise 22.

29. The matrix in Exercise 23.

30. The matrix in Exercise 24.

31. Suppose a Markov chain has a total of n states, of which m are absorbing. Express the dimensions of the four submatrices I, O, R, and Q in terms of m and n. (For example, I has dimension $m \times m$.)

32. (a) Assume there is a square matrix R and a matrix A such that the product AR exists. Refer to Appendix 1. Follow the steps to derive a formula for the sum $A + RA + R^2A \cdots + R^nA$.

(b) Suppose that n approaches infinity; that is, n gets larger and larger. Suppose, at the same time, that each component of R^n becomes less and less dis-

tinguishable from 0, so that R^n approaches a zero matrix. In this case, what will be the formula for $A + AR + AR^2 \cdots$?

33. (a) For an absorbing matrix $\mathbf{P}$, show that $\mathbf{P}^2$ will have the form $\begin{pmatrix} I & O \\ R + QR & Q^2 \end{pmatrix}$

(b) Use this result to show that $\mathbf{P}^n$ will have the form
$$\begin{pmatrix} I & O \\ R + QR + \cdots + Q^{n-1}R & Q^n \end{pmatrix}$$
Use the results of Exercise 32 to deduce the form of $\mathbf{P}^n$ as n approaches infinity, also under the assumption that Q^n approaches a zero matrix.

(c) Use this result to justify the assertion in the text concerning the number of times the process remains in each nonabsorbing state (on the average) prior to absorption.

Applications

34. *(Gambler's Ruin)* Solve the gambler's ruin problem if the probabilities for win and loss on each play are equal.

35. Solve the gambler's ruin problem if the probabilities for win and loss on each play are equal and if the gambler will stop when he or she acquires 5 dollars.

36. Solve the gambler's ruin problem if the probabilities of winning and losing are equal and the gambler will stop only when he loses all his money.

37. *(Higher Education)* The sons of Yale alumni invariably decide to go to Yale. The sons of Harvard men go to Harvard 50 percent of the time, and to Princeton and Yale 25 percent of the time. Sons of Princeton men go to Yale, Harvard, and Princeton one-third of the time. Which is the absorbing state? An exclusive Boston men's club consists of men, one-third of whom went to Harvard, one-third of whom went to Princeton, and one-third of whom went to Yale. What is the expected number of generations before all of their male descendants attend the absorbing college?

38. Solve Exercise 37 if sons of Harvard men go exclusively to Harvard.

39. *(Insect Control)* Normally, cockroaches are repelled by light. A roach attracted to light would be much easier to trap, control, and dispose of. In one experiment, a scientist noticed that a small portion of the roaches in a control group are attracted to light. All offspring of the light-attracted insects are themselves light-attracted, but only 10 percent of the offspring of the normal roaches are attracted to light. What is the expected number of generations after which the entire roach population in this experimental setup is of the light-attracted variety? Assume that 20 percent of the insects are light-attracted in the original experimental configuration.

 Calculator Exercises _____

40. What is the fundamental matrix F for

$$\begin{pmatrix} 1 & 0 & 0 \\ 0.5 & 0.36 & 0.14 \\ 0.33 & 0.33 & 0.34 \end{pmatrix}?$$

41. Consider the transition matrix

$$\begin{pmatrix} 1 & 0 & 0 \\ 0.35 & 0.35 & 0.30 \\ 0.45 & 0.26 & 0.29 \end{pmatrix}$$

(a) Compute F and FR for this matrix.

(b) What is the probability of ending up in state 1 when starting from either of the nonabsorbing states? How could this have been determined *without* doing the calculations?

(c) Suppose the process starts in state 2. On the average, how many steps will it go through before being absorbed?

(d) Suppose the process starts in state 3. On the average, how many steps will it go through before being absorbed?

CHAPTER REVIEW _____

Terms _____

steps	Markov chain	states
transition probability	transition matrix	stochastic matrix
probability matrix	Meaning of the components of P^n	regular matrix
fixed-point matrix	absorbing state	absorbing Markov chain
fundamental matrix		

Key Concepts _____

- A Markov chain is a process consisting of a chain or series of discrete steps. At each step, there are the same possible outcomes, only one of which may occur in each step. The outcome of any one step must depend only on the outcome of the previous step.

- The components p_{ij} of the transition matrix are the probabilities that the system proceeds from state i in one step to state j in the next.

- Suppose $\mathbf{A}^0$ is the probability row matrix describing the system at the outset and $\mathbf{A}^n$ describes it at any step n. Then the key Markov relations are $\mathbf{A}^{n+1} = \mathbf{A}^n\,\mathbf{P}$ or, equivalently, $\mathbf{A}^n = \mathbf{A}^0\,\mathbf{P}^n$.

- If the Markov transition matrix $\mathbf{P}$ is regular, then the long-term behavior of the Markov process can be quantified through the limit value of $\mathbf{P}$.

- Absorbing Markov process are those containing absorbing states; once in an absorbing state, the process remains in that state forever after. Eventually, every absorbing Markov process must wind up in one of the absorbing states if there is an absorbing state in the process. The fundamental matrix associated with an absorbing matrix yields important information regarding the expected number of steps before absorption takes place.

Review Exercises _____

Which of the following matrices could be used as transition matrices? If a matrix could not be used as such, state why not.

1. $\begin{pmatrix} -1 & 2 \\ 0.5 & 0.5 \end{pmatrix}$

2. $\begin{pmatrix} \frac{1}{3} & \frac{1}{3} & \frac{1}{3} \\ 0 & 0 & 1 \\ 0.5 & 0.25 & 0.25 \end{pmatrix}$

3. Let $\mathbf{P} = \begin{pmatrix} \frac{1}{6} & \frac{5}{6} \\ \frac{1}{2} & \frac{1}{2} \end{pmatrix}$ be a transition matrix for some Markov process.

(a) What is the probability that the process will proceed from state 2 to state 2 in a single step?

(b) What is the probability that the process will proceed from state 2 to state 2 in exactly two steps?

(c) What is the probability that the process will proceed from state 2 to state 2 in exactly two steps such that the process remains in the second state in the middle step as well?

Which of Exercises 4–7 are regular matrices? For all regular matrices, determine the limit value.

4. $\begin{pmatrix} 0.5 & 0.5 \\ 0.5 & 0.5 \end{pmatrix}$ **5.** $\begin{pmatrix} 0.7 & 0.3 \\ 0.3 & 0.7 \end{pmatrix}$

6. $\begin{pmatrix} 1 & 0 & 0 \\ \frac{1}{3} & \frac{1}{3} & \frac{1}{3} \\ 0 & \frac{1}{2} & \frac{1}{2} \end{pmatrix}$ **7.** $\begin{pmatrix} \frac{1}{2} & \frac{1}{4} & \frac{1}{4} \\ \frac{1}{4} & \frac{1}{2} & \frac{1}{4} \\ \frac{1}{4} & \frac{1}{4} & \frac{1}{2} \end{pmatrix}$

For Exercises 8–9, determine the fundamental matrix for each of the Markov matrices.

8. $\begin{pmatrix} 1 & 0 & 0 \\ 0.4 & 0.6 & 0 \\ 0.5 & 0.5 & 0 \end{pmatrix}$ **9.** $\begin{pmatrix} 1 & 0 & 0 \\ 0 & 1 & 0 \\ 0.3 & 0.35 & 0.35 \end{pmatrix}$

Applications

10. (Agriculture) A certain type of pea has plants that are either short or tall. Peas from short plants yield short pea plants $\frac{2}{3}$ of the time. Peas from tall plants yield tall plants $\frac{3}{4}$ of the time.

(a) Set up a transition matrix for this problem.

(b) Start with a batch of peas from which grow short and tall plants in equal proportions. If all these peas are planted, what proportion of the resulting plants will be tall and short?

(c) Suppose each plant gives rise to the same number of peas, and all these peas are again planted. What proportion of tall and short plants will we then see?

11. (The Stock Market) Investors have long sought models that will predict the behavior of stock markets. (As of this writing, no such model has ever been discovered.) One business student seeks to impose Markov behavior on the stock market. Observations on one major exchange seem to indicate that if a stock has risen today, it will rise tomorrow with probability $\frac{2}{3}$. If the stock falls today, it will fall tomorrow with probability $\frac{1}{2}$. (Assume no stock price ever remains unchanged in this exchange.) At the close of business on a Monday, 60 percent of all stocks had risen in price. What does a Markov model predict for this percentage at the close of business on Tuesday? at the close of business on Friday? at the close of business one month hence?

12. A more realistic stock market model takes into account stock prices that remain unchanged from one business day to the next. Preliminary data indicate that the following probabilities govern stock price changes from one day to the next. Here, the symbols $+$, $-$, and 0 refer to prices that rise, decrease, or remain unchanged from one day to the next

$$\begin{array}{c} \quad\quad + \quad\ \ 0 \quad\ \ - \\ \begin{array}{c} + \\ 0 \\ - \end{array} \begin{pmatrix} 0.5 & 0.2 & 0.3 \\ 0.7 & 0.2 & 0.1 \\ 0.3 & 0.1 & 0.6 \end{pmatrix} \end{array}$$

(a) On Friday night, financial analysts determine that half of all stocks have remained unchanged, while 30 percent have declined in price. Use the transition matrix to predict what these percentages will be at the close of business on Monday night. The stock markets in this country are closed on Saturdays and Sundays, so that no trading takes place these days.

(b) What does the Markov model predict about stock price changes for this stock exchange in the long term?

(c) What does the fact that the long-term matrix has a limit value indicate about the validity of the Markov chain model for stock price variation?

13. (Consumer Spending Habits) In one affluent Seattle suburb, the tendency of families to buy new cars in one year depends on whether they bought a new car last year. If not, then the family purchases a new car 75 percent of the time. If so, the family purchases a new car 90 percent of the time.

(a) One family bought a new luxury car in the summer of 1984. How likely are they to buy another one in 1985? in 1986?

(b) In the long-term, what percentage of the town's families will buy new cars in any given year?

Calculator Exercises (14, 15)

14. Suppose the transition matrix from Exercise 13 can be written as

New car   No new car

New car $\begin{pmatrix} 0.613 & 0.387 \\ 0.555 & 0.445 \end{pmatrix}$
No new car

Use a calculator or computer program to recompute the answers to (a) and (b) of that problem.

15. Suppose the transition matrix in Exercise 12 is

$$\begin{array}{c} \\ + \\ 0 \\ - \end{array} \begin{array}{ccc} + & 0 & - \\ \begin{pmatrix} 0.47 & 0.22 & 0.31 \\ 0.70 & 0.18 & 0.12 \\ 0.34 & 0.06 & 0.60 \end{pmatrix} \end{array}$$

Use a calculator or computer to obtain the answers to parts (a) and (b) of that problem.

16. *(Gambler's Ruin)* Solve the gambler's ruin problem if the gambler may go 1 dollar into debt before quitting. Assume further that there are equal probabilities of winning and losing and that the gambler quits if he acquires 4 dollars.

8

DECISIONS AND GAMES

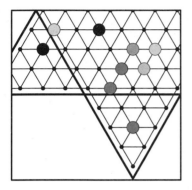

INTRODUCTION

A certain collection of mathematical techniques encourages the use of quantitative methods to make decisions in competitive situations. Planners use this *game theory* to analyze the strategies available to opposing sides in competitive situations like games, although these techniques are just as appropriate to business and war.

The hope is that through the analysis of game theory, we will be able to make our rewards as large as possible (or to minimize our losses).

First we establish the notation and conventions convenient to the theory and look at a class of games that always have a best strategy. Then, we learn how to decide on strategy in cases where the optimal strategy is less well defined.

Section 8.1 INTRODUCTION TO GAME THEORY

simple games   We shall be discussing *simple games*, competitive situations in which there are only *two* opposing sides, and each side has only a few well-defined options. Furthermore, each *play* of the game consists of a *single pair of moves*, one by each of the opponents.

play

At the end of the game, each player receives a payoff (which may be negative) according to rules laid down at the start. Games can be played over and over, and players seek to optimize their payoff over the long run. A positive payoff represents a reward, and a negative payoff is the penalty for having lost the game.

payoff matrices   *Payoff matrices* are useful in summarizing the rules of the game. These are charts or tables summarizing the payoffs for each pair of strategies of the two players. *Row labels* of the payoff matrix will summarize the alternatives of the first player, and *column headings* perform the same function for the competitor. We agree to present all payoffs *from the point of view of the row player*. That is,

- *Positive payoffs* indicate a gain to player R (the row player) and a loss to player C (the column player).

- *Negative payoffs* represent a loss to R and a gain to C.

zero-sum games   We may use this payoff matrix notation whenever one player's gain is the opponent's loss. Such conflicts are called *zero-sum games* because the sum of the payoffs to R and to C is exactly zero. Payoff matrices may also be used when the sum of the payoffs to *constant-sum games*   the two opponents is always the same (*constant-sum games*).

Historical Note

John von Neumann   *"I wonder whether a brain like von Neumann's does not indicate a species superior to that of man." This is the comment made by the Nobel laureate Hans Bethe of John von Neumann (NOY-mon; 1903–1957), the inventor of game theory. Although his first paper on this theory appeared in 1928, its full development was only set forth in the* Theory of Games and Economic Behavior *(1944), coauthored with economist Oskar Morgenstern. Von Neumann contributed prolifically to many scientific fields. To this day, for example, his theories on the organization of computer memory have shaped the manufacture of all high-speed digital computers.*

Example 1   The European children's game Morra is played by two children, who use stones from previously assembled piles as the payoffs in the game. Each child makes a fist and, at a signal, shows one or two fingers as in Figure 8.1. If the sum of the fingers is even, the first child gets that number of pebbles. If the sum is odd, the second child gets that number. The object of the game is to accumulate the most stones. Express this game as a payoff matrix.

Solution

Let the first player be R and the second be C. Represent a payoff to R as a positive number, and a payoff to C as negative. If both children show a single finger, a total

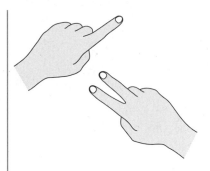

FIGURE 8.1
Morra.

of two fingers (an even number) are shown, so R collects two stones. But if C shows two fingers to R's one, then the sum is three (odd), and C collects three pebbles, which become a negative payoff from R's point of view. We use similar reasoning to complete the matrix when R shows two fingers.

		C	
		1	2
R	1	2	-3
	2	-3	4

Skill Enhancer 1   Once again, write the payoff matrix for Morra. This time, write the matrix from the point of view of C—payoffs to C are positive, and payoffs to R are negative.

Answer in Appendix E.

---

rational strategy

Game Strategy

Players in mathematical games are assumed to be *rational* at all times:

1. Player R chooses the alternative to *maximize* the payoff.

2. C attempts to *minimize* the payoff.

expected payoff

The payoff is always computed from R's point of view. Games can be played over and over again, so it makes sense to speak of *expected payoff*.

Example 2   A rock star R and a country-western singer C are committed to perform in the vicinity of one of the campuses of the University of Nevada on the same day, at the same time. Each star competes for the largest share of the concertgoing audience. Each may appear in either of two suburbs. If they both perform in the first suburb, R will draw 40 percent of the audience, with the remainder attending the country concert. If they both perform in the second town, C will draw 60 percent of the audience. However, if R performs in town 1 and C in town 2, then R will draw 80 percent of the concertgoers. If R performs in town 2 and C in town 1, then R will attract only 30 percent of the audience. How should the singers plan their engagements?

Solution

Set up the payoff matrix. The payoffs are expressed as percentages of the total audience, and are expressed from the point of view of R.

		C's Strategies	
		Town 1	Town 2
R's	Town 1	40%	80%
Strategies	Town 2	30%	40%

Since the sum of the payoffs to both performers is 100 percent of the audience, this is an example of a *constant-sum game.*

The rational player seeks to minimize losses. R might be tempted by the lure of town 1, but might pull back in fear that C will also choose town 1, and so steal R's thunder. C might choose to perform in town 1, hoping that R will not make the same choice, but nervous that he will. The best thing each can do is to implement a *maximin* strategy—plan for the *worst* case, and expect that any attendance over that is unlooked-for fortune.

Strategic planning. Here is how R should think: "Let me plan for the worst. The worst I can do by choosing town 1 is 40 percent, whereas the worst with town 2 is 30 percent. At least let me maximize the worst-case outcome, so I will plan my concert for town 1. Of these two worst cases, this is the better one for me. Should C not be as careful a planner as I am, I may do even better."

Here is how C's analysis should run. Remember, the *lower* the percentage in the table, the better it is for C. (Why?) The worst choices for C by choosing strategies 1 and 2 are 40 percent and 80 percent, so C might as well choose the strategy that guarantees the "best" worst case, which is 40 percent via the first town.

Both players adopt strategies that yield the same payoff—40 percent attendance for R's concert (and therefore 60 percent for C's). Of course, if one of the singers does not *value* take the trouble to plan carefully, then the other will reap the benefits. The *value* of the game is 40 percent, the payoff occurring when both players use this careful reasoning to determine the best course of action for each of them.

Skill Enhancer 2 Performers R and C are again committed to concerts at the same time. In this case, if they both perform in the first suburb, R will draw 50 percent of the audience. If they both perform in the second town, C will draw 65 percent of the audience. However, if R performs in town 1 and C in town 2, then R will draw 75 percent of the concertgoers. In the opposite case, R will attract only 30 percent of the audience.
(a) Construct the payoff matrix.   (b) What is R's maximin strategy?   (c) What is C's maximin strategy?   (d) What is the value of the game?

Answer in Appendix E.

fair   If the value of a game is 0, the game is said to be *fair*.

This example suggests one method for finding the best strategy in certain two-person, constant-sum games. (There is no guarantee that this algorithm will work for all games.)

Best Two-Person Strategy

1. Set up the payoff matrix.
2. Identify the *row minimum* in each row of the payoff matrix.
3. Identify the *column maximum* for each column.
4. If one or more elements are both the row minimum and the column maximum, then that payoff is the *optimal payoff* obtained from the optimal strategy.

   Rational players should choose a strategy (those rows and columns) to produce the optimal payoff. This optimal payoff is the *value* of the game.

---

pure strategy
strictly determined
saddle point

A game need not possess this *pure strategy*, but if it does, it is said to be *strictly determined*. The special payoff value when each player chooses the optimal strategy is the *saddle point* value. If there is more than one saddle point, all the saddle points will have the same value. (We do not prove this fact.)

Example 3   Find the saddle points (if they exist) for the following games. (Each matrix displays the payoffs for a particular game.)

(a) $\begin{pmatrix} \frac{1}{2} & 2 & 1 \\ 0 & -1 & -\frac{1}{2} \\ \frac{1}{4} & 1 & 2 \end{pmatrix}$ (b) $\begin{pmatrix} -1 & -2 & 3 \\ -1 & 4 & 0 \\ 0 & 0 & 1 \end{pmatrix}$

(c) $\begin{pmatrix} 8 & 7 & 8 & 7 \\ 9.5 & 3 & 9 & 5 \\ 8 & 2 & 8 & 4 \\ 11 & 7 & 9 & 7 \end{pmatrix}$ (d) $\begin{pmatrix} 0 & 1 & -1 \\ 1 & \frac{4}{3} & -\frac{1}{2} \\ -1 & 2 & 0 \end{pmatrix}$

Solution

We can identify a saddle point with the following bookkeeping method. Surround each row minimum with a right angle, "$\rfloor$," hugging the bottom of the element and each column maximum with an upward right angle, "$\lceil$." Saddle points will be surrounded by a box made of both brackets, and thus will be easy to identify.

(a) Using this convention, the individual row minima might be indicated as follows:

$$\begin{pmatrix} \frac{1}{2}\rfloor & 2 & 1 \\ 0 & -1\rfloor & -\frac{1}{2} \\ \frac{1}{4}\rfloor & 1 & 2 \end{pmatrix}$$

The largest elements in each column are

$$\begin{pmatrix} \lceil\frac{1}{2} & \lceil 2 & 1 \\ 0 & -1 & -\frac{1}{2} \\ \frac{1}{4} & 1 & \lceil 2 \end{pmatrix}$$

Superposing both marking types on the same matrix yields

$$\begin{pmatrix} \boxed{\dfrac{1}{2}} & \overline{2} & 1 \\ 0 & \underline{-1} & -\dfrac{1}{2} \\ \dfrac{1}{4} & 1 & \overline{2} \end{pmatrix}$$

Clearly, the saddle point is at row 1 and column 1. This game has a value of $\frac{1}{2}$.

(b) The saddle point's value is 0, at row 3 and column 1. This game is fair.

$$\begin{pmatrix} -1 & \underline{-2} & \overline{3} \\ \underline{-1} & \overline{4} & 0 \\ \boxed{0} & 0 & 1 \end{pmatrix}$$

(c) Rows may have two or more equal, minimum elements; columns may have more than one equal, maximum element; and games may have more than one saddle point! There are four saddle point values, all with the same value.

$$\begin{pmatrix} 8 & \boxed{7} & 8 & \boxed{7} \\ 9.5 & 3 & \overline{9} & 5 \\ 8 & 2 & 8 & 4 \\ \overline{11} & \boxed{7} & \overline{9} & \boxed{7} \end{pmatrix}$$

(d) Games need not possess saddle points, as this example illustrates. Optimum strategies for these games are determined using methods we discuss in the next two sections.

$$\begin{pmatrix} 0 & 1 & \underline{-1} \\ \overline{1} & \dfrac{4}{3} & \underline{-\dfrac{1}{2}} \\ \underline{-1} & \overline{2} & \overline{0} \end{pmatrix}$$

Skill Enhancer 3   What are the saddle points (if any) for these payoff matrices? If there is a saddle point, what is the value of the game?

(a) $\begin{pmatrix} 1 & 2 & 3 \\ 3 & 2 & 1 \\ 4 & 9 & 4 \end{pmatrix}$ (b) $\begin{pmatrix} 1 & 2 & 3 \\ 3 & 2 & 1 \\ 0 & -1 & 0 \end{pmatrix}$

Answer in Appendix E.

If one side in a game has more options than the opponent, the payoff matrix will not be square.

Example 4   Determine the saddle point(s) in the following games.

(a) $\begin{pmatrix} 1 & 4 & 1 \\ 1 & 2 & 1 \\ -1 & 5 & 6 \\ 0 & 5 & -3 \end{pmatrix}$ (b) $\begin{pmatrix} -2 & -3 & -5 & -2 & -4 \\ 1 & 0 & 1 & -1 & -1 \\ -2 & -4 & -9 & -9 & -8 \end{pmatrix}$

Solution

Here are these matrices with the row minima and column maxima bracketed.

(a) $\begin{pmatrix} \boxed{1} & 4 & 1 \rfloor \\ \boxed{1} & 2 & 1 \rfloor \\ -1 \rfloor & \lceil 5 & \lceil 6 \\ 0 & \lceil 5 & -3 \rfloor \end{pmatrix}$
(b) $\begin{pmatrix} -2 & -3 & -5 \rfloor & -2 & -4 \\ \lceil 1 & \lceil 0 & \lceil 1 & \boxed{-1} & \boxed{-1} \\ -2 & -4 & \underline{-9} \rfloor & \underline{-9} \rfloor & -8 \end{pmatrix}$

The matrix in (a) has two saddle points, both in column 1. The value of this game is 1. There are two saddle points in (b), and the value of this game is -1.

Skill Enhancer 4   What are the saddle points for these games? If there is a saddle point, what is its value?

(a) $\begin{pmatrix} 1 & 2 & 3 & 4 \\ 4 & 3 & 2 & 1 \end{pmatrix}$
(b) $\begin{pmatrix} 9 & 5 \\ 8 & 4 \\ 7 & 3 \\ 6 & 2 \end{pmatrix}$

Answer in Appendix E.

_____ *Exercises 8.1* _____

The matrices in Exercises 1–30 represent the payoff matrices for several games. Decide which games are strictly determined, and find the optimal strategy and value for such games. Locate all saddle points for the strictly determined games.

1. $\begin{pmatrix} \frac{1}{2} & \frac{1}{2} \\ \frac{1}{3} & \frac{1}{4} \end{pmatrix}$

2. $\begin{pmatrix} 0.7 & 0.3 \\ 0.3 & 0.7 \end{pmatrix}$

3. $\begin{pmatrix} 0.6 & 0.5 \\ 0.7 & 0.3 \end{pmatrix}$

4. $\begin{pmatrix} \frac{1}{3} & \frac{2}{3} \\ \frac{1}{3} & \frac{2}{3} \end{pmatrix}$

5. $\begin{pmatrix} 1 & -1 \\ -1 & 1 \end{pmatrix}$

6. $\begin{pmatrix} 1 & 1 \\ -1 & -1 \end{pmatrix}$

7. $\begin{pmatrix} 1 & 1 \\ 1 & 1 \end{pmatrix}$

8. $\begin{pmatrix} 1 & -1 \\ 1 & -1 \end{pmatrix}$

9. $\begin{pmatrix} 1 & 0 & -1 \\ -1 & 0 & 1 \\ 0 & -1 & 1 \end{pmatrix}$

10. $\begin{pmatrix} 0 & 1 & -1 \\ -1 & 0 & 1 \\ 1 & -1 & 0 \end{pmatrix}$

11. $\begin{pmatrix} \frac{1}{2} & \frac{1}{4} & \frac{1}{4} \\ \frac{1}{4} & \frac{1}{2} & \frac{1}{4} \\ \frac{1}{4} & \frac{1}{4} & \frac{1}{2} \end{pmatrix}$

12. $\begin{pmatrix} 0.3 & 0.6 & 0.1 \\ 0.6 & 0.4 & 0.4 \\ 0.4 & 0.2 & 0.4 \end{pmatrix}$

13. $\begin{pmatrix} 0 & -1 \\ -7 & -3 \end{pmatrix}$

14. $\begin{pmatrix} -1 & 2 \\ 2 & -1 \end{pmatrix}$

15. $\begin{pmatrix} 1 & 2 \\ -7 & -1 \end{pmatrix}$

16. $\begin{pmatrix} -0.01 & 0 \\ 0.03 & 0.2 \end{pmatrix}$

17. $\begin{pmatrix} 1 & 1 & 1 & 1 \end{pmatrix}$

18. $\begin{pmatrix} 10 & 1 \\ 15 & -1 \end{pmatrix}$

19. $\begin{pmatrix} 8 & 10 & 4 \\ 7 & 6 & 5 \end{pmatrix}$

20. $\begin{pmatrix} 3 & -3 \\ -2 & -2 \\ -1 & -5 \end{pmatrix}$

21. $\begin{pmatrix} 3 & 0 & 4 \\ 4 & -1 & 3 \\ 2 & 1 & 2 \end{pmatrix}$

22. $\begin{pmatrix} 4 & 5 & 4 \\ -2 & 10 & -1 \\ 0 & 12 & -7 \end{pmatrix}$

23. $\begin{pmatrix} -\frac{2}{3} & \frac{1}{4} & -\frac{1}{4} & -2 \\ \frac{1}{8} & -\frac{3}{4} & 0 & -1 \\ 0 & -1 & -\frac{5}{8} & -1 \end{pmatrix}$

24. $\begin{pmatrix} -\frac{1}{2} & \frac{1}{2} & -\frac{1}{12} \\ \frac{1}{6} & \frac{1}{4} & \frac{1}{8} \\ 0 & \frac{1}{3} & \frac{1}{5} \end{pmatrix}$

25. $\begin{pmatrix} -1 & -2 & -1 \\ 0 & 2 & 1 \\ -1 & -3 & -1 \end{pmatrix}$

26. $\begin{pmatrix} 0.03 & -0.1 & 0.02 \\ 0.005 & 0 & 0.01 \\ -0.1 & 0.2 & 0.015 \end{pmatrix}$

27. $\begin{pmatrix} 8.3 & 4.8 & -1.0 & -5.9 \\ 7.6 & 3.2 & 4.4 & 5.2 \\ 9.5 & -6.5 & -6.5 & 0.1 \\ 9.1 & -3.2 & 7.6 & 7.6 \end{pmatrix}$

28. $\begin{pmatrix} -1.9 & 0 & 4.6 & 2.0 \\ -3.1 & -1.1 & -2.4 & -1.1 \\ -0.1 & 0.7 & 0.8 & 1.6 \\ -2.3 & -1.1 & -1.2 & -1.1 \end{pmatrix}$

29. $\begin{pmatrix} 2 & 4 & 6 \\ 3 & 5 & 7 \\ 4 & 6 & 8 \end{pmatrix}$ **30.** $\begin{pmatrix} -1 & 0 & -1 \\ 3 & 3 & -2 \\ 6 & 9 & -1 \end{pmatrix}$

For what values of x (if any) will the games in Exercises 31–36 be strictly determined? What will the value of each game be when strictly determined?

31. $\begin{pmatrix} 1 & x \\ 2 & -1 \end{pmatrix}$ **32.** $\begin{pmatrix} -2 & 3 \\ x & 0 \end{pmatrix}$

33. $\begin{pmatrix} 1 & x \\ x & -1 \end{pmatrix}$

34. $\begin{pmatrix} 8+x & 10+x & 4+x \\ 7+x & 6+x & 5+x \end{pmatrix}$

35. $\begin{pmatrix} x-0.01 & x \\ x+0.03 & x+0.2 \end{pmatrix}$ **36.** $\begin{pmatrix} 3 & x & 4 \\ 5 & -1 & 3.5 \\ 2 & 1 & 1.5 \end{pmatrix}$

W 37. Why does it make sense to speak of a strictly determined game with value 0 as being fair?

W 38. Why do you suppose the saddle point is *called* the saddle point? (*Hint*: See Exercise 39.)

Applications

39. (*Science*) Dr. Rao, a geologist, and Professor Collings, a meteorologist, have received funding to build a building to house the Institute for Planetary and Atmospheric Sciences. The lab will be built in a hilly part of town, because the land is cheapest there. Geologist Rao wants the lab situated as low as possible to make earth measurements as easy to take as possible. Meteorologist Collings wants to be as high as possible, to make easy his measurements of weather conditions. The following chart gives the elevations (in feet above sea level) for each of nine possible building sites, which are at the intersections of the east-west roads and north-south streets that are there. Rao and Collings both realize they will never agree on the choice of location, so they agree on this compromise: Rao has to select an east-west road, Collings will select a north-south street, and they will build the lab at their intersection.

1,100	825	650
1,000	850	900
600	825	1,100

(a) Use game theory to help the researchers decide the best location for the lab.

(b) Sketch the surface of this portion of land, showing the relative heights of each of the nine possible building sites. Use your sketch to help explain why the term "saddle point" is an appropriate one.

40. (a) Solve Exercise 39 if the matrix giving the ground elevations is

950	1,100	925
1,000	1,000	1,100
900	825	950

(b) If the following matrix represents the elevations, what are the constraints on y if y is the saddle point of the problem?

1,000	1,200	1,050
950	1,000	1,100
y	1,100	1,100

41. (*Child's Play*) What is the game matrix for Morra if each child can show one of *three* fingers?

42. (*Restaurant Management*) Two fast-food outlets, Big Burger and Burger Czar, are separately contemplating locating in two neighboring suburbs. If they both locate in the first suburb, Big Burger will grab 65 percent of the business by virtue of its extensive national ad campaign. If both are in the second, Big Burger will net only 60 percent of the available business. On the other hand, if Big Burger locates in suburb 1 and Burger Czar chooses the other, Big Burger will attract only a 35 percent market share, whereas with Big Burger in suburb 2 and Burger Czar in suburb 1, Big Burger garners 85 percent of the fast-food market. Set this problem up as a game with a payoff matrix. Let Big Burger play the rôle of the row player R. Is there an optimal strategy? What is it?

43. (*History*) A history graduate student is attempting to clarify the issues surrounding a medieval French battle about which conflicting records exist. Adherents of two conflicting religious groups, *les Rondes* and *les Casse-têtes*, met and did battle on or near a pair of hills in Burgundy. Had both of them approached the battle line from the *different* hills, *les Casse-têtes* would have had a 50 percent chance of winning the battle. If they both approached from the first hill, the chances for *les Casse-têtes* increased to 75 percent, but their chances decreased to 25 percent if both groups approached from the second hill. Help the student decide what the optimal battle strategy would have been. Applying this best strategy, what would have been the chances that *les Rondes* won the battle?

44. Solve Exercise 43 again if the probabilities of the battle victories are different. If the two sides approach each other from the same hills, the probability of victory for *les Casse-têtes* is 66 percent. If the two sides approach from different hills, and *les Casse-têtes* approach from hill 1, their probability of winning is 25 percent; this rises to 75 percent if this group approaches from the second hill. What now is the optimal strategy for both sides?

45. *(Play)* One children's game consists of each of two children simultaneously showing one or two fingers. Robin and Carla play this game and, in honor of Robin's birthday, decide that Carla will give Robin as many dollars as there are total fingers showing. What is the best strategy for both Robin and Carla? What is the minimum that Robin can count on receiving providing she plays the game rationally? Show that this game is not fair.

Section 8.2   *GAMES OF MIXED STRATEGY*

Games with identifiable, unambiguous saddle point strategies are not common. Two-person games have payoff matrices similar to the one for Morra, which has no saddle point and is not strictly determined. (Why?) The Morra game matrix is

$$\begin{pmatrix} 2 & -3 \\ -3 & 4 \end{pmatrix}$$

Despite the absence of a saddle point, is there nevertheless a best strategy for this game, and, if so, what is it?

Whatever the optimal strategy is, it does *not* include choosing a single strategy and sticking with it, time after time. The opponent would soon notice the pattern and use it against us. For example if player R consistently plays row 2 (showing two fingers) in hopes of winning the payoff of $+4$, player C would soon start playing the column 1 strategy (showing a single finger), which would reduce R's payoff from $+4$ to -3. If there is a best strategy, *it must involve a random alternation among all available strategies.*

In a two-person game, suppose players R and C have m and n options available to them in each play. If the $m \times n$ payoff matrix contains no saddle points, then R and C *mixed strategy* will need an appropriate *mixed strategy*. That is, there will be some set of m optimal probabilities

$$r_1, \; r_2, \; \ldots, \; r_m \qquad r_1 + r_2 + \cdots + r_m = 1$$

such that R will choose strategy 1 with probability r_1, strategy 2 with probability r_2, and so on. Similarly, there will be a set of n probabilities

$$c_1, \; c_2, \; \ldots, \; c_n \qquad c_1 + c_2 + \cdots + c_n = 1$$

optimal strategy

that govern C's strategies. The two players make their strategic decisions *independently* of each other.

In games of mixed strategy, it is important to carefully state what we expect from an *optimal strategy*.

Rational Decision Making in Games of Mixed Strategy

Player R chooses probabilities r_i such that if alternative i is played with probability r_i, then R's expected payoff is as large as possible against the best counterstrategy that C can offer.

Similarly, if C plays option i with probability c_i, then the expected payoff will be a maximum against any and all counterstrategies that R chooses.

Three questions remain.

1. Given such probabilities for R and C, what are the expected payoffs for each opponent?

2. How may we determine the optimal probabilities r_i and c_i?

3. On a practical level, how do we set up a procedure to guarantee that alternative i for R occurs a fraction r_i of the time and that alternative j for C occurs a fraction c_i of the time?

Example 5   The payoff matrix for a certain game is

$$\begin{pmatrix} -1 & 3 \\ 2 & -4 \end{pmatrix}$$

C plays each column 50 percent of the time. On the other hand, R can choose between two game plans—either choosing the first row $\frac{1}{3}$ of the time and choosing the second row $\frac{2}{3}$ of the time, or vice versa, choosing the second row $\frac{1}{3}$ of the time and the first row $\frac{2}{3}$ of the time. Of these, what is R's best plan? (For the moment, ignore the issue of where these strategies came from; assume they are given.)

Solution

R and C do not collaborate, so their choices are *independent* in the sense of probability theory. There are four possible outcomes in any given game:

$$R1, C1 \qquad R1, C2 \qquad R2, C1 \qquad R2, C2$$

where the notation $R1, C2$ means R chooses option 1 and C chooses option 2, and so on. The probabilities of independent events are the product of the probabilities, so $P(R1, C1) = P(R1) \cdot P(C1)$, and so on.

Case 1:   For the case in which R chooses the first row $\frac{1}{3}$ of the time, we have this suggestive matrix:

$$\begin{array}{cc} & \begin{array}{cc} \frac{1}{2} & \frac{1}{2} \end{array} \\ \begin{array}{c} \frac{1}{3} \\ \frac{2}{3} \end{array} & \begin{pmatrix} -1 & 3 \\ 2 & -4 \end{pmatrix} \end{array}$$

which leads to this table:

	Probability	Payoff	Expectation
$R1, C1$	$\frac{1}{6}$	-1	$-\frac{1}{6}$
$R1, C2$	$\frac{1}{6}$	3	$\frac{1}{2}$
$R2, C1$	$\frac{1}{3}$	2	$\frac{2}{3}$
$R2, C2$	$\frac{1}{3}$	-4	$-\frac{4}{3}$
Totals	1		$-\frac{1}{3}$

The expected payoff to R in the long run is $-\frac{1}{3}$, a consistent loss.

Case 2:   The following table details the analysis for the case $P(R1) = \frac{2}{3}$. Again, if the matrix

$$\begin{array}{cc} & \begin{array}{cc} \frac{1}{2} & \frac{1}{2} \end{array} \\ \begin{array}{c} \frac{2}{3} \\ \frac{1}{3} \end{array} & \begin{pmatrix} -1 & 3 \\ 2 & -4 \end{pmatrix} \end{array}$$

suggests the strategy, then this table indicates R's expectation.

	Probability	Payoff	Expectation
$R1, C1$	$\frac{1}{3}$	-1	$-\frac{1}{3}$
$R1, C2$	$\frac{1}{3}$	3	1
$R2, C1$	$\frac{1}{6}$	2	$\frac{1}{3}$
$R2, C2$	$\frac{1}{6}$	-4	$-\frac{2}{3}$
Totals			$\frac{1}{3}$

In this second case, R makes out much better. In the long run, R *gains* $\frac{1}{3}$ for each play of the game. Of the available strategies, R should employ this one.

In both cases, the expected payoff is not one of the actual available payoffs.

Skill Enhancer 5   Consider the payoff matrix

$$\begin{array}{cc} & \begin{array}{cc} \frac{1}{2} & \frac{1}{2} \end{array} \\ \begin{array}{c} \alpha \\ 1-\alpha \end{array} & \begin{pmatrix} -2 & 2 \\ 2 & -3 \end{pmatrix} \end{array}$$

which suggests that C alternates between each strategy one-half of the time. R plays the first strategy α of the time, and the second strategy $1 - \alpha$ of the time.

(a) Does this game possess a saddle-point strategy?

(b) What is R's expected value for this game if $\alpha = \frac{1}{4}$?

(c) What is R's expected value for this game if $\alpha = \frac{1}{3}$?

Answer in Appendix E.

Example 5 illustrates three points.

1. We use probability to analyze mixed-strategy situations.

2. Some strategies are superior to others.

3. We need a better way to compute the expected payoffs for mixed strategies.

Matrix Methods for Computing Expected Payoff

It will be instructive to solve the last example again in greater generality. We may achieve that by replacing the specific game matrix with this more general one.

$$
\begin{array}{cc}
 & \begin{array}{cc} c_1 & c_2 \end{array} \\
\begin{array}{c} r_1 \\ r_2 \end{array} & \left(\begin{array}{cc} e & f \\ g & h \end{array} \right)
\end{array}
$$

The expected payoff for any game is the sum of the products in the right column of Figure 8.2. This is a difficult sum to remember, but is fortunately given by the easier-to-remember matrix product

(8.1) RGC

where G is the game matrix and the variables R and C are the matrices

$$ R = \begin{pmatrix} r_1 & r_2 \end{pmatrix} \qquad C = \begin{pmatrix} c_1 \\ c_2 \end{pmatrix} $$

Remember to

- *Pre*multiply (multiply on the left) by R, R's row strategy
- *Post*multiply (multiply on the right) by C, C's strategy column matrix

	Probability	*Payoff*	*Expectation*
$R1, C1$	$r_1 c_1$	e	$r_1 c_1 e$
$R1, C2$	$r_1 c_2$	f	$r_1 c_2 f$
$R2, C1$	$r_2 c_1$	g	$r_2 c_1 g$
$R2, C2$	$r_2 c_2$	h	$r_2 c_2 h$

FIGURE 8.2

The expected payoff for a 2×2 game—general case.

Although we presented Equation (8.1) in the context of a 2×2 game, it applies to the expected payoff for any two-person $m \times n$ game when the matrix and the probabilities r_i and c_i are defined appropriately. We will use the notation $E(R, C)$ to show that the expected payoff depends on the components of R and C. Of course, this payoff also depends on the components of G.

Example 6   The matrix

$$G = \begin{pmatrix} 2 & 4 & 0 \\ -3 & 1 & 1 \end{pmatrix}$$

describes the payoffs in a game between two players R and C. C never chooses the middle column, and plays the first column three times as often as the third. Player R decides among her two alternatives randomly (but R can choose only one strategy or the other, not both). Compute the expected payoff for this set of strategies.

Solution

These facts imply the existence of these probability matrices:

$$R = \begin{pmatrix} \frac{1}{2} & \frac{1}{2} \end{pmatrix} \quad \text{and} \quad C = \begin{pmatrix} \frac{3}{4} \\ 0 \\ \frac{1}{4} \end{pmatrix}$$

so the expected payoff $E(R, C)$ is the product RGC. By the associative rule of matrix multiplication, it does not matter which pair of matrices we multiply first. Performing the calculations,

$$RG = \begin{pmatrix} \frac{1}{2} & \frac{1}{2} \end{pmatrix} \begin{pmatrix} 2 & 4 & 0 \\ -3 & 1 & 1 \end{pmatrix} = \begin{pmatrix} -\frac{1}{2} & \frac{5}{2} & \frac{1}{2} \end{pmatrix}$$

$$(RG)C = \begin{pmatrix} -\frac{1}{2} & \frac{5}{2} & \frac{1}{2} \end{pmatrix} \begin{pmatrix} \frac{3}{4} \\ 0 \\ \frac{1}{4} \end{pmatrix} = -\frac{3}{8} + \frac{1}{8} = -\frac{1}{4}$$

so with this set of strategies, over the long run, R will steadily lose $\frac{1}{4}$ to C on each play.

Skill Enhancer 6   Let $R = \begin{pmatrix} \frac{1}{4} & \frac{3}{4} \end{pmatrix}$ and $C = \begin{pmatrix} \frac{1}{3} \\ \frac{1}{3} \\ \frac{1}{3} \end{pmatrix}$. Use these probability matrices to compute the value of the game in Example 6.

Answer in Appendix E.

This formalism is useful for discussing the *Fundamental Theorem of Game Theory*. (We omit the proof.) This theorem guarantees a best set of strategies (that is, probabilities for each alternative) for both R and C.

The Fundamental Theorem of Game Theory

For any valid matrices R and C whose elements are probabilities summing to 1, the expected payoff E is always given by the matrix product RGC:

$$E = RGC$$

For each player, there is an optimal set of probabilities described by the matrices R_{opt} and C_{opt}. It is optimal because if R uses this optimal strategy, then no matter what strategy the opponent uses, the expected payoff $R_{\text{opt}}GC$ will always be at least as great as some number v:

$$R_{\text{opt}}GC \geq v$$

In the same way, if C uses the optimal strategy C_{opt}, then C's expected payoff will *not* exceed the same number v,

$$RGC_{\text{opt}} \leq v$$

no matter what strategy R employs.

The number v is the *value* of the game.

This theorem fails to prescribe means for discovering either of the optimal strategies or the value v of the game. We delve into these important issues in the next section.

_____ *Exercises 8.2* _____

Use the payoff matrix $\begin{pmatrix} 6 & 5 \\ 7 & 3 \end{pmatrix}$ in each of Exercises 1–6 to find the expected payoff. Use the given strategy matrices R and C in each problem.

1. $R = \begin{pmatrix} \frac{1}{2} & \frac{1}{2} \end{pmatrix}$, $C = \begin{pmatrix} \frac{1}{2} \\ \frac{1}{2} \end{pmatrix}$

2. $R = \begin{pmatrix} 1 & 0 \end{pmatrix}$, $C = \begin{pmatrix} \frac{1}{3} \\ \frac{2}{3} \end{pmatrix}$

3. $R = \begin{pmatrix} \frac{2}{3} & \frac{1}{3} \end{pmatrix}$, $C = \begin{pmatrix} 0 \\ 1 \end{pmatrix}$

4. $R = \begin{pmatrix} 1 & 0 \end{pmatrix}$, $C = \begin{pmatrix} 0 \\ 1 \end{pmatrix}$

5. $R = \begin{pmatrix} \frac{1}{3} & \frac{2}{3} \end{pmatrix}$, $C = \begin{pmatrix} \frac{2}{3} \\ \frac{1}{3} \end{pmatrix}$

6. $R = \begin{pmatrix} \frac{1}{4} & \frac{3}{4} \end{pmatrix}$, $C = \begin{pmatrix} \frac{1}{3} \\ \frac{2}{3} \end{pmatrix}$

Use the payoff matrix $\begin{pmatrix} 1 & -2 \\ -1 & 2 \end{pmatrix}$ in each of Exercises 7–12 to find the expected payoff. Use the given strategy matrices R and C in each problem.

7. $R = \begin{pmatrix} \frac{1}{2} & \frac{1}{2} \end{pmatrix}$, $C = \begin{pmatrix} \frac{1}{2} \\ \frac{1}{2} \end{pmatrix}$

8. $R = \begin{pmatrix} 1 & 0 \end{pmatrix}$, $C = \begin{pmatrix} 0 \\ 1 \end{pmatrix}$

9. $R = \begin{pmatrix} \frac{2}{3} & \frac{1}{3} \end{pmatrix}$, $C = \begin{pmatrix} \frac{2}{3} \\ \frac{1}{3} \end{pmatrix}$

10. $R = \begin{pmatrix} \frac{1}{2} & \frac{1}{2} \end{pmatrix}$, $C = \begin{pmatrix} \frac{1}{4} \\ \frac{3}{4} \end{pmatrix}$

11. $R = \begin{pmatrix} \frac{1}{4} & \frac{3}{4} \end{pmatrix}$, $C = \begin{pmatrix} \frac{1}{4} \\ \frac{3}{4} \end{pmatrix}$

12. $R = \begin{pmatrix} 0 & 1 \end{pmatrix}$, $C = \begin{pmatrix} \frac{3}{4} \\ \frac{1}{4} \end{pmatrix}$

Suppose player R chooses each of the available row strategies an equal fraction of the time, and C chooses column strategies in the same way. In these conditions, what is the expected payoff $E(R, C)$ of the games in Exercises 13–25?

13. $\begin{pmatrix} 1 & -1 \\ 0 & 5 \end{pmatrix}$ **14.** $\begin{pmatrix} 0.5 & 0.3 \\ -0.3 & 0.8 \end{pmatrix}$

15. $\begin{pmatrix} 1 & 1 \\ 1 & 1 \end{pmatrix}$ **16.** $\begin{pmatrix} 1 & -1 \\ -1 & 1 \end{pmatrix}$

17. $\begin{pmatrix} 5 & 4 \\ -3 & 7 \end{pmatrix}$ **18.** $\begin{pmatrix} -2 & 3 \\ 4 & -1 \end{pmatrix}$

19. $\begin{pmatrix} 4 & -2 & 1 \\ -6 & 5 & -1 \end{pmatrix}$ **20.** $\begin{pmatrix} 3 & 4 \\ 5 & -1 \\ -3 & 3 \end{pmatrix}$

21. $\begin{pmatrix} 1 & 1 & 1 \\ 2 & 0 & -2 \\ -3 & 2 & -1 \end{pmatrix}$ **22.** $\begin{pmatrix} 2 & -1 & 3 \\ 3 & -2 & 0 \\ 0 & 1 & 4 \end{pmatrix}$

23. $\begin{pmatrix} 0 & 1 & -1 \\ -1 & 0 & 1 \\ 1 & -1 & 0 \end{pmatrix}$ **24.** $\begin{pmatrix} \frac{1}{2} & \frac{1}{4} & \frac{1}{4} \\ \frac{1}{4} & \frac{1}{2} & \frac{1}{4} \\ \frac{1}{4} & \frac{1}{4} & \frac{1}{2} \end{pmatrix}$

25. $\begin{pmatrix} -2 & 3 & -1 & 0 \\ 4 & -2 & 1 & -2 \\ -3 & 3 & -3 & 3 \\ 1 & -1 & 1 & -1 \end{pmatrix}$

26. Based on your work in the previous 13 problems, what is an easy way to compute the expected value of a game whenever R and C both have the same number of options (call this number n) and each chooses each option an equal fraction of the time?

Suppose C chooses strategies according to probabilities

$$C = \begin{pmatrix} \frac{1}{4} \\ \frac{3}{4} \end{pmatrix}$$

and R decides to play according to *either* of the probability row matrices $R_1 = \begin{pmatrix} 0.5 & 0.5 \end{pmatrix}$ or $R_2 = \begin{pmatrix} \frac{7}{8} & \frac{1}{8} \end{pmatrix}$. Use expected payoff to decide which of the R strategies is superior for the payoff matrices in Exercises 27–30.

27. $\begin{pmatrix} 1 & 0 \\ 0 & 1 \end{pmatrix}$ **28.** $\begin{pmatrix} \frac{1}{3} & -\frac{2}{3} \\ \frac{1}{6} & \frac{1}{3} \end{pmatrix}$

29. $\begin{pmatrix} 0.4 & -0.6 \\ -0.1 & -0.5 \end{pmatrix}$ **30.** $\begin{pmatrix} 10 & 1 \\ 2 & 9 \end{pmatrix}$

31. Use the Fundamental Theorem of Game Theory to de-

termine the expected payoff when *both* players use their optimal probability distributions for choosing strategies.

W 32. Describe the meaning of value for a game with a mixed strategy. How does this meaning differ from that of the value for a strictly determined game?

Applications

(Broadcasting) Two television stations serve the metropolitan area of Portland, Oregon—call them stations KCOL and KROW. These two stations are engaged in a ratings war over the Saturday morning "prime time" period, during which children watch TV for hours. There are many varieties of children's programming that can be shown, but local parents' groups have effectively pressured for restrictions to either cartoons or old, classic comedies between the hours of 10:00 and 11:00. An independent ratings service in the employ of KROW has determined the effect on the ratings of the program that KROW offers given the two types of programming that KCOL offers. This information is summarized in a payoff matrix. The elements in this matrix represent units by which ratings of KROW rise or fall. Use this information in Exercises 33–35.

33. Suppose the payoff matrix is

$$\begin{array}{cc} & \text{Cartoons} \quad \text{Comedies} \\ \begin{array}{c} \text{Cartoons} \\ \text{Comedies} \end{array} & \begin{pmatrix} -1 & 1 \\ 0 & 2 \end{pmatrix} \end{array}$$

and suppose further that the program director at KROW knows that the opposing station programs each of the two categories with equal probability.
   (a) What should the programming strategy of KROW be so as to maximize expected programming ratings?
   (b) Using 20-20 hindsight and common sense, how could you have predicted this result purely by considering the structure of the payoff matrix?

34. Suppose you know that the probability column matrix that KCOL uses to choose programming is

$$\begin{pmatrix} \frac{1}{4} \\ \frac{3}{4} \end{pmatrix}$$

Will KROW's strategy change? What will the expected change in KROW's ratings be?

35. Suppose the payoff matrix now has these components:

$$\begin{pmatrix} -1 & 2 \\ 0 & 1 \end{pmatrix}$$

and KCOL uses the column matrix

$$\begin{pmatrix} \frac{1}{2} \\ \frac{1}{2} \end{pmatrix}$$

Show that the expected payoff for KROW is independent of its strategy.

36. *(Advertising)* The Row Group and ColCorp are two large firms competing in the sale of personal products. Each has two new brands of toothpaste it is anxious to market. The following payoff matrix (the payoff is measured in percentages of the total market) shows the Row Group's payoffs when each of the toothpastes is paired against the competition.

	1	2
1	50%	60%
2	70%	30%

The Row Group has learned that ColCorp will advertise its products on a random basis, with each product to get the same amount of advertising. Row Group's advertising department has decided tentatively to allocate one-third of its advertising to Row Group's product 1, and two-thirds of the budget to the second product. Compute the expected payoff, and compare it to the expected payoff should the Row Group use two-thirds of the budget on product 1.

37. Solve Exercise 36 if the Row Group spends equal amounts on each of its products, but ColCorp spends one-third on its first product and two-thirds on its second product. Compare that with the case of ColCorp spending two-thirds of its budget on product 1.

38. *(Politics)* In the upcoming elections, the Republicans and Democrats have to focus on two special issues of interest to a particular small town in Utah. The following matrix gives the payoffs to the Republicans (rows) when each party takes a position on one of the issues against its opponents. The Democrats have the columns. The issues are identified by the numbers 1 and 2.

	1	2
1	40%	60%
2	70%	50%

What is the expected result of the election if the Republicans spend one-third of the time on issue 1 and two-thirds on issue 2, and the Democrats spend one-quarter of the time on the first issue and the remainder on the second issue?

Calculator Exercises

Use the payoff matrix $\begin{pmatrix} 0.55 & -0.45 \\ -0.77 & 1.11 \end{pmatrix}$ in each of Exercises 39–42 to find the expected payoff. Use the given strategy matrices R and C in each problem.

39. $R = \begin{pmatrix} \frac{1}{2} & \frac{1}{2} \end{pmatrix}, C = \begin{pmatrix} \frac{1}{2} \\ \frac{1}{2} \end{pmatrix}$

40. $R = \begin{pmatrix} 0.55 & 0.45 \end{pmatrix}, C = \begin{pmatrix} 0.37 \\ 0.63 \end{pmatrix}$

41. $R = \begin{pmatrix} 0.21 & 0.79 \end{pmatrix}, C = \begin{pmatrix} 0.38 \\ 0.62 \end{pmatrix}$

42. $R = \begin{pmatrix} 0.3491 & 0.6509 \end{pmatrix}, C = \begin{pmatrix} 0.2937 \\ 0.7063 \end{pmatrix}$

Section 8.3   OPTIMAL MIXED STRATEGIES

The material in this section will allow us to compute optimal game strategies—or rather, it will allow us to transform a problem in game theory into one in linear programming. The linear programming problem may then be tackled by the techniques of Chapters 3 and 4.

We begin, though, by considering an important set of special cases.

Dominant Rows and Recessive Columns

In special cases, general $m \times n$ games can be reduced to smaller matrices.

dominant row   One row is said to *dominate* another in a payoff matrix if each element in the dominating row is greater than or equal to the corresponding element in the dominated row. A rational row player will *always* choose the dominant row over a dominated row because these strategies will be at least as good as those of the dominated row.

Example 7 Analyze the game characterized by the following payoff matrix

$$\begin{pmatrix} 1 & 2 \\ 0 & 3 \\ -1 & 0 \end{pmatrix}$$

Solution

The first row dominates the third row because each of the components in the first row is greater than its counterpart in the third row. There will never be a reason for the rational row player to choose row 3 over row 1—no matter what C's strategy is, R will *always* do better choosing row 1 over row 3. (Of course, there may be times when R will opt for the second row.) The dominance of row 1 over row 3 means that we ignore that row in the payoff matrix. Therefore

$$\begin{pmatrix} 1 & 2 \\ 0 & 3 \end{pmatrix}$$

is entirely equivalent to the original payoff matrix, but has the advantage that we can analyze it for optimal strategies. (Note that row 2 also dominates row 3 in the original matrix.)

Using the methods presented earlier, we find that the saddle point for this game is the top left element. The value of the game is 1, and the first option is optimal for each player.

Skill Enhancer 7 Is there a 2×2 game equivalent to the following? What is it?

$$G = \begin{pmatrix} 1 & 2 \\ 3 & 4 \\ 4 & 3 \end{pmatrix}$$

Answer in Appendix E.

recessive column   In the same way, the elements in one column may all be less than or equal to the elements of some other column. The first column is called *recessive*. Although the row player's goal is to maximize payoff, the column player's goal is to *minimize* it. Therefore, a rational column player will *always* choose a recessive column where possible.

Example 8 Reduce the payoff matrix

$$\begin{pmatrix} 1 & 6 & 3 \\ 4 & 2 & 5 \end{pmatrix}$$

Solution

The first column is recessive to the third. Therefore, we may discard the third column; it will never play a part in strategy proposed by a rational column player. The resulting payoff matrix is therefore

$$\begin{pmatrix} 1 & 6 \\ 4 & 2 \end{pmatrix}$$

The reduced matrix has no saddle point, so we will be able to apply the mixed strategy results we present shortly.

Skill Enhancer 8   Reduce $G = \begin{pmatrix} 2 & 3 & 4 \\ 3 & 4 & 2 \end{pmatrix}$.

Answer in Appendix E.

Example 9   Show how the matrix

$$\begin{pmatrix} 1 & 2 & 3 \\ 2 & 3 & 4 \\ 0 & 0 & -1 \\ 3 & 0 & 8 \end{pmatrix}$$

can be reduced.

Solution

First, note that row 2 dominates both row 1 and row 3. We discard those rows immediately. The payoff matrix reduces to

$$\begin{pmatrix} 2 & 3 & 4 \\ 3 & 0 & 8 \end{pmatrix}$$

Finally, compare columns 2 and 3. The second is recessive to the third, so we discard the third column. We are left with

$$\begin{pmatrix} 2 & 3 \\ 3 & 0 \end{pmatrix}$$

which we can analyze using the formulas of this section.

Skill Enhancer 9   Consider the payoff matrix

$$G = \begin{pmatrix} 1 & 2 & 4 \\ 3 & 1 & 3 \\ 0 & 1 & 1 \\ 0 & 1 & 2 \end{pmatrix}$$

(a) Reduce the matrix by considering dominant rows only.
(b) Reduce this matrix by considering recessive columns.

Answers in Appendix E.

Now we begin the more general analysis.

Optimal Row Strategy

We proceed by considering a (fairly) general 2×2 game. That is, let

(8.2)
$$G = \begin{pmatrix} a_{11} & a_{12} \\ a_{21} & a_{22} \end{pmatrix}$$

subject only to the restriction that all components a_{ij} be strictly greater than 0. (As we will see, this is only a temporary restriction.)

We can compute the value of this game as

(8.3)
$$v = (RG)C = \begin{pmatrix} r_1 a_{11} + r_2 a_{21} & r_1 a_{12} + r_2 a_{22} \end{pmatrix} \begin{pmatrix} c_1 \\ c_2 \end{pmatrix}$$
$$= c_1(r_1 a_{11} + r_2 a_{21}) + c_2(r_1 a_{12} + r_2 a_{22})$$

The row player's analysis proceeds as follows: R makes a move, which C counters with a strategy that will minimize the game's value. The minimum of Equation (8.3) will occur when $c_1 = 0$ or $c_1 = 1$. That is, the minimum will be the smaller of $(r_1 a_{11} + r_2 a_{21})$ or $(r_1 a_{12} + r_2 a_{22})$. (See Exercise 26 to see why.) Let t be this minimum:

(8.4)
$$t = \min(r_1 a_{11} + r_2 a_{21}, r_1 a_{12} + r_2 a_{22})$$

The rational strategy for R will be one that *maximizes* t. Since the larger v is, the smaller will be $\frac{1}{t}$, we can equivalently state that R should attempt to minimize $\frac{1}{t}$.

From Equation (8.4), we must have

$$t \le r_1 a_{11} + r_2 a_{21}$$
$$t \le r_1 a_{12} + r_2 a_{22}$$

Since $v > 0$ (because r_1, r_2 are nonnegative and all components of G are > 0 by assumption), we may divide each equality by t without changing the sense of the inequalities.

$$1 \le \frac{r_1}{t} a_{11} + \frac{r_2}{t} a_{21}$$
$$1 \le \frac{r_1}{t} a_{12} + \frac{r_2}{t} a_{22}$$

These inequalities will look simpler if we define new variables:

(8.5)
$$x_1 = \frac{r_1}{t} \qquad x_2 = \frac{r_2}{t}$$

Since $r_1 + r_2 = 1$, we have

(8.6)
$$x_1 + x_2 = \frac{1}{t}$$

a quantity we seek to minimize. In terms of these new variables, the inequalities become

(8.7)
$$x_1 a_{11} + x_2 a_{21} \ge 1$$
$$x_1 a_{12} + x_2 a_{22} \ge 1$$

That is, the strategy for R is determined by finding the minimum of

$$x_1 + x_2$$

subject to Inequalities (8.7) and to $x_1 \ge 0$, $x_2 \ge 0$. But this is a special type of linear programming problem—it is a minimum problem subject to $\ge$ constraints. This special class of problem was a focus of our attention with regard to the simplex method, whose *dual* allowed us to solve maximum problems. In fact, the dual of this problem solves the problem of optimal strategy for C. (See Exercise 28.)

The solution to the linear programming problem yields a minimum value for $\frac{1}{t}$ and the values of x_1 and x_2 that produce this value. It remains to note how to retrieve the optimal strategy matrix R from the linear programming solution. Use the defining equations for the x_i:

$$r_1 = x_1 t \qquad r_2 = x_2 t$$

We state the generalization of this method for the case where G is an $m \times n$ matrix.

Determining Optimal Row Strategies

First, solve the linear programming problem defined by minimizing $\frac{1}{t} = x_1 + x_2 + \cdots + x_m$ subject to $x_i \geq 0$ and to

$$\begin{pmatrix} x_1 & x_2 & \cdots & x_m \end{pmatrix} G \geq \begin{pmatrix} 1 \\ 1 \\ \vdots \\ 1 \end{pmatrix}$$

where the right matrix is an $m \times 1$ column matrix all of whose components are 1s. Then, determine the r_i via

$$r_i = x_i t$$

Here,

G is any $m \times n$ game matrix, all of whose elements are positive.

Example 10   What are the probabilities for R's best strategy if $G = \begin{pmatrix} 1 & 10 \\ 11 & 2 \end{pmatrix}$?

Solution

We seek to minimize $x_1 + x_2$ subject to $x_1 \geq 0$, $x_2 \geq 0$ and to

$$\begin{pmatrix} x_1 & x_2 \end{pmatrix} \begin{pmatrix} 1 & 10 \\ 11 & 2 \end{pmatrix} \geq \begin{pmatrix} 1 \\ 1 \end{pmatrix}$$

(That is, each component of the matrix product is to be greater than or equal to 1.) This leads to the set of inequality constraints

$$x_1 + 11x_2 \geq 1$$
$$10x_1 + 2x_2 \geq 1$$

The graphical solution to this problem is indicated in Figure 8.3.

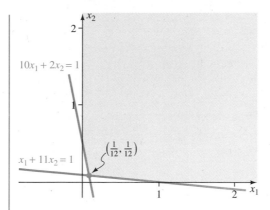

FIGURE 8.3
R's best strategy.

Examination of the three corners of this figure forces the conclusion that $x_1 = x_2 = \frac{1}{12}$ and therefore that $\frac{1}{t} = \frac{1}{6}$ and $t = 6$. Therefore

$$r_1 = x_1 t = \frac{6}{12} = \frac{1}{2} \qquad r_2 = x_2 t = \frac{6}{12} = \frac{1}{2}$$

The value of this game is 6.

Skill Enhancer 10   Consider the game

$$G = \begin{pmatrix} 2 & 11 \\ 10 & 1 \end{pmatrix}$$

(a) Is G strictly determined?
(b) What is the set of constraints we seek to satisfy?
(c) What are the values of the x_i at the optimum?
(d) What is the quantity we seek to minimize, and what is its minimum value?
(e) What is the optimum value for t? What are the optimal probabilities r_i determining R's strategy?

Answers in Appendix E.

Optimal Strategy for C

Write the value of the game as

$$v = R(GC) = r_1(c_1 a_{11} + c_2 a_{12}) + r_2(c_1 a_{21} + c_2 a_{22})$$

C realizes that R will attempt to maximize this expression. As before, this maximum will occur when either $c_1 = 0$ or $c_1 = 1$, and C will try to minimize this. Therefore, define w by

$$w = \min(c_1 a_{11} + c_2 a_{12}, c_1 a_{21} + c_2 a_{22})$$

Minimizing w is entirely equivalent to maximizing $\frac{1}{w}$. If we mirror the steps in the previous argument, this time choosing variables

$$y_1 = \frac{c_1}{w} \qquad y_2 = \frac{c_2}{w}$$

we conclude that the optimal column strategy is related to the solution of the linear programming maximization problem defined by maximizing $y_1 + y_2$ if $y_1, y_2 \geq 0$ and if

$$a_{11} y_1 + a_{12} y_2 \leq 1$$
$$a_{21} y_1 + a_{22} y_2 \leq 1$$

The relation between the two types of problems is via the connecting equations

$$c_1 = y_1 w \qquad c_2 = y_2 w$$

As before, we generalize this solution.

Optimal Column Strategy

First, solve the linear programming problem defined by maximizing $\frac{1}{w} = y_1 + y_2 + \cdots + y_n$ if all the y_i are greater than or equal to 0 and if

$$G \begin{pmatrix} y_1 \\ y_2 \\ \vdots \\ y_n \end{pmatrix} \leq \begin{pmatrix} 1 \\ 1 \\ \vdots \\ 1 \end{pmatrix}$$

where G contains only positive elements. The rightmost matrix is an $n \times 1$ column matrix all of whose components are 1s.

Next, this solution is related to column strategy via the equations

$$c_i = y_i w$$

Example 11   If the game matrix is $G = \begin{pmatrix} 1 & 10 \\ 11 & 2 \end{pmatrix}$, what is the optimal column strategy?

Solution

G defines the linear programming problem of maximizing $y_1 + y_2$ if

$$y_1 + 10 y_2 \leq 1 \qquad 11 y_1 + 2 y_2 \leq 1$$

for $y_1 \geq 0$, $y_2 \geq 0$. The graphical solution to this problem is presented in Figure 8.4.

We conclude that $y_1 = \frac{2}{27}$, $y_2 = \frac{5}{54}$, and the maximum value of $y_1 + y_2$ is therefore $\frac{2}{27} + \frac{5}{54} = \frac{1}{6}$. From this, $w = 6$, so that

$$c_1 = y_1 w = \frac{4}{9} \qquad c_2 = y_2 w = \frac{5}{9}$$

The value of this game is 6.

Skill Enhancer 11   Consider the game

$$G = \begin{pmatrix} 2 & 11 \\ 10 & 1 \end{pmatrix}$$

We want the optimal column strategy.
(a) What is the set of constraints we seek to satisfy?
(b) What are the values of the y_i at the optimum?
(c) What is the quantity we seek to minimize, and what is its minimum value?
(d) What is the optimum value for w?
(e) What is the value for this game? What are the optimal probabilities c_i determining C's strategy?

Answers in Appendix E.

Examples 10 and 11 taken together solve the game $\begin{pmatrix} 1 & 10 \\ 11 & 2 \end{pmatrix}$ by providing optimal strategies for both players and by providing the value of the game.

Negative Components in G

Suppose G is some $m \times n$ game matrix containing one or more negative components. The theory presented above will not work without some modifications. This is because crucial steps in the derivation involved division by t or w. We relied on the facts that neither quantity was ever zero, so that we were able to divide by t or w, and that these

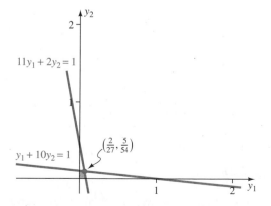

FIGURE 8.4
C's **best strategy.**

quantities were positive, so that the sense of the inequalities would not change upon division.

Associated with G is some positive number k and an $m \times n$ matrix E all of whose components are 1. Let a_{ij} be the negative component of G with the greatest absolute value, and define k as some number such that $k > |a_{ij}|$. Let us consider the game matrix G' where

$$G' = G + kE$$

G' is a matrix all of whose components are by definition greater than 0, so the previous analysis applies to G'.

First, note that the optimal strategies associated with G and with G' are identical. If every player's payoffs are all increased by the same amount, there is no advantage in switching strategies.

Next, what will the value of G' be? Suppose the value of G is v. We show in Exercise 25 that the value of G' is precisely $v + k$.

Therefore, to find strategies for G, simply construct a matrix G' whose solution is defined by the linear programming procedures laid out above. The strategies for G and G' are identical. Their values differ by the constant k used to construct G'.

Example 12 Determine the optimal strategies for the game defined by

$$G = \begin{pmatrix} -1 & 8 \\ 9 & 0 \end{pmatrix}$$

Solution

The component $a_{11} = -1$ is negative. A convenient value of k is $k = 2$, but any k such that $k > |-1|$ would do. If $E = \begin{pmatrix} 1 & 1 \\ 1 & 1 \end{pmatrix}$, then

$$G' = \begin{pmatrix} -1 & 8 \\ 9 & 0 \end{pmatrix} + 2 \begin{pmatrix} 1 & 1 \\ 1 & 1 \end{pmatrix} = \begin{pmatrix} 1 & 10 \\ 11 & 2 \end{pmatrix}$$

But this is precisely the game matrix of Examples 10 and 11. Consequently, the optimal strategies are as determined in those problems.

Since the value of G' must be 6, the value of G is $6 - 2 = 4$.

Skill Enhancer 12 Consider the game

$$G = \begin{pmatrix} -1 & 8 \\ 7 & -2 \end{pmatrix}$$

(a) If we add 3 to each payoff, what is the new game G'?
(b) What are the optimal strategies for G?
(c) What is the value of G?

Answers in Appendix E.

A useful special case involves 2×2 matrices. The linear programming procedures can be solved for a general game matrix to determine explicit formulas for such games. We will switch to a less formal notation for the components of G.

Optimal Strategies for 2×2 **Games**

If

$$G = \begin{pmatrix} e & f \\ g & h \end{pmatrix}$$

then

$$r_1 = \frac{h - g}{\Delta} \qquad c_1 = \frac{h - f}{\Delta}$$

$$r_2 = \frac{e - f}{\Delta} \qquad c_2 = \frac{e - g}{\Delta}$$

for

$$\Delta = e + h - f - g$$

The matrices $R = \begin{pmatrix} r_1 & r_2 \end{pmatrix}$ and $C = \begin{pmatrix} c_1 \\ c_2 \end{pmatrix}$ are the optimal row and column strategies. As before, the value v of the game is

$$v = RGC$$

Example 13   Use these formulas to solve $G = \begin{pmatrix} 1 & 10 \\ 11 & 2 \end{pmatrix}$.

Solution

Using the identifications $e = 1$, $f = 10$, $g = 11$, and $h = 2$, we have $\Delta = -18$. Therefore

$$r_1 = \frac{2 - 11}{-18} = \frac{1}{2} \qquad c_1 = \frac{2 - 10}{-18} = \frac{4}{9}$$

$$r_2 = \frac{1 - 10}{-18} = \frac{1}{2} \qquad c_2 = \frac{1 - 11}{-18} = \frac{5}{9}$$

These answers agree with the results of Examples 10 and 11, as they must. (The reader should also check that the formulas yield a value of 6 for the game.)

Skill Enhancer 13   Consider the game

$$G = \begin{pmatrix} 2 & 11 \\ 10 & 1 \end{pmatrix}$$

(a) What are the values for e, f, g, and h?
(b) What does Δ equal for this game?
(c) Using these values, determine the r_i and c_i for G.

Answers in Appendix E.

Example 14   Refer to the beginning of the chapter to review the rules for Morra. The game is played over and over, and the object is to amass as many pebbles as possible. What is each child's best strategy against the other? Which child has the advantage?

Solution

The payoff matrix is

$$
\begin{array}{cc}
 & \begin{array}{cc} 1 & 2 \end{array} \\
\begin{array}{c} 1 \\ 2 \end{array} & \begin{pmatrix} 2 & -3 \\ -3 & 4 \end{pmatrix}
\end{array} = \begin{pmatrix} e & f \\ g & h \end{pmatrix}
$$

Now, $\Delta = e + h - f - g = 2 + 4 - (-3) - (-3) = 12$, so that

$$
r_1 = \frac{4 - (-3)}{12} = \frac{7}{12}
$$

$$
r_2 = \frac{2 - (-3)}{12} = \frac{5}{12}
$$

For the second player, on the other hand (no pun),

$$
c_1 = \frac{\cdot\ 4 - (-3)}{12} = \frac{7}{12}
$$

$$
c_2 = \frac{2 - (-3)}{12} = \frac{5}{12}
$$

Each child should show one finger for $\frac{7}{12}$ of the time, and two fingers the remaining $\frac{5}{12}$ of the time. These are nonobvious results! The value of the game is

$$
v = \begin{pmatrix} \frac{7}{12} & \frac{5}{12} \end{pmatrix} \begin{pmatrix} 2 & -3 \\ -3 & 4 \end{pmatrix} \begin{pmatrix} \frac{7}{12} \\ \frac{5}{12} \end{pmatrix} = -\frac{1}{12}
$$

Over the long run, it is more advantageous to be the "column" child, who wins one-twelfth of a pebble per play on the average.

Skill Enhancer 14   In "Multiplication Morra," one child gets the *product* of the fingers shown, so the payoff matrix is

$$
G = \begin{pmatrix} 1 & -2 \\ -2 & 4 \end{pmatrix}
$$

(a) What are e, f, g, h, and Δ for this matrix?
(b) What are the row strategy probabilities r_i?
(c) What are the column strategy probabilities c_i?
(d) What is the value v of this game?

Answers in Appendix E.

_____ *Exercises 8.3* _____

Reduce the games in Exercises 1–8 by considering dominant rows and recessive columns.

1. $\begin{pmatrix} 1 & 2 \\ 3 & 3 \\ -2 & 1 \end{pmatrix}$

2. $\begin{pmatrix} 6 & -6 \\ 5 & 4 \\ 3 & 0 \end{pmatrix}$

3. $\begin{pmatrix} 9 & 6 & 8 \\ -1 & 2 & 0 \\ 0 & 0 & 1 \end{pmatrix}$

4. $\begin{pmatrix} 3 \\ 2 \\ 1 \end{pmatrix}$

5. $\begin{pmatrix} 1 & 3 \\ 2 & 4 \end{pmatrix}$

6. $\begin{pmatrix} 1 & 2 & 1.5 \\ 3 & -1 & -2 \\ 0 & 0 & -1 \end{pmatrix}$

7. $\begin{pmatrix} 1 & -1 & -10 \\ -1 & 1 & 1 \end{pmatrix}$ **8.** $\begin{pmatrix} 1 & 2 & 3 \\ -4 & -5 & -6 \\ 0 & 1 & 4 \end{pmatrix}$

Derive the optimal probabilities R_{opt} and C_{opt} for the matrices in Exercises 9–16. Use linear programming to find these answers. Reduce each matrix as much as possible before attempting the solution.

9. $\begin{pmatrix} 2 & 3 \\ 3 & 2 \end{pmatrix}$ **10.** $\begin{pmatrix} 11 & -1 \\ 0 & 10 \end{pmatrix}$

11. $\begin{pmatrix} 0 & 1 \\ 1 & 1 \end{pmatrix}$ **12.** $\begin{pmatrix} 0.5 & 0 \\ -0.1 & 0.75 \end{pmatrix}$

13. $\begin{pmatrix} -3 & -9 \\ -8 & -1 \end{pmatrix}$ **14.** $\begin{pmatrix} -100 & -15 \\ -10 & -150 \end{pmatrix}$

15. $\begin{pmatrix} 2 & 4 \\ 5 & 4 \\ 2 & 3 \end{pmatrix}$ **16.** $\begin{pmatrix} 1 & 1 & 3 \\ 4 & -1 & 3 \end{pmatrix}$

Use the special formulas for 2×2 matrices to solve the matrices in Exercises 17–22.

17. Exercise 9   **18.** Exercise 10

19. Exercise 11   **20.** Exercise 12

21. Exercise 13   **22.** Exercise 14

23. Show that if a 2×2 payoff matrix has no inverse, then it corresponds to a fair game.

24. In this problem, let G and E be $m \times n$ matrices. G is a game matrix, and E is a matrix all of whose components are 1s. Let R and C be optimal row and column strategy matrices for G. Let k be some positive number, and define a game $G' = G + kE$.
   (a) Show that $(kE)C$ is a column matrix of all ks. What is its dimension?
   (b) Compute RGC. Determine the value of G' in terms of the value of G, and verify the formula given in the text.

25. Show explicitly that adding a constant to each component of a 2×2 matrix does not affect the strategy of either player.

26. Let $y = hx + k(1 - x)$ be a function defined for $0 \le x \le 1$ and for $h > 0, k > 0$.
   (a) Show that the maximum or minimum of y occurs for either $x = 0$ or $x = 1$.

(b) Where does the maximum or minimum occur if $h = k$?
(c) Use these results to show that the minimum of the quantity $(r_1 a_{11} + r_2 a_{21})c_1 + (r_1 a_{12} + r_2 a_{22})c_2$ occurs for either $c_1 = 0$ or $c_1 = 1$. (*Hint:* Let $x = c_1$ and $1 - x = c_2$ and use the first part of this problem.)

27. Provide the details for the derivation of the results for optimal column strategies.

28. Show that the linear programming problems in Examples 10 and 11 are duals of each other.

W 29. Make up a simple game, and describe its rules thoroughly and in terms of the payoff matrix.

Applications

30. *(Farming)* Farmers in the midwestern Wheat Belt have to struggle against nature to bring in a good crop. In particular, will the growing season be normal or dry? Normal summers call for a wheat variety that does not do well if the weather turns out to be drier than normal. Conversely, if a dry-tolerant seed is planted, it will not do well in normal growing seasons. A payoff matrix showing the relative performances of the seed might look like this:

	Normal	Dry
Normal seed	10	1
"Dry" seed	3	10

   (a) Studies indicate that normal summers occur 60 percent of the time. What is the optimal strategy for the farmer?
   (b) The probabilistic answers to the first part might help an "ideal" farmer given many seasons to perfect a growing strategy. Unfortunately, real farmers have only a single growing season at their disposal, and they are unwilling to risk economic disaster this year for fortune the next—their farm may not last that long! What is a better interpretation of the answers to part (b)?

31. Solve Exercise 30 subject to the observation that of the last 25 summers, 20 have been observed to be normal. Revise the farmers' strategy in the light of these observations.

32. *(Textbook Selling)* The school bookstore often agrees to buy back school textbooks from students at the end of the school year. Many authors revise their texts frequently, rendering past editions worthless to both the bookstore and successive generations of students. For one particular math book, records indicate that a new edition is introduced in 3 years out of 10. If a new edition is to come out, the bookstore should *not* buy used copies of the old edition. The matrix showing the relative monetary profits is given as follows:

$$
\begin{array}{c}
\quad\quad\quad\quad\quad\quad\quad\quad\quad \text{Same ed.} \quad \text{New ed.} \\
\begin{array}{l}
\text{Order new books from publisher} \\
\text{Buy used texts from students}
\end{array}
\left(
\begin{array}{cc}
9 & 10 \\
10 & -3
\end{array}
\right)
\end{array}
$$

(a) If the store manager simply buys used books in alternate years, what is the expected bookstore profit or loss?

(b) What is the bookstore's optimal strategy against the textbook author?

(c) Assuming that bookstores always buy used books in alternate years, what is the *author's* optimal strategy against the stores?

33. Solve Exercise 32 again if the payoff matrix is
$$
\begin{pmatrix} 10 & 10 \\ 8 & -3 \end{pmatrix}.
$$

34. *(Fashion)* Clothing manufacturers are always trying to outguess the whims of fashion. Imagine one such manufacturer trying to decide to make men's pants with or without cuffs. A payoff matrix describing the result of good and bad guessing is as follows. The rows describe the manufacturer's choices, while the columns indicate the preferences of fashion.

$$
\begin{array}{c}
\quad\quad\quad\quad\quad\quad\quad \textit{Fashion} \\
\quad\quad\quad\quad\quad\quad \textbf{Cuffs} \quad \textbf{No cuffs} \\
\textit{Manufacturer} \;
\begin{array}{l}
\textbf{Cuffs} \\
\textbf{No cuffs}
\end{array}
\left(
\begin{array}{cc}
5 & -1 \\
-2 & 5
\end{array}
\right)
\end{array}
$$

(a) The manufacturer feels that fashion chooses cuffed pants one-third of the time. What is the manufacturer's best strategy in general, and what will be her expected gain or loss with this best strategy?

(b) Another manufacturer, a pessimist by nature, is convinced that "the fashion world is out to get him." That is, fashion prefers cuffs when he omits them, and vice versa. If this somehow were the case, what would fashion's optimal strategy be against this (or any) manufacturer?

35. Solve Exercise 34 again if the payoff matrix is
$$
\begin{pmatrix} 5 & -2 \\ -1 & 5 \end{pmatrix}.
$$

36. *(Politics)* The government of one particular Latin American government finds itself in constant and unremitting struggles against bands of insurgents. The insurgents aim their attacks either against the capital city itself, against a main water reservoir, or against a lone but symbolic military outpost. The government defends itself against these attacks as best it can with limited numbers of federal troops. A government applied mathematician has studied the situation, and has constructed the following payoff matrix. The elements in the matrix represent political gains by the regime. Rows represent the activities of the government, and columns the activities of the insurgents.

$$
\begin{array}{c}
\quad\quad\quad\quad\quad\quad\quad\quad\quad\quad\quad \textbf{Insurgents} \\
\quad\quad\quad\quad\quad\quad\quad\quad \textit{City} \quad \textit{Reservoir} \quad \textit{Outpost} \\
\textbf{Government} \;
\begin{array}{l}
\textit{City} \\
\textit{Reservoir} \\
\textit{Outpost}
\end{array}
\left(
\begin{array}{ccc}
8 & 3 & 0 \\
10 & 3 & 1 \\
1 & 2 & 2
\end{array}
\right)
\end{array}
$$

(a) Define the optimal government defense strategy.

(b) Define the optimal rebel strategy.

(c) What is the expected payoff if both sides use their optimal strategies?

37. Solve Exercise 36 if the matrix showing the gains of the regime against the insurgents is
$$
\begin{pmatrix} 8 & 2 & -1 \\ 9 & 3 & -1 \\ 2 & 1 & 2 \end{pmatrix}.
$$

▣ Calculator Exercises

Derive the optimal probabilities for the matrices in Exercises 38–39.

38. $\begin{pmatrix} 6.83 & 2.02 \\ 1.89 & 5.69 \end{pmatrix}$ **39.** $\begin{pmatrix} 0.07 & 1.12 \\ 0.98 & -0.05 \end{pmatrix}$

CHAPTER REVIEW

Terms

simple games	play	payoff matrices
zero-sum games	constant-sum games	rational strategy
expected payoff	value	fair

pure strategy   strictly determined   saddle point
mixed strategy   optimal strategy   dominant row
recessive column

Key Concepts

- Simple games involve confrontations between opposing players, each of which endeavors to optimize their gain from the game. Games are analyzed by setting up a *payoff matrix*, which is a chart listing the payoffs for each combination of plays that are open to the two players.

- If a payoff matrix possesses one or more *saddle points*, then these saddle points identify the optimal strategies for both sides. There is a well-defined algorithm for identifying any existing saddle point. If a game possesses more than one saddle point, they have the same payoff.

- Many games do not possess saddle points, so some prescription for *mixed strategies* is called for. In these cases,

strategies are devised to maximize the expected payoff of the game. One obtains these strategies by finding a linear programming problem which is equivalent to the game theory problem, and solving this problem.

- Probabilities for selecting available alternatives of a game depend only on the components of the payoff matrix.

- Sometimes, in cases where one side has more than 2 alternatives open to it, it is possible to eliminate certain rows or columns. For example, a dominant row will always be selected over the ones it dominates. In the same way, a rational column player should always reject a column in favor of its recessive mate.

Review Exercises

Determine the saddle point or points of the payoff matrices in Exercises 1–4.

1. $\begin{pmatrix} 10 & 12 & 6 \\ 9 & 8 & 7 \end{pmatrix}$

2. $\begin{pmatrix} 2 & 2.5 & 2 \\ -1 & 5 & -0.5 \\ 0 & 6 & -3.5 \end{pmatrix}$

3. $\begin{pmatrix} 1 & 2 & 1 \\ 0 & -2 & -1 \\ 1 & 3 & 1 \end{pmatrix}$

4. $\begin{pmatrix} 3 & -10 & 2 \\ 0.5 & 0 & 1 \\ -10 & 20 & 1.5 \end{pmatrix}$

Use payoff matrix $\begin{pmatrix} 2 & -3 \\ -2 & 1 \end{pmatrix}$ in each of Exercises 5–10 to find the expected payoff. Use the given strategy matrices R and C in each problem.

5. $R = \begin{pmatrix} \frac{1}{2} & \frac{1}{2} \end{pmatrix}, C = \begin{pmatrix} \frac{1}{2} \\ \frac{1}{2} \end{pmatrix}$

6. $R = \begin{pmatrix} 1 & 0 \end{pmatrix}, C = \begin{pmatrix} 0 \\ 1 \end{pmatrix}$

7. $R = \begin{pmatrix} \frac{2}{3} & \frac{1}{3} \end{pmatrix}, C = \begin{pmatrix} \frac{2}{3} \\ \frac{1}{3} \end{pmatrix}$

8. $R = \begin{pmatrix} \frac{1}{2} & \frac{1}{2} \end{pmatrix}, C = \begin{pmatrix} \frac{1}{7} \\ \frac{6}{7} \end{pmatrix}$

9. $R = \begin{pmatrix} \frac{1}{4} & \frac{3}{4} \end{pmatrix}, C = \begin{pmatrix} \frac{1}{4} \\ \frac{3}{4} \end{pmatrix}$

10. $R = \begin{pmatrix} 0 & 1 \end{pmatrix}, C = \begin{pmatrix} \frac{3}{4} \\ \frac{1}{4} \end{pmatrix}$

What are the expected payoffs for the payoff matrices in Exercises 11–14? Assume that all alternatives for each player are equally likely to be chosen.

11. $\begin{pmatrix} 1 & 3 \\ 4 & 2 \end{pmatrix}$

12. $\begin{pmatrix} -1 & 0.2 \\ 0.1 & -2 \end{pmatrix}$

13. $\begin{pmatrix} 12 & 1 & 1 \\ 1 & 1 & 2 \end{pmatrix}$

14. $\begin{pmatrix} 40 & 45 \\ 44 & 38 \end{pmatrix}$

Reduce the games in Exercises 15–20 by considering dominant rows and recessive columns.

15. $\begin{pmatrix} 1 & 2 & 3 \end{pmatrix}$

16. $\begin{pmatrix} 1 & 2 & -1 \\ -1 & 3 & 1 \end{pmatrix}$

17. $\begin{pmatrix} 1 & -1 \\ -1 & 1 \end{pmatrix}$

18. $\begin{pmatrix} 1 & -1 \\ -1 & 1 \\ 2 & -\frac{1}{2} \end{pmatrix}$

19. $\begin{pmatrix} 1 & -1 & 2 \\ -2 & 2 & -1 \\ 0 & -2 & 3 \end{pmatrix}$ **20.** $\begin{pmatrix} 3 \\ -1 \\ 0 \end{pmatrix}$

Derive the optimal probabilities R_{opt} and C_{opt} for the matrices in Exercises 21–24 and compute the expected payoffs.

21. $\begin{pmatrix} 8 & 9 \\ 9 & 8 \end{pmatrix}$ **22.** $\begin{pmatrix} -3 & -10 \\ -9 & -2 \end{pmatrix}$

23. $\begin{pmatrix} 110 & 80 \\ 109 & 79 \\ 82 & 112 \end{pmatrix}$ **24.** $\begin{pmatrix} 1 & 2 & 7 \\ 6 & 7 & 0 \end{pmatrix}$

Applications

25. *(Retailing)* An entrepreneur contemplates starting up an old-fashioned ice cream parlor. The market for old-fashioned ice cream is evenly divided among those who like soft ice cream, and those who prefer hard. The original business plan calls for serving only one kind of ice cream. If he serves soft ice cream, he will attract 75 percent of the "softies" but only 55 percent of the "hardies." If he serves hard ice cream, he will lure 60 percent of the softies and 50 percent of the hardies. What is his best business strategy?

26. Solve Exercise 25 if market research shows that if he carries hard ice cream, he will attract 65 percent of those addicted to hard ice cream, and 80 percent of soft ice cream fanatics. If he serves soft ice cream, he will lure 60 percent of the hard ice cream fanatics into his store, and 65 percent of the soft ice cream fans. What is his best business strategy in this case?

27. *(Politics)* A terrible scandal rocks the corridors of City Hall! The Republicans are trying to figure out how to exploit this situation, which threatens to upset the equilibrium between Republicans and Democrats. The Republicans feel they can either take an offensive or a defensive attitude. If they do, here's the matrix describing their relative projected performance in the upcoming election. Voters are assumed to be either Democratic or Republican.

		Democratic	
		offensive	defensive
Republican {	offensive stance	3	-2
	defensive stance	-1	4

From a game theory point of view, how should the Republicans handle the situation?

28. Solve Exercise 27 if 30 percent of the voters are Republican. Solve it if 80 percent are Republican.

9

STATISTICS

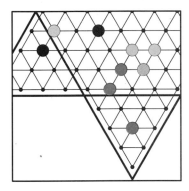

INTRODUCTION

So far, our methods have helped us derive *new* information from a *given* set of facts. The problem in statistical situations is subtly different—we are given many facts, and our aim is to hack away at these data so we can lay bare underlying trends.

Sometimes this "hatchet work" is straightforward. A judicious display of the data in a well-thought-out graph may be all that is needed. Perhaps a simple rearrangement to determine the *median* or *mode* may suffice.

Many people are familiar with the concept of averaging, and we can always compute the *average* (or *mean*) for any set of numbers. Will this average always be meaningful? We resolve this issue with our discussion of the *variation* of data around the average value.

Large groups of data are often distributed in distinct, recognizable ways. Two such common distributions are the *normal* and *binomial distributions*, and we devote two sections to them.

We need statistics because many situations generate more information than we can possibly deal with. We can determine the patterns and trends in such situations—census data, test grades, health records, and so on—only by running the mounds of data through the filter of a statistical technique.

Section 9.1   *DESCRIPTIVE STATISTICS*

data set Examine the following *data set*, or collection of numerical information. It represents sales figures for a medium-sized electrical supply firm.

1984	$3,900,000
1985	4,500,000
1986	4,750,000
1987	5,100,000
1988	4,750,000

It is difficult to see any trends or patterns in a table, and so we turn to a *graph* to improve the presentation.

Bar Graphs

graphs *Graphs* can display these data in any of several ways. In all cases, we first decide on the two different variables whose relationship we want to illustrate. Then we use the familiar Cartesian format for their display. In this example, the two variables are the year Y and the sales amount S for any particular year. We seek the relationship between S and Y, and will graph Y, the independent variable, as the horizontal quantity. S will be the vertical quantity.

Graphs of data differ from most graphs of mathematical functions in that the independent variable may take on only distinct, *discrete* values. It makes no sense to speak of the year "$1994\frac{1}{3}$" or "1997.68923." Our graph should consist exclusively of a series of individual, unconnected points. The discussion and examples that follow display some effective graphing techniques.

Figure 9.1 illustrates the first method. A series of heavy vertical bars connect each *bar graph* data point to the horizontal axis. For obvious reasons, we call this a *bar graph*. The graph quite clearly suggests that sales volume is rising slowly but consistently.

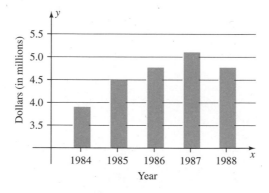

FIGURE 9.1
The bar graph representation of sales data.

Example 1 Display the sales data of Figure 9.1 with the bar graph drawn so that the bars are horizontal rather than vertical.

Solution

See Figure 9.2.

Skill Enhancer 1 Measure the lengths of the fingers of your right hand. Display the data as a bar graph.

Answer in Appendix E.

dual graphs   Dual graphs are often helpful for showing two sets of data on a single graph.

Example 2 Create a single display to help visualize the following hypothetical data from a study of learning in white rats.

| *Effect of Age on Learning in Rats* | | | |
| *2-Month-Old Rat* | | *24-Month-Old Rat* | |
Trial Number	Seconds	Trial Number	Seconds
1	24	1	22
2	22	2	20
3	15	3	20
4	10	4	17
5	11	5	18

Solution

See Figure 9.3. It is quite clear from the graph that young rats are better learners than old rats. This advantage of youth may not be quite so clear from the table.

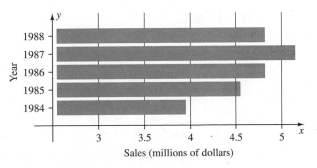

FIGURE 9.2
A variant bar graph representation.

Skill Enhancer 2   Now measure the lengths of the fingers of your left hand. Plot these data, together with the right-hand finger lengths, on a single dual graph.

Answer in Appendix E.

There will almost always be alternative methods for graphically displaying data. For one alternative view of the rat data in Example 2, see Figure 9.4.

The graph of Figure 9.5 displays *three* sets of data in a bar graph.

It is never appropriate to connect discrete data points with a continuous curve. Smooth, flowing curves incorrectly imply that for every value of the horizontal variable, there is a corresponding vertical data value. Nevertheless, in a *line graph* we do connect adjacent data points with straight line segments. The broken-line nature of these graphs keeps viewers from regarding the graph as a continuous curve. With line graphs, it is

line graph

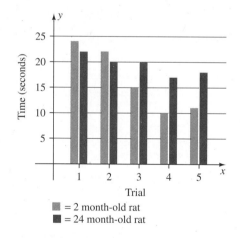

FIGURE 9.3
A double bar graph displays two sets of discrete data.

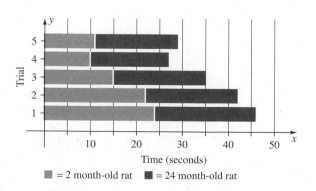

FIGURE 9.4
An alternative view of the double bar graph depicting rat learning data.

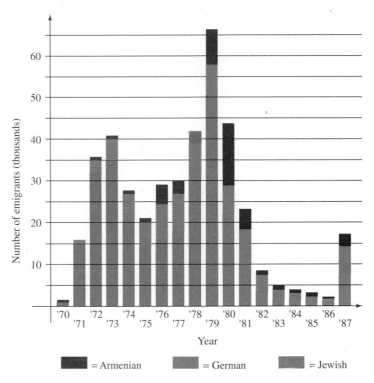

FIGURE 9.5
Fluctuating Soviet emigration. (Figure for 1987 is an estimate.) Source: *The New York Times*, **Dec. 6, 1987. Copyright © 1987 by The New York Times Company. Reprinted by permission.**

dual line graph   particularly easy to superimpose several sets of related data on a single graph to form a *dual line graph*.

	Mammals			Birds and Reptiles	
Length (cm)	Weight (gm)		Length (cm)	Weight (gm)	
8	100		10	100	
20	180		20	145	
25	200		30	150	
32	330		42	275	
50	350		50	250	
57	530		57	300	
70	900		70	350	
			83	320	
			100	450	

Example 3   Seven mammals were measured and weighed, as were nine reptiles and birds. Lengths are in centimeters and weights are in grams. Display these data as a dual line graph.

Solution

See Figure 9.6.

Skill Enhancer 3   Take the data on your lengths of fingers for both hands, and display this information as a dual line graph.

Answer in Appendix E.

pie charts   Although they are not quite graphs, *pie charts* are yet another useful device for the visual representation of data. Each wedge of the "pie" represents one data category. The size of the slice is proportional to the data value for that category. Pie charts are useful when the *absolute* value of the data is unimportant, whereas the *relative* value of the data is important.

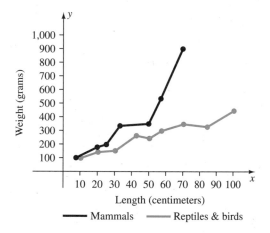

FIGURE 9.6

Relationship between body length and body weight for several classes of organisms (hypothetical data).

Example 4   Exhaustive research on the eating habits of American college students reveals that 20 percent of their diet is meat, 15 percent is vegetables (other than French fries), 18 percent is French fries, 25 percent is pizza, carbonated soft drinks form 13 percent, and the remaining 10 percent consists of all other foods and drinks. Display this information in a pie chart.

Solution

The pie chart appears in Figure 9.7. The importance of pizza in the diets of American college students is quite clear, while the total number of pizza pies consumed is unclear (and perhaps unimportant).

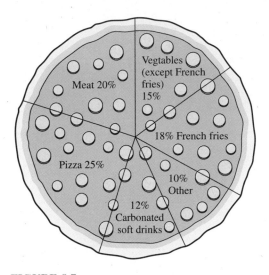

FIGURE 9.7
"Pizza" pie chart.

Skill Enhancer 4   For a certain club, the eating habits are very different—25 percent of their snacking is apples, 15 percent is plain yogurt, 20 percent is granola bars, 25 percent is carrots, and the remainder is peanut butter and jelly sandwiches (made with whole wheat bread, of course). Display these results as a pie chart.

Answers in Appendix E.

How to Best Display Data

Choosing the best way to display information is as much art as science. It is difficult to formulate precise rules that apply to all cases. The emphasis should always be on displaying the data in as effective and clear a way as possible. One virtuoso data display appears in Figure 9.8. This figure combines line graphs and bar graphs to summarize every aspect of the weather of New York City during 1986. Despite the vast quantity of data, the graph appears uncluttered and remains easy to read.

Another excellent example is the map in Figure 9.9, which also appears inside the back cover of this book. [E. R. Tufte brought this display to the author's attention in his book *The Visual Display of Quantitative Information* (Cheshire, CT: Graphics Press). This beautiful and relevant book explores the issue of proper presentation of data in great depth.] Here, the French engineer Charles Joseph Minard (1781–1870) shows six sets of data pertaining to the fate of Napoleon's army in Russia during the brutal winter of 1812–1813. The initial width of the band indicates the initial strength of 422,000 men. The decreasing width clearly shows the effects of war and weather. Although 100,000 men reached Moscow, only 10,000 troops returned back to Poland. The map maker had to develop several nonstandard graphing techniques. Altogether, six quantities are shown—the size of the army, the location of the army (two variables—latitude and longitude), the direction of movement (two variables—movements east-west and north-

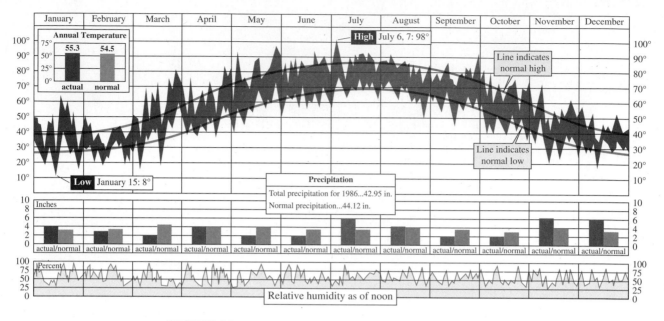

FIGURE 9.8

New York City weather for 1986. Source: *The New York Times,* **January 4, 1987. Copyright © 1987 by The New York Times Company. Reprinted by permission.**

south), and the temperature on various dates during the retreat from Moscow. (Minard uses the *Réaumar scale* for temperature. In this scale, 0° Réaumar represents 0° in the Celsius scale, and 80° Réaumar corresponds to 100° Celsius, the boiling point of pure water. The formula connecting Celsius and Réaumar is $C = \frac{5}{4}R$.)

Many of the graphs in this section were drawn from two consecutive issues of *The New York Times,* a graphic illustration (no pun) of the importance of graphic data in modern life.

_____ *Exercises 9.1* _____

1. Here is a table of data on the frequency with which crickets chirp as a function of Fahrenheit temperature. Graph the data as requested.

Temp. (°F)	Chirps per min.
60	4
70	6
75	8
80	8
85	10

(a) Display the data as a bar graph.

(b) Draw the data as a broken line graph.

W (c) Is it appropriate to represent these data as a pie chart? Why or why not?

2. The government of Ruritania spends its tax income on guns, butter, and shoes for the wife of the current dictator. Expenditures on these categories are $\overline{R}175,000$, $\overline{R}250,000$, and $\overline{R}100,000$, respectively. (The symbol $\overline{R}$ represents units of currency—the Ruritanian ruble.)

(a) Represent the data as a bar graph.

(b) Represent the Ruritanian spending budget as a pie chart. The angle of each wedge should represent the percentage of the total Ruritanian income.

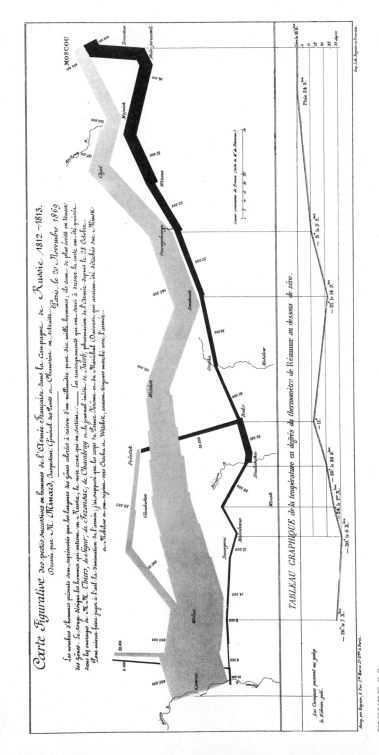

FIGURE 9.9

The defeat of Napoleon at the Battle of Borodin occurred in 1812. (Leo Tolstoy describes the battle in a wealth of memorable detail in the novel *War and Peace*.) Napoleon's army experienced immense suffering throughout this campaign, both before and after the actual battle. This graph masterfully summarizes these difficulties. The map maker displays six pieces of information. Source: Edward R. Tufte, *The Visual Display of Quantitative Information* (Cheshire, Connecticut: Graphics Press, 1983).

3. A certain chemistry experiment is performed three times, with minor variations differentiating the three experiments. The experiment consists of four steps, and the result of each step is that a certain measurable volume of radon gas is released. Here are the measurements for the three experiments.

Experiment 1		Experiment 2	
Step	Volume (ml)	Step	Volume (ml)
1	22	1	10
2	17	2	5
3	20	3	10
4	15	4	10

Experiment 3	
Step	Volume (ml)
1	15
2	0
3	10
4	12

(a) Display the data from the *three* experiments in a *single* compound line graph.
(b) Display the data from experiments 1 and 3 as a single bar graph, but in two different ways.
(c) Display the data from all three experiments in a single bar graph.
(d) Is it feasible to express the data from a single experiment as a pie chart?
(e) Display the data from the third experiment as a pie chart.
W (f) What problems do you foresee in displaying the data from the second experiment as a pie chart? Display these data in a pie chart and deal with this problem.

Applications

Refer to Figure 9.8 to answer Exercises 4–7.

4. What was the *second* lowest temperature recorded during 1986? What was the date of this occurrence?

5. January and August had about the same amount of pre-cipitation. What justification is there for regarding January as being "wetter" than August?

6. (a) What month *normally* has the highest precipitation? the lowest?
(b) What month *actually* had the highest precipitation? the lowest?

7. What month in Manhattan normally contains the coldest weather of the year? the hottest?

Answer Exercises 8–13 based on the map of Napoleon's 1812–1813 winter campaign displayed in Figure 9.9 on page 445. A larger view of the map appears in the inside back cover of this book.

W 8. During the retreat from Moscow, why would you call the crossing of the Berezina River a disaster?

9. When the advancing army reached Smolensk, what proportion of the original army remained with Napoleon?

10. Prepare a table showing the outside temperature *in degrees Celsius* at Dorogobonge, Smolensk, Stulienska, Moloderno, and the Nieman River. Use these data to prepare a line graph.

11. Prepare a table showing the strength of the army in retreat at each of the above locations.

12. Prepare a table displaying the distances of the above towns from Moscow in units of French leagues (*lieues communs* on the map).

13. Use the results of the previous two exercises to prepare a line graph relating the size (strength) of the army to distance from Moscow.

14. *(Fluctuating Soviet Emigration)* Refer to Figure 9.5.
(a) What was the peak year for Jewish emigration during the period summarized by this graph? What was the peak year for German emigration? for Armenian emigration?
W (b) Why is the bar graph presentation shown here superior to one in which each ethnic group has its own bar adjacent to other bars?
(c) Officials of the former Soviet Union clamped down on emigration during the early years of the 1980s, as Figure 9.5 clearly shows. Premier Gorbachev then began trumpeting a new era of *glasnost*—openness—between his country and the United States. One proof would be the relaxation of emigration restrictions. Judging from this figure, when would you suspect that the new *glasnost* era began?

15. *(Current Interest Rates)* Refer to Figure 9.10 to answer this question. At about what date were all long-term rates at a maximum? All short-term rates?

Long-term rates (%)

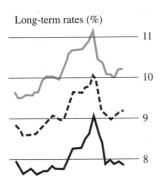

J J A S O N D
1987

— Aa-rated utility bonds
- - Treasury bonds
— Tax-exempt bonds

Short-term rates (%)

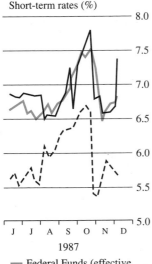

J J A S O N D
1987

— Federal Funds (effective weekly average)
— One-month commercial paper
- - Three-month treasury bills

FIGURE 9.10
Current interest rates. (a) Long-term rates. (b) Short-term rates. Source: *The New York Times,* **December 7, 1987. Copyright ©️ 1987 by The New York Times Company. Reprinted by permission.**

16. *(America's Top Customers)* Use the data from Figure 9.11 to prepare a line graph. What features does the pie chart reveal that are less clear in a line (or bar) graph?

17. *(Stock Performance)* There are many ways of displaying data besides those discussed in the text. One common example is shown in Figure 9.12, which displays the performance of leading stock price indicators over time. Study this figure to learn how to read it.
  (a) What week had the highest volume of shares traded? the lowest Dow Jones Industrial Average? the lowest New York Stock Exchange Composite Index?
  (b) From your knowledge of events, you may know why these lows occurred. Is it reasonable that a high volume might accompany a low selling price? Would any other data in this figure support your reasoning?

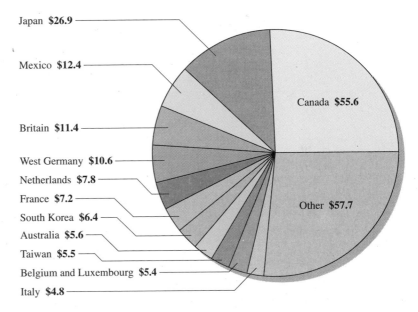

FIGURE 9.11
America's Top 12 Customers. Total dollar value of U.S. exports in 1986 was $217.3 billion. The 12 largest markets are shown, in billions of dollars. Source: *The New York Times,* **December 6, 1987. Copyright ©️ 1987 by The New York Times Company. Reprinted by permission.**

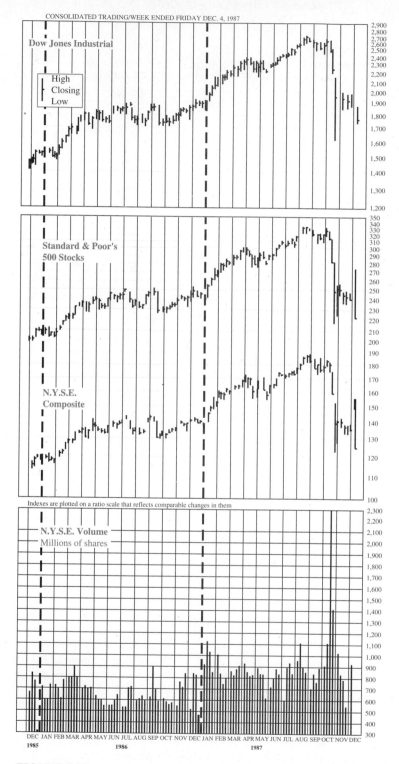

CONSOLIDATED TRADING/WEEK ENDED FRIDAY DEC. 4, 1987

Dow Jones Industrial

High
Closing
Low

Standard & Poor's
500 Stocks

N.Y.S.E.
Composite

Indexes are plotted on a ratio scale that reflects comparable changes in them

N.Y.S.E. Volume
Millions of shares

FIGURE 9.12

New York Stock Exchange indicators. Source: *The New York Times*, **December 6, 1987. Copyright © 1987 by The New York Times Company. Reprinted by permission.**

Section 9.2   *STATISTICS AND PROBABILITY; ORGANIZING DATA*

The easiest way to analyze any set of data starts with a careful rearrangement of all items. Below is a list of numbers representing the weights of the students in a college class. All readings on the scale were taken to the *nearest five pounds.*

110	115	95	135	120	120
95	110	105	115	120	120
140	130	125	105	95	125
100	120	130	125	120	120
90	130	140	115	125	115

These weights are listed in no special order. Observe that several individuals have the same weight (at least to the nearest five pounds). The first thing to do is rearrange these numbers in size order, making sure to *group* identical data items together.

group

Weight	Tally	Frequency
90	|	1
95	|||	3
100	|	1
105	||	2
110	||	2
115	||||	4
120	╫╫ ||	7
125	||||	4
130	|||	3
135	|	1
140	||	2

tally marks
frequency

The *tally marks* in the central column are a useful device for counting the frequency with which an item occurs. The *frequency* of a value is simply the number of times that value appears in the set of data. The value that occurs with the greatest frequency is the *mode*. Data sets may have more than one mode or no mode. (A fuller discussion of the mode appears in Section 9.3. The values we chose to lie at the boundary between one class and the next are the *class boundaries* or *class intervals*. In this example, the class boundaries are 90 pounds, 95 pounds, and so forth.

mode

class boundaries

The frequency of individual items often achieves central importance, and it is useful to graph the *frequency distribution* of the items. The data values appear on the horizontal axis, and the corresponding frequencies on the vertical axis.

frequency distribution

When graphed as a *bar graph*, as in Figure 9.13, we call the frequency distribution a *histogram*. A histogram strongly resembles a bar graph, but there are *never* any spaces

histogram

between the bars of a histogram. We expect the class boundaries to appear along the horizontal axis, and the values of the frequencies are plotted along the vertical axis.

frequency polygon   When we graph the frequency distribution as a *line graph*, as in Figure 9.14, we call it a *frequency polygon*. The shape of the frequency polygon in Figure 9.14, with its central peak surrounded by a series of lower and flatter hills, is characteristic of many frequency distributions for a wide variety of phenomena.

> **Example 5**   In one town in the Canadian province of Saskatchewan, social scientists are studying the incomes of the town's wage earners. For the sake of convenience, we categorize each wage earner into one of five Canadian income tax brackets. Here are

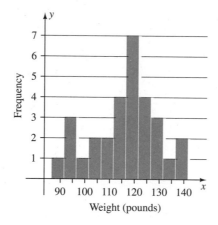

FIGURE 9.13

Frequency distribution for student weights as a bar graph.

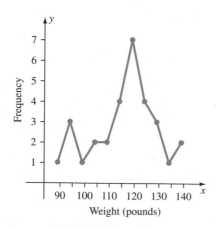

FIGURE 9.14

Frequency distribution as a line graph.

the raw data as gathered by one of the researchers. (Roman numerals indicate the tax bracket. Workers with the highest incomes fall into bracket V.)

II	I	I	III	III	IV
III	V	III	II	I	I
IV	I	III	II	IV	IV
I	III	II	II	III	III
III	III	III	IV	II	I
I	I	III	II	II	II

Organize these data by obtaining data frequencies and creating a histogram and frequency polygon.

Solution

First, obtain the frequencies for each tax bracket.

Bracket	I	II	III	IV	V
Frequency	9	9	12	5	1

The mode is bracket III, containing 12 persons. Having made this tabulation, we can present the histogram and frequency polygon as in Figures 9.15 and 9.16.

Skill Enhancer 5   Take a pile of 30 books, and classify them in length categories of 1 through 5, where category 1 is a book 100 pages or less, category 2 is 101 through 150 pages, category 3 is 151 through 200 pages, category 4 is 251 through 300 pages, and category 5 is 301 pages or longer. Obtain the data frequency for this collection of books, and create a histogram and frequency polygon.

Answer in Appendix E.

The frequency distribution depends strongly on the accuracy with which we make our original measurements and on the class boundaries of those original measurements. Refer to the beginning of this section, in which weights were given to the nearest five pounds. At that time, we saw that three students "weighed" 130 pounds, even though their actual weights might have been 128.0, 131.5, and 129.5 pounds. Now, suppose we recategorize the student weights, placing them this time in categories of students whose weights are the same to the nearest *half pound*. These three students now fall in different weight categories as a result of the finer classification. That is why the frequency polygon will change.

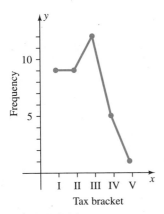

FIGURE 9.15
The histogram for Canadian tax bracket data.

FIGURE 9.16
Frequency polygon for Canadian tax bracket data.

Statistics and Probability

Frequency distributions provide a tie-in to probability theory. We earlier related the *empirical probability* of an event to the *relative* frequency with which we observed the event in some sample. Referring again to Example 5, we might ask the probability that a wage earner selected at random from this town in Saskatchewan is in the fifth of the Canadian tax brackets. Our data tell us that of the 36 wage earners sampled, only one is in that bracket. May we conclude that $\frac{1}{36}$ is a good estimate of the desired probability?

bias

That conclusion is valid only if the relatively small sample of 36 wage earners represents the entire town in miniature; that is, the conclusion is valid if our sample is not *biased* in any way. The sample might be biased, for example, if the scientists selected it by standing outside an exclusive restaurant and querying patrons as they left the premises. Presumably, few middle- or low-income wage earners eat at such a place, and they would be underrepresented in the statistician's sample. Often, in the absence of information to the contrary, we assume that samples presented to us for study will not be biased. In that case, the conclusion that the probability is $\frac{1}{36}$ is reasonable.

Example 6   Sam Johnson samples a paragraph from a popular novel, and observes that in one randomly chosen passage, there are 7 words which are two letters long, 7 of length three, 11 of length four, 10 of length five, and 15 containing more than five letters.
(a) If Sam selects a word in some other piece of written English, estimate the probability that it will contain three letters; five or more letters.
(b) Comment on the reliability of these estimates.

Solution

(a) There are a total of 50 words in the original paragraph. Of these, 7 are of length three and 25 have five or more letters. We estimate the two probabilities at $\frac{7}{50}$ and $\frac{25}{50} = \frac{1}{2}$.
(b) These are reliable estimates if Sam's original sample is not biased. Although written English contains prose with longer words than popular novels, there are also plenty of documents with shorter words. On the whole, these estimates are reasonable.

Skill Enhancer 6   Jim Boswell tallies the word length of sentences in some piece of writing. In one passage, 5 sentences have five or fewer words in them, 3 have six or seven words, 5 contain eight or nine words, and 11 comprise ten or more words. Jim then proposes to select a sentence at random from some other place in the document. What is a reasonable estimate of the probability that it will contain eight or nine words?

Answer in Appendix E.

Exercises 9.2

(Broadcast Advertising) In the New York City metropolitan region, there are four major television networks. Each network is concerned with the fraction of viewers tuned in to it, because the greater this fraction, the higher the rate it can charge to advertisers. In a survey of 200 "typical" households, an independent rating agency discovered the following statistics for the 8:00 and 8:30 time slots. Use this information for Exercises 1–4.

Station	8:00	8:30
WBCD	25	25
WCDE	50	80
WDEF	80	60
WEFG	45	35

1. Prepare a frequency polygon for the 8:00 time slot.

2. Prepare a frequency polygon for the 8:30 time slot.

3. What is the probability that a listener chosen at random at 8:15 is tuned in to station WBCD? Comment on the reliability of your answer.

4. What is the probability that a listener at 8:44 is tuned in to either WCDE or WDEF? Comment on the reliability of your answer.

(Fighting Forest Fires) Use this information to answer Exercises 5–10. Forest fires in a heavily forested portion of Montana are a severe problem during the summer. One area is divided into six patrol areas. A dispatcher keeps a log of all fires reported to her by simply marking the number of the district in which a reported fire occurs. Here is a copy of the dispatcher's log from July 1987:

4	4	1	5	6	3
5	1	2	6	4	5
2	2	4	1	6	4
1	6	2	5	3	4
3	3	1	2	4	

and here is the one from July 1988:

6	4	5	6	5	1
4	3	3	3	1	2
1	1	5	3	5	3
2	5	3	4	4	6

5. What is the mode for the 1987 data? Express the data from 1987 as a histogram.

6. What is the mode of the 1988 data? Express these data as a histogram.

7. Use the data from 1987 to compute the probability that a fire will occur in region 4. Use the data from 1988 for the same computation. Comment on the two answers.

8. Use the data from 1987 to compute the probability that a fire will occur in region 1. Use the data from 1988 for the same computation. Comment on the two answers.

9. Use the data from 1987 to compute the probability that a fire will occur in region 1 or 4. Use the data from 1988 for the same computation. Comment on the two answers.

10. Determine a good way to show the frequency polygons from two different data sets on the same graph, and express the data from 1987 and 1988 as a pair of frequency polygons.

(Biology) The following series of numbers refer to the litter size of mature female white rats in a particular study in which rats were exposed to a specific set of nutriments; use this information for Exercises 11–12.

6	6	8	7	3	7	8	4
4	4	5	6	7	8	8	6
6	6	6	8	9	6	4	3
5	6	6	6	7	6	3	4
7	8	8	2	5	6	6	6

11. Do these data have a mode? If so, what is it?

12. Express these data as a histogram.

(Culture and Literacy) For a sociological study of reading habits of suburban adolescent girls, a collection of the most well-read gothic romance novels was chosen from a local library. Here are the page lengths for these books. Use these data to answer Exercises 13–16.

179	180	180	182	175
181	181	178	178	180
201	165	210	180	178
178	177	181	171	179
188	182	178	178	180
180	184	183	179	

13. Do these data have a mode? If so, what is it?

14. Express these data as a histogram.

15. What is the most probable novel length? How probable is it? In your opinion, how reliable is this answer?

16. What is the probability that a gothic novel will be longer than 205 pages? shorter than 175 pages? In your opinion, how reliable are these answers?

Calculator Exercises

17. Computers make word counting much easier than it might have been for Sam Johnson back in Example 6 in the text. The following chart yields information about words and word lengths in a longer piece of text.

Length	Frequency
1	39
2	45
3	55
4	69
5	77
6	43
7 or more	96

A word is selected at random from this text.

(a) Estimate the probability that it will contain four or fewer letters.

(b) Estimate the probability that it will contain five letters.

W (c) Comment on the reliability of these estimates.

18. A contest at the Save-On Hardware Store requires customers to guess the number of jellybeans in a large glass jar. After the contest has closed, an employee counts the pieces of candy and records the amounts by color. Here are the results.

Color	Frequency
Red	444
Yellow	623
Green	298
Black	333
White	519

(a) If a single piece of candy is chosen, what is the probability that it will be white?

(b) If we choose two pieces, what is the probability that exactly one of them will be white?

Section 9.3   AVERAGES AND MEASURES OF CENTRAL TENDENCY

Data sets can be large. For example, the set of incomes for all Asian-American males in the 1995 calendar year might contain millions of elements. Is it possible to determine a few numbers that can somehow summarize the properties and trends of the larger, original set? Measures of the central tendency of the set are helpful in this regard. A *measure of the central tendency* is a number that reveals the "center" of the data. Many such measures are possible, depending on our precise interpretation of the "center" of the data.

measure of the central tendency

Computing Averages

average

Perhaps the best-known measure of central tendency is the *average* or *mean* of the data. We sometimes call this average the *arithmetic average* to distinguish it from other, less-often-used varieties of mean. We often use the terms *mean* and *average* interchangeably, and we use μ or $E(x)$ to symbolize it. The symbol μ is the Greek letter *mu* (pronounced *mew*). (The notation $E(x)$ should remind you of the *expectation* we studied in the context of game theory. Expectation is the average.)

Average

$$\mu = \frac{x_1 + x_2 + \cdots + x_N}{N}$$

where

N	is the total number of data items.
$x_1, \ldots, x_N$	are data items.
μ	is the average.

Informally, compute the average by totaling all numbers together, and then dividing by the number of items added.

Example 7

(a) In a small college seminar, the grades of the 5 students on a recent quiz were 85, 90, 65, 75, and 100. What was the class average?

(b) In another, larger class, 5 people received 75, 4 received 85, 3 obtained 95, and 2 received grades of 100. What was the average grade for this class?

Solution

(a) Since there are 5 grades, the average or mean grade is

$$\frac{85 + 90 + 65 + 75 + 100}{5} = \frac{415}{5} = 83$$

(b) The calculation here is similar but slightly more tedious, since we have to add all grades for all students. There are a total of 14 students, so the average is

$$\frac{75 + 75 + 75 + 75 + 75 + 85 + 85 + 85 + 85 + \cdots}{14}$$

A simpler computation is

$$\frac{5 \cdot 75 + 4 \cdot 85 + 3 \cdot 95 + 2 \cdot 100}{5 + 4 + 3 + 2} = \frac{1,200}{14} = 85\frac{5}{7} \approx 85.71$$

(Make sure you see why this works.)

Skill Enhancer 7

(a) In another small class, recent grades on an exam were 60, 85, 90, 100, 80, and 75. What was the class average?

(b) Suppose that in a somewhat larger class, exactly 6 people received each of these grades. What would the average be in this case?

Answers in Appendix E.

grouped data   The second part of this example suggests that a special formula for *grouped data* is worthwhile. The *frequency* f_i is the number of times the distinct data value x_i occurs.

Average Value of Grouped Data

$$\mu = \frac{f_1 x_1 + f_2 x_2 + \cdots + f_m x_m}{N}$$

where

m	is the number of *distinct* data values.
x_1 through x_m	are the distinct data values.
f_i	is the *frequency* of x_i.
$N = f_1 + f_2 + \cdots + f_m$	is the *total* number of all data items.

Example 8   Mike Wheelless is a salesman at the local car showroom. All of his income derives from commissions on the sale of new cars. During September, when his agency was clearing out last year's models, Mike sold 6 Blades, 8 Bullets, 4 Dune Buggies, 8 Cattlemen, and 5 pickup trucks. (These are the names of several popular automotive models.) His commission per vehicle on these models is $85, $90, $75, $95, and $115, respectively. What is his average commission per car in September?

Solution

The data have been grouped for us. The total number of vehicles he sold is

$$N = 6 + 8 + 4 + 8 + 5 = 31$$

so the calculations for the mean become

$$\frac{6 \cdot 85 + 8 \cdot 90 + 4 \cdot 75 + 8 \cdot 95 + 5 \cdot 115}{31} = \frac{2,865}{31} \approx \$92.42$$

Skill Enhancer 8   In another month, Mike sold 4, 4, 3, 6, and 8 of the Blades, Bullets, Dune Buggies, Cattlemen, and pickups.
(a) What is the total number of vehicles he sold in this month?
(b) What now is his average commission?

Answers in Appendix E.

sample mean   It is important to distinguish between the *sample mean* and the mean of the entire population. Typically, we have access to information relating only to some small
sample   subgroup—or *sample*—of the entire population. We compute the mean value only for the sample data. The entire population also possesses a mean value, but we are rarely in a position to measure it (nor will we want to, if the population contains many elements). To distinguish between these two quantities, statisticians use $\bar{x}$ to denote the measured, sample mean, with μ reserved for the mean of the entire population. It is our hope that $\bar{x}$ will not differ greatly from μ, and if the sample is not biased, this is generally true. Much statistical theory has been created to investigate the relationship between statistical quantities computed on the basis of sample data and the actual values of these quantities over the entire population.

Example 9   Six preschoolers line up in order by size. Their heights in centimeters are 85, 86, 86.5, 87, 87, and 90. Compute the sample mean $\bar{x}$ by choosing as a sample every other child starting with the first. Perform this same calculation starting with the second child. Compare these numbers with the population mean μ.

Solution

The first sample consists of the heights 85, 86.5, and 87 cm. The mean is

$$\bar{x}_1 = \frac{85 + 86.5 + 87}{3} \approx 86.2 \text{ cm}$$

For the second sample,

$$\bar{x}_2 = \frac{86 + 87 + 90}{3} \approx 87.7\,\text{cm}$$

The population mean is

$$\mu = \frac{85 + 86 + 86.5 + 87 + 87 + 90}{6} \approx 86.9\,\text{cm}$$

(All calculations were rounded to a single decimal place.) There are some differences among the three values. How significant are these differences? The answer requires further analysis, always bearing in mind the uses to which this information will be put.

Skill Enhancer 9   Using the same group of children, split them this time into two groups—the first three and the second three.
(a) Compute the sample mean for the first group.
(b) Compute the sample mean for the second group.

Answers in Appendix E.

Mode and Median

Two other simpler measures of central tendency are the *mode* and the *median*. The *mode* is the data value that occurs most often.

Example 10   What are the modes of the following sets of data?
(a) $\{4, 9, 32, -7.6, 9, 14, 8\}$
(b) $\{\pi, \pi, 0, -1, 0, \pi, -1, \sqrt{3}, -1\}$
(c) $\{99, 98, 97, 96, 94, 93\}$

Solution

(a) Since only 9 is repeated more than once, this value is the mode.
(b) Both π and -1 are repeated three times, so both values are modes.
(c) This set has *no* mode. (A set in which no element is repeated has *no* modes.)

Skill Enhancer 10   What are the modes?
(a) $\{1, 2, 3, 4, 5\}$
(b) $\{1, 2, 1, 2, 1\}$
(c) $\{1, 3, 6, 3, 1\}$

Answers in Appendix E.

Example 10 illustrates the weaknesses of the mode. Sometimes data have no mode, and sometimes the mode is not unique. A measure that resolves these difficulties is the *median*
median. Suppose we have arranged the elements in numerical order. If there are an *odd* number of items, the median value is that number above and below which lie an equal number of data values. If there is an *even* number, we choose the pair of central values, and define the median to be their average.

The Median

We arrange the data in numerical order. There are two cases, depending on whether there is an *odd* or an *even* number of elements.

Odd   The element lying in the middle of the data sequence is the median. There will be as many elements lying above it as below it.

Even   When data are in numerical order, there are *two* middle elements—let them have values x and y. The median is their average,

$$\frac{x+y}{2}$$

Example 11   Determine the median for each of the following sets of data.
(a) {10,000, 10,500, 10,900, 11,400, 45,000, 53,000, 72,000}
(b) {10,000, 10,500, 10,900, 11,400, 145,000, 153,000, 272,000}
(c) {10,500, 10,900, 11,400, 45,000, 53,000, 72,000}

Solution

In all cases, the data are already arranged in ascending numerical order.
(a) Here, there are seven items. The fourth item—11,400—has as many values above it as below it. This is the median value.
(b) As in part (a), this data set contains seven items. The first four items are the same as those in part (a). In this collection of data, the highest three values are much greater than the corresponding elements of the set in part (a). The median remains the same, 11,400.
(c) Here, there are six items—an even number. The two central items are the third and fourth, with values 11,400 and 45,000. Compute the midpoint—the median—as

$$\text{Median} = \frac{1}{2}(11,400 + 45,000) = \frac{56,400}{2} = 28,200$$

In this case, the median is *not* a member of the original data set.

Skill Enhancer 11   What are the medians?
(a) {1, 2, 3, 4, 5}
(b) {1, 2, 1, 2, 1}
(c) {1, 3, 6, 3, 1}

Answers in Appendix E.

The median is often an inadequate measure of central tendency. To see why, suppose that in Example 11, the given numbers stand for the current salaries of a small group of Vietnam veterans. By comparing the salaries in parts (a) and (b), we see that the median remains the same, even though the high salaries of the second group are substantially greater than the high salaries of the first group. We often expect a measure of central tendency to incorporate aspects of *all* data values. This is something that neither the median nor the mode does. On the other hand, the average is influenced by all data values, and this is one reason why the average is so important.

Dispersion from the Mean

Anyone can take the average of numerical data—but will this average always be meaningful? Study the following situation.

A friend is coming for a visit from far away. We receive a last-minute call concerning the local weather. What clothes should he bring with him? We explain that for the last five days the noontime Fahrenheit temperature readings have been 71, 70, 72, 72, and 70 degrees. The friend can decide on the basis of these data what type of clothing is necessary.

Now, imagine that several months later another old friend is about to visit. In response to a similar conversation, we remark that the noontime temperatures for the most recent five days were 84, 59, 64, 71, and 77 degrees.

Need we have quoted the five single temperatures in both cases? Would the *average* noontime temperature have sufficed?

In both cases, the average is 71 degrees. (Make sure you see why.) Clearly, the average *is* a useful summary of the *first* group of data. All the given temperatures cluster tightly about the average. The situation with the second group is less clear. Although they yield the same average, these temperature readings wandered much further from the mean. In fact, although the first friend could have made a decision on what clothes to bring on the basis of the average, the second friend could not.

The average by itself is not sufficient—we need a measure of the degree with which the given data cluster around or wander away from the mean value.

deviation   If μ is the mean of a set of values x_1 through x_N, then the *deviation about the mean* of x_i is $(\mu - x_i)$. Informally, the deviation of some number is the numerical distance between that number and the mean. Deviations may be positive, negative, or zero.

An *apparently* reasonable recipe for the deviation from the mean would be to use the sum of the deviations of the individual data values. Partly because some deviations are negative while others are positive, it is possible to show that all deviations always sum to zero, so this reasonable definition will not lead to a useful result. An improvement on this definition comes from observing that the square of any nonzero number is always positive, so that if we add the *squares* of the deviations, the terms will never cancel.

variance   The *variance* is the average of this sum of squared terms. Variance is denoted by σ^2, where σ is the Greek letter *sigma*. In the Greek alphabet, σ has the same sound as *s* does in the English alphabet.

Variance: Ungrouped Data

Compute the squares of the deviations, sum them, and divide by N.

$$\sigma^2 = \frac{(\mu - x_1)^2 + (\mu - x_2)^2 + \cdots + (\mu - x_N)^2}{N}$$

σ^2	is the variance.
μ	is the average data value of the population.
N	is the number of data items in the population.
x_1 through x_N	are the data items in the population.

TABLE 9.1   Steps in computing the variance. Note that the sums of the deviations are always zero.

x_i	$\mu = 71, N = 5$ $\mu - x_i$	$(\mu - x_i)^2$	x_i	$\mu = 71, N = 5$ $\mu - x_i$	$(\mu - x_i)^2$
71	0	0	84	−13	169
70	1	1	59	12	144
72	−1	1	64	7	49
72	−1	1	71	0	0
70	1	1	77	−6	36
Total	0	4	Total	0	398

Example 12   Compute the variances for the temperature readings discussed at the beginning of this section.

Solution

Normally, the first statistic computed for any data set is the mean. In this case that has already been done for us. The average temperature in both cases is 71 degrees. Table 9.1 shows the intermediate steps in the calculations. For the first readings,

$$\sigma^2 = \frac{4}{5}$$

For the second,

$$\sigma^2 = \frac{398}{5} = 79.6$$

The mathematics bears out our purely intuitive feelings. The variance of the second set of temperatures is much greater than that of the first set. The first set of values cluster much more tightly around the average. It makes more sense to use the average as a summary of the first set of data than it does for the second set.

Skill Enhancer 12   Here are two sets of data: $\{1, 2, 3, 4, 5\}$ and $\{-1, 0, 4, 5, 6\}$.
(a) What is the average for each set of data?
(b) What is the variance for the first set of data?
(c) What is the variance for the second set of data?

Answers in Appendix E.

Variance: Grouped Data

$$\sigma^2 = \frac{f_1 \cdot (\mu - x_1)^2 + \cdots + f_n \cdot (\mu - x_m)^2}{N}.$$

where
m is the number of *distinct* data items.
f_i is the frequency of the ith distinct data item x_i.
$N = f_1 + \cdots + f_m$ is the total number of data items.

The smaller the variance, the more closely the data huddle about the mean.

Example 13 Perhaps due to the effects of toxic industrial pollution, some small towns in coastal Louisiana have exceptionally high rates of liver cancer among their populations. By the time a doctor detects the disease, the prognosis is death within a few months. Here are data on the number of cancer patients in one small town and the numbers of months they lived *after* the cancer had been detected.

Number of patients	2	4	8	4	2
Months lived	5	4	3	2	1

What are the mean and variance of the survival times after detection?

Solution

Using the formula for grouped data, compute μ.

$$\mu = \frac{2 \cdot 5 + 4 \cdot 4 + 8 \cdot 3 + 4 \cdot 2 + 2 \cdot 1}{2 + 4 + 8 + 4 + 2}$$
$$= \frac{10 + 16 + 24 + 8 + 2}{20} = \frac{60}{20}$$
$$= 3$$

The average survival time for these patients is 3 months.

The calculation of the variance involves the mean and the frequencies.

$$\sigma^2 = \frac{2 \cdot (3-5)^2 + 4 \cdot (3-4)^2 + 8 \cdot (3-3)^2 + 4 \cdot (3-2)^2 + 2 \cdot (3-1)^2}{20}$$
$$= \frac{2 \cdot 4 + 4 \cdot 1 + 8 \cdot 0 + 4 \cdot 1 + 2 \cdot 4}{20} = \frac{8 + 4 + 0 + 4 + 8}{20} = \frac{24}{20}$$
$$= 1.2 \, \text{month}^2$$

Table 9.2 helps make the computation of the variance more systematic.

Skill Enhancer 13 In another year, the physicians of this town had more data. Here is the table now.

Number of patients	3	6	12	6	3
Months lived	5	4	3	2	1

What now are the mean and variance of the survival times after detection?

Answer in Appendix E.

The units of the variance in Example 13 are *months*^{2}, rather a curious unit. Variance has different units from the mean or the raw data themselves, so statisticians often work *standard deviation* with the *standard deviation*, σ. The standard deviation is the square root of the variance. The standard deviation and the mean have the same units as the raw data.

TABLE 9.2   Computing a variance using grouped data.

$\mu = 3, m = 5, N = 20$				
x_i	$(\mu - x_i)$	$(\mu - x_i)^2$	f_i	$f_i(\mu - x_i)^2$
5	-2	4	2	8
4	-1	1	4	4
3	0	0	8	0
2	1	1	4	4
1	2	4	2	8
Total	0			24

Standard Deviation

$$\sigma = \sqrt{\sigma^2}$$

Standard deviation $= \sqrt{\text{variance}}$

Example 14   Compute the standard deviation of the data of Example 13.

Solution

$$\sigma = \sqrt{1.2\,\text{month}^2} \approx 1.10\,\text{month}$$

Skill Enhancer 14   The variance of some experiment is $64\,\text{ohm}^2$. What is the standard deviation?

Answer in Appendix E.

In the same way we defined the sample mean $\bar{x}$ as an approximation to the population mean μ, we define a sample variance and standard deviation s^2 and s. Even in unbiased samples, a formula for s^2 based on the one for σ^2 is apt to consistently *underestimate* the true value of σ^2. It can be shown that the following expressions lead to better estimates.

Sample Variance and Standard Deviation

$$s^2 = \frac{(\bar{x} - x_1)^2 + \cdots + (\bar{x} - x_n)^2}{n - 1}$$

$$s = \sqrt{s^2}$$

Here,

s^2	is the sample variance.
s	is the sample standard deviation.
n	is the number of items in the sample.
x_1 through x_n	are the individual data items in the sample.
$\bar{x}$	is the sample mean.

We normally use N for the number of items in the entire population and n for the number of items in the sample. Remember to

Divide by	When dealing with
N	entire population
$n-1$	sample

when computing the variance or standard deviation.

Example 15   Compute the variances for the two samples of Example 9. Compute the variance for the entire population as well.

Solution

For the first sample,

$$s_1^2 = \frac{(85-86.2)^2 + (86.5-86.2)^2 + (87-86.2)^2}{3-1}$$

$$= \frac{(-1.2)^2 + (0.3)^2 + (0.8)^2}{2} = \frac{1.44 + 0.09 + 0.64}{2}$$

$$\approx 1.09\,\text{cm}^2$$

In the same way, for the second sample,

$$s_2^2 = \frac{(86-87.7)^2 + (87-87.7)^2 + (90-87.7)^2}{2}$$

$$= 4.34\,\text{cm}^2$$

You should show that

$$\sigma^2 = 2.37\,\text{cm}^2$$

for this population.

Skill Enhancer 15

(a) Compute the variances for the two samples of the enhancer to Example 9.
(b) Compute the variance for the entire population as well.

Answers in Appendix E.

Chebychev's Inequality

Chebychev inequality   We saw earlier the connection between probability and the frequency distribution in a collection of data. This connection is strengthened by the so-called *Chebychev inequality*, which provides a lower bound on the likelihood of a data value in terms of the parameters μ and σ^2 of the entire population.

Chebychev's Inequality

The probability that a data value x lies within k units of the mean is at least as great as $1 - (\sigma^2)/k^2$. That is,

$$P[(\mu - k) \leq x \leq (\mu + k)] \geq 1 - \frac{\sigma^2}{k^2}$$

In this inequality,

$P(\mu - k \leq x \leq \mu + k)$ is the probability that x lies within k units of μ.

μ is the mean value of the population.

σ^2 is the variance of the population.

Example 16   Certain data are distributed with mean 0.5 and a variance of 0.0015. Estimate the lower bound on the probability that an item chosen randomly from this population has a value between 0.4 and 0.6.

Solution

We use the Chebychev inequality, with $\mu = 0.5$ and $\sigma^2 = 0.0015$. We choose $k = 0.1$, because $\mu - 0.1 = 0.4$ and $\mu + 0.1 = 0.6$. The desired probability p must satisfy

$$p \geq 1 - \frac{0.0015}{(0.1)^2} = 1 - 0.15 = 0.85$$

That is, the probability that an item has a value between 0.4 and 0.6 is at least 0.85.

Skill Enhancer 16   Certain data are distributed with mean 0.4 and a variance of 0.002.
(a) Identify μ and σ^2.
(b) Let p be the lower bound on the probability that an item chosen randomly from this population has a value between 0.3 and 0.5. What is k and k^2? Estimate p.

Answers in Appendix E.

Example 17   It's known that the life span of a small laboratory animal is distributed with a mean of 6 days and a standard deviation of 0.5 day. An experiment will take between 5 and 7 days to conduct, so the researcher wants to know what is the smallest proportion of animals he can expect to survive for this length of time on the average.

Solution

Use the Chebychev inequality, with $\mu = 6$, $\sigma = 0.5$, and $k = 1$. If the proportion is p, then

$$p \geq 1 - \frac{(0.5)^2}{1^2} = 1 - 0.25 = 0.75$$

Skill Enhancer 17   A botanist asks a similar question about a species of tree. In this case, the mean life span is 75 years with a standard deviation of 6 months. A certain cycle in a forest ecosystem takes between 65 and 85 years to complete.
(a) For use in the Chebychev inequality, what are the mean, variance, and value of k and k^2 for this species?
(b) What is the smallest proportion of trees she can expect to survive for this length of time?

Answers in Appendix E.

Estimates of this probability can be sharpened if we know further details about the distribution of probability. Certain special distributions are explored in succeeding sections.

_____ Exercises 9.3 _____

1. Assemble a pile of coins of several types. What coin is the mode of this collection? What is the median coin? What is the average coin type? How appropriate is it to discuss these statistical quantities with respect to pocket change?

2. Pool the class schedules for yourself and two friends. Classify courses as science, math, English, foreign language and literature, social science, history, and other. What course category is the mode? What is the median? Is there an average course? Comment if any of these questions are inappropriate.

W 3. Will the median of a sample be less than, more than, or the same as the average? Why?

Give the mode or modes of the sets of data in Exercises 4–10. (If a data set has no mode, state that fact.)

4. Elephant, aardvark, aardvark, ant, elephant, tiger, elephant, aardvark, elephant

5. 1909, 1910, 1899, 1909, 1914, 1901

6. January, April, March, January, January, February, March, February, March

7. Seattle, Portland, San Diego, Portland, San Diego, Fresno, Davenport, Corvallis

8. 14, 77, −1, 0, 13.999

9. −0.001, 0.01, −0.001, −0.00001, −0.01

10. Suppose discrete data are graphed as a vertical bar graph

of frequency. What is an easy way to detect the mode or modes from this graph?

Determine the median values of the sets of data listed in Exercises 11–14.

11. 1, 3, 5, 6, 8, 10

12. 99, 100, 103, 107, 199, 201

13. $\frac{1}{2}, \frac{2}{3}, \frac{5}{8}, \frac{2}{3}, \frac{3}{4}$

14. −0.001, −0.001, −0.01, −0.1, 0, 0.0001

15. Determine the average value of the data in Exercise 11.

16. Determine the average value of the data in Exercise 12.

17. Determine the average value of the data in Exercise 13.

18. Determine the average value of the data in Exercise 14.

Compute the mean of the data in Exercises 19–24.

19. 2, 4, 6, 8, 10   20. −1, −2, −3, 1, 2, 3

21. 0, 1, −1, 3, −19, 12   22. −4, −2, 0, 2, 4

23. 17, 11, 12, 2, 0, 1   24. 1, $\frac{1}{2}, \frac{1}{4}, \frac{1}{3}$, 0

The tables in Exercises 25–30 give data values and their frequency. (The left column is the frequency.) Compute the average.

25. 1 : 1
2 : 1
3 : 1
4 : 1
5 : 1
6 : 1

26. 1 : 1
2 : 2
3 : 3
4 : 4
5 : 5
6 : 6

27. 1 : 6
2 : 5
3 : 4
4 : 3
5 : 2
6 : 1

28. 1 : 2
2 : 4
3 : 6
4 : 8
5 : 10
6 : 12

29. 1 : 1
2 : −2
3 : 3
4 : −4
5 : 5
6 : −6

30. 1 : 1
2 : $\frac{1}{2}$
3 : $\frac{1}{3}$
4 : $\frac{1}{4}$
5 : $\frac{1}{5}$
6 : $\frac{1}{6}$

31. In a nursery school class, 4 children are two years old, 4 are three years old, 7 are four years old, and 3 are five years old.
(a) What is the mean age? the median age?
(b) Compute the variance and standard deviation.

32. For a class experiment, each student was asked to count the paper currency he or she was carrying. The results were surprising: 4 people had no money at all, 7 had two dollars, 8 had three dollars, 5 had five dollars, 12 had exactly ten dollars, 3 had eleven dollars, and 1 student had ninety dollars!
(a) What is the median value? What is the mean?
(b) Compute the variance and standard deviation.

33. Consider the integers from 0 to 9 inclusive. What is their variance and standard deviation about the mean?

34. Consider the half-integers from $\frac{1}{2}$ to $9\frac{1}{2}$ inclusive.
(a) What are their variance and standard deviation about the mean?
(b) Compare these results with those of Exercise 33.

35. Consider the odd numbers between 0 and 9 inclusive as a sample of all whole numbers in that range. Compute the sample mean, standard deviation, and variance. Compare these results with the results of Exercise 33.

36. Consider the even numbers between 0 and 9 inclusive as a sample of all whole numbers in that range. Compute the sample mean, standard deviation, and variance. Compare these results with the results of Exercise 33.

37. Consider the odd half-integers between $\frac{1}{2}$ and $9\frac{1}{2}$ inclusive as a sample of all half integers in that range. Compute the sample mean, standard deviation, and variance. The odd half-integers are $1\frac{1}{2}$, $3\frac{1}{2}$, $5\frac{1}{2}$, and so on. Compare these results with the results of Exercise 34.

38. Consider the even half-integers between $\frac{1}{2}$ and $9\frac{1}{2}$ inclusive as a sample of all half integers in that range. Compute the sample mean, standard deviation, and variance. The even half-integers are $\frac{1}{2}$, $2\frac{1}{2}$, $4\frac{1}{2}$, and so on. Compare these results with the results of Exercise 34.

39. Determine the variance and standard deviation of the data in Exercise 11.

40. Determine the variance and standard deviation of the data in Exercise 12.

41. Determine the variance and standard deviation of the data in Exercise 13.

42. Determine the variance and standard deviation of the data in Exercise 14.

W43. Student B. C. Dull used (by mistake) a slightly different formula for variance, namely

$$\sigma^2 = \frac{(x_1 - \mu)^2 + \cdots + (x_N - \mu)^2}{N}$$

(Compare this with the formula for variance on page 459.) Nevertheless, all his computed variances were correct! Why?

44. Other means besides the arithmetic mean are possible. The *geometric mean* of n numbers is defined as $(x_1 \cdot x_2 \cdot \cdots \cdot x_n)^{\frac{1}{n}}$.
(a) Compare the geometric and arithmetic means of the two data sets 4, 9 and 9, 25.
(b) When will the geometric mean of two numbers be a whole number?
(c) When will the geometric mean of two numbers equal the arithmetic mean of two numbers?

45. Show that the sum of deviations (not standard deviations) about the mean must *always* be zero.

W46. Use the Chebychev inequality to compute the lower bound of probability that a data value lies within one-half unit of the mean when the mean is 5 with a variance of 1.

W47. Refer again to Exercise 46. What can you conclude about the usefulness of the inequality you obtained as your answer? What about the usefulness of the Chebychev inequality?

Use the Chebychev inequality in Exercises 48–53 to estimate the minimum probability that a randomly selected item from the population lies k units from the mean.

48. $\mu = 10$, $\sigma = 0.75$, $k = 1$

49. $\mu = 0.1$, $\sigma = 0.001$, $k = 0.002$

50. $\mu = 0$, $\sigma = 0.5$, $k = 0.25$

51. $\mu = -1$, $\sigma = 1$, $k = 0.5$

52. $\mu = 100$, $\sigma = 3$, $k = 5$ **53.** $\mu = 7$, $\sigma = 1.5$, $k = 3$

54. There are n numeric values, and their average is some number a. If each of the values is multiplied by the same constant k, what will the average be?

55. There are n numeric values, and their variance is some number σ^2. If each of the values is multiplied by the same constant k, what will the variance be?

Applications

56. *(Manufacturing)* A packaging machine is to create packages containing exactly 100 rubber bands per pack. Randomly selecting four packs, the plant manager discovers that one pack has 97 bands, two others contain 101, and the fourth contains 107! What is the median value? the mean? the sum of the deviations? the standard deviation about the mean?

57. You have asked your local lumberyard to cut 6 four-foot boards for some shelving. To your surprise, the planks aren't all quite the same length, and you measure them yourself. You find that 2 boards are $47\frac{1}{2}$ inches long, 1 is exactly 48 inches, and 3 are $48\frac{1}{4}$ inches long. What is the mean length? the mode? the median? What is the variance about the mean?

58. *(Zoology)* A famous naturalist returned from a six-month field expedition on the Amazon River, where she was fortunate enough to observe 6 specimens of a rare alligator. She weighed all 6, and performed some statistical calculations in the field, but in the course of returning home, some of her handwriting blurred. Here's what can be deciphered. (Variables represent smudged data values.)

Specimen	1	2	3	4	5	6
Weight (lb)	1	2	1.5	2	x	y

modes = 1.5, 2   mean = 2

Deduce the possible values of x and y. Compute the variance of all the weights.

59. Solve Exercise 58 if you now know that the median is 2 lb, and the mean is 2 lb. One of the smudged values is 2.5 lb. The second smudged value (y) is greater than 2. Recompute the variance.

60. *(College grades)* Damien needs to get an A in math. The professor awards this grade to anyone whose average on the five term quizzes is 85 or better. On the first four quizzes, Damien earned grades of 70, 95, 90, and 80. Is it still possible for him to obtain an A for the course? What is the minimum grade he needs on the final quiz? If he manages to achieve the minimum grade he needs for that A, what will the standard deviation of his grades be?

61. In another course, the final course grade is computed by counting the final exam as the equivalent of *two* term quizzes. Besides the final, there are three quizzes. Damien earned grades of 60, 90, and 95 on these.
  (a) If an A means an average of 85 or better, what is the minimum grade he needs on the final to guarantee this? What will the standard deviation of his grades be if he earns this minimum grade on the final?
  (b) Suppose that the professor decides to drop the lowest term quiz grade before computing the final course grade. What now is the minimum final grade that Damien needs?

62. *(Candy Manufacture)* The Happy Time Candy Company produces 1-lb bags of jelly beans. At least, the mean weight of the beans is 1 lb with a standard deviation of 1 oz. Bags that contain more than a pound detract from the company's profits, and bags that are too light generate customer complaints. What proportion of Happy Time's production can at least be expected to weigh between 14 and 18 ounces?

63. *(Freelancing)* Lon Noll is a freelance book designer. Based on his business for the last five years, his average annual income is $25,000 with a standard deviation of $1,000. In order to budget for a new car, he would like to know the minimum likelihood that his income in the upcoming year will be between $23,000 and $27,000. Can you help him out with this calculation?

Calculator Exercises

Compute the average and standard deviation for the lists of numbers in Exercises 64–69.

64. 3, 3, 9, 10, 11   **65.** 1, 2, 3, 6, 7, 8

66. 10, 11, 12, 13, 14, 15

67. 5, 6, 7, 7, 7, 7

68. 10.1, 10.2, 11.1, 12.2, 13.3, 14.4

69. 1.2, 2.2, 3.2, 4.2,
5.2, 6.2

The tables in Exercises 70–75 give data values and their frequency. (The left column is the frequency.) Compute the average and standard deviation.

70. 1 :   2
2 :   1
3 :   0
4 : −1

71. 1 :   1
2 :   2
3 :   3
4 :   4

72. 1 :   4
2 :   3
3 :   2
4 :   0.5

73. 1 :   2
2 :   1
3 : $\frac{2}{3}$
4 : $\frac{1}{2}$

74. 1 :   1.5
2 :   2.5
3 :   3.6
4 :   4.6

75. 1 :   91
2 :   92
3 :   94
4 :   95

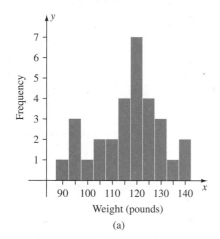

(a)

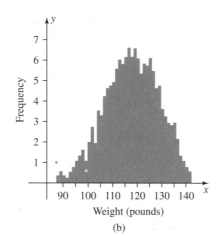

(b)

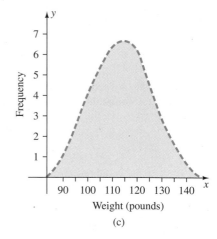

(c)

FIGURE 9.17
Successive histograms as the numbers of students increase, and as the sizes of the categories decrease.

Section 9.4   DISTRIBUTION FUNCTIONS; THE NORMAL DISTRIBUTION

At the beginning of Section 9.2 (page 449), we created a histogram by graphing weights of students. We gathered these weights by taking each weight to the nearest five pounds. Now, let us contemplate this graph as it might appear as we do two things:

1. *Increase* the number of students in the classroom

2. *Decrease* the size of the weight categories

Figure 9.17 illustrates several steps in this "thought experiment." With enough students and with a fine enough measure, the profile of the histogram becomes smoother and smoother, and more regular and pleasing in shape. This function appears to approach an *ideal shape*; one such is shown as Figure 9.17c. In real life, there are never an infinite number of students, nor can weight categories be infinitely close to one another, but statisticians often pretend that these assumptions are true. They then derive equations that describe this ideal shape and that reveal additional, important information. We call

distribution function

a function that we can use to determine the probability of a data value a *distribution function*.

Why are these distribution functions of any interest? After all, there can never be an infinity of anything, so the hypotheses under which we derive these functions are suspect. We make the assumption that *real-life distributions approximate an ideal distribution function*. These distributions help us answer important questions about probability.

normal distribution

The most important distribution function is the *normal distribution*, obeyed by many attributes, including the weights of college students. The graph of this function gives rise to the famous *bell-shaped curve*, several examples of which we see in Figure 9.18. (This curve is described by the equation

$$y = \frac{1}{\sigma\sqrt{2\pi}} e^{-(1/2)[(x-\mu)/\sigma]^2}$$

where π and e are mathematical constants—$\pi \approx 3.14159$, $e \approx 2.71828$—and μ and σ are the mean and standard deviation of the distribution. As μ and σ take on different values, the "bell" of the curve moves to the right or left, and becomes tall and skinny or short and squat.)

There are several important properties of these distributions, and we list them below. We will not discuss the theory behind these properties.

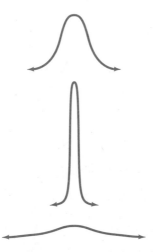

FIGURE 9.18
Bell-shaped, normal distribution curves.

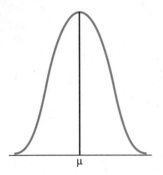

FIGURE 9.19
A vertical line bisects the normal curve at the x value corresponding to the mean. The curve is symmetric about this line.

Properties of the Normal Distribution

1. Normal curves are *symmetric about the vertical line* that intersects the highest point. See Figure 9.19.

2. The value along the horizontal axis that is the intersection with the vertical bisector is the *mean*. See Figure 9.19.

3. The total *area under the curve is always 1.* Furthermore, both shallow arms of the curve get straighter and straighter, and closer and closer to the horizontal axis. But no matter how far from the mean, they are never perfectly horizontal, nor do they ever intersect the horizontal axis.

4. The *probability* $P(a \leq x \leq b)$ that the result of some experiment lies between the values $x = a$ and $x = b$ is the area under the normal curve that lies between the vertical lines $x = a$ and $x = b$. See Figure 9.20.

5. The area between $x = \mu - \sigma$ and $x = \mu + \sigma$ is about 68.27 percent of the total area under the curve. The area between $\mu - 2\sigma$ and $\mu + 2\sigma$ is 95.45 percent of the total area, and the area between $\mu - 3\sigma$ and $\mu + 3\sigma$ is approximately 99.73 percent of the area. See Figure 9.21. We rarely encounter data located "farther" away than 3 standard deviation units from the mean.

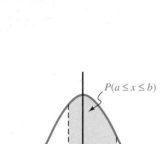

FIGURE 9.20
The probability of a result is related to the area under the curve.

Perhaps the most important of these properties are items 4 and 5, which relate *probability* to an *area* underneath the curve.

From the data presented in item 5 in the box and displayed in Figure 9.20, we can construct Figure 9.22. In this figure, we see the area of each "column" under the normal,

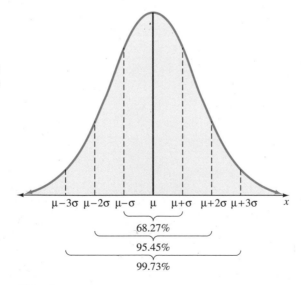

FIGURE 9.21
There is a relationship between the area under the normal curve and distance from the mean in units of standard deviation.

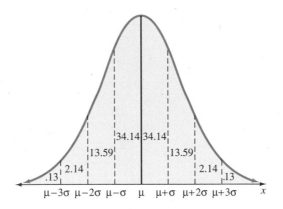

FIGURE 9.22
From the relationship between the areas under the normal curve and distance from the mean in units of standard deviation, we can determine the area of each standard deviation "column."

where vertical boundaries occur at $x = \mu - 3\sigma$, $\mu - 2\sigma$, $\mu - \sigma$, μ, $\mu + \sigma$, $\mu + 2\sigma$, and $\mu + 3\sigma$. This figure also shows the areas of the two tails of the curve.

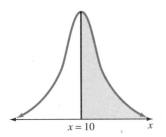

FIGURE 9.23
The probability that $x \geq 10$.

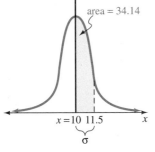

FIGURE 9.24
The probability that $10 \leq x \leq 11.5$.

Example 18   Certain data are distributed normally with a mean of 10 and a standard deviation of 1.5. An item is chosen at random from this population; let its value be x. What are the probabilities that x lies within the following ranges?
(a) $x \geq 10$ (b) $10 \leq x \leq 11.5$ (c) $7 \leq x \leq 11.5$ (d) $x \leq 5.5$

Solution

(a) By the symmetry of the normal distribution, data are as likely to have values above the mean as below. As Figure 9.23 shows, this probability is one-half the total area under the curve, which is 1. The answer is $\frac{1}{2}$.
(b) The difference between 11.5 and 10 is 1.5, the value of the standard deviation. The results shown in Figure 9.22 apply to *any* normal curve. This figure shows that the area under this portion of the curve is 34.14 percent. See also Figure 9.24.
(c) Divide the range between 7 and 11.5 into two portions, as in Figure 9.25. The area under the normal curve between 10 and 11.5 is 34.14 percent of the whole area. (Why? See the previous part.) Observe that the distance between 7 and 10 is 3, which is twice the standard deviation in this problem. This corresponds to the areas of the two segments to the left of $x = \mu$ in Figure 9.22 because 7 lies at a distance of 2σ to the left of the mean. The probability p is the sum of the areas under these three segments:

$$p = 0.3414 + 0.3414 + 0.1359 = 0.8187$$

which is 81.87 percent.
(d) The value 5.5 is significant because $\mu - 3\sigma = 10 - 3(1.5) = 5.5$ for this problem. Thus, the probability that x is less than 5.5 is represented by the left-hand "tail" in Figure 9.26. From Figure 9.22 we see that this area is 0.13 percent, the desired probability.

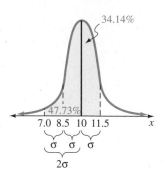

FIGURE 9.25
The probability that $7 \leq x \leq 11.5$.

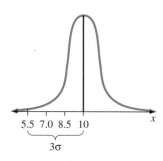

FIGURE 9.26
The probability that $x \leq 5.5$.

Skill Enhancer 18   Some data have a normal distribution with a mean of 0 and a standard deviation of 3. Let the value of an item selected at random from this population be z. With what probability does z lie within these intervals?
(a) $z \leq 0$
(b) $-3 \leq z \leq 6$
(c) $-6 \leq z \leq 0$

Answers in Appendix E.

Example 19   Instead of mowing her lawn, a statistics graduate student analyzes it. The heights of the grass plants are normally distributed with a mean of 5 cm and a standard deviation of 0.2 cm. There are 1,500 grass plants per square meter. Estimate how many of them have heights between 4.8 and 5 cm.

Solution

The difference $5 - 4.8$ is 0.2, one standard deviation. By property 5 above, 68.27 percent of the grass plants lie between 4.8 and 5.2 cm in height. By property 1, this area is symmetric about the mean value, so the percentage we want is *one-half* of 68.27

percent, or 34.14 percent. (See Figure 9.27.) Roughly $0.3414 \times 1,500$, or 512 plants lie within the range of heights between 4.8 and 5 cm.

Skill Enhancer 19   Paperback gothic romance novels have a mean length of 192 pages. The length is normally distributed with a standard deviation of 4 pages. We need the probability that a randomly selected novel will have a length of over 200 pages.
(a) What is $\mu + 2\sigma$ for this problem?
(b) Refer to Figure 9.22 on page 471 and say which of these segments of the normal curve correspond to the probability.
(c) What is the probability?

Answers in Appendix E.

Example 20   The distribution of body weights of college students is normal with a mean of 130 lb, and a standard deviation of 20 lb. There are 1,000 graduating seniors at Carson Community College. Estimate how many of them fit into the following weight categories.
(a) 110 to 130 lb
(b) 110 to 150 lb
(c) 90 to 110 lb
(d) 90 lb or under
(e) 190 lb or over

Solution
The reader will want to refer extensively to Figure 9.22 on page 471.
(a) Since the mean is 130 lb and the standard deviation is 20 lb, the interval between 110 and 130 lb is precisely one standard deviation unit in length ($130 - 110 = 20$). This interval lies to the left of the mean, and comprises an area that is 34.14 percent of the whole. A good estimate of the number of seniors in this range is $0.3414 \times 1,000 \approx 341$.
(b) The range of 110 to 150 lb is precisely one standard deviation unit in both directions from the mean, since $110 = 130 - 20$ and $150 = 130 + 20$. We estimate, therefore, that

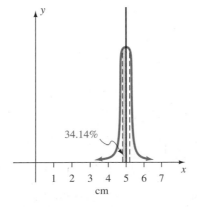

FIGURE 9.27
Analyzing grass.

68.27 percent of the seniors fall into this category. The final answer is $0.6827 \times 1{,}000 \approx$ 683 students.

(c) The range 90 to 110 lb includes students lying in the interval between $\mu - 2\sigma$ and $\mu - \sigma$. (Why?) From Figure 9.28, this range cuts off 13.59 percent of the area, so the best estimate is $0.1359 \times 1{,}000 \approx 136$ students.

(d) How many students weigh 90 lb or less? From Figure 9.28, note that $x = 90$ lies two standard deviations below the mean. In Figure 9.22 the two segments below $x = \mu - 2\sigma$ contain 2.28 percent of the total area under the curve. Our best estimate is that $0.0228 \times 1{,}000 \approx 23$ students weigh 90 lb or less.

(e) Note that $190 = \mu + 3\sigma$. From Figure 9.22 we know that this portion has an area of 0.13 percent. Therefore, the best estimate is $1{,}000 \times 0.0013 \approx 1$ student.

Skill Enhancer 20   As it happens, of these 1,000 students, exactly 500 are male. Their weights are distributed normally with a mean of 180 lb and a standard deviation of 25 lb.
(a) For this sample, what are $\mu - 3\sigma$, $\mu - 2\sigma$, $\mu - \sigma$, $\mu + \sigma$, $\mu + 2\sigma$, and $\mu + 3\sigma$?
(b) About how many men weigh more than 205 lb?
(c) About how many weigh between 155 and 180 lb?
(d) Estimate how many men weigh less than 130 lb.

Answers in Appendix E.

Suppose we have a set of bounds each of which is *not* necessarily a whole number of standard deviations from the mean. What then? Referring to Example 20, suppose we need an estimate of the number of students weighing between 110 and 140 lb. How shall we proceed?

In principle, the answer is straightforward. Construct the normal curve for this distribution, and compute the area under the curve bounded by the bounding numbers. (For Example 20, we need the area between x values of 110 and 140.) Unfortunately, it is not possible to compute areas under the normal curve in closed form, as we can the areas of simpler geometric figures.

Another approach might be to *make a table* of areas under normal curves in the same way as for square roots and trigonometric functions. This is a formidable task—we would need tables for normal curves with every conceivable pair of μ and σ.

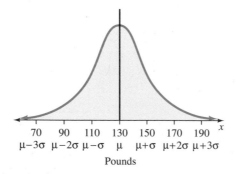

FIGURE 9.28
The normal curve applied to a student weight distribution.

A simple transformation on any data (as long as they follow a normal distribution) converts the data to transformed data with $\mu = 0$ and $\sigma = 1$. If we agree to perform this transformation wherever necessary, then we will not need an extensive series of tables of the normal distribution. In fact, we will need a table for only the single *standard* normal distribution, characterized by mean 0 and standard deviation 1.

To see why this transformation works, consider Example 20 once again. In solving this problem, what counted was *not* the actual data values, but rather the distance of these values from the mean measured in standard deviation units. What counted was not (say) that we needed the group of students weighing between 110 and 130 lb, but that these students had weights lying between the mean and a weight one standard deviation above it. If this transformation preserves these units (which it does), then we can expect that this technique will work.

Z score   The transformed data is called the *Z score*, and here is its formula.

Z **scores**

$$z = \frac{x - \mu}{\sigma}$$

z is the transformed Z score.
x is the original data value.
μ is the mean of the original data.
σ is the standard deviation.

In words, *divide the difference between any data item and the mean by the standard deviation* to get the Z score. Once we know the Z score, we need refer to a single table, such as Table 1 in Appendix 4. Figure 9.29 compares normal and standard normal curves. With respect to the lower, Z scale in these figures, the mean of both curves is 0 and the standard deviation is 1, even though the mean and standard deviation of the untransformed data are quite different.

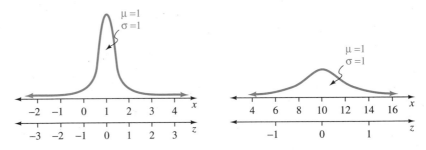

FIGURE 9.29
Normal curves with regular and Z-score scales.

Example 21   Certain data are distributed normally with a mean of 100 and a variance of 25. Compute the Z score for data with the following values:

(a) 90
(b) 110
(c) 95

Solution

For this problem, $\mu = 100$. Since the variance is 25, we have $\sigma = 5$.
(a) Take $x = 90$. Then

$$z = \frac{90 - 100}{5} = -2$$

(b) $z = \frac{110 - 100}{5} = 2$.
(c) $z = \frac{95 - 100}{5} = -1$.

Skill Enhancer 21   Compute these Z scores, assuming that the population mean is 50 and the standard deviation is 10.
(a) 0;   (b) 30;   (c) 55;   (d) 80.

Answers in Appendix E.

Example 22   Salaries of heads of households in a suburb of Houston are normally distributed with a mean of \$22,500 and standard deviation of \$1,500. Three heads of households chosen at random earn \$21,000, \$38,000, and \$22,500. What are the corresponding Z scores for these salaries?

Solution

$$z_1 = \frac{21,000 - 22,500}{1,500} = -\frac{1,500}{1,500} = -1$$

$$z_2 = \frac{38,000 - 22,500}{1,500} = 15,500/1,500 = 10.33$$

$$z_3 = \frac{22,500 - 22,500}{1,500} = 0$$

Skill Enhancer 22   Used car prices for a car of a particular make, model, and year are normally distributed with a mean of \$7,500 and a standard deviation of \$500. While shopping for just this kind of car, Darcy discovered prices of \$7,100, \$7,600, and \$7,995. What are the Z scores for these prices?
      Data values *below* the mean have a *negative* Z score.
      Data values *above* the mean have a *positive* Z score.

Answers in Appendix E.

Solving Normal Distribution Problems

1. Convert the data to the corresponding Z scores.
2. Exploit the properties of the standard normal curve to solve the problem.

The next example provides practice in reading Z-score tables.

Example 23 What areas under the standard normal curve correspond to Z-score values of:
(a) $z = 0.41$
(b) $z = -0.43$

Solution

A portion of the Z-score table from Appendix 4 appears as Table 9.3. The table is a series of pairs of columns; for a particular positive value of z, we read off the corresponding value of $A = A(z)$, where A is the area under the curve bounded on the *left* by the vertical axis and on the *right* by the vertical line located z units from the origin.
(a) A Z-score value of 0.41 corresponds to $A = 0.1591$. This area is the right shaded portion of Figure 9.30. The probability that the untransformed data value x has a Z-score value between 0 and 0.41 is therefore 0.1591, or about 16 percent. This is the left shaded portion of Figure 9.30.
(b) A Z-score value of $+0.43$ corresponds to $A = 0.1664\%$. A *negative* value means that the area lies to the *left* of the vertical axis, as in Figure 9.30. The probability that the untransformed data have a value between -0.43 and 0 is 0.1664, or about 16.6 percent.

TABLE 9.3 A fragment from a table of Z scores.

z	A
0.40	0.1554
0.41	0.1591
0.42	0.1628
0.43	0.1664
0.44	0.1700

Skill Enhancer 23 What areas under the standard normal curve correspond to Z-score values of
(a) $z = -0.4$
(b) $z = 0.44$

Answers in Appendix E.

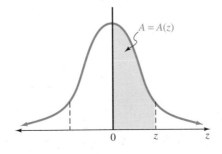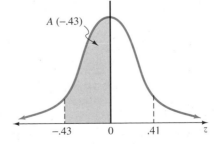

FIGURE 9.30
Interpreting $z = 0.41$ and $z = -0.43$.

Example 24   Richardson Construction Company orders hundreds of 2-by-4 pieces of wood precut to 4-ft lengths for their prefabricated houses. Because the wood is cut at the mill by hand, there is some slight variation in length. The actual lengths are distributed normally, with mean 4 ft and standard deviation 1/2 in.
(a) In any large batch of 2-by-4s, what proportion will be under 4 ft in length?
(b) What fraction will be between 47-7/8 and 48-1/8 in?
(c) What proportion will be over 49 in long?
(d) What proportion will be *under* 47.5 in in length?

Solution

(a) By the symmetry of *any* normal curve about the mean, as many pieces will be over 4 ft as under. The answer is therefore 50 percent.
(b) The Z score of 48-1/8 in is

$$\frac{48\frac{1}{8} - \text{mean}}{\text{standard deviation}} = \frac{48\frac{1}{8} - 48}{\frac{1}{2}} = \frac{1}{4}$$

By symmetry, the Z score of $47\frac{7}{8}$ in must be $-\frac{1}{4}$. (Verify this.)
    Referring to the table in Appendix 4, a Z score of 0.25 corresponds to an area of 0.0987. This is the area from $z = 0$ to $z = 0.25$. The area between $z = -0.25$ and $z = +0.25$ is twice this—0.1974, slightly less than 20 percent. See Figure 9.31.
(c) A length of 49 in corresponds to a Z score of

$$\frac{49 - 48}{\frac{1}{2}} = 2$$

From the table, the area under the normal curve between the mean and $Z = 2$ is 0.4772. The answer to the problem is the part of the curve—the right-hand tail—lying to the *right* of 2. By the symmetry of the curve, we know that the entire right half of the curve encloses an area of $\frac{1}{2}$. Therefore, the area of the tail is $0.5000 - 0.4772 = 0.0228 = 2.28$ percent. This is the proportion of cut pieces with length of 49 in or greater. See Figure 9.32.
(d) A Z score for 47.5 is

$$\frac{47.5 - 48}{\frac{1}{2}} = -1$$

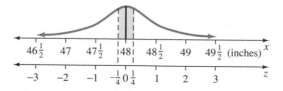

FIGURE 9.31
Computing fractions of wood cut to different lengths.

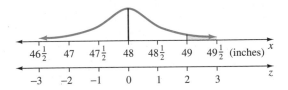

FIGURE 9.32
Cut wood for which $z > 2$.

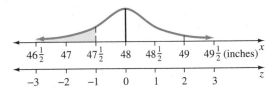

FIGURE 9.33
Cut wood for which $z < -1$.

with a corresponding area of 0.3414 under the normal curve. This area is to the left of the vertical axis, as in Figure 9.33. Since the desired proportion is the shaded area in this figure, we compute its area as $0.5 - 0.3414 = 0.1586$. That is, slightly less than 16 percent of the shipment will be 47.5 in long or shorter.

Skill Enhancer 24   Consider this same large batch of wood.
(a) What are the Z scores for $48\frac{1}{4}$ in and $48\frac{3}{4}$ in?
(b) What areas correspond to these scores?
(c) Estimate the fraction of wood between $48\frac{1}{4}$ and $48\frac{3}{4}$ in long.

Answers in Appendix E.

_____ *Exercises 9.4* _____

Assume that certain data are normally distributed with a mean of 2 and standard deviation of 1. Estimate the proportions of the data that lie within the ranges given in Exercises 1–12.

1. greater than 2

2. less than 2

3. between 1 and 3

4. between 1 and 4

5. between 0 and 4

6. between −1 and 5

7. between 2 and 5

8. between 1 and infinity

9. between −∞ and 3

10. between −∞ and 1

11. between 4 and ∞

12. between 3 and 4

Assume that certain data are normally distributed with a mean of −4 and standard deviation of 4. Estimate the proportions of the data that lie within the ranges given in Exercises 13–18.

13. greater than −4

14. between −4 and 0

15. between −4 and +8

16. between 0 and 2

17. value of exactly 0

18. value of exactly 0.5

Assume that some numerical data have a mean of 8.5 and a standard deviation of 2.5. Estimate the proportions of the data that lie within the ranges given in Exercises 19–24. The data are distributed normally.

19. between 1 and 3.5

20. less than 1

21. greater than 16

22. between 6 and 8.5

23. between −∞ and 11

24. between 1 and 8.5

Give the Z scores for the data items of Exercises 25–30, which have been drawn from normal distributions with the given mean and standard deviation.

25. $\mu = 1, \sigma = 0.5; 0, 1, 2$

26. $\mu = -1, \sigma = 2; -1, 1, 2$

27. $\mu = 0, \sigma = 10; -5, 1.3, 0.5$

28. $\mu = -\frac{3}{4}, \sigma = -\mu; 9, 11, 12$

29. $\mu = -0.12, \sigma = 0.42; -1, 1, 2$

30. $\mu = 0.43, \sigma = 0.28; -1, 1, 2$

Use Table 1 in Appendix 4 to determine the areas of the portions of the normal curves shaded in Exercises 31–36. All the curves are standard—$\mu = 0$ and $\sigma = 1$.

31.

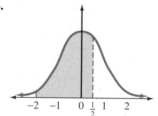

32.

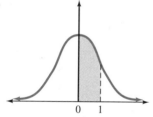

33.

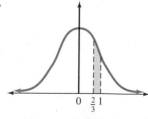

34.

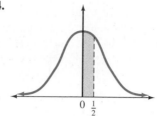

35.

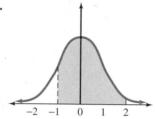

36.

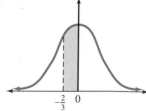

What fraction of the area under the standard normal curve lies between the intervals given in Exercises 37–42?

37. $0 \le z \le 1$ **38.** $-0.75 \le z \le -0.25$

39. $-0.2 \le z \le -0.1$ **40.** $-0.1 \le z \le 0.6$

41. $z \le 1.5$ **42.** $z \ge 2.3$

43. The area under the standard normal curve defined by the range $0 \le z \le q$ is $\frac{1}{2}$. What is q?

44. The area under the standard normal curve defined by the range $-0.1 \le z \le q$ is $\frac{1}{4}$. What is q?

45. The area under the standard normal curve defined by the range $0.1 \le z \le q$ is $\frac{1}{3}$. What is q?

46. The area under the standard normal curve defined by the range $-0.1 \le z \le q$ is $\frac{1}{8}$. What is q?

W 47. When solving problems involving the normal distribution, why is it more convenient to work with intervals that can be expressed as a whole number of standard deviation units from the mean?

W 48. The average length of time college students "go steady" has a distribution function with a mean of about six weeks. Do you think the distribution is normal? Why or why not?

Applications

49. (SAT Scores) SAT scores in the senior class of a certain high school have a mean of 500 with a standard deviation of 100 and are distributed normally. What is a good estimate of the fraction of these seniors with scores over 600? A certain college accepts only individuals whose SAT scores are 700 or better. What proportion of these seniors would be eligible for acceptance by this college?

50. (SAT Scores) The SAT scores of the freshman class at the same high school described in Exercise 49 have a normal distribution with a mean of 500 and a standard deviation of 150. Somehow, the identification for each test answer sheet has been lost. Is it possible to use the statistics of this chapter to decide whether a given test paper belongs to a freshman or a senior? Suppose that all the freshman papers remain together, as do the senior papers. Is it possible to decide which group of answer sheets belongs to which class? How so?

51. (Chemistry) A certain medical solution comes pre-mixed by the drug manufacturer. In principle, 2 milligrams (mg) of the drug are mixed into 1 L of fluid, but there is some variation in the amount of drug added. The actual amount added is normally distributed with a mean of 2 mg and $\sigma = 0.25$ mg. This solution is a vital part of the cure of a rare disease, but if the patient receives less than 1.6 mg, the drug is ineffective. On the other hand, if the patient receives more than 2.8 mg, the drug induces a fatal heart attack. What proportion of the patients receiving this treatment will receive an ineffective dose? What proportion will receive a fatal dose?

52. (Weather and Climate) Over the years, on average the first frost appears on October 10 in Bozeman, Mon-tana. The actual date of the first frost is normally distributed about October 10 with a standard deviation of 10 days. Estimate the number of years in a century when frost appears before September 15. Estimate the number of years in a century with a frost-free October. If the appearance of frost is *truly* normally distributed, can any year ever be without frost up till January 1?

53. (Agriculture) The yield per square meter of winter wheat is 0.75 bushel, normally distributed with a standard deviation of 0.15 bushel. How many square meters yield 1 bushel or more? What proportion of a large farm yields between 0.8 and 1.0 bushel per m^2? If the yield were *precisely* 0.75 bu/m^2, a farmer could easily estimate the yield by simply multiplying the size of the farm by the yield per square meter. Given the fact that the yield *is* normal, and given that the farm is large, how now shall a farmer estimate the yield?

54. (Retailing) (*Requires special thought*) On the average, 20 people a day walk into the offices of DriveUsCrazy Car Rentals, but the actual number is normally distributed with $\sigma = 6$ customers. How many cars shall the company keep on hand (at the beginning of the business day) to ensure that *all* the customers that day can be accommodated 75 percent of the time?

Calculator Exercises

55. What is the Z score for data values of 12.2 and 14.4 when it is known that the mean is 13.1 and the variance is 0.0625?

56. What is the Z score for data values of 29.007 and 44.619 when it is known that the mean is 35.126 and the variance is 2.012?

Section 9.5 THE BINOMIAL DISTRIBUTION

An important class of problems involves events with only two possible outcomes. For convenience, we refer to these distinct outcomes as *success* and *failure*, or *s* and *f*. Furthermore, we let

$$p = P(s) \quad \text{and} \quad q = P(f)$$

where the notation $P(s)$ represent the probability of an event *s*.
We must have

$$p + q = 1$$

(why?) or

$$q = 1 - p$$

Bernoulli trials   When it is possible to describe events within this framework, we call them *Bernoulli trials*, after Jacques Bernoulli, the mathematician who first investigated this problem extensively.

Historical Note

Bernoulli family   *The Bernoulli family (pronounced ber-NOO-ee) is celebrated in mathematics. Over the course of several generations, many of its members achieved fame by virtue of their achievements in mathematics and in the applications of mathematics to the natural sciences. This Bernoulli is the patriarch of the family, Jacques (1654–1705). Besides his work in probability, he is famous for his work with the then-new techniques of differential and integral calculus.*

Example 25   A student rolls a die. Discuss the event that 1 or a prime number shows.

Solution

Define s as the event that some member of the set $\{1, 2, 3, 5\}$ appears face up. The event f is the appearance of either 4 or 6. By our earlier work on probability of simple events, we know that

$$p = P(s) = \frac{4}{6} = \frac{2}{3}$$

so

$$q = 1 - p = \frac{1}{3}$$

Skill Enhancer 25   A child drops a piece of buttered bread. Discuss the event that the buttered side lands face down.

Answer in Appendix E.

We frequently need to consider possible outcomes in a *sequence* of Bernoulli trials. Suppose that in Example 25 we roll the die over and over again. We might ask questions like the following. Here, as in Example 25, we include 1 in the set of primes.

- In 5 rolls of the die, what is the probability that we observe 3 or fewer prime numbers?
- How likely is it that in 10 rolls of the die, we do not observe a prime number even once?
- In n rolls of the die, what is the probability that we observe success s in at least half of the outcomes?

We use results from our study of combinations and from our study of independent and mutually exclusive probabilities to derive an expression b for the probability that there will be *exactly k successes* in a sequence of *n Bernoulli trials*. This equation will depend on three quantities:

1. k, the number of successes
2. n, the number of trials
3. The probability p of success

We emphasize this dependence by writing $b(n, k; p)$.

binomial distribution   We can derive the form for $b(n, k; p)$ by examining a series of problems whose solutions provide the steps for this derivation. The function $b(n, k; p)$ is the *binomial distribution* function.

A Sequence of Bernoulli Trials

1. Each trial may have only *two outcomes*—success s and failure f.
2. The probability p of success must remain *constant* from trial to trial. (Therefore, the probability q of failure must also remain the same, since $q = 1 - p$.)
3. All trials are *independent*. (Recall what that means.)

Example 26   How can a woman about to give birth be said to undergo a Bernoulli trial?

Solution

The number of outcomes is *twofold*: Her child will be either male or female. If each outcome is *equiprobable*, and if we equate success with the birth of a daughter, then $p = q = \frac{1}{2}$.

There are families, though, where these probabilities are not equal. Suppose this woman's husband's family has records showing that over the course of four generations, wives give birth to sons 75 percent of the time. In this case, $p = 0.25$ and $q = 0.75$.

Skill Enhancer 26   Why is a flip of a coin a Bernoulli trial?

Answer in Appendix E.

Example 27   Refer again to Example 26 and consider the case of a couple where the husband belongs to the family in which sons predominate. The couple plan on two children, and would like to have exactly one son and one daughter. How likely is this if they would like the daughter to be born first? if the order of birth does not matter?

Solution

Having two children is a sequence of *two* Bernoulli trials. Since each is independent of the other, *the probability of a particular sequence is the product of the probabilities of each trial*. Thus, the probability of a girl followed by a boy ("success" followed by "failure") is $p \times q = 0.25 \times 0.75 = 0.1875$.

If the birth order does not matter, then there are two possible ways of having exactly one daughter: The girl followed by the son, or vice versa. These are two *mutually exclusive* possibilities. Probabilities of mutually exclusive events are the sums of the individual probabilities. In this case, we've computed the probability of girl-then-boy to be 0.1875. The probability of boy-then-girl is $q \times p = 0.75 \times 0.25$, also 0.1875. The likelihood of exactly one daughter is therefore $0.1875 + 0.1875 = 0.375$.

Skill Enhancer 27   If boys are as likely as girls, what is the probability of having exactly one daughter out of two children?

Answer in Appendix E.

The solution to the second part of Example 27 is equivalent to determining the value of $b(2, 1; \frac{1}{4})$. The couple wants two children; there are two "trials," so $n = 2$. We have decided that "success" means birth of a daughter, and in this case, $p = \frac{1}{4}$. We want exactly one daughter, so $k = 1$.

Example 28   A fair die is rolled 5 times. What is the probability of obtaining exactly one 6?

Solution

This experiment is a sequence of 5 Bernoulli trials, so we need to evaluate $b(5, 1; \frac{1}{6})$. If success is getting a 6, then $p = \frac{1}{6}$. There are precisely *five* ways in which precisely one 6 can occur in a sequence of 5 rolls. They are

$$sffff$$
$$fsfff$$
$$ffsff$$
$$fffsf$$
$$ffffs.$$

Since $s = P(6) = \frac{1}{6}$, then $q = P(\text{anything else}) = \frac{5}{6}$, and we can compute the probabilities of these 5 possibilities:

$$P(sffff) = \frac{1}{6} \times \frac{5}{6} \times \frac{5}{6} \times \frac{5}{6} \times \frac{5}{6} = \frac{5^4}{6^5}$$

$$P(fsfff) = \frac{5}{6} \times \frac{1}{6} \times \frac{5}{6} \times \frac{5}{6} \times \frac{5}{6} = \frac{5^4}{6^5}$$

$$P(ffsff) = \frac{5}{6} \times \frac{5}{6} \times \frac{1}{6} \times \frac{5}{6} \times \frac{5}{6} = \frac{5^4}{6^5}$$

$$P(fffsf) = \frac{5}{6} \times \frac{5}{6} \times \frac{5}{6} \times \frac{1}{6} \times \frac{5}{6} = \frac{5^4}{6^5}$$

$$P(ffffs) = \frac{5}{6} \times \frac{5}{6} \times \frac{5}{6} \times \frac{5}{6} \times \frac{1}{6} = \frac{5^4}{6^5}$$

The probabilities are all equal. The probability of obtaining exactly one 6, no matter how, is thus the sum of $\frac{5^4}{6^5}$ added 5 times, or simply $5 \times \frac{5^4}{6^5} = (\frac{5}{6})^5$.

Skill Enhancer 28   A Korean die has four faces instead of the usual six; the faces are numbered 1 through 4. Suppose the die is rolled 3 times. Let s be the roll of a 3; let f be the roll of any other number.
(a) How many ways can s occur? What are they?
(b) What is $P(s)$? $P(f)$?
(c) What is the probability that any of these ways occur?
(d) What is the probability of obtaining exactly one 3 in three rolls of the die?
(e) Using the b notation of the text, what quantity have we just computed?

Answers in Appendix E.

Determining Any Bernoulli Probability

These last two examples suggest a strategy for finding the Bernoulli probability.

1. Compute the probability for *one specific pattern* of exactly k successes in n trials. (It may be any specific pattern.)

2. Count the number of ways exactly k successes can be distributed among n trials.

3. The final probability $b(n, k; p)$ is the product of the probability in step 1 and the whole number of step 2.

Computing k Successes in n Trials: A Specific Case

A specific case we choose is one in which the first k trials are successes, whereas the final $(n - k)$ trials are failures. Because each trial is independent, the probability of this sequence is

$$(9.1) \qquad \underbrace{\underbrace{p \times p \times \cdots \times p}_{k \text{ times}} \times \underbrace{q \times q \times \cdots \times q}_{n-k \text{ times}}}_{\text{a total of } n \text{ factors}} = p^k q^{n-k}$$

There are a total of n factors.

How Many Ways of Achieving k Successes in n Trials?

A sequence of Bernoulli trials is like a sequence of n empty boxes that will contain an s if the corresponding trial is successful and will be empty otherwise. Determining the number of ways of achieving k successes is equivalent to deciding how many ways there are of distributing k s's among n otherwise empty boxes. (A box may hold only a single s.) But we know this number! It is the number of combinations of n things taken k at a time, $C(n, k)$. The formula for this number of combinations is

$$(9.2) \qquad C(n, k) = \frac{n!}{k!(n - k)!}$$

The binomial probability function is the product of Expressions (9.1) and (9.2).

> ## The Binomial Probability Function
>
> $$b(n, k; p) = \frac{n!}{k!(n-k)!} p^k q^{n-k}$$
>
> where
>
> $b(n, k; p)$ is the probability of obtaining exactly k successes in a sequence of n Bernoulli trials where the probability of success is p.
>
> n is the number of trials.
>
> k is the number of successes.
>
> p is the probability of success.
>
> $q = 1 - p$ is the probability of failure.

binomial probability function   Mathematicians call $b(n, k; p)$ the *binomial probability function* because the coefficients for the terms in the expansion of the binomial product $(p + q)^k$ are given by this function. (Refer back to our discussion of the Binomial Theorem; see Section 5.2 on page 253.) Specifically, the kth term in this expansion is $b(n, k; p)$.

Example 29   Use the definition of $b(n, k; p)$ to evaluate the following quantities to four-decimal-place accuracy. (Use a scientific calculator if necessary.)
(a) $b(4, 1; 0.5)$ (b) $b(5, 3; 0.75)$ (c) $b\left(5, 4; \frac{1}{8}\right)$ (d) $b\left(3, 3; \frac{1}{2}\right)$

Solution
(a) $b(4, 1; 0.5) = \frac{4!}{1!(4-1)!} 0.5^1 (1 - 0.5)^{4-1} = 4(0.5)(0.5)^3 = 0.2500$

(b) $b(5, 3; 0.75) = \frac{5!}{3!2!}(0.75)^3(0.25)^2 = 10(0.4219)(0.0625) = 0.2637$

(c) $b\left(5, 4; \frac{1}{8}\right) = \frac{5!}{4!1!} \left(\frac{1}{8}\right)^4 \left(\frac{7}{8}\right) = 5(0.0002)(0.875) = 0.0011$

(d) $b\left(3, 3; \frac{1}{2}\right) = \frac{3!}{3!0!} \left(\frac{1}{2}\right)^3 \left(\frac{1}{2}\right)^0 = \frac{1}{8} = 0.125$

Skill Enhancer 29   Evaluate, but leave the answers in exponential form. (a) $b\left(3, 2; \frac{1}{5}\right)$;
(b) $b(4, 1; 0.3)$;   (c) $b\left(5, 5; \frac{2}{3}\right)$.

Answers in Appendix E.

Having used probability theory to derive the form of $b(n, k; p)$, we may now use this function to determine the likelihoods of various sequences of Bernoulli trials. Study the next few examples carefully.

Example 30   The government plans a new space shuttle launch area for a region whose climate varies little throughout the year. Invariably, it is either rainy or sunny, and the

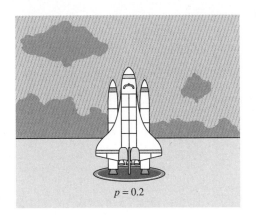 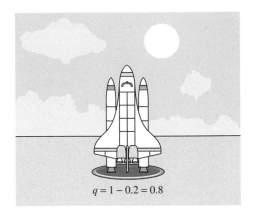

$p = 0.2$ $q = 1 - 0.2 = 0.8$

FIGURE 9.34

probability that it will rain on any given day is 0.2. What is the probability that it will rain exactly twice during a given, consecutive 7-day period? What is the probability that it will rain *not more than* 2 days during this 7-day period?

Solution

Weather at this site consists of a series of Bernoulli trials, each of which is the weather during a single day. The "successful" event is rain, with $p = 0.2$. The probability that it will rain *exactly* twice during a 7-day period is

$$b(7, 2; 0.2) = \frac{7!}{2!5!}(0.2)^2(0.8)^5 = 21 \times 0.04 \times 0.3277$$
$$= 0.2753$$

The event that it will rain *not more than 2 days* is the same as not raining at all *or* raining on exactly one day *or* raining on exactly two days. Consequently, the probability is

$$P = b(7, 0; 0.2) + b(7, 1; 0.2) + b(7, 2; 0.2)$$

Now, $b(7, 0, 0.2) = 0.8^7 = 0.2097$, $b(7, 1; 0.2) = \frac{7!}{1!6!}(0.2)(0.8)^6 = 1.4(0.2621) = 0.3670$, and we computed $b(7, 2; 0.2)$ above. Thus, the probability is

$$P = 0.2097 + 0.3670 + 0.2753 = 0.8520$$

Skill Enhancer 30   Tony flips a fair coin 5 times. Answer these questions, leaving the answers in factored, exponential form.
(a) What is the probability of getting exactly 1 head?
(b) What is the probability of getting no heads in this series?
(c) What is the probability of getting no more than 1 head in this series?

Answers in Appendix E.

Normal Approximation of the Binomial Distribution

The Russian mint is engaged in the manufacture of Russian *kopecks*, coins with the spending equivalent of much less than an American penny. One machine, responsible for 10,000 coins per day, produces an imperfect coin with probability 0.1. What is the probability that 1,050 or more kopecks will be imperfect on any given day?

This problem is simple in theory to solve, but quite tedious in practice. The probability that exactly k kopecks are defective is $b(10,000, k; 0.1)$. The event "1,050 or more defective kopecks" is equivalent to "exactly 1,050 defectives *or* exactly 1,051 defectives *or* exactly 1,052 defectives *or* ... *or* exactly 10,000 defectives." We would compute this probability by evaluating and then summing all terms of the form

$$b(10,000, k; 0.1) = \frac{10,000!}{k!(10,000 - k)!}(0.1)^k(0.9)^{10,000-k}$$

for $1,050 \leq k \leq 10,000$, a time-consuming and difficult chore. This probability is the sum of the shaded area in Figure 9.35.

But this figure is suggestive. The profiles of the verticals appear to form a bell-shaped curve. Might we be able to *approximate* the binomial distribution $b(n, k; p)$ by a normal distribution whenever n is suitably large?

The approximation is validated by the following fact, which we state without proof. Figures 9.36 through 9.39 provide a further, suggestive demonstration. (Each bar is centered about its whole number value on the horizontal axis, which is why the leftmost box extends to the left of the vertical axis in these figures.)

The Normal Approximation to the Binomial Distribution

For large values of n, the binomial distribution may be approximated by a normal distribution with parameters μ and σ, where

$$\mu = np$$
$$\sigma = \sqrt{npq}$$

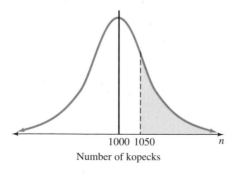

FIGURE 9.35
Computing the probability of producing defective kopecks.

Example 31   Refer again to the discussion of kopeck manufacture. One machine, responsible for 10,000 coins per day, produces an imperfect coin with probability 0.1. What is the probability that 1,050 or more kopecks will be imperfect on any given day?

Solution

Here, $p = 0.1$, so $q = 1 - 0.1 = 0.9$, and $n = 10,000$. Therefore, we may approximate this distribution by a normal distribution where

$$\mu = 10,000 \times 0.1 = 1,000$$
$$\sigma = \sqrt{10,000 \times 0.1 \times 0.9} = 30$$

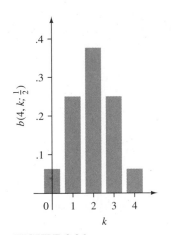

FIGURE 9.36
The binomial distribution for
$n = 4$, $p = \frac{1}{2}$.

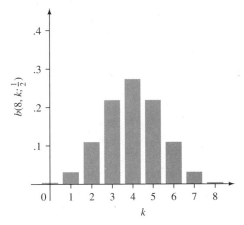

FIGURE 9.37
The binomial distribution for $n = 8$, $p = \frac{1}{2}$.

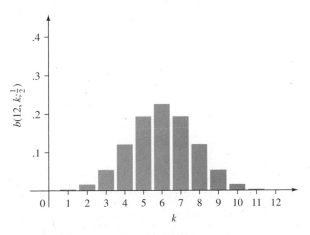

FIGURE 9.38
The binomial distribution for $n = 12$, $p = \frac{1}{2}$.

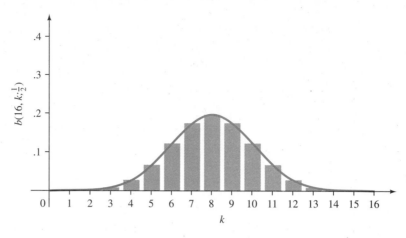

FIGURE 9.39
The binomial distribution for $n = 16$, $p = \frac{1}{2}$. On it is superimposed a normal distribution curve (the dotted curve) with $\mu = np = 8$, $\sigma = \sqrt{npq} = 2$.

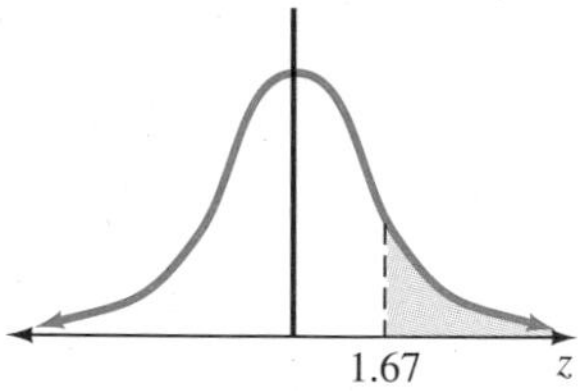

FIGURE 9.40
Defective kopecks and their probabilities.

A value of $k = 1,050$ therefore corresponds to a Z score of $\frac{1,050-1,000}{30} = 1.67$. As Figure 9.40 shows, we need the area to the *right* of this value.

From Table 1, Appendix 4, we see that the area corresponding to this value of z is 0.4525, so the answer to this example is

$$0.5 - 0.4525 = 0.0475$$

a slightly less than 5 percent probability.

Skill Enhancer 31   The Russians have installed a new and improved machine. It can produce 20,000 kopecks per day, but the probability of imperfection is 0.2.
(a) What are q, μ, and σ?
(b) Is it appropriate to use a normal approximation to the binomial distribution?
(c) What Z score corresponds to 4,100 imperfections?
(d) What is the probability p that 4,100 or more kopecks will be imperfect on any given day?

Answers in Appendix E.

As a *rule of thumb*, it is safe to apply this approximation for large n or where p is close to $\frac{1}{2}$.

Example 32   If a motorist "runs" a red light in Sleepy Springs, Montana, he or she runs a 20 percent risk of receiving a ticket. The town planners estimate that 100,000 people run one of Sleepy Springs's traffic lights. What is the probability that 20,000 people or less receive traffic tickets? that between 20,100 and 20,200 people receive tickets?

Solution

Here $n = 100,000$, so it is safe to solve this problem in binomial probability distribution using a normal approximation. To do so, take

$$\mu = np = 100,000 \times 0.2 = 20,000$$

$$\sigma = \sqrt{npq} = \sqrt{100,000 \times 0.2 \times 0.8} = \sqrt{16,000} \approx 126.5$$

To answer the first question, note that since 20,000 is the mean of the distribution, and since the normal distribution is symmetric about the mean, the probability that $\leq 20,000$ people receive tickets is 50 percent.

The Z score of 20,100 is

$$\frac{20,100 - 20,000}{126.5} = \frac{100}{126.5} = 0.79$$

and for 20,200 it is

$$\frac{20,200 - 20,000}{126.5} = \frac{200}{126.5} = 1.58$$

The desired probability is $A(1.58) - A(0.79)$. From Table 1 in Appendix 4, this is $0.4430 - 0.2852 = 0.1578$, or just under 16 percent.

Skill Enhancer 32   What is the probability that 20,000 or more people receive traffic tickets in this town?

Answer in Appendix E.

_____ *Exercises 9.5* _____

Evaluate the binomial probabilities in Exercises 1–6 using paper and pencil *only*.

1. $b(2, 1; 0.5)$ **2.** $b(3, 2; 0.4)$

3. $b\left(4, 3; \frac{1}{2}\right)$ **4.** $b\left(2, 0; \frac{1}{3}\right)$

5. $b\left(4, 2; \frac{1}{3}\right)$ **6.** $b(2, 2; 0.75)$

Calculator Exercises _____

Use calculators or statistical tables (if necessary) to compute the binomial probabilities in Exercises 7–16.

7. $b(10, 5; 0.5)$ **8.** $b\left(7, 4; \frac{1}{3}\right)$

9. $b(5, 4; 0.8)$ **10.** $b(6, 2; 0.4)$

11. $b(7, 3; 0.6)$ **12.** $b(8, 6; 0.4)$

13. $b(4, 1; 0.3)$ **14.** $b(9, 3; 0.1)$

15. $b\left(8, 5; \frac{1}{4}\right)$ **16.** $b(4, 4; 0.9)$

Someone tosses a fair coin 8 times. Compute the probabilities of the events listed in Exercises 17–22.

17. Exactly 4 heads occur.   **18.** Exactly 7 tails occur.

19. No heads occur.

20. Not more than 4 heads occur.

21. At least 4 heads occur.   **22.** At least 6 tails occur.

Of the voters in a certain town in Wyoming, 75 percent are Republican; the remainder are Democrats. A sample of 6 voters is chosen randomly. Use binomial probabilities to compute the probabilities that the samples chosen conform to the characteristics mentioned in Exercises 23–30.

23. Exactly half of these voters are Republicans.

24. Precisely 2 of them are Republicans.

25. Less than 3 of them are Democrats.

26. All of them are Republican.

27. All of them are Republican *or* all of them are Democratic.

28. At least half of them are Republican.

29. Two-thirds or more are Republican.

30. Less than half are Democrats.

A fair die is rolled 8 times. Use binomial probabilities and a calculator or tables to compute the probabilities of the events in Exercises 31–38.

31. A 5 appears exactly once.

32. A 3 appears precisely twice.

33. An odd number appears half of the time.

34. A 4 or a 5 appears at least once.

35. The sum of the rolls is 6.

36. The sum of the rolls is exactly 7.

37. The sum of the rolls is 7 or less.

38. All the rolls of the die yield the same number.

Use the formula for binomial probability to evaluate the expressions in Exercises 39–42.

39. $b(n + 1, n; \frac{1}{2})$

40. $b(n, 0; 10)$

41. $b(n, n - 1; q)$

42. $b(n, n - k; q)$

In Exercises 43–48, you are given values of n and of p and a range of k. Assume that these parameters describe a binomial distribution. Use the normal approximation to the binomial to determine the probability that a randomly selected value will lie within the given range of k.

43. $n = 120$, $p = 0.5$, $115 \le k \le 125$

44. $n = 144$, $p = 0.25$, $140 \le k$

45. $n = 200$, $p = 0.01$, $k \ge 211$

46. $n = 81$, $p = 0.4$, $0 \le k \le 80$

47. $n = 500$, $p = 0.6$, $k \ge 505$

48. $n = 360$, $p = 0.75$, $355 \le k \le 370$

49. (a) Evaluate $b(2, 1; \frac{1}{2})$, $b(4, 2; \frac{1}{2})$, and $b(8, 4; \frac{1}{2})$. Do you see an emerging trend in your answers?

   (b) What is the meaning of the expression $b(2n, n; \frac{1}{2})$? As n gets larger and larger (or, to use the language of calculus, as n goes to infinity), what value do you expect $b(2n, n; \frac{1}{2})$ to approach?

   (c) Substantiate your claim in the previous part by showing this mathematically. To do this, you will need the fact that for very large values of n,

$$n! \approx \sqrt{2\pi n}\; n^n e^{-n}$$

You need to know that e is a special mathematical constant, an irrational number with the value 2.7183 (to four decimal places). This number is important in much of higher mathematics. For this problem, ignore its possible significance and treat it like any other constant. Also, for very large positive values of n, the fraction $\frac{1}{n}$ gets closer and closer to zero.

50. Use the fact that $b(n, k; p)$ represents the terms in the expansions of $(p + q)^n$ to show that

$$C(n, 1) + C(n, 2) + \cdots + C(n, n) = 2^n.$$

Interpret this answer. (*Hint:* Let $p = q = \frac{1}{2}$.)

Applications

If necessary, you may use calculators or tables to solve the following problems.

51. **(Math Anxiety)** The student B. C. Dull is always well prepared for any mathematics exam. On the day of the exam, anxiety always plagues him. This anxiety reduces his chances of answering any short-answer question correctly from 90 percent to 65 percent. The passing grade at this university is 70 percent.

   (a) If the exam consists of 10 questions, what are his chances of passing the exam in a state of anxiety?

   (b) What would his chances of passing be were he not afflicted with math anxiety?

52. Solve Exercise 51 again if the number of questions on the exam is 20.

53. Solve Exercise 51 again if the passing grade is 65 percent.

54. **(Consumer Awareness)** Microcomputers are complicated pieces of equipment, and the chances that a brand new one will work properly are only 90 percent. A company needs to have 5 working microcomputers immediately. If the company buys 5, what are the chances that all 5 will work properly.

55. In Exercise 54, what are the chances that 5 will be working properly if the company purchases 6 new microcomputers?

56. In Exercise 54, how many microcomputers should the company purchase to guarantee with a probability of 95 percent that there will be 5 working computers? with a probability of 99 percent?

W 57. Refer again to Exercise 54. Is there a number that the company can buy to guarantee with 100 percent certainty that there will be 5 working microcomputers? Why or why not?

W 58. Do results computed with the help of the binomial distribution formula depend on which event is labeled "success" and which "failure"?

59. *(Psychological Studies)* A laboratory rat runs a T-maze with a 60 percent probability of turning to the left. In a run of 8 trials, what is the probability that the rat runs to the left 4 times *or more*?

60. *(Polls)* In order to check that its sampling methods are reliable, Cantor Polls asks the same question of the same individuals over the course of a week to get an idea of the reliability and consistency of members of the public. Suppose one individual changes his mind on the same issue from one poll to the next with probability $\frac{3}{4}$. If he is polled on the same topic once a day for 7 days, how likely is it that his answers will be entirely consistent over the course of this period?

61. *(Typesetting)* A telephone company contracts out the preparation of the telephone directory to a typesetting company in Singapore. This company can set a page of listings with only an 80 percent chance of avoiding all errors. If there are 10 pages in this directory, how likely is it that the directory is free from typographical errors? What if the directory contains 20 pages? 5 pages?

62. Solve Exercise 61 again if the chance of a perfectly typeset page is 95 percent.

63. *(Epic Poetry)* Epic poems such as *Iliad* and *Odyssey* are now thought not to have been formally written until hundreds of years after their composition, but rather to have been transmitted orally from one poet to another across the generations. If the probability that a poet remembers the poem perfectly from the time he or she learns it until the time the poem is taught to the successor poet is as high as 95 percent, what are the chances that we have the original poem after 5 generations of poets?

64. *(Manufacturing)* A lamp factory manufactures Art Deco floor lamp replicas. An individual lamp has a $\frac{2}{3}$ probability of passing the factory inspection before being shipped to the customer. A lot of 12 lamps is prepared. What is the probability that exactly 10 lamps are satisfactory? that 10 *or more* lamps are satisfactory?

65. *(Shipping)* Rara Avis Antiques in Chicago is preparing a consignment of antiques for a client in Salt Lake City. The material will be placed in 5 large crates, which will be shipped by a private carrier company. Unbeknownst to Rara Avis, the carrier has a success rate of only 80 percent with these goods. That is, the probability that one of these crates will be damaged in transit is 20 percent. What is the probability that none of the crates will be damaged? What is the probability that all will be damaged? What is the probability that at least 3 out of the 5 crates will be undamaged?

66. Referring to Exercise 65, the customer in Salt Lake City receives the goods and determines that 2 out of the 5 crates are damaged. Rara Avis complains to the company, which responds by citing a quote in its advertising literature—deliveries are made in undamaged condition 95 percent of the time. The carrier claims that the damage was a random, probabilistic event entirely consistent with its advertising. What are the chances that this damage would occur if the advertising is correct?

67. *(Messenger Services)* OnTime Delivery Service specializes in next-day delivery of packages with a success rate of 70 percent. If a company mails one parcel via this company, what is the probability that it will reach its destination the next day?

68. One businessperson absolutely must get some documents into the hands of a colleague, and decides to use the OnTime Delivery Service (see Exercise 67) despite its somewhat mediocre record. The businessperson decides to improve the chances of delivery by sending several copies of the documents to the colleague's address. If the businessperson mails 4 copies, what are the chances that at least one copy gets there on time?

69. *(ESP: Fact or fiction?)* One individual with self-proclaimed powers of extrasensory perception (ESP) claims the ability to identify shapes on which a trained investigator is concentrating in a different room with an accuracy of 90 percent. A test consists of 10 trials. For each trial, the investigator fixes her attention on one of four different geometric shapes. In the first room, the "mind reader" writes down the shapes.
(a) What is the probability that the mind reader scores 50 percent or better in this test, assuming that he possesses the mind reading abilities he claims?
(b) What is the probability of his achieving 50 percent

or better, assuming that the results are due to pure chance—that is, on any trial, the mind reader has a 1-in-4 chance of guessing correctly?

70. *(Biology)* Many sterile Petri dishes are needed for an important biology experiment. The procedure for sterilizing these dishes is not entirely perfect, and faultily prepared dishes can be screened by letting them sit for a few days and visually scanning them for the presence of opaque areas. On the average, 1 in 20 dishes are not sterile. What is the probability of obtaining 90 *or more* sterile dishes in a batch of 100?

71. *(Publishing)* A publisher has brought out a new college calculus text that is in competition with 19 other texts. If any school will choose any text with equal probability, and if the publisher makes contact with 600 schools, estimate the probability that 25 or fewer schools will adopt this text. Estimate the probability that 35 or more will adopt the text.

72. Solve Exercise 71 under the assumption that the publisher is so confident that the quality of the text makes up for its newness on the market that any school will choose it with probability 0.25.

73. Solve Exercise 71 if the publisher is pessimistic that the book's quality can compensate for its newness. Suppose now that its probability of adoption by any school is only 0.01.

74. *(Final Exams)* The final exam for a one-semester course in finite mathematics consists of 100 short-answer questions. The student B. C. Dull has not distinguished himself this semester, and assesses his chances of answering any question correctly at only 50 percent. The passing grade for the exam is 57 percent. What are his chances of passing?

75. **(a)** A colleague of B. C. Dull (see Exercise 74) is Penny Bright, a far more conscientious student. She estimates her chances of getting any individual question right at 92 percent. What are her chances of receiving an A on the exam, where any grade of 90 percent or more is an A?
    (b) What are her chances of receiving a B, where any grade between 84 percent and 89 percent becomes a B?

76. *(Botany)* Any grass seed will sprout with a probability of 75 percent. A square meter of lawn receives 1,500 grass seeds at sowing time in the spring. This small plot is said to be lush if 80 percent of the seeds sprout. What are the chances that such an area will grow a lush lawn?

77. *(Death and Taxes)* The Internal Revenue Service (IRS) reveals that any taxpayer has a 2 percent chance of having his or her tax returns audited in any given year. Assume that this probability remains constant from year to year. If a taxpayer plans on paying taxes from the time he is 21 until he is 70, estimate the probability that he will be subject to between 2 and 3 audits. Estimate the probability that he will be subject to *no* audits.

Calculator Exercises

78. Compute **(a)** $b(13, 5; 0.36)$ **(b)** $b(11, 7; 0.59)$

79. In one urban hospital, birth records show that the actual probability of a boy being born is 0.5211. Over several years, exactly 15,000 children were born here. The student B. C. Dull suspects that of that number, $0.5211 \times 15,000 = 7,816.5 \approx 7,817$ will have been boys. What is the probability that 7,817 or more of the children will be boys? What is the probability that more than 7,950 of the children will have been boys?

CHAPTER REVIEW

Terms

data set	graphs	bar graph
dual graphs	line graph	dual line graph
pie charts	group	tally marks
frequency	mode	class boundaries
frequency distribution	histogram	frequency polygon
bias	measure of the central tendency	average

grouped data   sample mean   sample
median   deviation   variance
standard deviation   Chebychev inequality   distribution function
normal distribution   Z score   Bernoulli trials
binomial distribution   binomial probability function

Key Concepts

- One way to begin analyzing raw data is by grouping identical data items together.
- The mode of a data set is the value or values that occur most often.
- To determine the median value, arrange the data items in numerical order. The median value of an odd number of data items is the item above and below which there are equal numbers of items. If there are an even number of items, the median is the average of the two middle values.
- There is a relationship between the frequency of a data value and the empirical probability that that value will occur in a randomly chosen data value from that sample.
- The most familiar mean value is that computed as the sum of the items divided by the number of items.

- The variance and standard deviation are ways to measure how closely the data huddle about the mean value.
- Distribution functions indicate the ideal distribution of items—an infinite number of items distributed continuously. The normal (bell-shaped) distribution is perhaps the most familiar, and we can infer the probabilities with which values of data lie within intervals from this distribution. These inferences are based on the number of standard deviation units away from the mean that the endpoints of the intervals are located.
- The binomial distribution function is a distribution giving probabilities for sequences of Bernoulli trials—sequences of events in which each event has only one of two outcomes (plus other requirements).

Review Exercises

The accompanying table presents data collected at 8 test sites within the estuary region where the Mississippi River enters the Gulf of Mexico. The numbers in the second row express the magnitude of the *salinity* (degree of saltiness) expressed in appropriate units of measurement. Use this chart to answer Exercises 1–3.

Site	1	2	3	4	5	6	7	8
Salinity	10.5	11.0	12.3	11.7	12.5	10.9	10.9	10.0

1. Display these data as a bar graph.
2. Display the data as a broken line graph.

W 3. Might it be appropriate to represent these data as a pie chart? Why or why not?

4. Use the data shown in Figure 9.41. During what years were the earnings per advertisement between $0.50 and $1.00? between $1.50 and $2.00? less than $0.25?

5. Refer to Figure 9.42. Use these data to create a pie chart. What does the pie chart help show that the bar graph does not?

6. What was the lowest value of the Dow Jones Industrial Average between October 29 and October 31, 1929? What was the highest value in the interval October 19 through December 4, 1987? Use the graph in Figure 9.43.

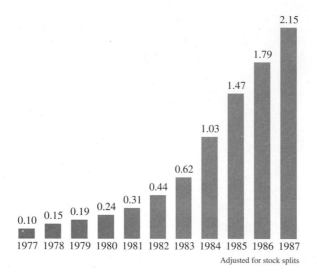

FIGURE 9.41
Company earnings. Source: *The New York Times*, **December 6, 1987. Copyright © 1987 by The New York Times Company. Reprinted by permission.**

Social scientists are studying the food shopping habits of families in various wage-earning categories. They begin by deciding upon 7 different price categories for items typically found in shopping baskets. Items in category 1 are the cheapest; those in category 7 are the most costly. The scientists inventory a "typical" shopping basket from one family by writing down the price category for each item. Here are the results:

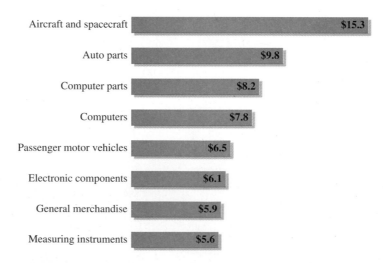

FIGURE 9.42
American exports. Source: *The New York Times*, **December 6, 1987. Copyright © 1987 by The New York Times Company. Reprinted by permission.**

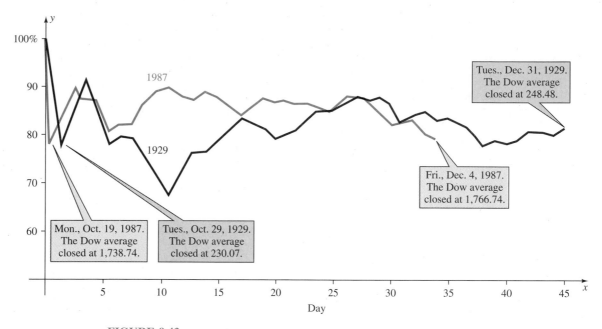

FIGURE 9.43

Stock market crashes. Source: *The New York Times*, **December 6, 1987. Copyright © 1987 by The New York Times Company. Reprinted by permission.**

5	4	7	7	1	3	4	4	7
5	6	1	1	4	6	2	3	1
7	7	4	3	2	3	5	1	1

Here are the tabulations for a second family:

2	3	7	7	2	6	7	2	2
1	3	2	5	1	6	7	5	3
3	3	4	2	1	4	5	2	3

7. Express the data from the first family as a histogram; as a frequency polygon.

8. Express the data from the second family as a histogram; as a frequency polygon.

9. Prepare a histogram for the data from *both* families using a single graph.

Use the following data items to answer Exercises 10–13.

$$1 \quad 1 \quad 9 \quad 4 \quad 5$$

10. What is the mode?

11. What is the median value?

12. What is the mean value of these data?

13. Compute the variance and standard deviation.

14. There are n items in one data set; call them a_1 through a_n. A second data set also contains n items, b_1 through b_n. Each item in the *second* data set is 9 times as great as the corresponding item in the *first* data set. What is the relationship between the *variances* of the two data sets? between the *standard deviations* of the two data sets?

15. The first 5 integers which are perfect squares are 1, 4, 9, 16, and 25. What are their median, mean, variance, and standard deviation?

Data are distributed normally with mean 0 and *variance* 4. On the basis of this, answer Exercises 16–21.

16. What is the probability that any data item will be positive? negative?

17. What is the probability that an item will lie between the values of -2 and 0?

18. What is the probability that an item will lie between the values of 2 and 4?

19. What is the probability that a number from this set of data will be greater than 4? less than -4?

20. What is the probability that a number from this set of data will have a value between -6 and 0?

21. What is the probability that a number drawn from this set will have a value of *exactly* 2?

Numbers are chosen from data distributed normally with mean 10 and standard deviation 3. Compute the corresponding Z scores.

22. 7   **23.** 11

24. 0   **25.** -1

26. 10.5   **27.** 8.5

Evaluate the quantities in Exercises 28–31.

28. $b(4, 3; \frac{3}{4})$ **29.** $b(4, 0; 0.25)$

30. $b(1, 1; 0.9)$ **31.** $b(4, 2; \frac{1}{3})$

In Exercises 32–35, assume that there is a binomial distribution with n and p as given. Estimate the probabilities that a data value chosen at random will have values of k within the given ranges.

32. $n = 400$, $p = 0.5$, $191 \le k \le 200$

33. $n = 400$, $p = 0.6$, $0 \le k \le 230$

34. $n = 225$, $p = 0.4$, $95 \le k \le 100$

35. $n = 300$, $p = 0.7$, $k \ge 220$

36. If a fair coin is tossed three times, what is the probability that exactly one head will occur? that at most one head will occur?

37. If a fair coin is tossed 7 times, what is the probability that exactly 3 heads will occur? that at least 3 heads will occur? that at most 3 heads will occur?

Applications

W 38. *(Animal Husbandry)* A certain prize pig gives birth to a litter of 7 piglets. What is the probability that there are exactly 3 males in this litter? that there are *at most* 3 males? Can you use the results of Exercise 37 to obtain the answers to this question? Why or why not?

W 39. A coin is tossed 50 times, and we observe 35 heads. Is this consistent with the supposition that the coin is fair? Why or why not?

W 40. A single fair die is rolled 60 times. We observe 12 1s, 9 2s, 9 3s, 13 4s, and 7 5s. Are these observations consistent with a fair die? Are these observations consistent with the hypothesis that a 4 is 10 percent more likely than the remaining, equally likely outcomes?

41. Dominic, the head chef at Dominic's Pizzeria, slices his pies into 8 wedges. He does his slicing by eye, and one of his customers determines that the areas of his slices are distributed normally with a mean of 20 in^2 and a standard deviation of 2 in^2.
   (a) What is the probability that a customer gets a wedge with an area between 22 and 24 in^2?
   (b) What is the probability that a customer gets a wedge with an area less than 18 in^2?

42. *(Garment Manufacture)* Sizes of clothes are supposed to be standard, but the actual dimensions of ready-made clothes are generally normally distributed. For one manufacturer of lady's pants, size 4 means that the pants will have a 20-in waist with a standard deviation of 1 in. A size 5 pair of pants has a 22-in waist with a standard deviation of 1 in. A woman with a 21-in waist feels comfortable only in pants with a waist size between 20 and 21 in.
   (a) If she chooses a pair of size 4 pants, what is the probability that she feels comfortable wearing them?
   (b) If she chooses a pair of size 5 pants, what is the probability that she feels comfortable wearing them?

43. Refer again to Exercise 41 (pizza pies). Use Z scores and the table of the standard normal distribution (Table 1 in Appendix D) to compute the probability that a customer gets a wedge with an area between 21 and 23 in^2. Compute the probability that a customer will get a wedge with an area less than 19 in^2.

44. Refer to Exercise 42 to compute the probability that a size 4 pair of pants will have a waist greater than 20.5 in. What is the probability that a pair of size 5 pants will have a waist between 20.5 and 21 in?

45. *(Canning Foods)* The New England Can Company cans cranberry sauce prepared from berries grown in the bogs of Cape Cod. Out of every 100 cans, 95 are perfect, and the remainder must be discarded. In a lot of 5 cans, what is the probability that they are all perfect? What is the probability that at least 4 are perfect and unblemished?

46. *(Postal Service)* At the time of the Civil War, letters mailed from one city would be received by recipients in a neighboring city within 7 days in 75 percent of all cases. Four letters were mailed from one city to addresses in the neighboring city. What is the probability that at least half of them reached their destinations within 7 days?

47. *(Advice to the Lovelorn)* If he asks a girl for a date, the student B. C. Dull estimates that his chance of being accepted by her are about 1 in 5. If he asks 70 girls for a date during one typical semester, what are his chances of having 17 or more dates?

48. *(Manufacturing)* The One-Tooth-Three Company manufactures toothpicks. Any single toothpick will pass the company's automatic inspection machines with a 90 percent probability. The daily quota is 500,000 acceptable toothpicks. If the company manufactures 550,000 toothpicks, what is the probability of meeting its quota?

49. Refer to Exercise 48. How many toothpicks must the company manufacture to meet its quota with a 75 percent probability?

10

FINANCIAL MATHEMATICS

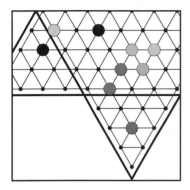

INTRODUCTION

Managers and leaders use the topics in the previous chapters to make the most of the raw materials available to their organizations. We have neglected until now that most important of resources—money. Using someone else's money (as when we borrow it from a bank) requires us to determine the "rental fee" for its use. We call this rental fee the *interest*, and we need some new methods to compute it.

These techniques play rôles in several related topics, such as mortgage loans, long-range financial planning, and financial planning in an inflationary economy. We will look at these topics in this chapter.

Section 10.1   *SIMPLE AND COMPOUND INTEREST*

interest   People often seek to *borrow* money. The borrower's promise to pay *interest* acts as the inducement for lenders to (temporarily) part with their money. This interest is a *rental fee* that the borrower pays for the use of the bank's money. (In the same way, people pay a rental fee—rent—to a landlord for the use of an apartment.) This definition applies even to those preparing to deposit money in a savings account. In this case, the depositor is the lender; the bank is *borrowing* this money from you.

principal   The borrowed money is the *principal*, and the interest depends on three quantities:

1. The *size* of the principal

2. The *length of time* that the borrower holds on to the borrowed money

rate of interest   **3.** A *rate of interest* per unit of time, which represents the portion of the principal that the borrower agrees to pay to the lender for the use of the money for each of the agreed-upon units of time

interest period   The *interest period* is the constant period of time after which interest is computed; interest is never computed continuously. If interest is computed *quarterly*, for example, then new interest is *posted* (computed and placed in the lender's account) only at the end of each three-month period; three months is a quarter of a year. When the borrower fully pays off the loan, the lender has his or her (or its) money back plus the interest.

In the simplest situation, the interest is proportional to the principal, the time, and *simple interest*   the interest rate. We call this *simple interest*.

Simple Interest

$$I = P \times r \times t$$

I   is the simple interest.
P   is the principal (amount borrowed).
r   is the interest rate per *year*.
t   is the time of the loan, measured in *years*.

We compute interest only at the end of each unit of time.

FIGURE 10.1

Example 1   Mr. Bradley lends his son Sam $8,000 to buy a new car. Sam promises to pay this money back to his father in 24 months, plus simple interest at 8.5 percent per year. How much interest will Sam pay?

Solution

Use the formula $I = Prt$. Here, the principal $P = \$8,000$, the interest rate is $r = 0.085$ per year, and the time is $t = 2$ years, since 24 months $= 2$ years. The interest I is

$$I = \$8,000 \times 0.085/\text{yr} \times 2 \text{ yr} = \$1,360.00$$

Figures 10.2 and 10.3 show the growth of interest with time and with the number of interest periods. Remember, 8.5 percent/yr is the same as 0.085/yr.

Skill Enhancer 1   What is simple interest computed on $1,800 at 12 percent for 18 months?

Beware of the common error of neglecting to convert a percentage to its equivalent decimal!

Answer in Appendix E.

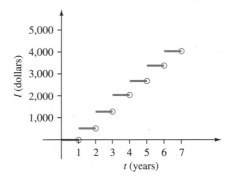

FIGURE 10.2
The growth of simple interest. The step-like nature of the graph reflects the fact that interest increases *only* at the end of each year.

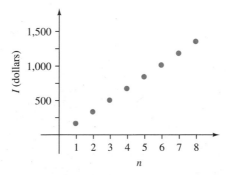

FIGURE 10.3
The growth of simple interest may also be graphed as a function of the number of interest periods.

Simple interest situations rarely occur, but in one way or another this concept forms the basis of all interest calculations.

Compound Interest

compound interest

In Example 1, Mr. Bradley did not demand any interest payments until the end of the period of the loan. Imagine slightly different conditions. In this variation, Mr. Bradley decides that the first year's interest will become his, even though Sam will not pay it to him for another twelve months. In this alternative version, Sam's total interest payment would be *greater*. For the final twelve months, Sam would pay interest on the original

period of compounding

$8,000 *and* on the interest earned in the first 12 months. The *period of compounding* refers to the time period after which interest is recomputed. Unless the borrower pays this interest to the lender, it becomes part of the lender's principal. The borrower will owe interest on this interest to the lender at the end of the next compounding period. In this alternative version of Example 1, interest is *compounded annually*.

Example 2   Solve Example 1 if Mr. Bradley requires annual compounding.

Solution

For the first year, the accumulated interest is

$$I = \$8,000 \times 0.085/\text{yr} \times 1 \text{ yr} = \$680$$

For the final 12 months of the loan, Sam is really using $8,680 of his father's money— $8,000 from the original loan, plus the $680 of interest that he now owes his father. For the final year, the interest is

$$I = \$8,680 \times 0.085/\text{yr} \times 1 \text{ yr} = \$737.80$$

so the *total* interest is

$$I_{\text{total}} = \$680.00 + \$737.80 = \$1417.80$$

In this example, interest compounded annually is greater than simple interest by $57.80.

Skill Enhancer 2   What is interest computed on $1,800 at 12 percent for 18 months? Interest is compounded annually.

Answer in Appendix E.

We need a formula for the accumulated interest under conditions of compounded interest. We can derive it by solving Example 2 in greater generality. Suppose there are N compounding periods per year. For annual compounding, $N = 1$; for quarterly compounding, $N = 4$, and so on. Let i be the interest rate *per compounding period*. For example, if the annual interest rate is $R = 0.12$ (12 percent), and interest is compounded quarterly, then the quarterly interest rate is $i = 0.12/4 = 0.03$. If we compound interest monthly, then $i = 0.12/12 = 0.01$, and so on. A simple formula connects N and the annual rate r with the rate i per period:

(10.1) $$i = \frac{r}{N}$$

Let the initial principal be P. We will use the notation P_i to represent the total *interest plus principal* that will have accumulated (accrued) to the lender at the *completion* of the ith interest period. In this notation, $P_0 = P$.

We use the formula for simple interest to compute i_1, the interest that accrues at the end of the *first* interest period. (In general, let I_j represent the interest that accrues at the end of the jth period.)

$$I_1 = \text{principal} \times \text{rate per interest period} \times 1 \text{ interest period}$$
$$= P \times i$$

The lender's current total is

$$P_1 = P + I_1 = P + Pi = P(1+i)$$

At the end of the second interest period,

$$I_2 = P_1 \times i$$
$$= P(1+i) \times i$$
$$= P(i + i^2)$$

so that

$$P_2 = P_1 + I_2$$
$$= P(1+i) + P(i + i^2) = P(1+i) + Pi(1+i)$$
$$= P(1+i)^2$$

In the same way, at the end of the nth interest period, the accumulated principal plus interest is $P_n = P(1+i)^n$. The growth of principal plus interest with n appears in Figure 10.4.

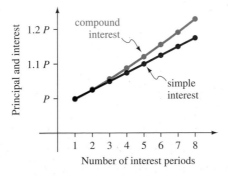

FIGURE 10.4
How interest and principal accumulate with n, the number of interest periods, in the compounding of interest. For comparison, the growth of interest with simple interest is shown by the points lying on the black line.

Compound Interest

$$S = P(1+i)^n$$

Here,

S is the accumulated principal and interest after n periods.
P is the principal lent to the borrower.
i is the rate of interest *per interest period*.
n is the number of interest periods.

Example 3   One bank offers 8 percent interest compounded quarterly on deposits in savings accounts. Homer Jackson deposits $2,000 in one such account in September of the fourth year before his daughter leaves for college. To how much will this money have grown by the time his daughter leaves for college?

Solution

Four years contain 16 quarterly interest periods. The quarterly interest rate is $i = \frac{0.08}{4} = 0.02$. The accumulated total is

$$S = P_{16} = P(1+i)^n$$
$$= 2,000(1+0.02)^{16} = \$2,745.57$$

Skill Enhancer 3   By the time Homer's second daughter prepares for college, the interest rate has fallen to 6 percent, but interest is still compounded quarterly. This time, he deposits $3,500 in an account in September of the fifth year before this daughter departs for college.
(a) What are n and i for this problem?
(b) How much will this sum have grown to?

Answers in Appendix E.

This example, like so many other problems in financial mathematics, requires calculators or computers for finding the answers.

Example 4   Good Bank is offering 8 percent interest compounded semiannually for deposits into regular savings accounts. Better Bank offers 7 percent compounded quarterly on its deposits. A person leaves money in an account for one year. Which account will contain the most money after one year?

Solution

For purposes of comparison, assume a depositor opens an account of $1,000 at each bank on January 1. We compute the balance one year later at each bank.

In the Good Bank, there are two interest periods in a year, with an interest rate of $\frac{0.08}{2} = 0.04$ per period. At the end of 2 periods, the accumulated total is

$$P_2 = 1,000(1.04)^2 = \$1,081.60$$

For Better Bank, the quarterly interest rate is $\frac{0.07}{4} = 0.0175$. Therefore, the total after 4 quarters (one year) is

$$P_4 = 1{,}000(1.0175)^4 = \$1{,}071.86$$

Good Bank provides a slightly better yield.

Skill Enhancer 4   B. C. Dull places $5,000 in an account in which money is left to accumulate at 8 percent compounded quarterly.
(a) How much is there in the account after 4 years?   (b) After 10 years?
(c) After 20 years?   (d) After 30 years (when Dull will retire)?

Answers in Appendix E.

Effective Interest

effective rate of interest

It is hard to compare two different accounts offering compound interest at different terms, as in the last example. In general, *the higher the interest and the more frequent the compounding, the better for the lender or depositor.* How may one compare one account with a higher rate to one with a more frequent compounding period? The answer lies with the *effective rate of interest*, an entirely artificial quantity. It is the rate that a loan would need to produce the same accrued interest *during one year* if interest accrued at simple interest. The effective rate is useful for consumers seeking the best terms from commercial and savings banks. A *borrower* should seek loans with the *lowest* effective rate. *Depositors* or *lenders* should seek to obtain the *highest* effective rate possible.

Example 5   Refer again to Example 4. What are the effective rates of interest for the two banks of that problem?

Solution

We have seen that a $1,000 deposit at the Good Bank grows to $1,081.60. The total earned interest is $81.60. The effective rate is the rate R such that

$$81.60 = 1{,}000 \times R \times 1$$

Clearly,

$$R = \frac{81.60}{1{,}000} = 8.160\%$$

A deposit in Better Bank earns $71.86, so the effective interest rate for this bank is

$$R = \frac{71.86}{1{,}000} = 7.186\%$$

Skill Enhancer 5   Hannah deposits $1,000 in an account bearing interest at 6 percent compounded monthly.
(a) What is i for this problem?   (b) How much interest accrues at the end of one year?   (c) What is the effective rate of interest for this account?

Answers in Appendix E.

The Power of Compounding

The next example shows how much faster compounded interest will grow than simple interest.

Example 6   A young, recently married couple receive $10,000 in cash as part of their wedding presents. In anticipation of retiring in 35 years, they deposit this money in an interest-bearing account and will leave it untouched until then. They have a choice of placing this money in several types of accounts. Decide which of the following they should choose.

Account A: 10% at simple interest
Account B: 10% compounded annually
Account C: 10% compounded quarterly
Account D: 8% compounded quarterly

Solution

The total interest in account *A* is simply

$$I = Prt = 10,000 \times 0.1 \times 35 = \$35,000$$

which when added to the original principal yields a total of

$$S_A = \$45,000$$

The remaining accounts require us to use the equation for compound interest.

For account *B*, there are 35 annual periods in which the interest is compounded. The total principal and interest S_B for this account is

$$S_B = 10,000(1 + 0.1)^{35} = \$281,024.37$$

Account *C* also provides compounding of interest, this time quarterly; 10 percent annually is equivalent to 2.5 percent quarterly, and 35 years contain 140 quarterly periods. The calculation becomes

$$S_C = 10,000(1.025)^{140} = \$317,205.83$$

For the final choice, account *D*, the quarterly interest rate is $\frac{0.08}{4} = 0.02$, so the total accumulation in the account after 140 interest periods (35 years) will be

$$S_D = 10,000(1.02)^{140} = \$159,964.66$$

A comparison of these results shows the young couple that they should choose account *C*. It is also clear, by comparing *A* and *D*, how important the effect of compounding can be. Even a lower interest rate when compounded, as in *D*, provides a much greater total return than the higher rate of *A* with no compounding.

Skill Enhancer 6   Consider two accounts: *A* is an account bearing 12 percent in simple interest, and *B* is an account bearing interest at 6 percent compounded quarterly. Assume you will be leaving $10,000 in one of the accounts for 30 years.
(a) How much does the sum in *A* grow to?
(b) What are *i* and *n* for account *B*?
(c) How much does the sum in *B* grow to?
(d) Which account is better for you?

Answers in Appendix E.

_____**Exercises 10.1**_____

In each of Exercises 1–6, you are given three quantities: a dollar amount, an annual interest rate, and a period. Compute the simple interest that accumulates for each sum at that rate for that time period. ($£$ = British pound)

1. $650; 7.5 percent; 24 months

2. $1,250; 6.5 percent; 18 months

3. $100.50; 8.8 percent; 13 months

4. $990.90; 8.2 percent; 1.7 years

5. £88; 10.1 percent; 15 months

6. £1,234; 11 percent; 9 months

In Exercises 7–12, solve for the indicated variable. Assume conditions of simple interest. P is the starting principal, t is the time of the loan, r is the annual interest rate, and I is the accumulated interest.

7. $P = \$1,000$; $r = 12$ percent; $I = \$12.50$; $t = ?$

8. $P = \$1,250$; $r = 8.5$ percent; $I = \$500$; $t = ?$

9. $r = 10.5$ percent; $t = 1.25$ yr; $I = \$150.00$; $P = ?$

10. $r = 8.8$ percent; $t = 10.5$ yr; $I = \$1,000$; $P = ?$

11. $P = \$750$; $I = \$80$; $t = 0.75$ yr; $r = ?$

12. $P = \$10,000$; $I = \$17,000$; $t = 5$ yr; $r = ?$

Use a calculator or a set of financial tables to determine the answers to Exercises 13–20.

13. What is the total when $10,000 is invested for 6.5 years at 8 percent compounded quarterly?

14. What is the total *interest* when $7,500 is invested at 10.5 percent compounded semiannually?

15. What is the total *interest* when $14,750 is invested for 6 years at 7 percent compounded quarterly?

16. What is the total when $11,000 is invested for 4 years at 10 percent compounded annually?

17. What is the total when $8,500 is invested at 9.6 percent for 30 months?

18. What is the total when $12,250 is invested for 30 years at 10.4 percent compounded *weekly*?

19. What is the total when $9,500 is invested for 2 years at 10.8 percent compounded *daily*? (Assume there are

only 360 days per year. This is a common assumption in financial calculations.)

20. What is the total when $7,600 is invested at 7.2 percent compounded daily for 90 months?

W 21. Refer to Example 2 in the text. What would have been the total interest due Mr. Bradley if Sam had paid the interest for the first 12 months at the end of *12* months rather than after 24 months? Compare this total with the answer to the first example in the text. Why are the two totals for interest now equal?

22. Compute the total accumulated money when 500,000 Japanese yen are invested for 18 months at 10.8 percent annual interest:
(a) at simple interest. (b) compounded annually.
(c) compounded quarterly. (d) compounded monthly.
(e) compounded daily.

23. Compute the total accumulated capital when 750,000 German marks are invested for 2.5 years at 9.6 percent annual interest
(a) compounded annually. (b) compounded quarterly.
(c) compounded monthly. (d) compounded weekly.
(e) compounded daily.

24. What is the effective rate of interest for a bank account that compounds an annual interest rate of 7.6 percent quarterly?

25. What is the effective rate of interest for a certificate of deposit that compounds semiannually an $8\frac{1}{2}$ percent annual rate of interest?

26. What is the effective rate of interest for a certificate of deposit that compounds weekly a 10.4 percent annual rate of interest?

27. What is the effective rate of interest for a bank that pays 9.6 percent compounded annually?

28. What is the effective rate of interest for a 12-month investment that pays 10 percent simple interest for the initial 6 months followed by 7 percent compounded quarterly for the final 6 months?

29. What is the effective rate of interest for a bank that pays 10 percent compounded quarterly for an initial 6-month period followed by 8.5 percent compounded semiannually for the subsequent 6 months?

W 30. Reread Example 6. By what factor will the original sum grow in account C of that problem? What does this imply for students recently graduated but still thinking about retirement planning?

(**d**) 12 percent?

31. Mrs. Kwamina deposits $10,000 in a bank account that computes interest quarterly, makes no further deposits, and withdraws the accumulated total exactly 10 years later. What is this total if the annual rate of interest is (**a**) 6 percent;   (**b**) 8 percent;   (**c**) 9 percent; or

32. Mrs. Lo deposits $10,000 in an account that pays 9 percent annual interest. No further deposits are made, and Mrs. Lo withdraws the accumulated amount exactly 10 years later. Compute this total if the interest is compounded (**a**) annually;   (**b**) semiannually;   (**c**) quarterly; or   (**d**) monthly.

Section 10.2   ANNUITIES AND THEIR VALUES

annuity
annuity period

ordinary annuity
annuity due

life of an annuity

future value

value
interest-free annuity

An *annuity* is a series of payments of money. In this text, we shall assume that payments are all the same, and are made at the end of equally spaced periods, called the *period* of an annuity. When the annuity interest is credited at the same time the annuity payment is made, we have an *ordinary annuity*. When the payment is made at the beginning of the period, we term this an *annuity due*. In this chapter, we work solely with ordinary annuities.

The *life of an annuity*, that period during which the payments are made, often extends over many years. The recipient of the payments usually places the payments in some interest-bearing account, so that the payments accrue interest. The *future value FV* of an annuity is the sum of these payments, plus any earned interest. Sometimes, future value is simply called *value*.

The simplest case is an *interest-free annuity*. Each payment is spent immediately and therefore earns no interest. If there are n payments of p dollars, the future value is

$$FV = np$$

See Figure 10.5.

> **Example 7**   After her father's death, 18-year-old Marcy begins receiving monthly checks from the Social Security Administration for $75.64. She receives these until

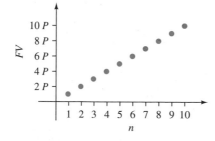

FIGURE 10.5
FV **as a function of** n **(no interest).**

she turns 22. If this annuity lasts exactly 4 years, and if Marcy spends all these checks upon receipt, what is the value of this annuity?

Solution

The total value is the number of months—48—times the monthly check amount.

$$FV = 48 \times 75.64 = \$3,630.72$$

Skill Enhancer 7   What is the value of this annuity if each monthly payment is $83.00 and if the annuity lasts for exactly 5 years? Each check is spent upon receipt.

Answer in Appendix E.

interest-bearing annuity   More often, each payment p rests in an account where it earns interest compounded in each period. The interest rate per period is i, and there are n periods. No one withdraws money from the account during the life of the annuity. This is an *interest-bearing annuity*.

Annuity payments are made at the *end* of each period. The first payment will grow to a value

$$p(1+i)^{n-1}$$

after n periods, the life of the annuity. (Since the payment is made at the end of the first period, there are only $n-1$ remaining periods during which the payment earns interest.) The second payment will grow to

$$p(1+i)^{n-2}$$

since it stays in the account one less period. The third period will grow to $p(1+i)^{n-3}$, and so on. The last period earns no interest, since receipt of this final payment ends the annuity. The value of an interest-bearing annuity is the sum

$$FV = p(1+i)^{n-1} + p(1+i)^{n-2} + \cdots + p(1+i) + p$$

geometric series   This sum forms a *geometric series* (see Appendix B). The quantities p and $(1+i)$ here play the rôles of a and r in that Appendix, which gives the sum FV as

$$FV = p\frac{(1+i)^n - 1}{(1+i) - 1}$$
$$= p\frac{(1+i)^n - 1}{i}$$

Future Value of an Interest-Bearing Annuity

$$FV = p\frac{(1+i)^n - 1}{i}$$

where

FV is the value at the end of the life of the annuity (includes principal plus all earned interest).

p is the amount of each payment.

i is the interest rate per period.

n is the number of periods in the life of the annuity.

In this case, the deposits are made at the end of each period.

Example 8   Mrs. Simms sets up an IRA (Individual Retirement Account) at age 30. This plan calls for her to deposit a maximum of $2,000 per year into an interest-bearing account. Suppose she contributes to this plan annually for 30 years, and the annual effective yield for the account is 8 percent. Compute the sum that will have accumulated by the time she retires at age 60. For purposes of comparison, calculate the value of a similar interest-free annuity, where money is deposited, but earns no interest.

Solution

The formula gives the value of Mrs. Simms's annuity. Take $p = \$2,000$, $n = 30$, and $i = 0.08$. The period is one year. Then,

$$FV = 2,000\frac{1.08^{30} - 1}{0.08}$$
$$= 2,000 \times 113.28 = \$226,566.42$$

By comparison, if the money earns no interest, the annuity's value is

$$FV = 2,000 \times 30 = \$60,000$$

By earning interest, and allowing the interest to earn interest, Mrs. Simms can retire with a nest egg almost four times the size of the sums she actually places in the account. This is another example of the power of compound interest.

Skill Enhancer 8   Calculate the value of the annuity once again. Assume that the effective rates are as indicated.
(a) 9 percent
(b) 10 percent

Answers in Appendix E.

Sinking Funds

Businesses are often able to foresee expenses they will undertake at some future time. They can prepare for this expense by depositing money in an interest-bearing account *sinking fund*   of some kind. This account is a *sinking fund*. In sinking fund problems, we seek to determine the amount of money p to be deposited in the sinking fund at the end of each period to accumulate to S, the desired sum. But S is simply the value of the payments

$$S = p\frac{(1+i)^n - 1}{i}$$

so that

$$p = S\left[\frac{(1+i)^n - 1}{i}\right]^{-1}$$

Financial analysts often use the notation $S(n, i)$ to stand for the quantity within the square brackets.

$$S(n, i) = \frac{(1+i)^n - 1}{i}$$

This curious quantity is read "S angle n at i" because it appears as $S_{\overline{n}\,i}$ in older texts. The notation $S_{n,i}$ may also appear from time to time. Tabulated values of $S(n, i)$ appear in Table 4, Appendix D.

Sinking Funds

$$P = \frac{S}{S(n, i)}$$

Here,

S is the value of the sinking fund;

p is the periodic payment to accumulate to S; and

$S(n, i) = \frac{(1+i)^n - 1}{i}$.   [Some texts use the notation $S_{\overline{n}\,i}$ or $S_{n,i}$ in place of $S(n, i)$.]

Example 9   The Asbestos Corporation of America is a large company that was sued in 1970 in a class action suit on behalf of the hundreds of individuals (allegedly) dying from exposure to asbestos. Although the company is defending itself vigorously in court, its lawyers have privately advised setting up a sinking fund to cover the cost of the settlement. The lawyers estimate that the litigation will continue for 15 years, and the amount of the settlement may be \$1 billion. If the company will make semiannual payments to a fund yielding 12 percent annually, what should the amount of those semiannual payments be?

Solution

In this problem, take

$$S = 1,000,000,000$$

$$i = \frac{0.12}{2} = 0.06$$

$$n = 2 \text{ payments per yr} \times 15 \text{ yr} = 30$$

Then

$$S(n, i) = \frac{(1+i)^n - 1}{i} = \frac{1.06^{30} - 1}{0.06} = 79.0582$$

and

$$p = \frac{1,000,000,000}{79.0582} = \$12,648,911.49$$

Each semiannual payment should be in the amount of $12,648,911.49.

Skill Enhancer 9 Remember Mrs. Simms, back in Example 8? Suppose her retirement accounts allow her to deposit as much as she would like each year. Assuming the same rate of return (an effective 8 percent annually) and the same life of the annuity, she wants to determine the size of her deposits to ensure a certain sum upon retirement.
(a) What are n and i for this problem?
(b) What is $S(n, i)$ for this problem?
(c) If she wants to retire with a sum of $350,000 in her account, what should her annual deposits be?

Answers in Appendix E.

An Introduction to Present Value

present value   How much should a rational investor pay *today* for the privilege of receiving $100 *one year from now*? That is, what is the *present value* of $100 to be received one year from now? We can deduce an answer provided we know the terms under which money earns interest. The *present value PV* will be that principal that grows to $100 after a specified period of time at a specified rate of interest.

Example 10 Jerry knows he can deposit money into a special savings account that pays 6 percent annually compounded monthly. He is saving up to pay for a deluxe trip around the world he wants to take in 18 months. This trip should cost him $10,000. How much must he deposit today to have that sum available to him at that time; that is, *what is the present value of $10,000?*

Solution

The monthly interest rate i is $\frac{0.06}{12} = 0.005$, and the number of interest periods is $n = 18$. Let PV be the present value. By the law of compound interest, this value must grow to equal $10,000 in 18 months:

$$10,000 = PV(1+i)^n = PV(1.005)^{18}$$

so that

$$PV = 10,000(1.005)^{-18} \approx \$9,141.36$$

Under these terms, *the present value of $10,000* is $9,141.36.

Skill Enhancer 10 How does the present value change if the method of compounding changes? Compute the present value for the trip if the compounding is done
(a) quarterly;   (b) twice a year.

Answers in Appendix E.

We can generalize this result.

Present Value of a Single Payment

$$PV = P(1+i)^{-n}$$

where

PV is the present value of a single payment.

P is the amount of a single payment to be received in the future.

n is the number of interest periods after which P will be received.

i is the *periodic* rate at which money can earn interest.

Future Value versus Present Value

Before launching into the full-scale discussion of present value in the next section, students would do well to make sure they understand the distinction between the value (or future value) of money and the present value of money. *Future value* is the sum of all future payments together with earned interest. Since money held today will earn interest between now and some future time, the *present value* is an attempt to make precise the relationship between a sum of money today and a sum valued at some future time. Except in the case of the interest-free annuity, the present value of some future sum of money is less than that sum because the present value can earn the necessary interest to make it equal in value to the future sum.

_____ *Exercises 10.2* _____

Compute the values of the annuities described in Exercises 1–6. Assume no payment earns interest.

1. $p = \$2{,}500$, $n = 33$ **2.** $p = \$175$, $n = 15$

3. $p = \$1{,}350$, $n = 11$ **4.** $p = \$100.25$, $n = 45$

5. $p = \$1{,}111.11$, $n = 111$

6. $p = \$1{,}200$, $n = 72$

Compute the values of the annuities in Exercises 7–18. Assume that all payments are placed into an interest-bearing account with the indicated interest rate per period.

7. $p = \$25$, $n = 25$, $i = 8$ percent

8. $p = 50$, $n = 52$, $i = 0.005$

9. $p = \$150$, $n = 25$, $i = 0.10$

10. $p = \$250$, $n = 20$, $i = 0.095$

11. $p - \$330$, $n - 75$, $i = 0.05$

12. $p = \$250$, $n = 33$, $i = 0.085$

13. $p = \$100$, $n = 36$, $i = 0.05$

14. $p = \$1{,}000$, $n = 10$, $i = 0.004$

15. $p = \$1{,}000$, $n = 15$, $i = 0.004$

16. $p = \$500$, $n = 50$, $i = 0.01$

17. $p = \$750$, $n = 66$, $i = 0.005$

18. $p = \$1{,}100$, $n = 12$, $i = 0.08$

A friend asks to borrow some money from you. She promises to make a single payment of $5,000 to you at the terms stated below in Exercises 19–26. In each case, what is the maximum amount you should be willing to lend; that is, what is the present value of $5,000 under these terms?

19. 20 years; 4 percent compounded annually

20. 20 months; 12 percent compounded monthly

21. 7 days; 24 percent compounded daily

22. 15 months; 8 percent compounded monthly

23. 30 months; 12 percent compounded semiannually

24. 2 years; 10 percent compounded annually

25. 18 months; 10.2 percent compounded semiannually

26. 36 weeks; 13 percent compounded weekly

Compute the payments in Exercises 27–32 for a sinking fund in which the goal is to accumulate $1,000,000 at the end of n periods, where the money will earn interest at a rate i per period.

27. $n = 12$, $i = 0.005$ **28.** $n = 10$, $i = 0.01$

29. $n = 18$, $i = 0.01$ **30.** $n = 20$, $i = 0.0075$

31. $n = 24$, $i = 0.08$ **32.** $n = 60$, $i = 0.05$

Applications

33. (Clipping Coupons) Wealthy Miss Pym owns many bonds. One bond's issuer distributes interest by attaching a sheet of coupons to the bond. Every six months, Miss Pym cuts one off and presents it to her banker, who redeems it for $50. Miss Pym spends this money immediately. These interest payments provide Miss Pym with a modest annuity. What is its value if this bond has coupons for a 25-year span?

34. (Book Royalties) Tim Kitchener is a writer of popular novels. His earnings from writing take the form of *royalties*, whereby he receives a small portion of the price of each book sold. Typically, his books sell 5,000 copies a month for 10 months, after which they are virtually forgotten by the reading public. If his royalty rate is $1.10 per book, and if all these royalties are spent on living expenses, what is the value to Kitchener of one novel?

35. (Financial Planning) Investigate the relative values of several different IRA plans. In each case, you will deposit $2,000 a year until age 60, and the deposited money will earn 7 percent annually. What will the values of the accounts be if you start the IRA at age 25? at age 35? at age 45?

36. Solve Exercise 35 again if the interest rate is 8.5 percent.

37. (College Financing) Mr. and Mrs. Charpentier need to plan for the college tuition for their newborn child, Mark. If Mark starts college at age 18, and if a year of college will cost $30,000 at that time, the Charpentiers will need to accumulate $4 \times 30,000 = \$120,000$ to send their son to school. If they deposit $2,000 annually in a 10 percent account, how much will they have by the time Mark is 18? (Assume annual compounding.)

38. Refer to Exercise 37. Use a calculator to help you estimate the minimum annual deposit the Charpentiers need to accumulate the cost of a college education.

39. Refer to Exercise 37. Estimate the minimum annual deposit needed if the deposited money earns 6 percent.

40. (Financial Planning) Acme Tool & Dye-Making Company foresees the need to replace a new metal-stamping machine in 5 years. Such a machine will cost $100,000. The company savings account yields 7 percent interest compounded quarterly. The company wants to determine the value p of a monthly deposit so there will be exactly $100,000 in the company account in 5 years. What is the proper value of p?

41. Solve Exercise 40 if the account yields 8 percent compounded semiannually. In this problem, p will refer to the size of a *semiannual* company deposit.

42. (Real Estate) Exactly 10 years from now, Estelle and Marvin Parsons have to pay the balance due on the mortgage. At that time, the balance will be $150,000. The Parsons decide to set up a sinking fund to raise this sum, and will deposit some money monthly in an account paying 10 percent annually. What should their monthly payments to this fund be?

43. (Financial Planning) A couple wishes to plan for their child's college education. The child is just 3, and will leave for college 15 years from now, at which point the parents estimate that the total cost of the college education will be $120,000. If the couple can deposit money monthly in an account paying 8 percent annually, how much should the monthly deposit be to ensure that the balance will be $120,000 in exactly 15 years?

W 44. Examine the examples we have covered so far in this chapter. When planning for retirement, evaluate the relative merits of high interest versus frequent period of compounding versus length of the life of the account.

Calculator Exercises

45. What is the value of an annuity whose life consists of 120 periods? The interest rate per period is 0.0123, and the amount of each payment is $1,333.34.

46. The Robles family has one daughter. When she was born, her parents estimated the cost of her college education to be $250,000 by the time she reaches the age of 18. The parents want to establish a fund to accumulate this total by the time the daughter is ready for college. What will be the amount of the payments to the fund if the parents make payments each month for exactly 18 years to an account paying 0.75 percent per month?

Section 10.3   *PRESENT VALUE; AMORTIZATION*

present value of an annuity   Each payment of an annuity has a present value. The *present value of an annuity* is the sum of the present values of all these payments. The next examples demonstrate a method that we can generalize to derive a formula for the present value of an annuity. In present value problems, it is helpful to examine the problem from the point of view of the party who will receive the payments (that is, the party who will receive the annuity).

Example 11   The Huntington Bay Savings and Loan Association loans money for home improvements at 10 percent per annum compounded annually. They will loan the money for 4 years at most. A homeowner can afford at most $3,000 a year in loan payments, and wants to know the maximum amount of money he can borrow in these circumstances. Show how this problem is solved by computing the present value of an annuity.

Solution

At the time that the loan is transacted, the homeowner receives the amount of the loan, and prepares to start making payments to the bank. From the *bank's* point of view, however, it has bought a 4-year annuity. The cost to the bank of the annuity is the amount it loans out, and its annuity payments are the monthly payments the homeowner makes to the bank. The bank is willing to purchase this annuity—$3,000 per year—for the present value of the annuity. This sum is the largest amount that the homeowner may borrow.

Skill Enhancer 11   Max Melchett needs to borrow money to buy a car. He can only afford $1,500 per year in payments to the bank. Discuss reasoning by which the bank can determine the largest amount Max may borrow. The loan will be repaid in 3 years.

Answer in Appendix E.

Example 12   Determine the maximum amount referred to in Example 11.

Solution

The money the homeowner receives is the sum that the bank uses to "purchase" a 4-year, $3,000 annuity. The maximum that the bank should pay for this annuity—the maximum

Homeowner borrows money:
traditional view

Bank buys an annuity

FIGURE 10.6

the homeowner can borrow—is the present value of the annuity at 10 percent. This present value is the sum of the present values of the four annual payments.

Let V_i be the present value of \$3,000 at the end of the ith interest period. For the first payment,

$$V_1 = 3,000(1.1)^{-1} = \$2,727.27$$

In the same way, the present values V_2 through V_4 of the other payments are

$$V_2 = 3,000(1.1)^{-2} = \$2,479.34$$
$$V_3 = 3,000(1.1)^{-3} = \$2,253.94$$
$$V_4 = 3,000(1.1)^{-4} = \$2,049.04$$

The present value PV to the bank of the loan payments is

$$PV = V_1 + V_2 + V_3 + V_4 = \$9,509.34$$

Thus, our homeowner cannot afford to borrow *more* than \$9,509.34 under these terms.

Skill Enhancer 12 We will want to know the maximum amount Max Melchett may borrow from a bank. The loan is to be completely repaid in 3 years, and Max's annual payments will be \$1,500 each year. Assume the interest rate of the loan is 8 percent.
(a) What is the present value of the first payment V_1?
(b) What are the present values of the remaining two payments V_2 and V_3?
(c) What is the maximum amount Max may borrow?

Answers in Appendix E.

May we use this reasoning in a more general form to discover a formula for the present value? Assume that n payments p are made at the end of each period and that there is some annual interest rate equivalent to a rate of i per interest period. The interest is compounded at the end of each payment period. The present value V_j of the jth payment is

$$V_j = p(1+i)^{-j}$$

for values of j ranging from 1 through n. PV is the sum of the V_j's.

$$PV = V_1 + V_2 + \cdots + V_n$$
$$= p[(1+i)^{-1} + (1+i)^{-2} + \cdots + (1+i)^{-n}]$$
$$= p(1+i)^{-n}[1 + (1+i)^1 + \cdots + (1+i)^{n-1}]$$

The sum within the square brackets in the last line forms a geometric series. We use the formula for the sum of a geometric series from Appendix B to find

$$PV = p(1+i)^{-n}[1 + (1+i)^1 + \cdots + (1+i)^{n-1}]$$
$$= \frac{p}{(1+i)^n}\frac{(1+i)^n - 1}{(1+i) - 1} = p\frac{(1+i)^n - 1}{i(1+i)^n}$$
$$= p\left[\frac{1 - (1+i)^{-n}}{i}\right]$$

For convenience, define a function of two variables $A(n, i)$ by

$$A(n, i) = \frac{1 - (1 + i)^{-n}}{i}$$

(Sometimes the notation $a_{\overline{n}\,i}$ is used for $A(n, i)$. This alternative notation is read "*a* angle *n* at *i*.") Tables of values for $A(n, i)$ (or $a_{\overline{n}\,i}$) appear in Table 3 of Appendix D.

The Present Value of an Annuity

$$PV. = p \cdot A(n, i)$$

where
PV is the present value.
p is the periodic payment of the annuity (we assume it has a constant value).
n is the number of payments.
i is the interest rate per period.
$A(n, i)$ $= 1 - (1 + i)^{-n}/i$.

In a loan, the amount borrowed is the PV for the income stream the bank will receive from the borrower. We can solve for p in this formula to compute the amount of the monthly payment.

$$p = \frac{PV}{A(n, i)}$$

Example 13   Ads selling a popular model kitchen range advertise these terms. The purchaser puts \$200 down and makes monthly payments of \$50 for 15 months, which represent an annual interest rate of 12 percent. Deduce the total cost of the stove from this information.

Solution

We begin by computing the present value of an annuity with these terms: $n = 15$, $i = 0.12/12 = 0.01$, and $p = 50$.

$$PV = 50 \cdot A(15, 0.01) = 50 \left[\frac{1 - (1.01)^{-15}}{0.01} \right]$$

$$= \$693.25$$

The present value $PV = \$693.25$ represents the equivalent value of money the financing institution would part with today in order to receive a stream of income in the future. That is, PV represents the money the purchaser borrows for the stove. This value plus the down payment, or $\$693.25 + \$200 = \$893.25$, is the retail cost of the stove.

Skill Enhancer 13   The same set of ads advertise a refrigerator under (apparently) more attractive terms. "With only \$50 down, and monthly payments of \$50 for 25 months, the 'fridge can be yours!" proclaim the ads. This represents financing at an annual rate of 18 percent.

(a) What are p, n, i, and $A(n, i)$ for this problem?

(b) What is the present value of the stream of payments?

(c) What is the true cost to a customer of the refrigerator?

(d) What is the total of all the payments made by the customer to the appliance store?

Answers in Appendix E.

In a loan, the present value of all loan payments represents the amount borrowed.

It may be helpful to summarize all the different values that a stream of money may have.

Future Value versus Present Value

It is important to distinguish between these concepts. Future Value or FV is the sum of the payments for the life of the annuity.

Value is another term for *future value*.

Present Value or PV represents the amount of money a rational person would pay *today* to receive the annuity payments in the future. It is easy to compute PV for a single future payment. PV for the entire stream of payments (the annuity) is the sum of the present values of each payment.

We have these formulas.

$$FV = p \frac{(1+i)^n - 1}{i}$$

$$PV = p \frac{1 - (1+i)^{-n}}{i}$$

Amortization

level payment

In loan problems, bankers make their calculations so the value of each loan payment remains constant—the payments are *level*. The formulas for present value ensure that the level periodic payments represent sums of which part represents interest due the bank and the remainder represents a portion of the principal. The varying portions of interest and principal shift slightly from payment to payment in such a way that

1. The payments remain level.

2. Once the final payment is made, there will be *no* more principal to pay off.

amortization
self-liquidating loans

This process whereby the principal is paid back to the lender during the life of the loan is called *amortization*. Amortized loans are also called *self-liquidating loans*.

mortgage

A variety of consumer loans are designed to be self-liquidating, but the most important example is the so-called *mortgage* loan used for the purchase of houses and land. The borrower pledges the land and house to the lender as security for the loan, so in

default

case the borrower cannot make the loan payments (the loan is in *default*), the banker can seize the pledged property. Then, the banker may sell it to recover the money owed to the bank. The vast majority of mortgages involve level monthly payments. The word

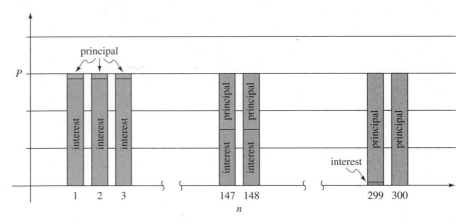

FIGURE 10.7
When level payments represent amortization of a loan principal, each payment represents a part that is principal and a part that is interest. The bar graph shows this variation for a *mortgage* loan with a life of 25 years.

mortgage itself comes from the Old French phrase meaning *dead pledge*. (The word *amortization* also contains the root word meaning "dead".)

Amortization loan problems are similar to other present value problems. Often, we need the value of the monthly payment given the remaining conditions of the loan. We merely solve for p in the formula for present value:

$$p = \frac{PV}{A(n, i)}$$

Example 14   The Rodriguez family has an opportunity to purchase a certain house for $195,000, well below its market value. If they put up a down payment of $45,000, their bank will finance the balance for them in the form of a 25-year mortgage loan at 10 percent. What will the monthly payments be for this loan? How much interest will the Rodriguezes pay over the life of the mortgage?

Solution

The difference between the purchase price and the down payment, $150,000, is the amount to finance. Under the bank's terms, $n = 25 \times 12 = 300$, and $i = 0.10/12 \approx 0.00833$. Therefore,

$$\frac{1}{A(300, 0.00833)} = 0.009087$$

and the monthly payments will be

$$150,000 \times 0.009087 = \$1,363.05$$

Over the course of 25 years, the Rodriguez family will have made total monthly payments to the value of

$$300 \times \$1,363.05 = \$408,915.00$$

of which only $150,000 is principal. Therefore, the total interest paid to the bank is

$$408,915 - 150,000 = \$258,915$$

Skill Enhancer 14 What would the monthly payments be for a $100,000 mortgage at 8 percent? The life of the mortgage is 25 years.

Answer in Appendix E.

In long-term self-liquidating loans, such as the mortgage of Example 14, the total interest paid the lender substantially exceeds the principal borrowed. This is certainly the case in Example 14.

_____ Exercises 10.3 _____

Exercises 1–6 describe terms for various annuities. Use present value analysis to decide upon the maximum price you would pay for the privilege of receiving these annuities. Assume all payments are $500.

1. Monthly payments for 5 years. Interest is 12 percent per annum.

2. Weekly payments for 26 weeks. Interest is 10 percent per annum.

3. Monthly payments for 20 years. Interest is 8 percent per annum.

4. Semiannual payments for 15 years. Interest is 8 percent per annum.

5. Monthly payments for 18 months. Interest is 10.5 percent per annum.

6. Weekly payments for 13 weeks. Interest is 10.4 percent per annum.

Each of Exercises 7–12 gives the monthly payment, the annual interest rate, and the term of several loans. In each case, what is the loan amount? Write the amount to the nearest penny.

7. 147.97; 9 percent; 60 months

8. 200.02; 6 percent; 60 months

9. 100.00; 9 percent; 48 months

10. 150.00; 9 percent; 48 months

11. 100.00; 6 percent; 60 months

12. 100.00; 9 percent; 72 months

Each of Exercises 13–18 presents an amount to be borrowed from a bank, the annual interest rate that is in effect, and the term of the loan. Calculate the monthly payments to the nearest penny.

13. $2,000; 9 percent; 48 months

14. $10,000; 8 percent; 60 months

15. $100,000; 8 percent; 60 months

16. $9,000; 12 percent; 48 months

17. $9,000; 9 percent; 48 months

18. $5,000; 9 percent; 36 months

Exercises 19–26 list the prices for various houses, followed by terms under which a bank will grant a mortgage. All interest rates given are *annual* rates. The amounts given are the amounts to be financed. All mortgage payments will be made on a monthly basis. Calculate the monthly mortgage payments.

19. $200,000; 15 years at 10 percent.

20. $100,000; 25 years at 12 percent.

21. $125,000; 25 years at 10.4 percent.

22. $225,000; 30 years at 8 percent.

23. $75,000; 12 years at 8 percent.

24. $150,000; 30 years at 12 percent.

25. $175,000; 30 years at 10 percent.

26. $200,000; 25 years at 15 percent.

Applications

27. *(Banking)* A bank loan officer is deciding on which of two loan applications to approve. The terms of loan *A* require the borrower to pay 48 monthly payments of

$250 at an annual interest rate of 12 percent. The terms of loan *B* call for the borrower to pay 24 bimonthly payments (every other month) of $500 at the same annual interest rate. All other things being equal, which one should the loan officer approve?

28. Solve Exercise 27 if the payments for loan *B* are only $450, but the annual interest rate is 13 percent.

29. *(Retirement Planning)* Carlos Goldstein began working for the Still River Police Department when he was 20 years old, and retired when he was 50. At that time, he had accumulated $250,000 in savings, which he used to purchase a 20-year annuity from the Foresightful Insurance Company. He expects monthly payments, and expects his money to work for him at 10 percent annually. If the monthly payments are constant, how large can he expect those payments to be?

30. *(Retirement Planning)* Sheila Rivera is planning for her financial future. She seeks to retire and live for 25 years with monthly income from an annuity that she will purchase upon retirement. If these annuities work at 10 percent per annum, what will the purchase price of the annuity be if she expects monthly payments of $500?

31. *(Home Economy)* A portable compact disk player costs $159.95 if purchased outright. However, the store allows customers to pay $40 down and make payments for 5 months. What should the payments be if the annual interest rate is 15 percent?

32. *(Home Economy)* A set of living room furniture costs $4,995 when purchased outright. If a customer puts down a deposit of $1,000, the balance can be paid off in 24 months. If the annual interest rate is 20 percent, what should the monthly payments be?

33. *(Home Economy)* A young couple buys a new refrigerator. They give the store a $250 down payment and make monthly payments of $100 for the next twelve months. This financing corresponds to an annual interest rate of 19.6 percent. What would the refrigerator cost if purchased outright?

34. *(Business Administration)* The United Copper Company needs a new smelter for its ore refining operations. The manufacturer requires a down payment of $100,000 together with monthly payments of $15,000 for the next 5 years. This corresponds to financing at an annual rate of 20 percent. What is the total cost of the smelter?

35. *(New York State Lottery)* The first prize in the New York State Lottery is often advertised as being "one million dollars" or "two million dollars." In actuality, this is not so, for the first prize represents a 20-year *annuity*

to the winner. For a first prize of $1,000,000, the state makes annual payments of $50,000.
  (a) Analyze this prize from the standpoint of the *state*. If the state is able to invest its money so as to earn 9.5 percent annually on its principal, what is the *present value* of this prize?
  (b) Now analyze this prize from the point of view of the winner. If the winner promptly invests each payment in a money market account that earns 8.5 percent annually on its deposits, what is the *value* of this annuity?

36. Solve Exercise 35 if both the state and the winner earn 10 percent annually on their money.

37. *(Real Estate)* The Ward family is buying a house for which they will finance $100,000 of the cost with a 25-year mortgage at 18 percent interest.
  (a) What will the monthly payments be?
  (b) What portion of the first month's payment represents interest, and what portion represents principal?

W **38.** Suggest a method whereby a real estate agent could *estimate* the monthly payment of a mortgage. This method would increase in accuracy as the interest rate becomes higher and higher. Use the results of the second part of Exercise 37 to guide your thinking.

39. *(Real Estate)* You will purchase a house by obtaining a mortgage loan for $100,000. The interest rate is 12 percent per annum, the payments will be monthly, but you have a choice of a 25-year or a 30-year term.
  (a) What will the monthly payments be if you choose a 25-year term?
  (b) What will the monthly payments be if you choose a 30-year term?
  (c) How much *less* interest will you pay over the life of the loan if you choose the 25-year term mortgage?

40. Solve Exercise 39 if the annual interest rate is 10 percent.

Calculator Exercises

41. Is it possible to arrange a loan in such a way that monthly payments are exactly $125.00? The amount of the loan is $5,000, and the annual interest rate is 9 percent.

42. B. C. Dull, a student, likes writing checks for even amounts of money (no pennies). He needs to borrow $7,500 for a new car, and would like to make monthly payments of exactly $150.00. The interest rate is 9 percent per annum. Is it possible to choose a term for the loan which would make this possible? Discuss the possibilities.

Section 10.4 APPLICATION: INFLATION FINANCIAL PLANNING

inflation Inflation is the loss of purchasing power of money, and we express it as an *annual inflation rate*. No one really knows all the underlying causes of inflation, but the results can be ruinous to savings. Even at a "modest" inflation rate of 3 percent per year, a family's nest-egg will decline in purchasing power by one-half in 24 years. At an inflation rate of 6 percent annually, this decline occurs in only 12 years, and at 8 percent per year, this steep decline occurs in about 9 years. We examined simple inflation problems in Chapter 1; in this section, we examine situations in which sums of money are subject both to erosion from inflation *and* to growth from interest.

The Inflated Value of Money

purchasing power We measure the value of money by its *purchasing power*. Gasoline cost about one dollar per gallon at the pump in 1987. Twenty years earlier, the same gallon cost about 30 cents. This suggests that 30 of the cents used in 1967 have the same *purchasing power* as one 1987 dollar. The only way to compare dollar amounts at different times is by comparing the purchasing power of each sum.

rate of inflation We measure the *rate of inflation* g as the relative difference in purchasing power between the value p of some money at some time and its inflated value q at some later time. Specifically,

$$(10.2) \qquad g = \frac{p - q}{q}$$

We express inflation as a percentage per time interval. The time interval should be as short as possible, but practically it is not possible to measure inflation in periods shorter than a month. Given some rate of inflation and a value of goods or money at the start of some time, we can solve Equation (10.2) for the inflated value q.

Inflated Value

$$q = \frac{p}{1 + g}$$

where

p is the original sum.

q is the inflated value of p at the end of some period of time.

g is the inflation rate per period.

Inflation does not affect all goods at the same rate. The rate of inflation is an average. That is, on the average, the value of goods and services seems to increase annually by the annual rate of inflation. Inflation will affect some goods more than others.

Example 15 On March 1, milk cost $1.90/gal. On July 1, a gallon of milk now costs $1.96. Use this data to estimate the rate of inflation g.

Solution

The price difference across the 4-month interval is $0.06. Equation (10.2) for g yields

$$g = \frac{0.06}{1.96} = 3.06 \text{ percent per 4 months}$$

When stating this result, make sure to say that it is for 4 months. Quote this rate as "the 4-month inflation rate," or "the rate per 4 months," or "the rate per third of a year," or something equivalent.

Skill Enhancer 15   The price of heating oil went up $0.10 per gallon in a 6-month period of time. It had cost $0.96 per gallon. What is the rate of inflation?

Answer in Appendix E.

Example 16
(a) Estimate the inflated value of $100 after one year if inflation persists at a rate of 4 percent per year.
(b) What is the inflated value after 2 years?
(c) What is the inflated value after n years?

Solution

(a) The inflated value q is given by

$$q = \frac{100}{1 + 0.04} = \$96.15$$

(b) Finding the inflated value of $100 after *two* years is equivalent to finding the inflated value of $96.15 after one year:

$$q = \frac{96.15}{1.04} = \$92.45$$

Alternatively, we may note that

$$q = \frac{100/1.04}{1.04} = \frac{100}{1.04^2}$$

which yields the same answer.
(c) An easy generalization of the previous part yields the result

$$q = \frac{100}{1.04^n}$$

for the inflated value of $100 after n years.

Skill Enhancer 16
(a) What is the inflated value of $500 after one year of inflation at 5 percent?
(b) What is the inflated value of $500 after two years of inflation at 5 percent?

Answers in Appendix E.

Example 17   It was always Tom's habit to put aside pennies that he received in change. At age 18 in 1986, he left home for college, having accumulated $400 in pennies. The pennies sat in rolls on the top shelf of his closet. He returned home 4 years later. For the first 2 years, the annual rate of inflation was measured to be 5 percent per year. During the third year, this rate increased to 7 percent per year. During the last year, inflation had moderated to 3 percent per year. Discuss the real value of his penny collection upon his return home from college at the end of his fourth year away.

Solution

We use the formula for inflated value to determine the decline in value. A chart such as the following helps keep track of the calculations.

Year	Annual Inflation Rate	Value at Year End
1986		$400.00
1987	5 percent	$\dfrac{400}{1.05} = \$380.95$
1988	5 percent	$\dfrac{380.95}{1.05} = \$362.81$
1989	7 percent	$\dfrac{362.81}{1.07} = \$339.07$
1990	3 percent	$\dfrac{339.07}{1.03} = \$329.20$

The collection's purchasing power has been eroded by

$$400 - 329.20 = \$70.80$$

its inflated value is only $329.80 expressed in 1986 dollars.

Skill Enhancer 17   What is the value of $500 after three years of inflation? The rates of inflation for these three years are 3 percent, 4 percent, and 5 percent.

Answer in Appendix E.

Simultaneous Growth and Inflation

real rate of growth

Money is often subject to growth and inflation at the same time. That is, money may earn interest during some time interval at the same time that the principal and interest both undergo inflation. A *real rate of growth* k depends on both interest and inflation to measure the growth in purchasing power of the money. If p and p' are the values of purchasing power at the beginning and end of this period, then we define k by

(10.3) $$p' = p(1 + k)$$

We can derive an alternate expression for p'. Suppose an investor places some amount p with some financial institution advertising an interest rate i. This constitutes a promise on the part of the institution to pay back an amount $p(1 + i)$ at the end of the period if it receives the investor's money p at the beginning. During this time, though, money is subject to inflation at a rate g. The purchasing power p' of $p(1+i)$ will be eroded by this inflation, so that

(10.4)
$$p' = \frac{p(1+i)}{1+g}$$

Equating Equations (10.3) and (10.4) yields an expression for growth:

(10.5)
$$1 + k = \frac{1+i}{1+g}$$

Newspapers and other media usually assert that real growth is the difference between inflation and interest, an assertion which is not precisely true. Equation (10.5) gives the proper relationship. In these problems, the interest rate i is more precisely an *actual* or *nominal rate of interest*.

actual rate of interest
nominal rate of interest

> **Example 18**   An investor wants her investments to grow in purchasing power by 10 percent per year. If the investor feels that inflation will persist at an annual rate of 4 percent per year, what nominal rate of interest must she demand?
>
> *Solution*
>
> Substitute $k = 0.1$ and $g = 0.04$ in Equation (10.5) and solve for i:
>
> $$1.1 = \frac{1+i}{1.04}$$
> $$1 + i = 1.1440$$
> $$i = 14.40 \text{ percent}$$
>
> She must demand a nominal rate of 14.4 percent.

> **Skill Enhancer 18**   Inflation seems to have leveled off at about 3 percent per year. I want a retirement account to grow in purchasing power by 12 percent per year. What nominal rate of interest must I demand?
>
> *Answer in Appendix E.*

The Annuity Equation

We seek the derivation of the present value of an annuity where the annuity payments are adjusted to reflect the effects of inflation. We want the present value PV' of such an annuity, where the primed accent mark $'$ on PV' reminds us we are expressing PV' in units of currency that are standard at the time the annuity begins. That is, the dollars we choose for PV' have the same purchasing power as those in force at the start of the annuity. (Due to inflation, the value of dollars, or British pounds sterling, or Japanese yen will most likely decline in purchasing power with time.)

Assume that inflation proceeds at a rate g per interest period, and assume that this rate remains constant. Suppose further that we *adjust* each annuity payment so each payment is constant in purchasing power. To maintain a level purchasing power, the actual value of the payment will increase bit by bit from payment to payment. Let p

represent this constant purchasing power value. Then the actual dollar amount of the ith payment p_i will be

$$p_i = p(1+g)^i$$

for values of i running between 1 and n (there are a total of n payments during the life of the annuity).

The present value V_i of each of these payments is

$$V_i = [p(1+g)^i](1+i)^{-i}$$
$$= p(1+k)^{-i}$$

That is, we can compute the present value for each payment subject to real growth k in the same way we compute the present value of each payment subject to nominal growth i by simply *replacing i by k in the present value formula.* In the same way, the formula for the value of an annuity will continue to hold *provided we replace i by k in that formula.*

Present Value of an Inflation-Adjusted Annuity

$$PV' = p \cdot A(n, k)$$

Here,

PV' is the present value of the annuity; each payment is adjusted to maintain the same purchasing power.

p is the actual value of the first payment.

k is the real rate of growth per payment period.

n is the number of payments.

$A(n, k)$ $= \frac{1-(1+k)^{-n}}{k}$.

Value of an Annuity

$$FV' = p\frac{(1+k)^n - 1}{k}$$

where

FV' is the future value of the annuity (purchasing power).

k is the real rate of growth per payment period.

n is the number of payments.

p is the value of the first payment (and consequently the purchasing power of each subsequent payment).

Equivalently, the primed quantities are the value and present value of annuities in an economy with *no* inflation and a nominal interest rate k.

Example 19   According to one simple model of the economy, inflation will persist at a rate of 5.5 percent annually for 20 years. One student of this model wants to buy a 20-year annuity that will pay out the equivalent of 12,000 current dollars per year.

What should he expect to pay for this annuity? The money will be invested in accounts guaranteeing the fixed (nominal) rate of 10 percent per year.

Solution

First compute the rate of growth from Equation (10.5).

$$1 + k = \frac{1.1}{1.055} = 1.042654$$

so

$$k = 4.2654 \text{ percent per year}$$

Next, compute the function $A(n, k)$:

$$A(20, 0.042654) = \frac{1 - 1.042654^{-20}}{0.042654} \approx 13.276$$

Then, from the annuity formula,

$$PV' = 12,000 \times A = (20, 0.042654) = \$159,317.40$$

is the price to pay.

Skill Enhancer 19   Another model predicts an inflation rate of 4 percent over the next quarter century. A student is interested in a 15-year annuity paying out the equivalent of $20,000 per year. The money will be invested in accounts guaranteeing the nominal rate of 10 percent per year.
(a) What is the rate of growth?
(b) Compute $A(n, k)$ for this problem. What is n?
(c) What is the present value of this annuity?
(d) What is the maximum this student should be willing to pay for this annuity?

Answers in Appendix E.

Example 20   One young couple plans to borrow as much money as they can as a mortgage loan, which they will use to buy a house. The monthly payments are $500. One bank has an innovative loan program—it will adjust the monthly payments so that the monthly payments have the same purchasing power throughout the life of the loan. This bank deals in rates of growth rather than in actual interest rates. It anticipates an annual rate of inflation of 3 percent. Assume that all these loans are 20-year loans. On March 1, the couple is told that the rate of growth for their mortgage loan will be 10.2 percent. If they wait a month, the rate will drop to 10.1 percent. They are anxious to buy their new house, and are tempted to forgo waiting, reasoning that 0.1 percent will have little effect on their borrowing power.
(a) Is it worth their while to wait the extra month?
(b) How would the amount they can borrow compare with a conventional mortgage loan, where the payments remain fixed for the life of the loan? Use the annual rate of 10.2 percent for this question.

Solution

(a) To assess these alternatives, we will compute the maximum amounts that the couple can borrow; these are the present values of the income stream of mortgage payments. In all cases, $n = 12 \times 20 = 240$. For the first loan, the monthly growth rate is $0.102/12 = 0.008500$. For the second, it is $0.101/12 = 0.008417$. We have

$$A(240, 0.008500) = 102.217$$

and

$$A(240, 0.008417) = 102.914$$

The amounts of the two loans are PV_1' and PV_2':

$$PV_1' = 500 \times 102.217 = \$51,108.50$$
$$PV_2' = 500 \times 102.917 = \$51,457.11$$

As the interest rate declines, so will the present value of the loan increase. In this case, the difference in values between the two rates is \$348.61. If the couple's living costs for the month before the interest rate decline goes into effect are greater than this amount, they should probably forgo the modest interest rate decline.

(b) Given an inflation rate of 3 percent and a growth rate of 10.2 percent, we can solve for the actual interest rate from the formula

$$1.102 = \frac{1+i}{1.03}$$

therefore

$$1 + i = 1.102 \times 1.03 = 1.1351$$

so that

$$i = 13.51\%$$

On a fixed payment scheme, the maximum amount the couple can borrow is the present value of an annuity with 240 monthly payments and an interest rate of $i = 0.1351/12$. Here,

$$A(240, 0.1351/12) = 82.7750$$

and the maximum amount they can borrow is

$$500 \times 82.7750 = \$41,387.50$$

about 20 percent less.

Skill Enhancer 20   Reexamine the facts of this example. This time, let's say the couple can afford \$750 per month. At the time the story opens, the rate of growth is 9.5 percent, but it will drop to 9 percent by the next month. The life of a mortgage is 25 years ($n = 300$).

(a) What are the two monthly growth rates?

(b) Compute the present values of income streams for the two interest rates.

(c) What is the difference between them? Do you think it is worth waiting the month for the interest rate drop?

(d) Using the current growth rate of 9.5 percent, what is the actual interest rate? What is the monthly interest rate i?

(e) What would be the maximum amount the couple could afford using a conventional mortgage?

Answers in Appendix E.

Example 21 What is the *value* to the bank in Example 20 at the start of the loan of the 20-year stream of mortgage payments? Each payment has a purchasing power of $500 expressed in the dollars current at the start of the loan. The bank invests each payment in such a way as to generate a real rate of growth of 10.2 percent on this money.

Solution

Here, $n = 240$, and the monthly growth rate is $k = 0.0085$, so the value becomes

$$FV' = 500 \cdot \frac{1.0085^{240} - 1}{0.0085}$$
$$= 500 \cdot 779.36 = \$389,680.00$$

The value (future value) of this annuity is $389,680.00. Remember, this calculation takes the purchasing power of money into account.

Skill Enhancer 21 Each payment of a mortgage computed like the one in Example 20 has a purchasing power of $750 expressed in the dollars current at the start of the loan. The bank invests each payment in such a way as to generate a real rate of growth of 9.6 percent.
(a) What is the monthly rate of growth?
(b) What is the value to the bank of this income stream?

Answers in Appendix E.

Interesting financial planning problems involve making deposits to accumulate a large fund for a specified period, at the end of which, payments will be withdrawn for another period of time.

Example 22 Mr. and Mrs. Smart have a newborn son in 1986 for whom they wish to plan a college education. At the time of the birth, yearly college costs are as high as $18,000 in many universities. The parents estimate that college costs will escalate by an amount sufficient to offset the effects of inflation. The Smarts also feel that if they invest an annual amount in a reasonable stock market portfolio, they can achieve a real rate of growth of 10 percent on the value of their investments for 18 years. At that time, they expect to begin withdrawing annual sums of 18,000 1986 dollars for 4 years. The money continues to earn interest at a 10 percent real rate of growth during that time. The Smarts will invest an annual sum with a constant value of p 1986 dollars. What is the smallest value of p consistent with these assumptions that will enable them to accomplish their goal?

Solution

Let FV' be the value of the 18 years worth of deposits by the Smarts. FV' depends on $n = 18$, $k = 0.1$, and the as-yet-unknown annual deposit p.

$$FV' = p\frac{1.1^{18} - 1}{0.1} = 45.5992p$$

Let PV' be the present value of the 4 years worth of tuition and other college costs. PV' depends upon the annual tuition $18,000, and the values $k = 0.1$, $n = 4$.

$$PV' = 18,000\frac{1 - 1.1^{-4}}{0.1} = \$57,057.58$$

The payment p is determined when the value of the deposits made while their son matures exactly equals the present value of all tuition payments and anticipated interest.

$$45.5992p = 57,057.58$$

$$p = \$1,251.28$$

The Smarts should make annual deposits whose value is $1,251.28 in 1986 dollars.

Skill Enhancer 22   By the time their daughter is born in 1990, college tuition has risen to about $21,000. Once again, the Smarts seek to determine the value of p so that if they deposit this amount annually for 18 years in an account with a real rate of growth of 10 percent, they will be able to withdraw an amount equal in purchasing power to $21,000 1990 dollars for four years.
(a) Express the future value FV' of 18 years of deposits in terms of p.
(b) Express the present value PV' of 4 years of tuition.
(c) Solve for p.

Answers in Appendix E.

_____ *Exercises 10.4* _____

In Exercises 1–6, find the indicated rate using the given information.

1. $i = 0.10$; $g = 0.03$; $k = ?$

2. $i = 0.075$; $g = 0.025$; $k = ?$

3. $g = 0.022$; $k = 0.085$; $i = ?$

4. $g = 0.096$; $k = 0.001$; $i = ?$

5. $k = 0.07$; $i = 0.100$; $g = ?$

6. $k = 0.09$; $i = 0.12$; $g = ?$

In Exercises 7–14, compute the present value of an annuity of 100 monthly payments, each of which varies so that the purchasing power of each remains constant. You are given the value p of the first payment, and the *monthly* rate of growth k.

7. $p = \$450$; $k = 0.10$ percent

8. $p = \$900$; $k = 0.3$ percent

9. $p = \$100$; $k = 0.8$ percent

10. $p = \$75$; $k = 0.45$ percent

11. $p = \$1,000$; $k = 0.01$

12. $p = \$750$; $k = 0.0075$

13. $p = \$1,250$; $k = 0.01667$

14. $p = \$667$; $k = 0.01667$

In Exercises 15–22, compute the values of the annuities. Each annuity consists of 72 monthly payments, each payment is level in terms of purchasing power, and k is the annual rate of growth.

15. $p = 1,000$; $k = 12$ percent

16. $p = \$1,250$; $k = 12$ percent

17. $p = \$1,666.66$; $k = 10$ percent

18. $p = \$900$; $k = 11$ percent

19. $p = \$700$; $k = 11$ percent

20. $p = \$850$; $k = 7.5$ percent

21. $p = \$1,250$; $k = 0.084$

22. $p = \$25.00$; $k = 0.072$

23. Show that an equivalent formula for k is $k = \dfrac{i - g}{1 + g}$.

24. (a) Compute $i - g$ and k for $i = 0.002$, $g = 0.001$.
    (b) Compute $i - g$ and k for $i = 0.02$, $g = 0.01$.
    (c) Compute $i - g$ and k for $i = 0.2$, $g = 0.1$.

W 25. The text asserted that the idea that k was simply the difference between i and g is a misconception. Use the results of Exercise 24 to express your opinion as to when the approximate formula $k = i - g$ will be useful.

W 26. Opinions are often heard to the effect that since wages rise at about the rate of inflation, we don't have to work any harder to buy goods at inflated prices, and so inflation is not so bad after all. Who (if anyone) would be hurt by the effects of inflation, and why?

Applications

27. Eunice and Fred Potts won the New York State "Million Dollar" Lottery. Official state propaganda leads them to believe that they have won one million dollars. What they actually have won is a 20-year annuity. Each annual payment is $50,000 in 1988 dollars, the year they won. The state's funds are invested to show an 8.4 percent annual real growth rate. The Potts take their winnings and invest in a stock mutual fund (a type of investment) that has historically shown a 12 percent real growth rate.
    (a) What is the actual present value of the Potts's prize?
    (b) What is the value of the Potts's investments?

28. Solve Example 22 if the real rate of growth is only 7 percent per annum.

29. Solve Example 22 if the couple starts the plan when their son is 3 years old.

30. Solve Example 22 if the real rate of growth is 7 percent annually *and* the couple begins the savings plan when their son is 3 years old.

31. *(Retirement Planning)* A couple starts a 30-year investment plan in 1975. Their first annual contribution was $1,000, and they increase each subsequent contribution by an amount proportional to the past year's inflation. Their contribution in 1985 was $2,000, and their goal is to accumulate a nest egg of 80,000 1970 dollars.
    (a) If inflation is constant, what is the real growth rate?
    (b) At the present rate of contribution, how large will the couple's nest egg be at the end of 30 years?
    (c) To what should they increase or decrease this contribution to meet their target?
    (d) Estimate how many actual dollars in the year 2000 the sum of 80,000 1970 dollars will be.

CHAPTER REVIEW

Terms

interest	principal	rate of interest
interest period	simple interest	compound interest
period of compounding	effective rate of interest	annuity
annuity period	ordinary annuity	annuity due
life of an annuity	future value	value
interest-free annuity	interest-bearing annuity	geometric series
sinking fund	present value	present value of an annuity
level payment	amortization	self-liquidating loans
mortgage	default	inflation
purchasing power	rate of inflation	real rate of growth
actual rate of interest	nominal rate of interest	

Key Concepts

- Interest is the rental fee borrowers pay for the use of some-one else's money. The easiest way to compute interest is via simple interest, although this is rarely encountered in real-life financial situations. Far more common is compound interest.
- The effective rate of interest is the equivalent simple rate of interest needed to earn in one year the interest that actually accrues as a result of compounding. Effective interest rates help consumers shop for the most economical loan terms.
- Problems in which the growth or decrease of one quantity depends on its value in a previous period of time can often be well modeled using difference equations.
- An annuity is a stream of payments. An annuity has a value and a present value. The present value can be thought of as the sum of money a person would be willing to pay

today for the privilege of receiving the annuity payments in the future. To evaluate the present value or the value of an annuity, one must know the conditions under which money earns interest.

- For a large and important class of loans, the periodic payments are constant and so computed that the payments amortize the principal of the loan. That is, each payment represents a portion of the principal and the interest due on the currently outstanding principal. The size of the payment is carefully adjusted so that the loan is completely paid off at the time the final loan payment is made.
- All of these concepts have to be adjusted slightly to take account of inflation, which is the tendency for money to have less and less purchasing power as time goes by.

Review Exercises

Compute simple interest for loans using the terms given in Exercises 1–4.

1. $725; 18 months at 6.5 percent
2. $1,800; 24 months at 10.7 percent
3. $6.25; 9 months at 12 percent
4. $1,400; 13 months at 11 percent
5. What is the total when $1,400 is invested at 10 percent compounded quarterly for 18 months?
6. What is the total when $900 is invested at 9.6 percent compounded monthly for 36 months?
7. What is the total *interest* when $400 is invested at 7.5 percent compounded semiannually for 24 months?
8. What is the total *interest* when $825 is invested at 6.8 percent compounded quarterly for 17 months?

Each of Exercises 9–12 describes the terms of an annuity. Compute the value of each annuity.

9. $p = \$300$; $n = 14$; $i = 0.01$
10. $p = \$625$; $n = 7$; $i = 0.0015$
11. $p = \$1$; $n = 4$; $i = 0.1$
12. $p = \$1,000$; $n = 22$; $i = 0.02$
13. Compute the present value of the annuity described in Exercise 9.

14. Compute the present value of the annuity described in Exercise 10.
15. Compute the present value of the annuity described in Exercise 11.
16. Compute the present value of the annuity described in Exercise 12.

In Exercises 17–20, compute the monthly payment for the following sinking funds if we need to accumulate $50,000 after n interest periods. The interest rate per period is i.

17. $n = 15$, $i = 0.01$
18. $n = 12$, $i = 0.005$
19. $n = 20$, $i = 0.0075$
20. $n = 12$, $i = 0.001$

In Exercises 21–24, assume the payments are graduated so that the payments are constant in terms of purchasing power. If you interpret the indicated rate as a real rate of growth, compute the inflation-adjusted present value of the annuities.

21. Compute the present value of the annuity described in Exercise 9.
22. Compute the present value of the annuity described in Exercise 10.
23. Compute the present value of the annuity described in Exercise 11.
24. Compute the present value of the annuity described in Exercise 12.

Applications

25. *(Financial Planning)* Reeney's Resume Service recently bought a microcomputer that the Service will use for word processing. The Service knows that it will need a new computer in 3 years to keep up with changing technology. It anticipates that this new system will cost $6,500 in 3 years. If the company savings account yields 8.5 percent compounded quarterly, compute the size of the payment the company will need to deposit quarterly so as to have exactly $6,500 in 3 years. What is the value of these payments?

W 26. Solve Exercise 25 if the company decides to wait 6 years before buying a new computer system. Why isn't this answer exactly one-half or exactly twice the answer to Exercise 25?

27. *(Retirement Planning)* Susanne Greengold places $1,000 in an IRA on an annual basis. The IRA is a money market fund that guarantees a minimum return of 11 percent compounded annually. If she plans on making deposits for 30 years, what will be the value of her IRA at the time of retirement?

28. *(Real Estate)* The Cleavers are saving for a house. They anticipate coming up with a $75,000 down payment, and feel that they can afford $1,250 a month in mortgage payments. If mortgages are currently being of-fered at 10 percent with 30-year payback periods, what is the highest price house that the Cleavers can afford?

29. *(Retirement Planning)* Ben Lee User began thinking of his retirement in 1990, when he was 30 years old. At that time, he began making payments into a special savings account that yields 8.4 percent compounded monthly. He plans on making monthly deposits for 35 years, until he is 65 years old, when he will retire, and withdraw $1,500 per month for 15 years. What should his monthly deposits be during his working career?

30. Solve Exercise 25 to take inflation into account. The current system the company would like to buy costs $5,000. It doesn't know what the equivalent system will cost in 3 years, but will assume that its price will rise in line with inflation, which the company assumes to be 1 percent per quarter for the next 3 years at least. Now, compute the quarterly payments the company will need to make to be able to purchase this system in 3 years. Assume that the quarterly payments are to be constant in terms of purchasing power.

31. Solve Exercise 29, under these additional assumptions. Inflation persists at a monthly rate of 0.4 percent during this entire period, and his deposits will be constant in terms of purchasing power. He expects his monthly retirement withdrawals to be $1,500 in terms of 1990 dollars. What should be the size of his deposits?

11

INTRODUCING DIFFERENCE EQUATIONS

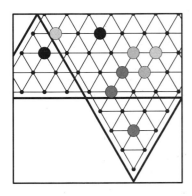

INTRODUCTION

One of the most important methods for analyzing a natural process involves examining the process at the times when the process undergoes change.

Difference equations provides one means for examining processes of growth and decay on the basis of change without requiring a background in calculus. It is the aim of this chapter to provide an introduction to these methods.

We will begin with a discussion of new concepts with appropriate new notation. What does *change to a system* really mean? We follow this with examples of the use of difference equations, largely in the field of finance. As we will see, the notation of difference equations needs some further manipulation to get the growth of a quantity in a more familiar form.

When working with difference equations, it is often important to deduce the *long-term* behavior of the equation. It is possible to do this without solving the equations explicitly.

Finally, we will examine applications of difference equations to finance and the natural sciences to demonstrate the power of these techniques.

We have implied that difference equations are useful for their own sake, but that alone does not justify their inclusion in this volume. Difference equations provide an alternative way to derive almost all the results of the preceding chapter on finance and money. They also provide another technique for examining quantities with a view towards predicting their long-term change and thereby optimizing their use, and that provides the ultimate justification for this material.

Section 11.1   DIFFERENCE EQUATIONS AND CHANGE

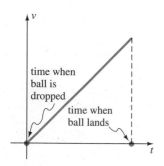

FIGURE 11.1
The speed of a falling ball changes at every instant between the time it is dropped and the time it lands.

When do things change? Many physical quantities change at each moment during the period when we study them. For example, if we drop a bowling ball from the roof of a building, we are most likely interested in the behavior of the ball from the moment we let it go to the moment it crashes on the ground. At each instant during its fall, the downward speed of the ball is different than it is at any other instant. Its speed changes constantly as it falls. Many of the quantities you may have studied in a physics course change constantly during the periods in which they are interesting.

In this text, though, we have seen a few examples of quantities that do *not* change at each instant. The balance in an interest-bearing account does not change all the time, but only at the times when the interest is posted to the account (or when a new deposit is made). Let us suppose that the depositor makes no further deposits after having opened the account. After the first posting of interest, we have a balance y_1. After the second, we have new balance y_2, and so on. Using this notation, it makes sense to refer to the original deposit as y_0. Thus, any account is completely described by stating the sequence of balances $y_0, y_1, \ldots, y_n$, and so on for that account. This point is made in Figure 10.2 on page 503, where we see that the value of the account is constant until it takes a sudden "leap" at the end of each interest period.

At the time interest is posted, how does the balance change? Suppose a bank account earns interest at a 12 percent annual rate (1 percent per month) compounded monthly. If the balance after 3 months is y_3, then next month's balance, y_4, is y_3 (last month's balance) plus whatever interest this money earns. But the earned interest is 1 percent of that—$0.01y_3$. The equation describing this change is

$$y_4 = y_3 + 0.01y_3$$

or

$$y_4 = 1.01y_3$$

There is nothing special about the change from y_3 to y_4. More generally, therefore, we may write

$$y_{n+1} = 1.01y_n$$

which applies to all values of n greater than zero. The equation that starts the process applies to the opening of the account:

$$y_0 = p$$

where p is the amount of the initial deposit. The balance after any month is completely specified by stating the equation relating successive monthly balances to each other and the amount used to open the account, y_0. In general, the *difference equation* for any system consists of a pair of equations—the general rule by which y_n becomes y_{n+1}, and the specific rule for y_0. The difference equation for this example is the pair of equations

difference equation

$$y_0 = p$$
$$y_{n+1} = 1.01y_n \qquad \text{for } n > 0$$

Example 1 The size of some process in one interval is 75 percent of its size in the previous interval less 5 percent of the original size h. Express this rule as a difference equation.

Solution

The difference equation is

$$y_{n+1} = 0.75y_n - 0.05h$$
$$y_0 = h$$

Skill Enhancer 1 The size of some process in one interval is 120 percent of its size in the previous interval plus 1 percent of the original size h. Express this rule as a difference equation.

Answer in Appendix E.

Example 2 In order to set up an account with the electric company in Somerville, Massachusetts, a new customer has to give the company a good faith deposit of $100. This money earns simple interest at an annual rate of 8 percent, credited quarterly. At the end of a year, provided the customer is in good standing, this money plus all accumulated interest is returned. Set up the difference equation for this system. The variables in this equation should represent the balance in the customer's account during each quarter.

Solution

Let y_0 be the initial deposit:

$$y_0 = 100 \qquad\qquad (11.1)$$

The accumulated total y_{n+1} in any quarter is the total from the previous quarter, plus the interest. Since it is simple interest, the amount earned is 2 percent (the interest rate per quarter) times the original amount y_0.

$$y_{n+1} = y_n + 0.02y_0 \qquad\qquad (11.2)$$

Equation (11.2) and the initial condition, Equation (11.1), determine the customer's balance. Note that Equation (11.2) is valid for $n = 1, 2, 3, 4$.

Skill Enhancer 2 After a rate change, the company demands a $250 deposit. This deposit now earns simple interest at 10 percent, credited semiannually. We want to set up a set of difference equations describing the customer's balance y_i in the account at the end of each 6-month period.
(a) What is the equation for y_0?
(b) What is the equation with y_n where $n > 0$?
(c) For what values of n do these solutions hold?

Answers in Appendix E.

Example 3 Harriet Fargo deposits $2,000 a year into an IRA (Individual Retirement Account) with a certain stock mutual fund. Historically, the fund earns interest as if it were an account earning 10.5 percent annually and compounded annually. Assuming this historical performance, what difference equation system describes the growth in value of Harriet's IRA?

Solution

Let y_n be the balance of her account after the nth year. Clearly,

$$y_0 = 2000$$

Assuming historical performance of interest earnings,

$$\begin{array}{ccccccc}
\text{Current} & = & \text{Previous} & + & \text{Earned} & + & \text{New} \\
\text{Balance} & & \text{Year's Balance} & & \text{Interest} & & \text{Deposit} \\
y_{n+1} & = & y_n & + & 0.105\,y_n & + & 2000
\end{array}$$

Upon simplification, this becomes

$$y_{n+1} = 1.105\,y_n + 2000$$

Skill Enhancer 3   Angel Gonzales is a big believer in saving. Each time he receives a paycheck (issued monthly), he deposits 10 percent of it immediately into an account that earns 9 percent compounded monthly. He opened the account with a deposit of $1,000. Let the variables y_i represent Angel's balance at the beginning of each month. His paycheck is for an amount of p dollars.
(a) What is the equation for y_0?
(b) What is the monthly interest rate on the account?
(c) Describe, in words, the contributions that cause the balance to change at the beginning of each month.
(d) What is the equation for y_{n+1}, $n > 0$?

Answers in Appendix E.

Example 4   Torus Tire Company has a matching company pension plan. An employee elects to belong to the plan by allowing the company to deduct a fixed sum from gross pay and depositing it directly into an account yielding interest at an annual rate of 6 percent compounded monthly. The company matches each deposit on a 50 percent basis. That is, for every dollar the employee contributes into the account, the company deposits 50 cents. What is the difference equation governing the employee's monthly balance? (Assume paychecks are distributed monthly.)

Solution

The initial balance is the first of the equal-sized monthly contributions. Let this amount be p.

$$y_0 = 1.5\,p$$

Thereafter, each monthly balance y_{n+1} is related to the previous balance by

$$\begin{array}{ccccccccc}
\text{Current} & = & \text{Previous} & + & \text{Interest} & + & \text{Employee's} & + & \text{Company's} \\
\text{balance} & & \text{month's} & & \text{earned} & & \text{contribution} & & \text{contribution} \\
& & \text{balance} & & & & & & \\
y_{n+1} & = & y_n & + & 0.005\,y_n & + & p & + & 0.5p
\end{array}$$

so that

$$y_{n+1} = 1.005y_n + 1.5p$$

is the difference equation we seek. Remember, 6 percent annually is 0.5 percent monthly.

Skill Enhancer 4   Consider the following pension plan from the stand point of difference equations. An employee chooses an amount p to be deducted from the paycheck and deposited automatically into the account. The account earns 9 percent compounded monthly, and paychecks are issued monthly. The employer supplements the employee's deposit by contributing one-third of the employee's deposit together with one-third of the interest earned since the last deposit.
(a) What is the equation for y_0?
(b) What is the monthly interest rate?
(c) What in words causes the increase in the balance at the beginning of each month?
(d) What is the difference equation describing the account balance?

Answers in Appendix E.

_____ *Exercises 11.1* _____

In Exercises 1–6, write the difference equations implied by the given rules.

1. A process is the same at all times.

2. One quantity is the negative of its value in the previous interval. Initially, it has a value of π.

3. Not only is one quantity one-half its numeric value from the previous interval, but it has the opposite sign. Its original value is 10.

4. A quantity is twice as large as it was in the previous interval, less 10 percent of this previous value. It originally had a value of -1.

5. The value of y_n is 95 percent of the value in the previous interval plus one-half of the original value (which was 4).

6. The absolute value of a quantity is 20 percent more than the absolute value of the quantity in the previous interval, to which was added the original value. The sign of the new quantity is the opposite of the value in the previous interval of time. The quantity began with a value of 1,000.

Applications

In Exercises 7–20, write the difference equation describing each situation. Make sure to specify y_0 and the general expression for y_{n+1}. Clearly state what y_n stands for in each problem.

7. A bank balance earns interest at a 15 percent annual rate compounded quarterly.

8. A bank balance earns interest at a 16 percent annual rate compounded monthly.

9. Seth borrowed money from Alan, and agrees to pay back the amount with interest. Simple interest at 10 percent per year is computed twice a month.

10. Beth borrowed money from Lisa, and agrees to pay back the amount with interest. Simple interest at 12 percent per year is computed monthly.

11. Jozefa deposits $1,000 yearly in a money market fund that earns 11 percent annually compounded annually.

12. Beatrice deposits $1,500 into a company bank account every year. The money earns 10 percent annual interest compounded quarterly.

13. *(GDP)* The gross domestic product (GDP) increases each year by an amount proportional to last year's GDP. Since 1950, that annual rate of increase has been 4 percent in the United States.

14. *(GDP)* The gross domestic product (GDP) increases each year by an amount proportional to last year's GDP. Before 1950, that annual rate of increase was 2 percent in the United States.

15. *(Radioactivity)* In a radioactive sample of material, the rate of radioactivity is proportional to the volume of material. Let this rate be k. As the material reacts, the

radioactive substance decays into inert substances. The rate in one sample is measured weekly.

16. (Radioactivity) In a radioactive sample of material, the rate of radioactivity is proportional to the volume of material. Let this rate be k. As the material reacts, the radioactive substance decays into inert substances. The rate in one sample is measured daily.

17. (Population Demographics) The population of Englermance increases each year. The annual increase is 2 percent of the previous year's population. Express the current population using difference equations.

18. (Wildlife Populations) The population of fish in New York harbor decreases each year by 4 percent of the previous year's population. (This decrease is due to the presence of harmful pollutants in the waters.) Express the current fish population using difference equations.

W 19. (Physics: Follow the Bouncing Ball) A pink handball is dropped from a 20-foot window. B. C. Dull suspects that the height on the nth bounce is 90 percent of the height of the previous bounce plus 10 percent of the original height. Express this height using difference equations assuming Dull's hypotheses are correct. Think about the behavior of bouncing balls (in particular, what must happen if you wait long enough?) and compare it to the long-term behavior predicted by this student. Why must his relationship be incorrect?

20. (Physics: Skipping Stones)
(a) A student "skips" a flat stone across a lake surface. Each skip is 25 percent shorter than the one before it. Use difference equations to express the distance traveled by the stone in the current skip.
(b) Using a new variable, express the total distance traveled by the stone.

Section 11.2   SOLVING DIFFERENCE EQUATIONS

The difference equations for the problems in the previous section give us the current balance (for example) in terms of the previous balance. Typically, though, we would like this information in terms of the initial balance, the interest rate, and any other constants that appear in the equation. That is, *instead* of information that (say)

$$y_5 = 1.1y_4 + 0.1$$

we would like to know how a general current balance variable depends on time (and the other parameters of the problem).

More formally, we observe that all the difference equations of the previous examples can be put in the general form

$$y_{n+1} = ay_n + b \qquad\qquad n = 1, 2, 3, \ldots$$

for various values of a and b. Can we instead write y_n exclusively in terms of the constants a and b and the initial value y_0?

To see how, it is instructive to examine the first few balances. For example,

$$y_1 = ay_0 + b$$

Similarly,

$$y_2 = ay_1 + b$$

But using the value for y_1, this becomes

$$y_2 = ay_1 + b$$

$$= a(ay_0 + b) + b$$
$$= a^2 y_0 + b(1 + a)$$

In the same way, we see that

$$y_3 = ay_2 + b$$
$$= a[a^2 y_0 + b(1+a)] + b$$
$$= a^3 y_0 + b(a^2 + a + 1)$$

This pattern is maintained for all y_n. That is,

$$y_n = a^n y_0 + b(a^{n-1} + a^{n-2} + \cdots + 1)$$

The series in parentheses is a *geometric series* (why?), and we may evaluate it using the formula given in Appendix B at the back of this book.

$$b(a^{n-1} + \cdots + 1) = \frac{b(a^n - 1)}{a - 1}$$

Thus, the solution to the general difference equation is

$$y_n = a^n y_0 + \frac{b(a^n - 1)}{a - 1} = a^n y_0 - \frac{a^n b}{1 - a} + \frac{b}{1 - a}$$
$$= \frac{b}{1 - a} + a^n \left(y_0 - \frac{b}{1 - a} \right)$$

which works, *provided* that $a \neq 1$. For when $a = 1$, the denominators of several of the fractions in this solution take the forbidden value of zero.

In case $a = 1$, we have to perform a similar (but simpler) analysis to learn that

$$y_n = y_0 + nb$$

Solving Difference Equations

The solution to the difference equation $y_{n+1} = ay_n + b$ for given values of a, b, and y_0 are

$$y_n = \frac{b}{1 - a} + a^n \left(y_0 - \frac{b}{1 - a} \right) \qquad\qquad a \neq 1$$

and

$$y_n = y_0 + nb \qquad\qquad a = 1$$

The intent of the next few examples is to generate some familiarity with this set of formulas. In each case, we also show the graph of y_n with n.

Example 5   The difference equations for some problem are given by

$$y_{n+1} = 0.9y_n + 10 \qquad n > 0$$
$$y_0 = p$$

(a) Solve the difference equation when $p = 110$.   (b) Solve the difference equation when $p = 100$. In both cases, graph y_n.

Solution

In this example, we have $a = 0.9$, $b = 10$, and $y_0 = p$. Since $a \neq 1$, we use the general form for the solution.

$$y_n = \frac{10}{1 - 0.9} + 0.9^n \left(p - \frac{10}{1 - 0.9} \right)$$
$$= 100 + (p - 100)0.9^n$$

(a) When $p = 110$, this solution becomes

$$y_n = 100 + (10)0.9^n$$

(b) When $p = 100$, the coefficient of 10^n vanishes identically. Therefore,

$$y_n = 100$$

for all n.

 The graphs of both solutions appear in Figure 11.2. For the first, 4_n starts at 110 and draws closer and closer to a value of 100, which it never actually reaches.

Skill Enhancer 5   For a certain difference equation, $a = 1.2$, $b = 2.4$, and $y_0 = 0$. Solve the equation.

Answer in Appendix E.

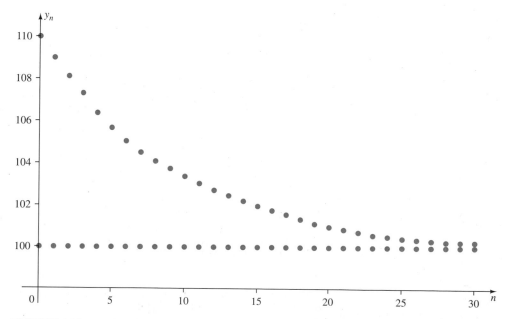

FIGURE 11.2
The graphs of y_n with n for $a = 0.9$ and $b = 10$. On the top, $y_0 = 110$; on the bottom, $y_0 = 100$.

This last example demonstrates that there may be values of a, b, and y_0 so that y_n remains constant.

Example 6 The parameters for a set of difference equations are $a = -0.9$, $b = 9.5$, and $y_0 = 10$. Solve this equation.

Solution

Again, since $a \neq 1$, we use the general form of the solution. We compute $\frac{b}{1-a} = \frac{9.5}{1.9} = 5$ so that

$$y_n = 5 + (10 - 5)(-0.9)^n = 5\left[1 + (-0.9)^n\right]$$

The graph of this solution appears in Figure 11.3. The values for y_n jump up and down, and appear to get ever closer to some value for increasing n.

Skill Enhancer 6 Solve the difference equation for which $a = -0.5$, $b = 9$, and $y_0 = -1$.

Answer in Appendix E.

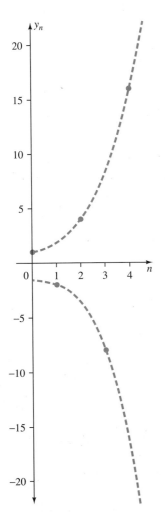

FIGURE 11.4
The graph for the difference equation given by $a = -2$, $b = 0$, and $y_0 = 1$.

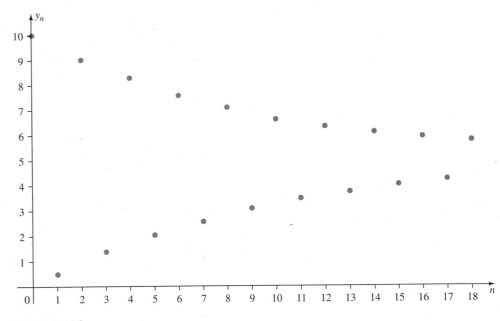

FIGURE 11.3
The solution for the difference equation given by $a = -0.9$, $b = 9.5$, and $y_0 = 10$.

Example 7 Solve a difference equation for which $a = -2$, $b = 0$, and $y_0 = 1$.

Solution

The solution is simply $y_n = (-2)^n$, and its graph appears in Figure 11.4. This graph illustrates yet another type of solution, one in which the values of y_n jump back and

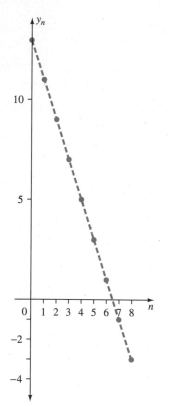

FIGURE 11.5
The solution for the
difference equation when
$a = 1, b = -2, y_0 = 13.$

forth about some value (in this case, they jump between positive and negative values), and yet the absolute values $|y_n|$ get ever larger for larger values of n.

Skill Enhancer 7   Solve the difference equation determined by $a = -3$ and $b = y_0 = 3$.

Answer in Appendix E.

Problems using the special form of the formula have a simpler solution.

Example 8   Solve the difference equation specified by $a = 1$, $b = -2$, $y_0 = 13$.

Solution
Since $a = 1$, we need the special case for the solution. Here,

$$y_n = 13 - 2n$$

whose graph appears in Figure 11.5.

Skill Enhancer 8   Write the solution for the difference equation given by $a = 1$, $b = 10$, and $y_0 = -1$.

Answer in Appendix E.

Later in this chapter we will examine many applications of difference equations.

_____ *Exercises 11.2* _____

Exercises 1–18 state parameters for a difference equation. Solve each equation.

1. $a = 1, b = 1, y_0 = 1$

2. $a = 1, b = -1, y_0 = -1$

3. $a = 2, b = 3, y_0 = 4$

4. $a = 1.2, b = 5, y_0 = 0$

5. $a = -1.5, b = 0, y_0 = -1$

6. $a = 1.3, b = 6, y_0 = 0$

7. $a = 1, b = 0, y_0 = -1$

8. $a = 1, b = 0.001, y_0 = 0.0001$

9. $a = \frac{1}{2}, b = 2, y_0 = 4$

10. $a = -\frac{1}{2}, b = 9, y_0 = 6$

11. $a = 1, b = \frac{1}{3}, y_0 = -\frac{2}{3}$

12. $a = \frac{7}{6}, b = \frac{1}{3}, y_0 = 0$

13. $a = 1.01, b = 1, y_0 = 100$

14. $a = -1.01, b = 0, y_0 = 0$

15. $a = 1,000, b = 1,000, y_0 = 1$

16. $a = 1.01, b = 2, y_0 = 200$

17. $a = 10,000, b = 10,000, y_0 = -1$

18. $a = -\frac{1}{2}, b = \frac{1}{2}, y_0 = 0$

19. For this problem, assume $y_0 = 0$. Solve these difference equations.

(a) $a = b = 10$;   (b) $a = b = 100$;

(c) $a = b = 1,000$;   (d) $a = b = 1,000,000$;

W (e) Examine your answers to these questions. Do you see a trend? What do you suppose the solution is for $a = b$ when a gets ever larger values (as a approaches infinity, for those readers who have studied the topic of limits)?

20. For the case of a difference equation where $a = 1$, derive the solution $y_n = y_0 + nb$ given in the text.

Section 11.3   *LONG-TERM BEHAVIOR OF DIFFERENCE EQUATIONS*

In the original difference equation approach, we expressed a current quantity in terms of the same quantity in the previous time period ($y_{n+1} = 1.1y_n + 10$, for example). When we solved the equations, we expressed a quantity in terms of time (together with other parameters). One advantage that the original approach *does* possess over the second approach is that long-term behavior of y_n may be more easily seen.

long-term behavior   What do we mean by *long-term behavior*? We are interested in the way y_n grows compared to y_0 after many ticks of the clock have passed. Will y_n be much bigger than y_0? Will it be much smaller? Will it be the same size? In Figure 11.6, we see the three possible cases, of convergence, divergence, and remaining constant. In addition, as Figure 11.7 shows, it is possible for values of y_n to *oscillate* about some value.

We can use the general solution to difference equations to determine simple rules to determine the long-term behavior of y_n. For the remainder of this section, we will assume that $a \neq 1$. Then, the general solution to the difference equation

$$y_{n+1} = ay_n + b \tag{11.3}$$

is

$$y_n = \frac{b}{1-a} + a^n \left(y_0 - \frac{b}{1-a} \right) \tag{11.4}$$

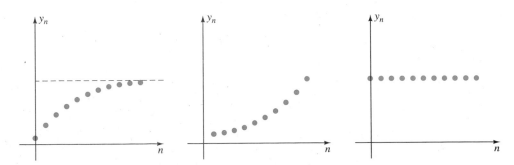

FIGURE 11.6
In the long run, y_n may converge, diverge, or stay the same.

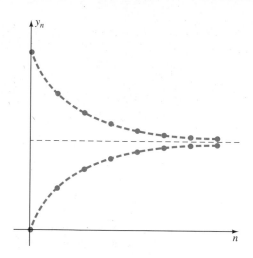

FIGURE 11.7
Values of y_n may oscillate about some value.

We also need to remind ourselves of several facts about quantities like a^n.

diverges   **1.** When $|a| > 1$, $|a^n|$ gets ever larger as n gets larger; it *diverges*.

converges   **2.** When $|a| < 1$, $|a^n|$ gets ever closer and closer to zero as n gets larger; it *converges* to zero.

         3. Of course, when $a = 1$, $a^n = 1$ (although we are not considering this case at this time).

oscillate   **4.** If $a < 0$, then the a^n *oscillate* in value about zero. That is, if $a^n > 0$, then $a^{n+1} < 0$ and vice versa.

Under what conditions will $y_n = y_0$ for all n? Let $y_{n+1} = y_n = y_0$ in Equation (11.3) to see that this condition implies:

$$y_0 = ay_0 + b$$

or

$$y_0 = \frac{b}{1 - a} \tag{11.5}$$

When this relationship is satisfied, $y_n = y_0$ for all n.

When will the y_n converge to a particular value? What value l do they converge to? In Equation (11.4), we know that a^n becomes smaller and smaller with larger values of n for $|a| < 1$. Eventually, the term

$$a^n \left(y_0 - \frac{b}{1 - a} \right)$$

becomes so close to zero that we could never measure it in any practical situation. At that time, y_n becomes so close in value to $\frac{b}{1-a}$ that it is hopeless to measure any deviation from it. We conclude that y_n converges to $\frac{b}{1-a}$ whenever $|a| < 1$.

When will the y_n diverge? The sequence y_n diverges whenever $|a| > 1$. In this case, the term $a^n \left(y_0 - \frac{b}{1-a}\right)$ (assuming that $y_0 \neq \frac{b}{1-a}$) grows larger and larger in magnitude because the coefficient a^n does.

Finally, we may assert the elements y_n will oscillate about the value $\frac{b}{1-a}$ value whenever $a < 0$. This follows from the behavior of a^n for $a < 0$ regardless of the individual values for a, b, and y_0.

Long-Term Behavior of Difference Equations

Convergence and divergence of the y_n are determined by values of a.

1. Whenever $|a| < 1$, they converge to the value $\frac{b}{1-a}$.
2. Whenever $|a| > 1$, they diverge.
3. Whenever a satisfies

$$y_0 = \frac{b}{1-a}$$

we have $y_n = y_0$ for all n.

In addition, we observe oscillatory behavior of the y_n about $\frac{b}{1-a}$ whenever $a < 0$.

The fraction

$$\frac{b}{1-a}$$

plays such an important role in these rules that it is a good idea to compute it as the first step in the following examples.

Typically, we need to know if a solution converges or diverges, and whether it oscillates or not. These facts are straightforward to ascertain—just compare the absolute value of a with 1 to determine convergence or divergence, and look at the sign of a for oscillatory behavior. But there is one important exception. Regardless of the results of these tests, if the parameters y_0, a, and b satisfy

$$y_0 = \frac{b}{1-a}$$

then all y_n have the same value.

Example 9   For one difference equation, $a = b = \frac{1}{2}$, $y_0 = 1$. Discuss the long-term behavior of the equation, and sketch a graph of y_n.

Solution

First we compute $\frac{b}{1-a}$.

$$\frac{b}{1-a} = \frac{\frac{1}{2}}{\frac{1}{2}} = 1$$

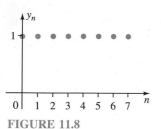

FIGURE 11.8
The difference equation given by $a = b = \frac{1}{2}$, $y_0 = 1$.

But this is the value of y_0. We conclude that $y_n = 1$ for all n. The graph appears in Figure 11.8.

Skill Enhancer 9   Consider the difference equation given by $a = \frac{1}{3}$, $b = 4a$, $y_0 = 2$.
(a) What is $\frac{b}{1-a}$?   (b) What is its relationship to y_0?   (c) Solve the equation.

Answers in Appendix E.

Example 10   Discuss the solution to the difference equation given by $a = 1.01$, $y_0 = b = 2,000$.

Solution

Since $a > 1$, the solution diverges. Since a is positive, it does not oscillate. Since $\frac{b}{1-a} = \frac{2,000}{-0.01} \neq y_0$, these conclusions stand.

Skill Enhancer 10   Consider the solution to the difference equation given by $a = 1.06$, $b = 1,000$, $y_0 = 5,000$.
(a) What is $\frac{b}{1-a}$?
(b) Is the equation $y_0 = \frac{b}{1-a}$ true?
(c) Comment on the convergence and oscillations of the solution.

Answers in Appendix E.

Example 11   Discuss the solution to the difference equation given by $a = -0.99$, $b = 0$, $y_0 = h > 0$.

Solution

Here $\frac{b}{1-a} = 0$, which can never equal y_0, for we know that whatever h is, it is strictly positive (greater than zero). Hence, we may apply our tests with no fear.

Since $|a| < 1$, the solution converges. It converges to $\frac{b}{1-a} = 0$. Because a is negative, the solution oscillates about zero. An informal sketch of the solution appears in Figure 11.9.

Skill Enhancer 11   Consider the difference equation given by $a = -0.7$, $b = 0.001$, and $y_0 = 10$.
(a) Compute $\frac{b}{1-a}$.
(b) Comment on the convergence and oscillations of the solution.
(c) If it converges, to what value does it converge?

Answers in Appendix E.

Example 12   Discuss the solution to the difference equation given by $a = -1.2$, $b = 22$, $y_0 = 10$.

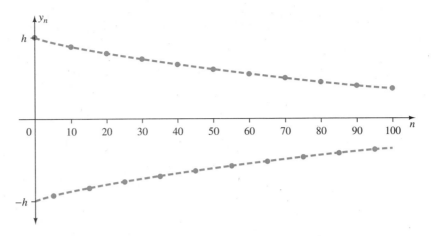

FIGURE 11.9
A convergent, oscillatory difference equation.

Solution

Based on an examination of a, it's tempting to conclude that the solution is divergent and oscillatory. However, the calculation

$$\frac{b}{1-a} = \frac{22}{2.2} = 10 = y_0$$

reveals a different ending. In fact, for all n, $y_n = 10$. The solution is *neither* divergent nor oscillatory.

Skill Enhancer 12   Consider the difference equation given by $a = -3$, $b = 12$, and $y_0 = 3$.
(a) What is $\frac{b}{1-a}$?
(b) Discuss the solution.

Answers in Appendix E.

———— *Exercises 11.3* ————————————————————————

In Exercises 1–6 the parameters a, b, and y_0 refer to a difference equation. Select the letter or letters of the values that answer the questions.

1. For which values of a will the solution diverge when $b = y_0 = 10$?
  (a) 0.9
  (b) 0.99
  (c) −0.99
  (d) −1.01

2. For which values of a will the solution oscillate when $b = 1$, $y_0 = 3$?

  (a) 1
  (b) −1
  (c) 2
  (d) −3

3. For which values of a will the solution converge and oscillate when $b = 23$, $y_0 = 10$?
  (a) −2
  (b) −1.3
  (c) 0
  (d) 1

4. For which values of b will the solution be constant when $a = 1.2$ and $y_0 = -2.4$?

(a) 1.4
(b) −1.4
(c) −2.4
(d) 2.4

5. For which values of y_0 will the solution be constant when $a = -1$ and $b = -2$?
(a) −1
(b) 1
(c) 2
(d) 3

6. For which values of a will the solution be convergent and oscillatory when $b = 3.9$ and $y_0 = 3$?
(a) −0.3
(b) −0.5
(c) −0.7
(d) −0.9

Discuss the nature of the solutions to the difference equations in Exercises 7–24 *without* explicitly solving the equations.

7. $a = 0$, $b = 0$, $y_0 = 0$ 8. $a = 0$, $b = 0$, $y_0 = -1$

9. $a = 0$, $b = 1$, $y_0 = 0$ 10. $a = -1$, $b = 0$, $y_0 = 1$

11. $a = 1.2$, $b = 1.2$, $y_0 = 1.2$

12. $a = 1.1$, $b = 2.2$, $y_0 = 3.3$

13. $a = -1.2$, $b = 3$, $y_0 = -2$

14. $a = -0.5$, $b = -0.5$, $y_0 = -0.5$

15. $a = -0.001$, $b = 0.002$, $y_0 = 0.003$

16. $a = -2$, $b = 3.4$, $y_0 = 5.6$

17. $a = -1.1$, $b = 0$, $y_0 = 3$

18. $a = \frac{3}{4}$, $b = 2$, $y_0 = 8$

19. $a = \frac{1}{9}$, $b = \frac{16}{3}$, $y_0 = 6$

20. $a = -\frac{1}{2}$, $b = \frac{1}{2}$, $y_0 = \frac{1}{3}$

21. $a = -\frac{4}{3}$, $b = \frac{2}{3}$, $y_0 = \frac{2}{7}$

22. $a = -3$, $b = -4$, $y_0 = -5$

23. $a = 1.2$, $b = 1,000$, $y_0 = 2,000$

24. $a = -1.2$, $b = -1,000$, $y_0 = -2,000$

W 25. From the discussion of the text, and from references to Figure 11.6, what do you think is meant by the terms "convergence" and "divergence"?

W 26. We can derive the criterion for constant values for y_n directly from the equation $y_{n+1} = ay_n + b$. What does this equation imply when $y_n = y_0$ for all n? Sketch (in words) the steps needed to draw the conclusion.

Section 11.4   USES AND APPLICATIONS OF DIFFERENCE EQUATIONS

We have left till now an examination of the uses to which difference equations may be put. It may seem most natural to apply them to financial problems, as the balance y_n in such a problem naturally conforms to the difference equations model:

1. The balance changes suddenly and only at certain well-defined time periods.

2. The current balance depends only on the balance in the previous time period.

Study the next examples carefully.

Example 13   Suppose that someone deposits money in an account. This principal P earns interest compounded periodically and earns interest at a rate of i per period. Use difference equations to obtain the total balance after n periods.

Solution

The initial deposit is P, so $y_0 = P$. Since the balance in period $(n + 1)$ is the balance in the nth period plus the earned interest, we have

$$y_{n+1} = y_n + i y_n$$
$$= (1 + i) y_n$$

This is a difference equation with $a = 1 + i$, $b = 0$, so the solution is

$$y_n = (1 + i)^n P$$

which agrees (as it must) with the formula we derived in Section 10.1.

Skill Enhancer 13   Ivan deposits \$1250 into an account that earns interest of 6 percent compounded quarterly. What is his balance after the nth quarter?

Answer in Appendix E.

Example 14   Solve Example 3.

Solution

The difference equation for the growth of Harriet's IRA is

$$y_{n+1} = 1.105 y_n + 2{,}000$$

so that $a = 1.105$, $b = 2{,}000$, and $y_0 = 2{,}000$. Its solution is

$$y_n = \frac{2{,}000}{1 - 1.105} + 1.105^n \times 2{,}000 \left(1 - \frac{1}{1 - 1.105} \right)$$

or

$$y_n = 2{,}000 \left(1.105^n + \frac{(1.105^n - 1)}{0.105} \right)$$

A graph of this solution as a function of n appears in Figure 11.10. In 20 years, total deposits of \$42,000 will have grown to almost \$136,000.

Skill Enhancer 14   Angel puts 10 percent of each monthly paycheck into a savings account that earns 9 percent compounded monthly. Angel started the account with \$1,000. Let y_n be the balance after the nth month.
(a) What is the monthly interest rate?
(b) What is the difference equation for y_n?
(c) What is the solution to this difference equation?

Answers in Appendix E.

Example 15   Simple interest is earned at 1 percent per month on \$100. Use difference equations to compute the balance after n months.

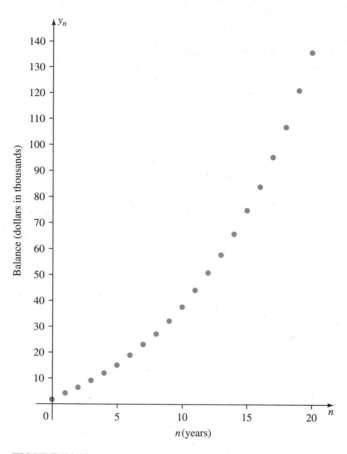

FIGURE 11.10
Harriet's IRA balance.

Solution

In this case $y_0 = 100$. The new month's balance equals last month's balance, plus the interest earned on the initial amount.

$$y_{n+1} = y_n + 0.01 y_0$$
$$y_0 = 100$$

The earned interest depends on y_0 only because we are assuming conditions of simple interest. In this difference equation, $a = b = 1$. The solution is therefore

$$y_n = 100 + n$$

A graph of y_n with n appears in Figure 11.11.

Skill Enhancer 15 A loans \$1,250 to B, who agrees to pay 12 percent annual simple interest. Interest is posted monthly.
(a) What is the difference equation for the balance that B owes A?

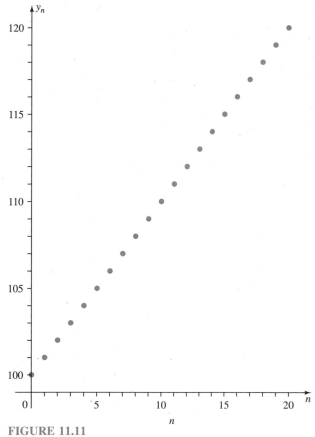

FIGURE 11.11
Simple interest via difference equations.

(b) What is the solution to this equation?

Answers in Appendix E.

Example 16   Solve Example 4.

Solution

The difference equation is $y_{n+1} = 1.005 y_n + 1.5p$, so that $a = 1.005$ and $b = 1.5p$. The condition of the problem requires that $y_0 = 1.5p$. The solution is

$$y_n = \frac{1.5p}{1 - 1.005} + 1.005^n \left(1.5p - \frac{1.5p}{1 - 1.005} \right)$$

$$= 1.5p \left[\frac{-1}{0.005} + \left(1 + \frac{1}{0.005} \right) 1.005^n \right]$$

$$= 1.5p \left[-200 + 201(1.005)^n \right]$$

Skill Enhancer 16   An employee chooses an amount p to be deducted from each paycheck and deposited automatically into the account. The account earns 9 percent compounded monthly, and paychecks are issued monthly. The employer supplements the employee's deposit by contributing one-third of the employee's deposit together with one-third of the interest earned since the last deposit.

(a) What is the equation for y_0?

(b) What is the difference equation describing the account balance? What are a and b?

(c) What is the solution to this difference equation?

Answers in Appendix E.

We may often use difference equations as an alternative means for deriving many of the formulas we worked with in the preceding chapter. Examine this next example carefully.

Example 17   A young couple wants to assume a mortgage with a face value of P. The monthly mortgage payments p assume that interest will be computed at a *monthly* interest rate i, and the mortgage will have a lifetime of N months. What must the monthly payment p be for this mortgage?

Solution

Here, we let y_n be the balance that the borrowers owe the bank at the completion of the nth month of the loan. Clearly, $y_0 = P$ and $y_N = 0$. If the homeowners owe y_n to the bank after the nth month, what will they owe the bank after the $(n + 1)$th month? Changes to the balance cause the balance to both grow and decrease.

1. It will *grow* by the amount of additional interest that the outstanding balance y_n earns during the month. This amount is iy_n.

2. It will *decrease* by the small amount of principal the homeowners pay as part of the monthly payment.

The monthly payment p is designed to be just enough to cover the additional interest together with the decrease in principal. Therefore, we can say that

$$y_{n+1} = y_n + iy_n - p$$

or

$$y_{n+1} = (1 + i)y_n - p$$

The solution to this difference equation, taking $a = 1 + i$ and $b = -p$, is

$$y_n = \frac{p}{i} + (1 + i)^n \left(P - \frac{p}{i} \right)$$

which can be rearranged to give

$$y_n = (1 + i)^n \left\{ p \left[\frac{(1 + i)^{-n} - 1}{i} \right] + P \right\} \tag{11.6}$$

At the end of the mortgage, when $n = N$, $y_N = 0$. Substituting these values into Equation (11.6) and solving for p yields

$$p = P\left[\frac{i}{1 - (1+i)^{-n}}\right]$$

which is equivalent to the result we obtained in the previous chapter.

Skill Enhancer 17   We will use difference equations to derive the future value of an interest-bearing annuity. Let the periodic rate of interest be i, and let y_n be the future value of the annuity after the nth interest period. Let p be the amount of each deposit.
(a) How much interest is earned in one interest period?
(b) By what additional amount does the future value change?
(c) What is the difference equation for this system? What are the values of a, b, and y_0?
(d) What is the solution to this system?

Answers in Appendix E.

Difference equations also appear in nonfinancial situations.

Example 18   If an object is placed between two mirrors, there would in theory be an infinite number of images of it reflected back and forth. In fact, though, about 4 percent of the light is lost in each reflection. If an object becomes "invisible" to the human eye when it loses 90 percent of its original intensity, how many visible images will there be? Will the images ever become truly invisible?

Solution

Let y_n be the amount of light present in the nth reflection. Clearly, $y_{n+1} = y_n - 0.04y_n = 0.96y_n$, a difference equation with $a = 0.96$, $b = 0$. Therefore,

$$y_n = 0.96^n y_0$$

If $y_n = 0.1y_0$ in the final reflection, then n is determined by

$$0.1 = 0.96^n$$

By trial-and-error experimentation with a calculator, we discover that $0.96^{56} = 0.1017$ and $0.96^{57} = 0.0976$. There will be, therefore, about 56 visible reflections.

The images will never truly vanish, because there is no finite value of n for which 0.96^n will ever equal zero.

Skill Enhancer 18   Sourdough bread is made slightly differently from other breads. A portion of dough is allowed to ferment, and this is the starter for dough for the loaves of bread. A baker starts with a volume v of the starter. This is divided into 10 portions, each of which is used to make another starter on the next day, and so on. Let y_n be the amount of the original starter present in the dough for a loaf baked on the nth day.
(a) Set up a difference equation for y_n.
(b) Solve this equation.

Answers in Appendix E.

_____ *Exercises 11.4* _____

In Exercises 1–14, set up and solve the problem using difference equations.

1. A bank balance earns interest at a 15 percent annual rate compounded quarterly.

2. A bank balance earns interest at a 16 percent annual rate compounded monthly.

3. Seth borrowed money from Alan, and agrees to pay back the amount with interest. Simple interest at 10 percent per year is computed twice a month.

4. Beth borrowed money from Lisa, and agrees to pay back the amount with interest. Simple interest at 12 percent per year is computed monthly.

5. Jozefa deposits $1,000 yearly in a money market fund that earns 11 percent annually compounded annually.

6. Beatrice deposits $1,500 into a company bank account every year. The money earns 10 percent annual interest compounded quarterly.

7. *(GDP)* The gross domestic product (GDP) increases each year by an amount proportional to last year's GDP. Since 1950, that annual rate of increase has been 4 percent in the United States.

8. *(GDP)* The gross domestic product (GDP) increases each year by an amount proportional to last year's GDP. Before 1950, that annual rate of increase was 2 percent in the United States.

9. *(Radioactivity)* In a radioactive sample of material, the rate of radioactivity is proportional to the volume of material. Let this rate be k. As the material reacts, the radioactive substance decays into inert substances. The rate in one sample is measured weekly.

10. *(Radioactivity)* In a radioactive sample of material, the rate of radioactivity is proportional to the volume of material. Let this rate be k. As the material reacts, the radioactive substance decays into inert substances. The rate in one sample is measured daily.

11. *(Population Demographics)* The population of Englermance increases each year. The annual increase is 2 percent of the previous year's population. Express the current population using difference equations.

12. *(Wildlife Populations)* The population of fish in New York harbor decreases each year by 4 percent of the previous year's population. (This decrease is due to the presence of harmful pollutants in the waters.) Express the current fish population using difference equations.

13. *(Physics: Follow the Bouncing Ball)* A pink handball is dropped from a 20-ft window. B. C. Dull suspects that the height on the nth bounce is 90 percent of the height of the previous bounce plus 10 percent of the original height. Express this height using difference equations assuming Dull's hypotheses are correct. Why must this relationship be incorrect?

14. *(Physics: Skipping Stones)* A student "skips" a flat stone across a lake surface. Each skip is 25 percent shorter than the one before it. Use difference equations to express the total distance travelled by the stone.

W 15. Set and solve Exercise 13 subject to slight changes. This time, B. C. Dull hypothesizes that the height of the ith bounce is 80 percent of the previous bounce plus 10 percent of the original height. Set up the difference equation and solve. What is distinctive about the solution?

W 16. The economic term *deflation* is the opposite of inflation. What do you think deflation is?

17. *(Economic Instability)* Refer to Exercise 16 for the term *deflation*. Suppose an economy undergoes deflation so that the price of one commodity is only 97 percent of what it was the month before. Use difference equations to express the current price in terms of last month's price. Solve this difference equation.

18. *(Biology)* The population of tiny one-celled marine life seems to be 23 percent more than it was the day before. A flask of sea water contains $1,000,000$ of these creatures at the start of the experiment. Use difference equations to express the daily growth of the population. How many of these creatures will exist on the fifth day of the experiment? on the nth day?

19. *(Solid Waste Accumulation)* The garbage in one Michigan community accumulates at a rate proportional to the size of the community. There are 15 percent more people in this town in one year than there were in the previous year. Use difference equations to express the total accumulated volume of solid waste as a function of the total of the previous year. Solve this equation in terms of the total amount present when this study began.

20. *(Animal Conservation)* The elephant population of a national park in Kenya is decreasing such that the current population is 3 percent less than last year's. What is the population in any given year in terms of the initial

population and the years elapsed? How long will it take for the elephants to decrease to the point where there are half as many as there were at the initial time?

21. *(Human Population Studies)* The population of

China is growing at a rate such that there are 2.5 percent more people in any given year than there were in the previous year. Express this growth as a difference equation. Solve this equation. How long will it take for China's population to double?

CHAPTER REVIEW

Terms

difference equation

converges

long-term behavior

oscillate

diverges

Key Concepts

- Difference equations may be an efficient means for describing quantities that change value only at regular, known, predictable instants of time. In order to use the difference equation tool kit developed in this chapter, we need to be able to relate the value y_{n+1} during one interval with its value y_n in the previous interval. This general relation should take the form $y_{n+1} = ay_n + b$ for suitable constants a and b. We also need to know y_0.

- The solutions to the difference equation $y_{n+1} = ay_n + b$ for given values of a, b, and y_0 are

$$y_n = \frac{b}{1-a} + a^n \left(y_0 - \frac{b}{1-a} \right) \qquad a \neq 1$$

and

$$y_n = y_0 + nb \qquad a = 1$$

- In case $a \neq 1$, it makes sense to discuss the long-term behavior—growth or decay—of the y_n. When $|a| < 1$, the y_n converge to the value $\frac{b}{1-a}$. When $|a| > 1$, the y_n diverge. When a is negative, successive values of y_n oscillate about $\frac{b}{1-a}$.

- When testing for the long-term behavior of the y_n, be sure to test whether the given values of a, b, and y_0 satisfy the equation $y_0 = \frac{b}{1-a}$. If they do, regardless of the values of a and b, all the y_n are equal to y_0.

- Difference equations are useful in applications involving finance, physics, and processes of growth and decay.

Review Exercises

In Exercises 1–6, write the difference equation describing each situation. Make sure to specify y_0 and the general expression for y_{n+1}. Clearly state what y_n stands for in each problem. Do not solve the problems.

1. A bank balance earns interest at a 12 percent annual rate compounded quarterly.

2. A bank balance earns interest at a 9 percent annual rate compounded monthly.

3. Pablo borrowed money from Felicia, and agrees to pay back the amount with interest. Simple interest at 10 percent per year is computed twice a month.

4. Nguyen borrowed money from Tran, and agrees to pay back the amount with interest. Simple interest at 12 percent per year is computed monthly.

5. *(Population Demographics)* The population of Gerolstein increases each year. The annual increase is 3.6 percent of the previous year's population. Express the current population using difference equations.

6. *(Wildlife Populations)* The population of fish in Seattle harbor decreases each year by 6 percent of the previous year's population. (This decrease is due to the presence of harmful pollutants in the waters.) Express the current fish population using difference equations.

7. *(Physics)* Graduate student B. C. Dull is testing a new rubberlike substance in the laboratory. He theorizes that the height on the nth bounce is 95 percent of the height of the previous bounce less a small fixed amount that represents the loss of height due to the presence of frictional forces. Express this height using difference equations assuming Dull's hypotheses are correct.

Exercises 8–13 state parameters for a difference equation. Solve each equation.

8. $a = -2$, $b = -2$, $y_0 = -2$

9. $a = 1.01$, $b = 1.01$, $y_0 = 1.01$

10. $a = 1$, $b = 0$, $y_0 = -1$

11. $a = 1$, $b = 3$, $y_0 = -2$

12. $a = -1.1$, $b = 7$, $y_0 = 3$

13. $a = -1.1$, $b = 3.15$, $y_0 = 1.5$

Discuss the nature of the solutions to the difference equations in Exercises 14–21 *without* explicitly solving the equations.

14. $a = 1.01$, $b = -2$, $y_0 = 3$

15. $a = -1.1$, $b = 2$, $y_0 = -2$

16. $a = -7$, $b = y_0 = 0.001$

17. $a = 0.99$, $b = 0$, $y_0 = 0$

18. $a = \frac{1}{2}$, $b = \frac{1}{2}$ $y_0 = \frac{1}{2}$ **19.** $a = -\frac{1}{3}$, $b = 1$ $y_0 = -1$

20. $a = 2$, $b = 3$ $y_0 = -3$ **21.** $a = -\frac{7}{8}$, $b = 1$ $y_0 = \frac{8}{15}$

22. Solve Exercise 1.   **23.** Solve Exercise 2.

24. Solve Exercise 3.   **25.** Solve Exercise 4.

26. Solve Exercise 5.   **27.** Solve Exercise 6.

W 28. *(Long-Term Behavior of Difference Equations)* In the text, we considered long-term behavior of difference equations only for the case $a \neq 1$. What happens when $a = 1$? How may we categorize long-term behavior in that case?

12

USING GRAPHS TO PLAN

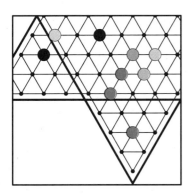

INTRODUCTION

Since the end of World War II, some projects have become so large that they are no longer comprehensive to a single manager or even to a small group of managers. Keeping track of the phases of such a project—whether it be building a skyscraper, highway, or submarine; launching a space flight; installing a new computer system; designing and marketing new products; or completing a corporate merger—has become as difficult as the undertaking itself. Great effort was spent in determining ways to effectively plan and control large-scale endeavors, and by the mid-1950s two closely related methods of analysis had been fully developed. Both were used to plan the development of the Polaris missile, and both were given credit for bringing in this project two years ahead of schedule.

CPM
critical path method

The two methods are identified by sets of initials. *CPM* stands for *critical path method* and works by representing the subtasks of a large project as components of a specially drawn network representation of the project on which are projected the times of completion for each subtask. CPM has become popular not only because of its effectiveness, but because it relies more on common sense than on sophisticated mathematical techniques. Anyone can understand CPM! (But do not, on that count, underestimate this method.)

PERT

The second method is *PERT*, or *probabilistic evaluation review technique*, and is really a variation on CPM analysis extended to the case where we are uncertain about the times of completion. What can we say about the probability of completing the task in less than a certain amount of time given information about the probabilities of completing the subtasks?

The aim of this chapter is to present the elements of these techniques to the reader.

Section 12.1   NETWORKS AND AN INTRODUCTION TO CPM

Throughout this book, we have emphasized the importance of using available resources to the best advantage when grappling with a problem. Most such methods concentrate on physical resources, and until now we have not looked at how to carry out tasks in the least amount of *time*. This network analysis helps address this lack. With elements of CPM and PERT, the conscientious manager can ensure that projects take the least amount of time, and identify potential bottlenecks in the task.

network   A *network* or *graph* is a drawing consisting of lines or simple curves that come
graph   together at special meeting points. The lines are called *arcs* whether they are straight or
arcs   curved, and the points at which arcs intersect are the *nodes*. (Note that we are using the
nodes   term *graph* with a different meaning than it had earlier in this text. For this reason, we will stick with the word *network*.) A spider web, a map of city streets, and the lines of grout in a tile floor are common examples of networks. (See Figure 12.1.) In the map, each street is an arc and each corner is a node.

The importance of networks lies not so much with the way they display geographical relationships, but with their ability to lay bare the relationships between various quantities (relationships that often have nothing whatever to do with geography). In this chapter, we will use networks to show relationships between subtasks of a project.

Consider the construction of a house. Here's a simplistic list of the individual tasks involved.

A. Dig the foundation.

B. Pour the foundation.

C. Erect the structural skeleton of the house.

D. Install the plumbing and electrical systems.

E. Put on the siding and roofing.

F. Interior plastering.

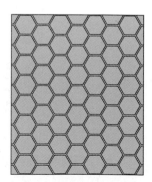

FIGURE 12.1
Some common networks.

G. Landscaping.

H. Final cleanup.

From a *planning* point of view, this list is not helpful for at least two reasons.

- We cannot perform any scheduling because we have no idea how long each task requires.

- Not only that, but we have no idea of the relation in which each subtask stands to the others. Obviously, we cannot begin the interior plastering until the basic structure has been erected and the pipes and wires put in place, but there may be other relationships that are not so obvious.

For that reason, after gathering more facts, we advance to the following chart.

	Task	*Time*	*Predecessors*
A	Dig the foundation.	1	—
B	Pour the foundation.	1	*A*
C	Erect the structural skeleton of the house.	2	*B*
D	Install the plumbing and electrical systems.	2	*C*
E	Put on the siding and roofing.	3	*C*
F	Interior plastering.	2	*D*
G	Landscaping.	1	*E*
H	Final cleanup.	1	*F, G*

immediate predecessors   The term *immediate predecessors* refers to what we must have finished before beginning the current job assignment. In this table, the times for each activity are in months.

Although the descriptions of the tasks are helpful and interesting, they are irrelevant to the planning process. Clearly, all we need to make a useful schedule is the list of tasks identified by letter, together with the times of completion and the immediate predecessors of each. A special network is the best way for visualizing the relationships between tasks.

As we have observed, a network consists of nodes connected by arcs. Let us agree that a node represents the point in time at which one or more tasks end and the successor tasks to these begin. Nodes are connected by arcs, and we will label each arc by the

activity on arc   activity we perform to cross from one node to the other. This convention is the *activity on arc* (or *aoa*) representation and seems to be most convenient when using computer techniques for generating CPM schedules. (An alternative scheme is the *aon* or *activity on node* representation.)

Figure 12.2 displays the network for building a house. We have embellished each activity label with the number indicating the length of time for that activity. Note finally that there is *no* relationship between the length of any arc and the time for that activity.

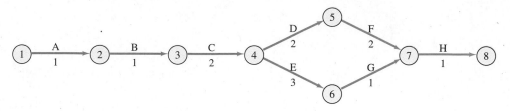

FIGURE 12.2
A network appropriate to building a house.

Example 1   Here is a list of tasks and their predecessors.

	Predecessors
A	—
B	*A*
C	*A*
D	*A*
E	*B*
F	*C*
G	*D*
H	*E, F, G*

What is the network for the project?

Solution

The solution appears in Figure 12.3.

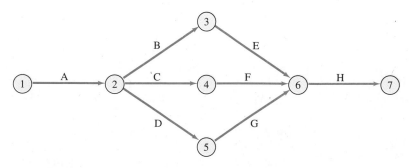

FIGURE 12.3
An example of a simple network.

Skill Enhancer 1   Given the short task list

	Predecessors
A	—
B	—
C	—
D	A
E	C
F	D, B, E

what is the associated network?

Answer in Appendix E.

Example 2   Study the network in Figure 12.4. What task list pertains to it?

Solution

It's best to proceed by considering the tasks in alphabetical order. Here is the table for this network.

	Predecessors
A	—
B	—
C	A
D	C
E	B, D
F	C
G	E, F

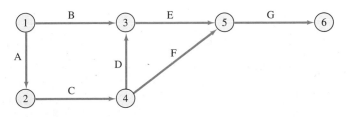

FIGURE 12.4
What task list pertains to this network?

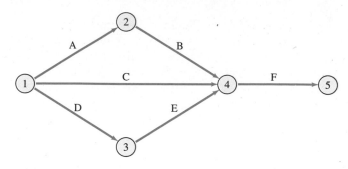

FIGURE 12.5
And what task list pertains to this network?

Skill Enhancer 2   What task list is appropriate for the network of Figure 12.5?

Answer in Appendix E.

There are some subtleties to appreciate with this method; not all networks are quite as straightforward as Figure 12.2 might lead us to believe. To understand these subtleties, we begin by formally stating the few rules governing arcs and nodes.

Rules for Nodes and Arcs

We label arcs by their activity. We label nodes with a number that has no connection with the activity.

1. We start each network with a beginning node, node 1, and draw a node indicating the finish of the project. (In Figure 12.5, for example, node 5 is the finishing node.)

2. The pattern by which we number nodes is generally irrelevant, but it proves useful to guarantee that the node label at the finish of any activity is greater than the node label at the start of the activity. (Generally, there will be more than one way to label the nodes and still satisfy this requirement.)

3. Any two nodes may be connected by *at most one arc*.

4. Any single activity may be represented by a single arc only.

It is the last two requirements that lead to the subtleties of graph drawing. Let us see why.

Dummy Activities

Consider this abbreviated task list.

	Predecessors
A	—
B	—
C	A, B

It is tempting to offer Figure 12.6 as the network for this project, but this violates item three of our network rules—more than one arc connects nodes 1 and 2.

Practitioners of CPM adopt an artificial labeling scheme to escape this difficulty. We agree to pretend there exists an imaginary node along one of the arcs, and that a *dummy activity* connects the new node with an existing node. Figure 12.7 indicates the two ways that dummy activities may be introduced to represent the given task and satisfy the network rules.

dummy activity

Example 3   What is the network for this task list?

	Predecessors
A	—
B	—
C	—
D	A, B, C

Solution

We need two dummy activities here. See Figure 12.8. (Beware the common error of drawing three arcs to connect nodes 1 and 4.)

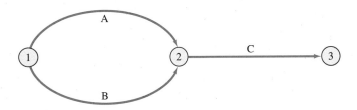

FIGURE 12.6
Not a correct network because two arcs may not connect nodes 1 and 2.

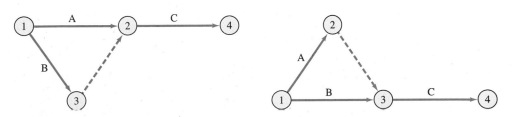

FIGURE 12.7
Two ways of introducing dummy activities to correct the network of Figure 12.6.

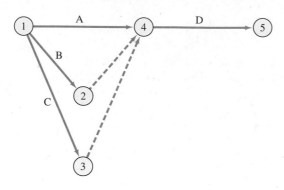

FIGURE 12.8
The solution to Example 3.

Skill Enhancer 3   The task list

	Predecessors
A	—
B	—
C	*A, B*
D	*A, B*
E	*C, D*

corresponds to what network?

Answer in Appendix E.

Dummy activities are also useful in "helping" a single activity span more than one pair of nodes. As example, consider this activity list.

	Predecessors
A	—
B	*A*
C	*B*
D	*A, C*

Beginners are sorely tempted by Figure 12.9, but this violates the last of our network rules, for a single activity (here, activity *A*) may not be represented by more than one arc.

The dilemma is resolved by imagining a dummy activity for *A* emanating from node 2. The correct network appears in Figure 12.10. The dashed arrow that indicates a dummy activity belongs to the predecessor activity.

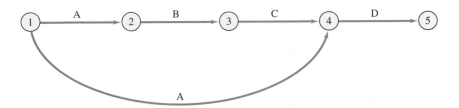

FIGURE 12.9
Another wrong network—a single activity may not be represented by more than one arc.

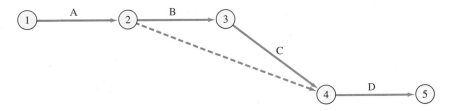

FIGURE 12.10
A dummy activity associated with activity _A_ corrects the network.

Example 4   Draw the network corresponding to this activity list.

	Predecessors
A	- - -
B	- - -
C	B
D	A
E	C, D
F	A, E

Solution
See Figure 12.11.

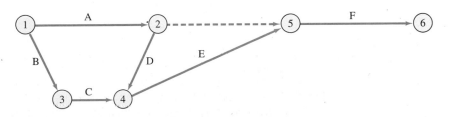

FIGURE 12.11
The solution to Example 4 involves a dummy activity.

Skill Enhancer 4   Draw the network corresponding to this task list.

	Predecessors
A	—
B	—
C	A
D	B, C
E	A, D

Answer in Appendix E.

Provided the reader keeps in mind the two rules for using dummy variables, any list of tasks and predecessor relationships can be translated into a network.

Example 5   Here is a task list. What is the network?

	Predecessors
A	—
B	—
C	—
D	A, B, C
E	D
F	E
G	D, F
H	E, G

Solution
See Figure 12.12.

Skill Enhancer 5   Given the task list

	Predecessors
A	—
B	A
C	B
D	A, B
E	C, D

determine the network.

Answer in Appendix E.

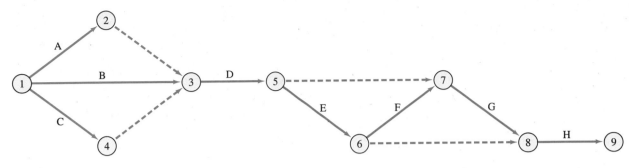

FIGURE 12.12

___Exercises 12.1___

Each of Exercises 1–8 displays a network of tasks. For each, provide the list of tasks showing the predecessors of each.

1.

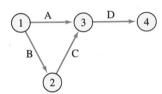

2.

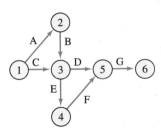

3.

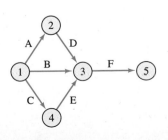

4.

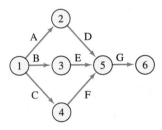

5.

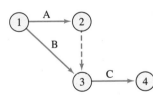

6.

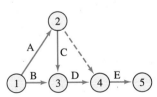

7.

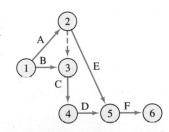

8.

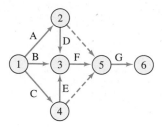

Each of Exercises 9–18 contains a task list. Create a network appropriate for each such list. Make sure that each arc and node is carefully labeled.

9.

	Predecessors
A	—
B	—
C	A
D	B, C

10.

	Predecessors
A	—
B	—
C	A
D	B, C
E	B, C
F	E
G	D, F
H	G
I	D, F
J	H, I

11.

	Predecessors
A	—
B	—
C	—
D	—
E	A
F	B
G	C
H	D
I	E, F, G, H

12.

	Predecessors
A	—
B	—
C	A
D	B, C
E	B, C
F	E
G	D
H	F, G

13.

	Predecessors
A	—
B	—
C	—
D	—
E	—
F	D
G	A, B, C, E, F

14.

	Predecessors
A	—
B	—
C	A, B
D	A, B
E	C, D

15.

	Predecessors
A	—
B	—
C	A
D	B, C
E	A, D

16.

	Predecessors
A	—
B	—
C	A
D	B
E	A, D
F	C, E
G	B
H	F, G

17.

	Predecessors
A	—
B	—
C	—
D	A
E	A
F	B
G	B
H	D, E
I	F, G
J	C
K	H, I, J

18.

	Predecessors
A	—
B	—
C	—
D	A, B
E	A, B
F	C, D, E

W 19. (a) Do you think it is possible to have more than one dummy activity proceed from a single node? Why or why not?

(b) Use your thoughts to create the network for this task list.

	Predecessors
A	—
B	—
C	—
D	A, B
E	B, C
F	D, E

20. Use the result of Exercise 19 to create a network for this task list.

	Predecessors
A	—
B	—
C	—
D	A, B
E	B
F	B, C
G	E
H	D, G
I	F, G
J	H, I

Section 12.2 THE CRITICAL PATH METHOD

The purpose of representing relationships between tasks by networks is to make possible the determination of the completion date of the project (as well as a few other things). To do this, we make one important assumption:

- The exact time of completion for each of the subtasks of the project is known with complete certainty.

Typically, the task list for a project includes not only the tasks and their predecessors, but the times for each subtask as well.

The network for the following task list appears in Figure 12.13. The numbers embellishing each task label are the units of time required for the completion of each task.

	Time	Predecessors
A	2	—
B	4	—
C	3	—
D	5	—
E	3	A, B
F	5	C, D
G	1	E, F

How Long Does the Project Take?

early start   The *early start* time is the earliest time we may begin a task. We determine the early start time for each task in a procedure that starts alphabetically from task A and goes to the final task.

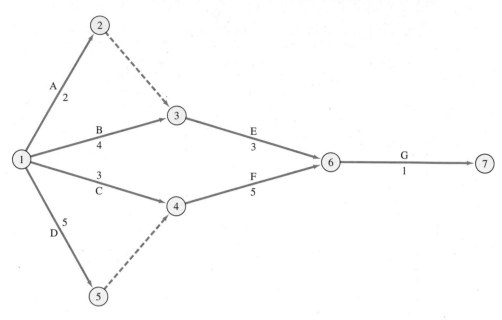

FIGURE 12.13
Computing the time of completion for this project.

early finish   Associated with the early start time is the *early finish* time of a task. It's easy to figure the early finish time of a task—simply add the time to complete a task to its early start time.

recursive technique   How do we compute the early start time? We use a *recursive technique*—the early start time of one task is based upon the early finish times of the predecessor tasks.

Computing the Early Start and Early Finish Times

1. For the initial tasks (the ones with no predecessors), the early start times are zero by definition.

2. Any task with predecessors cannot begin until *all* of these predecessors have finished. Consequently, the early start time of any task with predecessors is the *maximum* of the early finish times of the predecessors.

3. The early finish time of any task is the sum of its early start time and the time for completion of that task.

We use square bracket notation [a, b] for early start and early finish times to indicate that the early start time is a and the early finish time is b.

    The early finish time of the final task in the project is the minimum time needed to complete the project.

    Refer again to Figure 12.13. The early start time for task A is 0, and its early finish time is 2 + 0 = 2. We write A [0, 2].

In the same way, for B, C, and D (all tasks with no predecessors) we have $B\,[0,4]$, $C\,[0,3]$, and $D\,[0,5]$.

Next in our alphabetical circuit is E. It cannot begin until both A and B have finished—that is, until 4 four time units have passed. Therefore the early start time is 4, and the early finish time is $4+3=7$, since E takes 3 units of time.

In the same way, the early start time for F is 5, and (since F takes 5 units) the early finish time is $5+5=10$.

Make sure you understand why the times for G are $[10,11]$. The network, embellished with the early times, appears in Figure 12.14. Since the early finish time for the final task G is 11 units, we now know that this project will take at least 11 time units to complete.

Example 6   For the task list

	Time	Predecessors
A	1	—
B	2	—
C	3	A, B
D	5	A, B
E	3	A, B
F	3	C, D, E

(times are in weeks) determine the minimum time for the completion of the project.

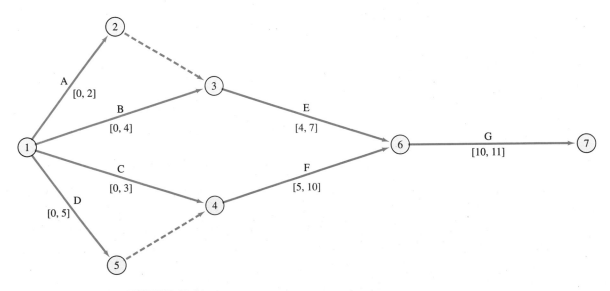

FIGURE 12.14
The network with early start and early finish times.

Solution

It's always a good idea to construct the network for the task list, even when that is not explicitly requested. The network with completion times appears in Figure 12.15.

Since A and B have no predecessors, we compute their early start and finish times as [0, 1] and [0, 2].

Tasks C, D, and E cannot begin until both A and B have finished, which happens 2 weeks into the project. Therefore, for C, D, and E we have times of [2, 5], [2, 7], and [2, 5].

Since F, the final task, cannot begin until C, D, and E finish, we have its start time as 7 weeks. It finishes in $7 + 3 = 10$ weeks, which is thus the completion time of the entire project under ideal circumstances.

We conclude by summarizing the early start and early finish times for the project.

Task	Early Times
A	[0, 1]
B	[0, 2]
C	[2, 5]
D	[2, 7]
E	[2, 5]
F	[7, 10]

Skill Enhancer 6   What are the early start and early finish times for these tasks? How long will the entire project take? (Times are measured in months.)

	Time	Predecessors
A	2	—
B	3	—
C	4	A
D	3	B, C
E	4	B, C
F	5	D, E

Answer in Appendix E.

Late Start and Late Finish Times

late start

The early start times let us know the earliest we may begin a task. But we far more frequently need to know the latest time we may begin a task. We mean by "latest time" the time beyond which a delay will delay the completion of the entire project. The *late start* time is the latest time we may start a task without delaying the entire project. (We know—in principle at least—the project completion time, because we compute it in the process of determining all the early start times.) Associated with each late start is the

late finish

late finish time, the time at which at a task is finished when it begins at the late start time.

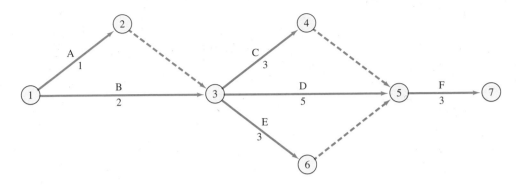

FIGURE 12.15
Computing early starts and finishes: the network.

We compute the late start times using methods similar to those used to compute the early start times. The important difference is this: We begin with the final task (instead of the initial task) and work right to left until we get to the tasks with no predecessors.

successor task   The opposite of a predecessor task is a *successor task*, a task that comes after another.

Late Start and Late Finish Times

1. Compute the completion time for the project by determining the early start and early finish times for each task.

2. The late finish time for the final task (or tasks) of the project is the project completion time.

3. Since any task comes before *all* its successors, the late finish time for any other task is the minimum of the late start times of its successors.

4. The late start time for a task is the difference between the late finish time and the time for completion.

   We will use curly bracket notation $\{a, b\}$ for the late start a and late finish b of any task.

   As we proceeded from left to right (or from start to finish) when analyzing a project with respect to early times, so must we proceed from right to left (or finish to start) when determining late start and late finish times.

Refer again to the network of Figure 12.14 and the task list on page 575. We learned that that project requires *11 units* of time for completion. What are the late start and finish times for each task?

The final task G finishes when $t = 11$; this is the late finish time. Since it takes one "click of the clock" to complete, the late start time is $11 - 1 = 10$. We write $G\{10, 11\}$.

E and F precede G. Since the late start of G is at 10, the latest that E and F can finish is also at $t = 10$. These tasks require 3 and 5 time units for their completion, so their late start times are $10 - 3 = 7$ and $10 - 5 = 5$.

In the same way, the late times for A and B are $\{5, 7\}$ and $\{3, 7\}$. That is, the absolute latest times we may begin these tasks without delaying the project are $t = 5$ for A and $t = 3$ for B. Make sure you see why the late times for tasks C and D are $\{2, 5\}$ and $\{0, 5\}$.

Figure 12.16 displays this network once again with both early and late times labeled.

slack   For some tasks, the early and late times are different. When there is this *slack* between early and late start times, as in tasks A, B, C, and E, then there is some choice in the

float   starting times. (Some texts refer to this slack time as *float*.)

In other tasks, such as D, F, and G, the early and late times are identical. In order to keep the project on schedule, we have no choice but to begin these tasks on the early and late start times (which are the same). These tasks lie in a continuous path from start to completion of the project's network, and they are *critical tasks* in the sense that failure to begin and end each at the proper time will delay the project. D, F, and G form the *critical path* for this network; hence the name of this technique (critical path method—CPM).

Example 7   Refer again to Example 6, where we see the task list

	Time	*Predecessors*
A	1	—
B	2	—
C	3	A, B
D	5	A, B
E	3	A, B
F	3	C, D, E

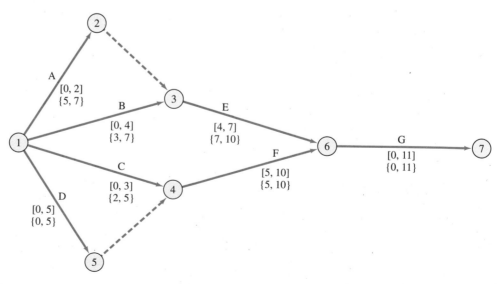

FIGURE 12.16
Early and late times for a project.

The network appears in Figure 12.15. This project lasts 10 weeks. What are the late start and finish times for the project? What are the critical tasks for the project?

Solution

Both the early and late times for this project appear in Figure 12.17. By inspection, the critical paths, along which the late and early times are identical, are *B*, *D*, and *F*.

Skill Enhancer 7   What are the late start and late finish times for these tasks? Times are in months, and we have seen in an earlier exercise that this project will last 15 months.

	Time	*Predecessors*
A	2	—
B	3	—
C	4	*A*
D	3	*B, C*
E	4	*B, C*
F	5	*D, E*

What are the critical tasks for this project? Draw the network for the project.

Answers in Appendix E.

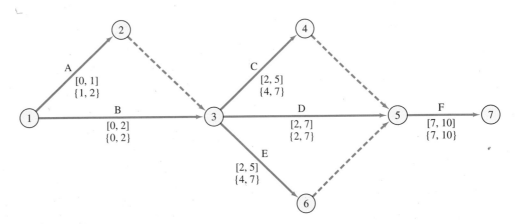

FIGURE 12.17
The early and late times for another project.

Example 8   The following table is one breakdown of some of the tasks involved in making a movie. Times are in months.

	Task	Time	Predecessors
A	Screen actors.	2	—
B	Arrange permissions for land where shooting will be done, etc.	3	—
C	Prepare sets.	3	B
D	Negotiate and sign contracts with actors.	6	A
E	Shoot the movie.	9	C, D
F	Prepare the advertising copy.	2	E
G	Edit the film.	3	E
H	Distribute advertising in newspapers, radios, TV, etc.	4	C, F
I	Reshoot some scenes.	5	G
J	Distribute film to theaters everywhere.	1	H, I

(a) What is the network for this project?

(b) What are the early start and finish times for each task?

(c) How many months does it take to make this movie from start to finish (getting the movie into theaters)?

(d) What are the late start and finish times for this project?

(e) What is the critical path for this project?

Solution

Here, in Figure 12.18, is the network with early and late times on it. The time for completion of the movie is 26 months, and the critical tasks seem to be A, D, E, G, I, and J.

Skill Enhancer 8 The following chart details some of the tasks involved in building a small bridge for a highway overpass. Times of completion are in months.

	Task	Time	Predecessors
A	Survey land.	1	—
B	Apply and receive permits.	2	—
C	Prepare and construct foundation and superstructure of bridge.	3	A, B
D	Observe environmental precautions.	8	A, B
E	Detail work on bridge.	4	C
F	Landscaping.	3	C
G	Final dedication.	1	D, E, F

Draw the network for this project. Determine the early start and finish times, and the total length of time needed for completion of this project. Determine the late start and finish times for the project. What are the tasks lying on the critical path of the project?

Answers in Appendix E.

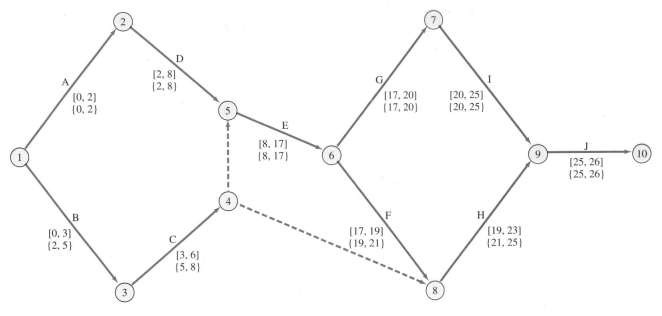

FIGURE 12.18
Making a movie.

Terror at the World Trade Center

A bombing occurred in the basement of one of the World Trade Center towers in New York City on Friday, February 26, 1993, shortly after noon. Although this act has been described as "the single most destructive act of terror ever committed on American soil," the authorities involved in maintaining the physical complex of the World Trade Center had to cope immediately with the problem of restoring the physical plant so tenants could return to work. The directors of the World Trade Center used CPM to manage this immense task.

There are thousands and thousands of subtasks that make up this task, from bracing girders to installing emergency lighting. These activities often affect each other. As we might expect, some activities are independent, others cannot begin until certain predecessors have completed, while groups of others may begin simultaneously. The CPM chart is generated by software running on personal computers, and as updates are made to this chart, the computer generates a new one several times a day.

Although this restoration project greatly resembles an ordinary construction project, certain important subtasks remind everyone that this project is different in important ways. The most important of these is clue-gathering by the Federal Bureau of Investigation and by the Bureau of Alcohol, Tobacco, and Firearms, over which the WTC managers have little control.

Tenants began returning to the WTC complex on April 1. That they could do so so soon after the damage occurred is a great testimony to both CPM and the dedication and hard work of the men and women involved. (For additional information on the use of CPM in the World Trade Center, see Matthew Wald's article "Aided by Computers,

Repairs are Charted" which appeared on page B5 of the Monday, March 8, 1993 edition of *The New York Times*.)

_____ *Exercises 12.2* _____

Each of Exercises 1–10 contains a task list with times of completion. Create a network appropriate for each such list including the early and late start and finish times. What is the time of completion for each project? What are the critical tasks for each project?

1.

	Time	Predecessors
A	1	—
B	4	—
C	2	A
D	2	B, C

2.

	Time	Predecessors
A	1	—
B	1	—
C	1	A
D	1	B, C
E	1	B, C
F	1	E
G	1	D, F
H	1	G
I	1	D, F
J	1	H, I

3.

	Time	Predecessors
A	1	—
B	2	—
C	2	—
D	1	—
E	3	A
F	4	B
G	5	C
H	6	D
I	1	E, F, G, H

4.

	Time	Predecessors
A	1	—
B	1	—
C	1	A
D	3	B, C
E	1	B, C
F	2	E
G	1	D
H	8	F, G

5.

	Time	Predecessors
A	2	—
B	3	—
C	3	—
D	2	—
E	2	—
F	1	D
G	2	A, B, C, E, F

6.

	Time	Predecessors
A	1	—
B	2	—
C	2	A, B
D	1	A, B
E	0.5	C, D

7.

	Time	Predecessors
A	2	—
B	1	—
C	1	A
D	1	B, C
E	1	A, D

8.

	Time	Predecessors
A	1	—
B	1	—
C	2	A
D	1	B
E	2	A, D
F	1	C, E
G	3	B
H	1	F, G

9.

	Time	Predecessors
A	1	—
B	1	—
C	4	—
D	1	A
E	2	A
F	2	B
G	1	B
H	2	D, E
I	1	F, G
J	3	C
K	11	H, I, J

10.

	Time	Predecessors
A	1	—
B	2	—
C	2	—
D	1	A, B
E	2	A, B
F	2	C, D, E

W **11.** Give at least one application for the CPM technique in these areas:
(a) corporate finance;   (b) real estate; and
(c) public works.

W **12.** Review the advantages of CPM. What are some disadvantages?

Applications

13. *(Marketing)* Mary Monohan is planning the annual sales meeting for the Moughton Hifflin Publishing Com-pany sales force. Here are some of the steps in the planning of the meeting. (Times are in weeks.)

	Task	Time	Predecessors
A	Develop theme.	2	—
B	Get editors' approval.	3	A
C	Prepare program.	2	—
D	Select location.	2	C
E	Firm up editors' travel plans.	1	B, D
F	Final double check with editors and upper management.	2	E
G	Prepare literature.	4	B, D
H	Make reservations.	4	G
I	Final details.	2	F, H

Draw the network for this project and compute early and late start and finish times. What is the time of completion for this project? What tasks comprise the critical path?

14. *(Experimental Biology)* The biology department at Yalvard University has received an immense grant of money to investigate a certain fatal childhood disease. It involves setting up some sophisticated and special-purpose equipment. In order to get the new laboratory up and running, the following project needs to be completed.

	Task	Time	Predecessors
A	Decide what equipment is needed.	4	—
B	Order equipment.	7	A
C	Install equipment.	8	B
D	Set up training laboratory.	5	A
E	Conduct the training course.	8	D
F	Test the new system.	5	C, E

(Times are in weeks.) How long will the whole project take? What are the critical tasks? What are the early start and finish and late start and finish times for these six tasks?

How often does it happen that we know with deadly certainty the precise time for the completion of a task? The answer must be for only a small fraction of tasks. The imprecision of weather, the mails, and the fact that we may not even be sure of what each task entails help contribute to this uncertain state.

It is more reasonable to specify optimistic and pessimistic times of completion and the most likely time of completion for each task. With this information, together with a few assumptions coupled with our work on probability and probability distributions, it becomes possible to estimate the likelihood that we can execute a project in some given period of time.

Optimism, Pessimism, and the Beta Distribution

Let us identify an optimistic and pessimistic time of completion a and b for a task. That means that no matter how lucky we are, no matter how smoothly things go, we can never complete a task in less than a units of time. At the same time, no matter how serious the delays, how daunting the circumstances, it will never take us longer than b units of time to complete the task.

Earlier in this text, we examined the *normal distribution* function, which helps us compute the probabilities for a wide variety of events. For any event that follows a normal distribution, it is always true that

$$P(t > T) > 0$$
$$P(t < T) > 0$$

where T is some given instant of time. To be sure, the probabilities may be very close to zero, but they are definitely *not* zero.

Evidently, if any event possesses an optimistic and pessimistic time of completion, it cannot follow a normal distribution. For we have just said that there are times a and b such that

(12.1) $\qquad\qquad\qquad\qquad P(t > b) = 0$

(12.2) $\qquad\qquad\qquad\qquad P(t < a) = 0$

beta distribution   Mathematicians have developed a number of probability distribution functions in addition to the normal distribution. If, in addition to a and b, we can identify the most likely time m for the tasks' completion, then it may be useful to assume that the probability of completion follows the *beta distribution*. Although further mathematical aspects of the beta distribution need not concern us, we need to point out that for this distribution, Equations (12.1) and (12.2) are true.

If we know a, b, and m for the task, then we may assert that the *mean* μ and *variance* σ^2 of the completion times are given by

(12.3)
$$\mu = \frac{a + 4m + b}{6}$$

(12.4)
$$\sigma^2 = \frac{(b - a)^2}{36}$$

Example 9   A certain task will never take more than four weeks to perform, and can never take less than four days. Most likely, though, it takes two weeks. If the time of completion is distributed according to a beta distribution, what is the average time of completion for this task? What is the variance? What is the standard deviation of completion?

Solution

We convert all units of time to days. According to the problem, we have

$$a = 4 \, \text{days}$$
$$m = 2 \, \text{weeks} = 14 \, \text{days}$$
$$b = 4 \, \text{weeks} = 28 \, \text{days}$$

using the same notation of Equations (12.3) and (12.4). By these equations, then, we have

$$\mu = \frac{4 + 4(14) + 28}{6} = 14\frac{2}{3}$$
$$\sigma^2 = \frac{(28 - 4)^2}{36} = 16$$

Since the standard deviation σ is the square root of the variance, we have

$$\sigma = 4$$

Skill Enhancer 9   In a laboratory, fruit flies can be trained to do a simple task. It will never take more than ten minutes to do, and can never take less than four minutes. Most likely, though, it takes seven minutes.
(a) What are a, b, and m for this problem?   (b) If the time of completion is distributed according to a beta distribution, what is the average time of completion for this task?
(c) What is the variance?   (d) What is the standard deviation of completion?

Answers in Appendix E.

Assumptions of PERT

We use the refinement of CPM called *PERT* to analyze project completion times under conditions of uncertainty. The method is founded upon the following assumptions.

1. The times of completion of all tasks are *independent* of one another.
2. The times of completion of each task follow a *beta distribution*.
3. If we were to analyze the project under CPM, we would identify the critical path of the project. We assume that the *same critical path* controls the project under the PERT analysis.
4. The completion time for the *entire* project follows a *normal distribution*.

In practice, all of these assumptions hold (or hold close enough) to ensure that the PERT method is useful in practice. The third assumption, though, is the most problematical. To see why, let's examine the trivial network of Figure 12.19. In this figure we adopt the notation

$$\boxed{a \mid m \mid b \mid \mu \mid \sigma^2}$$

to identify the optimistic, most likely, and pessimistic times and the mean and variance (which we derive from the first three values using Equations (12.3) and (12.4). Here, these numbers refer to days. In this figure, task *B* forms the critical path. (Make sure you see why.)

However, we don't know the times of completion with certainty. The pessimistic completion time is 7 days, and in the event that things go bad and task *A* requires that time, task *A* becomes the critical path (since 7 > 5). By our assumption, we will not allow this situation to arise.

In the next few examples, we examine the ways we can answer questions about the likelihoods of completion of a project using PERT.

Example 10   Refer to the project illustrated in Figure 12.20. With what likelihood will the project be completed in 10 days or less?

Solution

First, we need to identify the critical path. Figure 12.20 shows that for task *A*, the average time of completion will be 7 days. For *B*, the average is 9 days. Therefore, *B* defines the critical path. We assume that it remains the critical path under all conditions.

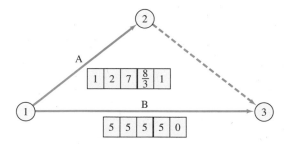

FIGURE 12.19
A trivial network for PERT analysis.

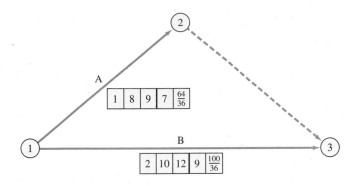

FIGURE 12.20
A simple project for PERT analysis. (Times are in days.)

Since the variance for this task is $\frac{100}{36}$, the standard deviation is

$$\sigma = \frac{10}{6}$$

The Z score for the data value of 10 is

$$z = \frac{10 - 9}{\frac{10}{6}} = 0.6$$

The area under the standard normal distribution function between the axis of symmetry (the y axis) and a Z-score value of 0.6 is, by Table 1 in Appendix 4, 0.2257. As Figure 12.21 shows, the probability we want is the *sum* of this area and the area under the left half of the curve. This probability is therefore

$$0.5 + 0.2257 = 0.7257$$

Skill Enhancer 10
(a) For this same project, what is the probability that it will take more than 10 days to complete?

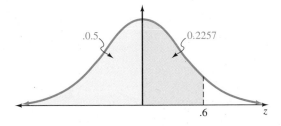

FIGURE 12.21
The probability for the problem posed by Example 10 is the sum of the areas of both shaded portions.

(b) What is the Z score corresponding to a completion date of 11 days? Referring to Table 1, what is $A(z)$ for this value?

(c) What part of the standard normal curve represents the probability that the project will require 11 or more days for completion? Describe this section in words.

(d) What is the probability that this project will require 11 days or more for completion?

Answers in Appendix E.

Typically, there will be more than one activity on the critical path. In that case, we appeal to two additional facts applied to tasks performed in sequence:

1. To compute *the average time for the project's completion, add* the averages for all the tasks on the critical path.

2. To compute *the variance for the project's completion, add* the individual variances for all the tasks on the critical path.

Example 11   A network for the tasks making up a project appears in Figure 12.22. The critical path is composed of the tasks C, D, and E; the estimates of completion times are indicated in the figure.

(a) What are the average and variance of the times of completion for these tasks?

(b) What are the average and variance of the time of completion for the project as a whole?

Solution

(a) For C, we have $a = m = b = 1$. Using Equations (12.3) and (12.4), we determine that

$$\mu_C = \frac{1 + 4 + 1}{6} = 1$$

$$\sigma_C^2 = (1 - 1)^2 \, (36) = 0$$

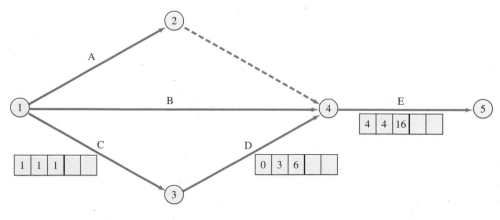

FIGURE 12.22
Using PERT to analyze a critical path containing three tasks. Times are in days.

In the same way, for D, since $a = 0$, $m = 3$, and $b = 6$, we have

$$\mu_D = \frac{0 + 3(6) + 6}{6} = 4$$

$$\sigma_D^2 = \frac{(6-0)^2}{36} = 1$$

Finally, for E, we use the facts $a = m = 4$ and $b = 16$ to compute

$$\mu_E = \frac{4 + 4(6) + 16}{6} = \frac{44}{6} = 7\frac{1}{3}$$

$$\sigma_E^2 = \frac{(16-4)^2}{36} = 4$$

(b) For the entire project, we add the averages and the variances.

$$\mu = \mu_C + \mu_D + \mu_E = 1 + 4 + 7\frac{1}{3} = 12\frac{1}{3}$$

$$\sigma^2 = \sigma_C^2 + \sigma_D^2 + \sigma_E^2 = 0 + 1 + 4 = 5$$

Skill Enhancer 11   Redo these calculations given that the times for the tasks are

C	2	2	2		
D	0	3	12		
E	2	4	8		

That is, complete this chart. What are the average and variance for completing the whole project?

Answer in Appendix E.

Example 12   Consider again the project illustrated in Figure 12.22.
(a) What is the probability that the task will be completed in 12 days or less?
(b) What is the probability of completing the project in 11 days or less?
(c) What is the probability of completing the project in more than 13 days?

Solution

In all cases, we compute the Z score of the given times.
(a) Here,

$$z = \frac{12 - 12\frac{1}{3}}{\sqrt{5}} = -0.149$$

and $A(z) = 44\%$, the probability for completing the task in 12 days or less.
(b) The Z score for $t = 11$ is

$$z = \frac{11 - 12\frac{1}{3}}{\sqrt{5}} \approx -0.60$$

with a corresponding value of $A(z) = 0.2251$. Because z is negative, we subtract the value of $A(z)$ from one-half (which is the area of the left half of the normal curve). The probability is $0.5 - 0.2251 = 0.2755$.

(c) For $t = 13$, the Z score is

$$z = \frac{13 - 12\frac{1}{3}}{\sqrt{5}} \approx 0.3$$

with the same corresponding value of $A(z) = 0.1179$. The probability of completing the project in 13 days or more is 0.3821.

Exercises 12.3

In Exercises 1–8, you are given the optimistic, most likely, and pessimistic estimates for times of completion of a task. Under the assumption that the time of completion follows the beta distribution, determine the average time of completion and the variance of the completion time. See Example 9.

1. 2 10 12
2. 2 2 2
3. 1 1 1
4. 6 8 10
5. 7 8 9
6. 14 16 18
7. 3 5 9
8. 1 3 7

The information in each of Exercises 9–14 consists of the time estimates for the tasks lying on the critical path for a certain project. Determine the mean and variance for each such task, and the mean and variance for the entire project. See Example 11.

9. A 1 2 3
　　 B 0 3 3

10. A 2 2 5
　　 B 5 6 9

11. A 9 9 9
　　 B 8 8 8
　　 C 7 7 7

12. A 3 6 8
　　 B 3 4 9
　　 C 2 8 9

13. A 2 2 9
　　 B 3 8 9
　　 C 3 4 8

14. A 2 3 4
　　 B 3 3 3
　　 C 5 5 5

15. (a) Suppose the times for completion of a task are dis-
tributed according to the beta distribution, and there are the estimates a, b, and m (using the notation from throughout the chapter). What is the effect on the average and variance μ and σ^2 if these parameters are each doubled?

(b) What is the effect if each parameter is tripled?

W **(c)** What is the effect if each parameter is multiplied by a general factor, call it n?

16. (a) Suppose 1 is added to each of the parameters a, b, and m that characterize some process governed by the beta distribution. What effect does this have on the average and variance?

(b) What effect does subtracting 5 have?

W **(c)** What effect does adding a general term T have on these parameters?

Applications

17. *(The Traveling Salesperson)* It takes a certain time for a salesperson on the road to visit a customer and make the sales pitch. Experience shows that this time is distributed according to the beta distribution. For the visit of one saleswoman to one particular customer, it can never take less than 2 hours or more than 8. The most likely time of visit is 3 hours. What are the average and variance for the times of these visits?

18. *(Manufacturing Quality Control)* Studies show that the time a light bulb lasts before burning out follows a beta distribution. The most likely life of a bulb is 1,200 h. The mean life of a bulb is 1,100 h with a standard deviation of 40 h. What are the shortest and longest bulb lifetimes?

19. *(House Construction)* Figure 12.23 displays the network for the construction of a house. Tasks B, E, F, and I form the critical path. (Times are in weeks.)

(a) What is the task list of predecessors for this project?

(b) What are the mean and variance for the time of completion of the house?

(c) The builders get a bonus if the house is finished in 12 weeks or less. With what likelihood will they earn this bonus?

(d) The builders have agreed to suffer a financial penalty in their fee if the house takes 15 weeks or longer to build. What is the probability that they will suffer the penalty?

20. *(A Data Processing Training Program)* The Torus Tire Company is converting its inventory procedures to computer. Figure 12.24 displays the network for the project of training the employees to learn and use this new system. The critical path consists of tasks A, C, E, and G. The times in the figure are in weeks.

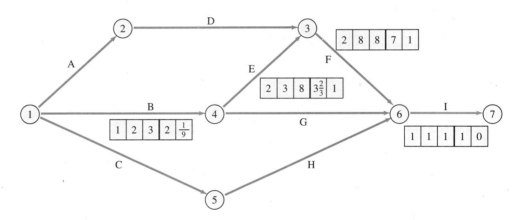

FIGURE 12.23
Building a house.

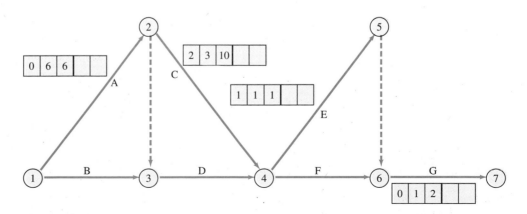

FIGURE 12.24
A data processing training program.

(a) What are the means and variances for the tasks on the critical path?

(b) What is the mean and variance for the completion of the entire project?

(c) Write the list of predecessors for each task in this project.

(d) What is the Z score corresponding to 9 weeks? corresponding to 12 weeks?

(e) The consulting company charged with executing this project gets a bonus if the training program is completed in 9 weeks or less. With what probability will they earn this bonus?

(f) The employees will suffer a decrease in their end-of-year bonus if they do not learn the new system within 12 weeks. What is the probability that they will suffer this penalty?

CHAPTER REVIEW

Terms

CPM	critical path method	PERT
network	graph	arcs
nodes	immediate predecessors	activity on arc
dummy activity	early start	early finish
recursive technique	late start	late finish
successor task	slack	float
beta distribution		

Key Concepts

- We may draw networks to represent the predecessor relationships pertaining to a set of tasks.

- We use the *critical path method* to determine the length of the project and the slack associated with each task of the project. The important assumption in this method is that the time for completing each separate subtask is fixed and known with certainty. Simple addition and subtraction enable us to compute the early start and finish times and the late start and finish times for each task, from which the other calculations follow.

- If the time of distribution of a task follows the *beta distribution,* then we can compute the average time of completion and the variance from the optimistic, most likely, and pessimistic times of completion.

- We use the *PERT* method to analyze times of completion when the time of completion of each of the subtasks follows the beta distribution. (Several additional assumptions are made also.) This method enables us to assess the likelihood (probability) that a project will or will not be completed within a certain period of time.

Review Exercises

Refer to Figure 12.25 for Exercises 1–2.

1. Write the task list for the network on the left.

2. Write the task list for the network on the right.

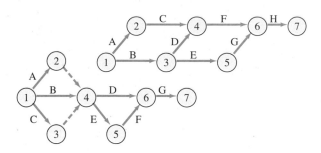

FIGURE 12.25

Determining task lists for these tasks.

For each of Exercises 3–4, create the network corresponding to the given task list.

3.

	Predecessors
A	—
B	—
C	—
D	A
E	B
F	C
G	D, E
H	F, G

4.

	Predecessors
A	—
B	—
C	A
D	A
E	B, C, D

In the left figure of Figure 12.25, we know that each time takes the number of days in the following table.

A	1
B	2
C	1.5
D	4
E	3
F	2
G	1

Use this information to answer Exercises 5–8.

5. Compute the early start and early finish times for each task.

6. Compute the late start and late finish times for each task.

7. How long will it take to complete this project?

8. What is the critical path for the project?

In each of Exercises 9–12 you are given the optimistic, most likely, and pessimistic estimates for completion of a task that follows the beta distribution. The times, in months, are given in the standard notation of this chapter. What are the means and variances of these tasks?

9. $\boxed{3\,|\,4\,|\,9\,|\;|\;}$ **10.** $\boxed{7\,|\,8\,|\,9\,|\;|\;}$

11. $\boxed{10\,|\,12\,|\,14\,|\;|\;}$ **12.** $\boxed{1\,|\,5\,|\,13\,|\;|\;}$

13. Refer again to the right-hand network of Figure 12.25. Let us suppose that this represents the network for a project to test a new nuclear submarine. The critical path comprises tasks *B*, *D*, *F*, and *H*. We have this table.

B	$\boxed{1\,|\,2\,|\,3\,|\;|\;}$
D	$\boxed{2\,|\,3\,|\,4\,|\;|\;}$
F	$\boxed{2\,|\,3\,|\,10\,|\;|\;}$
H	$\boxed{3\,|\,4\,|\,11\,|\;|\;}$

where times are in months.
(a) What are the means and variances for the completion times of the tasks on the critical path?
(b) What is the average time for completion of this project? What is the variance? What is the standard deviation?
(c) What is the probability of completing the project in 10 days or less?

APPENDIX A

ALGEBRA REVIEW

INTRODUCTION

This brief appendix contains brief descriptions of a few important topics from standard elementary and intermediate algebra.

We present it in the hope that students will find it useful and convenient to refer to "the back of the book" to refresh their memories. In no way, though, is it intended to be a complete and comprehensive discussion of all of college algebra.

Section A.1   ARITHMETIC OF SIGNED NUMBERS

absolute value   The *absolute value* of a number is the value of the number without its sign. For any quantity x, its absolute value is written $|x|$. Formally, the rule is

$$|a| = \left\{ \begin{array}{l} a, \; a > 0 \\ -a, \; a < 0 \end{array} \right\}$$

Example 1   Compute $|14.7|$, $|-0.003|$, and $|+1|$.

Solution

We have $|14.7| = 14.7$, $|-0.003| = 0.003$, and $|+1| = 1$.

Addition and Subtraction

To add two numbers with the *same* signs, add the absolute values of the numbers and affix the sign. Examples:

$$11 + 5 = 16$$
$$(-3) + (-7) = -10$$

To add two numbers with *different* signs, subtract the smaller absolute value from the larger absolute value, and affix the sign that accompanies the larger absolute value. For example, $7 + (-2) = 5$ and $3 + (-8) = -5$.

Formally, we transform subtraction problems into addition problems and use the rules for addition. To do this, change the sign of the number following the minus sign, and *add* this number. Examples:

$$-5 - (-2) = -5 + (+2) = -3$$
$$7 - 3 = 7 - (+3) = 7 + (-3) = 4$$

Example 2   Use the laws of signed numbers to compute $3 + 5$, $(-11) + (-12)$, $(-12) + 13$, $-3 - (-3)$, and $5 - (+5)$.

Solution

$3 + 5 = 8$, $(-11) + (-12) = -23$, $(-12) + 13 = 1$, $-3 - (-3) = 0$, and $5 - (+5) = 0$.

Multiplication and Division

Dividing or multiplying two numbers with the *same* sign always yields a positive answer. (Division by zero is *never* allowed.) Examples:

$$(-5) \times (-2) = 10$$
$$\frac{(-12)}{(-3)} = 4$$

Dividing or multiplying two numbers with different signs always results in a negative answer. Examples:

$$(-7) \times 3 = -21$$

$$\frac{100}{(-10)} = -10$$

Example 3   Compute $(+3)(-12)$, $(-1)(-1)$, $\frac{-36}{+6}$, and $\frac{-48}{-12}$.

Solution

$(+3)(-12) = -36$, $(-1)(-1) = +1$, $\frac{-36}{+6} = -6$, and $\frac{-48}{-12} = +4$ (or just 4).

We conclude this section with a few observations. Whenever we do arithmetic on purely positive numbers, the laws of signed numbers are identical to the laws of arithmetic familiar to us from elementary school.

Remember, any expression of the form $-x$ is equivalent to $-1 \times x$. Of course, we are always allowed to write $1 \times x$ for x and vice versa.

Finally, beginning algebraists may be confused at evaluating expressions similar to

$$3.1416 - 2.7183$$

Should this be interpreted as

$$3.1416 - (+2.7183)$$

or as

$$3.1416 + (-2.7183)$$

In all such cases, these are equivalent expressions. We may feel free to choose the most convenient expression to work with, or to choose the form we are most comfortable with.

Section A.2   FRACTIONS

numerator

denominator

We use fraction notation as a shorthand for division. For example, $\frac{1}{4}$ means the same as $1 \div 4$. The *numerator* is the upper number in a fraction. We call the bottom number the *denominator*. Generations of school children remember this because the d in "denominator" is like the d in "down" number (the bottom number).

Rewriting Fractions with Larger Denominators

We often need to transform a fraction of the form $\frac{a}{b}$ to one of the form $\frac{c}{d}$, where d is greater than b and is a multiple of d. Let n be the multiple:

$$d = nb$$

Then an equivalent fraction is $\frac{na}{d}$. This rule, confusing to state in this generality, is easy to apply.

Example 4   Convert $\frac{3}{4}$ to an equivalent fraction with 48 in the denominator.

Solution

The two denominators are 4 and 48. Since $4 \times 12 = 48$, we have $n = 12$. Therefore, the equivalent fraction is

$$\frac{12 \cdot 3}{12 \cdot 4} = \frac{36}{48}.$$

Adding or Subtracting Fractions

If two fractions have the *same* denominator, simply add or subtract the numerators and keep the same denominator. For example, $\frac{2}{9} - \frac{5}{9} = \frac{2-5}{9} = -\frac{3}{9} = \frac{1}{3}$.

If two fractions have *different* denominators, convert them to fractions with identical denominators and proceed as above. To do this, we find the lowest common multiple of the two denominators, but we don't plan to discuss this topic in this appendix. In any event, suppose the two unlike denominators are a and b. Then it is *always* true that both fractions can be converted to ones with denominators ab.

Example 5   (a) Determine $\frac{1}{3} - \frac{1}{2}$.   (b) What is $\frac{1}{12} + \frac{5}{18}$?

Solution

(a) We convert both fractions to sixths.

$$\frac{1}{3} = \frac{2}{6}$$
$$\frac{1}{2} = \frac{3}{6}$$

Therefore,

$$\frac{1}{3} - \frac{1}{2} = \frac{2}{6} - \frac{3}{6}$$
$$= \frac{2-3}{6} = -\frac{1}{6}$$

(b) The lowest common denominator (least common multiple) for these fractions is 36. It is quite possible, though, that a student may fail to notice this, or forget how to determine this. It is quite proper, since

$$12 \times 18 = 216$$

to convert both fractions to 216ths. 216 is a common denominator of both fractions; it is just not the smallest common denominator.

$$\frac{1}{12} = \frac{18 \times 1}{18 \times 12} = \frac{18}{216}$$
$$\frac{5}{18} = \frac{12 \times 5}{12 \times 18} = \frac{60}{216}$$

Therefore,

$$\frac{1}{12} + \frac{5}{18} = \frac{18}{216} + \frac{60}{216} = \frac{78}{216}$$

which can be reduced by cancelling 6 from numerator and denominator.

$$\frac{78}{216} = \frac{13}{36}$$

Multiplying and Dividing Fractions

The rules for multiplying and dividing fractions are much easier to state than those for adding and subtracting. To *multiply* fractions, form the product by multiplying the numerators together and the denominators together.

$$\frac{3}{4} \times \frac{1}{2} = \frac{3 \times 1}{4 \times 2} = \frac{3}{8}$$

Multiply a whole number n by considering it to be a fraction $\frac{n}{1}$.

$$3 \times \frac{5}{8} = \frac{3}{1} \times \frac{5}{8} = \frac{3 \times 5}{1 \times 8} = \frac{15}{8}$$

To divide one fraction by another, invert this other fraction and proceed as in multiplication ("invert and multiply").

$$\frac{1}{3} \div \frac{1}{2} = \frac{1}{3} \times \frac{2}{1} = \frac{2}{3}$$

Example 6   Compute (a) $\frac{7}{11} \times \frac{22}{49}$;   (b) $\frac{3}{8} \div \frac{3}{4}$.

Solution

(a) We have

$$\frac{7}{11} \times \frac{22}{49} = \frac{7 \times 22}{11 \times 49}$$

$$= \frac{2}{7}$$

(b) The division $\frac{3}{8} \div \frac{3}{4}$ is the same as

$$\frac{3}{8} \times \frac{4}{3} = \frac{1}{2}$$

Section A.3   MANIPULATING TERMS AND FACTORS IN EQUATIONS

commutative   Addition and multiplication are both *commutative* operations:

$$A + B = B + A$$

$$A \times B = B \times A$$

for any quantities A, B. When applied to numbers, these statements express the well-known facts that, for example, $4 + 7 = 7 + 4$ and $3 \times 9 = 9 \times 3$ or $x + 10 = 10 + x$ and $7y = y \cdot 7$.

associative   Addition and multiplication are both *associative*; that is, for any quantities A, B, and C, then

$$A + (B + C) = (A + B) + C$$
$$A(BC) = (AB)C$$

Examples: $x + (y + 2) = (x + y) + 2$
$0.5(5 \times 7) = (0.5 \times 5)(7) = 17.5$

For any quantities A, B, and C,

$$A(B + C) = AB + AC$$

distributive law   Use this *distributive law* to evaluate expressions of the form

$$4(3 + 5) = 4 \times 3 + 4 \times 5 = 32$$

and

$$x_1(x_1 + x_2 + 2) = x_1^2 + x_1 x_2 + 2x_1$$

Order of Operations

We need to agree on the *order* in which operations are to be executed in case we are confronted with a long expression to evaluate. In general, valid algebraic expressions are composed of simpler expressions separated by the operations $+$, $-$, $\times$, $\div$, or exponentiation; other, more advanced operations will not be treated here. The expressions on either side of one of these operations may be a constant, a variable (or variable raised to a power), or a valid expression enclosed in parentheses.

Evaluate parenthesized portions of the expression first. If there is more than one level of parentheses, evaluate the deepest level first.

Next, evaluate exponential expressions. Exponential quantities are reviewed in the next section.

Finally, evaluate the remaining operations—$+$, $-$, $\times$, and $\div$—according to a precise pecking order. Multiplication and division precede addition and subtraction. The four signs ($+$, $-$, $\times$, and $\div$) act like tiny magnets, each with different strengths. The magnets attract the operands on either side of them. $\times$ and $\div$ are magnets of equal strength, as are $+$ and $-$, but $\times$ and $\div$ are stronger than $+$ and $-$. When an expression contains both $\times$ and $\div$ or both $+$ and $-$, evaluate them from left to right as they are written. (I thank D. E. Knuth of Stanford University for the magnet analogy.)

Order of Operations

$$( ) \qquad x^n \qquad \begin{array}{c} \times \\ \div \end{array} \qquad \begin{array}{c} + \\ - \end{array}$$

Using these rules, we can easily evaluate any arithmetic expression.

$$12/(3+5) = \frac{12}{8} \qquad\qquad \text{evaluate parentheses first}$$
$$= \frac{3}{2}$$

$$12/3 + 5 = 4 + 5 \qquad\qquad \text{evaluate division first}$$
$$= 9$$

$$12/2 \times 3/4 = \left[\left(\tfrac{12}{2}\right) \times 3\right]/4 \quad \text{evaluate from left to right}$$
$$= 6 \times 3/4 = 4.5$$

The concepts in this section make it possible to evaluate and simplify many expressions.

Example 7   Simplify (and evaluate if possible):
(a) $2x + 3y - 2x - y$;   (b) $2 + 2 \times 2 + 2 \times 2$; and   (c) $(x + y)(2x - y)$.

Solution
(a) $2x + 3y - 2x - y = 2x - 2x + 3y - y = 2y$ (b) $2 + 2 \times 2 + 2 \times 2 = 2 + 4 + 4 = 10$
(c) We use the distributive law several times.

$$(x + y)(2x - y) = (x + y)2x - (x + y)y$$
$$= 2x(x + y) - y(x + y)$$
$$= 2x^2 + 2xy - xy - y^2 = 2x^2 + xy - y^2$$

Section A.4   EXPONENTS

exponential notation   For any positive integer n and any real number x, we agree to define *exponential notation* as a shorthand for repeated multiplication. By definition,

$$x^1 = x$$
$$x^n = x \cdot x \cdots\cdots x \quad (n \text{ factors}, n > 0)$$

$$x^{-n} = \frac{1}{x^n}$$

$$x^0 = 1$$

Here, x is called the *base*, and n is the *exponent* or *power*.

Examples:

$$100^1 = 100$$

$$9^2 = 9 \times 9 = 81$$

$$y^{-1} = \frac{1}{y}$$

$$2^{-2} = \frac{1}{2^2} = \frac{1}{4}$$

$$987,654,321^0 = 1$$

The following box summarizes the rules for manipulating factors with the same base.

Algebra of Exponents

$$x^m \cdot x^n = x^{m+n}$$

$$(x^m)^n = x^{mn}$$

$$(xy)^m = x^m y^m$$

$$\frac{x^n}{x^m} = x^{n-m}$$

$$\left(\frac{x}{y}\right)^n = \frac{x^n}{y^n}$$

Example 8   Use the laws of exponents to evaluate:
(a) $4^3 \times 4^{-1}$;   (b) $(2^2)^2$;   (c) $a^2 b^2$; and   (d) $\frac{a^2 b^3}{ab^{-2}}$.

Solution

(a) $4^3 \times 4^{-1} = 4^{3+(-1)} = 4^2 = 16$.
(b) $(2^2)^2 = 2^4 = 16$.
(c) Since (in general) $a \neq b$, the laws of exponents do not apply and we may not simplify this expression any further.
(d) We have

$$\frac{a^2 b^3}{ab^{-2}} = a^{2-1} b^{3-(-2)}$$

$$= a^1 b^5 = ab^5$$

Section A.5   *QUADRATIC EQUATIONS OF ONE VARIABLE*

quadratic equation   One common type of equation is the *quadratic equation* of one variable. The general form for such an equation is

$$ax^2 + bx + c = 0 \qquad\qquad \textit{Quadratic Equation}$$

where x is the variable and a, b, and c are three known constants. In terms of these constants, the solution is given by

$$x = \frac{-b \pm \sqrt{b^2 - 4ac}}{2a} \qquad \textit{The Quadratic Formula}$$

This represents *two* possible solutions, as one chooses the $+$ or $-$ answer.

Example 9   Solve $x^2 + 2x - 15 = 0$.

Solution

Upon comparison with the general form, we see that $a = 1$, $b = 2$, and $c = -15$. The solution is therefore

$$x = \frac{-2 \pm \sqrt{2^2 - 4 \cdot 1 \cdot (-15)}}{2 \cdot 1}$$
$$= \frac{-2 \pm \sqrt{64}}{2} = -1 \pm 4$$

The two roots x_1 and x_2 are

$$x_1 = -1 - 4 = -5$$
$$x_2 = -1 + 4 = 3$$

Exercises A.5

Use the laws of signed numbers to compute the answers to Exercises 1–12.

1. $-4 + 0$

2. $13 - (-12)$

3. $-6 - (3)$

4. $-17 - 12$

5. $0 + (+5)$

6. $2 + 2$

7. $+0 \times (-12)$

8. $\frac{0}{-1,000}$

9. $\frac{-3}{0}$

10. $-7 \times (-4)$

11. $\frac{-4}{2}$

12. $\frac{6}{3}$

Compute the answers to Exercises 13–24.

13. $\frac{1}{2} + \frac{1}{2}$

14. $\frac{3}{4} - \frac{1}{4}$

15. $\frac{1}{9} - \frac{1}{3}$

16. $\frac{1}{5} - \frac{1}{6}$

17. $\frac{7}{10} + \frac{2}{5}$

18. $\frac{1}{5} - \frac{2}{3}$

19. $\frac{1}{2} \times \frac{1}{2}$

20. $\frac{3}{4} \times 2$

21. $\frac{1}{2} \div \frac{1}{2}$

22. $\frac{2}{3} \div \frac{1}{5}$

23. $\frac{1}{9} \div 3$

24. $2\frac{2}{5} \div \frac{5}{6}$

Evaluate the expressions in Exercises 25–30.

25. 3^{-3}

26. $(-3)^{-3}$

27. $(2^3)^2$

28. $a^2 b^{-3} c a^{-2} bc$

29. $8^2 u^3$

30. $\frac{xyz}{(xyz)^2}$

Use the quadratic formula to solve the equations in Exercises 31–36.

31. $x^2 + x - 1 = 0$

32. $x^2 - 9 = 0$

33. $x^2 - 5x + 6 = 0$

34. $x^2 - x - 6 = 0$

35. $2x^2 - x - 1 = 0$

36. $3x^2 - x - 1 = 0$

APPENDIX B

GEOMETRIC SERIES

B.1 Working with Geometric Series

INTRODUCTION

Computations, calculations, and derivations frequently involve expressions that are long sequences of terms to be added. Indeed, sometimes such a long sequence may contain an infinite number of terms (whatever that may mean). This situation arises so often that it is helpful to study these sequences separately.

The most common such sequence of terms is called a *geometric series*. We have met it several times in the context of deriving certain formulas. The ability to replace long series with a short, equivalent expression means that it is easier to write certain results and that we will make fewer mistakes while writing these results. For this reason, and due to its inherent interest, we present certain elementary results in this appendix.

Section B.1   *WORKING WITH GEOMETRIC SERIES*

series   A *series* is a sequence of numbers to be added together. Each number in this sequence is
term   a *term* in the series. Interesting series are those in which each term relates to the terms
that precede it in some well-defined way. One such series might be

$$2 + 1 + \frac{1}{2} + \frac{1}{4} + \frac{1}{8}$$

geometric series   in which each term is $\frac{1}{2}$ of the term before it. This is a particular example of a *geometric series*, in which the ratio of one term to its successor (the term that follows it) is constant throughout the series. For any geometric series, there are numbers a and r such that the series may be written

$$a + ar + ar^2 + ar^3 + \cdots + ar^n$$

A natural question is to determine the *sum* of the series as a function of a, r, and n.

To do this, let this sum be denoted by S. That is,

$$S = a + ar + ar^2 + ar^3 + \cdots + ar^n \tag{B.1}$$

Multiplying this equation by r yields

$$rS = ar + ar^2 + ar^3 + \cdots + ar^n + ar^{n+1} \tag{B.2}$$

If we subtract Equation (B.2) from (B.1), we have

$$S = a + ar + ar^2 + ar^3 + \cdots + ar^n$$
$$\underline{-rS = \quad - ar - ar^2 - ar^3 - \cdots - ar^n - ar^{n+1}}$$
$$(1 - r)S = a - ar^{n+1}$$

so that

$$S = a\frac{r^{n+1} - 1}{r - 1} \tag{B.3}$$

There is one important special case. This occurs when the absolute value of r is strictly less than 1, and we sum an infinite number of terms in the series. In this case,

$$S = a\frac{1}{1 - r}$$

Example   Evaluate $1 + a^2 + \cdots + a^n$.

Solution

This is a geometric series, because the ratio of one term to its successor is a. The a in this series plays the role of r in Equation (B.3). The sum S is therefore

$$S = \frac{a^{n+1} - 1}{a - 1}$$

HYPOTHESIS TESTING: CHI-SQUARE METHODS

C.1 Chi-Square Analysis

INTRODUCTION

We have used statistics to find out something about data when we knew the nature of the model (see Chapter 9). A typical question: How probable is a certain combination of the dice when we know that all faces are equally likely? Can we reverse this pattern? Upon observing the results of an experiment, may we use these results to learn something about the underlying model?

Much of the time, we have only data and need to find the underlying patterns. In such circumstances, we will use statistical techniques like χ^2 to help resolve these issues.

Section C.1 *CHI-SQUARE ANALYSIS*

If we roll dice many times, can we use the results to conclude that the dice are fair? *Chi-square* analysis is helpful in this regard. *Chi* is the Greek letter χ; we often write *chi-square* as χ^2. Speakers of English often pronounce *chi* like *kie* (rhymes with *pie*), although it should be pronounced with a more gutteral initial consonant sound.

Identifying the Hypothesis

We use the term *experiment* in its probabilistic sense. *The first thing to do in any experiment is to identify the hypothesis that needs testing.* As an example, we have a certain coin. Is it "fair" in the statistical sense? (When we toss it, will we observe an equal number of heads and tails in the long run?) We will toss the coin many times in order to test the hypothesis that the coin is fair. If this hypothesis is true, then the probability of heads and tails is $P(H) = P(T) = \frac{1}{2}$.

Even if the coin obeys this hypothesis, we expect that in any experimental test sequence of tosses, there may be a substantial difference between what we *expect* and what we *observe*. How much deviation from the expected results is permissible before we must reject our hypothesis?

Contingency Table and Chi-Square

A statistical measure that helps answer this question contains a mixture of the observed and expected results. Begin by listing the observations and expectations systematically in a *contingency table*, a table with one row each for the observations and expectations, and one column for each possible outcome. A final column shows the total observations and expectations. The *expected results* are those that we would expect *assuming our hypothesis is true*. We see a contingency table for a sequence of 10 coin tosses in Table C.1.

contingency table

expected results

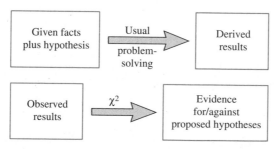

FIGURE C.1
Usual analysis versus chi-square analysis.

TABLE C.1   A typical contingency table
constructed for the tossing of coins *contingent* on
the hypothesis that each outcome is equally likely

| | Outcomes | | |
	H	T	Totals
Observed	6	4	10
Expected	5	5	10

chi-square

If there are n possible outcomes, then we *expect* to see outcomes E_1 through E_n; we *observe* outcomes O_1 through O_n. The following formula defines the *chi-square* or χ^2 measure of this experiment.

Chi-Square

$$\chi^2 = \frac{(O_1 - E_1)^2}{E_1} + \frac{(O_2 - E_2)^2}{E_2} + \cdots + \frac{(O_n - E_n)^2}{E_n} \qquad \text{(C.1)}$$

where

O_i is the *observed* data.
E_i is the *expected* data.

Example 1   Use the data from Table C.1 to compute χ^2.

Solution

$$\chi^2 = \frac{(6-5)^2}{5} + \frac{(4-5)^2}{5} = \frac{1}{5} + \frac{1}{5} = \frac{2}{5}$$

Example 2   A somewhat shady character gives us another coin to test. On the basis of some informal observations, we form the tentative hypothesis that the coin is weighted so that heads are twice as likely as tails. Now, we toss the coin 12 times, and observe 8 heads. Create a contingency table for this experiment, and compute χ^2.

Solution

Because heads are twice as likely as tails under our hypothesis, we *expect* 8 heads and 4 tails in 12 tosses, which (by coincidence) is what we observe. The contingency table is Table C.2. We compute χ^2 as

$$\chi^2 = \frac{(8-8)^2}{8} + \frac{(4-4)^2}{4} = 0$$

When the observed results match what we expect exactly, then $\chi^2 = 0$.

TABLE C.2   A contingency table for coin tossing *contingent* on heads being twice as likely as tails

	Outcomes		
	H	T	Totals
O	8	4	12
E	8	4	12

When Will Chi-Square Be Large?

Example 2 shows that χ^2 is exactly zero when the observations conform perfectly to the expectations. The larger the disagreement between the O_i and the E_i, the greater will be χ^2. We hope to relate the size of χ^2 to the likelihood that the observations are consistent with the hypothesis.

Two factors contribute to large values of χ^2:

1. The more individual observations *disagree* with expected results, the greater χ^2 will be.

2. The more *individual outcomes* there are in an experiment, the greater χ^2 will be (unless, of course, expected results consistently equal the observed results, a highly unlikely—and suspicious—circumstance).

Tables of χ^2 values relate the degree of disagreement with values of likelihood of the hypothesis, but these tables are subject to the number of possible outcomes, or rather to a closely related subject, *the number of degrees of freedom* of the experiment.

degrees of freedom   The *degrees of freedom* of an experiment refers to the number of ways it is possible to vary the results of an experiment if the number of trials remains fixed. Study these examples.

1. We toss a coin 10 times in a row. Let h be the number of heads we observe in this sequence. Although there are two outcomes (heads or tails), there is only a *single degree of freedom*. If we vary h, the number of tails is $10 - h$. There is no choice in the matter.

2. We roll a single die 42 times. There are 6 distinct outcomes, but only 5 degrees of freedom. Although it is possible to vary the number of times that any 5 of the 6 numbers turn up, the number of times the sixth number appears is fixed at

$$42 - \text{total of the other 5 numbers}$$

We usually represent the number of degrees of freedom by the Greek letter ν (nu), and there is a simple relationship between the outcomes and the degrees of freedom.

Degrees of Freedom

$$\nu = \text{number of possible outcomes} - 1$$

where ν is the number of degrees of freedom.

Confidence Levels

confidence level
critical level
Statisticians often agree to work with a 95 percent *confidence level* or (equivalently) with a 5 percent *critical level* when testing any hypothesis. That is, we agree to reject a hypothesis if the probability that we *incorrectly* reject the hypothesis is 5 percent or less. We can never specify a 100 percent confidence level because there is always a chance that experimental data will badly disagree with the expected results even when the hypothesis is true. χ^2 is important because for any value of ν, there is a value of χ^2 that identifies this confidence level.

Critical values of χ^2 for various values of ν appear in Table C.3.

TABLE C.3 Critical values of χ^2 at a confidence level of 95 percent (5 percent critical level) for various values of ν (degrees of freedom)

ν	1	2	3	4	5	6	7	8	9	10
χ^2	3.841	5.991	7.815	9.488	11.070	12.592	14.067	15.507	16.919	18.307

Testing a Hypothesis

Compute χ^2 and compare it with the critical value.

1. If the experimental value is *greater* than the critical value, we agree to *reject* the hypothesis.

2. If the experimental value is *less than or equal to* the critical value, we conclude that the hypothesis is *consistent* with the observed facts.

Example 3 An isolated village in the Scottish highlands contains people with only 5 different surnames. Rather than take a census of each of the 5 families, a sociologist supposes that each of the 5 are equally numerous, and tests that hypothesis by selecting 120 villagers at random, tabulating this small selection of names, and using chi-square techniques to evaluate this assumption. The numbers of people with the first 4 surnames are 30, 21, 24, and 19. What may the sociologist conclude?

Solution

First, note that since there are 5 outcomes (different surnames), there are only 4 degrees of freedom, and it is appropriate to specify the problem completely with the 4 numbers given. Let the surnames be A, B, C, D, and E. By our hypothesis, they are each equally numerous, so we expect each family will contribute $120/5 = 24$ members to the sample. The number of people in the sample with last name E is equal to $120 - (30 + 21 + 24 + 19) = 26$. A contingency table might appear as follows:

	A	B	C	D	E	Totals
O	30	21	24	19	26	120
E	24	24	24	24	24	120

and we compute

$$\chi^2 = \frac{(30-24)^2}{24} + \frac{(21-24)^2}{24} + \frac{(24-24)^2}{24} + \frac{(19-24)^2}{24} + \frac{(26-24)^2}{24}$$
$$= \frac{36+9+0+25+4}{24} = \frac{74}{24}$$
$$= 3\frac{1}{12}$$

Referring to Table C.3, we see that the critical value of chi-square with 4 degrees of freedom is 9.488, substantially greater than $3\frac{1}{12}$. Our hypothesis of equal family size is consistent with the data.

Example 4   A second researcher uses the data presented in Example 3. On the basis of talks with the villagers, she receives the impression that families A and E are equal sized, and that these two together have as many people as do the remaining families. These remaining families (namely B, C, and D) are also equally large. Use χ^2 to test this hypothesis.

Solution

If A and E constitute half the population and are groups of the same size, we expect that 60 people out of 120 have these surnames—30 and 30 for each family. The remaining 60 people should be equally divided among people with last names B, C, and D, that is, 20 apiece. The contingency table under this hypothesis takes this form:

	A	B	C	D	E	Totals
O	30	21	24	19	26	120
E	30	20	20	20	30	120

and

$$\chi^2 = 0 + \frac{1}{20} + \frac{16}{20} + \frac{1}{20} + \frac{16}{30}$$
$$= \frac{18}{20} + \frac{16}{30}$$
$$= 1\frac{13}{30}$$

This value is also well below the cutoff value of 9.488, so this second hypothesis is also consistent with the data.

Study Examples 3 and 4 carefully. Together, they show that one set of data may support several different hypotheses. For that reason, it is bad policy to use χ^2 analysis to conclude that a particular hypothesis is in fact true or false. The best a researcher can do is to conclude that hypotheses and data are consistent or inconsistent.

_____ Exercises C.1 _____

Determine the number of degrees of freedom of the experiments in Exercises 1–4.

1. Tossing a coin followed by rolling a single die.
2. Rolling a pair of dice.
3. Counting the offspring of fruit flies, and noting the sex of each insect.
4. Jelly beans come in any of 9 flavors. You select a piece of candy at random from a jar of mixed jelly beans.

The following six groups of data represent heads and tails observed when a presumably fair coin is tossed many times. Compute chi-square for each set of data. Determine whether the data are consistent with the hypothesis that the coin is fair. (For example, the pair _H40, T60_ represent observed results of 40 heads and 60 tails.)

5. H20, T20
6. H40, T60
7. H6, T4
8. H13, T15
9. H2, T0
10. H10, T0
11. Solve Exercises 9 and 10 again according to the hypothesis that the coin is four times as likely to come up heads as tails.
12. Solve Exercises 9 and 10 according to the hypothesis that the coin is ten times as likely to come up heads as tails.

Groups of data in Exercises 13–16 represent tabulated results when a single die is thrown many times. (For example, the sequence of 6 numbers "12, 11, 8, 8, 12, 10" means that 12 1s were observed, 11 2s, 8 3s and 4s, 12 5s, and 10 6s. In this experiment, there were a total of $12 + 11 + 8 + 8 + 12 + 10 = 61$ rolls of the die.) Calculate chi-square for each contingent on the hypothesis that each outcome is equally likely.

13. 12, 10, 8, 8, 12, 10
14. 2, 2, 3, 3, 2, 6
15. 6, 7, 2, 2, 7, 0
16. 10, 13, 9, 7, 11, 10
17. In this exercise, assume that a single coin is fair.
    (a) Contingent on that hypothesis, calculate chi-square if we observe 6 heads and 4 tails in a series of 10 tosses.
    (b) Calculate chi-square if the coin is tossed 20 times, with the same relative numbers of heads and tails as in part (a).
    (c) Recalculate chi-square if the coin is tossed 100 times, and you observe the same relative number of heads and tails as in part (a).

W 18. (_Requires special thought_) Examine the results of Exercise 17. Even though the relative numbers of heads and tails remain the same, the value of chi-square increases substantially as the number of trials increases. Wouldn't you expect chi-square to remain the same as long as the relative occurrences of the outcomes remain the same? How can you account for this trend from a _qualitative_ point of view? (That is, justify this result without the use of equations.)

Applications

19. **(Genetics)** According to the theory of genetics, when peas purebred for smooth skins are crossed with peas purebred for wrinkled skins, the offspring should show both wrinkled and smooth skins in a 1 to 3 ratio. This result is tested by an undergraduate biology class, who plant seeds from such a cross and obtain 108 plants. Of these, 35 are wrinkled. Is this result consistent with genetic theory?

20. **(Property Values)** An extension to the "Orange" subway line in Boston has been criticized for contributing to an excessive rise in surrounding property values, to the point where these will cause disruptions in the existing neighborhoods. The mayor's office seeks to verify or refute this charge. As part of a preliminary survey, researchers record the following price rises for 6 random parcels of property from 1984 (pre-subway) to 1987 (after the subway).

Parcel No.	1	2	3	4	5	6
Percent Increase	20%	75%	50%	60%	60%	40%

Specifically, this study tests this hypothesis: Property values in this period have risen by 55 percent or more in at least half of the surrounding properties. Use a chi-square analysis to test this hypothesis.

21. Refer again to Exercise 20. A group of merchants disputes these assertions, claiming that because of rising property values in all northeastern cities, values of properties have increased by at least 50 percent for two-thirds of all properties during this period. Do the data support this assertion?

22. *(Health and Nutrition)* Proponents of Vitamin C claim that large doses of this vitamin reduce the length of time people suffer from colds. According to their hypothesis, 75 percent of cold-sufferers who dose themselves with large amounts of this vitamin suffer from colds for 4 days or less. Detractors of this theory point out that 75 percent of cold sufferers will only keep colds for 6 days or less anyway regardless of any medications they take. In a survey of 100 people with colds who took massive doses of this vitamin, it was found that 15 kept their colds for 3 days, 25 for 4 days, 5 for 5 days, 15 for 6 days, and 40 for 7 days. Compute χ^2 contingent on both hypotheses.

23. *(Mind Reading: Skeptics)* Fury Teller is a self-proclaimed mind reader. One test of his abilities consists of a control subject concentrating on a sequence of 10 digits, each a 0 or a 1, one at a time. Fury attempts to read the control's mind, and writes down the number the control is thinking of. In one sequence, Fury "guessed" 7 correct answers, and proclaimed this result as evidence of his powers. Compute chi-square for this experiment under the assumption that his results were the result of pure luck. That is, there are two outcomes—right and wrong, with equal probabilities for each.

24. *(Mind Reading: Believers)* Recompute chi-square for Exercise 23 from the standpoint of one who believes in Teller's powers. In this case, there are two outcomes again—right and wrong—but you expect *no* wrong answers.

MATHEMATICAL TABLES

TABLE D.1   AREAS UNDER THE STANDARD NORMAL CURVE

z	$A(z)$	z	$A(z)$	z	$A(z)$	z	$A(z)$	z	$A(z)$
0.00	0.0000	0.25	0.0987	0.50	0.1915	0.75	0.2734	1.00	0.3413
0.01	0.0040	0.26	0.1026	0.51	0.1950	0.76	0.2764	1.01	0.3438
0.02	0.0080	0.27	0.1064	0.52	0.1985	0.77	0.2794	1.02	0.3461
0.03	0.0120	0.28	0.1103	0.53	0.2019	0.78	0.2823	1.03	0.3485
0.04	0.0160	0.29	0.1141	0.54	0.2054	0.79	0.2852	1.04	0.3508
0.05	0.0199	0.30	0.1179	0.55	0.2088	0.80	0.2881	1.05	0.3531
0.06	0.0239	0.31	0.1217	0.56	0.2123	0.81	0.2910	1.06	0.3554
0.07	0.0279	0.32	0.1255	0.57	0.2157	0.82	0.2939	1.07	0.3577
0.08	0.0319	0.33	0.1293	0.58	0.2190	0.83	0.2967	1.08	0.3599
0.09	0.0359	0.34	0.1331	0.59	0.2224	0.84	0.2995	1.09	0.3621
0.10	0.0398	0.35	0.1368	0.60	0.2257	0.85	0.3023	1.10	0.3643
0.11	0.0438	0.36	0.1406	0.61	0.2291	0.86	0.3051	1.11	0.3665
0.12	0.0478	0.37	0.1443	0.62	0.2324	0.87	0.3078	1.12	0.3686
0.13	0.0517	0.38	0.1480	0.63	0.2357	0.88	0.3106	1.13	0.3708
0.14	0.0557	0.39	0.1517	0.64	0.2389	0.89	0.3133	1.14	0.3729
0.15	0.0596	0.40	0.1554	0.65	0.2422	0.90	0.3159	1.15	0.3749
0.16	0.0636	0.41	0.1591	0.66	0.2454	0.91	0.3186	1.16	0.3770
0.17	0.0675	0.42	0.1628	0.67	0.2486	0.92	0.3212	1.17	0.3790
0.18	0.0714	0.43	0.1664	0.68	0.2517	0.93	0.3238	1.18	0.3810
0.19	0.0753	0.44	0.1700	0.69	0.2549	0.94	0.3264	1.19	0.3830
0.20	0.0793	0.45	0.1736	0.70	0.2580	0.95	0.3289	1.20	0.3849
0.21	0.0832	0.46	0.1772	0.71	0.2611	0.96	0.3315	1.21	0.3869
0.22	0.0871	0.47	0.1808	0.72	0.2642	0.97	0.3340	1.22	0.3888
0.23	0.0910	0.48	0.1844	0.73	0.2673	0.98	0.3365	1.23	0.3907
0.24	0.0948	0.49	0.1879	0.74	0.2704	0.99	0.3389	1.24	0.3925

TABLE D.1 AREAS UNDER THE STANDARD NORMAL CURVE (*cont.*)

z	A(z)	z	A(z)	z	A(z)	z	A(z)	z	A(z)
1.25	0.3944	1.75	0.4599	2.25	0.4878	2.75	0.4970	3.25	0.4994
1.26	0.3962	1.76	0.4608	2.26	0.4881	2.76	0.4971	3.26	0.4994
1.27	0.3980	1.77	0.4616	2.27	0.4884	2.77	0.4972	3.27	0.4995
1.28	0.3997	1.78	0.4625	2.28	0.4887	2.78	0.4973	3.28	0.4995
1.29	0.4015	1.79	0.4633	2.29	0.4890	2.79	0.4974	3.29	0.4995
1.30	0.4032	1.80	0.4641	2.30	0.4893	2.80	0.4974	3.30	0.4995
1.31	0.4049	1.81	0.4649	2.31	0.4896	2.81	0.4975	3.31	0.4995
1.32	0.4066	1.82	0.4656	2.32	0.4898	2.82	0.4976	3.32	0.4995
1.33	0.4082	1.83	0.4664	2.33	0.4901	2.83	0.4977	3.33	0.4996
1.34	0.4099	1.84	0.4671	2.34	0.4904	2.84	0.4977	3.34	0.4996
1.35	0.4115	1.85	0.4678	2.35	0.4906	2.85	0.4978	3.35	0.4996
1.36	0.4131	1.86	0.4686	2.36	0.4909	2.86	0.4979	3.36	0.4996
1.37	0.4147	1.87	0.4693	2.37	0.4911	2.87	0.4979	3.37	0.4996
1.38	0.4162	1.88	0.4699	2.38	0.4913	2.88	0.4980	3.38	0.4996
1.39	0.4177	1.89	0.4706	2.39	0.4916	2.89	0.4981	3.39	0.4997
1.40	0.4192	1.90	0.4713	2.40	0.4918	2.90	0.4981	3.40	0.4997
1.41	0.4207	1.91	0.4719	2.41	0.4920	2.91	0.4982	3.41	0.4997
1.42	0.4222	1.92	0.4726	2.42	0.4922	2.92	0.4982	3.42	0.4997
1.43	0.4236	1.93	0.4732	2.43	0.4925	2.93	0.4983	3.43	0.4997
1.44	0.4251	1.94	0.4738	2.44	0.4927	2.94	0.4984	3.44	0.4997
1.45	0.4265	1.95	0.4744	2.45	0.4929	2.95	0.4984	3.45	0.4997
1.46	0.4279	1.96	0.4750	2.46	0.4931	2.96	0.4985	3.46	0.4997
1.47	0.4292	1.97	0.4756	2.47	0.4932	2.97	0.4985	3.47	0.4997
1.48	0.4306	1.98	0.4761	2.48	0.4934	2.98	0.4986	3.48	0.4997
1.49	0.4319	1.99	0.4767	2.49	0.4936	2.99	0.4986	3.49	0.4998
1.50	0.4332	2.00	0.4772	2.50	0.4938	3.00	0.4987	3.50	0.4998
1.51	0.4345	2.01	0.4778	2.51	0.4940	3.01	0.4987	3.51	0.4998
1.52	0.4357	2.02	0.4783	2.52	0.4941	3.02	0.4987	3.52	0.4998
1.53	0.4370	2.03	0.4788	2.53	0.4943	3.03	0.4988	3.53	0.4998
1.54	0.4382	2.04	0.4793	2.54	0.4945	3.04	0.4988	3.54	0.4998
1.55	0.4394	2.05	0.4798	2.55	0.4946	3.05	0.4989	3.55	0.4998
1.56	0.4406	2.06	0.4803	2.56	0.4948	3.06	0.4989	3.56	0.4998
1.57	0.4418	2.07	0.4808	2.57	0.4949	3.07	0.4989	3.57	0.4998
1.58	0.4429	2.08	0.4812	2.58	0.4951	3.08	0.4990	3.58	0.4998
1.59	0.4441	2.09	0.4817	2.59	0.4952	3.09	0.4990	3.59	0.4998
1.60	0.4452	2.10	0.4821	2.60	0.4953	3.10	0.4990	3.60	0.4998
1.61	0.4463	2.11	0.4826	2.61	0.4955	3.11	0.4991	3.61	0.4998
1.62	0.4474	2.12	0.4830	2.62	0.4956	3.12	0.4991	3.62	0.4999
1.63	0.4484	2.13	0.4834	2.63	0.4957	3.13	0.4991	3.63	0.4999
1.64	0.4495	2.14	0.4838	2.64	0.4959	3.14	0.4992	3.64	0.4999
1.65	0.4505	2.15	0.4842	2.65	0.4960	3.15	0.4992	3.65	0.4999
1.66	0.4515	2.16	0.4846	2.66	0.4961	3.16	0.4992	3.66	0.4999
1.67	0.4525	2.17	0.4850	2.67	0.4962	3.17	0.4992	3.67	0.4999
1.68	0.4535	2.18	0.4854	2.68	0.4963	3.18	0.4993	3.68	0.4999
1.69	0.4545	2.19	0.4857	2.69	0.4964	3.19	0.4993	3.69	0.4999
1.70	0.4554	2.20	0.4861	2.70	0.4965	3.20	0.4993	3.70	0.4999
1.71	0.4564	2.21	0.4864	2.71	0.4966	3.21	0.4993	3.71	0.4999
1.72	0.4573	2.22	0.4868	2.72	0.4967	3.22	0.4994	3.72	0.4999
1.73	0.4582	2.23	0.4871	2.73	0.4968	3.23	0.4994	3.73	0.4999
1.74	0.4591	2.24	0.4875	2.74	0.4969	3.24	0.4994	3.74	0.4999

TABLE D.1 AREAS UNDER THE STANDARD NORMAL CURVE (*cont.*)

z	$A(z)$	z	$A(z)$	z	$A(z)$	z	$A(z)$	z	$A(z)$
3.75	0.4999	3.80	0.4999	3.85	0.4999	3.90	0.5000	3.95	0.5000
3.76	0.4999	3.81	0.4999	3.86	0.4999	3.91	0.5000	3.96	0.5000
3.77	0.4999	3.82	0.4999	3.87	0.4999	3.92	0.5000	3.97	0.5000
3.78	0.4999	3.83	0.4999	3.88	0.4999	3.93	0.5000	3.98	0.5000
3.79	0.4999	3.84	0.4999	3.89	0.4999	3.94	0.5000	3.99	0.5000

TABLE D.2 THE COMPOUND INTEREST FUNCTION $(1 + i)^n$

	i						
n	0.0050 (0.5%)	0.0075 (0.75%)	0.0100 (1%)	0.0125 (1.25%)	0.0150 (1.5%)	0.0175 (1.75%)	0.0200 (2%)
4	1.020 151	1.030 339	1.040 604	1.050 945	1.061 364	1.071 859	1.082 432
5	1.025 251	1.038 067	1.051 010	1.064 082	1.077 284	1.090 617	1.104 081
6	1.030 378	1.045 852	1.061 520	1.077 383	1.093 443	1.109 702	1.126 162
8	1.040 707	1.061 599	1.082 857	1.104 486	1.126 493	1.148 882	1.171 659
9	1.045 911	1.069 561	1.093 685	1.118 292	1.143 390	1.168 987	1.195 093
10	1.051 140	1.077 583	1.104 622	1.132 271	1.160 541	1.189 444	1.218 994
12	1.061 678	1.093 807	1.126 825	1.160 755	1.195 618	1.231 439	1.268 242
14	1.072 321	1.110 276	1.149 474	1.189 955	1.231 756	1.274 917	1.319 479
16	1.083 071	1.126 992	1.172 579	1.219 890	1.268 986	1.319 929	1.372 786
18	1.093 929	1.143 960	1.196 147	1.250 577	1.307 341	1.366 531	1.428 246
20	1.104 896	1.161 184	1.220 190	1.282 037	1.346 855	1.414 778	1.485 947
22	1.115 972	1.178 667	1.244 716	1.314 288	1.387 564	1.464 729	1.545 980
24	1.127 160	1.196 414	1.269 735	1.347 351	1.429 503	1.516 443	1.608 437
26	1.138 460	1.214 427	1.295 256	1.381 245	1.472 710	1.569 983	1.673 418
28	1.149 873	1.232 712	1.321 291	1.415 992	1.517 222	1.625 413	1.741 024
30	1.161 400	1.251 272	1.347 849	1.451 613	1.563 080	1.682 800	1.811 362
32	1.173 043	1.270 111	1.374 941	1.488 131	1.610 324	1.742 213	1.884 541
34	1.184 803	1.289 234	1.402 577	1.525 566	1.658 996	1.803 725	1.960 676
36	1.196 681	1.308 645	1.430 769	1.563 944	1.709 140	1.867 407	2.039 887
38	1.208 677	1.328 349	1.459 527	1.603 287	1.760 798	1.933 338	2.122 299
40	1.220 794	1.348 349	1.488 864	1.643 619	1.814 018	2.001 597	2.208 040
45	1.251 621	1.399 676	1.564 811	1.748 946	1.954 213	2.182 975	2.437 854
50	1.283 226	1.452 957	1.644 632	1.861 022	2.105 242	2.380 789	2.691 588
55	1.315 629	1.508 266	1.728 525	1.980 281	2.267 944	2.596 528	2.971 731
60	1.348 850	1.565 681	1.816 697	2.107 181	2.443 220	2.831 816	3.281 031
65	1.382 910	1.625 281	1.909 366	2.242 214	2.632 042	3.088 426	3.622 523
70	1.417 831	1.687 151	2.006 763	2.385 900	2.835 456	3.368 288	3.999 558
75	1.453 633	1.751 375	2.109 128	2.538 794	3.054 592	3.673 511	4.415 835
80	1.490 339	1.818 044	2.216 715	2.701 485	3.290 663	4.006 392	4.875 439
85	1.527 971	1.887 251	2.329 790	2.874 602	3.544 978	4.369 437	5.382 879
90	1.566 555	1.959 092	2.448 633	3.058 813	3.818 949	4.765 381	5.943 133
95	1.606 112	2.033 669	2.573 538	3.254 828	4.114 092	5.197 203	6.561 699
100	1.646 668	2.111 084	2.704 814	3.463 404	4.432 046	5.668 156	7.244 646
105	1.688 249	2.191 446	2.842 787	3.685 347	4.774 572	6.181 785	7.998 675
110	1.730 879	2.274 867	2.987 797	3.921 512	5.143 570	6.741 957	8.831 183

TABLE D.2   THE COMPOUND INTEREST FUNCTION $(1 + i)^n$ *(cont.)*

				i			
n	0.0050 (0.5%)	0.0075 (0.75%)	0.0100 (1%)	0.0125 (1.25%)	0.0150 (1.5%)	0.0175 (1.75%)	0.0200 (2%)
115	1.774 586	2.361 464	3.140 205	4.172 811	5.541 086	7.352 890	9.750 340
120	1.819 397	2.451 357	3.300 387	4.440 213	5.969 323	8.019 183	10.765 163
150	2.113 048	3.067 314	4.448 423	6.445 473	9.330 531	13.494 683	19.499 603
180	2.454 094	3.838 043	5.995 802	9.356 334	14.584 368	22.708 854	35.320 831
210	2.850 184	4.802 435	8.081 435	13.581 780	22.796 537	38.214 463	63.978 797
240	3.310 204	6.009 152	10.892 554	19.715 494	35.632 816	64.307 303	115.888 735
270	3.844 472	7.519 082	14.681 517	28.619 274	55.696 949	108.216 338	209.916 403
300	4.464 970	9.408 415	19.788 466	41.544 120	87.058 800	182.106 467	380.234 508
330	5.185 616	11.772 483	26.671 863	60.306 000	136.079 888	306.448 787	688.742 181
360	6.022 575	14.730 576	35.949 641	87.540 995	212.703 781	515.692 058	1247.561 128

TABLE D.3   THE FUNCTION $A(n, i)$

				i			
n	0.0050 (0.5%)	0.0075 (0.75%)	0.0100 (1%)	0.0125 (1.25%)	0.0150 (1.5%)	0.0175 (1.75%)	0.0200 (2%)
4	3.950 496	3.926 110	3.901 966	3.878 058	3.854 385	3.830 943	3.807 729
5	4.925 866	4.889 440	4.853 431	4.817 835	4.782 645	4.747 855	4.713 460
6	5.896 384	5.845 598	5.795 476	5.746 010	5.697 187	5.648 998	5.601 431
8	7.822 959	7.736 613	7.651 678	7.568 124	7.485 925	7.405 053	7.325 481
9	8.779 064	8.671 576	8.566 018	8.462 345	8.360 517	8.260 494	8.162 237
10	9.730 412	9.599 580	9.471 305	9.345 526	9.222 185	9.101 223	8.982 585
12	11.618 932	11.434 913	11.255 077	11.079 312	10.907 505	10.739 550	10.575 341
14	13.488 708	13.243 022	13.003 703	12.770 553	12.543 382	12.322 006	12.106 249
16	15.339 925	15.024 313	14.717 874	14.420 292	14.131 264	13.850 497	13.577 709
18	17.172 768	16.779 181	16.398 269	16.029 549	15.672 561	15.326 863	14.992 031
20	18.987 419	18.508 020	18.045 553	17.599 316	17.168 639	16.752 881	16.351 433
22	20.784 059	20.211 215	19.660 379	19.130 563	18.620 824	18.130 269	17.658 048
24	22.562 866	21.889 146	21.243 387	20.624 235	20.030 405	19.460 686	18.913 926
26	24.324 018	23.542 189	22.795 204	22.081 253	21.398 632	20.745 732	20.121 036
28	26.067 689	25.170 713	24.316 443	23.502 518	22.726 717	21.986 955	21.281 272
30	27.794 054	26.775 080	25.807 708	24.888 906	24.015 838	23.185 849	22.396 456
32	29.503 284	28.355 650	27.269 589	26.241 274	25.267 139	24.343 859	23.468 335
34	31.195 548	29.912 776	28.702 666	27.560 456	26.481 728	25.462 378	24.498 592
36	32.871 016	31.446 805	30.107 505	28.847 267	27.660 684	26.542 753	25.488 842
38	34.529 854	32.958 080	31.484 663	30.102 501	28.805 052	27.586 285	26.440 641
40	36.172 228	34.446 938	32.834 686	31.326 933	29.915 845	28.594 230	27.355 479
45	40.207 196	38.073 181	36.094 508	34.258 168	32.552 337	30.966 263	29.490 160
50	44.142 786	41.566 447	39.196 118	37.012 876	34.999 688	33.141 209	31.423 606
55	47.981 445	44.931 612	42.147 192	39.601 687	37.271 467	35.135 445	33.174 788
60	51.725 561	48.173 374	44.955 038	42.034 592	39.380 269	36.963 986	34.760 887

TABLE D.3   THE FUNCTION $A(n, i)$ (cont.)

	i						
n	0.0050 (0.5%)	0.0075 (0.75%)	0.0100 (1%)	0.0125 (1.25%)	0.0150 (1.5%)	0.0175 (1.75%)	0.0200 (2%)
65	55.377 461	51.296 257	47.626 608	44.320 980	41.337 786	38.640 597	36.197 466
70	58.939 418	54.304 622	50.168 514	46.469 676	43.154 872	40.177 903	37.498 619
75	62.413 645	57.202 668	52.587 051	48.488 970	44.841 600	41.587 478	38.677 114
80	65.802 305	59.994 440	54.888 206	50.386 657	46.407 323	42.879 935	39.744 514
85	69.107 505	62.683 836	57.077 676	52.170 060	47.860 722	44.065 005	40.711 290
90	72.331 300	65.274 609	59.160 881	53.846 060	49.209 855	45.151 610	41.586 929
95	75.475 694	67.770 377	61.142 980	55.421 127	50.462 201	46.147 933	42.380 023
100	78.542 645	70.174 623	63.028 879	56.901 339	51.624 704	47.061 473	43.098 352
105	81.534 058	72.490 703	64.823 247	58.292 409	52.703 809	47.899 110	43.748 964
110	84.451 795	74.721 851	66.530 526	59.599 704	53.705 500	48.667 149	44.338 245
115	87.297 670	76.871 181	68.154 944	60.828 269	54.635 330	49.371 374	44.871 974
120	90.073 453	78.941 693	69.700 522	61.982 847	55.498 454	50.017 087	45.355 389
150	105.349 975	89.864 247	77.520 123	67.588 188	59.521 664	52.908 385	47.435 845
180	118.503 515	98.593 409	83.321 664	71.449 643	62.095 562	54.626 532	48.584 405
210	129.829 103	105.569 641	87.625 960	74.109 756	63.742 246	55.647 537	49.218 491
240	139.580 772	111.144 954	90.819 416	75.942 278	64.795 732	56.254 267	49.568 552
270	147.977 248	115.600 671	93.188 715	77.204 681	65.469 713	56.614 814	49.761 810
300	155.206 864	119.161 622	94.946 551	78.074 336	65.900 901	56.829 069	49.868 502
330	161.431 778	122.007 487	96.250 731	78.673 432	66.176 758	56.956 389	49.927 404
360	166.791 614	124.281 866	97.218 331	79.086 142	66.353 242	57.032 049	49.959 922

TABLE D.4   THE FUNCTION $S(n, i)$

	i						
n	0.0050 (0.5%)	0.0075 (0.75%)	0.0100 (1%)	0.0125 (1.25%)	0.0150 (1.5%)	0.0175 (1.75%)	0.0200 (2%)
4	4.030 100	4.045 255	4.060 401	4.075 627	4.090 903	4.106 230	4.121 608
5	5.050 251	5.075 565	5.101 005	5.126 572	5.152 267	5.178 089	5.204 040
6	6.075 502	6.113 631	6.152 015	6.190 654	6.229 551	6.268 706	6.308 121
8	8.141 409	8.213 180	8.285 671	8.358 888	8.432 839	8.507 530	8.582 969
9	9.182 116	9.274 779	9.368 527	9.463 374	9.559 332	9.656 412	9.754 628
10	10.228 026	10.344 339	10.462 213	10.581 666	10.702 722	10.825 399	10.949 721
12	12.335 562	12.507 586	12.682 503	12.860 361	13.041 211	13.225 104	13.412 090
14	14.464 226	14.703 404	14.947 421	15.196 380	15.450 382	15.709 533	15.973 938
16	16.614 230	16.932 282	17.257 864	17.591 164	17.932 370	18.281 677	18.639 285
18	18.785 788	19.194 718	19.614 748	20.046 192	20.489 376	20.944 635	21.412 312
20	20.979 115	21.491 219	22.019 004	22.562 979	23.123 667	23.701 661	24.297 370
22	23.194 431	23.822 296	24.471 586	25.143 078	25.837 580	26.555 926	27.298 984
24	25.431 955	26.188 471	26.973 465	27.788 084	28.633 521	29.511 016	30.421 862
26	27.691 911	28.590 271	29.525 631	30.499 628	31.513 969	32.570 440	33.670 906
28	29.974 522	31.028 233	32.129 097	33.279 384	34.481 479	35.737 880	37.051 210
30	32.280 017	33.502 902	34.784 892	36.129 069	37.538 681	39.017 150	40.568 079
32	34.608 624	36.014 830	37.494 068	39.050 441	40.688 288	42.412 200	44.227 030
34	36.960 575	38.564 578	40.257 699	42.045 303	43.933 092	45.927 115	48.033 803
36	39.336 105	41.152 716	43.076 878	45.115 505	47.275 969	49.566 129	51.994 367
38	41.735 449	43.779 822	45.952 724	48.262 942	50.719 885	53.333 624	56.114 940

TABLE D.4   THE FUNCTION $S(n, i)$ *(cont.)*

	i						
n	0.0050 (0.5%)	0.0075 (0.75%)	0.0100 (1%)	0.0125 (1.25%)	0.0150 (1.5%)	0.0175 (1.75%)	0.0200 (2%)
40	44.158 847	46.446 482	48.886 373	51.489 557	54.267 894	57.234 134	60.401 983
45	50.324 164	53.290 112	56.481 075	59.915 691	63.614 201	67.598 584	71.892 710
50	56.645 163	60.394 257	64.463 182	68.881 790	73.682 828	78.902 225	84.579 401
55	63.125 775	67.768 834	72.852 457	78.422 456	84.529 599	91.230 163	98.586 534
60	69.770 031	75.424 137	81.669 670	88.574 508	96.214 652	104.675 216	114.051 539
65	76.582 062	83.370 852	90.936 649	99.377 125	108.802 772	119.338 614	131.126 155
70	83.566 105	91.620 073	100.676 337	110.871 998	122.363 753	135.330 758	149.977 911
75	90.726 505	100.183 314	110.912 847	123.103 486	136.972 781	152.772 056	170.791 773
80	98.067 714	109.072 531	121.671 522	136.118 795	152.710 852	171.793 824	193.771 958
85	105.594 297	118.300 130	132.978 997	149.968 153	169.665 226	192.539 280	219.143 939
90	113.310 936	127.878 995	144.863 267	164.705 008	187.929 900	215.164 617	247.156 656
95	121.222 430	137.822 495	157.353 755	180.386 232	207.606 142	239.840 185	278.084 960
100	129.333 698	148.144 512	170.481 383	197.072 342	228.803 043	266.751 768	312.232 306
105	137.649 787	158.859 454	184.278 652	214.827 734	251.638 126	296.101 986	349.933 735
110	146.175 867	169.982 280	198.779 720	233.720 931	276.237 994	328.111 820	391.559 160
115	154.917 242	181.528 515	214.020 489	253.824 843	302.739 039	363.022 275	437.516 992
120	163.879 347	193.514 277	230.038 689	275.217 058	331.288 191	401.096 196	488.258 152
150	222.609 504	275.641 853	344.842 290	435.637 828	555.368 701	713.981 881	924.980 138
180	290.818 712	378.405 769	499.580 198	668.506 759	905.624 513	1240.505 953	1716.041 568
210	370.036 893	506.991 356	708.143 519	1006.542 412	1453.102 444	2126.540 729	3148.939 852
240	462.040 895	667.886 870	989.255 365	1497.239 481	2308.854 370	3617.560 166	5744.436 758
270	568.894 351	869.210 884	1368.151 663	2209.541 903	3646.463 280	6126.647 868	10445.820 144
300	692.993 962	1121.121 937	1878.846 626	3243.529 615	5737.253 308	10348.940 980	18961.725 403
330	837.123 262	1436.331 126	2567.186 279	4744.479 992	9005.325 848	17454.216 378	34387.109 043
360	1004.515 042	1830.743 483	3494.964 133	6923.279 611	14113.585 393	29410.974 741	62328.056 387

Appendix E: Answers to Skill Enhancers

Chapter 1

1. (a) and (d) **2.** $x_1 + x_2 = 2$; $\frac{1}{3}x_1 - \frac{1}{4}x_2 + \frac{1}{5}x_3 + 10 = 0$. **3.** Let x_i stand for the number of hits of the ith player.
4. Let M_i be the number of subscriptions of the ith magazine she sells. **5.** $y = -2$ **6.** No solution.
7. All values of u satisfy this equation—there is no unique solution. This equation has an infinite number of solutions.
8. **9.** Choosing $x = 0$, $y = 0$, and $x = 1$ leads to points $(0, 6)$, $(2, 0)$, and $(1, 3)$.

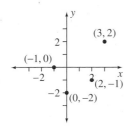

10. **11.** **12.** $(0, 2)$ and $(10, 0)$ **13.** $(0, 0)$

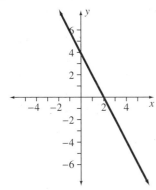

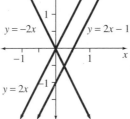

14. **15.** **16.** $y = -4$ and $x = -1$ **17.** $m = 1$; m is undefined; and $m = -15$.

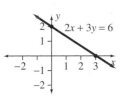

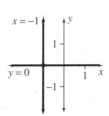

18. $y = -x - 1$ **19.** $3x + 4y = -2$ **20.** \$250,000 **21.** Here, $m = \frac{1}{2}$ and $b = 0$. The equation is $y = \frac{1}{2}x$.
22. $m = \frac{3}{2}$ **23.** They are all parallel. **24.** \$360; \$150 **25.** Approximately \$13,140 **26.** \$4200 **27.** \$75,000
per year; \$1,250,000 **28.** \$206,000 profit **29.** $p = .5x + 170$ **30.** \$2550 **31.** 60 days after March 1 **32.** 240
33. $(6, -2)$ **34.** Kenny has \$200 and Lenny has \$250. **35.** No solution possible. The equations represent parallel lines
with different slopes. Therefore, the lines, being distinct, can have no intersection. **36.** $(7\frac{1}{2}, 11)$ **37.** $(0, -1)$
38. Since these equations represent two different ways of writing the same equation, there is no unique solution possible. These
are dependent equations. **39.** $c = m = 100$ gal

Chapter 2

1. 2×4 and 3×2 **2.** 0, 0, and -2

3. $X + Y$ can't be computed, since these matrices have different dimensions. $Z + X = \begin{pmatrix} 0 \\ 0 \end{pmatrix}$, $X - Z = \begin{pmatrix} 4 \\ 6 \end{pmatrix}$, and $Y + W = \begin{pmatrix} 0 & 0 \end{pmatrix}$.

4. $10A = \begin{pmatrix} 10 & -20 & 40 \end{pmatrix}$ and $-\frac{1}{3}C = \begin{pmatrix} 0 & -\frac{1}{3} & -\frac{1}{3} \\ -\frac{1}{3} & 0 & -\frac{1}{3} \\ -\frac{1}{3} & -\frac{1}{3} & -\frac{1}{3} \end{pmatrix}$ **5.** $\begin{pmatrix} .5 & .25 \\ .35 & .35 \\ .25 & .55 \\ .4 & .45 \end{pmatrix}$ **6.** 0 and 40

7. -5; $\begin{pmatrix} 0 & -10 & -20 & -30 \end{pmatrix}$ **8.** $\begin{pmatrix} 8 \\ 3 \end{pmatrix}$

9. $DE = E$; $ED = E$. The fact that $DE = ED$ is due only to the very special form of matrix E. Normally, matrix multiplication is not commutative.

10. $\begin{pmatrix} 16 & 7 \end{pmatrix}$ **11.** p_{31} is undefined; $p_{11} = -7$ and $p_{24} = 18$.   **12.** $XY = \begin{pmatrix} -2 & -3 \\ 4 & 2 \end{pmatrix}$; $YX = \begin{pmatrix} 0 & -4 \\ 2 & 0 \end{pmatrix}$

13. $XY = \begin{pmatrix} -3 & 1 & -2 \\ -3 & -1 & -1 \end{pmatrix}$. YX is undefined.   **14.** $\begin{pmatrix} 2 & 1 & 2 \\ -1 & -2 & 0 \\ 1 & -1 & -1 \end{pmatrix}$ **15.** $\begin{pmatrix} -\frac{1}{4} & \frac{5}{4} \\ -2 & \frac{3}{2} \end{pmatrix}$

16. $S = 100 \begin{pmatrix} 15 & 11 & 10 \\ 12 & 12 & 12 \\ 11 & 12 & 15 \end{pmatrix}$; receipts $= \begin{pmatrix} \$99,000 & \$86,500 & \$87,500 \end{pmatrix}$; total weekly receipts are $273,000.

17. $x_1 = 1$, $x_2 = -2$, and $x_3 = -4$ **18.** $p = q = 0$

19. If $A = \begin{pmatrix} a & b & c \\ d & e & f \\ g & h & i \end{pmatrix}$ is an arbitrary 3×3 matrix, then we can use brute force to show that $A \times I_3 = I_3 \times A = A$.

20. Let $x_1 = u$, $x_2 = v$, $A = \begin{pmatrix} 1 & 2 \\ 3 & -2 \end{pmatrix}$, and $B = \begin{pmatrix} -1 \\ 7 \end{pmatrix}$. The augmented matrix is $A' = \left(\begin{array}{cc|c} 1 & 2 & -1 \\ 3 & -2 & 7 \end{array} \right)$.

21. $\left(\begin{array}{cc|c} 2 & -3 & 5 \\ 3 & 2 & -2 \end{array} \right)$

22. We get U if we multiply the first row of T by $\frac{1}{2}$ and the second by $-\frac{1}{2}$. We get V if we add the second row of T to the first row. We obtain W by subtracting the first row of T from the second row.

23. $u = \frac{3}{2}$, $v = -\frac{5}{2}$ **24.** $(x, y) = (3, -1)$ **25.** 2 of the first kind, 3 of the second

26. $\begin{pmatrix} 1 & 0 \\ 0 & 1 & 0 \\ 0 & 0 & 1 \end{pmatrix}$; $\begin{pmatrix} 1 & 0 & 0 \\ 0 & 1 & 1 \\ 0 & 0 & 0 \end{pmatrix}$; and $\begin{pmatrix} 1 & 0 \\ 0 & 1 \end{pmatrix}$ **27.** $\begin{pmatrix} 1 & 2 & 3 & 4 \\ 0 & -2 & -2 & -1 \\ 0 & 2 & 7 & 12 \end{pmatrix}$ **28.** $\begin{pmatrix} \frac{1}{2} & 1 & \frac{3}{2} \\ \frac{7}{2} & 0 & \frac{7}{2} \end{pmatrix}$

29. $\begin{pmatrix} 1 & 0 & 0 & 1 \\ 0 & 1 & 0 & -1 \\ 0 & 0 & 1 & 0 \end{pmatrix}$ **30.** $\begin{pmatrix} 1 & 0 & 0 & -1 \\ 0 & 1 & 0 & 2 \\ 0 & 0 & 1 & 1 \end{pmatrix}$ **31.** $x = 3$, $y = 1$, $z = -1$

32. Unique values of x, y, and z are not possible. The reduced matrix is $\begin{pmatrix} 1 & 0 & -1 & -2 \\ 0 & 1 & 2 & 5 \\ 0 & 0 & 0 & 0 \end{pmatrix}$. If we let $z = t$, where t is some arbitrary parameter, then $x = t - 2$ and $y = 5 - 2t$.

33. These two equations are inconsistent; there is no solution.   **34.** $x = 1$, $y = -1$, $z = 2$

35. The equations are $x + y + 2z = 4$, $2x - 2y = 4$, and $4y + z = 4$, with solutions $x = 3$, $y = 1$, and $z = 0$.

36. If we introduce the parameter t, then $x = 2 - \frac{t}{2}$, $y = 1 + \frac{3}{2}t$, and $z = t$.

37. The resulting set of equations has a solution. It is $a = 8000$, $b = -4000$, and $c = 4000$. Clearly, there is no physical solution corresponding to the mathematical one, since it is impossible to plant a negative number of bushels of variety B.

38. Verify that $bb^{-1} = B^{-1}B = I_2$.   **39.** $x_1 = 4$, $x_2 = 6$

40. 4 lb. peanuts, 6 lb. cashews; 8 lb. peanuts, 2 lb. cashews

41. $B^{-1} = \begin{pmatrix} -2 & 1 \\ \frac{3}{2} & -\frac{1}{2} \end{pmatrix}$ **42.** $B^{-1} = \begin{pmatrix} -\frac{1}{3} & 0 & \frac{2}{3} \\ 0 & \frac{1}{2} & 0 \\ \frac{2}{3} & 0 & -\frac{1}{3} \end{pmatrix}$

43. No inverse exists. The reduced augmented matrix is $\begin{pmatrix} 1 & \frac{1}{2} & | & \frac{1}{2} & 0 \\ 0 & 0 & | & 0 & 0 \end{pmatrix}$ and since the left part is not the identity matrix, B has no inverse.

44. Total production is $X = \begin{pmatrix} \$35,700 & \$41,300 \end{pmatrix}$.

45. The production matrix is given *exactly* by $X = \$1,000,000 \begin{pmatrix} \frac{20}{9} & 1 & \frac{76}{81} \end{pmatrix}$

Chapter 3

1. Only $3 > t$ is strict.   **2.**   **3.**   **4.**

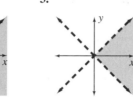

5.   **6.**   **7.** No change.   **8.** $(x_1, x_2) = (1, 1)$; $Q = 5$

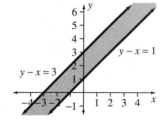

9. The minimum is $p = 1$ at the point $(1, 0)$.

10. $P = 12$ at $(x, y) = (\frac{8}{7}, \frac{15}{7})$.   **11.** Four shirts and six pants; total profit $26

12. Objective function: $P = 3x + 5y$. Constraint on white: $.5x + .2y \le 10$. Constraint on yellow: $.3x + .6y \le 12$. Non-negative constraints: $x \ge 0$, $y \ge 0$. The corners of the feasible region are $(0, 0)$, $(20, 0)$, $(15, 12\frac{1}{2})$, and $(0, 20)$. The optimal strategy is 15 gal of the first pastel, and $12\frac{1}{2}$ gal of the second. Under that strategy, the maximum profit is $107.50.

13. Defining x and y as in the example, $x = y = 500$ and the minimal shipping cost is $3100.

14. 13 refrigerators and 7 stoves; the minimum commission is $1175.

15. No feasible region is defined, so no minimum is possible.

16. We cannot find a definite maximum for P because the feasible region is unbounded. Feasible values of x and y will allow ever larger values of P; no one such value is the absolute maximum.

17. No, because the new feasible region is still unbounded.

18. The minimum is $p = 6$. Any point on the line segment joining the points $(3, 0)$ and $(0, 2)$ will give that minimum.

Chapter 4

1. (c)

2. We need the slack variables s_1 and s_2. With them, the system of equalities becomes

$$x_1 + x_2 + s_1 \qquad\qquad = 2$$
$$2x_1 + x_2 \qquad + s_2 \qquad = 3$$
$$-3x_1 - x_2 \qquad\qquad + P = 0$$

3. (a) There are 2 non-basic variables and 3 basic variables, including P. There are six ways of distributing zeros among the variables x_1, x_2, s_1 and s_2.

(b)

x_1	x_2	s_1	s_2	$P = 3x_1 + x_2$
0	0	2	3	0
0	2	0	1	2
0			0	unfeasible
	0	0		unfeasible
$1\frac{1}{2}$	0	$\frac{1}{2}$	0	$4\frac{1}{2}$
1	1	0	0	4

(c) The maximum value of P is $4\frac{1}{2}$ when $(x_1, x_2) = (1\frac{1}{2}, 0)$. Furthermore, at that point, $s_1 = s_2 = 0$.

4. (a)
$$x \quad + s_1 \qquad\qquad = 4$$
$$x + y \qquad + s_2 \qquad = 5$$
$$-3x + y \qquad\qquad + P = 0$$

(b)

x	y	s_1	s_2	$P = 3x - y$
0	0	4	5	0
0		0		inconsistent
0	5	4	0	-5
4	0	0	1	12
	0	0		unfeasible
4	1	0	0	11

(c) $P = 12$ at $(4, 0)$

5. (a)
$$x \quad + s_1 \qquad\qquad\qquad = 40$$
$$y \qquad + s_2 \qquad\qquad = 30$$
$$3x + 4y \qquad\qquad + s_3 = 180$$

(b)

x	y	s_1	s_2	s_3	$P = x + y$
0	0	40	30	180	0
0		0			inconsistent
0	30	40	0	60	30
0			0		unfeasible
40	0	0	30	60	40
	0		0		inconsistent
	0			0	unfeasible
		0	0		unfeasible
40	15	0	15	0	55
20	30	20	0	0	50

(c) 40 of the first, 15 of the second.

6. The basic variables are x_2, s_1, s_3, and P. The non-basic variables are x_1 and s_2.

7. There are three non-basic variables. The six basic variables include one corresponding to the objective function P.

8. The variables x_1, s_2, and (of course) P are basic.

9. The pivot variable is s_1. For the first row, we compute the pivot ratio as $2/2 = 1$. For the second row, the ratio is $6/3 = 2$. Remember, we never compute a pivot ratio for the final row.

10. The pivot column is s_1, and the smallest pivot ratio is at the first element of that column (the third 2 on the first row). Pivoting

about that element yields this matrix. $\begin{pmatrix} 1 & 1 & \boxed{1} & 0 & 0 & 1 \\ -2 & -3 & 0 & 1 & 0 & 3 \\ 9 & 10 & 0 & 0 & 1 & 21 \end{pmatrix}$

11. (a) $s_1 = s_2 = 0$. (b) $x_1 = 6$, $x_2 = 4$. (c) $P_{max} = 14$.

12. Upon introducing slack variables s_1 and s_2, we have an initial simplex tableau of $\begin{pmatrix} 1 & 1 & 1 & 0 & 0 & 3 \\ 2 & \boxed{3} & 0 & 1 & 0 & 6 \\ -3 & -4 & 0 & 0 & 1 & 0 \end{pmatrix}$ Upon

pivoting, we find that the pivot ratios for two rows are the same. Choose any one, and pivot. The final simplex tableau is

$$\begin{pmatrix} 1 & 0 & 3 & -1 & 0 & | & 3 \\ 0 & 1 & -2 & 1 & 0 & | & 0 \\ 0 & 0 & 1 & 1 & 1 & | & 9 \end{pmatrix}$$ from which we learn that the maximum value of P is 9, occurring when $x_1 = 3$ and when $x_2 = s_1 = s_2 = 0$.

13. $P_{max} = 7$. There are two variables, and when $x_1 = x_2 = 1$, then $P = 7$.

14. Let x_1 and x_2 be the number of 8% and 12% bonds. Now, $x_1 + x_2 \le 100$ and $4x_2 - 3x_1 \le 1$ are the constraints. The optimal strategy requires purchasing 57 low yield bonds and 43 high yield bonds. With that strategy, the maximum annual income is $9720.

15. $x_1 = 100$, $x_2 = 50$, and $P = \$140,000$.   **16.** There are 22 constraints and 23 unknowns.   **17.** (a)

18. $M^T = M$, $N^T = \begin{pmatrix} 1 & 1 \\ x & y \\ x^2 & y^2 \end{pmatrix}$, and $P^T = P$.

19. (a) Since there are two inequality constraints and two unknowns, the matrix M will have dimension 3×3.

(b) $M = \begin{bmatrix} 1 & 2 & 3 \\ 2 & 1 & 3 \\ 1 & 1 & 0 \end{bmatrix}$; $M^T = \begin{bmatrix} 1 & 2 & 1 \\ 2 & 1 & 1 \\ 3 & 3 & 0 \end{bmatrix}$.

(c) Subject to $y_1 \ge 0$ and $y_2 \ge 0$ and to $y_1 + 2y_2 \le 1$ we need the maximum value of $P = 3y_1 + 3y_2$.

$$2y_1 + y_2 \le 1$$

20. $P_{min} = 11$, $x_1 = 3$, $x_2 = 0$.

21. (a) We introduce two slack variables u_1 and u_2. $y_1 + 2y_2 + u_1 \qquad\qquad = 1$ (b) $\begin{pmatrix} 1 & 2 & 1 & 0 & 0 & | & 1 \\ 2 & 1 & 0 & 1 & 0 & | & 1 \\ -3 & -3 & 0 & 0 & 1 & | & 0 \end{pmatrix}$

$2y_1 + y_2 \qquad\quad + u_2 \qquad = 0$

$-3y_1 - 3y_2 \qquad\qquad\quad + P = 0$

(c) $\begin{pmatrix} 0 & 1 & \frac{2}{3} & -\frac{1}{3} & 0 & | & \frac{1}{3} \\ 1 & 0 & -\frac{1}{3} & \frac{2}{3} & 0 & | & \frac{1}{3} \\ 0 & 0 & 1 & 1 & 1 & | & 2 \end{pmatrix}$ (d) For the dual problem, $P_{MAX} = 2$ when $y_1 = y_2 = \frac{1}{3}$. (e) For the minimum problem, $P_{min} = 2$ for $x_1 = x_2 = 1$.

22. $\begin{bmatrix} 1 & 1 & 2 \\ 1 & 2 & 3 \\ 2 & 3 & 0 \end{bmatrix}$; $P = 2y_1 + 3y_2$ subject to $y_1 + y_2 \le 2$ and $y_1 + 2y_2 \le 3$;

y_1	y_2	u_1	u_2	P	
1	1	1	0	0	2
1	2	0	1	0	3
-2	-3	0	0	1	0

;

y_1	y_2	u_1	u_2	P	
1	0	2	-1	0	1
0	1	-1	1	0	1
0	0	1	1	1	5

; and $p = 5$ when $x_1 = x_2 = 1$.

23. $p = \frac{5}{2}$ when x_1, x_2, and x_3 take any values between $(0, \frac{5}{2}, 0)$ and $(0, \frac{3}{2}, 1)$

24. Fly 2 small and 5 large balloons (for a total cost of $7600).

25. Introduce a slack variable s, a surplus variable S, an artificial variable a, and the cost penalty M. The problem becomes

$$x_1 + x_2 + s \qquad\qquad\qquad = 2$$
$$x_1 - x_2 \qquad - S + a \qquad = 1$$
$$-3x_1 - 2x_2 \qquad\qquad + Ma + P = 0$$

26. When we introduce the auxiliary variables s, S, and a, the initial tableau matrix is $\begin{pmatrix} 1 & 1 & 1 & 0 & 0 & 0 & | & 2 \\ 1 & -1 & 0 & -1 & 1 & 0 & | & 1 \\ -3 & -2 & 0 & 0 & M & 1 & | & 0 \end{pmatrix}$.

Before pivoting, we need to make a a basic variable.
$$\left(\begin{array}{cccccc|c} 1 & 1 & 1 & 0 & 0 & 0 & 2 \\ \boxed{1} & -1 & 0 & -1 & 1 & 0 & 1 \\ -3-M & M-2 & 0 & M & 0 & 1 & -M \end{array}\right)$$
Now, we pivot about the

boxed element to find this sequence of augmented matrices.
$$\left(\begin{array}{cccccc|c} 0 & 1 & \frac{1}{2} & \frac{1}{2} & -\frac{1}{2} & 0 & \frac{1}{2} \\ 1 & -1 & 0 & -1 & 1 & 0 & 1 \\ 0 & -5 & 0 & -3 & M+3 & 1 & 3 \end{array}\right) \Rightarrow$$

$$\left(\begin{array}{cccccc|c} 0 & 1 & \frac{1}{2} & \frac{1}{2} & -\frac{1}{2} & 0 & \frac{1}{2} \\ 1 & 0 & \frac{1}{2} & -\frac{1}{2} & \frac{1}{2} & 0 & \frac{3}{2} \\ 0 & 0 & \frac{5}{2} & -\frac{1}{2} & M+\frac{1}{2} & 0 & \frac{11}{2} \end{array}\right) \Rightarrow$$
$$\begin{array}{cccccc} x_1 & x_2 & s & S & a & P \\ \left(\begin{array}{cccccc|c} 0 & 2 & 1 & 1 & -1 & 0 & 1 \\ 1 & 1 & 1 & 0 & 0 & 0 & 2 \\ 0 & 1 & 3 & 0 & M & 1 & 6 \end{array}\right) \end{array}$$

The maximum value is $P = 6$ when $x_1 = 2$ and $x_2 = 0$. Also, $s = a = 0$ and $S = 1$.

27. $P = 8$ when $x_1 = x_2 = 2$.

28. The strategy changes. The new optimum is $I = \$17,000$ when $x_1 = 3$ and $x_2 = 4$.

Chapter 5

1. Only (b) and (d) are true **2.** Both statements are true. **3.** 6 **4.** 14 **5.** $A' = \{a, 1, s, y, z\}$ **6.** Ø; U

7. There are four elements—hh, ht, th, and tt, where h = head and t = tail. **8.** $L \cap M = \{5, 6, 7, 8, 9, 10\}$

9. A **10.** $P \cap E = \{1, 2\}$, $P \cup E = \{1, 2, 3, 4, 5, 6, 7, 8, 9\}$, and $(P \cap E)' = \{0, 3, 4, 5, 6, 7, 8, 9\}$

11. (a) $N \cap C$; (b) $F \cap C'$; (c) $(F \cup N) \cap C$.

12. A; $(B' \cap A) \cup (A' \cap B)$ (but there are other ways of representing this)

13. **14.**

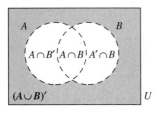

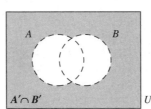

15. The theorem is proved by noting the equivalence of the diagrams in the previous skill enhancers.

16. A singular circular area represents both sets. In other words, the single area has two distinct names.

17. No. **18.** **19.** 10 **20.** 5 **21.** 16

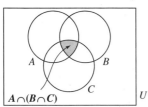

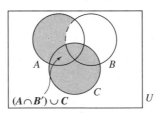

22. 4 cases **23.** (a) 3; (b) 10; (c) 5; and (d) 9. **24.** $3 \times 3 \times 2 = 18$. **25.** $5! = 120$; $7! = 5040$.

26. $6! = 720$. **27.** (a) 2; (b) 6; (c) 20. **28.** $P(10, 7) = \frac{10!}{3!}$ **29.** (a) 1; (b) 4; (c) 10. **30.** 462

31. (a) combinations; (b) combinations; and (c) permutations.

32. By the formula, $C(n, n) = \frac{n!}{(n-n)!n!} = \frac{n!}{0!n!} = 1$, as claimed. Because $0! = 1$, this formula applies even for $n = 0$.

33. $C(11, 5)C(6, 5) = \frac{11!}{5!5!1!} = 2772$. **34.** $C(20, 7)C(13, 7)C(6, 6) = \frac{20!}{7!7!6!}$

35. $(\{i\}, \{j\}, \{k\})$; $(\{i\}, \{k\}, \{j\})$; $(\{j\}, \{i\}, \{k\})$; $(\{j\}, \{k\}, \{i\})$; $(\{k\}, \{i\}, \{j\})$; and $(\{k\}, \{j\}, \{i\})$.

36. $\left(\begin{array}{ccc} & 20 & \\ 7, & 7, & 6 \end{array}\right) = \frac{20!}{7!}7!6! \approx 3.22 \times 10^{14}$. **37.** $C(4, 1) = 4$

38. Begin by observing that $(x + y)^n = (y + x)^n$. By the Theorem, the coefficient of $x^i y^{n-i}$ is $C(n, i)$. On the other hand, the coefficient of $y^{n-i} x^i$ is $C(n, n - i)$. But these coefficients must be the same, so we conclude that the statement is true.

Chapter 6

1. $S = \{v, i, b, g, y, o, r\}$ where each letter stands for a color.

2. The sample space is $S = \{$buttered side face up, buttered side face down$\}$, or $S = \{U, D\}$. There are four possible events in S—$\emptyset$, S, $\{U\}$, and $\{D\}$. (Which event is impossible? Which is certain?)

3. $S = \{hh, ht, th, tt\}$.   **4.** $= \{RR, RS, SR, SS\}$, where $R = $ rain and $S = $ sun.

5. $S_1 = \{0, 1, 2, 3\}$, where the number represents the number of heads observed; $S_2 = \{HHH, HHT, HTH, HTT, THH, THT, TTH, TTT\}$ where $H = $ heads and $T = $ tails.

6. $S = \{P_1 T_1, P_1 T_2, P_2 T_1, P_2 T_2, P_3 T_1, P_3 T_2\}$ where P_1 and so on are the presidential candidates and T_1, etc. are the candidates for Treasurer.

7. (a) $\{m, a, t, h, e, i, c, s\}$; (b) $\{a, e, i, o, u, l, s, y\}$; (c) $\{e, u, a\}$.

8. (a) $\{BBB\}$; (b) S_2; (c) $\{BBG, BGG, GBG, GGG\}$.   **9.** $\frac{15}{20} = \frac{3}{4}$.   **10.** $\frac{9}{12} = \frac{3}{4}$.

11. (a) valid; (b) invalid; (c) invalid; and (d) invalid.   **12.** (a) $3(c + d) = 1$; (b) c; (c) $c + 3d$; and (d) $2c + d$.

13. The relative frequency of germination is $\frac{74}{100} = 0.74$. This number is the empirical probability.   **14.** $\frac{211}{313} \approx 0.6741$.

15. $a = 3$, $b = 6$, $P(F) = \frac{3}{3+6} = \frac{3}{9} = \frac{1}{3}$.   **16.** $\frac{3}{8}$.   **17.** $a = 5$ and $a + b = 8$ so $a = 5$ and $b = 3$. Odds are 5 to 3.

18. 1 to 3; but 2 to 6, 3 to 9, and so on are also correct.

19. Yes. There are seven such events. The probability is $\frac{1}{7}$.   **20.** $\frac{1}{22}$

21. Two sample spaces are $S_1 = \{0, 1, 2, 3\}$ where each number represents the number of observed heads, and $S_2 = \{HHH, HTH, HHT, HTT, THH, TTH, THT, TTT\}$. Because all elements of S_2 are equiprobable, S_2 is preferred.

22. Use the sample space $S = \{HH, HT, TH, TT\}$. (a) $\frac{1}{4}$; (b) $\frac{1}{4}$; (c) $\frac{1}{2}$.

23. (a) Since there are 7 even numbers in this range, $P = \frac{7}{15}$. (b) Since there are eight odd numbers in this range, $P = \frac{8}{15}$.
(c) There are five distinct digits in the current date. Therefore, this time, $P = \frac{5}{15} = \frac{1}{3}$.

24. Use the same sample space. (a) $P = \frac{2}{36} = \frac{1}{18}$. (b) $P = \frac{4}{36} = \frac{1}{9}$. (c) Only four simple events are members of the event; this event is the same as saying that only a 1 or 2 appears on any die. $P = \frac{4}{36} = \frac{1}{9}$.

25. $P = \frac{34}{C(40,7)}$

26. Let E be the event that a player wins second prize in the New York lottery, and F the winning of the first prize in the modified state lottery. $P(E) \approx 0.0000323$. $P(F) = \frac{1}{C(20,6)} = 0.0000258$. F is the more likely event.

27. There are $C(6, 3) = 20$ equiprobable events in the sample space. (a) $P(3 \text{ reds}) = \frac{1}{20}$. (b) $P(3 \text{ reds or 3 blacks}) = \frac{1}{10}$.
(c) $P(\text{exactly 2 reds}) = \frac{C(3,2)}{C(6,3)} = \frac{3}{20}$. (d) Drawing at least two marbles is done by drawing all three reds or by drawing exactly two reds. $P(\text{at least 3 reds}) = \frac{1}{20} + \frac{3}{20} = \frac{1}{5}$.

28. (a) 'Vanity' and 'modesty' have a single letter in common (y), so these events are not mutually exclusive. (b) These words also have letters in common (e, i, l), so these events are also not mutually exclusive.

29. (a) This pair of events is not mutually exclusive; for example, the second day of the month is in the first two weeks of any month. (b) This pair of events is mutually exclusive. Only Monday begins with 'M', and this is not a day between Thursday and Saturday.

30. (a) Yes; (b) $P(G) = \frac{20}{81}$; (c) $P(H) = \frac{20}{81}$; (d) $G \cup H$; (e) $P(G \cup H) = \frac{40}{81}$.

31. (a) $P(TTT) = \frac{1}{8}$; (b) yes; (c) $P(HHH \cup TTT) = \frac{1}{8} + \frac{1}{8} = \frac{1}{4}$.

32. (a) $\frac{1}{2}$; $\frac{1}{2}$. (b) $P(\text{green and even}) = \frac{1}{6}$; (c) $P(\text{green or even}) = \frac{1}{2} + \frac{1}{2} - \frac{1}{6} = \frac{5}{6}$.

33. (a) No. (b) $P(F) = \frac{15}{20} = \frac{3}{4}$; $P(B) = \frac{7}{20}$. (c) $P(F \cap B) = \frac{5}{20} = \frac{1}{4}$. (d) $P(F \cup B) = \frac{3}{4} + \frac{7}{20} - \frac{1}{4} = \frac{17}{20}$.

34. $1 - \frac{1}{9} = \frac{8}{9}$.   **35.** $P(3 \text{ heads}) = \frac{1}{8}$. $P(\text{not 3 heads}) = 1 - \frac{1}{8} = \frac{7}{8}$.

36. (a) $P(R \cup H) = 0.7 + 0.4 - 0.3 = 0.8$; (b) $P((R \cup H)') = 1 - 0.8 = 0.2$.   **37.** the second event.   **38.** $\frac{1}{2}$

39. (a) $\frac{1}{2}$; (b) $\frac{2}{3}$.   **40.** (a) $P(F) = \frac{4}{10}$. (b) $P(G|F) = \frac{P(G \cap F)}{P(F)}$ (Note that $F \cap G$ is the same as $G \cap F$). (c) $\frac{1}{3}$.

41. $\frac{1}{6}$; $\frac{1}{3}$.   **42.** (a) $\frac{269}{500}$; (b) $\frac{64}{500}$; (c) $P(L|B) = \frac{P(L \cap B)}{P(B)}$; (d) $\frac{55}{269}$.

43. Not independent; the three given probabilities do not obey the law, for $0.6 \times 0.7 = 0.42 \neq 0.41$.

44. The probability of $IND \cap BMT$ would be 0.176 if 176 of the commuters took both lines. Since 150 take both already, an additional 26 would need to take both subway lines.

45. $\frac{1}{2}$.   **46.** (a) 8; $\frac{2}{3}$; (b) $\frac{2}{9} > 0.2$.   **47.** 0.41   **48.** $\frac{2}{3}$.   **49.** $P(S_1|B) = \frac{14}{41}$; $P(S_2|B) = \frac{27}{41}$.

50. They would remain the same.   **51.** (a) $P(A) = 0.5$ $P(B) = 0.5$ (b) $P(V) = 0.075$. (c) $P(A|V) = \frac{1}{3}$.

$P(V|A) = 0.05$ $P(V|B) = 0.1$

52. $P(W) = \frac{3}{7}$. $P(1|W) = \frac{2}{3}$; $P(2|W) = \frac{1}{3}$.   **53.** (a) $\frac{11}{28}$; (b) $\frac{2}{11}$; (c) $\frac{5}{11}$.

54. (a) $P(S|N) = \frac{1}{3}$; $P(S|C) = \frac{1}{4}$; $P(S|B) = 1$. (b) $\frac{3}{8}$. (c) $P(B|S) = P(C|S) = P(N|S) = \frac{1}{3}$.

55. (a) $P(T) = 0.33$, $P(T') = 0.67$; (b) $P(+) = (0.95)(0.33) + (0.55)(0.2) = 0.4475$; (c) $P(T|+) = \frac{(0.95)(0.33)}{0.4475} \approx 0.701$.

56. $(0.3)2 + (0.4)(-1) + (0.3)(-1) = -0.1$

57. $E = 2(\frac{1}{2}) + (-1)(\frac{1}{2}) = \frac{1}{2}$. That is, in the long run, you gain one-half dollar on each play. You should certainly want to play this game; on the long run, you pay thirty cents to win fifty cents, so you gain twenty cents per play on the average.

58. (a) $P(T) = 1 - 0.6 = 0.4$. (b) $(0.6)(1) + 0.4(-x) = -0.2$. (c) $x = 2$.

59. $E_{live} = \frac{1}{200,000} = 0.0005\%$; $P_u = \frac{1}{150}$; $E_{dead} = \frac{1}{150} \times \frac{15}{1000} = 0.0001 = 0.01\%$

60. (a) $\frac{1}{3}$; (b) $\frac{1}{6}$; (c) $4\frac{1}{6}$.   **61.** $E = 0(\frac{1}{3}) + 3(\frac{2}{3}) = 2$

62. (a) $E_I I = -150,000$. (b) Now, E_I and E_{II} are equal. Management will need to find another decision-making tool.

63. (a) \$2.04 million; (b) The first.

Chapter 7

1. (a) $\mathbf{A}^1 = \begin{pmatrix} 1 & 0 \end{pmatrix} \begin{pmatrix} .75 & .25 \\ .25 & .75 \end{pmatrix} = \begin{pmatrix} .75 & .25 \end{pmatrix}$, so the chance of sun on Saturday is 0.75.

(b) Now, $\mathbf{A}^1 = \begin{pmatrix} 0 & 1 \end{pmatrix} \begin{pmatrix} .75 & .25 \\ .25 & .75 \end{pmatrix} = \begin{pmatrix} .25 & .75 \end{pmatrix}$, and $\mathbf{A}^2 = \begin{pmatrix} .25 & .75 \end{pmatrix} \begin{pmatrix} .75 & .25 \\ .25 & .75 \end{pmatrix} = \begin{pmatrix} .375 & .625 \end{pmatrix}$, so the chance of sun on Sunday is now $0.375 = \frac{3}{8}$.

2. 0; 0.34; 0.236; 0.352.

3. (a) $\mathbf{P} = \begin{pmatrix} .5 & 0 & .5 \\ .5 & .5 & 0 \\ 0 & .5 & .5 \end{pmatrix} = \frac{1}{2}\begin{pmatrix} 1 & 0 & 1 \\ 1 & 1 & 0 \\ 0 & 1 & 1 \end{pmatrix}$. (b) $\mathbf{P}^2 = \frac{1}{4}\begin{pmatrix} 1 & 1 & 2 \\ 2 & 1 & 1 \\ 1 & 2 & 1 \end{pmatrix}$; $\mathbf{P}^3 = \frac{1}{8}\begin{pmatrix} 2 & 3 & 3 \\ 3 & 2 & 3 \\ 3 & 3 & 2 \end{pmatrix}$. (c) Since $(\mathbf{P}^2)_{23} = \frac{1}{4}$, there is a

probability of 0.25 (d) Since $(\mathbf{P}^3)_{11} = \frac{1}{8} \cdot 2 = \frac{1}{4}$, the likelihood is 0.25.

4. L is regular because every element of $L^2 = LL$ is greater than zero. M is regular because every one of its elements is greater than zero. N is not; to see why, compute N^2, N^3, and so on. For every power n, $N^n = N$, and since there are some zero elements in N, this matrix is not regular.

5. $\begin{pmatrix} u & v \end{pmatrix} = \begin{pmatrix} u & v \end{pmatrix} \begin{pmatrix} \frac{1}{3} & \frac{2}{3} \\ \frac{1}{2} & \frac{1}{2} \end{pmatrix}$ and $u + v = 1$. This leads to $3v = 4u$ together with $u + v = 1$. The solution is $\mathbf{w} = \begin{pmatrix} \frac{3}{7} & \frac{4}{7} \end{pmatrix}$.

6. $u = v = w = \frac{1}{3}$; $\mathbf{W} = \begin{pmatrix} \frac{1}{3} & \frac{1}{3} & \frac{1}{3} \\ \frac{1}{3} & \frac{1}{3} & \frac{1}{3} \\ \frac{1}{3} & \frac{1}{3} & \frac{1}{3} \end{pmatrix}$.

7. For $p = \frac{1}{2}$, we get $\mathbf{P}^2 = \mathbf{P}^3 = \begin{pmatrix} .5 & .5 \\ .5 & .5 \end{pmatrix}$. For $p = \frac{1}{3}$, $\mathbf{P}^2 = \begin{pmatrix} \frac{5}{9} & \frac{4}{9} \\ \frac{4}{9} & \frac{5}{9} \end{pmatrix}$, $\mathbf{P}^3 = \begin{pmatrix} \frac{14}{27} & \frac{13}{27} \\ \frac{13}{27} & \frac{14}{27} \end{pmatrix}$.   **8.** Only N is.

9. Here, $I = \begin{pmatrix} 1 & 0 \\ 0 & 1 \end{pmatrix}$, $O = \begin{pmatrix} 0 & 0 \\ 0 & 0 \end{pmatrix}$, $R = \begin{pmatrix} \frac{1}{3} & \frac{1}{6}\frac{1}{4} & \frac{1}{4} \end{pmatrix}$, and $Q = \begin{pmatrix} \frac{1}{6} & \frac{1}{3}\frac{1}{4} & \frac{1}{4} \end{pmatrix}$. Since I is 2×2, there are two absorbing states. Q is also of dimension 2×2, so there are two non-absorbing states.

10. Because there are m non-absorbing states, we know Q has dimension $m \times m$. By the rules of matrix multiplication, $Q' = Q^2$ also has dimension $m \times m$. Therefore, there are m absorbing states in $\mathbf{P}^{62}$.

11. We have previously seen that $Q = \begin{pmatrix} \frac{1}{6} & \frac{1}{3} \\ \frac{1}{4} & \frac{1}{4} \end{pmatrix}$. Therefore, $I - Q = \begin{pmatrix} \frac{5}{6} & -\frac{1}{3} \\ -\frac{1}{4} & \frac{3}{4} \end{pmatrix}$. The inverse is $F = \frac{1}{13}\begin{pmatrix} 18 & 8 \\ 6 & 20 \end{pmatrix}$.

12. (a) $R = \begin{pmatrix} \frac{1}{3} & \frac{1}{6} \\ \frac{1}{4} & \frac{1}{4} \end{pmatrix}$; $FR = \begin{array}{c} \\ 3 \\ 4 \end{array}\begin{pmatrix} \frac{8}{13} & \frac{5}{13} \\ \frac{7}{13} & \frac{6}{13} \end{pmatrix}$. (b) $p_{31} = \frac{8}{13}$, $p_{32} = \frac{5}{13}$; (c) $p_{41} = \frac{7}{13}$, $p_{42} = \frac{6}{13}$.

13. (a) $\frac{6}{13}$, $\frac{20}{13}$, 2. (b) 2.

Chapter 8

1.

		C	
		1	2
R	1	−2	+3
	2	+3	−4

2. (a) $\begin{array}{cc} 50\% & 75\% \\ 30\% & 35\% \end{array}$. (b) Choose town 1. (c) Choose town 1. (d) 50%.

3. (a) The first and third elements of the bottom row. Its value is 4. (b) No saddle points.

4. (a) No saddle point. (b) The saddle point is top right element; the game's value is 5.　　**5.** (a) No. (b) $-\frac{3}{8}$. (c) $-\frac{1}{3}$.

6. $\frac{1}{4}$.　　**7.** yes; $\begin{pmatrix} 3 & 4 \\ 4 & 3 \end{pmatrix}$.　　**8.** $\begin{pmatrix} 2 & 4 \\ 3 & 2 \end{pmatrix}$.　　**9.** (a) $\begin{pmatrix} 1 & 2 & 4 \\ 3 & 1 & 3 \end{pmatrix}$. (b) $\begin{pmatrix} 1 & 2 \\ 3 & 1 \end{pmatrix}$.

10. (a) No. (b) $2x_1 + 10x_2 \geq 1$, $11x_1 + x_2 \geq 1$. (c) $(x_1, x_2) = (\frac{1}{12}, \frac{1}{12})$. (d) Minimize x. Its minimum yields $\frac{1}{t} = \frac{1}{6}$.
(e) $t = 6$. $r_1 = r_2 = \frac{6}{12} = \frac{1}{2}$.

11. (a) $2y_1 + 11y_2 \leq 1$, $10y_1 + y_2 \leq 1$. (b) $(y_1, y_2) = (\frac{5}{54}, \frac{2}{27})$. (c) Maximize x. Its maximum yields $\frac{1}{w} = \frac{1}{6}$. (d) $w = 6$. $c_1 = 6(\frac{5}{54}) = \frac{5}{9}$, $c_2 = 6(\frac{2}{27}) = \frac{4}{9}$. (e) 6.

12. (a) $G' = \begin{pmatrix} 2 & 11 \\ 10 & 1 \end{pmatrix}$. (b) The strategies are as in Enhancers 10 and 11. (c) $v = 3$.

13. (a) $e = 2$, $f = 11$, $g = 10$, $h = 1$. (b) $\Delta = -18$. (c) $r_1 = r_2 = \frac{1}{2}$; $c_1 = \frac{10}{18} = \frac{5}{9}$; $c_2 = \frac{8}{18} = \frac{4}{9}$.

14. (a) $e = 1$, $f = g = -2$, $h = 4$, $\Delta = 9$. (b) $r_1 = \frac{2}{3}$, $r_2 = \frac{1}{3}$. (c) $c_1 = \frac{2}{3}$, $c_2 = \frac{1}{3}$. (d) $v = 0$.

Chapter 9

1. Do it!　　**2.** Do it!　　**3.** Do it!　　**4.**　　　　　　　　　　　　**5.** Do it!　　**6.** $\frac{5}{24}$.　　**7.** (a) $\frac{490}{6} = 81\frac{2}{3}$;

(b) $81\frac{2}{3}$.　　**8.** (a) 25; (b) $\frac{2415}{25} = \$96.60$.　　**9.** (a) $\bar{x} = \frac{85+86+86.5}{3} \approx 85.83$. (b) $\bar{x} = \frac{87+87+90}{3} = 88.0$.

10. (a) No mode. (b) 1. (c) 1 and 3.　　**11.** (a) 3; 2.8 (b) 1. (c) 3.　　**12.** (a) 3 for both. (b) $\sigma^2 = 2$. (c) $\sigma^2 = 7.76$.

13. $\mu = 3$; $\sigma^2 = 1.2$ mo^2.　　**14.** 8 ohms.　　**15.** (a) $s_1^2 = 0.58$; $s_2^2 = 3$. (b) $\sigma^2 = 2.37$ cm^2.

16. (a) $\mu = 0.4$, $\sigma^2 = 0.002$ (b) $k = 0.1$, $k^2 = 0.01$; 0.8.

17. (a) $\mu = 75$ yr, $\sigma^2 = \frac{1}{4}$ yr, $k = 10$ yr, and $k^2 = 100$ yr^2. (b) $1 - 0.0025$.　　**18.** (a) 0.5; (b) 0.8187; (c) 0.4772.

19. (a) 200; (b) The entire curve to the right of $x = \mu + 2\sigma$, which consists of one column plus the right-hand tail of the curve.
(c) 2.27%.

20. (a) 105, 130, 155, 205, 230, 255. (b) 79. (c) 171. (d) 11.   **21.** (a) -5; (b) -2; (c) $\frac{1}{2}$; (d) 3.

22. $z_{7100} = -0.8$, $z_{7600} = 0.2$, $z_{7995} = 0.99$.   **23.** (a) 0.1554; (b) 0.1700.

24. (a) $\frac{1}{2}$; $\frac{3}{2}$. (b) .1915; .4332. (c) .2417 $= .4332 - .1915$.

25. If s and f represent the butter landing down or up, then $p = P(s)$ is probably slightly greater than one-half (since butter adds some weight). $q = P(f) = 1 - p$ is slightly less than one-half.

26. There are only two outcomes, H or T, and the probability of each remains the same from trial to trial. Each flip is independent of any other.

27. $\frac{1}{2}$.   **28.** (a) 3; sff, fsf, ffs. (b) $P(s) = \frac{1}{4}$; $P(f) = \frac{3}{4}$. (c) $\frac{3^2}{4^3}$. (d) $3 \times \frac{3^2}{4^3}$. (e) $b(3, 1; \frac{1}{4})$.

29. (a) $\frac{3\cdot4}{5^5}$; (b) $4(0.3)(0.7)^4$; (c) $(\frac{2}{3})^5$.   **30.** (a) $\frac{5}{2^5}$; (b) $\frac{1}{2^5}$; (c) $\frac{6}{2^5}$.

31. (a) $q = 0.8$, $\mu = 4000$, $\sigma = \sqrt{3200} \approx 56.57$. (b) Yes. (c) $z = \frac{100}{56.57} = 1.77$. (d) $p = 0.5 - .4616 = .0384$.   **32.** 50%.

Chapter 10

1. $324.   **2.** $336.96.   **3.** (a) $n = 16$; $i = 0.015$. (b) $4441.45.

4. (a) $6863.93; (b) $11,040.20; (c) $24,377.20; (d) $53,825.82.   **5.** (a) 0.5% = 0.005. (b) $61.68. (c) 6.168%.

6. (a) $46,000.00; (b) $i = 0.015$, $n = 120$; (c) $59,693.23; (d) account B.   **7.** $4980.00

8. (a) $272,615.08; (b) $328,988.05   **9.** (a) $n = 30$, $i = 0.08$; (b) 113.283; (c) $3089.60.

10. (a) $9145.42; (b) $9151.42.

11. The money Max borrows is, from the bank's point of view, the purchase price to the bank of a 3-year annuity, during the life of which the bank receives $1500 per year. The value to the bank of this annuity is the present value of the annuity, which is the sum of the present values of the payments.

12. (a) $V_1 = 1500(1.08)^{-1} = \$1388.89$; (b) $V_2 = 1500(1.08)^{-2} = \$1286.01$; $V_3 = \$1190.75$; (c) $3865.65.

13. (a) $p = 50$, $n = 25$, $i = 0.015$, $A(25, .015) = 20.7196$; (b) $1035.98; (c) $1085.98; (d) $1300.00.

14. $771.82   **15.** $g = 10.4\%$ in this period.   **16.** (a) $476.19; (b) $453.51.

17. $444.54   **18.** 15.36%   **19.** (a) 5.7692% (b) $n = 15$; $A(15, 0.057692) = 9.8604$; (c) $197,208.13; (d) $197,208.13.

20. (a) $.095/12 = 0.00792$, $.09/12 = 0.00750$; (b) For 9.5%, $PV = 750A(300, .00792) = \$85,814.84$. For 9%, $PV = 750A(300, .00750) = \$89,371.22$. (c) $3,556.38. It is probably worth the wait. (d) $r = 12.785\%$, $i = 1.065\%$. (e) $67,465.65.

21. (a) 0.8% or 0.008; (b) $FV' = \$540,848.41$.   **22.** (a) $FV' = 45.5992p$. (b) $PV' = 66,567.17$. (c) $p = 1459.83$.

Chapter 11

1. $y_{n+1} = 1.2y_n + 0.001h$; $y_0 = h$.   **2.** (a) $y_0 = 250$; (b) $y_{n+1} = y_n + .05y_0$. (c) 1, 2

3. (a) $y_0 = 1000$; (b) $0.75\% = 0.0075$; (c) The balance increase is due to the earned interest for that month together with the new deposit. All this is added to the previous balance. (d) $y_{n+1} = 1.0075y_n + 0.1p$.

4. (a) $y_0 = 0$; (b) $0.75\% = 0.0075$; (c) The monthly increase is given by p from the employee, $\frac{p}{3}$ from the employer, the interest earned, and one-third of the earned interest $.0075y_n$ contributed by the employer. (d) $y_{n+1} = 1.001y_n + \frac{4}{3}p$.

5. $y_n = 12(1.2^n - 1)$.   **6.** $y_n = 6 - 7(-0.5)^n$.   **7.** $y_n = \frac{3}{4} + \frac{9}{4}(-3)^n$.   **8.** $y_n = -1 + 10n$.

9. (a) 2; (b) $\frac{b}{1-a} = y_0$; (c) $y_n = 2$ for all n.   **10.** (a) $-16,666\frac{2}{3}$; (b) no; (c) the solution diverges without oscillating.

11. (a) $\approx 5.88 \times 10^{-4}$; (b) the solution converges and oscillates; (c) $\approx 5.88 \times 10^{-4}$.   **12.** (a) 3. (b) $y_n = 3$ for all n.

13. $y_n = 1250(1.015)^n$.   **14.** (a) 0.0075; (b) $y_{n+1} = 1.0075y_n + .1p$; (c) $y_n = \frac{.1p}{.0075} + 1.0075^n \left(1000 - \frac{.1p}{.0075}\right)$.

15. (a) $y_{n+1} = y_n + .01y_0$, $y_0 = 1250$. (b) $y_n = 1250 + 12.5n$.

16. (a) $y_0 = 0$; (b) $y_{n+1} = 1.001y_n + \frac{4}{3}p$; $a = 1.001$, $b = \frac{4}{3}p$; (c) $y_n = \frac{-\frac{4}{3}p}{0.001} + (1.001)^n \left(\frac{\frac{4}{3}p}{.001}\right)$.

17. (a) iy_n; (b) $+p$, the amount of the next deposit; (c) $y_{n+1} = (1 + i)y_n + p$; $a = 1 + i$, $b = y_0 = p$; (d) $y_n = -\frac{p}{i} + (1 + i)^n \left[\frac{ip+p}{i}\right]$ which can be rearranged to be $y_n = \frac{p}{i}\left[(1 + i)^{n+1} - 1\right]$, the formula from the previous chapter.

18. (a) $y_{n+1} = \frac{1}{10}y_n$, $y_0 = v$; (b) $y_n = \frac{1}{10^n}v$.

Chapter 12

1.

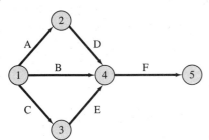

2.

	Predecessors
A	—
B	A
C	—
D	—
E	D
F	B, C, E

3.

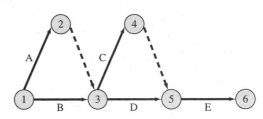

4.

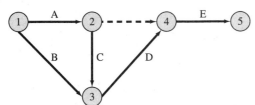

5.

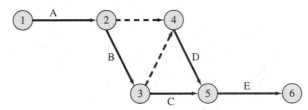

6. _____ The project will take 15 months.

A	[0, 2]
B	[0, 3]
C	[2, 6]
D	[6, 9]
E	[6, 10]
F	[10, 15]

7.

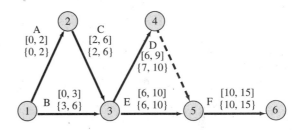

The critical tasks are A, C, E, and F.

8.

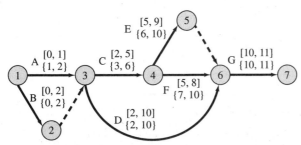

The time for building the bridge is 11 months, and the critical tasks are B, D, and G.

9. (a) $a = 4$, $b = 10$, and $m = 7$. (b) $\mu = 7$ min. (c) $\sigma^2 = 1$ min². (d) $\sigma = 1$ min.

10. (a) $1 - 0.7257 = 0.2743$. (b) $z = \frac{11-9}{\frac{10}{6}} = 1.2$; $A(1.2) = 0.3849$. (c) The "tail" of the right half of the curve, that is, the portion extending to the right of the vertical line drawn at $z = 1.2$. (d) $0.5 - 0.3849 = 0.1151$.

11.

C	2	2	2	2	0

D	0	3	12	4	4

E	2	4	8	$4\frac{1}{3}$	1

The average time for the project is $10\frac{1}{3}$; the variance is 5.

Appendix F: Answers to Odd-Numbered Exercises

Section 1.1

1. Use v_i for each vehicle type. **3.** $p = (.25n_1 - E_1) + \ldots + (.25n_i - E_i)$ **5.** b and c **7.** 4 **9.** No solution. **11.** 4 **13.** -6 **15.** $t_1 = 0$ **17.** $\frac{3}{20}$ **19.** $z = 72$ **21.** $x = 1$ **23.** $\frac{c-by}{a}$ **25.** $\frac{9(1+y)}{8}$ **27.** $\frac{10}{y}$

29. $\frac{3-2y}{y^2}$ **31.** **33.** **35.**

x	1	2	3
y	2	5	8

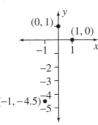

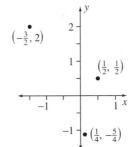

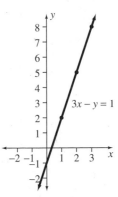

37.

x	1	2	3
y	4	6	8

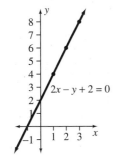

39.

x	1	2	3
y	-2	0	2

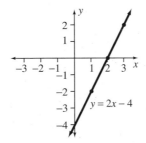

41.

x	0	2	1
y	$-2\frac{1}{2}$	$-1\frac{1}{2}$	-2

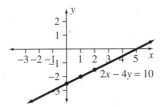

43.

x	6	0	-1
y	0	4	$4\frac{2}{3}$

45.

x	36	18	0
y	0	−36	−72

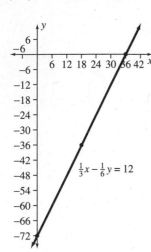

47.

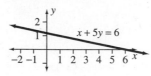

49.

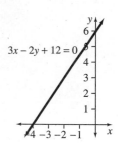

51.

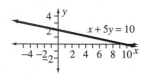

53.

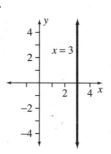

55.

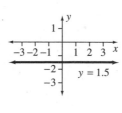

57.

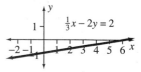

59.

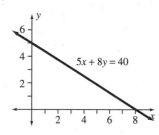

61.

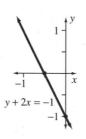

63.

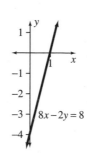

65.

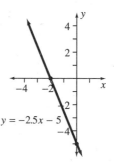

67.

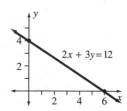

69.

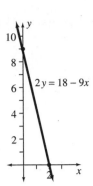

$2y = 18 - 9x$

71.

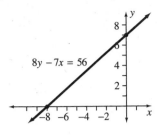

$8y - 7x = 56$

73.

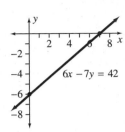

$6x - 7y = 42$

75.

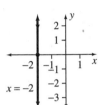

$x = -2$

77.

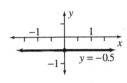

$y = -0.5$

79.

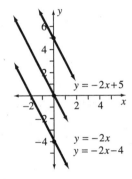

$y = -2x + 5$

$y = -2x$

$y = -2x - 4$

It affects their position above or below the origin.

81. Such lines have only a single intercept; two are needed for this method.

83.

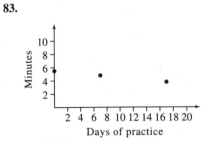

Minutes

Days of practice

85.

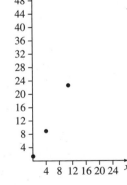

87.

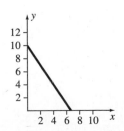

The fad dies out about six-and-two-thirds month after July 1.

89. 65.00

91. −4.11

Section 1.2

1. 1 **3.** undefined **5.** 0 **7.** 0 **9.** t **11.** $-\frac{9}{4}$ **13.** $\frac{25}{28}$ **15.** 5 **17.** The slope is $m = -3$ and it makes no difference which you choose for P_1 provided you remain consistent in the computation. **19.** $-x + y = 0$
21. $3x + 5y = -1$ **23.** $x = -1$ **25.** $-3x + 3y = 1$ **27.** $-x + y = 1$ **29.** $-7x + 2y = 20$
31. $-x + 2y = -9$ **33.** $6x + 12y = 5$ **35.** **37.**

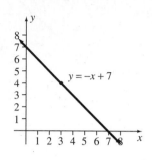

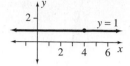

39.

41.

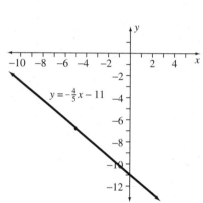

43.

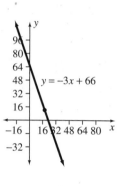

45.

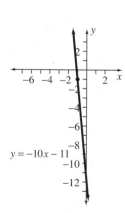

47.

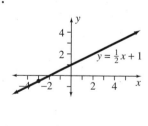

49. The form $y = mx + b$ is not appropriate here. The slope m is undefined because the line is vertical, and its equation is $x = 12$.

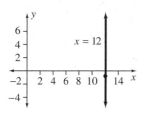

51. $3, 12, 0, \frac{3}{2}$ **53.** $y = -x$ **55.** $y = -x + \frac{1}{2}$ **57.** $y = -x$ **59.** $y = -1$ **61.** $x = 10$ **63.** $y = \frac{1}{2}x + \frac{1}{4}$
65. $y = 6x - 8$ **67.** $y = -3x + 5$ **69.** $-x + y = 0$ **71.** $y = 2$ **73.** $x = 2$ **75.** $x + y = 5$ **77.** $-10x +$
$4y = 1$ **79.** $x - y = 1$ **81.**   **83.**   **85.**

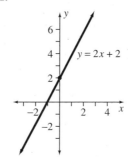 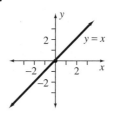 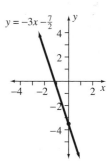

87.   **89.**   **91.**   **93.** -2 **95.** 0 **97.** $\frac{2}{9}$

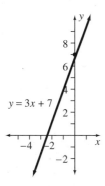 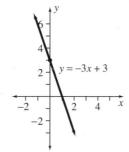 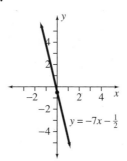

99. $\frac{2}{3}$ **101.** 4 **103.** $-\frac{1}{5}$ **105.** -1 **107.** 0
109. (a) Assume $m \neq 0$. Using m and $(a,0)$ in the point-slope form, we have $y - 0 = m(x - a)$ so that $y + ma = mx$. Divide through by m to prove the assertion. (b) We assume that $a \neq 0 \neq b$. Since the intercepts are $(0,b)$ and $(a,0)$, the slope of the line connecting them is $m = -b/a$. The slope-intercept formula implies that $y + (b/a)x = b$. Divide through by b to prove the assertion.
111. $y = .5x + 7.5$ **113.** $y = -\frac{3}{2}x$ **115.** (a) -2; (b) $y = -2x + 5$ **117.** \$740,000; $y = 60,000x + 380,000$
119. $R = 3.3$; yes; the units of resistance are volts/ampere (1 volt/ampere = 1 ohm)   **121.** The equation is $y = 75x + 160$ in
about 3.87 weeks.   **123.** -1.39 **125.** 0.53 **127.** $4.5x + y = 15.4$

Section 1.3

1. 547.5% per year (assuming 365 days per year)   **3.** 13.68%/year   **5.** $-273°$ **7.** If x is the number of regular cups and y is the number of large cups, the expression is $8.9 = .25x + .45y$.   **9.** Approximately 13,265   **11.** $200,000/year
13. (a) $5500; (b) cost is about $14,950; (c) $9450   **15.** 15%   **17.** $5F - 9C = 160$ **19.** 392 m/sec   **21.** $2,400
23. 2 coins; $16

Section 1.4

1. $x = 3$, $y = -2$ **3.** $p = q = 2$ **5.** $x_1 = 1$, $x_2 = 0$ **7.** $y_1 = \frac{145}{81}$, $y_2 = -\frac{26}{81}$ **9.** $u = \frac{3}{2}$, $t = \frac{1}{2}$
11. $(10, -2)$ **13.** $(0, 25)$ **15.** $(5, 4)$ **17.** $x = 3$, $y = -2$ **19.** $u = 1$, $t = 4$ **21.** $u = 3$, $v = 1$
23. $x = 0$, $y = 2$ **25.** $(1, 0)$ **27.** $(1, 1)$ **29.** inconsistent equations   **31.** $(3, 4)$ **33.** $x = \frac{-11}{7}$, $y = -\frac{1}{7}$
35. Not enough information present. Both equations are equivalent to $u = 2t + 7$.   **37.** $x = z = 3$ **39.** $x = 0$, $y = \frac{1}{4}$
41. inconsistent equations.   **43.** dependent equations.   **45.** $(1, -1)$ **47.** $(0, 0)$ **49.** No solutions.   **51.** $x_1 = 3$, $x_2 = 0$ **53.** Infinite number of solutions.   **55.** $t_1 = \frac{1}{4}$, $t_2 = \frac{1}{8}$ **57.** $x_1 = 0$, $x_2 = -1$ **59.** $(11, 13)$
61. $x = 2$ **63.** $-40°$ **65.** If x is the pounds of by-products, and y is the pounds of soybeans, then $x = 300$ pounds, $y = 100$ pounds.   **67.** $350,000 of 12% bonds, $400,000 of 7% bonds.   **69.** 1000 inexpensive seats, 1500 expensive seats.
71. 12 stamps from Sri Lanka, 10 from Burma.   **73.** $(1, -1.32)$

Answers to Review Questions, Chapter 1

1. 3   **3.** 5   **5.** -7 **7.**   **9.**   **11.**   **13.**

  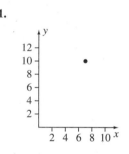 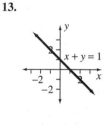

15.   **17.**   **19.**   **21.** 1   **23.** $\frac{1}{2}$ **25.** 0

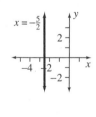 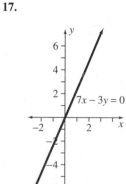

27. $x - y = -1$ **29.** $x - 2y = 4$ **31.** $y = q$ **33.** $y = 2x$ **35.** $y = \frac{1}{2}x + 3$ **37.** $y = -x + 2$ **39.** $\frac{17}{2}$
41. $\frac{9}{5}$ **43.** $x = 7.5$, $y = 2.5$ **45.** Infinite number of solutions.   **47.** $u = v = 2\frac{1}{4}$ **49.** $p = 2$, $q = 1$
51. $y = 7$, $z = -1$ **53.** About 25 degrees.   **55.** $17\frac{5}{6}\%$ **57.** Stretch about $\frac{2}{11}$ cm; shrink about .1 cm.   **59.** The faster train travels at 100 mph and the slower at 50 mph.   **61.** 40 ml of the 5% solution; 10 ml of the 15% solution.

Section 2.1

1. 2×3 **3.** 1×1; diagonal is 3   **5.** 5×1 **7.** 3×3; diagonal elements are 1, 1, and 1   **9.** (a) 1; (b) doesn't

exist; (c) 8; (d) 10; (e) 0; (f) -1; (g) 0; (h) 2; (i) 8   **11.** Not possible—the dimensions are not equal   **13.** $\begin{pmatrix} 7 & 14 \\ -7 & 0 \\ 7 & 7 \end{pmatrix}$

15. $\begin{pmatrix} 10\frac{1}{2} & -12 & -1 \end{pmatrix}$ **17.** $\begin{pmatrix} -9 & 0 & -3 \end{pmatrix}$ **19.** $\begin{pmatrix} 4\pi & -2\pi \\ 0 & \pi \\ 0 & 2\pi \end{pmatrix}$ **21.** Not possible—matrices have different dimen-

sions   **23.** $\begin{pmatrix} 14 & -14 \\ 7 & 7 \\ 14 & 14 \end{pmatrix}$ **25.** $\begin{pmatrix} -11 & -13 & -\frac{1}{2} \end{pmatrix}$ **27.** $\begin{pmatrix} -6 & 6 & -6 \end{pmatrix}$ **29.** $\begin{pmatrix} 16 & -16 \\ 8 & 8 \\ 16 & 16 \end{pmatrix}$ **31.** $\begin{pmatrix} 0 & -4 \\ 5 & 2 \end{pmatrix}$

33. $\begin{pmatrix} 3 & -8 \\ 3 & 4 \end{pmatrix}$ **35.** $\begin{pmatrix} -35 & 15 \\ -20 & -25 \end{pmatrix}$ **37.** $\begin{pmatrix} -8.5 & 6.5 \\ 0 & -.25 \\ 0.25 & -1.5 \end{pmatrix}$ **39.** $\begin{pmatrix} 1 & -14 \\ 7 & 0 \end{pmatrix}$ **41.** $\begin{pmatrix} 9 & 8 & 3 & -7 \\ \frac{1}{3} & 0 & -2 & 2 \end{pmatrix}$

43. $n \times m$

45. An arbitrary element of $A + B$ is $a_{ij} + b_{ij}$. An arbitrary element of $B + A$ is $b_{ij} + a_{ij} = a_{ij} + b_{ij}$. Since the two corresponding but arbitrary elements are equal, so are $A + B$ and $B + A$.

47. An element of the matrix $(c + d)A$ is $(c + d)a_{ij} = ca_{ij} + da_{ij}$. Since the right-hand side is the sum of cA and dA, the assertion is shown.

49. An arbitrary element of $c(dA)$ is $c(da_{ij}) = cda_{ij} = (cd)a_{ij}$. The last expression is a component of $(cd)A$, which completes the demonstration.

51. A typical matrix might be:
$$\begin{array}{c} \text{Math} \\ \text{English} \\ \text{History} \end{array} \begin{pmatrix} \begin{array}{cccc} \text{Quiz 1} & 2 & 3 & 4 \\ 100 & 95 & 98 & 89 \\ A & B & B- & C \\ 85 & 80 & 65 & 70 \end{array} \end{pmatrix}$$

53. (a) $S = \begin{pmatrix} 80 & 50 & 30 \\ 80 & 75 & 10 \\ 75 & 75 & 20 \\ 60 & 50 & 20 \end{pmatrix}$ is one such definition.

All components are percentages. $T = \begin{pmatrix} 80 & 75 & 80 \\ 80 & 75 & 80 \\ 80 & 75 & 80 \\ 80 & 75 & 80 \end{pmatrix}$; (b) $C = \begin{pmatrix} 500 \\ 500 \\ 500 \\ 500 \end{pmatrix}$

55. (a) $R_1 = \begin{pmatrix} 300 & 450 \\ 350 & 400 \\ 500 & 500 \end{pmatrix}$; $R_2 = \begin{pmatrix} 350 & 400 \\ 450 & 400 \\ 650 & 550 \end{pmatrix}$; (b) $R_2 - R_1 = \begin{pmatrix} 50 & -50 \\ 100 & 0 \\ 150 & 50 \end{pmatrix}$; (c) $R_2 + R_1 = \begin{pmatrix} 650 & 850 \\ 800 & 800 \\ 1150 & 1050 \end{pmatrix}$

57. $\begin{pmatrix} -3.43 & 0.00 \\ 0.00 & 1.15 \end{pmatrix}$ **59.** $\begin{pmatrix} -9.87 & 19.74 \\ -17.39 & -37.13 \end{pmatrix}$

Section 2.2

1. 1   **3.** Row times a row—not defined   **5.** Column times a row—not (yet) possible   **7.** -2 **9.** 18   **11.** $\frac{2}{3}$

13. Not compatible   **15.** $\begin{pmatrix} 2 \\ -2 \end{pmatrix}$ **17.** $\begin{pmatrix} -3 \\ -1 \end{pmatrix}$ **19.** $\begin{pmatrix} -3 \\ 2 \\ -\frac{5}{3} \end{pmatrix}$ **21.** $\begin{pmatrix} -3 & 26 \\ -6 & -238 \end{pmatrix}$ **23.** $\begin{pmatrix} -392 & 24 \\ -41.4 & -54 \end{pmatrix}$

25. $\begin{pmatrix} -34 & -7 & 30 \\ -28 & 6 & 30 \\ 12 & -6 & 18 \end{pmatrix}$ **27.** $\begin{pmatrix} 33 & 12 & 30 \\ 8 & 0 & 2 \end{pmatrix}$ **29.** $\begin{pmatrix} -9 & 18 \\ -2 & 19 \end{pmatrix}$ **31.** Matrix product is undefined.

33. $\begin{pmatrix} .027 & -.331 \\ .956 & .002 \\ .007 & .598 \end{pmatrix}$ **35.** $\begin{pmatrix} .004 & .254 & -.127 \\ .385 & -.993 & 0 \\ -.999 & -.004 & .925 \end{pmatrix}$ **37.** $\begin{pmatrix} \frac{1}{4} & \frac{21}{4} \\ \frac{7}{4} & \frac{7}{4} \end{pmatrix}$ **39.** $\begin{pmatrix} 1 & 3 \\ 0 & 1 \end{pmatrix}$ **41.** $\begin{pmatrix} 5 & 8 & -4 \\ 1 & 1 & 4 \\ 2 & 0 & 2 \\ 3 & 3 & -6 \\ 4 & 2 & -5 \end{pmatrix}$

43. $n = p$, $m = q$ **45.** (a) $\sqrt{13}$; (b) $\sqrt{13}$; (c) $\sqrt{13}$; (d) 5; (e) $\sqrt{\frac{1}{2}}$; (f) $\sqrt{2}$ **49.** $R = \begin{pmatrix} 1 & 0 \\ 0 & 1 \end{pmatrix}$

51. Notice that $A = A^T$. Thus $AA^T = AA = \begin{pmatrix} 5 & 2 \\ 2 & 4 \end{pmatrix}$, which is symmetric by inspection. In the same way, $A + A^T =$

$2A = \begin{pmatrix} 2 & 4 \\ 4 & 0 \end{pmatrix}$, which is also symmetric by inspection. Since $A = A^T$, $A - A^T = \begin{pmatrix} 0 & 0 \\ 0 & 0 \end{pmatrix}$, which is skew-symmetric by inspection.

53. The equation $a_{ij} + o_{ij} = o_{ij} + a_{ij} = a_{ij}$ holds since all the o_{ij} are zero. Therefore, the corresponding matrices $A = O$, $O + A$, and A are all equal.

55. Let us examine the (1-1) component of the matrix expression $A(B + C)$. It is $a_{11}(b_{11} + c_{11}) + a_{12}(b_{21} + c_{21}) = (a_{11}b_{11} + a_{12}b_{21}) + (a_{11}c_{11} + a_{12}c_{21})$. The right-hand side is the (1-1)-component of the matrix expression $AB + AC$. This follows for all components, and the assertion is true.

57. Look at the (1-1)-component of $A(cB)$. It is $a_{11}cb_{11} + a_{12}cb_{21}$, which also equals $(ca_{11})b_{11} + (ca_{12})b_{21}$ and equals $c(a_{11}b_{11} + a_{12}b_{21})$. These last expressions are the (1-1)-components of $(cA)B$ and $c(AB)$. This equality follows for all other components, as the reader may show, and the assertion follows.

59. (a) $\begin{array}{cc} & \text{To fad} \quad \text{Away from fad} \\ \begin{matrix} \text{Fad child} \\ \text{“No-fad” child} \end{matrix} & \begin{pmatrix} .75 & .25 \\ .75 & .25 \end{pmatrix} \end{array}$; (b) $(50 \quad 0)\begin{pmatrix} .75 & .25 \\ .75 & .25 \end{pmatrix} = (37.5 \quad 12.5)$; there are about 38 cult followers among the children and the remainder are unenthused; (c) this proportion (18 and 12) remains the same for all subsequent days.

61. (a) A typical matrix might be $G = \begin{array}{c} \\ \text{Math} \\ \text{English} \\ \text{History} \\ \text{Economics} \end{array} \begin{array}{c} \text{Quiz 1} \quad 2 \quad 3 \\ \begin{pmatrix} 100 & 75 & 80 \\ 85 & 85 & 80 \\ 75 & 85 & 95 \\ 90 & 90 & 90 \end{pmatrix} \end{array}$. Your grade for math (for example) would be computed

as $.25(100) + .25(75) + .5(80) = 83.75$. (b) Let $W = \begin{pmatrix} .25 \\ .25 \\ .50 \end{pmatrix}$. Then the product GW is a 4×1 matrix. (Why? Its first component is precisely $.25(100) + .25(75) + .5(80) = 83.75$. (Why?) In the same way, the ith component of GW is the average for the ith course.

63. $\begin{pmatrix} -11.39 \\ 43.53 \end{pmatrix}$ **65.** $\begin{pmatrix} 14.70 & -1.62 \\ 35.67 & -2.25 \end{pmatrix}$

Section 2.3

1. $\begin{pmatrix} 1 & 1 & | & 3 \\ -1 & 1 & | & 2 \end{pmatrix}$ **3.** $\begin{pmatrix} 1 & 1 & 1 & | & 2 \\ 3 & -3 & 0 & | & 12 \\ 0 & 0 & 1 & | & 7 \end{pmatrix}$ **5.** $\begin{pmatrix} 1 & 0 & 0 & 0 & | & 1 \\ 1 & 1 & 0 & 0 & | & 2 \\ 1 & 1 & 1 & 0 & | & 3 \\ 1 & 1 & 1 & 1 & | & 4 \end{pmatrix}$ **7.** $\begin{pmatrix} .03198 & -2.2201 & | & 6.33098 \\ 1.2983 & 1.8000 & | & 5.00159 \end{pmatrix}$

9. $\begin{aligned} x_1 + x_2 + x_3 &= 2 \\ 2x_1 + 2x_2 + 2x_3 &= 1 \end{aligned}$ **11.** $\begin{aligned} .93x_1 + .77x_2 &= -4.22 \\ -.34x_2 - .43x_3 &= 6.01 \\ .01x_1 + .04x_2 + .93x_3 &= -.92 \end{aligned}$ **13.** $\begin{aligned} x_1 - x_2 - x_3 &= 3 \\ -x_1 - 2x_2 &= 1 \\ 2x_1 + 5x_2 + .5x_3 &= 1 \end{aligned}$

15. $10x_1 + 9x_2 + 8x_3 = 7$
$6x_1 + 5x_2 + 4x_3 = 3$
$2x_1 + x_2 = 0$
   17. $R'_1 = -2R_1$ **19.** $R'_2 = R_2 + R_1$ **21.** Interchange rows 2 and 3 (that is, $R'_2 = R_3$, $R'_3 = R_2$)

23. $\begin{pmatrix} 1 & 0 & | & 3 \\ 0 & 1 & | & 2 \end{pmatrix}$ **25.** $\begin{pmatrix} 1 & 1 & | & 1 \\ 0 & 0 & | & 0 \end{pmatrix}$ **27.** $\begin{pmatrix} 1 & 0 & | & 1 \\ 0 & 1 & | & 0 \end{pmatrix}$ **29.** $\begin{pmatrix} 1 & 0 & | & -1 \\ 0 & 1 & | & 2 \end{pmatrix}$ **31.** $\begin{pmatrix} 1 & 0 & | & 2 \\ 0 & 1 & | & 2 \end{pmatrix}$ **33.** $\begin{pmatrix} 1 & 0 & | & \frac{1}{2} \\ 0 & 1 & | & \frac{1}{2} \end{pmatrix}$

35. $\begin{pmatrix} 1 & 0 & | & 0 \\ 0 & 1 & | & 1 \end{pmatrix}$ **37.** $(1, 1)$ **39.** $u = v = 1$ **41.** $u_1 = 1, u_2 = 0$ **43.** $x = 1, y = 3$ **45.** $s = 4, t = -2$

47. $x = y = -1$ **49.** Replace the word "row" by the word "column" in all rules.   **51.** $1\frac{2}{3}$ l of the 10% solution, $3\frac{1}{3}$ l of the 25% solution.   **53.** $6\frac{2}{3}$ l of the 10% solution, $3\frac{1}{3}$ l of the 25% solution.   **55.** Yes; choose 7 illusions and 1 sleight-of-hand.   **57.** 6 each of short-long selections.   **59.** $\begin{pmatrix} 4.6 & -3.7 & | & -3.1 \\ -4.2 & 7.9 & | & -4.2 \end{pmatrix}$ **61.** $\begin{pmatrix} 1 & 0 & | & 0.83 \\ 0 & 1 & | & 0 \end{pmatrix}$

Section 2.4

1. $\begin{pmatrix} 1 & 0 & 2 \\ -1 & 1 & -3 \end{pmatrix}$ **3.** $\begin{pmatrix} 0 & -6 & 0 \\ 1 & 3 & 2 \end{pmatrix}$ **5.** $\begin{pmatrix} 1 & 8 & 0 \\ 0 & -12 & 0 \end{pmatrix}$ **7.** $\begin{pmatrix} 1 & 2 & 1 & \frac{3}{2} \\ 0 & -2 & 1 & \frac{1}{2} \end{pmatrix}$ **9.** $\begin{pmatrix} 1 & 2 & 6 & 9 \\ 0 & -2 & -3 & -9 \\ 0 & 4 & 0 & 4 \end{pmatrix}$

11. $\begin{pmatrix} \frac{1}{3} & \frac{2}{3} & 1 & \frac{4}{3} \\ 1 & 0 & 0 & 3 \\ 2\frac{2}{3} & 1\frac{1}{3} & 0 & \frac{-4}{3} \end{pmatrix}$ **13.** $\begin{pmatrix} 1 & 0 & 2 & -2 \\ 0 & 1 & 6 & -3 \end{pmatrix}$ **15.** $\begin{pmatrix} -4 & 1 & -6 \\ 2\frac{1}{6} & 0 & 3 \end{pmatrix}$ **17.** $\begin{pmatrix} \frac{1}{3} & 0 & \frac{1}{6} \\ 0 & 1 & 0 \end{pmatrix}$

19. $\begin{pmatrix} 0 & -\frac{1}{2} & \frac{1}{2} & 2 \\ 1 & \frac{1}{2} & -\frac{3}{2} & 0 \\ 0 & 0 & 2 & 4 \\ 0 & -\frac{1}{2} & -\frac{3}{2} & 3 \end{pmatrix}$ **21.** $\begin{pmatrix} 1 & 0 & -1 \\ 0 & 1 & 3 \end{pmatrix}$ **23.** $\begin{pmatrix} 1 & 2 & 3 \\ 0 & 0 & 0 \end{pmatrix}$ **25.** $\begin{pmatrix} 1 & 0 & 0 \\ 0 & 1 & 0 \\ 0 & 0 & 1 \end{pmatrix}$ **27.** $\begin{pmatrix} 1 & 0 & 0 & 0 & 1 \\ 0 & 1 & 0 & 0 & 1 \\ 0 & 0 & 1 & 0 & 1 \\ 0 & 0 & 0 & 1 & 1 \end{pmatrix}$

29. $\begin{pmatrix} 1 & 20 \\ 0 & 0 \end{pmatrix}$ **31.** $\begin{pmatrix} 1 & 0 & -2 & -\frac{1}{3} \\ 0 & 1 & -\frac{8}{3} & -\frac{7}{9} \end{pmatrix}$ **33.** $\begin{pmatrix} 1 & 0 & 0 & \frac{1}{6} \\ 0 & 1 & 0 & \frac{5}{12} \\ 0 & 0 & 1 & \frac{5}{6} \end{pmatrix}$ **35.** $\begin{pmatrix} 1 & 0 & -7 \\ 0 & 1 & 5 \end{pmatrix}$ **37.** $\begin{pmatrix} 1 & 0 & 0 & 0 & 0 \\ 0 & 1 & 0 & 0 & 0 \\ 0 & 0 & 1 & 2 & 0 \\ 0 & 0 & 0 & 0 & 0 \end{pmatrix}$

39. $\begin{pmatrix} 1 & 0 & 0 & \frac{1}{6} \\ 0 & 1 & 0 & \frac{5}{12} \\ 0 & 0 & 1 & \frac{5}{6} \end{pmatrix}$ **41.** $x = -1, y = 1$ **43.** No unique solution, and we may write $x_1 = -\frac{1}{2}x_3 + 4$, $x_2 = \frac{1}{2}x_3 + 1$

45. $a = b = 0$ **47.** No unique solution. $\begin{pmatrix} 1 & 2 & 3 \\ 0 & 0 & 0 \end{pmatrix}$ **49.** $u_1 = u_2 = u_3 = 12$ **51.** $x_1 = x_2 = x_3 = x_4 = 1$

53. $a = b = \frac{1}{3}$, $c = 0$, $d = -1$ **55.** (a) $x_1 + x_2$, $x_2 + x_3$; (b) $x_1 = 20$, $x_2 = 25$, $x_3 = 15$ **57.** In this case, $x_1 = 28$, $x_3 = 48$ and the store will need a total of 113 workers.   **59.** 200, 175, and 125   **61.** $\begin{pmatrix} 0 & 0.8 & 5.79 \\ 1 & 0 & -0.9 \end{pmatrix}$

63. $\begin{pmatrix} 1 & 0 & | & 3.51 \\ 0 & 1 & | & -0.929 \end{pmatrix}$

Section 2.5

1. $\begin{pmatrix} 0 & 1 \\ 1 & -1 \end{pmatrix}$ **3.** $\begin{pmatrix} 4 & -1 \\ -4 & 2 \end{pmatrix}$ **5.** No inverse   **7.** $\begin{pmatrix} 1 & -1 \\ -1 & \frac{1}{2} \end{pmatrix}$ **9.** $\begin{pmatrix} \frac{1}{23} & \frac{2}{23} \\ -\frac{7}{46} & \frac{9}{46} \end{pmatrix}$ **11.** $\begin{pmatrix} .25 & .25 \\ .1 & -.1 \end{pmatrix}$

13. $\begin{pmatrix} \frac{1}{6} & -\frac{1}{6} \\ \frac{1}{8} & \frac{1}{8} \end{pmatrix}$ **15.** No inverse   **17.** $\begin{pmatrix} 0 & 0 & -1 \\ \frac{1}{3} & \frac{2}{3} & \frac{4}{3} \\ 0 & 2 & 2 \end{pmatrix}$ **19.** $\begin{pmatrix} .04 & -.08 & 0 \\ .08 & .04 & 0 \\ 0 & 0 & 1 \end{pmatrix}$ **23.** $\begin{pmatrix} \frac{1}{1-x} & -\frac{x}{1-x} \\ -\frac{1}{1-x} & \frac{1}{1-x} \end{pmatrix}$;

this is valid for all $x \neq 1$.   **25.** Since $ad - bc = 0$, its reciprocal won't exist, and neither will the matrix inverse given

by the formula.   **27.** (a) $\left(\frac{1}{6}\right)$; (b) 3; (c) 12   **29.** (a) $\begin{pmatrix} .2 & .4 \\ .1 & -.3 \end{pmatrix}$ (b) $u = .2, \ v = .1$; (c) $x_1 = .4, \ x_2 = .2$

31. (a) $\begin{pmatrix} \frac{2}{9} & \frac{1}{3} \\ \frac{1}{9} & -\frac{1}{3} \end{pmatrix}$ (b) $z_1 = z_2 = 1$; (c) $z_1 = -\frac{5}{9}, \ z_2 = \frac{2}{9}$ **33.** (a) $\begin{pmatrix} \frac{1}{2} & \frac{1}{2} \\ \frac{1}{2} & -\frac{1}{2} \end{pmatrix}$ (b) $x = 0, \ y = 2$; (c) $x = y = .5$

35. $\begin{pmatrix} 2 & 5 \\ 2 & -5 \end{pmatrix}$ (a) $(14, -6)$; (b) $u = -3, \ v = 7$; (c) $S = 12, \ T = -8$.   **37.** (a) No inverse; (b), (c) no solutions

39. (a) $\begin{pmatrix} \frac{1}{4} & -\frac{3}{4} & -\frac{1}{8} \\ 0 & 0 & \frac{1}{4} \\ \frac{1}{4} & \frac{1}{4} & -\frac{1}{8} \end{pmatrix}$; (b) $x = -1, \ y = 1, \ z = 3$; (c) $x = -3, \ y = 2, \ z = -1$ **41.** (a) No inverse; (b), (c) no

solutions   **43.** $\begin{pmatrix} \frac{1}{14} & -\frac{1}{2} & -\frac{2}{7} & \frac{6}{7} \\ -\frac{1}{7} & 0 & -\frac{3}{7} & \frac{2}{7} \\ \frac{3}{14} & \frac{1}{2} & \frac{1}{7} & -\frac{3}{7} \\ -\frac{1}{7} & 1 & \frac{4}{7} & -\frac{5}{7} \end{pmatrix}$ (a) $z_1 = \frac{9}{7}, \ z_2 + z_4 = -\frac{11}{7}, \ z_3 = \frac{13}{7}$ (b) $z_1 = -\frac{1}{2}, \ z_2 = 2, \ z_3 = -\frac{3}{2}, \ z_4 = 3$

45. Same number of rows and columns are necessary.   **47.** $166,666.67 apiece for salaries and supplies; $666,666.66 for
advertising.   **49.** beach lovers: 200,000; tennis lovers: 100,000   **51.** Beach lovers: 200,000; tennis lovers: 100,000
53. (a) 20 of first species; 40 apiece in second and third species (b) 20 of first species; 40 in second species; 30 in third species.
55. $x = 1.69, \ y = 0.08$

Section 2.6

1. $I - Q = \begin{pmatrix} -\frac{1}{4} & \frac{1}{4} \\ \frac{1}{6} & \frac{1}{6} \end{pmatrix}$; $(I - Q)^{-1} = \begin{pmatrix} -2 & 3 \\ 2 & 3 \end{pmatrix}$.   **3.** $\begin{pmatrix} .5 & -.5 \\ -.5 & .5 \end{pmatrix}$; no inverse   **5.** $\begin{pmatrix} p & -p \\ -p & p \end{pmatrix}$; no in-

verse   **7.** $\begin{pmatrix} \frac{3}{4} & -\frac{1}{2} \\ -\frac{1}{4} & 1 \end{pmatrix}$; $\begin{pmatrix} 1.6 & .8 \\ .4 & 1.2 \end{pmatrix}$ **9.** $\begin{pmatrix} .8 & -.1 & 0 \\ -.1 & .8 & 0 \\ 0 & 0 & 0 \end{pmatrix}$; no inverse   **11.** $\begin{pmatrix} .75 & -.5 & 0 & 0 \\ -.25 & 1 & 0 & 0 \\ 0 & 0 & .9 & -.1 \\ 0 & 0 & -.1 & .9 \end{pmatrix}$;

$\begin{pmatrix} 1.6 & .8 & 0 & 0 \\ .4 & 1.2 & 0 & 0 \\ 0 & 0 & 1.125 & .125 \\ 0 & 0 & .125 & 1.125 \end{pmatrix}$ **13.** $XQ = \begin{pmatrix} x_1 & x_2 \end{pmatrix} \begin{pmatrix} .15 & .20 \\ .25 & .10 \end{pmatrix} = \begin{pmatrix} .15x_1 + .25x_2 & .20x_1 + .10x_2 \end{pmatrix}$ **15.** Wheat:

$77,250; steel: $84,000   **17.** $D = \begin{pmatrix} \$37,000 & \$48,000 \end{pmatrix}$ **19.** $X = \begin{pmatrix} \$1,358,024.69 & \$1,111,111.11 & \$2,462,277.09 \end{pmatrix}$
21. $X = \begin{pmatrix} \$1,296,296.30 & \$833,333.33 & \$1,213,991.77 \end{pmatrix}$ **23.** A community can't use more resources that it has access to.
Thus the sum of the column elements for any industry must not be greater than 1, or the community or enterprise won't be able
to support its other industries. Profit-making companies derive their profit from an excess of goods that they market. Thus the
inequality must be strict to ensure that such a surplus exists.   **25.** $\begin{pmatrix} 1.1 & 1.9 \\ -0.1 & 2.2 \end{pmatrix}$; $\begin{pmatrix} 0.84 & -0.73 \\ 0.04 & 0.42 \end{pmatrix}$

Answers to Review Exercises, Chapter 2

1. 2×2 **3.** 3×1 **5.** 3×1 **7.** $\begin{pmatrix} 0 & 1 \\ 1 & 0 \end{pmatrix}$ **9.** Not possible—incompatible dimensions   **11.** $\begin{pmatrix} -\frac{11}{2} & -9' & -3 \\ 10 & -13 & -5 \end{pmatrix}$

13. $\left(-\frac{1}{3}\right)$

15. $\begin{pmatrix} \frac{1}{4} & \frac{1}{2} & -\frac{1}{4} & -1 \\ -3 & -2 & -1 & -8 \\ 6\frac{1}{8} & 6\frac{1}{4} & -\frac{1}{8} & 5\frac{1}{2} \\ 3 & 6 & -3 & -12 \end{pmatrix}$

17. $\begin{pmatrix} \frac{5}{4} & 2 \\ \frac{1}{4} & \frac{3}{2} \end{pmatrix}$

19. $\begin{pmatrix} -2.304 & 9.88 & 12.034 \\ .0003 & -.004 & 7.645 \end{pmatrix}$

21. $\begin{pmatrix} 10 & -7 & 12 \end{pmatrix}$

23. $\begin{pmatrix} 1 & 0 & 0 & 1 \\ 0 & 1 & 0 & 1 \\ 0 & 0 & 1 & 1 \end{pmatrix}$

25. $\begin{pmatrix} 1 & 0 & .5 \\ 0 & 1 & 2.5 \end{pmatrix}$

27. $\begin{pmatrix} 1 & 0 & 12 \\ 0 & 1 & 12 \end{pmatrix}$

29. $\begin{pmatrix} 1 & 0 & 0 & 3 \\ 0 & 1 & 0 & -2 \\ 0 & 0 & 1 & 1 \end{pmatrix}$

31. $\begin{pmatrix} 1 & 2 & 2 \\ 0 & 0 & 0 \\ 0 & 0 & 0 \end{pmatrix}$

33. $\begin{pmatrix} 1 & 0 & 10 \\ 0 & 1 & 10 \end{pmatrix}$

35. $\begin{pmatrix} 1 & 0 & 1 & 3\frac{1}{2} \\ 0 & 1 & 0 & \frac{1}{2} \end{pmatrix}$

37. $\begin{pmatrix} 1 & 0 & 0 & 4 \\ 0 & 1 & 0 & -3 \\ 0 & 0 & 1 & +3 \end{pmatrix}$

39. $\begin{pmatrix} 1 & 0 & 0 & 2 \\ 0 & 1 & -\frac{3}{2} & -\frac{3}{2} \end{pmatrix}$

41. $\begin{pmatrix} 1 & 0 & 5 & 11 \\ 0 & 1 & -8 & -15 \\ 0 & 0 & 0 & 0 \end{pmatrix}$

43. $\begin{pmatrix} 1 & 0 & 0 & -1 \\ 0 & 1 & 0 & 4 \\ 0 & 0 & 1 & 4 \end{pmatrix}$

45. $\begin{pmatrix} 1 & 0 \\ 0 & 1 \end{pmatrix}$

47. $\begin{pmatrix} -1 & 1 \\ 1 & 0 \end{pmatrix}$

49. $\begin{pmatrix} -\frac{20}{9} & \frac{1000}{9} \\ \frac{1}{3} & \frac{100}{3} \end{pmatrix}$

51. $\begin{pmatrix} \frac{3}{2} & -\frac{1}{4} & -\frac{1}{2} \\ -\frac{3}{4} & \frac{3}{8} & \frac{1}{4} \\ \frac{1}{4} & -\frac{1}{8} & \frac{1}{4} \end{pmatrix}$

53. $\begin{pmatrix} .4 & -.8 & .2 \\ .2 & .6 & -.4 \\ -.2 & .4 & .4 \end{pmatrix}$

55. (a) $x = y = 10$; (b) $x = 0$, $y = 15$

57. (a) $u = \frac{27}{4}$, $v = \frac{33}{2}$; (b) $u = -\frac{27}{4}$; $v = -\frac{33}{2}$

59. (a) $x = 11$, $y = 2$, $z = -9$; (b) $x = 9$, $y = 2$, $z = -8$

61. (a) $\begin{pmatrix} 48 & 51 \\ 52 & 50 \end{pmatrix}$ (b) $\begin{pmatrix} 60 & 63.75 \\ 65 & 62.5 \end{pmatrix}$ (c) $\begin{pmatrix} 48 & 50 \\ 58 & 59 \end{pmatrix}$ (d) $\begin{pmatrix} 12 & 13.75 \\ 7 & 35 \end{pmatrix}$

63. 4T, 6T

65.

	(a)	(b)
1st show	111,111	166,667
2nd show	88,889	133,333

Section 3.2

1.

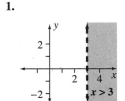

$x > 3$

3.

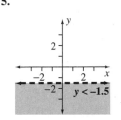

$y \geq 2x - y + 3$

5.

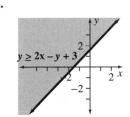

$y < -1.5$

7.

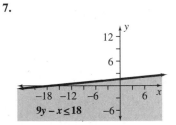

$9y - x \leq 18$

9.

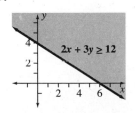

$2x + 3y \geq 12$

11.

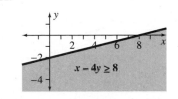

$x - 4y \geq 8$

13.

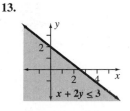

$x + 2y \leq 3$

15.

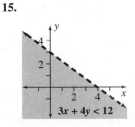

$3x + 4y < 12$

17.

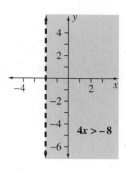

$4x > -8$

19.

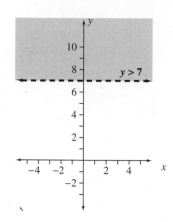

$y > 7$

21.

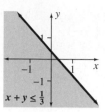

$x + y \leq \frac{1}{3}$

23.

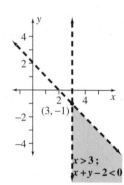

$(3, -1)$

$x > 3;$
$x + y - 2 < 0$

25.

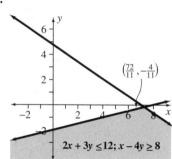

$\left(\frac{72}{11}, -\frac{4}{11}\right)$

$2x + 3y \leq 12; \, x - 4y \geq 8$

27.

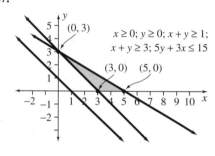

$(0, 3)$ $x \geq 0; \, y \geq 0; \, x + y \geq 1;$
$x + y \geq 3; \, 5y + 3x \leq 15$

$(3, 0)$ $(5, 0)$

29.

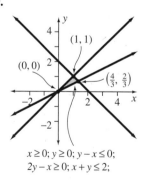

$(1, 1)$

$(0, 0)$ $\left(\frac{4}{3}, \frac{2}{3}\right)$

$x \geq 0; \, y \geq 0; \, y - x \leq 0;$
$2y - x \geq 0; \, x + y \leq 2;$

31.

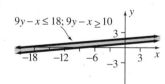

$9y - x \leq 18; \, 9y - x \geq 10$

33.   The solution is empty

35.   The solution is empty   **37.**   The solution is empty   **39.**

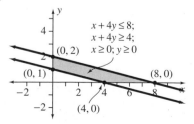

$x + 4y \leq 8;$
$x + 4y \geq 4;$
$(0, 2)$ $x \geq 0; \, y \geq 0$

$(0, 1)$ $(8, 0)$

-2 $(4, 0)$

41.

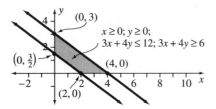

$x \geq 0; y \geq 0;$
$3x + 4y \leq 12; 3x + 4y \geq 6$

43.

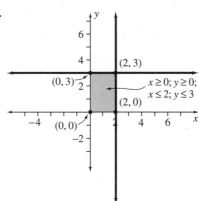

$x \geq 0; y \geq 0;$
$x \leq 2; y \leq 3$

45.

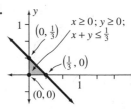

$x \geq 0; y \geq 0;$
$x + y \leq \frac{1}{3}$

47. Let y be the number of gold systems, and let x be the number of silver systems. The least number of gold systems to be sold is the smallest y-coordinate lying in the feasible region in the graph. This is the y-coordinate of the corner (3, 12); that is, at least 12 gold systems must be sold each week to satisfy the business constraints.

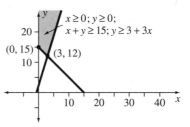

$x \geq 0; y \geq 0;$
$x + y \geq 15; y \geq 3 + 3x$

49. x and y represent the number of beads. $x + y$ is the sum of the beads. $x + y \geq 10$ means the total must be at least 10. $y \leq 6$ means the "y" beads cannot exceed 6.

51.

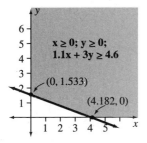

$x \geq 0; y \geq 0;$
$1.1x + 3y \geq 4.6$

Section 3.3
1. The maximum is 18 at (2; 0).

3. The maximum is 21 at $(2, \frac{3}{2})$.

5. The maximum is 9 at either (0, 4.5) or (1, 0) (or at any point on the straight-line segment connecting these two points).

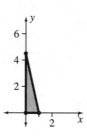

7. (3, 1) yields the max of 29.   **9.** (3, 8) yields the max of 43.   **11.** (2, 1) yields the max of 20.

13. The minimum is $\frac{1}{2}$ at the point (1, 0).   **15.** The minimum is $5\frac{1}{2}$ at the point (1, 1).

       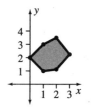

17. The minimum value is 0 at any point on the line segment connecting (0, 0) with (10, −1).

19. (2, 0) yields the min of 1.

21. (0, 1) yields the min of 5.   **23.** (1, 0) yields the min of $\frac{1}{2}$.   **25.** Max is 12 at (0, 4).   **27.** Min is 3 at (0, 1).   **29.** Max is 3 at (1, 6).   **31.** Min is −1 at (1, 0).   **33.** The maximum is 3 at the point (1, 2).

35. The minimum is −2 at the point (0, 2).   **37.** The maximum is 24 at the point (0, 8).

39. The minimum value is 0 at the origin (0, 0).   **41.** The maximum value is 12 at the point (6, 3).

43. (a) not allowed   (b) not allowed

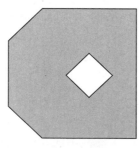   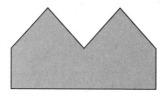

    (c) allowed   (d) not allowed

   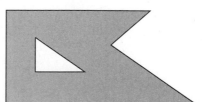

47. (1.3, 2.1); −4.690

49. 1.840; (−2, −0.8)

Section 3.4
1. No **3.** Now use 775 gal of each dessert. The profit is $6200. **5.** $112.50 by washing 5 each of cars and trucks.
7. $15,000 invested in each type of bond; annual yield is $3375. **9.** $10,000 at $7\frac{1}{2}\%$, $10,000 at 15%; yield is $2250.
11. Any combination of building on the "line segment" connecting $(0, 1.6)$ with $(\frac{32}{25}, 0)$ generates on average $6400 profit. Here, in $(0, 1.6)$, the first component refers to ranch houses, the second to colonials. **13.** From Tangiers: 50 animals to Liberia, none to Nairobi. From Capetown: 75 animals to Liberia, 125 to Nairobi. Transportation cost = $32,750. **15.** $174,000; 40 cars, 60 trucks. **17.** Interview 40 young men together with 20 older men, for a total of 60 interviewees. **19.** The minimum commissions totals $9100 (Carrie sells 5 cars, Barry sells 27). The maximum is $15,900 (Carrie sells 9, Barry sells 47). **21.** Deliver 200 bicycles from the north factory to the west showroom and 400 bicycles from the south factory to the east showroom.
23. Let x_A be the number of A floats, and so on. The spectator turnout will be $1000(100x_A + 5x_B + 3x_C)$, and this is subject to the constraints $x_A \geq 0$, $x_B \geq 0$, $x_C \geq 0$, $1000(100x_A + 40x_B + 20x_C) \leq 500,000$.
25. Computers work algebraically rather than geometrically. Although it is a straightforward task to instruct the computer to solve individual inequalities, it may be a problem to determine the feasible region, and its corners, of a *system* of inequalities.
27. No solution. **29.** $44.94

Section 3.5
1. The feasible region is empty. **3.** The feasible region is unbounded. **5.** The feasible set is empty. **7.** A constant cannot be optimized. **9.** The feasible region is semi-infinite opening up, so there is no absolute maximum. **11.** $P = 6$ at $(6, 0)$. **13.** $P = 6$ at any point on the line $x + y = 6$ between $(6, 0)$ and $(2, 4)$. **15.** The feasible region consists of the line between two points $-(-3.5, 4.5)$ and $(2.4, -1.4)$. The maximum is 13.08 at $(2.4, -1.4)$. **17.** P achieves a maximum value of 3 at a point on the line segment between (and including) the points $(1, 1)$ and $(2, 0)$. **19.** The feasible region is unbounded. The objective function achieves a minimum value of 1 at any point on the segment between and including the points $\left(\frac{2}{3}, \frac{4}{3}\right)$ and $\left(\frac{9}{5}, \frac{1}{5}\right)$. **21.** It is a crystallike space with plane faces, the region of overlap between all individual inequalities of the system. **23.** The minimum shipping cost is $1900. If x and y represent the quantity shipped from Brooklyn to Manhattan and from Brooklyn to Amityville, then any point on the line segment connecting $(500, 100)$ with $(200, 400)$ will achieve this minimum cost. $500 - x$ and $40 - y$ represent the quantities shipped from Baldwin to Manhattan and Amityville, respectively.

Answers to Review Exercises, Chapter 3.
1.

3.

5.

7. There are no corners; the feasible region is an infinite strip.

9. Corners are $(0, 0)$, $(0, 1)$, and $(1, 2)$.

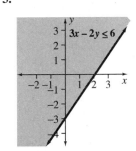

$x + 2y > 0; x + 2y < 4$

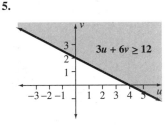

$x \geq 0; y \geq 0; y - x \leq 1; y \leq 2$

11. Corner coordinates are $(\frac{11}{3}, \frac{19}{3})$, $(\frac{20}{3}, \frac{10}{3})$, and $(\frac{2}{3}, \frac{1}{3})$.

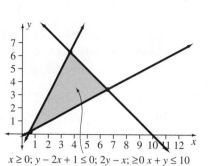

$x \geq 0;\ y - 2x + 1 \leq 0;\ 2y - x;\ \geq 0\ x + y \leq 10$

13. Min: $3\frac{1}{2}$ at $(1, 1)$;

max: 12 at $(4, 0)$.   **15.** Min: 3 at $(1, 0)$; max: 6 at $(2, 0)$.   **17.** Min: 3 at $(1, 0)$ or $(0, 6)$ or at any point on the line segment connecting these points; max: $15\frac{1}{2}$ at $(4, 7)$.   **19.** Max is $14\frac{1}{4}$ at $(3\frac{1}{2}, \frac{1}{2})$.   **21.** No solution—the feasible region is empty.   **23.** The feasible region is unbounded—no maximum.   **25.** 56 and 8 ladders; $1840.   **27.** Manufacture $22\frac{6}{7}$ ladders of the first type. The profit is $685\frac{5}{7}$. This finishing constraint is stronger than any other of the constraints of the problem. **29.** (a) The minimum cost is $5\frac{1}{7}$ when $\frac{16}{7}$ liter of the first freshener is mixed with $\frac{10}{7}$ liter of the second. (b) Mix the same amount of each freshener as does the first homeowner. This will put 30 ml of chemical C in the pool.

Section 4.1

1. No.   **3.** No.   **5.** No.   **7.** $x_1 + 2x_2 + s_1 = 3$
$2x_1 + x_2 + s_2 = 2$ **9.** $3x_1 + s_1 = 4$
$2x_1 + 2x_2 + s_2 = 9$ **11.** $x_1 + 2x_2 + s_1 = 1$
$x_1 + s_2 = \dfrac{3}{4}$

13. $3x_1 + 4x_2 + s_1 = 4$
$3x_1 + 3x_2 + s_2 = 3.25$ **15.** $x_1 + x_2 + s_1 = 3$
$x_2 + x_3 + s_2 = 4$

17. $x_1 + 2x_2 + s_1 = 6$ (a) $2x_1 + x_2 + s_2 = 6$; (b) 2 basic variables not including P; (c) 2 non-basic variables;
$-x_1 - x_2 + P = 0$
(d) 6 ways; (e) $x_1 = x_2 = P = 0$. This corresponds to the origin. (f) This solution implies that $s_1 = -6 < 0$. All the slack variables must not be negative.   **19.** $P = 6$ at $(x_1, x_2) = (0, 2)$.   **21.** $P = 4$ at $(2, 2)$.   **23.** $P = 8$ at $(1, 2)$.

25. $\left.\begin{array}{r} x_1 + s_1 = 3 \\ 2x_1 + x_2 + s_2 = 3 \end{array}\right\}$ max is 3 at $x_1 = 0,\ x_2 = 3$.   **27.** Max is $7\frac{1}{2}$ at $x_1 = \frac{3}{2},\ x_2 = 0$.

29. Max is 22 at $x_1 = 4$, $x_2 = 0$, $x_3 = 2$.   **31.** Nonbasic: x_1, s_1, Basic: P, x_2, s_2 **33.** Nonbasic: x_1, x_2, Basic: s_1, s_2, s_3, P **35.** Nonbasic: s_1, s_2, Basic: x_1, x_2, s_3, P **37.** Nonbasic: x_1, x_2, s_2, Basic: x_3, s_1, P **39.** Nonbasic: 3, Basic: 4.   **41.** Nonbasic: 4, Basic: 4.   **43.** Nonbasic: 3, Basic: 6.   **45.** Nonbasic: 1, Basic: 3.   **47.** Max income = $350; clean and repair 7 pieces each.   **49.** Basic: x_2, s_1, s_2, P; Nonbasic: x_1, s_3.

Section 4.2

1.
$$x_1 + 2x_2 + s_1 = 4$$
$$4x_1 + 3x_2 + s_2 = 4$$
$$3x_1 + x_2 + s_3 = 9$$
$$-10x_1 - 12x_2 + P = 0$$

3.
$$\tfrac{1}{2}x_1 + \tfrac{2}{3}x_2 + s_1 = \tfrac{7}{8}$$
$$x_1 + \tfrac{5}{4}x_2 + s_2 = \tfrac{4}{3}$$
$$\tfrac{3}{4}x_1 + x_2 + s_3 = 1$$
$$-\tfrac{3}{4}x_1 - \tfrac{3}{4}x_2 + P = 0$$

5.
$$2x_1 + 3x_2 + x_3 + s_1 = 7$$
$$x_1 + 3x_2 + 2x_3 + s_2 = 6$$
$$5x_1 + 3x_3 + s_3 = 4$$
$$-x_1 + x_2 - 2x_3 + P = 0$$

7. Maximum is 6 at $(1, 2)$.   **9.** Maximum is 6 at $(0, 2)$.

11. Maximum is 6 at $(0, 2)$.   **13.** Maximum is 5 at $(1, 2)$.   **15.** No.

Section 4.3

1. The max is $P = 25$ when $x_1 = 0$, $x_2 = 5$. There are two variables.   **3.** Two original variables; maximum is 3 at $(1, 2)$.

5. Two original variables; maximum is 10 at $x_1 = x_2 = 2$.   **7.** $\left(\begin{array}{cccc|c} 1 & \boxed{1} & 1 & 0 & 4 \\ \hline -5 & -6 & 0 & 1 & 0 \end{array} \right)$

9. $\left(\begin{array}{ccccc|c} 1 & 1 & 1 & 0 & 0 & 4 \\ 0 & \boxed{1} & 0 & 1 & 0 & 2 \\ \hline -2 & -5 & 0 & 0 & 1 & 0 \end{array} \right)$ **11.** $\left(\begin{array}{ccccc|c} \boxed{1} & 0 & 1 & 0 & 0 & 3 \\ 1 & 2 & 0 & 1 & 0 & 4 \\ \hline -6 & -1 & 0 & 0 & 1 & 0 \end{array} \right)$ **13.** $\left(\begin{array}{cccccc|c} 6.1 & \boxed{.1} & 1 & 0 & 0 & 0 & 4.5 \\ 5.5 & -4.4 & 0 & 1 & 0 & 0 & 3.3 \\ 1 & 0 & 0 & 0 & 1 & 0 & 3 \\ \hline -2.5 & -1.125 & 0 & 0 & 0 & 1 & 0 \end{array} \right)$

15. $\left(\begin{array}{cccccc|c} 1 & 2\frac{1}{5} & 1 & 0 & 0 & 0 & 2 \\ \frac{4}{5} & \boxed{2} & 0 & 1 & 0 & 0 & 2 \\ \frac{3}{5} & \frac{7}{10} & 0 & 0 & 1 & 0 & 3\frac{1}{5} \\ \hline -\frac{1}{5} & -\frac{2}{5} & 0 & 0 & 0 & 1 & 0 \end{array} \right)$ **17.** $\left(\begin{array}{ccccccc|c} 1 & 0 & -3 & 1 & 0 & 0 & 0 & 4 \\ 0 & \boxed{2} & 3 & 0 & 1 & 0 & 0 & 5 \\ 1 & 1 & 4 & 0 & 0 & 1 & 0 & 4 \\ \hline -2 & -3 & 0 & 0 & 0 & 0 & 1 & 0 \end{array} \right)$ **19.** $P_{\max} = 18\frac{1}{2}$ at $x_1 = 3$, $x_2 = \frac{1}{2}$

21. $P = 8$ for $x_1 = 4$, $x_2 = 0$ **23.** $P = 16$ for $x_1 = 4$, $x_2 = 0$ **25.** $P = 6$ for $x_1 = 0$, $x_2 = 3$ **27.** $P = 5$ for $x_1 = 1$, $x_2 = 2$, $x_3 = 1$ **29.** $P = 8$ for $x_1 = 4$, $x_2 = 0$, $x_3 = 2$ **31.** $P = 3$ for $x_1 = 2$, $x_2 = 1$, $x_3 = 0$

33. $P = 8$ for $x_1 = x_2 = 0$, $x_3 = 4$, $x_4 = 0$ **35.** $P = 28.4$ at $(3.2, 2.6)$ **37.** Most effective way: no newspaper ads or radio spots, 100 batches of fliers.

39. cake A: 64

cake B: 0

cake C: 32

All results should be interpreted as averages.

41. Study 10 type 1 galaxies, 60 type 2 galaxies.   **43.** These results should be interpreted as averages.   **45.** 1st row, 1st column   **47.** 1st element in third column   **49.** 2nd row, 1st column.

Section 4.4

1. (3)   **3.** $\begin{pmatrix} 1 & 3 \\ 2 & 4 \end{pmatrix}$ **5.** $\begin{pmatrix} 1 & 1 \\ 0 & 1 \end{pmatrix}$ **7.** $\begin{pmatrix} 0 & 0 \\ 0 & 0 \end{pmatrix}$ **9.** $\begin{pmatrix} 1 & -3 \\ 2 & -2 \\ 3 & -1 \end{pmatrix}$ **11.** $\begin{pmatrix} \frac{1}{2} & -1 & 0 \\ 2 & 1 & \frac{1}{2} \end{pmatrix}$ **13.** $\begin{pmatrix} -1 \\ 1 \\ 0 \\ 2 \\ 4 \\ 6 \\ 8 \end{pmatrix}$

15. $y_1 \le 2$

$y_2 \le 1$; $p = 3y_1 + y_2 + y_3$.

$y_3 \le 1$

17. $y_1 \le 1$ and $P = 3y_1$ **19.** $y_1 + 2y_2 + y_3 \le 2$

$2y_1 + y_2 + y_3 \le 1$ and $P = 6y_1 + 4y_2 + 2y_3$.

$3y_1 + y_2 \le 3$

21. Max $P = 2y_1 + 2y_2 + 3y_3$ subject to $y_1 \le 3$,

$y_2 \le 0$,

$y_3 \le 1$.

23. Max $P = 3y_1 + 3y_2$ subject to $y_1 + 2y_2 \le 3$,

$2y_1 \le 0$,

$y_2 \le 1$.

25. Minimize $P = 3x$ subject to $x_1 \ge 3$,

$x_2 \ge 4$.

27. Minimize $P = 4x_1 + 3x_2$ subject to $x_1 + 3x_2 \ge 3$

$2x_1 + 2x_2 \ge 4$.

29. (a) $M = \begin{pmatrix} 1 & 2 & 5 \\ 2 & 3 & 5 \\ 1 & 1 & 0 \end{pmatrix}$, $M^T = \begin{pmatrix} 1 & 2 & 1 \\ 2 & 3 & 1 \\ 3 & 5 & 0 \end{pmatrix}$. (b) Maximize $P = 3y_1 + 5y_2$ subject to $y_1 + 2y_2 \le 1$, $2y_1 + 3y_2 \le 1$, and to $y_1 \geq 0$, $y_2 \geq 0$. (c) $\begin{pmatrix} 1 & 2 & 1 & 0 & 0 & | & 1 \\ 2 & \boxed{3} & 0 & 1 & 0 & | & 1 \\ -3 & -5 & 0 & 0 & 1 & | & 0 \end{pmatrix}$ (d) $\begin{pmatrix} -\frac{1}{3} & 0 & 1 & -\frac{2}{3} & 0 & \frac{1}{3} \\ \frac{2}{3} & 1 & 0 & \frac{1}{3} & 0 & \frac{1}{3} \\ \frac{1}{3} & 0 & 0 & \frac{5}{3} & 1 & \frac{5}{3} \end{pmatrix}$ (e) The maximum value of P is

$\frac{5}{3}$ when $y_1 = 0$ and $y_2 = \frac{1}{3}$. (f) The minimum value of p is $\frac{5}{3}$ when $x_1 = 0$ and $x_2 = \frac{5}{3}$.   **33.** Minimum is 46 at (3, 4).   **35.** Minimum is 2 at (0, 2).   **37.** Minimum is 2 at $x_1 = x_3 = 0$, $x_2 = 2$.   **39.** (a) The cost is \$30 when $A = B = 1$, $C = 2$. (b) Same strategy as (a); there will be 6600 richness units per plate.   **41.** Use 2 ounces of α, 1 ounce each of β and γ.

43. When solving, let x_3 and x_4 be the numbers from Los Angeles to Albuquerque and Chicago. The number to Buffalo will be $600 - x_3 - x_4$. The minimum transportation cost is \$200,000 when 400 children travel from New York to Albuquerque, 100 children travel from New York to Buffalo, and 400 children travel from Los Angeles to Chicago. That is, $x_1 = 400$, $x_2 = x_3 = 0$, $x_4 = 400$.

45. The pivot element is 5.0 on the top row; $\begin{pmatrix} 0 & 1.18 & 1 & -0.5 & | & 0.5 \\ 1 & 0.95 & 0 & 0.45 & | & 0.45 \\ 0 & 0.91 & 0 & 1.91 & | & 1.91 \end{pmatrix}$

Section 4.5

1.
$$
\begin{array}{rcrcrcrcr}
x_1 & & & -\,S_1 & & +\,a_1 & & = & 1 \\
x_1 & +\,s_1 & & & & & & = & 4 \\
& & x_2 & -\,S_2 & & & +\,a_2 & = & 1 \\
& & x_2 & +\,s_2 & & & & = & 3 \\
-2x_1 & -\,4x_2 & & & & +\,Ma_1 & +\,Ma_2 & +\,P = & 0
\end{array}
$$
Here $x_1 = x$, $x_2 = y$.
$$
\begin{pmatrix}
1 & 0 & 0 & 0 & -1 & 0 & 1 & 0 & 0 & 1 \\
1 & 0 & 1 & 0 & 0 & 0 & 0 & 0 & 0 & 4 \\
0 & 1 & 0 & 0 & 0 & -1 & 0 & 1 & 0 & 1 \\
0 & 1 & 0 & 1 & 0 & 0 & 0 & 0 & 0 & 3 \\
-2 & -4 & 0 & 0 & 0 & 0 & M & M & 1 & 0
\end{pmatrix}
$$

3.
$$
\begin{array}{rcrcrcr}
x_1 & +\,x_2 & & -\,S_1 & +\,a_1 & = & 1 \\
& x_2 & +\,s_1 & & & = & 2 \\
x_1 & & +\,s_2 & & & = & 2 \\
-2x_1 & -\,4x_2 & & & +\,Ma_1 & +\,P = & 0
\end{array}
$$
$$
\begin{pmatrix}
1 & 1 & 0 & 0 & -1 & 1 & 0 & 1 \\
0 & 1 & 1 & 0 & 0 & 0 & 0 & 2 \\
1 & 0 & 0 & 1 & 0 & 0 & 0 & 2 \\
-2 & -4 & 0 & 0 & 0 & M & 1 & 0
\end{pmatrix}
$$

5.
$$
\begin{array}{rcrcrcrcr}
x_1 & +\,x_2 & +\,x_3 & & -\,S_1 & +\,a_1 & = & 3 \\
x_1 & +\,x_2 & & +\,s_1 & & & = & 4 \\
& x_2 & +\,x_3 & +\,s_2 & & & = & 4 \\
x_1 & & +\,x_3 & & +\,s_3 & & = & 4 \\
-x_1 & -\,2x_2 & -\,3x_3 & & & +\,Ma_1 & +\,P = & 0
\end{array}
$$
Here $x_1 = x$, $x_2 = y$, $x_3 = z$.
$$
\begin{pmatrix}
1 & 1 & 1 & 0 & 0 & 0 & -1 & 1 & 0 & 3 \\
1 & 1 & 0 & 1 & 0 & 0 & 0 & 0 & 0 & 4 \\
0 & 1 & 1 & 0 & 1 & 0 & 0 & 0 & 0 & 4 \\
1 & 0 & 1 & 0 & 0 & 1 & 0 & 0 & 0 & 4 \\
-1 & -2 & -3 & 0 & 0 & 0 & 0 & M & 1 & 0
\end{pmatrix}
$$

7.

$$
\begin{aligned}
x_1 + 2x_2 + 3x_3 \qquad\qquad -S_1 \quad + a_1 \qquad\qquad &= 4 \\
x_1 \qquad\qquad + s_1 \qquad\qquad\qquad\qquad &= 3 \\
x_2 \qquad + s_2 \qquad\qquad\qquad\qquad &= 2 \\
x_3 \qquad + s_3 \qquad\qquad\qquad\qquad &= 2 \\
-x_1 - 2x_2 - 3x_3 \qquad\qquad\qquad + Ma_1 \; + p \; &= 0
\end{aligned}
$$

where $x_1 = x$, $x_2 = y$, $x_3 = z$

$$
\begin{pmatrix}
1 & 2 & 3 & 0 & 0 & 0 & -1 & 1 & 0 & 4 \\
1 & 0 & 0 & 1 & 0 & 0 & 0 & 0 & 0 & 3 \\
0 & 1 & 0 & 0 & 1 & 0 & 0 & 0 & 0 & 2 \\
0 & 0 & 1 & 0 & 0 & 1 & 0 & 0 & 0 & 2 \\
-1 & -2 & -3 & 0 & 0 & 0 & 0 & M & 1 & 0
\end{pmatrix}
$$

9. (a) $x + y + s = 2$

$x + y - S + a = 1;$

$-x - 2y + Ma + P = 0$

$x = 0$, $y = 2$.

(b) $\begin{pmatrix} 1 & 1 & 1 & 0 & 0 & 0 & 2 \\ 1 & 1 & 0 & -1 & 1 & 0 & 1 \\ -1 & -2 & 0 & 0 & M & 1 & 0 \end{pmatrix}$; (c) $\begin{pmatrix} 0 & 0 & 1 & 1 & -1 & 0 & 1 \\ 1 & 1 & 1 & 0 & 0 & 0 & 2 \\ 1 & 0 & 2 & 0 & M & 1 & 4 \end{pmatrix}$ (d) $P = 4$ when

11. 20; $x = 4$, $y = 3$ **13.** 12; $x = y = 2$ **15.** 12 for $x = y = 0$, $z = 4$ **17.** 13; $x = 3$, $y = z = 2$ **21.** 6000 tennis balls; 1000 golf balls; $22,000.

23. Let x_1, x_2 be the number of long and short papers. Maximize $5x_1 + 3x_2$ when $\qquad x_1 \geq 3$, $x_2 \geq 3$,

$$150x_1 + 100x_2 \leq 1000,$$

$$x_2 \geq x_1 + 1$$

25. Problems with 3 or fewer variables are appropriate for the geometric method. Otherwise, we have no choice but to resort to the simplex method.

Answers to Review Exercises, Chapter 4

1. $x_1 \quad + s_1 \qquad = 4$
$x_1 + x_2 \quad + s_2 = 5$

3. $x_1 + x_2 + x_3 - S_1 \qquad + a_1 \qquad = 4$
$3x_1 + 2x_2 + x_3 \qquad - S_2 \quad + a_2 = 10$

5. $\begin{pmatrix} -1 & 1 \\ -1 & 1 \\ 0 & -2 \\ -1 & -1 \end{pmatrix}$ **7.** $\begin{pmatrix} .012 & -2.01 \\ -.444 & 0.79 \end{pmatrix}$

9. $\begin{pmatrix} 1 & 0 \\ .1 & 1 \end{pmatrix}$ **11.** Minimize $p = 2y_1 + 4y_2 + 2y_3$ subject to $2y_1 \geq 45$, $y_2 \geq 32$, $y_3 \geq 14$.   **13.** Minimize

$p = 4y_1 + y_2 + 5y_3 + 9y_4$ subject to $2y_1 + 7y_3 \geq 1$, $3y_2 + y_3 \geq 1$, $y_1 + 3y_4 \geq 1$, $y_2 + 4y_4 \geq 1$ and to all $y_i \geq 0$.

15. Maximize $P = 4y_1 + 7y_2 + 3y_3$ subject to all $y_i \geq 0$ and to $y_1 + 2y_2 \leq 2$, $2y_2 + 4y_3 \leq 1$, and $y_1 + 3y_2 + y_3 \leq 3$.   **17.** 20 at (4, 4).   **19.** 15 at (0, 3).   **21.** 10 at $x_1 = 2$, $x_2 = 4$, $x_3 = 0$.   **23.**

$$
\begin{aligned}
x_1 + x_2 \; + s_1 \qquad\qquad\qquad &= 2 \\
x_2 \qquad - S_1 + a_1 \qquad &= 1 \\
-2x_1 - \frac{7}{2}x_2 \qquad\qquad + Ma_1 + P &= 0
\end{aligned}
$$

25. 7 at (0, 2).   **27.** 100 Model C boards, none of the others. Profit is $1500.   **29.** None with the first or third species, 300 with the second species.   **31.** Maximum is 20 refugees. No Russians are interviewed, and 20 Taiwanese refugees are interviewed.

Section 5.1

1. T   **3.** F   **5.** T   **7.** T   **9.** F   **11.** F   **13.** F   **15.** F   **19.** M, I, S, P   **21.** F   **23.** T
25. T   **27.** Let the notation $(a \; b)$ represent the throw such that the red die shows a and the white die shows b. Then the outcomes are (1 1) (1 2) (1 3) (1 4) (1 5) (1 6) (2 1) (2 2) (2 3) (2 4) (2 5) (2 6) (3 1) (3 2) (3 3) (3 4) (3 5) (3 6) (4 1) (4 2) (4 3) (4 4) (4 5) (4 6) (5 1) (5 2) (5 3) (5 4) (5 5) (5 6) (6 1) (6 2) (6 3) (6 4) (6 5) (6 6)

29. Let h represent heads and t represent tails. Then the possible outcomes are ttt, tth, tht, thh, htt, hth, hht, hhh.
31. ϕ, $\{T\}$, $\{e\}$, $\{n\}$, $\{s\}$, $\{Te\}$, $\{Tn\}$, $\{Ts\}$, $\{en\}$, $\{es\}$, $\{ns\}$, $\{Ten\}$, $\{Tes\}$, $\{Tns\}$, $\{ens\}$, $\{Tens\}$ **33.** 6; $2^6 = 64$ **35.** 14
37. $2^8 - 1 = 255$ **39.** 63 **41.** 2 **43.** 11 **45.** (a) 2 (b) 4 (c) 8 (d) 1 (e) 2^i (f) 2^j
47. A could be a set like $A = \{e, \{e\}\}$, in which case the text's statement is incorrect.

Section 5.2

1. 0, 1, 2, 3, 4, 5, 6 **3.** 1, 3, 5, 7, 9 **5.** 7, 8, 9 **7.** {c, o, m, p, u, t, e } **9.** c, o, m, u, t, e, r
11. The entire alphabet **13.** A **15.** A **17.** A **19.**

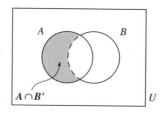

21.

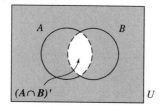

23.

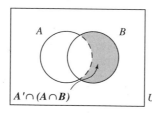

25.

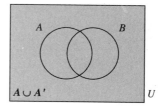

27. The set of all months not beginning with "A." **29.** u **31.** Empty set **33.** {a, b, c, d, e}
35.   **37.**   **39.**

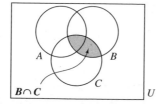

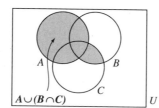

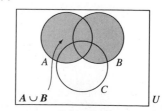

41. To get the left-hand side, note the Venn diagram (a), from which we can derive (b). To get the right-hand side, note that from Venn diagrams (c) and (d) we get diagram (e). Since the diagrams for both the right- and left-hand side are the same, we conclude the theorem is proven. (a)   (b)   (c)

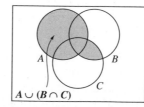

(d)   (e)

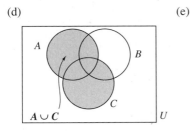 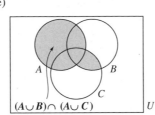

43. In the case that the two sets have a nonempty intersection and are not equal, the Venn diagrams are as follows. The left-hand set is clearly a subset of the right-hand set. In the same way, consider all other cases, such as $A = B$, A is completely contained within B, and so on, to demonstrate the assertion.

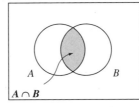 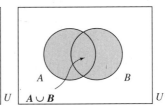

45. $1:\ A \cap (B \cup C)'$
     $2:\ A \cap B \cap C'$
     $3:\ B \cap (A \cup C)'$
     $4:\ A \cap B' \cap C$
     $5:\ A' \cap B \cap C$
     $6:\ C \cap (A \cup B)'$
     $7:\ (A \cup B \cup C)'$

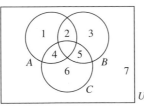

47. $S' \cup P$ **49.** $(P \cup R)' \cap S'$ **51.** $(P \cap R \cap S')'$ **53.** $M \cap W'$ **55.** $(C \cap M)'$ **57.** $(M \cap W') \cup (F \cap C)$
59. $P \cup E$ **61.** $(P \cup E)'$

Section 5.3
1. 14   **3.** 7   **5.** 50   **7.** 12   **9.** 17   **11.** 3   **13.** 19   **15.** 58   **17.** 0
19. We can prove the formula by noting first that for any two sets L and M with no elements in common, $c(L \cup M) = c(L) + c(M)$. Any sets $A \cup B$ can always be divided into two nonintersecting sets—the set A, and the set of all elements that are in B but not in A, $B \cap A'$. That is, since $A \cup B = A \cup (B \cap A')$,

$$c(A \cup B) = c(A) + c(B \cap A'). \tag{a}$$

(Take $L = A$ and $M = B \cap A'$.) In a similar fashion we can say that the elements in B consist of those elements which are in B and also in A and those which are in B and are not in A. That is, $B = (A \cap B) \cup (B \cap A')$. The sets on the right cannot have any elements in common, so we may conclude that

$$c(B) = c(A \cap B) + c(B \cap A'). \tag{b}$$

Combine equations (a) and (b) to prove the assertion.   **21.** $c(A) \geq 17$, $c(B) \geq 15$, $c(C) \geq 18$.   **23.** $c(A) \geq 300$, $c(B) \geq$ 300, $c(C) \geq 300$.   **25.** $c(A) \geq 12$, $c(B) \geq 11$, $c(C) \geq 12$.   **27.** 30   **29.** 1   **31.** 2   **33.** 6   **35.** 9

37. 7   **39.**

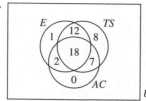

41. $\frac{20}{100} = \frac{1}{5}$ **43.** 18   **45.** The data in the

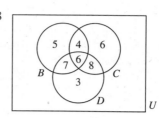

problem give rise to the accompanying Venn diagram where the labels B, C, D represent the children with blocks, eating cookies, and playing with blocks. The Venn diagram accounts for 39 children, but there are 41 in the Wee Care Center. Two children are unaccounted for.   **47.** It's hard to apply Venn diagrams to problems requiring more than 3 sets.   **49.** 10,170

Section 5.4

1. 120   **3.** 576   **5.** $\frac{1}{12}$ **7.** 495   **9.** 840   **11.** 120   **13.** $\frac{7!}{4!}$ **15.** 4!   **19.** If we tried to use problem **17.** we might compute $0! = 0 \cdot (-1)!$ which implies that $(-1)! = \frac{1}{0}$ an impossibility.   **21.** 1   **23.** 3   **25.** 1   **27.** $n+1$ **29.** $\frac{(3n)!}{n!}$ which is not the same as 3!   **31.** 20, 160   **33.** 35   **35.** 36   **37.** 8!   **39.** $\frac{26!}{18!}$ **41.** 120   **43.** 210 **45.** 40, 320; 604, 800   **47.** 87, 178, 291, 200   **49.** 39, 916, 800

Section 5.5

1. 3   **3.** 35   **5.** 21   **7.** 4   **9.** 20   **11.** $n+1$ **13.** 1   **15.** When $r = 0$ or $r = 1$.   **17.** 1   **19.** 7 **21.** 35   **23.** 3   **25.** 20   **27.** 30   **29.** $r = 0$ and $r = u$ **31.** 495   **33.** $\frac{4096!}{72!\,4084!}$ **35.** $\frac{14!}{(6!)^2\,2!}$ **37.** $\frac{52!}{(5!)^5\,27!}$ **39.** 4200   **41.** (a) $\frac{14!}{5!\,9!}$ (b) 60 (c) 3 (d) 60 (e) \$125; 1 way   **43.** $\frac{52!}{13!\,39!}$ bonds; $\frac{52!}{(13!)^4}$ deals **45.** $C(n, n)$ refers to the number of different ways that n item can be chosen from n items. This can happen only 1 way. **47.** 495

Answers to Review Exercises, Chapter 5

1. T   **3.** F   **5.** M, i, s, p   **7.** $\{a, c\}$ **9.** $\{b, d, f, z\}$ **11.** $\{a, b, c, d, e, f, g, h\}$
13.   **15.**   **17.**

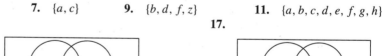

19.   **21.**   **23.** 5   **25.** 14   **27.** 24   **29.** 1   **31.** 6

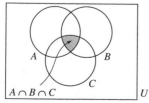

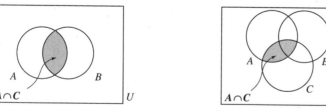

33. 60   **35.** 12   **37.** 25   **39.** 37   **41.** 169; 156   **43.** 4   **45.** 1632   **47.** 126

Section 6.1

1. The sample space is $S = \{l, o, q, u, a, c, i, s\}$.   **3.** The sample space is $S = \{b, a, n\}$.   **5.** The sample space is $S = \{aa, ap, al, ae, ar, pa, pp, pl, pe, pr, la, lp, ll, le, lr, ea, ep, el, ee, er, ra, rp, rl, re, rr\}$.   **7.** $\{uu, ud, du, dd\}$; u = tack lands point up; d = tack lands point down   **9.** $\{1\text{-}2, 1\text{-}3, 1\text{-}4, 2\text{-}3, 2\text{-}4, 3\text{-}4\}$; "1-2" means choosing books 1 and 2, etc.   **11.** $\{m, f, mm, mf, ff, mmm, mmf, mff, fff, mmmm, mmmf, mmff, mfff, ffff\}$; m = male puppy, f = female puppy   **13.** $\{\heartsuit\heartsuit, \diamondsuit\diamondsuit, \heartsuit\clubsuit, \heartsuit\spadesuit, \diamondsuit\heartsuit, \diamondsuit\diamondsuit, \diamondsuit\clubsuit, \diamondsuit\spadesuit, \clubsuit\heartsuit, \clubsuit\diamondsuit, \clubsuit\clubsuit, \clubsuit\spadesuit, \spadesuit\heartsuit, \spadesuit\diamondsuit, \spadesuit\clubsuit, \spadesuit\spadesuit\}$ **15.** $\{LLL, LRL, LLR, LRR, RLL, RRL, RLR, RRR\}$ **17.** $\{AAAA, AAAB, AABA, AABB, ABAA, ABAB, ABBA, ABBB, BAAA, BAAB, BABA, BABB, BBAA, BBAB, BBBA, BBBB\}$ **19.** $\{eggplant\}$ **21.** $\{bananas, figs\}$ **23.** $\emptyset$ **25.** $S_1 = \{(1, 1), (1, 2), \ldots, (1, 6), \ldots (6, 6)\}$ and $S_2 = \{1, 2, 3, \ldots, 12\}$. Each element of S_2 is the sum of the dice.   **27.** $\{1A, 2A, 3A, 1B, 2B, 3B\}$ **29.** Invalid   **31.** Valid   **33.** Valid   **35.** Valid   **37.** Valid

39. $\{(1\ 1)\ (1\ 2)\ (1\ 3)\ (1\ 4)\ (1\ 5)\ (1\ 6)\ (2\ 1)\ (2\ 2)\ (2\ 3)\ (2\ 4)\ (2\ 5)\ (2\ 6)\ (3\ 1)\ (3\ 2)\ (3\ 3)\ (3\ 4)\ (3\ 5)\ (3\ 6)\ (4\ 1)\ (4\ 2)\ (4\ 3)\ (4\ 4)\ (4\ 5)\ (4\ 6)\ (5\ 1)\ (5\ 2)\ (5\ 3)\ (5\ 4)\ (5\ 5)\ (5\ 6)\ (6\ 1)\ (6\ 2)\ (6\ 3)\ (6\ 4)\ (6\ 5)\ (6\ 6)\}$

41. $\{(4\ 6)\ (5\ 5)\ (6\ 4)\}$ **43.** $\{(1\ 1)\ (1\ 2)\ (1\ 3)\ (1\ 4)\ (1\ 5)\ (1\ 6)\ (2\ 1)\ (3\ 1)\ (4\ 1)\ (5\ 1)\ (6\ 1)\}$ **45.** Impossible event

47. $\{(1\ 3)\ (2\ 2)\ (3\ 1)\ (3\ 6)\ (4\ 5)\ (5\ 4)\ (6\ 3)\}$ **49.** $\{(5\ 5)\}$ **51.** $\{(1\ 2)\ (1\ 4)\ (1\ 6)\ (2\ 1)\ (2\ 3)\ (2\ 5)\ (3\ 2)\ (3\ 4)\ (3\ 6)\ (4\ 1)\ (4\ 3)\ (4\ 5)\ (5\ 2)\ (5\ 4)\ (5\ 6)\ (6\ 1)\ (6\ 3)\ (6\ 5)\}$ **53.** $\{ggg, gbg, ggb, gbb\}$ **55.** $\{ggg, ggb, gbb, bbb\}$

57. $\{ggg, bgg, gbg, ggb\}$ **59.** The entire sample space.   **61.** $\frac{1}{39}$; $\frac{1}{66}$.   **63.** $\frac{1}{13}$ **65.** $\frac{1}{31173}$.   **67.** $\frac{1}{66}$.

69. $\frac{104}{365}$.   **71.** $\frac{13}{365}$.   **73.** $\frac{3}{7}$ **75.** $\frac{5}{11}$ **77.** $\frac{1}{2}$ **79.** $\frac{2}{3}$ **81.** $\frac{4}{11}$ **83.** 1   **85.** 5   **87.** $\frac{10}{19}$

89. $\frac{3}{5}$ **91.** 7:3   **93.** 0 to anything   **95.** 4:1   **97.** 2:5   **99.** Invalid probability   **101.** 1:8   **103.** 4:5

105. 19:1   **107.** 5:3   **109.** The sum of the $P(E_i)$ is the probability for $E_1 \cup E_2 \cup \ldots \cup E_n$. $P(E_1 \cup \ldots \cup E_n) = 1$ means it's certain that at least one of the E_i will occur.

111. (a)

(b) E and $F \cap E'$ have no events in common, so that $P(E \cup F) = P(E) + P(F \cap E')$.

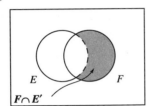

(c)

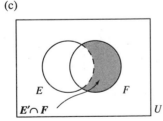
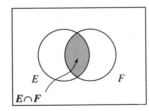
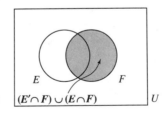

(d) $P(F) = P(E \cap F) + P(E' \cap F)$ so that $P(E' \cap F) = P(F) - P(E \cap F)$.   (e) Substituting the results of (d) into (b) yields $P(E \cup F) = P(E) + P(F \cap E') = P(E) + P(F) - P(E \cap F)$.

113. S1P1, S1P2, S2P1, S2P2, S3P1, S3P2   **115.** E_1M_1, E_1M_2, E_1M_3, E_2M_1, E_2M_2, E_2M_3, E_3M_1, E_3M_2, E_3M_3, E_4M_1, E_4M_2, E_4M_3 **117.** $\frac{3}{14}$ **119.** $\frac{4719}{10,723} = .4401$ **121.** $\frac{1404}{3000} = .468$ **123.** $\frac{1}{33}$

125. About 15 pieces   **127.** 0.5912

Section 6.2

1. $\frac{1}{30}$.   **3.** $\frac{2}{5}$.   **5.** $\frac{6}{26} = \frac{3}{13}$.   **7.** $\frac{2}{7}$.   **9.** $\frac{3}{12} = \frac{1}{4}$.   **11.** $\frac{1}{3}$.   **13.** (a) $S = \{H1, H2, \ldots, H36, T1, T2, \ldots, T36\}$; (b) Many possibilities. For example, $S = \{He, Ho, Te, To\}$.   **15.** $\frac{1}{36}$ **17.** $\frac{5}{18}$ **19.** $\frac{5}{36}$ **21.** $\frac{1}{72}$

23. $\frac{8}{27}$ **25.** $\frac{1}{216}$ **27.** $\frac{91}{216}$ **29.** $\frac{1}{8}$ **31.** .5   **33.** $\frac{1}{16}$ **35.** $\frac{1}{16}$ **37.** $\frac{1}{64}$ **39.** .25   **41.** $\frac{1}{52}$

43. $\frac{1}{52}$ **45.** $\frac{2}{52} = \frac{1}{26}$ **47.** $\frac{4}{C}(52, 13)$ **49.** $\frac{12}{26}$ **51.** $\frac{5}{26}$ **53.** $\frac{3}{13}$ **55.** 100   **57.** $\frac{99}{100}$

59. 0.25   **61.** 0.25   **63.** 0.09   **65.** $\frac{1}{45}$ **67.** $\frac{16}{45}$ **69.** 45   **71.** $\frac{44}{45}$ **73.** 10   **75.** $\frac{3}{10}$ **77.** $\frac{1}{10}$
79. $\frac{9}{10}$ **81.** $\frac{3}{14}$ **83.** $\frac{11}{14}$ **85.** $\frac{3}{14}$ **87.** $\frac{20}{21}$ **89.** 75%; $66\frac{2}{3}$%   **91.** $C(7,7)C(93,3)/C(100,10)$
93. $C(7,1)C(93,6)/C(100,10)$ **95.** $13,320/7,059,052 \approx 0.001887$
97. We would need to be explicitly given the probabilities for each simple event. Then, we express any outcome of interest as the union of simple events, and sum those probabilities to find the probability of the outcome.
99. 1287

Section 6.3

1. no   **3.** yes   **5.** yes   **7.** yes   **9.** no   **11.** no   **13.** $\frac{6}{26}$ **15.** $\frac{10}{26}$ **17.** 1   **19.** $\frac{6}{10}$ **21.** $\frac{4}{9}$
23. 0.05   **25.** $\frac{25}{26}$ **27.** $\frac{18}{26}$ **29.** $\frac{90}{99}$ **31.** $\frac{1}{108}$ **33.** $\frac{1}{108}$ **35.** $\frac{1}{54}$ **37.** $\frac{67}{72}$ **39.** $\frac{125}{216}$ **41.** 0.5
43. $\frac{7}{8}$ **45.** $\frac{1}{16}$ **47.** $\frac{3}{4}$ **49.** $\frac{5}{16}$ **51.** $\frac{9}{16}$ **53.** $\frac{2}{7}$ **55.** $\frac{83}{84}$ **57.** $\frac{43}{84}$ **59.** $\frac{1}{7}$ **61.** $\frac{1}{4}$
63. $\frac{4}{52}$ **65.** $\frac{3}{13}$ **67.** 0.5   **69.** $\frac{2}{7}$ **71.** $\frac{2}{11}$ **73.** $\frac{11}{144}$; $\frac{15}{144}$ **75.** 59; $\frac{26}{59}$; $\frac{45}{59}$ **77.** 55%   **79.** (a) $\frac{3}{10}$
(b) $\frac{13}{15}$ (c) $\frac{1}{2}$ **81.** (a) $\frac{7}{8}$ (b) $\frac{3}{5}$ (c) $\frac{19}{40}$ **83.** (a) $\frac{2}{5}$ (b) $\frac{1}{5}$ (c) $\frac{3}{20}$ (d) $\frac{1}{5}$ (e) $\frac{11}{20}$ (f) $\frac{9}{20}$ **85.** $P = 0.7236$ or $P = 0.2764$

Section 6.4

1. E is double sixes, F is the sum > 7; $E \cap F$ is the double sixes and a sum > 7, or just double sixes   **3.** 0.75   **5.** $\frac{2}{3}$
7. $\frac{5}{7}$ **9.** 0.25   **11.** $\frac{2}{5}$ **13.** $\frac{18}{19}$ **15.** $\frac{1}{6}$ **17.** $\frac{1}{3}$ **19.** $\frac{1}{6}$ **21.** 0.5   **23.** $\frac{3}{4}$ **25.** 0
27. $\frac{1}{2}$ **29.** $\frac{4}{11}$ **31.** $\frac{1}{9}$ **33.** $\frac{1}{9}$ **35.** (a) $\frac{1}{5}$ (b) $\frac{1}{2}$ **37.** $\frac{3}{4}$ **39.** 1   **41.** Not independent   **43.** Not
independent   **45.** Independent   **47.** no   **49.** no   **51.** yes   **53.** 0.07   **55.** 0.25   **57.** 0.2   **59.** 0.3
61. $\frac{1}{4}$ **63.** 0.44
65. By the definition of conditional probability, $P(A|B) = P(A \cap B)/P(B)$. By the definition of independent events, $P(A|B) = P(B)$. Equating the right-hand side of each equation proves the assertion.
67. For n topics, $P(C_1) = P(P_2) = 1/n$. Also, $P(C_1 \cap P_2) = 1/n(n-1) \neq 1/n^2$.   **69.** 25%   **71.** 40%   **73.** (a) $\frac{1}{25}$
(b) $\frac{1}{3}$ (c) $\frac{1}{4}$ **75.** 0.2035   **77.** $\frac{29}{40}$ **79.** $P(D/F_1)$ **81.** 0.6216

Section 6.5

1. 0.3   **3.** $\frac{1}{3}$ **5.** 0.7   **7.** 0.3   **9.** 0.1325   **11.** 0.2642   **13.** 0.7   **15.** 0.0789   **17.** $\frac{5}{12}$ **19.** $\frac{2}{5}$
21. 0.4   **23.** 0.25   **25.** 0.49   **27.** $\frac{6}{17} \approx 0.3529$ **29.** 0.8235   **31.** $\frac{1}{4}$ **33.** 0   **35.** $\frac{13}{648} \approx 2.01\%$
37. $\frac{1}{3}888 \approx 0.026\%$ **39.** $\frac{1}{13} \approx 0.0769$ **41.** $\frac{1}{4}$; $\frac{1}{3}$ **43.** 0.01   **45.** 0.5263   **47.** $\frac{109}{9000}$; $\frac{100}{109}$ **49.** The talk
occurred with a 9.17% likelihood.   **51.** (a) 0.4545; 0.4773 (b) 0.5018; 0.1745   **53.** 0.422; 0.802

Section 6.6

1. $\frac{1}{3}$ **3.** 0   **5.** $\frac{3}{4}$ **7.** $\frac{1}{3}$ **9.** -0.17 **11.** 2   **13.** 0   **15.** 0   **17.** $-\frac{1}{8}$ **19.** 1   **21.** 0
23. 3.5   **25.** \$3.50   **27.** (a) $-\frac{1}{19}$ dollars; (b) win \$36.84   **29.** \$2500   **31.** Large box   **33.** Site 1   **35.** Study!
37. $\frac{1}{3888} \approx 0.026\%$ **39.** -0.12 **41.** choose high royalties   **43.** 0.2294; 3.6034

Answers to Review Questions, Chapter 6

1. { Sun, clouds, rain, . . . }
3. {++++, +++−, ++−+, ++−−, +−++, +−+−, +−−+, +−−−, −+++, −++−, −+−+, −+−−, −−++, −−+−,
−−−+, −−−−}; + means a job offer; − means no job offer.
5. {1, 2, 3, 4}   **7.** {(4 4)}   **9.** {(1 2), (1 4), (3 2), (3 4), (2 1), (2 3), (4 1), (4 3)}   **11.** $\frac{1}{16}$ **13.** 0.5
15. $\frac{4}{25}$ **17.** $\frac{12}{25}$ **19.** 0.16   **21.** 0.48   **23.** $\frac{2}{5}$ **25.** $\frac{10}{19}$ **27.** 2:1   **29.** 11:14   **31.** $\frac{1}{3}$; $\frac{2}{3}$ **33.** 1:11

35. 5 people   **37.** $\frac{1}{52}$ **39.** $\frac{14}{52}$ **41.** 10   **43.** 0.3   **45.** 0.7   **47.** $\frac{1}{45}$ **49.** $\frac{6}{52}$ **51.** $\frac{10}{52}$ **53.** $\frac{2}{52}$

55. $\frac{1}{11}$ **57.** $\frac{6}{11}$ **59.** $\frac{3}{70}; \frac{3}{70}; \frac{9}{70}$ **61.** $\frac{1}{4}; \frac{7}{8}$ **63.** $\frac{1}{20}; \frac{1}{30}$ **65.** 0.6125   **67.** 0.8571   **69.** $\frac{9}{13}$

71. $\frac{3}{13}$ **73.** $\frac{12}{13}$ **75.** $\frac{1}{5}$ **77.** -1 **79.** Bird: $\frac{3}{25}$; reptile: $\frac{1}{5}$ **81.** (a) 144 (b) $\frac{1}{144}; \frac{3}{144}; \frac{6}{144}$ (c) identical

with cube   **83.** (a) $\frac{11}{20}$ (b) $\frac{1}{2}$ (c) $\frac{7}{10}$ **85.** (a) $\frac{13}{24}$ (b) $\frac{10}{13}$ **87.** (a) $\frac{3}{5}; \frac{2}{5}$ (b) $\frac{4}{7}$ **89.** (a) $\frac{31}{8}$ (b) $-\frac{1}{4}$

Section 7.1

1. $\begin{pmatrix} 1-p & p \\ p & 1-p \end{pmatrix}$ **3.** yes   **5.** No—now 1 does not sum to one.   **7.** yes   **9.** $\begin{pmatrix} \frac{1}{2} & \frac{1}{2} \end{pmatrix}$ **11.** $\begin{pmatrix} \frac{25}{42} & \frac{17}{42} \end{pmatrix}$

13. $\begin{pmatrix} \frac{19}{30} & \frac{11}{30} \end{pmatrix}$ **15.** $\begin{pmatrix} \frac{17}{30} & \frac{13}{30} \end{pmatrix}$ **17.** $\begin{pmatrix} \frac{1}{2}+\frac{1}{6}p & \frac{1}{2}-\frac{1}{6}p \end{pmatrix}$ **19.** $\begin{pmatrix} \frac{2}{3} & \frac{1}{3} \end{pmatrix}$ **21.** $\begin{pmatrix} \frac{1}{3} & \frac{2}{3} \end{pmatrix}$ **23.** $\begin{pmatrix} \frac{4}{7} & \frac{3}{7} \end{pmatrix}$

25. $\begin{pmatrix} \frac{3}{4} & \frac{1}{4} \end{pmatrix}$ **27.** $\begin{pmatrix} \frac{1}{3} & 0 & \frac{2}{3} \end{pmatrix}$ **29.** $\begin{pmatrix} \frac{2}{9} & \frac{1}{3} & \frac{4}{9} \end{pmatrix}$ **31.** $\begin{pmatrix} \frac{1}{4} & \frac{5}{12} & \frac{1}{3} \end{pmatrix}$ **33.** $\frac{1}{3}\begin{pmatrix} \frac{5}{6} & \frac{13}{12} & \frac{13}{12} \end{pmatrix}$

35. $\begin{pmatrix} \frac{4}{15} & \frac{11}{30} & \frac{11}{30} \end{pmatrix}$ **37.** $\begin{pmatrix} \frac{1}{3} & \frac{1}{3} & \frac{1}{3} \end{pmatrix}$ **39.** $\begin{pmatrix} \frac{1}{4} & \frac{5}{8} & \frac{1}{8} \end{pmatrix}$ **41.** $\begin{pmatrix} 0.25 & 0.5 & 0.25 \end{pmatrix}$

43. $\begin{pmatrix} 0.3875 & 0.3250 & 0.2875 \end{pmatrix}$ **45.** (a) 0.4 (b) Not defined—there are only two states in the Markov process.

47. (a) $p^2 = \begin{pmatrix} \frac{43}{64} & \frac{21}{64} \\ \frac{9}{32} & \frac{23}{32} \end{pmatrix}$; $p^3 = \begin{pmatrix} \frac{169}{512} & \frac{343}{512} \\ \frac{197}{256} & \frac{109}{256} \end{pmatrix}$ (b) $\frac{1}{8}; \frac{3}{4}$ (c) $\begin{pmatrix} \frac{7}{16} & \frac{9}{16} \end{pmatrix}$; $\begin{pmatrix} \frac{61}{128} & \frac{67}{128} \end{pmatrix}$ (d) $\begin{pmatrix} \frac{97}{256} & \frac{159}{256} \end{pmatrix}$

49. Sine P is the identity matrix, $A^n = A^0$ for all n.

51. p_{ij} is the probability of proceeding to the jth state from i. All states are mutually exclusive, so $p_{i1} + p_{i2} + \cdots + p_{in}$ is the probability that the process will be in some state in the next step. But the process will certainly be in some state, that is, this probability is 1. The assertion is proved.

53. (a) 0.6; (b) for the second exam, the probabilities of hard and easy exams are both equal to .5. For the third exam, the probability of the exam being difficult is 0.55.

55. (a) 0.6; (b) for the second exam the probabilities of hard and easy exams are both equal to 0.5. For the third exam, the probability of the exam being difficult is 0.55.

57. (a) $\begin{pmatrix} .2 & .8 \\ .6 & .4 \end{pmatrix}$; (b) in the second month, 40% of commuters take a car. In the third month, the proportion will be 44%.

59. (a) $\begin{pmatrix} 0.3 & 0.7 \\ 0.05 & 0.95 \end{pmatrix}$ (b) 95%; 93.75% (c) 13.33%; 8.33%   **61.** $\begin{pmatrix} 0.9 & 0.1 \\ 0.6 & 0.4 \end{pmatrix}$; 87%   **63.** 25%; 43.75%

67. $\begin{pmatrix} 0.4410 & 0.5590 \end{pmatrix}$

Section 7.2

1.

n	$p_{11} = p_{22}$	$p_{12} = p_{21}$
2	.6250	.3750
3	.4375	.5625
4	.5313	.4688
5	.4844	.5157
6	.5079	.4922
7	.4961	.5040
9	.4991	.5010
11	.4998	.5002

3. For all n, $p_{21} = 0$, $p_{22} = 1$.

n	p_{11}	p_{12}
2	.16	.84
3	.064	.936
4	.0256	.9744
5	.0102	.9898
6	.0041	.9959
7	.0016	.9984
8	.0007	.9993

5. Regular; $\begin{pmatrix} 0.4 & 0.6 \end{pmatrix}$ **7.** Not regular (not a transition matrix)   **9.** Not regular   **11.** Not regular   **13.** $\begin{pmatrix} 1 & 0 \\ 0 & 1 \end{pmatrix}$

15. $\begin{pmatrix} .5 & .5 \\ .5 & .5 \end{pmatrix}$ **17.** $\begin{pmatrix} \frac{1}{2} & \frac{1}{2} \\ \frac{1}{2} & \frac{1}{2} \end{pmatrix}$

19. $\begin{pmatrix} .4 & .3 & .3 \\ .4 & .3 & .3 \\ .4 & .3 & .3 \end{pmatrix}$ **21.** $\begin{pmatrix} \frac{1}{3} & \frac{1}{3} & \frac{1}{3} \\ \frac{1}{3} & \frac{1}{3} & \frac{1}{3} \\ \frac{1}{3} & \frac{1}{3} & \frac{1}{3} \end{pmatrix}$ **23.** $\begin{pmatrix} \frac{3}{7} & \frac{4}{7} \\ \frac{3}{7} & \frac{4}{7} \end{pmatrix}$ **25.** $\left(\frac{4}{9} \ \ \frac{5}{9} \right)$ **27.** $\left(\frac{3}{11} \ \ \frac{8}{11} \right)$ **29.** $\left(\frac{11}{19} \ \ \frac{8}{19} \right)$

31. $(.3 \ \ .7)$ **33.** $(.5 \ \ .5)$ **35.** $(.5 \ \ .5)$ **37.** All components in the product matrix are sums of products of components of P. Since all $p_{ij} > 0$, so are all sums of products of the p_{ij}.   **39.** Left-handed with probability 0.5716.

41. (a) 0.25 (b) $P(\text{Dem}) = 0.3192$

$$P(\text{Ind}) = 0.4256$$

$$P(\text{Rep}) = 0.2554 \quad \text{They remain the same.}$$

43. 0.4286   **45.** 37.5%   **47.** 40.625%,   43.75%,   15.625%   **49.** 40.625%,   43.75%,   15.625%

51. $\begin{pmatrix} 0.2940 & 0.3211 & 0.3849 \\ 0.2940 & 0.3211 & 0.3849 \\ 0.2940 & 0.3211 & 0.3849 \end{pmatrix}$

Section 7.3

1. Yes   **3.** Yes   **5.** Yes   **7.** No   **9.** No   **11.** No   **13.** $R = \left(\frac{1}{4} \right)$, $Q = \left(\frac{3}{4} \right)$ **15.** $Q = R = \left(\frac{1}{2} \right)$

17. $R = \begin{pmatrix} \frac{1}{6} \\ \frac{1}{4} \end{pmatrix}$, $Q = \begin{pmatrix} \frac{2}{3} & \frac{1}{6} \\ \frac{1}{2} & \frac{1}{4} \end{pmatrix}$ **19.** $R = (.25 \ \ .25)$, $Q = (0.4)$, $F = \left(\frac{5}{3} \right)$ **21.** $R = \begin{pmatrix} \frac{1}{3} \\ \frac{1}{4} \end{pmatrix}$, $Q = \begin{pmatrix} \frac{1}{3} & \frac{1}{3} \\ \frac{1}{4} & \frac{1}{2} \end{pmatrix}$, $F =$

$\begin{pmatrix} 2 & \frac{4}{3} \\ 1 & \frac{8}{3} \end{pmatrix}$ **23.** $R = \begin{pmatrix} \frac{1}{5} & \frac{1}{5} \\ \frac{1}{4} & \frac{1}{4} \end{pmatrix}$, $Q = \begin{pmatrix} \frac{3}{10} & \frac{3}{10} \\ \frac{1}{4} & \frac{1}{4} \end{pmatrix}$, $F = \frac{20}{9} \begin{pmatrix} \frac{3}{4} & \frac{3}{10} \\ \frac{1}{4} & \frac{7}{10} \end{pmatrix}$

25. Probability of ending in state 1 is 42% and in state 2 is 58%. On average, these will be $\frac{5}{3}$ of a step before absorption.

27. There is a certainty in ending up in the unique absorbing state. Starting from state 2, there are $3\frac{1}{3}$ steps before absorption. From state 3, there are $3\frac{2}{3}$ steps.

29. Starting from any nonabsorbing state, there is equal likelihood of ending up in any of the absorbing states. From states 3 and 4, there are $\frac{7}{3}$ and $\frac{19}{9}$ steps before absorption.

31. The dimensions of O, R and Q are $m \times (n - m)$, $(n - m) \times m$ and $(n - m) \times (n - m)$.

35. $R = \begin{pmatrix} .5 & 0 \\ 0 & 0 \\ 0 & 0 \\ 0 & .5 \end{pmatrix}$, $Q = \begin{pmatrix} 0 & .5 & 0 & 0 \\ .5 & 0 & .5 & 0 \\ 0 & .5 & 0 & .5 \\ 0 & 0 & .5 & 0 \end{pmatrix}$, $F = \begin{pmatrix} 1.6 & 1.2 & 0.8 & 0.4 \\ 1.2 & 2.4 & 1.6 & 0.8 \\ 0.8 & 1.6 & 2.4 & 1.2 \\ 0.4 & 0.8 & 1.2 & 1.6 \end{pmatrix}$. On the average, starting with 1, 2, 3, or 4 dollars, there will be 4, 6, 6, and 4 moves before the game terminates. The probabilities of win/loss starting with various sums

is given by the product matrix $FR = \begin{matrix} & \begin{matrix} 0 & \ \ 5 \end{matrix} \\ \begin{matrix} 1 \\ 2 \\ 3 \\ 4 \end{matrix} & \begin{pmatrix} .8 & .2 \\ .6 & .4 \\ .4 & .6 \\ .2 & .8 \end{pmatrix} \end{matrix}$. For example, starting with \$3, the player has a 40% chance of going broke and a 60% chance of winning.

37. Yale is the absorbing state. Within $3\frac{2}{3}$ generations (on the average), all Harvardian descendants will be blue (Yale graduates). It takes Princetonian descendants $3\frac{1}{3}$ generations to achieve the same fate.

39. $P = \begin{pmatrix} 1 & 0 \\ 0.1 & 0.9 \end{pmatrix}$, $F = (10)$ **41.** (a) $F = \begin{pmatrix} 1.8514 & 0.7823 \\ 0.6780 & 1.6949 \end{pmatrix}$, $FR = \begin{pmatrix} 1 \\ 1 \end{pmatrix}$ (b) 100% (c) 2.6337 (d) 2.3729

Answers to Review Questions, Chapter 7

1. No negative probabilities are forbidden.   **3.** (a) $\frac{1}{2}$ (b) $\frac{2}{3}$ **5.** Regular, $\begin{pmatrix} \frac{1}{2} & \frac{1}{2} \\ \frac{1}{2} & \frac{1}{2} \end{pmatrix}$ **7.** Regular, $\begin{pmatrix} \frac{1}{3} & \frac{1}{3} & \frac{1}{3} \\ \frac{1}{3} & \frac{1}{3} & \frac{1}{3} \\ \frac{1}{3} & \frac{1}{3} & \frac{1}{3} \end{pmatrix}$

9. $F = \left(\frac{20}{13} \right)$ **11.** 60% risen on Tuesday, on Friday, and in the long term.   **13.** (a) 90%, 88.5% (b) $\frac{15}{17}$ **15.** (a) 54.6% up, 15.2% no change, 30.2% down (b) 45.3% up, 15.1% no change, 39.6% down

Section 8.1

1. Value $= \frac{1}{2}$; optimal strategies are row 1 and any column.   **3.** Value $= \frac{1}{2}$; row 1, col 2   **5.** Not strictly determined.
7. Not strictly determined.   **9.** Not strictly determined.   **11.** Not strictly determined.   **13.** -1 **15.** 1   **17.** All components   **19.** 5   **21.** 1   **23.** -1 in rightmost column   **25.** 0   **27.** Not strictly determined.   **29.** Value is 4; optimal strategy is row 3, column 1.   **31.** $-1 < x < 1$, value is x; $x \leq -1$, value is -1.   **33.** $-1 < x < 1$; value is 1.
35. All x; value is $x + 0.03$.   **37.** No player is favored in the long run, provided each plays rationally.   **39.** At 850'. The drawing shows that the area in question resembles a saddle, hence the term "saddle point."

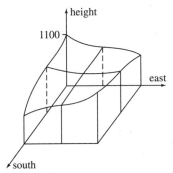

41.

2	-3	4
-3	4	-5
4	-5	6

43. Les Rondes should use hill 2; les Casse-têtes should approach from hill 1. Under this strategy, les Rondes have a 50% chance of winning.

45. Carla should show one finger; Robin should show two. Robin's minimum gain per play is \$3 if she plays rationally. This is the value of the game; since it is not zero, the game is not fair.

Section 8.2

1. $5\frac{1}{4}$ **3.** $\frac{13}{5}$ **5.** $\frac{17}{3}$ **7.** 0   **9.** 0   **11.** $\frac{5}{8}$ **13.** $\frac{5}{4}$ **15.** 1   **17.** $3\frac{1}{4}$ **19.** $\frac{1}{6}$ **21.** $\frac{1}{9}$
23. 0   **25.** $\frac{1}{16}$ **27.** Strategy 1   **29.** Strategy 2   **31.** $R_{opt} \, G \, C_{opt}$ **33.** (a) Choose row 2 all the time; value $= \frac{3}{2}$
(b) Each payoff is greater in row 2 than the payoff in the element above it.   **35.** Payoff is $\frac{1}{2}(x + y)$; since $x + y = 1$, the strategy is irrelevant.   **37.** $\frac{1}{2}$; $\frac{11}{20}$ **39.** 0.1100   **41.** 0.2978

Section 8.3

1. (3) **3.** (6) **5.** (2) **7.** $\begin{pmatrix} 1 & -10 \\ -1 & 1 \end{pmatrix}$ **9.** $R_1 = R_2 = C_1 = C_2 = \frac{1}{2}$ **11.** $R_1 = C_2 = 0$, $R_2 = C_1 = 1$

13. $R_1 = \frac{7}{13}$, $R_2 = \frac{6}{13}$, $C_1 = \frac{8}{13}$, $C_2 = \frac{5}{13}$ **15.** $R_2 = C_1 = \frac{1}{4}$, $R_2 = C_2 = \frac{3}{4}$, $R_3 = 0$ **17.** $R_1 = R_2 = C_1 = C_2 = \frac{1}{2}$
19. $R_1 = C_2 = 0$, $R_2 = C_1 = 1$ **21.** $R_1 = \frac{7}{13}$, $R_2 = \frac{6}{13}$, $C_1 = \frac{8}{13}$, $C_2 = \frac{5}{13}$

23. The value for a general payoff matrix is q/Δ, $q = eh - fg$. The inverse to a 2×2 matrix is $\dfrac{1}{q}\begin{pmatrix} h & -f \\ -g & e \end{pmatrix}$ so it won't have

an inverse whenever the game is fair ($q = 0$).   **25.** Define $G' = \begin{pmatrix} e' & f' \\ g' & h' \end{pmatrix} = \begin{pmatrix} e+k & f+k \\ g+k & h+k \end{pmatrix}$ Then $\Delta^1 = e^1 + h^1 - g^1 - f^1 =$

$e + h - g - f = \Delta$ (all the K's cancel). Then $r_1^1 = \dfrac{h^1 - g^1}{\Delta} = \dfrac{h - g}{\Delta} = r_1$ and so on.   **31.** (a) $r_1 = \frac{7}{16}$, $r_2 = \frac{9}{16}$ (b) $\frac{7}{16}$ to the

normal variety $\frac{9}{16}$ to the drought resistant variety.   **33.** (a) 7.7 (b) $r_1 = \frac{13}{14}$, $r_2 = \frac{1}{14}$ **35.** (a) $r_1 = \frac{7}{13}$, $r_2 = \frac{6}{13}$; $\frac{23}{13}$

(b) $c_1 = \frac{6}{13}$, $c_2 = \frac{7}{13}$ **37.** (a) $r_1 = 0$, $r_2 = \frac{1}{5}$, $r_3 = \frac{4}{5}$ (b) $c_1 = 0$, $c_2 = \frac{3}{5}$, $c_3 = \frac{2}{5}$ (c) $\frac{12}{5}$ **39.** $R = \begin{pmatrix} 0.4952 & 0.5048 \end{pmatrix}$

$C = \begin{pmatrix} 0.5625 \\ 0.4375 \end{pmatrix}$

Answers to Review Questions, Chapter 8

1. 7   **3.** All four 1's   **5.** $-\frac{1}{2}$ **7.** $-\frac{1}{9}$ **9.** $-\frac{1}{4}$ **11.** 2.5   **13.** 3   **15.** $\begin{pmatrix} 1 \end{pmatrix}$ **17.** No reduction

possible.   **19.** $\begin{pmatrix} 1 & -1 \\ -2 & 2 \\ 0 & -2 \end{pmatrix}$ **21.** $r_1 = r_2 = c_1 = c_2 = \frac{1}{2}$; $v = 8.5$ **23.** $r_1 = r_3 = \frac{1}{2}$; $r_2 = 0$; $C_1 = \frac{8}{15}$, $C_2 = \frac{7}{15}$, $V = 96$

25. Serve soft ice cream.   **27.** Republicans assume a "middle of the road" strategy—Let $r_1 = r_2 = \frac{1}{2}$.

Section 9.1

1. (a)   (b)

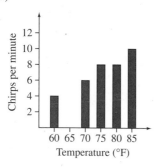
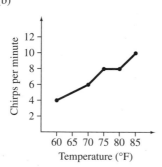

(c) No. If a pie chart were appropriate, we might conclude that the total number of chirps/minute forms a quantity in which we have some interest. This is not the case—adding the chirps in the right-hand column is not meaningful.

3. (a)

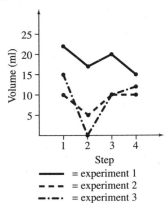

= experiment 1
= experiment 2
= experiment 3

(b)

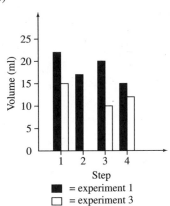

= experiment 1
= experiment 3

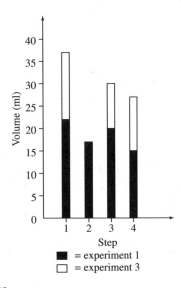

= experiment 1
= experiment 3

(c)

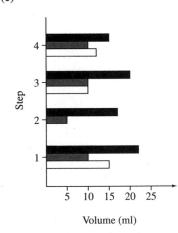

= experiment 1
= experiment 2
= experiment 3

(d) yes (e)

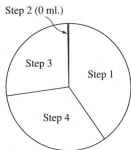

Step 2 (0 ml.)
Step 3
Step 1
Step 4

(f)

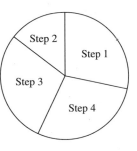

Step 2
Step 1
Step 3
Step 4

5. January is "wetter" weather because the actual precipitation is relatively much greater than the normal amount in January than in August.

7. January; July

9. About 50%

11.

City	Strength in retreat
Dorogobonge	55,000
Smolensk	37,000
Stulienska	50,000
Moloderno	12,000
Nieman River	10,000

13.

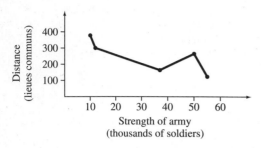

15.   October 20; October 20

17.   (a) The third week of October, 1987; December, 1985. (b) It is reasonable. The graph lends tentative support to the hypothesis that relative dips in the price (the Dow-Jones average) are associated with relative peaks in volume. See, for example, mid-September, 1986.

Section 9.2

1.

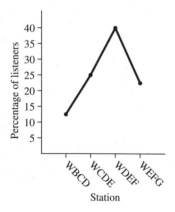

3.   25%   **5.**   Region 4;

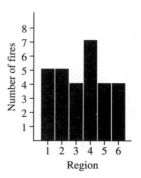

7. $\frac{7}{29}$, $\frac{1}{6}$ **9.** $\frac{12}{29}$, $\frac{1}{3}$

11.   Yes; mode is 6.   **13.**   Yes; 178 and 180 pages.   **15.**   178 or 180 pages; both have empirical probabilities of $\frac{6}{29}$.
17.   0.4906; 0.1816

Section 9.3

3.   Insufficient information to answer the question.   **5.**   1909   **7.**   Portland and San Diego
9. -0.001, -0.01 **11.**   5.5   **13.** $\frac{2}{3}$ **15.**   5.5   **17.** $\frac{77}{120}$ **19.**   6   **21.** $-\frac{2}{3}$ **23.** $7\frac{1}{6}$ **25.**   1
27. $2\frac{2}{3}$ **29.** -1 **31.**   (a) mean is 3.5, median is 4. (b) $\sigma^2 = 1.0278$, $\sigma = 1.0138$ **33.** $\sigma^2 = 8.25, \sigma = 2.8723$
35.   5, $\sqrt{10} = 3.1623$, 10   **37.**   5, $9\frac{1}{6}$, 3.0277   **39.**   8.917; 2.986   **41.**   0.0067, 0.0816
43.   Each term of Dull's formula is the negative of the corresponding term of the original formula. When squared, the "negative" becomes "positive."
45.   The deviation $\Delta_i = x_i - \mu$. If there are n deviations, then $\Delta_1 + \Delta_2 + \ldots + \Delta_n = x_1 - \mu + x_2 - \mu + \ldots + x_n - \mu$

$$= x_1 + x_2 + \ldots + x_n - n\mu$$

$$= x_1 + x_2 + \ldots + x_n - (x_1 + x_2 + \ldots + x_n)$$

$$= 0$$

where we used the definition of μ.

47. Not a useful equality here since any probability is always greater than any negative number.   **49.** 0.75   **51.** 0 (The Chebychev inequality is not useful here).   **53.** 0.75   **55.** $k^2\sigma^2$ **57.** 47.9583, 48.25, 48.125, 0.1128   **59.** The missing weights are 2.5 and 3 pounds; $\sigma^2 = 0.4167$ pounds^2.   **61.** (a) 90, 12.65 (b) 77.5   **63.** 0.75   **65.** 4.5, 2.63   **67.** 6.5, 0.7638   **69.** 3.7, 1.7078   **71.** 3.1   **73.** 8, 0.4397   **75.** 93.7, 1.4177

Section 9.4

1. 50%   **3.** 68.27%   **5.** 95.46%   **7.** 43.32%   **9.** 0.8414   **11.** 2.28%   **13.** 50%   **15.** 43.32%   **17.** 0%
19. 2.14%   **21.** 0.14%   **23.** 84.14%   **25.** −2, 0, 2   **27.** −$\frac{1}{2}$, 0.13, 0.05   **29.** −2.10, 2.07, 5.05   **31.** 0.3413
33. 0.0927   **35.** 0.8185   **37.** 0.3413   **39.** 0.0395   **41.** 0.9332   **43.** Infinity   **45.** About 1.14   **49.** 0.1587,
0.0228   **51.** 0.0548, 0.0007   **53.** 0.0478, 0.3217   **55.** −3.6, 5.2

Section 9.5

1. $\frac{1}{2}$ **3.** $\frac{1}{4}$ **5.** $\frac{8}{27}$ **7.** $\frac{252}{2^{10}} = 0.2461$ **9.** 0.4096   **11.** 0.1935   **13.** 0.4116   **15.** 0.0231   **17.** 0.2734
19. $\frac{1}{256}$ **21.** $\frac{163}{256}$ **23.** 0.1318   **25.** 0.8306   **27.** 0.1782   **29.** 0.8306   **31.** 0.3721   **33.** 0.2734
35. 0   **37.** 0   **39.** $\frac{n+1}{2^{n+1}}$ **41.** npq^{n-1}, where $q = 1 - p$ **43.** 0   **45.** 0   **47.** 0   **49.** (a) $\frac{1}{2} = 0.5$,
$\frac{3}{8} = 0.375$, $\frac{70}{256} = 0.2734$; the probabilities are decreasing. (b) This is the probability of observing exactly one-half of the tosses of a fair coin to be heads. This probability goes to zero as the number of tosses gets very large. (c) We conclude that $(n!)^2 = 2\pi n n^{2n} e^{-2n}$, $(2n)! = 2\sqrt{\pi n}\, 2^{2n} n^{2n} e^{-2n}$ so that $b(2n,\ n;\ \frac{1}{2}) = \frac{1}{\sqrt{\pi n}}$, which approaches zero for large n. We can use this expansion to estimate $b(16,\ 8;\ \frac{1}{2})$ by taking $n = 8$ so this expression is approximately $\frac{1}{\sqrt{8\pi}} \approx 0.20$.   **51.** (a) 0.5138 (b) 0.9872
53. (a) 0.7516 (b) 0.9984   **55.** 0.8857   **57.** No   **59.** 0.8270   **61.** 0.1074, 0.0115, 0.3277   **63.** 0.7738
65. 0.3277, 0.00032, 0.9421   **67.** 0.7   **69.** (a) 1.0 (b) 0.0781   **71.** 0.1736, 0.1736   **73.** Certain; virtually zero.
75. (a) 0.7704 (b) 0.1319   **77.** 0.2465; 0.3642   **79.** 0.5; 0.0146

Answers to Review Exercises, Chapter 9

1.

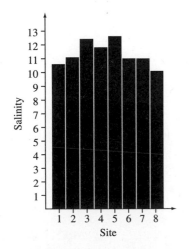

3. No   **5.**

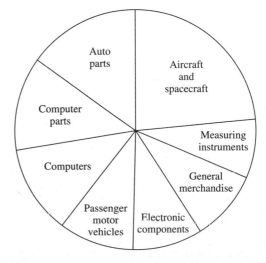

7.

9.

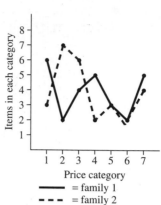

11.   4

13. 8.8, 2.9665   **15.** 9, 11, 74.80, 8.65   **17.** 34.14%   **19.** −2.28%, 2.28%   **21.** 0   **23.** $\frac{1}{3}$ **25.** $-\frac{11}{3}$
27. $-\frac{1}{2}$ **29.** $\frac{81}{256}$ **31.** $\frac{8}{27}$ **33.** 0.1539   **35.** 0.1038   **37.** $\frac{35}{128}$, $\frac{99}{128}$, $\frac{1}{2}$ **39.** Yes   **41.** (a) 0.1359
(b) 0.1587   **43.** (a) 0.2417 (b) 0.3085   **45.** 0.7737; 0.9774   **47.** 0.1841   **49.** Approximately 555,390.

Section 10.1

1. $97.50   **3.** $9.58   **5.** £11.11   **7.** 0.1042 years   **9.** $1142.86   **11.** 14.22%/yr   **13.** $16,734.18
15. $7616.53   **17.** $10,745.47   **19.** $11,790.09
21. $1360. Since the interest is given to the lender as soon as it is posted, it does not itself accumulate interest. The calculation now reduces to the same mathematics as for simple interest.
23. (a) DM 943,164.94 (b) DM 950,737.95 (c) DM 952,526.85 (d) DM 953,225.92 (e) DM 953,406.36 (DM = German Marks)
(f) The more frequent interest is computed, the greater the yield. We assume 360 days per year in this problem.
25. 8.68%   **27.** 9.6%   **29.** 9.53%   **31.** $18,140.18; $22,080.40; $24,351.89; $32,620.38

Section 10.2

1. $82,500   **3.** $14,850   **5.** $123,333.21   **7.** $1827.65   **9.** $14,752.06   **11.** $249,695.73   **13.** $9583.63
15. $15,427.37   **17.** $58,473.73   **19.** $2281.93   **21.** $4976.73   **23.** $3736.29   **25.** $4306.87   **27.** $81,066.43
29. $50,982.05   **31.** $14,977.96   **33.** $2500.00   **35.** $276,473.76; $126,498.08; $50,258.04   **37.** $91,198.35
39. $3882.78   **41.** $8329.09   **43.** $346.78   **45.** $361,648.55

Section 10.3

1. $22,477.52   **3.** $59,777.15   **5.** $8293.59   **7.** $7128.21   **9.** $4018.48   **11.** $5172.56   **13.** $49.77
15. $2027.64   **17.** $223.97   **19.** $2149.21   **21.** $1171.31   **23.** $811.84   **25.** $1535.75   **27.** Loan A
29. $2412.55   **31.** $24.90   **33.** $1331.75   **35.** (a) $440,619.11 (b) $2,418,850.66   **37.** (a) $1517.43 (b) $1500
interest, $17.43 principal   **39.** (a) $1053.22 (b) $1028.61 (c) $54,333.60   **41.** In case the loan is for 48 months, monthly
payments of $125 will correspond to a loaned amount of $5023.10. This seems to be the best we can do. At 47 months, for example,
the principal would be $4935.77.

Section 10.4

1. 6.8%   **3.** 10.89%   **5.** 2.80%   **7.** $42,802.82   **9.** $6,865.48   **11.** $63,028.88   **13.** $60,630.98
15. $104,709.93   **17.** $163,518.20   **19.** $73,296.95   **21.** $116,504.55   **23.** $K = \frac{1+i}{1+g} - 1 = \frac{1+i-(1+g)}{1+g} = \frac{i-g}{1+g}$
27. (a) $476,632.53 (b) $3,602,622.12   **29.** $1795.82   **31.** (a) 7.18% (b) $97,522.99 (c) decrease; $820.27 is the payment.

Answers to Review Exercises, Chapter 10

1. $70.69   **3.** $0.56   **5.** $1623.57   **7.** $63.46   **9.** $4484.23   **11.** £4.64   **13.** $3901.11   **15.** £3.17
17. $3106.19   **19.** $2326.53   **21.** $3901.11   **23.** £3.17   **25.** $481.24; $6500.00   **27.** $199,020.88   **29.** $60.52

Section 11.1

1. $y_{n+1} = y_n = y_0$ **3.** $y_{n+1} = -\frac{1}{2}y_n$; $y_0 = 10$ **5.** $y_{n+1} = 0.95y_n + 2$; $y_0 = 4$ **7.** $y_{n+1} = \left(1 + \frac{0.15}{4}\right)y_n$
9. $y_{n+1} = \frac{0.1}{24}y_0$; y_n is the amount of interest earned in each period.   **11.** $y_{n+1} = 1.11y_n + 1000$ **13.** $y_{n+1} = 1.04y_n$
15. $y_{n+1} = (1-k)y_n$ **17.** $y_{n+1} = 1.02y_n$ **19.** $y_{n+1} = 0.9y_n + 0.1y_0$

Section 11.2

1. $y_n = 1 + n$ **3.** $y_n = 7(2n) - 3$ **5.** $y_n = -(-1.5)^n$ **7.** $y_n = -1$ **9.** $y_n = 4$ **11.** $y_n = \frac{n-2}{3}$
13. $y_n = 200(1.01)^n - 100$ **15.** $y_n = \left(1 + \frac{1000}{999}\right)1000^n - \frac{1000}{999}$ **17.** $y_n = \frac{10{,}000^n - 10{,}000}{9{,}999}$ **19.** (a) $y_n = \frac{10n+1-10}{9}$
(b) $y_n = \frac{100^{n+1} - 100}{99}$ (c) $y_n = \frac{1000^{n+1} - 1000}{999}$ (d) $y_n = \frac{1{,}000{,}000^{n+1} - 1{,}000{,}000}{999{,}999}$ (e) For $a = b$, as a gets very very large, the solution gets
closer to $y_n = a^n$.

Section 11.3

1. d   **3.** none   **5.** a   **7.** All y_n are trivially zero.   **9.** $y_n = 1$ for $n > 0$.   **11.** divergent   **13.** divergent and
oscillatory   **15.** divergent and oscillatory   **17.** divergent and oscillatory   **19.** constant   **21.** constant   **23.** divergent

Section 11.4

1. $P = p(1+i)^n$, $n =$ number of quarters, $i = \frac{0.15}{4}$, $p =$ original balance, $P =$ current balance.   **3.** $i = Prn$, $R = \frac{0.1}{24}$, $P =$
amount borrowed. i represents the accumulated interest. $n =$ number of periods   **5.** $P_n = 1.11^n \left(1000 + \frac{1000}{0.11}\right) - \frac{1000}{0.11}$ where
P_n represents the total at the $(n+1)^{\text{st}}$ year.   **7.** $g_n = g_0(1.04)^n$ is the value of the GNP after the nth year. g_0 is the
original value of GNP.   **9.** $a_n = (1-k)^n a_0$ where a_0 is the original amount, and a_n is the amount remaining after n weeks.
11. $P_n = p(1.02)^n$ where p is the original population and P_n is the population of the nth year.   **13.** If h_n represents the
height after the nth bounce, then $h_n = 20$.   **15.** $a_n = (0.8)^n \left(\frac{h_0}{2}\right) + \frac{h_0}{2}$. Now, this equation predicts that as the number of
bounces gets large, the ball bounces to one-half the original height h_0. In real life, we expect h_n to approach zero for large n.
17. $y_{n+1} = 0.97y_n$; $y_0 = p$. $y_n = (0.97)_p^n$.   **19.** $g_{n+1} = 1.15g_n$; $g_n = (1.15)_{g_0}^n$ **21.** $y_{n+1} = 1.025y_n$; $y_n = (1.025)^n y_0$;
between 28 and 29 years.

Answers to Review Questions, Chapter 11

1. $y_{n+1} = 1.03y_n$ **3.** $P_{n+1} = A + ni$, $i = A \times \frac{0.1}{24}$ is the simple interest earned each period, A is the amount of the loan.
5. $P_{n+1} = 1.036p_n$ **7.** $h_{n+1} = 0.95h_n - \alpha$ **9.** $y_n = (1.01)^n (102.01) - 101$ **11.** $y_n = 3n - z$ **13.** Since the 3
parameters satisfy $y_0 = \frac{b}{1-a}$, $y_n = 1.5$ for all n.   **15.** Divergent and oscillatory   **17.** Convergent to 0.   **19.** Oscillatory,
and convergent to $\frac{3}{4}$.   **21.** All $y_n = \frac{8}{15}$ since $y_0 = \frac{b}{1-a}$.   **23.** $p_n = p(1.0075)^n$, n the number of months.
25. Let A represent the amount borrowed, let p_n represent the amount owed Tran after n months. Then

$$p_n = A\left(1 + \frac{n}{100}\right)$$

27. $p_n = 0.94_p^n$; p the original population, p_n the population after the nth year.

Section 12.1

1.
A –
B –
C B
D A, C

3.
A –
B –
C –
D A
E C
F D, B, E

5.
A –
B –
C A, B

7.
A –
B –
C A, B
D C
E A
F D, E

9.

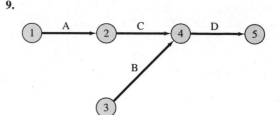

11.

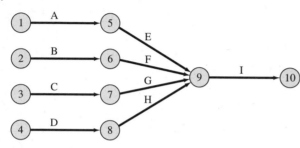

13.

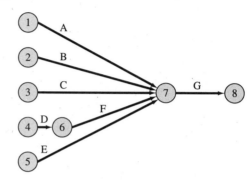

15.

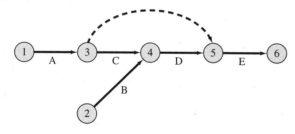

17.

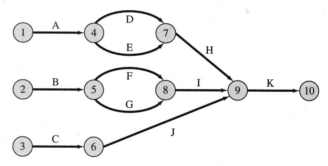

19.

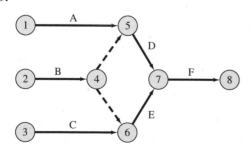

Section 12.2

1.   Completion time: 6; critical path: B-D

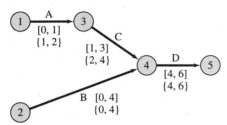

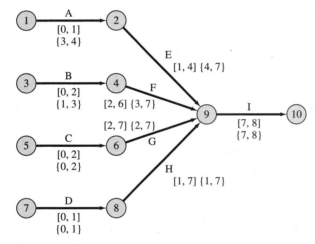

3.   Time of completion: 8; 2 critical paths: C-G-J and D-H-I.

5.   Completion time = 5; there are three critical paths: B-G, C-G, D-F-G.

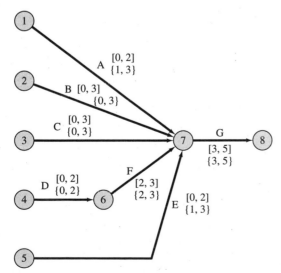

7.

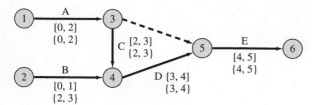

Completion time = 5; critical path is ACDE.

9.

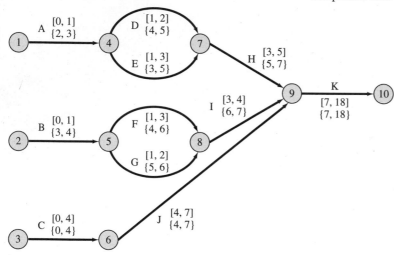

Completion time = 18; critical path is C-J-K

11. In corporate finance, CPM is useful for achieving merger of two corporations. In both real estate and public works, CPM is useful for proper scheduling of the many simultaneous activities that take place in large building projects, such as skyscrapers and bridges.

13.

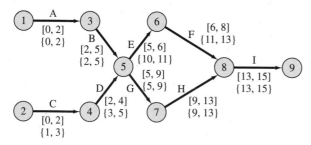

Completion time is 15 weeks; the critical path is ABGHI.

Section 12.3

1. $\mu = 9$, $\sigma^2 = \frac{25}{9}$ **3.** $\mu = 1$, $\sigma^2 = 0$ **5.** $\mu = 8$, $\sigma^2 = \frac{1}{9}$ **7.** $\mu = 5\frac{1}{3}$, $\sigma^2 = 1$ **9.** $\mu_a = 2$, $\sigma_a^2 = \frac{1}{9}$; $\mu_B = 2\frac{1}{2}$, $\sigma_B^2 = \frac{1}{4}$; $\mu = 4\frac{1}{2}$, $\sigma^2 = \frac{13}{36}$ **11.** $\mu_A = 9$, $\mu_B = 8$, $\mu_C = 7$, $\mu = 24$; $\sigma_a^2 = \sigma_B^2 = \sigma_c^2 = \sigma^2 = 0$

13. $\mu_a = \frac{19}{6}$, $\sigma_a^2 = \frac{49}{36}$; $\mu_B = \frac{22}{3}$, $\sigma_B^2 = 1$; $\mu_\infty = \frac{9}{2}$, $\sigma_c^2 = \frac{25}{36}$; $\mu = 15$, $\sigma^2 = \frac{55}{18}$ **15.** Doubling each parameter doubles μ and quadruples the variance. Tripling each parameter triples the average and multiplies the variance by nine. In general, multiplying each factor by n increases the average by a factor of n and increases the variance by a factor of n^2.   **17.** $\mu = 3\frac{2}{3}$ hr; $\sigma^2 = 1$ hr^2.

19. (a)

	Predecessors
A	—
B	—
C	—
D	A
E	B
F	D, E
G	B
H	C
I	F, G, H

(b) $\mu = 13\frac{2}{3}$, $\sigma^2 = 2\frac{1}{9}$. (c) $0.1251 \approx \frac{1}{8}$. (d) 0.1788.

Answers to Review Exercises, Chapter 12

1.

	Predecessors
A	—
B	—
C	—
D	A, B, C
E	A, B, C
F	E
G	D, F

3.

5.

A	[0,1]
B	[0,2]
C	[0,1.5]
D	[2,6]
E	[2,5]
F	[5,7]
G	[7,8]

7. 8 days

9. $4\frac{2}{3}$, 1　　**11.** 12, $\frac{4}{9}$　　**13.** (a) B: 2, $\frac{1}{9}$; D: 3, $\frac{1}{9}$; F: 4, $\frac{16}{9}$; H: 5, $\frac{16}{9}$. (b) $\mu = 14$, $\sigma^2 = \frac{34}{9}$, $\sigma = \sqrt{\frac{34}{9}} \approx 1.9437$.
(c) 0.0207.

Appendix A, Section A.5

1. -4　　**3.** -9　　**5.** 5　　**7.** 0　　**9.** undefined　　**11.** -2　　**13.** 1　　**15.** $-\frac{2}{9}$　　**17.** $\frac{11}{10}$　　**19.** $\frac{1}{4}$　　**21.** 1
23. $\frac{1}{27}$　　**25.** $\frac{1}{27}$　　**27.** 64　　**29.** $64u^3$　　**31.** $\frac{-1\pm\sqrt{5}}{2}$　　**33.** 2, 3　　**35.** $-\frac{1}{2}, 1$

Appendix C, Section C.5

1. 11　　**3.** 1　　**5.** $\chi^2 = 0$; data is consistent　　**7.** $\chi^2 = \frac{2}{5}$; data is consistent　　**9.** $\chi^2 = 1$; data is consistent
11. problem 9 not computable; $\chi^2 = 2.5$ and data is consistent　　**13.** 1.6　　**15.** 11.5　　**17.** (a) $\frac{2}{5}$ (b) $\frac{4}{5}$ (c) 4　　**19.** no
21. yes　　**23.** $\frac{8}{5}$

Index

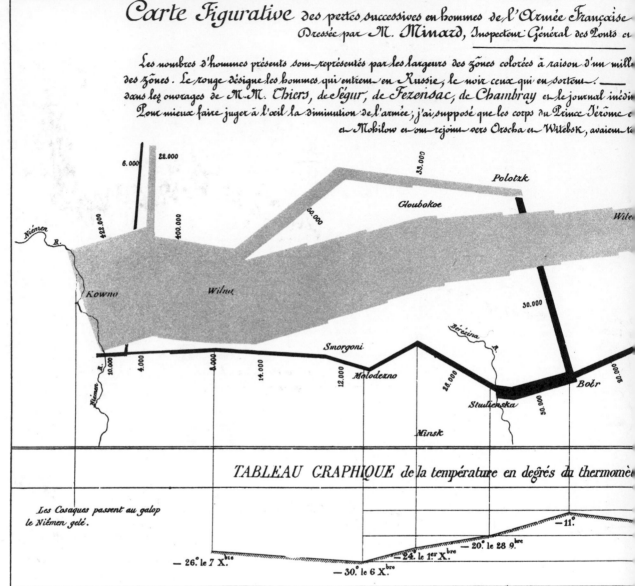

Napoleon's March to Moscow   The War of 1812

This classic of Charles Joseph Minard (1781–1870), the French engineer, shows the terrible fate of Napoleon's army in Russia. Describe
by E. J. Marey as seeming to defy the pen of the historian by its brutal eloquence, this combination of data map and time-series, drawn
1861, portrays the devastating losses suffered in Napoleon's Russian campaign of 1812. Beginning at the left on the Polish-Russian bord
near the Niemen River, the thick band shows the size of the army (422,000 men) as it invaded Russia in June 1812. The width of the bar
indicates the size of the army at each place on the map. In September, the army reached Moscow, which was by then sacked and deserte
with 100,000 men. The path of Napoleon's retreat from Moscow is depicted by the darker, lower band, which is linked to a temperatu